ESE 2024

UPSC ENGINEERING SERVICES EXAMINATION

Main Examination

Civil Engineering

Topicwise *Conventional* Solved Papers

Paper-II

25 Years SOLVED PAPERS

- ✓ Thoroughly revised and updated
- ✓ Fully solved with explanations
- ✓ Also useful for State Engineering Services Exams

www.madeeasypublications.org

MADE EASY Publications Pvt. Ltd

Corporate Office: 44-A/4, Kalu Sarai (Near Hauz Khas Metro Station), New Delhi-110016
E-mail: infomep@madeeasy.in
Contact: 0-9021300500
Visit us at: www.madeeasypublications.org

ESE-2024 : Main Examination
Civil Engineering : Paper-II | Conventional Solved Questions : (1999-2023)

1st Edition : 2010
2nd Edition : 2011
3rd Edition : 2012
4th Edition : 2013
5th Edition : 2014
6th Edition : 2015
7th Edition : 2016
8th Edition : 2017
9th Edition : 2018
10th Edition : 2019
11th Edition : 2020
12th Edition : 2021
13th Edition: 2022 (Feb)
14th Edition: 2022 (Aug)

15th Edition : 2023

Director's Message

During the last few decades of engineering academics, India has witnessed geometric growth in engineering graduates. It is noticeable that the level of engineering knowledge has degraded gradually, while on the other hand competition has increased in each competitive examination including GATE and UPSC examinations. Under such scenario higher level efforts are required to take an edge over other competitors.

The objective of **MADE EASY books** is to introduce a simplified approach to the overall concepts of related stream in a single book with specific presentation. The topic-wise presentation will help the readers to study & practice the concepts and questions simultaneously.

The efforts have been made to provide close and illustrative solutions in lucid style to facilitate understanding and quick tricks are introduced to save time.

Following tips during the study may increase efficiency and may help in order to achieve success.

- Thorough coverage of syllabus of all subjects
- Adopting right source of knowledge, i.e. standard reading text materials
- Develop speed and accuracy in solving questions
- Balanced preparation of Paper-I and Paper-II subjects with focus on key subjects
- Practice online and offline modes of tests
- Appear on self assessment tests
- Good examination management
- Maintain self motivation
- Avoid jumbo and vague approach, which is time consuming in solving the questions
- Good planning and time management of daily routine
- Group study and discussions on a regular basis
- Extra emphasis on solving the questions
- Self introspection to find your weaknesses and strengths
- Analyze the exam pattern to understand the level of questions
- Apply shortcuts and learn standard results and formulae to save time

B. Singh (Ex. IES)
CMD, MADE EASY Group

ESE 2024 Main Examination

Civil Engineering Conventional Solved Questions

Paper-II

Contents

1 Fluid Mechanics, Hydraulic Machines and OCF

Revised Syllabus of ESE : *Fluid Mechanics, Open Channel Flow, Pipe Flow: Fluid properties; Dimensional Analysis and Modeling; Fluid dynamics including flow kinematics and measurements; Flow net; Viscosity, Boundary layer and control, Drag, Lift, Principles in open channel flow, Flow controls. Hydraulic jump; Surges; Pipe networks.*

Hydraulic Machines and Hydro power: Various pumps, Air vessels, Hydraulic turbines – types, classifications & performance parameters; Power house – classification and layout, storage, pondage, control of supply.

1. Fluid Properties

1.1 A plate with surface area of 0.4 m^2 and weight of 500 N slides down on an inclined plane at 30° to the horizontal at a constant speed of 4 m/s. If the inclined plane is lubricated with an oil of dynamic viscosity 2 poise, find the thickness of lubricant film.

[10 marks : 2006]

Solution:

Assuming linear relationship between shear stress developed in the lubricant and velocity gradient.

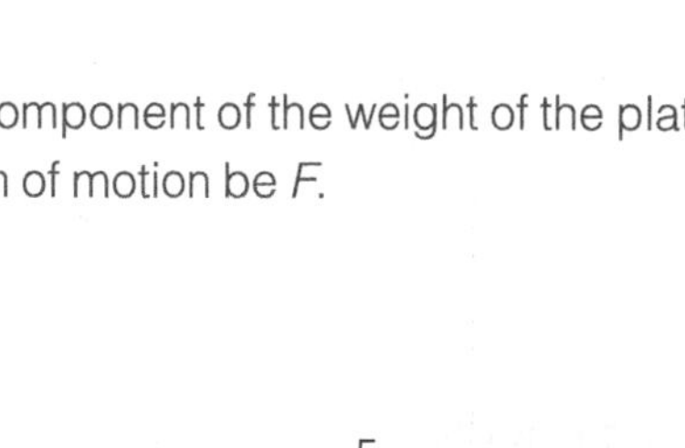

Let the thickness of the lubricating film be y

Surface Area of plate, $A = 0.4\ m^2$

Weight of plate, $W = 500$ N

Speed of sliding of plate, $V = 4$ m/s

Dynamic viscosity, $\mu = 2$ poise $= 0.2$ kg/m-s

The shear stress will be developed in the lubricant due to the component of the weight of the plate in the direction of motion. Let the component of weight in the direction of motion be F.

$\therefore \quad F = W\sin 30° = 500 \sin 30° = 250$ N

According to Newton's law of viscosity,

$$F = \frac{\mu A V}{y} \qquad \left[\because F = \tau A; \quad \tau = \mu\frac{du}{dy}\right]$$

$$\Rightarrow \quad 250 = \frac{0.2 \times 0.4 \times 4}{y}$$

$$\Rightarrow \quad y = 1.28 \times 10^{-3}\ \text{m} = 1.28\ \text{mm}$$

1.2 **A rotating viscometer has two cylinders. The radius of inner fixed cylinder is R_1 and the radius of the outer rotating cylinder is R_2. This viscometer is used for the measurement of viscosity. Derive an expression for the viscosity in terms of the torque acting on the inner cylinder of height L, gap between the bottoms of the two cylinders b, and the angular speed ω (omega).**

[9 marks : 2007]

Solution:

It consists of two co-axial cylinders, having radius R_1 and R_2 as shown in the figure. The very small space $(R_2 - R_1)$ is left in between the two. The space between them is filled with the liquid whose viscosity is to be determined.

The inner cylinder is suspended by a torsion wire on spring and it is held stationary. The outer cylinder is then rotated at a constant angular velocity. When the outer cylidner rotates, the torque generated by such rotation is transmitted by the thin liquid film to the inner stationary cylinder, which causes rotation of torsion wire. The rotation of wire can be measured by means of a circular dial attached to the wire and a fixed pointer.

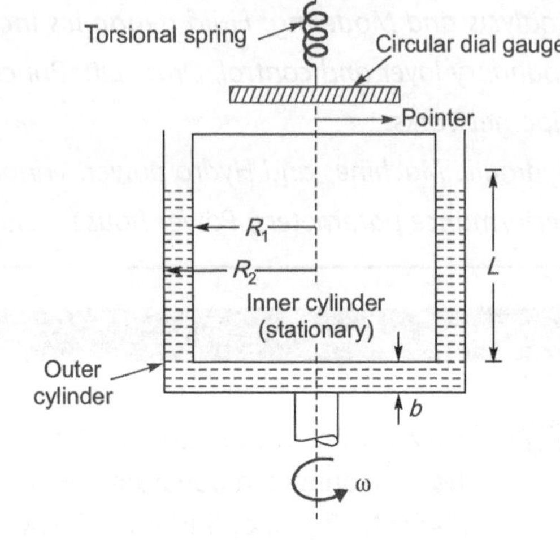

From the previously obtained calibration curve between the torque and the rotation of torsion wire, the torque exerted on wire and hence on the inner cylinder, corresponding to the measured rotation of wire can be known.

Total torque, $$T = T_1 + T_2 \quad \text{...(i)}$$

T_1 = Torque due to side

T_2 = Torque due to bottom

Case-1:

Torque contributed from the sides, T_1

Circumferential velocity of the outer cylinder

$$V = \omega R_2$$

Clearance between the cylinders, $h = R_2 - R_1$

Assuming linear variation of velocity across the gap,

Velocity gradient $$\frac{du}{dr} = \frac{V}{r} = \frac{\omega R_2}{R_2 - R_1}$$

Shear stress, $$\tau = \mu \frac{du}{dr} = \frac{\mu \omega R_2}{R_2 - R_1}$$

Shear force, $$F_s = \tau \times 2\pi R_1 \times L$$

$$\therefore \quad T_1 = F_s \times R_1$$

$$\Rightarrow \quad T_1 = \tau \times 2\pi R_1 \times L \times R_1$$

$$\Rightarrow \quad T_1 = \frac{\mu \omega R_2}{(R_2 - R_1)} \times 2\pi R_1^2 L$$

$$= \frac{2\pi \mu \omega R_1^2 R_2 L}{R_2 - R_1} \quad \text{...(ii)}$$

Case-2:

Torque contributed from the bottom (T_2)

Consider an element of inner cylinder of width 'dr' at a radial distance r.

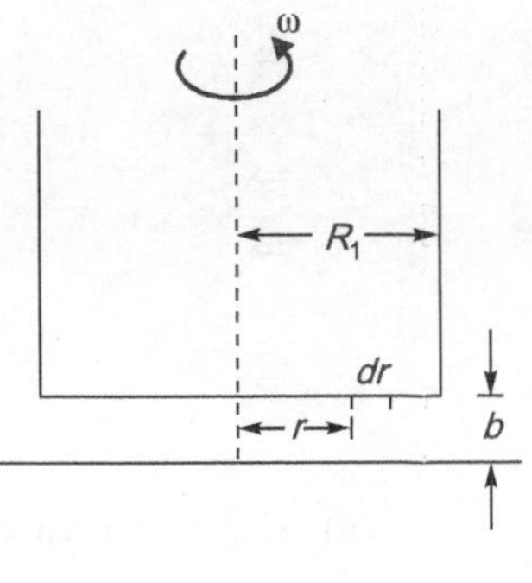

Velocity at this radius, $v = r\omega$

Assuming linear variation of velocity with depth in the gap 'b'

Shear stress, $$\tau = \frac{\mu v}{b} = \frac{\mu r\omega}{b}$$

Torque of the element, $$dT_2 = \frac{\mu r\omega}{b}(2\pi r\, dr)r = \frac{\mu\omega}{b}2\pi r^3 dr$$

Total torque on the cylinder, $$T_2 = \int_0^{R_1}\frac{\mu\omega}{b}2\pi r^3 dr$$

$$\Rightarrow \quad T_2 = \frac{\mu\omega}{b}2\pi\left[\frac{r^4}{4}\right]_0^{R_1} = \frac{\mu\omega}{b}\frac{2\pi R_1^4}{4} = \frac{\pi\mu\omega R_1^4}{2b} \quad \text{... (iii)}$$

Total torque, $$T = T_1 + T_2$$

$$T = \frac{2\pi\mu\omega R_1^2 R_2 L}{R_2 - R_1} + \frac{\pi\mu\omega R_1^4}{2b}$$

$$T = \left(\frac{2\pi R_1^2 . R_2 L}{R_2 - R_1} + \frac{\pi R_1^4}{2b}\right)\omega\mu$$

$$\mu = \frac{T}{\omega\left(\frac{2\pi R_1^2 R_2 L}{R_2 - R_1} + \frac{\pi R_1^4}{2b}\right)}$$

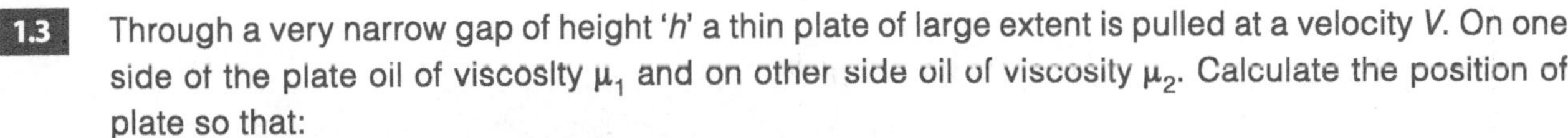

1.3 **Through a very narrow gap of height 'h' a thin plate of large extent is pulled at a velocity V. On one side of the plate oil of viscosity μ_1 and on other side oil of viscosity μ_2. Calculate the position of plate so that:**

(i) The shear force on two sides of plate is equal

(ii) The pull required to drag the plate is minimum

[10 marks: 2008]

Solution:

Let y be the distance of the thin plate from the top surface. Assuming linear relationship between shear stress developed and the velocity gradient.

(i) **Shear stress developed on the top portion** is given by,

$$\tau_1 = \mu_1\frac{du}{dy}$$

$$\Rightarrow \quad \tau_1 = \mu_1 \times \frac{V}{y}$$

Shear stress developed on the bottom portion is given by

$$\tau_2 = \mu_2 \times \frac{V}{h-y}$$

If A is the area of thin plate, then shear force on the top and bottom portion,

$$F_1 = \tau_1 \times A \text{ and } F_2 = \tau_2 \times A$$

But $\qquad F_1 = F_2$

$$\Rightarrow \quad \tau_1 \times A = \tau_2 \times A$$

$$\Rightarrow \quad \mu_1 \times \frac{V}{y} = \mu_2 \times \frac{V}{h-y}$$

$$\Rightarrow \quad \mu_1(h-y) = \mu_2 y$$

$$\Rightarrow \quad y(\mu_1 + \mu_2) = \mu_1 h$$

$$\Rightarrow \quad y = \frac{\mu_1 h}{\mu_1 + \mu_2} \text{ (Ans.)}$$

(ii) The pull required to drag the plate = Total shear force

$$\Rightarrow \quad F = F_1 + F_2$$

$$\Rightarrow \quad F = \tau_1 A + \tau_2 A$$

$$\Rightarrow \quad F = \mu_1 \times \frac{V}{y} \times A + \mu_2 \times \frac{V}{h-y} \times A = \left[\frac{\mu_1}{y} + \frac{\mu_2}{h-y}\right] VA$$

For F to be minimum, $\qquad \frac{dF}{dy} = 0$

$$\Rightarrow \quad -\frac{\mu_1}{y^2} + \frac{\mu_2}{(h-y)^2} = 0$$

$$\Rightarrow \quad \frac{y^2}{(h-y)^2} = \frac{\mu_1}{\mu_2}$$

$$\Rightarrow \quad \frac{y}{h-y} = \frac{\sqrt{\mu_1}}{\sqrt{\mu_2}}$$

$$\Rightarrow \quad \left(\sqrt{\mu_1} + \sqrt{\mu_2}\right) y = \sqrt{\mu_1} h$$

$$\Rightarrow \quad y = \frac{\sqrt{\mu_1} h}{\sqrt{\mu_1} + \sqrt{\mu_2}} \text{ (Ans.)}$$

1.4 The velocity distribution for flow over a plate is given by

$$u = 2y - y^2$$

in which u is the velocity in ms^{-1} at a distance y metres from the plate. Determine the shear stress in Nm^{-2} at the boundary and at 0.2 m from it. Dynamic viscosity of fluid is 0.9 Ns/m^2.

[4 marks : 2013]

Solution:

Given $\qquad u = 2y - y^2$

and $\qquad \mu$ = Dynamic viscosity of fluid = 0.9 Ns/m^2

$$\text{Shear stress } (\tau) = \mu \frac{\partial u}{\partial y} = \mu(2 - 2y)$$

$$\therefore \quad \tau|_{y=0.2m} = 0.9(2 - 2 \times 0.2) = 1.44 \text{ N/m}^2$$

and $\qquad \tau_{1_{y=0}} = 0.9 \times 2 = 1.8 \text{ N/mm}^2$

1.5 A rectangular plate of 0.50 m × 0.50 m dimensions weighing 500 N slides down an inclined plane making 30° angle with the horizontal, at a velocity of 1.75 m/s. If the 2 mm gap between the plate and the inclined surface is filled with a lubricating oil, find its viscosity and express it in poise as well as in Ns/m^2.

[4 marks : 2014]

Solution:

Area of plate, $A = 0.50 \times 0.50 = 0.25\,\text{m}^2$

Weight of plate, $W = 500\,\text{N}$

$$W\sin\theta = F_{drag}$$

$$500\sin 30^\circ = \tau \cdot A$$

$$\Rightarrow \quad \mu\frac{du}{dy}A = 500\sin 30^\circ$$

$$\Rightarrow \quad \mu\frac{(V-0)}{2\times10^{-3}}\times 0.25 = 500\sin 30^\circ$$

$$\mu = \frac{500\sin 30^\circ \times 2\times10^{-3}}{1.75\times 0.25} = 1.143\ \text{N-s/m}^2$$

Since; $1\ \text{Poise} = 10^{-1}\ \text{N-s/m}^2$

$$\Rightarrow \quad 1\frac{\text{N-s}}{\text{m}^2} = 10\ \text{Poise}$$

$\therefore$ $\mu = 11.43$ poise or $1.143\ \text{N-s/m}^2$

1.6 A rectangular plate of 0.5 m × 0.5 m dimensions, weighing 500 N slides down an inclined plane making 30° angle with the horizontal at a velocity of 1.75 m/s. If the 2 mm gap between the plate and inclined surface is filled with a lubricating oil, find its viscosity in poise.

[6 marks : 2020]

Solution:

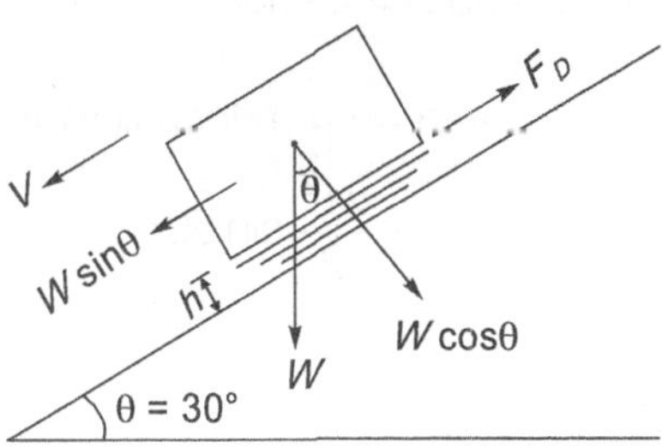

Force analysis in direction of motion

$$F_D = W\sin\theta$$

$$\tau A = 500\sin 30^\circ \qquad \text{...(i)}$$

$\because$ Shear stress, $\tau = \mu\dfrac{du}{dy}$

{Since the gap is very-very small so velocity variation is considered as linear.}

$$\tau = \mu\frac{V-0}{h}$$

$$\tau = \mu\frac{V}{h}$$

By eq. (i) $\mu \frac{V}{h} A = 500 \sin 30°$

$$\mu \frac{(1.75)}{0.002} \times 0.5 \times 0.5 = 500 \sin 30°$$

$$\mu = 1.143 \text{ Ns/m}^2$$

$$\mu = 11.43 \text{ Poise}$$

2. Manometry and Hydrostatic Forces

2.1 **Find the depth of water required to topple the rectangular flashboard and reaction at the hinge of the flashboard shown in figure.**

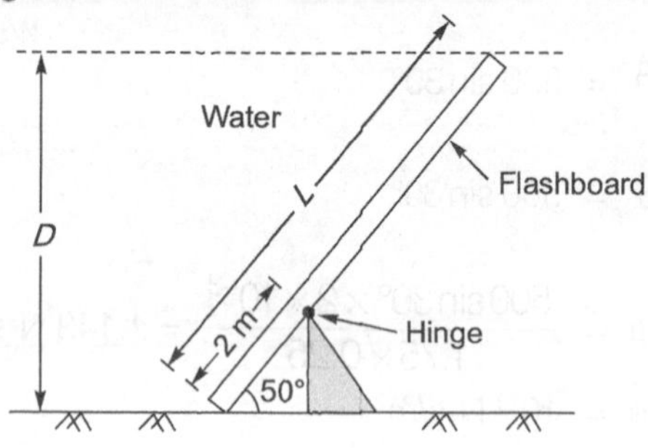

[10 marks : 2006]

Solution:

Let the centre of gravity of the flashboard be at a distance $\bar{x}$ from the free surface.

Assuming unit width of the flashboard perpendicular to the plane of paper

Hydrostatic force on the flashboard is given by

$$F = wA\bar{x} = w(L \times 1) \times \bar{x} = wL\bar{x}$$

$$\sin 50° = \frac{D}{L} = \frac{\bar{x}}{L/2}$$

$$\therefore \quad \bar{x} = \frac{L}{2} \sin 50° \text{ and } D = L \sin 50°$$

$$\therefore \quad F = wL\bar{x} \quad [\because \text{Area} = L \times 1, \text{ for unit width}]$$

$$\Rightarrow \quad F = wL \times \frac{L}{2} \sin 50° = \frac{wL^2}{2} \sin 50°$$

The hydrostatic force F will act at the centre of pressure $(\bar{h})$.

$$\therefore \quad \bar{h}_{cp} = \bar{x} + \frac{I_G}{A\bar{x}} \sin^2\theta$$

$$\bar{h}_{cp} = \frac{L}{2} \sin 50° + \frac{1 \times L^3 \times 2}{12 \times (L \times 1) \times L \sin 50°} \sin^2 50° \qquad \left[\because I_G = \frac{bd^3}{12} = \frac{1 \times L^3}{12}\right]$$

$$= \frac{L}{2} \sin 50° + \frac{L}{6} \sin 50° = \frac{2}{3} L \sin 50°$$

Now, we have

$$\sin 50° = \frac{\bar{h}_{cp}}{y}$$

$$\Rightarrow \quad y = \frac{\frac{2}{3} L \sin 50°}{\sin 50°} = \frac{2}{3} L$$

Thus the perpendicular distance of the line of action of the hydrostatic force F from the hinge is given by

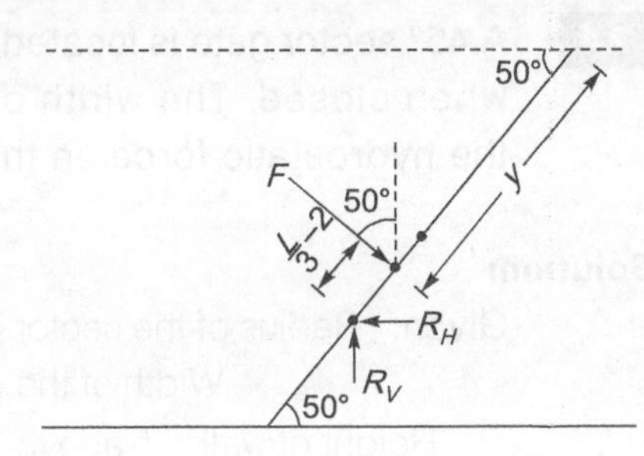

$$\text{Lever Arm} = L - \frac{2}{3}L - 2 = \frac{L}{3} - 2$$

Taking the moment of all the forces about the hinge, we get

$$F\left(\frac{L}{3} - 2\right) = 0$$

$$\Rightarrow \quad \frac{wL^2}{2}\sin 50°\left(\frac{L}{3} - 2\right) = 0$$

$$\Rightarrow \quad L = 6 \text{ m}$$

$$\therefore \quad D = L \sin 50° = 6 \sin 50° = 4.6 \text{ m}$$

Now for equilibrium,

$$R_H = F\sin 50° = \frac{wL^2}{2}\sin 50° \times \sin 50° = \frac{9810 \times (6)^2}{2} \times \sin^2 50° = 103.62 \text{ kN}$$

$$R_V = F\cos 50° = \frac{wL^2}{2}\sin 50° \times \cos 50° = \frac{9810 \times (6)^2}{2} \times \sin 50° \times \cos 50° = 86.95 \text{ kN}$$

Resultant Reaction, $R = \sqrt{R_H^2 + R_V^2} = \sqrt{(103.62)^2 + (86.95)^2} = 135.27 \text{ kN}$

2.2 **Determine the total pressure on a plane rectangular plate 1 m wide and 3 m deep when its upper edge is horizontal and coincides with water surface and plate is held perpendicular to water surface.**

[2 marks : 2010]

Solution:

Let the width and depth of the rectangular plate be b and d respectively.

Total pressure on the rectangular plate will be given as

$$P = \gamma A\bar{x}$$

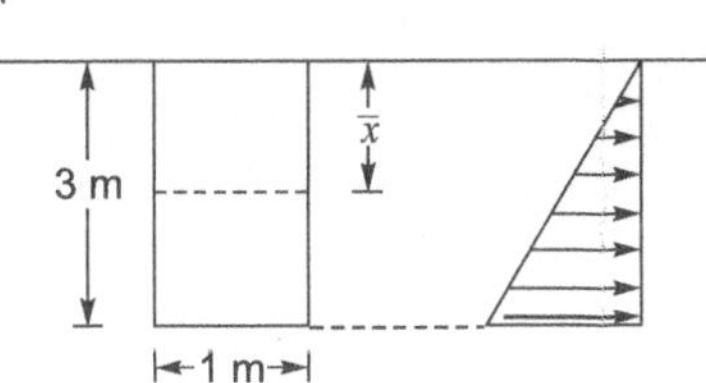

where γ is the unit weight of water, A is the area of rectangular plane surface and $\bar{x}$ is the distance of centre of gravity from water surface.

$$\therefore \quad P = \gamma \times b \times d \times \bar{x}$$

$$= 9810 \times 1 \times 3 \times \frac{3}{2} \qquad \left[\because \bar{x} = \frac{d}{2}\right]$$

$$= 44145 \text{ N} = 44.145 \text{ kN}$$

2.3 **Show that the hydraulic pressure remains invariant in a horizontal plane parallel to free surface.**

[4 marks : 2010]

Solution:

Consider an element of area dA, is y height below the free surface level, in a fluid of density ρ, hence for equilibrium

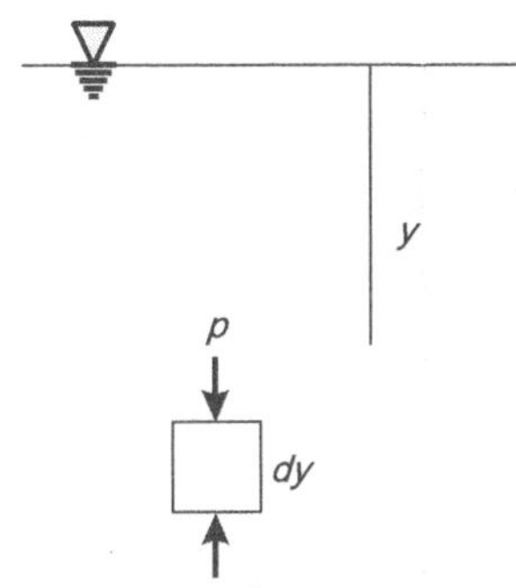

pdA + Weight of liquid in a volume of $dA \cdot dy = (p + dp)dA$

$$pdA + \rho d(dAdy) = (p + dp)dA$$

$$pg \cdot dAdy = dp \cdot dA$$

$$\frac{dp}{dy} = \rho g$$

$$p = \rho g y + \text{constant}$$

$$p \propto y$$

Hydrostatic pressure $\propto$ Depth

Hence, hydrostatic pressure varies only in vertical direction. Hence at a particular depth below the free surface hydrostatic pressure will remain same in a horizontal plane.

2.4 **A 45° sector gate is located on the crest of spillway. The water is upto the mid-point of the gate when closed. The width of the gate is 10 m. The radius of the sector gate is 2 m. Determine the hydrostatic force on the gate. Mass density 1000 kg/m^3, g = 9.79 ms^{-2}.**

[10 marks : 2011]

Solution:

Given: Radius of the sector gate = R = 2 m

Width of the gate = L = 10 m

Height of water, h above the bottom tip of the gate

$$\Rightarrow \quad \sin\theta = \frac{h}{R}$$

$$\Rightarrow \quad h = R\sin\theta = 2\sin 22.5 = 0.765 \text{ m}$$

Hydrostatic force, $P = \sqrt{P_H^2 + P_V^2}$ ...(i)

$$P_H = \gamma_w A\bar{x}$$

Here A is area of vertical projection of the gate and $\bar{x}$ is the C.G. of the vertical projection from top.

$$P_H = \gamma_w \cdot (h \times L) \times \frac{h}{2} = (9.79 \times 1000)(0.765 \times 10) \times \frac{0.765}{2}$$

$$= 28646.76 \text{ N} = 28.65 \text{ kN}$$

P_v = Vertical component of the water pressure

= Weight of imaginary volume of water ABC.

Area ABC = Area of sector AOC – Area of triangle AOB

$$= \frac{\pi \times 2^2}{360} \times 22.5 - \frac{1}{2} \times 0.765 \times \frac{0.765}{\tan 22.5}$$

$$= 0.0794 \text{ m}^2$$

$$\therefore \quad P_v = (0.0794 \times 10 \times 9.79 \times 1000)$$

$$= 7773.26 \text{ N} = 7.773 \text{ kN}$$

From (i) Hydrostatic Force,

$$P = \sqrt{28.65^2 + 7.773^2} = 29.69 \text{ kN}$$

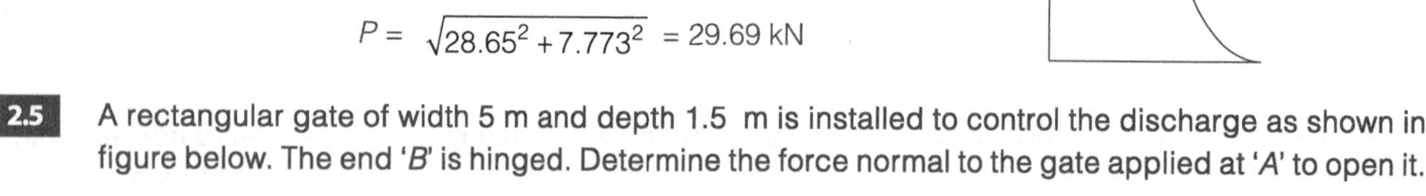

2.5 **A rectangular gate of width 5 m and depth 1.5 m is installed to control the discharge as shown in figure below. The end 'B' is hinged. Determine the force normal to the gate applied at 'A' to open it.**

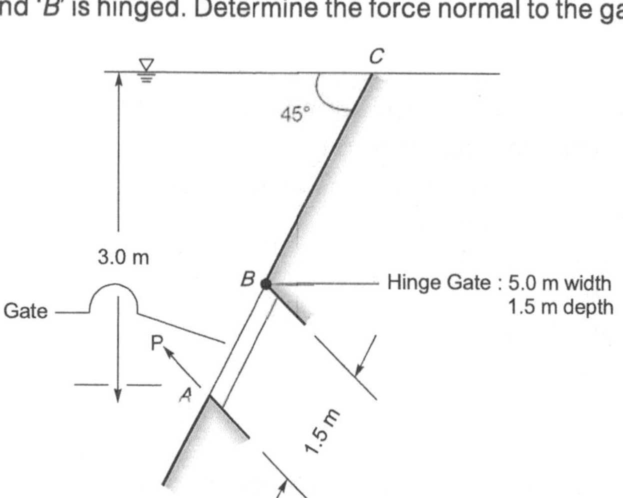

[6 marks : 2012]

Solution:

Given: A = Area of gate = 1.5 × 5 = 7.5 m^2

Depth of C.G. of gate from free surface of water = $\bar{h}$

$$= DG = AC - AE$$
$$= 3 - AG \sin 45°$$
$$= 3 - 0.75 \times \frac{1}{\sqrt{2}}$$
$$= 2.4697 \text{ m}$$

The total pressure force (F) acting on the gate,

$$F = \rho g A\bar{h} = 1000 \times 9.81 \times 7.5 \times 2.4697 = 181708.18 \text{ N} = 181.71 \text{ kN}$$

This force is acting at point H where depth of H from free surface is given by

$$h^* = \frac{I_G \sin^2\theta}{A\bar{h}} + \bar{h}$$

where, $$I_G = \text{M.O.I of gate} = \frac{bd^3}{12} = \frac{5 \times 1.5^3}{12} = 1.40625 \text{m}^4$$

∴ **Depth of centre of pressure** $$h^* = \frac{1.40625 \times \sin^2 45°}{7.5 \times 2.4697} + 2.4697$$
$$= 0.03796 + 2.4697 = 2.508 \text{ m}$$

As from figure $$\sin 45° = \frac{h^*}{OH}$$

∴ Distance, $$OH = \frac{h^*}{\sin 45°} = \frac{2.508}{\frac{1}{\sqrt{2}}} = 3.547 \text{m}$$

Distance, $$AO = \frac{3}{\sin 45°} = 4.243 \text{m}$$

∴ Distance, $$AH = AO - OH = 4.243 - 3.547 = 0.696 \text{ m}$$

∴ Distance $$BH = AB - AH = 1.5 - 0.696 = 0.804 \text{ m}$$

Taking the moments about the hinge B

$$P \times AB = F \times (BH)$$

∴ $$P \times 1.5 = 181708.18 \times 0.804$$

⇒ $$P = \frac{181708.18 \times 0.804}{1.5} = 97395.585 \text{ N} = 97.396 \text{ kN}$$

2.6 The gates of lock are 5 cm wide by 6 m and when closed, at an angle of 120°. Each gate is held on by two hinges placed at the top and bottom of the gate. If the water levels are 6 m and 4.5 m on the upstream and downstream sides respectively, determine the magnitude of the forces on the hinges due to the water pressure.

[20 marks : 2013]

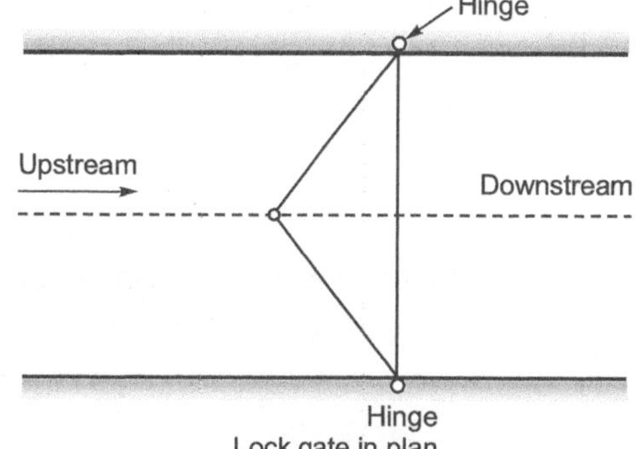

Lock gate in plan

Solution:

Width of the gates of lock = 5 cm

Although 5 cm width is practically not possible, it may be due to printing error in exam. Assuming data given is correct.

$\therefore$ Width, $W = 5$ cm

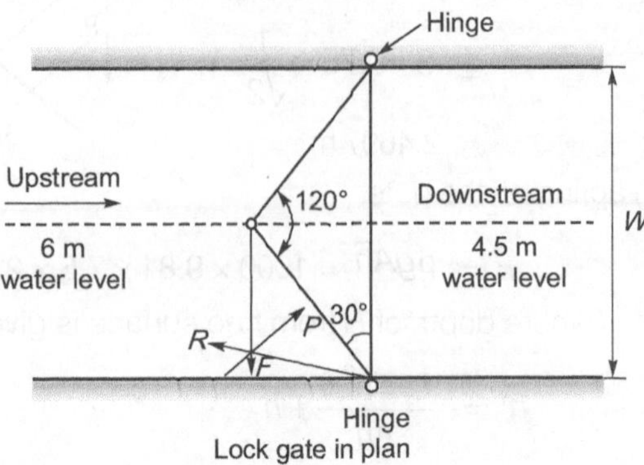

Fig. (i)

$$\therefore \quad \text{Width of each gate} = \frac{5\times 10^{-2}}{2\cos 30^\circ} = 0.02887 \text{ m}$$

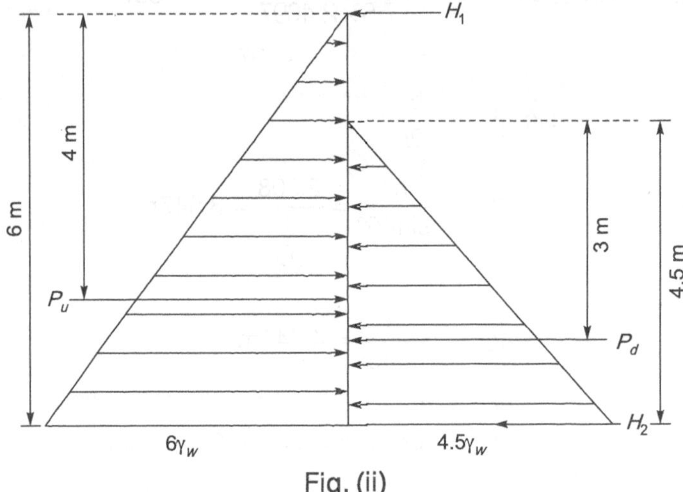

Fig. (ii)

Total pressure on the upstream face of the gate is

$$P_u = \rho g\bar{h}A = 1000\times 9.81\times\frac{6}{2}\times[6\times 0.02887]$$

$$= 5097.8646 \text{ N} = 5.0978 \text{ kN}$$

The depth of the centre of pressure on the upstream face is given by

$$\bar{h}_u = \bar{h}+\frac{I_{CG}}{A\bar{h}} = 3+\frac{\frac{1}{12}\times 0.02887\times 6^3}{(0.02887\times 6)\times 3} = 4 \text{ m}$$

Total pressure on the downstream face of the gate is

$$P_d = \rho g\bar{h}A = 1000\times 9.81\times\frac{4.5}{2}\times(0.02887\times 4.5)$$

$$= 2867.5488 \text{ N} = 2.86755 \text{ kN}$$

Depth of centre of pressure on downstream face is

$$\bar{h}_d = \bar{h} + \frac{I_{CG}}{A\bar{h}} = 2.25 + \frac{\frac{1}{12} \times 0.02887 \times 4.5^3}{(0.02887 \times 4.5) \times 2.25} = 3 \text{ m}$$

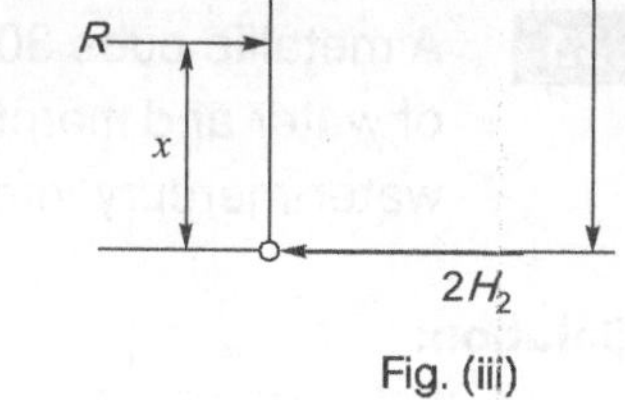

Fig. (iii)

Resultant water pressure on each gate

$$P = P_u - P_d = 5.0978 - 2.8675 = 2.2303 \text{ kN}$$

If x is the height of the point of application ofthe resultant water pressure on the gate, then

$$P \times x = P_u \times (6 - \bar{h}_u) - P_d \times (4.5 - \bar{h}_d)$$

$$\therefore \quad 2.2303\, x = 5.0978 \times (6 - 4) - 2.8675\,(4.5 - 3)$$

$$\therefore \quad x = 2.643 \text{ m}$$

Consider Free Body Diagram from fig. (i), we get

$$F = R$$

Also, $\quad F \sin 30 + R \sin 30 = P$

$$\therefore \quad R = \frac{P}{2 \sin 30} = P = 2.2303 \text{ kN}$$

Now, from fig. (iii)

$\Sigma F_x = 0;$ $\quad R = 2H_1 + 2H_2$

$\Sigma M_{H_2} = 0;$ $\quad R \times x = 2H_1 \times 6$

$\therefore$ Force on one top hinge, $H_1 = \dfrac{2.2303 \times 2.643}{6} = 0.4912$ kN

$$\therefore \quad 2H_2 = R - 2H_1 = 2.2303 - 0.4912 = 1.7391 \text{ kN}$$

Force on on one bottom hinge, $\quad H_2 = 0.6239$ kN

2.7 **A 9 m deep tank contains 6 m of water and 3 m of oil of relative density 0.88. Determine the pressure at the bottom of the tank. What is the pressure at the bottom of the tank if the entire tank is filled with water? What is the water thrust in this case? Draw the pressure distribution diagram in both the cases.**

[8 marks : 2015]

Solution:

Case 1: Pressure at top = 0

Pressure at interface $= s_l \gamma_w z_1$

$= 0.88 \times 9.81 \times 3$

$= 25.90$ kN/m^2

Pressure at bottom $= s_l \gamma_w z_1 + \gamma_w z_2$

$= 0.88 \times 9.81 \times 3 + 9.81 \times 6$

$= 84.76$ kN/m^2

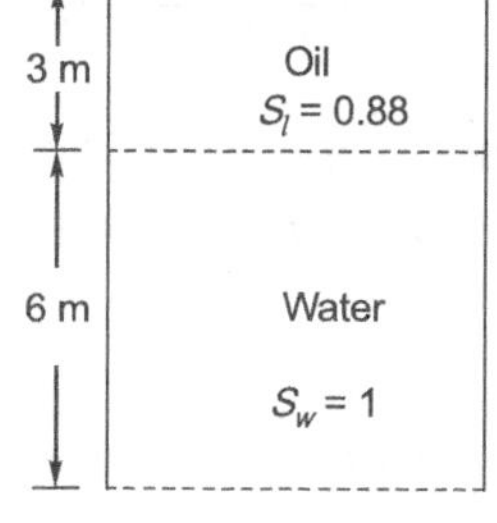

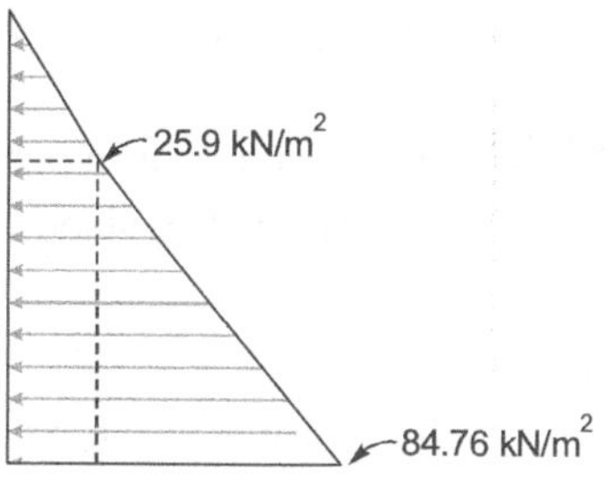

Case 2:

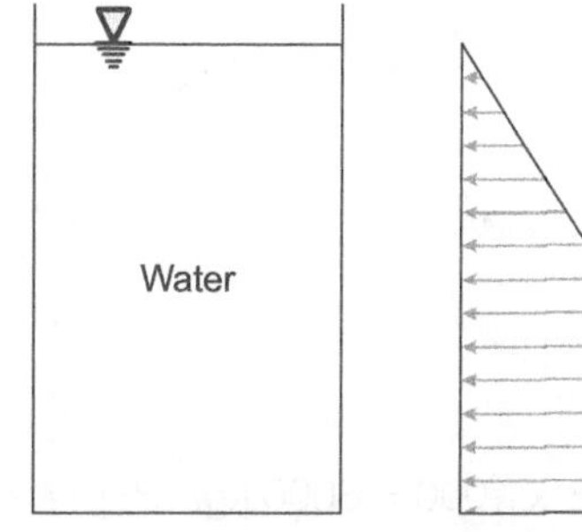

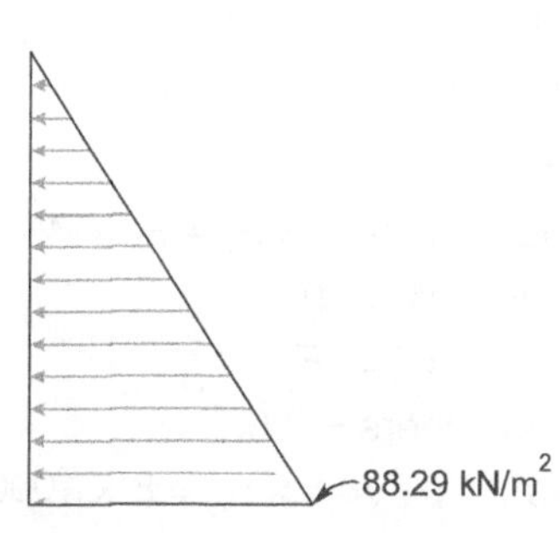

$$\text{Pressure at top} = 0$$

$$\text{Pressure at bottom} = \gamma_w H = 9 \times 9.81 = 88.29 \text{ kN/m}^2$$

$$\text{Water thrust} = \frac{1}{2}(\gamma_w H) \times H = \frac{1}{2} \times 88.29 \times 9 = 397.305 \text{ kN per meter width of wall}$$

3. Buoyancy & Floatation

3.1 **A metallic cube 30 cm side and weighing 450 N is lowered into a tank containing a two fluid layer of water and mercury. Top edge of the cube is at water surface. Determine the position of block at water-mercury interface when it has reached equilibrium.**

[10 marks : 2009]

Solution:

Let the top edge of the cube be at a distance x from the water mercury interface.

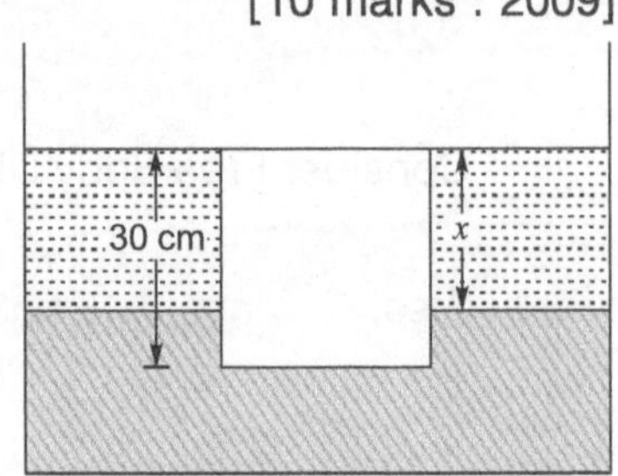

As per Archimedes principle, when a body is immersed in a fluid either wholly or partially, it is lifted up by a force which is equal to the weight of the fluid displaced by the body. Thus, the force of buoyancy is given by

$$F_B = \text{Weight of water and mercury displaced by cube}$$

$$= \text{Volume of cube in water} \times \rho_w g + \text{Volume of cube in mercury} \times \rho_{Hg} g$$

$$= \left(\frac{30}{100} \times \frac{30}{100} \times x\right) \times 1000 \times 9.81 + \frac{30}{100} \times \frac{30}{100} \times \left(\frac{30}{100} - x\right) \times 13600 \times 9.81$$

$$= 0.3 \times 0.3 \times 9810 \times x + 0.3 \times 0.3 \times 13600 \times 9.81(0.3 - x)$$

$$\Rightarrow \quad F_B = 882.9x + 12007.44(0.3 - x) \quad \text{...(i)}$$

As per the principle of floatation the weight of a body floating in a fluid is equal to the buoyant force at equilibrium

$$\therefore \quad F_B = 450 \text{ N} \quad \text{...(ii)}$$

From (i) and (ii), we have

$$450 = 882.9x + 12007.44\,(0.3 - x)$$

$$\Rightarrow \quad 12007.44x - 882.9x = 3602.232 - 450$$

$$\Rightarrow \quad 11124.54x = 3152.232$$

$$\Rightarrow \quad x = \frac{3152.232}{11124.54} = 0.283 \text{ m} = 28.3 \text{ cm}$$

3.2 **A metallic sphere of specific gravity 8.0 falls in an oil of density 800 kg/m^3. The diameter of the sphere is 10 mm. The viscosity of oil is 7.848 N-s/m^2. Determine the terminal velocity of metallic sphere.**

[4 marks : 2010]

Solution:

If a body drops in a fluid, at the instant it has acquired terminal velocity the net force acting on the body will be zero. The forces acting on the body at this state will be

(i) Weight of body (W) acting downward

(ii) Drag force (F_D) acting opposite to the direction of motion of body

(iii) Buoyant force (F_B) acting vertically up.

$\therefore$ Net force on the metallic sphere $= F_B + F_D - W$

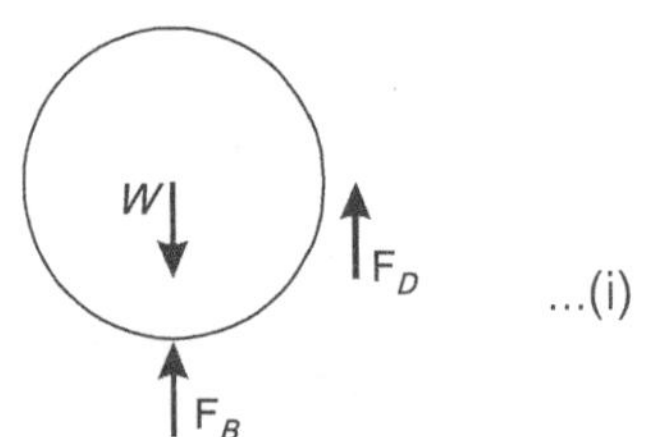

But $\quad F_B + F_D - W = 0$

$$\Rightarrow \quad W = F_B + F_D \quad \text{...(i)}$$

Specific gravity of metallic sphere = 8.0

Density of metallic sphere, $\rho_s = 8 \times \rho_w = 8 \times 1000 = 8000 \text{ kg/m}^3$

Diameter of sphere, $D = 10 \text{ mm} = 10 \times 10^{-3}$ m

Viscosity of oil, $\mu = 7.848$ N-s/m^2

$$\therefore \quad W = \rho_s \times g \times \text{volume of sphere}$$

$$= 8000 \times 9.81 \times \frac{\pi}{6} \times (10 \times 10^{-3})^3 = 0.0411 \text{ N}$$

and

$$F_B = \rho_o \times g \times \text{volume of sphere}$$

$$= 800 \times 9.81 \times \frac{\pi}{6} \times (10 \times 10^{-3})^3 = 0.00411 \text{ N}$$

Assuming Stoke's law to be valid in this case, we get

$$F_D = 3\pi\mu VD \qquad \text{where } V \text{ is the terminal velocity}$$

From (i), we get

$$0.0411 = 0.00411 + 3\pi\mu VD$$

$$\Rightarrow \quad 3\pi\mu VD = 0.0411 - 0.00411$$

$$\Rightarrow \quad V = \frac{0.03699}{3\pi \times 7.848 \times 10 \times 10^{-3}} = 0.05 \text{ m/s}$$

But Stoke's law is valid only upto Reynolds number less than 0.2

$$\therefore \quad \text{Re} = \frac{\rho_o VD}{\mu} = \frac{800 \times 0.05 \times 10 \times 10^{-3}}{7.848} = 0.051 < 0.2 \text{ (ok)}$$

Thus the terminal velocity of the sphere is 0.05 m/s

3.3 **A vertical gap 23.5 mm wide of infinite extent contains oil of specific gravity 0.9 and viscosity 2.5 N-s/m^2. A metal plate 1.5 m × 1.5 m × 1.5 mm weighing 50 N is to be lifted through the gap at a constant speed of 0.1 m/sec. Estimate the force required to lift the plate.**

[6 marks : 2012]

Solution:

Width of gap = 23.5 mm

Viscosity, μ = 2.5 Ns/m^2

Specific gravity oil = 0.9

$\therefore$ Weight density of oil = 0.9 × 1000 = 900 kgf/m^3

= 900 × 9.81 N/m^3 ($\because$ 1 kgf = 9.81 N)

Assuming that the plate lies in the middle of the gap

Volume of plate = 1.5 m × 1.5 m × 1.5 mm

= 1.5 × 1.5 × 0.0015 m^3

= 0.003375 m^3

Thickness of plate = 1.5 mm

Velocity of plate = 0.1 m/sec

Weight of plate = 50 N

When the plate is in the middle of the gap, the distance of plate from vertical surface of the gap

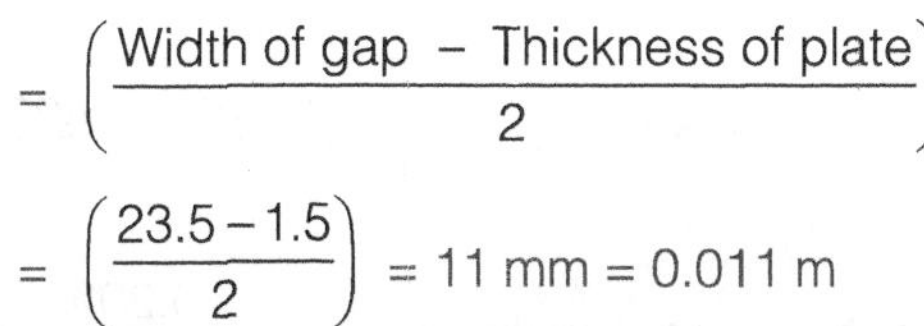

$$= \left(\frac{\text{Width of gap} - \text{Thickness of plate}}{2}\right)$$

$$= \left(\frac{23.5 - 1.5}{2}\right) = 11 \text{ mm} = 0.011 \text{ m}$$

Now, shear force on left side of the metallic plate

$$F_1 = \text{Shear stress} \times \text{Area}$$

$$= \mu\left(\frac{du}{dy}\right)_1 \times (1.5\times1.5) = 2.5\times\left(\frac{0.1}{0.011}\right)\times1.5\times1.5 = 51.136 \text{ N}$$

Similarly, the shear force on the right side of the metallic plate

$$F_2 = \text{Shear stress} \times \text{Area}$$

$$= 2.5\times\left(\frac{0.1}{0.011}\right)\times(1.5\times1.5) = 51.136 \text{ N}$$

∴ **Total shear force** $= F_1 + F_2 = 51.136 + 51.136 = 102.272$ N

In this case the weight of plate (which is acting downward) and upward thrust is also to be taken into account.

∴ The upward thrust = weight of fluid displaced = ρvg

= (unit weight of fluid) × Volume of fluid displaced

= 9.81 × 900 × 0.003375

= 29.80 N

The net force acting in the downward direction due to the weight of the plate and upward thrust

= weight of plate – upward thrust = 50 – 29.80 = 20.20 N

∴ Total force required to lift the plate up

= Total shear force + 20.20 = 102.272 + 20.20 = 122.472 N

3.4 **The weights of a cube (side = 1.2 m) and a sphere (diameter = 1.25 m) are 20 kN and 5 kN respectively. Both cube and sphere are connected together by a short rope in a water reservoir. Computer the tension in the rope and percentage of sphere that will be above water surface.**

[8 marks : 2016]

Solution:

Let subscript 's' denotes sphere and subscript 'c' denotes cube and T be tension in the rope.

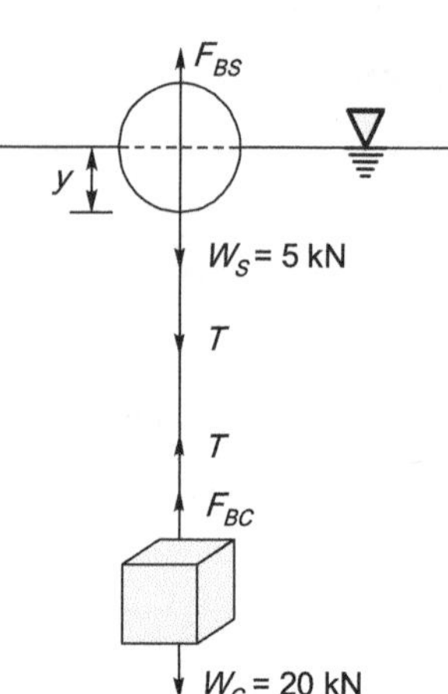

y = Depth of immersion of sphere

Given: Diameter of sphere (d_s) = 1.25 m

Side of cube (a_c) = 1.2 m

For cube, $T = W_C - F_{BC}$

$= 20 - (a_c)^3 9.81$

$= 20 - (1.2)^3 9.81$

∴ Tension in the rope, $T = 3.048$ kN

For sphere $T + W_S = F_{BS}$

⇒ $3.048 \times 10^3 + 5 \times 10^3 = 1000 \times V_{immersed} \times 9.81$

⇒ $V_{immersed} = 0.8204 \text{ m}^3$ = Immersed volume sphere in water.

∵ Volume of sphere in air $= \left[\frac{\pi}{6}(1.25)^3 - 0.8204\right]\text{m}^3 = 0.2023 \text{ m}^3$

∴ Percentage of sphere above water surface $= \dfrac{0.2023}{\frac{\pi}{6}(1.25)^3} \times 100 = 19.78\%$

3.5 A cylindrical gate of 3.0 m dia. and having 4.0 m length is made of 60 mm thick steel sheet of 7.8 ton/m^3 density. The gate has water on both its sides.

(i) Water level is up to 2.0 m depth from bottom of gate on the left side and up to 1.5 m depth on the right side of the gate. Determine the magnitude, location and direction of the resultant hydrostatic force. Also determine whether the gate will float in this case or not.

(ii) If the water level rises up to 3.0 m on the left of the gate (up to the top of the gate), determine whether the gate will float or not. If yes, what should be the thickness of the gate sheet to prevent it from floating?

[20 Marks : 2021]

Solution:

(i) Length, $L = 4$ m

Diameter, $D = 3$ m

$t = 60$ mm

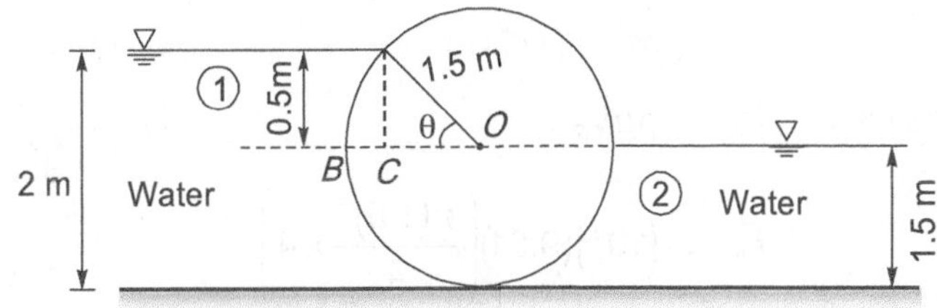

For horizontal force,

Due to lift side water, $F_{H_1} = \rho_w . g\bar{h}_1 A_{V_1}$

$$= (10^3)(9.81)\left(\frac{2}{2}\right)(2 \times 4)$$

$$= 78.48 \text{ kN } (\rightarrow)$$

Due to right side water, $F_{H_2} = \rho_w g\bar{h}_2 A_{V_2}$

$$= (10^3)(9.81)\left(\frac{1.5}{2}\right)(1.5 \times 4)$$

$$= 44.145 \text{ kN } (\leftarrow)$$

Net horizontal force, $F_H = F_{H_1} - F_{H_2}$

$$= 34.335 \text{ kN } (\rightarrow)$$

For vertical force,

Due to left side water, $F_{V_1} = \rho g \forall_1$

$\forall_1$ = (Area of sector OBD + Area of sector OBA + Area of ΔOAE) × 4

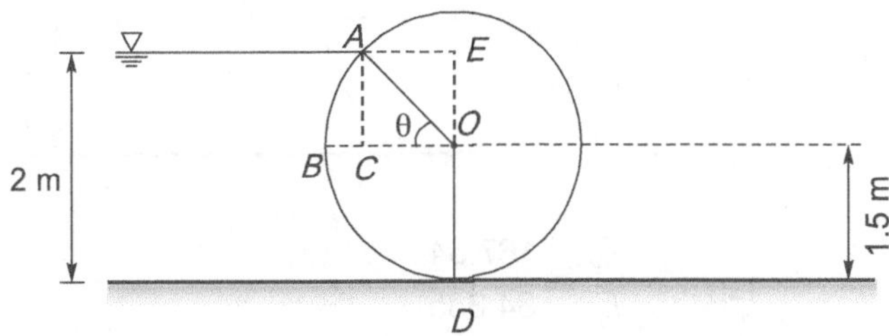

$$AE = \sqrt{1.5^2 - 0.5^2} = 1.414 \text{ m}$$

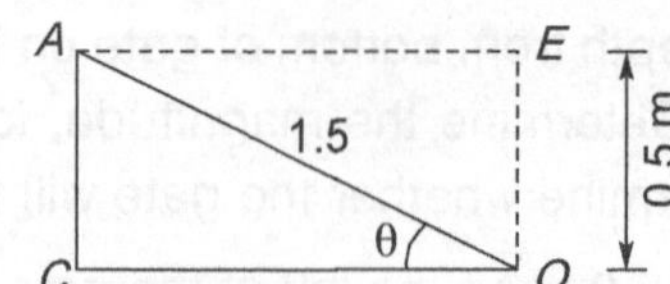

$$\sin\theta = \frac{AC}{OA} = \frac{0.5}{1.5}$$

$$\theta = 19.47$$

$$\forall_1 = \left[\frac{\pi(1.5)^2}{4} + \frac{19.47°}{360°} \times \pi(1.5)^2 + \left(\frac{1}{2} \times 1.414 \times 0.5\right)\right]4$$

$$= 10.01 \text{ m}^3$$

When, $F_{V_1} = (10^3)(9.81)(10.01)$

$$= 98.2 \text{ kN } (\uparrow)$$

Due to right side water, $F_{V_2} = \rho g \forall_2$

$$F_{V_2} = (10^3)(9.81)\left(\frac{\pi(1.5)^2}{4} \times 4\right)$$

$$= 69.34 \text{ kN } (\uparrow)$$

$\bar{x}_2 = F_{v2}$ act at $\left(\frac{4 \times 1.5}{3\pi}\right)$ m from DE i.e. 0.64 m.

Net vertical force due to water on gate

$$F_V = F_{V_1} + F_{V_2}$$

$$= 98.2 + 69.34$$

$$= 167.54 \text{ kN } (\uparrow)$$

Resultant force due to water on gate

$$F_R = \sqrt{{F_H}^2 + {F_V}^2}$$

$$= \sqrt{34.335^2 + 167.54^2}$$

$$= 171.02 \text{ kN}$$

Direction of resultant force

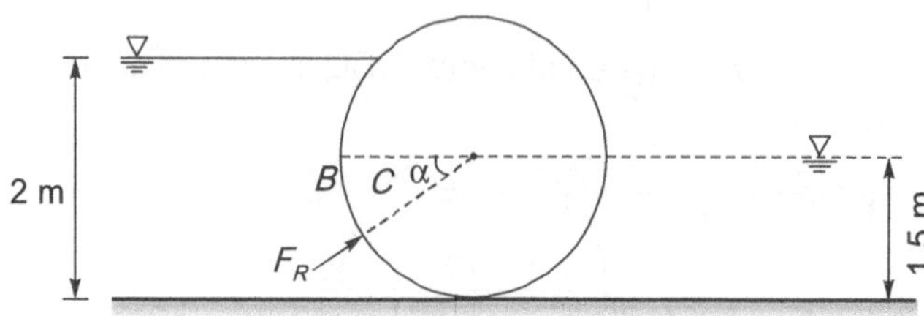

$$\tan\alpha = \frac{F_V}{F_H} = \frac{167.54}{34.335}$$

$$\alpha = 78.418°$$

Resultant force due to water acts normal to the cylindrical surface and passes through the centre O.

$$\text{Weight of gate} = \rho_s \forall g$$
$$= \rho_s(\pi DLt)g$$
$$= (7.8\times10^3)\pi(3)(4)(0.06)(9.81)$$
$$= 173.08 \text{ kN}$$

Since weight of gate is higher than net vertical force due to water on gate so gate will not float.

(ii)

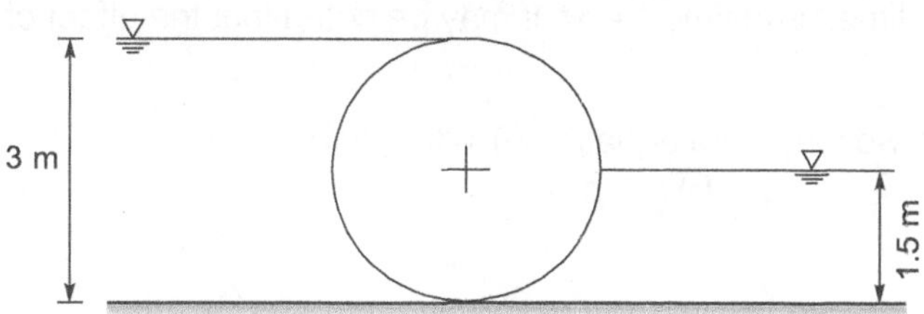

Net vertical force due to water on gate

$$F_V = \rho g\left(\frac{3}{4}\pi R^2\right)\times 4$$
$$= (10^3)(9.81)\left(\frac{3}{4}\pi(1.5)^2\times4\right)$$
$$= 208.03 \text{ kN } (\uparrow)$$

Weight of cylindrical gate = 173.08 kN

Since F_V is greater than weight of cylindrical gate so gate will float.

Thickness of gate sheet to prevent floating

$$F_V = \text{Weight of gate}$$
$$10^3\times208.03 = (7.8)(\pi DLt)\times9.81\times10^3$$
$$208.03 = 7.8\times\pi(3)(4)t(9.81)$$
$$t = 0.072116 \text{ m}$$
$$= 72.12 \text{ mm}$$

4. Fluid Kinematics

4.1 A small rocket with an initial weight of 3600 N is to be launched vertically. It consumes 5 kg/sec of fuel and exhausts it at atmospheric pressure with a velocity of 1100 m/s relative to the rocket. What is the initial acceleration of the rocket and what is its velocity 20 seconds after launch?

[10 marks : 2002]

Solution:

As per **Newton's second law of motion**, the resultant of all the external forces acting on a body is equal to the rate of change of linear momentum of the body i.e.

$$F = \frac{dp}{dt}$$
$$\Rightarrow \quad Fdt = dp$$
$$\Rightarrow \quad \int Fdt = p_2 - p_1$$

[Integrating both sides]

$$\Rightarrow \qquad p_1 + \int F dt = p_2 \qquad ...(i)$$

The integral is a vector known as linear impulse of the force F during the time interval considered. The equation expresses that, when a body is acted upon by a force F during a given time interval, the final momentum p_2 of the body may be obtained by adding its initial momentum p_1, and the impulse of the force F during the interval of time

Let the rocket has a initial mass M and it is launched vertically at time $t = 0$. The fuel is consumed at a constant rate 'q' and is expelled at a constant speed V_e relative to the rocket. At time t, the mass of the rocket and remaining fuel is $M - qt$ and the velocity 'v'. During the time interval Δt, a mass of fuel $q\Delta t$ is expelled. Denoting by 'u', the absolute velocity of expelled fuel, we apply the principle of impulse and momentum between time t and time $t + \Delta t$. It may be noted that the effect of air resistance is neglected here.

For a time interval Δt, we may write equation (i) in the form

$$p_1 + F\Delta t = p_2 \qquad ...(ii)$$

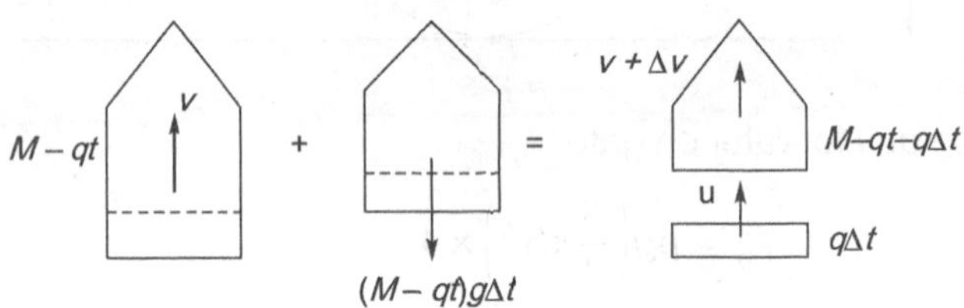

From (ii), we get

$$(M - qt)v - (M - qt)g\Delta t = [(M - qt - q\Delta t)(v + \Delta v) + q\Delta t u]$$

$$\Rightarrow \quad (M - qt)\, v - (M - qt)g\, \Delta t = (M - qt)v + (M - qt)\Delta v - q\Delta t(v + \Delta v) + q\Delta tu$$

$$\Rightarrow \quad -(M - qt)g\, \Delta t = (M - qt)\Delta v + q\Delta t[u - (v + \Delta v)]$$

Dividing both sides by Δt, we get

$$\Rightarrow \quad -(M - qt)g = (M - qt)\frac{\Delta v}{\Delta t} + q[u - (v + \Delta v)]$$

Replacing $u-(v + \Delta v)$ with V_e, the velocity of the expelled fuel relative t_0 the rocket, we get

$$-(M - qt)g = (M - qt)\frac{\Delta v}{\Delta t} - qV_e$$

As Δt approaches zero, we get

$$-(M - qt)g = (M - qt)\frac{dv}{dt} - qV_e$$

Dividing both sides by $M-qt$, we get

$$\frac{dv}{dt} = \frac{qV_e}{M - qt} - g \qquad ...(iii)$$

But

$$\frac{dv}{dt} = a$$

$$\therefore \quad a = \frac{qV_e}{M - qt} - g$$

Integrating both sides of equation (iii), we get from $v = 0$ at $t = 0$ to $v = v$ at $t = t$

$$\therefore \quad \int_o^v dv = \int_o^t \frac{qV_e}{M - qt} dt - \int_o^t g dt$$

$$\Rightarrow \quad v = \frac{qV_e}{-q} \times [\log_e(M - qt)]_0^t - g \times [t]_o^t$$

$$\Rightarrow \quad v = -V_e \times [\log_e(M - qt) - \log_e M] - gt$$

$$\Rightarrow \quad v = V_e \log_e\left(\frac{M}{M - qt}\right) - gt$$

∴ Initial acceleration of the rocket i.e. at $t = 0$

$$a = \frac{qV_e}{M - qt} - g$$

at $t = 0$

$$a = \frac{qV_e}{M - q \times 0} - g \qquad \left[q = 5\text{ kg/s } V_e = 1100\text{ m/s}, M = \frac{3600}{g}\text{kg}\right]$$

$$\Rightarrow \quad a = \frac{5 \times 1100}{3600/9.81} - 9.81 = 15 - 9.81$$

$$\Rightarrow \quad a = 5.19\text{ m/s}^2 \textbf{ (Ans.)}$$

Velocity at any time t is given by

$$v = V_e \log_e\left(\frac{M}{M - qt}\right) - gt$$

at $t = 20$ seconds.

$$v_{20} = 1100 \times \log_e\left(\frac{3600/9.81}{\frac{3600}{9.81} - 5 \times 20}\right) - 9.81 \times 20$$

$$\Rightarrow \quad v_{20} = 1100 \times \log_e\left(\frac{3600}{2619}\right) - 196.2$$

$$\rightarrow \quad v_{20} = 1100 \times 0.3181 - 196.2$$

$$\Rightarrow \quad V_{20} = 350 - 196.2$$

$$\Rightarrow \quad V_{20} = 153.8\text{ m/s} \quad \textbf{(Ans.)}$$

4.2 In a three dimensional flow, the components of velocity are
$u = xy$, $v = 4yz^3$ and $w = -(yz + z^4)$
Test whether the continuity equation for incompressible fluid flow is satisfied.
Determine the acceleration vector at point (1,1,1).

[9 marks : 2007]

Solution:

The continuity equation for a 3–D steady incompressible flow is given by

$$\frac{\partial u}{\partial x} + \frac{\partial v}{\partial y} + \frac{\partial w}{\partial z} = 0$$

Here,

$$u = xy \Rightarrow \frac{\partial u}{\partial x} = y; \frac{\partial u}{\partial y} = x; \frac{\partial u}{\partial z} = 0$$

$$v = 4yz^3 \Rightarrow \frac{\partial v}{\partial x} = 0; \frac{\partial v}{\partial y} = 4z^3; \frac{\partial v}{\partial z} = 12yz^2$$

$$w = -(yz + z^4) \Rightarrow \frac{\partial w}{\partial x} = 0; \frac{\partial w}{\partial y} = -z; \frac{\partial w}{\partial z} = -(y + 4z^3)$$

$$\therefore \quad \frac{\partial u}{\partial x} + \frac{\partial v}{\partial y} + \frac{\partial w}{\partial z} = y + 4z^3 - (y + 4z^3)$$

$$\Rightarrow \quad \frac{\partial u}{\partial x} + \frac{\partial v}{\partial y} + \frac{\partial w}{\partial z} = 0 \quad \text{(Hence proved)}$$

For steady flow, $a_x = u\frac{\partial u}{\partial x} + v\frac{\partial u}{\partial y} + w\frac{\partial u}{\partial z}$

$\Rightarrow \quad a_x = (xy \times y) + (4yz^3 \times x) + [-(yz + z^4) \times 0]$

$\Rightarrow \quad a_x = xy^2 + 4xyz^3$ ...(i)

Also, $a_y = u\frac{\partial v}{\partial x} + v\frac{\partial v}{\partial y} + w\frac{\partial v}{\partial z}$

$\Rightarrow \quad a_y = (xy \times 0) + (4yz^3 \times 4z^3) - (yz + z^4)(12\, yz^2)$

$\Rightarrow \quad a_y = 16\, yz^6 - 12y^2z^3 - 12yz^6$

$\Rightarrow \quad a_y = 4yz^6 - 12y^2z^3$ ...(ii)

and $a_z = u\frac{\partial w}{\partial x} + v\frac{\partial w}{\partial y} + w\frac{\partial w}{\partial z}$

$\Rightarrow \quad a_z = (xy \times 0) + 4yz^3 \times (-z) + (yz + z^4)(y + 4z^3)$

$\Rightarrow \quad a_z = -4yz^4 + y^2z + 4yz^4 + yz^4 + 4z^7$

$\Rightarrow \quad a_z = y^2z + yz^4 + 4z^7$ ...(iii)

We know that acceleration vector $\vec{a}$ is given by

$$\vec{a} = a_x\hat{i} + a_y\hat{j} + a_z\hat{k}$$

$$\Rightarrow \quad \vec{a} = (xy^2 + 4xyz^3)\hat{i} + (4yz^6 - 12y^2z^3)\hat{j} + (y^2z + yz^4 + 4z^7)\hat{k}$$

$\therefore$ Acceleration vector $\vec{a}$ at point (1, 1, 1) will be

$$\vec{a} = (1 \times 1^2 + 4 \times 1 \times 1 \times 1^3)\hat{i} + (4 \times 1 \times 1^6 - 12 \times 1^2 \times 1^3)\hat{j} + (1^2 \times 1 + 1 \times 1^4 + 4 \times 1^7)\hat{k}$$

$$\Rightarrow \quad \vec{a} = 5\hat{i} - 8\hat{j} + 6\hat{k} \text{ (Ans.)}$$

4.3 **The velocity components in x and y directions are given as $u = x + y$ and $v = x - y$ respectively. Find whether these velocity components satisfy possible two dimensional incompressible flow or not?**

[2 marks : 2010]

Solution:

For 2-D steady incompressible flow, the continuity equation should be satisfied i.e.

$$\frac{\partial u}{\partial x} + \frac{\partial v}{\partial y} = 0$$

Here $u = x + y; \quad \frac{\partial u}{\partial x} = 1$

$v = x - y; \quad \frac{\partial v}{\partial y} = -1$

$$\therefore \quad \frac{\partial u}{\partial x} + \frac{\partial v}{\partial y} = 1 + (-1) = 0 \text{ (Hence ok)}$$

Thus the velocity components satisfy the possible two dimensional incompressible flow.

4.4 **A three dimensional flow is given by $\vec{V} = (y^2 + z^2)i + (x^2 + z^2)j + (x^2 + y^2)k$**
Determine the components of acceleration at a point (2, 3, 4).

[4 marks : 2010]

Solution:

In Cartesian co-ordinates, the velocity vector is expressed as

$$\vec{V} = ui + vj + wk$$

$$\therefore \quad u = y^2 + z^2;\ v = x^2 + z^2;\ w = x^2 + y^2$$

At point (2, 3, 4), we have

$$u = 3^2 + 4^2 = 25$$

$$v = 2^2 + 4^2 = 20$$

$$w = 2^2 + 3^2 = 13$$

Let acceleration in x, y, z directions be a_x, a_y and a_z respectively.

$$\therefore \quad a_x = u\frac{\partial u}{\partial x} + v\frac{\partial u}{\partial y} + w\frac{\partial u}{\partial z}$$

$$= 25 \times \frac{\partial}{\partial x}(y^2 + z^2) + 20 \times \frac{\partial}{\partial y}(y^2 + z^2) + 13 \times \frac{\partial}{\partial z}(y^2 + z^2)$$

$$= 25 \times 0 + 20 \times 2y + 13 \times 2z = 20 \times 2 \times 3 + 13 \times 2 \times 4 = 224$$

Also,

$$a_y = u\frac{\partial v}{\partial x} + v\frac{\partial v}{\partial y} + w\frac{\partial v}{\partial z}$$

$$= 25 \times \frac{\partial}{\partial x}(x^2 + z^2) + 20 \times \frac{\partial}{\partial y}(x^2 + z^2) + 13 \times \frac{\partial}{\partial z}(x^2 + z^2)$$

$$= 25 \times 2x + 20 \times 0 + 13 \times 2z = 25 \times 2 \times 2 + 13 \times 2 \times 4 = 204$$

Also,

$$a_z = u\frac{\partial w}{\partial x} + v\frac{\partial w}{\partial y} + w\frac{\partial w}{\partial z}$$

$$= 25 \times \frac{\partial}{\partial x}(x^2 + y^2) + 20 \times \frac{\partial}{\partial y}(x^2 + y^2) + 13 \times \frac{\partial}{\partial z}(x^2 + y^2)$$

$$= 25 \times 2x + 20 \times 2y + 13 \times 0 = 25 \times 2 \times 2 + 20 \times 2 \times 3 = 220$$

4.5 **The x and y components in a three dimensional flow are given by**

$$u = x^2 + z^2;\ v = y^2 + z^2$$

Find the simplest z-component of velocity that satisfies the continuity equation.

[4 marks : 2012]

Solution:

The **continuity equation for incompressible fluid** is given by

$$\frac{\partial u}{\partial x} + \frac{\partial v}{\partial y} + \frac{\partial w}{\partial z} = 0$$

Now,

$$u = x^2 + z^2 \qquad \therefore \frac{\partial u}{\partial x} = 2x$$

$$v = y^2 + z^2 \qquad \therefore \frac{\partial v}{\partial y} = 2y$$

Substituting the value of $\frac{\partial u}{\partial x}$ and $\frac{\partial v}{\partial y}$ in the continuity equation, we have

$$2x + 2y + \frac{\partial w}{\partial z} = 0$$

$$\Rightarrow \quad \frac{\partial w}{\partial z} = -(2x + 2y)$$

Integrating both sides, we get

$$\int \partial w = \int -(2x+2y)\partial z$$

$$\Rightarrow \quad w = -(2xz + 2yz) + \text{constant of integration}$$

where constant of integration cannot be a function of z but it can be a function of x and y or constant i.e., $f(x, y, c)$

∴ **The *z*-component of velocity**

$$w = -(2xz + 2yz) + f(x, y, c)$$

4.6 **If $\phi = \frac{-A}{2\pi}\log r$, where *A* is a positive constant, determine Ψ and plot the typical equipotential lines and stream-lines. Identify the flow pattern.**

[8 marks : 2014]

Solution:

$$\phi = -\frac{A}{2\pi}\log r$$

Above velocity potential function shows radial flow like source or sink flow.

For such flows, the radial and tangential components in terms of ϕ are given by

$$u_r = \frac{\partial \phi}{\partial r}$$

and

$$u_\theta = \frac{1}{r}\frac{\partial \phi}{\partial \theta} = \frac{\partial}{\partial r}\left[-\frac{A}{2\pi}\log r\right]$$

$$\Rightarrow \quad u_r = -\frac{A}{2\pi}\cdot\frac{1}{r} \quad \text{...(i)}$$

Also

$$u_\theta = \frac{1}{r}\frac{\partial}{\partial \theta}\left[-\frac{A}{2\pi}\log r\right] = 0$$

Also the radial and tangential component in tems of ψ are given by

$$u_r = \frac{1}{r}\frac{\partial \psi}{\partial \theta}$$

and

$$u_\theta = -\frac{\partial \psi}{\partial r}$$

$$\therefore \quad \frac{1}{r}\frac{\partial \psi}{\partial \theta} = -\frac{A}{2\pi r} \quad \text{[From (i)]}$$

$$\Rightarrow \quad \frac{\partial \psi}{\partial \theta} = -\frac{A}{2\pi}$$

$$\Rightarrow \quad \partial \psi = -\frac{A}{2\pi}d\theta$$

$$\Rightarrow \quad \psi = -\frac{A}{2\pi}\theta + C_1$$

Let $\psi = 0°$, when $\theta = 0°$,

Then $\quad 0 = 0 + C_1 \quad (\therefore\ C_1 = 0)$

Hence the equation of stream function becomes

$$\psi = -\frac{A}{2\pi}\theta$$

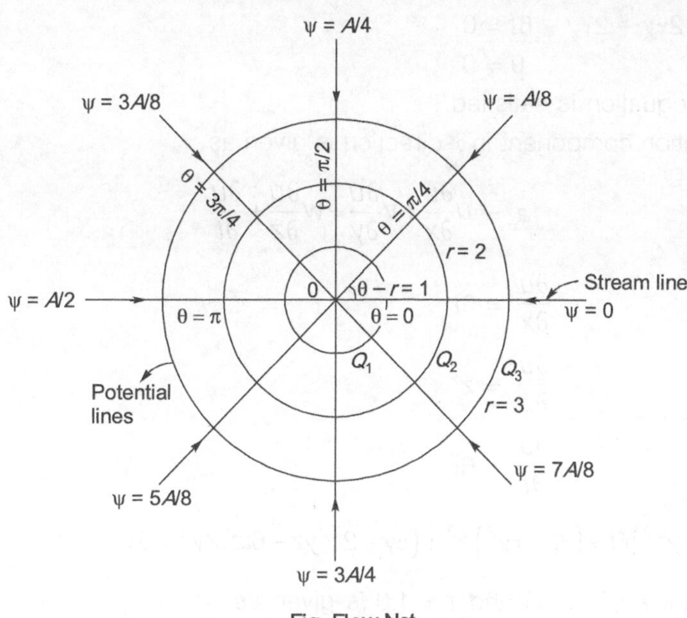

Fig. Flow Net

The flow net shown above represents typical sink flow.

4.7 The velocity vector in an incompressive flow is given by

$$V = \left(6xt + yz^2\right)i + \left(3t + xy^2\right)j + \left(xy - 2xyz - 6tz\right)k$$

(i) Verify whether the continuity equation is satisfied.

(ii) Determine the acceleration in x direction at point $A(1, 1, 1)$ and $t = 1.0$.

[12 marks : 2019]

Solution:

(i) For the given flow field

$$u = 6xt + yz^2$$
$$v - 3t + xy^2$$
$$w = xy - 2xyz - 6tz$$

$$\therefore \quad \frac{\partial u}{\partial x} = 6t$$

$$\frac{\partial v}{\partial y} = 2xy$$

$$\frac{\partial v}{\partial y} = -2xy - 6t$$

For an incompressible flow, continuity equation is given as,

$$\frac{\partial u}{\partial x} + \frac{\partial v}{\partial y} + \frac{\partial w}{\partial z} = 0 \quad ...(i)$$

$\therefore$ Substituting values of $\frac{\partial u}{\partial x}, \frac{\partial v}{\partial y}$ & $\frac{\partial w}{\partial z}$ in equation (i)

$$6t + 2xy - 2xy - 6t = 0$$
$$0 = 0$$

Hence continuity equation is satisfied.

(ii) The acceleration component in x-direction is given as

$$a_x = u\frac{\partial u}{\partial x} + v\frac{\partial u}{\partial y} + w\frac{\partial u}{\partial z} + \frac{\partial u}{\partial t}$$

$$\frac{\partial u}{\partial x} = 6t$$

$$\frac{\partial u}{\partial z} = z^2$$

$$\frac{\partial u}{\partial t} = 6x$$

$$\therefore \quad a_x = \left(6\times t + yz^2\right)6t + \left(3t + xy^2\right)z^2 + \left(xy - 2\times yz - 6tz\right)2yz + 6x$$

Value of a_x at point A (1, 1, 1) and t = 1.0 is given as

$$a_x = (6 + 1)\cdot 6 + (3 + 1)\cdot 1 + (1 - 2 - 6)\cdot 2 + 6 = 38 \text{ units}$$

4.8 **If the velocity components for two dimensional flow are given by**

$$u = \frac{x}{x^2 + y^2} \text{ and } v = \frac{y}{x^2 + y^2}$$

determine the acceleration components in X and Y direction and rotation in Z direction at two points in the flow field (i) (2, 3) and (ii) (4, 5). Coordinates are in meters.

[12 Marks : 2021]

Solution:

Given: For 2D flow,

Velocity components,

$$u = \frac{x}{x^2 + y^2},\ v = \frac{y}{x^2 + y^2}$$

Acceleration in x-direction:

$$a_x = u\frac{\partial u}{\partial x} + v\frac{\partial u}{\partial y}$$

$$\frac{\partial u}{dx} = \frac{\partial}{\partial x}\left(\frac{x}{x^2 + y^2}\right) = \frac{\left(x^2 + y^2\right) - x(2x)}{\left(x^2 + y^2\right)^2} = \frac{y^2 - x^2}{\left(x^2 + y^2\right)^2}$$

$$\frac{\partial u}{\partial y} = \frac{\partial}{\partial y}\left(\frac{x}{x^2 + y^2}\right) = \frac{-2xy}{\left(x^2 + y^2\right)^2}$$

$$a_x = \left(\frac{x}{x^2 + y^2}\right)\left[\frac{y^2 - x^2}{\left(x^2 + y^2\right)^2}\right] + \left(\frac{y}{x^2 + y^2}\right)\left[\frac{-2xy}{\left(x^2 + y^2\right)^2}\right]$$

$$= \frac{xy^2 - x^3 - 2xy^2}{\left(x^2 + y^2\right)^3} = \frac{-x^3 - xy^2}{\left(x^2 + y^2\right)^3} = \frac{-x\left(x^2 + y^2\right)}{\left(x^2 + y^2\right)^3} = \frac{-x}{\left(x^2 + y^2\right)^2}$$

in y-direction: $a_y = u\dfrac{\partial v}{\partial x} + v\dfrac{\partial v}{\partial y}$

$$\frac{\partial v}{\partial x} = \frac{\partial}{\partial x}\left(\frac{y}{x^2+y^2}\right) = -\frac{2xy}{\left(x^2+y^2\right)^2}$$

$$\frac{\partial v}{\partial y} = \frac{\partial}{\partial y}\left(\frac{y}{x^2+y^2}\right)$$

$$= \frac{\left(x^2+y^2\right)-y(2y)}{\left(x^2+y^2\right)^2} = \frac{x^2-y^2}{\left(x^2+y^2\right)^2}$$

Now, $a_y = u\dfrac{\partial v}{\partial x} + v\dfrac{\partial v}{\partial y}$

$$= \left(\frac{x}{x^2+y^2}\right)\left[\frac{-2xy}{\left(x^2+y^2\right)^2}\right] + \left(\frac{y}{x^2+y^2}\right)\left[\frac{x^2-y^2}{\left(x^2+y^2\right)^2}\right]$$

$$= \frac{-2x^2y + x^2y - y^3}{\left(x^2+y^2\right)^3} = \frac{-y\left(x^2+y^2\right)}{\left(x^2+y^2\right)^3}$$

$$a_y = \frac{-y}{\left(x^2+y^2\right)^2}$$

$$a_x = \frac{-x}{\left(x^2+y^2\right)^2} \text{ and } a_y = \frac{-y}{\left(x^2-y^2\right)^2}$$

At point (2, 3) $a_x = \dfrac{-2}{\left(2^2+3^2\right)^2},\ a_y = \dfrac{-3}{\left(2^2+3^2\right)^2}$

$$a_x = \frac{-2}{169}\text{ m/s}^2,\ a_y = \frac{-3}{169}\text{ m/s}^2$$

At point (4, 5) $a_x = \dfrac{-4}{\left(4^2+5^2\right)^2},\ a_y = \dfrac{-5}{\left(4^2+5^2\right)^2}$

$$a_x = \frac{-4}{1681}\text{ m/s}^2,\ a_y = \frac{-5}{1681}\text{ m/s}^2$$

Rotation: $\omega_z = \dfrac{1}{2}\left(\dfrac{\partial v}{\partial x} - \dfrac{\partial u}{\partial y}\right)$

$$= \frac{1}{2}\left[\left\{\frac{-2xy}{\left(x^2+y^2\right)^2}\right\} - \left\{\frac{-2xy}{\left(x^2+y^2\right)^2}\right\}\right]$$

$$\omega_z = 0$$

Hence, rotation is zero in z-direction for all values of *x* and *y*.

5. Fluid Dynamics, Flow Measurements and Vortex Flow

5.1 A steel pipe of 15 cm diameter carries water at a rate of 30 liters per second from point *A* to *B* along the pipe, the point *B* being 20 m higher than point A and 600 m apart along the pipe. If the pressure at *B* is to be 2.8 kg/cm^2, what pressure must be maintained at *A*, if friction factor for the pipe is 0.024? What will be the capacity of the pipe after 15 years of service, when the friction factor is tripled? Assume that the same pressures are to be maintained at *A* and *B*, (kg = unit of force in metric system).

[15 marks : 1999]

Solution:

The brief schematic given below shows the arrangement clearly.

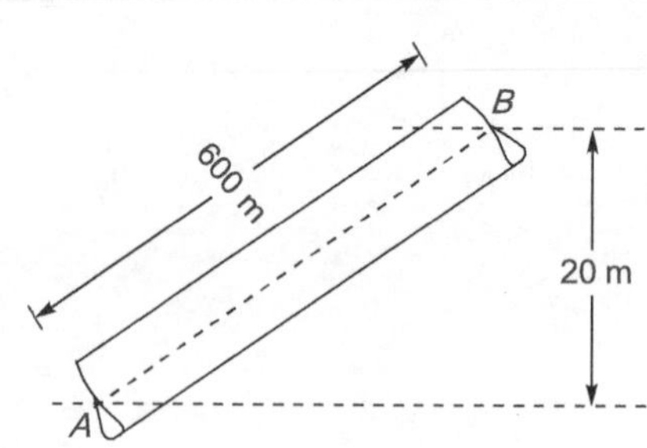

Given data:

$$Q = 30 \text{ litres per second} = 30 \times 10^{-3} = 0.03 \text{ m}^3/\text{s}$$

$$d = 15 \text{ cm} = 0.15 \text{ m}$$

$$p_B = 2.8 \text{ kg/cm}^2 = \frac{2.8 \times 9.81}{(10^{-2})^2} = 2.75 \times 10^5 \text{ N/m}^2$$

$$L = 600 \text{ m};\ f = 0.024$$

Assuming point *A* as datum, we have $Z_A = 0$ and $Z_B = 20$ m

Applying Bernoulli's equation between *A* and *B*, we get,

$$\frac{p_A}{w} + \frac{V_A^2}{2g} + Z_A = \frac{p_B}{w} + \frac{V_B^2}{2g} + Z_B + h_f$$

$$\Rightarrow \quad \frac{p_A}{w} + \frac{V^2}{2g} + 0 = \frac{2.75 \times 10^5}{9810} + \frac{V^2}{2g} + 20 + \frac{fLQ^2}{12.1d^5}$$

$$\Rightarrow \quad \frac{p_A}{w} = 28 + 20 + \frac{0.024 \times 600 \times (0.03)^2}{12.1 \times (0.15)^5}$$

$$\Rightarrow \quad p_A = 62.10 \times 9810 = 6.1 \times 10^5 \text{ N/m}^2$$

When friction factor is tripled

$$f = 3 \times 0.024 = 0.072$$

$$p_A = p_B$$

Again applying Bernoulli's equation between *A* and *B*, we get

$$\frac{p_A}{w} + \frac{V_A^2}{2g} + Z_A = \frac{p_B}{w} + \frac{V_B^2}{2g} + Z_B + h_f \qquad [V_A = V_B]$$

$$\frac{6.096 \times 10^5}{9.81 \times 1000} + 0 + \frac{V_A^2}{2g} = \frac{2.75 \times 10^5}{9.81 \times 1000} + 20 + \frac{V_B^2}{2g} + \frac{8fLQ^2}{\pi^2 g D^5}$$

$$62.141 = 28.033 + 20 + \frac{0.072 \times 8 \times 600 Q^2}{\pi^2 \times 9.81 \times 0.15^5}$$

$$Q = 0.0173 \text{ m}^3/\text{s}$$

5.2 A venturimeter having a throat diameter of 100 mm is fitted into a pipeline which has a diameter of 200 mm and through which oil of specific gravity 0.9 is flowing. The difference in pressure between the entry and throat tappings is measured by a *U*-tube manometer, containing mercury of specific gravity 13.6 and the connections are filled with the oil flowing in the pipeline. If the difference of level indicated by the mercury in the *U*-tube is 0.6 m, calculate the theoretical volume rate of flow through the meter.

[12 marks : 2004]

Solution:

The brief schematic of the arrangement is shown below:

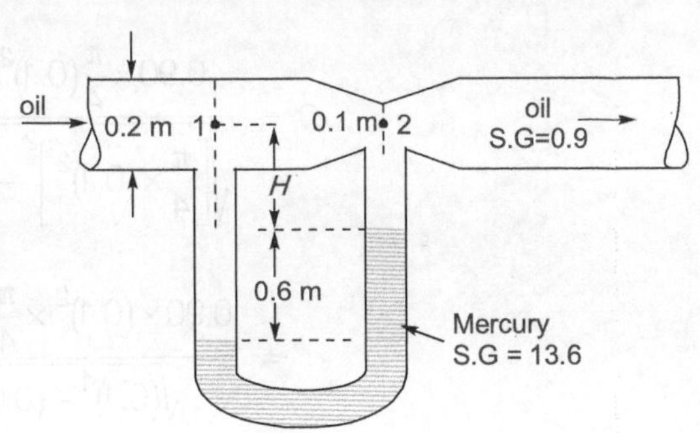

Let the distance between point (1) and upper meniscus of mercury be H.

Applying pressure equation between (1) and (2), we get

$$\frac{p_1}{w} + H + 0.6 - 0.6\left(\frac{13.6}{0.9}\right) - H = \frac{p_2}{w}$$

$$\Rightarrow \quad \frac{p_1}{w} - \frac{p_2}{w} = 8.47 \qquad \text{...(i)}$$

Now applying Bernoulli's equation between (1) and (2), we get

$$\frac{p_1}{w} + \frac{V_1^2}{2g} + Z_1 = \frac{p_2}{w} + \frac{V_2^2}{2g} + Z_2 \qquad \text{(Neglecting losses)}$$

Assuming venturimeter as horizontal, i.e., $Z_1 = Z_2$

$$\therefore \quad \frac{p_1}{w} - \frac{p_2}{w} = \frac{V_2^2 - V_1^2}{2g}$$

$$\Rightarrow \quad \frac{Q^2}{2g}\left[\frac{1}{A_2^2} - \frac{1}{A_1^2}\right] = \frac{p_1}{w} - \frac{p_2}{w}$$

Substituting value of $\frac{p_1}{w} - \frac{p_2}{w}$ from (i), we get

$$\Rightarrow \quad \frac{Q}{\sqrt{2g}} = \frac{A_1 A_2 \sqrt{8.47}}{\sqrt{A_1^2 - A_2^2}}$$

$$\Rightarrow \quad Q = \frac{\frac{\pi}{4}\times(0.2)^2 \times \frac{\pi}{4}\times(0.1)^2 \times \sqrt{2\times 9.81\times 8.47}}{\frac{\pi}{4}\sqrt{(0.2)^4 - (0.1)^4}} = 0.1046 \text{ m}^3/\text{s}$$

5.3 A flow nozzle is used for measuring the discharge in the laboratory. The approach flow pipe has a 100 mm diameter pipe and the throat diameter of the flow nozzle is 60 mm. The coefficient of velocity is 0.90. The differential mercury manometer shows 13.5 cm of Hg. Obtain the discharge in litres per second.

[10 marks : 2005]

Solution:

Formula is same as in venturimeter

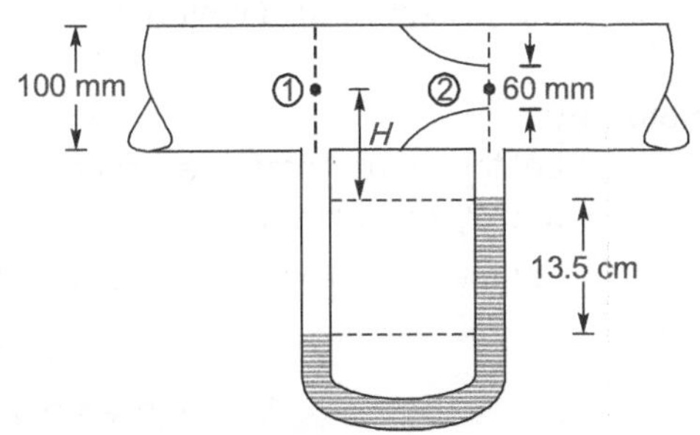

Here, $c_d = c_v \quad \{\because c_c = 1\}$

$$Q = \frac{c_v A_1 A_2}{\sqrt{A_1^2 - A_2^2}}\sqrt{2gh} \qquad \left\{\begin{aligned} h &= 13.5\left[\frac{13.6}{1} - 1\right] \\ &= 170.1 \text{ cm} \\ &= 1.7 \text{ m}\end{aligned}\right\}$$

$$Q = \frac{0.90\times\frac{\pi}{4}(0.1)^2\times\frac{\pi}{4}(0.06)^2}{\sqrt{\left[\frac{\pi}{4}\times(0.1)^2\right]^2-\left[\frac{\pi}{4}(0.06)^2\right]^2}}\times\sqrt{2\times9.81\times1.7}$$

$$= \frac{0.90\times(0.1)^2\times\frac{\pi}{4}(0.06)^2}{\sqrt{(0.1)^4-(0.06)^4}}\times\sqrt{2\times9.81\times1.7}$$

$$= 0.015752\ \text{m}^3/\text{sec or } 15.752\ \text{litre/sec}$$

5.4 What is fluid transient?

[2 marks : 2005]

Solution:

Fluid transient is a condition where velocity and pressure of a fluid changes over time. These changes may be due to starting or stopping of a pump, opening or closing of valves or fluctuations in the supply pressure from reservoir or tanks.

Transient flow events can be extremely destruct if the magnitude and velocity of the pressure wave exceeds the capacity of the system in which it takes place.

5.5 **A conical tube of length 2 m fixed vertically with its smaller end upwards. The velocity head and pressure head at upper ends are 1.27 m and 2.50 m respectively. The velocity head and pressure head at lower ends are 0.203 m and 5.407 m respectively. Find the direction of flow.**

[2 marks : 2010]

Solution:

Let us denote upper and lower ends of the tube by 1 and 2 respectively

Assuming section 2 as datum.

Total energy at section 1 = Pressure head + velocity head + datum head

= 2.50 + 1.27 + 2 = 5.77 m

Total energy at section 2 = Pressure head + velocity head + datum head

= 5.407 + 0.203 + 0 = 5.61 m

Since the total energy at the lower end is less than the total energy at the upper end, therefore the direction of flow will be from upper end to lower end i.e. from section 1 to section 2.

5.6 **A pipe carrying water tapers from cross-section 0.3 m² at A to 0.14 m² at *B*. The average velocity at *A* is 1.8 m/sec and pressure is 441 kN/m² (gauge). If the frictional effects are negligible, determine the pressure at *B* which is 5.5 m above the level of *A*.**

[4 marks : 2010]

Solution:

Let the area of cross section at *A* and *B* be A_A and A_B respectively. Also let the velocity at *A* and *B* be V_A and V_B respectively.

For the flow to occur, the **continuity equation** must be satisfied i.e.

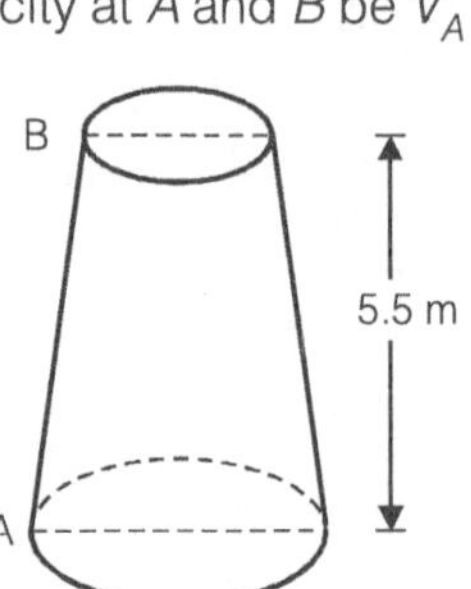

$$A_AV_A = A_BV_B$$

$$\Rightarrow \quad 0.3\times1.8 = 0.14\times V_B$$

$$\Rightarrow \quad V_B = 3.86\ \text{m/sec}$$

Assuming *A* as datum and applying **Bernoulli's equation** between *A* and *B*, we have

$$\frac{p_A}{\gamma_w}+\frac{V_A^2}{2g}+Z_A = \frac{p_B}{\gamma_w}+\frac{V_B^2}{2g}+Z_B+h_f$$

Since frictional losses are zero, therefore $h_f = 0$

$$\therefore \quad \frac{441}{9.81}+\frac{(1.8)^2}{2\times 9.81}+0 = \frac{p_B}{9.81}+\frac{(3.86)^2}{2\times 9.81}+5.5+0$$

$$\Rightarrow \quad \frac{p_B}{9.81} = \frac{441}{9.81}+\frac{(1.8)^2}{2\times 9.81}-\frac{(3.86)^2}{2\times 9.81}-5.5$$

$$\Rightarrow \quad p_B = 38.86\times 9.81 = 381.21 \text{ kN/m}^2$$

5.7 **A 10 cm diameter orifice discharges water at 45 litres per second under a head of 2.75 m. A flat plate is held normal to the jet just downstream from the vena contracta requiring a force of 310 N to resist the impact of the jet. Find coefficients of contraction C_c, velocity C_v and discharge C_d.**

[4 marks : 2012]

Solution:

Given data:

Head, $H = 2.75$ m

Diameter of orifice $D = 10$ cm $= 0.1$ m

$\therefore$ Area, $a = \frac{\pi}{4}\times(0.1)^2 = 0.007854\,\text{m}^2$

Actual discharge, $Q = 45$ liters/sec $= 0.045$ m^3/sec

Theoretical velocity, $V_{th} = \sqrt{2gH} = \sqrt{2\times 9.81\times 2.75} = 7.345$ m/s

$\therefore$ Theoretical discharge, $Q_{th} = V_{th}\times$ Area of orifice $= 7.345\times 0.007854 = 0.05769$ m^3/s

$\therefore$ Coefficient of discharge, $C_d = \frac{\text{Actual discharge}}{\text{Theoretical discharge}} = \frac{0.045}{0.05769} = 0.78$

Now, **According to impulse – momentum principle**

Force, $F = \frac{WV}{g}$

Now, force required $= 310$ N

Where W = weight of water in N/s

and V = Actual velocity

Actual discharge, $Q = 0.045$ m^3/s

$\therefore$ Weight of water, $W = 0.045\times 1000\times 9.81$ N/s $= 441.45$ N/sec

$$\therefore \quad 310 = 441.45\times\frac{V}{9.81}$$

$$\Rightarrow \quad V = 6.889 \text{ m/s}$$

Hence, $C_v = \frac{\text{Actual velocity}}{\text{Theoretical velocity}} = \frac{6.889}{7.345} = 0.9379$

Since, $C_d = C_c\times C_v$

$$\therefore \quad C_c = \frac{C_d}{C_v} = \frac{0.78}{0.9379} = 0.832$$

5.8 A 30 mm diameter vertical pipe conveys oil of dynamic viscosity of 1 poise and mass density of 0.85 gm/cc. The pressures measured at two points P and Q located 25 m apart are 1882 cm and 4706 cm. If the flow is laminar, determine the direction of the flow and the discharge.

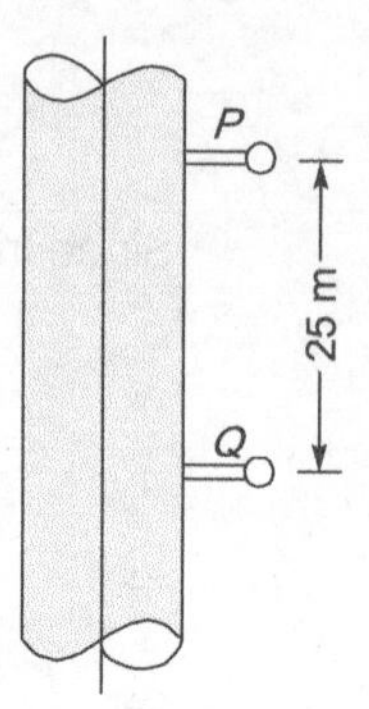

[4 marks : 2014]

Solution:

Given: Diameter of pipe (D) = 30 mm, Dynamic viscosity (μ) = 1 Poise, Mass density (ρ) = 0.85 g/cc

$$\text{Piezometric head at } Q = \frac{P_Q}{\rho g} + Z_1 = 4706 \text{ cm} + 0 = 4706 \text{ cm}$$

$$\text{Piezometric head at } P = \frac{P_P}{\rho g} + Z_2 = 1882 \text{ cm} + 2500 = 4382$$

Since Piezometric head at P is less than Q. Hence direction of flow will be from Q to P.

Let velocity of flow is V.

Applying **Bernoulli's equation** between P and Q

$$\frac{P_Q}{\rho g} + Z_1 + \frac{V^2}{2g} = \frac{P_P}{\rho g} + Z_2 + \frac{V^2}{2g} + h_L$$

$$\left(\frac{P_Q}{\rho g} + Z_1\right) - \left(\frac{P_P}{\rho g} + Z_2\right) = h_L$$

$$\therefore \quad (4706) - (4382) = h_L$$

$$\Rightarrow \quad h_L = 324 \text{ cm}$$

$$\Rightarrow \quad \frac{32\mu VL}{\rho g D^2} = 324$$

$$\Rightarrow \quad V = \frac{324 \times \rho g D^2}{32\mu L} = \frac{324 \times 0.85 \times 981 \times 3^2}{32 \times 1 \times 2500} = 30.40 \text{ cm/s}$$

Discharge,

$$Q = AV = \frac{\pi \times 3^2}{4} \times 30.40 = 214.88 \text{ cm}^3/\text{s}$$

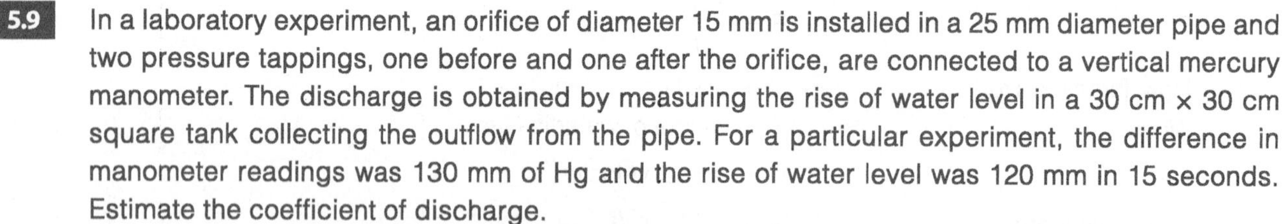

5.9 In a laboratory experiment, an orifice of diameter 15 mm is installed in a 25 mm diameter pipe and two pressure tappings, one before and one after the orifice, are connected to a vertical mercury manometer. The discharge is obtained by measuring the rise of water level in a 30 cm × 30 cm square tank collecting the outflow from the pipe. For a particular experiment, the difference in manometer readings was 130 mm of Hg and the rise of water level was 120 mm in 15 seconds. Estimate the coefficient of discharge.

[8 marks : 2014]

Solution:

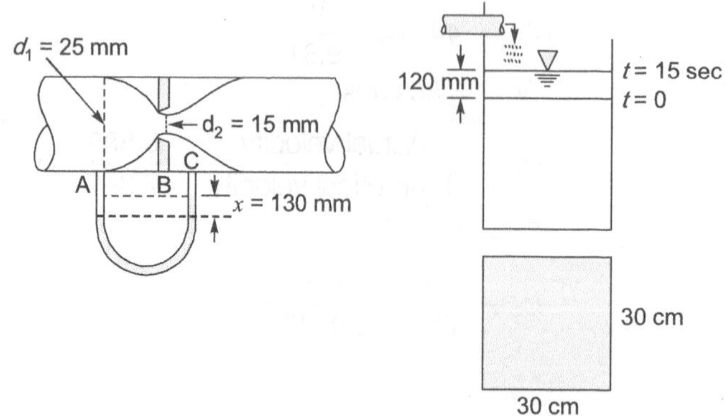

Discharge of pipe, $Q = \frac{\text{Volume stored in tank}}{\text{Time taken}}$

$$= \frac{0.30 \times 0.30 \times 0.120}{15} = 7.2 \times 10^{-4}\ \text{m}^3/\text{s}$$

Reading of manometer, $x = 130$ mm

Assume density of mercury, $\rho_m = 13600$ kg/m³

$$h = x\left(\frac{\rho_m}{\rho_w} - 1\right) = 0.130\left(\frac{13600}{1000} - 1\right) = 1.638\ \text{m}$$

For orifice, flow rate is given by

$$Q = C_d A_1 A_2 \sqrt{\frac{2gh}{A_1^2 - A_2^2}}$$

where, A_1 = Area cross-section of pipe

A_2 = Area cross-section of orifice

thus, $A_1 = \frac{\pi}{4}(0.025)^2 = 4.908 \times 10^{-4}\ \text{m}^2$

$$A_2 = \frac{\pi}{4}(0.015)^2 = 1.767 \times 10^{-4}\ \text{m}^2$$

$$\therefore \quad Q = C_d \times \left(4.908 \times 10^{-4}\right) \times \left(1.767 \times 10^{-4}\right) \sqrt{\frac{2 \times 9.81 \times 1.638}{\left(4.908 \times 10^{-4}\right)^2 - \left(1.767 \times 10^{-4}\right)^2}}$$

$$7.2 \times 10^{-4} = C_d \times 1.0737 \times 10^{-3}$$

or $C_d = 0.67057$

5.10 **An open cylindrical tank, 0.50 m in diameter and 1.0 m in height, is completely filled with water and rotated about is axis at 240 rpm. Find the radius up to which bottom will be exposed and the volume of water spilled out of the tank.**

[8 marks : 2014]

Solution:

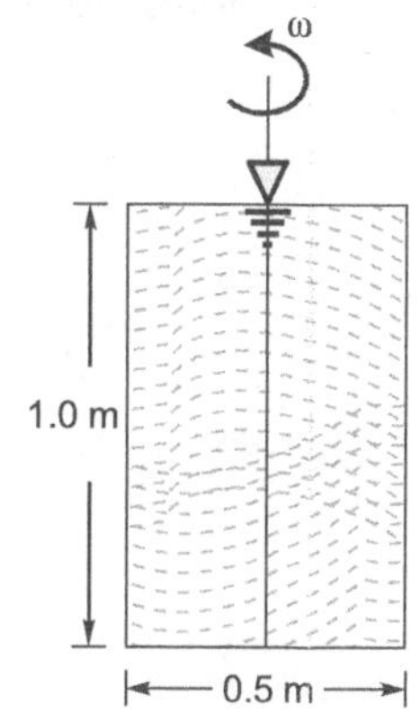

Angular rotation, $\omega = \frac{2\pi N}{60} = \frac{2\pi \times 240}{60} = 25.13\ \text{rad/sec}$

Depth of cavity formed when cylinder is rotated about its axis.

$$z = \frac{R^2\omega^2}{2g} = \frac{(0.25)^2 \times (25.13)^2}{2 \times 9.81} = 2.01\ \text{m}$$

Let r be the radius upto which bottom will be exposed.

$$z = \frac{\omega^2 r^2}{2g}$$

$$\Rightarrow \quad 1.01 = \frac{(25.13)^2 \cdot r^2}{2 \times 9.81}$$

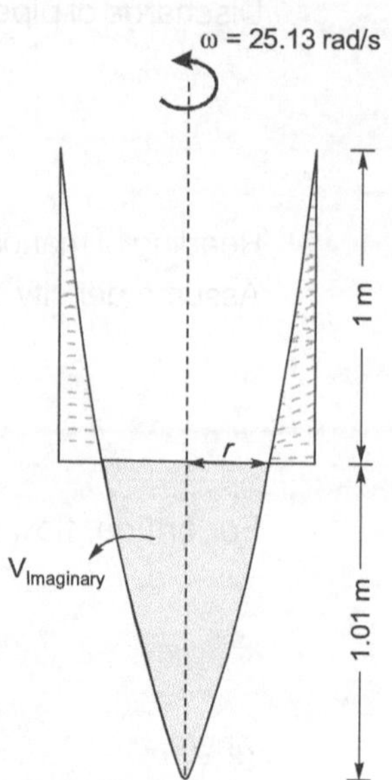

$$\Rightarrow \quad r^2 = \frac{1.01 \times 2 \times 9.81}{(25.13)^2} = 0.0314$$

$$\Rightarrow \quad r = 0.177\text{ m}$$

Initial volume, $V_i = \pi \times 0.25^2 \times 1 = 0.1964\text{m}^3$

Final volume, $V_f = (V_i) - [V_{\text{cavity}} - V_{\text{imaginary}}]$

$$V_{\text{cavity}} = \frac{1}{2}\pi(0.25)^2 \times 2.01 = 0.1973\text{m}^3$$

$$V_{\text{imaginary}} = \frac{1}{2} \times \pi(0.177)^2 \times 1.01 = 0.0497\text{m}^3$$

$$\therefore \quad V_f = [0.1964] - [0.1973 - 0.0497] = 0.049\text{m}^3$$

Spilled volume, $V_{\text{spilled}} = V_i - V_f = 0.1964 - 0.049 = 0.1473\text{ m}^3$

5.11 **A horizontal venturi meter which has an inlet diameter of 120 mm and throat diameter of 60 mm is connected to a pipeline. The coefficient of discharge is 0.95. The inlet pressure is 10 kPa (gauge) and local atmospheric pressure is 96 kPa (absolute). Determine the maximum discharge of water that can be allowed without causing cavitation. Assume vapour pressure of water as 4 kPa.**

[8 marks : 2015]

Solution:

Given: $C_d = 0.95$

$p_1 = 10\text{ kPa} = 10\text{ kN/m}^2$ (gauge)

$p_1 = 10 + 96\text{ kPa} = 106\text{ kPa} = 106\text{ N/m}^2$ (absolute)

$p_2 = 4\text{ kPa} = 4\text{ kN/m}^2$ (absolute)

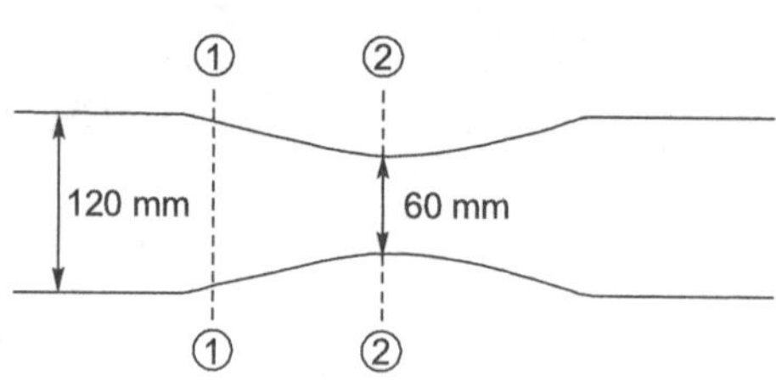

$$a_1 = \frac{\pi}{4} \times 0.12^2 = 11.31 \times 10^{-3}\text{ m}^2$$

$$a_2 = \frac{\pi}{4} \times 0.06^2 = 2.83 \times 10^{-3}\text{ m}^2$$

Applying Bernoulli's equation between section (1) and (2)

$$\frac{p_1}{\gamma} + \frac{v_1^2}{2g} + z_1 = \frac{p_2}{\gamma} + \frac{v_2^2}{2g} + z_2$$

$$\Rightarrow \quad \frac{1}{2g}(v_2^2 - v_1^2) = \frac{p_1}{\gamma} - \frac{p_2}{\gamma}$$

$$\Rightarrow \quad \frac{Q^2}{2g}\left(\frac{1}{a_2^2} - \frac{1}{a_1^2}\right) = \frac{p_1 - p_2}{\gamma}$$

$$\Rightarrow \quad Q^2 = \frac{2g\, a_1^2 a_2^2}{\gamma(a_1^2 - a_2^2)}(p_1 - p_2)$$

$$\Rightarrow \quad Q = \frac{a_1 a_2 \sqrt{2g(p_1 - p_2)}}{\sqrt{\gamma\left(a_1^2 - a_2^2\right)}}$$

$$\Rightarrow \qquad Q = \frac{11.31\times10^{-3}\times2.83\times10^{-3}\times\sqrt{2\times9.81\times(106-4)}}{\sqrt{9.81\times\left[\left(11.31\times10^{-3}\right)^2-\left(2.83\times10^{-3}\right)^2\right]}}$$

$$\Rightarrow \qquad Q = 0.04175 \text{ m}^3/\text{s} = 41.75\ l/\text{s}$$

5.12 Alignment of a water pipeline has been changed by 45° in horizontal plane using reducing bend having 40 cm and 20 cm dia. at the beginning and end of the bend respectively. A discharge of 400 lit/sec is flowing in the pipeline. The pressure in the pipeline at the starting end of the bend is 3000 bar. The frictional head loss due to the bend may be assumed as 5% of the kinetic energy at the exit off the bend. Determine the force exerted on the bend. Also determine the direction of resulting force.

[20 Marks : 2021]

Solution:

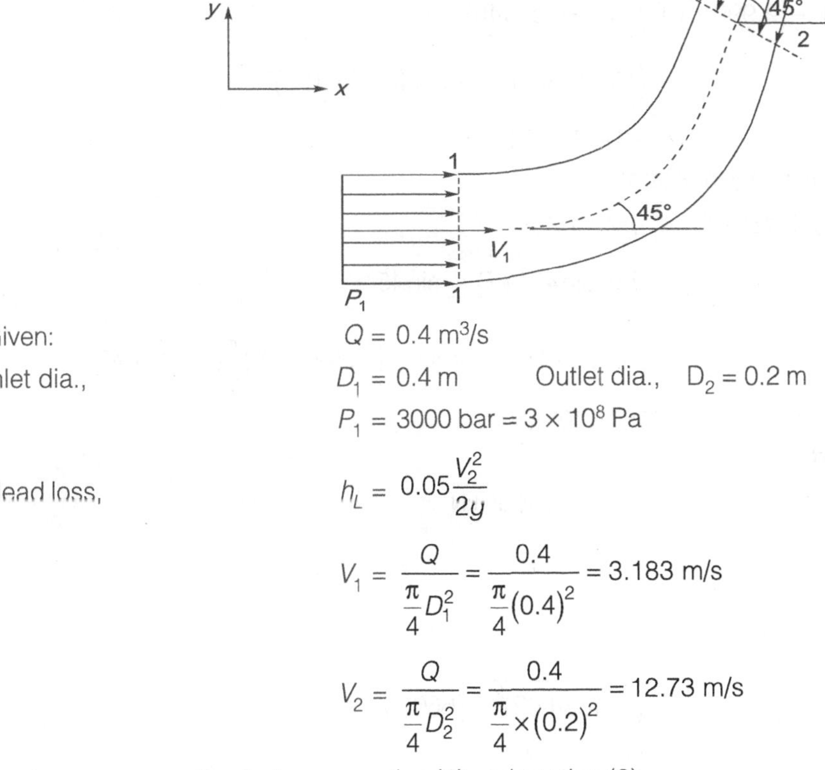

Given: $Q = 0.4 \text{ m}^3/\text{s}$

Inlet dia., $D_1 = 0.4$ m Outlet dia., $D_2 = 0.2$ m

$P_1 = 3000 \text{ bar} = 3\times10^8$ Pa

Head loss, $h_L = 0.05\dfrac{V_2^2}{2g}$

$$V_1 = \frac{Q}{\frac{\pi}{4}D_1^2} = \frac{0.4}{\frac{\pi}{4}(0.4)^2} = 3.183 \text{ m/s}$$

$$V_2 = \frac{Q}{\frac{\pi}{4}D_2^2} = \frac{0.4}{\frac{\pi}{4}\times(0.2)^2} = 12.73 \text{ m/s}$$

Apply energy equation between section (1) and section (2)

$$\frac{P_1}{\rho g}+\frac{V_1^2}{2g}+Z_1 = \frac{P_2}{\rho g}+\frac{V_2^2}{2g}+Z_2+h_L \qquad (Z_1 = Z_2)$$

$$\frac{P_1}{\rho g}+\frac{V_1^2}{2g} = \frac{P_2}{\rho g}+\frac{V_2^2}{2g}+0.05\frac{V_2^2}{2g}$$

$$= \left(3\times10^8\right)+\left(\frac{10^3}{2}\right)\left[3.183^2-1.05(12.73)^2\right] = P_2$$

$$P_2 = 2.999\times10^8 \text{ Pa}$$

In x-y plane
Water system

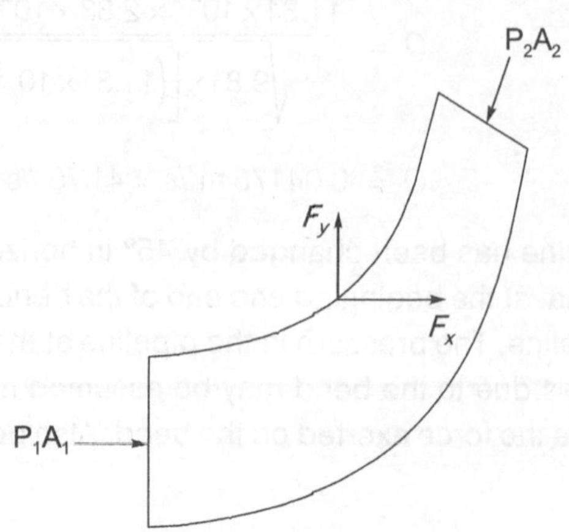

F_x and F_y = Force exerted by bend on water in x and y direction respectively.

Apply Newton's 2nd law in x-direction

$$P_1A_1 - P_2A_2\cos 45° + F_x = \dot{m}V_2\cos 45° - \dot{m}V_1$$

$$\left(3\times10^8\right)\frac{\pi}{4}(0.4)^2 - \left(2.999\times10^8\right)\frac{\pi}{4}(0.2)^2\cos 45° + F_x$$

$$= \left(10^3\right)(0.4)(12.73\cos 45° - 3.183)$$

$$= -31034.68 \text{ kN}$$

Apply Newton's 2nd law in y-direction

$$-P_2A_2\sin 45° + F_y = \dot{m}V_2\sin 45°$$

$$F_y = \dot{m}V_2\sin 45° + P_2A_2\sin 45°$$

$$= \left(10^3\right)(0.4)12.73\sin 45° + \left(2.999\times10^8\right)\frac{\pi}{4}(0.2)^2\sin 45°$$

$$= 6667.703 \text{ kN}$$

Force exerted on bend

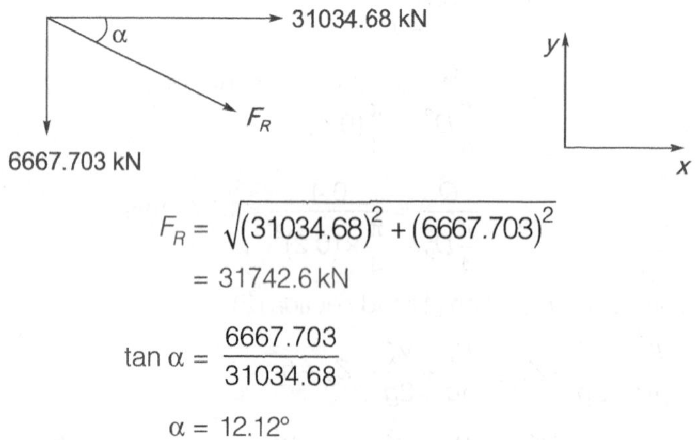

$$F_R = \sqrt{(31034.68)^2 + (6667.703)^2}$$

$$= 31742.6 \text{ kN}$$

$$\tan\alpha = \frac{6667.703}{31034.68}$$

$$\alpha = 12.12°$$

5.13 An open tank of diameter D containing water to depth h_0 is emptied by a smooth orifice at the bottom. Derive an expression for the time taken to reduce the height to h.
Also deduce the expression for time if $d \ll D$. Then estimate the time if D is 0.5 m, diameter of orifice is 0.025 m with water level as 0.5 m.

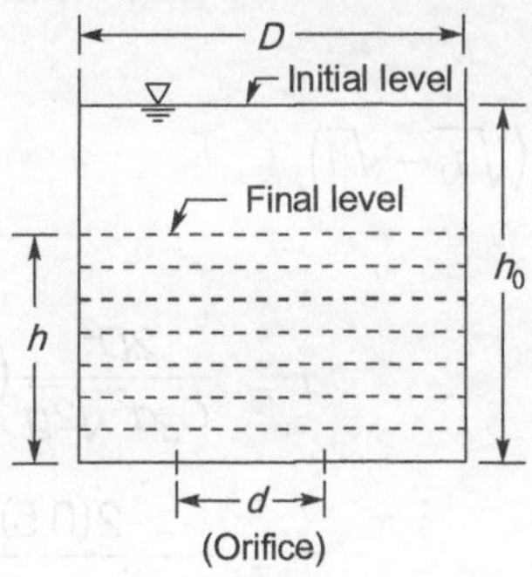

[20 marks : 2022]

Solution:

Given, $D = 0.5$ m, $d = 0.025$ m, $h_0 = 0.5$ m

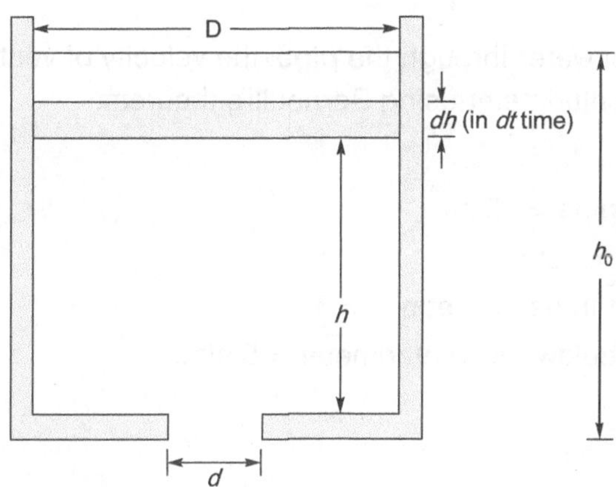

$$\text{Diameter of tank} = D$$

$$\text{Area of tank} = A = \frac{\pi}{4}D^2$$

$$\text{Diameter of orifice} = d$$

$$\text{Area of orifice} = a = \frac{\pi}{4}d^2$$

Let time for water level to fall from h_0 to $h = T$

and Coefficient of discharge $= C_d$

$\therefore$
$$-A.dh = C_d a\sqrt{2gh}\cdot dt$$

$\Rightarrow$
$$\frac{-A}{C_d a\sqrt{2g}}\frac{dh}{\sqrt{h}} = \text{dt}$$

Integrating,

$$\frac{-A}{C_d a\sqrt{2g}}\int_{h_0}^{h}\frac{dh}{\sqrt{h}} = \int_0^T dt$$

$\Rightarrow$
$$\frac{A}{C_d a\sqrt{2g}}\int_{h}^{h_0}\frac{dh}{\sqrt{h}} = \text{T}$$

$\Rightarrow$
$$\frac{2A}{C_d\cdot a\cdot\sqrt{2g}}\left(\sqrt{h_0} - \sqrt{h}\right) = T$$

$$\Rightarrow \quad \frac{2\frac{\pi}{4}D^2}{C_d \frac{\pi}{4} d^2 \sqrt{2g}}\left(\sqrt{h_0} - \sqrt{h}\right) = T$$

$$\Rightarrow \quad T = \frac{2D^2}{C_d d^2 \sqrt{2g}}\left(\sqrt{h_0} - \sqrt{h}\right)$$

$$\Rightarrow \quad T = \frac{2(0.5)^2}{C_d (0.025)^2 \sqrt{2g}}(\sqrt{0.5} - \sqrt{h})$$

$$\Rightarrow \quad T = \frac{180.609}{C_d}(\sqrt{0.5} - \sqrt{h}) \text{ seconds}$$

where h is in meter.

5.14 **Find the discharge of water through the pipe, the velocity of venturimeter throat and the pressure 600 cm above the venturimeter using Bernoulli's theorem.**

Given:
Diameter of vertical pipe = 15 cm
Venturiimeter throat = 7 cm
Absolute pressure at throat = 1 atm
Pressure at 600 cm below the venturimeter = 5 atm.

[20 marks : 2023]

Solution:

In the figure shown, section (1) , (2) and (3) are at inlet of pile, throat of venturimeter and 6 m above throat respectively.

Diameter at venturimeter inlet, $D_1 = 0.15$ m
Diameter at venturimeter throat, $D_2 = 0.07$ m
Pressure at venturimeter throat, P_2 = 1 atm = 101325 Pa
Pressure at venturimeter inlet, $P_1 = 5 \times 101325 = 506625$ Pa

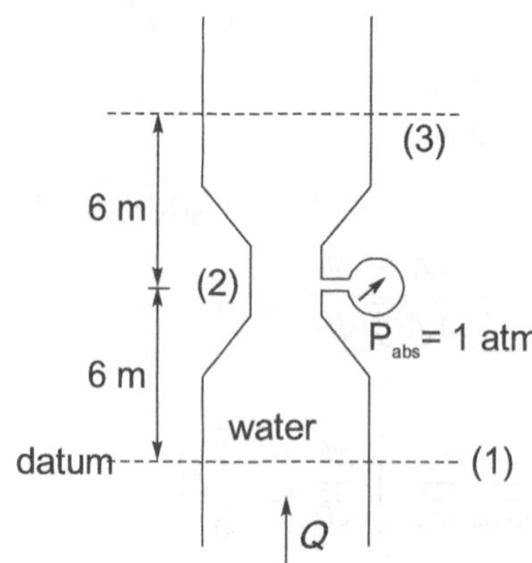

Considering section (1) as datum and applying energy equation between section (1) and section (2)

$$\frac{P_1}{\rho g} + \frac{V_1^2}{2g} + z_1 = \frac{P_2}{\rho g} + \frac{V_2^2}{2g} + z_2$$

where V_1 and V_2 are velocity at (1) and (3)

$$\Rightarrow \quad \left(\frac{P_1}{\rho g}+z_1\right)-\left(\frac{P_2}{\rho g}+z_2\right) = \frac{V_2^2-V_1^2}{2g} \quad \text{(Apply continuity: } Q = A_1V_1 = A_2V_2)$$

$$\Rightarrow \quad \frac{P_1-P_2}{\rho g}+z_1-z_2 = \frac{1}{2g}\left(\frac{Q^2}{A_2^2}-\frac{Q^2}{A_1^2}\right)$$

$$\Rightarrow \quad \frac{(5-1)\times 101325}{(10^3)(9.81)}-6 = \frac{1}{2(9.81)}\frac{Q^2}{A_1^2}\left[\left(\frac{A_1}{A_2}\right)^2-1\right] \quad [\because z_1 - z_2 = -6]$$

$$\Rightarrow \quad 692.88\times\left[\frac{\pi}{4}(0.15)^2\right]^2 = Q^2\left[\left(\frac{0.15}{0.07}\right)^4-1\right]$$

$$\therefore \quad Q = 0.1038 \text{ m}^3\text{/s}$$

Now, Velocity at throat, $V_2 = \dfrac{Q}{A_2}$

$$= \frac{0.1038}{\frac{\pi}{4}(0.07)^2}$$

$$\therefore \quad V_2 = 26.97 \text{ m/s}$$

Also, Pressure at (3) is given by applying energy equation between (1) and (3)

$$\frac{P_1}{\rho g}+\frac{V_1^2}{2g}+z_1 = \frac{P_3}{\rho g}+\frac{V_3^2}{2g}+z_3$$

Now, $V_1 = V_3$ and $z_1 - z_3 = -12$ m

$$\therefore \quad P_3 = P_1 - 12\rho g = 5\text{atm}-\frac{12\times 9.81\times 1000}{101325} = 3.838\text{ atm}$$

6. Dimensional Analysis & Model Analysis

6.1 **A river stretch 10 km long and 850 m wide is to be modelled for flood studies in a hydraulics laboratory, where the available flume size is of 2 m width, 20 m length and 70 cm height. It is decided to use a depth scale of 10. If the peak flood in the river is 3000 cumec at a depth of 6 m, find the suitable horizontal scale to be adopted for the construction of the model. Maximum available flow is 150 litres per second in the laboratory. The horizontal scale should be in the multiple of 10 and such that the model is as big as possible under the given conditions. Determine the model dimension, depth and discharge. Assume that hydraulic depth is same as depth of flow in both the model and prototype.**

[15 marks : 2001]

Solution:

Given $Q_p = 3000 \text{ m}^3\text{/s}; h_p = 6 \text{ m};$ Depth scale, $(L_r)_V = 10$

But $(L_r)_V = \dfrac{\text{Vertical dimension of prototype}}{\text{Vertical dimension of model}}$

$$\Rightarrow \quad (L_r)_V = \frac{h_p}{h_m}$$

$$\Rightarrow \quad 10 = \frac{6}{h_m}$$

$$\Rightarrow \quad h_m = 0.6 \text{ m}$$

$$\Rightarrow \quad h_m = 60 \text{ cm}$$

The available height of model is 70 cm, therefore the calculated height of model i.e. 60 cm is ok
Now, maximum available flow in laboratory i.e. Q_m = 150 litres per second.

But
$$\frac{Q_p}{Q_m} = \frac{V_p \times A_p}{V_m \times A_m}$$

$$\Rightarrow \quad \frac{Q_p}{Q_m} = \left(\frac{h_p}{h_m}\right)^{1/2} \times \frac{B_p \times h_p}{B_m \times h_m}$$

$$\Rightarrow \quad \frac{Q_p}{Q_m} = \left[(L_r)_V\right]^{1/2} \times \frac{B_p}{B_m} \times \frac{h_p}{h_m}$$

$$\Rightarrow \quad \frac{Q_p}{Q_m} = \left[(L_r)_V\right]^{1/2} \times (L_r)_H \times (L_r)_V = (L_r)_H \times (L_r)_V^{3/2}$$

$$\Rightarrow \quad \frac{3000}{150 \times 10^{-3}} = (L_r)_H \times (10)^{3/2}$$

$$\Rightarrow \quad (L_r)_H = 632.46$$

But the horizontal scale should be in the multiple of 10, therefore the horizontal scale for the biggest model will be 640

But
$$(L_r)_H = \frac{B_p}{B_m}$$

$$\Rightarrow \quad 640 = \frac{850}{B_m}$$

$$\Rightarrow \quad B_m = \frac{850}{640} = 1.33 \text{ m}$$

The available width in laboratory is 2 m, therefore the calculated width of the model i.e. 1.33 m is OK

Also,
$$(L_r)_H = \frac{L_p}{L_m}$$

$$\Rightarrow \quad 640 = \frac{10 \times 10^3}{L_m}$$

$$\Rightarrow \quad L_m = 15.625 \text{ m}$$

The available length in laboratory is 20 m, therefore the calculated length of model i.e. 15.625 m is OK.

Now,
$$\frac{Q_p}{Q_m} = (L_r)_H \times [(L_r)_V]^{3/2}$$

$$\Rightarrow \quad \frac{3000 \times 10^3}{Q_m} = 640 \times (10)^{3/2}$$

$$\Rightarrow \quad Q_m = 148.23 \text{ litres per second (OK)}$$

Thus the model will be of 1.33 m width, 15.625 m length and 60 cm height. The discharge in the model will be 148.23 litres per second.

6.2 It is proposed to improve the river channel and it is expected that the water surface profile will be altered. Hence, it is proposed to study using the scale model and can be written as

$$\frac{y}{y_0} = f\left(\frac{x}{y_b}, \frac{v}{\sqrt{gy_0}}\right)$$

in which y is depth at a distance x, y_0 is the depth at the reference location, v is the velocity at the reference location. Froude similitude is proposed for modelling. Scale of model is 1 : 20.

(i) If the prototype flow is 50 m^3s^{-1}. What is the model flow rate?

(ii) If the reference depth is 5 m, what is model depth?

(iii) If the model depths are given below, what are the corresponding depths in the prototype?

x/y_0	0	1	10	100	1000
y/y_0	1.0	0.96	0.90	0.71	0.83

[20 marks : 2005]

Solution:

(i)
$$Q_p = 50 \text{ m}^2\text{s}^{-1}$$

By Froude law,

$$\left(\frac{v}{\sqrt{gy_0}}\right)_m = \left(\frac{v}{\sqrt{gy_0}}\right)_p$$

$$\frac{v_m}{v_p} = \sqrt{\frac{(y_0)_m}{(y_0)_p}}$$

$$v_r = \sqrt{L_r}$$

$$Q_r = A_r \cdot v_r$$

$$Q_r = L_r^2 \times \sqrt{L_r} = L_r^{2.5}$$

$$\frac{Q_m}{Q_p} = \left(\frac{L_m}{L_p}\right)^{2.5}$$

$$\frac{Q_m}{50} = \left(\frac{1}{20}\right)^{2.5}$$

$$Q_m = 0.028 \text{ m}^3\text{/sec}$$

(ii) If the reference depth for prototype is 5 m i.e.

$$(y_0)_p = 5 \text{ m}$$

As given relationship,

$$\frac{y}{y_0} = f\left[\frac{x}{y_0}, \frac{v}{\sqrt{gy_0}}\right]$$

$$\left(\frac{y}{y_0}\right)_m = \left(\frac{y}{y_0}\right)_p$$

$$\frac{y_m}{(y_0)_m} = \frac{y_p}{(y_0)_p}$$

$$(y_0)_m = (y_0)_p \times \frac{y_m}{y_p}$$

$$(y_0)_m = 5 \times \frac{1}{20}$$

$$(y_0)_m = 0.25 \text{ m}$$

Hence, reference depth for model is 0.25 m.

(iii) As per given relationship

$$\left(\frac{y}{y_0}\right) = f\left[\frac{x}{y_0}, \frac{V}{\sqrt{gy_0}}\right]$$

$$\left(\frac{y}{y_0}\right)_m = \left(\frac{y}{y_0}\right)_p$$

and

$$\left(\frac{x}{y_0}\right)_m = \left(\frac{x}{y_0}\right)_p$$

$\left(\frac{x}{y_0}\right)_m$	0	1	10	100	1000
$\left(\frac{y}{y_0}\right)_m$	1.0	0.96	0.9	0.71	0.83

Hence,

$\left(\frac{x}{y_0}\right)_p$	0	1	10	100	1000
$\left(\frac{y}{y_0}\right)_p$	1.0	0.96	0.9	0.71	0.83

and put $(y_0)_p = 5 \text{ m}$

then prototype depths are

x_p	0	5	50	500	5000
y_p	5	4.8	4.5	3.55	4.15

6.3 **The boundary layer thickness δ (delta) in flows with pressure gradient depends on distance x, mass density of fluid ρ (rho), viscosity μ (mu), free stream velocity U and the pressure gradient $\frac{\partial p}{\partial x}$. Perform the dimensional analysis to prove that**

$$\frac{\delta}{x} = \phi\left\{\frac{U.x.\rho}{\mu}, \frac{\frac{\partial p}{\partial x}.x}{\rho U^2}\right\}$$

[8 marks : 2007]

Solution:

Given that: $\delta = fn\left(x, \rho, \mu, U, \frac{\partial p}{\partial x}\right)$

For using **Buckingham's π theorem**, the dimensions of each variable are listed below:

$$\delta = [L];\ x = [L];\ \rho = [ML^{-3}];\ \mu = [ML^{-1}T^{-1}];\ U = [LT^{-1}];\ \frac{\partial p}{\partial x} = [ML^{-2}T^{-2}]$$

There are a total of six variables ($n = 6$) and three primary dimensions ($m = 3$). Thus there should be ($n - m$) dimensionless π terms.

$$\therefore \quad n - m = 6 - 3 = 3\ \pi \text{ terms}$$

Using x, ρ and U as repeating variables, we get

$$\pi_1 = \delta\, x^a \rho^b U^c$$

$$\Rightarrow \quad [M^0 L^0 T^0] = [L] \times [L]^a \times [ML^{-3}]^b \times [LT^{-1}]^c$$

Hence by equating powers of M, L and T, we get

$$b = 0;\ a - 3b + c + 1 = 0;\ c = 0$$

Solving above equations, we get

$$a = -1;\ b = 0;\ c = 0$$

$$\therefore \quad \pi_1 = \delta x^{-1}$$

$$\Rightarrow \quad \pi_1 = \frac{\delta}{x} \qquad \text{...(i)}$$

and

$$\pi_2 = \mu\, x^a \rho^b U^c$$

$$\Rightarrow \quad [M^0 L^0 T^0] = [ML^{-1}T^{-1}]\,[L]^a\,[ML^{-3}]^b\,[LT^{-1}]^c$$

Equating powers of M, L and T, we get

$$b + 1 = 0;\ a - 3b + c - 1 = 0;\ c + 1 = 0$$

Solving the above equations, we get

$$a = -1;\ b = -1;\ c = -1$$

$$\therefore \quad \pi_2 = \mu\, x^{-1} \rho^{-1} U^{-1}$$

$$\Rightarrow \quad \pi_2 = \frac{\mu}{x.\rho.U}$$

or

$$\pi_2 = \frac{x.\rho.U}{\mu} \qquad \text{...(ii)}$$

and

$$\pi_3 = \frac{\partial p}{\partial x} x^a \rho^b U^c$$

$$\Rightarrow \quad M^0 L^0 T^0 = [ML^{-2}T^{-2}]\,[L]^a\,[ML^{-3}]^b\,[LT^{-1}]^c$$

Equating powers of M, L and T, we get

$$b + 1 = 0;\ a - 3b + c - 2 = 0;\ c + 2 = 0$$

Solving above equations, we get

$$a = 1;\ b = -1;\ c = -2$$

$$\therefore \quad \pi_3 = \frac{\partial p}{\partial x}.x.\rho^{-1}.U^{-2}$$

$$\Rightarrow \quad \pi_3 = \frac{\frac{\partial p}{\partial x}.x}{\rho U^2} \qquad \text{...(iii)}$$

Thus, from (i), (ii) and (iii) we get

$$\frac{\delta}{x} = \phi\left\{\frac{x.\rho.U}{\mu}, \frac{\frac{\partial p}{\partial x}.x}{\rho U^2}\right\}$$

6.4 On the apron of spillway model built to a scale of 1 : 30, a hydraulic jump of 100 mm is formed and time required to open the spillway gate is 100 seconds. If the discharge in the prototype spillway is 1500 cumecs, make calculations for the corresponding

(i) height of the jump in prototype

(ii) time in the prototype to produce the same surge effects in model.

(iii) discharge over the model

[10 marks : 2008]

Solution:

In case of hydraulic jump over a spillway, gravity forces are predominant, hence Froude model law is applied.

(i) For **geometric similarity,**

$$L_r = d_r$$

$$\Rightarrow \quad \frac{L_m}{L_p} = \frac{d_m}{d_p}$$

$$\Rightarrow \quad \frac{1}{30} = \frac{100}{d_p}$$

$$\Rightarrow \quad d_p = 3000 \text{ mm} = 3 \text{ m}$$

Thus a hydraulic jump of 3 m will be formed in the prototype.

(ii) For **kinematic similarity, velocity scale ratio** is given by

$$V_r = \frac{L_r}{T_r}$$

But according to **Froude model law,**

$$V_r = \sqrt{L_r}$$

$$\therefore \quad \sqrt{L_r} = \frac{L_r}{T_r}$$

$$\Rightarrow \quad T_r = \sqrt{L_r}$$

$$\Rightarrow \quad \frac{T_m}{T_p} = \sqrt{\frac{L_m}{L_p}}$$

$$\Rightarrow \quad \frac{100}{T_p} = \sqrt{\frac{1}{30}}$$

$$\Rightarrow \quad \frac{(100)^2}{T_p^2} = \frac{1}{30}$$

$$\Rightarrow \quad T_p^2 = 30 \times 10^4$$

$$\Rightarrow \quad T_p = 547.72 \text{ seconds.}$$

Time in the prototype to produce the same surge is 547.72 seconds.

(iii) We know that

$$Q = AV$$

$$\therefore \quad Q_r = A_r \times V_r$$

$$\Rightarrow \quad Q_r = (L_r)^2 \times V_r$$

But according to **Froude model law,**

$$V_r = \sqrt{g_r L_r}$$

But $g_r = 1$

$\therefore$ $V_r = \sqrt{L_r}$

$\Rightarrow$ $Q_r = (L_r)^2 \times (L_r)^{1/2}$

$\Rightarrow$ $Q_r = (L_r)^{5/2}$

$$\Rightarrow \quad \frac{Q_m}{Q_p} = \left(\frac{L_m}{L_p}\right)^{5/2}$$

$$\Rightarrow \quad \frac{Q_m}{1500} = \left(\frac{1}{30}\right)^{5/2}$$

$\Rightarrow$ $Q_m = 1500 \times 2.029 \times 10^{-4}$

$\Rightarrow$ $Q_m = 0.3043$ cumecs

Thus, the discharge over the model is 0.3043 cumecs.

6.5 **In 1 : 30 model spillway, the velocity and discharges are 2.0 m/sec and 2.0 m³/sec respectively. Find the corresponding velocity and discharge in the prototype.**

[4 marks : 2010]

Solution:

In hydraulic structures such as spillway, **Froude model law** is applicable i.e.

$$(F_r)_{model} = (F_r)_{prototype}$$

$$\Rightarrow \quad \frac{V_m}{\sqrt{g_m L_m}} = \frac{V_p}{\sqrt{g_p L_p}}$$

$$\Rightarrow \quad \frac{V_p}{V_m} = \sqrt{\frac{g_p}{g_m} \times \frac{L_p}{L_m}} \qquad [\because g_p = g_m]$$

$$\Rightarrow \quad \frac{V_p}{V_m} = \left(\frac{L_p}{l_m}\right)^{1/2} \qquad ...(i)$$

$$\Rightarrow \quad V_p = V_m \times \left(\frac{L_p}{L_m}\right)^{1/2}$$

$$\Rightarrow \quad V_p = 2 \times \left(\frac{30}{1}\right)^{1/2}$$

$\Rightarrow$ $V_p = 10.95$ m/sec

We know that discharge is given by

$$Q = AV$$

$$\Rightarrow \quad \frac{Q_m}{Q_p} = \frac{A_m}{A_p} \times \frac{V_m}{V_p}$$

$$\Rightarrow \quad \frac{Q_m}{Q_p} = \left(\frac{L_m}{L_p}\right)^2 \times \left(\frac{L_m}{L_p}\right)^{1/2} \qquad \text{[from (i)]}$$

$$\Rightarrow \quad \frac{Q_m}{Q_p} = \left(\frac{L_m}{L_p}\right)^{5/2}$$

$$\Rightarrow \quad \frac{2}{Q_p} = \left(\frac{1}{30}\right)^{5/2}$$

$$\Rightarrow \quad Q_p = 2 \times 4929.50$$

$$\Rightarrow \quad Q_p = 9859 \text{ m}^3/\text{sec}$$

6.6 The shear stress τ in open channel depends on depth of flow y, velocity v, density ρ, surface tension σ and acceleration due to gravity g. Using Buckingham's Pi Theorem, make out the dimensional analysis of the problem.

[4 marks : 2010]

Solution:

Given that $\tau = fn(y, v, \rho, \sigma, g)$

For using **Buckingham's Pi theorem**, the dimensions of each variable are given below:

$$\tau = [ML^{-1}T^{-2}];\ y = [L];\ v = [LT^{-1}];\ \rho = [ML^{-3}];\ \sigma = [MT^{-2}];\ g = [LT^{-2}]$$

There are a total of six variables ($n = 6$) and three primary dimensions ($m = 3$). Thus, there should be $(n - m)$ dimensionless pi terms.

$$\therefore \quad n - m = 6 - 3 = 3 \ pi \text{ terms}$$

using y, v and ρ as repeating variables, we get

$$\pi_1 = \tau y^a v^b \rho^c$$

$$\Rightarrow \quad M^0L^0T^0 = [ML^{-1}T^{-2}]\,[L]^a\,[LT^{-1}]^b\,[ML^{-3}]^c$$

On equating powers of M, L and T on both sides, we get

$$c + 1 = 0$$
$$a + b - 3c - 1 = 0$$
$$b + 2 = 0$$

Solving above equations, we get, $a = 0$, $b = -2$, $c = -1$

$$\therefore \quad \pi_1 = \tau y^0 v^{-2} \rho^{-1}$$

$$\Rightarrow \quad \pi_1 = \frac{\tau}{\rho v^2} \qquad \text{...(i)}$$

Now, $\pi_2 = \sigma y^a v^b \rho^c$

$$\Rightarrow \quad M^0L^0T^0 = [MT^{-2}]\,[L]^a\,[LT^{-1}]^b\,[ML^{-3}]^c$$

On equating powers of M, L and T we get

$$c + 1 = 0$$
$$a + b - 3c = 0$$
$$b + 2 = 0$$

Solving the above equations, we get $a = -1$, $b = -2$, $c = -1$

$$\therefore \quad \pi_2 = \sigma y^{-1} v^{-2} \rho^{-1}$$

$$\Rightarrow \quad \pi_2 = \frac{\sigma}{y \rho v^2} \qquad \text{...(ii)}$$

Also, $\pi_3 = g y^a v^b \rho^c$

$$\Rightarrow \quad M^0L^0T^0 = [LT^{-2}]\,[L]^a\,[LT^{-1}]^b\,[ML^{-3}]^c$$

On equating powers of M, L and T we get

$$c = 0$$
$$a + b - 3c + 1 = 0$$
$$b + 2 = 0$$

Solving the above equations, we get $a = 1$, $b = -2$, $c = 0$

$\therefore$ $$\pi_3 = gyv^{-2}\rho^0$$

$\Rightarrow$ $$\pi_3 = \frac{gy}{v^2} \quad \text{...(iii)}$$

Thus from (i), (ii) and (iii), we get

$$\frac{\tau}{\rho v^2} = \phi\left[\frac{\sigma}{y\rho v^2}, \frac{gy}{v^2}\right]$$

$\Rightarrow$ $$\tau = \rho v^2\phi\left[\frac{\sigma}{y\rho v^2}, \frac{gy}{v^2}\right]$$

6.7 **The measured loss of head in a 50 mm diameter pipe conveying water at 0.6 ms^{-1} is 800 mm of water per 100 m length. Calculate the loss of head in mm of water per 400 m length when air flows through a 200 mm diameter pipe at the corresponding speed. Assume that the pipes have geometrically similar roughness and assume the densities of air and water as 1.23 kg/m^3 and 1000 kg/m^3 respecitvely and the absolute viscosities as 1.8 × 10^{-5} Pa.s and 0.12 Pa.s respectively.**

[10 marks : 2011]

Solution:

Given:

Dia. of pipe = $d_w = 50$ mm

Velocity of flow of water = $V_w = 0.6$ m/s

Loss of head = $h_{fw} = 800$ mm/100 m

Length of pipe = $l_w = 100$ m

Also, for flow of air

Dia. of pipe = $d_a = 200$ mm

Velocity of flow of air = V_a

Length of pipe = $l_a = 400$ m

Applying Reynolds model law of dynamic similarity.

$$\frac{\rho_w \cdot V_w \cdot d_w}{\mu_w} = \frac{\rho_a \cdot V_a \cdot d_a}{\mu_a}$$

or $$V_a = \left(\frac{\rho_w}{\rho_a}\right)\left(\frac{\mu_a}{\mu_w}\right)\left(\frac{d_w}{d_a}\right)\cdot V_w$$

$\Rightarrow$ $$V_a = \left(\frac{1000}{1.23}\right)\left(\frac{1.8\times10^{-5}}{0.12}\right)\left(\frac{50}{200}\right)\times 0.6 = 0.0183 \text{ m/s}$$

For water pipe

Head loss, $$h_{f_w} = \frac{f_w l_w V_w^2}{2gd_w}$$

$\Rightarrow$ $$800\times10^{-3} = \frac{f_w\times100\times0.6^2}{2\times9.81\times0.05}$$

$\Rightarrow$ $$f_w = 0.0218$$

For geometrically similar roughness and same reynolds number $f_w = f_a$

$\therefore \quad h_{fa} = \dfrac{f_a l_a v_a^2}{2gd_a} = \dfrac{0.0218 \times 400 \times 0.0183^2}{2 \times 9.81 \times 0.2} = 0.0007442 \text{ m} = 0.744 \text{ mm}$

Now, $\quad \rho_a g h_{fa} = \rho_w g h_{fw}$

$$1.23 \times 10^{-3} \times 0.744 = 10^3 \times h_{fw}$$

$$h_{fw} = 0.915 \times 10^{-3} \text{ mm of } H_2O \text{ column}$$

6.8 **In 1 : 20 model of a spillway, the velocity and discharge are 1.5 m/sec and 2.0 m³/sec respectively. Find the corresponding velocity and discharge in the prototype.**

[4 marks : 2012]

Solution:

As we know for free surface flows such as flows over spillways **Froude model law** is applicable in which the Froude number for both model and prototype should be equal

$\therefore \quad (Fr)_{model} = (Fr)_{Prototype}$

$\Rightarrow \quad \dfrac{V_m}{\sqrt{g_m L_m}} = \dfrac{V_p}{\sqrt{g_p L_p}}$

Now, $\quad g_m = g_p$

$\therefore \quad \dfrac{V_m}{\sqrt{L_m}} = \dfrac{V_p}{\sqrt{L_p}}$

$\Rightarrow \quad \dfrac{V_p}{V_m} = \sqrt{\dfrac{L_p}{L_m}} = \sqrt{L_r} \qquad \left(\because \dfrac{L_p}{L_m} = L_r\right)$

Where L_r = scale ratio for length

$\dfrac{V_p}{V_m} = V_r$ = scale ratio for velocity

$\therefore \quad \dfrac{V_p}{V_m} = V_r = \sqrt{L_r}$

For discharge $\quad Q = A \times V = L^2 \times \dfrac{L}{T} = \dfrac{L^3}{T}$

$\therefore \quad Q_r = \dfrac{Q_p}{Q_m} = \dfrac{\left(L^3/T\right)_p}{\left(L^3/T\right)_m} = \left(\dfrac{L_p}{L_m}\right)^3 \times \left(\dfrac{T_m}{T_p}\right)$

$$= L_r^3 \times \frac{1}{\sqrt{L_r}} = L_r^{2.5}$$

$$\left[\because \frac{T_r}{T_m} = \frac{(L/V)_p}{(L/V)_m} = \frac{L_p/V_p}{L_m/V_m} = \frac{L_p}{L_m} \times \frac{V_m}{V_p} = L_r \times \frac{1}{\sqrt{L_r}} = \sqrt{L_r}\right]$$

Given data

Scale ratio of length, $\quad L_r = 20$

Velocity in model, $\quad V_m = 1.5$ m/s

Discharge in model, $\quad Q_m = 2$ m³/sec

As; $\quad \dfrac{V_p}{V_m} = \sqrt{L_r} = \sqrt{20}$

$\therefore \quad V_p = V_m \times \sqrt{20} = 1.5 \times \sqrt{20} = 6.708$ m/s

For discharge $\quad \dfrac{Q_p}{Q_m} = L_r^{2.5} = (20)^{2.5}$

$\therefore \quad Q_p = Q_m \times (20)^{2.5} = 2 \times 20^{2.5} = 3577.71$ m³/sec

6.9 Prove that the scale ratio for discharge for a distorted model is given as

$$\frac{Q_p}{Q_m} = (L_r)_H \times [(L_r)_V]^{3/2}$$

where,

Q_p = discharge through prototype
Q_m = discharge through model
$(L_r)_H$ = horizontal scale ratio
$(L_r)_V$ = vertical scale ratio

[4 marks : 2014]

Solution:

Distorted models follows Froude Law's

(i) Velocity scale ratio:

$$\frac{V_m}{V_p} = \sqrt{\frac{2gh_m}{2gh_p}} = \sqrt{\frac{h_m}{h_p}} = \sqrt{(L_r)_v}$$

$$V_r = \sqrt{(L_r)_v}$$

(ii) Area scale ratio:

$$A_r = \frac{A_m}{A_p} = \frac{b_m \cdot h_m}{b_p \cdot h_p} = \left(\frac{b_m}{b_p}\right) \times \left(\frac{h_m}{h_p}\right)$$

or $\quad A_r = (L_r)_H \cdot (L_r)_v$

$\therefore$ Discharge scale ratio, $\quad Q_r = A_r \cdot V_r$

$$\frac{Q_p}{Q_m} = (L_r)_H \cdot (L_r)_v \cdot \sqrt{(L_r)_v} = (L_r)_H \times \left[(L_r)_v\right]^{3/2}$$

6.10 The pressure drop per unit length, $\Delta p/L$, in a pipe depends on the diameter D, velocity V, roughness height ε, density ρ and viscosity μ. The Moody's friction factor chart uses the non-dimensional parameters f, Re and ε/D, and is convenient in finding out the pressure drop in a pipe when the other parameters are known. However, it cannot be directly used to find the velocity for a given pressure drop. Obtain the set of non-dimensional parameters whose plot will directly provide the velocity if other parameters are known.

[8 marks : 2014]

Solution:

$\dfrac{\Delta P}{L}$ is a function of $D, V, \in, \rho, \mu$

$$\frac{\Delta P}{L} = f[D, V, \in, \rho, \mu]$$

$\therefore$ Total no. of variables, $\quad m = 6$

Choosing D, V, ρ as fundamental variables.

$\therefore$ No. of fundamental variables $n = 3$

$\therefore \quad$ No. of π-groups $= m - n = 6 - 3 = 3$

S.N.	Variable	Dimension
1	$\Delta P/L$	$ML^{-2}T^{-2}$
2	D	L
3	V	LT^{-1}
4	$\in$	L
5	ρ	ML^{-3}
6	μ	$ML^{-1}T^{-1}$

We have the three π-group as

$$\pi_1 = D^{a_1} \cdot V^{b_1} \rho^{c_1} \cdot \frac{\Delta P}{L}$$

$$\pi_2 = D^{a_2} \cdot V^{b_2} \rho^{c_2} \cdot \in$$

$$\pi_3 = D^{a_3} \cdot V^{b_3} \rho^{c_3} \cdot \mu$$

First π-group:

$$\pi_1 = D^{a_1} \cdot V^{b_1} \rho^{c_1} \cdot \frac{\Delta P}{L}$$

$$M^o L^o T^o = L^{a_1} \left[LT^{-1}\right]^{b_1} \cdot \left[ML^{-3}\right]^{c_1} \left[ML^{-2}T^{-2}\right]$$

Equating the powers of M, L, T on both side

Power of M: $0 = c_1 + 1$ $\therefore c_1 = -1$

Power of L: $0 = a_1 + b_1 - 3c_1 - 2$ ...(i)

Power of T: $0 = -b_1 - 2$ $\therefore b_1 = -2$

From (ii), $a_1 - 2 + 3 \times 1 - 2 = 0$ $\therefore a_1 = 1$

Hence,

$$\pi_1 = D^1 V^{-2} \rho^{-1} \frac{\Delta P}{L} = \frac{D \cdot \Delta P}{\rho L V^2}$$

Second π-group:

$$\pi_2 = D^{a_2} V^{b_2} \rho^{c_2} \in$$

$$M^o L^o T^o = L^{a_2} \left[LT^{-1}\right]^{b_2} \left[ML^{-3}\right]^{c_2} \cdot L$$

Equating the powers of M, L, T on both side

Power of M: $0 = c_2$ $\therefore c_2 = 0$

Power of L: $0 = a_2 + b_2 - 3c_2 + 1, \quad a_2 + b_2 + 1 = 0$ ...(ii)

Power of T: $0 = -b_2$ $\therefore b_2 = 0$

From (ii), $a_2 + 0 + 1 = 0$ $\therefore a_2 = -1$

Hence,

$$\pi_2 = D^{-1} V^0 \rho^0 \in = \frac{\in}{D}$$

Third π-group:

$$\pi_3 = D^{a_3} V^{b_3} \rho^{c_3} \mu$$

$$M^o L^o T^o = L^{a_3} \left[LT^{-1}\right]^{b_3} \left[ML^{-3}\right]^{c_3} \cdot \left[ML^{-1}T^{-1}\right]$$

Equating the powers of M, L, T on both side

Power of M: $0 = c_3 + 1$ $\therefore c_3 = -1$

Power of L: $0 = a_3 + b_3 - 3c_3 - 1$ ...(iii)

Power of T: $0 = -b_3 - 1$ $\therefore b_3 = -1$

From (iii), $a_3 - 1 + 3 - 1 = 0$ $\therefore a_3 = -1$

Hence,

$$\pi_3 = D^{-1} V^{-1} \rho^{-1} \mu = \frac{\mu}{\rho V D}$$

According to Buckingham's π-theorem

$$f(\pi_1, \pi_2, \pi_3) = 0$$

$$f\left(\frac{\Delta P.D}{\rho L V^2}, \frac{\epsilon}{D}, \frac{\mu}{\rho V D}\right) = 0$$

$$\frac{\Delta P.D}{\rho L V^2} = k\phi\left[\frac{\epsilon}{D}, \frac{\mu}{\rho V D}\right]$$

or

$$\frac{\Delta P}{L} = \frac{\rho V^2}{D} \cdot k\phi\left[\frac{\epsilon}{D}, \frac{\mu}{\rho V D}\right]$$

where, k is a constant.

6.11 **Estimate for 1 : 20 model of a spillway (i) prototype velocity corresponding to a model velocity of 2 m/sec, (ii) prototype discharge per unit width corresponding to a model discharge per unit width of 0.3 m³/sec/m, (iii) pressure head in the prototype corresponding to a model head of 5 cm of mercury at a point and (iv) the energy dissipated per second in the model corresponding to a prototype value of 1.5 kW.**

[8 marks : 2015]

Solution:

Length ratio, $L_r = \frac{L_m}{L_p} = \frac{1}{20}$

(i) Prototype velocity for model velocity of 2 m/sec.

Velocity ratio, $V_r = \sqrt{L_r}$

$$\Rightarrow \quad \frac{V_m}{V_p} = \sqrt{\frac{1}{20}}$$

$$\Rightarrow \quad V_p = V_m\sqrt{20} = 2\sqrt{20}$$

$$V_p = 8.944 \text{ m/s}$$

(ii) Prototype discharge per unit width for model discharge per unit width of 0.3 m³/s/m.

Discharge intensity ratio, $q_r = \frac{V_r A_r}{L_r} = \frac{\sqrt{L_r} \times L_r^2}{L_r} = L_r^{3/2}$

$$\frac{q_m}{q_p} = \left(\frac{1}{20}\right)^{3/2}$$

$$q_p = q_m(20)^{3/2} = 0.3 \times (20)^{3/2}$$

$$q_p = 26.83 \text{ m}^3\text{/s/m}$$

(iii) Prototype pressure head for model pressure head of 5 cm of Hg.

$$\left(\frac{p}{\rho g}\right)_r = L_r$$

$$\Rightarrow \quad \frac{(p/\rho g)_m}{(p/\rho g)_p} = \frac{1}{20}$$

$$\Rightarrow \quad (p/\rho g)_p = 20 \times (p/\rho g)_m$$

$$= 20 \times 5 = 100 \text{ cm of Hg}$$

(iv) Energy dissipated per second in model for 1.5 kW. Energy dissipated in prototype.

$$\text{Energy dissipated/sec.} = \text{Work done/sec.} = \text{Power}$$

$$P_r = F_r V_r$$

$$= \rho_r L_r^3 \times g_r \times \sqrt{L_r}$$

$$= L_r^{7/2} \qquad (\because \rho_r = 1 \text{ and } g_r = 1)$$

$$\Rightarrow \quad \frac{P_m}{P_P} = \left(\frac{1}{20}\right)^{7/2}$$

$$\Rightarrow \quad P_m = 1.5 \times \left(\frac{1}{20}\right)^{7/2}$$

$$\Rightarrow \quad P_m = 4.19 \times 10^{-5} \text{ kW} = 0.042 \text{ W}$$

6.12 **An aircraft propeller of diameter *d*, rotating at a speed N r.p.m. and advancing with velocity *V* in air of density ρ and viscosity μ, develops a thrust *T*. Using Buckingham's Pi theorem, show that**

$$\frac{T}{\rho d^2 V^2} = f\left(\frac{Vd\rho}{\mu}, \frac{dN}{V}\right)$$

[20 marks : 2017]

Solution:

The functional relationship for the thrust, T may be written as

$$T = f(d, N, V, \rho, \mu)$$

which in its most general from may be written as

$$f(T, d, N, V, \rho, \mu) = C$$

Thus, in present case the number of variables, $n = 6$ and these can be completely described by three fundamental dimensions. Hence, $m = 3$.

$$\therefore \quad (n - m) = (6 - 3) = 3$$

Thus, there are 3-dimensionless π-groups in terms of which the above functional relationship may be expressed.

Thus, taking ρ, d and V as repeating variables, we have

$$\pi_1 = \rho^{a_1} V^{b_1} d^{c_1} T \qquad \text{...(i)}$$

$$\pi_2 = \rho^{a_2} V^{b_2} d^{c_2} N \qquad \text{...(ii)}$$

and

$$\pi_2 = \rho^{a_3} V^{b_3} d^{c_3} \mu \qquad \text{...(iii)}$$

Inserting fundamental dimensions in eq. (i)

$$\left[M^0 L^0 T^0\right] = \left[\frac{M}{L^3}\right]^{a_1} \left[LT^{-1}\right]^{b_1} [L]^{c_1} \left[MLT^{-2}\right]$$

Equating the exponents of M, L and T, we obtain

For $\quad M : 0 = a_1 + 1 \Rightarrow a_1 = -1$

For $\quad L : 0 = -3a_1 + c_1 + 1 + b_1$

$\Rightarrow \quad 0 = -3 \times (-1) + c_1 + b_1 + 1 = c_1 + 4 + b_1$

$\Rightarrow \quad 0 = c_1 - 2 + 4$

$\Rightarrow \quad c_1 = -2$

For $\quad T : 0 = -b_1 - 2$

$\Rightarrow \quad b_1 = -2$

$$\therefore \quad \pi_1 = \frac{T}{\rho V^2 d^2} \quad \text{...(i)}$$

Again we have,

$$\pi_2 = \left[ML^{-3}\right]^{a_2}\left[LT^{-1}\right]^{b_2}[L]^{C_2}\left[T^{-1}\right]$$

$$\left[M^0L^0T^0\right] = \left[M^{a_2}L^{-3a_2+b_2+c_2}T^{-b_2-1}\right]$$

Equating exponents of M, L and T

$$\Rightarrow \quad a_2 = 0$$

$$\Rightarrow \quad -b_2 - 1 = 0$$

$$\Rightarrow \quad b_2 = -1$$

$$\Rightarrow \quad -3a_2 + b_2 + c_2 = 0$$

$$\Rightarrow \quad -3 \times 0 - 1 + c_2 = 0$$

$$\Rightarrow \quad c_2 = 1$$

$$\therefore \quad \pi_2 = \rho^0 v^{-1} dN = \frac{dN}{V} \quad \text{...(ii)}$$

Again we have $\quad \pi_3 = \left[M^0L^0T^0\right] = \left[ML^{-3}\right]^{a_3}\left[LT^{-1}\right]^{b_3}[L]^{c_3}\left[ML^{-1}T^{-1}\right]$

Equating exponents of M, L and T, we have

$$\Rightarrow \quad a_3 + 1 = 0$$

$$\Rightarrow \quad a_3 = -1$$

$$\Rightarrow \quad -b_3 - 1 = 0$$

$$\Rightarrow \quad b_3 = -1$$

$$\Rightarrow \quad -3a_3 + b_3 + c_3 - 1 = 0$$

$$\Rightarrow \quad +3 + b_3 + c_3 - 1 = 0$$

$$\Rightarrow \quad b_3 + c_3 = -2$$

$$\Rightarrow \quad c_3 = -1$$

$$\therefore \quad \pi_3 = \frac{\mu}{\rho d V}$$

$$\text{or} \quad \pi_3 = \frac{\rho d V}{\mu}$$

$$\Rightarrow \quad \pi_1 = f(\pi_2, \pi_3)$$

$$\therefore \quad \frac{T}{\rho d^2 V^2} = f\left(\frac{dN}{V}, \frac{\rho d V}{\mu}\right)$$

6.13 **A circular cylinder is placed in a uniform flow with its axis perpendicular to the flow direction. The drag force on the cylinder per unit length, F_D, depends on the flow velocity, V, fluid density, ρ, fluid viscosity, μ' and the cylinder diameter, D. Obtain the non-dimensional sets (Pi numbers) which could be used to analyse this problem.**

[12 marks : 2018]

Solution:

According to the question

Mathematically, $\quad F_D = f(V, D, \rho, \mu)$

Total no. of variables, $\quad n = 5$

No. of fundamental dimensions, $\quad m = 3$

$\therefore$ π-terms, $n - m = 2$

Let the repeating variables be V, D and ρ.

And, each π-term contains $m + 1$ variables.

We have,

$$\pi_1 = F_D V^{a_1} D^{b_1} \rho^{c_1} \quad ...(i)$$

Substituting dimensions on both sides,

$$[M^0 L^0 T^0] = [ML^0T^{-2}][LT^{-1}]^{a_1}[L]^{b_1}, [ML^{-3}]^{c_1}$$

Equating the power of M, L, T on both sides,

Power of M, $0 = 1 + c_1 \Rightarrow c_1 = -1$

Power of L, $0 = 0 + a_1 + b_1 - 3c_1 \Rightarrow a_1 + b_1 = -3$

$$a_1 = -2$$

$\therefore$ $b_1 = -1$

So, $\pi_1 = F_D V^{-2} D^{-1} \rho^{-1}$

$$\pi_1 = \frac{F_D}{\rho V^2 D}$$

Similarly, we have

$$\pi_2 = \mu V^{a_2} D^{b_2} \rho^{c_2} \quad ..(ii)$$

Substituting dimensions on both sides,

$$[M^0L^0T^0] = [ML^{-1}T^{-1}][LT^{-1}]^{a_2}[L]^{b_2}[ML^{-3}]^{c_2}$$

Equating the power of M, L, T on both sides

Power of M, $0 = 1 + c_2$ $\therefore c_2 = -1$

Power of T, $0 = -1 - a_2$ $\therefore a_2 = -1$

Power of L, $0 = -1 + a_2 + b_2 - 3c_2$ $\therefore b_2 = -1$

Subsituting the values of a_2, b_2, c_2 in π_2,

$$\pi_2 = \mu V^{-1} D^{-1} \rho^{-1}$$

or $\pi_2 = \left(\frac{\mu}{\rho VD}\right)$

So, non-dimensional sets are

$$\pi_1 = \left(\frac{F_D}{\rho V^2 D}\right) \text{ and } \pi_2 = \left(\frac{\mu}{\rho VD}\right)$$

6.14 What are the different forces acting in a moving fluid? Also discuss different similarity model laws used to study the dynamic similarity of a prototypes.

[12 Marks : 2021]

Solution:

Forces acting on a moving fluid are:

(i) Inertia force
(ii) Viscous force
(iii) Gravity force
(iv) Pressure force
(v) Elastic force
(vi) Surface tension force

For dynamically similar model:

(i) **Reynold's model law:** For the flow where in addition to inertia, viscous force is the only other predominant force, the similarity of flow in the model and its prototype can be established if the Reynold's number is same for both the system, thus,

$$(\text{Re})_{\text{model}} = (\text{Re})_{\text{prototype}}$$

or $$\frac{\rho_m V_m L_m}{\mu_m} = \frac{\rho_p V_p L_p}{\mu_p}$$

or $$\frac{\rho_r V_r L_r}{\mu_r} = 1$$

or $$\frac{V_r L_r}{\nu_r} = 1$$

Reynold's model law in valid for:

- Flow of incompressible fluid is closed pipes.
- Motion of submarines completely under water.

(ii) **Froude model law:** The gravity force is the only predominant force in addition to the inertia force, which controls the motion

$$(\text{Fr})_{\text{model}} = (\text{Fr})_{\text{prototype}}$$

or $$\frac{V_m}{\sqrt{g_m L_m}} = \frac{V_p}{\sqrt{g_p L_p}}$$

or $$\frac{V_r}{\sqrt{g_r L_r}} = 1$$

$$V_r = \sqrt{g_r L_r}$$

Since, in most of the cases $g_r = 1$,then

$$V_r = \sqrt{l_r}$$

Froude model is valid for:

- Free surface flows such as flow over spillways, sluices etc.
- Flow of jet from an orifice or nozzle.

(iii) **Euler Model Law:** This model is applicable where pressure force controls flow in addition to inertial force. Thus,

$$(Eu)_m = (Eu)_p$$

or $$\frac{V_m}{\sqrt{\frac{p_m}{\rho_m}}} = \frac{V_p}{\sqrt{\frac{p_p}{\rho_p}}}$$ or $$\frac{V_r}{\sqrt{\frac{p_r}{\rho_r}}} = 1$$

Euler model is applicable in case of cavitation.

(iv) **Weber Model Law:** When surface tension effects predominates in addition to inertia force, this model is used in order to establish similarity between model and prototype.

$$(\text{We})_{\text{model}} = (\text{We})_{\text{prototype}}$$

or $$\frac{V_r}{\sqrt{\frac{\sigma_r}{\rho_r L_r}}} = 1$$

Weber model law is applicable for capillary rise in narrow passages

(v) **Mach Model Law:** When elastic compression forces are significant in addition to inertia force then this model is used.

$$(M_a)_{model} = (M_a)_{prototype}$$

or

$$\frac{V_r}{\sqrt{\frac{k_r}{\rho_r}}} = 1$$

This model is applicable for hydraulic model testing in unsteady flow (water hammer).

6.15 **State Buckingham's π-theorem. Write the procedure for selecting the repeating variables.**

[15 marks : 2022]

Solution:

Buckingham π–Method

- The theorem states that if there are n dimensional variables involved in a phenomenon, which can be completely described by m fundamental quantities or dimensions (such as mass, length, time etc.), and are related by dimensionally homogeneous equation, then the relationship among the n-quantities can always be expressed in terms of exactly $(n - m)$ dimensionless and independent π terms.
- In a general dimensional analysis problem, there is one π that we call the dependent π, giving it the notation π. The parameter π, is in general a function of several other π's, which we call independent π's. Thus,

$$\pi_1 = f(\pi_2, \pi_3, \ldots . \pi_k)$$

where k is the total number of π's.
- If a quantity is dimensionless, it is a π-term.
- If any two quantities have same dimension, then their ratios will be a π-term.
- Any π-term can be replaced by any power of that term, it can also be replaced by multiplying or dividing it by another π-term.

Procedure for Selecting Repeating Variables

(i) Select the first repeating variable from those describing the geometry of flow e.g. size and shape of fluid passage or of the moving body such as diameter, length, height, etc.

(ii) Select the second repeating variable from those representing the fluid properties such as density, viscosity, surface tension, elasticity, vapour pressure, etc.

(iii) Select the third repeating variable from those characterising the fluid motion such as velocity, acceleration, discharge, pressure, force, power.

7. Flow Through Pipes

7.1 **Water flows from a reservoir through a pipe of 0.15 m diameter and 180 m long to a point 13.5 m below the open surface of the reservoir. Here it branches into two pipes each of 0.1 m diameter one of which is 48 m long discharging to atmosphere at a point 18 m below reservoir level, and the other 60 m long discharging to atmosphere 24 m below reservoir level. Assuming a constant friction coefficient of 0.032, calculate the discharge from each pipe. Neglect any losses at the junction**

[20 marks : 2000]

Solution:

The figure below shows the layout of the reservoir and pipes.

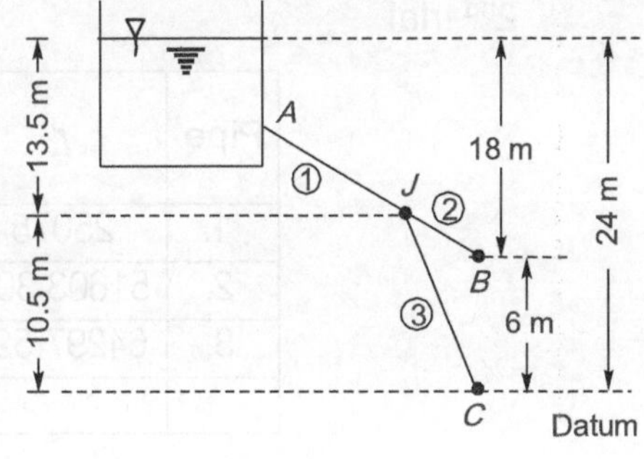

Given data:

$d_1 = 0.15$ m $d_2 = 0.1$ m $d_3 = 0.1$ m

$L_1 = 180$ m $L_2 = 48$ m $L_3 = 60$ m

Friction factor, $f = 0.032 \times 4 = 0.128$ for all pipes

As per **Darcy-Weisbach equation** head loss is expressed as

$$h_f = rQ^n = \left(\frac{fL}{12.1d^5}\right)Q^2$$

If open to atm. then head loss is expressed as

$$h_f = \left(\frac{fL}{12.1d^5}\right)Q^2 + \frac{V^2}{2g} = \left(\frac{fL}{12.1d^5}\right)Q^2 + \frac{(Q/A)^2}{2g}$$

$$= \left[\frac{fL}{12.1d^5} + \frac{1}{2g\left(\frac{\pi}{4}d^2\right)}\right]Q^2 = \left[\frac{fL}{12.1d^5} + \frac{1}{12.1d^4}\right]Q^2$$

$$\therefore \quad r_1 = \frac{fL_1}{12.1(d_1)^5} = \frac{0.128 \times 180}{12.1 \times (0.15)^5} = 25075$$

$$r_2 = \frac{fL_2}{12.1(d_2)^5} + \frac{1}{12.1(d_2)^4} = \frac{0.128 \times 48}{12.1 \times (0.1)^5} + \frac{1}{12.1 \times (0.1)^4} = 51603.306$$

$$r_3 = \frac{fL_3}{12.1(d_3)^5} + \frac{1}{12.1(d_3)^4} = \frac{0.128 \times 60}{12.1 \times (0.1)^5} + \frac{1}{12.1 \times (0.1)^4} = 64297.521$$

Taking flow towards junction J as positive and away from junction J as negative. Assuming piezometric head at J as $H_j = 10.5$ m

1st trial

Pipe	r	Estimated h_f	$Q = \left(\frac{h_f}{r}\right)^{1/2}$	$\left\lvert\frac{Q}{h_f}\right\rvert$
1.	25075	24 – 10.5 = 13.5	0.0232	0.00172
2.	51603.306	6 – 10.5 = – 4.5	– 0.00934	0.00207
3.	64297.521	0 – 10.5 = – 10.5	– 0.0128	0.00122
		Total	ΔQ = 0.0011	0.00501

$$\therefore \quad \Delta H_j = \frac{n \times \Delta Q}{\Sigma\left\lvert\frac{Q}{h_f}\right\rvert} = \frac{2 \times 0.0011}{0.00501} = +0.44 \text{ m}$$

H_j for 2nd trial = 10.5 + 0.44 = 10.94 m

2nd trial

Pipe	r	Estimated h_f	$Q=\left(\frac{h_f}{r}\right)^{1/2}$	$\left\|\frac{Q}{h_f}\right\|$
1.	25075	24 – 10.94 = 13.06	0.0228	0.00174
2.	51603.306	6 – 10.94 = – 4.94	– 0.0098	0.001984
3.	64297.521	0 – 10.94 = – 10.94	– 0.013	0.00119
		Total	ΔQ = 0	0.004914

$$\therefore \quad \Delta H_j = \frac{n \times \Delta Q}{\Sigma\left|\frac{Q}{h_f}\right|} = \frac{2 \times 0}{0.004914} = 0$$

Hence, the refined value of H_j = 10.94 m

Applying Bernoulli's equation between *A* and *J*, we get

$$H_A = H_J + r_1 Q_1^2$$

$$\Rightarrow \quad 24 = 10.94 + 25075\, Q_1^2$$

$$\Rightarrow \quad Q_1^2 = 5.2084 \times 10^{-4}$$

$$\Rightarrow \quad Q_1 = 0.0228 \text{ m}^3/\text{s}$$

Applying Bernoulli's equation between *J* and *B*, we get

$$H_J = H_B + r_2 Q_2^2$$

$$\Rightarrow \quad 10.94 = 6 + 51603.306\; Q_2^2$$

$$\Rightarrow \quad Q_2^2 = 9.573 \times 10^{-5}$$

$$\Rightarrow \quad Q_2 = 0.00978 \text{ m}^3/\text{s}$$

By continuity equation, we get

$$Q_3 = Q_1 - Q_2 = 0.0228 - 0.00978 = 0.01302 \text{ m}^3/\text{s}$$

$$Q_1 = 0.0228 \text{ m}^3/\text{s (towards junction)}$$

$$Q_2 = 0.00978 \text{ m}^3/\text{s (away from junction)}$$

$$Q_3 = 0.01302 \text{ m}^3/\text{s (away from junction)}$$

Alternate Method

Applying Bernoulli's between point *A* and *J* (considering J as datum and gauge pressures)

$$\frac{P_A}{\rho g} + \frac{V_A^2}{2g} + Z_A = \frac{P_J}{\rho g} + \frac{V_J^2}{2g} + Z_J + h_f$$

$$\Rightarrow \quad 0 + 0 + 13.5 = \frac{P_J}{\rho g} + 0 + 0 + \frac{fL_1V_1^2}{2gd_1}$$

$$13.5 = \frac{P_J}{\rho g} + 0 + 7.828V_1^2 \qquad \text{...(i) } [V_A = 0]$$

Similarly, applying Bernoulli's between *J* and *B*

$$\frac{P_J}{\rho g} + \frac{V_J^2}{2g} + Z_J = \frac{P_B}{\rho g} + \frac{V_2^2}{2g} + Z_B + h_f$$

$$\frac{P_J}{\rho g} + 0 + 10.5 = 0 + \frac{V_2^2}{2g} + 6 + \frac{fL_2V_2^2}{2gd_2}$$

$$\frac{P_J}{\rho g} = \left[\frac{1}{2g} + \frac{fL_2}{2gd_2}\right]V_2^2 - 4.5$$

$$\frac{P_J}{\rho g} = 3.182\,V_2^2 - 4.5 \quad \text{...(ii)}$$

and applying Bernoulli's between J and C

$$\frac{P_J}{\rho g} + \frac{V_J^2}{2g} + Z_J = \frac{P_C}{\rho g} + \frac{V_3^2}{2g} + Z_C + h_f$$

$$\frac{P_J}{\rho g} + 0 + 10.5 = 0 + \frac{V_3^2}{2g} + 0 + \frac{fL_3V_3^2}{2gd_3}$$

$$\frac{P_J}{\rho g} = \left[\frac{1}{2g} + \frac{fL_3}{2gd_3}\right]V_3^2 - 10.5$$

$$\frac{P_J}{\rho g} = 3.965\,V_3^2 - 10.5 \quad \text{...(iii)}$$

From eq. (i) and (ii)

$$18 = 3.182\,V_2^2 + 7.828\,V_1^2 \quad \text{...(iv)}$$

Similarly from eq. (i) and (iii)

$$24 = 3.965\,V_3^2 + 7.828\,V_1^2 \quad \text{...(v)}$$

Applying continuity equation at junction J

$$A_1V_1 = A_2V_2 + A_3V_3$$

$$\frac{\pi}{4} \times 0.15^2 V_1 = \frac{\pi}{4} \times 0.1^2 \times V_2 + \frac{\pi}{4} \times 0.1^2 \times V_3$$

$$2.25V_1 = V_2 + V_3 \quad \text{...(vi)}$$

Consider $V_2 = mV_1$

From eq. (vi)

$$2.25V_1 - mV_1 = V_3$$

$$V_3 = (2.25 - m)\,V_1$$

Solving eq. (iv), (v) and (vi)

From eq. (iv) $\quad 18 = 3.182\,m^2V_1^2 + 7.828\,V_1^2$

$$V_1^2 = \frac{18}{3.182m^2 + 7.828}$$

From eq. (v) $\quad 24 = [3.965\,(2.255 - m)^2 + 7.828]V_1^2$

$$24 = \left[3.965(2.25 - m)^2 + 7.828\right]\left[\frac{18}{3.182m^2 + 7.828}\right]$$

$$m = 0.964$$

$$V_1 = 1.292 \text{ m/s} \Rightarrow Q_1 = 0.0228 \text{ m}^3\text{/s}$$

$$V_2 = 1.245 \text{ m/s} \Rightarrow Q_2 = 0.00978 \text{ m}^3\text{/s}$$

$$V_3 = 1.662 \text{ m/s} \Rightarrow Q_3 = 0.01305 \text{ m}^3\text{/s}$$

(direction of flow as shown in fig.)

7.2 Water is pumped through a 600 mm diameter pipe with a head loss of 20 m. It is intended to reduce the power consumption by laying another parallel pipe of appropriate diameter laid by the side of the existing pipeline with the same overall length and the same friction factor by reducing the head loss to 10 m, but still delivering the same discharge jointly through both pipes. What should be the diameter of this addition at pipe?

[10 marks : 2002]

Solution:

Let the discharge through the pipe be Q.

The head loss through the pipe may be given as

$$h_L = \frac{fLQ^2}{12.1D^5}$$

where f is friction factor, L is length of pipe and D is the diameter of the pipe.

Given that $h_L = 20$ m, $D = 600$ mm

$$\therefore \quad 20 = \frac{fLQ^2}{12.1 \times (0.6)^5}$$

$$\Rightarrow \quad fLQ^2 = 18.82 \qquad \text{... (i)}$$

Let the diameter of the additional pipe be 'd'. The friction factor and length of this pipe are same as that of the existing pipe.

Let discharge in the existing pipe and additional pipe be Q_1 and Q_2 respectively

But $Q = Q_1 + Q_2$ [Given]

Since both these pipes are laid parallel to each other the head loss remains same

$$\therefore \quad h_{L_1} = h_{L_2}$$

$$\Rightarrow \quad \frac{fLQ_1^2}{12.1D^5} = \frac{fLQ_2^2}{12.1d^5} \qquad [\because Q = Q_1 + Q_2]$$

$$\Rightarrow \quad \frac{Q_1^2}{(0.6)^5} = \frac{(Q - Q_1)^2}{d^5} \qquad \text{... (ii)}$$

Also $h_{L_1} = 10$ m

$$\Rightarrow \quad \frac{fLQ_1^2}{12.1(0.6)^5} = 10$$

$$\Rightarrow \quad fLQ_1^2 = 9.41 \qquad \text{... (iii)}$$

Dividing (i) by (iii), we get

$$\frac{fLQ^2}{fLQ_1^2} = \frac{18.82}{9.41}$$

$$\Rightarrow \quad \frac{Q^2}{2} = Q_1^2$$

$$\Rightarrow \quad Q_1 = \frac{Q}{\sqrt{2}}$$

$$\Rightarrow \quad Q_1 = 0.707\, Q$$

$$\therefore \quad Q_2 = Q - Q_1 = Q - 0.707\, Q = 0.293\, Q$$

From (ii), we have

$$\frac{Q_1^2}{(0.6)^5} = \frac{(Q-Q_1)^2}{d^5}$$

$$\Rightarrow \quad \frac{(0.707Q)^2}{(0.6)^5} = \frac{(0.293Q)^2}{d^5}$$

$$\Rightarrow \quad d^5 = \left(\frac{0.293}{0.707}\right)^2 \times (0.6)^5$$

$$\Rightarrow \quad d = 0.4218 \text{ m} = 421.8 \text{ mm (Ans.)}$$

7.3 **Three reservoirs *A*, *B* and *C* supply water to a common junction J through pipes *AJ*, *BJ* and *CJ*, respectively and the characteristics of flow through these pipes are given respectively by $AJ - 4Q^2$; $BJ\text{–}7Q^2$ and $CJ - 9Q^2$. The water surface elevations in the three reservoirs are *A*-150 m, *B*-120 m and *C*-40 m. Determine the flows through each of the pipes *AJ*, *BJ* and *CJ*.**

[10 marks : 2003]

Solution:

Let H_j be the piezometric head at the common junction *J*.

for all pipes, $h_f = rQ^n$ with different values of *r* and *n*

Average value of *n* is given by

$$\bar{n} = \frac{2+2+2}{3} = 2$$

A trial and error method is adopted by assuming piezometric head as $H_J = 115$ m

Flow towards the junction is taken positive and flow away from the junction is taken negative.

$$h_f = rQ^n \qquad Q = \left(\frac{h_f}{r}\right)^{1/n}$$

First trial

Pipe	r	n	Estimated h_f	$Q=\left(\frac{h_f}{r}\right)^{1/n}$	$\left\lvert\frac{Q}{h_f}\right\rvert$
AJ	4	2	150 – 115 = 35	2.958	0.0845
BJ	7	2	120 – 115 = 5	0.845	0.1690
CJ	9	2	40 – 115 = – 75	– 2.887	0.0385
			Total	ΔQ = +0.916	0.292

$$\therefore \quad \Delta H_J = \frac{\bar{n} \times \Delta Q}{\Sigma\left\lvert\frac{Q}{h_f}\right\rvert} = +\frac{2 \times 0.916}{0.292} = +6.27 \text{ m}$$

New H_J for 2nd trial = 115 + 6.27 = 121.27 m

Second trial

Pipe	r	n	Estimated h_f	$Q=\left(\frac{h_f}{r}\right)^{1/n}$	$\left\lvert\frac{Q}{h_f}\right\rvert$
AJ	4	2	150 – 121.27 = 28.73	2.68	0.0933
BJ	7	2	120 – 121.27 = –1.27	–0.426	0.3354
CJ	9	2	40 – 121.27 = –81.27	–3.005	0.037
			Total	ΔQ = –0.751	0.4657

Case (i)

$$m = 0.925$$

$$Q_1 = 2.75 \text{ m}^2/\text{s} \qquad \text{(From } A \text{ to } J\text{)}$$

$$Q_2 = 0.22 \text{ m}^3/\text{s} \qquad \text{(From } B \text{ to } J\text{)}$$

$$Q_3 = 2.975 \text{ m}^3/\text{s} \qquad \text{(From } J \text{ to } C\text{)}$$

Case (ii)

$$m = 2.496$$

$$Q_1 = 4.493 \text{ m}^3/\text{s} \qquad \text{(From } A \text{ to } J\text{)}$$

$$Q_2 = -2.698 \text{ m}^3/\text{s} \qquad \text{(–ve means from } J \text{ to } B\text{)}$$

$$Q_3 = 1.8 \text{ m}^3/\text{s} \qquad \text{(from } J \text{ to } B\text{)}$$

7.4 Derive the following relationship for the pressure wave velocity for the fluid transient

$$a = \sqrt{\frac{k/\rho}{1 + \frac{kD}{Ee} c_1}}$$

[8 marks : 2005]

Solution:

In situations, where the velocity can change suddenly, the elastic properties of the pipe and liquid becomes significant. For ex. sudden valve closure causes an increase in pressure head, ΔH to occur which propagates at a speed 'a'.

Let us consider a section of pipe of incremental length 'ΔL'. The pressure wave propagates at a speed 'a'. The wave speed is the speed relative to the observer at rest rather than the speed relative to work velocity. For relatively rigid pipes, choosing any reference would give same result.

But, in unsteady flow situation, we cannot apply linear momentum equation for a steady flow. However, we can translate the coordinate system so that the flow appears steady.

If we move our reference system to the left at a speed 'a', we have for all appearances a steady for

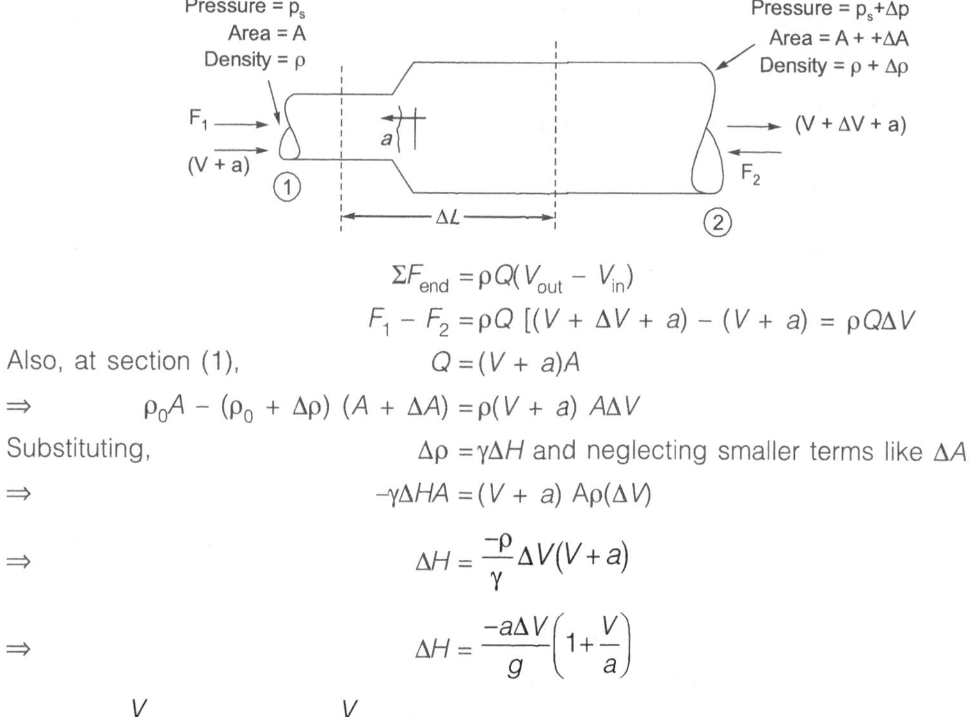

$$\Sigma F_{end} = \rho Q(V_{out} - V_{in})$$

$$F_1 - F_2 = \rho Q\,[(V + \Delta V + a) - (V + a) = \rho Q \Delta V$$

Also, at section (1), $\quad Q = (V + a)A$

$$\Rightarrow \quad \rho_0 A - (\rho_0 + \Delta\rho)\,(A + \Delta A) = \rho(V + a)\,A\Delta V$$

Substituting, $\quad \Delta\rho = \gamma\Delta H$ and neglecting smaller terms like ΔA

$$\Rightarrow \quad -\gamma \Delta H A = (V + a)\,A\rho(\Delta V)$$

$$\Rightarrow \quad \Delta H = \frac{-\rho}{\gamma}\Delta V(V + a)$$

$$\Rightarrow \quad \Delta H = \frac{-a\Delta V}{g}\left(1 + \frac{V}{a}\right)$$

Normally, $\frac{V}{a}$ is very small, $\frac{V}{a} < 0.01$

$\therefore$ ΔH can be written as, $$\Delta H = -\frac{a}{g}\Delta V \quad \text{...(i)}$$

Expression for wave speed:

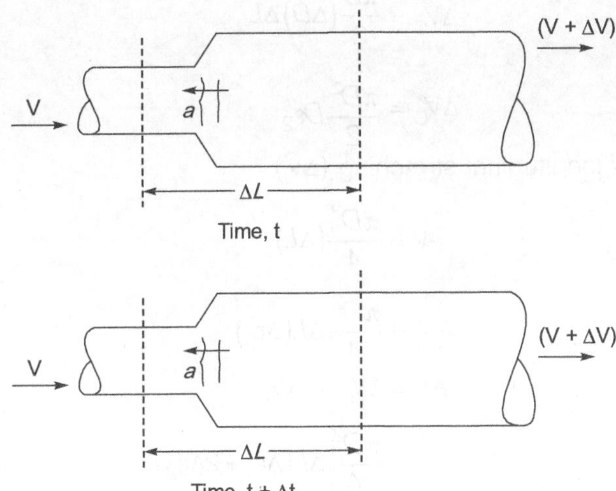

During the time interval required for the wave to pass through the control volume, mass has accumulated in the section in the amount,

$$\Delta M = VA\rho\Delta t - (V + \Delta V)(A + \Delta A)(\rho + \Delta\rho)\Delta t$$

Expanding and neglecting smaller terms we get

$$\Delta M = A\rho\Delta V\Delta t$$

$$\Rightarrow \quad \Delta M = -A\rho\Delta V\left(\frac{\Delta L}{a}\right)$$

This extra liquid is stored in the C.V. partly by being compressed slightly to a larger density, and partly by occupying additional space provided by stretching the pipe cross-section by a small amount.

Change in liquid volume due to compressibility

$$K = \frac{-\Delta\rho}{\frac{\Delta\forall}{V}} = \Delta\forall = -\Delta\rho\left(\frac{\forall}{k}\right) \quad \text{...(ii)}$$

Change in liquid volume due to elasticity:

We know that, $$\Delta\sigma_1 = \left(\frac{\Delta\varepsilon_1 + \mu\Delta\varepsilon_2}{1-\mu^2}\right)E$$

or $$\Delta\varepsilon_1 = \frac{\Delta\sigma_1 - \mu\Delta\sigma_2}{E}$$

$$\Delta\sigma_2 = \left(\frac{\Delta\varepsilon_2 + \mu\Delta\varepsilon_1}{1-\mu^2}\right)E$$

or $$\Delta\varepsilon_2 = \frac{\Delta\sigma_2 - \mu\Delta\sigma_1}{E} \quad \text{...(iii)}$$

Where, $\Delta\sigma_1$ and $\Delta\varepsilon_1$ are the stress and strain respectively in the direction along the pipe axis, while $\Delta\sigma_2$ and $\Delta\varepsilon_2$ are the values in circumferential direction.

For water hammer, pressure waves, there is usually a stress and strain already resident in the pipe caused by steady state flow. Hence above equations are written in incremental form.

Change in volume due to circumferential stretching (ΔV_C)

$$V = \frac{\pi D^2}{4}(\Delta L)$$

$$\Delta V_C = \frac{\pi D}{2}(\Delta D)\Delta L$$

$\Rightarrow$ $$\Delta V_C = \frac{\pi D^2}{2} D\varepsilon_2 \qquad \left(\because \frac{\Delta D}{D} = \Delta\varepsilon_2\right)$$

Change in volume by longitudinal stretching (ΔV_l)

$$V = \frac{\pi D^2}{4}(\Delta L)$$

$$\Delta V_l = \frac{\pi D^2}{4}\Delta L(\Delta\varepsilon_1)$$

Total volume change, $$\Delta V = \Delta V_C + \Delta V_l$$

$$\Delta V = \frac{\pi D^2}{4}\Delta L(\Delta\varepsilon_1 + 2\Delta\varepsilon_2) \qquad \text{...(iv)}$$

Also, we know that, $$\Delta\sigma_2 = \frac{(\Delta p)D}{2e}$$

Where e is pipe wall thickness from equation (iii)

$$\frac{(\Delta p)D}{2e} = \left(\frac{\Delta\varepsilon_2 + \mu\Delta\varepsilon_1}{1-\mu^2}\right)E \qquad \text{...(v)}$$

While the relation between circumferential stress and pressure is valid for all types of restrain the relation between longitudinal stress and strain varies with restraint type.

For ex.

(i) If pipe is anchored at one end and free to stretch longitudinally, longitudinal stress would be

$$\Delta\sigma_1 = \frac{(\Delta\rho)D}{4e}$$

(ii) If the pipe is rigidly anchored to prevent any axial strain, then

$$\Delta\sigma_1 = \mu\sigma_2 \text{ and } \Delta\varepsilon_1 = 0$$

(iii) However, if pipe has functional expansion joints then $\Delta\sigma_1 = 0$ and $\Delta\varepsilon_1$ is of no significance.

As per Wylie and Steeter, three cases are observed:

Case-1: Pipe anchorage only at upstream end

Case-2: Pope restraint from axial movement

Case-3: Longitudinal expansion joints along the pipeline

However, buried pipelines are more common and might be expected to develop full restraint axially by soil friction, so we will examine Case-2.

∴ Wave speed for Case-2 restraint

$$\Delta\sigma_1 = \frac{\mu(\Delta\varepsilon_2)}{(1-\mu^2)}E \qquad (\because \Delta\varepsilon_1 = 0)$$

Evaluation (v) becomes, $$\frac{(\Delta\rho)D}{2e} = \left(\frac{\Delta\varepsilon_2}{1-\mu^2}\right)E$$

$$\Rightarrow \qquad \Delta\varepsilon_2 = \frac{(\Delta\rho)D}{2e}\frac{(1-\mu^2)}{E}$$

Substituting values of $\Delta\varepsilon_2$ and $\Delta\varepsilon_1$ in equation (iv)

$$\Delta V = \frac{\pi D^2}{4}\Delta L\left(\frac{(\Delta\rho)D}{e}\right)\frac{(1-\mu^2)}{E} \qquad \text{...(vi)}$$

$$\Delta M = (\rho+\Delta\rho)(A\Delta L+\Delta\forall)-\rho A\Delta L$$

Equation this expression of ΔM with previously obtained ΔM and dropping smaller terms

$$\Delta\rho A\Delta L + \rho\Delta V = A\rho\Delta V\left(\frac{\Delta L}{a}\right) \qquad \text{...(vii)}$$

Also, $$\frac{\Delta\rho}{\rho} = -\frac{\Delta\forall}{\forall}$$

From eq. (ii), we have, $$\frac{\Delta\forall}{\forall} = -\frac{\Delta\rho}{k}$$

Replacing $\Delta\rho$ by $\gamma\Delta H$ in preceding equation and solving with eq. (vi) and (vii) we get

$$\gamma\Delta H\left[\frac{1}{k}+\left(\frac{1-\mu^2}{E}\right)\frac{D}{e}\right] = \frac{\Delta V}{a}$$

Combining above equation with equation (i), we get

$$\Rightarrow \qquad a^2\rho\left[\frac{1}{k}+\frac{D}{e}\left(\frac{1-\mu^2}{E}\right)\right] = 1$$

$$\Rightarrow \qquad a = \sqrt{\frac{\frac{k}{\rho}}{1+\frac{kD}{eE}(1-\mu^2)}} \qquad \text{(for Case-2)}$$

Above expression is conveniently expressed as,

$$\sigma = \sqrt{\frac{\frac{k}{\rho}}{1+\frac{kD}{eE}(1-\mu^2)}}$$

where, $C_1 = \frac{5}{4} - \mu$ is (Case-1) restraint

$C_1 = 1 - \mu^2$ is (Case-2) restraint

$C_1 = 1$ is (Case-3) restraint

7.5 Water hammer occurs in an elastic pipe due to rapid closure of the wall. Derive an expression for the rise of pressure in the pipeline.

[8 marks : 2007]

Solution:

Consider a pipe AB in which water is flowing as shown in figure.

Let the pipe thickness 't' is small compared to the diameter D of the pipe. It is given that pipe is elastic.

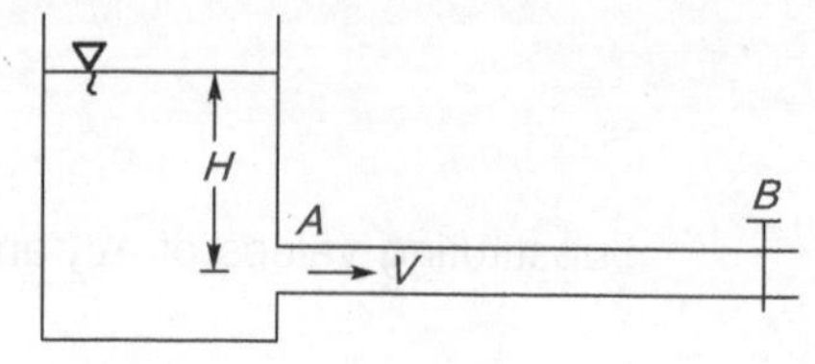

Let A = Area of cross-section of pipe

E = Modulus of elasticity of pipe material

$\frac{1}{m} = \mu$ = Poisson's ratio for pipe material

p = Increase in pressure due to water hammer

L = Length of pipe

When the valve is closed rapidly, a wave of high pressure of intensity 'p' will be set up in the water. Due to this high pressure 'p' circumferential and longitudinal stresses in the pipe wall will be produced.

Let σ_1 = circumferential stress in pipe

σ_2 = longitudinal stress in pipe

$$\therefore \quad \sigma_1 = \frac{pD}{2t} \quad \text{and} \quad \sigma_2 = \frac{pD}{4t}$$

The strain energy stored in pipe material per unit volume is given by

$$= \frac{1}{2E}\left[\sigma_1^2 + \sigma_2^2 - \frac{2\sigma_1\sigma_2}{m}\right] = \frac{1}{2E}\left[\frac{p^2D^2}{4t^2} + \frac{p^2D^2}{16\,t^2} - \frac{p^2D^2}{4\,mt^2}\right]$$

$$\Rightarrow \quad = \frac{1}{2E} \times \frac{p^2D^2}{4t^2}\left[\frac{1}{4} + 1 - \frac{1}{m}\right] = \frac{p^2D^2}{8Et^2}\left(\frac{5}{4} - \frac{1}{m}\right)$$

Total volume of pipe material = $\pi D\,tL$

$\therefore$ Total strain energy stored in pipe material,

$$U = \frac{p^2D^2}{8Et^2}\left(\frac{5}{4} - \frac{1}{m}\right)\pi D\,t\,L = \frac{p^2ADL}{2Et}\left(\frac{5}{4} - \frac{1}{m}\right)$$

Strain energy stored in water $= \frac{1}{2} \times \frac{p^2}{K} \times AL$

Loss of kinetic energy of water $= \frac{1}{2} \times \rho AL \times V^2$

Thus, loss of kinetic energy of water = Gain of strain energy in water + strain energy stored in pipe material

$$\Rightarrow \quad \frac{1}{2} \times \rho AL \times V^2 = \frac{1}{2} \times \frac{p^2}{K} \times AL + \frac{p^2ALD}{2Et}\left(\frac{5}{4} - \frac{1}{m}\right)$$

$$\Rightarrow \quad \rho V^2 = \frac{p^2}{K} + \frac{p^2D}{Et}\left(\frac{5}{4} - \frac{1}{m}\right)$$

$$\Rightarrow \quad p^2 = \frac{\rho V^2}{\frac{1}{K} + \frac{D}{Et}\left(\frac{5}{4} - \frac{1}{m}\right)} = \frac{\rho^2 K V^2}{\rho K\left[\frac{1}{K} + \frac{D}{tE}\left(\frac{5}{4} - \frac{1}{m}\right)\right]} = \rho^2V^2 \times \frac{K}{\rho} \times \frac{1}{1 + \frac{KD}{tE}\left(\frac{5}{4} - \frac{1}{m}\right)}$$

$$\Rightarrow \quad p = \rho V \times \sqrt{\frac{K/\rho}{1 + \frac{KD}{tE}\left(\frac{5}{4} - \frac{1}{m}\right)}}$$

If $$C = \sqrt{\frac{K/\rho}{1 + \frac{KD}{tE}\left(\frac{5}{4} - \frac{1}{m}\right)}}, \text{ then}$$

Rise in pressure is given by

$$p = \rho VC \quad \text{where } C \text{ is velocity of pressure wave in the elastic pipe}$$

Generally Poisson's ratio for pipe materials = $\frac{1}{4}$ i.e. $\frac{1}{m} = \frac{1}{4}$

$$\therefore \qquad C = \sqrt{\frac{K/\rho}{1+\frac{KD}{tE}}}$$

Further if pipe is rigid ($E \rightarrow \infty$), then the term $\frac{KD}{tE}$ can be neglected.

$$\therefore \qquad C = \sqrt{\frac{K}{\rho}} \qquad \left[\because \text{velocity of sound} = \sqrt{\frac{K}{\rho}}\right]$$

Thus, when the pipe is rigid the velocity of pressure wave is equal to velocity of sound.

7.6 **Derive Darcy-Weisbach equation for the head loss due to friction in circular pipe. Assume flow is steady and incompressible.**

[10 marks : 2008]

Solution:

Let
p_1 = Pressure intensity at section 1-1
V_1 = Velocity of flow at section 1-1
L = Length of pipe between sections 1-1 and 2-2
d = Diameter of pipe
f' = Frictional resistance per unit wetted area per unit velocity
h_f = Head loss due to friction
p_2, V_2 = Pressure intensity and velocity at section 2-2

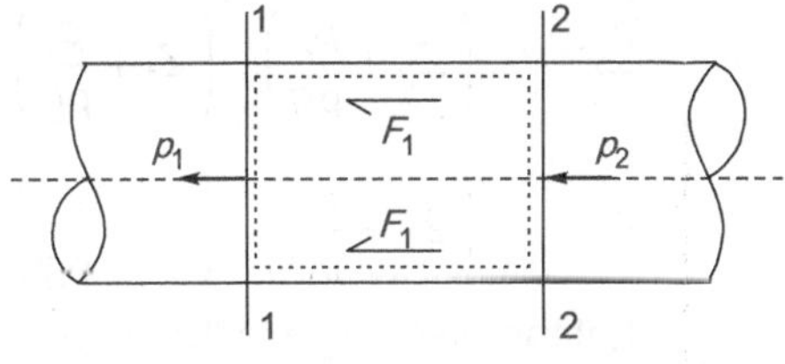

Applying Bernoulli's equation between section 1-1 and 2-2

$$\frac{p_1}{\gamma}+\frac{V_1^2}{2g}+Z_1 = \frac{p_2}{\gamma}+\frac{V_2^2}{2g}+Z_2+h_f$$

Assuming the pipe as horizontal, thus $Z_1 = Z_2$

The diameter of pipe is same at both the sections, therefore the velocities at both the sections are equal i.e. $V_1 = V_2 = V$

$$\frac{p_1}{\gamma}+\frac{V_1^2}{2g}+Z_1 = \frac{p_2}{\gamma}+\frac{V_2^2}{2g}+Z_2+h_f$$

$$\Rightarrow \qquad \frac{p_1}{\gamma}-\frac{p_2}{\gamma} = h_f$$

$$\Rightarrow \qquad p_1 - p_2 = \gamma h_f \qquad \text{...(i)}$$

Now frictional resistance, F_1 = frictional resistance per unit wetted area per unit velocity × wetted area × (velocity)^2

$$\Rightarrow \qquad F_1 = f'\pi dL \times V^2 \qquad \text{...(ii)}$$

Resolving all the forces acting between section 1-1 and 2-2 in the horizontal direction, we get

$$p_1A - p_2A - F_1 = 0$$

$$\Rightarrow \qquad (p_1 - p_2)A = F_1$$

$$\Rightarrow \qquad (p_1 - p_2) = \frac{f' \times \pi d \times L \times V^2}{A} \qquad \text{[from (ii)]}$$

$$\Rightarrow \qquad \gamma h_f = \frac{f' \times \pi d \times L \times V^2}{A} \qquad \text{[from (i)]}$$

$$\Rightarrow \quad h_f = f' \times \frac{4\pi d}{\pi d^2} \times \frac{LV^2}{\gamma} \qquad \left[\because A = \frac{\pi d^2}{4}\right]$$

$$\Rightarrow \quad h_f = \frac{4f'}{\gamma} \times \frac{LV^2}{d} \qquad [\because \gamma = \rho g]$$

$$\Rightarrow \quad h_f = \frac{4f'}{\rho} \times \frac{LV^2}{gd} = \frac{8f'}{\rho} \times \frac{LV^2}{2gd} = \frac{fLV^2}{2gd} \qquad \text{where } f = \frac{8f'}{\rho}$$

and it is called **friction factor**.

7.7 **There is a sudden enlargement of a water pipeline from 200 mm to 400 mm. The hydraulic gradient rises by 10 mm. Estimate the discharge in the pipe.**

[4 marks : 2010]

Solution:

Diameter of smaller pipe, $D_1 = 200 \text{ mm} = 0.2 \text{ m}$

Diameter of larger pipe, $D_2 = 400 \text{ mm} = 0.4 \text{ m}$

Area of smaller pipe, $A_1 = \frac{\pi D_1^2}{4} = \frac{\pi \times (0.2)^2}{4} = 0.01\,\pi \text{ m}^2$

Area of larger pipe, $A_2 = \frac{\pi D_2^2}{4} = \frac{\pi \times (0.4)^2}{4} = 0.04\,\pi \text{ m}^2$

Rise of hydraulic gradient is given as

$$\left(Z_2 + \frac{p_2}{\rho g}\right) - \left(Z_1 + \frac{p_1}{\rho g}\right) = 10 \text{ mm}$$

$$\Rightarrow \quad \left(Z_2 + \frac{p_2}{\rho g}\right) - \left(Z_1 + \frac{p_1}{\rho g}\right) = \frac{10}{1000} \text{m}$$

$$\Rightarrow \quad \left(Z_2 + \frac{p_2}{\rho g}\right) - \left(Z_1 + \frac{p_1}{\rho g}\right) = \frac{1}{100} \text{m} \qquad ...(i)$$

Let the discharge in the pipe be Q.

From continuity equation, we get

$$A_1 V_1 = A_2 V_2$$

$$\Rightarrow \quad V_1 = V_2 \times \left(\frac{A_2}{A_1}\right)$$

$$\Rightarrow \quad V_1 = V_2 \times \left(\frac{0.04\,\pi}{0.01\,\pi}\right)$$

$$\Rightarrow \quad V_1 = 4V_2 \qquad ...(ii)$$

Applying Bernoulli's equation between smaller and larger ends of the pipe, we get

$$\frac{p_1}{\rho g} + \frac{V_1^2}{2g} + Z_1 = \frac{p_2}{\rho g} + \frac{V_2^2}{2g} + Z_2 + \text{Head loss due to enlargement} \qquad ...(iii)$$

But head loss due to enlargement is given by

$$h_e = \frac{(V_1 - V_2)^2}{2g}$$

$$\Rightarrow \quad h_e = \frac{(4V_2 - V_2)^2}{2g} = \frac{9V_2^2}{2g} \qquad ...(iv)$$

Substituting values of rise in hydraulic gradient from (i), velocity V_1 from (ii) and head loss from (iv) in (iii), we get

$$\Rightarrow \quad \frac{p_1}{\rho g}+\frac{(4V_2)^2}{2g}+Z_1 = \frac{p_2}{\rho g}+\frac{V_2^2}{2g}+Z_2+\frac{9V_2^2}{2g}$$

$$\Rightarrow \quad \frac{16V_2^2-9V_2^2-V_2^2}{2g} = \left(\frac{p_2}{\rho g}+Z_2\right)-\left(\frac{p_1}{\rho g}+Z_1\right)$$

$$\Rightarrow \quad \frac{6V_2^2}{2g} = \frac{1}{100}$$

$$\Rightarrow \quad V_2 = \sqrt{\frac{9.81}{300}}$$

$$\Rightarrow \quad V_2 = 0.181 \text{ m/s}$$

$\therefore$ Discharge, $Q = A_2V_2 = 0.04\pi \times 0.181 = 0.023 \text{ m}^3/\text{s}$

7.8 **Two pipes of lengths 2500 m each and diameters of 80 cm and 60 cm respectively are connected in parallel. The friction factor for each pipe is $4f = 0.024$. Total flow is equal to 250 litres per second. Find the discharge in each pipe.**

[4 marks : 2010]

Solution:

Length of pipe 1, $L_1 = 2500$ m

Length of pipe 2, $L_2 = 2500$ m

Diameter of pipe 1, $D_1 = 80$ cm

Diameter of pipe 2, $D_2 = 60$ cm

Friction factor for each pipe, $4f = 0.024$

Total discharge, $Q = 250$ litres per second

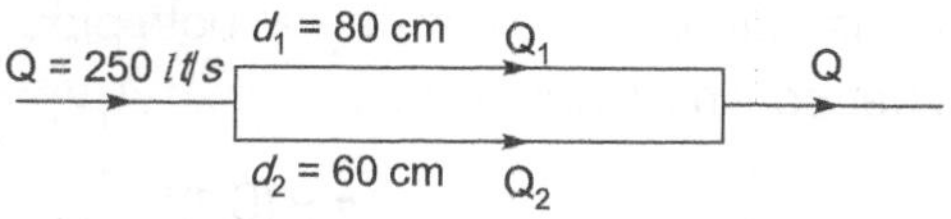

$d_1 = 80$ cm Q_1
Q = 250 l/s
Q
$d_2 = 60$ cm Q_2

When two pipes are connected in parallel head loss through each one of them is equal.

$\therefore$ **Head loss due to friction,** $h_f = \dfrac{4fLV^2}{2gD}$

Let the discharge in pipe 1 and pipe 2 be Q_1 and Q_2 respectively.

$$\therefore \quad Q = Q_1 + Q_2 \quad \text{...(i)}$$

Also, $h_{f1} = h_{f2}$

$$\Rightarrow \quad \frac{4fL_1V_1^2}{2gD_1} = \frac{4fL_2V_2^2}{2gD_2} \quad [\text{where } L_1 = L_2]$$

$$\Rightarrow \quad \frac{V_1}{V_2} = \sqrt{\frac{D_1}{D_2}}$$

$$\Rightarrow \quad \frac{V_1}{V_2} = \sqrt{\frac{80}{60}}$$

$$\Rightarrow \quad V_1 = 1.155\, V_2 \quad \text{...(ii)}$$

But, $V_1 = \dfrac{Q_1}{A_1} = \dfrac{4Q_1}{\pi D_1^2}$

$$V_2 = \frac{Q_2}{A_2} = \frac{4Q_2}{\pi D_2^2}$$

∴ From (ii), we get

$$\frac{4Q_1}{\pi D_1^2} = \frac{4Q_2}{\pi D_2^2} \times 1.155$$

$$\Rightarrow \quad Q_1 = 1.155 \times \left(\frac{80}{60}\right)^2 \times Q_2$$

$$\Rightarrow \quad Q_1 = 2.05\, Q_2$$

Substituting in (i) we get

$$Q_1 + Q_2 = Q$$

$$\Rightarrow \quad 2.05Q_2 + Q_2 = 250$$

$$\Rightarrow \quad Q_2 = \frac{250}{3.05}$$

$$\Rightarrow \quad Q_2 = 81.97 \text{ litres per second}$$

$$\therefore \quad Q_1 = 250 - 81.97 = 168.03 \text{ litres per second}$$

7.9 Two reservoirs are connected by a pipe 100 m long and 100 mm in diameter followed by another pipe 60 m long and 50 mm in diameter. The total head loss between the reservoirs is 10.3 m. Given f = 0.03. Compute discharge neglecting minor losses.

[4 marks : 2010]

Solution:

The discharge will be constant in both pipes as they are connected in series.

As per continuity equation, $A_1V_1 = A_2V_2$

$$\Rightarrow \quad \frac{\pi \times (0.1)^2}{4} \times V_1 = \frac{\pi \times (0.05)^2}{4} \times V_2$$

$$\Rightarrow \quad V_2 = 4V_1 \qquad \text{...(i)}$$

As per Darcy-Weisbach equation, head loss is given by

$$\frac{fL_1V_1^2}{2gD_1} + \frac{fL_2V_2^2}{2gD_2} = h_L$$

$$\Rightarrow \quad \frac{f}{2g}\left[\frac{L_1V_1^2}{D_1} + \frac{L_2V_2^2}{D_2}\right] = h_L$$

$$\Rightarrow \quad \frac{0.03}{2 \times 9.81}\left[\frac{100 \times V_1^2}{0.1} + \frac{60 \times (4V_1)^2}{0.05}\right] = 10.3$$

$$\Rightarrow \quad 1000\, V_1^2 + 19200\, V_1^2 = 6736.2$$

$$\Rightarrow \quad V_1^2 = 0.3335$$

$$\Rightarrow \quad V_1 = 0.58 \text{ m/s}$$

$$\therefore \quad \text{Discharge, } Q = A_1V_1$$

$$= \frac{\pi \times (0.1)^2}{4} \times 0.58$$

$$= 4.5553 \times 10^{-3} \text{ m}^3\text{/s} = 4.553 \text{ litres per second}$$

7.10 Estimate the maximum water hammer pressure generated in a rigid pipe of a 3 m dia pipe with an initial velocity of 3.0 ms^{-1} and the pipe is 8 km long. The downstream valve at the pipe end is closed in 4 seconds. The bulk modulus of water is 2.25×10^6 kPa and the mass density is 995.7 kg/m^3. Also determine the critical time of closure.

[4 marks : 2011]

Solution:

Given: Dia of pipe = d = 3 m, Length of pipe = l = 8000 m, Time of closure = t = 4 sec

Velocity of flow = V = 3 m/s

Velocity of wave, $$C = \sqrt{\frac{k}{\rho}} = \sqrt{\frac{2.25 \times 10^6 \times 10^3}{995.7}} = 1503.24 \text{ m/s}$$

Critical time of closure, $$t_c = \frac{2L}{C} = \frac{2 \times 8000}{1503.24} = 10.645 \text{ s}$$

Since the valve is closed in 4 sec which is less than critical time of closure therefore the value closure is rapid.

Pressure generated in pipe, $P = \rho CV = 995.7 \times 1503.24 \times 3 = 4.49 \times 10^6$ N/m^2

7.11 Water flows at a velocity of 1 ms^{-1} in a 150 mm dia new ductile iron pipe. Estimate the head loss over 500 m using Darcy-Weisbach formula for the given k_S = 0.26mm. Assume viscosity = 1.00×10^{-6} m^2s^{-1}.

$$\left[\text{Hint}: \frac{1}{\sqrt{f}} = -2\log\left[\frac{k_S}{3.7D} + \frac{2.51}{\text{Re}\sqrt{f}}\right]\right]$$

[6 marks : 2011]

Solution:

Given: Dia. of pipe = D = 0.15 m, Velocity of flow = V = 1 m/s, Viscocity, $\nu = 10^{-6}$ m^2/s, k_s = 0.26 mm

Head loss as per **Darcy-Weisbach formula**

$$h_f = \frac{flV^2}{2gD} \quad \text{...(i)}$$

Now, Reynolds number of flow, $$Re = \frac{VD}{\nu}$$

or $$Re = \frac{1 \times 0.15}{10^{-6}} = 15 \times 10^4$$

Using formula given in the Hint:

$$\frac{1}{\sqrt{f}} = -2\log\left[\frac{0.26 \times 10^{-3}}{3.7 \times 0.15} + \frac{2.51}{15 \times 10^4 \times \sqrt{f}}\right]$$

$$\Rightarrow \quad \frac{1}{\sqrt{f}} = -2\log\left[4.68 \times 10^{-4} + \frac{16.73 \times 10^{-6}}{\sqrt{f}}\right]$$

Using trial and error method:

$$f = 0.02383$$

Substituting in (i): $$h_f = \frac{0.02383 \times 500 \times 1^2}{2 \times 9.81 \times 0.15} = 4.048 \text{ m} \simeq 4.05 \text{ m}$$

7.12 Determine the maximum time for rapid closure of pipe 60 cm diameter, 450 m long made of steel ($E = 2.07 \times 10^{11}$ N/m^2) with a wall thickness of 1.25 cm. The pipe carries a liquid of specific gravity 0.88 and bulk modulus $K = 1.03 \times 10^9$ N/m^2 flowing at 1 m^3/sec. E = Young's modulus of steel.

[4 marks : 2012]

Solution:

Let us consider the pipe to be elastic.

The velocity of pressure wave in the elastic pipe is given by,

$$C = \sqrt{\frac{k}{\rho_w}\left(\frac{1}{\sqrt{1+\left(\frac{k}{E}\right)\left(\frac{d}{t}\right)}}\right)}$$

Given data:

$L = 450$ m, $D = 60$ cm $= 0.6$ m, $k = 1.03 \times 10^9$ N/m^2, $t = 1.25$ cm $= 0.0125$ m, $E = 2.07 \times 10^{11}$ N/m^2

Specific gravity of liquid = 0.88

$\therefore$ Density of liquid, $\rho = 0.88 \times 1000 = 880$ kg/m^3

$$\therefore \quad C = \sqrt{\frac{\frac{1.03\times10^9}{880}}{1+\frac{1.03\times10^9\times0.6}{0.0125\times2.07\times10^{11}}}}$$

$$\Rightarrow \quad C = \sqrt{\frac{1.17045\times10^6}{1.2388}} = 972.007 \text{ m/s}$$

The closure will be rapid if time required to close the pipe i.e.,

$$T < \frac{2L}{C}$$

$\therefore$ **Maximum, time for rapid closure of pipe**

$$= \frac{2L}{C} = \frac{2\times450}{972.007} = 0.926 \text{ sec}$$

7.13 **A pipeline of length 26.5 km is used for transmission of water. If the 100 kW power is to be transmitted through the pipe in which water is having 490.5 N/cm^2 pressure at inlet of the pipe, find the diameter of the pipe corresponding to maximum efficiency of transmission. Use head loss due to friction = $\frac{fLU^2}{2gd}$ where $f = 0.026$.**

[4 marks : 2012]

Solution:

Given: Length of pipe, $L = 26.5$ km $= 26500$ m

Pressure at inlet, $P = 490.5$ N/cm^2 $= 490.5 \times 10^4$ N/m^2

$\therefore$ Pressure head at inlet, $H = \frac{P}{\rho g} = \frac{490.5\times10^4}{1000\times9.81} = 500\text{m}$

Let the diameter corresponding to maximum efficiency $(\eta_{max}) = d$

As we know that $\eta = \frac{H-h_f}{H}$

Also for maximum efficiency of transmission

$$h_f = \frac{H}{3}$$

$$\therefore \qquad \eta_{max} = \frac{H - H/3}{H} = \frac{2}{3}$$

$$\therefore \qquad h_f = \frac{500}{3} = 166.67\,\text{m}$$

Now, Power transmitted = 100 kW

Power transmitted is given by the relation

$$P = \frac{\rho g \times Q \times (H - h_f)}{1000}$$

$$\Rightarrow \qquad 100 = \frac{1000 \times 9.81 \times Q \times (500 - 166.67)}{1000}$$

$$\Rightarrow \qquad Q = \frac{100}{9.81 \times (500 - 166.67)} = 0.03058\ \text{m}^3/\text{s}$$

$$= \text{Discharge corresponding to maximum efficiency}$$

But $\qquad Q = \text{Area of pipe} \times \text{Velocity of flow} = \frac{\pi}{4} d^2 \times U$

$$\therefore \qquad 0.03058 = \frac{\pi}{4} d^2 \times U$$

$$\Rightarrow \qquad U = \frac{0.03058 \times 4}{\pi \times d^2} = \frac{0.03894}{d^2}$$

Now, the head loss due to friction

$$h_f = \frac{f \times L \times U^2}{2g \times d}$$

But $\qquad h_f = 166.67\ \text{m}$

$$\therefore \qquad 166.67 = \frac{0.026 \times 26500 \times U^2}{2 \times 9.81 \times d}$$

$$\Rightarrow \qquad 166.67 = \frac{35.117 U^2}{d} = \frac{35.117}{d} \times \left(\frac{0.03894}{d^2}\right)^2 \qquad \left(\because U = \frac{0.03894}{d^2}\right)$$

$$\Rightarrow \qquad 166.67 = \frac{0.05325}{d^5}$$

$$\Rightarrow \qquad d^5 = \frac{0.05325}{166.67} = 0.0003195$$

$$\Rightarrow \qquad d = (0.0003195)^{1/5} = 0.199\ \text{m} \approx 0.2\ \text{m}$$

$\therefore$ Diameter of pipe corresponding to maximum efficiency = 0.2 m or 200 mm

7.14 **A pipeline of 0.6 m diameter is 1.5 km long. To augment the discharge another pipeline of the same diameter is introduced parallel to the first in the second half of the length. Neglecting minor losses find the increase in discharge if $f = 0.04$ and head above the outlet is 30 m.**

[6 marks : 2012]

Solution:

Given : Diameter of pipe line, $\qquad D = 0.6$ m

Length of pipe line, $\qquad L = 1.5\ \text{km} = 1.5 \times 1000 = 1500$ m

$$f = 0.04$$

Head above the outlet $= 30$ m

$\therefore$ Head at inlet $h = 30$ m

Head at outlet = atmospheric head $= 0$

$\therefore$ Head lost, $h_f = 30$ m

Length of another parallel pipe, $L_1 = \dfrac{1500}{2} = 750\text{m}$

Diameter of another parallel pipe,

$d_1 = 0.6$ m

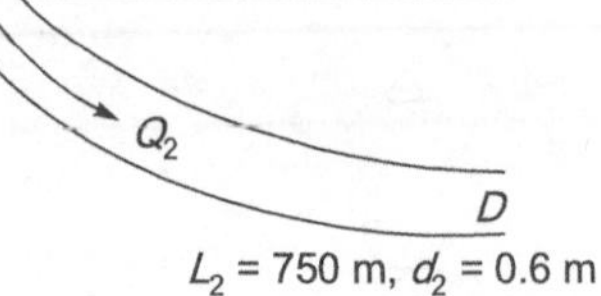

Figure shows the arrangement of pipe systems.

Case-1:

Discharge for a single pipe of length 1500 m and diameter = 0.6 m

The head lost due to friction in single pipe is

$$h_f = \frac{f \times L \times V^{*2}}{d \times 2g}$$

where V^* = Velocity of flow for single pipe

$$\therefore \quad 30 = \frac{0.04 \times 1500 \times V^{*2}}{0.6 \times 2 \times 9.81}$$

$$\Rightarrow \quad V^* = \sqrt{\frac{30 \times 0.6 \times 2 \times 9.81}{0.04 \times 1500}} = 2.426 \text{ m/s}$$

$\therefore$ **Discharge,**

$$Q^* = V^* \times \text{Area} = 2.426 \times \frac{\pi}{4} \times 0.6^2$$

$$= 0.6859 \text{ m}^3/\text{sec} \quad \text{...(i)}$$

Case-2:

When an additional pipe of length 750 m and diameter 0.6 m is connected in parallel with the last half length of the pipe.

Let Q_1 = discharge in 1st parallel pipe

Q_2 = discharge in 2nd parallel pipe

$\therefore \quad Q = Q_1 + Q_2$

where, Q = Discharge in main pipe when pipes are parallel

But as the length and diameters of each parallel pipe is same

$\therefore \quad Q_1 = Q_2 = Q/2$

Consider the flow through *ABC* or *ABD*.

Head loss through *ABC* = Head lost through *AB* + head lost through *BC* ... (ii)

Head lost due to friction through *ABC* = 30 m (given)

Now, Head lost due to friction through,

$$AB = \frac{f \times 750 \times V^2}{0.6 \times 2 \times 9.81}$$ Where, V = Velocity of flow through *AB*.

$$V = \frac{Q}{\text{Area}} = \frac{Q}{\frac{\pi}{4} \times (0.6)^2} = \frac{4Q}{\pi \times 0.36}$$

$\therefore$ **Head lost due to friction through *AB***

$$= \frac{0.04 \times 750}{0.6 \times 2 \times 9.81} \times \left(\frac{4Q}{\pi \times 0.36}\right)^2 = 31.87Q^2$$

Head lost due to friction through *BC*

$$= \frac{f \times L_1 \times V_1^2}{d_1 \times 2g} = \frac{0.04 \times 750}{0.6 \times 2 \times 9.81} \times \left[\frac{Q}{2 \times \frac{\pi}{4} \times 0.6^2}\right]^2$$

$$\left[\because V_1 = \frac{Q_1}{\frac{\pi}{4} \times (0.6)^2} = \frac{Q}{2 \times \frac{\pi}{4} \times 0.6^2}\right]$$

$$= \frac{0.04 \times 750}{0.6 \times 2 \times 9.81} \times \frac{16}{4 \times \pi^2 \times 0.36^2} \times Q^2 = 7.969\ Q^2$$

Subsituting these values in equation (ii), we get

$$30 = 31.87Q^2 + 7.969Q^2 = 39.839Q^2$$

$$\Rightarrow \quad Q = \sqrt{\frac{30}{39.839}} = 0.8677 \text{m}^3/\text{sec}$$

$\therefore$ Increase in discharge = $Q - Q^*$

$$= 0.8677 - 0.6859 = 0.1818 \text{m}^3/\text{sec.} = 26.51\% \text{ increase in discharge.}$$

7.15 **A new 5 km long pipeline connects two reservoirs. The water surface elevation of the upper reservoir is 1100 m and that of the lower reservoir is 835 m. The pipeline is 400 mm nominal diameter of welded steel (Hazan Williams Coefficient C = 130) with a square mouth inlet (14.32 m long) and including the following fittings:**

1.	Gate Valves	10 numbers	Each gate valve of equivalent length	0.98 m
2.	Standard radius 90° elbow	19 numbers	Each elbow of equivalent length	6.4 m
3.	Standard radius 45° elbow	37 numbers	Each elbow of equivalent length	4.57 m
4.	Straight tees	8 numbers	Each Tee length	2.19 m

What is the water flow rate between the reservoirs? Use the concept of equivalent pipe.

Given : $h_f = \dfrac{10.7Q^{1.85}L}{C^{1.85}D^{4.87}}$

[8 marks : 2013]

Solution:

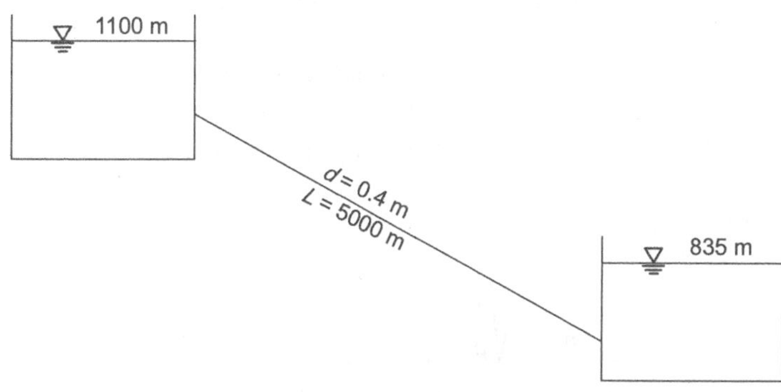

Equivalent length for gate valves $= 0.98 \times 10 = 9.8$ m

Equivalent length for 90° elbow $= 6.4 \times 19 = 121.6$ m

Equivalent length for 45° elbow = $4.57 \times 37 = 169.09$ m

Equivalent length for straight tees = $2.19 \times 8 = 17.52$ m

Total = 318.01 m

Length of square mouth inlet = 14.32 m

∴ Total length of pipe (L) = $5000 + 318.01 + 14.32$ = 5332.33 m

Hazen William's coefficient (C) = 130

$$\therefore \quad h_f = \frac{10.7 Q^{1.85} L}{C^{1.85} D^{4.87}}$$

Where $h_f = 1100 - 835 = 265$ m

$$\therefore \quad 265 = \frac{(10.7) Q^{1.85} (5332.33)}{(130)^{1.85} (0.4)^{4.87}}$$

$$\Rightarrow \quad Q = 0.63868 \text{ m}^3/\text{s} = 638.68 \text{ lt./s}$$

7.16 **A 300 mm diameter pipe of mild steel having 6 mm thickness carries water at the rate of 200 litres/s. What will be the rise in pressure if the valve at the downstream end is closed instantaneously? Compare the results assuming the pipe to be rigid as well as elastic. What should be the maximum closing time for the computed results to be valid? Take pipe length as 8.0 km. The ratio of modulus of elasticity and bulk modulus of water may be taken as 100. Assume any other data only if required and state the same clearly.**

[8 marks : 2014]

Solution:

Given: Diameter of pipe, $D = 300$ mm, Thickness of pipe material,

t = 6 mm

Q = 200 l/sec

Length of pipe = 8 km

Ratio of $\frac{E}{k}$ = 100

Area of cross-section of pipe, $A = \frac{\pi D^2}{4} = \frac{\pi \times (0.3)^2}{4} = 0.071 \text{m}^2$

Initial velocity of flow, $V_i = \frac{Q}{A} = \frac{200 \times 10^{-3}}{0.071} = 2.82 \text{m/s}$

For sudden closure, the pressure rise due to water hammer is given by

$$p = \rho V_i C_p$$

where ρ = density of fluid i.e. water

V_i = initial velocity of flow

C_p = velocity of propagation of wave which is velocity of sound

Case I: If pipe is rigid:

Velocity of sound in rigid medium,

$$C_p = \sqrt{\frac{k}{\rho}}$$

Given $\frac{E}{k} = 100$ and assuming $E = 2 \times 10^5$ N/mm² for mild steel

$$\Rightarrow \quad k = \frac{2 \times 10^5}{100} \text{N/mm}^2 = 20 \times 10^8 \text{ N/m}^2$$

Assume density of water, $\rho = 1000 \text{ kg/m}^3$

$$\therefore \quad C_p = \sqrt{\frac{20\times10^8}{10^3}} = 1414.2 \text{ m/s}$$

Rise in pressure, $P = \rho V_i C_p = 1000 \times 2.82 \times 1414.2 \text{ Pa} = 3.99 \text{ MPa} \simeq 4 \text{ MPa}$

Case II: If pipe is elastic

Velocity of sound in elastic medium

$$C_p = \sqrt{\frac{k}{\rho}}\left[\frac{1}{1+\frac{Dk}{tE}}\right]^{1/2} = 1414.21\times\left[\frac{1}{1+\frac{300}{6}\times\frac{1}{100}}\right]^{1/2} = 1154.70 \text{ m/s}$$

Hence rise in pressure, $p = \rho V_i C_p = 1000 \times 2.83 \times 1154.70 \text{ Pa} = 3.26 \text{ MPa}$

$$\text{Critical closure time, } t_{cr} = \frac{2L}{C}$$

where, L = length of pipe, C = velocity of sound

(i) **For rigid pipe medium,** $t_{cr} = \dfrac{2\times8000}{1414.21} = 11.31 \text{ sec}$

(ii) **For elastic pipe medium,** $t_{cr} = \dfrac{2\times8000}{1154.70} = 13.86 \text{ sec}$

Above results are valid if the valve is closed before 11.31 sec

7.17 **Two pipelines have been connected to a large water reservoir. One of the pipes is 15 cm × 300 m length with an outlet 4 m below the reservoir water level, while the other pipe is 20 cm × 600 m length. Both the pipes have free discharge at outlet end and the total discharge rate is 50 *L/s*. Calculate the difference of elevation between the reservoir water level and the outlet of 20 cm pipe. Neglect entrance losses, and for all pipes $f = 0.08$ in Darcy's formula**

$$h_f = \frac{flv^2}{2gd}$$

[8 marks : 2014]

Solution:

Total discharge in both pipes, $Q_0 = 50 \text{ } l/\text{s} = 0.05 \text{ m}^3/\text{s}$

Assume discharge in pipe 1 (15 cm diameter) is Q_1 and discharge in pipe 2 (20 cm diameter) is Q_2

$$\therefore \quad Q_0 = Q_1 + Q_2$$

$$\therefore \quad Q_2 = Q_0 - Q_1$$

$$Q_2 = 0.05 - Q_1 \quad \text{....(i)}$$

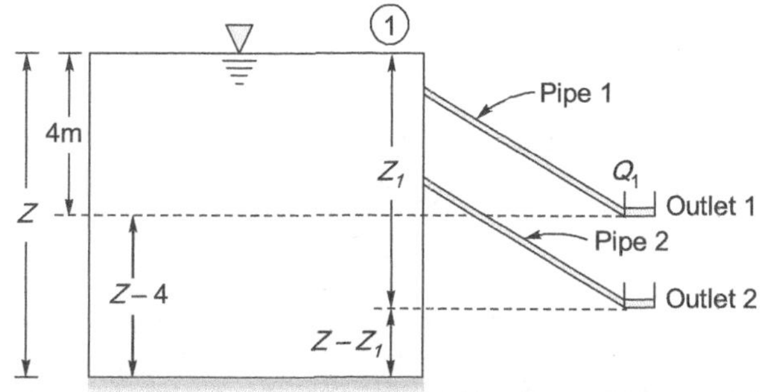

Let velocity of flow in pipe 1 is V_1 m/s and pipe 2 is V_2 m/s.

Applying Bernoulli's equation between free surface and outlet 1.

$$Z+\frac{p_{atm.}}{\rho g}+\frac{V^2}{2g} = (Z-4)+\frac{p_{atm}}{\rho g}+\frac{V_1^2}{2g}+h_{L_1}$$

Minor losses are neglected.
At free surface of reservoir,

$\therefore$ $$V = 0$$

$\therefore$ $$0 = -4+\frac{V_1^2}{2g}+\frac{fL_1V_1^2}{2gD_1}$$

$\therefore$ $$4 = V_1^2\left[\frac{1}{2g}+\frac{fL_1}{2gD_1}\right] = V_1^2\left[\frac{1}{2\times 9.81}+\frac{0.08\times 300}{2\times 9.81\times 0.15}\right]$$

$\therefore$ $$V_1 = 0.69818 \text{ m/s}$$

$\therefore$ $$Q_1 = A_1V_1$$

$$= \frac{\pi}{4}D_1^2\times V_1 = \frac{\pi}{4}\times 0.15^2\times 0.69818 = 0.01234 \text{ m}^3/\text{s}$$

$\therefore$ eq. (i)

$\Rightarrow$ $$Q_2 = 0.05-0.01234$$

$$Q_2 = 0.03766 \text{ m}^3/\text{s}$$

$\therefore$ $$V_2 = \frac{Q_2}{A_2} = \frac{0.03766}{\frac{\pi}{4}\times D_2^2} = \frac{0.03766}{\frac{\pi}{4}\times 0.2^2} = 1.19875 \text{ m/s}$$

Now, **applying Bernoulli's equation between free surface and outlet 2.**
Where Z_1 is the difference of elevation between the reservoir water level and outlet 2.

$$Z+\frac{P_{atm}}{\rho g}+\frac{V^2}{2g} = Z-Z_1+\frac{P_{atm}}{\rho g}+\frac{V_2^2}{2g}+h_{L2}$$

$\therefore$ $$Z_1 = \frac{V_2^2}{2g}+\frac{fL_2V_2^2}{2gD_2}$$

$$= \frac{1.19875^2}{2\times 9.81}+\frac{0.08\times 600\times 1.19875^2}{2\times 9.81\times 0.2} = 17.65124 \text{ m}$$

7.18 **Three reservoirs *A*, *B*, *C* are interconnected by pipes as shown in the figure.**

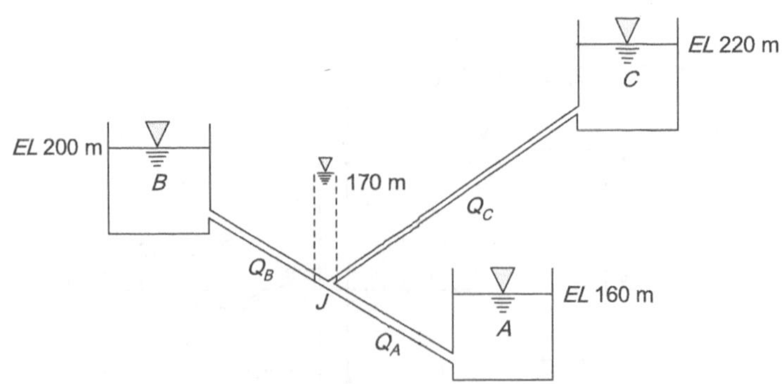

The water surface elevations in the reservoirs and the piezometric head at junction *J* are indicated in the figure. Write the mass balance equation at junction *J* and justify the answer.

[4 marks : 2014]

Solution:

$$\text{Piezometric head at } B = 200 \text{ m}$$
$$\text{Piezometric head at } C = 220 \text{ m}$$
$$\text{Piezometric head at } J = 170 \text{ m}$$
$$\text{Piezometric head at } A = 160 \text{ m}$$

Since Piezometric head at J is less than B. Hence, direction of flow will be towards J from B.

Also the Piezometric head at J is less than C. Hence, direction of flow will be towards J from C.

But Piezometric head at A is less than the Piezometric head at J. Hence, direction of flow will be towards A from J.

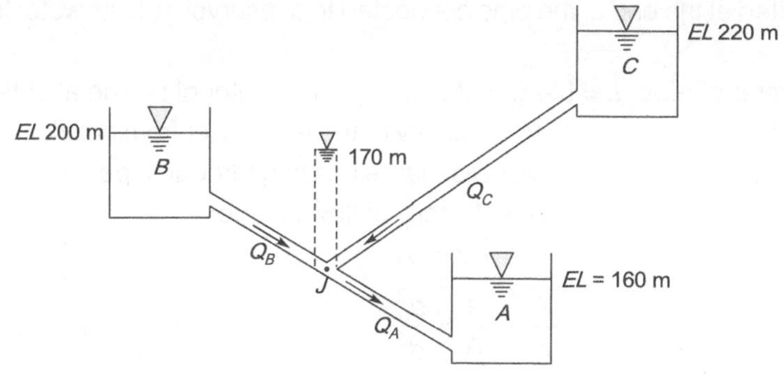

By the continuity equation at junction J,

$$\text{Total mass in} = \text{Total mass out}$$
$$Q_B + Q_C = Q_A$$

7.19 **Water (bulk modulus of elasticity = 2.2 × 10³ MPa) is flowing with a velocity of 2 m/s through an elastic pipe (modulus of elasticity = 2.1 × 10⁵ MPa) of diameter = 500 mm, thickness = 5 mm and length = 2 km. If a valve provided at the end of the pipe is closed in 2s, what will be the rise in pressure due to the resulting water hammer?**

[8 marks : 2016]

Solution:

Bulk modulus of elasticity, $k_W = 2.2 \times 10^3 \text{ MPa} = 2.2 \times 10^9 \text{ Pa}$

Flow velocity, $V = 2 \text{ m/sec}$

Modulus of elasticity, $E = 2.1 \times 10^5 \text{ MPa} = 2.1 \times 10^{11} \text{ Pa}$

Diameter of pipe, $D = 500 \text{ mm} = 0.5 \text{ m}$

Thickness of pipe, $t = 5 \text{ mm} = 0.005 \text{ m}$

Length of pipe, $L = 2 \text{ km} = 2000 \text{ m}$

$$\text{Critical Time} = \frac{2L}{C}$$

$$\because \quad C = \sqrt{\frac{k_w}{\rho\left[1+\frac{Dk_w}{Et}\right]}} = \sqrt{\frac{2.2\times10^9}{10^3\left[1+\frac{0.5(2.2\times10^9)}{(2.1\times10^{11})(0.005)}\right]}} = 1036.54 \text{ m/sec}$$

$$\therefore \quad \text{Critical time} = \frac{2\times2000}{1036.54} = 3.86 \text{ sec}$$

Since time taken to close the valve is 2 sec which is less than critical time. Hence, it is a case of sudden closure.

Since, $T < \frac{2L}{C}$ where T is time taken to close the valve.

$$\therefore \quad \text{Rise in pressure, } P = \rho.V.C = 10^3 \times 2 \times 1036.54 \times 10^{-3} = 2073.1 \text{ kPa}$$

7.20 **A pipe of diameter *d* and length *L* leads from a reservoir. At the discharge end a nozzle of diameter d_n is fitted. At the inlet of pipe head available is *H*. The nozzle is discharging into atmosphere. Prove that for maximum power transmitted by the nozzle $d_n = \sqrt[4]{\frac{d^5}{2fL}}$. Head loss in pipe is given by $h_f = \frac{fLV^2}{2gd}$ where f = Darcy's friction factor.**

[20 marks : 2017]

Solution :

Let a nozzle be fitted at the end of the pipe connected to a reservoir with its water level at a height *H* above centreline

Given: d = Diameter of pipe, L = Length of pipe, d_n = Diameter of nozzle at outlet

Let v = Velocity of the issuing jet from nozzle

Q = Discharge passing through pipe

V = Velocity of flow in pipe

$\therefore \quad Q = AV = av$

$$\Rightarrow \quad \frac{V}{v} = \frac{a}{A} = \frac{d_n^2}{d^2} \quad \text{...(i)}$$

Further neglect minor losses and the loss of energy in nozzle

We have;

$$H = \frac{fLV^2}{2d \times g} + \frac{v^2}{2g} \quad \text{...(ii)}$$

Maximum power available from a nozzle:

Power transmitted by the jet issuing from a nozzle is

$$P = \rho g(av) \times \frac{v^2}{2g}$$

Substituting value of $\frac{v^2}{2g}$ from equation (ii), we get

$$P = \rho g(av)\left[H - \frac{fLV^2}{2gd}\right]$$

$$\Rightarrow \quad P = \rho g(av)\left[H - \frac{fLa^2v^2}{dA^2 \times 2g}\right]$$

$$\text{or} \quad P = \rho g(a)\left[Hv - \frac{fLa^2v^3}{dA^2 \times 2g}\right]$$

The condition of maximum power transmitted by the nozzle may be obtained by differentiating above equation w.r.t. *v* and equating to zero; thus

$$\frac{dP}{dv} = \rho ga\left[H - \frac{3fLa^2v^2}{d \times A^2 \times 2g}\right] = 0$$

$$\text{or} \quad H - \frac{3fLa^2v^2}{d \times A^2 \times 2g} = 0$$

$$\text{or} \quad H = \frac{3fLV^2}{2g \times d} = 3h_f$$

Substituting H from equation (ii) in above equation;

$$\frac{fLV^2}{2dg}+\frac{v^2}{2g} = \frac{3fLV^2}{2g\times d}$$

$$\frac{fLV^2}{d\times 2g} = \frac{1}{2}\times\frac{v^2}{2g}$$

or
$$\frac{V^2}{v^2} = \frac{d}{2}\times\frac{1}{fL}$$

Put value of $\frac{V^2}{v^2}$ from equation (i) in above equation;

$$\Rightarrow \quad \left(\frac{d_n^2}{d^2}\right)^2 = \frac{d}{2fL}$$

$$\Rightarrow \quad d_n^4 = \frac{d^5}{2fL}$$

or
$$d_n = \sqrt[\frac{1}{4}]{\frac{d^5}{2fL}}$$

Hence proved.

7.21 **There are two reservoirs, A and B, with the water surface elevation in A 10 m higher than that in B. These are connected through two pipes in series with pipe 1 starting from A and pipe 2 ending in B. Lengths of both the pipes are 100 m, diameter of pipe 1 is 30 cm and diameter of pipe 2 is 25 cm. At the junction of these pipes, water is being withdrawn at the rate of 0.02 m³/s. Friction loss in the pipes is given by $h_f = f\frac{L}{D}\frac{v^2}{2g}$ and the Darcy friction factor, f, for both pipes is 0.02. Neglecting the head losses at the entrance and at the junction, estimate the total water withdrawal from reservoir A.**
[12 marks : 2018]

Solution:

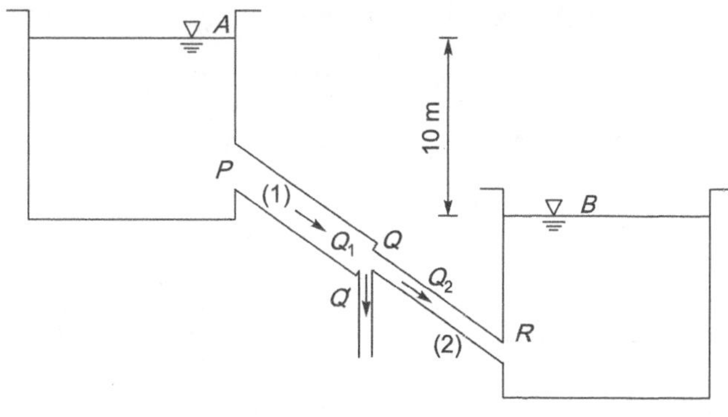

Given:

Pipe 1: L_1 = 100 m, d_1 = 30 cm

Pipe 2: L_2 = 100, d_2 = 25 cm

Let Q_1 = Discharge through pipe (1)

and Q_2 = discharge through pipe (2)

Also, Q' = discharge withdrawn at junction = 0.02 m³/s

Now, from continuity equation, at junction

$$Q_1 = Q_2 + Q'$$

or
$$Q_2 = (Q_1 - 0.02) \quad \text{...(i)}$$

Applying Bernoulli's equation to points A and B,

$$\frac{p_A}{\rho g}+\frac{V_A^2}{2g}+Z_A = \frac{p_B}{\rho g}+\frac{V_B^2}{\rho g}+Z_B+\left(h_{f_1}+h_{f_2}\right)$$

We know,

$$P_A = P_B = P_{atm} = 0$$
$$V_A = V_B = 0$$

$\therefore$
$$Z_A - Z_B = h_{f_1} + h_{f_2}$$

or
$$10 = \frac{fL_1V_1^2}{d_1\times 2g}+\frac{fL_2V_2^2}{d_2\times 2g}$$

$$10 = \frac{fL_1Q_1^2}{d_1\times 2g\left(\frac{\pi}{4}d_1^2\right)^2}+\frac{fL_2Q_2^2}{d_2\times 2g\left(\frac{\pi}{4}d_2^2\right)^2}$$

$$10 = \frac{0.02\times 100\times Q_1^2}{12.1d_1^5}+\frac{0.02\times 100\times Q_2^2}{12.1\times d_2^5} \quad \text{...(ii)}$$

Substituting the value of Q_2 from equation (i) to (ii),

$$60.5 = \frac{Q_1^2}{0.3^5}+\frac{(Q_1-0.02)^2}{0.25^5}$$

On solving this quadratic equation,

$$Q_1 = 0.2194 \text{ m}^3\text{/s}$$

$\therefore$ Water withdrawal from reservoir A, $Q_1 = 0.2194$ m/^{3}s

7.22 **A straight 20 cm diameter pipeline 4 km long is laid between two reservoirs having a difference of levels of 40 m. To increase the capacity of the system, an additional 2 km long, 20 cm diameter pipe is laid parallel from the upper reservoir to the mid-point of the original pipe. Find the increase in discharge due to installation of the new pipe. Assume f as 0.00625.**

[12 marks : 2022]

Solution:

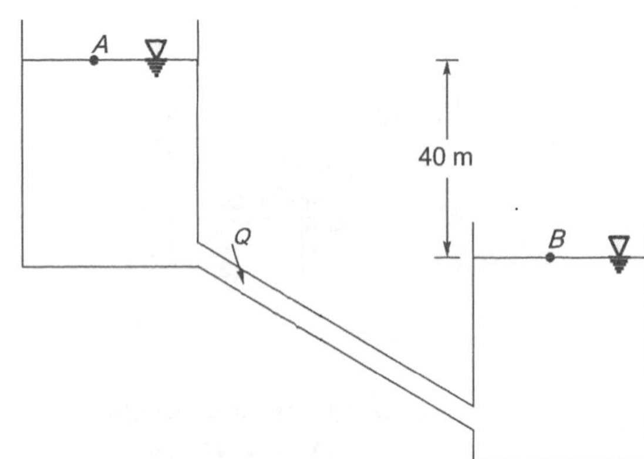

Diameter, $d = 0.2$ m

Length, $L = 4000$ m, $f = 0.00625$

Apply energy equation between (A) and (B)

$$\frac{P_A}{\rho g}+\frac{V_A^2}{\rho g}+Z_A = \frac{P_B}{\rho g}+\frac{V_B^2}{\rho g}+Z_B+h_f$$

$$\Rightarrow \quad Z_A - Z_B = \frac{8Q^2}{\pi^2 g} \cdot \frac{fL}{D^5}$$

$$\Rightarrow \quad 40 = \frac{8 \times Q^2}{\pi^2 g}\left[\frac{0.00625 \times 4000}{(0.2)^5}\right]$$

$$\Rightarrow \quad Q = 0.078718 \text{ m}^3\text{/sec.} = 78.718 \text{ lps}$$

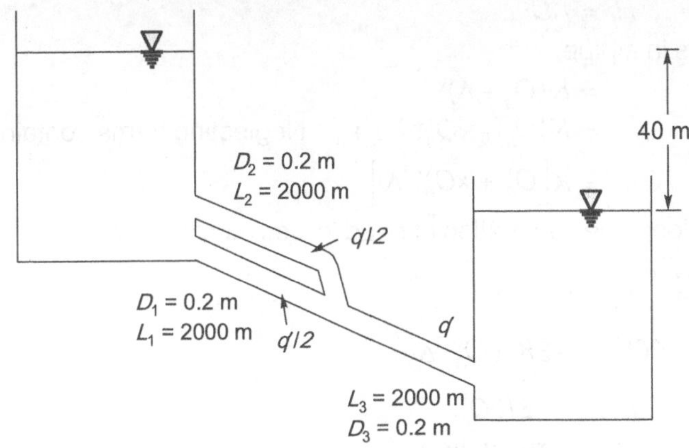

Apply energy equation between (A) and (B)

$$\frac{P_A}{\rho g} + \frac{V_A^2}{2g} + Z_A = \frac{P_B}{\rho g} + \frac{V_B^2}{2g} + Z_B + h_f$$

$$\Rightarrow \quad 40 = \frac{8(q'/2)^2}{\pi^2 g} \times \frac{f \cdot L_1}{D_1^5} + \frac{8 \times [q']^2}{\pi^2 g} \cdot \frac{f(L/2)}{D^5}$$

$$\Rightarrow \quad 40 = \frac{8(q')^2}{\pi^2 g} \times \frac{f \cdot L_1}{D^5}\left[\frac{1}{2} \times \frac{1}{4} + \frac{1}{2}\right]$$

$$\Rightarrow \quad 40 = \frac{8(q')^2}{\pi^2 \times 9.81} \times \frac{0.00625 \times 4000}{(0.2)^5} \times 0.625$$

$$\Rightarrow \quad q' = 0.099571 \text{ m}^3\text{/sec.} = 99.57 \text{ lps}$$

$\therefore$ Increase in discharge: $= 99.57 - 78.718$

$= 20.852$ lps

7.23 **Explain the Hardy Cross method used for pipe network analysis in the water distribution system. Also derive the expression to apply correction to assumed flow successively for each pipe loop in the network.**

[20 marks : 2022]

Solution:

In any pipe network, the following two conditions must be satisfied:

(i) The algebraic sum of the pressure drops around a closed loop must be zero, i.e. there can be no discontinuity in pressure.

(ii) The flow entering a junction must be equal to the flow leaving the same junction; i.e. the law of continuity must be satisfied.

Hardy-Cross Method: The flow in each pipe is assumed by the designer (in magnitude as well as in direction) in such a way that the principle of continuity is satisfied at each junction (i.e. the inflow at any junction becomes equal to the outflow at that junction).

A correction to these assumed flows is then computed successively for each pipe loop in the network, until the correction is reduced to an acceptable magnitude.

If Q_a is the assumed flow and Q is the actual flow in the pipe, then that correction Δ is given by

$$\Delta = Q - Q_a$$

or

$$Q = Q_a + \Delta$$

Now, expressing the head loss (H_L) as

$$H_L = KQ^x$$

We have, the head loss in a pipe,

$$= K(Q_a + \Delta)^x$$

$$= K[Q_a^x + xQ_a^{x-1}\Delta + ... \text{ Neglecting terms containing higher powers of } \Delta]$$

$$= K\left[Q_a^x + xQ_a^{x-1}\Delta\right]$$

Now, around a closed loop, the summation of head losses must be zero.

$$\therefore \quad \Sigma K\left[Q_a^x + xQ_a^{x-1}\cdot\Delta\right] = 0$$

$$\Rightarrow \quad \Sigma K\cdot Q_a^x = -\Sigma K.xQ_a^{x-1}\Delta$$

$$\Rightarrow \quad \Delta = -\frac{\Sigma K\cdot Q_a^x}{\Sigma x\cdot Q_a^{x-1}K}$$

Since Δ is given the same sign (or direction) in all pipes of the loop, the denominator of the above equation is taken as the absolute sum of the individual items in the summation. Hence

$$\Rightarrow \quad \Delta = \frac{\Sigma K\cdot Q_a^x}{\Sigma\left|x\cdot KQ_a^{x-1}\right|}$$

$$\Rightarrow \quad \Delta = \frac{-\Sigma H_L}{x\cdot\Sigma\left|\frac{H_L}{Q_a}\right|} \quad (\because KQ^x = H_L \text{ and } x = \text{constant})$$

where H_L = head loss for the assumed flow Q_a.

8. Viscous Flow of Incompressible Fluid

8.1 Laminar flow of oil takes place down the slope between parallel plates, 8 mm apart held inclined at 30° to the horizontal. The oil is of viscosity 0.92 kg/m-s and of mass density 1320 kg/m^3. The pressures of two points, 1 m vertically apart are 40 kN/m^2 and 170 kN/m^2, when the upper plate moves at 0.3 m/s velocity relative to the lower plate in the direction opposite to the oil flow. Determine:

(i) The maximum velocity of the flow within the oil medium,

(ii The shear stress on the top plate, and

(iii) The rate of oil flow per meter width.

[15 marks : 2003]

Solution:

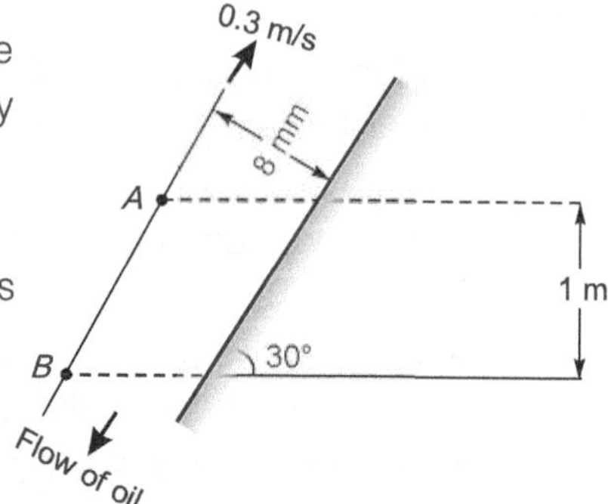

We know that for a laminar flow between two plates with one plate moving and other plate stationary, the velocity distribution is given by

$$u = \frac{U}{B}\times y + \frac{1}{2\mu}\times\frac{\partial p}{\partial x}\times(y^2 - By)$$

But in this case the moving plate and oil are moving in different directions

$$\therefore \quad u = -\frac{U}{B}\times y + \frac{1}{2\mu}\times\frac{\partial p}{\partial x}\times(y^2 - By)$$

The vertical distance between A and B is 1 m.

$$\therefore \quad \sin 30° = \frac{1}{AB}$$

$$\Rightarrow \quad AB = 2 \text{ m}$$

Taking B as datum, we get

$$p_B = 40 \times 10^3 \text{ N/m}^2$$

$$p_A = 170 \times 10^3 + \rho_{oil} \times g \times h$$

$$= 170 \times 10^3 + 1320 \times 9.81 \times 1 = 170 \times 10^3 + 13 \times 10^3 = 183 \times 10^3 \text{ N/m}^2$$

$$\therefore \quad \frac{\partial p}{\partial x} = \frac{p_B - p_A}{AB} = \frac{40 \times 10^3 - 183 \times 10^3}{2} = -71.5 \times 10^3 \text{ N/m}^3$$

(i) **For maximum velocity of flow,** $\frac{du}{dy} = 0$

$$\Rightarrow \quad \frac{d}{dy}\left[-\frac{U}{B} \times y + \frac{1}{2\mu} \times \frac{\partial p}{\partial x} \times \left(y^2 - By\right)\right] = 0$$

$$\Rightarrow \quad -\frac{U}{B} + \frac{1}{2\mu} \times \frac{\partial p}{\partial x} \times (2y - B) = 0$$

$$\Rightarrow \quad 2y - B = \frac{U}{B} \times \frac{2\mu}{\partial p/\partial x}$$

$$\Rightarrow \quad 2y = B + \frac{U}{B} \times \frac{2\mu}{\partial p/\partial x}$$

$$\Rightarrow \quad y = \frac{1}{2}\left[B + \frac{U}{B} \times \frac{2\mu}{\partial p/\partial x}\right]$$

$$\therefore \quad y = \frac{1}{2}\left[8 \times 10^{-3} + \frac{0.3}{8 \times 10^{-3}} \times \frac{2 \times 0.92}{(-71.5 \times 10^3)}\right]$$

$$= \frac{1}{2}\left[8 \times 10^{-3} - 9.65 \times 10^{-4}\right] = \frac{7.035}{2} \times 10^{-3} \times 10^3 = 3.52 \text{ mm}$$

The velocity of flow is maximum at $y = 3.52$ mm

$$\therefore \quad u = -\frac{U}{B} \times y + \frac{1}{2\mu}\frac{\partial p}{\partial x}\left(y^2 - By\right)$$

$$\Rightarrow \quad u_{max} = -\frac{0.3}{8 \times 10^{-3}} \times 3.52 \times 10^{-3} + \frac{1}{2 \times 0.92} \times 71.5 \times 10^3$$

$$[8 \times 10^{-3} \times 3.52 \times 10^{-3} - (3.52 \times 10^{-3})^2]$$

$= 0.48$ m/s **(Ans.)**

(ii) As per **Newton's law of viscosity**

$$\tau = \mu \times \frac{du}{dy}$$

$$\Rightarrow \quad \tau = \mu \times \left[-\frac{U}{B} + \frac{1}{2\mu} \times \frac{\partial p}{\partial x} \times (2y - B)\right]$$

At $\quad y = B, \tau = \tau_{top}$

$$\therefore \quad \tau_{top} = \mu \times \left[-\frac{U}{B} + \frac{1}{2\mu} \times \frac{\partial p}{\partial x} \times (2B - B) \right] = -\frac{\mu U}{B} + \frac{B}{2} \times \frac{\partial p}{dx}$$

$$= -\frac{0.92 \times 0.3}{8 \times 10^{-3}} + \frac{8 \times 10^{-3}}{2} \times (-71.5 \times 10^3) = -320.5 \text{ N/m}^2 \textbf{ (Ans.)}$$

(iii) **The discharge q of the flow per unit width of the plate** can be obtained by integrating a small discharge 'dq' passing through strip 'dy'

$$\therefore \quad dq = udy$$

$$\Rightarrow \quad q = \int_0^B \left[-\frac{U}{B} y + \frac{1}{2\mu} \times \frac{\partial p}{\partial x} \times (y^2 - By) \right] dy$$

$$= \frac{1}{2\mu} \times \frac{\partial p}{\partial x} \times \left[\frac{y^3}{3} - \frac{By^2}{2} \right]_0^B - \frac{U}{B} \times \left[\frac{y^2}{2} \right]_0^B = \frac{1}{2\mu} \times \frac{\partial p}{\partial x} \times \left[\frac{B^3}{3} - \frac{B^3}{2} \right] - \frac{UB}{2}$$

$$= -\frac{B^3}{12\mu} \times \frac{\partial p}{\partial x} - \frac{UB}{2} = -\frac{(8 \times 10^{-3})^3}{12 \times 0.92} \times (-71.5 \times 10^3) - \frac{0.3 \times 8 \times 10^{-3}}{2}$$

$$= 2.11 \times 10^{-3} \text{ m}^2\text{/sec/m } \textbf{(Ans.)}$$

8.2 Consider steady uniform flow of a Newtonian liquid of dynamic viscosity 'μ' and density 'ρ' between two parallel plates, kept at a distance B apart under a pressure gradient $\frac{\partial p}{\partial x}$ with the top plate moving at a uniform velocity U in the x-direction and with the bottom plate stationary. Develop the general expression for $\frac{u}{U}$ where 'u' is the velocity of the liquid flow at height 'y' above the bottom plate as a function of (y/B).

[5 marks : 2003]

Solution:

The flowing fluid is Newtonian, hence **Newton's law of viscosity** can be applied. As per this law, the shear stress developed between the plates is given by

$$\tau = \mu \frac{du}{dy}$$

$$\Rightarrow \quad \frac{\partial \tau}{\partial y} = \mu \frac{d^2 u}{dy^2} \quad \text{...(i)}$$

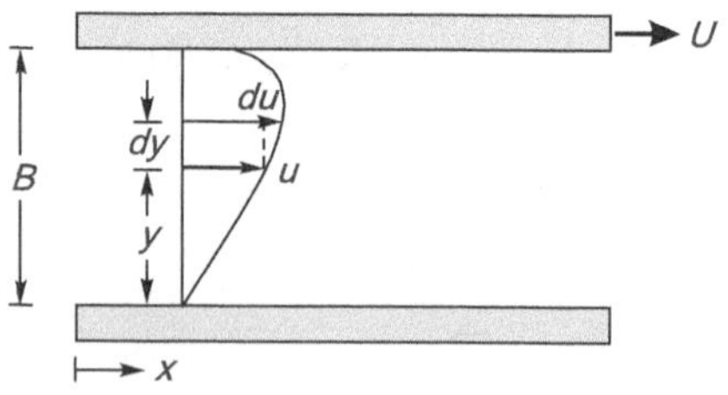

But for a steady and uniform laminar flow, the relation between pressure gradient and shear gradient is given by

$$\frac{\partial \tau}{\partial y} = \frac{\partial p}{\partial x} \quad \text{...(ii)}$$

From (i) and (ii), we get

$$\mu \frac{d^2 u}{dy^2} = \frac{\partial p}{\partial x}$$

$$\Rightarrow \quad \frac{d^2 u}{dy^2} = \frac{1}{\mu} \times \frac{\partial p}{\partial x}$$

Integrating both sides w.r.t y, we get

$$\frac{du}{dy} = \frac{1}{\mu} \times \frac{\partial p}{\partial x} \times y + C_1 \quad \text{...(iii)}$$

Integrating both sides w.r.t. y again, we get

$$u = \frac{1}{\mu} \times \frac{\partial p}{\partial x} \times \frac{y^2}{2} + C_1 y + C_2 \quad \text{...(iv)}$$

But at $y = 0$, $u = 0$

Thus, from (iv) we get

$$u = \frac{1}{\mu} \times \frac{\partial p}{\partial x} \times \frac{y^2}{2} + C_1 y + C_2$$

$$\Rightarrow \quad 0 = \frac{1}{\mu} \times \frac{\partial p}{\partial x} \times 0 + C_1 \times 0 + C_2$$

$$\Rightarrow \quad C_2 = 0$$

Also at $y = B$, $u = U$

Thus from (iv), we get

$$u = \frac{1}{\mu} \times \frac{\partial p}{\partial x} \times \frac{y^2}{2} + C_1 y$$

$$\Rightarrow \quad U = \frac{1}{\mu} \times \frac{\partial p}{\partial x} \times \frac{B^2}{2} + C_1 B$$

$$\Rightarrow \quad \frac{U}{B} = \frac{1}{\mu} \times \frac{\partial p}{\partial x} \times \frac{B}{2} + C_1$$

$$\Rightarrow \quad C_1 = \frac{U}{B} - \frac{B}{2\mu} \times \frac{\partial p}{\partial x}$$

Substituting value of C_1 in (iv), we get

$$u = \frac{1}{\mu} \times \frac{\partial p}{\partial x} \times \frac{y^2}{2} + U \times \frac{y}{B} - \frac{By}{2\mu} \times \frac{\partial p}{\partial x}$$

$$\Rightarrow \quad u = U \times \frac{y}{B} + \frac{1}{2\mu} \times \frac{\partial p}{\partial x} \times \left(y^2 - By\right)$$

$$\Rightarrow \quad u = U \times \frac{y}{B} + \frac{1}{2\mu} \times \frac{\partial p}{\partial x} \times y \times (y - B)$$

$$\Rightarrow \quad u = U \times \frac{y}{B} + \frac{B}{2\mu} \times \frac{\partial p}{\partial x} \times y \times \left(\frac{y}{B} - 1\right)$$

$$\Rightarrow \quad \frac{u}{U} = \frac{y}{B} + \frac{1}{2\mu} \times \frac{By}{U} \times \frac{\partial p}{\partial x} \times \left(\frac{y}{B} - 1\right)$$

If

$$K = \frac{1}{2\mu} \times \frac{By}{U}$$

Then

$$\frac{u}{U} = \frac{y}{B} + K \times \frac{\partial p}{\partial x} \times \left(\frac{y}{B} - 1\right)$$

8.3 Assuming the following velocity distribution in a circular pipe:

$u = U_{max}\left(1-\frac{r}{R}\right)^{1/7}$, where U_{max} = maximum velocity; R = radius of pipe; u = velocity at a radius 'r' from the centre of pipe.

(a) calculate the ratio of the mean and maximum velocities.

(b) calculate the radius at which the actual velocity equals the mean velocity.

[20 marks : 2004]

Solution:

(a) Given that

$$u = U_{max}\left(1-\frac{r}{R}\right)^{1/7}$$

The discharge Q passing through any cross-section of a circular pipe can be obtained by integrating a small discharge (dQ) passing through an elementary ring of thickness dr, considered at a radial distance r, out of the cross-sectional area of the pipe.

$$\therefore \quad dQ = udA$$

$$\Rightarrow \quad dQ = u \times 2\pi r \times dr$$

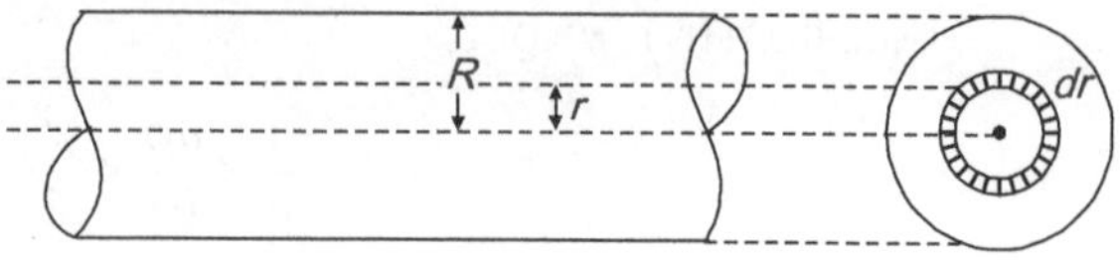

$$\Rightarrow \quad Q = \int_0^R U_{max}\left(1-\frac{r}{R}\right)^{1/7} \times 2\pi r \times dr$$

$$= 2\pi U_{max}\int_0^R r\left(1-\frac{r}{R}\right)^{1/7} dr$$

$$= 2\pi U_{max}\left[r \times \int\left(1-\frac{r}{R}\right)^{1/7} dr - \int\left\{\frac{dr}{dr} \times \int\left(1-\frac{r}{R}\right)^{1/7} dr\right\}dr\right]_0^R$$

$$= 2\pi U_{max}\left[-rR \times \frac{7}{8} \times \left(1-\frac{r}{R}\right)^{8/7} + \int R \times \frac{7}{8} \times \left(1-\frac{r}{R}\right)^{8/7} dr\right]_0^R$$

$$= 2\pi U_{max}\left[-rR \times \frac{7}{8} \times \left(1-\frac{r}{R}\right)^{8/7} - R^2 \times \frac{7}{8} \times \frac{7}{15} \times \left(1-\frac{r}{R}\right)^{15/7}\right]_0^R$$

$$= 2\pi U_{max}\left[0 - 0 + 0 + \frac{49}{120} \times R^2\right]$$

$$\Rightarrow \quad Q = \frac{49}{60} \times U_{max} \times \pi R^2$$

Mean velocity, $$U = \frac{\text{Total discharge}}{\text{Total area of flow}} = \frac{\frac{49}{60} \times U_{max} \times \pi R^2}{\pi R^2} = \frac{49}{60} \times U_{max}$$

$$\therefore \quad \frac{U}{U_{max}} = \frac{\frac{49}{60} \times U_{max}}{U_{max}} = \frac{49}{60}$$

(b) $$u = U_{max}\left(1-\frac{r}{R}\right)^{1/7} \quad \ldots(i)$$

But $$U = \frac{49}{60} \times U_{max} \quad \text{...(ii)}$$

As per the problem (i) is equal to (ii) i.e.

$$U_{max}\left(1-\frac{r}{R}\right)^{1/7} = \frac{49}{60} \times U_{max}$$

$$\Rightarrow \quad 1-\frac{r}{R} = \left(\frac{49}{60}\right)^7$$

$$\Rightarrow \quad \frac{r}{R} = 1-\left(\frac{49}{60}\right)^7$$

$$\Rightarrow \quad r = 0.758\,R$$

8.4 **An oil having viscosity 0.08 N-s/m², specific weight 8829 N/m³, density 900 kg/m³ flows at the rate of 5.4×10^{-3} m³/s through a horizontal circular pipe of 0.12 m diameter and length 150 m. Find,**
(i) pressure difference in 150 m length in kN/m²
(ii) wall shear stress in N/m² and
(iii) average and maximum velocity

[4 marks : 2010]

Solution:

Viscosity (μ) = 0.08 N-s/m², Specific weight (γ) = 8829 N/m³, Density (ρ) = 900 kg/m³, Discharge (Q) = 5.4×10^{-3} m³/s, Diameter (D) = 0.12 m, Length of pipe (L) = 150 m

$\therefore$ Velocity of flow,
$$V = \frac{Q}{\pi D^2/4} = \frac{4 \times 5.4 \times 10^{-3}}{\pi \times (0.12)^2} = 0.48 \text{ m/s}$$

Reynolds number for the flow may be given as

$$Re = \frac{\rho VD}{\mu} = \frac{900 \times 0.48 \times 0.12}{0.08} = 648 < 2000$$

Thus the flow in the pipe is laminar.

(i) **Pressure difference** in the pipe is given by

$$\Delta p = \frac{32\,\mu VL}{D^2} = \frac{32 \times 0.08 \times 0.48 \times 150}{(0.12)^2} \times 10^{-3} = 12.8 \text{ kN/m}^2$$

(ii) **Wall shear stress** is given by

$$\tau_0 = \frac{8\mu V}{D} = \frac{8 \times 0.08 \times 0.48}{0.12} = 2.56 \text{ N/m}^2$$

(iii) **Average velocity,** $V = 0.48$ m/s

Maximum velocity, $V_{max} = 2V = 2 \times 0.48 = 0.96$ m/s

8.5 **The velocity distribution in a stream is generally governed by one by seventh law, i.e., $\frac{V}{V_0} = \left(\frac{Y}{Y_0}\right)^{1/7}$, where V and V_0 are point velocities at heights Y and Y_0 from the stream bed. Show that point velocity at 0.6 depth from free surface is equal to average velocity.**

[5 marks : 2013]

Solution:

$$\frac{v}{v_0} = \left(\frac{y}{y_0}\right)^{1/7}$$

$\Rightarrow$
$$v = v_0\left(\frac{y}{y_0}\right)^{1/7}$$

Assuming unit width of stream

$\therefore$
$$A = (D \times 1) = D$$

Discharge in elementary strip $(dQ) = v(dy.1) = vdy$

$$= v_0\left(\frac{y}{y_0}\right)^{1/7} dy$$

$\therefore$ Total discharge,
$$Q = \int dQ = \int_0^D v_0\left(\frac{y}{y_0}\right)^{1/7} dy$$

$$= \frac{v_0}{y_0^{1/7}}\frac{7}{8}\left[y^{8/7}\right]_0^D = \frac{7v_0}{8y_0^{1/7}}D^{8/7}$$

$\therefore$ Average velocity,
$$v = \frac{Q}{A} = \frac{Q}{D} = \frac{7}{8}\frac{v_0}{y_0^{1/7}}D^{1/7} \quad \text{...(i)}$$

Velocity at 0.6 depth from free surface= Velocity at 0.4 depth from stream bed

Hence,
$$y = 0.4D$$

$\therefore$
$$v_{0.4D} = v_0\left(\frac{0.4D}{y_0}\right)^{1/7} = \frac{v_0}{y_0^{1/7}}(0.87731)D^{1/7}$$

Hence,
$$y = 0.4D$$

$$\simeq \frac{7}{8}\frac{v_0 D^{1/7}}{y_0^{1/7}} \quad \text{...(ii)}$$

From eq. (i) and (ii)

Average velocity = Velocity at 0.6 m depth from free surface.

8.6 What are the functions of a surge tank?

[5 marks : 2020]

Solution:

Surge tank: It is a device introduced within a hydropower water conveyance system having a rather long pressure conduit to absorb the excess pressure rise in case of a sudden value closure thus eliminating water hammer effect.

- It also acts as a small storage from which water may be supplied in case of a sudden valve opening of the turbine.
- It reduces the distance between turbine and free surface.

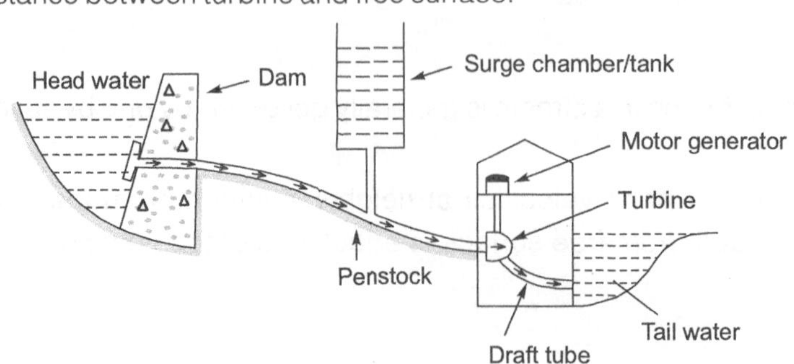

9. Turbulent Flow in Pipes

9.1 **A pipeline carrying water has surface roughness of average height 1.165 mm. If the shear stress developed is 3.6 N/m^2, what is the nature of the pipe surface?**

[4 marks : 2017]

Solution:

Given:

Height of surface roughness/protrusions = k = 1.165 mm,

Boundary shear stress, $\tau_0 = 3.60\text{ N/m}^2$

We know,

Kinematic viscosity of water, $\nu = 1 \times 10^{-6}\text{ m}^2\text{/sec at }20°\text{C}$

∴ Shear velocity, $$u_* = \sqrt{\frac{\tau_0}{\rho}} = \sqrt{\frac{3.60}{1000}} = 0.06\text{ m/sec}$$

Hence, $$\frac{u_* k}{\nu} = \frac{0.06 \times 1.165 \times 10^{-3}}{1 \times 10^{-6}} = 69.90 \simeq 70$$

Check:

$$\frac{k}{\delta'} = \frac{u_* k}{11.6\nu} = \frac{0.06 \times 1.165 \times 10^{-3}}{11.6 \times 10^{-6}} = 6.026 > 6$$

∴ The pipe surface is hydrodynamically rough.

9.2 **The velocity of flow in a badly corroded 8.0 cm pipe is found to increase 30% as a Pitot tube is moved from a point 1.0 cm from the wall to 2.0 cm from the wall. Assuming pipe to be a rough pipe, estimate the height of roughness elements.**

[8 marks : 2017]

Solution:

The flow is rough turbulent and as such $\dfrac{u}{u_*} = 5.75\log_{10}\left(\dfrac{y}{k}\right) + 8.50$ ...(i)

where, u_* = Shear velocity, k = Height of protrusions/roughness

Hence, $$\frac{u_2 - u_1}{u_*} = 5.75\log\left(\frac{y_2}{y_1}\right)$$

Given: $u_2 = 1.30u_1$, y_1 = 1.0 cm and y_2 = 2.0 cm from pipe wall.

$$\frac{0.30u_1}{u_*} = 5.75\log\left(\frac{2.0}{1.0}\right) = 1.731$$

$$\Rightarrow \quad \frac{u_1}{u_*} = 5.77 \quad \text{...(ii)}$$

Put this value in equation (i);

$$5.77 = 5.75\log\left(\frac{0.01}{k}\right) + 8.50$$

$$\Rightarrow \quad \log\left(\frac{0.01}{k}\right) = -0.475$$

$$\Rightarrow \quad k = 2.984 \times 10^{-2}\text{ m} = 29.84\text{ mm} \simeq 29.8\text{ mm}$$

10. Boundary Layer Theory

10.1 Explain: (a) Boundary layer thickness (b) Displacement thickness (c) Form Drag

The velocity distribution in a laminar boundary layer is given by $\frac{u_x}{U_\infty} = 3\left(\frac{y}{\delta}\right) - 2\left(\frac{y}{\delta}\right)^2$

where

u_x = velocity in x-direction at a distance y from the boundary

δ = boundary layer thickness

U_∞ = free stream velocity.

Find the ratio of the displacement thickness to the boundary layer thickness.

[15 marks : 1999]

Solution:

(i) **Boundary layer Thickness:** It is defined as the distance from the boundary surface to the point where the velocity reaches 99% of the free stream velocity. It is also known as nominal boundary thickness and is denoted by (δ). In other words, the boundary layer thickness may be considered equal to the distance y from the boundary surface at which $u = 0.99\ U_\infty$

(ii) **Displacement Thickness:** It is defined as the distance by which the boundary would have to be displaced in order to compensate for the reduction in mass flow rate on account of boundary layer growth. It is denoted by (δ^*). Mathematically,

$$\delta^* = \int_0^\delta \left(1 - \frac{u}{U_\infty}\right) dy$$

(iii) **Form drag:** If the surface of a immersed object along which the boundary layer forms is such that it curves away from the flow, there exists a tendency for the flowing fluid to leave the boundary. This phenomenon is known as separation of flow, which usually occurs at higher value of Reynolds number. On the downstream side of the body on account of the separation a region of low pressure is developed which is known as 'wake'. Since on the upstream side of the body the pressure being considerably high, there exists a pressure difference between the upstream and downstream sides. The pressure difference so created results in producing a drag on the body, which is known as form drag or pressure drag.

Now we have the velocity distribution as

$$\frac{u_x}{U_\infty} = 3\left(\frac{y}{\delta}\right) - 2\left(\frac{y}{\delta}\right)^2$$

Displacement thickness is given by

$$\delta^* = \int_0^\delta \left(1 - \frac{u_x}{U_\infty}\right) dy$$

$$\therefore \quad \delta^* = \int_0^\delta \left(1 - \frac{u_x}{U_\infty}\right) dy = \int_0^\delta \left(1 - 3\left(\frac{y}{\delta}\right) + 2\left(\frac{y}{\delta}\right)^2\right) dy$$

$$= \int_0^\delta \left(1 - \frac{3y}{\delta} + \frac{2y^2}{\delta^2}\right) dy = \left(y - \frac{3y^2}{2\delta} + \frac{2y^3}{3\delta^2}\right)_0^\delta = \delta - \frac{3\delta}{2} + \frac{2\delta}{3} = \frac{6\delta - 9\delta + 4\delta}{6}$$

$$\delta^* = \frac{\delta}{6}$$

The ratio of displacement thickness (δ^*) to boundary layer thickness (δ) will be

$$\frac{\delta^*}{\delta} = \frac{\delta/6}{\delta} = \frac{1}{6}$$

10.2 Calculate the friction drag on a plate 0.15 m wide and 0.45 m long placed longitudinally in a stream of oil having a free stream velocity of 6 m/sec. Also find the thickness of the boundary layer and shear stress at the trailing edge. Specific gravity of oil is 0.925 and kinematic viscosity is 0.9×10^{-4} m²/sec.

[15 marks : 2000]

Solution:

Length of plate, $L = 0.45$ m

Width of plate, $B = 0.15$ m

Free stream velocity, $U_\infty = 6$ m/s

Specific gravity of oil, $S = 0.925$

Kinematic viscosity of oil $v = 0.9 \times 10^{-4}$ m²/s

We know that Reynolds number at the trailing edge is given by

$$Re_L = \frac{U_\infty L}{v} = \frac{6\times0.45}{0.9\times10^{-4}} = 30000 < 5 \times 10^5$$

Hence the boundary layer is laminar over the entire length of plate.

Average drag coefficient, $$C_D = \frac{1.328}{\sqrt{Re_L}} = \frac{1.328}{\sqrt{30000}} = 7.667 \times 10^{-3}$$

Friction drag on both sides of the plate $= C_D \times \rho \times A \times U_\infty^2$

$$= 7.667 \times 10^{-3} \times 0.925 \times 10^3 \times 0.15 \times 0.45 \times (6)^2 = 17.23 \text{ N}$$

Thickness of boundary layer may be given by

$$\delta = \frac{5x}{\sqrt{Re_x}}$$

∴ Thickness of boundary layer at the trailing edge will be

$$\delta_L = \frac{5L}{\sqrt{Re_L}} = \frac{5\times0.45}{\sqrt{30000}} = 0.013 \text{ m} = 13 \text{ mm}$$

Shear stress at the trailing edge will be given by

$$\frac{\tau_0}{\rho U_\infty^2/2} = \frac{0.664}{\sqrt{Rc_L}}$$

$$\Rightarrow \quad \frac{\tau_0 \times 2}{0.925\times10^3\times(6)^2} = \frac{0.664}{\sqrt{30000}}$$

$$\Rightarrow \quad \tau_o = 63.83 \text{ N/m}^2$$

10.3 Consider the basic equations wherein (i) for an entirely laminar boundary layer on a flat plate, the average drag coefficient is $k_1 Re_L^{-1/2}$ and (ii) for an entirely turbulent boundary layer on the flat plate, the average drag coefficient is $k_2 Re_L^{-1/5}$. Based on these with adopting the known values of k_1 and k_2, derive an expression for the average drag coefficient on the plate when a part of the plate is covered by laminar boundary layer and the rest by turbulent boundary layer with the transition occurring where the local Reynolds number is 5×10^5.

[15 marks : 2002]

Solution:

Let the average drag coefficient be C_f.

When the boundary layer is entirely laminar for the total length of the flat plate, the average drag coefficient is then given by Blasius equation i.e.

$$C_d = \frac{1.328}{(\text{Re}_L)^{1/2}} \quad \text{where } \text{Re}_L = \frac{VL}{\nu}$$

But as per the problem,

$$C_f = \frac{k_1}{(\text{Re}_L)^{1/2}}$$

$$\therefore \quad k_1 = 1.328$$

Similarly, when the boundary layer is entirely turbulent for the total length of the flat plate, then

$$C_d = \frac{0.074}{(\text{Re}_L)^{1/5}}$$

$$\therefore \quad k_2 = 0.074$$

Given that the critical Reynolds number = 5×10^5

$$\therefore \quad \text{Re}_{x,cr} = \frac{Vx_{cr}}{\nu} = 5 \times 10^5$$

where x_{cr} is the distance from the leading edge of the flat plate where the transition from laminar to turbulent boundary layer occurs.

Also $$\frac{\text{Re}_{x,cr}}{\text{Re}_L} = \frac{Vx_{cr}/\nu}{VL/\nu} = \frac{x_{cr}}{L}$$

Let the local drag coefficient be $C_{f,x}$.

The average drage coeficient for the entire plate of length L is determined by integration as

$$C_f = \frac{1}{L}\int_0^L C_{f,x}\, dx$$

$$\Rightarrow \quad C_f = \frac{1}{L}\left[\int_0^{x_{cr}} C_{f,x,\text{laminar}}\, dx + \int_{x_{cr}}^{L} C_{f,x,\text{turbulent}}\, dx\right]$$

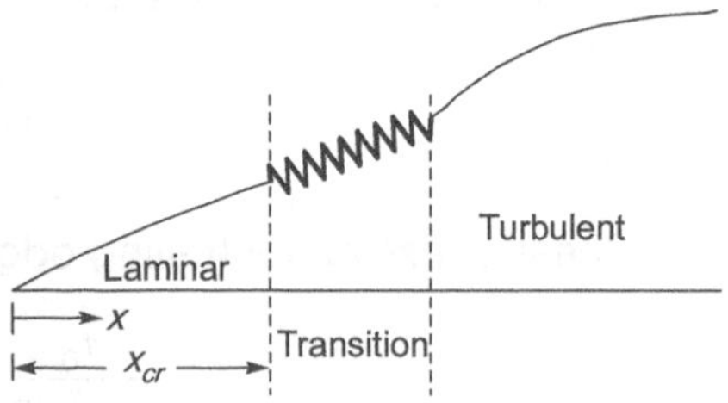

As per Blasius equation,

$$C_{f,x,\text{laminar}} = \frac{0.664}{(\text{Re}_x)^{1/2}} \text{ and } C_{f,x,\text{turbulent}} = \frac{0.059}{(\text{Re}_x)^{1/5}}$$

Laminar
Turbulent
Transition
x
x_{cr}

$$\therefore \quad C_f = \frac{1}{L}\left[\int_0^{x_{cr}} \frac{0.664}{(\text{Re}_x)^{1/2}}dx + \int_{x_{cr}}^{L}\frac{0.059}{(\text{Re}_x)^{1/5}}dx\right]$$

$$= \frac{1}{L}\left[\int_0^{x_{cr}} 0.664 \times \left(\frac{Vx}{\nu}\right)^{-1/2} dx + \int_{x_{cr}}^{L} 0.059 \times \left(\frac{Vx}{\nu}\right)^{-1/5} dx\right]$$

$$= \frac{1}{L}\left[0.664 \times \left(\frac{V}{\nu}\right)^{-1/2} \times 2 \times \left[x^{1/2}\right]_0^{x_{cr}} + 0.059 \times \left(\frac{V}{\nu}\right)^{-1/5} \times \frac{5}{4} \times \left[x^{4/5}\right]_{x_{cr}}^{L}\right]$$

$$= \frac{1}{L}\left[1.328 \times \left(\frac{Vx_{cr}}{\nu}\right)^{-1/2} \times x_{cr} + 0.074 \times \left(\frac{V}{\nu}\right)^{-1/5} \times \left(L^{4/5} - x_{cr}^{4/5}\right)\right]$$

$$= \frac{1}{L}\left[0.074 \times \left(\frac{VL}{\nu}\right)^{-1/5} \times \frac{L^{4/5}}{L^{-1/5}} - 0.074 \times \left(\frac{Vx_{cr}}{\nu}\right)^{-1/5} \times \frac{x_{cr}^{4/5}}{x_{cr}^{-1/5}} + 1.328 \times \left(\frac{Vx_{cr}}{\nu}\right)^{-1/2} \times x_{cr}\right]$$

$$= \frac{1}{L}\left[\frac{0.074}{(\text{Re}_L)^{1/5}} \times L - 0.074 \times (5\times10^5)^{-1/5} \times x_{cr} + 1.328 \times (5\times10^5)^{-1/2} \times x_{cr}\right]$$

$$= \frac{1}{L}\left[\frac{0.074}{(\text{Re}_L)^{1/5}} \times L - 3.485 \times 10^{-3} \times x_{cr}\right]$$

$$= \frac{0.074}{(\text{Re}_L)^{1/5}} - 3.485 \times 10^{-3} \times \frac{x_{cr}}{L}$$

$$= \frac{0.074}{(\text{Re}_L)^{1/5}} - 3.485 \times 10^{-3} \times \frac{5\times10^5}{\text{Re}_L}$$

$$\Rightarrow \quad C_f = \frac{0.074}{(\text{Re}_L)^{1/5}} - \frac{1742}{\text{Re}_L}$$

10.4 **Wind is blowing with a velocity of 120 km per hour over a smooth two dimensional flat plate. If the laminar boundary layer exists up to a value of Reynolds number Re_x equal to 4×10^5, find the maximum distance upto which laminar boundary layer persists and its maximum thickness. Assume kinematic viscosity of air as 1.49×10^{-5} m²/s.**

[10 marks : 2006]

Solution:

Velocity of air, $U_\infty = 120$ kmph

Reynolds number, $\text{Re}_x = 4 \times 10^5$

Kinematic viscosity of air = 1.49×10^{-5} m²/s.

We know that

$$\text{Re}_x = \frac{U_\infty x}{\nu}$$

where x is the distance upto which laminar boundary layer persists.

$$\therefore \quad 4 \times 10^5 = \frac{120}{3.6} \times \frac{x}{1.49 \times 10^{-5}}$$

$$\Rightarrow \quad x = \frac{3.6 \times 4 \times 1.49}{120} = 0.1788 \text{ m}$$

Since the velocity profile for the laminar boundary layer is not known, Blasius equation may be used to determine the thickness of boundary layer.

As per Blasius equation, the thickness of laminar boundary layer is given by

$$\delta = \frac{5x}{\sqrt{\text{Re}_x}} = \frac{5 \times 0.1788}{\sqrt{4 \times 10^5}}$$

$$= 1.41 \times 10^{-3} \text{ m} = 1.41 \text{ mm}$$

10.5 **If the velocity distribution in a pipe is given by $u = u_{max}\left(1 - \frac{r}{r_0}\right)^n$ where u is the velocity at any radius r, r_0 is the radius of pipe and u_{max} is the maximum velocity, find (i) average velocity and (ii) kinetic energy correction factor.**

[10 marks : 2009]

Solution:

(i) Let the average velocity of flow in the pipe be U.

Let the discharge through an elementary area dA of the pipe be dQ.

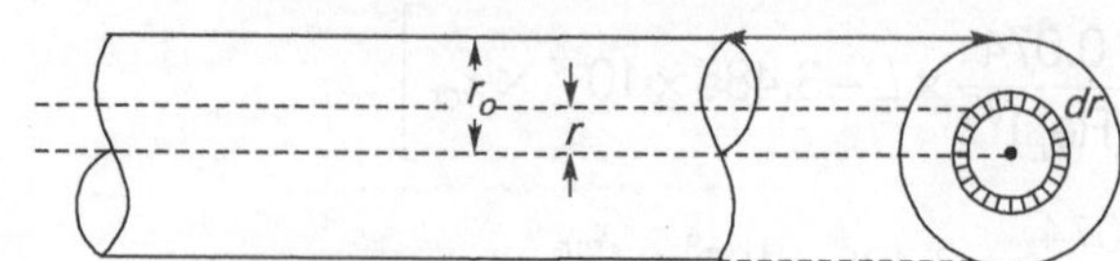

$\therefore \quad dQ = udA$

$$\Rightarrow \quad dQ = u_{max}\left(1-\frac{r}{r_o}\right)^n 2\pi r\, dr$$

$$\Rightarrow \quad Q = \int_0^{r_o} 2\pi r \cdot u_{max}\left(1-\frac{r}{r_o}\right)^n dr = 2\pi u_{max}\int_0^{r_o} r\left(1-\frac{r}{r_o}\right)^n dr$$

$$\Rightarrow \quad Q = 2\pi u_{max}\left[-\frac{r \times r_o}{n+1}\times\left(1-\frac{r}{r_o}\right)^{n+1} - \int\left[\frac{dr}{dr}\times\int\left(1-\frac{r}{r_o}\right)^n dr\right]dr\right]_0^{r_o}$$

$$\Rightarrow \quad Q = 2\pi u_{max}\left[-\frac{rr_o}{n+1}\times\left(1-\frac{r}{r_o}\right)^{n+1} + \int\frac{r_o}{n+1}\left(1-\frac{r}{r_o}\right)^{n+1} dr\right]_0^{r_o}$$

$$\Rightarrow \quad Q = 2\pi u_{max}\left[-\frac{rr_o}{n+1}\times\left(1-\frac{r}{r_o}\right)^{n+1} - \frac{r_o^2}{(n+1)(n+2)}\times\left(1-\frac{r}{r_o}\right)^{n+2}\right]_0^{r_o}$$

$$\Rightarrow \quad Q = 2\pi u_{max}\left[0-0+0+\frac{r_o^2\times(1)^{n+2}}{(n+1)(n+2)}\right] = \frac{2\pi r_o^2 u_{max}}{(n+1)(n+2)}$$

But $\quad Q = AU = \pi r_o^2 U$

$$\therefore \quad U = \frac{2\pi r_o^2 u_{max}}{(n+1)(n+2)\times\pi r_o^2}$$

$$\Rightarrow \quad U = \frac{2u_{max}}{(n+1)(n+2)} \text{ (Ans.)}$$

(ii) We know that kinetic energy correction factor is given by

$$\alpha = \frac{1}{AU^3}\int u^3 dA = \frac{1}{\pi r_o^2}\int_0^{r_o}\left(\frac{u}{U}\right)^3 2\pi r \cdot dr$$

$$= \frac{1}{\pi r_o^2}\int_0^{r_o}\left(\frac{u_{max}\left(1-\frac{r}{r_o}\right)^n}{\frac{2u_{max}}{(n+1)(n+2)}}\right)^3 2\pi r dr = \frac{2\pi}{8\pi r_o^2}(n+1)^3(n+2)^3\int_0^{r_o} r\left(1-\frac{r}{r_o}\right)^{3n} dr$$

$$= \frac{(n+1)^3(n+2)^3}{4r_o^2}\left[r\int\left(1-\frac{r}{r_o}\right)^{3n} dr - \int\left[\frac{dr}{dr}\int\left(1-\frac{r}{r_o}\right)^{3n} dr\right]dr\right]_0^{r_o}$$

$$= \frac{(n+1)^3(n+2)^3}{4r_o^2}\left[-\frac{rr_o}{3n+1}\times\left(1-\frac{r}{r_o}\right)^{3n+1} + \int \frac{r_o}{3n+1}\times\left(1-\frac{r}{r_o}\right)^{3n+1} dr\right]_0^{r_o}$$

$$= \frac{(n+1)^3(n+2)^3}{4r_o^2}\left[-\frac{rr_o}{3n+1}\times\left(1-\frac{r}{r_o}\right)^{3n+1} - \frac{r_o^2}{(3n+1)(3n+2)}\times\left(1-\frac{r}{r_o}\right)^{3n+2}\right]_0^{r_o}$$

$$= \frac{(n+1)^3(n+2)^3}{4r_o^2}\left[0-0+0+\frac{r_o^2}{(3n+1)(3n+2)}\right] = \frac{(n+1)^3(n+2)^3}{4r_o^2}\times\frac{r_o^2}{(3n+1)(3n+2)}$$

$$\alpha = \frac{(n+1)^3(n+2)^3}{4(3n+1)(3n+2)} \text{ (Ans.)}$$

10.6 Show that if a plate of size 10 m x 5 m is towed through a fluid so that the boundary layer is entirely laminar, the ratio of towing speeds so that the drag force remains constant regardless of whether 10 m or 5 m side is in the flow direction is given by $\frac{U_{10m}}{U_{5m}} = 1.2599$

If the boundary layer is entirely turbulent and Reynolds number is less than 10^7, $\frac{U_{10m}}{U_{5m}} = 1.08$.

[4 marks : 2010]

Solution:

(i) When the flow is laminar i.e. Re < 5 × 10^5

When 10 m side is in the flow direction,

$$Re_{10} = \frac{\rho \times U_{10} \times 10}{\mu} \quad ...(i)$$

Drag force,
$$F_{D_{10}} = \frac{C_{D_{10}}\rho A U_{10}^2}{2} \quad ...(ii)$$

But for laminar flow,
$$C_{D_{10}} = \frac{1.328}{(Re_{10})^{1/2}} \quad ...(iii)$$

Substituting value of $C_{D_{10}}$ from (iii) in (ii), we get

$$F_{D_{10}} = \frac{1.328}{(Re_{10})^{1/2}} \times \frac{\rho A U_{10}^2}{2}$$

Similarly, drag force on the plate when 5 m side is in the flow direction.

$$F_{D_5} = \frac{1.328}{(Re_5)^{1/2}} \times \frac{\rho A U_5^2}{2}$$

But as per the question, the drag force is constant whether 10 m or 5 m side is in the flow direction

$$\therefore \quad F_{D_{10}} = F_{D_5}$$

$$\Rightarrow \quad \frac{1.328}{(Re_{10})^{1/2}} \times \frac{\rho A U_{10}^2}{2} = \frac{1.328}{(Re_5)^{1/2}} \times \frac{\rho A U_5^2}{2}$$

$$\Rightarrow \quad \left(\frac{U_{10}}{U_5}\right)^2 = \left(\frac{Re_{10}}{R_5}\right)^{1/2}$$

$$\Rightarrow \quad \left(\frac{U_{10}}{U_5}\right)^4 = \frac{Re_{10}}{Re_5} = \frac{\rho \times U_{10} \times 10/\mu}{\rho \times U_5 \times 5/\mu} = \frac{U_{10}}{U_5} \times 2$$

$$\Rightarrow \quad \frac{U_{10}}{U_5} = (2)^{1/3}$$

$$\Rightarrow \quad \frac{U_{10}}{U_5} = 1.2599 \text{ (Hence proved)}$$

(ii) When the flow is entirely turbulent i.e. Re < 10^7

Coefficient of drag, $\quad C_D = \dfrac{0.074}{[Re_1]^{1/5}}$

Since the drag force is constant, therefore

$$F_{D_{10}} = F_{D_5}$$

$$\Rightarrow \quad \frac{0.074}{[Re_{10}]^{1/5}} \times \frac{\rho A U_{10}^2}{2} = \frac{0.074}{[Re_5]^{1/5}} \times \frac{\rho A U_5^2}{2}$$

$$\Rightarrow \quad \left(\frac{U_{10}}{U_5}\right)^2 = \left(\frac{Re_{10}}{Re_5}\right)^{1/5}$$

$$\Rightarrow \quad \left(\frac{U_{10}}{U_5}\right)^{10} = \left[\frac{\rho \times U_{10} \times 10/\mu}{\rho \times U_5 \times 5/\mu}\right]$$

$$\Rightarrow \quad \left(\frac{U_{10}}{U_5}\right)^{10} = 2 \times \frac{U_{10}}{U_5}$$

$$\Rightarrow \quad \left(\frac{U_{10}}{U_5}\right)^{9} = 2$$

$$\Rightarrow \quad \frac{U_{10}}{U_5} = (2)^{1/9}$$

$$\Rightarrow \quad \frac{U_{10}}{U_5} = 1.08$$

(Hence proved)

10.7 The velocity distribution within the boundary layer is given by $\dfrac{u}{U} = \left(\dfrac{y}{\delta}\right)^{1/n}$. Obtain the ratio of displacement thickness to momentum thickness.

[4 marks : 2010]

Solution:

Displacement thickness is expressed as

$$\delta^* = \int_0^\delta \left(1 - \frac{u}{U}\right) dy$$

$$\Rightarrow \quad \delta^* = \int_0^\delta \left[1-\left(\frac{y}{\delta}\right)^{1/n}\right] dy$$

$$= \left[y - \frac{1}{\delta^{1/n}} \cdot \frac{y^{\frac{1}{n}+1}}{\frac{n+1}{n}}\right]_0^\delta$$

$$= \left[y - \left(\frac{n}{n+1}\right)\frac{y^{\frac{1}{n}+1}}{\delta^{\frac{1}{n}}}\right]_0^\delta$$

$$= \delta - \frac{n}{n+1} \times \delta$$

$$= \frac{\delta}{n+1}$$

Momentum thickness (θ)

$$\theta = \int_0^\delta \frac{u}{u_\infty}\left(1-\frac{u}{u_\infty}\right) dy$$

$$= \int_0^\delta \left(\frac{y}{\delta}\right)^{1/n}\left[1-\left(\frac{y}{\delta}\right)^{1/n}\right] dy$$

$$= \int_0^\delta \left[\left(\frac{y}{\delta}\right)^{1/n} - \left(\frac{y}{\delta}\right)^{2/n}\right] dy$$

$$= \left[\frac{y^{\frac{1}{n}+1}}{\delta^{\frac{1}{n}}\left[\frac{1}{n}+1\right]} - \frac{y^{\frac{2}{n}+1}}{\delta^{\frac{2}{n}}\left[\frac{2}{n}+1\right]}\right]_0^\delta$$

$$= \frac{\delta^{\frac{1}{n}+1}}{\delta^{\frac{1}{n}}\left[\frac{n+1}{n}\right]} - \frac{\delta^{\frac{2}{n}+1}}{\delta^{\frac{2}{n}}\left[\frac{2+n}{n}\right]}$$

$$= \frac{n.\delta}{n+1} - \frac{n\delta}{n+2}$$

$$= \frac{n\delta[n+2-n-1]}{(n+1)(n+2)}$$

$$= \frac{n\delta}{(n+1)(n+2)}$$

Thus from (i) and (ii), we get

$$\frac{\delta^*}{\theta} = \frac{\delta/(n+1)}{n\delta/(n+1)(n+2)} = \frac{n+2}{n} \quad \text{(Ans.)}$$

10.8 The velocity in a boundary layer over a horizontal flat plate held in a free stream with a velocity U is given as

$$\frac{u}{U} = a + b\eta + c\eta^2$$

where u is velocity at y and U at δ and $\eta = y/\delta$; y and δ are measure of normal to the flat plate. Determine the value of coefficients a, b and c using appropriate boundary conditions. If the shear stress on the plate is given $\tau_0 = K\frac{\mu U}{\delta}$, find the value of '$K$'.

[4 marks : 2012]

Solution:

We know that boundary conditions which must be satisfied by any velocity profile whether it is in laminar layer zone, or in turbulent boundary layer zone are:

(i) At $y = 0$, $u = 0$

(ii) At $y = \delta$, $u = U$

(iii) At $y = \delta$, $\frac{du}{dy} = 0$

Given: $$\frac{u}{U} = a + b\left(\frac{y}{\delta}\right) + c\left(\frac{y}{\delta}\right)^2$$

From (i) $$u = 0 \text{ at } y = 0$$

$\therefore$ $$0 = a + 0 + 0$$

$\Rightarrow$ $$a = 0$$

$\therefore$ $$\frac{u}{U} = b\left(\frac{y}{\delta}\right) + c\left(\frac{y}{\delta}\right)^2$$

From (ii) $$u = U \text{ at } y = \delta$$

$\therefore$ $$\frac{U}{U} = b\left(\frac{\delta}{\delta}\right) + c\left(\frac{\delta}{\delta}\right)^2$$

$\Rightarrow$ $$1 = b + c$$

or $$b + c = 1 \quad \text{...(iv)}$$

From (iii) $$\frac{du}{dy} = 0 \text{ at } y = \delta$$

Now, $$u = U\left[b\left(\frac{y}{\delta}\right) + c\left(\frac{y}{\delta}\right)^2\right]$$

$\therefore$ $$\frac{du}{dy} = U\left[b\left(\frac{1}{\delta}\right) + \frac{c}{\delta^2}2y\right]$$

$\therefore$ $$0 = U\left[\frac{b}{\delta} + \frac{c}{\delta^2}2\delta\right] \quad [\because U \text{ is constant}]$$

$\Rightarrow$ $$\frac{b}{\delta} + \frac{2c}{\delta} = 0$$

$\Rightarrow$ $$b + 2c = 0 \quad \text{...(v)}$$

Solving eq. (iv) and (v) we get

$$b = 2$$
$$c = -1$$

$$\therefore \qquad \frac{u}{U} = 2\left(\frac{y}{\delta}\right) - \left(\frac{y}{\delta}\right)^2$$

The shear stress, on the plate is given by Newton's law of viscosity as

$$\tau_0 = \mu\left(\frac{du}{dy}\right)_{y=0}$$

$$u = U\left[\frac{2y}{\delta} - \frac{y^2}{\delta^2}\right]$$

$$\left(\frac{du}{dy}\right)_{y=0} = U\left[\frac{2}{\delta} - \frac{2\times(0)}{\delta^2}\right] = \frac{2U}{\delta}$$

$$\therefore \qquad \tau_0 = \mu \times \frac{2U}{\delta} = \frac{2\mu U}{\delta} \qquad \text{...(vi)}$$

Given $\qquad \tau_0 = k\frac{\mu U}{\delta} \qquad$ ...(vii)

Comparing (vi) and (vii), $\qquad k = 2$

10.9 A flat plate of 2 m width and 4 m length is kept parallel to air flowing at 5 m/s velocity at 15°C. Determine the length of the plate over which boundary layer is laminar, shear at the location where boundary layer ceases to be laminar and total force on both sides on that portion of plate the boundary layer is laminar.

Take ρ = 1.208 kg/m^3 and ν = 1.47 × 10^{-5} m^2/s.

[15 marks : 2020]

Solution:

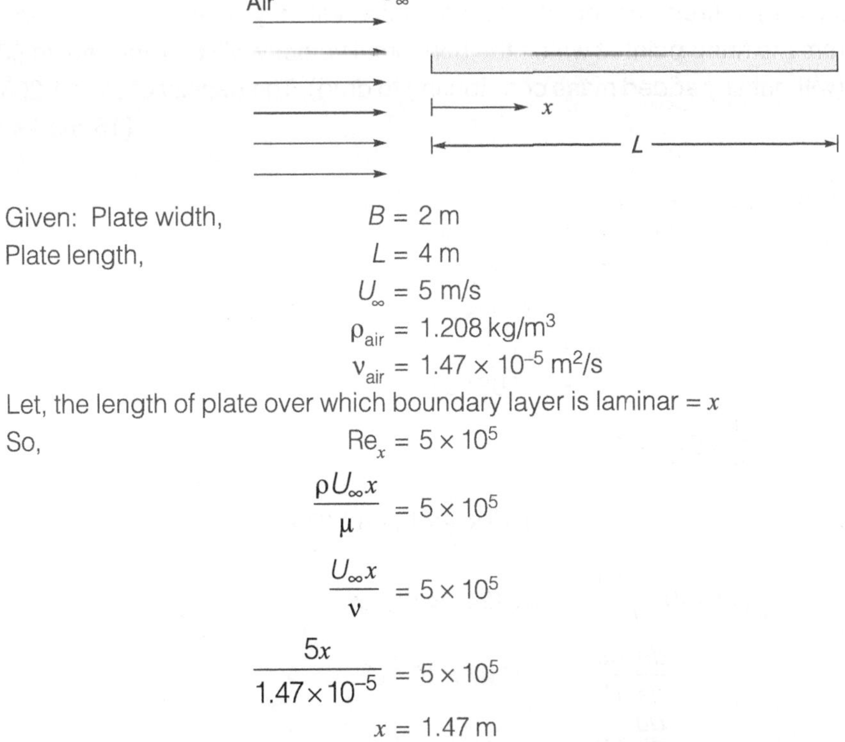

Given: Plate width, $\qquad B = 2$ m

Plate length, $\qquad L = 4$ m

$$U_\infty = 5 \text{ m/s}$$

$$\rho_{air} = 1.208 \text{ kg/m}^3$$

$$\nu_{air} = 1.47 \times 10^{-5} \text{ m}^2/\text{s}$$

Let, the length of plate over which boundary layer is laminar = x

So, $\qquad Re_x = 5 \times 10^5$

$$\frac{\rho U_\infty x}{\mu} = 5 \times 10^5$$

$$\frac{U_\infty x}{\nu} = 5 \times 10^5$$

$$\frac{5x}{1.47\times 10^{-5}} = 5 \times 10^5$$

$$x = 1.47 \text{ m}$$

Shear stress at the location where boundary layer ceases to be laminar

$$\frac{\tau_{0,x}}{\frac{1}{2}\rho U_\infty^2} = C_{f,x} \qquad \left[\text{According to Blasius result } C_{f,x} = \frac{0.664}{\sqrt{\text{Re}_x}}\right]$$

$$\tau_{0,x} = C_{f,x} \times \frac{1}{2} \times \rho \times U_\infty^2$$

$$= \frac{0.664}{\sqrt{\text{Re}_x}} \times \frac{1}{2} \times \rho \times U_\infty^2$$

$$= \frac{0.664}{\sqrt{5 \times 10^5}} \times \frac{1}{2} \times 1.208 \times 5^2$$

$$\tau_{0,x} = 1.418 \times 10^{-2}\ \text{N/m}^2$$

Drag force on that portion of plate where the boundary layer is laminar considering both the sides of plate.

$$= 2 \times \left[\bar{C}_{f,x} \times \frac{1}{2} \times \rho \times A \times U_\infty^2\right]$$

$$= 2 \times \left[\frac{1.328}{\sqrt{\text{Re}_x}} \times \frac{1}{2} \times 1.208 \times (1.47 \times 2) \times 5^2\right]$$

$$= 2 \times \frac{1.328}{\sqrt{5 \times 10^5}} \times \frac{1}{2} \times 1.208 \times 1.47 \times 2 \times 25$$

$$= 0.1668\ \text{N}$$

11. Drag and Lift Force

11.1 **A bullet of mass 12 gm and cross sectional area 27 mm^2 and drag coefficient of 0.43 (assumed to be an invariant throughout) is fired horizontally with an initial velocity of 90 m/sec. Determine the horizontal distance from the firing point at which the horizontal velocity will be reduced to 50 m/sec due to air resistance (without any added mass contributing to drag). The density of air is 1.208 kg/m^3.**

[15 marks : 2003]

Solution:

Mass of bullet, $m = 12$ gm $= 12 \times 10^{-3}$ kg

Cross-sectional area of bullet, $A = 27$ mm^2 $= 27 \times 10^{-6}$ m^2

Coefficient of drag, $C_D = 0.43$

Initial velocity, $u_i = 90$ m/s

Final velocity, $u_f \approx 50$ m/s

Density of air, $\rho = 1.208$ kg/m^3

Drag force on the bullet, $F_D = \frac{1}{2} \times C_D A \rho u^2$

$$\text{ma} = 0.43 \times \frac{1}{2} \times 1.208 \times 27 \times 10^{-6} u^2$$

$$\left(12 \times 10^{-3}\right)\frac{du}{dt} = 7.01244 \times 10^{-6}\ u^2$$

or

$$\frac{du}{ds} \cdot \frac{ds}{dt} = 584.37 \times 10^{-6}\ u^2$$

$$\frac{du}{ds} \times u = 584.37 \times 10^{-6}\ u^2$$

$$\frac{du}{u} = 584.37 \times 10^{-6}\, ds$$

Integrate it $$\int_{50}^{90} \frac{du}{u} = 584.37 \times 10^{-6} \int_{0}^{s} ds$$

$$s = 1005.85 \text{ m}$$

11.2 **A metallic sphere of specific gravity 8.0 falls in an oil of density 800 kg/m³. The diameter of the sphere in 10 mm. The viscosity of oil is 7.848 N-sec/m². Determine the terminal velocity of metallic sphere.**

[4 marks : 2010]

Solution:

Given data: Diameter of metallic sphere, D = 10 mm; Density of fluid medium, ρ = 800 kg/m²; Viscosity of fluid medium, μ = 7.848 N-sec/m²; Density of sphere, $\rho_s = 8 \times 1000 = 8000$ kg/m³

The forces acting on the sphere are,

Weight of sphere,
$$W = \text{Density of sphere} \times g \times \text{Volume of sphere}$$
$$= 8000 \times 9.81 \times \frac{\pi}{6} \times D^3$$
$$= 8000 \times 9.81 \times \frac{\pi}{6} \times \left(10 \times 10^{-3}\right)^3$$
$$= 0.041092 \text{ N}$$

Buoyant force,
$$F_B = \text{Density of fluid medium} \times g \times \text{Volume of sphere}$$
$$= 800 \times 9.81 \times \frac{\pi}{6} \times D^2$$
$$= 800 \times 9.81 \times \frac{\pi}{6} \times \left(10 \times 10^{-3}\right)^3$$
$$= 0.0041092 \text{ N}$$

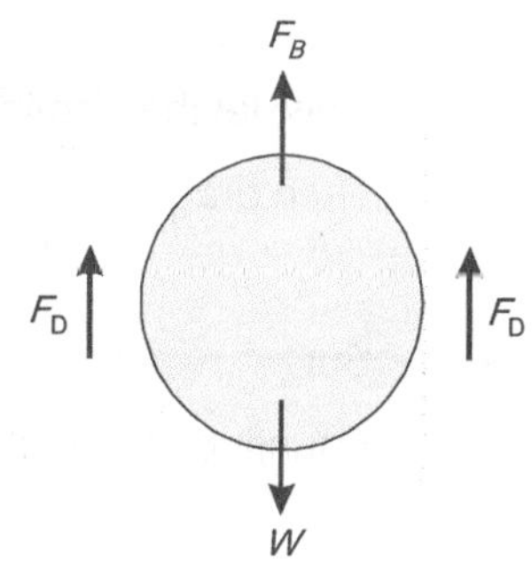

When the metallic sphere reaches the terminal velocity,

∴ We have

Hence, equilibrium of forces

$$W = F_B + F_D$$
$$F_D = W \quad F_B$$
$$F_D = 0.0410192 - 0.0041092$$
$$= 0.0369828 \text{ N}$$

Assuming Stoke's law is valid (Re < 0.2)

Hence,
$$F_D = 3\pi\mu UD$$
$$3\pi\mu UD = 0.0369828$$
$$3\pi \times 7.848 \times U \times 10 \times 10^{-3} = 0.0369828$$
$$U = 0.04999 \text{ m/sec}$$

Hence, check:

$$Re = \frac{\rho UD}{\mu} = \frac{800 \times 0.04999 \times 10 \times 10^{-3}}{7.848}$$
$$= 0.0509 \quad (< 0.2)$$

Hence, the Reynold's number is less than 0.2, so the expression $F_D = 3\pi\mu UD$ for calculating terminal velocity is valid.

Hence, terminal velocity of metallic sphere = 0.04999 m/sec.

11.3 **A metallic ball of diameter 5 mm drops in a fluid of density 800 kg/m³ and viscosity 3N-s/m². The specific gravity of metallic ball is 9.0. Find the terminal fall velocity of the ball, and the drag experienced by the ball.**

[4 marks : 2012]

Solution:

Given data:

Dia of metallic ball, $D = 5 \times 10^{-3}$ m Density of fluid, $\rho_0 = 800$ kg/m³

Viscosity of fluid, $\mu = 3$ N-s/m² Specific gravity of ball = 9

$\therefore$ Density of ball, $\rho_s = 9 \times 1000 = 9000$ kg/m³

The forces acting on the ball are

Weight of ball, W = Density of ball × g × Volume of ball

$$= 9000 \times 9.81 \times \frac{\pi}{6} D^3 = 9000 \times 9.81 \times \frac{\pi}{6} \times \left(5 \times 10^{-3}\right)^3 \text{ N}$$

$$= 0.0057785 \text{ N}$$

Buoyant force, F_B = Density of fluid × g × Volume of ball

$$= 800 \times 9.81 \times \frac{\pi}{6} \times \left(5 \times 10^{-3}\right)^3 \text{ N} = 0.0005136 \text{ N}$$

When the metallic ball reaches the terminal velocity, we have

$$W = F_D + F_B$$

$\therefore$ **Drag experienced by the ball**

$$F_D = W - F_B = 0.0057785 - 0.0005136$$

$$= 0.0052649 \text{ N}$$

Now, let the terminal fall velocity of the ball = U

Then drag force (F_D) is given by equation as

$$F_D = 3\pi\mu DU$$

But $F_D = 0.0052694$N

Equating the two values of F_D, we have

$$3\pi\mu DU = 0.0052649$$

or $3\pi \times 3 \times 5 \times 10^{-3} \times U = 0.0052649$

$$\Rightarrow \quad U = \frac{0.0052649}{3\pi \times 3 \times 5 \times 10^{-3}} = 0.03724 \text{ m/s}$$

CHECK:

Reynold's number, $R_e = \dfrac{\rho UD}{\mu} = \dfrac{800 \times 0.03724 \times 5 \times 10^{-3}}{3} = 0.049 < 0.2$

Hence the Reynold's number is less than 0.2, so the expression $F_D = 3\pi\mu DU$ for calculating terminal velocity is valid.

11.4 **What is the difference between form drag and friction drag? For a flat plate kept perpendicular to the flow direction in a uniform flow field, the coefficient of pressure is constant at –1.2 on the back face and varies parabolically on the front face from a value of 1.0 at the center to –1.2 at both edges. Find the drag coefficient.**

[8 marks : 2014]

Solution:

The force exerted by the fluid on the body in the direction of free stream is known as drag force.

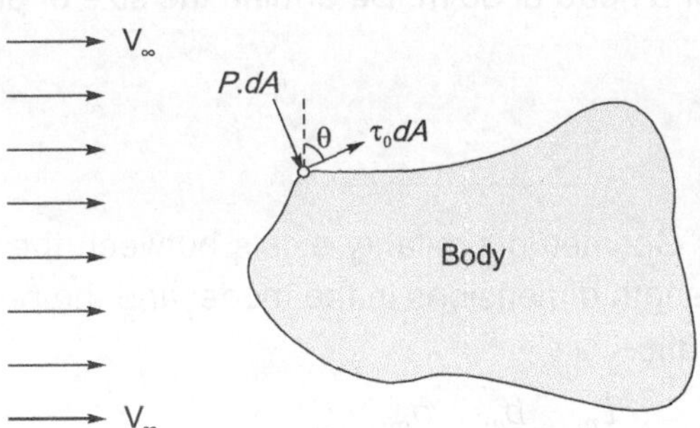

$$F_D = \oint PdA\cos\theta + \oint \tau_0 dA\sin\theta$$

The term $\oint \tau_0 dA \sin\theta$ is known as 'Skin friction drag', it is the component of total drag due to surface shear or skin friction. It is maximum when body is truly stream line and zero when body is normal to the stream line.

Whereas the total drag force, when an object has a high relative flow velocity, resulting situation of irregular fluid flow in which stream lines may be broken is called as form drag.

Assuming dimension of plate 10 cm × 10 cm,

Also assuming profile of pressure coefficients frontal area as

$$x = Ay^2 + B$$

at $\quad y = 0, \quad x = 1.0$

$\therefore \quad 1 = 0 + B$ or $B = 1$

at $\quad y = 5$ cm, $\quad x = -1.2$

$\therefore \quad -1.2 = A(5)^2 + 1$ or $A = -0.088$

thus the profile of pressure coefficient is

$$x = -0.088y^2 + 1$$

	Frontal Area		Rear Area	
Point	Coefficient of Pressure	Average Coefficient of Pressure	Coefficient of Pressure	Average Coefficient of Pressure
$y = 0\ (y_0)$	1.0	$C_p = \dfrac{y_0 + 2y_1 + 2y_2 + 2y_3 + 2y_4 + 2y_5}{11}$ $= 0.12$	1.2	1.2
$y = 1\ (y_1)$	0.912			
$y = 2\ (y_2)$	0.648			
$y = 3\ (y_3)$	0.208			
$y = 4\ (y_4)$	−0.408			
$y = 5\ (y_5)$	−1.20			

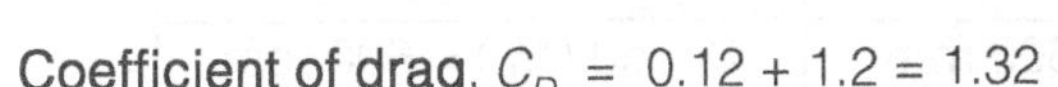

$\therefore \quad$ Coefficient of drag, $C_D = 0.12 + 1.2 = 1.32$

11.5 **Explain geometric similarity, kinematic similarity and dynamic similarity. Two homologous pumps are to run at the same speed of 600 r.p.m. Pump A has an impeller of 50 cm diameter and discharges 0.4 m³/sec of water under a head of 50 m. Determine the size of pump B and its net head if it is to discharge 0.3 m³/sec.**

[20 marks : 2019]

Solution:

Geometrical Similarity : Geometric similarity exists between the model and the prototype if the ratios of corresponding length dimensions in the model and the prototype are equal. Such a scale ratio is defined as scale ratio.

Length scale ratio : $L_r = \frac{L_m}{L_p} = \frac{b_m}{b_p} = \frac{d_m}{d_p}$ etc.

Area scale ratio : $A_r = \frac{A_M}{A_P} = \left(\frac{L_m \times b_m}{L_p \times b_p}\right) = L_r^2$

Volume scale ratio : $V_r = \frac{V_m}{V_p} = \left(\frac{L_m \times b_m \times d_m}{L_p \times b_p \times d_p}\right) = L_r^3$

Kinematic Similarity : Kinematic parameters are velocity, acceleration and discharge. This similarity exists if

(a) At all corresponding points in the model and prototype, the ratio of velocity as well as acceleration are same (both in magnitude and direction).

(b) The path of the corresponding moving particles are geometrically similar.

Time scale ratio : $T_r = \frac{T_m}{T_p}$

Velocity ratio : $V_r = \frac{V_m}{V_p} = \frac{L_r}{T_r}$

Acceleration ratio : $a_r = \frac{a_m}{a_p} = \frac{L_r}{T_r^2}$

Discharge ratio : $Q_r = \frac{Q_m}{Q_p} = \frac{L_r^3}{T_r}$

Dynamic Similarity : For dynamic similarity to exist between model and prototype, identical type of forces (viscous, pressure, elastic, etc.) must be parallel and must bear the same ratio at all corresponding sets of points. Dynamic parameters are force and power.

Force ratio : $\frac{F_m}{F_p} = \rho_r \times Q_r \times a_r = \frac{\rho_r L_r^4}{T_r^2}$

Power ratio : $\frac{P_m}{P_P} = F_r \,.\, V_r = \frac{\rho_r L_r^5}{T_r^3}$

Pump A	Pump B
Diameter (D_A) = 50 cm	Diameter (D_B) = ?
Discharge (Q_A) = 0.4 m³/s	Discharge (Q_B) = 0.3 m³/s
Head (H_A) = 50 m	Head (H_B) = ?
Speed (N_A) = 600 rpm	Speed (N_B) = 600 rpm

For geometrical, kinematic and dynamic similarity,

$$\left(\frac{Q}{D^3N}\right)_A = \left(\frac{Q}{D^3N}\right)_B$$

$$\frac{0.4}{50^3 \times 600} = \frac{0.3}{(D_B)^3 \times 600}$$

$$D_B = 45.428 \text{ m}$$

$$\left(\frac{H}{D^2N^2}\right)_A = \left(\frac{H}{D^2N^2}\right)_B$$

$$\frac{H_A}{H_B} = 1.2114$$

$$\therefore \quad H_B = 41.274 \text{ m}$$

12. Notches and Weirs

12.1 **A stream has a width of 25 m and depth of flow 2.5 m. The mean velocity of the flow is 1.10 m/s. Find the height of a weir to be built in the stream to raise the water level by 1.2 m. Assume coefficient of discharge as 0.90.**

[10 marks : 2006]

Solution:

Width of stream = 25 m

Depth of flow = 2.5 m

∴ Area of flow section, $A = 25 \times 2.5 = 62.5 \text{ m}^2$

Mean velocity of flow, $V = 1.10$ m/s

∴ Discharge, $Q = AV = 62.5 \times 1.10 = 68.75 \text{ m}^3/\text{s}$

The weir is to be constructed to raise the water level by 1.2 m, the depth of flow on the upstream of the weir will be

$$= 2.5 + 1.2 = 3.7 \text{ m}$$

∴ Velocity of approach is given by

$$V_a = \frac{Q}{25 \times 3.7} = \frac{68.75}{25 \times 3.7} = 0.74 \text{ m/s}$$

and
$$h_a = \frac{V_a^2}{2g} = \frac{(0.74)^2}{2 \times 9.81} = 0.0279 \text{ m}$$

Assuming that the weir is discharging free, then

$$Q = \frac{2}{3} C_d L \sqrt{2g}\left[(H + h_a)^{3/2} - (h_a)^{3/2}\right]$$

$$\Rightarrow \quad 68.75 = \frac{2}{3} \times 0.90 \times 25 \times \sqrt{2 \times 9.81}\left[(H + 0.0279)^{3/2} - (0.0279)^{3/2}\right]$$

$$\Rightarrow \quad H = 0.9982 = 1 \text{ m}$$

∴ Height of weir, $Z = 3.7 - 1 = 2.7$ m

12.2 **A discharge of 150 liters per second is to be measured by a V-notch of vertex angle 60°. What would be the head over the vertex? If the accuracy of the reading of the head is 1 mm, what error in discharge can be expected at this level? Assume $C_d = 0.62$.**

[8 marks : 2015]

Solution:

$$Q = 150 \, l/s = 0.15 \, m^3/s$$

$$\theta = 60°, \quad dH = 1 \text{ mm}, \quad C_d = 0.62$$

Discharge from a triangular notch is given by

$$Q = \frac{8}{15} C_d \sqrt{2g} H^{5/2} \tan\left(\frac{\theta}{2}\right)$$

$$0.15 = \frac{8}{15} \times 0.62 \times \sqrt{2 \times 9.81} \times H^{5/2} \times \tan\left(\frac{60}{2}\right)$$

$$H^{5/2} = 0.17738 \text{ m}$$

$$\Rightarrow \quad H = 0.5 \text{ m} = 500 \text{ mm}$$

$$\frac{dQ}{Q} = \frac{5}{2} \frac{dH}{H}$$

$$\Rightarrow \quad dQ = \frac{5}{2} \times \frac{1}{500} \times (0.15 \times 1000) = 0.75 \, l/s$$

12.3 During a major flood event, water overflows a roadway during which the roadway acts as a broad-crested weir. A vehicle stalled on the roadway will be washed off the roadway into the downstream channel when water depth exceeds 0.6 m.

(i) Determine the head (H) on the broad-crested weir for a discharge rate of 400 m^3/s if the overflow section of the roadway is horizontal and 200 m long. Assume a weir coefficient of 0.33.

(ii) Also, compute the water depth and velocity on the roadway.

(iii) Will a stalled vehicle on the roadway remain on the road during the flood or get washed off? Take $g = 9.81 \, m/s^2$. Give reasons for your answer.

[12 marks : 2017]

Solution:

Given: Discharge rate, $Q = 400 \, m^3/sec$

Weir coefficient, $C_d = 0.33$

Length of crest/weir = 200 m

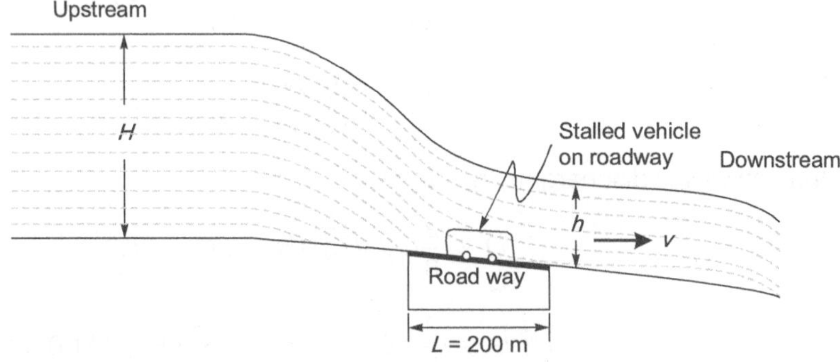

Neglect velocity of approach.

Let H = Head over the weir

h = Depth of water at the weir middle

v = Velocity of flow over the weir (roadway)

(i) We know,

For maximum discharge, $Q_{max} = 1.705 C_d \times L \times H^{3/2}$

$\therefore \quad 400 = 1.705 \times 0.33 \times 200 \times H^{3/2}$

$\Rightarrow$ $$H = 2.329 \text{ m}$$

(ii) Water depth, $$h = \frac{2}{3}H = \frac{2}{3} \times 2.329 = 1.553 \text{ m}$$

Using Bernoulli's equation, on upstream and downstream sides,

$$0 + 0 + H = h + \frac{v^2}{2g}$$

$\Rightarrow$ $$v = \sqrt{2g(H-h)} = \sqrt{2 \times 9.81(2.329 - 1.553)} = 3.902 \text{ m/sec}$$

(iii) $\because$ $(h = 1.553 \text{ m}) > 0.60 \text{ m}$

$\Rightarrow$ The stalled vehicle on the roadway during flood will be washed off.

12.4 The circular base of a cylindrical tank is of 1 m diameter and has an orifice of 10 cm diameter. The discharge coefficient, C_d, for the orifice is 0.6. Initially the tank is empty and then it is filled from the top using a pipe discharging 0.025 m³/s. How long will it take to fill the tank up to a height of 1 m? What will be the maximum height to which the tank can be filled?

[20 marks : 2018]

Solution:

Let A = Area of tank, Diameter of tank, D = 1 m
Area of orifice = a, Orifice of diameter, d = 10 cm
Height to be filled, H = 1 m.
Let at any time, height of liquid from orifice is h, and liquid surface rises by small height dh in time dT.

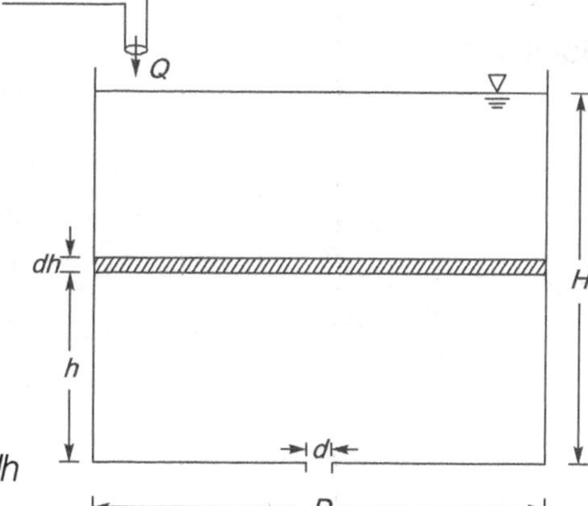

Theoretical velocity through orifice $= \sqrt{2gh}$

According to the question,

$$\left(Q - C_d a\sqrt{2gh}\right)dT = Adh$$

$$dT = \left(\frac{A}{Q - C_d . a\sqrt{2gh}}\right)dh$$

Integrating the above equation,

$$\int_0^T dT = \int_0^H \left(\frac{A}{Q - C_d a\sqrt{2gh}}\right)dh$$

Let, $$h = x^2$$

$$T = \int_0^{\sqrt{H}} \frac{A2x}{Q - C_d a\sqrt{2g}\,x}dx \qquad \therefore dh = 2xdx$$

$$T = \frac{-2A}{C_d a\sqrt{2g}} \int_0^{\sqrt{H}} \left(\frac{-C_d a\sqrt{2g}\,x + Q - Q}{Q - C_d a\sqrt{2g}\,x}\right)dx$$

$$T = \frac{-2A}{C_d a\sqrt{2g}} \left[x + \frac{Q}{C_d a\sqrt{2g}} \ln\left(Q - C_d a\sqrt{2g}x\right)\right]_0^{\sqrt{H}}$$

$$T = \frac{-2A}{C_d a\sqrt{2g}} \left[\sqrt{H} + \frac{Q}{C_d a\sqrt{2g}} \ln\left(\frac{\left(Q - C_d a\sqrt{2gH}\right)}{Q}\right)\right]$$

Here, area of tank, $A = \frac{\pi}{4} \times 1^2 \text{ m}^2$

Area of orifice, $a = \frac{\pi}{4} \times 0.1^2 \text{ m}^2$

Discharge coefficient, $C_d = 0.6$

Discharge, $Q = 0.025 \text{ m}^3/\text{s}$

On substituting these values

Time to fill the tank upto 1 m height, $T = 87.11$ seconds

Now, for maximum height upto which tank can be filled

$$Q_{in} = Q_{out}$$

$$\therefore \quad Q = C_d a\sqrt{2gh_{max}}$$

$$\Rightarrow \quad 0.025 = 0.6 \times \frac{\pi}{4} \times 0.1^2 \times \sqrt{2 \times 9.81 \times h_{max}}$$

$$\Rightarrow \quad h_{max} = 1.434 \text{ m}$$

12.5 **A stream has a width of 25 m, depth of 2.5 m and mean velocity of 1.5 m/sec. Find the height of weir to be constructed on the stream floor to raise the water level by 1 m. Assume value of discharge coefficient as 0.95.**

[12 Marks : 2021]

Solution:

Given,

$$B = 25 \text{ m}$$

$$C_D = 0.95$$

$$y = 2.5 \text{ m}$$

$$v = 1.5 \text{ m/s}$$

Discharge,

$$Q = A \times v = (B \times y) \times v$$

$$Q = 25 \times 2.5 \times 1.5 \text{ m}^3/\text{s}$$

$$Q = 93.75 \text{ m}^3/\text{s}$$

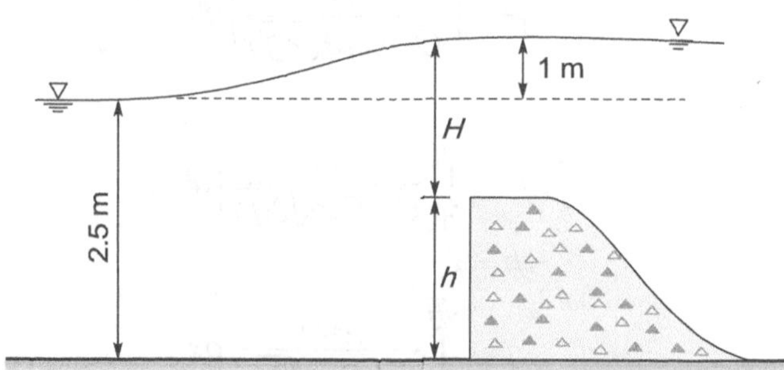

Assuming the weir to be rectangular crested,

$$Q = \frac{2}{3} C_D \sqrt{2g}\, BH^{3/2}$$

$$93.75 = \frac{2}{3} \times 0.95 \times \sqrt{2 \times 9.81} \times 25 \times (3.5 - h)^{3/2}$$

$$h = 2.2865 \text{ m}$$

13. Impact of Jets and Turbines

13.1 An inward flow reaction turbine works under a head of 30 m and discharge of 10 m^3/s. The speed of the runner is 300 rpm. At inlet tip of runner vane, the peripheral velocity of wheel is $0.9\sqrt{2gH}$ and the radial velocity of flow is $0.3\sqrt{2gH}$, where H is the head on the turbine. If the overall efficiency and the hydraulic efficiency of the turbine are 80% and 90% respectively, determine

(i) the power developed in kW.
(ii) diameter and width of runner at inlet.
(iii) guide blade angle at inlet.
(iv) inlet angle of runner vane.
(v) diameter of runner at outlet.

Assume that the discharge at outlet is radial.

[15 marks : 1999]

Solution:

Given data:

Discharge, $Q = 10$ m^3/s

Head, $H = 30$ m

Speed of runner, $N = 300$ rpm

Peripheral velocity at inlet, $u_1 = 0.9\sqrt{2gH}$

Radial velocity of flow at inlet, $V_{f_1} = 0.3\sqrt{2gH}$

Overall efficiency, $\eta_0 = 80\%$

Hydraulic efficiency, $\eta_h = 90\%$

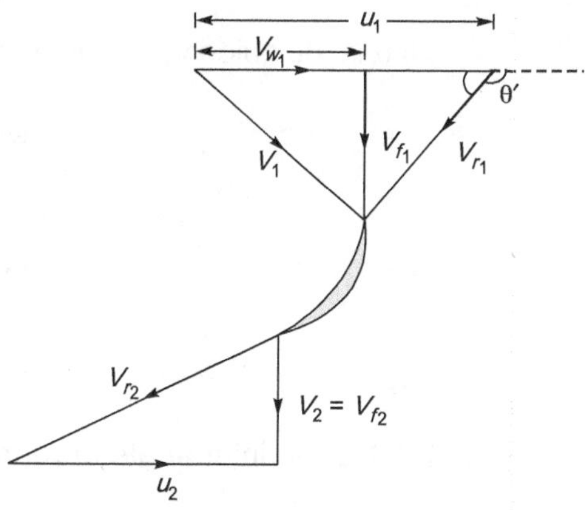

(i) The power developed by the turbine is given by

$$P = \gamma Q H \times \eta_0$$

$$= 9.81 \times 1000 \times 10 \times 30 \times \frac{80}{100} = 2354.40 \text{ kW } \textbf{(Ans.)}$$

(ii) Peripheral velocity at inlet is given by

$$u_1 = 0.9\sqrt{2gH}$$

But, $$u_1 = \frac{\pi D_1 N}{60}$$

where D_1 is the diameter of runner at inlet

$$\Rightarrow \quad 0.9\sqrt{2gH} = \frac{\pi D_1 N}{60}$$

$$\Rightarrow \quad 0.9\sqrt{2 \times 9.81 \times 30} = \frac{\pi \times D_1 \times 300}{60}$$

$$\Rightarrow \quad D_1 = 1.39 \text{ m } \textbf{(Ans.)}$$

Also, $Q = \pi D_1 B_1 V_{f_1}$ where B_1 is the width of runner at inlet

$$\Rightarrow \quad 10 = \pi \times 1.39 \times B_1 \times 0.3\sqrt{2gH}$$

$$\Rightarrow \quad 10 = \pi \times 1.39 \times B_1 \times 0.3\sqrt{2 \times 9.81 \times 30}$$

$$\Rightarrow \quad B_1 = 0.315 \text{ m } \textbf{(Ans.)}$$

(iii) Hydraulic efficiency is given by

$$\eta_h = \frac{u_1 V_{w_1} - u_2 V_{w_2}}{gH}$$

where V_{w_1} is whirl velocity at inlet and V_{w_2} is whirl velocity at outlet

But it is given that discharge at outlet is radial.

$$\therefore \quad V_{w_2} = 0$$

$$\Rightarrow \quad \eta_h = \frac{u_1 V_{w_1} - u_2 \times 0}{gH}$$

$$\Rightarrow \quad \eta_h = \frac{u_1 V_{w_1}}{gH}$$

$$\Rightarrow \quad \frac{90}{100} = \frac{0.9\sqrt{2\times 9.81\times 30}\times V_{w_1}}{9.81\times 30}$$

$$\Rightarrow \quad V_{w_1} = 12.13 \text{ m/s}$$

If α is the guide vane angle at inlet, then

$$\tan\alpha = \frac{V_{f_1}}{V_{w_1}}$$

$$\Rightarrow \quad \tan\alpha = \frac{0.3\sqrt{2\times 9.81\times 30}}{12.13}$$

$$\Rightarrow \quad \alpha = \tan^{-1}(0.6)$$

$$\Rightarrow \quad \alpha = 30.96^\circ \approx 31^\circ \textbf{ (Ans.)}$$

(iv) If θ is the inlet angle at runner vane, then

$$\tan\theta = \frac{V_{f_1}}{u_1 - V_{w_1}}$$

$$\Rightarrow \quad \tan\theta = \frac{0.3\sqrt{2\times 9.81\times 30}}{0.9\sqrt{2\times 9.81\times 30} - 12.13}$$

$$\Rightarrow \quad \theta = 36.87^\circ \textbf{ (Ans.)}$$

(v) Diameter of runner at outlet is given by

$$D_2 = \frac{D_1}{2} = \frac{1.39}{2} = 0.695 \text{ m}$$

13.2 **The following data were obtained from a test on Pelton wheel:**

(i) Head at the base of nozzle = 32 m

(ii) Discharge of the nozzle = 0.18 m³/s

(iii) Area of the jet = 7500 mm²

(iv) Power available at the shaft = 44 kW

(v) Mechanical efficiency = 94%

Calculate the power lost:

(a) in the nozzle,

(b) in the runner and

(c) in the mechanical friction.

[15 marks : 2000]

Solution:

Given data: $H = 32$ m; $Q = 0.18$ m³/s; $A = 7500$ mm²; S.P. = 44 kW; $\eta_m = 94\%$

(i) **Power** at the base of nozzle = $\frac{\gamma QH}{1000} = \frac{9.81 \times 0.18 \times 32 \times 10^3}{1000}$ = 56.51 kW

Kinetic energy of jet per second, KE = $\frac{1}{2}\rho Q v^2$

where V is the velocity of jet

$$\therefore \quad KE = \frac{1}{2} \times 1000 \times 0.18 \times \left(\frac{0.18}{7500 \times 10^{-6}}\right)^2 \times \frac{1}{1000} = 51.84 \text{ kW}$$

∴ Power lost in the nozzle = Power at the base of nozzle – KE

= 56.51 – 51.84 = 4.67 kW

(ii)
$$\eta_m = \frac{\text{Power available at the shaft}}{\text{Power delivered by water to runner}}$$

$$\Rightarrow \quad \frac{94}{100} = \frac{44}{\text{Runner Power}}$$

⇒ Runner Power = 46.81 kW

Power supplied to the runner is equal to the kinetic energy of the jet

i.e. KE = 51.84 kW

∴ Power lost in runner = KE – Runner power

= 51.84 – 46.81 = 5.03 kW

(iii) **Power lost in mechanical friction** = Runner power – shaft power

= 46.81 – 44 = 2.81 kW

13.3 **An inward flow reaction turbine works under an average head of 160 m with a discharge of 80 m³/s. The inlet and outlet diametres of the turbine are 4 m and 2 m respectively. The runner blade angle at the inlet is 120°. Radial discharging velocity at the outlet is 15 m/s. Assuming constant breadth of the wheel and 90% hydraulic efficiency, determine the HP produced in MW and the rpm of the machine.**

[15 marks : 2001]

Solution:

Given data: Head (H) = 160 m, Discharge (Q) = 80 m³/s, Inlet diameter (D_1) = 4 m

Outlet diameter (D_2) = 2 m, Velocity of flow at outlet, V_{f_2} = 15 m/s, Hydraulic efficiency, η_h = 90%

The inlet and outlet velocity triangle for inward flow reaction turbine is given here:

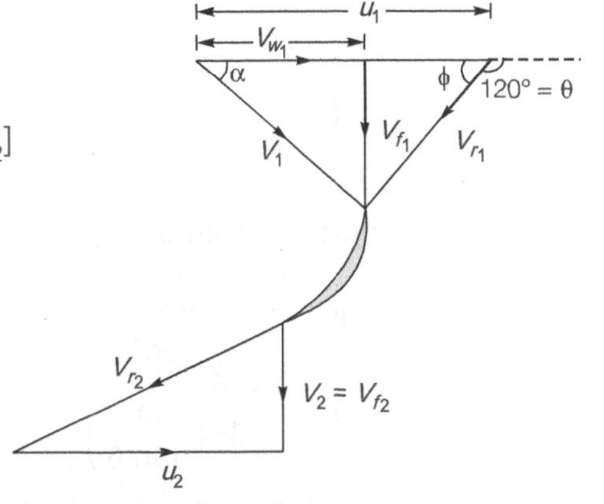

As per continuity equation, we have

$$Q = \pi D_1 B_1 V_{f_1} = \pi D_2 B_2 V_{f_2} \quad [\because B_1 = B_2]$$

$$\Rightarrow \quad 4 \times V_{f_1} = 2 \times V_{f_2}$$

$$\Rightarrow \quad V_{f_1} = \frac{2 \times 15}{4}$$

$$\Rightarrow \quad V_{f_1} = 7.5 \text{ m/s}$$

Now,
$$\tan(180 - \theta) = \frac{V_{f_1}}{u_1 - V_{w_1}}$$

$$\Rightarrow \quad \tan 60° = \frac{7.5}{u_1 - V_{w_1}}$$

$$\Rightarrow \quad \sqrt{3}u_1 - \sqrt{3}V_{w_1} = 7.5 \qquad \text{...(i)}$$

Also,

$$\eta_h = \frac{u_1 V_{w_1}}{gH}$$

$$\Rightarrow \quad \frac{90}{100} = \frac{u_1 V_{w_1}}{9.81 \times 160}$$

$$\Rightarrow \quad u_1 = \frac{1412.64}{V_{w_1}} \qquad \text{...(ii)}$$

Substituting (ii) in (i), we get

$$\sqrt{3} \times \frac{1412.64}{V_{w_1}} - \sqrt{3} \times V_{w_1} = 7.5$$

$$\Rightarrow \quad \sqrt{3}\, V_{w1}^2 + 7.5\, V_{w_1} - 2446.76 = 0$$

$$\Rightarrow \quad V_{w_1} = 35.48 \text{ m/s}$$

$$\therefore \quad u_1 = \frac{1412.64}{35.48} = 39.81 \text{ m/s}$$

Power developed is given by

$$P = \rho Q\left(u_1 V_{w_1} - u_2 V_{w_2}\right) \qquad \left[\because V_{w_2} = 0\right]$$

$$= 1000 \times 80 \times (39.81 \times 35.48 - u_2 \times 0)$$

$$= 80000 \times 39.81 \times 35.48 = 112.99 \text{ MW} \approx 113 \text{ MW}$$

Peripheral velocity at inlet is given by

$$u_1 = \frac{\pi D_1 N}{60}$$

$$\Rightarrow \quad 39.81 = \frac{\pi \times 4 \times N}{60}$$

$$\Rightarrow \quad N = 190 \text{ rpm}$$

13.4 **An inward flow reaction turbine discharges radially, the velocity of flow through the runner being constant. Show that the hydraulic efficiency is given by**

$$\eta_h = \frac{1}{1 + \dfrac{\frac{1}{2}\tan^2\alpha}{1 - \left(\dfrac{\tan\alpha}{\tan\theta}\right)}}$$

Where α is the guide blade angle and θ is the inlet vane angle of the runner and there is no friction on the blades.

[10 marks : 2003]

Solution:

The inlet and outlet velocity triangles for an inward flow reaction turbine is given here.

Head extracted by the turbine is given by

$$H_e = \frac{u_1 V_{w_1} - u_2 V_{w_2}}{g}$$

$$\Rightarrow \quad H_e = \frac{u_1 V_{w_1}}{g} \qquad \left[\because V_{w_2} = 0\right] \quad \text{...(i)}$$

But $\quad H_e = H - \frac{V_2^2}{2g}$ where H is head available ...(ii)

From (i) and (ii), we get

$$\frac{u_1 V_{w_1}}{g} = H - \frac{V_2^2}{2g} \qquad \left[\because V_{f_1} = V_{f_2} = V_2\right]$$

$$\Rightarrow \quad H = \frac{(V_{f_1})^2}{2g} + \frac{u_1 V_{w_1}}{g}$$

$$\Rightarrow \quad gH = \frac{(V_{f_1})^2}{2} + u_1 V_{w_1} \quad \text{...(iii)}$$

Also, $\quad \eta_h = \frac{u_1 V_{w_1} - u_2 V_{w_2}}{gH} \qquad \left[\text{Where } V_{w_2} = 0\right]$

$$\Rightarrow \quad \eta_h = \frac{u_1 V_{w_1}}{\frac{V_{f_1}^2}{2} + u_1 V_{w_1}} \qquad \text{[From (iii)]}$$

$$\Rightarrow \quad \eta_h = \frac{1}{1 + \frac{(V_{f_1})^2}{2u_1 V_{w_1}}}$$

but $\quad \tan\alpha = \frac{V_{f_1}}{V_{w_1}}$

and $\quad \tan\theta = \frac{V_{f_1}}{V_{w_1} - u_1}$

$$\Rightarrow \quad \tan\theta \times (V_{w_1} - u_1) = V_{f_1}$$

$$\Rightarrow \quad u_1 = V_{w_1} - \frac{V_{f_1}}{\tan\theta} = V_{w_1} - \frac{V_{w_1}\tan\alpha}{\tan\theta} = V_{w_1}\left(1 - \frac{\tan\alpha}{\tan\theta}\right)$$

$$\therefore \quad \eta_h = \frac{1}{1 + \frac{(V_{w_1})^2 \tan^2\alpha}{2(V_{w_1})^2\left(1 - \frac{\tan\alpha}{\tan\phi}\right)}}$$

$$\Rightarrow \quad \eta_h = \frac{1}{1 + \frac{\frac{1}{2}\tan^2\alpha}{1 - \frac{\tan\alpha}{\tan\theta}}} \qquad \textbf{(Hence proved)}$$

13.5 The turbine shown in the figure below, extracts 45 mhp (metric horse power) from the water flowing through it. The 30 cm diameter 100 m long pipe has a friction factor of 0.02. Neglecting minor losses, determine the flow rate through the pipe and the turbine.

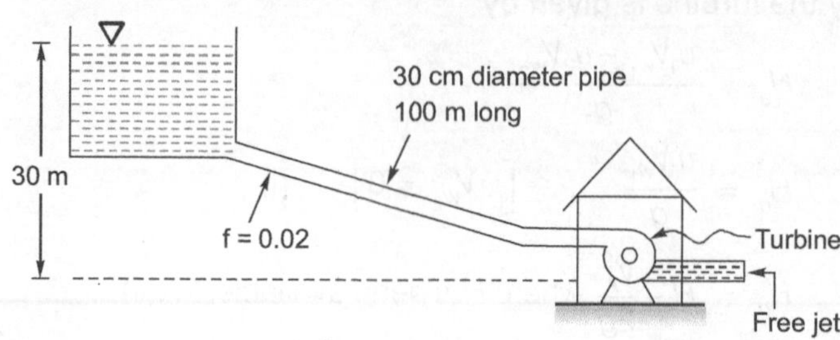

[20 marks : 2004]

Solution:

Power extracted by the turbine, $P = 45$ mhp

Diameter of pipe $(D) = 30$ cm, Length of pipe $(L) = 100$ m, Friction factor $(f) = 0.02$

Head available to the turbine, $H = 30$ m

Let the discharge through the pipe and turbine be Q.

The head loss due to friction is given by Darcy-weisbach equation as

$$h_f = \frac{fLV^2}{2gD}$$

where V is the velocity of flow in the pipe

But
$$V = \frac{\text{Discharge through pipe}}{\text{Area of cross-section of pipe}}$$

$$\Rightarrow \quad V = \frac{4 \times Q}{\pi \times (0.3)^2}$$

$$\Rightarrow \quad V = \frac{4Q}{0.09\pi} \quad \text{..(i)}$$

$$\therefore \quad h_f = \frac{0.02 \times 100}{2 \times 9.81 \times 0.3} \times \left[\frac{4Q}{0.09\pi}\right]^2 \quad \text{[From (i)]}$$

$$\Rightarrow \quad h_f = 68Q^2 \quad \text{...(ii)}$$

Head extracted by the turbine,

$$H_e = H - h_f$$

$$\Rightarrow \quad H_e = 30 - 68\,Q^2 \quad \text{[from (ii)]}$$

Power developed by the turbine is given by

$$P = \gamma Q H_e \quad \text{[1 mhp = 735 W]}$$

$$\Rightarrow \quad 45 \times 735 = 9810 \times Q \times (30 - 68\,Q^2)$$

$$\Rightarrow \quad 68Q^3 - 30\,Q + 3.371 = 0$$

$$\therefore \quad Q = 0.598 \text{ m}^3\text{/s and } 0.116 \text{ m}^3\text{/s}$$ **(Ans.)**

13.6 A turbine develops power of 150 kW at 250 rpm under a head of 20 m. Determine the scale ratio and speed of the similar machine which will generate 750 kW when working under a head of 30 m.

[10 marks : 2006]

Solution:

Given Data:

Power developed by 1st turbine, $P_1 = 150$ kW

Speed of 1st turbine, $N_1 = 250$ rpm

Head available for 1st turbine, $H_1 = 20$ m

Similarly, $P_2 = 750$ kW; $H_2 = 30$ m; $N_2 = ?$

Let the diameter of runners of 1st and 2nd turbine be D_1 and D_2 respectively.

If the scale effect is considered in turbines, then unit power is given by

$$P_u = \frac{P}{D^2 H^{3/2}}$$

$$\Rightarrow \quad \frac{P_1}{D_1^2 (H_1)^{3/2}} = \frac{P_2}{D_2^2 (H_2)^{3/2}}$$

$$\Rightarrow \quad \frac{150}{D_1^2 (20)^{3/2}} = \frac{750}{D_2^2 (30)^{3/2}}$$

$$\Rightarrow \quad \left(\frac{D_1}{D_2}\right)^2 = \frac{150}{750} \times \left(\frac{30}{20}\right)^{3/2} = 0.3674$$

$$\Rightarrow \quad \frac{D_1}{D_2} = 0.6061$$

$$\Rightarrow \quad \left(\frac{D_2}{D_1}\right) = 1.65 \textbf{ (Ans.)}$$

Also, the unit speed is given by

$$N_u = \frac{ND}{\sqrt{H}}$$

$$\Rightarrow \quad \frac{N_1 D_1}{\sqrt{H_1}} = \frac{N_2 D_2}{\sqrt{H_2}} \qquad \left[\because \frac{D_2}{D_1} = 1.65\right]$$

$$\Rightarrow \quad 250 \times \sqrt{\frac{30}{20}} = N_2 \times 1.65$$

$$\Rightarrow \quad N_2 = 185.57 \text{ rpm } \textbf{(Ans.)}$$

13.7 Give the classification of turbines on the basis of:

(i) Head available

(ii) Action of water

(iii) Specific speed in SI units

Also derive an expression for unit speed, unit discharge and unit power.

[9 marks : 2007]

Solution:

(i) Classification of turbines on the basis of head available:

(a) High head turbines

- These are capable of working under very high heads ranging from several hundred metres to few thousand metres
- These turbines require relatively less quantity of water.
- Pelton wheel is an example of high head turbines.

(b) Medium head turbines

- These are capable of working under medium heads ranging from about 60 m to 250 m.

- These turbines require relatively large quantity of water.
- Modern Francis turbines may be classified as medium head turbines.

(c) **Low head turbines**

- These are capable of working under the heads less than 60 m.
- Kaplan turbine is the example of low head turbine.

(ii) Classification of turbines on the basis of action of water:

(a) **Impulse turbine**

- In an impulse turbine, all the available energy of water is converted into kinetic energy.
- As the water flows over the vanes, the pressure is atmospheric from inlet to outlet of the turbine.
- Some of the impulse turbines are Pelton wheel, Girard turbine, Banki turbine, Jonval turbine, etc.

(b) **Reaction turbine**

- In a reaction turbine, at the entrance to the runner, only a part of the available energy of water is converted into kinetic energy and a substantial part remains in the form of pressure energy.
- As the water flows through runner, the water is under pressure and the pressure energy goes on changing into kinetic energy.
- The runner is completely enclosed in an air-tight casing and the runner and casing is completely full of water.
- Thomson, Francis, Propeller, Kaplan, etc. are some examples of reaction turbines.

(iii) Classification of turbines on the basis of specific speed in SI units:

(a) **Low specific speed turbines**

- Specific speed varies from 8.5 to 30.
- Example is Pelton wheel with single jet.

(b) **Medium specific speed turbines**

- Specific speed varies from 50 to 340.
- Example is Francis turbine.

(c) **High specific speed turbines**

- Specific speed varies from 255 to 860.
- Example is Kaplan turbine.

Derivation of the expression for unit speed

It is defined as the speed of the turbine working under a unit head. It is denoted by N_u.

Let N is the speed of a turbine under a head H and 'u' is the tangential velocity.

The tangential velocity, absolute velocity of water and head on the turbine are related as

$$u \propto V \text{ and } V \propto \sqrt{H}$$

$\therefore$ $$u \propto \sqrt{H}$$

But $$u = \frac{\pi DN}{60} \text{ where } D \text{ is diameter of turbine}$$

For a given turbine D is constant

$\therefore$ $$u \propto N$$

$\Rightarrow$ $$N \propto \sqrt{H}$$

$\Rightarrow$ $$N = k_1\sqrt{H} \text{ where } k_1 \text{ is a constant of proportionality}$$

If head on the turbine becomes unity, the speed becomes unit speed

i.e. when $H = 1, N = N_u$

$\therefore \quad N_u = k_1\sqrt{1}$

$\Rightarrow \quad k_1 = N_u$

$\therefore \quad N = N_u\sqrt{H}$

$\Rightarrow \quad N_u = \dfrac{N}{\sqrt{H}}$ (Ans.)

Derivation of expression for unit discharge

It is defined as the discharge passing through a turbine, which is working under a unit head. It is denoted by Q_u.

Let H = Head of water on turbine

Q = Discharge passing through turbine when head is H

a = Area of flow of water

The discharge passing through a given turbine under a head 'H' is given by

$$Q = a \times V$$

If a is constant, then

$$Q \propto V$$

But $V \propto \sqrt{H}$

$\therefore \quad Q \propto \sqrt{H}$

$\Rightarrow \quad Q = k_2\sqrt{H}$

If $H = 1$ then $Q = Q_u$

$\therefore \quad Q_u = k_2\sqrt{1}$

$\Rightarrow \quad k_2 = Q_u$

$\therefore \quad Q = Q_u\sqrt{H}$

$\Rightarrow \quad Q_u = \dfrac{Q}{\sqrt{H}}$ (Ans.)

Derivation of expression for unit power

It is defined as the power developed by a turbine, working under a unit head. It is denoted by P_u.

Let H = Head of water on the turbine

P = Power developed by the turbine under a head of H

Q = Discharge through turbine under a head H

The overall efficiency (η_o) is given by

$$\eta_o = \frac{\text{Power developed}}{\text{Water power}} = \frac{P}{\dfrac{\rho \times g \times Q \times H}{1000}}$$

$\Rightarrow \quad P = \eta_o \times \dfrac{\rho \times g \times Q \times H}{1000}$

$\Rightarrow \quad P \propto Q \times H$

But $Q \propto \sqrt{H}$

$\therefore \quad P \propto \sqrt{H} \times H$

$\Rightarrow \quad P \propto H^{3/2}$

$\Rightarrow \quad P = k_3 H^{3/2}$

If $\quad H = 1$ then $P = P_u$

$\therefore \quad P_u = k_3(1)^{3/2}$

$\Rightarrow \quad k_3 = P_u$

$\therefore \quad P = P_u \times H^{3/2}$

$\Rightarrow \quad P_u = \frac{P}{H^{3/2}}$ (Ans.)

13.8 **A Francis turbine working under a head of 40 m has a wheel diameter of 1.0 m at the entrance and 0.5 m at exit. The vane angle at the entrance is 90° and guide blade angle is 15°. The water at exit leaves the vane without shock and without any tangential velocity. Velocity of flow in the runner is constant. Neglecting the effect of draft tube and losses in the guide and runner passages, determine the speed of wheel in revolutions per minute and vane angle at exit.**

[10 marks : 2008]

Solution:

Given data: Head, $H = 40$ m, Inlet diameter, $D_1 = 1$ m, Outlet diameter, $D_2 = 0.5$ m

Vane angle at entrance, $\theta = 90°$, Guide blade angle, $\alpha = 15°$

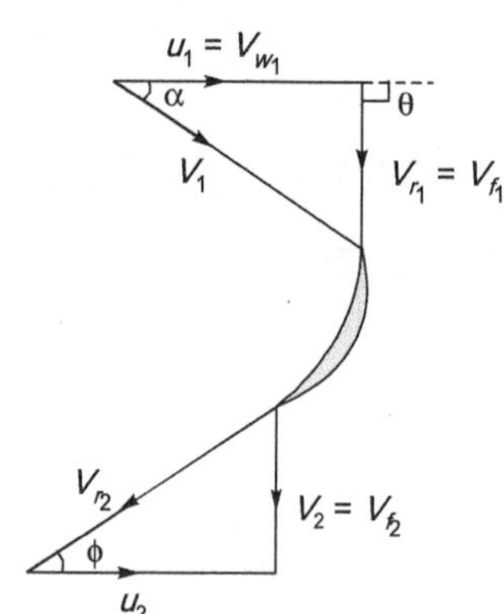

The velocity of flow in the runner is constant i.e. $V_{f_1} = V_{f_2}$

When $\theta = 90°$, the peripheral velocity at inlet u_1 is equal to velocity of whirl at inlet V_{w_1}.

The water at exit leaves the vane without shock and without any tangential velocity i.e. velocity of whirl at outlet V_{w_2} is zero.

From inlet velocity triangle, we have

$$\tan\alpha = \frac{V_{f_1}}{V_{w_1}}$$

$$\Rightarrow \quad \tan 15° = \frac{V_{f_1}}{u_1} \qquad [\because u_1 = V_{w_1}]$$

$$\Rightarrow \quad V_{f_1} = u_1 \tan 15° \qquad ...(i)$$

We know that, head utilized by the turbine is given by

$$H_e = \frac{u_1 V_{w_1} - u_2 V_{w_2}}{g}$$

$$\Rightarrow \quad H_e = \frac{u_1 V_{w_1}}{g} \qquad [\because V_{w_2} = 0]$$

But

$$H_e = H - \frac{(V_2)^2}{2g}$$

$$\Rightarrow \quad \frac{u_1 V_{w_1}}{g} = H - \frac{(V_2)^2}{2g}$$

$$\Rightarrow \quad \frac{(u_1)^2}{g} + \frac{(V_{f_2})^2}{2g} = H \qquad [\because u_1 = V_{w_1} \text{ and } V_2 = V_{f_2}]$$

$$\Rightarrow \quad (u_1)^2 + \frac{(V_{f_1})^2}{2} = gH \qquad [\because V_{f_1} = V_{f_2}]$$

$$\Rightarrow \quad 2 \times (u_1)^2 + (u_1 \tan 15°)^2 = 2gH$$

$$\Rightarrow \quad [2 + (\tan 15°)^2] \times (u_1)^2 = 2gH \qquad \text{[From (i)]}$$

$$\Rightarrow \quad (u_1)^2 = \frac{2 \times 9.81 \times 40}{2.072}$$

$$\Rightarrow \quad u_1 = 19.46 \text{ m/s}$$

But

$$u_1 = \frac{\pi D_1 N}{60}$$

$$\Rightarrow \quad 19.46 = \frac{\pi \times 1 \times N}{60}$$

$$\Rightarrow \quad N = 371.66 \text{ rpm}$$ **(Ans.)**

From outlet velocity triangle we have

$$\tan\phi = \frac{V_{f_2}}{u_2}$$

$$\Rightarrow \quad \tan\phi = \frac{u_1 \tan 15°}{u_2} \qquad [\because V_{f_2} = V_{f_1} = u_1 \tan 15°]$$

$$\Rightarrow \quad \tan\phi = \frac{D_1}{D_2} \times \tan 15° \qquad \left[\because \frac{u_1}{u_2} = \frac{\pi D_1 N/60}{\pi D_2 N/60} = \frac{D_1}{D_2}\right]$$

$$\Rightarrow \quad \tan\phi = \frac{1}{0.5} \times \tan 15°$$

$$\Rightarrow \quad \phi = 28.19°$$ **(Ans.)**

13.9 **A Pelton wheel is producing 300 kW working under a head of 180.0 m with discharge of 0.2 m³/s. Compute hydraulic efficiency and velocity of whirl at inlet and outlet and the mean bucket speed. Take coefficient of velocity = 0.985, angle of deflection of jet = 165° and relative velocity at exit = relative velocity at inlet.**

[10 marks : 2009]

Solution:

The inlet and outlet velocity triangles for the Pelton wheel is shown below:

Mean bucket speed, $u = u_1 = u_2$

$$\therefore \quad u = 0.46\sqrt{2gH} = 0.46\sqrt{2 \times 9.81 \times 180} = 27.34 \text{ m/s}$$

Velocity of jet at inlet, $V_1 = C_V\sqrt{2gH} = 0.985\sqrt{2 \times 9.81 \times 180}$

$= 58.53$ m/s

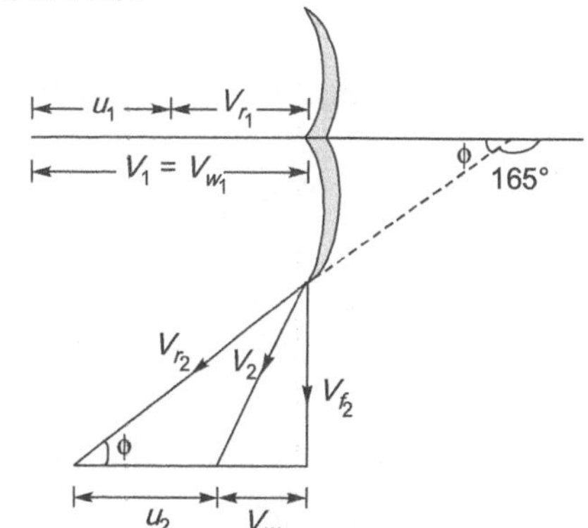

Velocity of whirl at inlet, $V_{w_1} = V_1 = 58.53$ m/s

$$\phi = 180° - 165° = 15°$$

Given that $V_{r_1} = V_{r_2}$

But $V_{r_2} = kV_{r_1}$

$$\therefore \quad V_{r_1} = kV_{r_1}$$

$\Rightarrow \quad k = 1$

Also, $\quad V_{r_1} = V_1 - u_1 = 58.53 - 27.34 = 31.19$ m/s

$\therefore \quad V_{r_2} = 31.19$ m/s

Velocity of whirl at outlet is given by

$$V_{w_2} = V_{r_2}\cos\phi - u_2 = 31.19\cos 15^\circ - 27.34 = 2.79 \text{ m/s}$$

Hydraulic efficiency, $$\eta_h = \frac{u(V_1 - u)(1 + k\cos\phi)}{gH} = \frac{27.34(58.53 - 27.34)(1+\cos 15^\circ)}{9.81 \times 180}$$

$$= 0.9494 = 94.94\% \text{ (Ans.)}$$

Overall efficiency, $$\eta_o = \frac{P}{\gamma QH} = \frac{300 \times 10^3}{9810 \times 0.2 \times 180} = 0.8495 = 84.95\%$$

Mechanical efficiency, $$\eta_m = \frac{\eta_o}{\eta_H} = \frac{0.8495}{0.9494} = 0.8948 = 89.48\%$$

13.10 **Following data are observed during testing of a Kaplan turbine: Power developed = 2500 kW, Head = 60 m and speed = 350 rpm. What will be the dimensionless specific speed of the turbine?**

[2 marks : 2010]

Solution:

Given Data: $\quad P = 2500$ kW; $H = 60$ m; $N = 350$ rpm

The non-dimensional specific speed is given by

$$N_s = \frac{N\sqrt{P}}{\rho^{1/2}(gH)^{5/4}} = \frac{350 \times \sqrt{2500 \times 10^3}}{(1000)^{1/2} \times (9.81 \times 60)^{5/4}} = 6.04$$

13.11 **A power house has 5 impulse turbines. Each turbine has two runners. Each runner is installed with 4 nozzles. Total discharge is 40 m³/s. Find the diameter of jet. Take coefficient of velocity as 0.985 and head as 250 m.**

[4 marks : 2010]

Solution:

Number of impulse turbines, $\quad n_1 = 5$

Number of runners in each turbine, $\quad n_2 = 2$

Number of nozzles in each runner, $\quad n_3 = 4$

$\therefore$ Total number of nozzles, $\quad n = n_1 \times n_2 \times n_3 = 5 \times 2 \times 4 = 40$

Discharge through single nozzle, $$Q = \frac{\text{Total discharge}}{n} = \frac{40}{40} = 1 \text{ m}^3/\text{s}$$

Velocity of jet, $$V = C_v\sqrt{2gH} = 0.985\sqrt{2 \times 9.81 \times 250} = 68.99 \text{ m/s}$$

$$\text{Area of jet} = \frac{Q}{V}$$

$$\Rightarrow \quad \frac{\pi d^2}{4} = \frac{1}{68.99}$$

$$\Rightarrow \quad d = 0.13585 \text{ m} = 135.85 \text{ mm}$$

13.12 A reaction turbine having inlet diameter as 1.0 m and rotational speed 400 rpm has flow area 0.25 m^2 at the inlet and is working under a head of 65 m. Flow is radial at the outlet. Compute the hydraulic efficiency and power developed by the wheel taking velocity of flow at inlet as 8.0 m/s and velocity of whirl at inlet as 25.0 m/s.

[4 marks : 2010]

Solution:

Inlet diameter, $D_1 = 1.0$ m

Rotational speed, $N = 400$ rpm

Area of flow, $A = 0.25$ m^2

Net available head, $H = 65$ m

Velocity of flow at inlet, $V_{f1} = 8.0$ m/s

Velocity of whirl at inlet, $V_{w1} = 25.0$ m/s

Flow is radial at outlet i.e. velocity of whirl at outlet, $V_{w2} = 0$

Let the peripheral velocity at inlet and outlet be u_1 and u_2 respectively

$$\therefore \quad u_1 = \frac{\pi D_1 N}{60} = \frac{\pi \times 1 \times 400}{60} = 20.94 \text{ m/s}$$

Discharge, $Q = A \times V_{f1} = 0.25 \times 8 = 2$ m^3/s

Power developed by the wheel is expressed as

$$P = \rho Q(u_1 V_{w1} - u_2 V_{w2})$$

$$= 1000 \times 2 \times (20.94 \times 25 - u_2 \times 0) \times 10^{-3} = 1047 \text{ kW}$$

Hydraulic efficiency,

$$\eta_h = \left[\frac{u_1 V_{w1} - u_2 V_{w2}}{gH}\right] \times 100 = \left[\frac{20.94 \times 25 - u_2 \times 0}{9.81 \times 65}\right] \times 100 = 82.1\%$$

13.13 An inward flow reaction turbine is supplied water at the rate of 0.36 m^3/s. Outlet pipe of the turbine is 380 mm in diameter. Turbine operates under a head of 55 m. Radial velocity of wheel is the same as the velocily of flow in the outlet pipe. Thc tangential velocity of wheel at inlet is 20 m/s. Compute guide vane angle and vane tip angle at inlet.

[4 marks : 2010]

Solution:

Discharge, $Q = 0.36$ m^3/s

Outlet diameter, $D_2 = 380$ mm

Net available head, $H = 55$ m

Tangential velocity of wheel at inlet, $u_1 = 20$ m/s

Radial velocity of wheel is the same as the velocity of flow in the outlet pipe i.e. velocity of whirl at outlet (V_{w2}) is zero. Let the guide vane angle and vatic tip angle at inlet be α and θ respectively

Absolute velocity in the outlet pipe,

$$V_2 = \frac{Q}{\pi D_2^2 / 4} = \frac{4 \times 0.36}{\pi \times (0.38)^2} = 3.17 \text{ m/s}$$

Now $V_{f_1} = V_{f_2} = V_2 = 3.17$ m/s

From energy balance, Head supplied = Work done + Kinetic head at exit

$$\Rightarrow \quad H = \frac{V_{w_1} u_1}{g} + \frac{V_2^2}{2g}$$

$$\Rightarrow \quad \frac{V_{w_1} u_1}{g} = H - \frac{V_2^2}{2g}$$

$$\Rightarrow \quad V_{w_1} = \left[55 - \frac{(3.17)^2}{2 \times 9.81}\right] \times \frac{9.81}{20} = 26.73 \text{ m/s}$$

But $$\tan \alpha = \frac{V_{f_1}}{V_{w_1}}$$

$$\Rightarrow \quad \alpha = \tan^{-1}\left(\frac{3.17}{26.73}\right) = 6.76° \textbf{ (Ans.)}$$

Also, $$\tan \theta = \frac{V_{f_1}}{V_{w_1} - u_1} = \frac{3.17}{26.73 - 20} = 0.4717$$

$$\Rightarrow \quad \theta = \tan^{-1}(0.4717) = 25.22° \textbf{ (Ans.)}$$

13.14 **A turbine for the following site condition is to be designed. Head is 120 m, power produced = 5 × 120 MW. No. of turbines 5, Head loss due to friction as 3% the gross head. Length of the penstock 1300 m each. Overall efficiency 0.87. Number of pairs of poles 18. Speed 166.7 rpm. Breadth to diameter 0.3. Darcy weisback f = 0.02. Determine the diameter of the penstock. Runner and the type of turbine ϕ = 0.75.**

[10 marks : 2011]

Solution:

Gross head, $H_G = 120$ m

Head loss due to friction $= 3\%$ of gross head $= \frac{3}{100} \times 120 = 3.6$ m

Net head, $H = 120 - 3.6 = 116.4$ m

Shaft power of each turbine, $SP = 120$ MW

Overall efficiency, $\eta_0 = 0.87$

$$\therefore \quad \text{Water power} = \frac{SP}{\eta_0} = \frac{120}{0.87}$$

$$\therefore \quad \dot{m}gH = 137.931 \text{ MW}$$

$$\therefore \quad \dot{m} = \frac{137.931 \times 10^6}{9.81 \times 116.4}$$

$$\therefore \quad Q = \frac{137.931 \times 10^6}{9.81 \times 116.4 \times 1000} = 120.792 \text{ m}^3/\text{s}$$

Length of penstock, $L = 1300$ m

Head loss due to friction, $h_f = 3.6$ m

$$h_f = \frac{fLQ^2}{12.1 D^5}$$

where D = diameter of penstock

$$\therefore \quad 3.6 = \frac{0.02 \times 1300 \times 120.792^2}{121.1 \times D^5} \quad \text{...[f = 0.02]}$$

$$\therefore \quad D = 6.1375 \text{ m}$$

Now, speed ratio, $\phi = 0.75$

$$\therefore \quad \frac{u}{\sqrt{2gH}} = 0.75$$

$$\therefore \quad u = 0.75\sqrt{2 \times 9.81 \times 116.4} = 35.8416 \text{ m/s}$$

Now, $u = \dfrac{\pi D_1 N}{60}$

where, D_1 = diameter of runner

N = speed = 166.7 rpm

$$\therefore \quad 35.8416 = \frac{\pi \times D_1 \times 166.7}{60}$$

$$\therefore \quad D_1 = 4.1063 \text{ m}$$

Type of turbine = ?

Specific speed, $N_S = \dfrac{N\sqrt{P}}{H^{5/4}}$

$$P = \text{power in kW} = \frac{120 \times 10^6}{10^3} = 120 \times 10^3 \text{kW}$$

$$\therefore \quad N_S = \frac{166.7\sqrt{120 \times 10^3}}{(116.4)^{5/4}} = 151.037 \text{ (SI units)}$$

∴ Turbine can be classified as Francis turbine based on specific speed.

13.15 **(i) Explain the function of governor in Impulse type of turbine.**

(ii) A Pelton wheel develops 3500 kW under a head of 120 m with an overall efficiency of 85%. Find the diameter of the nozzle if the coefficient of velocity is 0.98.

[2 + 8 = 10 marks : 2011]

Solution:

(i) A **hydraulic turbine** is required to run at the same constant speed as that of the electric generator to which it is directly coupled. This synchronous speed must be maintained for all condition of load on the generator. But since the demand for the power fluctuates with time, the hydro unit must adjust itself to such variations in load in such a way that the turbine speed remains constant/unaffected. Whenever the load on the turbine decrease, if the discharge entering the turbine is not adjusted accordingly,its speed will increase. On the other hand, increase in load will tend to reduce the turbine speed unless the turbine discharge is increased in accordance with the new load. Such regulation of discharge with a view to maintain a synchronous speed of the turbine runner is called governing of turbine and is done automatically by means of a governor.

(ii)

P = 3500 kW

H = 120 m

η_0 = 85%

c_v = 0.98

d = diameter of nozzle

$$\eta_0 = 0.85 = \frac{P}{\gamma Q H} = \frac{3500 \times 10^3}{9810 \times Q \times 120}$$

$$\Rightarrow \quad Q = 3.50 \text{ m}^3\text{/sec}$$

Also, $V_1 = c_v\sqrt{2gH}$

$$= 0.98\sqrt{2\times9.81\times120} = 47.55 \text{ m/s}$$

Discharge through nozzle, $Q = \frac{\pi d^2}{4}\cdot V_1$

$$\Rightarrow \quad \frac{\pi d^2}{4}\times 47.55 = 3.50$$

$$\Rightarrow \quad d = 0.3061 \text{ m}$$

$$= 30.61 \text{ cm}$$

13.16 **What are the advantageous of differential surge tanks over simple and restricted orifice surge tanks?**

[4 marks : 2011]

Solution:

A surge tank is a storage reservoir fitted at some opening made on a long penstock to receive the rejected flow when the penstock is suddenly closed by a valve fitted at its steep end. It also serves the purpose of supplying initial water for an increasing load on the turbine.

These are three main types of surge tank:

(i) Simple surge tank
(ii) Restricted orifice type surge tank
(iii) Differential surge tank

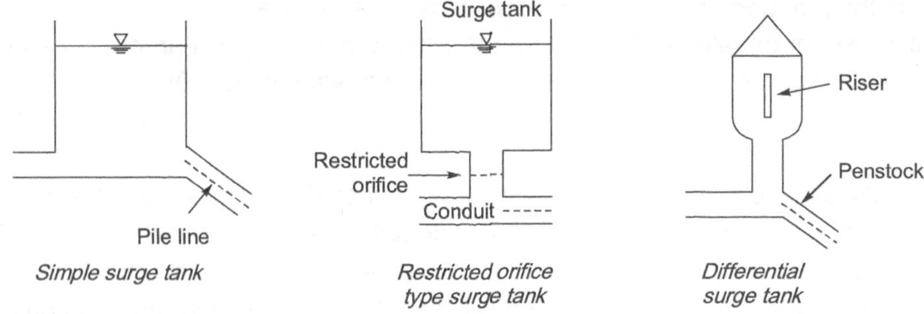

Simple surge tank *Restricted orifice type surge tank* *Differential surge tank*

Advantages of differential surge tank over simple and restricted orifice type surface tank:

A differential surge tank is a combinational of simple and restricted orifice type, with an addition of internal riser.

The superior merit of the differential tank lies in its capacity to separate its function as an accelerator or retarder or conduit velocity from its function as a storage tank. In the simple and restricted orifice type surge tank, the corrective action on the conduit velocity accumulates only as fast as the water level changes in the tank, and hence the duration of the surge is much prolonged and the tank has to be made much larger because it must supply or receive water for a longer period.

13.17 **A Pelton wheel has a mean bucket speed of 10 m/sec with a jet of water flowing at the rate of 0.7 m^3/sec under a head of 30 m. The bucket deflects the jet through an angle of 160°. Calculate power given by water to the runner and the hydraulic efficiency of the turbine. Assume efficiency of the nozzle as 0.98.**

[4 marks : 2012]

Solution:

Given:

Speed of bucket, $u = u_1 = u_2 = 10$ m/s

Discharge,	$Q = 0.7\ m^3/sec$
Head of water,	$H = 30\ m$
Angle of deflection	$= 160°$
∴ Angle,	$\phi = 180° - 160° = 20°$
Efficiency of nozzle	$= 0.98$
∴ Coefficient of velocity	$C_v = 0.98$

The velocity of jet,
$$V_1 = C_v\sqrt{2gH} = 0.98 \times \sqrt{2 \times 9.81 \times 30} = 23.78\ m/s$$

∴
$$V_{r_1} = V_1 - u_1 = 23.78 - 10 = 13.78\ m/s$$
$$V_{w_1} = V_1 = 23.78\ m/s$$

From outlet velocity triangle

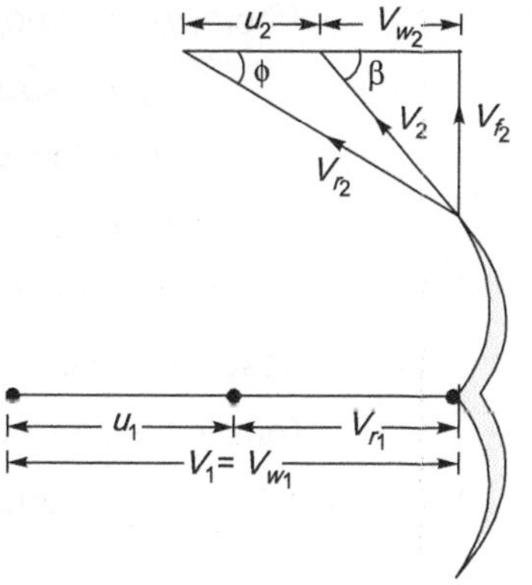

$$V_{r_2} = V_{r_1} = 13.78\ m/s$$
$$V_{w_2} = V_{r_2}\cos\phi - u_2$$
$$= 13.78 \cos 20° - 10 = 2.95\ m/s$$

Work done by the jet per second on the runner is given by

$$= \rho a V_1\left[V_{w_1} + V_{w_2}\right] \times u$$
$$= 1000 \times 0.7 \times [23.78 + 2.95] \times 10$$
$$[\because aV_1 = Q = 0.7\ m^3/s]$$
$$= 187110\ Nm/s$$

∴ Power given to turbine
$$= \frac{187110}{1000} = 187.11 kW$$

The hydraulic efficiency of the turbine is given by

$$\eta_h = \frac{\text{Runner power}}{\text{Water power}} = \frac{\rho Q\left[V_{w_1} + V_{w_2}\right] \times u}{\frac{1}{2}\rho Q V_1^2} = \frac{\left[V_{w_1} + V_{w_2}\right] \times u}{\left(\frac{V_1^2}{2}\right)}$$

∴
$$\eta_h = \frac{[23.78 + 2.95] \times 10}{\left(\frac{23.78^2}{2}\right)} = 0.94537 = 94.54\%$$

13.18 Define: (i) Shape number (ii) NPSH in pumps
Describe the occurrence of cavitation in propeller turbine of Kaplan turbine.

[4 marks : 2013]

Solution:

(i) **Shape Number:** Efficiency of centrifugal pumps depends on pump size, speed and shape number. Shape number (N_s) defines the shape of the pump as

$$N_s = \frac{\text{Revolutions/min}}{[\text{Head(in m)}]^{3/4}}$$

Alternatively it can be defined as speed at which the pump will run if reduced geometrically to give

1 lt/sec at 1 m head. It is also regarded as the ratio defining the shape of the pump.

(ii) NPSH in pumps: Quantitatively it is defined as summation of absolute pressure head at the pump inlet, less the absolute vapour pressure head of the liquid corresponding to the temperature at which it is being pumped, plus the velocity head at this point.

$$\therefore \quad \text{NPSH} = \left(\frac{p_a}{\rho g} + \frac{p_s}{\rho g}\right) - \frac{p_v}{\rho g} + \frac{V_s^2}{2g} \quad \text{...(i)}$$

But absolute pressure head at pump inlet is,

$$\frac{p_a}{\rho g} + \frac{p_s}{\rho g} = \frac{p_a}{\rho g} - \left[\frac{V_s^2}{2g} + h_s + h_f\right] \quad \text{...(ii)}$$

Where, p_a = Atmospheric pressure

p_s = Gauge pressure at pump inlet

V_s = flow velocity in suction pipe

h_s = suction lift

h_f = head loss in value, strainer, suction pipe

Absolute pressure lower than absolute vapour pressure of liquid is impossible to create at pump inlet.

Thus in the limiting case $\quad p_a + p_a = p_v$

$\therefore$ Limiting value of suction lift

$$(h_s) = \left(\frac{p_a - p_v}{\rho g}\right) - \frac{V_s^2}{2g} - h_f$$

Thus if h_s exceeds the limiting value, then rapid vaporization of liquid takes place thereby leading to cavitation.

Substituting the value of $\left(\frac{p_s}{\rho g} + \frac{p_s}{\rho g}\right)$ from (ii) in (i)

$$\text{NPSH} = \frac{p_a}{\rho g} - \frac{p_v}{\rho g} - h_s - h_f$$

Thus NPSH is the head required to make the liquid to flow through the suction pipe of the pump.

Cavitation in turbines: When in any part of the flow passage, the pressure reaches the vapour pressure of liquid. It starts vapourizing and formation of bubbles of vapour take place. These bubbles are carried away along with the flow and in high pressure zones. These gets collapse due to condensation of vapours to liquid. Due to sudden collapse of bubbles, surrounding liquid rushes to fill them. Liquid moving from all directions into the bubble collide at the bubble centre thus giving rise to very high localized pressures (~700 MN/m^2). Thus the solid surface in the vicinity is subjected to very high pressures. Alternate formation and collapse of bubbles result in severe damage to the surface which may lead to fatigue failure and surface becomes severely eroded.

In Kaplan turbine, cavitation may occur at the runner exit or inlet of draft tube where pressures are very much reduced.

To predict the occurrence of cavitation in any part of the turbine, a dimensionless parameter called "THOMA'S CAVITATION FACTOR" is defined which for Kaplan turbine is expressed as

$$\sigma = 1.1\left[0.28 + \frac{2}{15}\left(\frac{N_s}{444}\right)^2\right]$$

13.19 **An inward flow reaction turbine has an inlet and outlet diameter of 1.5 m and 0.75 m respectively. The breadth at the inlet is 0.30 m and at the outlet is 0.40 m. At a speed of rotation of 300 rpm, the relative velocity at the entrance is 5.25 m/sec and is radial. Calculate the**

(i) absolute velocity at entrance and inclination to the tangent of the runner;

(ii) discharge;

(iii) velocity of flow at the outlet

[8 marks : 2015]

Solution:

$D_1 = 1.5$ m; $D_2 = 0.75$ m, $B_1 = 0.3$ m; $B_2 = 0.4$ m

$$N = 300 \text{ rpm}; \quad V_{r_1} = V_{f_1} = 5.25 \text{ m/sec}$$

(i)
$$u_1 = \frac{\pi D_1 N}{60} = \frac{\pi \times 1.5 \times 300}{60} = 23.56 \text{ m/sec}$$

$$\because \quad \text{Absolute velocity } (V_1) = \sqrt{V_{r_1}^2 + u_1^2}$$

$$= \sqrt{23.56^2 + 5.25^2} = 24.14 \text{ m/sec}$$

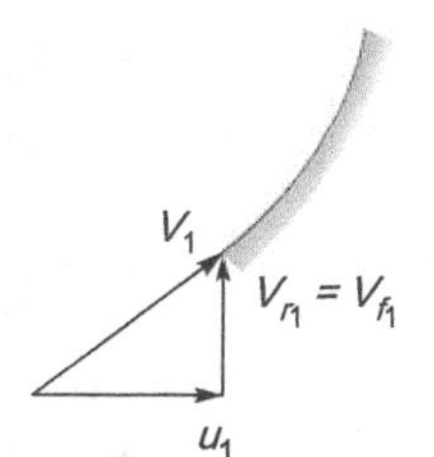

(ii)
$$\text{Discharge } (Q) = \pi D_1 B_1 V_{f_1}$$

$$= \pi \times 1.5 \times 0.3 \times 5.25 = 7.42 \text{ m}^3/\text{sec}$$

(iii)
$$\text{Velocity of flow at outlet } (V_{f_2}) = \frac{Q}{\pi D_2 B_2}$$

$$\therefore \quad V_{f_2} = \frac{7.42}{\pi \times 0.75 \times 0.4} = 7.875 \text{ m/sec}$$

13.20 **What are hydraulic turbines? Describe different types of hydraulic turbines with their characteristics and suitability of use. Differentiate between 'Pelton' and 'Francis' turbines.**

[8 marks : 2016]

Solution:

Hydraulic turbine: It is a device that converts potential energy of water (owing to storage at an elevation) into mechanical energy. The mechanical energy generated is used to rotate the shaft of generator in hydro-electric power plants.

Type of hydraulic turbines: Turbines are classified on various bases which are enumerated below:

(i) Based on change in pressure (i.e. type of energy at inlet)

- **Impulse turbine:** In impulse turbine, the total head of the incoming fluid is converted into a large kinetic head at the exit of nozzle. This is suitable for large head and low discharge conditions, e.g. Pelton turbine.

- **Reaction turbine:** In reaction turbine, pressure of the liquid changes as it flows through the rotor. Change in flow velocity and decrease in pressure imparts a reaction on the blades of the turbine. The penstock feeds water through a row of fixed blades through the casing. The fixed blades convert a part of pressure energy into kinetic energy before water enters the runner. Thus, water entering the runner has both pressure and kinetic energies, e.g. Francis turbine and Kaplan turbine.

(ii) Based on flow path

- **Axial flow turbine:** In this type of turbine, the flow path of the fluid remains parallel to the axis of rotation. e.g. Kaplan turbine.

- **Radial flow turbine:** In this type of turbine, the flow path is perpendicular to the axis of rotation.

–**Inward flow turbine** where water enters from periphery and leaves through the center of turbine. e.g. Old type of Francis turbine.

–**Outward flow turbine** where water enter from center and leaves from periphery of the turbine.

- **Mixed flow turbine:** It is a combination of axial flow and radial flow turbines where a significant proportion of both the flow paths exist in the fluid. e.g. Modern Francis turbine wherein fluid enters radially and exits in the axial direction.
- **Tangential flow turbine:** Here water strikes the runner in tangential direction to the wheel. e.g. Pelton turbine

(iii) Based on head available at inlet of turbine:

- **High head turbine:** In this turbine, the head available at turbine inlet varies from 150 m to 200 m or even more in certain cases. This turbine is associated with small amount of discharge. e.g. Pelton turbine.
- **Low head turbine:** In low head turbine, the net head available at turbine inlet is less than 30 m. This type of turbine requires large discharge. e.g. Kaplan turbine.
- **Medium head turbine:** In this turbine, the net head available at turbine inlet varies from 30 m to 150 m. This type of turbine requires moderate amount of discharge. e.g. Francis turbine.

(iv) Based on specific speed of turbine

- **High specific speed turbine:** In this, the specific speed is more than 250. e.g. Kaplan turbine.
- **Low specific speed turbine:** In this, the specific speed is less than 50. e.g. Pelton turbine.
- **Medium specific speed turbine:** In this, the specific speed of turbine varies from 50 to 150. e.g. Francis turbine

Table : *Pelton turbine vs. Francis turbine*

S.No.	Pelton turbine	Francis turbine
1.	It is an impulse turbine.	It is a reaction turbine.
2.	It is suitable for high heads (> 200 m).	It is suitable for medium heads (30 m to 150 m).
3.	It requires low discharge.	It requires high discharge.
4.	It has low specific speed.	It has medium specific speed.
5.	This turbine is always installed above the tail race and hence as such no draft tube is required.	This turbine is connected to the tail race through a draft tube.
6.	Flow is regulated though a needle valve fitted to the nozzle.	Flow is regulated by means of guide vane assembly.
7.	It has more hydraulic efficiency.	It has low hydraulic efficiency.
8.	No water casing is required.	Here since the flow is under pressure and thus casing is required to maintain the pressure.

13.21 Neglecting losses, compute the power produced by the turbine shown in figure below. Take η_0 = 100%.

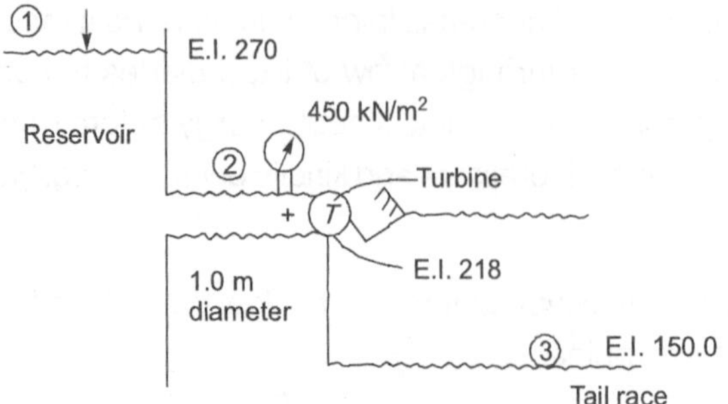

[20 marks : 2017]

Solution:

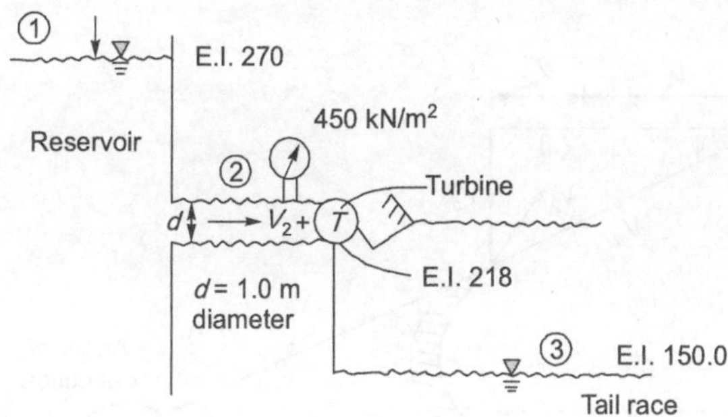

Water friction loss neglected.

Let Q be discharge and velocity V_2 in the pipe entering in turbine.

Applying Bernoulli's equations to point (1) and (2);

$$\frac{P_1}{\rho g}+\frac{V_1^2}{2g}+z_1 = \frac{P_2}{\rho g}+\frac{V_2^2}{2g}+z_2$$

$$\Rightarrow \quad 0+0+270 = \frac{450\times10^3}{9810}+\frac{V_2^2}{2g}+218$$

$$\Rightarrow \quad \frac{V_2^2}{2g} = (270-218)-\frac{450\times10^3}{9810} = 6.13 \text{ m}$$

$$\Rightarrow \quad V_2 = \sqrt{2\times9.81\times6.13} = 10.97 \text{ m/sec}$$

Let H_T be head extracted by the turbine

$$\frac{P_1}{\rho g}+\frac{V_1^2}{2g}+z_1 = H_T+\frac{P_3}{\rho g}+\frac{V_3^2}{2g}+z_3$$

$$\Rightarrow \quad 0+0+270 = H_T+0+\ 0+150$$

$$H_T = 120 \text{ metre}$$

Now, Discharge, $Q = A\times V_2$

$$= \frac{\pi}{4}\times(d)^2\times V_2$$

$$= \frac{\pi}{4}\times(1)^2\times10.97 = 8.62 \text{ m}^3\text{/sec}$$

$\therefore$ Power extracted by turbine $= \gamma Q H_T$

$= 9810\times8.62\times120$

$= 10.15$ MW

$\because$ η_0 = Overall efficiency of turbine = 100%

Hence, Power output of turbine $= \eta_0\times10.15$

$\Rightarrow$ $P_n = 10.15$ MW

13.22 A pelton wheel turbine bucket deflects the water jet by an angle θ. Show that if friction losses are ignored, the maximum power will be developed when the bucket speed is half of the jet speed. If friction on the bucket surface reduces the relative velocity of the jet by 10% at the exit, find the ratio of the bucket speed and jet speed for maximum power.

[14 marks : 2018]

Solution:

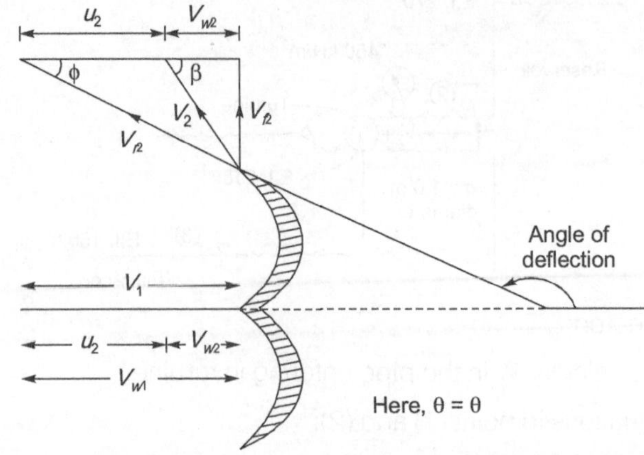

Let H = Net head acting on Pelton wheel

(i) Frictional losses are ignored

$$V_{r_1} = V_1 - u$$

$$V_{r_2} = V_{r_1} \qquad \ldots\ u_1 = u_2 = u = \frac{\pi DN}{60}$$

$$V_{w_2} = V_{r_2}\cos\phi - u$$

Force exerted by jet in the direction of motion,

$$F_x = \rho a V_1\left(V_{w_1} + V_{w_2}\right) \qquad \text{(where, } a = \text{area of jet)}$$

Now,Work done by the jet on the runner per seconds,

$$= F_x \times u = \rho a V_1\left[V_{w_1} + V_{w_2}\right] \times u \text{ Nm/s}$$

Power given to the runner by the jet

$$P = \frac{\rho a V_1\left[V_{w_1} + V_{w_2}\right] \times u}{1000} \text{kW} \qquad \ldots(i)$$

Here,

$$V_{w_1} = V_1$$

$$V_{w_2} = V_{r_2}\cos\phi - u$$

$$= V_{r_1}\cos\phi - u$$

$$= (V_1 - u)\cos\phi - u$$

So,

$$P = \frac{\rho a V_1\left[V_1 + (V_1 - u)\cos\phi - u\right] \times u}{1000}$$

$\Rightarrow$

$$P = \frac{\rho a V_1\left[(V_1 - u)(1 + \cos\phi)\right]u}{1000}$$

or

$$P = \frac{\rho a u(1 + \cos\phi)}{1000}\left(V_1^2 - uV_1\right)$$

So, for maximum power

$$\frac{dP}{du} = 0$$

$$\Rightarrow \quad \frac{d}{du}\left[u\left(V_1^2 - uV_1\right)\right] = 0$$

$$\therefore \quad \frac{d}{du}\left(uV_1 - u^2\right) = 0$$

or $$2u = V_1$$

or $$u = \left(\frac{V_1}{2}\right)$$ (Hence proved)

(ii) Frictional losses are considered

$$V_{r_2} = 0.9V_{r_1}$$

So, $$V_{r_2} = 0.9\,(V_1 - u)$$

From eq. (i) $$P = \frac{\rho a V_1\left[V_{w_1} + V_{w_2}\right] \times u}{1000} \text{ kW}$$

Here, $$V_{w_1} = V_1$$

$$V_{w_2} = V_{r_2}\cos\phi - u$$

$$= 0.9V_{r_1}\cos\phi - u$$

$$= 0.9\,(V_1 - u)\cos\phi - u$$

So, $$P = \frac{\rho a V_1\left[V_1 + 0.9(V_1 - u)\cos\phi - u\right] \times u}{1000}$$

$$\Rightarrow \quad P = \frac{\rho a V_1\left[(V_1 - u)(1 + 0.9\cos\phi)\right]u}{1000}$$

$$\Rightarrow \quad P = \frac{\rho a V_1(1 + 0.9\cos\phi)}{1000}\left(V_1 u - u^2\right)$$

For maximum power,

$$\frac{dP}{du} = 0$$

$$\Rightarrow \quad \frac{d}{du}\left(V_1 u - u^2\right) = 0$$

$$\Rightarrow \quad V_1 = 2u$$

or $$\frac{u}{V_1} = 0.5$$

Here u = bucket speed

and V_1 = jet speed

13.23 **An outward flow turbine running at 200 rpm, works on a discharge of 5 m^3/s under a head of 40 m. Internal and external diameters of the wheel are 2 m and 2.5 m respectively while the width at the inlet and outlet is 200 mm. Assuming the discharge to be radial at the outlet, determine angles of the turbine at the inlet and outlet. Also draw the velocity triangles for outward flow turbine.**

[20 marks : 2020]

Solution:

Outward flow turbine

$N = 200$ rpm, $Q = 5$ m^3/s, $H = 40$ m, $D_1 = 2$ m, $D_2 = 2.5$ m, $B_1 = B_2 = 200$ mm

$$\beta = 90°, V_{w2} = 0 \text{ (Discharge is radial at output)}$$

$$u_1 = \frac{\pi D_1 N}{60} = \frac{\pi(2)(200)}{60} = 20.94 \text{ m/s}$$

$$u_2 = \frac{\pi D_2 N}{60} = \frac{\pi(2.5)(200)}{60} = 26.18 \text{ m/s}$$

$$Q = \pi D_2 B_2 V_{f_2}$$

$$5 = \pi(2.5)(0.2)V_{f_2}$$

$$V_{f_2} = 3.183 \text{ m/s}$$

Also,

$$Q = \pi D_1 B_1 V_{f_1}$$

$$5 = \pi(2)(0.2)V_{f_1}$$

$$V_{f_1} = 3.98 \text{ m/s}$$

$$H = \frac{RP}{\dot{w}g} + \frac{V_2^2}{2g} \qquad (V_2 = V_{f_2})$$

$$40 = \frac{V_{w_1}u_1}{g} + \frac{V_{f_2}^2}{2g}$$

$$40 = \frac{V_{w_1}(20.94)}{g} + \frac{(3.183)^2}{2g}$$

$$40 = \frac{V_{w_1}u_1}{g} + \frac{(3.183)^2}{2g}$$

$$V_{w_1} = 18.5 \text{ m/s}$$

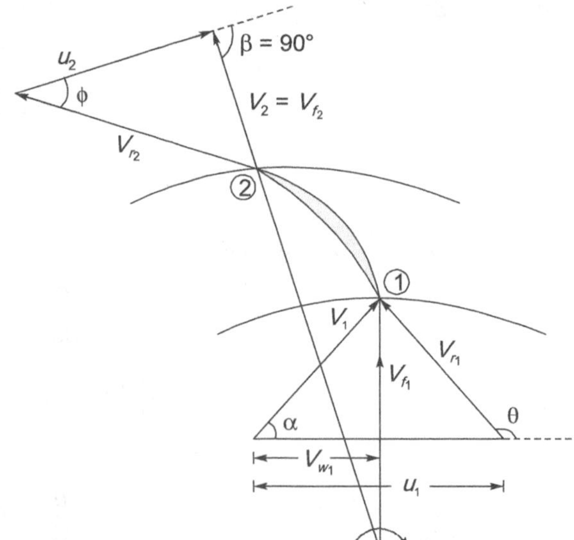

At outlet

$$\tan\phi = \frac{V_{f_2}}{u_2} = \frac{3.183}{26.18}$$

$$\phi = 6.93°$$

At inlet

$$\tan(180° - \theta) = \frac{V_{f_1}}{u_1 - V_{w_1}} = \frac{3.98}{20.94 - 18.5}$$

$$\theta = 121.51°$$

13.24 An inward flow reaction turbine works under a head of 40.0 m and discharge of 12 cu.m/sec. The speed of the runner is 350 rpm. At the inlet tip of runner vane, the peripheral velocity of wheel is 25.20 m/s and the radial velocity of flow is 8.40 m/s. If the overall efficiency and hydraulic efficiency of the turbine are 80% and 90% respectively, determine
(i) power developed by the turbine in kW, (ii) diameter and width of runner at inlet, (iii) guide blade angle at inlet, (iv) inlet angle at runner vane, and (v) diameter of runner at outlet.
Assume flow is radial at outlet. Also assume no blade friction and radial velocities at inlet and outlet are equal.

[20 Marks : 2021]

Solution:

Inward flow reaction turbine

$H = 40$ m	Determine
$Q = 12\ m^3/s$	(i) S.P.
$N = 350$ rpm	(ii) D_1 & B_1
$u_1 = 25.20$ m/s	(iii) α
$V_{f_1} = 8.4$ m/s	(iv) Q
$\eta_0 = 80\%$	(v) D_2
$\eta_h = 90\%$	

$V_2 = V_{f_2},\ \beta = 90°$

$V_{f_1} = V_{f_2} = 8.4$ m/s

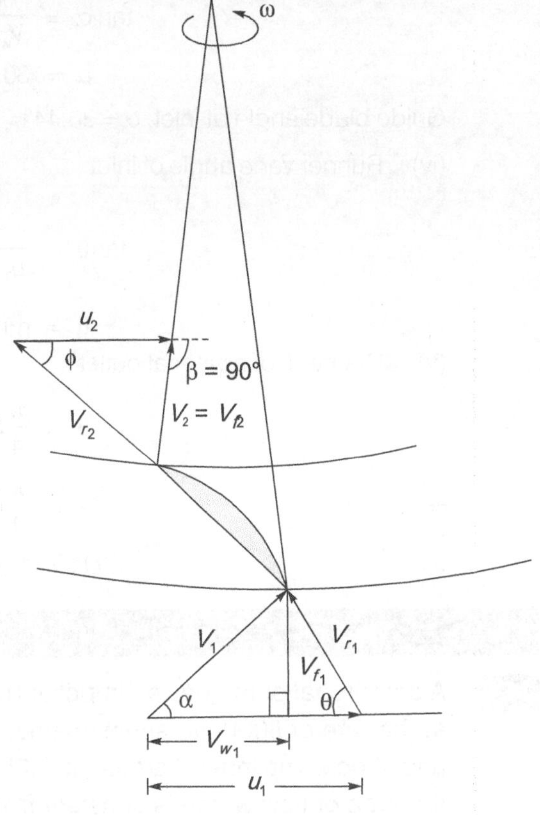

(i) Power developed by the turbine:

$$\eta_0 = \frac{S.P.}{W.P.}$$

$$0.8 = \frac{S.P.}{\dot{m}gH}$$

$$0.8 = \frac{S.P.}{\rho QgH}$$

$$0.8 = \frac{S.P.}{(10^3)(12)(9.81)(40)}$$

$$SP = 3767.04 \text{ kW}$$

(ii) Dia. and width at inlet

$$u_1 = \frac{\pi D_1 N}{60}$$

$$25.20 = \frac{\pi D_1 (350)}{60}$$

$$D_1 = 1.375 \text{ m}$$

Also,

$$Q = \pi D_1 B_1 V_{f_1}$$

$$12 = \pi(1.375) B_1 (8.4)$$

$$B_1 = 0.331 \text{ m}$$

Dia. at inlet = 1.375 m

Width at inlet = 0.331 m

(iii) Also, since flow is radial at outlet, hydraulic efficiency:

$$\eta_h = \frac{V_{w_1} u_1}{gH}$$

$$0.9 = \frac{V_{w_1} u_1}{gH}$$

$$0.9 = \frac{V_{w_1} \times 25.2}{9.81 \times 40}$$

$$V_{w_1} = 14.014 \text{ m/s}$$

$$\tan\alpha = \frac{V_{f_1}}{V_{w_1}} = \frac{8.4}{14.014}$$

$$\alpha = 30.94°$$

Guide blade angle at inlet, $\alpha = 30.44°$

(iv) Runner vane angle of inlet

$$\tan\theta = \frac{V_{f_1}}{u_1 - V_{w_1}} = \frac{8.4}{25.2 - 14.014}$$

$$\theta = \text{runner vane angle at inlet} = 36.91°$$

(v) Diameter of runner at outlet:

$$\theta = \frac{\pi}{4} D_2^2 V_{f_2}$$

$$\Rightarrow \quad 12 = \frac{\pi}{4} D_2^2 \times 8.4$$

$$\Rightarrow \quad D_2 = 1.35 \text{ m}$$

14. Pumps

14.1 **A centrifugal pump has an impeller 0.5 m outer diameter and when running at 600 rpm discharges at the rate of 8000 litres/min against a head of 8.5 m. The water enters the impeller without whirl and shock. The inner diameter is 0.25 m and the vanes are set back at outlet at an angle of 45° and the area of flow which is constant from inlet to outlet of the impeller is 0.06 m^2. Determine:**

(i) The manometric efficiency of the pump

(ii) The vane angle at inlet

(iii) The least speed at which the pump commences to work

[15 marks : 2000]

Solution:

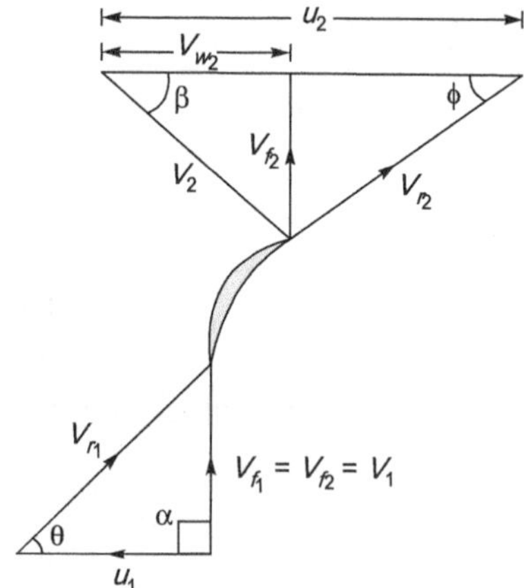

Impeller diameter at outlet, $D_2 = 0.5$ m

Impeller diameter at inlet, $D_1 = 0.25$ m

Speed of impeller, $N = 600$ rpm

Discharge through the pump, $Q = 8000$ lit/min

Head, $H_m = 8.5$ m

Vane angle at outlet, $\phi = 45°$

Area of flow $= 0.06$ m^2

Velocity of flow at inlet, $V_{f_1} = \dfrac{\text{Discharge}}{\text{Area of flow}}$

$$V_{f_1} = \frac{8000 \times 10^{-3}}{60 \times 0.06} = 2.22 \text{ m/s}$$

Tangential velocity at outlet, $u_2 = \dfrac{\pi D_2 N}{60} = \dfrac{\pi \times 0.5 \times 600}{60} = 15.71$ m/s

From velocity triangle at outlet, we have

$$\tan\phi = \frac{V_{f_2}}{u_2 - V_{w_2}}$$

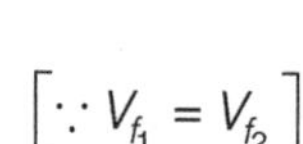
$\left[\because V_{f_1} = V_{f_2}\right]$

$$\Rightarrow \quad \tan 45° = \frac{2.22}{15.71 - V_{w_2}}$$

$$\Rightarrow \quad V_{w_2} = 15.71 - 2.22$$

$$\Rightarrow \quad V_{w_2} = 13.49 \text{ m/s}$$

(i) Manometric efficiency,

$$\eta_m = \frac{gH_m}{V_{w_2} u_2} \times 100 = \frac{9.81 \times 8.5}{13.49 \times 15.71} = 39.35\%$$

(ii) Vane angle at inlet is given by

$$\tan\theta = \frac{V_{f_1}}{u_1}$$

where
$$u_1 = \frac{\pi D_1 N}{60} = \frac{\pi \times 0.25 \times 600}{60} = 7.85 \text{ m/s}$$

$$\therefore \quad \tan\theta = \frac{2.22}{7.85}$$

$$\Rightarrow \quad \theta = 15.791°$$

(iii) For minimum starting speed, we have

$$\frac{(u_2)^2}{2g} - \frac{(u_1)^2}{2g} \geq H_m$$

$$\Rightarrow \quad \left(\frac{\pi N}{60}\right)^2 \times \left(D_2^2 - D_1^2\right) = H_m \times 2g$$

$$\Rightarrow \quad \frac{\pi^2}{3600} \times N^2 \times \left(0.5^2 - 0.25^2\right) = 8.5 \times 2 \times 9.81$$

$$\Rightarrow \quad N^2 = \frac{3600 \times 8.5 \times 2 \times 9.81}{\pi^2 (0.5^2 - 0.25^2)}$$

$$\Rightarrow \quad N \simeq 569.59 \text{ rpm}$$

14.2 A single acting reciprocating pump has the following details:

(i) Length of stroke = 180 mm
(ii) Speed of rotation = 72 rpm
(iii) Piston diameter = 120 mm
(iv) Diameter of suction pipe = 90 mm
(v) Length of suction pipe = 6 m
(vi) Suction head = 2.5 m
(vii) Suction pipe friction factor = 0.022

Determine the pressure head in the cylinder at the beginning, middle and end of stroke.

[10 marks : 2002]

Solution:

Given data: Length of stroke (L) = 180 mm, Speed of rotation (N) = 72 rpm, Piston diameter (D) = 120 mm
Diameter of suction pipe (d_s) = 90 mm, Length of suction pipe (L_s) = 6 m
Suction head (H_s) = 2.5 m, Suction pipe friction factor (f_s) = 0.022

The general expression for pressure head due to acceleration in suction pipe is given by

$$H_{as} = \frac{L_s}{g} \times \frac{A}{a_s} \times \omega^2 r\cos\theta$$

where A = area of cross-section of piston,

a_s = area of cross-section of suction pipe

ω = angular velocity = $\frac{2\pi N}{60}$

r = radius of crank = $\frac{L}{2} = \frac{180}{2}$ = 90 mm

θ = angular distance travelled by crank.

$$\therefore \quad H_{as} = \frac{L_s}{g} \times \frac{\pi D^2}{4} \times \frac{4}{\pi (d_s)^2} \times \omega^2 r\cos\theta$$

$$\Rightarrow \quad H_{as} = \frac{L_s}{g} \times \left(\frac{D}{d_s}\right)^2 \times \omega^2 r\cos\theta$$

The head loss due to friction in suction pipe is given by

$$H_{fs} = \frac{f_s L_s}{2gd_s} \times \left(\frac{A}{a_s}\omega r\sin\theta\right)^2$$

$$\Rightarrow \quad H_{fs} = \frac{f_s L_s}{2gd_s} \times \left(\frac{D}{d_s}\right)^4 (\omega r\sin\theta)^2$$

Pressure head at the beginning of stroke

$$H_{ts} = H_{atm} - H_{as} - H_{fs} - H_s$$

At the beginning, $\theta = 0°$

$$\therefore \quad H_{as} = \frac{6}{9.81} \times \left(\frac{120}{90}\right)^2 \times \left(\frac{2\pi \times 72}{60}\right)^2 \times 90 \times 10^{-3} \times \cos 0°$$

$$\Rightarrow \quad H_{as} = 5.56 \text{ m}$$

Also,

$$H_{fs} = \frac{0.022 \times 6}{2 \times 9.81 \times 90 \times 10^{-3}} \times \left(\frac{120}{90}\right)^4 \times \left(\frac{2\pi \times 72}{60} \times 90 \times 10^{-3} \times \sin 0°\right)^2$$

$$\Rightarrow \quad H_{fs} = 0$$

In terms of water, H_{atm} = 10.3 m of water

$$\therefore \quad H_{ts} = 10.3 - 5.56 - 0 - 2.5 = 2.24 \text{ m}$$

Pressure head at the middle of stroke,

$$\theta = 90°$$

$$\Rightarrow \quad \cos 90° = 0$$

and $\sin 90° = 1$

$$\therefore \quad H_{as} = 0$$

$$H_{fs} = \frac{0.022 \times 6}{2 \times 9.81 \times 90 \times 10^{-3}} \times \left(\frac{120}{90}\right)^4 \times \left[\frac{2\pi \times 72}{60} \times 90 \times 10^{-3}\right]^2 = 0.1088$$

$$\therefore \quad H_{ts} = 10.3 - 0 - 0.1088 - 2.5 = 7.6912 \text{ m}$$

Pressure head at the end of stroke

$$\theta = 180°$$

$$\Rightarrow \quad \sin 180° = 0$$

and $\cos 180° = -1$

$$\therefore \quad H_{as} = 5.56 \times \cos 180° = -5.56 \text{ m}$$

$$H_{fs} = 0$$

$$\therefore \quad H_{ts} = 10.3 - (-5.56) - 0 - 2.5 = 13.36 \text{ m}$$

14.3 A centrifugal pump discharges 1800 litres of water developing a head of 22 m when running at 315 rpm. The impeller diameter at the outlet is 1.5 m and the exiting velocity there is 2.47 m/sec. If the vanes are set back at an angle of 30° at the outlet, determine:
(i) The manometric efficiency of the pump.
(ii) The power required by the pump.
What will be the minimum starting speed? Take the impeller eye is at half size.

[15 marks : 2003]

Solution:

Given Data:

Discharge, $Q = 1800$ lit/sec (Assume)

Manometric head, $H_m = 22$ m

Speed of impeller, $N = 315$ rpm

Impeller diameter at outlet, $D_2 = 1.5$ m

Velocity of flow at outlet, $V_{f_2} = 2.47$ m/s

Vane angle at outlet, $\phi = 30°$

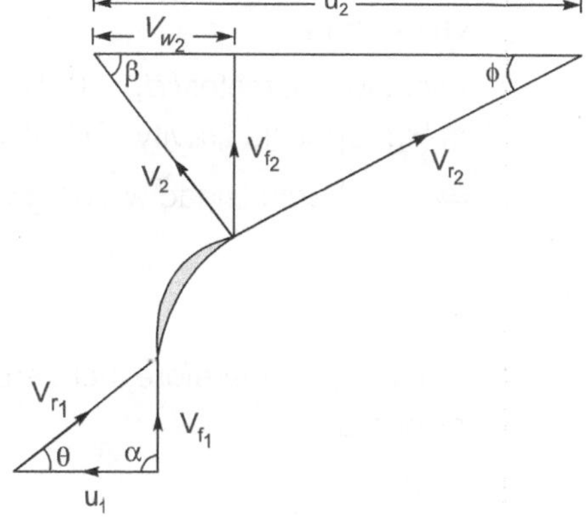

Tangential velocity at outlet, $u_2 = \dfrac{\pi D_2 N}{60}$

$$= \frac{\pi \times 1.5 \times 315}{60} = 24.74 \text{ m/s}$$

From velocity triangle at outlet, we have

$$\tan\phi = \frac{V_{f_2}}{u_2 - V_{w_2}}$$

$$\Rightarrow \quad \tan 30° = \frac{2.47}{24.74 - V_{w_2}}$$

$$\Rightarrow \quad V_{w_2} = 24.74 - \frac{2.47}{\tan 30°} = 24.74 - 4.28$$

$$\Rightarrow \quad V_{w_2} = 20.46 \text{ m/s}$$

(i) Manometric efficiency $\quad \eta_m = \dfrac{gH_m}{V_{w_2} u_2} \times 100 = \dfrac{9.81 \times 22}{20.46 \times 24.74} \times 100 = 42.64\%$

(ii) Power required by the pump, $\quad P = \dfrac{\gamma Q H_m}{1000 \times \eta_o}$ [Assume $\eta_o = 65\%$]

$$= \frac{9810 \times 1800 \times 10^{-3} \times 22}{1000 \times 0.65} = 597.66 \text{ kW}$$

Minimum starting speed is given by

$$\frac{\pi^2}{3600} \times N^2 \times (D_2^2 - D_1^2) = 2gH_m \qquad \left[\because D_1 = \frac{D_2}{2}\right]$$

$$\frac{\pi^2}{3600} \times N^2 \times (1.5^2 - 0.75^2) = 2 \times 9.81 \times 22$$

$$\Rightarrow \quad N^2 = \frac{3600 \times 2 \times 9.81 \times 22}{\pi^2 \times (1.5^2 - 0.75^2)}$$

$$\Rightarrow \quad N = 305.45 \text{ rpm}$$

14.4 A pump is installed in a pipeline of diameter 0.2 m. Close to the pump, gauges are installed on both sides. The gauge installed on the suction side indicates a pressure of 0.25 m vacuum of mercury and the gauge installed on the delivery side indicates a pressure of 40 m of water. A venturimeter 0.2 m × 0.1 m is also installed to measure the discharge of water flowing through the pipeline. The differential manometer installed between the main and the throat shows a deflection of 0.3 m of mercury (13.6). Taking coefficient of discharge for venturimeter as 1.0, compute the pump power in kW.

[8 marks : 2007]

Solution:

Given data:

Inlet pipe diameter (d_1) = 0.2 m, Throat diameter (d_2) = 0.1 m, Differential manometer reading (x) = 0.3 m of Hg, Specific gravity of (Hg) = 13.6, Coefficient of discharge (C_d) = 1.0

⇒ Venturi head will be given as,

$$h = 0.3\,(13.6 - 1)$$

$$\Rightarrow \quad h = 3.78 \text{ m of water}$$

Since $C_d = 1$, therefore there are no losses which means the theoretical discharge will also be the actual discharge

$$\Rightarrow \quad Q = C_d \frac{A_1 A_2 \sqrt{2gh}}{\sqrt{A_1^2 - A_2^2}} = \frac{1.0 \times \frac{\pi}{4} \times (0.2)^2 \times \frac{\pi}{4} \times (0.1)^2 \times \sqrt{2 \times 9.81 \times 3.78}}{\sqrt{\left[\frac{\pi}{4} \times (0.2)^2\right]^2 - \left[\frac{\pi}{4} \times (0.1)^2\right]^2}}$$

$$= \frac{\frac{\pi}{4} \times (0.2)^2 \times \frac{\pi}{4} \times (0.1)^2 \times \sqrt{74.1636}}{\frac{\pi}{4}\sqrt{(0.2)^4 - (0.1)^4}}$$

$$\Rightarrow \quad Q = 0.0699 \text{ m}^3\text{/s}$$

Applying Bernoulli's equation between 1 and 2,

$$\frac{p_1}{w} + \frac{V_1^2}{2g} + Z_1 + H_p = \frac{p_2}{w} + \frac{V_2^2}{2g} + Z_2$$

H_p is added as some work is done by the pump.

Assuming the points 1 and 2 are at the same level

i.e. $Z_1 = Z_2$.

Also, the diameter of both the pipes is same, hence the velocities V_1 and V_2 are same.

$$\therefore \quad \frac{p_1}{w} + H_p = \frac{p_2}{w}$$

$$-\frac{13.6 \times 1000 \times 9.81 \times 0.25}{1000 \times 9.81} + H_p = 40$$

$$\Rightarrow \quad H_p = 40 + 3.4 = 43.4 \text{ m}$$

Now pump power is given by

$$P = wQH_p$$
$$= 1000 \times 9.81 \times 0.0699 \times 43.4$$
$$= 29.76 \text{ kW}$$

14.5 Classify centrifugal pumps on the basis of design of casing and fluid handled. Define net positive suction head (NPSH) and obtain the maximum permissible static suction head.

[8 marks : 2007]

Solution:

On the basis of design of casing the centrifugal pumps are classified as volute pump and diffuser pump.

Volute Pump:

- In a volute pump the impeller is surrounded by a spiral shaped casing which is known as volute chamber.
- The area of flow increases gradually in the volute chamber resulting in decrease in the velocity of flow.
- The decrease in velocity increases the pressure of water flowing through the casing.
- It has been observed that in case of volute casing, the efficiency of the pump decreases slightly as a large amount of energy is lost due to the formation of eddies in this type of casing.

Diffuser Pump:

- In the diffuser pump, the impeller is surrounded by a series of guide blades (vanes) mounted on a ring called diffuser ring.
- The diffuser ring and guide vanes are fixed in position.
- The adjacent guide vanes provide gradually enlarged passages for the flow of liquid
- The liquid after leaving the impeller passes through these passages of increasing area, wherein the velocity of flow decreases and the pressure increases.
- Diffuser pumps are more costly than volute pumps.

On the basis of fluid handling centrifugal pumps are classified as radial flow pump, mixed flow pump and axial flow pump.

Radial Flow Pump:

- In radial flow pump the liquid flows through the impeller in the radial direction only.

Mixed Flow Pump:

- In mixed flow pumps, the liquid flows through the impeller axially as well as radially.
- A mixed flow impeller discharges more than the radial impeller.
- These are used where a large quantity of liquid is to be discharged to low heights.

Axial Flow Pump:

- In axial flow pumps, the flow of liquid through the impeller is in axial direction only
- The axial flow pumps are designed to deliver large quantities of liquid at relatively low heads.

- However, it is not justified to call axial flow pumps as centrifugal pumps as there is hardly any centrifugal action in their operation.

Net Positive Suction Head (NPSH)

It is defined as the absolute pressure head at the inlet to the pump minus the vapour pressure head (in absolute units) corresponding to the temperature of the liquid pumped, plus the velocity head at this point. Thus,

$$\text{NPSH} = \left(\frac{p_a}{\gamma}+\frac{p_s}{\gamma}\right)-\frac{p_v}{\gamma}+\frac{V_s^2}{2g}$$

But
$$\frac{p_s}{\gamma} = -\left(\frac{V_s^2}{2g}+h_s+h_{fs}\right)$$

$\therefore$
$$\text{NPSH} = \frac{p_a}{\gamma}-\frac{V_s^2}{2g}-h_s-h_{fs}-\frac{p_v}{\gamma}+\frac{V_s^2}{2g}$$

$\Rightarrow$
$$\text{NPSH} = \frac{p_a}{\gamma}-\frac{p_v}{\gamma}-h_s-h_{fs}$$

The right hand side of the above equation represents the total suction head (H_{sv})

i.e.
$$H_{sv} = \frac{p_a}{\gamma}-\frac{p_v}{\gamma}-h_s-h_{fs}$$

$\therefore$
$$\text{NPSH} = H_{sv}$$

Where p_a is the atmospheric pressure acting on the free liquid surface in the sump, p_s is the gauge pressure at the inlet to the pump, V_s is the velocity of flow in the suction pipe, h_s is the suction lift and h_{fs} is the head lost in foot valve, strainer and the suction pipe and p_v is the vapour pressure of liquid in absolute units.

14.6 **The internal and external diameters of the impeller of a centrifugal pump are 150 mm and 300 mm respectively. The pump is running at 1200 rpm. The vane angles at inlet and outlet are 20° and 30° respectively. The water enters radially and velocity of flow is constant. Determine the work done by the impeller per unit weight of water.**

[10 marks : 2008]

Solution:

Given data:

Impeller diameter at inlet, D_1 = 150 mm

Impeller diameter at outlet, D_2 = 300 mm

Speed of Pump, N = 1200 rpm

Vane angle at inlet, $\theta = 20°$

Vane angle at outlet, $\phi = 30°$

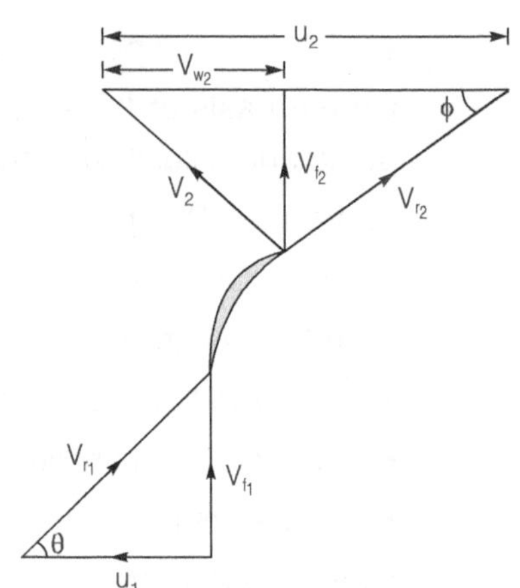

Tangential velocity at inlet,

$$u_1 = \frac{\pi D_1 N}{60} = \frac{\pi\times150\times10^{-3}\times1200}{60}$$
$$= 9.42 \text{ m/s}$$

From velocity triangle at inlet, we have

$$\tan\theta = \frac{V_{f_1}}{u_1}$$

$$\Rightarrow \quad V_{f_1} = 9.42 \times \tan 20°$$

$$\Rightarrow \quad V_{f_1} = 3.43 \text{ m/s}$$

The velocity of flow at outlet is given by

$$V_{f_2} = V_{f_1} = 3.43 \text{ m/s}$$

Tangential velocity at outlet is given by

$$u_2 = \frac{\pi D_2 N}{60} = \frac{\pi \times 300 \times 10^{-3} \times 1200}{60} = 18.85 \text{ m/s}$$

From velocity triangle at outlet, we have

$$\tan \phi = \frac{V_{f_1}}{u_2 - V_{w_2}} \qquad \left[\because V_{f_2} = V_{f_1}\right]$$

$$\Rightarrow \quad \tan 30° = \frac{3.43}{18.85 - V_{w_2}}$$

$$\Rightarrow \quad V_{w_2} = 18.85 - \frac{3.43}{\tan 30°}$$

$$\Rightarrow \quad V_{w_2} = 12.91 \text{ m/s}$$

Work done by the impeller on water per unit weight of water per sec is given by

$$P = \frac{V_{w_2} u_2 - V_{w_1} u_1}{g} \qquad \left[\text{where } V_{w_1} = 0\right]$$

$$\Rightarrow \quad P = \frac{V_{w_2} u_2}{g} = \frac{12.91 \times 18.85}{9.81} = 24.81 \text{ m}$$

14.7 **The diameters of the impeller of a centrifugal pump at inlet and outlet are 20 cm and 40 cm respectively. Determine the minimum starting speed of the pump against a head of 25 m.**

[2 marks : 2010]

Solution:

Given data: D_1 = 20 cm; D_2 = 40 cm; H_m = 25 m

The minimum starting speed required for a centrifugal pump to commence the flow is given as

$$\left(\frac{\pi N}{60}\right)^2 \times \left(D_2^2 - D_1^2\right) \geq 2gH_m$$

$$\Rightarrow \quad \left(\frac{\pi N}{60}\right)^2 \times \left[\left(\frac{40}{100}\right)^2 - \left(\frac{20}{100}\right)^2\right] \geq 2 \times 9.81 \times 25$$

$$\Rightarrow \quad \frac{\pi N}{60} \geq \sqrt{4087.5}$$

$$\Rightarrow \quad N \geq 1221.041 \text{ rpm}$$

Hence to start the pump we require minimum speed = 1222 rpm.

14.8 **A five cylinders reciprocating pump raises the water level by 150 m and the theoretical discharge is 0.20 m^3/s. The velocity in the delivery pipe is 2.0 m/s. Total head loss in pipes is 20 m. What is the input power, if the efficiency of the pump is 0.87?**

[4 marks : 2010]

Solution:

Theoretical discharge, $Q_{th} = 0.20\ m^3/s$

Velocity in delivery pipe, $V = 2.0\ m/s$

Total head loss in pipes, $h_{sd} = 20\ m$

Delivery head, $H_d = 150\ m$

Efficiency of pump, $\eta = 0.87$

Assuming coefficient of discharge of pump as 0.98, we get

$$Q_a = 0.98 \times Q_{th} = 0.98 \times 0.20 = 0.196\ m^3/s$$

Total head generated by pump, $$H = H_d + h_{sd} + \frac{V^2}{2g} = 150 + 20 + \frac{(2)^2}{2 \times 9.81} = 170.2\ m$$

$$\text{Input power, } P = \frac{\gamma_w Q_a H}{\eta} = \frac{9.81 \times 0.196 \times 170.2}{0.87} = 376.15\ kW$$

14.9 What is priming?

[1 mark : 2011]

Solution:

Priming is the process in which the impeller of a centrifugal pump will get fully submerged in liquid without any air trapped inside.This is especially required when there is a first start up.

All centrifugal pumps must be primed by filling them with water before they can operate. The objective of priming is to remove a sufficient amount of air from the pump and suction line to permit atmospheric pressure and submergence pressure to cause water to flow into the pump when pressure at the eye of the impeller is reduced below atmospheric as the impeller rotates.

14.10 **A centrifugal pump having an impeller of 35 cm outside diameter rotates at 1050 rpm. The vanes are radial at exit and are 7.0 cm wide. The velocity of radial flow through the impeller is 3 m/sec. The velocity in the suction and delivery pipes are 2.5 m/sec and 1.5 m/sec respectively. Neglecting frictional losses, determine the height through which pump lifts and the horse-power of the pump.**

[8 marks : 2012]

Solution:

Given data:

Outside diameter of impeller, $D_1 = 35\ cm = 0.35\ m$

Speed, $N = 1050\ rpm$

Width at outlet, $B_1 = 7\ cm = 0.07\ m$

Velocity of flow, $V_{f_1} = V_{f_2} = 3\ m/sec$

Velocity in suction pipe, $= 2.5\ m/sec$

Velocity in delivery pipe, $V_d = 1.5\ m/sec$

Now, tangential velocity of impeller at outlet

$$u_1 = \frac{\pi D_1 N}{60} = \frac{\pi \times 0.35 \times 1050}{60} = 19.24\ m/sec$$

As the vanes are radial at exit

$$\therefore \quad V_{w_1} = u_1 = 19.24\ m/s$$

As we know, the head developed will be given by

$$H_m = \frac{V_{w_1} u_1}{g} \eta_{man}$$

Assuming losses of the pump to be zero, head developed will be

$$H_m = \frac{V_{w_1} u_1}{g} = \frac{u_1^2}{g} = \frac{(19.24)^2}{9.81} = 37.735 \text{ m} \qquad [\text{As } V_{w_1} = u_1]$$

Now,
$$H_m = h_s + h_d + h_{f_s} + h_{f_d} + \frac{V_d^2}{2g}$$

Where $h_s + h_d = H_s$

which is known as the static head which is the net vertical height through which the liquid is lifted by the pump.

$\therefore$ Neglecting frictional losses, $H_s = H_m - \frac{V_d^2}{2g} = 37.735 - \frac{1.5^2}{2 \times 9.81} = 37.62 \text{ m}$

The horse power of the pump is given by

$$HP = \frac{WQH_m}{75}$$

Where
$$Q = \pi \times 0.35 \times 0.07 \times 3 \text{ m}^3/\text{sec} = 0.231 \text{ m}^3/\text{sec}$$

$$HP = \frac{1000 \times 0.231 \times 37.735}{75} = 116.224 \text{ h.p.}$$

14.11 Define specific speed of a pump. Where is it useful? A centrifugal pump runs at a speed of 250 rpm for 50 m³/sec discharge at a head of 50 m. Find the specific speed.

[4 marks : 2015]

Solution:

Specific speed: The specific speed of a pump may be defined as the speed in revolution per minute of a geometrically similar pump of such a size that under corresponding conditions it would deliver unit discharge against a unit head.

Specific speed,
$$N_s = \frac{N\sqrt{Q}}{(H_m)^{3/4}}$$

$N = 250$ rpm, $Q = 50 \text{ m}^3/\text{s}$, $H = 50$ m

$$\Rightarrow \quad N_s = \frac{250\sqrt{50}}{(50)^{3/4}} = 94 \text{ rpm}$$

14.12 What is centrifugal pump efficiency? Describe the important factors influencing centrifugal pump efficiency.

[4 marks : 2016]

Solution:

Centrifugal pump efficiency: Centrifugal pump efficiency or the overall efficiency of a centrifugal pump is defined as the ratio of power output from the centrifugal pump to the power input from the prime mover driving the pump i.e.

$$\eta_0 = \frac{\rho Q g H_m}{\text{Power delivered to the shaft}}$$

Overall efficiency is the product of three efficiencies viz. manometric efficiency, mechanical efficiency and the volumetric efficiency i.e.

$$\eta_0 = \eta_{mano} \cdot \eta_{mech} \cdot \eta_{vol}$$

Factors influencing centrifugal pump efficiency:

(i) Manometric head (H_m) required to be developed by the pump.

(ii) Discharge through the pump.

(iii) Density of the fluid to be pumped.

(iv) Speed of the pump (u_1).

(v) Velocity of whirl (V_{w1}) at inlet.

14.13 **A single-acting reciprocating pump discharges 0.018 m³ of water per second when running at 60 rpm. The stroke length is 50 cm and the diameter of piston is 22 cm. If the total lift is 15 m, determine the following:**
(i) Theoretical discharge of the pump.
(ii) Slip and percentage slip of the pump.

[8 marks : 2016]

Solution:

(i) Theoretical discharge of pump $(Q_{Th}) = \frac{ALN}{60} = \frac{\frac{\pi}{4}(0.22)^2 \times 0.5 \times 60}{60} = 0.019 \text{ m}^3/\text{s}$

(ii) Slip of the pump $= (Q_{Th} - Q_{Actual}) = (0.019 - 0.018) = 0.001 \text{ m}^3/\text{s}$

Percentage slip of pump $= \frac{Q_{Th} - Q_{Actual}}{Q_{Th}} \times 100 = \frac{0.001}{0.019} \times 100 = 5.26\%$

14.14 **When does cavitation occur in a pump and what are its harmful effects? How is the available net positive suction head defined and used in the analysis of cavitation?**

[6 marks : 2018]

Solution:

Cavitation is defined as the phenomenon of formation of vapour bubbles of a flowing liquid in a region where the pressure of a liquid falls below its vapour pressure and sudden collapsing of these vapour bubbles in a region of higher pressure.

Effects: When the vapour bubbles collapse, a very high pressure is created.

- The metallic surfaces, above which these vapour bubbles collapse, is subjected to these high pressure, which causes pitting action on the surface. Thus cavities are formed on the surface.
- There in considerable noise and vibrations also.
- Due to sudden collapse on the metallic surface, high pressure is produced and metallic surfaces are subjected to high local stresses. Thus surfaces are damaged.
- Efficiency of turbine decreases due to cavitation.

Available net positive suction head (NPSH) is very commonly used in pump industry. It is defined as the absolute pressure head at the inlet of the pump, minus the vapour pressure head (in absolute units) plus the velocity head.

$$\therefore \quad \text{NPSH} = \frac{p_1}{\rho g} - \frac{p_v}{\rho g} + \frac{V_s^2}{2g} \quad \text{...(i)}$$

Applying Bernoulli's principle at free surface of liquid and section 1 in the section pipe just at the inlet of pump.

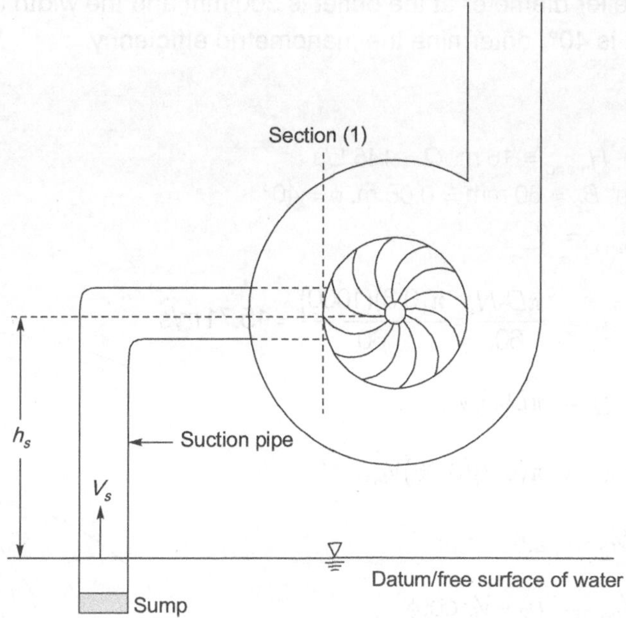

$$\frac{p_a}{\rho g} + \frac{V_a^2}{2g} + Z_a = \frac{p_1}{\rho g} + \frac{V_1^2}{2g} + Z_1 + h_L$$

$$\frac{p_a}{\rho g} = \frac{p_1}{\rho g} + \frac{V_s^2}{2g} + h_s + h_{fs}$$

$$\frac{p_1}{\rho g} = \frac{p_a}{\rho g} - \left(\frac{V_s^2}{2g} + h_s + h_{fs}\right) \quad \text{...(ii)}$$

Using eq. (ii) in (i)

$$\text{NPSH} = \frac{p_a}{\rho g} - \left(\frac{V_s^2}{2g} + h_s + h_{fs}\right) - \frac{p_v}{\rho g} + \frac{V_s^2}{2g}$$

$$\text{NPSH} = \frac{p_a}{\rho g} - \frac{p_v}{\rho g} - h_s - h_{fs}$$

$$\text{NPSH} = \left[H_a - H_v - h_s - h_{fs}\right] = \text{Total suction head}$$

Thus NPSH may also be defined as the total head required to make the liquid flow through the suction pipe to the impeller.

Thoma's cavitation factor is used to indicate whether cavitation will occur in pumps. The value of Thoma's cavitation factor for pump is given by:

$$\sigma = \frac{(H_{atm} - H_v) - H_s - H_{fs}}{H_m}$$

$$\sigma = \frac{NPSH}{H_m} \quad \text{...(iii)}$$

If the value of σ [calculated from (iii)] is less than the critical value. σ_c then cavitation will occur in the pumps.

σ_c depends on specific speed of the pump, N_s.

The following empirical relation may be used to calculate σ_c.

$$\sigma_c = 1.03 \times 10^{-3} (N_s)^{4/3}$$

14.15 A centrifugal pump runs at 1000 rpm against a head of 16 m and carries 145 liters/s of water discharge. The impeller diameter at the outlet is 300 mm and the width there is 60 mm. If the vane angle ϕ at the outlet is 40°, determine the manometric efficiency.

[15 marks : 2020]

Solution:

Given: $N = 1000$ rpm, $H_{mano} = 16$ m, $Q = 145$ L/s
$D_2 = 300$ mm $= 0.3$ m, $B_2 = 60$ mm $= 0.06$ m, $\phi = 40°$

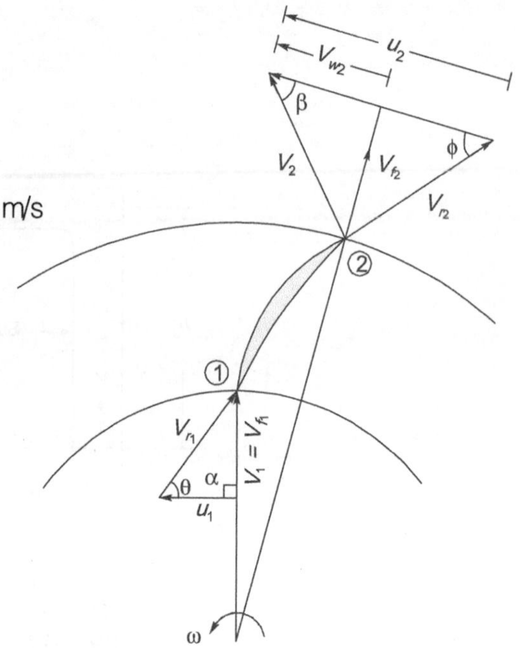

$$\eta_{mano} = ?$$

$$u_2 = \frac{\pi D_2 N}{60} = \frac{\pi(0.3)(1000)}{60} = 15.71 \text{ m/s}$$

$$Q = \pi D_2 B_2 V_{f_2}$$

$$0.145 = \pi(0.3)(0.06)V_{f_2}$$

$$V_{f_2} = 2.564 \text{ m/s}$$

At outlet.

$$V_{w2} = u_2 - V_{f_2}\cot\phi$$

$$= 15.71 - 2.564 \cot 40°$$

$$V_{w2} = 12.654 \text{ m/s}$$

$$\eta_{mano} = \frac{H_{mano}}{\frac{V_{w2}u_2}{g}} = \frac{16}{\frac{(12.654)(15.71)}{9.81}} = 0.7896$$

$$\eta_{mano}(\%) = 78.96$$

14.16 What is cavitation? How does it affect the performance of hydraulic machines? Also mention the significance of Thoma's cavitation number.

[12 marks : 2022]

Solution:

When pumping liquids, it is possible for the local pressure inside the pump to fall below the vapor pressure of the liquid. P_v(P_v, is also called the saturation pressure P_{sat} and is listed in thermodynamics tables as a function of saturation temperature.) When $P < P_v$, vapour-filled bubbles called cavitation bubbles appear. The liquid boils locally typically on the suction side of the rotating impeller blades where the pressure is lowest. After the cavitation bubbles are formed, they are transported through the pump to regions where the pressure is higher, causing rapid collapse of the bubbles. It is this collapse of the bubbles that is undesirable, since it causes noise, vibration, reduced efficiency, and most importantly, damage to the impeller blades. Repeated bubble collapse near a blade surface leads to pitting or erosion of the blade and eventually leads to catastrophic blade failure.

For Turbine:

Thoma's Cavitation Number (σ):

$$\sigma = \frac{(H_a - h_s) - H_v}{H} = \frac{\text{NPSH}}{H}$$

H_a = Atmospheric pressure head, H_v = Vapour pressure head, h_s = Suction head (at the outlet of reaction turbine), H = Working head, NPSH = Net Positive Suction Head

Critical value of Thoma's cavitation number (σ_c) depends upon N_s and is given by:

(i) For Francis turbine, $\sigma_c = 0.044\left(\frac{N_s}{100}\right)^2$

(ii) For Propeller turbine, $\sigma_c = 0.3 + 0.0032\left(\frac{N_s}{100}\right)^{2.73}$

(iii) For Kaplan turbine, σ_c is 10% higher than σ_c for propeller turbine.

Condition for no cavitation

$$\sigma \geq \sigma_c$$

$$\therefore \quad \text{NPSH} \geq \sigma_c \times H$$

For pump:

For no cavitation $\quad \text{NPSH} \geq \sigma_c H$

σ_c = Critical Thoma's number = $1.042 \times 10^{-3} (N_s)^{4/3}$

where N_s is specific speed of pump.

$$\text{NPSH} = (H_a - h_s) - H_v$$

where,
H_a = atmospheric pressure head
H_v = vapour pressure head
h_s = suction head

14.17 **A centrifugal pump having outer diameter equal to two times the inner diameter and running at 1200 rpm, works against a total head of 75 m. The velocity of flow through the impeller is constant and is equal to 3.0 m/s. The vanes are set back at an angle of 30° at the outlet. If the outlet diameter of the impeller is 600 mm and width at the outlet is 50 mm, determine the following:**

(i) Vane angle at the inlet.

(ii) Work done per second by the impeller.

(iii) Manometric efficiency.

(iv) Loss of head at inlet to impeller when the discharge is reduced by 40% with changing the speed.

[20 marks : 2022]

Solution:

Given: Manometric head, H_m = 75 m

Flow velocity, $V_{f1} = V_{f2}$ = 3 m/s

N = 1200 rpm

ϕ = 30°

Outer diameter of impeller, D_2 = 0.6 m

Width of outlet, B_2 = 0.05 m

Inner diameter of impeller, $D_1 = \frac{D_2}{2} = \frac{0.6}{2} = 0.3\,\text{m}$

$$\text{Discharge, } Q = \pi D_2 B_2 \times V_{f2}$$

$$= \pi \times 0.6 \times 0.05 \times 3 = 0.2827\ \text{m}^3/\text{s}$$

$$u_1 = \frac{\pi \times 0.3 \times 1200}{60} = 18.85\,\text{m/s}$$

$$u_2 = \frac{\pi \times 0.6 \times 1200}{60} = 37.7\,\text{m/s}$$

(i) Vane angle at inlet, $\tan\theta = \frac{V_{f1}}{u_1} = \frac{3}{18.85}$

$\Rightarrow$ $\theta = 9.043°$

$$V_{w2} = u_2 - \frac{V_{f2}}{\tan\phi} = 37.7 - \frac{3}{\tan 30°} = 32.504\,\text{m/s}$$

(ii) Work done per second,

$$W = \rho Q V_{w2} u_2$$
$$= 1000 \times 0.2827 \times 32.504 \times 37.7$$
$$= 346420.81\ \text{Nm/s}$$

(iii) Manometric efficiency,

$$\eta_m = \frac{gH_m}{V_{w2} \times u_2} = \frac{9.81 \times 75}{32.504 \times 37.7}$$

$\Rightarrow$ $\eta_m = 0.6004 \simeq 60\%$

14.18 **Find out the power required to drive the centrifugal pump if it lifts water to a height of 22 m. The overall efficiency of the pump is 70%. The pipe diameter is 10 cm and length is 100 m. It delivers 1500 *l*/min as discharge with the coefficient of friction of pipe as 0.004 and γ as ρ*g* (1000 × 9.81).**
[12 marks : 2023]

Solution:

Given: Static head, $H_{stat} = 22$ m

Overall efficiency, $\eta_0 = 0.70$

Diameter of pipe, $D = 0.1$ m

Length of pipe, $L_s + L_d = L = 100$ m

Discharge, $Q = 1500$ L/min

$$= \frac{1.5}{60}\ \text{m}^3/\text{s}$$

Friction factor, $f' = 0.004$

Coefficient of friction, $f = 4f' = 4 \times 0.004 = 0.016$

Now, Velocity, $V_d = \frac{Q}{A}$

$$= \frac{\frac{1.5}{60}}{\frac{\pi}{4}(0.1)^2} = 3.1831\ \text{m/s}$$

$\therefore$ Velocity head, $\frac{V_d^2}{2g} = 0.5164$ m

Since, $h_{fs} + h_{fd} = h_f$

$$= \frac{8Q^2}{\pi^2 g} \cdot \frac{f.L}{D^5}$$

$$= \frac{8\left(\frac{1.5}{60}\right)^2}{\pi^2 g} \frac{(0.016)(100)}{(0.1)^5}$$

$\therefore$ $$h_{fs} + h_{fd} = 8.2627 \text{ m}$$

Now, $$H_{mano} = H_{stat} + h_{fs} + h_{fd} + \frac{V_d^2}{2g}$$

$$= 22 + 8.2627 + 0.5164$$

$$= 30.7791 \text{ m}$$

$\therefore$ Overall efficiency, $$\eta_0 = \frac{H_{mano}}{S.P/\dot{m}g}$$

$\Rightarrow$ $$0.7 = \frac{(10)^3\left(\frac{1.5}{60}\right)(9.81)(30.7791)}{S.P}$$

$\therefore$ $$S.P. = 10.7837 \text{ kW}$$

Hence, power required to drive the pump = 10.7837 kW

15. Open Channel Flow

15.1 **A rectangular channel 2.4 m wide carries uniform flow of water at a rate of 7 cum/sec at a depth of 1.5 m. If there is a local rise of 15 cm in the bed, calculate the change in water level. What is the maximum rise in bed that will be permissible so that there is no change in the upstream depth of flow?**

[15 marks : 1999]

Solution:

Discharge per unit width, $$q = \frac{Q}{B} = \frac{7}{2.4} = 2.92 \text{ m}^3\text{/s/m}$$

Volocity, $$V_1 = \frac{q}{y_1} = \frac{2.92}{1.5} = 1.95 \text{ m/s}$$

Froude number, $$F_1 = \frac{V_1}{\sqrt{gy_1}} = \frac{1.95}{\sqrt{9.81 \times 1.5}} = 0.508 < 1$$

Hence the upstream flow is subcritical and the local rise will cause a drop in the water surface elevation.

Specific energy, $$E_1 = y_1 + \frac{V_1^2}{2g} = 1.5 + \frac{(1.95)^2}{2 \times 9.81} = 1.694 \text{ m}$$

Specific energy in the downstream may be given as

$$E_2 = E_1 - \Delta Z \text{ rise}$$

$\Rightarrow$ $$E_2 = 1.694 - \frac{15}{100} = 1.544 \text{ m}$$

Critical depth, $$y_c = \left(\frac{q^2}{g}\right)^{1/3} = \left[\frac{(2.92)^2}{9.81}\right]^{1/3} = 0.954 \text{ m}$$

Specific energy at the critical depth will be

$\therefore$ $$E_c = \frac{3}{2}y_c = \frac{3}{2} \times 0.954 = 1.431 \text{ m}$$

The minimum specific energy E_c is less than E_2, the available specific energy at the downstream. Hence $y_2 > y_c$ and the upstream depth y_1 will remain unchanged.

Now $E_2 = y_2 + \frac{V_2^2}{2g}$ $\left[V_2 = \frac{q}{y_2}\right]$

$\Rightarrow$ $E_2 = y_2 + \frac{q^2}{2gy_2^2}$

$\Rightarrow$ $1.544 = y_2 + \frac{(2.92)^2}{2\times 9.81\times y_2^2} = y_2 + \frac{0.4346}{y_2^2}$

$\Rightarrow$ $1.544\,y_2^2 = y_2^3 + 0.4346$

$\Rightarrow$ $y_2^3 - 1.544\,y_2^2 + 0.4346 = 0$

$\Rightarrow$ $y_2 = 1.278\text{ m}$

and $y_2 = 0.7312\text{ m}$ $(Fr_2 > 1)$

Thus, the change in water level is 1.5 – 1.278 = 0.222 m fall.

So $y_2 = 1.278\text{ m}$

The rise in the bed may be increased till the specific energy at the downstream becomes minimum i.e. $E_2 = E_c$

$\therefore$ $E_2 = E_1 - \Delta Z_{max}$

$\Rightarrow$ $E_c = E_1 - \Delta Z_{max}$

$\Rightarrow$ $\Delta Z_{max} = E_1 - E_c = 1.694 - 1.431$

$\Rightarrow$ $\Delta Z_{max} = 0.263\text{ m} = 26.3\text{ cm}$

15.2 **A rectangular channel having 30 m width carries a discharge of 200 cumecs at a normal depth of 4 m. It is to be smoothly contracted at the bridge site so that the Froude number of flow at the contracted rectangular section under the bridge is 0.5. Assuming no loss in energy, find the clear water-way and the depth of flow under the bridge.**

[15 marks : 2001]

Solution:

Discharge, $Q = 200\text{ m}^3/\text{s}$

Normal depth, $y_1 = 4\text{ m}$

Width, $B_1 = 30\text{ m}$

Downstream Froude Number, $F_2 = 0.5$

$\therefore$ $A_1 = B_1 y_1 = 30 \times 4 = 120\text{ m}^2$

$$V_1 = \frac{Q}{A_1} = \frac{200}{120} = 1.67\text{ m/s}$$

$$F_1 = \frac{V_1}{\sqrt{gy_1}} = \frac{1.67}{\sqrt{9.81\times 4}} = 0.266 < 1.$$ Hence flow is subcritical.

Now, assuming no loss in energy, the specific Energy at the upstream as well as downstream will be equal.

$\therefore$ $E_1 = E_2$

$\Rightarrow$ $y_1 + \frac{V_1^2}{2g} = y_2 + \frac{V_2^2}{2g}$

$\Rightarrow$ $4 + \frac{(1.67)^2}{2\times 9.81} = y_2 + \frac{V_2^2}{2g}$

$$\Rightarrow \quad y_2 + \frac{V_2^2}{2g} = 4.142 \quad \text{... (i)}$$

But Froude number at the downstream is given as $F_2 = 0.5$

$$\Rightarrow \quad \frac{V_2}{\sqrt{gy_2}} = 0.5$$

$$\Rightarrow \quad V_2^2 = 0.25gy_2$$

$\therefore$ Substituting in (i), we get

$$\Rightarrow \quad y_2 + \frac{0.25gy_2}{2g} = 4.142$$

$$\Rightarrow \quad y_2 = \frac{4.142}{1.125} = 3.682 \text{ m}$$

$$\therefore \quad V_2^2 = 0.25 \times 9.81 \times 3.682$$

$$\Rightarrow \quad V_2 = 3 \text{ m/s}$$

But
$$V_2 = \frac{Q}{A_2} = \frac{Q}{B_2 y_2}$$

$$\Rightarrow \quad B_2 = \frac{Q}{V_2 y_2} = \frac{200}{3 \times 3.682} = 18.11 \text{ m } \textbf{(Ans.)}$$

15.3 **Water is flowing at a depth of 1.3 m and velocity of 0.95 m/s when controlled by a sluice gate at a downstream location. The gate is abruptly dropped 0.14 m. Find the resulting depth on the upstream side of the gate. The coefficient of contraction under the gate is 0.61 m.**

[15 marks : 2002]

Solution:

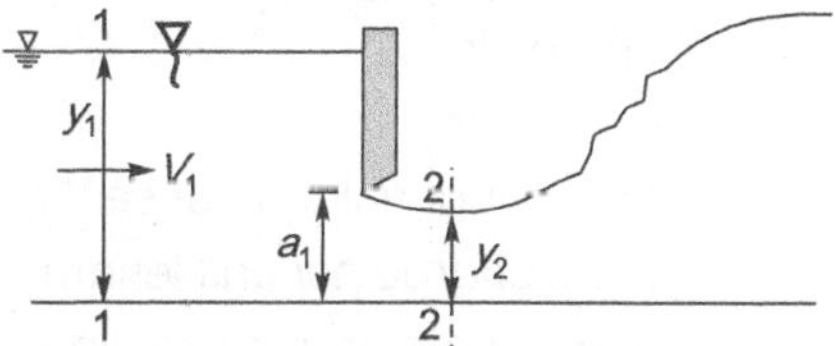

$$y_1 = 1.3 \text{ m}$$

$$V_1 = 0.95 \text{ m/s}$$

$$q = a_c\sqrt{2gy_1}$$

$$V_1 y_1 = 0.61 \times a \times \sqrt{2 \times 9.81 \times 1.3}$$

$$1.3 \times 0.95 = 0.61 \times a \times \sqrt{2 \times 9.81 \times 1.3}$$

$$a = 0.40 \text{ m}$$

Gate is dropped so a positive surge will more upstream and negative downstream (since negative surge can't be soled so we will consider positive surge)

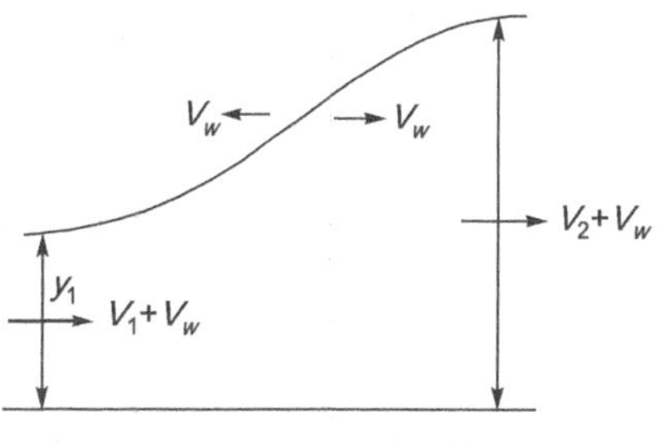

By continuity equation

$$V_2 y_2 = C_c(a - 0.14)\sqrt{2gy_2}$$

$$V_2 y_2 = 0.61 \times (0.40 - 0.14) \times \sqrt{2gy_2}$$

$$V_2 = \frac{0.61 \times (0.40 - 0.14) \times \sqrt{2 \times 9.81 \times y_2}}{y_2}$$

$$V_2 = \frac{0.705}{\sqrt{y_2}}$$

$$(V_1 + V_w)\,y_1 = (V_2 + V_w)\,y_2$$

$$(0.95 + V_w) \times 1.3 = (V_2 y_2 + V_w y_2)$$

$$1.235 + 1.3\,V_w = 0.705\sqrt{y_2} + V_w y_2$$

$$V_w = \frac{1.235 - 0.705\sqrt{y_2}}{y_2 - 1.3}$$

By Momentum equation

$$\gamma A_1 \bar{y}_1 - \gamma A_2 \bar{y}_2 = \rho Q\left[(V_2 + V_w) - (V_1 + V_w)\right]$$

$$\frac{g}{2}\left(y_1^2 - y_2^2\right) = (V_1 + V_w)\,y_1\left[V_2 - V_1\right]$$

$$4.905\,[1.3^2 - y_2^2] = \left[0.95 + \frac{1.235 - 0.705\sqrt{y_2}}{y_2 - 1.3}\right] \times 1.3 \times \left[\frac{0.705}{\sqrt{y_2}} - 0.95\right]$$

$$\Rightarrow \qquad y_2 = 1.435 \text{ m}$$

15.4 At an energy dissipater structure below a low spillway, the discharge is 19 m³/s and the energy loss is 1 m at the hydraulic jump forming therein. Determine the depths of flow at both ends of the jump.

[15 marks : 2002]

Solution:

Depths of flow at both ends of jump

$$y_1 = ?,\ y_2 = ?$$

Energy loss, $E_L = 1$ m

Discharge, $Q = 19$ m³/s

Assuming rectangular section and consider unit length

∴ Discharge per unit length, $q = 19$ m³/s/m

Now, for rectangular section,

$$\frac{2q^2}{g} = y_1 y_2 (y_1 + y_2)$$

$$\therefore \qquad y_1 y_2 (y_1 + y_2) = \frac{2 \times 19^2}{9.81} = 73.598 \qquad \text{...(i)}$$

$$E_L = \frac{(y_2 - y_1)^3}{4 y_1 y_2}$$

$$\therefore \qquad \frac{(y_2 - y_1)^3}{y_1 y_2} = 4 \times 1 = 4 \qquad \text{...(ii)}$$

Assume, $y_1 = x y_2$

$$\therefore \text{ eq. (i)} \Rightarrow \qquad x y_2^2 (x y_2 + y_2) = 73.598$$

$$\therefore \qquad x(x + 1) y_2^3 = 73.598 \qquad \text{...(iii)}$$

$$\text{eq. (ii)} \Rightarrow \qquad \frac{(y_2 - x y_2)^3}{x y_2^2} = 4$$

$$\therefore \quad y_2 = \frac{4x}{(1-x)^3} \quad \text{...(iv)}$$

$$\text{eq. (iii)} \Rightarrow \quad x(x+1)\left[\frac{4x}{(1-x)^3}\right]^3 = 73.598$$

From trial and error, on solving we get

$$x = 0.35623$$

$$\text{eq. (iv)} \Rightarrow \quad y_2 = \frac{4 \times 0.35623}{(1-0.35623)^3}$$

$$y_2 = 5.3407 \text{ m}$$

$$y_1 = 1.9025 \text{ m}$$

15.5 **Derive the correlation between roughness factor 'f' and Manning's coefficient 'n' in open channel flow.**

[5 marks : 2003]

Solution:

Let τ_o is the average shear stress at the channel boundary and it may be written as

$$\tau_o = \gamma RS \quad \text{... (i)}$$

where γ is unit weight of water, R is hydraulic radius and S ls the slope of channel bottom.

But the shear stress (τ_o) has been related to mass density (ρ) and average velocity (V) by the equation

$$\tau_o = \frac{f}{8}\rho V^2 \quad \text{... (ii)}$$

Equating (i) and (ii), we get

$$\gamma RS = \frac{f}{8}\rho V^2$$

$$\Rightarrow \quad V^2 = \left(\frac{\gamma}{\rho}\right) \times \frac{8}{f} \times RS$$

$$\Rightarrow \quad V = \sqrt{\frac{8g}{f}}\sqrt{RS} \quad [\because \gamma = \rho g]$$

$$\Rightarrow \quad V = C\sqrt{RS} \quad \text{... (iii)}$$

where C is called Chezy's coefficient and it is given by

$$C = \sqrt{\frac{8g}{f}}$$

Now from **Manning's equation**, we have

$$V = \frac{1}{n}R^{2/3}S^{1/2}$$

$$V = \frac{1}{n} \times R^{2/3} \times \frac{R^{1/2}}{R^{1/2}} \times S^{1/2}$$

$$\Rightarrow \quad V = \frac{1}{n} \times R^{1/6} \times (RS)^{1/2} = \frac{1}{n} \times R^{1/6}\sqrt{RS} \quad \text{... (iv)}$$

Comparing equation (iii) and (iv) we get

$$C = \frac{1}{n} \times R^{1/6} \quad \left[\because C = \sqrt{\frac{8g}{f}}\right]$$

$$\Rightarrow \quad \sqrt{\frac{8g}{f}} = \frac{1}{n} \times R^{1/6}$$

$$\Rightarrow \quad \frac{8g}{f} = \frac{1}{n^2} \times R^{1/3}$$

$$\Rightarrow \quad f = \frac{8gn^2}{R^{1/3}} \text{ (Ans.)}$$

15.6 **Compute the average shear stress at the sides and bottom of a rectangular flume, 3.2 m wide, flowing 1.1 m deep and laid on a slope of 1.3 m per 1000 m.**

[10 marks : 2003]

Solution:

Let us take a control volume encompassing sections 1 and 2 for the rectangular flume as shown below

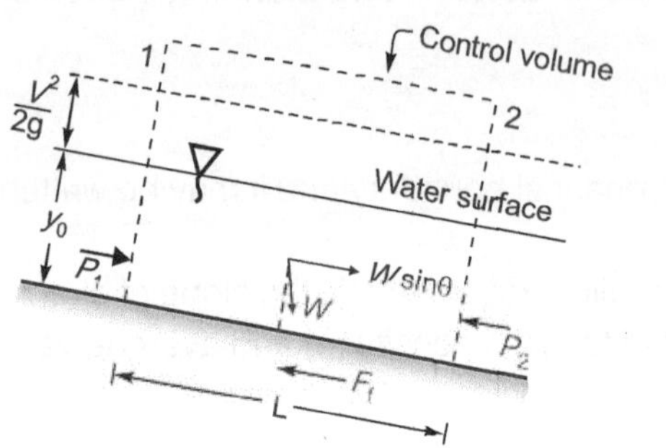

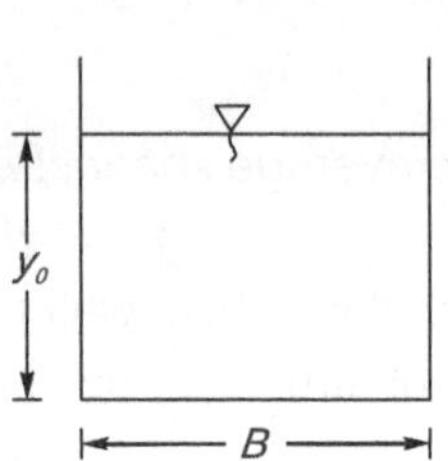

Applying momentum equation between section 1 and 2, distance L apart, we get

$$P_1 + W\sin\theta - F_f - P_2 = M_2 - M_1 \quad \text{... (i)}$$

where P_1 and P_2 are pressure forces and M_1 and M_2 are the momentum fluxes at section 1 and 2 respectively, W = Weight of fluid in the control volume and F_f = shear force at the boundary.

Now, for a uniform flow acceleration is constant.

$$\therefore \quad P_1 = P_2 \text{ and } M_1 = M_2$$

Also $\quad W = \gamma AL$ and $F_f = \tau_0 PL$

where τ_o = average shear stress on the wetted perimeter P and γ = unit weight of water. For very small values of θ, $\sin\theta \simeq \tan\theta$ and $\tan\theta = S$ = bed slope

Thus from (i) we get

$$P_1 + \gamma ALS - \tau_o PL - P_2 = M_2 - M_1$$

$$\Rightarrow \quad \tau_o PL = \gamma ALS$$

$$\Rightarrow \quad \tau_o = \gamma \times \frac{A}{P} \times S = \gamma RS$$

For rectangular flume, $\quad B = 3.2$ m; $y_0 = 1.1$ m

$$S = \frac{1.3}{1000} = 1.3 \times 10^{-3}$$

$$A = By_0 = 3.2 \times 1.1 = 3.52 \text{ m}^2$$

$$P = B + 2y_0 = 3.2 + (2 \times 1.1) = 5.4 \text{ m}$$

$$\therefore \quad R = \frac{A}{P} = \frac{3.52}{5.4} = 0.652 \text{ m}$$

The **average shear stress** will be

$$\tau_o = \gamma RS = 9810 \times 0.652 \times 1.3 \times 10^{-3} = 8.31 \text{ N/m}^2$$

15.7 **A trapezoidal channel with bottom width 3.5 m and side slopes 1 H : 1V on the left and 1.5 H: 1V on the right, with n = 0.016 and a bed slope of 2.6 in 10,000 carries a discharge of 8 m³/sec. Determine the normal depth and average shear stress on the channel bed.**

[15 marks : 2003]

Solution:

Bottom width, $B = 3.5 \text{ m}$

Left side slope, $z_1:1 = 1:1$

Right side slope, $z_2:1 = 1.5:1$

Manning's coefficient, $n = 0.016$

Bed slope, $S = \dfrac{2.6}{10000} = 2.6 \times 10^{-4}$

Discharge, $Q = 8 \text{ m}^3/\text{s}$

Let the normal depth be y

Area, $A = By + \dfrac{z_1 y^2}{2} + \dfrac{z_2 y^2}{2} = 3.5\,y + \dfrac{1}{2}y^2 + \dfrac{1.5 \times y^2}{2} = (3.5 + 1.25\,y)y$

Wetted Perimeter, $P = B + y\sqrt{z_1^2+1} + y\sqrt{z_2^2+1} = 3.5 + y\left[\sqrt{(1)^2+1} + \sqrt{(1.5)^2+1}\right]$

$P = 3.5 + 3.22\,y$

∴ Hydraulic Radius, $R = \dfrac{A}{P} = \dfrac{(3.5+1.25y)y}{3.5+3.22y}$

By **Manning's equation**, we have

$$Q = \frac{1}{n} A R^{2/3} S^{1/2}$$

$$\Rightarrow \quad 8 = \frac{1}{0.016} \times (3.5+1.25y)y \times \left[\frac{(3.5+1.25y)y}{3.5+3.22y}\right]^{2/3} \times \left[\frac{2.6}{10,000}\right]^{1/2}$$

$$\Rightarrow \quad 8 = \frac{1}{0.016} \times \frac{\left[(3.5+1.25y)y\right]^{5/3}}{(3.5+3.22y)^{2/3}} \times 0.016$$

$$\Rightarrow \quad 8 \times (3.5 + 3.22y)^{2/3} = \left[(3.5+1.25y)y\right]^{5/3}$$

Taking cube on both sides, we get

$$\Rightarrow \quad 512 \times (3.5 + 3.22\,y)^2 = \left[(3.5+1.25y)y\right]^5$$

$$\Rightarrow \quad (3.5 + 3.22y)^2 = \frac{\left[(3.5+1.25y)y\right]^5}{512}$$

Solving the above equation by hit and trial, we get

$$y = 1.511 \text{ m}$$

$$R = \frac{(3.5+1.25y)y}{3.5+3.22y} = \frac{(3.5+1.25 \times 1.511) \times 1.511}{3.5+(3.22 \times 1.511)} = 0.9733 \text{ m}$$

The average shear stress on the channel bed may be given as

$$\tau_o = \gamma RS = 1000 \times 9.81 \times 0.9733 \times \frac{2.6}{10000} = 2.48 \text{ N/m}^2$$

15.8 **A concrete lined trapezoidal irrigation canal has a bottom width of 10 m, side scope of 1H : 1V and longitudinal bottom slope of 0.0005. If the channel is several kilometers long, what is the flow depth near the downstream end for a flow of 60 m³/sec under free fall condition?**
[Take α = 1 (energy correction coefficient)]

[12 marks : 2004]

Solution:

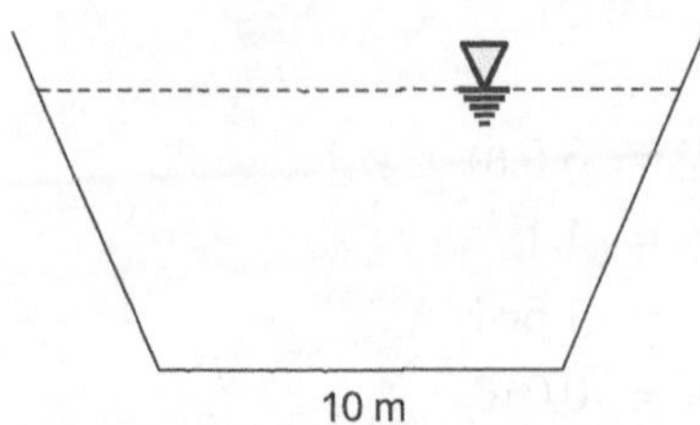

Given data: Longitudinal slope, $S_0 = 0.0005$; Discharge, $Q = 60$ m³/sec; Bottom width, $B = 10$ m; Side slope = 1H : 1V

$$\therefore \quad Z = 1$$

Under free fall condition depth of flow at downstream section will be critical depth (y_c).

Hence, for critical flow, $F_r = 1$

For non rectangular section,
$$F_r^2 = \frac{Q^2T}{gA^3} = 1$$

$$T = B + 2zy_c$$
$$= 10 + 2y_c$$

Area of cross-section $= A$

$$A = (B + zy_c)y_c$$
$$= (10 + y_c)y_c$$

Hence,
$$\frac{Q^2T}{gA^3} = 1$$

$$\frac{(60)^2(10+2y_c)}{9.81\times\left[(10+y_c)y_c\right]^2} = 1$$

By trial and error approach,

$$y_c = 1.46 \text{ m}$$

Hence, flow depth near the downstream end will be 1.46 m.

15.9 **(i) Show that hydraulically efficient trapezoidal channel is half hexagon.**
(ii) What is the difference between the economical section and hydraulically efficient channel section?

[5 + 5 marks : 2005]

Solution:

(i) Let the bottom width of the trapezoidal section be B, depth of flow be y and side slope be z horizontal to 1 vertical.

$\therefore$ Area, $$A = (B + zy)y \quad \text{... (i)}$$

Wetted perimeter, $$P = B + 2y\sqrt{1+z^2} \quad \text{... (ii)}$$

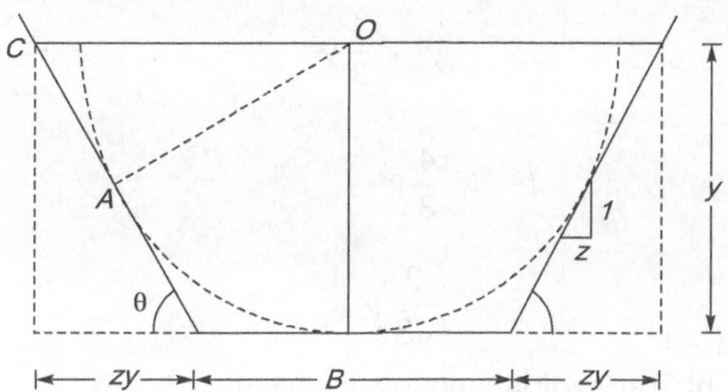

The conveyance of a channel section of a given area increases with a decrease in its perimeter. With the slope, roughness coefficient and area of flow fixed, a minimum perimeter section will represent the hydraulically efficient section as it conveys maximum discharge.

From (i), we get,

$$B = \frac{A}{y} - zy$$

Substituting value of B in (ii), we get,

$$P = \frac{A}{y} - zy + 2y\sqrt{1+z^2} \quad \text{... (iii)}$$

Assuming A and z to be constant equation (iii) can be differentiated w.r.t y and equated to zero for obtaining the condition of minimum P.

$$\therefore \quad \frac{dP}{dy} = -\frac{A}{y^2} - z + 2\sqrt{1+z^2} = 0$$

$$\Rightarrow \quad \frac{A}{y^2} + z = 2\sqrt{1+z^2}$$

Substituting value of A from (i), we get

$$\frac{(B+zy)y}{y^2} + z = 2\sqrt{1+z^2}$$

$$\Rightarrow \quad \frac{B+2zy}{y} = 2\sqrt{1+z^2}$$

$$\Rightarrow \quad \frac{B+2zy}{2} = y\sqrt{1+z^2}$$

$$\Rightarrow \quad B + 2zy = 2y\sqrt{1+z^2} \quad \text{...(iv)}$$

Now, for most economical section, differentiating P wrt. z and equated to zero by keeping A and y as constant, we get,

$$\frac{dP}{dz} = 0 - y + \frac{2y}{2\sqrt{1+z^2}} \times (2z) = 0$$

$$\Rightarrow \quad y\left(\sqrt{1+z^2}\right) = 2yz$$

$$\Rightarrow \quad 1 + z^2 = 4z^2$$

$$\Rightarrow \quad z = \frac{1}{\sqrt{3}}$$

If L is the length of the sloping side of the trapezoidal section, then

$$L^2 = y^2 + z^2y^2$$

$$\Rightarrow \quad L^2 = y^2 + \frac{1}{3}y^2 \qquad \left[\text{for } z = \frac{1}{\sqrt{3}}\right]$$

$$\Rightarrow \quad L^2 = \frac{4}{3}y^2$$

$$\Rightarrow \quad L = \frac{2}{\sqrt{3}}y$$

But for **most efficient Trapezoidal section**, we have

$$B + 2zy = 2y\sqrt{1+z^2}$$

$$\Rightarrow \quad B + 2\times\frac{1}{\sqrt{3}}y = 2\times y\times\sqrt{1+\frac{1}{3}}$$

$$\Rightarrow \quad B + \frac{2y}{\sqrt{3}} = \frac{2y\times 2}{\sqrt{3}}$$

$$\Rightarrow \quad B = \frac{2y}{\sqrt{3}}$$

Thus, length of the sloping side is equal to the bottom width of the trapezoidal section which is a property of hexagon. Thus the hydraulically most efficient trapezoidal section is half of a regular hexagon.

(ii) The conveyance of a channel section of a given area increases with a decrease in its perimeter. Hence a channel section having minimum perimeter for a given area of flow provides the maximum value of the conveyance. With the slope, roughness coefficient and area of flow fixed, a minimum perimeter section will represent the hydraulically-efficient section as its conveys the maximum discharge.

The channel section is said to be economical when the cost of construction of the channel is minimum. But the cost of construction of the channel depends on wheather the channel is in filling or cutting and its lining. To keep the cost down the wetted perimeter is kept minimum for a given discharge. However in daily practice these two terms are used synonymously.

15.10 **Design a hydraulically efficient trapezoidal channel for conveying 100 m³/sec. The roughness coefficient is 0.015. The bed slope of the channel is 1:5000. Add a free board of 10% of the depth of flow. Compute the Froude number. Sketch the designed cross section.**

[12 marks : 2005]

Solution:

For a trapezoidal channel to be most efficient hydraulically the following three criteria should be satisfied:

(i) **Hydraulic radius,** $R = \frac{y}{2}$

(ii) **Area of channel section,** $A = \frac{y^2}{z}$

(iii) **Side slope of the channel section,** $z = \frac{1}{\sqrt{3}}$

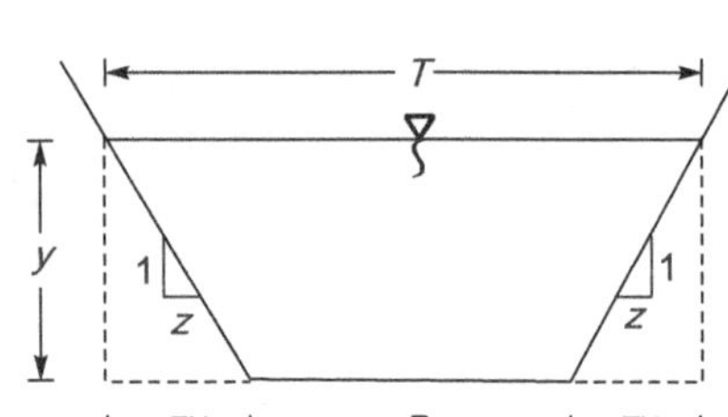

We know that

$$A = (B + zy)y$$

$$\Rightarrow \quad \frac{y^2}{z} = (B + zy)y \qquad \left[\because A = \frac{y^2}{z}\right]$$

$$\Rightarrow \quad \sqrt{3}y = B + \frac{y}{\sqrt{3}} \qquad \left[\because z = \frac{1}{\sqrt{3}}\right]$$

$$\Rightarrow \quad B = \frac{2y}{\sqrt{3}}$$

$$\therefore \quad A = \left(\frac{2y}{\sqrt{3}} + \frac{1}{\sqrt{3}} \times y\right)y = \sqrt{3}y^2$$

Now, by **Manning's equation**, we have

$$Q = \frac{1}{n} \times A \times (R)^{2/3} \times (S)^{1/2}$$

$$\Rightarrow \quad 100 = \frac{1}{0.015} \times \sqrt{3} \times y^2 \times \left(\frac{y}{2}\right)^{2/3} \times \left(\frac{1}{5000}\right)^{1/2}$$

$$\Rightarrow \quad y^{8/3} = \frac{100 \times 0.015 \times (2)^{2/3} \times (5000)^{1/2}}{\sqrt{3}}$$

$$\Rightarrow \quad y = 5.56 \text{ m}$$

$$\therefore \quad B = \frac{2y}{\sqrt{3}} = \frac{2 \times 5.56}{\sqrt{3}} = 6.42 \text{ m}$$

Hydraulic depth $\quad D = \frac{A}{T} = \frac{(B+zy)y}{B+2zy} = \frac{\sqrt{3}y^2}{4y/\sqrt{3}} = \frac{3}{4} \times y = \frac{3}{4} \times 5.56 = 4.17 \text{ m}$

Velocity of flow, $\quad V = \frac{Q}{A} = \frac{100}{\sqrt{3} \times (5.56)^2} = 1.87 \text{ m/s}$

Froude number, $\quad F = \frac{V}{\sqrt{gD}} = \frac{1.87}{\sqrt{9.81 \times 4.17}} = 0.292$

Free board $= \frac{10}{100} \times y = \frac{10}{100} \times 5.56 = 0.556 \text{ m} \approx 0.6 \text{ m}$

(10% of flow depth is kept as free board)

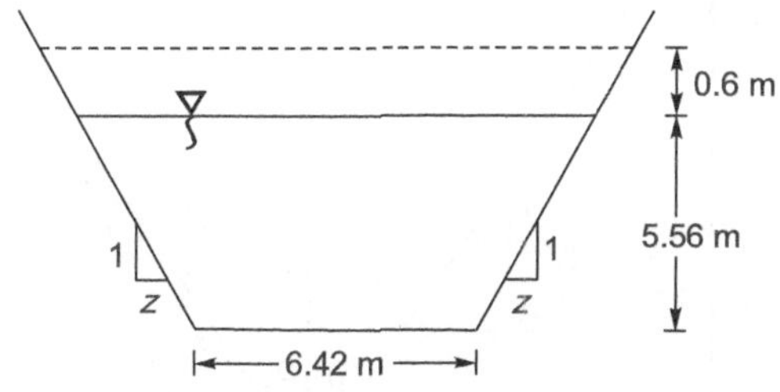

15.11 **Prove that the most efficient triangular cross-section channel is half of a square with its diagonal horizontal.**

[8 marks : 2006]

Solution:

For a **triangular channel section**, if θ is the angle of inclination of each of the sloping sides with the vertical and *y* is the depth of flow, then the wetted area '*A*' and wetted perimeter '*P*' can be respectively given by

$$A = zy^2 \qquad [\because z = \tan\theta]$$

$$\Rightarrow \quad A = y^2 \tan\theta$$

$$\Rightarrow \quad y = \sqrt{A/\tan\theta} \qquad \text{...(i)}$$

and

$$P = 2y\sec\theta \qquad \text{...(ii)}$$

Substituting value of *y* from (i) in (ii), we get

$$P = 2\times\sqrt{\frac{A}{\tan\theta}}\times\sec\theta \qquad \text{...(iii)}$$

Assuming wetted area, *A* to be constant, equation (iii) can be differentiated with respect to θ and equated to zero for obtaining the condition for minimum wetted perimeter.

$$\therefore \quad \frac{dP}{d\theta} = 2\sqrt{A}\left[\frac{\sec\theta\tan\theta}{\sqrt{\tan\theta}} - \frac{\sec^3\theta}{2(\tan\theta)^{3/2}}\right] = 0$$

$$\Rightarrow \quad \sec\theta(2\tan^2\theta - \sec^2\theta) = 0 \qquad [\because \sec\theta \neq 0]$$

$$\therefore \quad 2\tan^2\theta - \sec^2\theta = 0$$

$$\Rightarrow \quad \sqrt{2}\tan\theta = \sec\theta$$

$$\Rightarrow \quad \theta = 45^\circ$$

$$\Rightarrow \quad z = 1$$

Hence a triangular channel will be most efficient when each of its sloping sides makes an angle of 45° with the vertical.

Hydraulic radius,

$$R = \frac{A}{P} = \frac{y^2\tan\theta}{2y\sec\theta} \qquad [\because \theta = 45^\circ]$$

$$\Rightarrow \quad R = \frac{y^2\tan 45^\circ}{2y}\times\cos 45^\circ = \frac{y}{2\sqrt{2}} = R$$

Thus it can be seen that the most efficient triangular channel section will be half the square in scribed on a diagonal and having equal sloping sides.

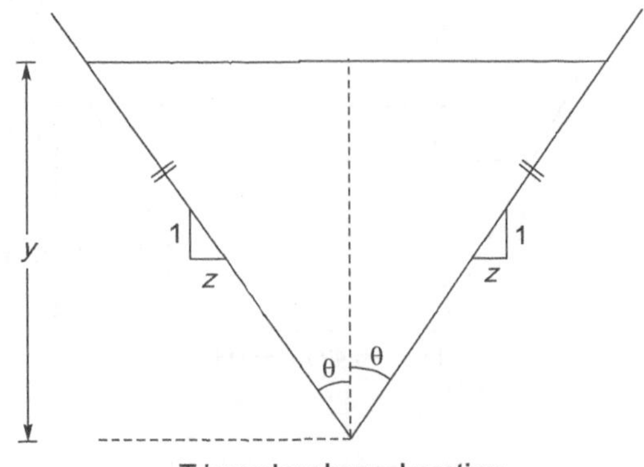

Triangular channel section

15.12 Show that the relation between the alternate depths in a rectangular channel is

$$h_c^3 = \frac{2h_1^2 h_2^2}{h_1 + h_2}$$

where h_c = critical depth, h_1 and h_2 are alternate depths. [7 marks : 2006]

Solution:

Let the discharge intensity of the rectangular channel be q.

Thus, $q = V_1h_1 = V_2h_2$

where V_1 and V_2 are velocities at alternate depths h_1 and h_2 respectively.

Now, the specific energy may be given as

$$E = h_1 + \frac{V_1^2}{2g} = h_2 + \frac{V_2^2}{2g}$$

$$\Rightarrow \quad h_1 + \frac{q^2}{2gh_1^2} = h_2 + \frac{q^2}{2gh_2^2}$$

$$\Rightarrow \quad h_1 - h_2 = \frac{q^2}{2g}\left[\frac{1}{h_2^2} - \frac{1}{h_1^2}\right]$$

$$\Rightarrow \quad 2(h_1 - h_2) = \frac{q^2}{g}\left[\frac{h_1^2 - h_2^2}{h_1^2 h_2^2}\right]$$

But for a **rectangular channel,**

$$h_c^3 = \frac{q^2}{g}$$

$$\therefore \quad \frac{2(h_1 - h_2)h_1^2 h_2^2}{(h_1 - h_2)(h_1 + h_2)} = h_c^3$$

$$\Rightarrow \quad h_c^3 = \frac{2h_1^2 h_2^2}{h_1 + h_2}$$

15.13 Hydraulic jump occurs on a horizontal rectangular channel. Prove that post jump depth will be given by:

$$\frac{y_2}{y_1} = \frac{1}{2}\left[-1 + \sqrt{1 + 8F_{r_1}^2}\right]$$

where y_1 and y_2 are pre jump and post jump depths respectively, and F_{r_1} is the Froude number for the supercritical flow.

Also show that the relative pre jump depth and the relative post jump depth are functions of F_{r_1} only.

[8 marks : 2007]

Solution:

Let the horizontal rectangular channel carries a discharge Q.

Let the depth of flow before the jump at section 1 be y_1 and depth of flow after the jump at section 2 be y_2.

The symbols A_1, V_1 and $\bar{z}_1$ represent the area of cross-section, mean velocity of flow and the depth of the centroid of area A_1 below the free surface respectively at section 1 and A_2, V_2 and $\bar{z}_2$ are the corresponding quantities at section 2 after the jump.

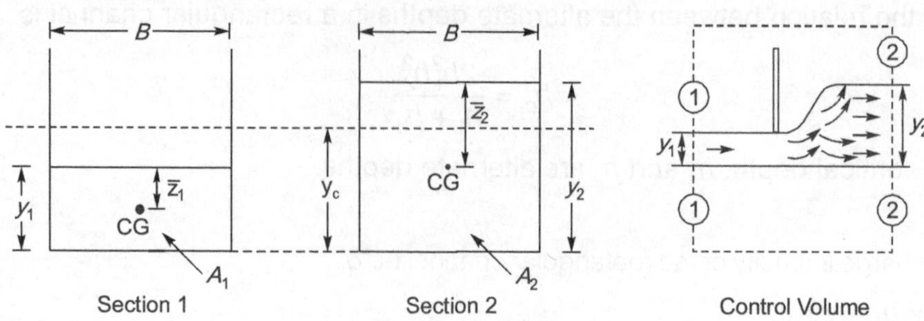

The only external forces acting on the mass of water between the section 1 and 2 are the hydrostatic pressure forces P_1 and P_2 at sections 1 and 2 respectively as the frictional loss has been assumed to be negligible.

Thus, on applying momentum equation, we get

$$P_2 - P_1 = \rho Q(V_1 - V_2)$$

$$\Rightarrow \quad wA_2\bar{z}_2 - wA_1\bar{z}_1 = \frac{wQ}{g}\left(\frac{Q}{A_1} - \frac{Q}{A_2}\right) \qquad [\because Q = A_1V_1 = A_2V_2]$$

$$\Rightarrow \quad \frac{Q^2}{gA_1} + A_1\bar{z}_1 = \frac{Q^2}{gA_2} + A_2\bar{z}_2$$

$$\Rightarrow \quad \frac{Q^2}{g(By_1)} + By_1\left(\frac{y_1}{2}\right) = \frac{Q^2}{g(By_2)} + By_2\left(\frac{y_2}{2}\right)$$

Dividing both sides by B, we get

$$\Rightarrow \quad \frac{Q^2}{gB^2y_1} - \frac{Q^2}{gB^2y_2} = \frac{1}{2}\left(y_2^2 - y_1^2\right)$$

$$\Rightarrow \quad \left(\frac{Q}{B}\right)^2 \times \frac{1}{g}\left[\frac{1}{y_1} - \frac{1}{y_2}\right] = \frac{1}{2}\left(y_2^2 - y_1^2\right)$$

$$\Rightarrow \quad \frac{q^2}{g}\left(\frac{y_2 - y_1}{y_1y_2}\right) = \frac{1}{2}(y_2 - y_1)(y_2 + y_1) \qquad \left[\because q = \frac{Q}{B}\right]$$

$$\Rightarrow \quad \frac{2q^2}{g} = y_1y_2\,(y_1 + y_2) = y^2\,y_2 + y_1y_1^{\,2}$$

$$\frac{2q^2}{gy_1} = y_2^2 + y_1y_2$$

$$y_2^2 + y_1y_2 - \frac{2q^2}{gy_1} = 0$$

This equation is the momentum equation for hydraulic jump in rectangular channels. This equation can be considered as quadratic equation in terms of y_1 and y_2 and its solution gives the relations between the conjugate depths y_1 and y_2 as

$$y_2 = -\frac{y_1}{2} \pm \sqrt{\left(\frac{y_1}{2}\right)^2 + \frac{2q^2}{gy_1}}$$

$$y_1 = -\frac{y_2}{2} \pm \sqrt{\left(\frac{y_2}{2}\right)^2 + \frac{2q^2}{gy_2}}$$

The negative sign before the square root has not been used since it gives negative values.

$$y_2 = -\frac{y_1}{2} + \sqrt{\left(\frac{y_1}{2}\right)^2 + \frac{2q^2}{gy_1}}$$

$$\Rightarrow \quad \frac{y_2}{y_1} = \frac{1}{2}\left[-1 + \sqrt{1 + \frac{8q^2}{gy_1^3}}\right]$$

But for a rectangular channel, $\frac{q^2}{g} = y_c^3$

$$\therefore \quad \frac{y_2}{y_1} = \frac{1}{2}\left[-1 + \sqrt{1 + 8\left(\frac{y_c}{y_1}\right)^3}\right]$$

But $$F_{r_1} = \frac{q}{\sqrt{gy_1^3}}$$

and $$F_{r_2} = \frac{q}{\sqrt{gy_2^3}}$$

$$\therefore \quad \frac{y_2}{y_1} = \frac{1}{2}\left[-1 + \sqrt{1 + 8F_{r_1}^2}\right]$$

Similarly it can be shown that, $$\frac{y_1}{y_2} = \frac{1}{2}\left[-1 + \sqrt{1 + 8F_{r_2}^2}\right]$$

It is clear from the two equations that relative pre and post jump depths are functions of Froude number only.

15.14 **(i) Show that for given discharge Q, the specific energy 'E' is minimum when $\frac{Q^2T}{gA^3}$ is unity. Here T is the top surface width and A is cross-sectional area of the flow.**

(ii) Obtain an expression for critical depth and minimum specific energy for a rectangular channel. Sketch the specific energy diagram showing
Variation of E with y
Minimum E and critical depth y_c
Alternate depths y_1 and y_2
for rectangular channel.

[10 marks : 2008]

Solution:

(i) The specific energy of flow at any channel section is defined as the energy per unit weight of water measured with respect to the channel bottom as the datum. Thus, **specific energy E at any section** is the sum of depth of flow at that section and velocity head i.e.

$$E = y + \frac{V^2}{2g}$$

But $$V = \frac{Q}{A}$$

$$\Rightarrow \quad E = y + \frac{Q^2}{2gA^2}$$

For a given discharge Q, the specific energy will be minimum when

$$\frac{dE}{dy} = 0$$

$$\Rightarrow \quad \frac{d}{dy}\left[y + \frac{Q^2}{2gA^2}\right] = 0$$

$$\Rightarrow \quad 1 - \frac{2Q^2}{2gA^3} \times \frac{dA}{dy} = 0$$

If T is the top width of flow, then differential water area dA near the free surface is equal to Tdy i.e.

$$Tdy = dA$$

$$\Rightarrow \quad T = \frac{dA}{dy}$$

$$\therefore \quad 1 - \frac{Q^2}{gA^3} \times T = 0$$

$$\Rightarrow \quad \frac{Q^2 T}{gA^3} = 1 \text{ (Ans.)}$$

(ii) We know that for any channel section, the **critical state of flow** is given by,

$$\frac{Q^2 T}{gA^3} = 1$$

$$\Rightarrow \quad \frac{Q^2}{g} = \frac{A^3}{T}$$

If it is assumed that 'q' represents the discharge per unit width of the rectangular section, then the total discharge will be

$$Q = q \times B$$

Corresponding to the critical depth of flow y_c, the area of rectangular section,

$$A = B \times y_c$$

$$\therefore \quad \frac{(q \times B)^2}{g} = \frac{(B \times y_c)^3}{B} \qquad \text{[For a rectangular channel, } T = B\text{]}$$

$$\Rightarrow \quad \frac{q^2 \times B^2}{g} = B^2 \times y_c^3$$

$$\Rightarrow \quad y_c = \left(\frac{q^2}{g}\right)^{1/3}$$

But specific energy, $E = y + \dfrac{Q^2}{2gA^2}$

Specific energy is minimum at critical depth of flow.

$$\therefore \quad E_{min} = y_c + \frac{q^2 \times B^2}{2g \times B^2 \times y_c^2}$$

$$\Rightarrow \quad E_{min} = y_c + \frac{1}{2} \times \frac{q^2}{g} \times \frac{1}{y_c^2}$$

$$\Rightarrow \quad E_{min} = y_c + \frac{y_c}{2} \qquad \left[\because \frac{q^2}{g} = y_c^3\right]$$

$$\Rightarrow \quad E_{min} = \frac{3}{2} y_c$$

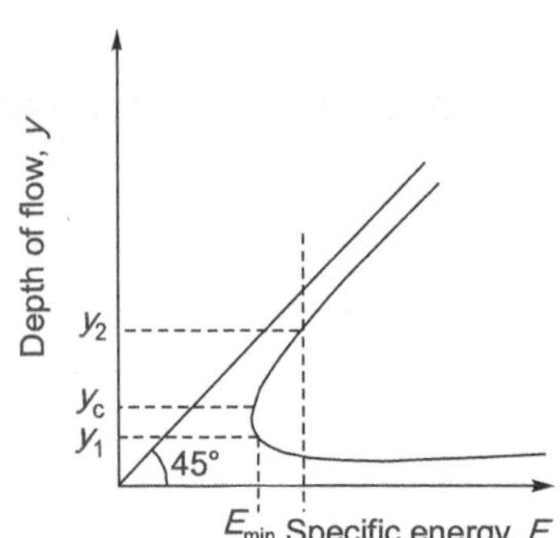

15.15 A 10.0 m wide rectangular channel with bottom slope of 0.00016 carries a discharge of 22.92 m^3/s at a normal depth of 2.0 m and critical depth of 0.8 m. The depth immediately upstream of dam is 10.0 m. Compute the length of the surface profile between 10.0 m and 6.0 m using Chow's or step method. Take step of 2.0 m and assume $M = 3.0$ and $N = 3.33$ and Manning's $n = 0.015$. The length of surface profile by Chow's method between the depths is given by

$$x_2 - x_1 = \frac{y_n}{S_o}\left[(u_2 - u_1) - \{F(u_2, N) - F(u_1, N)\} + \left(\frac{y_c}{y_n}\right)^M \frac{J}{N}\{F(v_2, J) - F(v_1, J)\}\right]$$

u	F(u, N)	v	F(v, J)
5	0.01	9	0.027
4	0.017	8	0.031
3	0.034	7	0.038
		6	0.048
		5	0.062
		4	0.087

[10 marks : 2009]

Solution:

(i) Step method

Given data:

$$y_n = 2.0 \text{ m}; B = 10.0 \text{ m}; Q = 22.92 \text{ m}^3/\text{s}$$
$$y_c = 0.8 \text{ m}; n = 0.015; S_0 = 0.00016$$

Discharge per unit width,

$$q = \frac{Q}{B} = \frac{22.92}{10} = 2.292 \text{ m}^3/\text{s/m}$$

Velocity, $$V = \frac{Q}{By}$$

Friction slope, $$S_f = \frac{V^2 n^2}{R^{4/3}}$$

$$\Delta x = \frac{\Delta E}{S_0 - \bar{S}_f}$$

where $$\bar{S}_f = \frac{1}{2}(S_{f1} + S_{f2}) \text{ and } E = y + \frac{V^2}{2g}$$

The calculations are tabulated below:

y(m)	V(m/s)	E(m)	ΔE(m)	S_F (× 10^{-6})	$\bar{S}_F$ (× 10^{-6})	Δx(m)	L(m)
10	0.2292	10.0027		2.3738			
			1.9985		3.2502	12749.62	12749.62
8	0.2865	8.0042		4.1267			
			1.9968		6.3716	12997.60	25747.22
6	0.382	6.0074		8.6166			

(ii) Chow's method:

$$M = 3.0; N = 3.33$$

$$u = \frac{y}{y_n}$$

$$\frac{y_n}{S_0} = \frac{2}{0.00016} = 12500$$

$$J = \frac{N}{N-M+1} = \frac{3.33}{3.33-3+1} = 2.5$$

$$\frac{N}{J} = N - M + 1 = 3.33 - 3 + 1 = 1.33$$

$$\frac{J}{N} = \frac{1}{1.33} = 0.75$$

$$\left(\frac{y_c}{y_n}\right)^M = \left(\frac{0.8}{2}\right)^3 = 0.064$$

$$v = u^{N/J} = \left(\frac{y}{y_n}\right)^{1.33}$$

$$x_2 - x_1 = \frac{y_n}{S_0}\left[(u_2-u_1) - \{F(u_2,N) - F(u_1,N)\} + \left(\frac{y_c}{y_n}\right)^M \frac{J}{N}\{F(v_2,J) - F(v_1,J)\}\right]$$

$$x_2 - x_1 = 12500[(5-4) - (0.017-0.01) + 0.064 \times 0.75\,(0.045-0.029)]$$

$\Rightarrow$ $\Delta x_1 = 12422.10$

Also $\Delta x_2 = 12500[(4-3) - (0.034-0.017) + 0.064 \times 0.75\,(0.079-0.045)]$

$\Rightarrow$ $\Delta x_2 = 12307.90$

The length of GVF profile by chow's method is 24730 m. The channel slope is mild because $y_n > y_c$ and the water surface profile will be M_1 because $y > y_n > y_c$.

The calculations are tabulated below:

y(m)	u	v	$F(u, 3.33)$	$F(v, 2.5)$	Dx(m)	L(m)
10	5	8.504	0.01	0.029	0	0
8	4	6.320	0.017	0.045	12422.10	12422.10
6	3	4.311	0.034	0.079	12307.90	24730.00

15.16 **Water emerges from a spillway with a velocity of 15 m/sec and a depth of 0.5 m. Calculate the necessary subcritical depth at the toe of the spillway for the occurrence of a hydraulic jump. Calculate the associated energy loss.**

[10 marks : 2009]

Solution:

Let the velocity of flow, depth and specific energy are V_1, y_1 and E_1 respectively before the hydraulic jump and V_2, y_2 and E_2 after the hydraulic jump.

The discharge per unit width is given by

$$q = V_1 y_1 = 15 \times 0.5 = 7.5 \text{ m}^2/\text{s}$$

The sequent depth ratio is given by

$$\frac{y_2}{y_1} = \frac{1}{2}\left[-1 + \sqrt{1 + 8F_1^2}\right]$$

where F_1 is the Froude number of the approach flow $= \dfrac{V_1}{\sqrt{gy_1}}$

$$\therefore \quad F_1 = \frac{15}{\sqrt{9.81 \times 0.5}} = 6.773$$

$$\therefore \quad y_2 = \frac{y_1}{2}\left[-1+\sqrt{1+8F_1^2}\right]$$

$$\Rightarrow \quad y_2 = \frac{0.5}{2}\left[-1+\sqrt{1+8\times(6.773)^2}\right] = 0.25[-1+19.18] = 4.55 \text{ m}$$

The energy loss is given by

$$E_L = E_1 - E_2$$

$$\Rightarrow \quad E_L = \left(y_1 + \frac{q^2}{2gy_1^2}\right) - \left(y_2 + \frac{q^2}{2gy_2^2}\right) = (y_1 - y_2) + \frac{q^2}{2g}\left(\frac{y_2^2 - y_1^2}{y_1^2 y_2^2}\right)$$

$$\Rightarrow \quad E_L = (0.5-4.55) + \frac{(7.5)^2}{2\times 9.81} \times \left[\frac{(4.55)^2-(0.5)^2}{(0.5\times 4.55)^2}\right] = 7.3 \text{ m}$$

Alternatively, for a rectangular channel

$$E_L = \frac{(y_2-y_1)^3}{4y_1y_2} = \frac{(4.55-0.5)^3}{4\times 0.5\times 4.55} = 7.3 \text{ m}$$

15.17 Draw surface profiles in the following cases:

(i) Upstream and downstream of sluice gate on a mild sloped channel when critical depth line is above the gate opening.

(ii) A steep sloped channel ending in an abrupt drop having reservoir level on the downstream side above the critical depth line.

(iii) Steep sloped channel followed by a mild sloped channel.

(iv) Mild sloped channel followed by a steep sloped channel.

[4 marks : 2010]

Solution:

The different surface profiles are shown below:

(i) Upstream and downstream of sluice gate on a mild sloped channel when critical depth line is above the gate opening.

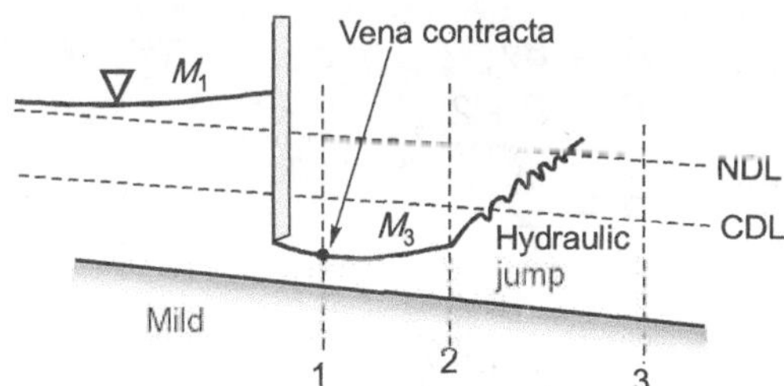

(ii) A steep sloped channel ending in an abrupt drop having reservoir level on the downstream side above the critical depth line.

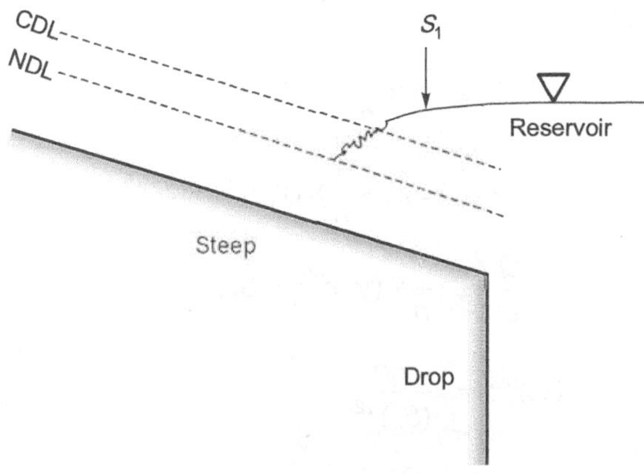

(iii) Steep sloped channel followed by a mild sloped channel.

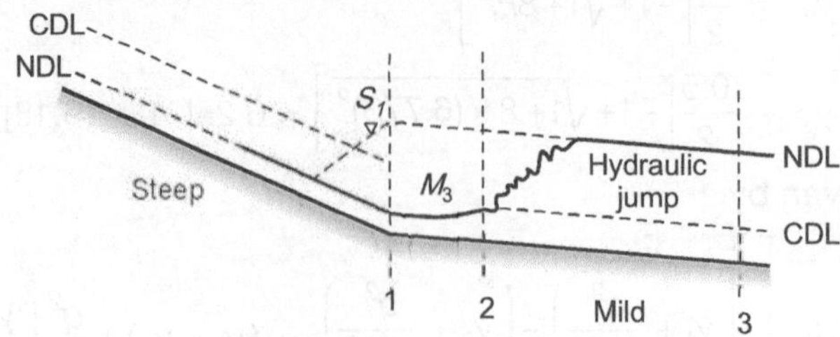

(iv) Mild sloped channel followed by a steep sloped channel.

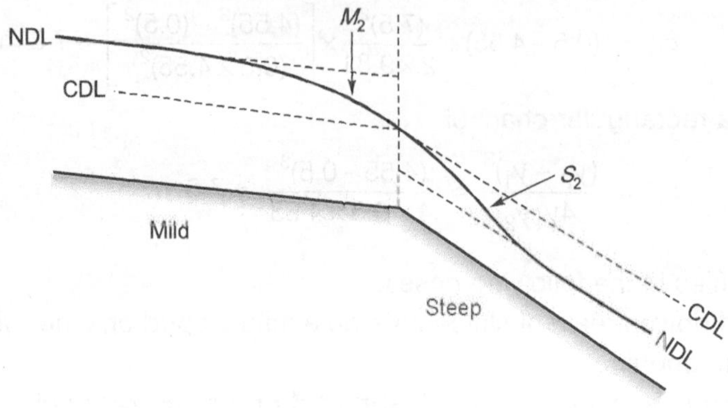

15.18 **Show that for wide rectangular channel the bed slope 'S_o' is mild or steep according S_o being less than or greater than $\dfrac{n^2g^{10/9}}{q^{2/9}}$.**

[4 marks : 2010]

Solution:

For a wide rectangular channel, width (B) is very large as compared to normal depth (y_o).

Area, $A = By_o$

Wetted perimeter, $P = B + 2y_o$

Hydraulic radius, $R = \dfrac{A}{P}$

$$= \frac{By_0}{B+2y_0} \qquad [\because y_o << B]$$

$$\Rightarrow \quad = \frac{By_0}{B} = y_o \qquad [\because B + 2y_o \approx B]$$

As per Manning's equation discharge is given by

$$Q = A \times \frac{1}{n} R^{2/3} S_0^{1/2}$$

$$\Rightarrow \quad Q = By_0 \times \frac{1}{n} \times (y_0)^{2/3} \times (S_0)^{1/2}$$

$$\Rightarrow \quad \frac{Q}{B} = \frac{1}{n} \times (y_0)^{5/3} \times (S_0)^{1/2}$$

$$\Rightarrow \quad (y_0)^{5/3} = \frac{qn}{(S_0)^{1/2}} \qquad \left[\because \frac{Q}{B} = q\right]$$

$$\Rightarrow \quad y_o = \left[\frac{qn}{(S_o)^{1/2}}\right]^{3/5} \quad \text{...(i)}$$

We also know that critical depth for a rectangular channel is expressed as

$$y_c = \left(\frac{q^2}{g}\right)^{1/3} \quad \text{...(ii)}$$

We know that for a mild slope $y_o > y_c$ and for a steep slope $y_o < y_c$.

$$\therefore \text{ For mild slope,} \quad \left[\frac{qn}{(S_o)^{1/2}}\right]^{3/5} > \left(\frac{q^2}{g}\right)^{1/3}$$

$$\Rightarrow \quad \frac{qn}{(S_o)^{1/2}} > \frac{q^{10/9}}{g^{5/9}}$$

$$\Rightarrow \quad (S_o)^{1/2} < \frac{ng^{5/9}}{q^{1/9}}$$

$$\Rightarrow \quad S_o < \frac{n^2 g^{10/9}}{q^{2/9}} \quad \text{(Hence proved)}$$

Similarly for **steep slope,** $\quad y_o < y_c$

$$\Rightarrow \quad S_o > \frac{n^2 g^{10/9}}{q^{2/9}} \quad \text{(Hence Proved)}$$

15.19 **A 5.0 m wide rectangular channel carries 15 m³/s of water with a velocity of 6 m/s. State whether hydraulic jump is a possibility. If yes compute height of the jump and power dissipated.**

[4 marks : 2010]

Solution:

Width of channel, $B = 5.0$ m

Discharge, $Q = 15$ m³/s

Velocity of flow before jump, $V_1 = 6$ m/s

Discharge per unit width, $q = \frac{Q}{B} = \frac{15}{5} = 3$ m³/s/m

Height before jump, $y_1 = \frac{q}{V_1} = \frac{3}{6} = 0.5$ m

Froude number before jump, $F_1 = \frac{V_1}{\sqrt{gy_1}} = \frac{6}{\sqrt{9.81 \times 0.5}} = 2.71 > 1$ i.e. flow is super critical

The ratio of the sequent depth in a horizontal frictionless rectangular channel is given by

$$\frac{y_2}{y_1} = \frac{1}{2}\left[\sqrt{1+8F_1^2} - 1\right]$$

$$\Rightarrow \quad y_2 = \frac{0.5}{2}\left[\sqrt{1+8\times(2.71)^2} - 1\right]$$

$$\Rightarrow \quad y_2 = 1.68 \text{ m}$$

Velocity of flow after the jump, $V_2 = \frac{q}{y_2} = \frac{3}{1.68} = 1.79$ m/s

Froude number after jump, $F_2 = \frac{V_2}{\sqrt{gy_2}} = \frac{1.79}{\sqrt{9.81 \times 1.68}} = 0.44 < 1$

∴ Flow is subcritical

It is obvious that flow before the jump is supercritical and after the jump is subcritical. Therefore hydraulic jump formation is a possibility.

$$\text{Height of jump} = y_2 - y_1 = 1.68 - 0.5 = 1.18 \text{ m}$$

Energy dissipated,
$$E_L = \frac{(y_2 - y_1)^3}{4y_1y_2} = \frac{(1.68 - 0.5)^3}{4 \times 0.5 \times 1.68} = 0.49 \text{ m}$$

Power dissipated,
$$P = \gamma Q E_L = 9810 \times 15 \times 0.49 \times 10^{-3} = 72.1035 \text{ kW}$$

15.20 **Find the critical depth for a discharge of 4 m³/sec for flow in right angled triangular channel.**

[4 marks : 2010]

Solution:

Let the area of flow be A.

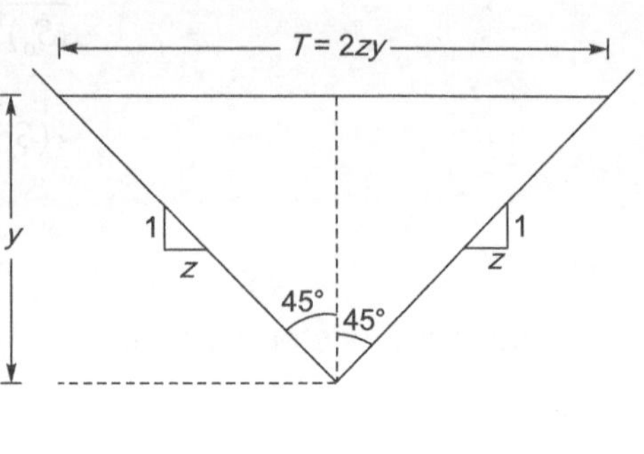

$$\therefore \quad A = zy^2$$

The basic equation governing the critical flow conditions in a channel is

$$\frac{Q^2}{g} = \frac{A^3}{T} \quad \text{...(i)}$$

For triangular channel having a side slope of z horizontal to 1 vertical

$$A = zy^2; \quad T = 2zy$$

But for a right angled triangular channel, $2\theta = 90°$ i.e. $\theta = 45°$

Also, $\tan\theta = z$

$\Rightarrow \tan 45° = z$

$\Rightarrow z = 1$

From (i), we have
$$\frac{Q^2}{g} = \frac{(zy_c^2)^3}{2zy_c}$$

$$\Rightarrow \frac{Q^2}{g} = \frac{z^3y_c^6}{2zy_c} \quad [\because z = 1]$$

$$\Rightarrow y_c = \left(\frac{2Q^2}{g}\right)^{1/5}$$

$$\Rightarrow y_c = \left(\frac{2 \times 4^2}{9.81}\right)^{1/5} \quad [\because Q = 4 \text{ m}^3/\text{s}]$$

$$\Rightarrow y_c = 1.27 \text{ m}$$

15.21 **Find the magnitude of Froude number after the hydraulic jump. Given the Froude number before the jump is $\sqrt{6}$.**

[4 marks : 2010]

Solution:

Assuming the channel section as rectangular

Sequent depth ratio before the jump is given by

$$\frac{y_2}{y_1} = \frac{1}{2}\left[\sqrt{1 + 8F_1^2} - 1\right] \quad \left[\because F_1 = \sqrt{6}\right]$$

$$\Rightarrow \frac{y_2}{y_1} = \frac{1}{2}\left[\sqrt{1 + 8 \times 6} - 1\right] = 3$$

$$\Rightarrow \quad \frac{y_1}{y_2} = \frac{1}{3} \quad ...(i)$$

Sequent depth ratio after the jump is given by

$$\frac{y_1}{y_2} = \frac{1}{2}\left[\sqrt{1+8F_2^2} - 1\right]$$

$$\Rightarrow \quad \frac{1}{3} = \frac{1}{2}\left[\sqrt{1+8F_2^2} - 1\right] \quad \text{[from (i)]}$$

$$\Rightarrow \quad \frac{2}{3}+1 = \sqrt{1+8F_2^2}$$

$$\Rightarrow \quad \left(\frac{5}{3}\right)^2 = 1+8F_2^2$$

$$\Rightarrow \quad 8F_2^2 = \frac{25}{9}-1$$

$$\Rightarrow \quad F_2^2 = \frac{16}{9\times 8}$$

$$\Rightarrow \quad F_2 = 0.47$$

Thus, the magnitude of Froude number after the hydraulic jump is 0.47.

Alternate **Solution:**

$$\because \quad \frac{y_2}{y_1} = \frac{1}{2}\left[\sqrt{1+8F_1^2} - 1\right] \quad ...(i)$$

$$\frac{y_1}{y_2} = \frac{1}{2}\left[\sqrt{1+8F_2^2} - 1\right] \quad ...(ii)$$

By multiplying eq. (i) and (ii)

$$4 = \left[\sqrt{1+8F_1^2} - 1\right]\times\left[\sqrt{1+8F_2^2} - 1\right]$$

$$\because \quad F_1 = \sqrt{6}$$

$$\Rightarrow \quad F_2 = 0.47$$

15.22 Show that the normal depth of flow in a triangular channel having side slope zH : 1 V is given by

$$y_n = 1.189\left[\frac{Q\cdot n}{\sqrt{S_o}}\right]^{3/8}\left(\frac{z^2+1}{z^5}\right)^{1/8}$$

[4 marks : 2010]

Solution:

The figure of the triangular channel is given below:

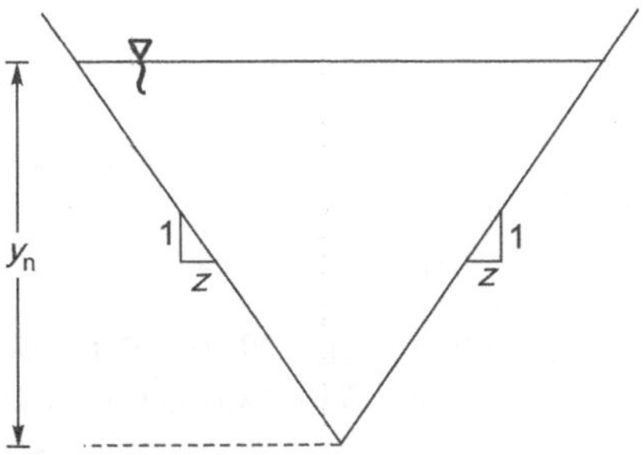

Area of flow, $A = zy_n^2$

Wetted perimeter, $P = 2y_n\sqrt{z^2+1}$

Hydraulic radius, $R = \dfrac{A}{P} = \dfrac{zy_n^2}{2y_n\sqrt{z^2+1}} = \dfrac{zy_n}{2\sqrt{z^2+1}}$

y_n 1 z 1 z

As per Manning's equation, discharge through the triangular channel is given by

$$Q = A \times \frac{1}{n} \times R^{2/3} \times \sqrt{S_0}$$

$$\Rightarrow \quad \frac{Qn}{\sqrt{S_0}} = zy_n^2 \times \left[\frac{zy_n}{2\sqrt{z^2+1}}\right]^{2/3}$$

$$\Rightarrow \quad \frac{Qn}{\sqrt{S_0}} = \frac{z^{5/3} \times y_n^{8/3}}{(2)^{2/3} \times (z^2+1)^{1/3}}$$

$$\Rightarrow \quad y_n^{8/3} = (2)^{2/3} \times \frac{Qn}{\sqrt{S_0}} \times \frac{(z^2+1)^{1/3}}{(z^5)^{1/3}}$$

$$\Rightarrow \quad y_n = (2)^{1/4} \times \left[\frac{Qn}{\sqrt{S_0}}\right]^{3/8} \times \left[\frac{z^2+1}{z^5}\right]^{1/8}$$

$$\Rightarrow \quad y_n = 1.189\left[\frac{Qn}{\sqrt{S_0}}\right]^{3/8} \times \left[\frac{z^2+1}{z^5}\right]^{1/8}$$

15.23 Show that the most efficient trapezoidal channel section is half regular hexagon given the side slope $z = \frac{1}{\sqrt{3}}$

[4 marks : 2010]

Solution:

Let the bottom width of the trapezoidal section be B, depth of flow be y and side slope be z horizontal to 1 vertical.

$\therefore$ Area, $\quad A = (B + zy)y \quad$ ... (i)

Wetted perimeter, $\quad P = B + 2y\sqrt{1+z^2} \quad$ ... (ii)

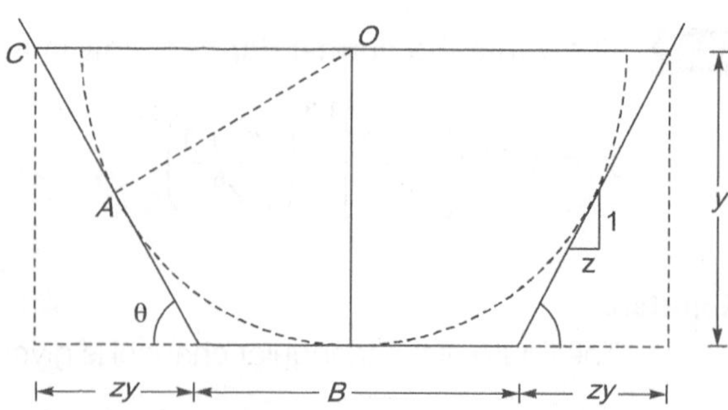

The conveyance of a channel section of a given area increases with a decrease in its perimeter. With the slope, roughness coefficient and area of flow fixed, a minimum perimeter section will represent the hydraulically efficient section as it conveys maximum discharge.

From (i), we get,

$$B = \frac{A}{y} - zy$$

Substituting value of B in (ii), we get,

$$P = \frac{A}{y} - zy + 2y\sqrt{1+z^2} \quad \text{... (iii)}$$

Assuming A and z to be constant equation (iii) can be differentiated w.r.t y and equated to zero for obtaining the condition of minimum P.

$$\therefore \quad \frac{dP}{dy} = -\frac{A}{y^2} - z + 2\sqrt{1+z^2} = 0$$

$$\Rightarrow \quad \frac{A}{y^2} + z = 2\sqrt{1+z^2}$$

Substituting value of A from (i), we get

$$\frac{(B+zy)y}{y^2} + z = 2\sqrt{1+z^2}$$

$$\Rightarrow \quad \frac{B+2zy}{y} = 2\sqrt{1+z^2}$$

$$\Rightarrow \quad \frac{B+2zy}{2} = y\sqrt{1+z^2}$$

$$\Rightarrow \quad B + 2zy = 2y\sqrt{1+z^2} \qquad \text{...(iv)}$$

If L is the length of the sloping side of the trapezoidal section, then

$$L^2 = y^2 + z^2y^2$$

$$\Rightarrow \quad L^2 = y^2 + \frac{1}{3}y^2 \qquad \left[\because z = \frac{1}{\sqrt{3}}\right]$$

$$\Rightarrow \quad L^2 = \frac{4}{3}y^2$$

$$\Rightarrow \quad L = \frac{2}{\sqrt{3}}y$$

But for **most efficient trapezoidal section**, we have

$$B + 2zy = 2y\sqrt{1+z^2}$$

$$\Rightarrow \quad B + 2 \times \frac{1}{\sqrt{3}}y = 2 \times y \times \sqrt{1+\frac{1}{3}}$$

$$\Rightarrow \quad B + \frac{2y}{\sqrt{3}} = \frac{2y \times 2}{\sqrt{3}}$$

$$\Rightarrow \quad B = \frac{2y}{\sqrt{3}}$$

Thus, length of the sloping side is equal to the bottom width of the trapezoidal section which is a property of hexagon. Thus the hydraulically most efficient trapezoidal section is half of a regular hexagon.

15.24 A trapezoidal channel with side slope 2H : 1 V is carrying a discharge of 25 m^3/s. The slope of the channel bed is 1 in 800. Take Chezy's C = 45 and design the channel.

[4 marks : 2010]

Solution:

Let the depth of flow in the channel be y and width at the bottom be B.

Designing the section as **most efficient trapezoidal channel section.**

The condition for most efficient trapezoidal section for constant area and side slope is given as

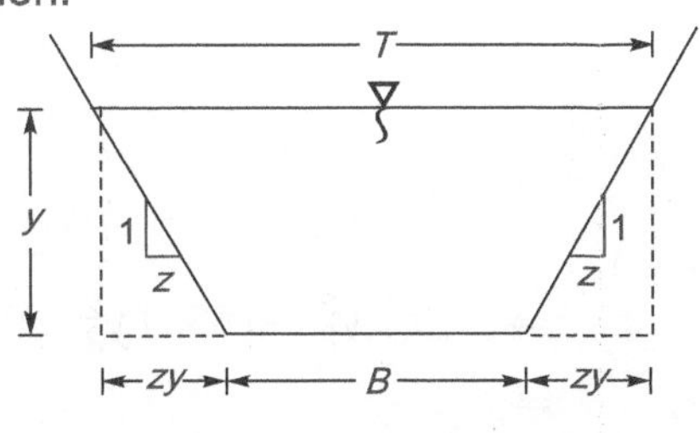

$$\frac{B+2zy}{2} = y\sqrt{1+z^2}$$

$$\Rightarrow \quad B = 2y\sqrt{1+z^2} - 2zy$$

$$\Rightarrow \quad B = 2y\left(\sqrt{1+z^2} - z\right) \qquad \text{...(i)}$$

Hydraulic radius for most efficient trapezoidal section for any value of z is given as

$$R = \frac{y}{2} \quad \text{...(ii)}$$

Given that $Q = 25\ \text{m}^3/\text{s}$; $z = 2$, $S = 1$ in 800, $C = 45$

As per Chezy's equation, velocity of flow is given by

$$V = C(R)^{1/2}(S)^{1/2} \quad \text{...(iii)}$$

But, $$Q = AV \quad [\text{where } A = (B + zy)y]$$

$\Rightarrow$ $$Q = (B + zy)y \times C(R)^{1/2}(S)^{1/2}$$

$\Rightarrow$ $$Q = \left[2y\left(\sqrt{1+z^2} - z\right) + zy\right]y \times C \times \left(\frac{y}{2}\right)^{1/2} \times (S)^{1/2}$$

$\Rightarrow$ $$25 = \left[2y\left(\sqrt{1+(2)^2} - 2\right) + 2y\right]y \times 45 \times \left(\frac{y}{2}\right)^{1/2} \times \left(\frac{1}{800}\right)^{1/2}$$

$\Rightarrow$ $$25 = \left[2y\left(\sqrt{5} - 2\right) + 2y\right]y \times 1.125 \times (y)^{1/2}$$

$\Rightarrow$ $$25 = 2.472y^2 \times 1.125 \times (y)^{1/2} = 2.781\ y^{5/2}$$

$\Rightarrow$ $$y^{5/2} = 8.9896$$

$\Rightarrow$ $$y = 2.41$$

From (i), we have

$$B = 2y\left(\sqrt{1+z^2} - z\right) = 2 \times 2.41\left(\sqrt{1+(2)^2} - 2\right) = 1.14\ \text{m}$$

15.25 **A rectangular channel 15 m wide has a normal depth of 0.8. The discharge carried is 10 m³/s, what is the alternate depth?**

[4 marks : 2010]

Solution:

Discharge, $$Q = 10\ \text{m}^3/\text{s}$$

Width, $$B = 15\ \text{m}$$

$\therefore$ Velocity, $$V_1 = \frac{Q}{By} = \frac{10}{15 \times 0.8} = 0.83\ \text{m/s}$$

Specific energy, $$E = y + \frac{V_1^2}{2g} = 0.8 + \frac{(0.83)^2}{2 \times 9.81} = 0.8351\ \text{m}$$

For alternate depth y_2, we have

$$E = y_2 + \frac{V_2^2}{2g}$$

$\Rightarrow$ $$E = y_2 + \frac{Q^2}{2gB^2y_2^2}$$

$\Rightarrow$ $$0.8351 = y_2 + \frac{(10)^2}{2 \times 9.81 \times 15^2 \times y_2^2}$$

$\Rightarrow$ $$0.8351 = y_2 + \frac{0.02265}{y_2^2}$$

$\Rightarrow$ $$y_2^3 - 0.8351\,y_2^2 + 0.02265 = 0$$

$\Rightarrow$ $$y_2 = 0.187\ \text{m} \quad \textbf{(Ans.)}$$

15.26 A hydraulic jump is formed in a 5 m wide outlet at a short distance downstream of control gate. If the flow depths are 10 m and 2 m in the u/s and d/s respectively of the sluice gate and the $Q = 150\ m^3s^{-1}$, determine

(a) Flow depth downstream of the jump

(b) Thrust on the gate

(c) Head losses in the jump

[10 marks : 2011]

Solution:

$y_1 = 10$ m, $y_2 = 2$ m, $Q = 150\ m^3/s$, $B = 5$ m, y_3 = Depth downstream of the jump.

Discharge per unit width

$$= \frac{Q}{B} = q = \frac{150}{5} = 30\ m^3/s/m$$

(a) Using,

$$\frac{y_3}{y_2} = \frac{1}{2}\left[-1+\sqrt{1+8F_{r2}^2}\right]$$

where

$$F_{r2} = \frac{q}{\sqrt{gy_2^3}} = \frac{30}{\sqrt{9.81\times 2^3}} = 3.39$$

$$\Rightarrow \quad \frac{y_3}{2} = \frac{1}{2}\left[-1+\sqrt{1+8\times 3.39^2}\right]$$

$$\Rightarrow \quad y_3 = 8.64\ m$$

(b) Thrust on the gate is given by

Force per unit width

$$\frac{F}{B} = \frac{1}{2}\gamma\frac{(y_1-y_2)^3}{(y_1+y_2)}$$

$$\frac{F}{B} = \frac{1}{2}\times(9.81\times 1000)\frac{(10-2)^3}{(10+2)} = 209280\ N/m = 209.28\ kN/m$$

Thrust on the gate $= 209.28 \times 5 = 1046.4$ kN

(c) Head loss in jump

$$E_L = \frac{(y_3-y_2)^3}{4y_2y_3} = \frac{(8.64-2)^3}{4\times 2\times 8.64} = 4.24\ m$$

15.27 A 25 cm pipe is compressed by a tree root into an elliptical section until it's inside height is 18 cm. Determine the hydraulic mean radius when the pipe is half full.

[5 marks : 2011]

Solution:

Diameter of circular section = 25 cm

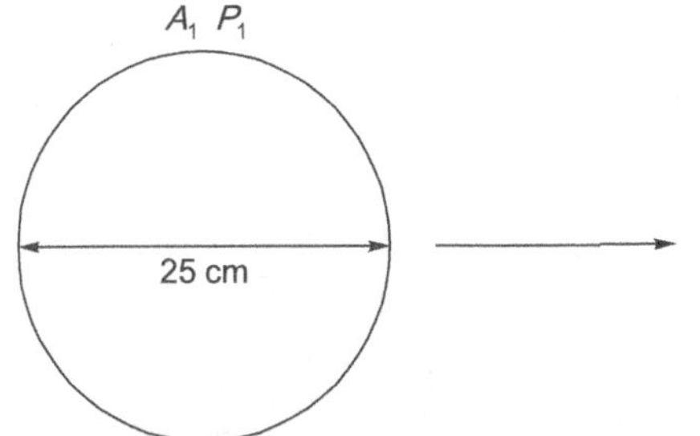

Initial Section

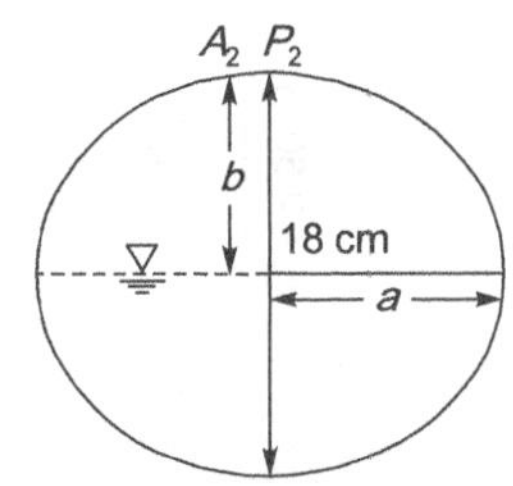

Final Section

a = ?

$$b = \frac{18}{2} = 9\ cm$$

We have to calculate the hydraulic mean radius of elliptical section while flowing half full.

Under this condition, $R_2 = \dfrac{\frac{A_2}{2}}{\frac{P_2}{2}} = \dfrac{A_2}{P_2}$

A_2 is area of elliptical section

P_2 is perimeter of elliptical section

Calculation of A_2:

Cross-sectional area of the circular and elliptical section will be same.

$\therefore$ Area of elliptical section $A_2 = \dfrac{\pi}{4}(25)^2 = 490.874 \text{ cm}^2$

Calculation of perimeter of elliptical section (P_2)

$$\pi r^2 = \pi ab$$

$$\left(\frac{25}{2}\right)^2 = \left(\frac{18}{2}\right) \times a \qquad (a = 17.361 \text{ cm})$$

We know that perimeter of ellipse is given by

$$P_2 = 2\pi\sqrt{\frac{a^2+b^2}{2}} = 2\pi\sqrt{\frac{(17.361)^2+(9)^2}{2}} = 86.88 \text{ cm}$$

Hydraulic mean radius, $R_2 = \dfrac{A_2}{P_2} = \dfrac{490.874}{86.88} = 5.65 \text{ cm}$

Hydraulic mean radius of the elliptical section while flowing half-full = 5.65 cm.

15.28 An overflow spillway is 50 m high. At the design the energy head is 2.5 m over the spillway. Find the sequent depth and energy loss in a hydraulic jump formed on a horizontal apron at the toe of the spillway. Neglect energy loss due to flow over the spillway face. Assume coefficient of discharge $C_d = 0.735$ and velocity at the toe, before the jump can be approximated as $V = \{2\,g\,(\text{total head})\}^{1/2}$. Where g is acceleration due to gravity.

[6 marks : 2012]

Solution:

The discharge per metre width of the spillway is

$$q = \frac{2}{3} C_d \sqrt{2g}\, H_d^{3/2}$$

Now, $H_d = 2.5 \text{ m}$

$C_d = 0.735$

$$\therefore \quad q = \frac{2}{3} \times 0.735 \times \sqrt{2 \times 9.81} \times (2.5)^{3/2} = 8.579 \text{ m}^3/\text{s/m}.$$

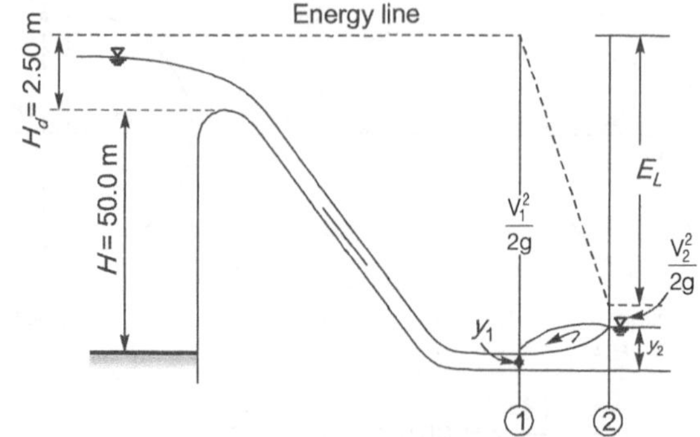

As given, velocity at toe before the jump,

$$V_1 = \{2g(\text{total head})\}^{1/2}$$

Now, Total head $= 50 + 2.5 = 52.5$ m

∴ Velocity, $V_1 = \sqrt{2\times 9.81\times 52.5} = 32.094\text{m/sec}$

$$q = 8.579 \text{ m}^3/\text{s/m}$$

∴ $$y_1 = \frac{q}{V_1} = \frac{8.579}{32.094} = 0.267\text{m}$$

∴ $$F_1 = \frac{V_1}{\sqrt{gy_1}} = \frac{32.094}{\sqrt{9.81\times 0.267}} = 19.83$$

Now, as we know, $$\frac{y_2}{y_1} = \frac{1}{2}\left(-1+\sqrt{1+8F_1^2}\right)$$

$$\frac{y_2}{0.267} = \frac{1}{2}\left[-1+\sqrt{1+8\times(19.83)^2}\right]$$

⇒ $$y_2 = 0.267 \times 27.548 = 7.355 \text{ m}$$

Energy loss $$E_L = \frac{(y_2-y_1)^3}{4y_1y_2} = \frac{(7.355-0.267)^3}{4\times 0.267\times 7.355} = 45.333 \text{ m}$$

$$E_1 = \text{Energy at section 1} = 52.5\text{m}$$

∴ **Percentage of initial energy lost** $= \frac{E_L}{E_1}\times 100 = \frac{45.333}{52.5}\times 100 = 86.35\%$

15.29 **A triangular channel of apex angle 90° and a rectangular channel of the same material have the same bed slope. If the rectangular channel has the depth of flow equal to the width of the channel and flow areas in both channels are the same, find the ratio of discharges in the rectangular and triangular channels respectively. Use Manning's roughness equation for estimation of velocity.**

[6 marks : 2012]

Solution:

Let width of rectangular channel = B

∴ Depth of flow = B

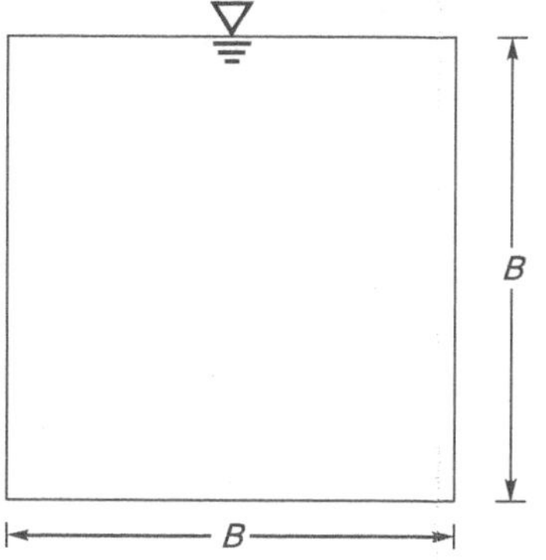

∴ Area, $A_1 = B^2$

Wetted perimeter, $P_1 = B + 2B = 3B$

Hydraulic radius, $$R_1 = \frac{A_1}{P_1} = \frac{B^2}{3B} = \frac{B}{3}$$

∴ Discharge in the rectangular channel

$$Q_1 = A_1\frac{1}{n}(R_1)^{2/3}S^{1/2}$$

⇒ $$Q_1 = B^2\frac{1}{n}\left(\frac{B}{3}\right)^{2/3}S^{1/2}$$

⇒ $$Q_1 = \frac{1}{n}B^2\left(\frac{B}{3}\right)^{2/3}S^{1/2}$$

Now, for a triangular channel of apex angle 90°, we have

Let depth of flow $= y_0$

Now $\cos 45^\circ = \dfrac{BD}{AB}$

$\Rightarrow$ $AB = \dfrac{BD}{\cos 45^\circ} = \dfrac{y_0}{\cos 45^\circ} = \sqrt{2} y_0$

Also, $\tan 45^\circ = \dfrac{AD}{BD}$

$\Rightarrow$ $AD = BD \times \tan 45^\circ = y_0 \times \tan 45^\circ = y_0$

A, D, C, B, 45°, 90°, y_0

$\therefore$ **Area,** $A_2 = \dfrac{1}{2} \times BD \times AC$ where $AC = 2AD = 2y_0$

$$= \frac{1}{2} \times y_0 \times 2 \times y_0 = y_0^2$$

Wetted perimeter $P_2 = AB + BC$

$$= 2\,AB \quad \text{(As AB = BC)}$$

$$= 2 \times \sqrt{2} y_0 = 2\sqrt{2} y_0$$

$\therefore$ $R_2 = \dfrac{A_2}{P_2} = \dfrac{y_0^2}{2\sqrt{2} y_0} = \dfrac{y_0}{2\sqrt{2}}$

Now, flow areas in both channel as same

$\therefore$ $A_1 = A_2$

$\Rightarrow$ $B^2 = y_0^2$

$\Rightarrow$ $B = y_0$

Now, $Q_2 = A_2 \dfrac{1}{n}(R_2)^{2/3} S^{1/2} = y_0^2 \dfrac{1}{n}\left(\dfrac{y_0}{2\sqrt{2}}\right)^{2/3} (S)^{1/2}$

$\therefore$ Ratio of discharge $= \dfrac{Q_1}{Q_2}$

$$\Rightarrow \quad \frac{Q_1}{Q_2} = \frac{B^2 \frac{1}{n}\left(\frac{B}{3}\right)^{2/3} (S)^{1/2}}{y_0^2 \frac{1}{n}\left(\frac{y_0}{2\sqrt{2}}\right)^{2/3} (S)^{1/2}}$$

As $B = y_0$

$$\therefore \quad \frac{Q_1}{Q_2} = \frac{(B/3)^{2/3}}{\left(B/2\sqrt{2}\right)^{2/3}}$$

$$\Rightarrow \quad \frac{Q_1}{Q_2} = \left(\frac{2\sqrt{2}}{3}\right)^{2/3} = 0.961$$

15.30 A trapezoidal channel (slope = 0.006, $n = 0.014$) has a bed width of 4 m and is with a side slope of 1 : 2 (V : H). The flow rate is 100 m^3s^{-1}. At a particular point canal confluences with a stream of 3.5 m depth. The invert elevations of the stream and the canal are same.

Determine whether the flow in the channel is subcritical or supercritical flow.

What is the type of control and stream represents the channel?

[20 marks : 2013]

Solution:

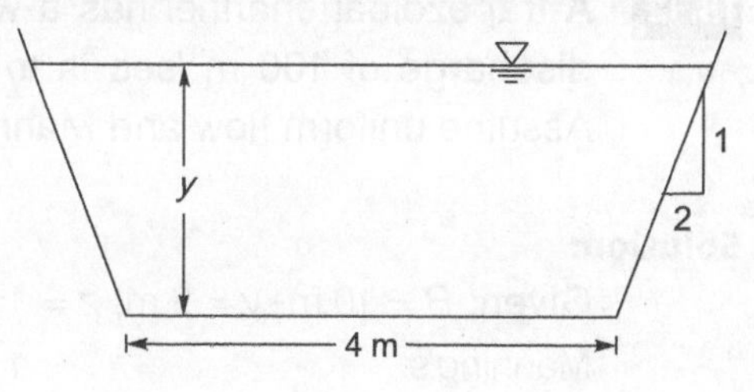

$S_0 = 0.006$, $n = 0.014$, $Q = 100\ m^3/s$, $A = (B + my)y$

Where $B = 4$ m and $m = 2$, y = normal depth of flow.

$$A = (4 + 2y)y$$

For $y = 3.5$ m

$\Rightarrow$ $A = (4 + 2 \times 3.5) \times 3.5 = 38.5\ m^2$

$$T = B + 2\,my$$

$$T = 4 + 2 \times 2 \times 3.5$$

$$T = 18\ m$$

$\because$ $$Fr^2 = \frac{Q^2T}{gA^3} = \frac{100^2 \times 18}{9.81 \times 38.5^3}$$

$$Fr = 0.567 < 1$$

So the flow is sub-critical at given depth.

Therefore, control section will be on downstream side.

15.31 **A hydraulic jump takes place in a horizontal, frictionless triangular channel with a bottom angle of 90°. Find the discharge if the pre-jump an post-jump depths are 5 cm and 15 cm respectively.**

[5 marks : 2014]

Solution:

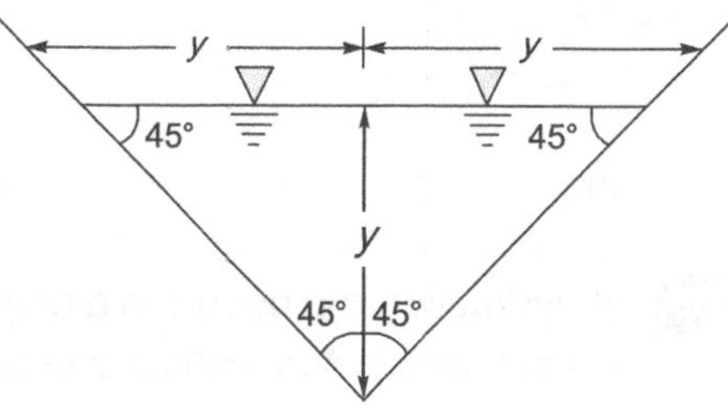

Given:

Pre-jump hydraulic depths, $y_1 = 5$ cm

Post-jump hydraulic depths, $y_2 = 15$ cm

Top width, $T = 2y$

Area of flow, $A = \frac{1}{2} \times 2y \times y = y^2$

In hydraulic jump, specific force remain constant at conjugate depths.

$$(F_s)_1 = (F_s)_2$$

$$\frac{Q^2}{A_1 g} + A_1\bar{z}_1 = \frac{Q^2}{A_2 g} + A_2\bar{z}_2$$

$$\frac{Q^2}{y_1^2 g} + y_1^2 \cdot \frac{y_1}{3} = \frac{Q^2}{y_2^2 g} + y_2^2 \cdot \frac{y_2}{3}$$

Here, $\bar{z}_1 = \frac{y_1}{3}$ and $\bar{z}_2 = \frac{y_2}{3}$

$$\frac{Q^2}{gy_1^2} + \frac{y_1^3}{3} = \frac{Q^2}{gy_2^2} + \frac{y_2^3}{3}$$

$$\frac{Q^2}{g}\left[\frac{1}{y_1^2} - \frac{1}{y_2^2}\right] = \frac{1}{3}\left[y_2^3 - y_1^3\right]$$

$$Q^2\left[\frac{y_2^2 - y_1^2}{y_1^2 y_2^2}\right] = \frac{g}{3}\left[y_2^3 - y_1^3\right]$$

Substituting $y_1 = 5$ cm and $y_2 = 15$ cm

$$Q^2\left[\frac{0.15^2 - 0.05^2}{(0.05 \times 0.15)^2}\right] = \frac{9.81}{3}\left[0.15^3 - 0.05^3\right]$$

$$355.55\,Q^2 = 0.0106$$

or $$Q^2 = 2.981 \times 10^{-5}$$

or $$Q = 5.46 \times 10^{-3}\ m^3/s$$

15.32 A trapezoidal channel has a width of 10 m and a side slope of 1.5 horizontal to 1.0 vertical. A discharge of 100 m³/sec is to be passed at a depth of 3 m. Design the slope of the channel. Assume uniform flow and Manning's n as 0.020.

[4 marks : 2015]

Solution:

Given: $B = 10$ m, $y = 3$ m, $z = 1.5$, $Q = 100$ m³/s

Manning's, $n = 0.02$

Area, $A = (B + zy)y$

$= (10 + 1.5 \times 3) \times 3 = 43.5 \text{ m}^2$

Wetted perimeter, $P = B + 2y\sqrt{z^2+1}$

$= 10 + 2 \times 3 \times \sqrt{1.5^2+1} = 20.82 \text{ m}$

Hydraulic Radius, $R = \frac{A}{P} = \frac{43.5}{20.82} = 2.09 \text{ m}$

Velocity, $V = \frac{1}{n} R^{2/3} S^{1/2}$

Discharge, $Q = \frac{1}{n} A R^{2/3} S^{1/2}$

$\Rightarrow \quad S = \left[\frac{Qn}{AR^{2/3}}\right]^2 = \left[\frac{100 \times 0.02}{43.5 \times 2.09^{2/3}}\right]^2$

$S = 7.91 \times 10^{-4}$

or $\quad S = \frac{1}{1264}$

15.33 A hydraulic jump occurs in a horizontal rectangular channel. Froude number before the jump is 12 and energy loss is 4 m. Estimate sequent depths, discharge intensity and Froude number after the jump.

[8 marks : 2015]

Solution:

Given, $F_{r_1} = 12$, $E_L = 4$ m

$$\frac{y_2}{y_1} = 0.5\left[-1+\sqrt{1+8F_{r_1}^2}\right]$$

$$\Rightarrow \quad \frac{y_2}{y_1} = 0.5\left[-1+\sqrt{1+8\times 12^2}\right] = 16.478$$

$$E_L = \frac{(y_2-y_1)^3}{4y_1y_2} = \frac{y_1^3\left(\frac{y_2}{y_1}-1\right)^3}{4y_1y_2} = \frac{y_1\left(\frac{y_2}{y_1}-1\right)^3}{4\left(\frac{y_2}{y_1}\right)}$$

$$\Rightarrow \quad 4 = \frac{y_1(16.478-1)^3}{4\times 16.478}$$

$$\Rightarrow \quad y_1 = 0.0711 \text{ m}$$

$$\Rightarrow \quad y_2 = y_1 \times 16.478 = 0.0711 \times 16.478 = 1.172 \text{ m}$$

$$y_c^3 = \frac{y_1y_2(y_1+y_2)}{2} = \frac{0.0711\times 1.172\times(0.0711+1.172)}{2} = 0.052$$

$$y_c = 0.373 \text{ m}$$

Discharge intensity, $q = \sqrt{y_c^3 \times g} = \sqrt{0.052\times 9.81} = 0.714 \text{ m}^3\text{/s/m}$

Froude number after jump, $\frac{y_1}{y_2} = 0.5\left[-1+\sqrt{1+8F_{r_2}^2}\right]$

$$\frac{0.0711}{1.172} = 0.5\left[-1+\sqrt{1+8F_{r_2}^2}\right]$$

$$\Rightarrow \quad F_{r_2} = 0.179 \simeq 0.18$$

15.34 **A trapezoidal channel has bed width = 4 m, side slope = 1.5, bed slope = 0.01 and Manning's n = 0.012. If it is carrying a discharge = 40 m³/s at normal depth = 1.5 m, then classify its slope by computing critical depth.**

[4 marks : 2016]

Solution:

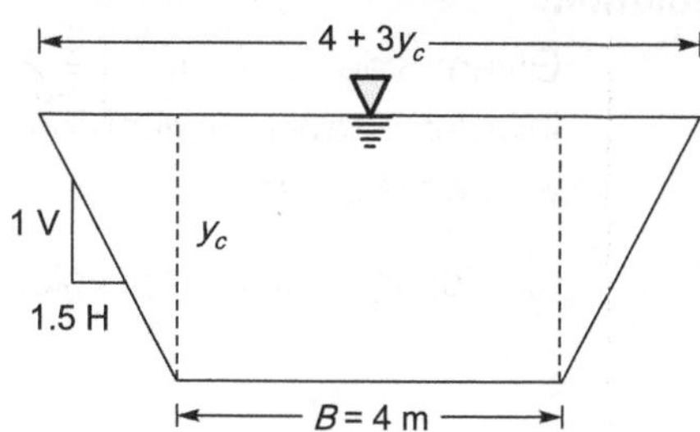

$$T = 4 + 3y_c$$

$$A_c = \frac{1}{2}[4+4+3y_c]y_c = \left(\frac{8+3y_c}{2}\right)y_c$$

$$\frac{Q^2}{g} = \frac{A_c^3}{T}$$

$$\frac{40^2}{9.81} = \frac{\left[\left(\frac{8+3y_c}{2}\right)y_c\right]^3}{4+3y_c}$$

$$1304.791 = \frac{[(8+3y_c)y_c)]^3}{4+3y_c}$$

By hit and trial $\quad y_c = 1.73$ m

Since y_0 (= 1.5 m) is less than y_c (= 1.73 m), it is Steep slope.

15.35 **A hydraulic jump is formed in a horizontal rectangular channel. If post-jump flow depth = 5 m for a discharge per unit width = 5 m³/s/m, then compute the pre-jump depth of flow and energy dissipated in the hydraulic jump.**

[8 marks : 2016]

Solution:

Given, Post jump flow depth, $\quad y_2 = 5$ m

Discharge per unit width, $\quad q = 5$ m³/s/m

Square of Froude's number after jump, $F_2^2 = \frac{q^2}{gy_2^3} = \frac{5^2}{9.81\times 5^3} = 0.0204$

$\therefore \quad F_2 = \sqrt{0.0204} = 0.143 < 1$

Thus the flow is sub-critical.

Now $\quad \frac{y_1}{y_2} = \frac{1}{2}\left[-1+\sqrt{1+8F_2^2}\right]$

$\Rightarrow \quad \frac{y_1}{5} = \frac{1}{2}\left[-1+\sqrt{1+8(0.0204)}\right]$

$\Rightarrow \quad y_1 = 0.196$ m

Check: Pre-jump Froude's number, $F_1 = \frac{q}{(gy_1^3)^{1/2}} = \frac{5}{(9.81\times 0.196^3)^{1/2}}$

$= 18.4 > 1$ (O.K.) (Super-critical flow)

Energy lost in hydraulic jump, $E_L = \dfrac{(y_2 - y_1)^3}{4y_1y_2} = \dfrac{(5-0.196)^3}{4\times 0.196\times 5} = 28.28$ m

15.36 **A trapezoidal channel with side slopes of 1 : 2 has to be designed to convey 8.0 m³/s at a velocity of 1.0 m/s, so that the amount of concrete lining for the bed and sides is the minimum.**
(i) Calculate the area of lining required for one metre length of the channel.
(ii) If the roughness coefficient is n = 0.015, calculate the bed slope of the channel for uniform flow.

[12 marks : 2017]

Solution:

Given: Side slope, m : 1 = 2 : 1, Discharge capacity, $Q = 8.0$ m³/sec, Velocity of flow, $v = 1$ m/sec

In order to have minimum concrete lining for bed and sides, the perimeter is minimum and hence discharge will be maximum.

(i) For an efficient trapezoidal section,

$$A_e = \left(2\sqrt{1+m^2} - m\right)y_e^2 = \left(2\sqrt{1+(2)^2} - 2\right)y_e^2 = 2.47y_e^2$$

Using Manning's formula, $Q = \dfrac{1}{n}A_eR_e^{2/3}S_0^{1/2}$

$\Rightarrow$ $A_e = \dfrac{Q}{V} = \dfrac{8.0}{1.0} = 8 \text{ m}^2$

Also, $8 = 2.47\,y_e^2$

$\Rightarrow$ $y_e = 1.80$ m

$\therefore$ $P_e = 2y_e\left\{2\sqrt{1+m^2} - m\right\}$

$= 2\times 1.80\left\{2\sqrt{1+(2)^2} - 2\right\}$

$= 8.8997 \text{ m} \simeq 8.9$ m

$\therefore$ Area of lining in one meter $= (8.9 \times 1)$ m² $= 8.9$ m²

(ii) Roughness coefficient, $n = 0.015$

$\therefore$ $V = \dfrac{1}{n}R_e^{2/3}s_0^{1/2}$

$\Rightarrow$ $1 = \dfrac{1}{0.015}\times\left(\dfrac{1.80}{2}\right)^{2/3}s_0^{1/2}$

$\Rightarrow$ $s_0 = 2.5894 \times 10^{-4}$

$\Rightarrow$ $s_0 = 1$ in 3862

15.37 **Design a most economical trapezoidal canal section for carrying discharge of 45 cumecs and side slope of $1\frac{1}{2}$: 1 at a velocity of 0.9 m/sec. Assume value of n = 0.022 (Manning's coefficient).**

[12 marks : 2018]

Solution:

Given: Discharge, $Q = 45$ m³/s, Velocity, $V = 0.9$ m/s, $n = 0.022$

$$\text{Side slope} = 1\frac{1}{2}:1 \quad (\therefore m = 1.5)$$

For economical canal section,

$$\text{Top width} = 2 \times \text{Length of each slopping side}$$

$$B + 2my = 2y\sqrt{1+m^2}$$

$$B + 2(1.5\,y) = 2y\sqrt{1+1.5^2}$$

So, $\qquad B = 0.6056\,y \qquad$ ...(i)

Now, $\qquad Q = AV$

$$\Rightarrow \quad V = \frac{Q}{A} \Rightarrow \frac{45}{A} = 0.9 \text{ m/s}$$

$$\Rightarrow \quad A = 50 \text{ m}^2$$

or $\qquad (B + my)y = 50$

Substituting value of B from equation (i)

We get, $\qquad (0.6056y^2 + 1.5y^2) = 50$

$$\Rightarrow \quad y = 4.87 \text{ m}$$

And, $\qquad B = 0.6056 \text{ y} = 2.95 \text{ m}$

Now, using Manning's equation, $V = \frac{1}{n}R^{2/3}\,S^{1/2},\ R = \frac{A}{P}$

Here, $\qquad A = (B + my)\,y$ and $P = B + 2y\sqrt{1+m^2} = 20.51 \text{ m}$

So, $$0.9 = \frac{1}{0.022} \times \left(\frac{A}{P}\right)^{2/3} S^{1/2}$$

On solving, $\qquad S = 1.1964 \times 10^{-4}$

15.38 **A hydraulic jump occurs in a horizontal rectangular channel, which is 1 m wide. The pre-jump depth is 5 cm and the post-jump depth is 20 cm. Assuming the channel to be frictionless, estimate the discharge of water. If the friction is not neglected, and the friction force is estimated to be 20 N over the jump-length, what will be the estimated discharge?**

[20 marks : 2018]

Solution:

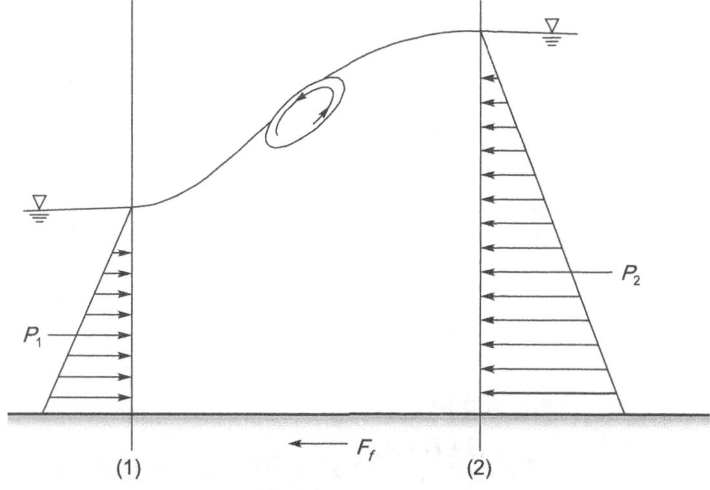

Given, Width, $\qquad B = 1$ m

Pre-jump depth, $\qquad y_1 = 5$ cm

Post-jump depth, $\qquad y_2 = 20$ cm

Channel is rectangular and horizontal.

Hence, for horizontal channel, no component of weight acts in the longitudinal direction.

Now, by momentum equation, in the direction of flow,

$$P_1 - F_f - P_2 = M_2 - M_1$$

$$\frac{1}{2}\gamma y_1^2 - F_f - \frac{1}{2}\gamma y_2^2 = \rho Q(V_2 - V_1)$$

$$= \rho Q\left(\frac{Q}{A_1} - \frac{Q}{A_1}\right)$$

$$= \frac{\rho Q^2}{B}\left(\frac{1}{y_2} - \frac{1}{y_1}\right) \quad ...(i)$$

(i) For frictionless channel $F_f = 0$

Substituting the values in eq. (i)

So, $$\frac{\gamma}{2}\left(y_1^2 - y_2^2\right) = \frac{\rho Q^2}{B}\left(\frac{1}{y_2} - \frac{1}{y_1}\right)$$

On substituting the values, $y_1 = 5$ cm, $y_2 = 20$ cm, $B = 1$ m

$$\frac{\rho g}{2}\left(0.05^2 - 0.2^2\right) = \frac{\rho Q^2}{1}\left(\frac{1}{0.2} - \frac{1}{0.05}\right)$$

On solving, $Q = 0.1107 \text{ m}^3/\text{s}$

(ii) When frictional force is considered

$$F_f = 20 \text{ N}$$

Using eq. (i) $$\frac{\gamma}{2}\left(y_1^2 - y_2^2\right) - F_f = \frac{\rho Q^2}{B}\left(\frac{1}{y_2} - \frac{1}{y_1}\right)$$

$$\frac{1000 \times 9.81}{2}\left(0.05^2 - 0.2^2\right) - 20 = \frac{1000Q^2}{1}\left(\frac{1}{0.2} - \frac{1}{0.05}\right)$$

On solving, $Q = 0.1166 \text{ m}^3/\text{s}$

15.39 A trapezoidal channel is to be designed to convey a discharge of 50 m³/sec at a velocity of 2 m/sec. The bed width to depth ratio is 0.8. The side slopes are 1H : 1V. Calculate the bed width, depth of flow and bed slope of the channel. Assume Manning's coefficient, $n = 0.02$.

[20 marks : 2019]

Solution:

Given data: Discharge = 50 m³/s

Velocity = 2 m/s

$$\frac{B}{y} = 0.8$$

$$n = 0.02$$

y, 1, 1, B

We have, Discharge = Velocity × Area

$$50 = 2 \times A$$

$\therefore$ $$\text{Area} = 25 \text{ m}^2 \quad ...(i)$$

$\therefore$ $$A = \frac{B + B + 2y}{2} \times y = (B + y)y = (0.8y + y)y$$

$$A = 1.8y^2 \quad ...(ii)$$

$\therefore$ Equating (i) and (ii)

$$1.8y^2 = 25$$
$$y = 3.726 \text{ m}$$
$$B = 0.8 \times y = 2.9814 \text{ m}$$
$$V = \frac{1}{n}R^{2/3}S^{1/2}$$
$$R = \frac{A}{P}$$
$$A = 1.8y^2$$
$$P = B + 2\times\sqrt{y^2+y^2} = 0.8y + 2\times\sqrt{2}y = 3.628y$$
$$\therefore \quad R = \frac{1.8y^2}{3.628y} = 0.496y$$
$$\therefore \quad 2 = \frac{1}{0.02}\times(0.496\times 3.726)^{2/3}\times S^{1/2}$$
$$S = 7.0548\times 10^{-4} \text{ i.e. } \frac{1}{1417.46}$$

$\therefore$ Bed width = 2.9814 m

Depth of flow = 3.726 m

Bed slope = $\frac{1}{1417.46}$

15.40 In a hydraulic jump occurring in a horizontal channel, the Froude's number before the jump is 10.0 and energy loss is 3.2 m. Estimate sequent depths, discharge intensity and Froude's number after the jump.

[10 marks : 2019]

Solution:

$$F_1 = 10.0 \text{ m } E_L = 3.20 \text{ m}$$

The sequent depth ratio $\frac{y_2}{y_1} = \frac{1}{2}\left[-1+\sqrt{1+8F_1^2}\right] = \frac{1}{2}\left(-1+\sqrt{1\times 8\times(10)^2}\right) = 13.651$

Energy loss, $E_L = \frac{(y_2 - y_1)^3}{4y_1y_2}$

$$\frac{E_L}{y_1} = \frac{\left(\frac{y_2}{y_1}-1\right)^3}{4\left(\frac{y_2}{y_1}\right)}$$

$$\therefore \quad \frac{3.20}{y_1} = \frac{(13.651-1)^3}{4\times 13.651} = 37.08$$

- y_1 = depth before the jump = $\frac{3.20}{37.08} = 0.0863$ m

  y_2 = depth after jump = $13.651 \times 0.0863 = 1.178$ m

- $F_1 = \frac{V_1}{\sqrt{g\,y_1}}$

$$10.0 = \frac{V_1}{\sqrt{9.81 \times 0.0863}}$$

$$V_1 = 9.201 \text{ m/s}$$

Discharge intensity, $q = V_1 y_1 = 9.201 \times 0.0863 = 0.7941 \text{ m}^3\text{/s/m}$

- Froude number after the jump $= F_2$

$$F_2 = \frac{V_2}{\sqrt{g\, y_2}} = \frac{q}{y_2\sqrt{g\, y_2}} = \frac{0.7941}{1.178\sqrt{9.81 \times 1.178}} = 0.1983$$

15.41 A channel has two sides vertical and semi-circular bottom of 2 m diameter. Calculate the discharge of water through the channel, when depth of flow is 2 m. Take $C = 70$ and slope of bed as 1 in 1000.

[6 marks : 2020]

Solution:

Given, Chezy's constant $(C) = 70$

Bed slope, $S = \frac{1}{1000}$

From figure,

Wetted Area, $A = \text{Area } ABCD + \text{Area } BEC$

$$= (1 \times 2) + \left(\frac{\pi \times 1^2}{2}\right) = 3.57 \text{ m}^2$$

Wetted perimeter, $P = AB + BEC + CD = 1 + (\pi \times 1) + 1 = 5.14 \text{ m}$

$\therefore$ Hydraulic mean depth 'or' Hydraulic radius

$$(R) = \frac{A}{P} = \frac{3.57}{5.14} = 0.695 \text{ m}$$

$\therefore$ Discharge, $'Q' = AC\sqrt{R \cdot S}$ (From Chezy's equation)

$$= 3.57 \times 70\sqrt{0.695 \times \frac{1}{1000}} = 6.58 \text{ m}^3\text{/s}$$

15.42 The depth of flow of water at a certain section of a rectangular channel 3 m wide is 0.45 m. The discharge through the channel is 3.1 m³/s. Determine whether a hydraulic jump will occur, and if so, determine its height and loss of energy per kg of water.

[10 Marks : 2021]

Solution:

Rectangular channel,

$$\text{Width } (B) = 3 \text{ m}$$

$$\text{depth } (y_1) = 0.45 \text{ m}$$

$$\text{Discharge } (Q) = 3.1 \text{ m}^3\text{/s}$$

For critical depth of flow,

$$y_c = \left(\frac{q^2}{g}\right)^{\frac{1}{3}} = \left(\frac{\left(\frac{3.1}{3}\right)^2}{9.81}\right)^{\frac{1}{3}} = 0.477 \text{ m}$$

Hence, $y_c > y_1$

So, flow will be supercritical and therefore, hydraulic jump will occur.

For sequent depth of H.J,

$$\frac{y_2}{y_1} = \frac{-1+\sqrt{1+8F_{r1}^2}}{2}$$

For a R.C,
$$F_{r1}^2 = \frac{q^2}{gy_1^3} = \frac{\left(\frac{3.1}{3}\right)^2}{9.81\times0.45^3} = 1.194$$

$$\frac{y_2}{y_1} = \frac{-1+\sqrt{1+8\times1.194}}{2} = 2.248$$

$$y_2 = 0.506 \text{ m}$$

Height of the jump, $H_j = y_2 - y_1 = (0.506 - 0.45) \text{ m} = 0.056 \text{ m}$

Energy loss (ΔE),
$$\Delta E = \frac{(y_2-y_1)^3}{4y_1y_2} = \frac{(0.056)^3}{4\times0.45\times0.506} = 1.928\times10^{-4} \text{ m}$$

$$\text{Energy loss per kg of water} = \frac{\rho Qg(\Delta E)}{\rho Q} = 1.892\times10^{-3} \text{ N.m/kg}$$

15.43 **A rectangular channel of 5 m width discharge water at 2 m³/s into a 5 m wide apron with 1/3500 slope at a velocity of 6 m/s. Determine the height of hydraulic jump and energy loss.**

[10 marks : 2022]

Solution:

Given, width, $B = 5 \text{ m}$

Bed slope, $S_0 = \frac{1}{3500}$

Discharge, $Q = 2 \text{ m}^3\text{/s}$

Velocity, $V_1 = 6 \text{ m/s}$

As $Q = AV_1$

$$2 = B + y_1 \times V_1$$
$$2 = 5\times y_1\times 6$$
$$y_1 = 0.066 \text{ m}$$

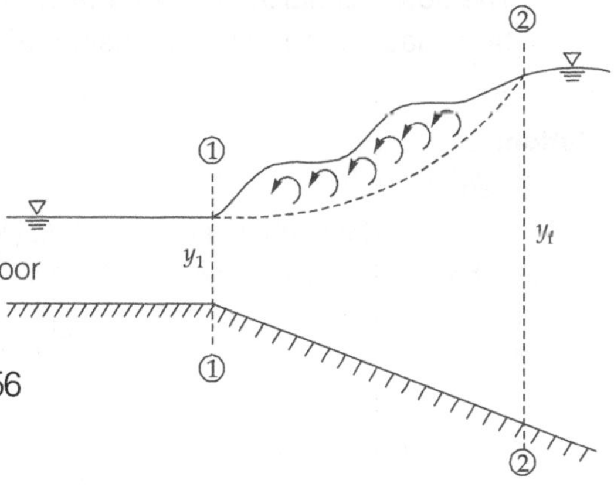

Let, y_2 = Sequent depth corresponding to y_1 on horizontal floor

y_f = Sequent depth on inclined floor

Froude no. at I.
$$Fr_1 = \frac{V_1}{\sqrt{gy_1}} = \frac{6}{\sqrt{9.81\times0.066}} = 7.456$$

$$\frac{y_2}{y_1} = \frac{-1+\sqrt{1+8Fr_1^2}}{2}$$

$$\frac{y_2}{0.066} = \frac{-1+\sqrt{1+8\times(7.456)^2}}{2}$$

$$y_2 = 0.663 \text{ m}$$

As per experiment conducted by U.S. Bureau of Reclamation (USBR))

$$\frac{y_f}{y_2} = 1.0071\exp(3.2386\tan\theta)$$

$$\frac{y_f}{0.663} = 1.0071\exp\left(3.2386\times\frac{1}{3500}\right)$$

$$y_f = 0.668 \text{ m}$$

$$\text{Height of jump} = y_f - y_1 = 0.668 - 0.066$$

$$H_j = 0.602 \text{ m}$$

Energy loss on slopping floor,

$$E_L = (E_1 + L_j \tan\theta) - E_f$$

$$E_1 = y_1 \cos\theta + \frac{V_1^2}{2g}$$

As bed slope in small, so cosθ = 1

$$E_1 = 0.066 + \frac{6^2}{2\times\pi.81} = 1.90 \text{ m}$$

As per USBR, $\frac{L_f}{y_2} = 6.1 + 4 \tan\theta$

$$\frac{L_j}{0.663} = 6.1 + 4\times\frac{1}{3500}$$

$$L_j = 4.045 \text{ m}$$

$$E_f = y_f \cos\theta + \frac{V_f^2}{2g} = 0.668\times 1 + \frac{Q^2}{2g(By_f)^2} = 0.668 + \frac{2^2}{2\times 9.81\times(5\times 0.668)^2}$$

$$E_f = 0.686 \text{ m}$$

So, $E_L = 1.9 + 4.045\times\frac{1}{3500} - 0.686$

$$E_L = 1.215 \text{ m}$$

15.44 **A 1.3 m wide rectangular channel had 0.35 m depth of water at a certain section of the channel. The flow discharge through the channel is 20 cumecs. Determine whether the hydraulic jump will take place or not. Find the height of jump and loss of energy.**

[12 marks : 2023]

Solution:

Given:

Width of channel, B = 1.3 m, Depth of flow, y_1 = 0.35 m

Discharge through channel, Q = 2 m^3/s

Now, discharge per unit width, $q = \frac{Q}{B} = \frac{2}{1.3}$ = 1.538 m^3/s/m

Froude number, $Fr_1 = \frac{V}{\sqrt{gy_1}}$ Where V is velocity in channel.

$$\Rightarrow \quad Fr_1 = \frac{Q}{A\sqrt{gy_1}}$$

$$\Rightarrow \quad Fr_1^2 = \frac{q^2}{gy_1^3} \qquad [\because A = B\cdot y]$$

$$= \frac{(1.538)^2}{9.81\times(0.35)^3} = 5.624$$

$$\Rightarrow \quad Fr_1 = \sqrt{5.624} = 2.37$$

i.e. $Fr_1 > 1$. Hence, flow is supercritical.

Now, the sequent depth ratio is given as

$$\frac{y_2}{y_1} = \frac{1}{2}\left[-1+\sqrt{1+8Fr_1^2}\right]$$

$$\frac{y_2}{0.35} = \frac{1}{2}\left[-1+\sqrt{1+8\times 5.624}\right]$$

$$y_2 = 1.012 \text{ m}$$

$\therefore$ Sequent depth, $y_2 = 1.012$ m

Now,

Froude number corresponding to y_2

$$Fr_2^2 = \frac{q^2}{gy_2^3}$$

$$= \frac{(1.538)^2}{9.81\times(1.012)^3} = 0.233$$

$\Rightarrow$ $Fr_2 = \sqrt{0.233} = 0.48$

i.e. $Fr_2 < 1$ hence flow is subcritical.

Thus, flow is taking place from supercritical to subcritical. Hence, hydraulic jump will takes place.

$$\text{Height of jump} = y_2 - y_1$$

$$= 1.012 - 0.35 = 0.662 \text{ m}$$

Now, loss of energy, E_L is given as:

$$E_L = \frac{(y_2-y_1)^3}{4y_1 \cdot y_2} = \frac{(1.012-0.35)^3}{4\times(1.012)(0.35)}$$

$$E_L = 0.205 \text{ m}$$

■■■■

2 Engineering Hydrology

Revised Syllabus of ESE : *Hydrological cycle, Ground water hydrology, Well hydrology and related data analysis; Streams and their gauging; River morphology; Flood, drought and their management; Capacity of Reservoirs.*

1. Precipitation and General Aspects of Hydrology

1.1 For a drainage basin of 640 km^2, isohyetals based on a storm event yield the following data:

Isohyetal Interval (cm)	Inter-isohyetal area (km^2)
14 - 12	90
12 - 10	140
10 - 8	125
8 - 6	140
6 - 4	85
4 - 2	40
2 - 0	20

Estimate the average depth of precipitation over the basin.

[10 marks : 2002]

Solution:

The calculations are tabulated below:

Isohyetal Interval (cm)	Average value of precipitation (cm)	Inter-Isohyetal area (km^2)	Fraction total area col. 3/640	Weighted p(cm) col. 2 × col. 4
(1)	(2)	(3)	(4)	(5)
14-12	13	90	0.1406	1.8278
12-10	11	140	0.2187	2.4057
10-8	9	125	0.1953	1.7578
8-6	7	140	0.2187	1.5309
6-4	5	85	0.1328	0.664
4-2	3	40	0.0625	0.1875
2-0	1	20	0.0312	0.0312
		ΣA = 640		Σp = 8.4049

Thus average depth of precipitation over the basin is 8.4049 cm.

1.2 A watershed has five non-recording rain gauges, installed in its area. The amount of rainfall recorded for one of the years is given below:

Rain gauge station	Annual rainfall in cm
I	100
II	120
III	190
IV	95
V	125

Find the required optimum number of non-recording and recording rain gauges for this watershed. Assume an error of 10% in the estimation of mean rainfall.

[15 marks : 2004]

Solution:

$$\Sigma p = 100 + 120 + 190 + 95 + 125 = 630 \text{ cm}$$

$$\therefore \quad \bar{p} = \frac{\Sigma p}{5} = \frac{630}{5} = 126 \text{ cm}$$

Optimum number of stations can be calculated as

$$N = \left(\frac{C_v}{E}\right)^2$$

where C_v is coefficient of variation of rainfall and E is allowable degree of error in %

$$\text{Now,} \quad C_v = \frac{\sigma_{m-1} \times 100}{\bar{p}}$$

$$\text{where} \quad \sigma_{m-1} = \sqrt{\frac{\sum_{i=1}^{m}(p_i - \bar{p})^2}{m-1}}$$

$$\therefore \quad \sigma_{m-1} = \sqrt{\frac{(100-126)^2 + (120-126)^2 + (190-126)^2 + \left(95-126^2\right) + (125-126)^2}{5-1}}$$

$$\Rightarrow \quad \sigma_{m-1} = 37.98 \text{ cm}$$

$$\therefore \quad C_v = \frac{37.98 \times 100}{126} = 30.14\%$$

$$\Rightarrow \quad N = \left(\frac{30.14}{10}\right)^2 = 9.0842 \approx 10 \text{ stations}$$

So, 9 non-recording type and 1 recording type rain gauge may be provided for this watershed.

1.3 What are the limitations of Theissen polygon method when compared to isohyetal method?

[4 marks : 2005]

Solution:

Theissen polygon method is only a mechanical and mathematical process and do not require any special skill, but on the contrary isohyetal method requires a lot of extra special judgement for drawing the contours. Theissen polygon method is inferior to isohyetal method in terms of accuracy and precision. **Isohyetal method** is the most accurate method, if the completion of contours is best done. **Isohyetal method** is superior to **arithmetical mean method** as well as Theissen **polygon method** especially when the stations are large in number.

1.4 **There are five rain gauge sections viz. *P, Q, R, S, T*. Theissen polygon network details are given in table below. Calculate the equivalent uniform depth (EUD).**

Rain gauge	Area %	Rainfall (mm)
P	24	45
Q	21	57
R	37	65
S	08	67
T	10	78

[6 marks : 2005]

Solution:

The calculations for determining the equivalent uniform depth are tabulated below:

Station	Area %	Fraction of total area	Rainfall (mm)	Weighted Rainfall (mm) (col. 3 × col. 4)
(1)	(2)	(3)	(4)	(5)
P	24	0.24	45	10.80
Q	21	0.21	57	11.97
R	37	0.37	65	24.05
S	08	0.08	67	5.36
T	10	0.10	78	7.80
	$\Sigma A = 100$	1.00		59.98 mm

Therefore the equivalent uniform depth of rainfall = 59.98 mm

1.5 The annual rainfall at station *X* and the average annual rainfall at 18 surrounding stations during 1952 to 1970 are as follows:

Annual rainfall in cm at *X*:

30.5, 38.9, 43.7, 32.2, 27.4, 32.0, 49.3, 28.4, 24.6, 21.8, 28.2, 17.3, 22.3, 28.4, 24.1, 26.9, 20.6, 29.5 and 28.4.

18 stations average annual rainfall in cm:

22.8, 35.0, 30.2, 27.4, 25.2, 28.2, 36.1, 18.4, 25.1, 23.6, 33.3, 23.4, 36.0, 31.2, 23.1, 23.4, 23.1, 33.2, and 26.4.

Explain how the consistency of the record at station X can be verified and how to determine the year in which a change in regime has occurred.

[8 marks : 2007]

Solution:

The given data is arranged in a reverse chronological order and their cumulative values are worked out as shown in the table below:

Year	Annual rainfall at X in cm	Cumulative annual rainfall at X in cm	18 stn. avg. annual rainfall in cm	18stn.cum.avg. annual rainfall in cm
1970	28.4	28.4	26.4	26.4
1969	29.5	57.9	33.2	59.6
1968	20.6	78.5	23.1	82.7
1967	26.9	105.4	23.4	106.1
1966	24.1	129.5	23.1	129.2
1965	28.4	157.9	31.2	160.4
1964	22.3	180.2	36.0	196.4
1963	17.3	197.5	23.4	219.8
1962	28.2	225.7	33.3	253.1
1961	21.8	247.5	23.6	276.7
1960	24.6	272.1	25.1	301.8
1959	28.4	300.5	18.4	320.2
1958	49.3	349.8	36.1	356.3
1957	32.0	381.8	28.2	384.5
1956	27.4	409.2	25.2	409.7
1955	32.2	441.4	27.4	437.1
1954	43.7	485.1	30.2	467.3
1953	38.9	524.0	35.0	502.3
1952	30.5	554.5	22.8	525.1

A graph is now plotted, by taking cumulative annual rainfall at *X* on *Y*-axis and cumulative average annual rainfall of 18 stations on *X*-axis respectively. This technique is called **double mass curve technique**.

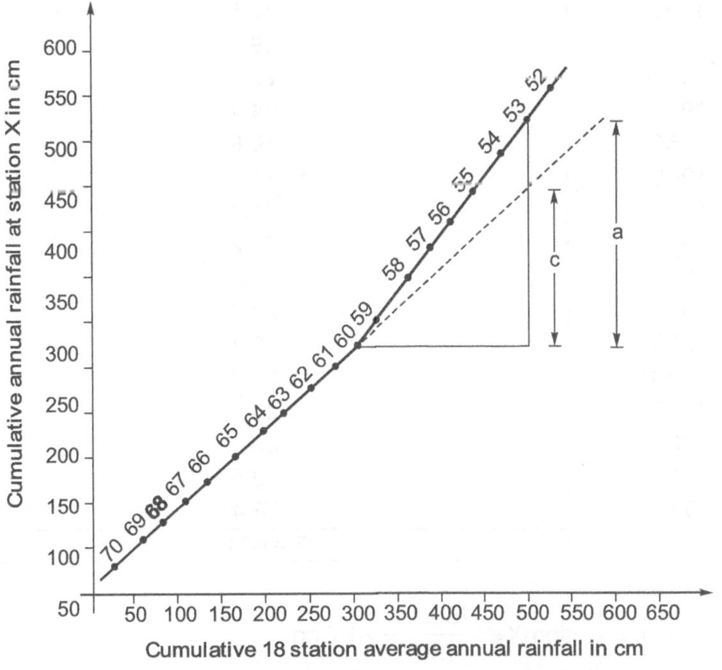

A perusal of this graph shows that inconsistency has occurred from 1960. Hence the present data, since 1960 to 1970 will be treated as correct and the previous data prior to the year 1960 will be corrected. In other words, the rainfall values of year 1959 to 1952 will be corrected.

The correction ratio is read out from the graph, and is given by

$$\text{Correction factor} = \frac{c}{a} = \frac{173}{248} = 0.6976$$

The yearly rainfall values of station *X* between the years 1959 to 1952 are thus corrected by multiplying the original values by 0.6976 as shown in table below:

Year (1)	Annual rainfall at station X in cm (2)	Corrected annual rain fall at station X = 0.6976 × col(2) (3)
1959	28.4	19.8
1958	49.3	34.4
1957	32.0	22.3
1956	27.4	19.1
1955	32.2	22.5
1954	43.7	30.5
1953	38.9	27.1
1952	30.5	21.3

The final adjusted values of annual rainfall at station *X* can be tabulated and their mean value is determined as shown below:

Year	Corrected annual rainfall values at station X in cm
1952	21.3
1953	27.1
1954	30.5
1955	22.5
1956	19.1
1957	22.3
1958	34.4
1959	19.8
1960	24.6
1961	21.8
1962	28.2
1963	17.3
1964	22.3
1965	28.4
1966	24.1
1967	26.9
1968	20.6
1969	29.5
1970	28.4
	Σ = 469.1

$$\text{Average annual rainfall at station } X = \frac{469.1}{19} = 24.7 \text{ cm}$$

1.6 The recorded annual rainfall from five raingauge stations in a catchment and the corresponding Thiessen polygon areas are as follows:

Thiessen polygon areas (cm^2)	Rainfall (cm)
25	125
30	175
30	225
10	275
5	325

The scale of the map is 1 : 50,000. Estimate the volume and the mean depth of the rainfall. Estimate the average annual discharge at the outlet, if the runoff coefficient of the catchment is 0.3.

[10 marks : 2009]

Solution:

The scale of map is 1 : 50,000. It means that 1 cm on the map represents 50,000 cm on the ground.

$$\therefore \quad 1 \text{ cm on map} = 50{,}000 \text{ cm on ground}$$
$$= 500 \text{ m on the ground}$$
$$= 0.5 \text{ km on the ground}$$
$$\therefore \quad 1 \text{ cm}^2 \text{ on the map} = 0.5 \times 0.5 \text{ km}^2 \text{ on the ground}$$
$$= 0.25 \text{ km}^2 \text{ on the ground}$$

The calculations are tabulated below:

Map Area (cm^2)	Ground Area(km^2) col.1 × 0.25	Fraction total area col.2/25	Rainfall (cm)	Weighted P(cm) col.3 × col.4
1	2	3	4	5
25	6.25	0.25	125	31.25
30	7.5	0.30	175	52.50
30	7.5	0.30	225	67.50
10	2.5	0.10	275	27.50
5	1.25	0.05	325	16.25
Total	25	1.00		195

$$\text{Mean depth of the rainfall} = 195 \text{ cm}$$
$$\text{Volume of rainfall} = 1.95 \times 100 \times (50000)^2 \times 10^{-4} \text{ m}^3 = 48.75 \text{ Mm}^3$$

Average annual discharge at the outlet is given by

$$Q = \frac{\text{Run off coefficient} \times \text{Volume of rainfall}}{365 \times 24 \times 60 \times 60}$$

$$\Rightarrow \quad Q = \frac{0.3 \times 487500}{31536000} \times 100 = 0.464 \text{ m}^3/\text{s}$$

1.7 (i) The areas between the isohyets are given in table below. Obtain the equivalent uniform depth. What is the depth of flow if the coefficient of runoff is 0.4? Find the volume of runoff.

Isohyete mm	75	90	100	125	140	150	165	180
Area between Isohyets (sq. km)	60.0	275	260	150	380	215	120	

[3 marks : 2011]

(ii) What are the limitations of the isohyetal method?

[1 marks : 2011]

(iii) What is coefficient of variation? [1 marks : 2011]

Solution:

(i) The equivalent uniform depth is given by

$$P = \frac{\left\{\begin{array}{l}60\times\left(\frac{75+90}{2}\right)+275\left(\frac{90+100}{2}\right)+260\left(\frac{100+125}{2}\right)+150\left(\frac{125+140}{2}\right)\\+380\left(\frac{140+150}{2}\right)+215\left(\frac{150+165}{2}\right)+120\left(\frac{165+180}{2}\right)\end{array}\right\}}{(60+275+260+150+380+215+120)}$$

$$= \frac{189862.5}{1460} = 130.04$$

If the coefficient of runoff, $c = 0.4$ then.

Depth of flow $= cP = 0.4 \times 130.04 = 52.016$ mm

Volume of runoff $= 52.016 \times$ Area of the catchment $\times 10^{-3}$

$= 52.016 \times (60 + 275 + 260 + 150 + 380 + 215 + 120) \times 10^6 \times 10^{-3}$

$= 75943.36 \times 10^3$ m^3

$= 75.943 \times 10^6$ m^3

(ii) **Limitations of Isohyetal method are as follows:**

(a) It is a time consuming process.

(b) Drawing isohyets is a difficult and cumbersome process.

(c) New isohyets have to be drawn to account for changes in rainfall patterns over time.

(iii) **Coefficient of variation** calculated as the standard deviation divided by the mean is often used to make comparisons of rainfall variability. Coefficient of variation counts the variation of a rainfall with respect to the long period average (LPA) of rainfall.

1.8 In a watershed four non-recording raingauges have been installed to record rainfall data. The annual rainfall record for one of the years is furnished below:

Location site of rainguage station	*A*	*B*	*C*	*D*
Recorded annual rainfall in '*cm*'	100	120	140	80

Assuming an error of 10 per cent in the estimation of mean rainfall find out the optimum number of non-recording and recording raingauges for this watershed.

[10 marks : 2012]

Solution:

The optimum number of station can be calculated as

$$N = \left(\frac{C_v}{\epsilon}\right)^2$$

where C_v is coefficient of variation of rainfall and $\in$ is allowable degree of error in %

Now, $$C_v = \frac{100 \times \sigma_m - 1}{\overline{P}}$$

where $\sigma_{m-1} = \sqrt{\sum_{i=1}^{m} \frac{(P_i - \bar{P})^2}{m-1}}$

Now, $\Sigma\bar{P} = 100 + 120 + 140 + 80 = 440$ cm

$\therefore$ $\bar{P} = \frac{\Sigma P}{4} = \frac{440}{4} = 110\text{cm}$

$\therefore$ $\sigma_{m-1} = \sqrt{\frac{(100-110)^2 + (120-110)^2 + (140-110)^2 + (80-110)^2}{4-1}}$

$\Rightarrow$ $\sigma_{m-1} = 25.82$ cm

$\therefore$ $C_v = \frac{100 \times 25.82}{110} = 23.47\%$

$\therefore$ $N = \left(\frac{23.47}{10}\right)^2 = 5.51 \simeq 6$ stations

So 5 non-recording type and 1 recording type rain gauge may be provided for this water shed.

1.9 **Five raingauge stations namely *A*, *B*, *C*, *D* and *E* are located on a circular shape basin of diameter 20 km as shown in the below figure. Compute the mean areal rainfall over the basin using Thiessen Polygon method, if the rainfall at stations, *A*, *B*, *C*, *D* and *E* are 100 cm, 90 cm, 110 cm, 120 cm and 80 cm, respectively.**

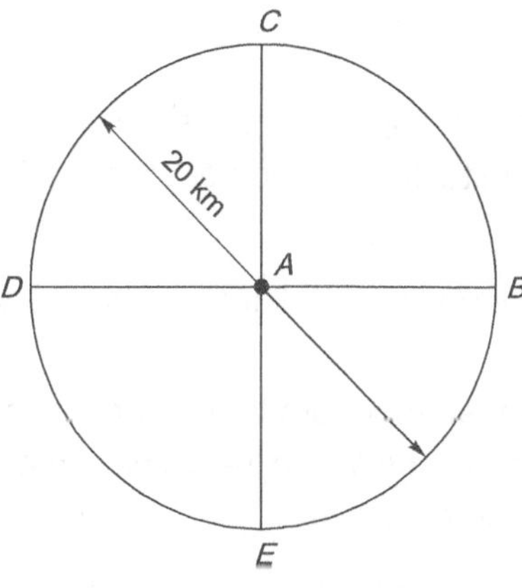

[5 marks : 2013]

Solution:

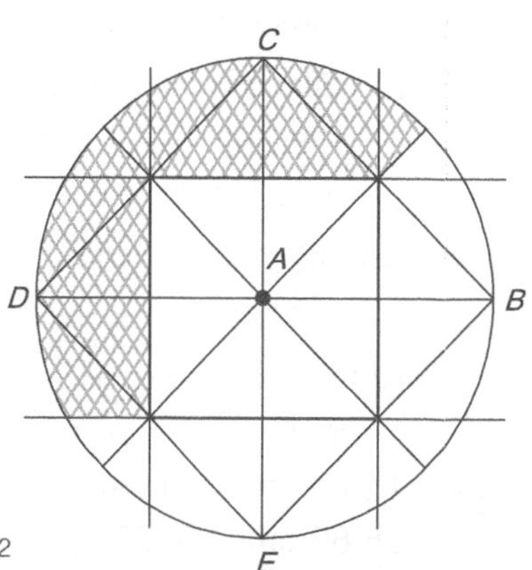

Basin dia = 20 km

Side of square around station A = 10 km × 10 km

Area of station A = 10 × 10 = 100 km^2

Total area of circular basin $= \frac{\pi}{4}(20)^2 = 314.159 \text{ km}^2$

$\therefore$ Area of station, B = Area of station C
= Area of station D
= Area of station E
$= \frac{(314.159 - 100)}{4} \text{ km}^2 = 53.54 \text{ km}^2$

$\therefore$ Mean rainfall $= \frac{100(100) + 90(53.54) + 110(53.54) + 120(53.54) + 80(53.54)}{314.159}$
= 100 cm

$\therefore$ Mean rainfall over the circular basin = 100 cm

1.10 The isohytes due to a storm in a catchment were drawn and the areas of the catchment bounded by the isohytes were tabulated as below:

Determine the uniform flow depth over the catchment.

[4 marks : 2014]

Isohyets (cm)	Areas (km^2)
Station-12.0	30
12.0-10.0	140
10.0-8.0	80
8.0-6.0	180
6.0-4.0	20

Solution:

$$\text{Uniform flow depth} = \frac{\sum (P_{avg})_i \times A_i}{\sum A_i}$$

Isohyets (cm)	$P_{avg} = \frac{P_1+P_2}{2}$	Areas (km^2)
12.0	12	30
12.0 - 10.0	11	140
10.0 - 8.0	9	80
8.0 - 6.0	7	180
6.0 - 4.0	5	20
		$\Sigma A = 450$

$$\text{Uniform flow depth} = \frac{(12\times 30)+(11\times 140)+(9\times 80)+(7\times 180)+(5\times 20)}{450} = 8.84\,\text{cm}$$

1.11 **What is precipitation in hydrology? Describe briefly various forms of precipitation.**

[4 marks : 2016]

Solution:

Precipitation: The term '*precipitation*' implies all forms of water that reaches the earth's surface is referred to as precipitation. Rain is the most common form of precipitation. Other forms of precipitation are hail, sleet, snow, drizzle, glaze etc.

(i) **Rain :** It is the most common form of precipitation. In this the water droplet size are larger than *0.5 mm and smaller than 6 mm.*

(ii) **Snow Fall :** Snow fall is another important form of precipitation. Snow are made up of ice crystals which usually combine to form *flakes*. When these flakes reaches to the earth's surface then snow fall occur. *Average density of snow 0.1 g/cm*^{3}

(iii) **Drizzle :** A fine sprinkle of numerous water droplets which have *size less than 0.5 mm* and intensity of these water *droplet is less than 1 mm/hr*, then it is referred as drizzle.

(iv) **Glaze:** When the water droplet (rain or drizzle) come in contact with ground which have the temperature near about 0°C or less than it, these water drops freeze and form ice coating on ground surface which is called glaze or freezing rain.

(v) **Sleet:** It is frozen rain drops which have *transparent grains*. These are formed when rainfall is at subfreezing temperature.

(vi) **Hail:** It is a large size of snow which have *size greater than 8 mm*. They are in the form of irrigator pellets or lumps of ice.

1.12 **The annual rainfall of six different raingauge stations of a river basin is 136.70 cm, 102.90 cm, 98.80 cm, 180.30 cm, 82.60 cm and 110.30 cm.**

Determine:

(i) The standard error in the estimation of average rainfall of the six raingauge stations.

(ii) The optimum number of raingauge stations in the river basin for a 10% error in the estimation of average rainfall.

[12 marks : 2023]

Solution:

Mean rainfall, $\bar{P}$ is given as

$$\bar{P} = \frac{\Sigma P}{n} \quad \text{where } n \text{ is number of raingauges.}$$

$$= \frac{136.70 + 102.90 + 98.80 + 180.30 + 82.60 + 110.30}{6}$$

$\Rightarrow$ $$\bar{P} = 118.6 \text{ cm}$$

$$\Sigma P^2 = (136.70)^2 + (102.90)^2 + (98.80)^2 + (180.30)^2 + (8260)^2 + (110.30)^2$$

$$= 90533.68 \text{ cm}^2$$

$$\overline{P^2} = \frac{\Sigma P^2}{n} = \frac{90533.68}{6} = 15088.946 \text{ cm}^2$$

Standard deviation, σ is given as

$$\sigma = \sqrt{\frac{n}{n-1} \times \left[\overline{P^2} - (\bar{P})^2\right]}$$

$$\sigma = \sqrt{\frac{6}{(6-1)}\left[15088.946 - (118.6)^2\right]}$$

$$\sigma = 35.037 \text{ cm}$$

Now, the coefficient of variation, C_v is given as

$$C_v = \frac{\sigma}{\bar{P}} \times 100$$

$$C_v = \frac{35.037}{118.6} \times 100$$

$$C_v = 29.542\%$$

(i) The standard error in the estimation of average rainfall of the six raingauge stationsis given as

$$N = \left(\frac{C_v}{\varepsilon}\right)^2 \quad \text{where } \varepsilon \text{ is the standard error.}$$

$$\varepsilon = \frac{C_v}{\sqrt{N}}$$

$$\varepsilon = \frac{29.542}{\sqrt{6}}$$

$$\varepsilon = 12.06\%$$

Hence, the standard error is 12.06% in the estimation of average rainfall of the six raingauge stations.

(ii) The optimum number of raingauge stations in the river basin for a 10% error in the estimation of average rainfall is given as

$$N = \left(\frac{C_v}{\varepsilon}\right)^2$$

Coefficient of variation, $C_v = 29.542$ [Calcualted above]

Now, $$N = \left(\frac{29.542}{10}\right)^2$$

$$N = 8.72$$

So, total 9 stations will be required to have 10% error in estimation of average rainfall.

2. Evaporation, Transpiration, Evapotranspiration and Stream Flow Measurement

2.1 In order to compute the flood discharge in a stream by the slope area method, the following data have been obtained:

	Upstream section	Middle section	Downstream section
Area(m^2) →	108.6	103.1	99.80
Wetted perimeter (m) →	65.3	60.7	59.40
Gauge reading (m) →	316.8	-	316.55

Determine the flood discharge assuming Manning's n = 0.029 and length between downstream section and upstream section as 250 m.

[15 marks : 2000]

Solution:

Let us use subscripts 1, 2 and 3 for U/S section, D/S section and middle section

Now, hydraulic mean depth, $R_1 = \frac{A_1}{P_1} = \frac{108.6}{65.3} = 1.663 \text{ m}$

$$R_2 = \frac{A_2}{P_2} = \frac{99.80}{59.40} = 1.680 \text{ m}$$

$$R_3 = \frac{A_3}{P_3} = \frac{103.1}{60.7} = 1.699 \text{ m}$$

Conveyance,
$$K_1 = \frac{1}{n}A_1R_1^{2/3} = \frac{1}{0.029} \times (108.6) \times (1.663)^{2/3} = 5256.46$$

$$K_2 = \frac{1}{n}A_2R_2^{2/3} = \frac{1}{0.029} \times (99.80) \times (1.680)^{2/3} = 4863.39$$

$$K_3 = \frac{1}{n}A_3R_3^{2/3} = \frac{1}{0.029} \times (103.10) \times (1.699)^{2/3} = 5062.01$$

Average conveyance is given by

$$K_{avg} = (K_1 \times K_2 \times K_3)^{1/3} = (5256.46 \times 4863.39 \times 5062.01)^{1/3} = 5058.07 \text{ m}$$

(i) 1st iteration

Assuming, $V_1 = V_2$

$$h_f = (h_1 - h_2) + \left(\frac{V_1^2}{2g} - \frac{V_2^2}{2g}\right) - K\left(\frac{V_1^2}{2g} - \frac{V_2^2}{2g}\right)$$

Friction loss,
$$= (h_1 - h_2) = 316.80 - 316.55 = 0.25 \text{ m}$$

Now,
$$Q = K_{avg}\sqrt{\frac{h_f}{L}} = 5058.07\sqrt{\frac{0.25}{250}} = 159.95 \text{ m}^3\text{/sec}$$

$$V_1 = \frac{Q}{A_1} = \frac{159.95}{108.6} = 1.47 \text{ m/s}$$

$$V_2 = \frac{Q}{A_2} = \frac{159.95}{99.8} = 1.602 \text{ m/s}$$

(ii) 2nd iteration

Take K = 0.1 for gradual contraction

$$h_f = (h_1 - h_2) + \left(\frac{V_1^2}{2g} - \frac{V_2^2}{2g}\right) - K\left(\frac{V_1^2}{2g} - \frac{V_2^2}{2g}\right)$$

$$= (316.8 - 316.55) + \left(\frac{1.47^2}{2\times9.81} - \frac{1.602^2}{2\times9.81}\right) - 0.1\left(\frac{1.47^2}{2\times9.81} - \frac{1.602^2}{2\times9.81}\right)$$

$$= 0.25 - 0.021 - 0.0021 = 0.2269 \text{ m}$$

$$Q = K_{avg}\sqrt{\frac{h_f}{L}} = 5058.07 \times \sqrt{\frac{0.2269}{250}} = 152.38 \text{ m}^3\text{/sec}$$

$$V_1 = \frac{Q}{A_1} = \frac{152.38}{108.6} = 1.403 \text{ m/s}$$

$$V_2 = \frac{Q}{A_2} = \frac{152.38}{99.8} = 1.53 \text{ m/s}$$

(iii) 3rd iteration

$$h_f = 0.25 + \left(\frac{1.403^2}{2\times9.81} - \frac{1.53^2}{2\times9.81}\right) - 0.1\left(\frac{1.403^2}{2\times9.81} - \frac{1.53^2}{2\times9.81}\right)$$

$$= 0.25 - 0.01898 - 0.001898 = 0.229 \text{ m}$$

$$Q = K_{avg}\sqrt{\frac{h_f}{L}} = 5058.07\sqrt{\frac{0.229}{250}} = 153.085 \text{ m}^3\text{/sec}$$

$$V_1 = \frac{Q}{A_1} = \frac{153.085}{108.6} = 1.41 \text{ m/s}$$

$$V_2 = \frac{Q}{A_2} = \frac{153.085}{99.8} = 1.534 \text{ m/s}$$

(iv) 4th iteration

$$h_f = 0.25 + \left(\frac{1.41^2}{2\times9.81} - \frac{1.534^2}{2\times9.81}\right) - 0.1\left(\frac{1.41^2}{2\times9.81} - \frac{1.534^2}{2\times9.81}\right)$$

$$\Rightarrow \quad h_f = 0.25 - 0.0186 - 0.00186 = 0.2295$$

$$Q = K_{avg}\sqrt{\frac{h_f}{L}} = 5058.07\sqrt{\frac{0.2295}{250}} = 153.252 \text{ m}^3\text{/sec}$$

2.2 During the passage of a flood, the following data were estimated at two sections; 500 m apart:

Section	Water surface elevation (m)	Area of flow section (m²)	Hydraulic mean depth (m)
Upstream *P*	85.233	91.746	2.835
Downstream, *R*	85.176	84.354	2.917

The eddy loss coefficient for gradual contraction is to be taken as 0.1 and for gradual expansion as 0.35. Estimate the flood discharge passing through the reach, Manning's $n = 0.022$.

[15 marks : 2003]

Solution:

Given data:

$A_P = 91.746 \text{ m}^2$, $A_R = 84.354 \text{ m}^2$

$R_P = 2.835$ m, $R_R = 2.917$ m, $L = 500$ m

$h_P = 85.233$ m, $h_R = 85.176$ m, $n = 0.022$

$K = 0.1$ ($\because A_P > A_R$, the stream is contracting)

Conveyance, $K_P = \frac{1}{n} A_P R_P^{2/3} = \frac{1}{0.022} \times 91.746 \times (2.835)^{2/3} = 8353.46$

Similarly, $K_R = \frac{1}{0.022} \times 84.354 \times (2.917)^{2/3} = 7827.82$

∴ Average conveyance, $K_{avg} = (K_P \times K_R)^{1/2} = (8353.46 \times 7827.82)^{1/2} = 8086.37$

Friction loss, $h_f = (h_P - h_R) + \left(\frac{v_P^2}{2g} - \frac{v_R^2}{2g}\right) - K\left|\frac{v_P^2}{2g} - \frac{v_R^2}{2g}\right|$ where K is 0.1

(i) 1st iteration

Assuming $v_P = v_R$, we get

$$h_f = h_P - h_R = 85.233 - 85.176 = 0.057 \text{ m}$$

But $Q = K_{avg}\sqrt{\frac{h_f}{L}} = 8086.37 \times \sqrt{\frac{0.057}{500}} = 86.34 \text{ m}^3/\text{sec}$

∴ $v_P = \frac{Q}{A_P} = \frac{86.34}{91.746} = 0.94 \text{ m/s}$

$$v_R = \frac{Q}{A_R} = \frac{86.34}{84.354} = 1.02 \text{ m/s}$$

(ii) 2nd iteration

$$v_P = 0.94 \text{ m/s}, \; v_R = 1.02 \text{ m/s}$$

∴ $h_f = 0.057 + \left(\frac{0.94^2}{2\times9.81} - \frac{1.02^2}{2\times9.81}\right) - 0.1\left|\frac{0.94^2}{2\times9.81} - \frac{1.02^2}{2\times9.81}\right|$

$$= 0.057 - 0.0079918 - 0.00079918 = 0.0482 \text{ m}$$

∴ $Q = K_{avg}\sqrt{\frac{h_f}{L}} = 8086.37 \times \sqrt{\frac{0.0482}{500}} = 79.39 \text{ m}^3/\text{sec}$

∴ $v_P = \frac{Q}{A_P} = \frac{79.39}{91.746} = 0.86 \text{ m/s}$

$$v_R = \frac{Q}{A_R} = \frac{79.39}{84.354} = 0.94 \text{ m/s}$$

(iii) 3rd iteration

$$v_P = 0.86 \text{ m/s}, \; v_R = 0.94 \text{ m/s}$$

∴ $h_f = 0.057 + \left(\frac{0.86^2}{2\times9.81} - \frac{0.94^2}{2\times9.81}\right) - 0.1\left|\frac{0.86^2}{2\times9.81} - \frac{0.94^2}{2\times9.81}\right|$

$$= 0.057 - 0.0073 - 0.00073 = 0.04897 \text{ m}$$

∴ $Q = K_{avg}\sqrt{\frac{h_f}{L}} = 8086.37 \times \sqrt{\frac{0.04897}{500}} = 80.0264 \text{ m}^3/\text{sec}$

∴ $v_P = \frac{Q}{A_P} = \frac{80.0264}{91.746} = 0.87 \text{ m/s}$

$$v_R = \frac{Q}{A_R} = \frac{80.0264}{84.354} = 0.95 \text{ m/s}$$

(iv) 4th iteration

$$v_P = 0.87 \text{ m/s}, \; v_R = 0.95 \text{ m/s}$$

$$\therefore \quad h_f = 0.057 + \left(\frac{0.87^2}{2\times 9.81} - \frac{0.95^2}{2\times 9.81}\right) - 0.1\left|\frac{0.86^2}{2\times 9.81} - \frac{0.95^2}{2\times 9.81}\right|$$

$$= 0.057 - 0.0074 - 0.00074 = 0.04886 \text{ m}$$

$$\therefore \quad Q = K_{avg}\sqrt{\frac{h_f}{L}} = 8086.37 \times \sqrt{\frac{0.04886}{500}} = 79.94 \text{ m}^3/\text{sec}$$

$$\therefore \quad v_P = \frac{Q}{A_P} = \frac{79.94}{91.746} = 0.87 \text{ m/s}$$

$$v_R = \frac{Q}{A_R} = \frac{79.94}{84.354} = 0.9476 \text{ m/s}$$

Hence the flood discharge, $Q = 79.94 \text{ m}^3/\text{sec}$

2.3 **Using the equation V(m/s) = 0.65 N + 0.03, obtain the velocity at 0.6 times the depth from the free surface. Here *N* stands for revolutions/s. Data on current meter observations are given below in tabular form.**

Distance from one bank (m)	Depth (m) (y)	Current meter observation at 0.6 y	
		No. of revolutions	Time in seconds
3.0	0.4	30	150
6.0	0.8	50	130
9.0	1.2	70	100
12.0	2.0	100	80
15.0	3.0	150	60
18.0	2.5	200	50
21.0	2.2	130	40
24.0	1.0	90	130

Also compute the discharge through the section :

[10 marks : 2008]

Solution:

The characteristic equation is given as

$$V(\text{m/s}) = 0.65\,N + 0.03$$

where N is the number of revolutions per second.

Assuming depth at a distance of 27 m from one bank is zero.

The total discharge is calculated by method of mid sections.

For the first and last section average width, $$\bar{W} = \frac{\left(W_1 + \frac{W_2}{2}\right)^2}{2W_1} = \frac{\left(3 + \frac{3}{2}\right)^2}{2\times 3} = 3.375 \text{ m}$$

For the rest of segments, $$\bar{W} = \left(\frac{3}{2} + \frac{3}{2}\right) = 3 \text{ m}$$

Distance from one bank(m)	Depth(m) y	Average width (m) $\bar{W}$	V	Segmental discharge $\Delta Q = y \times V \times \bar{W}$
3.0	0.4	3.375	0.16	0.216
6.0	0.8	3.0	0.28	0.672
9.0	1.2	3.0	0.485	1.746
12.0	2.0	3.0	0.843	5.058
15.0	3.0	3.0	1.655	14.895
18.0	2.5	3.0	2.63	19.725
21.0	2.2	3.0	2.14	14.124
24.0	1.0	3.375	0.48	1.62
27.0	0	—	—	$\Sigma\Delta Q_i = 58.056\ m^3/s$

$\therefore$ Total discharge, $Q = 58.056\ m^3/sec$

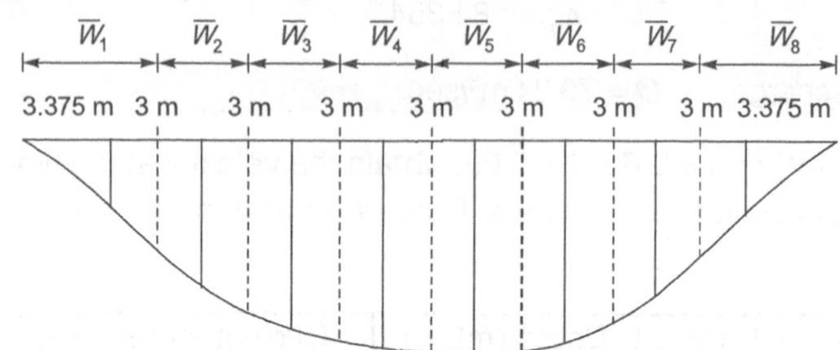

2.4 Following observations were made for conducting a water budget of a reservoir over a period of one month:
Average surface area = 10 km^2, Mean surface inflow rate = 10 cumec, Mean surface outflow rate = 15 cumec, Rainfall = 10 cm, Fall in the reservoir level = 1.5 m, Pan evaporation = 20 cm. Assuming the pan-factor as 0.7, estimate the average seepage discharge from the reservoir during the month.

[10 marks : 2009]

Solution:

In a time period Δt, the water budget for a reservoir can be written as

Mass inflow – Mass outflow = Change in mass storage

Mean surface inflow rate, $I = 10\ m^3/s$

Mean surface outflow rate, $Q = 15\ m^3/s$

Rainfall, $p = 10\ cm$

Average surface area, $A = 10\ km^2$

Change in reservoir storage volume is given by

$$\Delta S = \text{Fall in reservoir level} \times A$$
$$= 1.5 \times 10 \times 10^6\ m^3 = 1.5 \times 10^7\ m^3$$

Since there is a fall in reservoir level, therefore $\Delta S = -1.5 \times 10^7\ m^3$

Evaporation, $E = \text{Pan factor} \times \text{Pan evaporation} = 0.7 \times 20 \times 10^{-2} = 0.14\ m$

Time period, $\Delta t = \text{One month} = 30 \times 24 \times 60 \times 60 = 2.592 \times 10^6$ seconds

Let the average seepage discharge be Q_s.

Thus, the water budget equation will be

$$(I\Delta t + pA) - (Q\Delta t + EA + Q_s\Delta t) = \Delta S$$

$\Rightarrow (10 \times 2.592 \times 10^6 + 10 \times 10^{-2} \times 10 \times 10^6) - (15 \times 2.592 \times 10^6 + 0.14 \times 10 \times 10^6 + Q_s \times 2.592 \times 10^6$
$= -1.5 \times 10^7$

$\Rightarrow \quad 2.692 \times 10^7 - 4.028 \times 10^7 - Q_s \times 2.592 \times 10^6 = -1.5 \times 10^7$

$$\Rightarrow \quad Q_s = \frac{0.164 \times 10^7}{2.592 \times 10^6} = 0.6327 \text{ m}^3/\text{s}$$

Thus the average seepage discharge is 0.6327 m^3/s from the reservoir.

2.5 **Determine the amount of evapotranspiration from an area if the total rainfall precipitated during a storm is 10.0 mm. Given the antecedent moisture at the root in the soil was 5.00 mm, the loss of water due to seepage 2.5 mm, losses due to percolation 2.00 mm, surface run-off 3.00 mm and the moisture retained in the soil is 1.00 mm.**

[4 marks : 2012]

Solution:

As we know, from water budget equation,

$$P - R - G - E - T = \Delta S$$

Total rainfall, $P = 10$ mm

Antecedent moisture at root in the soil = 5 mm

Loss of water due to seepage = 2.5 mm

Loss due to percolation = 2 mm

Surface run-off = 3 mm

Moisture retained in the soil = 1 mm

∴ The water budget equation can be written as,

$$10 + 5 - 2.5 - 2 - 3 - T = 1$$

$$\Rightarrow \quad 7.5 - T = 1$$

$$\Rightarrow \quad T = 6.5 \text{ mm}$$

∴ Amount of evapotranspiration = 6.5 mm

2.6 **What is hydrologic cycle? Describe with equation that is used to quantify water going through various individual paths of the cycle. Give a neat diagram too.**

[8 marks : 2016]

Solution:

Hydrologic cycle: On planet Earth, water is stored in following reservoirs viz. oceans, rivers, soils, glaciers and as groundwater. Hydrologic cycle is a sort of conceptual model that describes the movement of water within the Earth traversing through the hydrosphere (oceans), lithosphere (ground), atmosphere and biosphere. Thus water moves from one reservoir to another by various processes like evaporation, condensation, transpiration, precipitation, runoff, infiltration, melting of ice, ground water flow etc. All this is described by the Hydrologic cycle.

Hydrologic Cycle

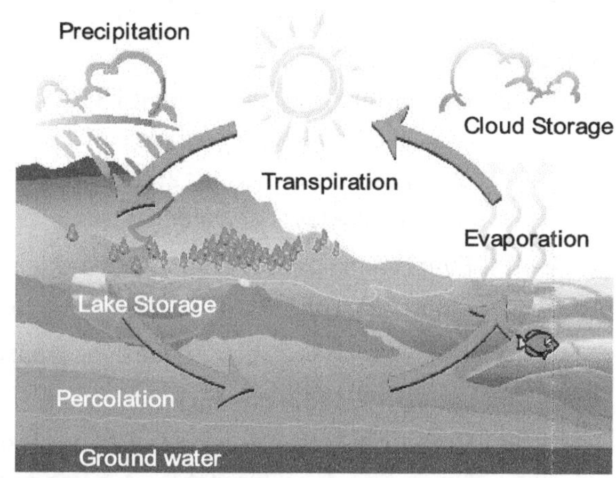

The various processes involved in the hydrologic cycle are as follows:

Evaporation: Evaporation from the oceans and other water bodies like lakes, ponds, streams etc. goes on throughout the day and night but its rate varies from time of the day. Evaporation rate is highest somewhere around mid-noon and least at night. Thus liquid water gets converted to gaseous state which leads to formation of clouds. Evaporation is measured by Pan evaporation method as

$$\text{Evaporation (in mm)} = \text{Pan coefficient} \times \text{Pan evaporation (in mm)}$$

Transpiration: It is the loss of water from the plant leaves. It also leads to formation of clouds. Transpiration rate varies from plant to plant and from time to time in a day and in the year too.

Evaporation and transpiration (i.e. evapotranspiration) are quantified by the use of *Blaney Criddle* equation as

$$ET_o = kp$$

where $$k = 0.46T_{avg} + 8$$

Condensation: This gaseous form of water when rises up gets cooler and becomes denser where gaseous water gets converted into tiny water droplets. This commences the formation of clouds. These clouds get carried away due to wind.

Precipitation: When clouds become sufficiently heavy due to condensation of water, they cannot retain the weight of water and thus it comes back to earth in the form of precipitation. Precipitation can occur in various forms like rain, hail, sleet, mist etc. depending of characteristics of the location and weather conditions. Precipitation is measured by rain gauges.

Infiltration: Once precipitation has started, some part of it flows on the surface called as *overland flow* which joins the streams. A part of this precipitation infiltrates through the ground and flows below ground called as base flow. It is quantified by Horton's infiltration equation as:

$$f_t = f_c + (f_o - f_c)e^{-kt}$$

where f_t = Infiltration rate at any time 't'

f_c = Final or equilibrium infiltration capacity of the soil

f_o = Infiltration capacity at time $t = 0$.

k = A constant (Horton's constant) that represents a decrease in infiltration capacity.

2.7 Describe briefly the following methods for estimation of evaporation:

(a) Evaporation-Pan Method

(b) Empirical Method

(c) Analytical Method

[10 marks : 2017]

Solution:

(a) **Evaporation-Pan Method:** ISI Standard Pan evaporimeter specified by IS : 5973-1970, also known as modified class A pan consists of a pan 1220 mm in diameter with 255 mm of depth. The pan is made of copper sheet of 0.9 mm thickness, tinned inside and pointed white outside. A fixed point gauge indicate the level of water.

Evaporation pans are not exact models of the large reservoirs and hence evaporation observed from a pan has to corrected to get the evaporation from lake or reservoir.

$$\text{Lake evaporation} = C_P \times \text{Pan evaporation}$$

where, C_P = Pan coefficient

(b) **Empirical Method:** A large number of empirical equations are available to estimate lake evaporation using commonly available metrological data. Most formulae are based on the Dalton type equation and can be expressed as

$$E_L = kf(u)(e_w - e_a)$$

E_L = Lake evaporation in mm/day

e_W = Saturated vapour pressure at water surface temperature in mm of mercury.

e_a = Actual vapour pressure of overlying air at a specific height in mm of mercury.

$f(u)$ = Wind-speed correction function
k = a coefficient

Two commonly used formulae are:

1. **Meyer's formula:**

$$E_L = k_m(e_w - e_a)\left(1+\frac{u_9}{16}\right)$$

where, u_9 = monthly mean wind velocity in km/h at about 9.0 m above ground.

2. **Rhower's formula:**

(c) **Analytical Methods:** The analytical method for the determination of lake evaporation can be broadly classified as:

1. **Water budget method:** It involves writing the hydrological continuity equation for lake and determining the evaporation from a knowledge or estimation of other variables.

   Continuity equation is written as

$$P + V_{is} + V_{ig} = V_{as} + V_{ag} + E_L + \Delta s + T_L$$

   P = Daily precipitation
   V_{is} = Daily surface inflow into the lake
   V_{ig} = Daily groundwater inflow
   V_{os} = Daily surface outflow from lake
   V_{og} = Daily seepage outflow
   E_L = Daily lake evaporation
   Δs = Increase in lake storage in a day
   T_L = Daily transpiration loss

2. **Energy Budget Method:** The energy budge method is an application of the law of conservation of energy. The energy available for evaporation is determined by considering the incoming energy, outgoing energy and energy stored in water body over a known time interval.

$$E_L = \frac{H_n - H_g - H_c - H_i}{\rho L(1+\beta)}$$

   where,

   $H_n = H_g + H_e + H_g + H_s + H_i$
   $H_n = H_c(1-x) - H_g$ = Net heat received by the water surface.
   H_g = Heat flux into the ground
   H_s = Heat stored in water body
   H_i = Net heat conducted out of system of water flow

3. **Mass Transfer Method:** This method is based on theories of turbulent mass transfer in boundary layer to calculate mass vapour transfer from surface to the surrounding atmosphere.

2.8 **(i) Define the "dilution method" of flow measurement by sudden injection and constant injection of chemicals in flowing water with diagram and governing equations.**

[10 marks : 2023]

(ii) A Rhodamine dye solution was discharged in a river section at a constant rate. Estimate the discharge if the dye is found to reach an equilibrium of 5 parts per billion (ppb).

Given : Amount of Rhodamine Dye = 25 g/*l*

Constant rate of flow : 10 cm^3/s

Assume $C_0 = 0$

[10 marks : 2023]

Solution:

(i)

Dilution method: The dilution method of flow measurement, also known as the chemical method depends upon the continuity principle applied to a tracer which is allowed to mix completely with the flow.

Sudden injection method: Consider a tracer in amount of C_1 and introduce it in upstream of river at section 1. Take an another section (2) at downstream which is far away from section (1) such that the complete mixing of tracer takes place. If the stream also contain some amount of tracer material of concentration C_0 and at section (2) the tracer concentration is C_2 as shown in figure

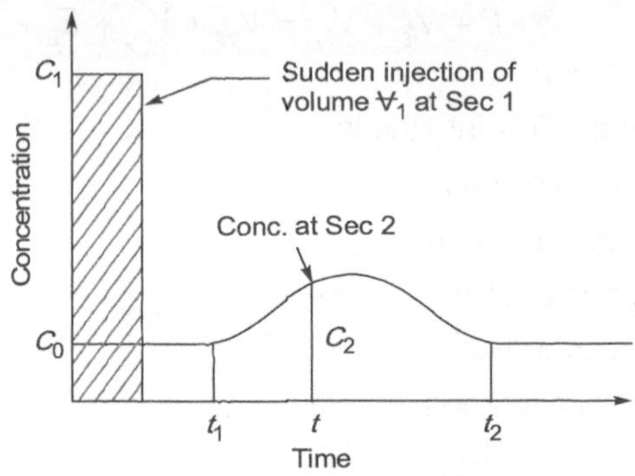

Fig. Injection Method

Now, M_1 = mass of tracer added at section (1) = V_1C_1

where V_1 is volume of tracer

Also, M_1 = mass of tracer in river discharge + mass of tracer in volume 1

$$= \int_{t_1}^{t_2} Q(C_2 - C_0)dt + \int_{t_1}^{t_2} \frac{V_1(C_2 - C_0)dt}{t_2 - t_1}$$

where Q is discharge of stream.

Neglecting the 2nd term on right hand side of above equation as it is very small, we get

$$V_1C_1 = \int_{t_1}^{t_2} Q(C_2 - C_0)dt$$

$$Q = \frac{V_1C_1}{\int_{t_1}^{t_2} Q(C_2 - C_0)dt}$$

Constant injection Method: It is another way of dilution method. In this method, tracer of concentration C_1 is injected at section 1 with a constant rate Q_1. If concentration at section 2 is gradually increased from C_0 at time t_1 to a constant value C_2 as shown in figure below, then, by continuity equation for tracer at steady state, we get

$$Q_1C_1 + QC_0 = (Q_1 + Q)C_2$$

where Q is discharge of stream

$$Q = \frac{Q_1(C_1 - C_2)}{(C_2 - C_0)}$$

Note: This method is also known as plateau gauging.

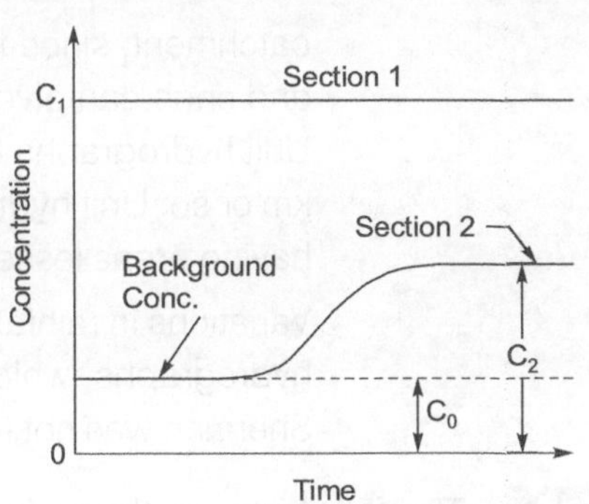

Fig. Constant Rate Injection Method

(ii)

Using constant rate injection method

$$Q = \frac{Q_1(C_1 - C_2)}{C_2 - C_0}$$

$$C_0 = 0 \qquad [\because \text{Given}]$$

$$Q_1 = 10 \text{ cm}^3/\text{s} = 10 \times 10^{-6} \text{ m}^3/\text{s}$$

$$C_1 = 25 \text{ g}/l = 25000 \text{ mg}/l = 25000 \text{ ppm}$$

$$C_2 = 5 \text{ ppb} = 5 \times 10^{-3} \text{ ppm}$$

Putting values, we get

$$Q = \frac{10 \times 10^{-6} \times (25000 - 5 \times 10^{-3})}{5 \times 10^{-3} - 0} = 50 \text{ m}^3/\text{s}$$

3. Infiltration, Runoff and Hydrograph

3.1 What are the assumptions and limitations of Sherman's Unit Hydrograph Theory?

[10 marks : 1999]

Or

What are the assumptions made in unit hydrograph theory?

[4 marks : 2005]

Solution:

Assumptions of Unit Hydrograph Theory are:

(i) **Time invariance:** The runoff produced from a given drainage basin due to a given effective rainfall shall always be the same irrespective of the time of its occurrence. In other words, the runoff response of a basin to a given effective rainfall is assumed to be time invariant.

(ii) **Linear response:** The runoff response of a drainage basin to the excess rainfall is assumed to be linear. It means that if the excess rainfall of D cm occurs in a duration of T hour, then the resulting runoff hydrograph will have its ordinates equal to D times the ordinates of a unit hydrograph.

Limitations of Unit Hydrograph Theory:

The basic assumptions made in defining a UH were that:

(a) the excess rain should occur uniformly over the entire basin and

(b) that its intensity should be constant during the entire duration.

In actual practice, however these two conditions are never strictly satisfied, since storms do have non-uniform areal distribution, and their intensities also vary during the specified duration.

Even under non-uniform areal distribution of rainfall, unit hydrographs can be used, if the variations in areal extent are consistent within the different storms, which is reasonably so for intense rains falling in smaller catchments. This however imposes a limit on the size of the catchment, since for very large basins, the centre of the storm can vary from storm to storm, and each can give different surface runoff hydrograph under otherwise identical conditions. Unit hydrographs therefore cannot give reliable results for basins exceeding about 5000 sq. km or so. Unit hydrographs are also generally not developed and used for very small basins, having areas lesser than 2 km^2.

Variations in rainfall intensity during the specified duration can also affect the accuracy of unit hydrographs, which can be minimized by choosing a smaller duration which according to Sherman was not to exceed the time of concentration.

3.2 **The following are the ordinates for a flood hydrograph resulting from an isolated storm of 6 hours duration.**

Time (hr)	0	12	24	36	48	60	72	84	96
Ordinates of flood hydrograph (cumec)	5	15	40	80	60	50	25	15	5

Determine the ordinates of 1 cm-6 hr unit hydrograph if the catchment area is 450 km^2.

[15 marks : 2001]

Solution:

The calculations are tabulated below:

Time (hr) (1)	FH (m^3/s) (2)	Base Flow (m^3/s) (3)	DRH (m^3/s) (4)	1 cm UH = $\frac{\text{col.4}}{\text{DRD}}$ (5)
0	5	5	0	0
12	15	5	10	4.17
24	40	5	35	14.58
36	80	5	75	31.25
48	60	5	55	22.92
60	50	5	45	18.75
72	25	5	20	8.33
84	15	5	10	4.17
96	5	5	0	0
			$\Sigma O = 250$	$\Sigma U = 104.17$

Assuming base flow = 5 m^3/sec

Now, Direct Runoff Depth, $\text{DRD} = \frac{0.36 \times \Sigma O \times t}{A}$

where $\Sigma O = 250$ m^3/s; $t = 12$ hr; $A = 450$ km^2

$$\therefore \quad \text{DRD} = \frac{0.36 \times 250 \times 12}{450} = 2.4 \text{ cm}$$

Check: DRD of unit hydrograph $= \frac{0.36\, \Sigma U \times t}{A}$

$$= \frac{0.36 \times 104.17 \times 12}{450} = 1.00 \text{ cm}$$

3.3 **Given below are the monthly rainfall, *P*, and the corresponding runoff values, *R*, for a period of 10 months for a catchment. Develop a correlation between *R* and *P*.**

Month No.	*P*(cms)	*R*(cms)
1	4	0.2
2	22	7.1
3	28	10.9
4	15	4.0
5	12	3.0
6	8	1.3
7	4	0.4
8	15	4.1
9	10	2.0
10	5	0.3

[15 marks : 2003]

Solution:

The relationship between rainfall and the resulting runoff is quite complex and is influenced by a host of factors relating the catchment and climate. One method is to fit a linear regression line between *R* and *P*. The calculations are tabulated below:

Month No.	P	R	P^2	PR
1	4	0.2	16	0.8
2	22	7.1	484	156.2
3	28	10.9	784	305.2
4	15	4.0	225	60
5	12	3.0	144	36
6	8	1.3	64	10.4
7	4	0.4	16	1.6
8	15	4.1	225	61.5
9	10	2.0	100	20.0
10	5	0.3	25	1.5
	$\Sigma P = 123$	$\Sigma R = 33.3$	$\Sigma P^2 = 2083$	$\Sigma PR = 653.2$

The **equation for straight-line regression** between runoff *R* and rainfall *P* is

$$R = aP + b$$

Then,

$$\Sigma R = a\Sigma P + nb$$

$$33.3 = 123a + 10b \quad \text{...(i)}$$

$$\Sigma PR = a\Sigma P^2 + b\Sigma P$$

$$653.2 = 2083a + 123b \quad \text{...(ii)}$$

From eq. (i) and (ii)

$$a = 0.43$$

$$b = -1.93$$

$$R = 0.43P - 1.93$$

3.4 **In a catchment, the average rainfall for a storm at two successive 6 hour intervals was 3.0 cm and 6.0 cm respectively. The abstraction losses ϕ-index were estimated to be 0.20 cm/hr. For the same catchment, the calculated data for a 6-hr unit hydrograph is available and is given below. Find the direct runoff hydrograph due to the storm.**

Time (hr)	Unit hydrograph ordinate (m³/s)
0	0
6	10
12	25
18	40
24	100
30	150
36	100
42	75
48	25
54	15
60	0

[20 marks : 2004]

Solution:

Let R_1 and R_2 be the runoffs at two successive 6 hour intervals.

Similarly, Let P_1 and P_2 be the average rainfall in cm.

Given $\phi_{index} = 0.20$ cm/hr

$$R_1 = P_1 - (\phi\text{-index} \times 6) = 3 - (0.2 \times 6) = 1.8 \text{ cm}$$

$$R_2 = P_2 - (\phi\text{-index} \times 6) = 6 - (0.2 \times 6) = 4.8 \text{ cm}$$

Now, the respective values of R_1 and R_2 are multiplied by the ordinates of the unit hydrograph with a lag of 6 hr. When these ordinates are added with the respective lag, they will give the direct runoff hydrograph. This is *S*-curve method.

The calculations are tabulated below:

Time (hr)	Unit hydrograph ordinates, U (m³/s)	R_1U	R_2U	R_1U+R_2U (DRH)
0	0	0	—	0
6	10	18	0	18
12	25	45	48	93
18	40	72	120	192
24	100	180	192	372
30	150	270	480	750
36	100	180	720	900
42	75	135	480	615
48	25	45	360	405
54	15	27	120	147
60	0	0	72	72
			0	0

3.5 A 30 minute unit hydrograph for a catchment is given in Table-1. The ϕ-index is 4 mm/hr. The storm details are given in Table-2. Obtain the runoff hydrograph for the storm.

Table-1

Time (min)	Runoff (m³/s)
0	0
30	1.2
60	2.8
90	1.7
120	1.4
150	1.2
180	1.1
210	0.91
240	0.74
270	0.61
300	0.5
330	0.28
360	0.17
390	0

Table-2

Time (min)	Runfall (cm)
0-30	3.3
30-60	2.7
60-90	1.9

[16 marks : 2005]

Solution:

Calculation of effective rainfall:

Given: $P_1 = 3.3$ cm, ϕ-index = 4 mm/hr, $P_2 = 2.7$ cm, $P_3 = 1.9$ cm

$t_1 = 30$ min = 0.5 hr, $t_2 = 30$ min, $t_3 = 30$ min

$R_1 = 3.3 - \frac{4}{10} \times 0.5 = 3.1$ cm; $R_2 = 2.7 - \frac{4}{10} \times 0.5 = 2.5$ cm; $R_3 = 1.9 - \frac{4}{10} \times 0.5 = 1.7$ cm

The values of R_1, R_2, R_3 are multiplied with the ordinates of unit hydrograph and they are added with a lag of 30 minutes. The calculations are tabulated below:

Time (min)	Runoff (m^3/s)	R_1U	R_2U	R_3U	$R_1U + R_2U + R_3U$ (DRH)
0	0	0	—	—	0
30	1.2	3.72	0	—	3.72
60	2.8	8.68	3	0	11.68
90	1.7	5.27	7	2.04	14.31
120	1.4	4.34	4.25	4.76	13.35
150	1.2	3.72	3.50	2.89	10.11
180	1.1	3.41	3.00	2.38	8.79
210	0.91	2.821	2.75	2.04	7.611
240	0.74	2.294	2.275	1.87	6.439
270	0.61	1.891	1.850	1.547	5.288
300	0.50	1.55	1.525	1.258	4.333
330	0.28	0.868	1.250	1.037	3.155
360	0.17	0.527	0.70	0.85	2.077
390	0	0	0.425	0.476	0.901
			0	0.289	0.289
				0	0

3.6 The run-off data at a stream gauging station for a flood in m^3/sec at three hourly intervals are as follows: 50, 50, 75, 125, 225, 290, 270, 145, 110, 90, 80, 70, 60, 55, 51, 50.

The drainage area is 40 km^2. The duration of the rainfall is 3 hr. Derive the three hour unit hydrograph for the basin. Assume a constant base flow of 50 m^3/sec throughout the duration.

[15 marks : 2007]

Solution:

The calculations for unit hydrograph are tabulated below:

Time (h) (1)	Ordinate of storm hydrograph (2)	Base flow (3)	DRH Col.(2) – Col.(3) (4)	3h unit hydrograph col. 4 ÷ 26.9 (5)
0	50	50	0	0
3	50	50	0	0
6	75	50	25	0.9
9	125	50	75	2.8
12	225	50	175	6.5
15	290	50	240	8.9
18	270	50	220	8.2
21	145	50	95	3.5
24	110	50	60	2.2
27	90	50	40	1.5
30	80	50	30	1.1
33	70	50	20	0.7
36	60	50	10	0.4
39	55	50	5	0.2
42	51	50	1	0.04
45	50	50	0	—
		Total	996	

$$\text{Direct runoff depth} = \frac{\text{Volume of DRH}}{\text{Drainage area}}$$

$$\text{Volume of DRH} = (\text{Sum of DRH ordinates}) \times \text{time of rainfall}$$
$$= 996 \times 3 \times 60 \times 60 = 10756800 \text{ m}^3$$
$$\text{Drainage area} = 40 \text{ km}^2 = 40 \times 10^6 \text{ m}^2$$
$$\therefore \quad \text{Direct runoff depth} = \frac{10756800}{40 \times 10^6} = 0.2689 \text{ m} = 26.9 \text{ cm}$$

Thus the ordinates of DRH (Col. 4) are divided by 26.9 cm to obtain the ordinates of 3h unit hydrograph.

3.7 Define unit hydrograph what are it's limitations? When do you recommend to use triangular unit hydrograph? Write down the expressions for peak discharge, base width and period of rise for triangular unit hydrograph?

[10 marks : 2008]

Solution:

A unit hydrograph is defined as "the hydrograph of direct runoff resulting from one unit depth of rainfall excess occurring uniformly over the basin and at a uniform rate for a specific duration."

Limitation of unit hydrograph:

(i) Unit hydrograph assumes uniform distribution of rainfall over the catchment, and the intensity of rainfall is assumed constant for the duration of rainfall excess. In practice these two conditions are never strictly satisfied.

(ii) Unit hydrograph method can not be used for a catchment area greater than 5000 km^2 and less than 2 km^2.

(iii) Precipitation must be from rainfall only. Snow melt runoff can not be satisfactory represented by unit hydrograph.

(iv) The catchment should not have large storage (Ex. tank, pond etc.) which affects the linear relationship between storage and discharge.

Use of triangular unit hydrograph:

Natural resources conservation services (NRCS) developed a dimensionless unit hydrograph (DUH) based on analysis of large number of watersheds. The DUH is very useful for constructing a synthetic unit hydrograph for a wide variety of watersheds.

Triangular unit hydrograph is used to solve the dimensionless unit hydrograph (DUH).

The soil conservation services (SSC), DUH consists of 37.5% of total runoff volume before the peak discharge and remaining volume after the peak discharges occurs.

Assuming 37.5% of the volume on left of T_p for the triangular unit hydrograph.

Therefore base with $T_b = \dfrac{T_p}{0.375} = 2.67 T_p$

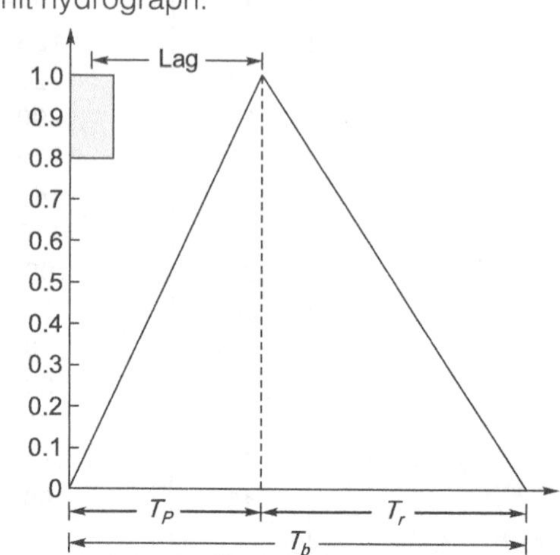

Q = Runoff volume

q_P = Peak flow rate

T_p = Time to peak

T_b = Time base of triangular unit hydrograph

T_L = Time lag

T_C = Time of concentration

D = Duration of rainfall

$$Q = \frac{1}{2} \times T_P \times q_P + \frac{1}{2} T_r \times q_P$$

$$Q = \frac{1}{2} \times q_P \times (T_P + T_r)$$

$$q_P = \frac{2Q}{T_P + T_r}$$

$$q_P = \frac{2Q}{T_b} = \frac{2Q}{2.67 T_P} \simeq \frac{0.75Q}{T_P}$$

Peak discharge: $$q_P = \frac{0.75Q}{T_P}$$

Time to peak (or Period of rise): $$T_P = \frac{D}{2} + T_L$$

where lag time (T_L) can be computed as

$$T_L = 0.6 \times T_C$$

3.8 **The ordinates of 6 hour unit hydrograph of a catchment are as follows:**

Time (hr)	0	6	12	18	24	30	36	42
Discharge (m^3/sec)	0	10	40	55	45	30	7	0

The unit depth of the unit hydrograph is 1 cm. Arrive at the direct runoff hydrograph resulting from the following excess-rainfall hyetograph occurring over the catchment.

Time (hr)	0-6	6-12
Effective Rainfall intensity (cm/hr)	1	0.5

[10 marks : 2009]

Solution:

Excess rainfall in first 6 hours = 1 × 6 = 6 cm

Excess rainfall in next 6 hours = 0.5 × 6 = 3 cm

It may be noted that 6 cm DRH occurs before 3 cm DRH. Thus the ordinates of the 3 cm DRH are lagged by 6 hours. The DRHs due to 6 cm and 3 cm rainfall excess are calculated by multiplying the ordinates of the unit hydrograph by 6 and 3 respectively. Using the method of superposition, the ordinates of the resulting DRH are obtained by combining the ordinates of the 6 cm and 3 cm DRH at any instant.

The calculations are tabulated below:

Time	Ordinate of 6-h UH (m^3/s)	Ordinate of 6 cm DRH col.(2) × 6	Oridnate of 3 cm DRH (col. 2 lagged by 6 h) × 3	Ordinate of 9 cm DRH (col.3 + col.4) (m^3/s)
1	2	3	4	5
0	0	0	0	0
6	10	60	0	60
12	40	240	30	270
18	55	330	120	450
24	45	270	165	435
30	30	180	135	315
36	7	42	90	132
42	0	0	21	21
45	0	0	0	0

3.9 (i) The watershed at a particular site on a river has 77000 ha area. The mean annual precipitation is 950 mm. About 25% of precipitation reaches the basin outlet as stream flow. Estimate the mean flow rate Q at this site in $m^3\ s^{-1}$?

[3 marks : 2011]

(ii) If the head available at the site is 150 m and assuming the transmission loss of 10%, what would be the power that can be generated in MW?

[2 marks : 2011]

Solution:

(i) Mean volume of precipitation over the catchment

$$V = 77000 \times 10^4 \times 0.95 = 731.5 \times 10^6\ m^3$$

if only 25% of precipitation reaches the basin outlet then volume of runoff that reaches outlet

$$= \frac{25}{100} \times 731.5 \times 10^6$$

$$\Rightarrow \quad \text{Mean flow rate } Q = \frac{\left(\frac{25}{100} \times 731.5 \times 10^6\right)}{365 \times 24 \times 60 \times 60} = 5.79\ m^3/s$$

(ii) Given:

Head available at site, H = 150 m

and Transmission loss = 10%

or Efficiency = 90%

$$\text{Power generated} = (\gamma_w QH)\eta = \frac{(9810 \times 5.79 \times 150)}{10^6} \times 0.9 = 7.67\ MW$$

3.10 Rainfall over a basin in three consecutive hours are 4 cm, 5 cm and 3 cm respectively. Estimate the surface runoff from the basin assuming negligible surface retention and evaporation losses. The infiltration loss can be estimated using the following Horton's equation

$$f = 1.2 + 4.2e^{-2.5t}$$

Here f = infiltration in cm/hr and t = time in hour from start of rainfall.

[8 marks : 2013]

Solution:

$$f(\text{cm/hr}) = 1.2 + 4.2\ e^{-2.5t(\text{hr})}$$

Surface retention and evaporation losses are negligible

$\therefore$ Runoff = Rainfall – Infiltration

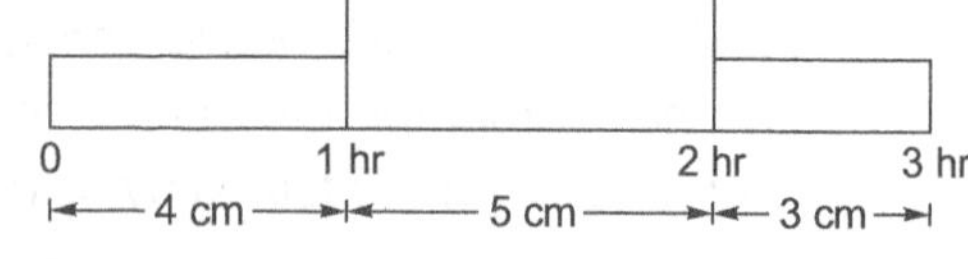

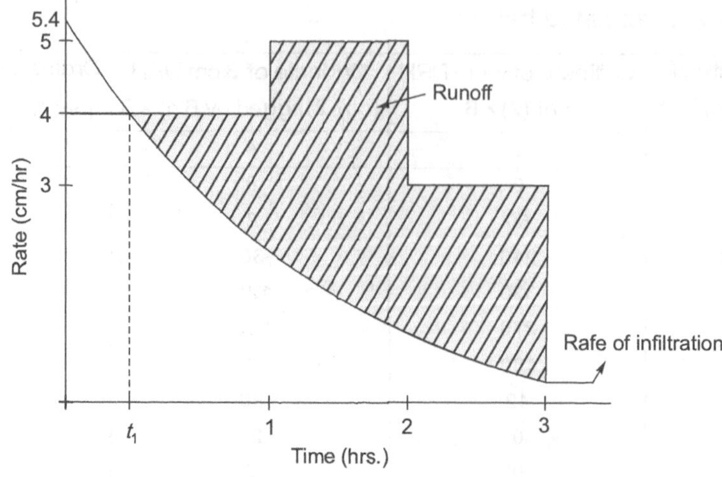

$$f_{at}\ t = 0 \text{ hr} = 1.2 + 4.2\, e^{-2.5\times 0}$$
$$= 1.2 + 4.2 = 5.4 \text{ cm/hr}$$

For t_1 $\Rightarrow$ $f = 1.2 + 4.2\, e^{-2.5\,t} = 4$

$\Rightarrow$ $t = t_1 = 0.162$ hr

Calculating total infiltration

= (infiltration from 0 to 0.162 hr) + (Infiltration from 0.162 hr to 3 hrs)

$$= 4\times(0.162-0)+\int_{0.162}^{3}\left(1.2+4.2e^{-2.5t}\right)dt$$

$$= (0.648\ cm)+\left[1.2t+\frac{4.2e^{-2.5t}}{(-2.5)}\right]_{0.162}^{3}$$

$$= (0.648) + 1.2\,(3-0.162) - 1.68(e^{-2.5\times 3} - e^{-2.5\times 0.162})$$
$$= 0.648 + 3.406 - 1.68(5.531\times 10^{-4} - 0.667)$$
$$= 0.648 + 3.406 + 1.12 = 5.174 \text{ cm}$$

Total rainfall = $4\times 1 + 5\times 1 + 3\times 1 = 12$ cm

Surface runoff = $12 - 5.174 = 6.826 \text{ cm} \approx 6.83$ cm

3.11 **A 12 hour rainfall with uniform intensity of 4 cm/hr produces a storm hydrograph of peak discharge 1000 m³/s. The abstractions of rainfall is at the rate of 1 cm/hr and base flow is 30 m³/s. Compute the peak discharge of 12 hour unit hydrograph.**

[4 marks : 2013]

Solution:

Duration of rainfall excess = 12 hours

Peak discharge = 1000 m^3/s

Rainfall intensity (i) = 4 cm/hr

$\therefore$ Total depth of rainfall $= \dfrac{4\text{ cm}}{\text{hr}}\times 12\text{hr} = 48\text{cm}$

Abstractions @ 1 cm/hr $= \dfrac{1\text{ cm}}{\text{hr}}\times 12\text{ hr} = 12\text{ cm}$

$\therefore$ Rainfall excess = 48 – 12 = 36 cm

Base flow = 30 m^3/s

$\therefore$ Peak of DRH = 1000 – 30 = 970 m^3/s

$\therefore$ Peak of 12 hr unit hydrograph $= \dfrac{970}{36} = 26.944\ m^3/s$

3.12 **(i) A storm with a 15.0 cm precipitation produced a direct runoff 8.7 cm. The time distribution of the storm is as follows:**

Time from start (in hour)	1	2	3	4	5	6	7	8
Incremental rainfall in each hour (in cm)	0.6	1.35	2.25	3.45	2.7	2.4	1.5	0.75

Estimate the ϕ index of the storm.

(ii) What is the difference in ϕ index and *w* index?

[6 + 2 = 8 marks : 2014]

Solution:

(i)

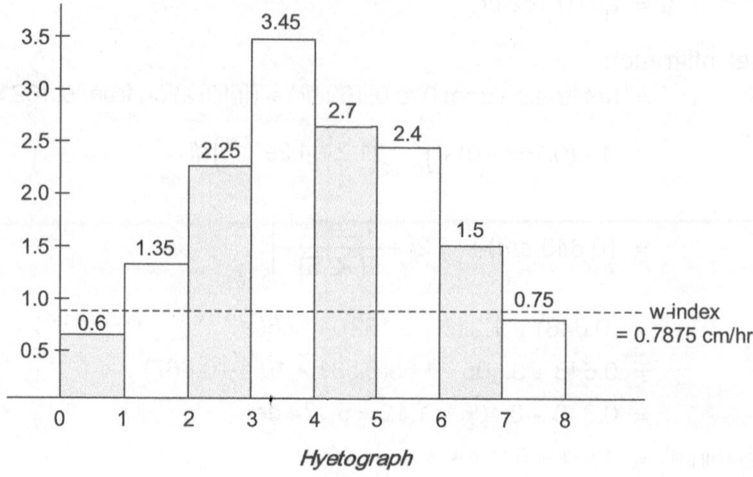

Hyetograph

$$P = 15.0 \text{ cm}$$

$$Q = 8.7 \text{ cm}$$

$$w\text{-index} = \text{Avg. Infiltration Rate during storm}$$

$$= \frac{P-Q-\Delta}{t} = \frac{15.0-8.7-0}{8} = 0.7875 \text{ cm/hr}$$

We know ϕ-index $\geq w$ - index

$$\phi\text{-index} = \text{Avg. Infiltration rate during rainfall excess}$$

Note: During railfall excess 0.6 cm/hr and 0.75 cm/hr does not contribute to runoff.

$$\phi\text{-index} = \frac{P-Q-(\text{Rainfall intensity below } w\text{-index} \times \text{duration})}{t-(\text{time of rainfall below } w\text{-index})}$$

$$= \frac{15.0-8.7-(0.6\times 1+0.75\times 1)}{8-1-1} = 0.825 \text{ cm/hr}$$

(ii) **Difference between ϕ and w-index**

W-index: It is the ratio of total infitration to the total time period of storm.

ϕ-index: It is the average rate of loss such that total rainfall volume in excess of that rate is equal total direct run-off volume.

3.13 **Given below are the observed flows from a storm of 4 h duration on a stream with a drainage area of 2000 km^2. Derive 4 h unit hydrograph. Assume a constant base flow of 150 m^3/sec**

Time (day)	Flow (m^3/sec)	Time (day)	Flow (m^3/sec)
1.	150	8	330
2.	1050	9	268
3.	880	10	230
4.	680	11	205
5.	570	12	180
6.	470	13	160
7.	400	14	150

[8 marks : 2015]

Solution:

Time (day)	Flow (m^3/s)	Base flow (m^3/s)	DRH (m^3/s)	UH (m^3/s)
1	150	150	0	0
2	1050	150	900	57.51
3	880	150	730	46.64
4	680	150	530	33.86
5	570	150	420	26.84
6	470	150	320	20.45
7	400	150	250	15.97
8	330	150	180	11.50
9	268	150	118	7.54
10	230	150	80	5.11
11	205	150	55	3.51
12	180	150	30	1.92
13	160	150	10	0.64
14	150	150	0	0
			$\Sigma = 3623$	

$$\text{Depth of runoff} = \frac{3623 \times 24 \times 60 \times 60}{2000 \times 10^6} \times 100 = 15.65 \text{ cm}$$

$$\text{Ordinate of UH} = \frac{\text{Ordinate of DRH}}{15.65}$$

3.14 The surface runoff from a flood on a drainage basin amounted to 5.0 cm. The area of the basin is 250 km^2. The equivalent uniform depth of rainfall on the drainage basin was 15 cm and the time distribution of the rainfall is given as follows. Calculate the ϕ-index:

Hour	08-09	09-10	10-11	11-12	12-13	13-14	Total
ppt. (mm)	15	20	50	20	30	20	155

[8 marks : 2015]

Solution:

Time for start (in hour)	1	2	3	4	5	6
Rainfall (in cm)	1.5	2	5	2	3	2
Intensity (cm/hr)	1.5	2	5	2	3	2

Total precipitation, $P = 155 \text{ mm} = 15.5 \text{ cm}$

Total runoff, $Q = 5 \text{ cm}$

$$W_{index} = \frac{P - Q}{t} = \frac{15.5 - 5}{6} = 1.75 \text{ cm/hr}$$

Since ϕ_{index} has to be somewhat more than W_{index}

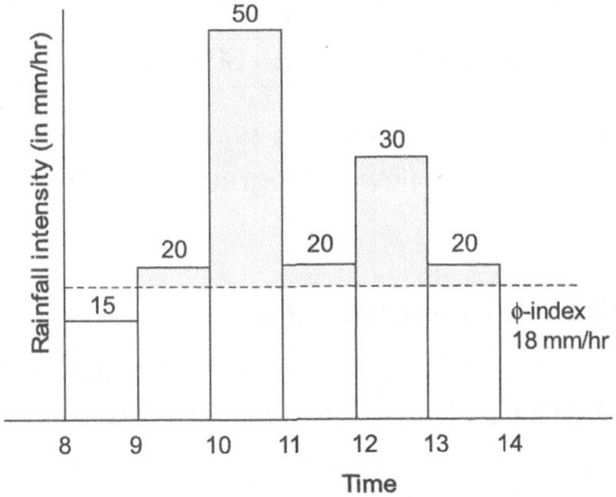

Note: During rainfall excess 1.5 cm/hr rainfall does not contribute to runoff

$$\phi_{index} = \frac{P - Q - (\text{Rainfall intensity below W-index} \times \text{Duration})}{t - (\text{Time of rainfall below W-index})}$$

$$\phi_{index} = \frac{15.5 - 5 - 1.5 \times 1}{6 - 1}$$

$$\phi_{index} = 1.8 \text{ cm/hr}$$

3.15 **What is hydrograph? Explain different types of hydrograph depending upon unit of time involved and their corresponding applications.**

[8 marks : 2016]

Solution:

Hydrograph: It is a continuous plot of instantaneous discharge versus time at a point in the catchment. It results from the combined interaction of various catchment and meteorological characteristics.

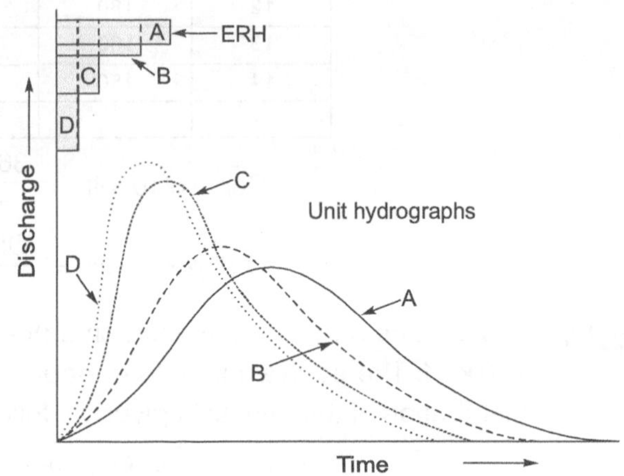

Unit hydrograph and different duration

Depending upon the unit of time involved, we have

1. **Annual Hydrograph:** It shows variation of daily, weekly or mean flow of any no. of successive days over a year.
2. **Monthly Hydrograph:** It shows the variation of daily mean flow over a month.
3. **Seasonal Hydrograph:** It shows the variation of the discharge in a particular season such as the monsoon season or dry season.
4. **Flood hydrograph or hydrographs due to storm:** It shows stream flow due to a storm over a catchment hydrograph 1, 2 and 3 as above are called long term hydrograph and used for long term studies like
   (a) Calculating the surface water potential of stream
   (b) Reservoir studies and
   (c) Drought studies

3.16 **What are the basic assumptions of a unit hydrograph? Six hour unit hydrograph of a watershed having a drainage area equal to 393 km² is as follows:**

Time (hour)	0	6	12	18	24	30	36	42	48
Unit hydrogrraph (cumec/cm)	0	1.8	30.9	85.6	41.8	14.6	5.5	1.8	0

For a storm over the watershed having excess rainfall of 5 cm for first six hours and 15 cm for second six hours, compute the stream flow hydrograph assuming constant base flow of 100 m³/sec.

[20 marks : 2018]

Solution:

The basic assumptions of a unit hydrograph are:

(i) **Time invariance:** The direct-runoff response to a given effective rainfall in a catchment is time invariant. This implies that the DRH for a given ER in a catchment is always the same irrespective of when it occurs.

(ii) **Linear Response:** The direct runoff response to the rainfall is assumed to be linear. It means principle of superposition may apply. So that we can find hydrograph of different duration from a given hydrograph.

Linear response means that if an input $x_1(t)$, causes an output $y_1(t)$ and an input $x_2(t)$, cause an output $y_2(t)$, then an input $x_1(t) + x_2(t)$ gives an output $y_1(t) + y_2(t)$

Also, if $x_2(t) = rx_1(t)$

Then $y_2(t) = ry_1(t)$

Hence, this assumption enable the method of superposition to be used to derive DRHs.

(iii) Rainfall is uniformly distributed thgoughout the area within its specified period of time.

(iv) The rainfall intensity is uniform over entire period of storm.

Now, given data:

Excess rainfall for first six hours,

$$ER_1 = 5 \text{ cm}$$

Excess rainfall for second six hours,

$$ER_2 = 15 \text{ cm}$$

Base flow, $BF = 100 \text{ m}^3/\text{s}$

6-hr UH is also given.

To find: Stream flow hydrograph.

Time (hrs)	UH (cumec/cm)	DRH_1 for $ER_1 = 5$ cm	DRH_1 for $ER_2 = 15$cm Lagged by 6hrs	$DRH_1 + DRH_2$	BF (m^3/s)	Stream flow hydrograph $= DRH_1 + DRH_2 + BF(m^3/s)$
0	0	0	—	0	100	100
6	1.8	9	0	9	100	109
12	30.9	154.5	27	181.5	100	281.5
18	85.6	428	463.5	891.5	100	991.5
24	41.8	209	1284	1493	100	1593
30	14.6	73	627	700	100	800
36	5.5	27.5	219	246.5	100	346.5
42	1.8	9	82.5	91.5	100	191.5
48	0	0	27	27	100	127
			0	0	100	100

3.17 **Define a unit hydrograph. Explain two basic assumptions made in the derivation of unit hydrograph. Following are the ordinates of a 4-hr unit hydrograph. Using this, derive the ordinates of a 12-hr unit hydrograph (do not plot the graph):**

Time (hr)	0	4	8	12	16	20	24	28	32	36	40	44
Ordinate of 4-hr UH	0	20	80	130	150	130	90	52	27	15	05	0

What are the uses and limitations of unit hydrograph?

[20 marks : 2019]

Solution:

Unit hydrograph : A unit hydrograph is defined as the hydrograph of direct runoff resulting from one unit depth of rainfall excess occurring uniformly over the basin and at a uniform rate for a specific duration.

The basic assumptions of a unit hydrograph are:

(i) **Time invariance:** The direct-runoff response to a given effective rainfall in a catchment is time invariant. This implies that the DRH for a given effective rainfall in a catchment is always the same irrespective of when it occurs.

(ii) **Linear Response:** The direct runoff response to the rainfall is assumed to be linear. It means principle of superposition may apply. So that we can find hydrograph of different duration from a given hydrograph.

Linear response means that if an input $x_1(t)$, causes an output $y_1(t)$ and an input $x_2(t)$, cause an output $y_2(t)$, then an input $x_1(t) + x_2(t)$ gives an output $y_1(t) + y_2(t)$

Also, if $x_2(t) = rx_1(t)$

Then $y_2(t) = ry_1(t)$

Hence, this assumption enable the method of superposition to be used to derive DRHs.

Time (h)	Ordinate of 4-h UH (m^3/s)	S-curve addition (m^3/s)	S-curve ordinate (m^3/s) (Col. 2+ Col. 3)	S-curve lagged by 12 h (m^3/s)	(Col. 4 - Col. 5)	Col. 6 / (12/4) = 12-h UH ordinates (m^3/s)
1	2	3	4	5	6	7
0	0	–	0	–	0	0
4	20	0	20	–	20	6.7
8	80	20	100	–	100	33.3
12	130	100	230	0	230	76.7
16	150	230	380	20	360	120.0
20	130	380	510	100	410	136.7
24	90	510	600	230	370	123.3
28	52	600	652	380	272	90.7
32	27	652	679	510	169	56.3
36	15	679	694	600	94	31.3
40	5	694	699	652	47	15.7
44	0	699	699	679	20	6.7
48		699	699	694	5	1.7
52			699	699	0	0

Uses of Unit Hydrographs : The unit hydrograph establish a relationship between *effective rainfall* and *direct runoff* for a catchment. This relationship is very useful in study of the hydrology of a catchment, as

(i) In the development of flood hydrograph for extreme rainfall magnitude

(ii) In extension of flood-flow records based on rainfall records

(iii) In development of flood forecasting and warning system based on rainfall

Limitation of unit hydrograph:

(i) Unit hydrograph assumes uniform distribution of rainfall over the catchment, and the intensity of rainfall is assumed constant for the duration of rainfall excess. In practice these two conditions are never strictly satisfied.

(ii) Unit hydrograph method can not be used for a catchment area greater than 5000 km^2 and less than 2 km^2.

(iii) Precipitation must be from rainfall only. Snow melt runoff can not be satisfactory represented by unit hydrograph.

(iv) The catchment should not have large storage (Ex. tank, pond etc.) which affects the linear relationship between storage and discharge.

3.18 What is ϕ index? How is it estimated? What are the factors that affect ϕ index?

[8 marks : 2020]

Solution:

ϕ-index: It is defined as the average rate of loss such that the volume of rainfall in excess of that rate will be equal to volume of direct runoff.

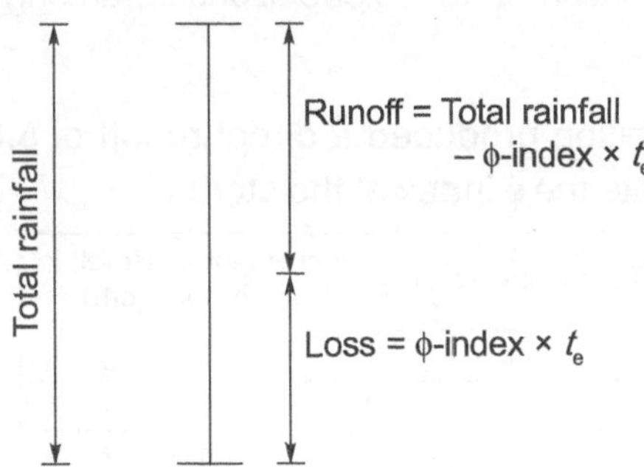

t_e = Time period of rainfall excess

OR

ϕ-index is defined as the ratio of total infiltration in which rainfall excess occur to the time period in which rainfall excess occur.

Procedure for calculation of ϕ-index:

Let us consider an example of a storm by means of hyetograph as given below:

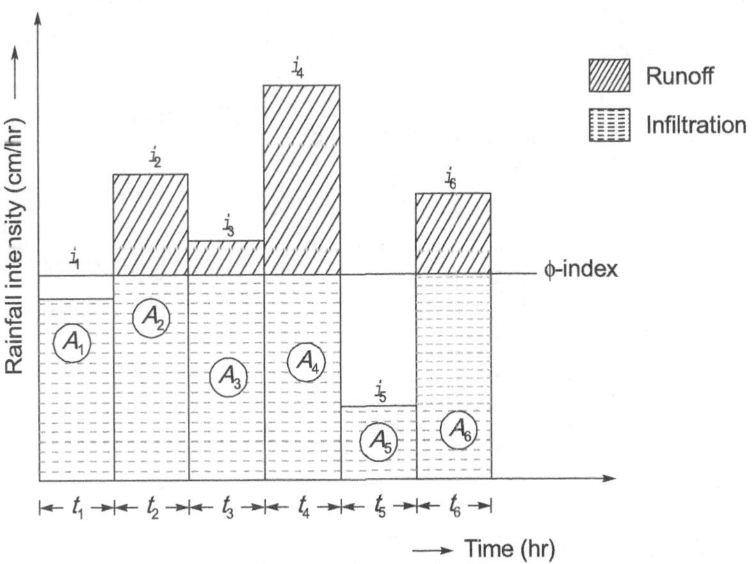

Mathematically,

$$\phi\text{-index} = \frac{\text{Total infiltration in which rainfall excess occur}}{\text{Time period in which rainfall excess occur}}$$

$$= \frac{(A_2 + A_3 + A_4 + A_6)}{(t_2 + t_3 + t_4 + t_6)} \text{ cm/hr}$$

Factors affecting φ-index:

(i) Vegetal coverage
(ii) Initial moisture conditions
(iii) Storm durations
(iv) Intensity of storm
(v Type of soil

To obtain the complete information on the interrelationship among above factors, a detailed study of catchment is necessary.

3.19 **A storm with 10 cm precipitation produced a direct runoff of 5.8 cm. The time distribution of the storm is given below. Estimate the φ index of the storm.**

Time from start (*h*)	Incremental rainfall in each hour (cm)
1	0.4
2	0.9
3	1.5
4	2.3
5	1.8
6	1.6
7	1.0
8	0.5

[12 marks : 2020]

Solution:

$$\text{Total infiltration} = \text{Total rainfall} - \text{Runoff}$$
$$= 10 - 5.8$$
$$= 4.2 \text{ cm}$$

$$\therefore \quad \text{w-index} = \frac{\text{Total infiltration}}{\text{Total duration of storm}} = \frac{4.2}{8} = 0.525 \text{ cm/hr}$$

Hyetograph for the given storm is as shown below:

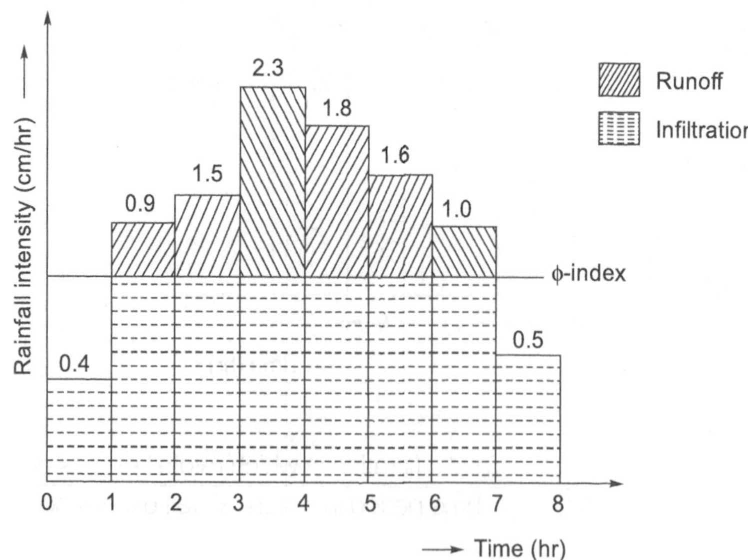

As $\quad$ φ-index > w-index

Hence storm of intensities 0.4 cm/hr and 0.5 cm/hr will not produce rainfall excess or runoff.

Let us consider the storms after 0.4 cm/hr and 0.5 cm/hr that will produce rainfall excess.

$$\therefore \quad \phi\text{-index} = \frac{\text{Total infiltration in which rainfall excess occur}}{\text{Time period in which rainfall excess occur}}$$

$$= \frac{\text{Total infiltraion} - \text{Infiltration in which no rainfall excess occur}}{t_e}$$

$$= \frac{4.2\text{ cm} - (0.4\times1 + 0.5\times1)}{6}$$

$$= \frac{4.2-0.9}{6} = \frac{3.3}{6} = 0.55 \text{ cm/hr}$$

Now, **check:**

Total runoff = Area of hatched portion

$\Rightarrow$ Total runoff = $(0.9-\phi\text{-index})\times1 + (1.5-\phi\text{-index})\times1 + (2.3-\phi\text{-index})\times1$
$+ (1.8-\phi\text{-index})\times1 + (1.6-\phi\text{-index})\times1 + (1-\phi\text{-index})\times1$
$= (0.9-0.55)\times1 + (1.5-0.55)\times1 + (2.3-0.55)\times1$
$+ (1.8-0.55)\times1 + (1.6-0.55)\times1 + (1.0-0.55)\times1$
$= 5.8$ cm = given runoff

Hence all right.

Our assumption is correct.

$\therefore \quad \phi\text{-index} = 0.55$ cm/hr

3.20 Calculate the infiltration rate and cumulative infiltration after 1.50 hours of rainfall of 4.0 cm/hr intensity on a silt loam soil with an initial effective saturation of 25%. Value of saturated hydraulic conductivity is 0.65 cm/hr, effective porosity of silt loam soil is 0.486, and wetting front soil suction head is 16.70 cm.

[20 Marks : 2021]

Solution:

For the silt loam

Effective saturation, $\sigma_e = 0.25$

Hydraulic conductivity, $k = 0.65$ cm/hr

Effective porosity, $\theta_e = 0.486$

Suction head, $\psi = 16.70$ cm

$\therefore \quad \Delta\theta = \theta_e(1-\sigma_e) = 0.486(1-0.25) = 0.3645$

$\therefore \quad |\psi|\Delta\theta = 16.70 \times 0.3645 = 6.0871$

From Green Ampt. equation,

Cumulative infiltration is given by

$$F - |\psi|\Delta\theta \ln\left(1+\frac{F}{|\psi|\Delta\theta}\right) = kt$$

$$\Rightarrow \quad F - 6.0871\ln\left(1+\frac{F}{6.0871}\right) = 0.65\,t$$

$$\therefore \quad F - 6.0871\ln\left(1+\frac{F}{6.0871}\right) = 0.65 \times 1.50$$

$$\Rightarrow \quad F - 6.0871 \ln\left(1 + \frac{F}{6.0871}\right) - 0.975 = 0$$

$$\Rightarrow \quad \frac{(F - 0.975)}{6.0871} = \ln\left(1 + \frac{F}{6.0871}\right)$$

Take $x = \dfrac{F}{6.0871}$

$$\therefore \quad x - 0.184 = \ln(1 + x)$$

$$\Rightarrow \quad x = 0.735$$

$$\therefore \quad F = 0.735 \times 6.0871$$

$$= 4.744 \text{ cm}$$

Infiltration capacity is given by, $f = k\left[\dfrac{|\psi|\Delta\theta}{F} + 1\right] = 0.65\left[\dfrac{6.0871}{F} + 1\right]$

$$= 0.65\left[\frac{6.0871}{3.8940} + 1\right]$$

$$= 1.666 \text{ cm/hr}$$

4. Floods, Flood Routing & Flood Channel

4.1 The amounts of water flowing from a certain catchment area at the proposed dam site are as tabulated below:

Determine:

(i) The minimum capacity of reservoir if water is to be used to feed the turbines of hydropower plant at an uniform rate and no water is to be spilled over.

(ii) The initial storage required to maintain the uniform demand as above.

[15 marks : 1999]

Month	Inflow ($\times 10^5$ m^3)
Jan	2.83
Feb	4.25
Mar	5.66
Apr	18.40
May	22.64
June	22.64
July	19.81
Aug	8.49
Sep	7.10
Oct	7.10
Nov	5.66
Dec	5.66

Solution:

Total inflow in the year = Summation of the given monthly inflows

$$= (2.83 + 4.25 + 5.66 + 18.40 + 22.64 + 22.64 + 19.81 + 8.49 + 7.10 + 7.10 + 5.66 + 5.66) \times 10^5$$

$$= 130.24 \times 10^5 \text{ m}^3$$

Since the hydropower plant is providing water at a uniform rate, therefore average monthly rate at which water can be withdrawn to avoid any wastage i.e. max. average monthly rate

$$= \frac{130.24 \times 10^5}{12} = 10.8533 \times 10^5 \text{ m}^3$$

Now, to calculate, the storage capacity, the computations are tabulated as follows:

Month	Monthly inflow (x 10^5 m^3)	Monthly outflow (x 10^5 m^3)	Monthly deficit (x 10^5 m^3) col.3 – col.2	Monthly surplus (x 10^5 m^3) col.2 – col.3	Cumulative deficit (x 10^5 m^3)	Cumulative surplus (x 10^5 m^3)	Net water available in reservoir (x 10^5 m^3)
(1)	(2)	(3)	(4)	(5)	(6)	(7)	(8)
Jan	2.83	10.8533	8.0233	—	—	—	—
Feb	4.25	10.8533	6.6033	—	—	—	—
Mar	5.66	10.8533	5.1933	—	19.82	—	(–) 19.82
Apr	18.40	10.8533	—	7.5467	—	—	—
May	22.64	10.8533	—	11.7867	—	—	—
June	22.64	10.8533	—	11.7867	—	—	—
July	19.81	10.8533	—	8.9567	—	40.08	(+) 20.26
Aug	8.49	10.8533	2.3633	—	—	—	—
Sep	7.10	10.8533	3.7533	—	—	—	—
Oct	7.10	10.8533	3.7533	—	—	—	—
Nov	5.66	10.8533	5.1933	—	—	—	—
Dec	5.66	10.8533	5.1933	—	20.26	—	0

(i) The highest value in col.6 and col.7 is 40.08×10^5 m^3 which represents the minimum storage capacity required to meet the demand without any spilling.

(ii) To compute the minimum initial storage, the net storage left in the reservoir with the above inflows and outflows is calculated. The maximum negative storage here works out to be 19.82×10^5 m^3. In order the reservoir fully meet the demand, there should be no negative storage in it, and in the limiting case, the maximum initial storage should be equal to zero. Hence the minimum initial storage in the reservoir should be 19.82×10^5 m^3.

4.2 **The regression analysis of a 30 year flood data at a point on a river yielded sample mean $\bar{x}$ = 1200 m^3/sec and standard deviation S_x = 650 m^3/sec. For what discharge would you design the structure at this point to provide 95% assurance that the structure would not fail in the next 50 years? Use Gumbel's method. The value of mean and standard deviation of the reduced variate for n = 30 are 0.53622 and 1.11238 respectively.**

[15 marks : 2000]

Solution:

Given

$$n = 30$$

$$\bar{y}_n = \text{reduced mean} = 0.53622$$

$$S_n = \text{reduced standard deviation} = 1.11238$$

$$\text{Reliability} = 95\%$$

$$\bar{x} = \text{mean of sample} = 1200 \text{ m}^3/\text{sec}$$

$$\sigma = \text{standard deviation of sample} = 650 \text{ m}^3/\text{sec}$$

$$\text{Reliability} = \left(1-\frac{1}{T}\right)^n$$

where n is number of years for which reliability is given i.e. 50 years.

$$\Rightarrow \quad \frac{95}{100} = \left(1-\frac{1}{T}\right)^{50}$$

$$\Rightarrow \quad T = 975.29 \text{ years}$$

We also know that

$$x_T = \bar{x} + K\sigma$$

and $$K = \frac{y_T - \bar{y}_n}{S_n}$$

But $$y_T = y_{975.29} = -\log_e \log_e \left(\frac{975.29}{975.29 - 1}\right) = 6.88222$$

$\therefore$ $$K = \frac{6.88222 - 0.53622}{1.11238} = 5.7049$$

$\therefore$ $$x_T = 1200 + (5.7049 \times 650) = 4908.19 \text{ m}^3/\text{sec}$$

4.3 The daily flows in a river for three consecutive years are given in the table by class intervals along with the number of days the flow belonged to this class. What are the 50% and 75% dependable flows for the river?

Daily mean discharge (m^3/s)	Number of days in each class interval		
	1981	1982	1983
100 - 90.1	0	6	10
90 - 80.1	16	19	16
80 - 70.1	27	25	38
70 - 60.1	21	60	67
60 - 50.	43	51	58
50 - 40.1	59	38	38
40 - 30.1	64	29	70
30 - 20.1	22	48	29
20 - 10.1	59	63	26
10 - negligible	54	26	13

[15 marks : 2002]

Solution:

With the given daily flows, we will calculate the annual yields for the given 3 years. Since daily discharge figures are given in range, mid value of range can be assumed to the flow discharge, which when multiplied by number of days will give yield during those days. Summation of yield values over 365 days will give annual yield.

Daily mean discharge (m^3/s)	Mean value of discharge	No. of days 1981	Yield (Mm^3)	No. of days 1982	Yield (Mm^3)	No. of days 1983	Yield (Mm^3)
(1)	(2)	(3)	(4)	(5)	(6)	(7)	(8)
100-90.1	95.05	0	0	6	49.27	10	82.12
90-80.1	85.05	16	117.57	19	139.62	16	117.57
80-70.1	75.05	27	175.08	25	162.11	38	246.40
70-60.1	65.05	21	118.03	60	337.22	67	376.56
60-50.1	55.05	43	204.52	51	242.57	58	275.87
50-40.1	45.05	59	229.65	38	147.91	38	147.91
40-30.1	35.05	64	193.81	29	87.82	70	211.98
30-20.1	25.05	22	47.61	48	103.89	29	62.76
20-10.1	15.05	59	76.72	63	81.92	26	33.81
10-Negligible	5.05	54	23.56	26	11.34	13	5.67
	Total	365	1186.55	365	1363.67	365	1560.65

$$\text{Yield (Mm}^3) = \text{Mean value of discharge} \times \text{Number of days for which the flow equalled} \times \frac{24 \times 60 \times 60}{10^6}$$

$$\text{Annual Yield (Mm}^3) = \text{Summation of Yield of each year}$$

Thus, the yields are arranged in descending order.

Year	Annual Yield (Mm^3)	Order (m)
1983	1560.65	1
1982	1363.67	2
1981	1186.55	3

Now, Total data points, $N = 3$

But
$$P = \frac{m}{N} \times 100$$

(i) **For 50% dependable flow, we get**

$$50 = \frac{m}{3} \times 100$$

$$\Rightarrow \quad m = 1.5$$

Since order number of 1.5 is not an integer, the mean value of yield corresponding to $m = 1$ and $m = 2$ is taken as the yield of 50% dependability

$$= \frac{1560.65 + 1363.67}{2} = 1462.16 \text{ Mm}^3$$

Hence annual flow with 50% dependability = 1462.16 Mm^3

$$\therefore \text{Daily discharge with 50\% dependability} = \frac{1462.16 \times 10^6}{365 \times 24 \times 60 \times 60} = 46.36 \text{ m}^3/\text{sec}$$

(ii) **For 75% dependable flow**

$$m = \frac{75}{100} \times 3$$

$$m = 2.25$$

2.25 is not an integer, thus yield is taken by interpolation

$$Q - 1363.67 = \frac{1186.55 - 1363.67}{3 - 2}(2.25 - 2) = 1319.39 \text{ Mm}^3$$

$$\therefore \text{Daily discharge with 75\% dependability} = \frac{1319.39 \times 10^6}{365 \times 24 \times 60 \times 60} = 41.837 \text{ m}^3/\text{sec}$$

4.4 Flood frequency computations for a flashy river at a point 50 km upstream of a bund site indicated the following:

Return Period T(years)	Peak Period (m^3/sec)
50	20,600
100	22,150

Estimate the flood magnitude in the river with a return period of 500 years through use of Gumbel's method.

[15 marks : 2003]

Solution:

We know that

$$x_T = \bar{x} + K\sigma$$

where $$K = \frac{y_T - \bar{y}_n}{S_n}$$

where $$y_T = -\log_e \log_e\left(\frac{T}{T-1}\right)$$

It may be noted that $\bar{y}_n$ and S_n remains same for one analysis.

Given data: $X_{50} = 20{,}600$ and $X_{100} = 22{,}150$

$$X_{500} = ?$$

$$y_{50} = -\log_e \log_e\left(\frac{50}{50-1}\right) = 3.90194$$

$$y_{100} = -\log_e \log_e\left(\frac{100}{100-1}\right) = 4.60015$$

$$y_{500} = -\log_e \log_e\left(\frac{500}{500-1}\right) = 6.21361$$

Now, we have

$$\bar{x} + \left(\frac{y_{50}}{S_n} - \frac{\bar{y}_n}{S_n}\right)\sigma = 20{,}600 \quad \text{... (i)}$$

$$\bar{x} + \left(\frac{y_{100}}{S_n} - \frac{\bar{y}_n}{S_n}\right)\sigma = 22{,}150 \quad \text{... (ii)}$$

Subtracting (i) from (ii), we get

$$\left(\frac{y_{100} - y_{50}}{S_n}\right)\sigma = 1550$$

$$\Rightarrow \quad (4.60015 - 3.90194)\frac{\sigma}{S_n} = 1550$$

$$\Rightarrow \quad \frac{\sigma}{S_n} = 2219.9625$$

Also, we have

$$x_{500} = \bar{x} + \left(\frac{y_{500} - \bar{y}_n}{S_n}\right)\sigma \quad \text{... (iii)}$$

Subtracting (ii) from (iii), we get

$$x_{500} - 22150 = (y_{500} - y_{100})\frac{\sigma}{S_n}$$

$$\Rightarrow \quad x_{500} = 22150 + (6.21361 - 4.60015) \times 2219.9625 = 25731.82 \text{ m}^3/\text{sec}$$

4.5 A hydraulic structure is sized for a 50 year recurrence interval design discharge. What is the risk that the flow capacity will be exceeded during any future 20 year period? What is the probability that the 50 year recurrence interval peak flow rate will be exceeded in the next 50 years?

[6 marks : 2005]

Solution:

Given recurrence time = 50 years = T

$\therefore$ The probability that the peak flow will be exceeded $= P = \frac{1}{T} = \frac{1}{50} = 0.02$

$\therefore$ Probability that the peak flow will not be exceeded = $q = 1 - P = 1 - 0.02 = 0.98$

Now, **risk that the flow capacity is exceeded in 20 years** is given by

$$R = 1 - q^{20} = 1 - (0.98)^{20} = 0.3324$$

Again **risk that the flow capacity is exceeded in 50 years** is given by

$$R = 1 - q^{50} = 1 - (0.98)^{50} = 0.6358$$

4.6 **Define recurrence interval or return period.**

[4 marks : 2005]

Solution:

Recurrence interval may be defined as the time interval during which any given storm is likely to be equalled or exceeded.

Thus, if we say that at a given station, the maximum precipitation of 20 cm has got a recurrence interval of 5 years, we mean that on this station, the chances of rainfall are such that once in five years, rain is likely to equal or exceed 20 cm. This does not mean that a rain equal to 20 cm or more will occur after every five year. It may occur twice in one set of five years and may not occur at all for consecutive eight years or so.

The recurrence interval is also known as return period. Its magnitude can be given by $T = 1/P$

where T is return period in years and P is the probability of occurrence of an event whose magnitude is equal to or in excess of a specified magnitude.

4.7 **The catchment area of a drainage basin is 2100 km^2. The length of main stream (L) is equal to 80 km. The distance along the main stream from the basin outlet to a point on the stream which is nearer to the centroid of the basin (L_c) is equal to 50 km. Compute the widths of the 3-hr synthetic unit hydrograph at 50% and 75% of peak discharge using Snyder's method.**

Take the coefficients

$C_t = 1.85$

$C_p = 0.45$

C_t = coefficient accounting catchment slope

C_p = dimensionless coefficient

[10 marks : 2006]

Solution:

Basin lag, $t_P = C_t(L \,.\, Lc)^{0.3}$

Here $C_t = 1.85, \quad L = 80 \text{ km}, \quad L_c = 50 \text{ km}$

$\therefore$ $t_P = 1.85 \times (80 \times 50)^{0.3}$

$\Rightarrow$ $t_P = 22.27 \text{ hrs}$

Now, a standard duration of effective rainfall in hrs is given by

$$t_r = \frac{t_P}{5.5}$$

$$\Rightarrow \quad t_r = \frac{22.27}{5.5} = 4.05 \text{ hrs}$$

When a non-standard rainfall duration t_R hours is adopted, instead of the standard value, t_r to derive a unit hydrograph the value of basin lag is affected. The modified basin lag is given by

$$t_P' = \frac{21}{22} t_P + \frac{t_R}{4}$$

Here $t_R = 3 \text{ hrs}$

$$\therefore \quad t_P' = \frac{21}{22} \times 22.27 + \frac{3}{4}$$

$$\Rightarrow \quad t_P' = 22.01 \text{ hrs}$$

Therefore, the peak discharge for a non-standard *ER* of duration t_R in m³/s is given by

$$Q_P = \frac{2.78\, C_P A}{t_P'}$$

where $C_P = 0.45, \quad A = 2100 \text{ km}^2, \quad t_P' = 22.01 \text{ hrs}$

$$\Rightarrow \quad Q_P = \frac{2.78 \times 0.45 \times 2100}{22.01} = 119.36 \text{ m}^3/\text{sec}$$

Now, the peak discharge per unit catchment area in m³/s/km² is given by

$$q = \frac{Q_P}{A} = \frac{119.36}{2100} = 0.0568 \text{ m}^3/\text{s/km}^2$$

∴ **The width (in time units) are given as**

$$W_{50} = \frac{5.87}{q^{1.08}} = \frac{5.87}{(0.0568)^{1.08}} = 129.9996 \text{ hrs.}$$

and
$$W_{75} = \frac{W_{50}}{1.75} = \frac{129.9996}{1.75} = 74.285 \text{ hrs.}$$

where W_{50} = width of unit hydrograph in hrs at 50% peak discharge

W_{75} = width of unit hydrograph in hrs at 75% peak discharge

Also, time base, $t_b = 72 + 3t_P' = 72 + 3 \times 22.01 = 138.03$ hrs

4.8 **The inflow hydrograph readings for a stream reach in m³/sec at 12 hourly intervals are as follows: 42, 45, 88, 272, 342, 288, 240, 198, 162, 133, 110, 90, 79, 68, 61, 56, 54, 51, 45, 45 and 42. The Muskingum coefficients for the stream reach are k = 36 hr and x = 0.15. Determine the attenuation in peak flow discharge and the time of peak outflow.**

[8 marks : 2007]

Solution:

Since k = 36 h and $2kx = 2 \times 36 \times 0.15 = 10.8$ h, Δt should be such that $k > \Delta t > 2kx$ i.e. 36 h > Δt > 10.8 h

In the present case Δt = 12 hr is selected to suit the given inflow hydrograph ordinate interval.

We know that
$$O_n = C_o I_n + C_1 I_{n-1} + C_2 O_{n-1}$$

where
$$C_0 = \frac{-kx + 0.5\,\Delta t}{k - kx + 0.5\,\Delta t} = \frac{-36 \times 0.15 + 0.5 \times 12}{36 - 36 \times 0.15 + 0.5 \times 12} = 0.0164$$

$$C_1 = \frac{kx + 0.5\,\Delta t}{k - kx + 0.5\,\Delta t} = \frac{36 \times 0.15 + 0.5 \times 12}{36 - 36 \times 0.15 + 0.5 \times 12} = 0.3115$$

$$C_2 = \frac{k - kx - 0.5\,\Delta t}{k - kx + 0.5\,\Delta t} = \frac{36 - 36 \times 0.15 - 0.5 \times 12}{36 - 36 \times 0.15 + 0.5 \times 12} = 0.6721$$

For the first time interval, 0 to 12 h

$I_1 = 42$ $\quad C_1 I_1 = 0.3115 \times 42 = 13.08$

$I_2 = 45$ $\quad C_0 I_2 = 0.0164 \times 45 = 0.74$

$C_0 = I_1$ $\quad\quad C_2 O_1 = 0.6721 \times 42 = 28.23$

$\therefore \quad\quad O_2 = C_0 I_2 + C_1 I_1 + C_2 O_1 = 0.74 + 13.08 + 28.23 = 42.05$

Peak of inflow hydrograph = 342 m^3/sec

Peak of outflow hydrograph = 231.13 m^3/sec

$\therefore$ Attenuation in peak flow discharge = 342 – 231.13 = 110.87 m^3/s.

The inflow hydrograph has a peak value at t = 48 h

The outflow hydrograph has a peak value at t = 84 h

$\therefore \quad\quad$ Time lag = 84 – 48 = 36 hrs.

Time(h)	I(m^3/s)	0.0164 I_2	0.3115 I_1	0.6721 O_1	O(m^3/s)
0	42				42
12	45	0.74	13.08	28.23	42.05
24	88	1.44	14.02	28.26	43.72
36	272	4.46	27.41	29.38	61.25
48	342	5.61	84.73	41.17	131.51
60	288	4.72	106.53	88.39	199.64
72	240	3.94	89.71	134.18	227.83
84	198	3.25	74.76	153.12	231.13
96	162	2.66	61.68	155.34	219.68
108	133	2.18	50.46	147.65	200.29
120	110	1.80	41.43	134.61	177.84
132	90	1.48	34.26	119.53	155.27
144	79	1.30	28.03	104.36	133.69
156	68	1.12	24.61	89.85	115.58
168	61	1.00	21.18	77.68	99.86
180	56	0.92	19.00	67.12	87.04
192	54	0.89	17.44	58.50	76.83
204	51	0.84	16.82	51.64	69.3
216	48	0.79	15.89	46.58	63.26
228	45	0.74	14.95	42.52	58.21
240	42	0.69	14.02	39.12	53.83

4.9 Determine the design flood discharge (allowing an increase of one third) for a bridge side with the following data: catchment area = 2×10^5 hectares, duration of storm = 8 hr, storm precipitation = 3 cm, time of concentration = 2 hr, gauged discharge for a past flood with average maximum daily rainfall of 18 cm was 3400 m^3/sec.

[8 marks : 2007]

Solution:

Given data: Catchment area = 2×10^5 hectares = $2 \times 10^5 \times 10^4$ m^2

Duration of storm = 8 hr

Storm precipitation = 3 cm

Time of concentration = 2 hr

For a past flood gauged discharge = 3400 m^3/sec

Intensity corresponding to past flood $= \dfrac{18 \text{ cm}}{1 \text{ day}} = \dfrac{18\times10^{-2} \text{ m}}{24\times60\times60 \text{ sec}} = 2.0833\times10^{-6}$ m/sec

Runoff formula, $Q = CiA$

Where, i = Intensity of rainfall

C = Runoff coefficient

For past flood, $Q = CiA$

$$3400 = C\times 2.0833\times10^{-6}\times2\times10^5\times10^4$$

$$C = 0.816013$$

Hence flood discharge for storm

$$Q = CiA$$

$$Q = 0.816013\times\frac{3\times10^{-2}}{8\times60\times60}\times2\times10^5\times10^4$$

$$= 1700.027 \text{ m}^3/\text{sec}$$

Hence design flood discharge $= 1700.027+\dfrac{1}{3}\times1700.027 = 2266.70 \text{ m}^3/\text{sec}$

4.10 **A flood series have mean equal to 300 m³/s and standard deviation as 50 m³/s. Compute the magnitude of 50 years flood using Gumbel and Chow methods.**

[10 marks : 2008]

Solution:

As per **Gumbels method**, the magnitude of flood is

$$x_T = \bar{x} + k\sigma$$

where $\bar{x}$ is mean value of hydrological data, σ is standard deviation and k is frequency factor

But for infinite sample size, $k = \dfrac{y_T - 0.577}{1.2825}$

Reduced variate, $y_T = -\log_e \log_e\left[\dfrac{T}{T-1}\right]$

where T is the return period or recurrence interval

$\therefore$ $y_T = -\log_e \log_e\left[\dfrac{50}{50-1}\right] = 3.902$

Thus, $k = \dfrac{y_T - 0.577}{1.2825} = \dfrac{3.902-0.577}{1.2825} = 2.5926$

$\therefore$ $x_T = 300 + 2.5926\times50 = 429.63 \text{ m}^3/\text{s}$

4.11 **Flood frequency computation yields the following results:**

Return Period (years)	Peak flood (m³/s)
50.0	20,500
100.0	25,400

Using Gumbel's method, estimate the flood for a return period of 150 years.

[4 marks : 2010]

Solution:

We know that, $x_T = \bar{x} + K\sigma$

and $$K = \frac{y_T - \bar{y}_n}{S_n}$$

where $$y_T = -\log_e \log_e\left(\frac{T}{T-1}\right)$$

But $\bar{y}_n$ and S_n are same for one analysis because total number of observations are fixed in one analysis.

$\therefore$ $$x_{50} = \bar{x} + \left(\frac{y_{50}}{S_n} - \frac{\bar{y}_n}{S_n}\right)\sigma$$

$\Rightarrow$ $$20500 = \bar{x} + \left(\frac{y_{50}}{S_n} - \frac{\bar{y}_n}{S_n}\right)\sigma \quad ...(i)$$

Similarly $$25400 = \bar{x} + \left(\frac{y_{100}}{S_n} - \frac{\bar{y}_n}{S_n}\right)\sigma \quad ...(ii)$$

Subtracting (i) from (ii), we get

$$\left(\frac{y_{100}}{S_n} - \frac{y_{50}}{S_n}\right)\sigma = 4900 \quad ...(iii)$$

But $$y_{100} = -\log_e \log_e\left(\frac{100}{100-1}\right) = 4.6$$

$$y_{50} = -\log_e \log_e\left(\frac{50}{50-1}\right) = 3.9$$

Substituting values of y_{50} and y_{100} in (iii), we get

$$(4.6 - 3.9) \times \frac{\sigma}{S_n} = 4900$$

$\Rightarrow$ $$\frac{\sigma}{S_n} = 7000 \quad ...(iv)$$

Now $$x_{150} = \bar{x} + \left(\frac{y_{150}}{S_n} - \frac{y_n}{S_n}\right)\sigma \quad ...(v)$$

Subtracting (ii) from (v), we get

$$x_{150} - 25400 = (y_{150} - y_{100}) \times \frac{\sigma}{s_n}$$

But $$y_{150} = -\log_e \log_e\left(\frac{150}{150-1}\right) = 5$$

$\therefore$ $$x_{150} = 25400 + (5 - 4.6) \times 7000 \quad \text{[from (iv)]}$$

$\Rightarrow$ $$x_{150} = 25400 + 2800$$

$\Rightarrow$ $$x_{150} = 28200 \text{ m}^3/\text{s}$$ **(Ans.)**

4.12 **Analyses of the maximum annual flood over the past 150 years in a small river indicates the following cumulative distribution**

Number	1	2	3	4	5	6	7	8	9	10	11
Q	0	25	50	75	100	125	150	175	200	225	250
$P(x < x_n)$	0	0.19	0.35	0.52	0.62	0.69	0.88	0.92	0.95	0.98	1.00

Estimate 10, 50, 100 years flood.

[5 marks : 2011]

Solution:

Probability of excedence $P(x > x_n)$ of:

$$10 \text{ year flood} = \frac{1}{10} = 0.1$$

$$50 \text{ year flood} = \frac{1}{50} = 0.02$$

$$100 \text{ year flood} = \frac{1}{100} = 0.01$$

Probability of non exceedence $P(x < x_n)$ of

$$10 \text{ year flood} = 1 - 0.1 = 0.9$$

$$50 \text{ year flood} = 1 - 0.02 = 0.98$$

$$100 \text{ year flood} = 1 - 0.01 = 0.99$$

from the given table using interpolation

$$10 \text{ year flood} = 150 + \frac{(175-150)}{(0.92-0.88)} \times (0.9 - 0.88) = 162.5 \text{ m}^3/\text{s}$$

$$50 \text{ year flood} = 225 \text{ m}^3/\text{s}$$

$$100 \text{ year flood} = 225 + \frac{(250-225)}{(1-0.98)} \times (0.99 - 0.98) = 237.5 \text{ m}^3/\text{s}$$

4.13 The working life of dam built to store irrigation requirement is expected to be 100 yrs. The spillway capacity is designed to accommodate the peak flood having a return period of 500 yrs. Calculate the risk of failure of the dam.

[4 marks : 2012]

Solution:

Risk is given by

$$\bar{R} = 1 - (1 - P)^n$$

$$\bar{R} = 1 - \left(1 - \frac{1}{T}\right)^n$$

where P = Probability = $\frac{1}{T}$

T = Return period

Given: n = 100 years

T = 500 years

$$\therefore \quad \bar{R} = 1 - \left(1 - \frac{1}{500}\right)^{100} = 0.1814$$

$\therefore$ The risk of failure of dam = 18.14%

4.14 Annual maximum flood data in a river at a station have been processed to estimate the maximum flood for different return periods using the Gumbel's method. If the estimated maximum flood for return periods 100 and 50 years are 450 m³/s and 400 m³/s, respectively, estimate the flood discharge for return period of 500 years.

[10 marks : 2013]

Solution:

Flood discharge (x_{100}) = 450 m³/s for 100 years return period.

and Flood discharge (x_{50}) = 400 m³/s for 50 years return period

$$\therefore \quad x_{100} = \bar{x} + k_{100}\sigma \quad \text{...(i)}$$

$$x_{100} = \bar{x} + (y_1 - \bar{y}_n)\left(\frac{\sigma}{s_n}\right)$$

$$x_{50} = \bar{x} + k_{50}\sigma \quad \text{...(ii)}$$

$$\therefore \quad (k_{100} - k_{50})\sigma = x_{100} - x_{50} = 450 - 400 = 50 \quad \text{...(iii)}$$

Also $k_T = \dfrac{y_T}{s_n} - \dfrac{\bar{y}_n}{s_n}$ Where s_n and $\bar{y}_n$ are constants.

$$\therefore \quad \left(\frac{y_{100}}{s_n} - \frac{\bar{y}_n}{s_n} - \frac{y_{50}}{s_n} + \frac{\bar{y}_n}{s_n}\right)\sigma = 50 \quad \text{(Substuting } k_T \text{ in (iii)}$$

$$\Rightarrow \quad (y_{100} - y_{50})\frac{\sigma}{s_n} = 50 \quad \text{...(iv)}$$

But $y_T = -\left[ln.ln\left(\dfrac{T}{T-1}\right)\right]$

$$\therefore \quad y_{100} = -\left[ln.ln\frac{100}{100-1}\right] = -\left[ln.ln\frac{100}{99}\right] = 4.60015$$

and $y_{50} = -\left[ln.ln\dfrac{50}{50-1}\right] = 3.90194$

Substuting y_{100} and y_{50} in (iv)

$$(4.60015 - 3.90194)\frac{\sigma}{s_n} = 50$$

$$\Rightarrow \quad \frac{\sigma}{s_n} = 71.6117$$

For return period $T = 500$ years

$$y_{500} = \left[-ln.ln\left(\frac{500}{500-1}\right)\right] = 6.21361$$

$$\therefore \quad (y_{500} - y_{100})\frac{\sigma}{s_n} = x_{500} - x_{100}$$

$$\Rightarrow \quad [6.21361 - 4.60015](71.6117) = x_{500} - 450$$

$$\Rightarrow \quad x_{500} = 565.54 \text{ m}^3/\text{s}$$

Thus discharge corresponding to return period of 500 years

$$= 565.54 \text{ m}^3/\text{s} = 566 \text{ m}^3/\text{s (say)}$$

Alternative Solution:

$$x_T = u + \alpha y_T$$

$$\therefore \quad x_{100} = u + \alpha y_{100}$$

$$\Rightarrow \quad 450 = u + \alpha(4.60015) \quad \text{...(i)}$$

Similarly $x_{50} = u + \alpha y_{50}$

$$\Rightarrow \quad 400 = u + \alpha(3.90194) \quad \text{...(ii)}$$

Solving eq. (i) and (ii)

$$\alpha = 71.6117$$

and $u = 120.5754$

$$\therefore \quad x_{500} = u + \alpha y_{500} = 120.5754 + 71.6117\,(6.21361) = 565.54 \text{ m}^3/\text{s}$$

4.15 Inflow and outflow hydrographs of a channel reach are triangular in shape and are plotted simultaneously as shown in the figure. The peak of inflow hydrograph is 10000 m^3/hr and occurs after one hour from the starting. The base is 96 hours. Similarly the peak of the outflow hydrograph is 8000 m^3/hr and falls on recession of inflow hydrograph.

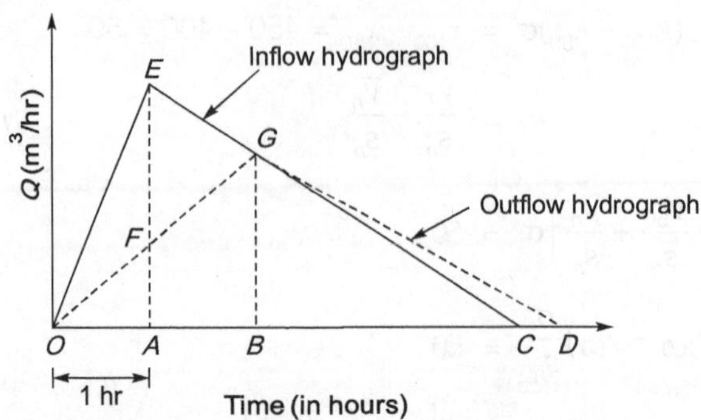

Determine the channel storage after one hour (i.e. under the peak of inflow hydrograph) and the maximum storage and the time at which it occurs along with the principle on which these are computed.

[8 marks : 2014]

Solution:

(i) Channel storage after one hour

By the property of similar triangle,

In ΔCAE and ΔCBG

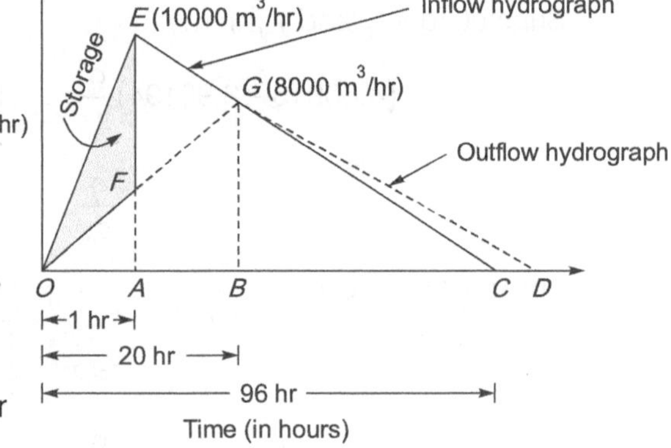

$$\frac{AC}{AE} = \frac{BC}{BG}$$

$$\Rightarrow \quad \frac{95}{10000} = \frac{BC}{8000}$$

$$\therefore \quad BC = \frac{95 \times 8000}{10000} = 76\,\text{hr}$$

Also in similar triangles, ΔOBG and ΔOAF

$$\frac{BG}{OB} = \frac{AF}{OA}$$

$$\Rightarrow \quad \frac{8000}{(96-76)} = \frac{AF}{1}$$

$$\therefore \quad AF = \frac{8000}{20} = 400\,\text{m}^3/\text{hr}$$

Channel storage after one hour,

$$\Delta S = \text{Area of } \Delta OFE$$

$$= \text{Area of } \Delta OAE - \text{Area of } \Delta OAF$$

$$= \left[\frac{1}{2} \times 1 \times 10000\right] - \left[\frac{1}{2} \times 1 \times 400\right] = 4800\,\text{m}^3$$

(ii) **In case of channel reach,** storage is increasing till peak inflow occurs and it continuous until outflow rises to equal to the receding inflow. At this time the flood crest occupies a central position in the reach. At the instant of maximum storage, the net change in storage equals zero and inflow equal to outflow.

At t = 20 hr (Point F)

Inflow equal to outflow, thus maximum storage occurs at 20 hr duration

$$\text{Storage} = \text{Inflow} - \text{Outflow}$$
$$= \text{Area of } OAE + \text{Area of } ABG - \text{Area of } OBG$$
$$= \frac{1}{2}\times 1\times 10000 + \frac{1}{2}\times 19(10000+8000) - \frac{1}{2}\times 8000\times 20$$
$$= 5000 + 171000 - 80000 = 96000 \text{ m}^3$$

4.16 **The inflow hydrograph a river reach is given in the table below. Determine the outflow hydrograph from this reach if coefficient K = 2.4 h, X = 0.15. Initial outflow in the reach in 85 m^3/s. Routing time step Δt = 1 h.**

S.No.	Routing Period (hr)	Inflow (m^3/s)
1	1	93
2	2	137
3	3	208
4	4	320
5	5	442
6	6	546
7	7	630
8	8	678
9	9	691
10	10	675
11	11	634
12	12	571
13	13	477
14	14	390
15	15	329
16	16	247
17	17	184
18	18	134
19	19	108
20	20	90

[10 Marks : 2021]

Solution:

Given: Muskingum coefficient

k = 2.4 hr, x = 0.15

$$C_0 = \frac{-kx+0.5\Delta t}{k(1-x)+0.5\Delta t} = \frac{-2.4\times 0.15+0.5\times 1}{2.4(1-0.15)+0.5\times 1} = \frac{0.14}{2.54} = 0.055$$

$$C_1 = \frac{kx+0.5\Delta t}{k(1-x)+0.5\Delta t} = \frac{2.4\times 0.15+0.5\times 1}{2.4(1-0.15)+0.5\times 1} = \frac{0.86}{2.54} = 0.339$$

$$C_2 = \frac{k(1-x)0.5\Delta t}{k(1-x)+0.5\Delta t} = \frac{2.4(1-0.15)-0.5\times 1}{2.4(1-0.15)+0.5\times 1} = \frac{1.54}{2.54} = 0.606$$

Time (hr)	Inflow (m³/sec)	Outflow (m³/sec)
1	$I_1 = 93$	$Q_1 = 85$
2	$I_2 = 137$	$Q_2 = 90.57$
3	$I_3 = 208$	$Q_3 = 112.77$
4	$I_4 = 320$	$Q_4 = 156.45$
5	$I_5 = 442$	$Q_5 = 227.59$
6	$I_6 = 546$	$Q_6 = 317.79$
7	$I_7 = 630$	$Q_7 = 412.32$
8	$I_8 = 678$	$Q_8 = 500.73$
9	$I_9 = 691$	$Q_9 = 571.29$
10	$I_{10} = 675$	$Q_{10} = 617.58$
11	$I_{11} = 634$	$Q_{11} = 637.95$
12	$I_{12} = 571$	$Q_{12} = 632.93$
13	$I_{13} = 477$	$Q_{13} = 603.36$
14	$I_{14} = 390$	$Q_{14} = 548.79$
15	$I_{15} = 329$	$Q_{15} = 482.87$
16	$I_{16} = 247$	$Q_{16} = 417.74$
17	$I_{17} = 184$	$Q_{17} = 347$
18	$I_{18} = 134$	$Q_{18} = 280.03$
19	$I_{19} = 108$	$Q_{19} = 221.06$
20	$I_{20} = 90$	$Q_{20} = 175.52$

$$Q_2 = C_0I_2 + C_1I_1 + C_2Q_1$$
$$= (0.055\times 137)+(0.339\times 93)+(0.606\times 85)$$
$$= 90.57 \text{ m}^3/\text{sec}$$

$$Q_3 = C_0I_3 + C_1I_2 + C_2Q_2$$
$$= (0.055\times 208)+(0.339\times 137)+(0.666\times 90.57)$$
$$= 112.77 \text{ m}^3/\text{sec}$$

Similarly we can identify, Q_4, Q_5 ... Q_{20}

■■■■

3 Water Resources Engineering

Revised Syllabus of ESE : *Multipurpose uses of Water, River basins and their potential; Irrigation systems, water demand assessment; Resources - storages and their yields; Water logging, canal and drainage design, Gravity dams, falls, weirs, Energy dissipaters, barrage Distribution works, Cross drainage works and head-works and their design; Concepts in canal design, construction & maintenance; River training, measurement and analysis of rainfall.*

---

1. Water Requirement of Crops

1.1 A sandy loam soil holds water at 140 mm/m depth between field capacity and permanent wilting point. The root depth of the crop is 30 cm and the allowable depletion of water is 35%. The daily water use by the crop is 5 mm/day. The area to be irrigated is 60 ha and water can be delivered at 28 litres per second. The surface irrigation application efficiency is 40%. There are no rainfall and ground water contribution.

Determine,

(i) allowable depletion depth between irrigations

(ii) frequency of irrigation

(iii) net application depth of water

(iv) volume of water required

(v) time to irrigate 4 ha plot

[15 marks : 1999]

Solution:

Given,

Moisture storing capacity of soil = 140 mm per m root zone depth

Root depth of crop = 30 cm

Allowable depletion of water = 35%

Consumptive use of crop = 5 mm per day

Area to be irrigated = 60 ha = 60×10^4 m^2

Rate at which water can be delivered = 28 litres per second

Irrigation application efficiency = 40%

(i) **Moisture storing capacity of soil** = Depth of water stored in root zone

∴ Depth of water stored in root zone = 140 mm per m root zone depth

$$= 140 \times \frac{30}{100} = 42 \text{ mm}$$

∴ Allowable depletion depth between irrigation $= \frac{35}{100} \times 42 = 14.7$ mm

(ii) Frequency of irrigation $= \dfrac{\text{Allowable depletion depth between irrigations}}{\text{Consumptive use of crop}}$

$$= \frac{14.7}{5} = 2.94 \text{ days}$$

(iii) Irrigation application efficiency $= \dfrac{\text{Water stored in root zone during irrigation}}{\text{Water delivered to the field}}$

$$\Rightarrow \quad \frac{40}{100} = \frac{14.7}{\text{Water delivered to the field}}$$

$\Rightarrow$ Water delivered to the field = 36.75 mm

This water delivered to the field is in terms of depth only i.e. it is the net application depth of water.

(iv) **Volume of water required** = Area to be irrigated × net application depth of water

$$= 60 \times 10^4 \times \frac{36.75}{1000} \times 10^3 \times 10^{-6} = 22.05 \text{ M litres}$$

(v) Volume of water required to irrigate 4 ha plot

$$= 4 \times 10^4 \times \frac{36.75}{1000} \times 10^3 \times 10^{-6} = 1.47 \text{ M litres}$$

$\therefore$ Time to irrigate 4 ha plot $= \dfrac{\text{Volume of water required}}{\text{Rate at which water is delivered}}$

$$= \frac{1.47 \times 10^6}{28} = 52{,}500 \text{ seconds} = 14.583 \text{ hrs.}$$

1.2 **For border strip method of irrigation, determine the time required to irrigate a strip of land of 0.04 hectares in area from a tubewell with a discharge of 0.02 cumec. The infiltration capacity of the soil may be taken as 5 cm/hr and the average depth of flow on the field as 10 cm. Also determine the maximum area that can be irrigated from this tube well.**

[15 marks : 2000]

Solution:

We know that the approximate time required to cover a given area with water can be calculated as

$$t = 2.303 \times \frac{y}{f} \times \log_{10}\left(\frac{Q}{Q - fA}\right)$$

where
Q = Discharge
y = Depth of water flowing over border strip
f = Rate of infiltration
A = Area of land strip to be irrigated

Given data:
$A = 0.04 \text{ ha} = 0.04 \times 10^4 \text{ m}^2 = 400 \text{ m}^2$
$Q = 0.02 \text{ cumecs} = 0.02 \times 60 \times 60 = 72 \text{ m}^3/\text{hr}$
$f = 5 \text{ cm/hr} = 0.05 \text{ m/hr}$
$y = 10 \text{ cm} = 0.1 \text{ m}$

$$\therefore \quad t = 2.303 \times \frac{y}{f} \times \log_{10}\left(\frac{Q}{Q - fA}\right)$$

$$= 2.303 \times \frac{0.1}{0.05} \times \log_{10}\left(\frac{72}{72 - 0.05 \times 400}\right) = 0.651 \text{ hrs} = 39.06 \text{ minutes}$$

Now, maximum area that can be irrigated, $A_{max} = \frac{Q}{f} = \frac{72}{0.05} = 1440 \text{ m}^2 = 0.144 \text{ ha}$

It may be noted that after irrigating this much of area, surface flow will stop and deep percolation will start.

1.3 **A loam soil has a field capacity of 25% and wilting coefficient of 10%. The dry unit weight of soil is 1.5 gm/cc. If the root zone depth is 60 cm, determine the storage capacity of the soil. Irrigation water is applied when moisture content falls to 15%. If the water application efficiency is 75%, determine the water depth required to be applied in the field.**

[10 marks : 2004]

Solution:

Given Data:

Field capacity, $F = 25\%$ Wilting coefficient, $\phi = 10\%$

Dry unit weight of soil, $\rho_s = 1.5$ gm/cc Root zone depth, $d = 60$ cm

Moisture storage capacity of the soil in the root zone depth, d is given by

$$= \frac{\gamma_s}{\gamma_w} \times d[F-\phi] = \frac{\rho_s}{\rho_w} \times d[F-\phi] = \frac{1.5}{1} \times 60 \times \left[\frac{25}{100} - \frac{10}{100}\right] = 13.5 \text{ cm}$$

Now, when the moisture content falls to 15%, the deficiency of water depth created will be given by

$$= \frac{\rho_s}{\rho_w} \times d \times [F - \text{fall in moisture content}] = \frac{1.5}{1} \times 60 \times \left[\frac{25}{100} - \frac{15}{100}\right] = 9 \text{ cm}$$

Hence, 9 cm depth of water is the net irrigation requirement.

$$\therefore \quad \text{Water application frequency} = \frac{\text{Net irrigation requirement}}{\text{Water depth required in the field}}$$

$$\Rightarrow \quad \frac{75}{100} = \frac{9}{\text{Water depth required in the field}}$$

$$\Rightarrow \quad \text{Water depth required in the field} = \frac{9 \times 100}{75} = 12 \text{ cm}$$

1.4 **Define duty, delta and base period of a crop and express the relationships connecting them. Explain briefly with the help of a diagram the concept of 'frequency of irrigation'.**

[10 marks : 2004]

Solution:

Duty: It is the area which can be irrigated in hectares when one cumec of discharge is available throughout the base period of a crop.

For example if 2 cumecs of discharge is available throughout the base period of a crop and 2000 hectares can be irrigated by this discharge, then duty of water for the given crop is

$$D = \frac{2000}{2} = 1000 \text{ ha/cumecs}$$

Delta: It is the total depth of water required by a crop during its entire base period.

For example if fodder requires 10 waterings in its base period and in every watering 10 cm depth of water is supplied, then total depth of water required is given by

Delta, $\Delta = 10 \times 10 = 100$ cm

Base Period: The time between the first watering of a crop at the time of its sowing to its last watering before harvesting is called the base period or the base of the crop.

Relation between duty, delta and Base Period of a crop:

Let 1 cumec discharge is supplied for entire base period of B days, then total volume of water applied to this crop

$$= B \times 24 \times 60 \times 60 \times 1 = 86400\, B \text{ m}^3$$

If this volume is used to irrigate D hectares of land and Δ is the total depth due to 1 cumec supply, then

$$86400\, B = D \times 10^4 \times \Delta$$

$$\therefore \quad D = \frac{8.64\, B}{\Delta}$$

where B is base period of crop in days, D is duty of crop in ha/cumec and Δ is delta of crop in m

Concept of frequency of irrigation: During irrigation the amount of water applied should be such that the moisture content of the soil is raised to its field capacity. After the irrigation is discontinued the moisture content of the soil is gradually reduced due to transpiration by the plants growing on the soil and evaporation from the soil surface. The total water (moisture) lost from the soil due to transpiration and evaporation is termed as evapotranspiration or consumptive use. The moisture content of the soil should however not be allowed to get depleted below the amount of the readily available moisture. As such when the soil moisture content reaches the lower limit of the readily available moisture, it should be replenished by irrigation and again raised to the field capacity of the soil as shown in figure below.

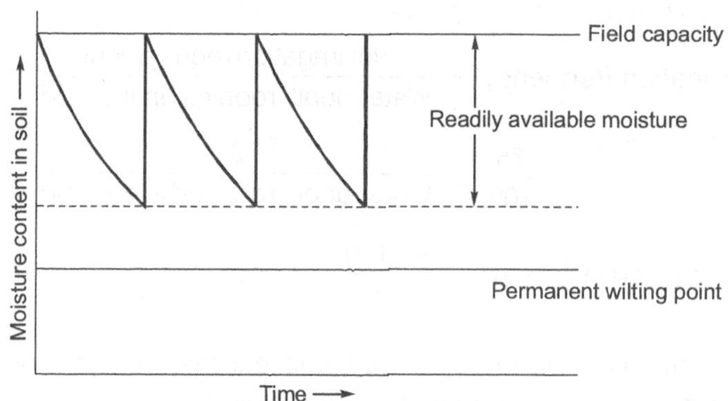

The frequency of irrigation depends on the amount of readily available moisture in the root zone of the soil and the rate of consumptive use. Thus if C_u is the rate of consumptive use expressed in terms of depth of moisture lost from the soil per day then the frequency of irrigation (f_w) is given by

$$f_w = \frac{d_w}{C_u} \text{ (days)}$$

where d_w is the minimum depth of water to be applied during irrigation.

1.5 **The irrigated area of a water course is 800 hectares. The intensity of rice in this area is 65%. The duration for transplantation of rice crop is 15 days and the total depth of water required by the crop is 60 cm, on the field, during transplantation period. The useful rain falling during the transplantation period is 15 cm. Find the duty of the irrigation water for the crop on the field during transplantation at the head of the field channel and at the head of distributary, if the loss of water in the water course is 25%. Also find the discharge in the water course.**

[10 marks : 2006]

Solution:

Give data: Total irrigated area of water course = 800 ha

Intensity of rice in this area = 65%

Transplantation period, B = 15 days

Depth of water required by the crop = 60 cm

Useful rainfall during transplantation period = 15 cm

Transit losses = 25%

$$= \frac{65}{100} \times 800 = 520 \text{ ha}$$

Irrigation required for transplantation of rice is given by

$$\Delta = \text{Depth of water required by the crop} - \text{Useful rainfall}$$
$$= 60 - 15 = 45 \text{ cm}$$

$\therefore$ **Duty at the head of distributary,**
$$D = \frac{8.64\,B}{\Delta} = \frac{8.64 \times 15}{45/100} = 288 \text{ ha/cumec}$$

Duty at the head of field channel (water course) $= 288 \times \left(1 - \frac{25}{100}\right) = 288 \times 0.75 = 216$ ha/cumec

Also, **Discharge in the water course** $= \dfrac{\text{Area required for rice transplantation}}{\text{Duty at the head of water course}}$

Area of rice transplantation = Intensity of rice × Area of water course

$$= \frac{520}{216} = 2.41 \text{ cumecs or } 2.41 \text{ m}^3/\text{sec}$$

1.6 **Determine the evapotranspiration and irrigation requirement for wheat, if the water application efficiency is 65% and the consumptive use coefficient for the growing season is 0.8 from the following data:**

Month	Mean Monthly Temp., °C	Monthly (%) of sunshine hrs	Effective rainfall, (cm)
November	18	7.20	2.60
December	15	7.15	2.80
January	13.5	7.30	3.50
February	14.5	7.10	2.00

[8 marks : 2007]

Solution:

The monthly consumptive use (in cm) is calculated by Blaney-Criddle formula i.e.

$$C_u = k \times \frac{p}{40} \times (1.8t + 32)$$

where k is consumptive use coefficient, p is monthly per cent of annual day light hours and t is mean monthly temperature in °C

If $\frac{p}{40}(1.8t + 32) = f$, then

$$C_u = kf$$

Seasonal consumptive use, $C_u = k\Sigma f$

The calculations are tabulated below:

Month	t	p	R_e	$f = \frac{p}{40}(1.8\,t + 32)$
November	18	7.20	2.60	11.6
December	15	7.15	2.80	10.5
January	13.5	7.30	3.50	10.3
February	14.5	7.10	2.00	10.3
			ΣR_e = 10.9 cm	Σf = 42.7 cm

∴ Seasonal consumptive use or evapotranspiration,

$$C_u = k \times \Sigma f = 0.8 \times 42.7 = 34.16 \text{ cm}$$

$$\text{Consumptive irrigation requirement} = C_u - \Sigma R_e = 34.16 - 10.9 = 23.26 \text{ cm}$$

$$\text{Field irrigation requirement} = \frac{CIR}{\eta_a} = \frac{23.26}{0.65} = 35.8 \text{ cm}$$

1.7 **For the data given in table and taking kor depth and kor period for wheat as 13.5 cm and 4 weeks respectively and for rice as 19.0 cm and 2.5 weeks respectively, compute the average discharge requirement and peak demand :**

Crop	Area under crop (hectare)	Total depth (cm)	Base period (day)	Average duty ha/cumec
Wheat	5000	37.50	140.0	3225.60
Rice	2500	120.00	120.0	864.00

[10 marks : 2008]

Solution:

Average discharge required for wheat

$$= \frac{\text{Area under wheat}}{\text{Average duty for wheat}} = \frac{5000}{3225.60} = 1.55 \text{ cumec}$$

Average discharge required for rice

$$= \frac{\text{Area under rice}}{\text{Average duty for rice}} = \frac{2500}{864.00} = 2.89 \text{ cumec}$$

Thus, the average discharge requirement is maximum of the two values i.e. 2.89 cumec **(Ans.)**

$$\text{Kor period for wheat} = 4 \text{ weeks} = 4 \times 7 = 28 \text{ days}$$

$$\text{Kor period for rice} = 2.5 \text{ weeks} = 2.5 \times 7 = 17.5 \text{ days}$$

Duty for wheat (for peak demand) $= \frac{864 \times \text{Kor period for wheat}}{\text{Kor water depth for wheat}} = \frac{864 \times 28}{13.5} = 1792$ ha/cumec

Duty for rice (for peak demand) $= \frac{864 \times \text{Kor period for rice}}{\text{Kor water depth for rice}} = \frac{864 \times 17.5}{19.0} = 795.8$ ha/cumec

Outlet discharge required for wheat (for kor demand)

$$= \frac{\text{Area under wheat}}{\text{Duty for wheat}} = \frac{5000}{1792} = 2.79 \text{ cumec}$$

Outlet discharge required for rice (for kor demand)

$$= \frac{\text{Area under rice}}{\text{Duty for rice}} = \frac{2500}{795.8} = 3.14 \text{ cumec}$$

Thus, the peak demand is the maximum of the two values i.e. 3.14 cumec **(Ans.)**

1.8 **What is a furrow? Give names of some row crops. State the advantages of sprinkler irrigation method over surface irrigation method.**

[5 marks : 2008]

Solution:

Furrows are narrow field ditches, excavated between rows of plants and carry irrigation water through them. Furrows vary from 8 cm to 30 cm in depth and may be as much as 400 m long.

Sugarcane, garden fruits, vegetables are some examples of row crops.

Advantages of sprinkler systems are as follows:

(i) Seepage losses which occur in earthen channels of surface irrigation methods, are completely eliminated.

(ii) Land levelling is not required, and thus avoiding removal of top fertile soil as happens in other surface irrigation methods.

(iii) No cultivation area is lost for making ditches as happens in surface irrigation methods.

(iv) In sprinkler system, the water is to be applied at a rate lesser than the infiltration capacity of the soil and thus avoiding surface runoff and its bad effects such as loss of water, washing of top soil, etc.

(v) Fertilizers can be uniformly applied, because they are mixed with irrigation water itself.

(vi) This method leaches down salts and prevents water logging or salinity.

1.9 **The depth of moisture in root zone at field capacity and permanent wilting point per m depth of soil are 0.5 m/m and 0.2 m/m respectively. Compute the field capacity and permanent wilting point. Take dry weight of soil as 13.73 kN/m³.**

[4 marks : 2010]

Solution:

Depth of water stored in the root zone in filling the soil upto field capacity is given as

$$h = \frac{\gamma_d dF}{\gamma_w}$$

$$\Rightarrow \quad 0.5 = \frac{13.73 \times 1 \times F}{9.81}$$

$$\Rightarrow \quad F = \frac{0.5 \times 9.81}{13.73}$$

$$\Rightarrow \quad F = 0.3572$$

$$\Rightarrow \quad F = 35.72\%$$

Thus, field capacity is 35.72%.

Depth of water stored in the root zone in filling the soil upto permanent wilting point is given as

$$h_1 = \frac{\gamma_d dP}{\gamma_w}$$

$$\Rightarrow \quad 0.2 = \frac{13.73 \times 1 \times P}{9.81}$$

$$\Rightarrow \quad P = \frac{0.2 \times 9.81}{13.73}$$

$$\Rightarrow \quad P = 0.1429$$

$$\Rightarrow \quad P = 14.29\%$$

Thus, the permanent wilting point is 14.29%.

1.10 Define the following :

1. Field irrigation requirement
2. Water application efficiency
3. Water storage efficiency
4. Water distribution efficiency
5. Consumptive use

[5 marks : 2011]

Solution:

1. **Field Irrigation requirement (FIR):** It is defined as the amount of water required to meet the net irrigation requirements plus the amount of water lost as surface runoff and through deep percolation.

$$\text{FIR} = \frac{\text{NIR}}{\eta_a}$$

η_a =Factor accounting for the loss of irrigation water by surface runoff and through deep percolation.

2. **Water application efficiecy (η_a):** It is defined as the ratio of the quantity of water stored in the root zone of the plants to the quantity of water delivered to the field.

$$\eta_a = \frac{W_s}{W_f} \times 100$$

3. **Water Storage Efficiency (η_s):** It is defined as the ratio of the quantity of water stored in the root zone during irrigation to the quantity of water needed to bring the moisture content of the soil to the field capacity. Thus if W_s is the quantity of water stored in the root zone during irrigation and W_n is the quantity of water needed to bring the moisture content of the soil to the field capacity.
(i.e., W_n = Field capacity – Available moisture in the soil prior to irrigation), then

$$\eta_s = \frac{W_s}{W_n} \times 100$$

4. **Water Distribution Efficiency (η_d):** Water distribution efficiency evaluates the degree to which water is uniformly distributed throughout the root zone during irrigation and hence it is also known as uniformity coefficient. The water distribution efficiency provides a measure for comparing different methods of irrigation.

   It is determined from the following expression in which y is the average numerical deviation in depth of water stored from the average depth of water d stored in the root zone during irrigation.

$$\eta_d = \left[1 - \frac{y}{d}\right] \times 100$$

5. **Consumptive use:** Consumptive use of water or evapotranspiration is defined as the total quantity of water used by the vegetative growth of a given area in transpiration and building of plant tissue, and that evaporated from the adjacent soil in the area in any specified time. Consumptive use is expressed in hectare-metres per hectare or depth of water in metres for specified periods such as days, months or crop growing season.

1.11 **What is crop calendar? Illustrate the same.** **[3 marks : 2011]**

Solution:

A Crop Calendar provides information about sowing and planting seasons and agronomic practices of the crops grown by farmers in a particular agro-ecological zone. It is a tool developed to assist farmers, extension workers, civil society and the private sector to be able to access and make available quality seeds of specific crop varieties for a particular agro-ecological zone at the appropriate sowing/planting season. It can be used by development-aid personnels in the planning and implementation of seed relief and rehabilitation activities following natural or human-led disasters. Furthermore, the Crop Calendar can serve as a quick reference tool in selecting crop varieties to adapt to changing weather patterns accelerated by climate change.

1.12 **Determine the discharge of a distributary at the tail end from the following data:**
Gross Command Area (GCA) = 20000 ha
Cultivable Command Area (CCA) = 70%
Losses beyond the tail end = 1 m^{3}s^{-1}

Kharif	Rice	Irrigation intensity 15%	Kor depth 19 cm	Kor period 2.5 weeks
Rabi	Wheat	Irrigation intensity 30%	Kor depth 13.5 cm	Kor period 4 weeks
	Sugarcane	Irrigation intensity 10%	Kor depth 16.5 cm	Kor period 4 weeks

[7 marks : 2011]

Solution:

Given:

Gross command area = 20000 ha

Cultivable command area i.e. area on which actual cultivation is done

$$= \frac{70}{100} \times 20000 = 14000 \text{ ha}$$

The discharge of the distributary at tail end will depend upon the type of crop i.e. kharif plus sugarcane or Rabi plus sugarane (Sugarcane is grown throughout the year). The maximum of the discharges required for two combination should be the discharge capacity of the tributary.

Kharif:

Kor depth = 19 cm

Kor period = 2.5 weeks

Duty of water for kor period of 2.5 weeks; using

$$D = \frac{8.64\,B}{\Delta}$$

$$D_k = \frac{8.64 \times 2.5 \times 7}{0.19} = 795.79 \text{ ha/cumec}$$

Area to be irrigated at 15% irrigation intensity = 0.15 × 14000 = 2100 ha

$$\text{Outlet discharge required for kharif} = \frac{2100}{795.79} = 2.64 \text{ cumec}$$

Outlet discharge including losses beyond tail end = Q_k = 2.64 + 1 = 3.64 cumec

Rabi:

$$\text{Kor depth} = 0.135 \text{ m}$$
$$\text{Kor period} = 4 \text{ weeks}$$
$$\text{Area to be irrigated} = \frac{30}{100} \times 14000 = 4200 \text{ ha}$$
$$\text{Duty of water required} = D_R = \frac{8.64 \times (4 \times 7)}{0.135} = 1792 \text{ ha/cumec}$$
$$\text{Outlet discharge required} = \frac{4200}{1792} = 2.34 \text{ cumec}$$
$$\text{Outlet discharge including losses} = Q_R = 2.34 + 1 = 3.34 \text{ cumec}$$

Sugarcane:

$$\text{kor depth} = 0.165 \text{ m}$$
$$\text{kor period} = 4 \text{ weeks}$$
$$\text{Area to be irrigated} = \frac{10}{100} \times 14000 = 1400 \text{ ha}$$
$$\text{Duty of water required} = D_s = \frac{8.64 \times (4 \times 7)}{0.165} = 1466.18 \text{ ha/cumec}$$
$$\text{Outlet discharge required} = Q_s = \frac{1400}{1466.18} = 0.955 \text{ cumec}$$

Excluding losses consideration here since sugarcane will be grown in combination with either Kharif or Rabi.

For, Sugarcane + Kharif : Discharge = 4.595 cumec

Sugarcane + Rabi : Discharge = 4.295 cumec

So, **Discharge of distributary** = 4.595 cumec

1.13 **The intensity of irrigation is 20% for wheat and 10% for rice, where cultural commanded area of a distributary is 6000 ha. The Kor period for wheat is 3 weeks and for rice 2 weeks. Find the total outlet discharge required excluding losses in the channel. Assume depth for Kor watering as 9 cm and 25 cm for wheat and rice respectively.**

[8 marks : 2014]

Solution:

Given:

Cultural commanded area (CCA) = 6000 ha

Area to be irrigated under wheat, A_w = Intensity of Irrigation for wheat × CCA

= 0.20 × 6000 = 1200 ha

Area to be irrigated under rice, A_R = Intensity of irrigation for Rice × CCA

= 0.1 × 6000 = 600 ha

Duty for Wheat and Rice:

Given: Kor period for wheat = 3 Weeks or 21 days

Depth of kor watering for wheat = 9 cm

Kor period for rice = 2 weeks or 14 days

Depth of Kor watering for rice = 25 cm

Using relation, $\Delta = 8.64\, B/D$

where, Δ = Kor water depth (in m)

B = days

D = Duty (hectare/ cumecs)

For Wheat:

$$D_w = \frac{8.64 \times 21}{0.09\text{m}} = 2016\,\text{ha/cumecs}$$

For Rice:

$$D_R = \frac{8.64 \times 14}{0.25} = 483.84\,\text{ha/cumecs}$$

Water required at the head of distributary to irrigate,

(i) Wheat: $$Q_W = \frac{A_W}{D_W} = \frac{1200}{2016} = 0.595\,\text{m}^3/\text{s or cumecs}$$

(ii) Rice: $$Q_R = \frac{A_R}{D_R} = \frac{600}{483.84} = 1.240\,\text{m}^3/\text{s or cumecs}$$

The required total outlet discharge is the maximum of the two i.e.

Q = 1.240 cumecs

1.14 **Three distributaries are used for irrigation. The details are given below. Find which one is less efficient.**

Discharge	15 m³/sec	20 m³/sec	25 m³/sec
CCA	15000 ha	25000 ha	30000 ha
Intensity of irrigation	60%	80%	60%
Base period	200 days (cotton)	120 days (wheat)	360 days (sugarcane)

[8 marks : 2015]

Solution:

Distributary (1)

$$Q_1 = 15\,\text{m}^3/\text{s}, \quad \text{CCA}_1 = 15000\text{ ha}, \quad I_1 = 60\%$$

$$B_1 = 200\text{ days}$$

$$\text{Area irrigated} = 15000 \times 0.6 = 9000\text{ ha}$$

$$\text{Duty, } D_1 = \text{Area irrigated per cumec of discharge}$$

$$= \frac{9000}{15} = 600\,\text{ha/cumec}$$

Distributary (2)

$$Q_2 = 20\,\text{m}^3/\text{s}, \quad \text{CCA}_2 = 25000\text{ ha}, \quad I_2 = 80\%$$

$$B_2 = 120\text{ days}$$

$$\text{Area irrigated} = 25000 \times 0.8 = 20000\text{ ha}$$

$$\text{Duty, } D_2 = \frac{20000}{20} = 1000\,\text{ha/cumec}$$

Distributary (3)

$$Q_3 = 25\,\text{m}^3/\text{s}, \quad \text{CCA}_3 = 30000\text{ ha}, \quad I_3 = 60\%$$

$$B_3 = 360\text{ days}$$

$$\text{Area irrigated, } A_3 = 30000 \times 0.6 = 18000\text{ ha}$$

$$\text{Duty, } D_3 = \frac{18000}{25} = 720\,\text{ha/cumec}$$

The distributary with least duty is least efficient. Therefore distributary (3) is least efficient.

1.15 **Define the following:**

1. **Potential evapotranspiration**
2. **Actual evapotranspiration**
3. **Field capacity**
4. **Permanent wilting point**

How can evapotranspiration be measured for a given vegetation type?

[4 marks : 2016]

Solution:

1. **Potential evapotranspiration:** When there is free availability of sufficient moisture to meet the needs of vegetation fully covering an area is called as potential evapotranspiration (PET).
2. **Actual evapotranspiration:** The real/actual evapotranspiration occurring in an area in a specific condition is called as actual evapotranspiration (AET).
3. **Field capacity:** Immediately after the rainfall, gravity water drains down through the soil and only certain water is retained within the soil due to molecular attraction offered by the soil particles. This water which retains in the soil against gravity expressed in terms of depth of water (assuming water to be spread on the catchment) is called as field capacity.
4. **Permanent wilting point:** It is that water content in the soil at which plant is no longer able to draw water from the soil and this is called as permanent wilting. Thus available moisture is the difference of field capacity and permanent wilting point i.e.

Available moisture = Field capacity – Permanent wilting point

Evapotranspiration need of a plant can be measured by an instrument called as lysimeter.

1.16 **A watercourse is designed to irrigate an area of 960 ha of rice. The transplantation of rice takes 15 days and during this the total depth of water required is 40 cm. Find the duty of irrigation water on the field if there is an effective rainfall of 12 cm. Also, find the duty and discharge at the head of the watercourse assuming losses of water as 20% in the watercourse.**

[10 marks : 2017]

Solution:

Given: Area of rice to be irrigated = Intensity of rice × Area of water course = 960 hectares

Useful rainfall during transplantation period = 12 cm

Depth of water required by the crop = 40 cm

Transplantation period, B = 15 days

Irrigation required for transplantation of rice is

Δ = Depth of water required by the crop-useful rainfall = 40 – 12 = 28 cm

$\therefore$ Duty at the head of distributary,

$$D = \frac{8.64B}{D} = \frac{8.64 \times 15}{28/100} = 462.86 \text{ ha/cumecs}$$

Duty at head of field channel (water course)

$$= 462.86 \times \left(1 - \frac{20}{100}\right) = 370.29 \text{ ha/cumecs}$$

Also, Discharge in the water course

$$= \frac{\text{Area required for rice transplantation}}{\text{Duty at head of water course}}$$

$$= \frac{960}{370.29} = 2.593 \text{ cumecs or m}^3\text{/sec}$$

1.17 Find the capacity of a reservoir for irrigating an area having GCA = 40000 ha. The cropping pattern consists of mainly the following crops:

Crop	Period	Base period (days)	Outlet factor (ha/cumec)	Intensity of cropping (%)	Crop ratio
Sugarcane	Oct-Sept	360	800	60	4
Cotton	May-Nov	180	1200	60	3
Wheat	Dec-April	120	1800	70	2
Gram	Dec-April	120	2000	70	3

The area has 25% non-culturable area.
Assume

(i) Time factor = 7/10

(ii) Extra allowance for period of peak water use = 20%

(iii) Conveyance losses = 20%

(iv) Reservoir losses = 10%

Also calculate design discharge and capacity factor of the main canal.

[20 marks : 2018]

Solution:

Monthly cropping pattern is shown below:

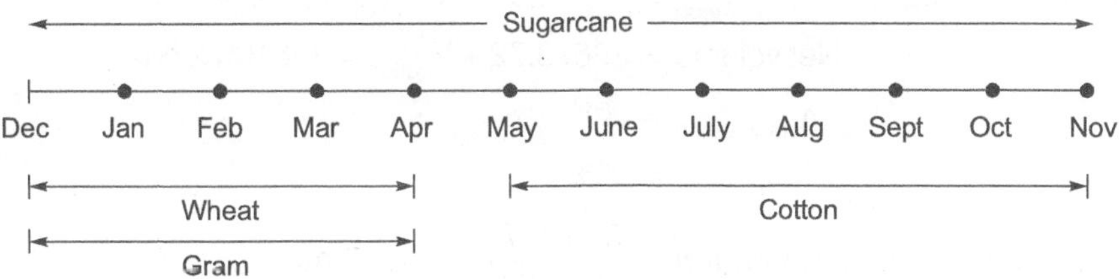

Given that:

$$GCA = 40000 \text{ hac}$$

$$CCA = \frac{100-25}{100} \times GCA = 30000 \text{ hac}$$

Crop-1: Sugarcane

$$\text{Area under irrigation} = \text{Intensity} \times \text{Crop ratio} \times \text{CCA}$$

$$= \frac{60}{100} \times \frac{4}{12} \times 30000$$

$$\text{Area} = 6000 \text{ hac}$$

We know,

$$\Delta = \left(\frac{8.64B}{D}\right)$$

where,

Δ = Delta in m

B = Base period in days

D = Duty (outlet factor) hac/cumec

$$\therefore \quad \Delta_S = \frac{8.64 \times 360}{800} = 3.888 \text{ m}$$

$$\therefore \quad \text{Volume required} = \text{Area}_S \times \Delta_S = 23328 \text{ hac-m}$$

Crop-2: Cotton

$$\text{Area under irrigation} = \frac{60}{100}\times\frac{3}{12}\times 30000 = 4500 \text{ hac}$$

$$\Delta_C = 8.64\times\frac{B}{D} = 8.64\times\frac{180}{1200} = 1.296 \text{ m}$$

$$\therefore \quad \text{Volume}_C = \text{Area}_C\times\Delta_C = 5832 \text{ hac-m}$$

Crop-3: Wheat

$$\text{Area}_W = \frac{70}{100}\times\frac{2}{12}\times 30000 = 3500 \text{ hac}$$

$$\Delta_W = 8.64\frac{B}{D} = 8.64\times\frac{120}{1800} = 0.576\text{m}$$

$$\therefore \quad \text{Volume}_W = \text{Area}_W\times\Delta_W = 2016 \text{ hac-m}$$

Crop-4: Gram

$$\text{Area}_G = \frac{70}{100}\times\frac{3}{12}\times 30000 = 5250 \text{ ha}$$

$$\Delta_G = 8.64\times\frac{120}{2000} = 0.5184\text{m}$$

$$\therefore \quad \text{Volume}_G = \text{Area}_G\times\Delta_G = 2721.6 \text{ ha}$$

$$\text{Period of peak water use} = \text{December to April}$$

$$= V_S + V_W + V_G = 28065.6 \text{ ha}$$

Due to 20% extra allowance $\quad V_{reqd} = 1.2\times 28065.6 = 33678.72 \text{ ha}$

$$\text{Net volume} = 33678.72 + V_{cotton} = 39510.72 \text{ ha}$$

Given, $\quad \eta_{conveyance} = 80\%$

$$\eta_{reservoir} = 90\%$$

$$\text{Gross reservoir capacity needed} = \frac{39510.72}{0.8\times 0.9} = 54876 \text{ ha}$$

Design discharge calculations

$$q_{sugarcane} = \frac{6000}{800 \text{ ha/cum}} = 7.5 \text{ cumec}$$

$$q_{cotton} = \frac{4500}{1200 \text{ ha/cum}} = 3.75 \text{ cumec}$$

$$q_{wheat} = \frac{3500}{1800 \text{ ha/cum}} = 1.94 \text{ cumec}$$

$$q_{gram} = \frac{5250}{2000 \text{ ha/cum}} = 2.625 \text{ cumec}$$

$$Q_{design} = \{\text{Max of (Dec-Apr, May-Nov}\}$$

$$= \text{Max of } (12.07, 11.25)$$

$$= 12.07 \text{ cumec}$$

$$\text{Capacity factor} = \frac{\bar{Q}_{mean}}{Q_{max}} = \frac{11.66}{12.07} = 0.966$$

$$\bar{Q}_{mean} = \frac{12.07+11.25}{2} = 11.66 \text{ cumec}$$

1.18 **Define field capacity, permanent wilting point and average moisture content. Explain how these will be useful in deciding the frequency of irrigation. (A schematic diagram showing less and more frequent irrigation is to be drawn for clarity).**

[3 + 4 + 3 = 10 marks : 2019]

Solution:

Field Capacity (F.C.)

- It is the amount of water content held in the soil after excess water gets drained off due to gravity.
- Physically, it can be defined as the bulk water content retained in the soil at particular hydraulic head.
- Field capacity is characterized by measuring water content after wetting a soil profile, covering it (to prevent evaporation) and monitoring the changing soil moisture in the profile.
- When the rate of changing soil moisture is very small, the water content at that point in said to be field capacity.

  Field Capacity is expressed in percentage

$$\text{Field Capacity of Soil} = \frac{\text{Weight of maximum amount of moisture held by soil against gravity}}{\text{Unit weight of the dry soil}}$$

- In medium textured soils, field capacity is about 50% of pore volume.
- Soil Moisture tension at field capacity ranges between one-tenth to one-third atmospheres.

Permanent Wilting Point (PWP)

- Water content at which plant is no longer able to extract water from the soil for its growth.
- At permanent wilting point, films of water are held very tightly and hence plant roots are not able to extract sufficient water for their growth.

$$\text{PWP} = \frac{\text{Weight of moisture held by the soil when plants gets permanently wilted}}{\text{Weight of dry soil}}$$

- Soil moisture tension of soil at permanent wilting point is in the range of 7 to 32 atm.

Average moisture content

- Depending upon the depth of irrigation water applied and frequency the average moisture content in the soil may lie between the lower limit wilting point or upper limit field capacity.
- Average moisture content of soil is moisture present in the soil that is available for plant.

Frequenoy of irrigation

Soil moisture in the root zone varies between field capacity (upper limit) and permanent wilting point (lower limit). When the soil moisture is consumed by plants through their roots, the soil moisture depletes. However, it is not allowed to deplete upto the permanent Wilting point, as it would result in considerable fall in crop yield. The optimum level up to which the soil moisture is allowed to deplete is called optimum moisture content.

The irrigation water should be supplied as soon as the moisture falls upto this optimum falls upto this optimum level, and its quantity should be sufficient enough to bring the moisture content upto its field capacity.

Water will be utilized by the plants after the fresh irrigation dose is given, and the soil moisture will start falling. It is then again recouped by a fresh irrigation dose as soon as the soon as the soil moisture reaches the optimum level.

The time interval between two successive irrigation dose is termed as frequency of irrigation.

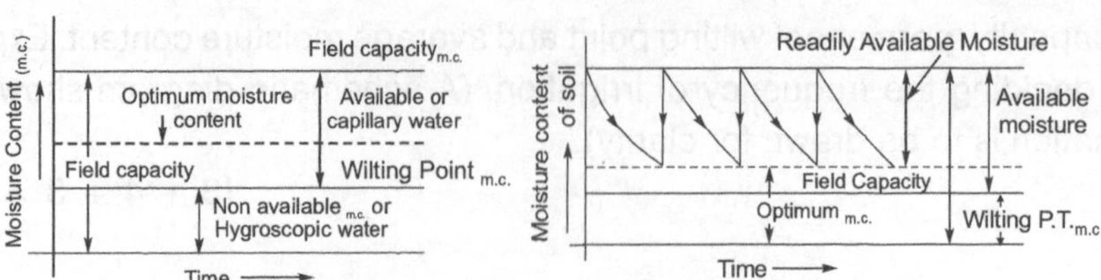

Available moisture for the plant = F.C. – ϕ

Readily available moisture for the plant = F.C. – M_O

Here, F.C. = Field capacity

ϕ = Wilting point or wilting coefficients below which plant can't survive.

M_O = Readily available moisture content.

$$\text{Frequency of Irrigation} = \frac{\text{Available/ Readily available moisture depth}}{\text{Consumptive use rate}}$$

1.19 **After how many days will you supply water to soil (clay loam) in order to ensure efficient irrigation of the given crop, if:**

Field capacity of soil = 27%

Permanent wilting point = 14%

Density of soil = 1.5 g/cm³

Effective depth of root zone = 75 cm

Daily consumptive use of water for the given crop = 11 mm

[12 marks : 2020]

Solution:

Given, Field capacity, (FC) = 27%

Permanent wilting point, (PWP) = 14%

Density of soil, (ρ_d) = 1.5 g/cm³ (Assuming dry density)

Effective depth of root zone, (d) = 75 cm

Daily consumptive use of water for the given crop (C_u) = 11 mm/day

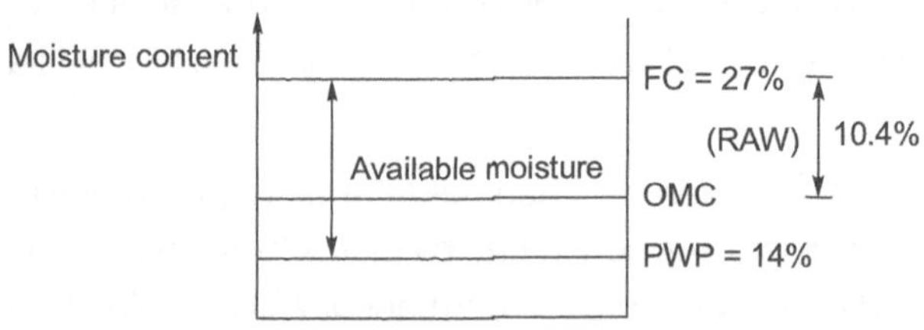

OMC = Optimum moisture content

RAW = Readily available water

The available moisture = FC – PWP = 27 – 14 = 13%

Let us assume that the readily available moisture 'or' the optimum soil moisture is 80% of available moisture.

i.e. Readily Available Moisture (RAW) = 0.80 × 13 = 10.4%

∴ Optimum moisture content (OMC) = 27 – 10.4 = 16.6%

It means that the moisture will be filled by irrigation between 16.6% and 27%.

Depth of water stored in root zone between these two limits:

$$d' = \frac{\rho_d}{\rho_w} \cdot [\text{FC} - \text{OMC}]$$

$$d' = \frac{1.5}{1} \times 0.75\,[0.27 - 0.166] = 0.117 \text{ m}$$

$$d' = 117 \text{ mm}$$

Now, frequency of irrigation

$$(\text{FOI}) = \frac{d'}{C_u} = \frac{117}{11} = 10.63 \text{ days} \simeq 10 \text{ days}$$

Hence, after 10 days, water should be applied to the given crop.

2. Design of Stable Channels and Canals

2.1 A river discharges 1000 m^3/sec of water at high flood level of *RL*= 103. A weir is constructed for flow diversion with a crest length of 255 m and total length of concrete floors as 40 m. The weir has to sustain the under seepage at a maximum static head of 2.4 m. The silt factor and the safe exit gradient for the river bed material are 1.1 and 1/6 respectively. Determine the depth of cut-off required at the downstream end of the concrete floor. Take the level of downstream concrete floor as *RL* = 100. Check for exit gradient.

[15 marks : 1999]

Solution:

Given,

High flood discharge, $Q = 1000$ m^3/sec

Length of weir, $L = 255$ m

∴ Discharge per unit length of weir is given by

$$q = \frac{Q}{L} = \frac{1000}{255} = 3.92 \text{ m}^3/\text{s}$$

Now for a discharge intensity, q, the normal depth of scour is given by Lacey's equation i.e.

$$R = 1.35\left[\frac{q^2}{f}\right]^{1/3} = 1.35\left[\frac{(3.92)^2}{1.1}\right]^{1/3} = 3.25 \text{ m}$$

Adopting downstream cut off upto 1.5 *R* below the downstream water level

∴ $1.5\,R = 1.5 \times 3.25 = 4.875$ m

Now, upstream water level = Upstream HFL = 103 m (given)

H = maximum static head causing seepage = 2.4 m (given)

∴ Downstream water level = Upstream HFL – H = 103 – 2.4 = 100.6 m

Hence, *RL* of bottom of downstream cut off = 100.6 – 4.875 = 95.725 m

Now *RL* of downstream floor = 100 m (given)

∴ Depth of downstream cut off = 100 – 95.725 = 4.275 m

Hence $d = 4.275$ m

Check for exit gradient

$$G_E = \frac{H}{d} \times \frac{1}{\pi\sqrt{\lambda}} \qquad \text{where } \lambda = \frac{1+\sqrt{1+\alpha^2}}{2}$$

where

$$\alpha = \frac{b}{d} = \frac{\text{Length of weir floor}}{\text{Depth of downstream cutoff}}$$

$$\therefore \qquad \alpha = \frac{b}{d} = \frac{40}{4.275} = 9.357$$

$$\therefore \qquad \lambda = \frac{1+\sqrt{1+(9.357)^2}}{2} = 5.2$$

$$\therefore \qquad G_E = \frac{2.4}{4.275} \times \frac{1}{\pi\sqrt{5.2}}$$

$$= \frac{2.4}{30.64} = \frac{1}{12.77} < \frac{1}{6} \text{ (Safe exit gradient)}$$

Hence, the weir is safe from exit gradient considerations with bottom of downstream cut off at *RL* of 95.725 m.

2.2 **A lined irrigation canal with trapezoidal cross-section has 5 m bed width, 2.5 m depth and 1.5(H) : 1(V) side slope. Longitudinal bed slope of the canal is 1 in 1000 and Manning's $n = 0.016$. What is the maximum carrying capacity of the canal? What area of land in hectares the canal can irrigate if the crop has 150 mm field irrigation requirement in a core period of 10 days?**

[10 marks : 2001]

Solution:

Given data:

Bed width, $B = 5$ m

Depth, $y = 2.5$ m

Side slope $= NH : 1V = 1.5\,H : 1V$

$\Rightarrow \quad N = 1.5$

Longitudinal slope, $S = \frac{1}{1000}$

Manning's, $n = 0.016$

Now, for a lined trapezoidal channel,

Area, $A = By + y^2 (\theta + \cot\theta)$

Wetted Perimeter, $P = B + 2y (\theta + \cot\theta)$

where θ is in radians

Bu $N = \cot\theta$

$$\Rightarrow \qquad \theta = \cot^{-1}(N) = \tan^{-1}\left(\frac{1}{N}\right) = \tan^{-1}\left(\frac{1}{1.5}\right) = 0.5880 \text{ radians}$$

$$\therefore \qquad A = By + y^2(\theta + \cot\theta) = 5 \times 2.5 + [(2.5)^2 \times (0.5880 + 1.5)] = 25.55 \text{ m}^2$$

$$\text{and} \qquad P = B + 2y(\theta + \cot\theta) = 5 + [2 \times 2.5 \times (0.5880 + 1.5)] = 15.44 \text{ m}$$

$$\text{But,} \qquad R = \frac{A}{P} = \frac{25.55}{15.44} = 1.655 \text{ m}$$

According to Manning's equation,

$$V = \frac{1}{n} R^{2/3} S^{1/2}$$

$$\text{But} \qquad Q = AV = \frac{1}{n} \times A \times R^{2/3} \times S^{1/2} = \frac{1}{0.016} \times 25.55 \times (1.655)^{2/3} \times \left(\frac{1}{1000}\right)^{1/2}$$

$$= 70.65 \text{ m}^3/\text{sec}$$

Hence maximum carrying capacity of canal = 70.65 m^3/sec.

During the core period of 10 days, volume of water which can be supplied by the channel

$$= 70.65 \times 10 \times 24 \times 60 \times 60 = 61.04 \times 10^6 \text{ m}^3$$

Volume of water available = Area which can be irrigated × Depth of irrigation

$\therefore$ Area which can be irrigated $= \dfrac{\text{Volume of water available}}{\text{Depth of irrigation}} = \dfrac{61.04 \times 10^6}{150 \times 10^{-3} \times 10^4}$ ha = 40693.33 ha

2.3 **An unlined irrigation canal has its bed and side composed of cohesionless material having mean diameter 6 mm. Angle of repose of the material is 40°. The bed width of the canal is 5 m and the side slope 1.5(H): 1(V). Determine the minimum discharge that can be admitted into the canal without any sediment movement. Longitudinal slope of the canal is 1 in 5000 and Manning's *n* = 0 .025.**

[15 marks : 2001]

Solution:

Given data:

Particle mean diameter, $d = 6 \text{ mm} = 6 \times 10^{-3}$ m

Angle of repose, $\phi = 40°$

Bed width of canal, $B = 5$ m

Side slope $= NH : 1\,V = 1.5\,H : 1\,V$

$\Rightarrow$ $N = 1.5$

Longitudinal slope, $S = \dfrac{1}{5000}$

Manning's $n = 0.025$

But $N = \cot\theta$

$$\Rightarrow \quad \theta = \tan^{-1}\left(\frac{1}{N}\right) = \tan^{-1}\left(\frac{1}{1.5}\right) = 33.69°$$

According to Shield's equation, the critical tractive stress acting on bottom of canal is given by

$$\tau_c = 0.056\,\gamma_w d (S_s - 1)$$

where, γ_w is unit weight of water, d is mean diameter of particles and S_s is specific gravity of sediment particles

For soils having inorganic sediments $S_s = 2.65$

$$\therefore \quad \tau_c = 0.056\,\gamma_w d (2.65 - 1) = \frac{\gamma_w d}{11}$$

Now, if τ_o is the average unit tractive force acting on the bottom of canal, then for no sediment movement from bottom

$$\tau_0 \le \tau_c$$

$$\Rightarrow \quad \gamma_w RS \le \frac{\gamma_w d}{11}$$

$$\Rightarrow \quad R \times \frac{1}{5000} \le \frac{6 \times 10^{-3}}{11}$$

$$\Rightarrow \quad R \le 2.73 \text{ m} \qquad \text{...(i)}$$

Also, if τ_0' is the average unit tractive force acting on the sides of the canal, then for no sediment movement from the sides

$$\tau_0' \le \tau_c'$$

where τ_c' is critical tractive stress acting on side slopes of canal

Now, τ_c and τ_c' are related as

$$\frac{\tau_c'}{\tau_c} = \sqrt{1 - \frac{\sin^2\theta}{\sin^2\phi}}$$

$$\Rightarrow \quad \frac{\tau'_c}{\tau_c} = \sqrt{1-\left(\frac{\sin 33.69°}{\sin 40°}\right)^2}$$

$$\Rightarrow \quad \tau'_c = 0.5053\,\tau_c$$

But $$\tau_c = \frac{\gamma_w d}{11}$$

$$\therefore \quad \tau'_c = 0.5053 \times \frac{\gamma_w d}{11}$$

Now, we have $$\tau'_0 = 0.75\,\gamma_w RS$$

$$\therefore \quad 0.75\,\gamma_w RS \le 0.5053 \times \frac{\gamma_w d}{11}$$

$$\Rightarrow \quad 0.75 \times R \times \frac{1}{5000} \le \frac{0.5053 \times 6 \times 10^{-3}}{11}$$

$$\Rightarrow \quad R \le 1.837 \text{ m} \quad \text{...(ii)}$$

Lesser value of R is adopted from (i) and (ii)

$$\therefore \quad R = 1.837 \text{ m}$$

Now, area of a trapezoidal channel, $A = (B + Ny)y$

Wetted perimeter of trapezoidal channel, $P = B + 2y\sqrt{N^2+1}$

$$\therefore \quad R = \frac{(B+Ny)y}{B+2y\sqrt{N^2+1}}$$

$$\Rightarrow \quad 1.837 = \frac{(5+1.5y)y}{5+2\times y\times\sqrt{(1.5)^2+1}}$$

$$\Rightarrow \quad 1.837 = \frac{5y+1.5y^2}{5+3.61y}$$

$$\Rightarrow \quad 1.5\,y^2 + 5y = 9.185 + 6.632\,y$$

$$\Rightarrow \quad 1.5\,y^2 - 1.632\,y - 9.185 = 0$$

Solving above equation, we get, $y = 3.08$ m

$$\therefore \quad A = (B + Ny)\,y = (5 + 1.5 \times 3.08) \times 3.08 = 29.63 \text{ m}^2$$

Now, the minimum discharge, $Q = AV$

$$= \frac{1}{n} AR^{2/3}S^{1/2}$$

$$= \frac{1}{0.025} \times 29.63 \times (1.837)^{2/3} \times \left(\frac{1}{5000}\right)^{1/2} = 25.14 \text{ m}^3\text{/sec}$$

2.4 Design a regime channel for a discharge of 35 m^3/sec with silt factor of 0.9 by Lacey's theory, taking side slopes as 1H : 2V.

[15 marks : 2002]

Solution:

The design steps to be followed to design a regime channel using Lacey's theory are:

(a) Calculation of velocity

$$V = \left[\frac{Qf^2}{140}\right]^{1/6}$$

where Q is discharge in m³/sec and f is silt factor

$$\therefore \quad V = \left[\frac{35 \times (0.9)^2}{140}\right]^{1/6} = 0.766 \text{ m/s}$$

(b) Calculation of hydraulic mean depth

$$R = \frac{5}{2}\left[\frac{V^2}{f}\right] = \frac{5}{2} \times \left[\frac{(0.766)^2}{0.9}\right] = 1.63 \text{ m}$$

(c) Calculation of area of channel

$$A = \frac{Q}{V} = \frac{35}{0.766} = 45.69 \text{ m}^2$$

But for a trapezoidal, channel

$$A = (B + Ny)\,y$$

$$\text{where } N \text{ is side slope} = 1\text{ H} : 2\text{ V} = \frac{1}{2}\text{ H} : 1\text{ V}$$

$$\Rightarrow \quad N = 0.5$$

$$\therefore \quad (B + 0.5\,y)\,y = 45.69$$

$$\Rightarrow \quad By + 0.5\,y^2 = 45.69 \quad \text{... (i)}$$

(d) Calculation of wetted perimeter

$$P = 4.75\sqrt{Q} = 4.75\sqrt{35} = 28.10 \text{ m}$$

But for a trapezoidal channel, $P = B + 2y\sqrt{N^2 + 1}$

$$\therefore \quad B + 2y\sqrt{(0.5)^2 + 1} = 28.10$$

$$\Rightarrow \quad B + 2.2361\,y = 28.10 \quad \text{... (II)}$$

Substituting value of B from (ii) in (i), we get

$$(28.10 - 2.2361\,y)\,y + 0.5\,y^2 = 45.69$$

$$\Rightarrow \quad 28.10\,y - 2.2361\,y^2 + 0.5\,y^2 - 45.69 = 0$$

$$\Rightarrow \quad 1.7361\,y^2 - 28.10\,y + 45.69 = 0$$

Solving above equation we get, $y = 1.834$ m

Substituting value of y in (ii), we get

$$B = 28.10 - 2.2361 \times 1.834$$

$$\Rightarrow \quad B = 23.999 \text{ m} \approx 24 \text{ m}$$

where B is width of channel and y is the depth of channel.

(e) Calculation of bed slope

$$S = \frac{f^{5/3}}{3340\,Q^{1/6}} = \frac{(0.9)^{5/3}}{3340 \times (35)^{1/6}}$$

$$= 1.3888 \times 10^{-4} = \frac{1}{7200}$$

2.5 A channel is 40 m wide, 2.6 m deep with slope of 1 in 5000. The representative bed material size d_{65} is 0.64 mm and mean diameter of the bed material is 0.38 mm. Manning's $n = 0.021$. Neglecting the side friction, find the bed load transported by the channel in tonnes per day by Meyer-Peter equation, considering $\frac{B}{D}$ as large and $\tau_c = 0.075d$.

[10 marks : 2003]

Solution:

Given data:

Width of channel (B) = 40 m, Depth of channel (D) = 2.6 m

Bed slope (S) = $\frac{1}{5000}$

Mean diameter of bed material (d_a) = 0.38 mm

Representative size (d_{65}) of bed material, $d = 0.64 \text{ mm} = 0.64 \times 10^{-3}$ m

Manning's $n = 0.021$

2.6 m

40 m

The critical shear stress required to move the grain is given by

$$\tau_c = 0.075\, d_a \quad \text{where } d_a \text{ is in mm.}$$

It is clearly understood from the above equation that τ_c is in kg/m²

$$\therefore \quad \tau_c = 0.075 \times 9.81 \times 0.38 = 0.2796 \text{ N/m}^2$$

Unit tractive force produced by the flowing water on the bed is given by

$$\tau_o = 0.97\, \gamma_w RS$$

where γ_w is unit weight of water in N/m³

$$R = \frac{A}{P} = \frac{B \times D}{B} = D\left(\text{Assuming } \frac{B}{D} \text{ as large}\right) = 2.6 \text{ m}$$

$$\therefore \quad \tau_o = 0.97 \times 9.81 \times 1000 \times 2.6 \times \frac{1}{5000} = 4.9482 \text{ N/m}^2$$

Now, the rate of bed load transport by Meyer-Peter equation is given by

$$g_b = 0.417\left[\tau_0\left(\frac{n'}{n}\right)^{3/2} - \tau_c\right]^{3/2}$$

where g_b is in N per metre width of channel per second

n' = Manning's coefficient pertaining to grain size on an unrippled bed and given by Strickler's formula i.e.

$$n' = \frac{1}{24} d^{1/6}$$

where d is the representative size of the bed material in meters usually takes as median size (d_{50}) or d_{65}

$$n' = \frac{1}{24} \times (0.64 \times 10^{-3})^{1/6} = 0.01223$$

$$\therefore \quad g_b = 0.417\left[4.9482\left(\frac{0.01223}{0.021}\right)^{3/2} - 0.2796\right]^{3/2} = 1.109 \text{ N/m/s}$$

∴ Bed load transported by 40 m wide channel

$$= \frac{1.109}{9.81} \times 40 \text{ kg/sec} = \frac{1.109}{9.81} \times 40 \times \frac{1}{1000} \times 24 \times 60 \times 60 \text{ tonnes/day}$$

= 390.70 tonnes per day.

2.6 Design a Lacey channel to carry 5 m³/sec through 0.5 mm sand.

[8 marks : 2007]

Solution:

Design procedure for Lacey's theory is as follows:

(a) Calculation of velocity

$$V = \left[\frac{Qf^2}{140}\right]^{1/6}$$

where Q is discharge in cumecs, V is velocity in m/s and f is silt factor, it is given by $f = 1.76\sqrt{d_{mm}}$ and d_{mm} is average particle size in mm

Thus,
$$f = 1.76\sqrt{0.5} = 1.24$$

$$V = \left(\frac{5 \times (1.24)^2}{140}\right)^{1/6}$$

$$\Rightarrow \quad V = 0.617 \text{ m/s}$$

(b) Calculation of hydraulic mean depth

$$R = \frac{5}{2} \times \frac{V^2}{f} = \frac{5}{2} \times \frac{(0.617)^2}{1.24} = 0.768 \text{ m}$$

(c) Calculation of area of channel section

$$A = \frac{Q}{V} = \frac{5}{0.617} = 8.104 \text{ m}^2$$

(d) Calculation of wetted perimeter

$$P = 4.75\sqrt{Q} = 4.75\sqrt{5} = 10.62 \text{ m}$$

For a trapezoidal channel with side slope of $\frac{1}{2}H$: 1V, we get

$$P = B + 2y\sqrt{N^2 + 1} \text{ and } A = (B + Ny)y$$

$$\therefore \quad (B + 0.5y)y = 8.104 \quad \text{...(i)}$$

$$B + 2y\sqrt{(0.5)^2 + 1} = 10.62 \quad \text{...(ii)}$$

Solving (i) and (ii), we get $y = 0.9$ m and $B = 8.6$ m

(e) Calculation of bed slope

$$S = \frac{f^{5/3}}{3340\, Q^{1/6}} = \frac{(1.24)^{5/3}}{3340 \times (5)^{1/6}} = \frac{1}{3051.68} \approx \frac{1}{3052}$$

2.7 Design a transition using Mitra's hyperbolic transition given by $B_x = \dfrac{B_c B_f L_f}{L_f B_c - (B_c - B_f)x}$ and compare the results using Chaturvedi's semi-cubical parabolic transition given by

$$x = \frac{L_f B_c^{3/2}}{B_c^{3/2} - B_f^{3/2}}\left(1 - \left(\frac{B_f}{B_x}\right)^{3/2}\right)$$

Given nominal bed width = 25 m, width of flumed section = 10.0 m and total length of transition = 15.0 m.

[4 marks : 2010]

Solution:

The design of transition as per **Mitra's hyperbolic transition equation** is given as:

$$B_x = \frac{B_c B_f L_f}{L_f B_c - (B_c - B_f)x}$$

where B_c is nominal bed width of the normal channel section, B_f is the width of flumed section, B_x is the bed width at any distance x from the flumed section and L_f is the total length of transition.

Here B_c = 25 m; B_f = 10.0 m; L_f = 15 m

$$\therefore \quad B_x = \frac{25 \times 10 \times 15}{15 \times 25 - (25 - 10)x}$$

$$\Rightarrow \quad B_x = \frac{3750}{375 - 15x}$$

$$\Rightarrow \quad B_x = \frac{250}{25 - x} \quad \text{...(i)}$$

For various values of x lying between 0 and 15 m, values of B_x are worked out and tabulated below:

x(in m)	0	3	6	9	12	15
$B_x = \frac{250}{25-x}$	10	11.36	13.16	15.63	19.23	25

As per Chaturvedi's semicubical parabolic transition equation various distances from the flumed section can be calculated by choosing various convenient values of B_x i.e.

$$x = \frac{L_f B_c^{3/2}}{B_c^{3/2} - B_f^{3/2}}\left[1 - \left(\frac{B_f}{B_x}\right)^{3/2}\right]$$

B_x as per Mitra's transition equations	10	11.36	13.16	15.63	19.23	25
Corresponding x value as per Chaturvedi's transition equation	0	3.5	6.78	9.80	12.55	15

2.8 The velocity distribution $v(y)$ in a wide channel is given by $v(y) = 5.75\sqrt{g y_0 S_0}\log\left(\frac{30y}{k}\right)$ in which y is the distance from the bottom and y_0 is the depth of flow, S_0 is the bed slope of the channel, k is the roughness height at the bed. Estimate the shear stress at 1 cm above the bottom of a channel on a slope of 0.5%. The depth of flow is 2.5 m and the characteristic roughness is 1 cm. If the critical shear stress required to move the bed material in the channel is 1 N/m^2, find whether the channel is in stable condition The viscosity of the water is 0.001 N s m^{-2}. [Assume log = $\log_e$ (natural log)]

[4 marks : 2011]

Solution:

Given velocity distribution $v(y) = 5.75\sqrt{g y_0 S_0}.\log\left(\frac{30y}{k}\right)$

$S_0 = 0.5\% = 0.005$

$k = 1 \text{ cm} = 0.01 \text{ m}$

$y_0 = 2.5 \text{ m}$

$\mu = 0.001 \text{ N-sm}^{-2}$

Substituting values in the equation of velocity profile

$$\Rightarrow \qquad v(y) = 5.75\sqrt{9.81\times 2.5\times 0.005}.\log\left(\frac{30y}{0.01}\right)$$

$$= 2.01\,(\log(3000\,y) = 2.01\,[\log 3000 + \log y]$$

$$= 6.989 + 2.01\log y$$

Shear stress,
$$\tau = \frac{\mu\partial v(y)}{\partial y} = \mu\left(\frac{2.01}{y}\right)$$

or τ at 1 cm above the bottom

$$\tau = 0.001\times\frac{2.01}{0.01} = 0.201\,\text{N/m}^2$$

Since, τ < critical shear stress at bottom = 1 N/m^2

$\therefore$ **Channel is stable.**

2.9 **Derive the following Lacey's regime equations**

$$V = \left[\frac{Qf^2}{140}\right]^{1/6}$$

and $P = 4.75\sqrt{Q}$

with the use of following basic equations of Lacey

$$V = \sqrt{\frac{2}{5}fR}$$

$$Af^2 = 140\,V^5$$

Here V = **velocity;**
Q = **discharge;**
f = **silt factor;**
P = **wetted perimeter, and**
R = **hydraulic mean radius**

[10 marks : 2013]

Solution:

To prove: In Lacey's regime channel

$$v = \left(\frac{Qf^2}{140}\right)^{1/6}$$

and
$$P = 4.75\sqrt{Q}$$

Given:
$$v = \sqrt{\frac{2}{5}fR} \qquad \text{...(i)}$$

and
$$Af^2 = 140\,v^5 \qquad \text{...(ii)}$$

Proof: $\because$
$$Af^2 = 140\,v^5$$

Substuting v from (i) in (ii)

$$Af^2 = 140\left[\frac{2}{5}fR\right]^{5/2}$$

$$\Rightarrow \qquad Af^{-1/2} = 140\left[\frac{2}{5}R\right]^{5/2}$$

$$\Rightarrow \quad \frac{A}{\sqrt{f}} = 140\frac{v^5}{f^{5/2}} \qquad \text{(from (i)}$$

$$\Rightarrow \quad \frac{Av}{\sqrt{f}} = \frac{140v^6}{f^{5/2}} \qquad \text{(multiplying both side by 'v')}$$

$$\Rightarrow \quad \frac{Q}{\sqrt{f}} = \frac{140v^6}{f^{5/2}} \qquad [\because Q = Av]$$

$$\Rightarrow \quad Q = \frac{140v^2}{f^2}$$

$$\Rightarrow \quad v = \left[\frac{Qf^2}{140}\right]^{1/6} \qquad \text{Hence Proved.}$$

Now, $\because$

$$v = \left(\frac{2}{5}fR\right)^{1/2}$$

$$\Rightarrow \quad v^2 = \frac{2}{5}f\left(\frac{A}{P}\right) \qquad \left[\because R = \frac{A}{P}\right]$$

$$\Rightarrow \quad P = \frac{2}{5}\frac{fA}{v^2} = \frac{2}{5v^2}\left(\frac{140v^5}{f}\right) \qquad \text{[Substuting 'Af' from (ii)]}$$

$$\Rightarrow \quad P = \frac{2\times 140}{5f}v^3$$

$$\Rightarrow \quad P = \frac{2\times 140}{5f}\left(\frac{Qf^2}{140}\right)^{3/6} \qquad \text{[Putting the value of 'v' from (i)]}$$

$$\Rightarrow \quad P = \frac{2\times(140)^{1/2}}{5}\sqrt{Q} = 4.73286\sqrt{Q}$$

$$\Rightarrow \quad P = 4.75\sqrt{Q} \qquad \textbf{Hence Proved.}$$

2.10 **Design a rigid boundary canal carrying 50 m³/sec. Take Manning's n = 0.014 and longitudinal slope of the canal as 12 per ten thousand. Channel is rectangular section.**

[8 marks : 2015]

Solution:

Given:

$$Q = 50 \text{ m}^3/\text{s}$$

$$n = 0.014$$

$$S = \frac{12}{10000}$$

Designing canal as hydraulically efficient section.

$$B = 2y$$

where, B = Width of section and y = Depth of flow

Area, $A = By = 2y^2$

Wetted perimeter, $P = B + 2y = 4y$

Hydraulic radius, $R = \frac{A}{P} = \frac{2y^2}{4y} = \frac{y}{2}$

Using Manning's equation, $Q = \frac{1}{n}AR^{2/3}S^{1/2}$

$$\Rightarrow \quad 50 = \frac{1}{0.014} \times 2 \times y^2 \times \left(\frac{y}{2}\right)^{2/3} \times \left(\frac{12}{10000}\right)^{1/2}$$

$$\Rightarrow \quad y^{8/3} = \frac{50 \times 0.014 \times 2^{2/3}}{2} \times \sqrt{\left(\frac{10000}{12}\right)}$$

$$\Rightarrow \quad y = 2.83 \text{ m}$$

$$\Rightarrow \quad B = 2y = 5.66 \text{ m}$$

Thus, provide a rigid boundary canal of width 5.66 m and depth of flow 2.83 m and an additional free board of 0.75 m.

2.11 **Using Lacey's theory, design an irrigation channel carrying 30 m³/sec. Take silt factor as 1.0.**

[8 marks : 2015]

Solution:

The design steps to be followed to design irrigation channel using **Lacey's theory** are:

(a) Calculation of velocity, $V = \left(\frac{Qf^2}{140}\right)^{1/6}$

where Q is discharge in m³/s and f is silt factor

$$\therefore \quad V = \left(\frac{30 \times 1.0^2}{140}\right)^{1/6} = 0.774 \text{ m/s}$$

(b) Calculation of hydraulic mean radius

$$R = \frac{5}{2}\left(\frac{V^2}{f}\right) = \frac{5}{2} \times \frac{0.774^2}{1} = 1.50 \text{ m}$$

(c) Area of channel, $A = \frac{Q}{V} = \frac{30}{0.774} = 38.76 \text{ m}^2$

For a trapezoidal channel, $A = (B + zy)y$

Assuming side slope to be 1H : 2V (z will be 1/2)

$$\Rightarrow \quad A = (B + 0.5y)y$$

$$\Rightarrow \quad 38.76 = By + 0.5y^2 \quad \text{...(i)}$$

(d) Wetted perimeter

$$P = 4.75\sqrt{Q} = 4.75\sqrt{30} = 26.02 \text{ m}$$

For a trapezoidal channel, $P = B + 2y\sqrt{z^2 + 1}$

$$\Rightarrow \quad 26.02 = B + 2y\sqrt{0.5^2 + 1}$$

$$\Rightarrow \quad B + 2.2361y = 26.02 \quad \text{...(ii)}$$

Substituting value of B from (ii) in (i), we get

$$(26.02 - 2.2361y)y + 0.5y^2 = 38.76$$

$$1.7361y^2 - 26.02 + 38.76 = 0$$

$$\Rightarrow \quad y = 1.68 \text{ m}$$

Substituting value of y in (ii), we get

$$B = 26.02 - 2.2361 \times 1.68 = 22.27 \text{ m}$$

Where B is width of channel and y is depth of channel.

(e) Calculation of bed slope, $S = \frac{f^{5/3}}{3340Q^{1/6}} = \frac{1^{5/3}}{3340 \times 30^{1/6}} = 1.698 \times 10^{-4} = \frac{1}{5888}$

2.12 Using Lacey's theory, design an irrigation canal to carry a discharge of 30 m^3/s with silt factor = 1.05.

[8 marks : 2016]

Solution:

Given: Discharge (Q) = 30 m^3/s, Silt factor (f)= 1.05

$\therefore$ Flow velocity, $$v = \left(\frac{Qf^2}{140}\right)^{1/6} = \left(\frac{30 \times 1.05^2}{140}\right)^{1/6} = 0.786 \text{ m/s}$$

Hydraulic radius, $$R = \frac{5}{2}\left(\frac{v^2}{f}\right) = \frac{5}{2}\left(\frac{0.786^2}{1.05}\right) = 1.471 \text{ m}$$

Area of canal section, $$A = \frac{Q}{v} = \frac{30}{0.786} = 38.168 \text{ m}^2$$

Wetted perimeter, $$P = 4.75\sqrt{Q} = 4.75\sqrt{30} = 26.017 \text{ m}$$

Let side slopes of trapezoidal canal section is $\frac{1}{2}$H : 1V. i.e., $m = \frac{1}{2}$

$$\therefore \quad P = b + 2y\sqrt{1+\left(\frac{1}{2}\right)^2} = b + y\sqrt{5} = 26.017$$

and $$A = (b+my)y = \left(b+\frac{y}{2}\right)y$$

$$\Rightarrow \quad 38.168 = \left(b+\frac{y}{2}\right)y = \left(26.017 - y\sqrt{5} + \frac{y}{2}\right)y$$

$$\Rightarrow \quad 38.168 = -1.736y^2 + 26.017y$$

$$\Rightarrow \quad y^2 - 14.987y + 21.986 = 0$$

$$\therefore \quad y = 13.34 \text{ m}, 1.65 \text{ m}$$

Neglecting $y = 13.34$ m (because of very high flow depth)

$\therefore$ Flow depth $y = 1.65$ m

$$\therefore \quad b = 26.017 - y\sqrt{5} = 26.017 - 1.65\sqrt{5} = 22.33 \text{ m}$$

Bed slope of canal, $$S = \frac{f^{5/3}}{3340\, Q^{1/6}} = \frac{1.05^{5/3}}{3340(30)^{1/6}} = 1.842 \times 10^{-4}$$

$$= \frac{1}{5427.725} = 1 \text{ in } 5427.725$$

2.13 An unlined channel having bed slope 1 in 4000, carries a certain discharge at flow depth = 5 m with hydraulic radius = 3.5 m. Determine the concentration of the suspended sediment load at 1.5 m above the channel bed if the concentration at 0.25 m above the bed is 400 ppm and the fall/settling velocity of the suspended particle is 0.3 m/s. Assume that Von Karman universal constant = 0.4.

[8 marks : 2016]

Solution:

The sediment concentration at any distance 'y' is determined by **Rouse equation** as below:

$$\frac{C}{C_a} = \left[\frac{a(D-y)}{y(D-a)}\right]^{w_0/kV_*}$$

where, $V^* =$ Shear velocity $= \sqrt{\frac{\tau_0}{\rho_w}}$

$\tau_0 =$ Shear stress at the bottom

$\rho_w =$ Density of water

$D =$ Total water depth

$C_a =$ Sediment concentration at distance 'a'

$k =$ Von-Karman universal constant = 0.4

$w_0 =$ Velocity of free fall of the grain

Here, $y = 1.5$ m

$D = 5$ m

$S = \text{bed slope} = \frac{1}{4000}$

Hydraulic radius, $R = 3.5$ m

$C_a = 400$ ppm

$a = 0.25$

$w_0 = 0.3$ m/s

$$V^* = \sqrt{\frac{\tau_0}{\rho_w}} = \sqrt{\frac{\gamma_w RS}{\rho_w}} = \sqrt{gRS}$$

$$\Rightarrow \quad V^* = \sqrt{9.81 \times 3.5 \times \frac{1}{4000}}$$

$$\Rightarrow \quad V^* = 0.0926 \text{ m/sec}$$

Now,

$$\frac{C}{400} = \left[\frac{0.25(5-1.5)}{1.5(5-0.25)}\right]^{\frac{0.3}{0.4\times0.0926}}$$

$$\Rightarrow \quad \frac{C}{400} = \left[\frac{0.875}{7.125}\right]^{8.099}$$

$$C = 16.81 \times 10^{-6} \text{ ppm}$$

2.14 **Explain the terms 'initial regime' and 'final regime' as explained in Lacey's regime theory of stable channels. Design a stable channel for carrying a discharge of 30 m^3/sec using Lacey's method assuming a silt factor equal to 1.0.**

[20 marks : 2019]

Solution:

Initial regime

- If only the bed slope of the channel varies and its wetted perimeter remains unaffected then a condition of non-silting and non scouring may exit which is called initial regime.
- Under initial regime condition Lacey's theory is not valid.

Final regime

- If all the variables such as perimeter, depth, flow are free to vary and can adjust according to the discharge there occurs no effective silting and scouring then such channels are said to have achieved permanent stability and said to be in final regime condition.
- Lacey's design theory is valid for such a regime condition.

Given:

$$Q = 30 \text{ m}^3/\text{s}, \ f = 1$$

$$V = \left(\frac{Qf^2}{140}\right)^{1/6} = \left(\frac{30\times 1^2}{140}\right)^{1/6} = 0.77 \text{ m/s}$$

$$A = \frac{Q}{V} = \frac{30}{0.77} = 38.96 \text{ m}^2$$

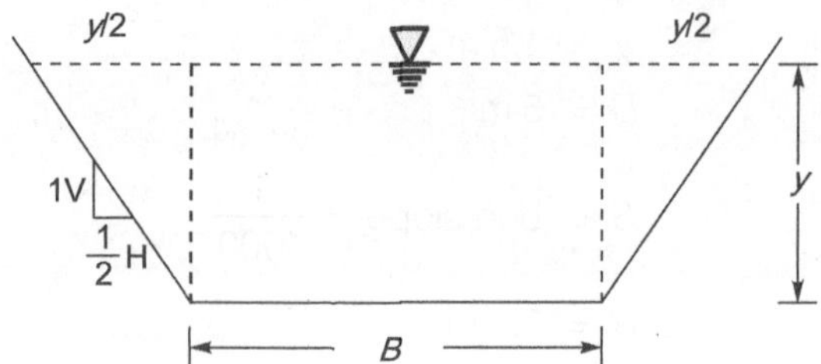

$$R = \frac{5}{2}\left(\frac{V^2}{f}\right) = \frac{5}{2}\left(\frac{0.77^2}{1}\right) = 1.48 \text{ m}$$

$$P = 4.75\sqrt{Q} = 4.75\sqrt{30} = 26.0168 \text{ m}$$

$$R = \frac{A}{P} = \frac{\frac{1}{2}(2B+y)y}{26.0168}$$

$$\Rightarrow \quad (2B + y)y = 77 \quad \text{...(i)}$$

$$P = 26.0168 = B + 2y\sqrt{\frac{1}{4}+1}$$

$$26.0168 = B + 2.236y$$

$$\Rightarrow \quad B = 26.0168 - 2.236y \quad \text{...(ii)}$$

From equation (i) and (ii),

$$3.472y^2 - 52.03y + 77 = 0$$

$$y = 13.32 \text{ m}, \ 1.66 \text{ m}$$

$$B = 22.3 \text{ m}$$

$$S = \frac{f^{5/3}}{3340Q^{1/6}} = \frac{(1)^{5/3}}{3340\times(30)^{1/6}} = \frac{1}{5887.53}$$

3. Design and Construction of Gravity Dams

3.1 The figure below shows the section of a gravity dam (non-overflow portion) built of concrete. Calculate the maximum vertical stresses at the heel and toe of the dam. Assume weight of concrete 23.5 kN/m^3. Neglect earthquake effects.

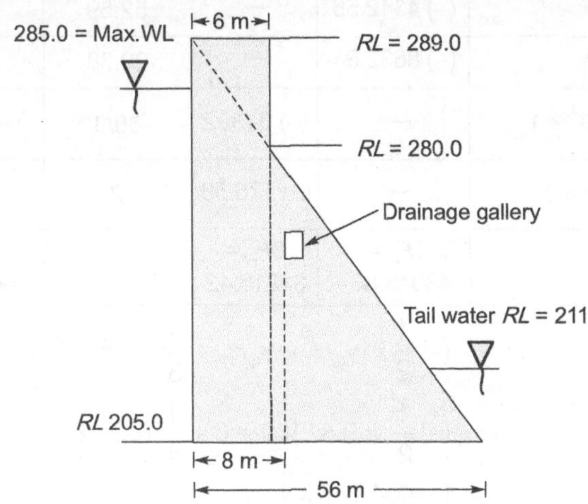

[20 marks : 2000]

Solution:

The cross section of the dam is shown below:

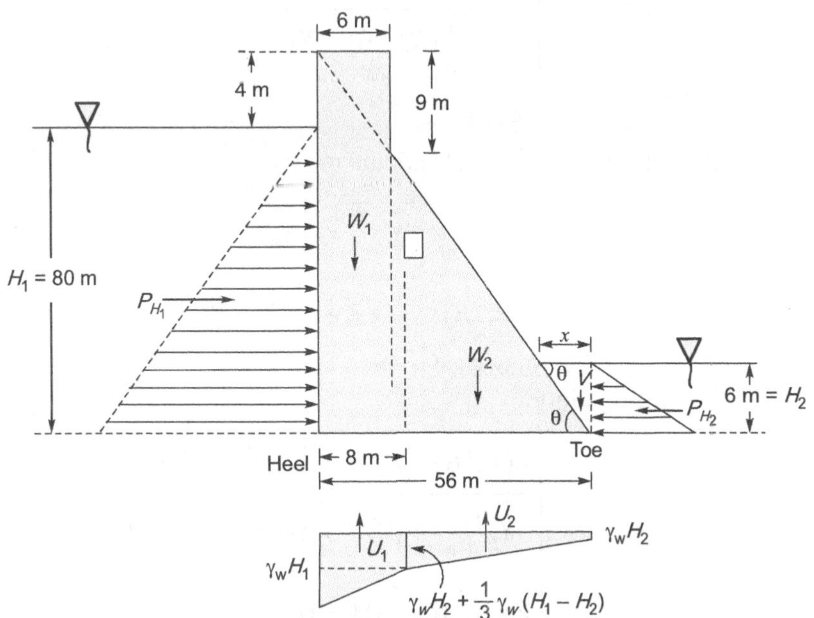

Height of maximum water level be H_1 = 80 m and height of tail water be H_2 = 6 m. The calculation of various forces is tabulated below.

Force	Description	F_V	F_H	LA from toe	M_R	M_O
W_1	(+) 84 × 6 × 1 × 23.5	+11844	—	53	627732	—
W_2	$(+)\frac{1}{2} \times 75 \times 50 \times 1 \times 23.5$	+44062.5	—	100/3	1468750	—
V	$(+)\frac{1}{2} \times 4 \times 6 \times 1 \times 9.81$	+117.72	—	4/3	156.96	—
U_1	(–) 4342.56 kN	(–) 4342.56	—	52.59	—	228375.23
U_2	(–) 8632.8 kN	(–) 8632.8	—	29.38	—	253631.66
P_{H_1}	$(-)\frac{1}{2} \times 9.81 \times 80^2 \times 1$	—	(–) 31392	80/3	—	837120
P_{h_2}	$(-)\frac{1}{2} \times 9.81 \times 6^2 \times 1$	—	(+) 176.58	2	353.16	—
		$\Sigma F_V =$ 43048.86	$\Sigma F_H =$ 31215.42		$\Sigma M_R =$ 2096992.12	$\Sigma M_O =$ 1319126.89

$$U_1 = (-)\frac{1}{2} \times [\gamma_w H_1 + \gamma_w H_2 + \frac{1}{3}\gamma_w (H_1 - H_2)] \times 8$$

$$= (-)\frac{1}{2} \times 9.81\,[80 + 6 + \frac{1}{3}(80 - 6)] \times 8$$

$$= (-)\ 4342.56 \text{ kN}$$

$$CG \text{ of } U_1 \text{ from heel} = \left[\frac{2\left(\gamma_w H_2 + \frac{1}{3}\gamma_w (H_1 - H_2)\right) + \gamma_w H_1}{\gamma_w H_2 + \frac{1}{3}\gamma_w (H_1 - H_2) + \gamma_w H_1}\right] \times \frac{8}{3}$$

$$= \left[\frac{2\left(6 + \frac{1}{3}(80 - 6)\right) + 80}{6 + \frac{1}{3}(80 - 6) + 80}\right] \times \frac{8}{3}$$

$$= 3.41 \text{ m}$$

$$CG \text{ of } U_1 \text{ from toe} = 56 - 3.41 = 52.59 \text{ m}$$

$$U_2 = \frac{1}{2} \times [\gamma_w H_2 + \gamma_w H_2 + \frac{1}{3}\gamma_w (H_1 - H_2)] \times (56 - 8)$$

$$= \frac{1}{2} \times 9.81 \times [6 + 6 + \frac{1}{3}(80 - 6)] \times 48$$

$$= 8632.8 \text{ kN}$$

CG of U_2 from drainage gallery

$$= \left[\frac{2(\gamma_w H_2) + \gamma_w H_2 + \frac{1}{3}\gamma_w (H_1 - H_2)}{\gamma_w H_2 + \gamma_w H_2 + \frac{1}{3}\gamma_w (H_1 - H_2)}\right] \times \left(\frac{56 - 8}{3}\right)$$

$$= \left[\frac{2 \times 6 + 6 + \frac{1}{3}(80 - 6)}{6 + 6 + \frac{1}{3}(80 - 6)}\right] \times \left(\frac{48}{3}\right) = 18.62 \text{ m}$$

$$CG \text{ of } U_2 \text{ from toe} = 48 - 18.62 = 29.38 \text{ m}$$

Now the location of resultant force from toe is given by

$$\bar{x} = \frac{\Sigma M_R - \Sigma M_0}{\Sigma F_V} = \frac{2096992.12 - 1319126.89}{43048.86} = 18.07 \text{ m}$$

∴ Eccentricity (e) of the resultant force

$$= \frac{B}{2} - \bar{x} = \frac{56}{2} - 18.07 = 9.93 \text{ m}$$

∴ Resultant force is nearer to the toe and $e > \frac{B}{6}$; **Hence tension is developed**

$$\sigma_{toe}/\sigma_{heel} = \frac{\Sigma F_V}{B}\left[1 \pm \frac{6e}{B}\right] = \frac{43048.86}{56}\left[1 \pm \frac{6 \times 9.93}{56}\right] = \frac{43048.86}{56}[1 \pm 1.064]$$

∴ $$\sigma_{toe} = \frac{43048.86}{56} \times 2.064 = 1586.66 \text{ kN/m}^2$$

and $$\sigma_{heel} = \frac{43048.86}{56} \times (-0.064) = -49.2 \text{ kN/m}^2$$

3.2 **Classify different types of dams. What type of analysis is to be carried out from the point of view of its breaching/breaking?**

[15 marks : 2005]

Solution:

Dams can be broadly classified as ancient type of dams and modern type of dams. Ancient type of dams are earthen dams, rock fill dams and solid masonry gravity dams. Modern type of dams are hollow masonry gravity dams, timber dams, steel dams and arch dams.

Some general information on all these dams is as follows:

(i) **Earthen dams:** These are made of soil that is pounded down solidly. They are built in areas where the foundation is not strong enough to bear the weight of a concrete dam and where earth is more easily available as a building material compared to concrete, stone or rock. Example is Hirakund dam in India.

(ii) **Rockfill dams:** Rockfill are formed of loose rocks and boulders piled in the river bed. A slab of RCC is often laid across the upstream face of a rockfill dam to make it water tight. Example is Tehri dam in India.

(iii) **Solid masonry gravity dam:** These are built with solid blocks of concrete which hold back the flow of water by shear weight. They can be constructed on any dam site, where there is a natural foundation strong enough to bear the great weight of the dam. Example is Bhakra dam in India.

(iv) **Hollow masonry gravity dam:** These are essentially designed on the same principles on which solid masonry gravity dams are designed, But they contain less concrete or masonry; about 35 to 40% or so. Generally the weight of water is carried by a deck of RCC or by arches that share the weight burden. They are difficult to build and are adopted only if very skilled labour is easily available, otherwise the labour cost is too high to build its complex structure.

(v) **Steel dams:** These are used as temporary coffer dams needed for the construction of permanent dams. These are usually reinforced with timber or earthfill.

(vi) **Timber dams:** These are short lived since in a few years time, rotting sets in. Their life is not more than 30 to 40 years and must have regular maintenance during that time.

(vii) **Arch dams:** These are very complex and complicated. They make use of horizontal arch section in place of weight to hold back the water. They are best suited where the dam is extremely high and narrow. Example is Idukki dam in Kerala across Periyar which is the only arch dam in India.

3.3 **What is an elementary profile of a gravity dam? Derive the expressions for determining base width of such dam based on stress and sliding criterion. Also derive the expressions for normal, principal and shear stresses at the base of the dam.**

[10 marks : 2006]

Solution:

Elementary profile of a gravity dam is a right angled triangle, having zero width at the water level and a base width (B) at the bottom i.e. the point where maximum hydrostatic pressure acts. It is subjected only to the external water pressure on the upstream side. In other words, the shape of such a profile is similar to the shape of the hydrostatic pressure distribution.

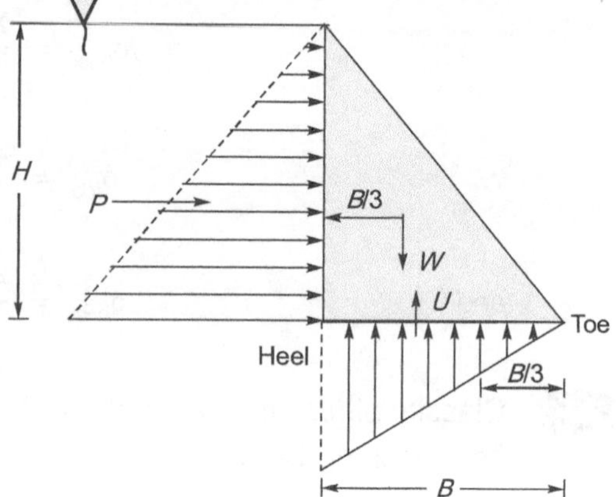

Let
P = Hydrostatic water pressure
W = Self weight
U = Uplift pressure
B = Base width of the dam

Base width (B) of elementary profile for no tension:

For no tension to be developed, the resultant of all the forces P, W and U must pass through **the outermost middle third point**. Thus taking moments of all the forces about the **lower middle third point** (i.e.) the point which resultant is passing, we get

$$W\left(\frac{B}{3}\right) - U\left(\frac{B}{3}\right) - P\left(\frac{H}{3}\right) = 0$$

$$\Rightarrow \quad (W-U)\left(\frac{B}{3}\right) - \frac{PH}{3} = 0$$

But
$$W = \frac{1}{2} \times B \times H \times 1 \times S_c \times \gamma_w$$

where, S_c is Specific gravity of concrete and γ_w is Unit weight of water

$$U = \frac{1}{2} C\gamma_w H \times B$$

where C is a constant which is taken equal to 1 in calculation and will be equal to zero when no uplift is considered.

$$P = \frac{1}{2}\gamma_w H^2$$

$$\therefore \quad \left[\frac{1}{2} B H S_c \gamma_w - \frac{1}{2} C\gamma_w H B\right] \times \frac{B}{3} - \frac{1}{2}\gamma_w H^2 \times \frac{H}{3} = 0$$

$$\Rightarrow \quad \frac{1}{2} \times \frac{1}{3} H\gamma_w (B^2 S_c - B^2 C) = \frac{1}{2}\gamma_w H^2 \times \frac{H}{3}$$

$$\Rightarrow \quad B^2 (S_c - C) = H^2$$

$$\Rightarrow \quad B = \frac{H}{\sqrt{S_c - C}}$$

Hence, if B is taken equal to or greater than $\frac{H}{\sqrt{S_c - C}}$, no tension will be developed at the heel with full reservoir.

Base width of elementary profile for no sliding:

For no sliding, the frictional resistance $\mu\,(W - U)$ should be equal to or more than the horizontal forces i.e. P

$$\therefore \quad \mu(W - U) \geq P$$

$$\Rightarrow \quad \mu\left[\frac{1}{2} BH\, S_c\, \gamma_w - \frac{1}{2} C\gamma_w\, HB\right] \geq \frac{1}{2}\gamma_w H^2$$

$$\Rightarrow \quad \mu BS_c - CB \geq H$$

$$\Rightarrow \quad B \geq \frac{H}{\mu\,(S_c - C)}$$

Hence B should be greater than or equal to $\frac{H}{\mu\,(S_c - C)}$ for no sliding.

Stresses developed in the elementary profile:

The normal stress at the toe and heel is given by

$$P_{max}/P_{min} = \frac{\Sigma F_V}{B}\left[1 \pm \frac{6e}{B}\right]$$

In case of elementary profile

$$\Sigma F_V = W - U$$

$$= \frac{1}{2} BH\gamma_w S_c - \frac{1}{2} C\gamma_w HB = \frac{1}{2} BH\gamma_w (S_c - C)$$

Further in the limiting case of no tension at any point in the base of the dam $e = \frac{B}{6}$ and hence

$$P_{max}/P_{min} = \frac{\frac{1}{2} BH\gamma_w (S_c - C)}{B}\left[1 \pm \frac{6 \times B/6}{B}\right]$$

$$P_{max} \text{ at toe} = H\gamma_w (S_c - C)$$

$$P_{min} \text{ at heel} = 0$$

The principal stress at the toe of the dam is given by

$$\sigma_1 = P_{max} \sec^2\phi \qquad \left[\text{where } \tan\phi = \frac{B}{H}\right]$$

$$\Rightarrow \quad \sigma_1 = H\gamma_w (S_c - C)(1 + \tan^2\phi)$$

$$\Rightarrow \quad \sigma_1 = H\gamma_w (S_c - C)\left(1 + \frac{B^2}{H^2}\right)$$

But

$$B = \frac{H}{\sqrt{S_c - C}}$$

$$\therefore \quad \sigma_1 = H\gamma_w (S_c - C)\left[1 + \frac{1}{S_c - C}\right]$$

$$\Rightarrow \quad \sigma_1 = H\gamma_w (S_c - C + 1)$$

$$\sigma_2 = 0$$

The shear stress at the toe of the dam is given by

$$\tau_o = P_{max}\tan\phi = \gamma_w H(S_c - C)\times\frac{B}{H}$$

$$= \gamma_w H(S_c - C)\times\frac{1}{\sqrt{S_c - C}}$$

$$\Rightarrow \qquad \tau_0 = \gamma_w H\sqrt{S_c - C}$$

$$\text{Shear stress at heel} = 0$$

3.4 **Find the width of elementary gravity dam whose height is 100 m. Specific gravity of dam material 2.2 and seepage coefficient at the base $C = 0.8$.**

[2 marks : 2010]

Solution:

Given,

Height of dam; $H = 100\,\text{m}$

Specific gravity of dam material, $S_c = 2.2$

Seepage coefficient at the base, $C = 0.8$

The width of elementary gravity dam is given by

$$B = \frac{H}{\sqrt{S_c - C}} = \frac{100}{\sqrt{2.2-0.8}} = \frac{100}{1.1832} = 84.52\text{ m}$$

3.5 **Determine the maximum base width of the elementary profile of a gravity dam with the following data:**

Specific gravity of the dam material = 2.4

Uplift intensity factor = 1.0

Coefficient of static friction = 0.80

Height of the dam = 20.0 m

[4 marks : 2015]

Solution:

Elementary profile of gravity dam is given as

Given: $S_c = 2.4,\quad c = 1.0$

$\mu = 0.80,\quad H = 20.0\text{ m}$

(i) For no tension failure

$$B = \frac{H}{\sqrt{S_c - c}} = \frac{20}{\sqrt{2.4-1}} = 16.9\text{m}$$

(ii) For no sliding failure

$$B = \frac{H}{\mu(S_c - c)} = \frac{20}{0.8(2.4-1)} = 17.86\text{ m}$$

Therefore width of dam will be greater of the two value calculated above i.e. $B = 17.86$ m

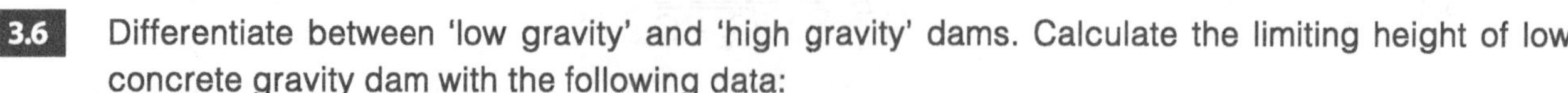

3.6 **Differentiate between 'low gravity' and 'high gravity' dams. Calculate the limiting height of low concrete gravity dam with the following data:**

Compressive strength of the concrete = 2500 kN/m^2

Unit weight of water = 9.81 kN/m^3

Specific gravity of the concrete = 2.4

[8 marks : 2016]

Solution:

Low gravity and high gravity dams:

The maximum normal stress in the dam i.e. principal stress near the toe computed for an elementary profile of gravity dam is given by,

$$\sigma = \gamma_w H(S_c - C + 1)$$

where
H = Height of dam
S_c = Specific gravity of dam material
C = A constant in the expression for uplift at heel

Let, f = Allowable compressive stress of the dam material.

$$\therefore \quad \sigma \leq f$$

$$\Rightarrow \quad \gamma_w H(S_c - C + 1) \leq f$$

$$\Rightarrow \quad H \leq \frac{f}{\gamma_w(S_c - C + 1)}$$

In case of no uplift pressure, $C\gamma_w H = 0$ then $C = 0$ and $H = H_{max}$

$$\therefore \quad H_{max} = \frac{f}{\gamma_w(S_c + 1)} \quad ...(i)$$

If $H > H_{max}$, then compressive stress in dam material will exceed allowable compressive stress (f) leading to its failure.

Thus, when height of dam (H) < H_{max} (as per eq. (i)) then it is called as low gravity dam else high gravity dam.

Given
$f = 2500 \text{ kN/m}^2$
$\gamma_w = 9.81 \text{ kN/m}^3$
$S_c = 2.4$

$\therefore$ **Limiting height of low concrete gravity dam,**

$$H_{max} = \frac{f}{\gamma_w(S_c + 1)}$$

$$_{max} = \frac{2500}{9.81(2.4 + 1)} = 74.95 \text{ m}$$

3.7 **Briefly describe various geological considerations for the site selection of the dam.**

[5 marks : 2017]

Solution:

Geological considerations in the selection of dam site: The important geological requirements that should be considered in the selection of a dam site are as follows:

- **Narrow River Valley :** If the proposed dam site contains a narrow river valley, only a small dam is required which means cost of dam construction will also be less. On the other hand, if the valley is wider, then construction cost will be very high and maintenance cost of dam will also be high.
- **Occurrence of bed rock at shallow depths :** If the dam rests on very strong and stable rocks, the stability and safety of dam will be very high.

- **Competent rocks to offer a stable foundation** : If the dam site consists of igneous rocks, they will offer a safe basis. If sedimentary rocks, particularly shales, poorly cemented sand stones and coverneous limestones are present then they are undesirable.

3.8 **Draw the schematic diagram of a gravity dam and indicate the major forces acting on it. Draw the diagram of the uplift force when (i) drain is not provided and (ii) drain is provided.**

[12 marks : 2019]

Solution:

Forces Acting on Gravity Dam

(i) Water Pressure

$$P = \frac{1}{2}\gamma_w H^2$$

Acting at $\frac{H}{3}$ from base

where γ_w = Unit weight of water.

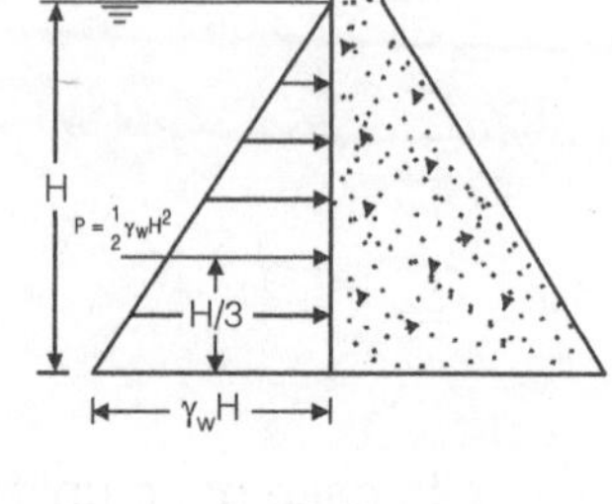

(ii) Uplift Pressure

(a) When Drainage Gallery is not Provided

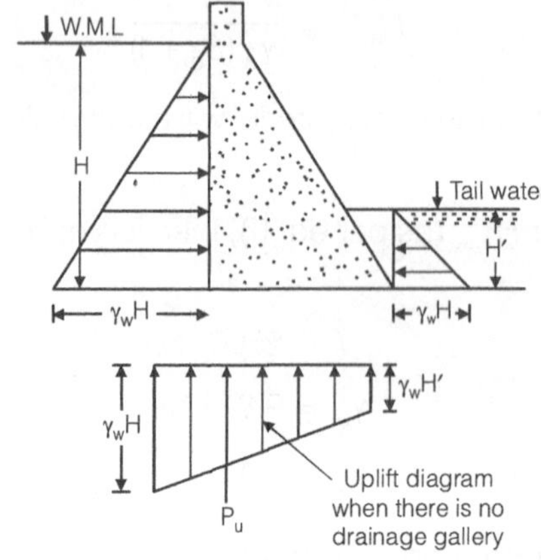

(b) When Drainage Gallery is Provided

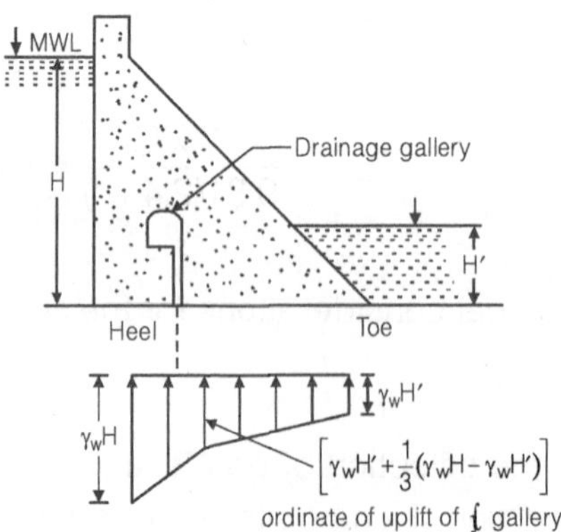

(iii) Earthquake Force

$$\alpha_H = 0.1g \text{ to } 0.2g$$

$$\alpha_V = 0.75\alpha_H$$

$$\alpha = \beta I \alpha_0$$

where, α_H = Horizontal acceleration, α_v = Vertical acceleration
α = Seismic coefficient, β = Soil foundation system factor
I = Importance factor
α_0 = Basic seismic coefficient which depends upon seismic zone of country.

$$F_g = \frac{W}{g}(g \pm \alpha_V)$$

where, F_g = Body force
g = Acceleration due to gravity, +ve for upward and −ve for downward.

(a) Hydrodynamic force

(i) $F_H = \left(\frac{W}{g}\right)\alpha_H$ F_H = Horizontal Inertia force.

(ii) It effect of water pressure due to earthquake distribution of pressure is parabolic.

$$P_e = 0.555\alpha\gamma_w H^2$$

at 0.424 H from base.
α = 0.1g to 0.2g

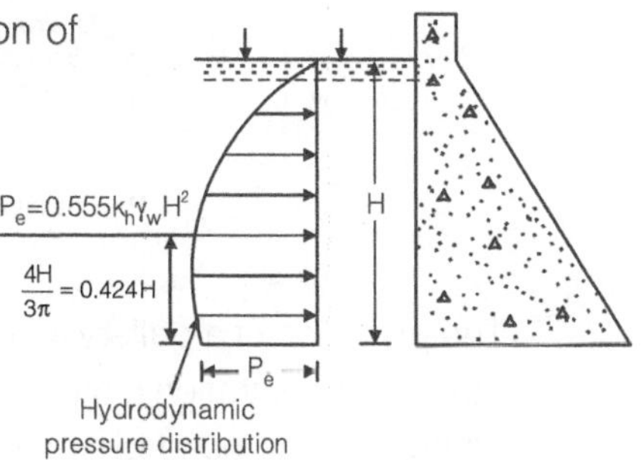

Hydrodynamic pressure distribution

(iv) Silt Pressure

$$F_{silt} = \frac{1}{2}k_a \gamma_{sub} \cdot h_s^2$$

where, h_s = Height of silt from the base.

k_a = Coefficient of active earth pressure = $\frac{1-\sin\phi}{1+\sin\phi}$

γ_{sub} = Submerged unit weight.

According to U.S.B.R

$$F_{silt} = \frac{360 h_s^2}{2}(kgf)$$

(v) Wave Pressure

$$P_w = 2 \cdot 4\gamma_w h_w$$

Acts at $\frac{h_w}{8}$ from still water level.

where, P_w = Resultant wave pressure.

$F_w = 2\gamma_w h_w^2$ acts at $\frac{3h}{8}$ from still water surface.

where F_w = Total wave force.

$h_w = 0.032\sqrt{VF} + 0.763 - 0.271(F)^{3/4}$ if $F < 32$ km.

$h_w = 0.032\sqrt{VF}$ when $F > 32$ km.

where, F = Length of reservoir in km
h_w = Height of wave in meter
V = Wind velocity in km/hr.

(vi) Self Weight of Dam

$$W = \gamma_c V$$

where, γ_c = Unit weight of concrete
V = Volume of dam body per unit length

3.9 **What do you understand by galleries and shafts and why are they provided in gravity dams?**

[12 marks : 2020]

Solution:

Galleries are the horizontal or sloping openings or passages left in the body of dam. They may run longitudinally (i.e. parallel to the dam axis) or transversely (i.e. normal to the dam axis) and are provided at various elevations. All the galleries are interconnected by steeply sloping passages or by vertical shafts fitted with stairs or mechanical lifts. The size of gallery will depend upon the size of the dam and the function of the gallery. Certain important shapes of the commonly used galleries are shown in figure below.

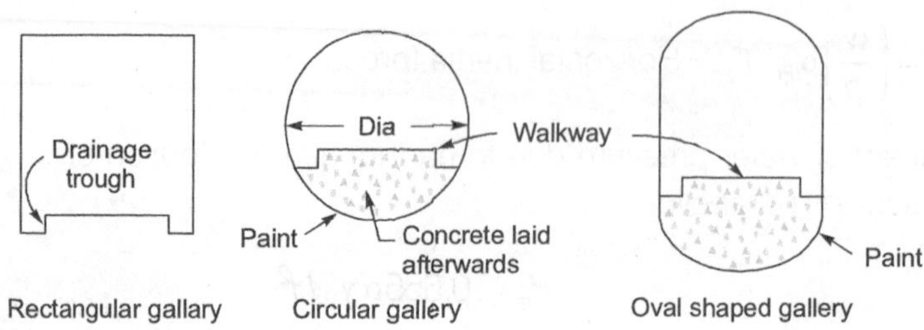

The provision of a gallery in a dam body, changes the normal pattern of stresses in the body of the dam. Stress concentration may occur at corners and hence in order to minimise this stress concentrations, the corner must be rounded smoothly. Tension and compression zones may be workout and proper reinforcement etc. are provided to counteract them.

Function of a gallery:

1. To provide drainage for the dam section.
2. To provide space for equipment required for drilling holes and grouting the hole to form a grout curtain in the foundation.
3. To provide space for header and return pipes for post cooling of concrete.
4. A gallery provides an access to the interior of the dam for inspection and maintenance.
5. A gallery also provides space for installing various instruments in the dam to study its structural behaviour.
6. A gallery can provide space for the mechanical and electrical equipment required for the operation of gates for spillways and outlets.

Shaft in gravity dam: A shaft is a vertical opening provided in a dam. Shafts are required for locating headers of the post cooling system and for locating measuring devices.

- Shafts are also required for the movement of elevators and the hoisting equipment. Sometimes, shafts are constructed inclined to connect two galleries or the same gallery at two different elevations.
- A plumb line shaft is constructed at the maximum section of the dam to make observations of the deflection of the dam under loads.
- A stilling well shaft is a special shaft used to record fluctuations of the water level in the reservoir. The shaft is connected to the reservoir at a point below the minimum reservoir level.

3.10 **A gravity dam is 18 m high, in triangular shape. The specific gravity of the dam material is 2.25. Find the minimum safe width of the dam. Use uplift factor K as 0.45. Also calculate the principal and shear stress at the toe of the dam. Consider safety against sliding when the reservoir is full.**

[20 marks : 2022]

Solution:

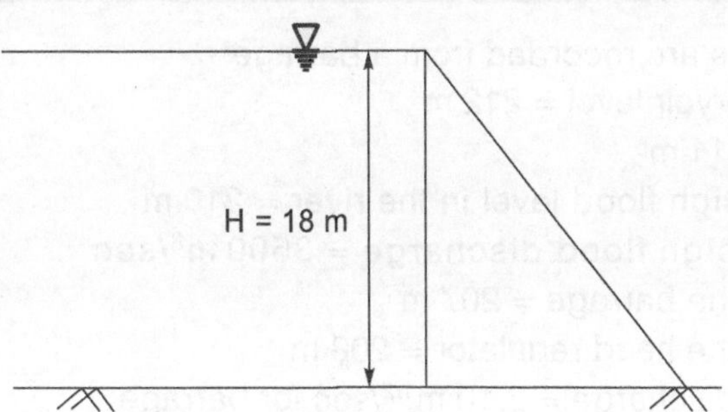

$$G_c = \text{Specific gravity of dam material} = 2.25$$

$$\text{Uplift factor, } K = 0.45$$

For safety against tension,

Let B_{min} = Minimum width of the dam

$$\therefore \quad B_{min} = \frac{H}{\sqrt{G_c - K}} = \frac{18\text{m}}{\sqrt{2.25 - 0.45}} = 13.4164 \text{ m}$$

$$= 13.42 \text{ m}$$

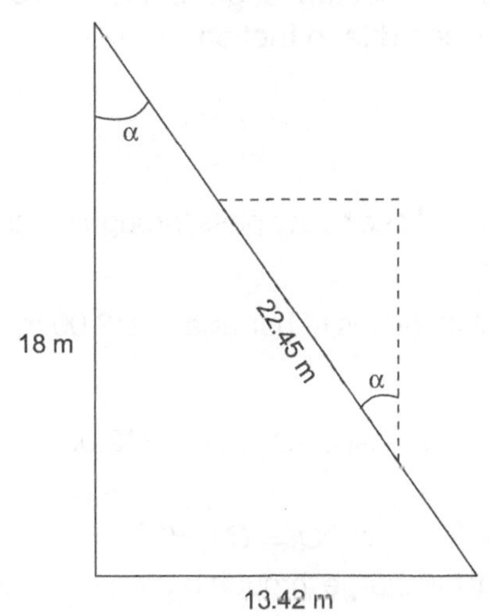

Principal stress at the toe of the dam

$$\sigma = \gamma_w H(G_c - K + 1)$$

$$= 9.81 \text{ kN/m}^3 \times 18 \text{ m} \times (2.25 - 0.45 + 1)$$

$$= 494.424 \text{ kN/m}^2$$

Maximum normal stress at the toe of the dam,

$$p_{\text{max toe}} = \gamma_w H(G_c - K)$$

$$= 9.81 \text{ kN/m}^3 \times 18 \text{ m}(2.25 - 0.45)$$

$$= 317.844 \text{ kN/m}^2$$

Shear stress at the toe,

$$\tau_{toe} = p_{\text{max toe}} \tan\alpha$$

$$= 317.844 \times \frac{13.42}{18} = 236.97 \text{ kN/m}^2$$

4. Water Logging, Theories of Seepage and Spillway

4.1 Following particulars are recorded from a barrage:
(i) Maximum reservoir level = 212 m
(ii) Pond level = 211 m
(iii) Downstream high flood level in the river = 210 m
(iv) Maximum design flood discharge = 3500 m^3/sec
(v) Crest level of the barrage = 207 m
(vi) Crest level of the head regulator = 208 m
(vii) Coefficient of discharge = 2.10 $m^{1/2}$/sec for barrage
(viii) Coefficient of discharge = 1.50 $m^{1/2}$/sec for head regulator
(ix) River bed level = 205 m
(x) Design discharge of main canal = 500 m^3/sec

Determine the number of gates required for the barrage and the head regulator if each gate has 10 m clear span. Neglect
(a) End contractions due to piers and abutments
(b) Velocity of approach

If a stilling basin is provided downstream of the barrage for the energy dissipation, find the length and RL of the basin floor. Assume that the length of basin is 5 times the conjugate depth required for hydraulic jump. Neglect losses due to friction.

[20 marks : 2001]

Solution:

1. Waterway for Barrage

High flood discharge of 3500 m^3/sec has to pass through the gated openings provided at the barrage site, with the following data:

Maximum water level at high flood u/s of barrage = 212.00 m

Crest level of Barrage = 207.00 m

∴ Opening available above the crest of barrage = 212.00 – 207.00 = 5.0 m

Now, we know that

$$Q = C\,L\,H^{3/2}$$

where C = coefficient of discharge through barrage = 2.10
L = effective waterway of barrage
H = 5.00 m
Q = 3500 m^3/sec

∴ $$3500 = 2.10 \times L \times (5)^{3/2}$$

⇒ $$L = \frac{3500}{2.10 \times (5)^{3/2}} = 149.1 \text{ m say } 150 \text{ m}$$

Effective waterway = Total waterway (ignoring end contractions)

Hence providing 15 bays, each of 10 m clear span.

∴ Number of gates required for barrage = 15

2. Waterway for canal head regulator

Discharge through main canal = 500 m^3/s

Coefficient of discharge for head regulator = 1.50

Head over the crest of regulator at pond level = 211 – 208 = 3 m

We know that $$Q_1 = C_1\,L_1\,H_1^{3/2}$$

$$\Rightarrow \quad 500 = 1.5 \times L_1 \times (3)^{3/2}$$

$$\Rightarrow \quad L_1 = \frac{500}{1.5 \times (3)^{3/2}} = 64.15 \text{ m say } 64 \text{ m}$$

Hence, providing 7 bays each of 10 m clear span, thereby providing 70 m waterway. So, number of gates required for canal head regulator each of 10 m span = 7

3. **Design of stilling basin**

$$\text{U/S HFL} = 212.00$$

$$\text{D/S HFL} = 210.00$$

$$\text{U/S TEL} = 212.00 \text{ (Ignoring velocity of approach)}$$

$$\text{D/S TEL} = 210.00 \text{ (Ignoring velocity of approach)}$$

$$H_L = 212 - 210 = 2 \text{ m}$$

Discharge intensity through barrage bays $= \frac{3500}{150} = 23.33 \text{ m}^3\text{/sec/m}$

Critical depth, $\quad y_c = \left(\frac{q^2}{g}\right)^{1/3} = \left[\frac{(23.33)^2}{9.81}\right]^{1/3} = 3.81 \text{ m}$

Now let $\quad Y = y_2/y_c$

and $\quad z = \frac{H_L}{y_c}$

$\therefore \quad z = \frac{2}{3.81} = 0.5249$

Computing value of Y using $\quad Y = 1 + 0.93556\, z^{0.368}$ for $z < 1$

$\Rightarrow \quad Y = 1 + 0.93556\,(0.5249)^{0.368}$

$\Rightarrow \quad Y = 1.738$

Now, $\quad Y = \frac{y_2}{y_c}$

$\Rightarrow \quad y_2 = Y \times y_c = 1.738 \times 3.81 = 6.62 \text{ m}$

$\therefore \quad 5y_2 = \text{length of cistern} = 5 \times 6.62 = 33.10 \text{ m}$

Now E_{f_2} can be calculated as $\quad \frac{E_{f_2}}{y_c} = Y + \frac{1}{2Y^2}$

$$E_{f_2} = 3.81 \times \left[1.738 + \frac{1}{2 \times (1.738)^2}\right] = 7.25 \text{ m}$$

Hence level at which jump will form = D/S TEL − E_{f_2} = 210 − 7.25 = 202.75 m

Hence provide cistern at RL = 202.75 m

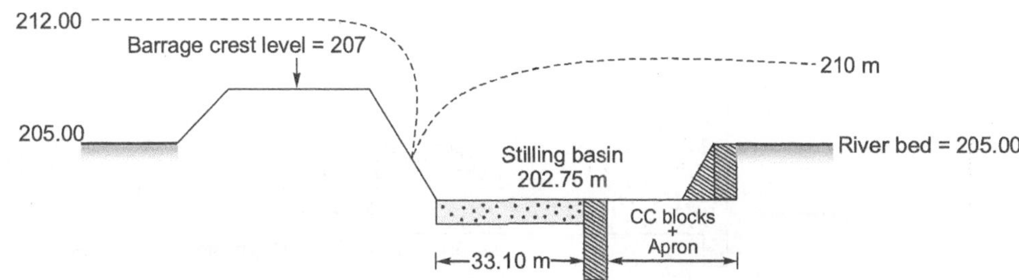

4.2 **Water discharges at the rate of 298 m³/sec over a spillway 14 m wide into a stilling basin of same width. The water level behind the spillway is 61 m over datum. The river water level in the downstream is 29.2 m over datum. The river bed is paved flush with the bed level to just adequate length to form the jump basin. Assuming that no energy is dissipated in the flow down the spillway, find the basin invert level required for a hydraulic jump to be formed within the basin.**

[15 marks : 2003]

Solution:

Given: Discharge (Q) = 298 m³/sec, Width of spillway (B) = 14 m, Water level behind spillway = 61 m (above datum), River water level in D/S = 29.2 m (above datum)

$\therefore$ Head loss, $H_L = 61 - 29.2 = 31.8$ m

Also, discharge intensity, $q = \frac{Q}{B} = \frac{298}{14} = 21.29$ m³/s/m

We know that critical depth,

$$y_c = \left(\frac{q^2}{g}\right)^{1/3} = \left[\frac{(21.29)^2}{9.81}\right]^{1/3} = 3.59 \text{ m}$$

Now Let $z = H_L/y_c = \frac{31.8}{3.59} = 8.86$

and $Y = 1 + 0.93556\, z^{0.240}$ for $z > 1$

$\therefore$ $Y = 1 + 0.93556\,(8.86)^{0.240}$

$\Rightarrow$ $Y = 2.58$

But $Y = \frac{y_2}{y_c}$

$\Rightarrow$ $y_2 = Y \times y_c = 2.58 \times 3.59 = 9.26$ m

$\therefore$ The basin invert level $= D/S$ water level $- y_2 = 29.2 - 9.26 = 19.94$ m above datum

4.3 **Explain the various relative positions of Tail Water-rating Curve (TWC) and Jump Height Curve (JHC), and the types of energy dissipating devices which shall be recommended below the spillway for each condition.**

[10 marks : 2006]

Solution:

In relation to the relative positions of tail water rating curve (TWC) and Jump height rating curve (JHC) **following types of energy dissipation devices** may be recommended:

(i) When TWC coinciding with JHC at all discharges

This is the most ideal condition for jump formation. The hydraulic jump will form at the top of the spillway at all discharges. In such a case a simple concrete apron of length 5 ($y_2 - y_1$) is generally sufficient to provide protection in the region of hydraulic jump.

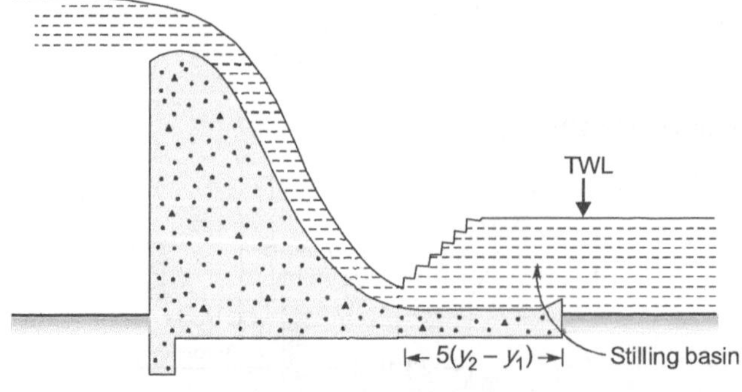

Simple horizontal apron

(ii) When TWC is lying above the JHC at all discharges

In this case, when y_2 is always below the tail water, the jump forming at toe will be drowned out by the tail water and little energy will be dissipated. This problem can be solved by providing a sloping apron above the river bed level or by providing a roller bucket type of energy dissipator.

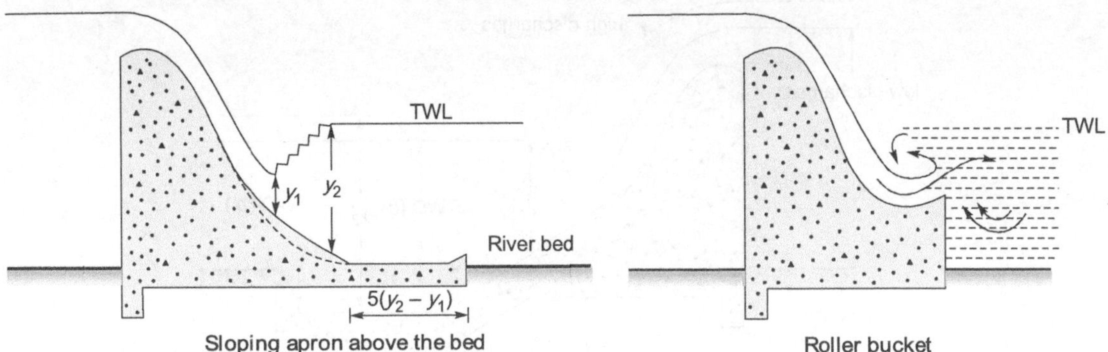

Sloping apron above the bed

Roller bucket

(iii) When TWC lies below the JHC at all discharges

If the tail water depth is very low, the water may shoot out of the above bucket and fall harmlessly into the river at some distance downstream of the bucket. This bucket is then known as Ski jump.

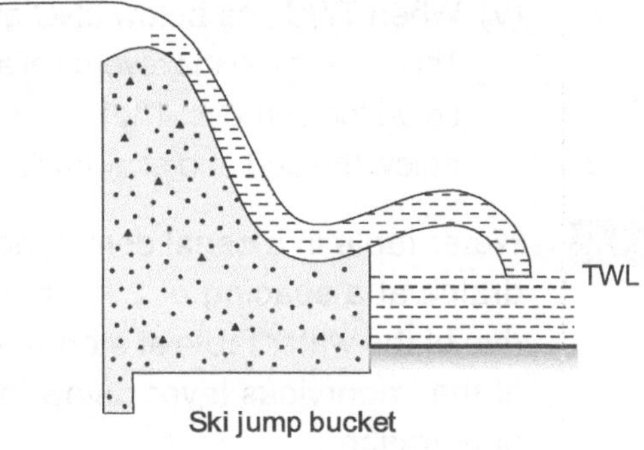

Ski jump bucket

The second solution to this problem can be the provision of providing a sloping apron but below the river bed. The required depth y_2 which is greater than TW depth can thus be made available by letting the jump form on this sloping apron as shown.

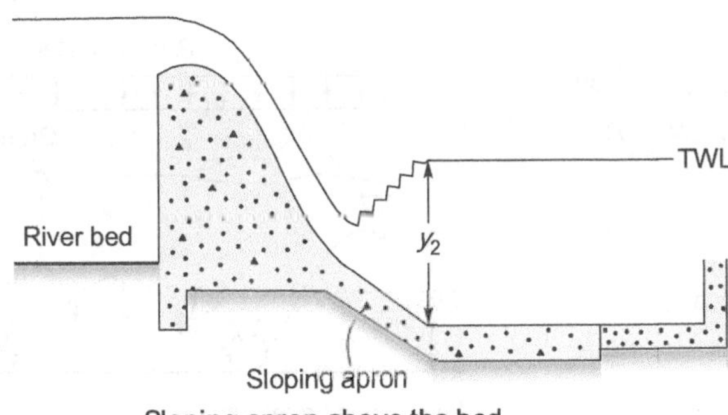

Sloping apron above the bed

The third solution to this problem may be the construction of a subsidiary dam below the main dam, so as to increase the tail water depth and cause a jump to form at the toe of the main dam.

If the tail water deficiency is small, a baffle wall or a row of friction blocks may be provided so as to dissipate the residual energy.

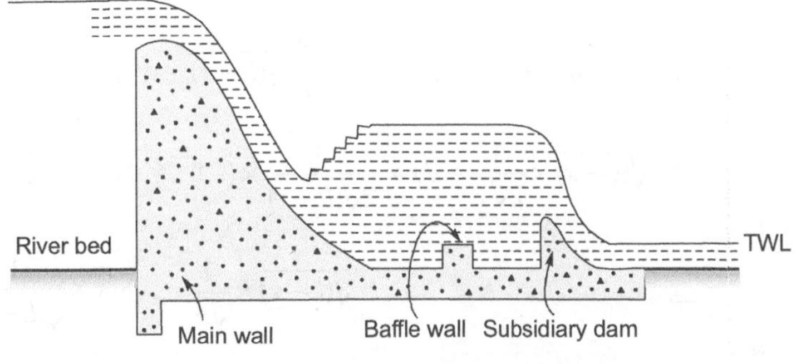

(iv) When TWC lies above JHC at low discharges and lies below JHC at high discharges.

In this case at low discharges, the jump will be drowned and at high discharges tail water is insufficient. The solution to this problem lies in providing a sloping apron partly above and party below the river bed.

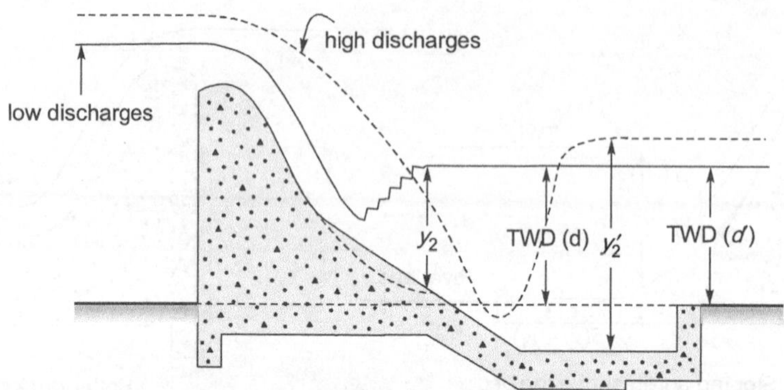

Sloping apron partly above and partly below the ground level

(v) When TWC lies below JNC at low discharges and lies above at high discharges.

This case is just the reverse of above case and hence the same arrangement which is made above will serve the purpose. The only difference will be that at low discharges, the jump will form on the apron below the bed and at high discharges, the jump will form on the apron at a point above the bed.

4.4 Water table in a canal command receives a recharge at the rate 2.5 mm/day. Sub-surface ditch drains at a spacing of 2 km are provided for the sub-surface drainage. Estimate the maximum rise of the water table at steady state. Given : hydraulic conductivity of the soil = 10 m/day, depth of the impervious layer below initial water table position = 20 m. Assume the ditches to be fully penetrating.

[10 marks : 2009]

Solution:

The water table profile for one dimensional **Dupit's flow** with recharge in the case of ditch drains is given by

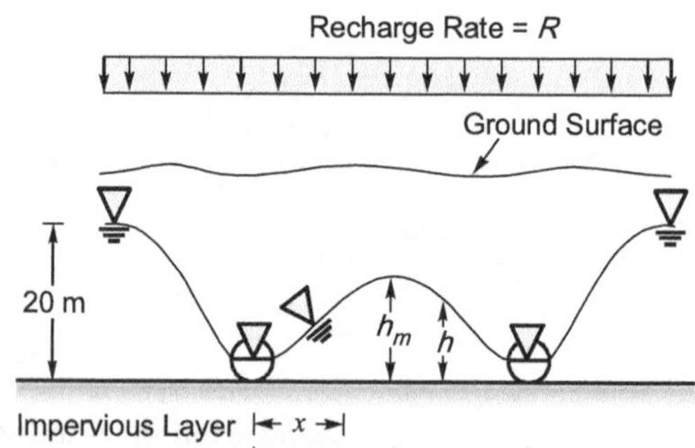

$$h^2 = \frac{R}{k}(L-x)x \qquad \text{...(i)}$$

where k is the hydraulic conductivity of the aquifer.

For maximum rise of the water table at the steady state,

$$\frac{dh}{dx} = 0$$

$$\Rightarrow \quad \frac{d}{dx}\sqrt{\frac{R}{k}(L-x)x} = 0$$

$$\Rightarrow \quad \sqrt{(L-x)}\frac{d}{dx}\sqrt{x} + \sqrt{x}\frac{d}{dx}\sqrt{(L-x)} = 0$$

$$\Rightarrow \quad \sqrt{(L-x)} \times \frac{1}{2\sqrt{x}} + \sqrt{x} \times \frac{1}{2\sqrt{(L-x)}} \times (-1) = 0$$

$$\Rightarrow \quad \sqrt{(L-x)} \times \frac{1}{\sqrt{x}} = \sqrt{x} \times \frac{1}{\sqrt{(L-x)}}$$

$$\Rightarrow \quad L - x = x$$

$$\Rightarrow \quad x = \frac{L}{2}$$

Substituting, $x = \frac{L}{2}$ in (i), we get $\quad (h_m)^2 = \frac{R}{k}\left(L - \frac{L}{2}\right) \times \frac{L}{2}$

$$\Rightarrow \quad h_m = \frac{L}{2} \times \sqrt{\frac{R}{k}}$$

Here L = 2 km; R = 2.5 mm per day; k = 10 m per day

$$\therefore \quad h_m = \frac{2 \times 10^3}{2} \times \sqrt{\frac{2.5 \times 10^{-3}}{10}} = 15.81 \text{ m}$$

4.5 **Drawdown from multiple observation well in a confined aquifer under steady state flow condition when 100 m^3/day is pumped is given below. Determine the transmissivity assuming the linear variation semi log graph on the given data :**

r (m) distance	30	40	50	100	200	400
s (m) drawdown	2.79	2.505	2.38	1.83	1.28	0.73

[10 marks : 2011]

Solution:

Using **Jacob method** to determine transmissivity

$$s = \frac{2.3Q}{4\pi T}\log_{10}\left(\frac{2.25\,Tt}{r^2 S}\right) \quad \text{...(i)}$$

s = Drawdown

r = Distance of measuring of drawdown from well

Q = Constant discharge = 100 m^3/d

T = Transmissivity, constant for the aquifer

S = Storage coefficient, constant for the aquifer

t = Time of measurement of drawdown

At r = 30 m, s = 2.79 m

From (i)
$$2.79 = \frac{2.3 \times 100}{4\pi T}\log_{10}\left(\frac{2.25T\,t}{30^2 \times S}\right) \quad \text{...(ii)}$$

At r = 40 m, s = 2.505
$$2.505 = \frac{2.3 \times 100}{4\pi T}\log_{10}\left(\frac{2.25T\,t}{40^2 \times S}\right) \quad \text{...(iii)}$$

Equation (ii) – (iii)

$$\Rightarrow \quad 0.285 = \frac{2.3 \times 100}{4\pi T}\log_{10}\left[\frac{2.25T\,t}{30^2 \times S} \times \frac{40^2 S}{2.25\,Tt}\right]$$

$$\Rightarrow \quad 0.285 = \frac{230}{4\pi T}\log_{10}\left[\left(\frac{40}{30}\right)^2\right]$$

$$\Rightarrow \quad T = 16.05 \text{ m}^2\text{/day} = 1.86 \text{ cm}^2\text{/s}$$

4.6 Check the safety of a barrage shown below against the piping action. The safe exit gradient is 1/6.

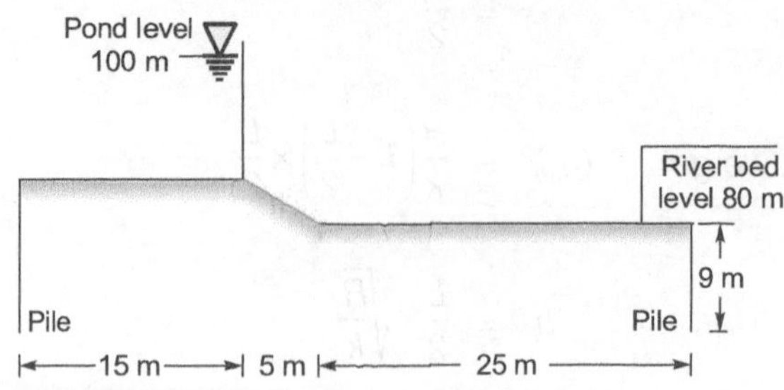

[4 marks : 2013]

Solution:

Max. seepage head available

$$H = 100 - 80 = 20 \text{ m}$$

Depth of downstream cut-off = 9 m

Total flow length, $b = 15 + 5 + 25 = 45$ m

$$\therefore \qquad \alpha = \frac{b}{d} = \frac{45}{9} = 5$$

$$\therefore \qquad \lambda = \frac{1+\sqrt{1+\alpha^2}}{2} = 3.04951$$

$$\therefore \qquad \frac{1}{\pi\sqrt{\lambda}} = 0.1823$$

$$\therefore \quad \text{Exit gradient} \quad (G_E) = \frac{H}{d}\cdot\frac{1}{\pi\sqrt{\lambda}} = \frac{20}{9}(0.1823) = 0.40511$$

$$\text{Given safe exit gradient} = \frac{1}{6} = 0.16778 < 0.40511$$

Thus barrage is not safe against piping.

4.7 An over flow ogee spillway has six bays each of 8 m clear length and five rounded nose piers ($K_p = 0.01$) each of 2.5 m thick. Both the abutments are sharp edged ($K_a = 0.2$). The upstream face of the spillway is vertical. Compute the discharge that will pass over the spillway at the design head, $H_d = 3$ m. The discharge coefficient for design head $C_o = 2.1$. Also compute the passing discharge over spillway for head over the crest $H = 3.45$ m.

Neglect the approach velocity to the spillway. Discharge coefficients for head other than design head are given below:

H/H_d	0.6	0.85	1.15	1.50
C/C_O	0.94	0.98	1.02	1.06

C is discharge coefficient for head *H*.

[10 marks : 2013]

Solution:

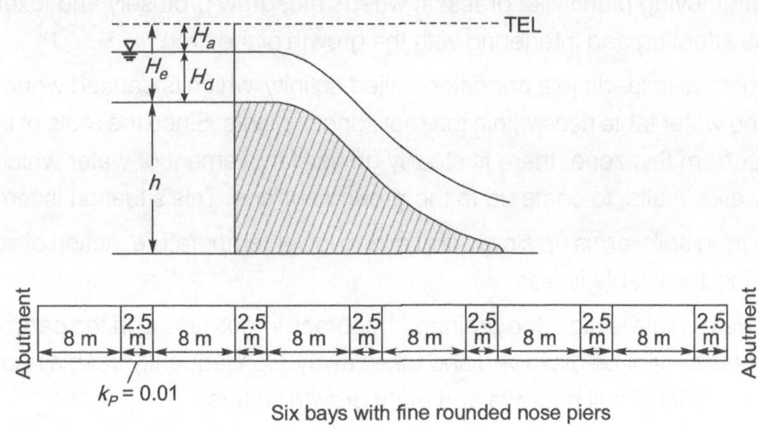

Six bays with fine rounded nose piers

Design head, $H_d = 3$ m

Design head discharge coefficient, $C_o = 2.1$

$$H_e = H_d + H_a$$

Where H_a = Head due to velocity of approach $= \dfrac{v^2}{2g}$

Spillway discharge, $Q = CL_eH_e^{3/2}$

Where $L_e = L - 2[NK_p + K_a]H_e$

Initially neglecting the approach velocity, $H_a = 0$

$\therefore$ $H_e = H_d = 3$ m

$\therefore$ $Q = C_oL_eH_e^{3/2} = (2.1)L_e(3)^{3/2}$

Where $L_e = 6 \times 8 - 2[5 \times 0.01 + 0.2]3 = 46.5$ m

$= (2.1)(46.5)(3)^{3/2} = 507.404$ m³/s

$\therefore$ $$V_a = \frac{507.404}{3(6\times8+5\times2.5)} = 2.7956 \text{ m/s}$$

$$\frac{H}{H_d} = \frac{3.45}{3} = 1.15$$

$\Rightarrow$ $$\frac{C}{C_o} = 1.02 \text{ (from table)}$$

Hence $C = (1.02)(2.1) = 2.142$

$$Q = CL_eH_e^{3/2} = (2.142)(46.5)(3.45)^{3/2}$$

$= 638.27$ m³/s

4.8 What are the effects of water logging?

[5 marks : 2020]

Solution:

Effect of water logging are as follows:

- Normal cultivation operations, such as tiling, ploughing etc. cannot be easily carried out in wet soils. In extreme cases, the free water may rise above the ground level making agricultural operations impossible.

- Certain water loving plants like grasses, weeds etc. grow profusely and luxuriantly in waterlogged lands, thus affecting and interfering with the growth of the crop.
- Water logging also leads to a condition called salinity, which is caused when the capillary fringe of the elevated water table rises within the root zone of plants. Since the roots of the plants continuously draw water from this zone, there is steady upward movement of water which causes rise of salts, especially alkali salts, to come up to the ground surface. This situation is termed as salinity.
- A water logged soil warms up slowly and due to lower temperature, action of soil bacteria is sluggish and plant food available is less.
- When the water table is high, the drainage becomes impossible and the carbon dioxide liberated by the plant roots cannot be dissolved and taken away. Consequently fresh air containing oxygen is not drawn and activity of soil bacteria and plant growth suffers.
- In waterlogged soils, natural flora such as water hyacinth grows profusely. This reduces the crop yield.
- The climate of a waterlogged area becomes damp. Formation of stagnant pools may become breeding places for mosquitoes. The climate thus becomes extremely detrimental to the health of community.

5. River Training Work, Diversion Headwork, Cross-drainage Work and Miscellaneous

5.1 **Explain the purpose of providing groynes. State the factors which affect the selection of type and performance of groyne. Mention about the orientation, desirable length and spacing of groyne in a groyne system.**

[10 marks : 2001]

Solution:

Groynes are the embankment type structures constructed transverse to the river flow, extending from the bank into the river. That is why they may also be called transverse dykes. These are constructed in order to protect the bank from which they are extended by deflecting the current away from the bank. As the water is unable to take a sharp turn, the bank gets protected for certain distance upstream and downstream of the groyne. However, the nose of the groyne is subjected to tremendous action of water and has to be heavily protected by pitching, etc. The action of eddies reduces from the head towards the bank and therefore the thickness of slope pitching and apron can be reduced accordingly.

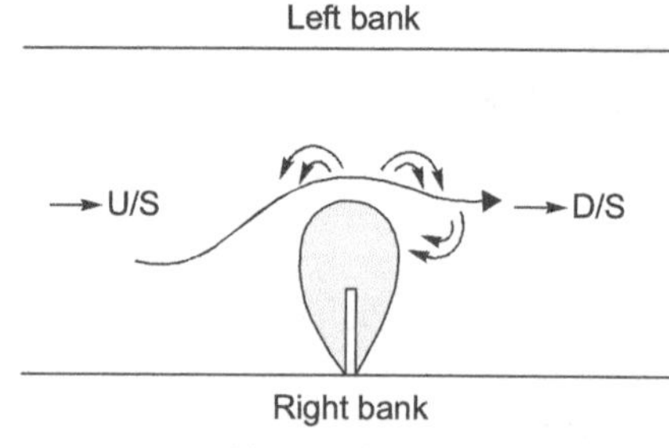

Normal Groyne

Factors affecting the selection of type of performance of groyne:

(i) River discharge
(ii) Angle of attack
(iii) Sediment load
(iv) Meander length
(v) Curvature of river
(vi) U/S and D/S river training measures

Based on the orientation or alignment groynes may be built either perpendicular to the bank line or they may be inclined U/S or D/S. Thus, **they may be classified as normal groyne, repelling groyne and attracting groyne.**

The length of groynes depend upon the position of the existing bank line and the designed or expected bank line for trained river. Though long groynes on easily erodible rivers are susceptible to damage and failure. In such cases, the best results can be obtained by starting with a shorter length and to extend the groyne gradually as silting between them proceeds. Groyne length is usually restricted to less than 20% of the river width and at the same time the groyne length is kept longer than 1.5 to 2 times the depth of flow.

Spacing of groynes depends on length of groyne, type of bank, width of river and type of groyne. Large spacing is required for locating groynes on convex banks and a smaller one for concave banks. A spacing of 2 to 2.5 times the length of groyne is generally adopted at convex banks while a spacing equal to the length of the groyne is most adopted for concave banks. Larger spacing is preferred for wider rivers than for narrower rivers. A higher value of spacing may be used for permeable groynes as compared to that required for impermeable groynes.

5.2 **Explain briefly the functions of the following for a typical canal head works:**
(i) Divide wall (ii) Guide bank (iii) Sediment ejector.
Explain briefly how does seepage endanger the safety of a hydraulic structure on permeable foundation.

[15 marks : 2004]

Solution:

(i) Divide wall :

The main functions served by the divide wall are:

(a) It separates the under sluice from weir proper. Since the crest level of the under sluices is lower than of the weir proper, the two must be separated and this is being done by the divide wall.

(b) It helps in providing a comparatively less turbulent pocket near the canal head regulator, resulting in deposition of silt in this pocket and thus to help in the entry of silt free water into the canal.

(c) Divide wall keeps the cross currents, if at all they are formed, away from the weir because cross currents leads to vortices and deep scours.

(ii) **Guide banks :** The functions of guide banks are:

(a) Guide banks confine the flood water of alluvial rivers within reasonable waterway and provide a straight non-tortuous approach towards a work constructed across the river.

(b) Guide banks tends to prevent the river from changing its course and out flanking the work.

(iii) Sediment ejector : Sediment ejector are those silt control devices which remove the silt which has already entered the canal from the head. These are provided in a canal a little distance downstream from the head of the canal.

The seepage may endanger the safety of a hydraulic structure on permeable foundation in the following two ways:

(a) By piping or undermining: If the water percolating through the foundation has sufficient force when it emerge at the downstream end of the floor it may lift up the soil particles at the end of the floor. With the removal of the surface soil there is further concentration of flow into the resulting depression and more soil is removed. This process of erosion progressively extends backwards towards the upstream side and results in the removal of soil and developing pipe like formation beneath the floor. The floor may subside in the hollows so formed and fail which is known as failure due to piping or undermining. This problem can be solved by providing sufficient length of the floor and by providing piles at the U/S and D/S ends of the floor.

(b) By uplift pressure: The water percolating through the foundation exerts an upward pressure on the floor which is known as uplift pressure. It the uplift pressure is not counter balanced by the weight of the structure, it may fail by rupture. This can be prevented by providing sufficient thickness of the floor or by providing pile at the U/S end of the floor so that the uplift pressure is reduced on the D/S side.

5.3 **Discuss the structural and non-structural measures for flood management and with a specific bias to levee.**

[10 marks : 2005]

Solution:

The commonly adopted measures for flood management are:

(i) Reduction of peak flow by reservoirs.
(ii) Reduction of peak stage by increased velocity resulting from channel improvement.
(iii) Diversion of flood waters through flood ways or by passes.
(iv) Temporary or permanent evacuation of the flood plain.
(v) Flood proofing of specific properties.
(vi) Reduction of flood runoff by land management.
(vii) Confinement of the flow within a predetermined channel by levees, flood walls or closed conduits.

Levees are one of the oldest and most widely used methods of protecting land from flood water. These are essentially longitudinal dams erected roughly parallel to a river, to serve as artificial high banks of the river and thus prevent the river water from spilling over the country side during floods. The flood water is confined between the levees and is made to flow down the river without causing any damage to the country side of the levees.

Levees are most frequently used for flood control because they can be built at relatively low cost of materials available at the site. Levees are generally built of material excavated from borrow pits parallel to the levee line. The material should be placed in layers and compacted, with the least pervious material being placed along the river side of the levee. Usually, there is no suitable material for a core, and hence most of the levees are homogeneous embankments.

5.4 **Explain the causes of meandering and describe the various types of groynes used for river training.**

[10 marks : 2006]

Solution:

Formation of successive bends of reverse order may lead to the formation of a complete S-curve called **meander**. When consecutive curves of reverse order connected with a short straight reaches are developed in a river reach, the river is said to be a meandering river. The cause of meandering is the excess turbulence generated by excess of river sediment during floods. During floods, the river carries tremendous amount of silt charge. It has been established that when the silt charge is in excess of the quantity required for stability, the river starts building up its slope by depositing the silt on its bed. In other words, the river reach becomes an aggrading or of an accreting type. This accretion is the primary process, which consequently leads to meandering.

Various types of groynes can be grouped under three heads:

(i) Based on type of alignment
(ii) Based on material of construction
(iii) Special groynes

There are three types of groynes based on type of alignment which are as follows:

(i) **Normal groyne:** This groyne is aligned perpendicular to the bank line and is also known as ordinary groyne.

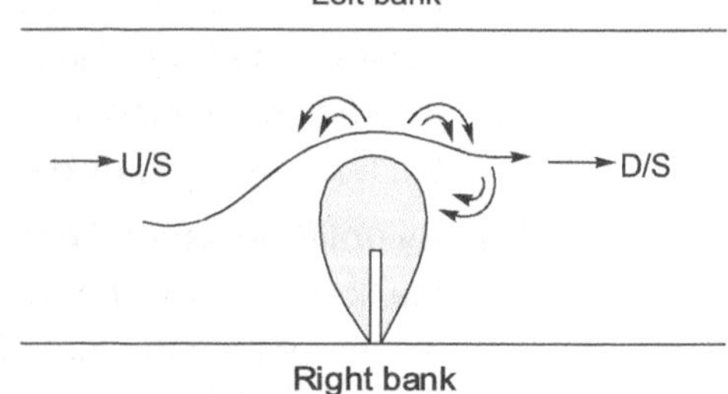

(ii) **Repelling groyne:** This groyne always point upstream and has the property of repelling the flow away from it and scour holes caused by the formation of vertical eddies are developed away from the bank and near the head of the groyne.

(iii) **Attracting groyne:** This groyne points downstream and has the property of attracting the flow towards it. In such a groyne scour holes are developed near the bank as compared to those in a repelling groyne.

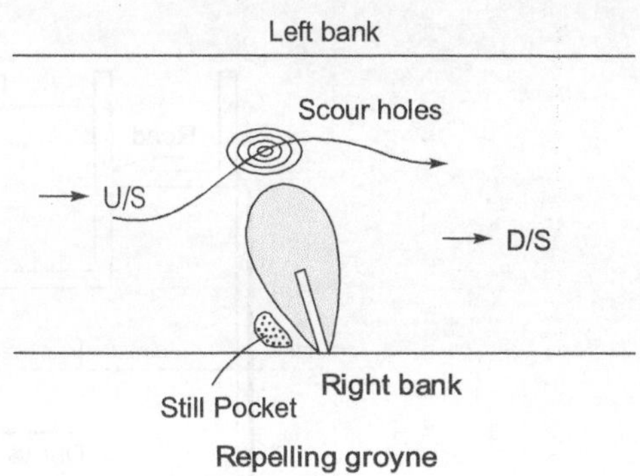

Repelling groyne

There are two types of groynes based on material of construction:

(i) **Impermeable groyne:** These are also called solid groynes or embankment groynes. These groynes may be rockfill embankments or earthen embankments armoured with stone pitching, concrete blocks, etc.

(ii) **Permeable groynes:** These groynes do permit restricted flow through them. These groynes are more or less temporary structures and are susceptible to damage by floating debris, etc. The common materials used are tree spurs or balli spurs.

There are two type of special groynes:

(i) **T-shaped groyne:** Denehey's *T*-shaped groyne is a special type of groyne developed in India. It is an ordinary groyne provide with an extra cross groyne at the head given it a *T*-shape. The longer arm of the *T* is provided on U/S and shorter one on D/S.

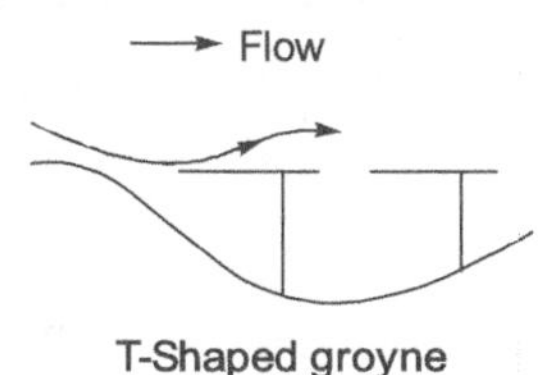

T-Shaped groyne

(ii) **Hockey shaped groyne:** These groynes are shaped like hockey sticks at their lower end. These groynes exert an attracting type of influence on the flow and hence are not useful for bank protection for repelling the current away from it.

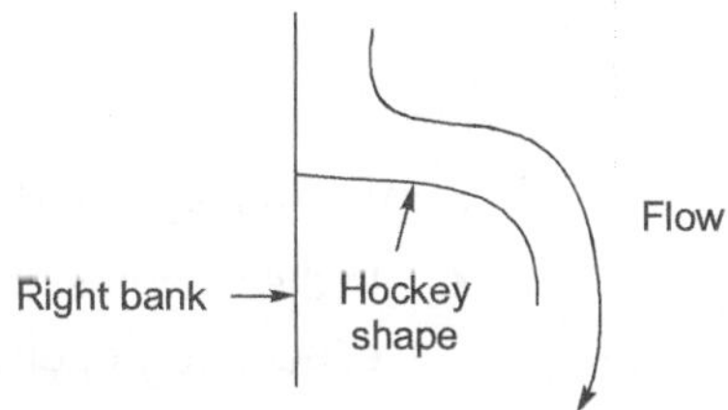

5.5 **What is a cross drainage work? What are the various types of cross drainage works?**

[5 marks : 2008]

Solution:

A cross drainage work is a structure which is constructed at the crossing of a canal and a natural drain, so as to dispose of drainage water without interrupting the continuous canal supplies. In whatever way the canal is aligned, such cross drainage works generally becomes unavoidable. A cross-drainage work is generally a costly construction and must be avoided as far as possible. Since a watershed canal crosses minimum number of drains such an alignment is preferred to a contour canal which crosses maximum number of drains.

Types of cross drainage works are as follows:

(i) **Aqueduct and syphon aqueduct:**

- In these works, the canal is taken over the natural drain, such that the drainage water runs below the canal either freely or under syphoning pressure.
- When the HFL of the drain is sufficiently below the bottom of the canal, so that the drainage water flows freely under gravity, the structure is known as an aqueduct.

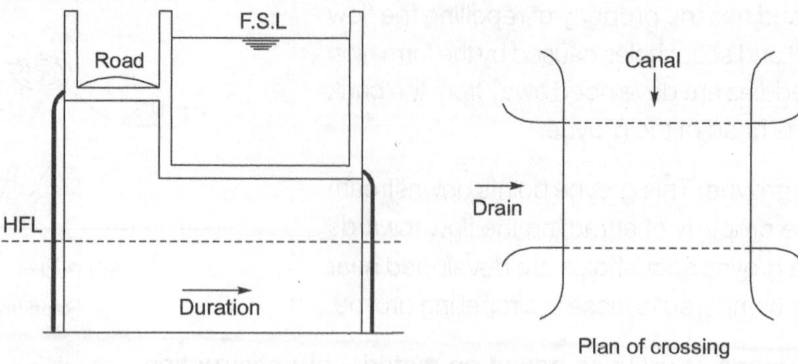

- If the HFL of the drain is higher than the canal bed and the water passes through the aqueduct barrels under syphonic action, the structure is known as syphon aqueduct.

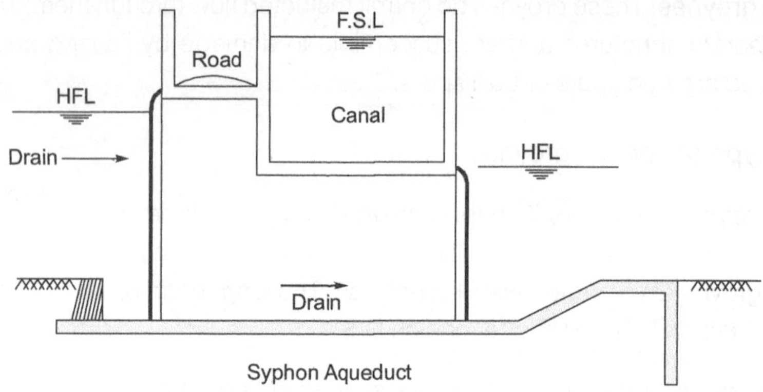

Syphon Aqueduct

(ii) Super Passage and Syphon:

- In these works, the drain is taken over the canal such that the canal water runs below the drain either freely or under syphoning pressure
- When FSL of the canal is sufficiently below the bottom of the drain trough, so that the canal water flows freely under gravity, the structure is known as a super passage.

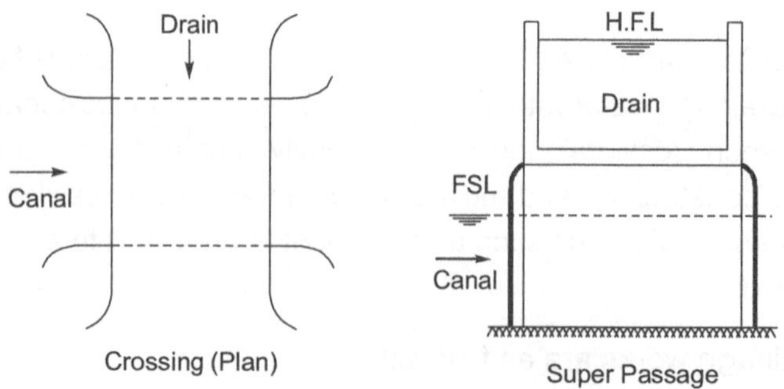

- If the FSL of the canal is sufficiently above the bed level of the drainage trough, so that the canal flows under syphonic action under the trough, the structure is known as a canal syphon or a syphon.

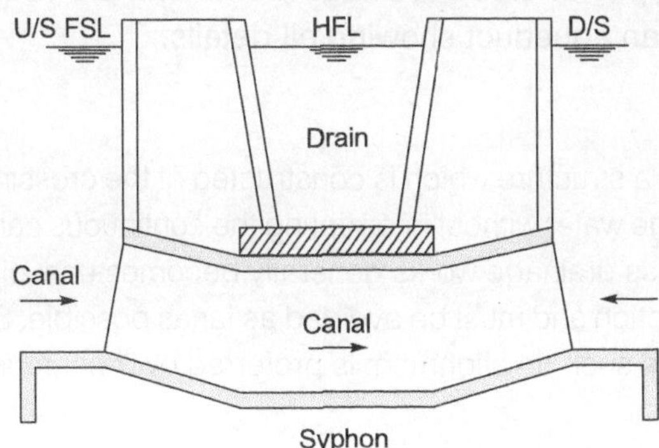

(iii) Level Crossing:

- In this type of cross-drainage work, the canal water and drain water are allowed to intermingle with each other.
- A level crossing is provided generally when a large canal and a high drain approach each other practically at the same level.

(iv) Inlets and Outlets:

- An inlet is a structure constructed in order to allow the drainage water to enter the canal and get mixed with the canal water and thus to help in augmenting canal supplies.
- Such a structure is generally adopted when the drainage discharge is small and the drain crosses the canal with its bed level equal to or slightly higher than the canal FSL.
- When the drainage discharge is high or if the canal is small, so that the canal cannot take the entire drainage water, an outlet may sometimes be constructed to escape out the additional discharge at a suitable site, a little downstream along the canal.

5.6 **Find the probable life of a reservoir with an initial capacity of 4×10^6 m^3 if the annual sediment inflow into reservoir is 4×10^4 m^3. The average trap efficiency is 0.9 and life of the reservoir is terminated when 90 per cent of initial capacity is occupied by sediment.**

[2 marks : 2010]

Solution:

Given, Initial capacity of reservoir $= 4 \times 10^6$ m^3

Annual sediment inflow $= 4 \times 10^4$ m^3

Average trap efficiency $= 0.9$

Life of reservoir is terminated when 90% of initial capacity is occupied by sediment.

$\therefore$ 90% of initial capacity $= 0.9 \times 4 \times 10^6 = 36 \times 10^5$ m^3

Volume of sediment deposited annually till 90% of initial capacity is filled

= Annual sediment inflow × trap efficiency

$= 4 \times 10^4 \times 0.9$

$= 36 \times 10^3$ m^3

$\therefore$ Number of years during which 90% of the initial capacity i.e. 36×10^5 m^3 shall be filled up is given by

$$\text{Probable life of reservoir} = \frac{36 \times 10^5}{36 \times 10^3} = 100 \text{ years}$$

5.7 What are the different types of cross drainage works that are required when a canal crosses a natural drainage? Draw an aqueduct showing all details.

[8 marks : 2012]

Solution:

A **cross drainage work** is a structure which is constructed at the crossing of a canal and a natural drain, so as to dispose of drainage water without interrupting the continuous canal supplies. In whatever way the canal is aligned, such cross drainage works generally becomes unavoidable. A cross-drainage work is generally a costly construction and must be avoided as far as possible. Since a watershed canal crosses minimum number of drains such an alignment is preferred over contour canal which crosses maximum number of drains.

Types of cross drainage works are as follows:

(i) Aqueduct and Syphon Aqueduct:

- In these works, the canal is taken over the natural drain, such that the drainage water runs below the canal either freely or under syphoning pressure.
- When the HFL of the drain is sufficiently below the bottom of the canal, so that the drainage water flows freely under gravity, the structure is known as an aqueduct.

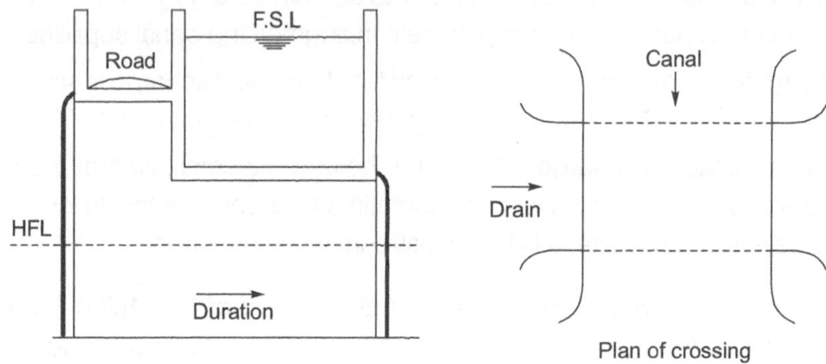

Plan of crossing

- If the HFL of the drain is higher than the canal bed and the water passes through the aqueduct barrels under syphonic action, the structure is known as syphon aqueduct.

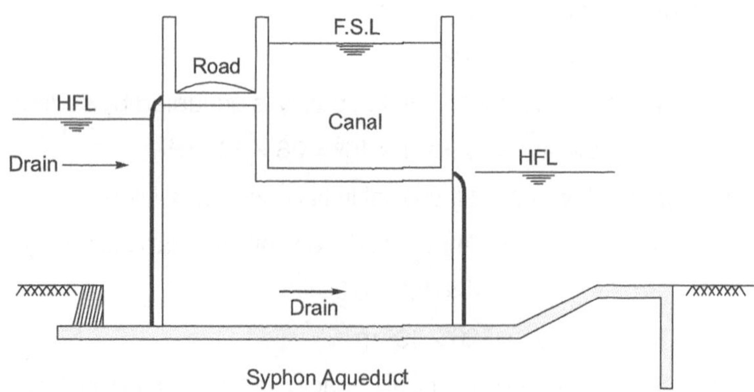

Syphon Aqueduct

(ii) Super Passage and Syphon:

- In these works, the drain is taken over the canal such that the canal water runs below the drain either freely or under syphoning pressure.

- When FSL of the canal is sufficiently below the bottom of the drain trough, so that the canal water flows freely under gravity, the structure is known as a super passage.

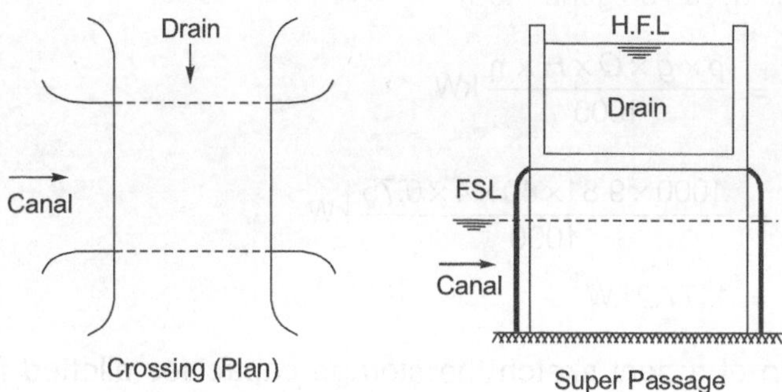

- If the FSL of the canal is sufficiently above the bed level of the drainage trough, so that the canal flows under syphonic action under the trough, the structure is known as a canal syphon or a syphon.

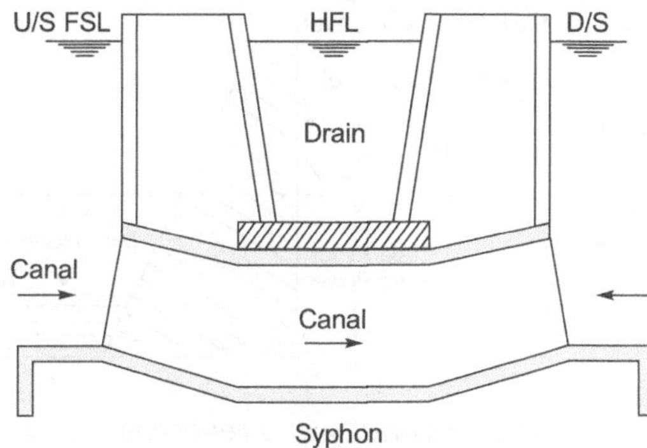

(iii) Level Crossing:

- In this type of cross-drainage work, the canal water and drain water are allowed to intermingle with each other.
- A level crossing is provided generally when a large canal and a high drainage approach each other practically at the same level.

(iv) Inlets and Outlets:

- An inlet is a structure constructed in order to allow the drainage water to enter the canal and get mixed with the canal water and thus to help in augmenting canal supplies.
- Such a structure is generally adopted when the drainage discharge is small and the drain crosses the canal with its bed level equal to or slightly higher than the canal FSL.
- When the drainage discharge is high or if the canal is small, so that the canal cannot take the entire drainage water, an outlet may sometimes be constructed to escape out the additional discharge at a suitable site, a little downstream along the canal.

5.8 **The main canal from the headwork of a dam has been designed to carry a discharge of 40 m^3/s. At a certain location along the course of the canal a drop of 4.0 m head is available. It is proposed to utilize the drop for generation of hydropower. Estimate how much kW (kilowatt) of energy can be generated assuming efficiency of the machinery used as 75 per cent.**

[4 marks : 2012]

Solution:

Given: Discharge (Q) = 40 m³/s, Head available (H) = 4 m, Efficiency of machinery (η) = 75%

$\therefore$ Amount of energy that can be generated

$$= \frac{\rho \times g \times Q \times H \times \eta}{1000} \text{ kW}$$

$$= \frac{1000 \times 9.81 \times 40 \times 4 \times 0.75}{1000} \text{ kW}$$

$$= 1177.2 \text{ kW}$$

5.9 **Show with the help of a neat sketch the storage capacities allotted for various purposes in a multipurpose reservoir.**
Briefly explain what is meant by useful life of a reservoir and how it is estimated.

[6 marks : 2012]

Solution:

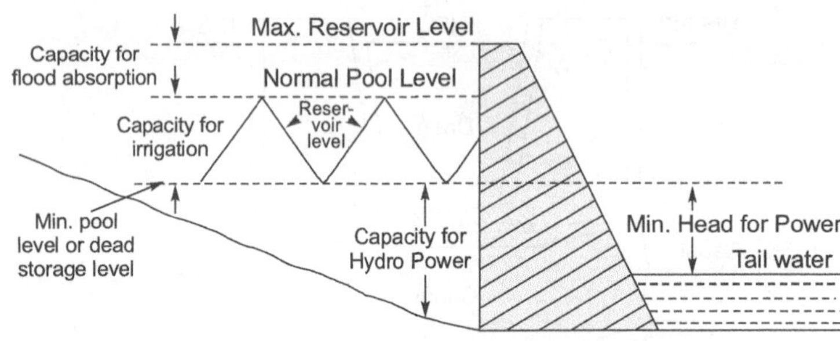

Storage capacity in a reservoir

Useful Life of Reservoir: It is the period during which the capacity occupied by sediments does not prevent the reservoir from serving its intended primary purpose. It is said to be over when the reservoirs cannot meet the intended commitments.

1. **By working out capacity inflow ratio percent and finding the annual loss of reservoir capacity.** For example, the capacity inflow ratio of Bhakra reservoir is 50% which gives annual loss of reservoir capacity as 0.5% which agrees closely with the figure of 0.57% actually worked out.
2. **By working out trap efficiency ratio for different reserovir capacities.** Divide the reservoir capacity into intervals of 10%. Find trap efficiencies for each capacities.

   Multiply the total annual sediment transported by the trap efficiency. Convert this sediment load into volume deposited per year. Divide the volume interval (i.e., 10% of the capacity) by the sediment deposited to get the number of years to fill this volume intervals of 10% capacity. Repeat the procedure and add the number of years for 80%, 70% ... 20% reservoir capacities.

5.10 **Obtain the diameter of the pipe outlet of a non-modular outlet:**
Discharge through the outlet = 0.02 m³ s⁻¹.
Length of the outlet = 15 m
FSL of the distributary = 200.00 m
Available working head = 5 cm
Coefficient of discharge = 0.51

[4 marks : 2014]

Solution:

Given: Discharge through the outlet = $0.02 \text{ m}^3 \text{s}^{-1}$

Length of the outlet = 15 m

FSL of the distributary = 200.00 m

Available working head or head loss, $H_L = 5.0 \text{ cm} = 0.05 \text{ m}$

Coefficient of discharge, $C_d = 0.51$

The discharge in non modular submerged pipe outlet is given by

$$Q = C_d \cdot A \cdot \sqrt{2gH_L}$$

where, A = Area cross section of pipe

$$0.02 = 0.51 \cdot \left(\frac{\pi D^2}{4}\right) \cdot \sqrt{2 \times 9.81 \times 0.05}$$

$$0.02 = 0.39673\, D^2$$

$$\therefore \quad D = 0.2245 \text{ m or } 224.5 \text{ mm}$$

5.11 **There are four important theories with regard to formation of meanders (i) disturbance theory, (ii) helicoidal flow theory (iii) excess energy theory and (iv) instability theory. Friedkin and Werner subscribed to the idea of disturbance theory. In brief, explain their ideas.**

[4 marks : 2015]

Solution:

Friedkin (in 1945) gave the most complete sets of laboratory meander data and said that meandering results primarily from load bank erosion and consequent load overloading and deposition by the river of the heavier sediments which move along the bed.

Once a local disturbance in the bank is initiated, it will start to grow into a bend. Further disturbances will develop as a result of impingement and deflection of the current which keeps on going until a sequence of bends exists, this is referred to as stream meander.

Werner (in 1951) has presented an analytical approach to the local disturbance hypothesis. It was suggested that the transverse waves play on important part in the formation of meanders. At the slightest disturbance a gravity wave is created, which is swept downstream under the characteristic standing wave pattern.

5.12 **A guide bank is required for a bridge on a river. For a design flood discharge of 50000 m³/sec, compute the following:**

(i) Lacey waterway

(ii) Total length between banks

(iii) u/s length of guide bund and d/s length of guide bund

(iv) Radius of u/s curved head and d/s curved head with angles

[8 marks : 2015]

Solution:

Given: $Q = 50000 \text{ m}^3/\text{s}$

(i) Lacey waterway $= 4.75\sqrt{Q} = 4.75\sqrt{50000} = 1062.13\text{m}$

(ii) Allowing 20% extra for piers etc, the **net spacing between the two guide bunds** will be,

$$L = 1.2 \times 1062.13 = 1274.56 \text{ m}$$

(iii) u/s length of guide bund $= 1.1 \text{ L}$

$$= 1.1 \times 1274.56 \simeq 1402 \text{ m}$$

d/s length of guide bund = 0.1 L to 0.2L
= 127.46 m to 254.9 m

(iv) Radius of u/s curve = 0.45 L
= 0.45 × 1274.56 = 573.55 m

Upstream curve is extended to subtend an angle of 120° to 140°.

Radius of d/s curve is equal to half the radius at the u/s side i.e. $\frac{573.55}{2} = 286.8\text{m}$ and sweep angle of 45° to 60° may be provided.

5.13 **Explain the use of stilling basin as an energy dissipater and describe the various components of a stilling basin with their functions with the help of sketches. List the four types of Indian standard basins.**

[20 marks : 2017]

Solution:

Stilling basin : Stilling basins are concrete aprons provided with auxiliary devices which help in dissipating the energy of flow by offering resistance to flow and stabilise the flow in a shorter length of basin, thus affecting economy.

In general, stilling basin may be defined as a structure in which the energy dissipation action is confined. If the phenomenon of hydraulic jump is basically used for dissipating this energy, it may be called a hydraulic jump type of stilling basin. Auxiliary devices may be used as additional measure for controlling the jump etc.

The various auxiliary devices with their functions are as follows:

(a) **Chute Blocks :** Chute blocks are kind of serrated device (row of small projections like teeth of saw) and provided at the entrance of the stilling basin. The dentated still diffuses the residual portion of high velocity of jet of water is forward and partly lifted from the floor, producing a shorter length of jump. They also help in stabilizing the flow and thus improve the jump performance.

(b) **Sills and Denoted Sills :** Sills or more preferably denoted sill is generally provided at the end of stilling basin. The denoted sill diffuses the residual portion of high velocity of jet reaching the end of basin. They therefore help in dissipating residual energy and to reduce the length of the jump or the basin.

(c) **Baffle Piers :** They are blocks placed within the basin across the basin floor. They help in breaking the flow and dissipate energy mostly by impact. These baffle piers sometimes called friction blocks are very useful in small structures such as low spillways and weirs etc.

List of standardized ISI stilling basin : A full description of such basins is available in IS : 4997-1968.

1. Stilling basin I [Froude's number < 4.5] with horizontal aprons
2. Stilling basin II [Froude's number > 4.5] with horizontal aprons
3. Stilling basin III with sloping aprons
4. Stilling basin IV with sloping aprons

U.S.B.R stilling basin IV : This type of stilling basin is shown below. It is used for Froude number varying between 2.5 and 4.50 which generally occurs in canal weirs, canal falls, diversion dams etc. This basin is applicable only to rectangular cross sections. Since oscillating waves are generated in this range of Froude number, they are tried to be controlled at source by providing large chute blocks.

The length of basin is generally kept equal to $5(y_2 - y_1)$.

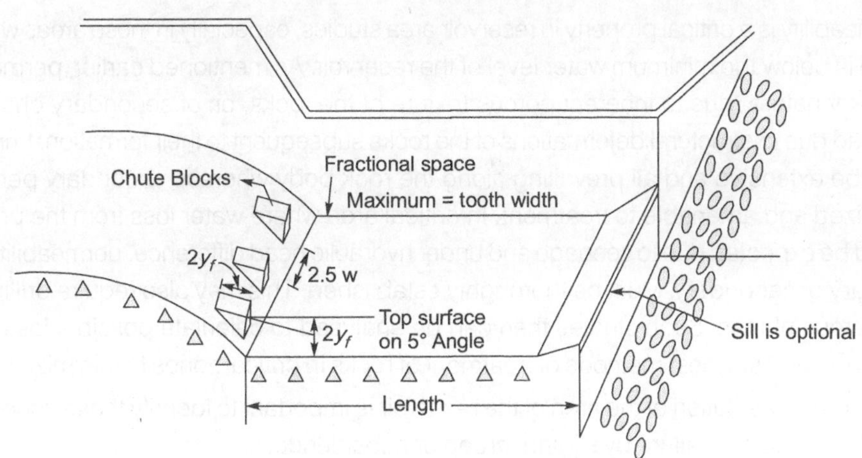

U.S.B.R. stilling basin IV [F_1 lies between 2.5 and 4.5]

5.14 **Discuss the geological characteristics necessary for the design and construction of reservoirs.**

[10 marks : 2020]

Solution:

Reservoirs may broadly be defined as artificially created water storage basins with storage capacity that may range from a few thousand cubic meters to thousands of million cubic meters. Depending on the purpose of storage, reservoirs are classified into three main categories :

(i) Storage and conservation reservoirs, where river water is stored by creating barriers or dams in its path and is then released from gated or ungated outlets. These feed the canal systems for irrigation and power generation.

(ii) Flood control reservoirs which have as their main function accommodating large volumes of surplus water during peak flow times of a river. The surplus water is released after the flood abates. Such reservoirs are provided with large sluice ways to discharge the inflow received by the reservoir during a flood up to a volume which could be safely accommodated in the channel downstream. Excess or surplus inflow is retained back till a desirable time.

(iii) Distribution reservoirs are actually small storage reservoirs which hold water supplies in a water-supply system for short spells of time. Water is constantly pumped into these reservoirs, from where it is distributed for drinking and other purposes in a regulated manner.

Since reservoirs are essentially water storage basins, the area should possess such geological characteristics that favour holding of water in the basin so created. Such characters are : topographic suitability, ground water conditions, permeability and structural stability.

(a) Topographically, the area should be a broad natural valley preferably ending in a narrow gorge where a barrier could be placed. The valley could be a U-shaped glacial valley, or V-shaped river valley or a broad synclinal valley. Flatlands and plains cannot make convenient places for reservoirs. In alluvial plains, the reservoir area has to be an extensive low-lying stretch bordered by high-lands on flanks.

(b) Groundwater conditions in the proposed reservoir area must be thoroughly established for an approximately accurate evaluation of storage capacity of the reservoir. Any one of the three in view of such a close relationship between water table and storage capacity of the proposed reservoir, it is absolutely essential that water table position all along the reservoir areas, especially along the lanks, must be thoroughly established. This may necessitate drilling test holes upto required depths and numbering many hundreds or even thousands.

(c) Permeability is a critical property in reservoir area studies, especially in those areas where groundwater table is below the minimum water level of the reservoir. As mentioned earlier, permeability may be of primary nature, due to inherent porous texture of the rocks, or. of secondary character, where it is caused due to structural deformations of the rocks subsequent to their formation. Primary permeability may be extensive and all prevailing along the rock body whereas secondary permeability is often localized and amenable to treatment. In critical area where water loss from the proposed reservoir could be expected due to seepage and under hydraulic head difference, permeability-values, whether primary or secondary. Must be thoroughly established. This may also require drilling of test-wells at numerous places. Such studies then can be analysed to calculate possible loss due to expected seepage and suggest methods of treatment of rocks in critical zones to minimize the expected loss.

(d) Structural constitution of the area of the reservoir is important to identify those zones along the flanks which are liable to failure by sliding, creep or subsidence.

(e) Trend and rate of weathering in catchment area is of considerable importance in determining the age of the reservoir. If the catchment happens to be made up predominantly of weathered, barren slopes which are broken, falling, shattered and jointed profusely, the run off and slope wash would contribute heavy load of sediments to the reservoir after every rain. This would obviously decrease the effective storage capacity of the reservoir on the one hand and its total life on the other hand. However, stable rock slopes covered with thick forests and vegetation would make ideal catchment areas for reservoirs.

5.15 **Calculate (i) total installed capacity of turbo generators, (ii) load factor, (iii) plant factor, and (iv) utilization factor of three generators of a hydel power station, each having a capacity of 10000 kW. The load of the plant varies from 12000 kW to 26000 kW.**

[20 marks : 2023]

Solution:

(i)
$$\text{Total installed capacity} = \text{Number of generators} \times \text{capacity of each generator}$$
$$= 3 \times 10000 \text{ kW}$$
$$= 30000 \text{ kW}$$

(ii)
$$\text{Load factor} = \frac{\text{Average load over a Certain period}}{\text{Peak load during that period}}$$

$$= \frac{\frac{(12000+26000)\text{kW}}{2}}{26000\text{kW}} = \frac{19}{26} = 0.73$$

(iii)
$$\text{Plant factor} = \frac{\text{Energy actually produced in time 't'}}{\text{Max.energy that can be produce in time 't'}}$$

$$= \frac{\left(\frac{12000+26000}{2}\right)\times t}{30000\times t} = \frac{19}{30} = 0.63$$

(iv)
$$\text{Utilization factor} = \frac{\text{Maximum power utilised}}{\text{Maximum power available}}$$

$$= \frac{26000}{30000} = 0.87$$

■■■■

4 Environmental Engineering

Revised Syllabus of ESE : *Water Supply Engineering:* *Sources, Estimation, quality standards and testing of water and their treatment; Rural, Institutional and industrial water supply; Physical, chemical and biological characteristics and sources of water, Pollutants in water and its effects, Estimation of water demand; Drinking water Standards, Water Treatment Plants, Water distribution networks.*

Waste Water Engineering: *Planning & design of domestic waste water, sewage collection and disposal; Plumbing Systems. Components and layout of sewerage system; Planning & design of Domestic Waste-water disposal system; Sludge management including treatment, disposal and re-use of treated effluents; Industrial waste waters and Effluent Treatment Plants including institutional and industrial sewage management.*

Solid Waste Management: *Sources & classification of solid wastes along with planning & design of its management system; Disposal system, Beneficial aspects of wastes and Utilization by Civil Engineers.*

Air, Noise pollution and Ecology: *Concepts & general methodology*

---

1. Water Demand

1.1 The census record of a particular town shows the population figures as follows:

Years	1960	1970	1980	1990
Population	55,500	63,700	71,300	79,500

Estimate the population for the year 2020 by Decreasing Rate of Growth.

[10 marks : 1999]

Solution:

The decreasing rate of growth method is as follows:

Year	Population	Increase in Population	% Increase in Population (r)	Decrease in% Increase (r′)
1960	55,500			
		8,200	14.77	
1970	63,700			2.84
		7,600	11.93	
1980	71,300			0.43
		8,200	11.50	
1990	79,500			

Avg. value of $r' = \dfrac{2.84 + 0.43}{2} = 1.635$

$r = 11.50$

$$P_{2000} = P_{1990} + \left(\frac{r-r'}{100}\right) \times P_{1990} = 79500 + \left(\frac{11.50-1.635}{100}\right) \times 79500 = 87,343$$

$$P_{2010} = P_{2000} + \left(\frac{r-2r'}{100}\right) \times P_{2000} = 87343 + \left(\frac{11.50-2\times 1.635}{100}\right) \times 87343 = 94,531$$

$$P_{2020} = P_{2010} + \left(\frac{r-3r'}{100}\right) \times P_{2010} = 94531 + \left(\frac{11.50-3\times 1.635}{100}\right) \times 94531 = 1,00,765$$

The population for the year 2020 is 1,00,765.

1.2 Compute the 'fire demand' for a city of 2 lac population by any two formulae including that of the National Board of Fire Underwriters.

[8 marks : 2007]

Solution:

(i) The rate of fire demand as per **National Board of Fire Underwriters formula** for a central congested city whose population is less than or equal to 2 lakh is given by

$$Q = 4637\sqrt{P}(1-0.01\sqrt{P})$$

where Q is amount of water required in litres per minute and P is population in thousands

$$Q = 4637\sqrt{200}\left[1-0.01\sqrt{200}\right]$$

$$= 56303.08 \text{ litres per minute}$$

$$= \frac{56303.08 \times 60 \times 24}{10^6}\text{MLD} = 81.08 \text{ MLD}$$

(ii) Kuichling's formula,

$$Q = 3182\sqrt{P} = 3182\sqrt{200}$$

$$= 45000.28 \text{ litres per minute} = 64.8 \text{ MLD}$$

(iii) Freeman Formula,

$$Q = 1136\left[\frac{P}{10}+10\right]$$

$$= 1136\left[\frac{200}{10}+10\right] = 34080 \text{ litres per minute}$$

$$= 49.0752 \text{ MLD}$$

(iv) Buston's formula,

$$Q = 5663\sqrt{P} = 5663\sqrt{200}$$

$$= 80086.91 \text{ litres per minute} = 115.33 \text{ MLD}$$

1.3 A city has the following recorded population:

Year 1971 : 60000

Year 1991 : 120000

Year 2011 : 180000

Estimate (i) the saturation population and (ii) expected population in the year 2031 by logistic curve method.

[12 marks : 2019]

Solution:

Year	Population
1971	60000
1991	120000
2011	180000

Saturated population:

$$P_s = \frac{2P_0P_1P_2 - P_1^2(P_0 + P_2)}{P_0P_2 - P_1^2}$$

$$= \frac{2 \times 60000 \times 120000 \times 180000 - (120000)^2(60000 + 180000)}{60000 \times 180000 - 120000^2}$$

$= 240000$

$$m = \frac{P_s - P_0}{P_0}$$

$m = 3$

$$n = \frac{1}{t_1}\ln\left[\frac{P_0(P_S - P_1)}{P_1(P_S - P_0)}\right]$$

$$= \frac{1}{20}\ln\left[\frac{60000(240000 - 120000)}{120000(240000 - 60000)}\right]$$

$= -0.0549 \text{ year}^{-1}$

$$P = \frac{P_S}{1 + m\log_e^{-1}(nt)}$$

$$P = \frac{240000}{1 + 3 \times \log_e^{-1}(-0.0549 \times 60)}$$

$= 215960.2945$

$= 215961$

2. Conduits for Transporting Water and Distribution Systems

2.1 A town of 200,000 population is to be supplied water from a source 2500 m away. The lowest water level in the source is 15 m below the water works of the town. The demand of water is estimated as 150 lit/capita/day. A pump of 300 HP is operated for 15 hours. If the maximum demand is 1.5 times the average demand, the velocity of flow through the rising main is 1.3 m/sec and the pump efficiency is 70%, determine

(i) hydraulic gradient and (ii) friction factor for the pipe.

[15 marks : 1999]

Solution:

Maximum demand = Population × 1.5 × average demand

= 200,000 × 1.5 × 150 = 45 MLD

The pumping is done for 15 hours a day, hence maximum discharge required for pumping,

$$Q = \frac{45 \times 10^6}{10^3 \times 15 \times 60 \times 60} = 0.8333 \text{ m}^3/\text{sec}$$

Power of pump = 300

But, we know that $$P(HP) = \frac{\gamma_w Q H}{0.7457 \times \eta}$$

$$\Rightarrow \quad 300 = \frac{9.81 \times 0.8333 \times H}{0.7457 \times 0.70}$$

$$\Rightarrow \quad H = 19.16 \text{ m}$$

But total head (H) = head difference between source and water works + head loss due to friction in rising main

$$\Rightarrow \quad 19.16 = 15 + h_f$$

$$\Rightarrow \quad h_f = 19.16 - 15$$

$$\Rightarrow \quad h_f = 4.16 \text{ m}$$

Now length of pipe, $L = 2500$ m

(i) **Hydraulic gradient** $= \frac{h_f}{L} = \frac{4.16}{2500} = \frac{1}{601}$

(ii) **We know that** $$h_f = \frac{fLV^2}{2gd}$$

Now $$\frac{\pi d^2}{4} = \frac{Q}{V} = \frac{0.8333}{1.3}$$

$$\therefore \quad d = \sqrt{\frac{0.8333}{1.3} \times \frac{4}{\pi}}$$

$$\Rightarrow \quad d = 0.9 \text{ m}$$

$$\because \quad h_f = \frac{fLV^2}{2gd}$$

$$\Rightarrow \quad f = \frac{4.16 \times 2 \times 9.81 \times 0.9}{2500 \times (1.3)^2} = 0.0174$$

2.2 **Following mean monthly flows were observed at a site on a stream in a typical year:**

Month	Jan.	Feb.	March	April	May	June	July	Aug.	Sep.	Oct.	Nov.	Dec.
Mean Monthly flow (cumec)	15	10	8	6	5	12	25	40	71	60	40	20

Assuming that the stream flow is to be fully utilised for delivering water to a city to meet its fixed monthly demand by diversion of flow from storage reservoir through a conduit, find the capacity of the conduit in cumec for which it is to be designed. Determine the minimum storage capacity in ha-m of the storage reservoir to ensure that all the available flow is fully utilised to meet the constant monthly demand of the city.

[20 marks : 2001]

Solution:

Since the stream flow is to be fully utilized for delivering water to the city to meet its fixed monthly demand

$\therefore$ Average fixed monthly demand = Average monthly Inflow

$$= \frac{821.9232}{12} = 68.4936$$

S.No.	Month	Monthly Inflow (cumec)	Monthly inflow (Mm^3)	Monthly Outflow (Mm^3)	Monthly deficit	Monthly Surplus	Cum. deficit	Cum. Sul.
(1)	(2)	(3)	(4)	(5)	(6)	(7)	(8)	(9)
1.	Jan	15	40.176	68.4936	28.3176	—	—	—
2.	Feb	10	24.192	68.4936	44.3016	—	—	—
3.	March	8	21.4272	68.4936	47.0664	—	—	—
4.	Apr	6	15.552	68.4936	52.9416	—	—	—
5.	May	5	13.392	68.4936	55.1016	—	—	—
6.	June	12	31.104	68.4936	37.3896	—	—	—
7.	July	25	66.960	68.4936	1.5336	—	266.652	
8.	August	40	107.136	68.4936	—	38.6424	—	—
9.	Sep	71	184.032	68.4936	—	115.5384	—	—
10.	Oct	60	160.704	68.4936	—	92.2104	—	—
11.	Nov	40	103.680	68.4936	—	35.1864	—	281.5776
12.	Dec	20	53.568	68.4936	14.9256	—	14.9256	—
Total		312	821.9232					

The minimum storage is the maximum of cumulative surplus and cumulative deficit. Thus the minimum storage in storage reservoir $= 281.5776 \text{ Mm}^3 = 28157.76 \text{ ha-m}$

$$\text{Design discharge in m}^3\text{/s for the conduit} = \frac{821.9232 \times 10^6}{24 \times 3600 \times 365} = 26.06 \text{ m}^3\text{/s}$$

2.3 **Discuss the functioning and applicability of the following:**
(i) Air valve (ii) Reflux valve (iii) Flanged Joint (iv) Expansion Joint

[8 marks : 2007]

Solution:

(i) Air valve:

- These are the special kind of valves which are generally placed along the pipe line at "summits" on both sides of the sluice valves and also on the downstream side of all other sluice valves.
- When placed on summits which are very near the HGL or sometimes above the HGL (especially during negative water hammers) they ensure the safety of the pipe against collapse.
- Similarly when placed below the ordinary sluice valves, they will protect the pipe against the negative pressures.
- A valve which will open out automatically as soon as the pressure in the pipe falls below a certain fixed predetermined value, and thus allowing air to enter the pipe is known as an air inlet valve.
- Sometimes air gets accumulated at high points when the supply is restored and the pipe is refilled after repairs. The accumulated air obstruct the free flow of water and the pipe may get air locked. Air relief valves are therefore required to be provided at all summits to remove the accumulated air under such circumstances.
- The most commonly used air valve is single orifice air valve. However, sometimes double acting orifice air valve is also provided.

(ii) Reflux valve:

- These are sometimes called non-return valves because they prevent water to flow back in the opposite direction.
- They may be installed on the delivery side of the pumping set, so as to prevent the back flow of stored or pumped water, when the pump is stopped.

- These are also installed on pump discharges to reduce water hammer forces on the pump.
- These are also required at inter-connections between a polluted water system and a potable water system, so as to prevent the entry of pollution into the pure water.
- The simplest type of reflux valve or check valve is a flap shutter hinged at the outlet end of pipe, which opens out in the direction of flow but closes of its own weight when the flow in the permissible direction ceases.

(iii) Flanged joint:

- These are used for pumping stations, filter plants, and at other locations where it may be necessary to occasionally disjoint the pipe.
- Two flanged pipes are brought together, keeping a rubber washer in between so as to make them water tight. They are then fixed by means of nuts and bolts.
- These joints are strong but rigid and hence cannot be used where deflections or vibrations are expected.
- They are expensive and mostly used for indoor works such as pumping stations, filter plants, etc.

(iv) Expansion joint:

- These are provided at suitable intervals in the pipelines, so as to counteract the thermal stresses produced due to temperature variations.
- The socket end is cast flanged and the spigot end is kept plain to provide an expansion joint in cast iron pipe.
- While making this joint, a small space is kept between the face of the spigot and the inner face of the socket and the spigot is filled up by means of a rubber gasket.
- The flanges are then tightened by means of nuts and bolts.
- When the pipe expand, the socket end moves forward and the gap left just closed.
- Similarly, when the pipes contract the socket moves backward creating the gap.

2.4 Population of a town is 50000 and area is 90 hectares. Calculate the discharge for which the sewers of a proposed combined system will be designed for the town, if the average coefficient of run-off for this area is 0.65 and the time of concentration of the design rain is 20 minutes. Assume any data required.

[8 marks : 2007]

Solution:

Assuming that the town is provided with a planned water supply at an average rate of 150 Lpcd and 80% of this water supply is converted into sewage

$\therefore$ Quantity of sewage produced per day $= \dfrac{80}{100} \times 150 \times 50000 \times 10^{-3} = 6000 \text{ m}^3$

Quantity of sewage produced per second $= \dfrac{6000}{24 \times 60 \times 60} = 0.0694 \text{ m}^3/\text{s}$

$\therefore$ Average sewage discharge $= 0.0694 \text{ m}^3/\text{s}$

Maximum sewage discharge $= 3 \times$ Average sewage discharge

$= 3 \times 0.0694 = 0.2083 \text{ m}^3/\text{s}$

The storm water discharge $= KpA$

Here $K = 0.65$; $T = 20$ minutes

$A = 90 \text{ ha} = 90 \times 10^4 \text{ m}^2$

$p =$ Design rainfall intensity in cm/hr

The value of p may be calculated as

$$p = \frac{75}{T+10} \text{ where } T \text{ is time of concentration in minutes}$$

$$= \frac{75}{20+10} = 2.5 \text{ cm/hr}$$

$\therefore$ Storm water discharge $= 0.65 \times \frac{2.5 \times 10^{-2}}{60 \times 60} \times 90 \times 10^4 = 4.0625 \text{ m}^3/\text{s}$

Hence, the **peak discharge** for which the sewers of the proposed combined system should be designed

= Max. sewage discharge + Max. storm discharge

$= 0.2083 + 4.0625 = 4.2708 \text{ m}^3/\text{s}$.

2.5 **A pipe network in the form of a triangle *ABC* has inflows of 5 m³/s and 4 m³/s at *A* and *B* respectively. The outflow at *C* is 9 m³/s. Given $K_{AB} = 10$, $K_{BC} = 50$, $K_{AC} = 20$, compute discharges in each pipeline [$h_f = KQ^2$].**
[10 marks : 2009]

Solution:

First of all, the magnitudes as well as directions of the possible flows in each pipe are assumed keeping in consideration the law of continuity at each junction. The assumed flows are given in the figure shown below.

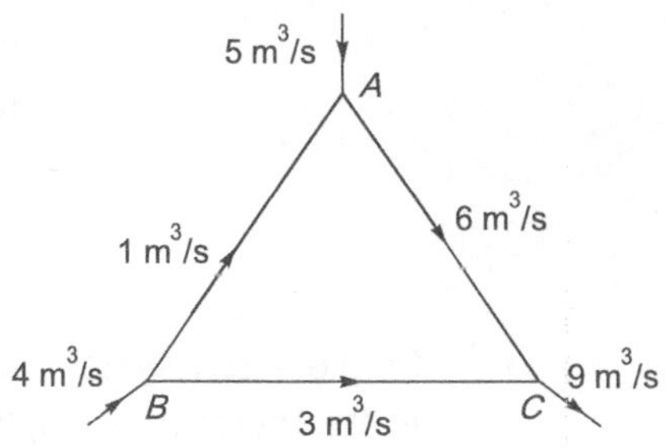

It is given that $h_f = KQ^2$

The given pipe network can be analysed by Hardy Cross method.

1st iteration for the loop BACB

Pipe	Assumed flows Q_a	K	$H_L = KQ_a^2$	$\left\|\frac{H_L}{Q_a}\right\|$	Corrected Q after first correction $Q_{a_1} = Q_a + \Delta$
BA	1	10	10	10	0.5
AC	6	20	720	120	5.5
CB	–3	50	–450	150	–3.5
			$\Sigma H_L = 280$	$\sum\left\|\frac{H_L}{Q_a}\right\| = 280$	

$$\Delta = -\frac{\Sigma H_L}{2\sum\left|\frac{H_L}{Q_a}\right|} = -\frac{280}{2 \times 280} = -0.5 \text{ m}^3/\text{s}$$

2nd iteration for the loop BACB

Pipe	Assumed flows Q_a	K	$H_L = KQ_a^2$	$\left\|\frac{H_L}{Q_a}\right\|$	Corrected Q after second correction $Q_{a_1} = Q_a + \Delta$
BA	0.5	10	2.5	5	0.5086
AC	5.5	20	605	110	5.5086
CB	–3.5	50	–612.5	175	–3.4914
			$\Sigma H_L = -5$	$\sum\left\|\frac{H_L}{Q_a}\right\| = 290$	

$$\Delta = -\frac{\Sigma H_L}{2\sum\left|\frac{H_L}{Q_a}\right|} = -\frac{-5}{2\times 290} = 0.0086\ \text{m}^3/\text{s}$$

Thus the discharges in pipe *BA*, *AC* and *CB* are 0.5086 m^3/s, 5.5086 m^3/s and 3.4914 m^3/s respectively.

2.6 **Draw and discuss the single stack system of plumbing?**

[10 marks : 2009]

Solution:

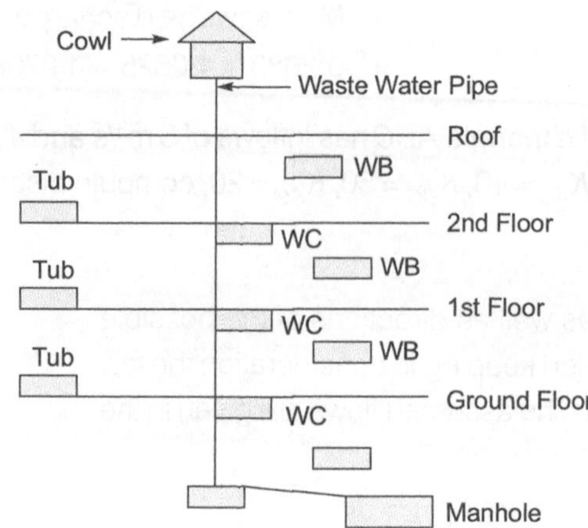

- Single stack system is one of the four principle system of plumbing adopted in a building.
- In this system, a single pipe in provided without any separate ventilation pipe.
- Since there is no separate ventilation pipe hence there is always the danger of four smell.
- To partially avoid this problem, the pipe is extended for about 2 m higher than roof level and provided with a cowl for removal of foul gases.

2.7 **For a population of 2 lac having per capita daily demand of water of 150 litres, determine the water horse power to raise the water from a river of RL 120 m to treatment plant of RL 140 m through a rising main 1 m in diameter. Assume 2 m as the total head loss due to friction, etc. and pumping efficiency of 70%.**

[2 marks : 2010]

Solution:

Population $= 2 \times 10^5$

Per capita daily demand = 150 litres

Discharge required for the pump will be

$$Q = \frac{2\times 10^5 \times 150 \times 10^{-3}}{24\times 60\times 60} = 0.3472\ \text{m}^3/\text{s}$$

Delivery head, $H_d = 140 - 120 = 20$ m

Loss of head due to friction, $h_f = 2$ m

Total head against which the pump has to work will be

$$H = H_d + h_f = 20 + 2 = 22\ \text{m}$$

Water horse power of the pump is given by

$$\text{WHP} = \frac{\gamma_w QH}{0.7457} = \frac{9.81\times 0.3472\times 22}{0.7457} \qquad [\gamma_w = 9.81\ \text{kN/m}^3]$$

$[\because 1HP = 745.7\ W = 0.7457\ \text{kW}]$

$$= 100.486 \approx 100.5\ \text{HP}$$

If η is the efficiency of pump, then **Brake horse power** is given as

$$BHP = \frac{WHP}{\eta} = \frac{100.5}{0.70} = 143.57\ \text{HP}$$

2.8 Explain briefly with neat sketches different systems of layout of water distribution network. Also mention for which town or city, each of the above distribution networks is suitable with their advantages/disadvantages.

[8 marks : 2016]

Solution:

Various layouts of water distribution network: The various water distribution networks differ only in the layout of pipes conveying water.

In general there are four types of water distribution network which are as follows:

(i) **Dead end system:** This network system is adopted for cities which have developed in a haphazard manner with no definite pattern of growth like old cities.

Dead End or Tree System

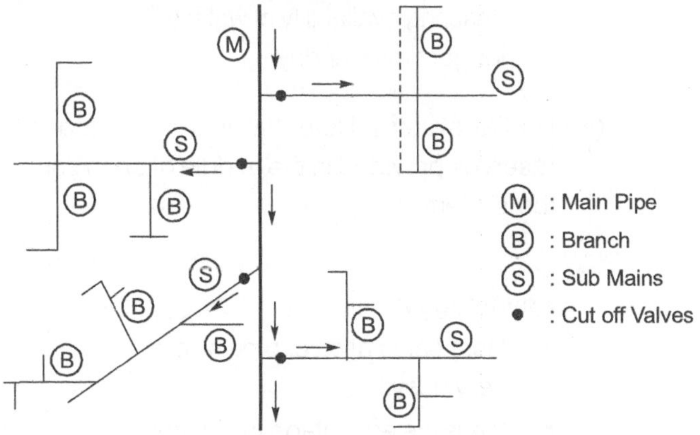

Ⓜ : Main Pipe
Ⓑ : Branch
Ⓢ : Sub Mains
● : Cut off Valves

Advantages:

- It is relatively a cheap system.
- Determination of discharge and pressure is easy due to less number of valves being employed in the network.
- Lying of pipes is easy.
- Pipes used are of lesser diameter and hence economical.

Disadvantages:

- Owing to a large number of dead ends, stagnation of water occurs which cannot be utilized at locations where demand is high.
- During repairs and maintenance, entire downstream portion below the valve has to be cut-off.

(ii) **Grid iron system:** It is suitable for cities with rectangular layout where water mains and branches are laid in a rectangular fashion.

Grid-iron System

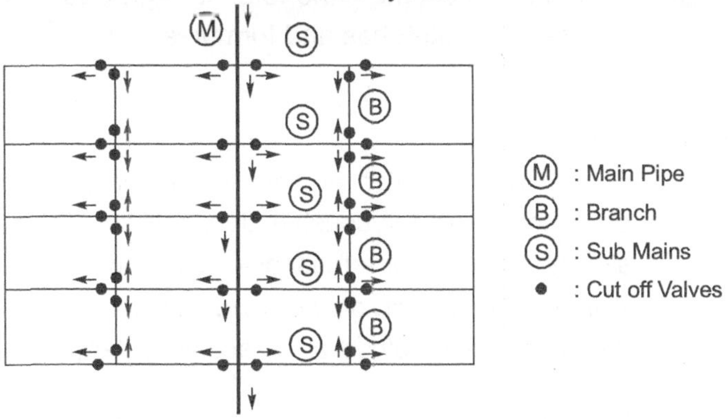

Ⓜ : Main Pipe
Ⓑ : Branch
Ⓢ : Sub Mains
● : Cut off Valves

Advantages:

- Due to the absence of dead ends, water remains in circulation.
- Due to continuously circulating water, there are less chances of water getting contaminated as compared to still water.
- In case of repairs and breakdown, water is available for other locations and only a part of network needs to be isolated.

Disadvantages:

- Due to large number of valves need to be provided at all branches, it is difficult to compute discharge and pressure and thus exact calculation of pipe size required is difficult.

(iii) Circular/Ring system: In this system of water distribution, the main line is laid all along the periphery roads and sub-mains branch out from the main system of pipes.

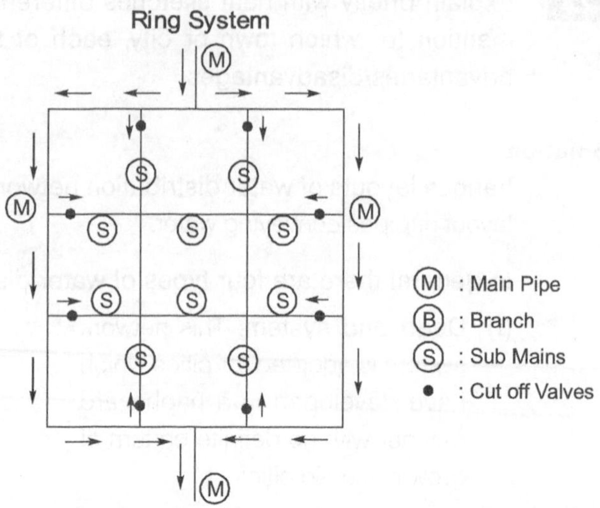

Advantages:

- The biggest advantage of this system is that water can be supplied to any point from at least two directions thereby ensuring availability of water for longer time per day.

(iv) Radial system: Here the area is divided into different zones and water is kept in the distribution reservoir provided in the middle of each zone. Supply pipes are laid radially towards the periphery of the system.

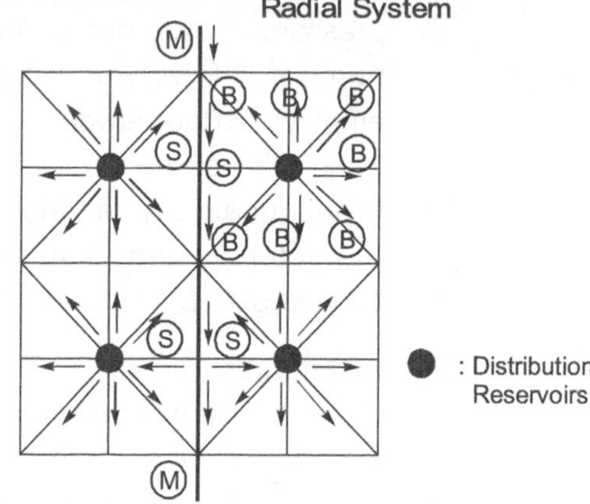

Advantages:

- Here calculation of pipe size is very easy.
- It has the advantage that water is made available in a short time as compared to other distribution systems.
- It is the most economical system if combined pumping and gravity flow is adopted.

2.9 **How will you estimate the total storage capacity of a distribution reservoir? Support your answer with suitable sketches and formulae.**

[15 marks : 2019]

Solution:

The quantity of water required to be stored in the reservoir for equalising or balancing this variable demand against the constant supply is known as the balancing reservoir or balancing storage or the storage capacity of a balancing reservoir.

This storage capacity of balancing reservoirs is worked out with help of hydrographs of inflow and outflow by mass curve method or by analytical tabular solution.

Mass Curve Method

- A mass curve diagram is the plot of accumulated supply or demand versus time.
- The supply is also called as inflow and demand as outflow. First mass curve of supply known as supply line is drawn and over this demand curve is superimposed.
- The amount of balancing storage is determined by adding the maximum ordinates between the demand and supply lines.

- To construct such diagram for a particular water supply project, we have to proceed as follows.
  - (a) From the past records, determine the hourly demand (or draft) for all 24 hours for typical days (maximum, average and minimum).
  - (b) Calculate and plot the cumulative demand against time and then plot the mass curve of demand.
  - (c) Draw the cumulative supply also against time, which is a straight line if the supply is constant.
  - (d) The storage required is calculated as the sum of the two maximum ordinates between demand and supply lines.
  - (e) Repeat the procedures for all the typical days (maximum, average and minimum) and determine the maximum storage required for the worst days.

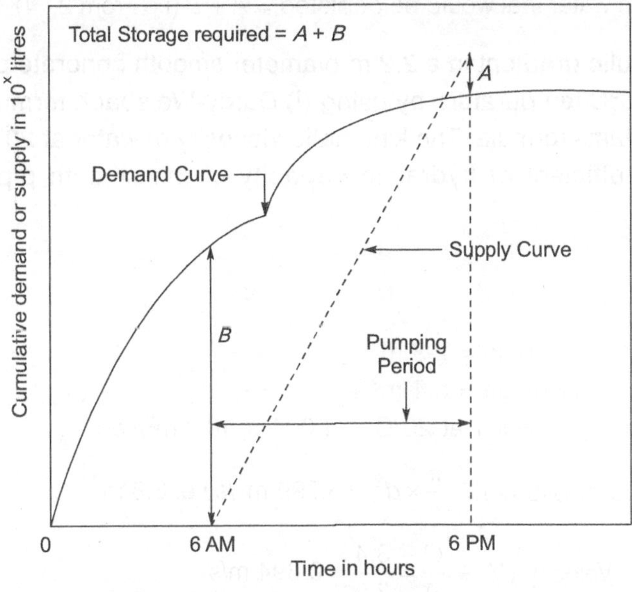

Limited Hour Supply

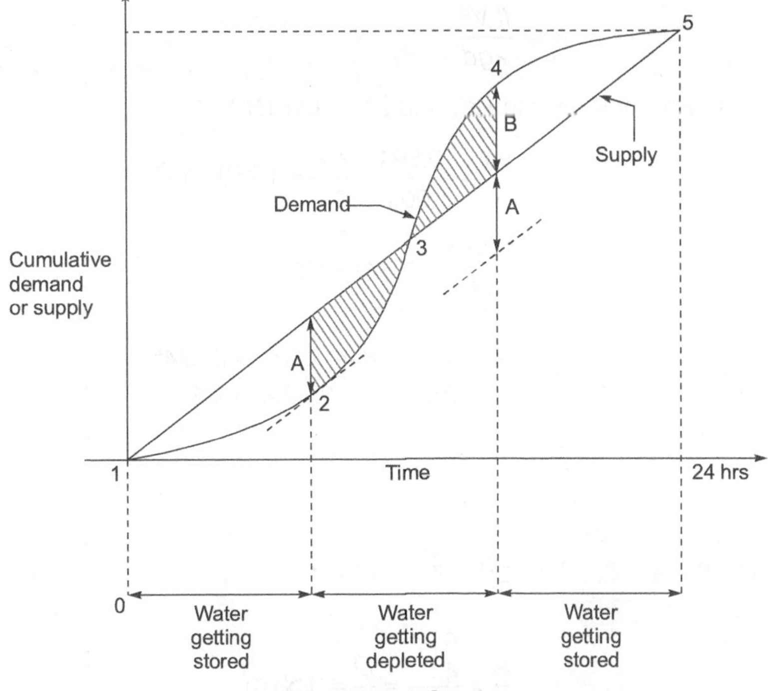

Limited Hour Supply

From (2 - 4) → Demand rate > supply rate
→ Water getting depleted
From (4 - 5 - 1 - 2) → Demand rate < supply rate
→ Water getting stored
From (4 - 5) → Accumulation = B
From (1 - 2) → Accumulation = A
From (2 - 3) → Depletion = A
From (3 - 4) → Depletion = B

⇒ Maximum water that would be accumulated = $A + B$ (i.e From 4 - 2)

⇒ Maximum water that would be depleted = $A + B$ (i.e From 2 - 4)

2.10 Estimate the hydraulic gradient in a 2.2 m diameter smooth concrete pipe carrying a discharge of 3.4 cumecs at 20°C temperature by using (i) Darcy-Weisbach formula, (ii) Manning's formula and (iii) Hazen-Williams formula. The kinematic viscosity of water at 20°C = 1.004 × 10^{-6} m²/sec, Hazen-Williams coefficient of hydraulic capacity of the smooth pipe = 130 and Manning's coefficient = 0.013.

[20 marks : 2019]

Solution:

Diameter of pipe = 2.2 m

Discharge = 3.4 m³/s

Kinematic viscosity of water at 20°C = 1.004 × 10^{-6} m²/sec

$$\text{Area of pile } (A) = \frac{\pi}{4} \times d^2 = 3.799 \text{ m}^2\text{/s i.e. } 3.8 \text{ m}^2$$

$$\text{Velocity } (V) = \frac{Q}{A} = \frac{3.4}{3.8} = 0.894 \text{ m/s}$$

(i) Darcy-weisbach formula,

$$h_f = \frac{fLV^2}{2gd}$$

where, F = friction factor which Reynold's number (R_e)

$$R_e = \frac{vd}{\nu} = \frac{0.894 \times 2.2}{1.004 \times 10^{-6}} = 1.958 \times 10^6$$

$$\therefore \quad f = \frac{0.316}{R_e^{1/4}} = 8.447 \times 10^{-3}$$

$$\text{Hydraulic gradient} = \frac{h_f}{L} = \frac{fv^2}{2gd} = \frac{8.44 \times 10^{-3} \times 0.894^2}{2 \times 9.81 \times 2.2}$$

$$\text{i.e.} \quad \frac{h_f}{L} = \frac{1}{6398.88}$$

(ii) Manning's formula, $V = \frac{1}{n} R^{2/3} S^{1/2}$

$$R = \frac{A}{P} = \frac{\frac{\pi}{4}D^2}{\pi D} = \frac{D}{4} = 0.55 \text{ m}$$

$$\therefore \quad 0.894 = \frac{1}{0.013} \times 0.55^{2/3} \times (S)^{1/2}$$

$$S = \frac{h_f}{L} = 2.9974 \times 10^{-4} = \frac{1}{3336.22}$$

(iii) Hazen William's formula

$$V = 0.849 \times C_H \times R^{0.63} \times S^{0.54}$$

$$0.894 = 0.849 \times 130 \times 0.55^{0.63} \times S^{0.54}$$

$$S = 2.6899 \times 10^{-4} = \frac{1}{3717.60}$$

2.10 **EA city with population of 5 lakhs is to be supplied water @ 150 lpcd. Using the data given below, determine the storage capacity of the reservoir assuming**

(i) Continuous pumping

(ii) Pumping for 9 hours from 6 PM - 3 AM

Also assume fire demand @ 2 lpcd and power breakdown for 2 hours.

0 - 3 AM	:	5% of total
3 - 6 AM	:	5% of total
6 - 9 AM	:	40% of total
9 - 12 PM	:	10% of total
12 - 3 PM	:	5% of total
3 - 6 PM	:	10% of total
6 - 9 PM	:	20% of total
9 - 12 AM	:	5% of total

[20 marks : 2023]

Solution:

Population of city, P = 500000

Per capita supply, q = 150 lpcd

So, total water supplied per day = 75×10^6 litre/day = 75 MLD

(i) When pumping is continuous

$$\text{Water supplied per hour} = \frac{75 \times 10^6 \text{ litre}}{24 \text{hours}} = 3.125 \times 10^6 \text{ L/hour}$$

Water supplied in 3 hours = 3.125×10^6 L/hour × 3 hour

= 9.375 million litres

Now, calculations are done below in table:

Time	Demand (Million litres)	Cumulative demand (Million litres)	Supply (Million litres)	Cumulative supply (Million litres)	Excess demand (Million litres)	Excess Supply (Million litres)
0–3A.M.	0.05 × 75 = 3.75	3.75	9.375	9.375	Nil	5.625
3–6A.M.	0.05 × 75 = 3.75	7.5	9.375	18.75	Nil	11.25
6–9A.M.	0.4 × 75 = 30	37.5	9.375	28.125	9.375	Nil
9–12P.M.	0.1 × 75 = 7.5	45	9.375	37.5	7.5	Nil
12–3P.M.	0.05 × 75 = 3.75	48.75	9.375	46.875	1.875	Nil
3–6P.M.	0.1 × 75 = 7.5	56.25	9.375	56.25	0	0
6–9P.M.	0.2 × 75 = 15	71.25	9.375	65.625	5.625	Nil
9–12A.M.	0.05 × 75 = 3.75	75	9.375	75	0	0

From the table,

Maximum value of excess demand = 9.375 ML

Maximum value of excess supply = 11.25 ML

So, Balancing storage required = 9.375 ML + 11.25 ML

= 20.625 Million litres

Now, Fire demand = 2 lpcd × 500000 persons

= 1 million litres per day

$$\text{Breakdown demand} = \frac{75MLD}{24\text{hours}} \times 2\text{hours}$$

= 6.25 million litre per day

Therefore, storage capacity of reservoir.

= (20.625 + 1 + 6.25)ML

= 27.875 Million litres

(ii) When pumping is done for 9 hours

$$\text{Water supplied per hour} = \frac{75 \times 10^6 \text{ litre}}{9 \text{ hours}}$$

$$= \frac{75}{9} \text{Million litre/hour}$$

$$\text{Water supplied in 3 hours} = \frac{75}{9}\frac{ML}{hour} \times 3\text{hour}$$

= 25 million litres

Now, calculations are done in table below:

Time	Demand (Million litres)	Cumulative demand (Million litres)	Supply (Million litres)	Cumulative supply (Million litres)	Excess demand (Million litres)	Excess Supply (Million litres)
0–3A.M.	0.05 × 75 = 3.75	3.75	25	25	Nil	21.25
3–6A.M.	0.05 × 75 = 3.75	7.5	0	25	Nil	17.5
6–9A.M.	0.4 × 75 = 30	37.5	0	25	12.5	Nil
9–12A.M.	0.1 × 75 = 7.5	45	0	25	20	Nil
12–3P.M.	0.05 × 75 = 3.75	48.75	0	25	23.75	Nil
3–6P.M.	0.1 × 75 = 7.5	56.25	0	25	31.25	Nil
6–9P.M.	0.2 × 75 = 15	71.25	25	50	21.25	Nil
9–12A.M.	0.05 × 75 = 3.75	75	25	75	0	0

From the table,

Maximum value of excess demand = 31.25 ML

Maximum value of excess supply = 21.25 ML

So, Balancing storage = (31.25 + 21.25) ML = 52.5 ML

Fire demand = 1 million litre per day

Breakdown demand = 6.25 Million litre per day

So, storage capacity of reservoir = 52.5 + 1 + 6.25 = 59.75 million litres

3. Development of Ground Water and Well Hydraulics

3.1 **During recuperation test, the water level in an open well was depressed by pumping upto 2.5 m The water level was raised by 1 m within 60 minutes just after stopping the pumping. Determine the yield from the well of 2 m diameter when the depression head is 3 m.**

[10 marks : 2001]

Solution:

Let at any time 't' the drawdown is 'h' and the well recuperates by 'dh' in time 'dt'. If at this stage 'q' is the discharge, then we know that

$$q \propto h$$

$$\Rightarrow \quad q = Ch$$

where C is a constant of proportionality

Also, if the cross-section area of the well is A, then

$$qdt = A(-dh)$$

$$\Rightarrow \quad Ch\,dt = -A\,dh$$

$$\Rightarrow \quad \frac{C}{A}dt = -\frac{dh}{h}$$

Integrating both sides, we get

$$\Rightarrow \quad \frac{C}{A}\int_0^T dt = \int_{h_1}^{h_2} -\frac{dh}{h}$$

$$\Rightarrow \quad \frac{C}{A} \times T = \left[-\log_e h\right]_{h_1}^{h_2} = \log_e\left(\frac{h_1}{h_2}\right)$$

$$\Rightarrow \quad \frac{C}{A} = \frac{1}{T}\log_e\left(\frac{h_1}{h_2}\right) = \frac{2.303}{T}\log_{10}\left(\frac{h_1}{h_2}\right)$$

where $\frac{C}{A}$ is called specific capacity or specific yield of the open well.

But, we know that

$$q = \frac{C}{A} \times \text{volume of water for given depression head}$$

$$\Rightarrow \quad q = \frac{C}{A} \times AH = \left[\frac{2.303}{T}\log_{10}\left(\frac{h_1}{h_2}\right)\right] \times AH$$

Here,

$$T = 60 \text{ min} = 60 \times 60 = 3600 \text{ sec}$$

$$h_1 = 2.5 \text{ m}; \quad h_2 = 2.5 - 1 = 1.5 \text{ m}$$

$$A = \frac{\pi}{4}d_w^2 = \frac{\pi}{4} \times 2^2 = 3.14 \text{ m}^2$$

$$H = 3 \text{ m}$$

$$\therefore \quad q = \left[\frac{2.303}{3600}\log_{10}\left(\frac{2.5}{1.5}\right)\right] \times 3.14 \times 3$$

$$\Rightarrow \quad q = 1.337 \times 10^{-3} \text{ m}^3/\text{sec} = 4.81 \text{ m}^3/\text{hr}$$

3.2 **Explain the reasons for the functional and physical failures of tube wells. What are the methods for reducing these failures?**

[10 marks : 2001]

Solution:

Normally tube wells last for about 15 to 20 years.

They may fail due to incrustation or corrosion:

(a) Incrustation: It occurs due to deposition of alkali salts on the inside walls of the pipe. The important alkali salt causing incrustation is calcium carbonate. The incrustation of the well pipe reduces the effective diameter of the well pipe and hence reduces the discharge of the tube well.

The incrustation can be reduced by:

(i) Reducing the drawdown and hence pumping rate.

(ii) Using screens having larger area of openings so as to keep some allowance for the future incrustation.

(iii) Using such materials for the strainers (screens) that may easily permit the removal of incrustating material by chemical action without effecting that strainer material.

(iv) Properly maintaining and periodically cleaning the well screens.

(b) **Corrosion:** The well pipe is gradually destroyed by corrosion due to the action of acidic waters on the pipe material. When chlorides and sulphates or carbon dioxide are present in the water, the well pipe will definitely get corroded. The aquifer sand, surrounding the well pipe finds a way out into the corroded pipe through the worn out pipe walls, thus bringing sand along with water. Hence, corrosion results in excessive withdrawl of sand along with well water. Thicker pipe may be used to avoid corrosion.

Other measures which can help in reducing corrosion and thereby increasing the life of the tube well as are:

(i) By reducing the drawdown and pumping rate.

(ii) By reducing corrosion resistant materials for pipes.

(iii) By using corrosion resistant coatings on the pipes.

(iv) By using thicker pipes.

3.3 Explain: (a) Aquifer, (b) Aquitard and (c) Aquiclude.

[6 marks : 2005]

Solution:

(a) **Aquifer:** It is a saturated formation of earth material which not only stores water but yields in sufficient quantity. Thus, an aquifer transmits water relatively easily due to its high permeability. Unconsolidated deposits of sand and gravel form good aquifers.

The availability of ground water from an aquifer at a place depends on the rates of withdrawal or replenishment. Aquifers play the roles of both a transmission conduit and a storage. Aquifers are classified as unconfined aquifers and confined aquifers on the basis of their occurrence and field situation. An unconfined aquifer (water table aquifer) is one in which a free water surface i.e. a water table exists. A confined aquifer (artesian aquifer) is an aquifer which is confined between two impervious beds such as aquicludes or aquifuges.

(b) **Aquitard:** An aquitard is a formation through which only seepage is possible and thus the yield is insignificant compared to an aquifer. It is partly permeable but highly porous. A sandy clay unit or silty units are examples of aquitard. Through an aquitard appreciable quantities of water may leak to an aquifer below it.

(c) **Aquiclude:** Aquiclude is a geological formation which is essentially impermeable to the flow of water. It may be considered as close to water movement even though it may contain large amount of water due to its high porosity. Clay is an example of aquiclude.

3.4 What is "*well function*"?

[2 marks : 2005]

Solution:

Well function is a special integral which is used to calculate the drawdown from a well for an unsteady flow in a confined aquifer.

The flow in a confined aquifer in unsteady state can be given as

$$\frac{\partial^2 h}{\partial x^2} + \frac{\partial^2 h}{\partial y^2} + \frac{\partial^2 h}{\partial z^2} = \frac{S}{T}\frac{\partial h}{\partial t}$$

$$\nabla^2 h = \frac{S}{T}\frac{\partial h}{\partial t}$$

The equation above is basic differential equation governing unsteady ground water flow in a homogeneous isotropic confined aquifer. This is known as diffusion equation. In polar coordinates, to represent the radial flow in a well, this equation takes the form

$$\frac{\partial^2 h}{\partial r^2} + \frac{1}{r}\frac{\partial h}{\partial r} = \frac{S}{T}\frac{\partial h}{\partial t}$$

The solution of this equation is given as

$$S = H - h = \frac{Q}{4\pi T}\int_u^{\infty}\frac{e^{-u}}{u}du \qquad ...(i)$$

Here $\int_u^{\infty}\frac{e^{-u}}{u}du$ is called as well function and denoted by W(u)

$$\therefore \qquad W(u) = \int_u^{\infty}\frac{e^{-u}}{u}du = -0.577216 - \log_e u + u - \frac{u^2}{2\underline{|2}} + \frac{u^3}{3\underline{|3}}.....$$

The value of W(u) can easily be calculated by the series above to the required number of significant digits which rarely exceeds 4. For small vales of $u(u \leq 0.01)$ only first two terms of the series are adequate. From equation (i)

$$s = \frac{Q}{4\pi T}\int_u^{\infty}\frac{e^{-u}}{u}du = \frac{Q}{4\pi T}W_{(u)}$$

$$S = \frac{Q}{4\pi T}\left[\ln\frac{4Tt}{r^2 s} - 0.5772\right]$$

Where, s = drawdown in observation well, δ = storativity, t = time, T = transmissivity, r = distance of observational well

3.5 **A 0.5 m diameter gravity well is being pumped at a steady rate of 1500 liters per minute. The drawdowns of 6 m and 2 m were observed in nearby test wells at distance of 6 m and 16 m from the well being pumped respectively, after steady state has been reached. Assume the well to be fully penetrating. The bottom of the well is 100 m below the undisturbed ground water table assuming that all observed points lie on Dupit's curve, compute the drawdown in the well being pumped.**

[8 marks : 2007]

Solution:

Given data: Q = 1500 litres per minute = $\frac{1500 \times 10^{-3}}{60}$ m³/s = 0.025 m³/s

r_1 = 6 m, s_1 = 6 m, r_2 = 16 m, s_2 = 2 m, H = 100 m

$$\therefore \qquad h_1 = H - s_1 = 100 - 6 = 94 \text{ m}$$

$$h_2 = H - s_2 = 100 - 2 = 98 \text{ m}$$

Using **Thiem's equation for unconfined aquifers,** we have

$$Q = \frac{\pi k\left(h_2^2 - h_1^2\right)}{2.303\log_{10}\left(\frac{r_2}{r_1}\right)}$$

$$\Rightarrow \quad 0.025 = \frac{\pi k\left(98^2 - 94^2\right)}{2.303 \log_{10}\left(\frac{16}{6}\right)}$$

$$\Rightarrow \quad \pi k = 3.193 \times 10^{-5}$$

$$\Rightarrow \quad k = 1.016 \times 10^{-5} \text{ m/s}$$

Thus the coefficient of permeability, k = 1.016×10^{-5} m/s

Radius of gravity well, $r_w = \frac{0.5}{2} = 0.25$ m

Using Theim's equation again, we get

$$Q = \frac{\pi k(h_1^2 - h_w^2)}{2.303 \log_{10}\left(\frac{r_1}{r_w}\right)}$$

$$\Rightarrow \quad 0.025 = \frac{3.193 \times 10^{-5} \times (94^2 - h_w^2)}{2.303 \log_{10}\left(\frac{6}{0.25}\right)}$$

$$\Rightarrow \quad 2488.3 = 94^2 - h_w^2$$

$$\Rightarrow \quad h_w = 79.67 \text{ m}$$

∴ **Drawdown in the pumped well = $H - h_w$ = 100 – 79.67 = 20.33 m**

3.6 **Two tube wells each of 200 mm diameter penetrate fully in a confined aquifer and are spaced at 150 m. Find the ratio of discharge from one well to the discharge from both wells under the same drawdown. Take radius of influence as 200 m.**

[5 marks : 2008]

Solution:

Given: Diameter of tube well = 200 mm, Spacing between tube wells = 150 m, Radius of influence, $R = 200$ m

∴ Radius of tube well, $r_w = \frac{200}{2} = 100$ mm

The discharge from a fully penetrating tubewell in a confined aquifer is given by

$$Q_1 = \frac{2\pi kHs}{2.3 \log_{10}\left(\frac{R}{r_w}\right)}$$

where H = thickness of confined aquifer

s = drawdown

k = coefficient of permeability

The discharge from two tubewells at a distance B apart is given by

$$Q_2 = Q_3 = \frac{2\pi kHs}{2.3 \log_{10}\left(\frac{R^2}{r_w B}\right)}$$

∴ **Total discharge from both wells,**

$$Q_4 = Q_2 + Q_3 = \frac{2 \times 2\pi\, kHs}{2.3 \log_{10}\left(\frac{R^2}{Br_w}\right)}$$

$$\frac{Q_1}{Q_4} = \frac{\log_{10}\left(\frac{R^2}{Br_w}\right)}{2\log_{10}\left(\frac{R}{r_w}\right)}$$

$$\Rightarrow \quad \frac{Q_1}{Q_4} = \frac{\log_{10}\left(\frac{200^2}{150\times 100\times 10^{-3}}\right)}{2\log_{10}\left(\frac{200}{100\times 10^{-3}}\right)} = 0.52$$

3.7 **A gravity well has a radius of 150 mm and the depth of water in the well is 30.5 m when pumping has not been started. If the radius of influence is twenty times the yield and the coefficient of permeability = 0.003125 m/min, compute the specific capacity of the well and the maximum rate at which water can be pumped.**

[5 marks : 2008]

Solution:

Specific capacity of a well is defined as the well yield per unit drawdrawn.

$$\therefore \quad s = 1 \text{ m}$$

When the pumping has not been started, the depth of water in the well is equal to the saturated thickness of the unconfined aquifer.

$$\therefore \quad d = 30.5 \text{ m}$$

Depth of water in well after pumping has been started is given by

$$h_w = d - s = 30.5 - 1 = 29.5 \text{ m}$$

Radius of well, $r_w = 150 \text{ mm} = 0.15 \text{ m}$

Coefficient of permeability, $k = 0.003125 \text{ m/min}$

Radius of influence, $R = 20 \times 1 = 20 \text{ m}$

$\therefore$ Discharge,

$$Q = \frac{\pi k(d^2 - h_w^2)}{2.3\log_{10}\left(\frac{R}{r_w}\right)} = \frac{\pi\times 0.003125\times(30.5^2 - 29.5^2)}{2.3\log_{10}\left(\frac{20}{0.15}\right)}$$

$$= 120.52\times 10^{-3} \text{ m}^3\text{/min} = 120.52 \text{ litre/min} = 2 \text{ litre/sec}$$

$$\therefore \quad \text{Specific capacity} = \frac{Q}{s} = \frac{2}{1} = 2 \text{ litre/sec}$$

For maximum discharge, $h_w = 0$ and $R = 20 \times 30.5 = 610$ m

$$\therefore \quad Q_{max} = \frac{\pi k(d^2 - 0^2)}{2.3\log_{10}\left(\frac{R}{r_w}\right)} = \frac{\pi\times 0.003125\times 30.5^2}{2.3\log_{10}\left(\frac{610}{0.15}\right)} = 1.1 \text{ m}^3\text{/min}$$

$$= \frac{1.1\times 1000}{60} \text{ litres/sec} = 18.34 \text{ litres/sec}$$

3.8 **Water is pumped from a well tapping an unconfined aquifer at a rate of 2400 m³/day. A no-drawdown boundary exists at a distance of 5 km from the well centre. Assuming the well to be fully penetrating, compute the steady state drawdown at the well face. Given : Initial saturated thickness = 50 m, hydraulic conductivity = 20 m/day, effective well radius = 1 m.**

[10 marks : 2009]

Solution:

Given data: Discharge (Q) = 2400 m^3/day, Radius of influence (R) = 5 km, Initial saturated thickness (d) = 50 m, Hydraulic conductivity (k) = 20 m/day, Effective well radius (r_w) = 1 m, The drawdown in the steady state can be calculated by using Dupit's formula for unconfined aquifer. The formula can be derived as shown below.

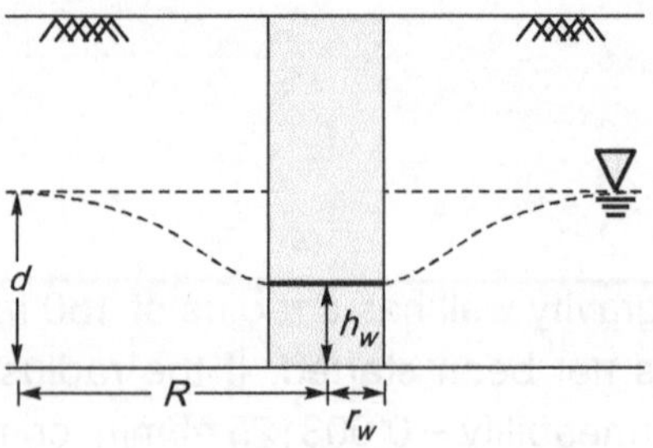

Ground level

Water table

d, h_w, R, r_w

As per Darcy's law, the discharge may be given as

$$Q = kiA$$

or
$$Q = k\frac{dh}{dr}2\pi rh$$

or
$$\frac{dr}{r} = \frac{2\pi k}{Q}\times hdh$$

Integrating between limits r_w and R, we get

$$\int_{r_w}^{R}\frac{dr}{r} = \frac{2\pi k}{Q}\times\int_{h_w}^{d} hdh$$

$$\Rightarrow \quad [\log_e r]_{r_w}^{R} = \frac{2\pi k}{Q}\times\left(\frac{h^2}{2}\right)_{h_w}^{d}$$

$$\Rightarrow \quad \log_e\frac{R}{r_w} = \frac{\pi k}{Q}(d^2 - h_w^2)$$

$$\Rightarrow \quad 2.303\log_{10}\frac{R}{r_w} = \frac{\pi k}{Q}(d^2 - h_w^2)$$

$$\Rightarrow \quad k = \frac{2.303Q\log_{10}\frac{R}{r_w}}{\pi\left(d^2 - h_w^2\right)}$$

$$\Rightarrow \quad d^2 - h_w^2 = \frac{2.303Q\log_{10}\frac{R}{r_w}}{\pi k}$$

$$\Rightarrow \quad h_w^2 = d^2 - \frac{2.303\,Q\log_{10}\frac{R}{r_w}}{\pi k}$$

$$= (50)^2 - \frac{2.303\times 2400\,\log_{10}\left(\frac{5000}{1}\right)}{\pi\times 20}$$

$$\Rightarrow \quad h_w = \sqrt{2174.6085} = 46.63\text{m}$$

Thus the steady state drawdown,

$$s = d - h_w = 50 - 46.63 = 3.37 \text{ m}$$

3.9 An aquifer of 25 m average thickness is overlain by an impervious layer of 30 m thickness. A test well 0.4 m in diameter and two observation wells at a distance of 12 m and 48 m are located in the aquifer. After pumping at a rate of 0.2 m^3/s for a long time, drawdowns in the wells were observed to be 3.5 m and 1.5 m respectively. Determine the coefficient of permeability in m/day.

[2 marks : 2010]

Solution:

Theim's formula for confined aquifer case is applicable here

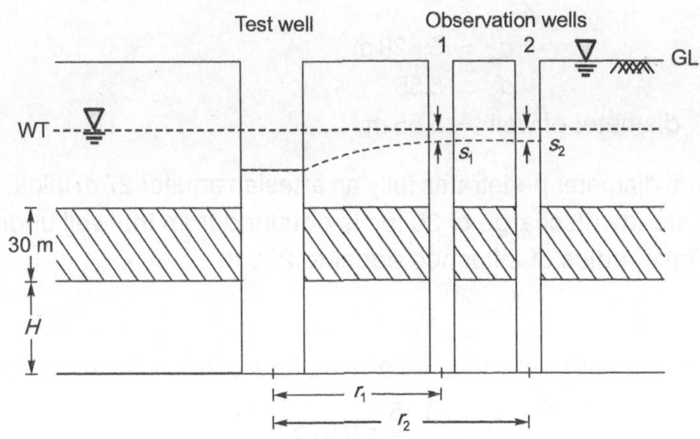

Given data: Discharge, $Q = 0.2\ m^3/s$

Radial distance of observation well 1 from test well, $r_1 = 12$ m

Radial distance of observation well 2 from test well, $r_2 = 48$ m

Drawdown in observation well 1, $s_1 = 3.5$ m

Drawdown in observation well, 2, $s_2 = 1.5$ m

Thickness of confined aquifer, $H = 25$ m

$$\text{Coefficient of permeability, } k = \frac{Q}{2\pi H(s_1 - s_2)} \times \log_e\left(\frac{r_2}{r_1}\right)$$

$$= \frac{0.2}{2\pi \times 25 \times (3.5 - 1.5)} \times \log_e\left(\frac{48}{12}\right) \text{ m/s}$$

$$= 8.8254 \times 10^{-4} \times 60 \times 60 \times 24 = 76.25 \text{ m/day}$$

3.10 **In fine sand aquifer region, design a well to get 10 liters/sec yield, under depression head of 2.5 meters. Value of $\frac{C'}{A}$ in cum./hr. per sq. meter of the area may be taken as 0.60.**

[10 marks : 2012]

Solution:

We know that, $q = \left(\frac{C'}{A}\right) \times$ Volume of water for given depression head

$\Rightarrow$ $q = \left(\frac{C'}{A}\right) \times AH$

where $\frac{C'}{A}$ is called specific capacity or specific yield of the open well in cum/hr/m^2 of area

Given, $\frac{C'}{A}$ for Fine sand = 0.6

Also, Depression head = 2.5 m

q = 10 liters/sec = 0.01 m^3/sec = 0.01 × 3600 m^3/hr = 36 m^3/hr

$\therefore$ 36 = 0.6 × A × 2.5

$$\Rightarrow \quad A = \frac{36}{0.6 \times 2.5} = 24\text{m}^2$$

$$\therefore \quad \frac{\pi d_w^{\ 2}}{4} = 24$$

$$\Rightarrow \quad d_w = 5.528 \text{ m}$$

$$\Rightarrow \quad d_w \simeq 5.55$$

$$\therefore \quad \text{diameter of well} = 5.55 \text{ m}$$

3.11 **A tube well of 15 cm diameter penetrates fully an artesian aquifer 27 m thick. Determine permeability of the aquifer if a steady discharge of 30 lps is obtained from the well under a drawdown of 3.0 m at the well face. Take radius of influence equal to 200 m.**

[6 marks : 2012]

Solution:

Given: Diameter of tube well (d_w) = 15 cm = 0.15 m, Radius of influence (R) = 200 m

$\therefore$ Radius of tube well, $r_w = \frac{0.15}{2} = 0.075\text{m}$

Discharge, Q = 30 lps = 0.03 m³/sec

Thickness of aquifer, H = 27 m

Drawdown, s = 3 m

The **discharge from a fully penetrating tube-well in a confined aquifer** is given by,

$$Q = \frac{2\pi kHS}{2.3\log_{10}\left(\frac{R}{r_w}\right)}$$

$$\therefore \quad 0.03 = \frac{2 \times \pi \times k \times 27 \times 3}{2.3\log_{10}\left(\frac{200}{0.075}\right)}$$

$$\Rightarrow \quad 2 \times \pi \times k \times 27 \times 3 = 0.2364$$

$$\Rightarrow \quad k = 4.645 \times 10^{-4} \text{ m/sec}$$

3.12 **What is a water bearing stratum? On what basis ground water flows in it? What is the difference between Specific Yield and Specific Retention?**

[8 marks : 2012]

Solution:

Water in the soil mantle is called subsurface water and is considered in two zones.

1. Saturated zone, and 2. Aeration zone

Saturated Zone: This zone, also known as groundwater zone, is the space in which all the pores of the soil are filled with water. The water table forms its upper limit and marks a free surface, i.e. a surface having atmospheric pressure.

Zone of Aeration: In this zone the soil pores are only partially saturated with water. The space between the land surface and the water table marks the extent of this zone.

All earth materials, from soils to rocks have pore spaces. Although these pores are completely saturated with water below the water tables, from the ground water utilization aspect only such material through which water moves easily and hence can be extracted with ease are significant. On this basis the saturated formations are classified into four categories. These are discussed below:

Aquifer: An aquifer is a saturated formation of earth material which not only stores water but yields it in sufficient quantity. Thus an aquifer transmits water relatively easily due to its high permeability. Unconsolidated deposits of sand and gravel form good aquifers.

Aquitard: It is a formation through which only seepage is possible and thus the yield is insignificant compared to an aquifer. It is partly permeable. A sandy clay unit is an example of aquitard.

Aquiclude: It is a geological formation which is essentially impermeable to the flow of water. It may be considered as closed to water movement even though it may contain large amounts of water due to its high porosity. Clay is an example of an aquiclude.

Aquifuge: It is a geological formation which is neither porous nor permeable. Massive compact rock without any fractures is an aquifuge. Usually the perched water table is of limited extent and the yield from such a situation is very small. In groundwater exploration a perched water table is quite often confused with a general water table.

The availability of groundwater from an aquifer at a place depends upon the rate of withdrawal and replenishment (recharge). Aquifers play the roles of both a transmission conduit and a storage. Aquifers are classified as unconfined aquifers and confined aquifers on the basis of their occurrence and field situation. An unconfined aquifer is one in which a free water surface i.e., a water table exists. Only the saturated zone of this aquifer is important in groundwater studies. Recharge of this aquifer takes place through infiltration of precipitation from the ground surface.

A water table is the free water surface in unconfined aquifer. The static level of a well penetrating an unconfined aquifer indicates the level of the water table at that point. The water table is constantly in motion adjusting its surface to achieve a balance between the recharge and outflow from the subsurface storage. Fluctuations in the water level in a dug well during various seasons of the year, lowering of the groundwater table in a region due to heavy pumping of the wells and the rise in the water table of an irrigated area with poor drainage, are some common examples of the fluctuation of the water table. In a general sense, the water table follows the topographic features of the surface. If the water table intersects the land surface the groundwater comes out to the surface in the form of springs or seepage.

The position of the water table relative to the water level in a stream determines whether the stream contributes water to the groundwater storage or the other way about. If the bed of the stream is below the groundwater table, during periods of low flows in the stream, the water surface may go down below the general water table elevation and the groundwater contributes to the flow in the stream. Such streams which receive groundwater flow are called effluent streams. Perennial rivers and streams are of this kind. If, however, the water table is below the bed of the stream, the stream-water percolates to the groundwater storage and a hump is formed in the groundwater table. Such streams which contribute to the ground water are knows as influent streams. Intermittent rivers and streams which go dry during long periods of dry spell (i.e. no rain periods) are of this kind

Sometimes a lens or localised patch of impervious stratum can occur inside an unconfined aquifer in such a way that it retains a water table above the general water table. Such a water table retained around the impervious material is known as perched water table.

The porosity gives a measure of the water-storage capability of a formation, not all the water held in the pores is available for extraction by pumping or draining by gravity. The pores hold back some water by molecular attraction and surface tension. The actual volume of water that can be extracted by the force of gravity from a unit volume of aquifer material is known as specific yield, S_y. The fraction of water held back in the aquifer is known as specific retention, S_r. Thus porosity

$$n = S_y + S_r$$

3.13 **Describe briefly the following in a well with neat sketches:**

(i) Cone of depression

(ii) Area of influence

(iii) Drawdown

(iv) Radius of influence

[8 marks : 2016]

Solution:

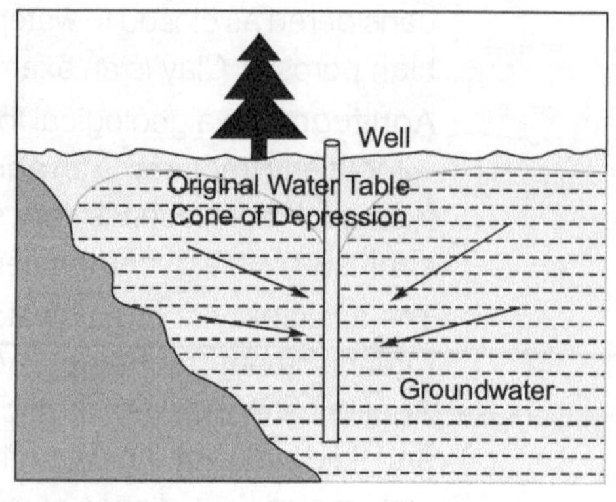

(i) **Cone of depression:** When ground water is pumped out from a well, a cone of depression is formed. In case of **unconfined aquifers**, this cone of depression is the actual water level in ground but in case of **confined aquifers**, this is the reduction in the pressure head of water. As one moves farther away from the well, the effect of this depression goes on decreasing.

(ii) **Area of influence:** Under steady state conditions of pumping from the well, the drawdown of water level gets stabilized. This altered water profile spreads outwards giving rise to a pattern called as cone of depression and the area under the influence of this depression is called as **area of influence**.

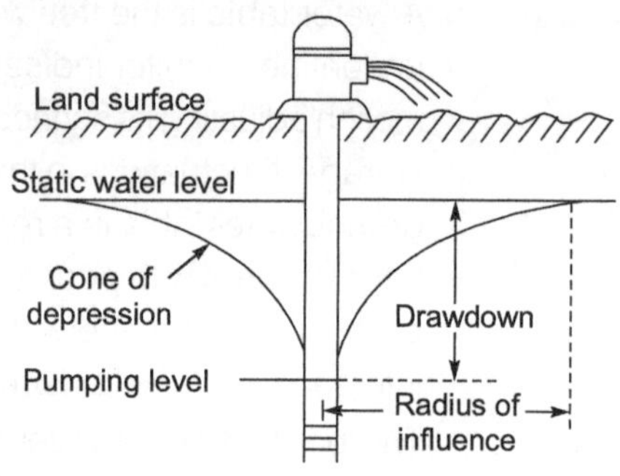

(iii) **Drawdown:** The decline in the water surface level (in case of unconfined aquifers) and pressure head (in case of confined aquifers) from the original level due to pumping from the well is called as drawdown. This drawdown is maximum at the well location and decreases as one moves away from the well.

(iv) **Radius of influence:** It is the radial distance from the center line of well up to which the effect of withdrawal of water from the well persists. Beyond this distance, the effect is negligible.

3.14 **What is the purpose of intake structure? What factors should be considered in locating an intake structure and what are the main considerations in the design of an intake structure?**

[8 marks : 2016]

Solution:

Intake structure: Intake structure is used to withdraw water from the surface sources and then to supply it to the water treatment plant.

Factors to be considered in locating the intake structure:

(i) Intake structure, as far as possible, must be located near to the water treatment plant so as to reduce cost on infrastructure of water conveyance.

(ii) The intake must be located in the purer zone of source of water so as to draw the best quality of water requiring less treatment.

(iii) The intake site must remain accessible at all the times round the year.

(iv) The intake must be located at the point in the source of water from where water can be drawn even during the dry periods of the year.

(v) The intake must be located at a point where it does not gets flooded during high floods.

(vi) Wherever possible, intake must be located in the straight part of the river or can so as to avoid silting and scouring.

(vii) On the upstream of intake structure, there must not be any point of disposal of waste into the river in the vicinity of the intake structure.

Factors to be considered in the design of intake structure:

(i) Type of intake structure.

(ii) The intake structure must be sufficiently heavy so that it does not get lifted up by the up-thrust of water.

(iii) Fluctuations in the water level.

(iv) Distance of water treatment plant from the intake structure.

3.15 **A 0.5 m diameter well fully penetrates an unconfined aquifer whose bottom is 150 m below the undisturbed groundwater table. The water is pumped out from the well at a constant rate of 0.1 m³/s. After equilibrium is reached, the drawdown observed in two observation wells at radial distances of 10 m and 50 m are respectively 10 m and 5 m. Determine**

(i) the coefficient of permeability (in m/s)

(ii) the drawdown in the well (in m)

[20 marks : 2017]

Solution:

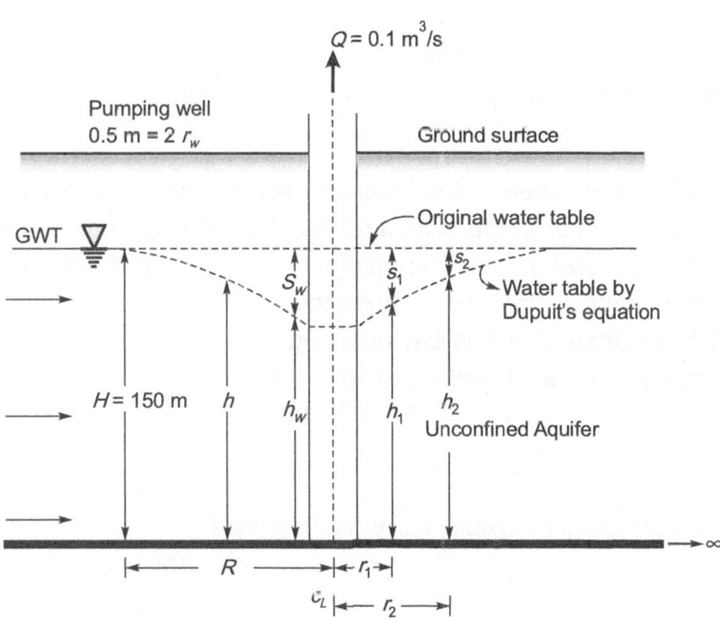

Radial flow to well in an unconfined aquifer

(i) Coefficient of permeability (k in m/s);

Given: $Q = 0.1\ m^3/sec$, $s_1 = 10$ m and $s_2 = 5$ m

$$h_2 = 150 - 5 = 145 \text{ m}$$

$$h_1 = 150 - 10 = 140 \text{ m}$$

We know that,

For unconfined aquifer, discharge,

$$Q = \frac{\pi k\left(h_2^2 - h_1^2\right)}{\ln\left(\frac{r_2}{r_1}\right)}$$

Substituting the respective values, we get

$$\Rightarrow \quad 0.1 = \frac{\pi \times k\left(145^2 - 140^2\right)}{\ln\left(\frac{50}{10}\right)} \Rightarrow \frac{\pi k \times 1425}{1.6094}$$

$$\therefore \qquad k = \frac{0.1 \times 1.6094}{\pi \times 1425} = 3.595 \times 10^{-5} \text{ m/s}$$

(ii) Let the drawdown at the well be s_w.

$$s_w = ?$$

At the pumping well, $\qquad r_w = \frac{0.50}{2} = 0.25 \text{ m}$

Using Dupuit's equation;

$$\therefore \qquad Q = \frac{\pi k\left(h_1^2 - h_w^2\right)}{\ln\left(\frac{r_1}{r_w}\right)}$$

$$\Rightarrow \qquad 0.10 = \frac{\pi \times 3.595 \times 10^{-5} \times \left[(140)^2 - h_w^2\right]}{\ln\left(\frac{10}{0.25}\right)}$$

$$\Rightarrow \qquad (140)^2 - h_w^2 = 3266.22$$

$$\Rightarrow \qquad h_w^2 = 16333.78$$

$$\Rightarrow \qquad h_w = 127.80$$

Drawdown at the well, $s_w = H - h_w = (150 - 127.8)$ m

$$s_w = 22.20 \text{ m}$$

3.16 **Give the characteristics of different formations in which groundwater exists. A fully penetrating artesian well is discharging at a rate of 25 litres/sec. The storage coefficient and transmissivity of the aquifer are 4.5×10^{-4} and 0.15 m²/min respectively. Find the drawdown at**

(i) A radius of 5 m distance after 2 hours pumping.

(ii) A radius 150 m distance after one day pumping

Using the following approximation well function

$$W(u) = -0.5772 - \ln(u) - u$$

[20 marks : 2018]

Solution:

Different formations in which groundwater exists are as follows:

1. Aquifer
2. Aquitard
3. Aquiclude
4. Aquifuge

There characteristics are mentioned below:

1. **Aquifer:** An aquifer is a saturated formation of earth material which not only stores water but yields it in sufficient quantity. Thus an aquifer transmits relatively easily due to its high permeability. Unconsolidated deposits of sand and gravel from good aquifer.

2. **Aquitard:** It is a formation through which only seepage is possible and thus the yield is insignificant compared to an aquifer. It is partly permeable. A sandy clay unit is an example of aquitard.

3. **Aquiclude:** It is a geological formation which is essentially impermeable to the flow of water. It may be considered as closed to water movement even though it may contain large amount of water due to high porosity. Clay is an example of acquiclude.

4. **Aquifuge:** It is a geological formation which is neither porous, nor permeable. There are no interconnected openings and hence cannot transmit water. Massive compact rock without any fractures in an acquifuge.

Given $\qquad T = 0.15 \text{ m}^2\text{/min}, S = 4.5 \times 10^{-4}$

(i)
$$u = \frac{r^2 S}{4Tt} = \frac{(5\text{m})^2 \times (4.5 \times 10^{-4})}{4 \times 0.15\ \text{m}^2/\text{min} \times (2 \times 60)\text{min}} = 1.5625 \times 10^{-4}$$

Using these method and calculating $W(u)$,

$$W(u) = -0.5772 - \ln(u) - u$$
$$= -0.5772 - \ln(1.5625 \times 10^{-4}) - (1.5625 \times 10^{-4})$$
$$= 8.1867$$

∴ Drawdown at 5 m after 2 hours pumping

$$S_{5\text{m}} = \frac{Q}{4\pi T} W(u) = \frac{0.025\ \text{m}^3/\text{s} \times 60}{4\pi \times 0.15\text{m}^2/\text{min}} \times 8.1867 = 6.518\ \text{m}$$

(ii)
$$u = \frac{r^2 S}{4Tt} = \frac{(150\text{m})^2 \times (4.5 \times 10^{-4})}{4 \times 0.15\ \text{m}^2/\text{min} \times (1440)\text{min}} = 0.011718$$

Using this method and calculating $W(u)$

$$W(u) = -0.5772 + 2 - \ln(u) - u = 3.8576$$

∴ Drawdown at 150 m after 1 day pumping

$$S_{150\text{ m}} = \frac{Q}{4\pi T} W(u) = \frac{0.025\ \text{m}^3/\text{s} \times 60}{4\pi \times 0.15\ \text{m}^2/\text{min}} \times 3.8576 = 3.071\ \text{m}$$

3.17 **Three tube wells of 25 cm, diameter each are located at the three vertices of an equilateral triangle of side 100 m. Each tube well penetrates fully in a confined aquifer of thickness 25 m. Assume the radius of influence for these wells and the coefficient of permeability of the aquifer as 300 m and 40 m/day respectively.**

(i) Calculate the discharge when only one well is pumping with a drawdown of 3 m.

(ii) What will be the percent change in discharge of this well if all the three wells were to pump such that the drawdown is 3 m in all the wells?

[12 marks : 2019]

Solution:

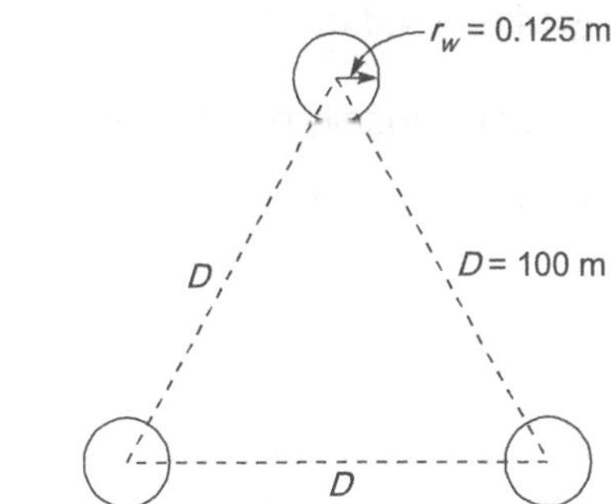

(i) When only one well is pumping

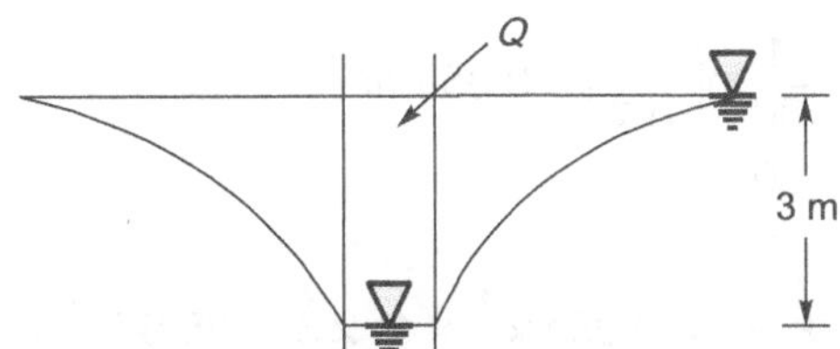

Discharge without interference,

$$Q = \frac{2\pi \times K \times Z \times s}{\ln\left(\frac{R}{r_w}\right)} = \frac{2\pi \times 40 \times 25 \times 3}{\ln\left(\frac{300}{0.125}\right)} = 2421.81 \text{ m}^3/\text{d}$$

(ii) When all three wells are pumping,

When all three wells interfere, discharge through each well would be

$$Q' = \frac{2\pi \times K \times z \times (s)}{\ln\left(\frac{R}{r_w} \times \frac{R^2}{D^2}\right)}$$

$$= \frac{2\pi \times 40 \times 25 \times 3}{\ln\left(\frac{300}{0.125} \times \frac{300^2}{100^2}\right)} = 1888.648 \text{ m}^3/\text{d}$$

$$\text{Percentage change in discharge} = \frac{\text{Discharge without interference} - \text{Discharge with interfence}}{\text{Discharge without interference}} \times 100$$

$$= \frac{2421.8 - 1888.64}{2421.8} \times 100 = 22.01\%$$

3.18 **During a recuperation test, the water in an open well was depressed by 2.5 m by pumping and it recuperated 1.8 m in 80 minutes. Find yield from a well of 4 m diameter under a depression head of 3 m.**

[8 marks : 2020]

Solution:

Recuperation test:

Yield from a well is given by

$$Q = k_s A h$$

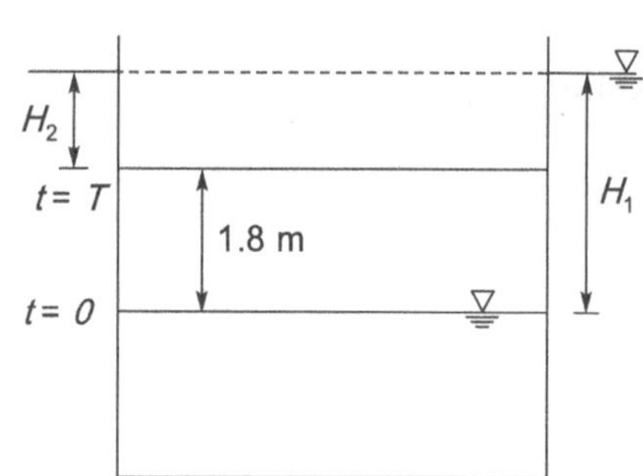

Given: $A = \frac{\pi}{4} \times 4^2$, $H = 3$ m

k_s = Specific capacity per unit well area

$$k_s = \frac{1}{T} \ln \frac{H_1}{H_2}$$

T = Recuperation time = 80 min

H_1 = Drawdown at ($t = 0$) = 2.5 m

H_2 = Drawdown at ($t = T$) = 2.5 – 1.8 = 0.7 m

$$k_s = \frac{1}{\frac{80}{60}} \ln\left(\frac{2.5}{0.7}\right) = 0.954 \text{ hr}^{-1}$$

$$Q = k_s A h = 0.954 \times \frac{\pi}{4} \times 4^2 \times 3 = 35.97 \text{ m}^3/\text{hr}$$

Therefore yield from the well is 35.97 m^3/hr

3.19 A 20 cm diameter well fully penetrates a confined aquifer of thickness 25 m when the will is pumped at a rate of 200 litres/minute. The steady state drawdown in two observation well located at 10 m and 100 m distance from the pumping well are found to be 3.5 m and 0.05 m, respectively. Calculate the permeability and transmissivity of the aquifer.

[10 marks : 2022]

Solution:

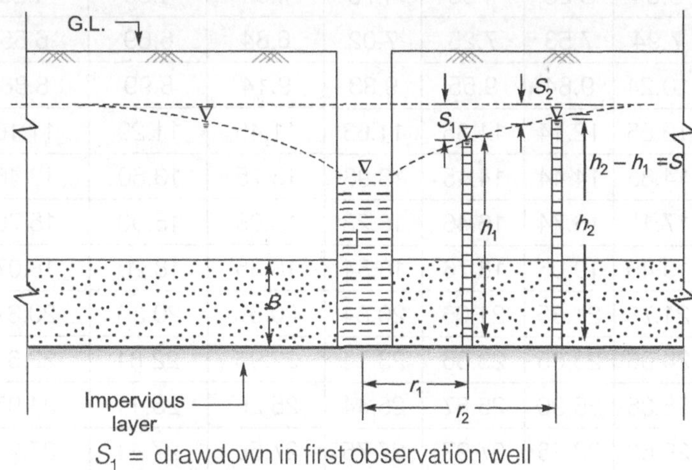

S_1 = drawdown in first observation well

$\Rightarrow$ $S_1 = 3.5$ m

S_2 = drawdown in second observation well

$\Rightarrow$ $S_2 = 0.05$ m

Applying Thiem's equation:

$$Q = \frac{2\pi kB(h_2 - h_1)}{ln\left(\frac{r_2}{r_1}\right)} = \frac{2\pi kB(S_1 - S_2)}{ln\left(\frac{r_2}{r_1}\right)}$$

$$\Rightarrow \quad \frac{200 \times 10^{-3}}{60} \text{m}^3/\text{s} = \frac{2\pi \times k \times 25\text{m}(3.5 - 0.05)\text{m}}{\ln\left(\frac{100}{10}\right)}$$

$\Rightarrow$ $k = 1.4163 \times 10^{-5}$ m/s

Transmissibility, $T = kB = 1.4163 \times 10^{-5}$ m/s $\times$ 25 m

$= 3.541 \times 10^{-4}$ m²/s

3.20 A well is pumping near a barrier boundary at a rate of 0.04 m³/s from a confined aquifer 20 m thick. The hydraulic conductivity of the aquifer is 3.5×10^{-4} m/s and its storactivity is 3×10^{-5}. Determine the drawdown in the observation well after 15 hours of continuous pumping. What is the fraction of the drawdown attributable to the impermeable barrier boundary?

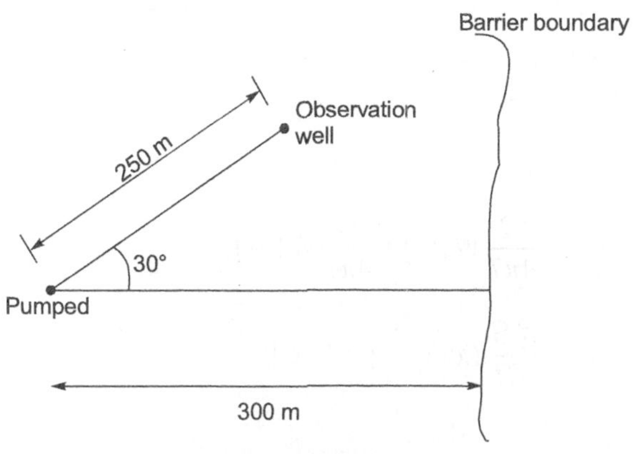

u	1.0	2.0	3.0	4.0	5.0	6.0	7.0	8.0	9.0
×1	0.219	0.049	0.013	0.0038	0.0011	0.00036	0.00012	0.000038	0.000012
$\times 10^{-1}$	1.82	1.22	0.91	0.70	0.56	0.45	0.37	0.31	0.26
$\times 10^{-2}$	4.04	3.35	2.96	2.68	2.47	2.30	2.15	2.03	1.92
$\times 10^{-3}$	6.33	5.64	5.23	4.95	4.73	4.54	4.39	4.26	4.14
$\times 10^{-4}$	8.63	7.94	7.53	7.25	7.02	6.84	6.69	6.55	6.44
$\times 10^{-5}$	10.94	10.24	9.84	9.55	9.33	9.14	8.99	8.86	8.74
$\times 10^{-6}$	13.24	12.55	12.14	11.85	11.63	11.45	11.29	11.16	11.04
$\times 10^{-7}$	15.54	14.85	14.44	14.15	13.93	13.75	13.60	13.46	13.34
$\times 10^{-8}$	17.84	17.15	16.74	16.46	16.23	16.05	15.90	15.76	15.65
$\times 10^{-9}$	20.15	19.45	19.05	18.76	18.54	18.35	18.20	18.07	17.95
$\times 10^{-10}$	22.45	21.76	21.35	21.06	20.84	20.66	20.50	20.37	20.25
$\times 10^{-11}$	24.75	24.06	23.65	23.36	23.14	22.96	22.81	22.67	22.55
$\times 10^{-12}$	27.05	26.36	25.96	25.67	25.44	25.26	25.11	24.97	24.86
$\times 10^{-13}$	29.36	28.66	28.26	27.97	27.75	27.56	27.41	27.28	27.16
$\times 10^{-14}$	31.66	30.97	30.56	30.28	30.05	29.87	29.71	29.58	29.46
$\times 10^{-15}$	33.96	33.27	32.86	32.58	32.35	32.17	32.02	31.88	31.76

[12 marks : 2022]

Solution:

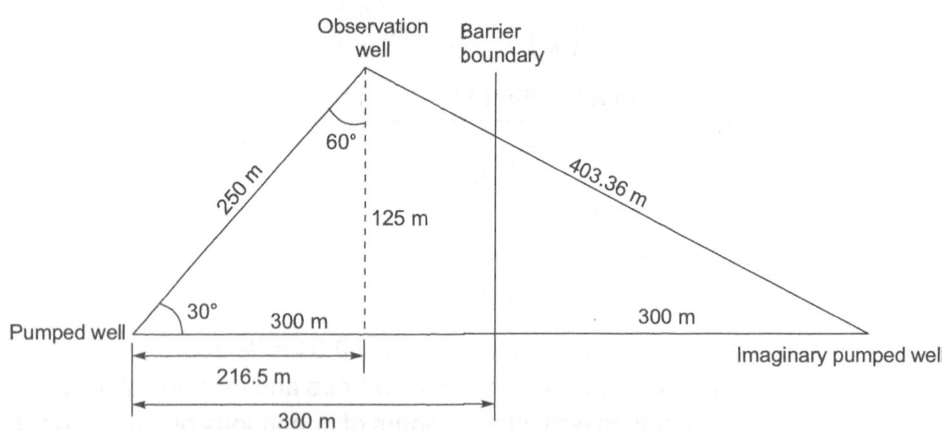

Discharge through a pumped well, $Q = 0.04 \text{ m}^3/\text{s}$

Thickness of confined aquifer, $B = 20$ m

Hydraulic conductivity of aquifer, $k = 3.5 \times 10^{-4}$ m/s

Storativity, $s = 3 \times 10^{-5}$

Drawdown in the observation well is

$$s = s_1 + s_2$$

$$= \frac{Q}{4\pi T} W(u_1) + \frac{Q}{4\pi T} W(u_2)$$

where, $$u_1 = \frac{r_1^2 S}{4Tt} \text{ (for pumped well)}$$

and $u_2 = \frac{r_2^2 S}{4Tt}$ (for imaginary pumped well)

$$u_1 = \frac{250^2 \times 3 \times 10^{-5}}{4 \times 3.5 \times 10^{-4} \times 20 \times (15 \times 3600)} = 1.24 \times 10^{-3}$$

$$u_2 = \frac{403.36^2 \times 3 \times 10^{-5}}{4 \times 3.5 \times 10^{-4} \times 20 \times 15 \times 3600} = 3.228 \times 10^{-3} \simeq 3.23 \times 10^{-3}$$

The value of $W(u_1)$ from the table is:

For $u = 1 \times 10^{-3}$, $W(u) = 6.33$

For $u = 2 \times 10^{-3}$, $W(u) = 5.64$

By interpolation, for $u_1 = 1.24 \times 10^{-3}$:

$$W(u_1) - 6.33 = \frac{5.64 - 6.33}{2 \times 10^{-3} - 1 \times 10^{-3}}(1.24 \times 10^{-3} - 1 \times 10^{-3})$$

$$\Rightarrow \quad W(u_1) = 6.1644$$

The value of $W(u_2)$ from the table is

For $u = 3 \times 10^{-3}$, $W(u) = 5.23$

For $u = 4 \times 10^{-3}$, $W(u) = 4.95$

By interpolation, for $u_2 = 3.23 \times 10^{-3}$:

$$W(u_2) - 5.23 = \frac{4.95 - 5.23}{(4 \times 10^{-3}) - (3 \times 10^{-3})}(3.23 \times 10^{-3} - 3 \times 10^{-3})$$

$$W(u_2) = 5.1656$$

$$\therefore \quad S = \frac{0.04\,m^3/s}{4\pi \times (3.5 \times 10^{-4} \times 20)\,m^2/s}[6.1644 + 5.1656] = 5.15\ m$$

If the impermeable barrier boundary would not exist, then drawdown in the well,

$$S' = \frac{0.04\,m^3/s}{4\pi \times (3.5 \times 10^{-4} \times 20)\,m^2/s} \times 6.1644 = 2.80\ m$$

Drawdown additionally caused due to impermeable barrier

$$S - S' = 5.15 - 2.80 = 2.35\ m$$

$\therefore$ Fraction of total drawdown due to barrier $= \frac{2.35}{5.15} = 0.46$

3.21 **Referring to figure given below, calculate the discharge from the well in steady state condition. The well completely penetrates the confined aquifer.**

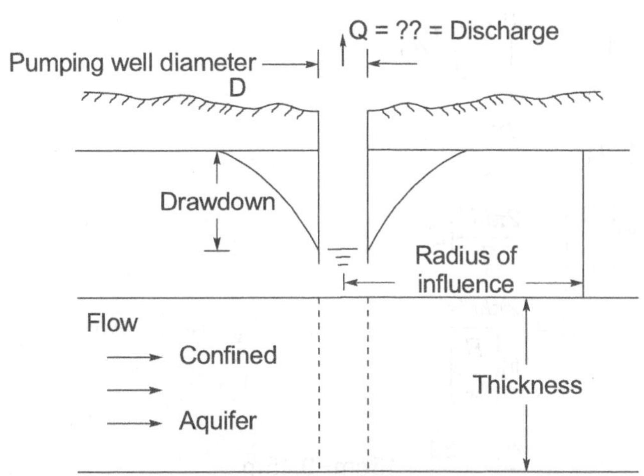

Given:
Diameter of pumping well = 30 cm
Permeability = 45 m/day
Length of the strainer = 20 m
Drawdown = 3.0 m
Radius of influence = 300 m
Discharge = ??

[20 marks : 2023]

Solution:

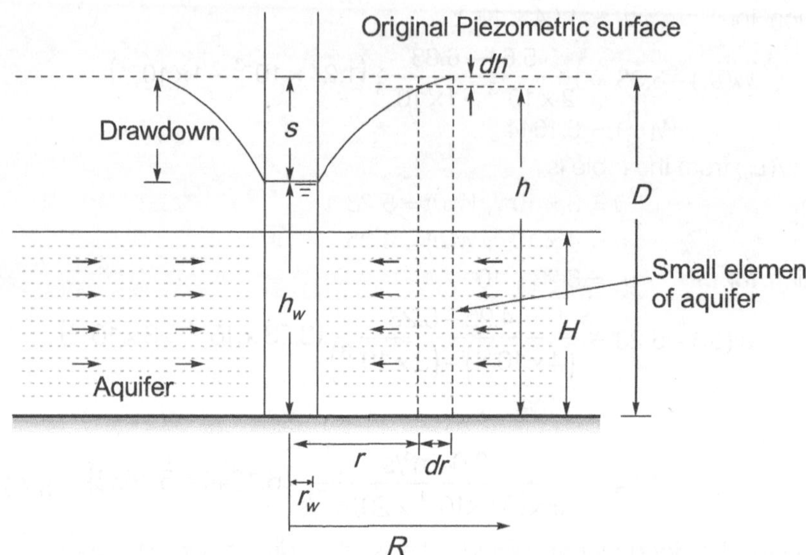

Consider a radial element of thickness '*dr*' at radius '*r*' from the main well and let '*dh*' be its drawdown.
Now, discharge through element, $Q = kiA$

where k is permeability

$$\Rightarrow \quad Q = k\frac{dh}{dr}\cdot 2\pi r\cdot H$$

$$\Rightarrow \quad \frac{dr}{r} = \frac{2\pi kH}{Q}dh$$

Integrating between r_w and R, we get

$$\Rightarrow \quad \int_{r_w}^{R}\frac{dr}{r} = \int_{h_w}^{D}\frac{2\pi kH}{Q}dh$$

$$\Rightarrow \quad [\ln r]_{r_w}^{R} = \frac{2\pi kH}{Q}[h]_{h_w}^{D}$$

$$\Rightarrow \quad \ln\left[\frac{R}{r_w}\right] = \frac{2\pi kH}{Q}[D-h_w]$$

$$\therefore \quad Q = \frac{2\pi kHs}{\ln\left[\frac{R}{r_w}\right]} \qquad [D-h_w = \text{drawdown} = s]$$

Now, $$r_w = \frac{d_w}{2} = \frac{30}{2} = 15\text{cm} = 0.15\text{m}$$

where d_w is diameter of main well.

$$k = 45 \text{ m/d} = 5.208 \times 10^{-4} \text{ m/s}$$
$$s = 3 \text{ m}$$
$$H = 20 \text{ m}$$
$$R = 300 \text{ m}$$

$$\therefore \quad Q = \frac{2\times\pi\times 5.208\times 10^{-4}\times 20\times 3}{\ln\left[\frac{300}{0.15}\right]} = 0.0258 \text{ m}^3\text{/s}$$

So, Discharge from well = 25.8 litre/sec.

4. Quality Control of Water Suppliers

4.1 Define MPN and CI. Explain membrane filtration technique.

[10 marks : 1999]

Solution:

Most Probable Number (MPN):

- It is an index to suggest quantity of coliforms present in a water sample
- In this test, 3 sample sizes, with each size having 5 samples and volume 10 mL, 1 mL and 0.1 mL respectively are tested for positive presumptive test and afterwards confirmatory test.
- This index is calculated by multiple tube fermentation technique.

Coliform Index (CI)

- It is an index to rate purity of water based on coliform bacteria present in water sample.
- It is defined as the reciprocal of the smallest quantity of a sample which would give a positive coliform test.

Membrane Filter Technique

- This technique gives a direct count of coliform bacteria.
- The water sample is filtered through a sterile membrane of pore size 5 to 10 μm.
- Bacteria retained an membrane is then put is contact with nutrients (M-Endo broth) and incubated at 35°C for 20-22 hours.
- Coliform bacteria will produce characteristic colonies which are pink to dark red.
- The final result is then reported in number of organism per 100 mL of water.

4.2 **Explain their qualitative differences in respect of total disolved solids, turbidity and bacterial quality, suggset the units needed for the treatment of water drawn from (a) ground (b) lake (c) river.**

[8 marks : 2010]

Solution:

(a) Ground

- The total dissolved solids, turbidity and bacterial quality are generally low in ruler drawn from ground.
- The treatment process recommended is

Aeration → Disinfection → Supply

(b) Lake

- Lake water has large bacterial activity.
- Turbidity and TDS are also generally high.
- The treatment process recommended is
  Prechlorination → Aeration → Coagulation → Rapid sand filter → Disinfection → Supply

(c) River

- Since sufficient air available to mix the total dissolved solids.
- The treatment process recommended is

Screening → Coagulation aided sedimentation → Filteration → Disinfection → Supply

4.3 **What are the sources and impact of hardness in water supplies? Also mention standard of hardness in water supply.**

[4 marks : 2013]

Solution:

Hardness: It is the concentration of multivalent metallic cations (M^{n+}) in solution. At super saturated conditions, these hardness causing cations react with anions in water resulting in the formation of solid precipitate. Depending upon the associating anions, hardness is classified as carbonate hardness or non-carbonate hardness. Hardness which is equivalent to alkalinity of water is termed as carbonate hardness and remaining hardness being the non-carbonate hardness.

Source: Multivalent metallic cations like Calcium (Ca^{+2}) and Magnesium (Mg^{+2}) are abundantly present in natural waters. Other cations that are found in smaller proportions are iron (Fe^{+2}), Manganese (Mn^{+2}), Aluminium (Al^{+3}).

Impact of hardness: Hard water leads to excessive consumptions of soap thereby posing an economic loss to the user. Usually soaps containing sodium ions (Na^{+}) react with multivalent cations to form precipitate. This precipitate gets adhered to tanks, clothes, dishes etc. Magnesium hardness has a laxative effect on certain persons.

Acceptable limit of total hardness is 200 mg/L and cause for rejection limit is 600 mg/L.

4.4 **Give suitable reasons why the following are important parameters for drinking water quality:**
(i) Nitrate (ii) Fluoride (iii) Total coliforms (iv) Iron

[8 marks : 2015]

Solution:

Nitrate

- Nitrate is not harmful because it is fully oxidised. But too much of nitrate affects infants. The reason behind this is it causes blue baby disease or mathemoglobinemia. Nitrate concentration should not be more than 45 mg/*l*.
- Nitrate concentration is measured by colour matching technique. Colour is formed by phenol-di-sulphonic + Potassium hydroxide.
- Free Ammonia + Organic Ammonia = Kjeldahl Ammonia

Fluoride: Fluorides occur sometimes naturally in water. If not, it should be added in controlled quantity during treatment process.

- Fluoridation of water supplies to a level of a 1 mg/*l* is safe and effective in reducing dental cavities.
- The greatest advantage against teeth decay is when water is drunk in childhood during the period of tooth formation.
- However, optimum concentration of fluoride has to be controlled because excessive amount leads to fluorosis, which causes decolouration or mottling of teeth and sometime bone damage both in children and adults.
- Greater than 5 mg/*l* cause deformation of bones called bone fluorosis and other skeleton abnormalities.

Permissible Limit: Acceptable limit is upto 1 mg/*l* and greater than 1.5 mg/*l* is cause for rejection.

Total Coliform

- Testing and counting of pathogenic bacteria can be done but only with great difficulty. Hence, these tests are not performed generally in routine.

- The usual routine tests are generally conducted to detect and count the presence of coliforms, which themselves are harmless aerobic lacteal fermentery organisms but their presence or absence indicates the presence and absence of pathogenic bacteria.
- Coliforms also known as bacteria coli [B-coli] or Escherichia coli (E-coli) are important harmless aerobic microorganisms, which are found residing in the intestine of all warm blooded animals including human beings, therefore, are excreted with their faces.
- Since they are harmless organisms to coliform group live longer in water than the pathogenic bacteria. It is generally presumed that the water will be safe and free from pathogenic bacteria if no coliform bacteria are present in it.

The tests for coliforms are:

1. Membrance filter technique
2. MPN test (Most probable number)
3. Coliform index

Iron: Iron in concentration greater than 0.3 ppm is undesirable as they may cause colouration of clothes washed in such waters.

- They may cause incrustation in water mains due to deposition of ferric hydroxides ($Fe(OH)_3$).
- Some bacteria use iron compound for an energy source and results in lime formation which may produce taste and odour.

$$Fe(HCO_3)_2 \xrightarrow{\text{Oxidation}} Fe(OH)_3 \downarrow$$

$$FeCl_3 \xrightarrow{\text{Oxidation}} Fe(OH)_3 \downarrow$$

$$FeSO_4 \xrightarrow{\text{Oxidation}} Fe(OH)_3 \downarrow$$

- Iron posses problems in ground water and bottom layer of lakes but not in surface water. Because surface water has sufficient O_2 and hence precipitate of $Fe(OH)_3$ will occur.

Permissible Limit: Acceptable limit for Fe is 0.1 – 1.0 mg/*l*.

4.5 Describe different causes responsible for water-borne diseases. Name the diseases caused by presence of various impurities. What are the preventive measures that can be taken to stop the spreading of these diseases?

[8 marks : 2016]

Solution:

Causes of water borne diseases: Water borne disease is caused due to drinking water getting contaminated by the presence of microbes and pathogens. Apart from this, many pathogens are acquired due to the consumption of contaminated foods and beverages.

Diseases caused by impurities present in water:

(i) **Fluorine:** Excess of fluorine in water causes **fluorosis**.
(ii) **Arsenicosis:** Even a minute amount of arsenic in water causes **arsenicosis**.
(iii) **Lead poisoning:** It is caused due to the presence of lead in potable water.
(iv) **Cholera:** This disease is cause due to the presence of 'vibrio chlorae' bacterium in water.
(v) **Amoebiasis:** Water contaminated by amoeba causes this disease.
(vi) **Typhoid:** This disease is caused by the presence of 'bacillus typhosus' bacterium.

Preventive measures to curb spreading of water borne diseases:

(i) Use treated and filtered water only.

(ii) Wash hands properly before and after the meals.

(iii) Eat washed fruits and cooked food only.

(iv) Avoid consumption of cut-fruits and juices from outside.

(v) Always keep food and beverages well closed.

(vi) Wash utensils and containers properly before using them.

(vii) Avoid flies by proper disposal of house hold wastes.

4.6 **Present the permissible drinking water quality standards of the following parameters. Also, explain the effects of presence of these parameters in water bodies:**

(i) Fluorides **(ii) Total Hardness**

(iii) Iron **(iv) Nitrates**

[20 marks : 2018]

Solution:

Effects of presence of these parameters and their permissible water quality standard limits are presented below:

1. **Fluorides:** Upto 1 mg/L, it helps to prevents dental cavities. During formation of permanent teeth it combines chemically with tooth enamel, resulting in harder, stronger teeth that are more resistant to decay.

   Excess value greater than 1.5-2.0 mg/L results in decoloration of teeth called mottling of teeth.

   Greater than 5 mg/L causes deformation of bones called bone fluorosis.

   Acceptable limit is upto 1 mg/L and greater than 1.5 mg/L is cause for rejection.

2. **Total hardness:** It is defined as concentration of multivalent metallic cations in the solution. For e.g., (Ca^{2+}, Mg^{2+}, Fe^{2+}, Al^{3+}, Mn^{2+} etc.)

   Hardness can be divided in two parts: Carbonate hardness noncarbonate hardened.

   Carbonate and bicarbonates of calcium and magnesium cause carbonate hardness, also called temporary hardness.

   Sulphate, chloride and nitrate of calcium and magnesium gives permanent hardness, also called non-carbonate hardness.

   Hard water leads to lesser foam formation, hence soap consumption is more.

   - It leads to scaling of boilers.
   - It causes of corrosion and incrustation of pipes.
   - It makes the food tasteless.
   - Magnesium hardness with sulphate ion have laxative effects.
   - Calcium hardness however does not cause any health problems.

   Acceptable limits of total hardness = 200 mg/L

   And cause of rejection limit = 600 mg/L

(iii) **Iron:** Iron poses colour problems, if greater than 0.3 mg/L.

Some, bacteria use iron compounds as an energy source and resulting into slime growth which may produce taste and odour.

- $Fe(HCO_3)_2 \xrightarrow{O_2} Fe(OH)_3 \downarrow$

  $FeCl_3 \xrightarrow{O_2} Fe(OH)_3 \downarrow$

$$FeSO_4 \xrightarrow{O_2} Fe(OH)_3 \downarrow$$

- Iron poses problems in ground water and bottom layers of lakes but not in surface water, because surface water has sufficient O_2.
- Acceptable limit for iron is 0.1 mg/L
  Rejection limit for iron is 1.0 mg/L

(iv) Nitrates: The presence of nitrates in water indicates complete decomposition of organic matter. It signifies old pollution in water.

- The presence of nitrate in water is not harmful as it is fully oxidized, but in some cases if it is present in large amount of affects the infants (new-born children).
- $NO_3^- \xrightarrow{\text{Acids in Body of Infants}} NO_2^-$

  $NO_2^- + \underbrace{HbO_2}_{\text{Haemoglobic}} \longrightarrow HbNo_2$
- Nitrates is converted into nitrite due to presence of acids in the intestine of infants which has very high affinity for haemoglobin hence replaces oxygen from it.
- Thereby leading to deficiency of oxygen in body. The disease caused is known as Blue Baby disease or Methemoglobinemia.
- Acceptable limit/cause for rejection limit = 45 mg/L

	Parameter	Cause for Rejection	Acceptable limit
1.	Fluoride	1.5 mg/L	1 mg/L
2.	Total hardness	600 mg/L	200 mg/L
3.	Iron	1 mg/L	0.1 mg/L
4.	Nitrates	45 mg/L	45 mg/L

4.7 Find out the pH of a mixture formed by mixing the following two water solutions:
Solution A : Volume 450 mL; pH = 7.5
Solution B : Volume 550 mL; pH = 6.5

[8 marks : 2020]

Solution:

Solution A : Volume 450 mL; pH = 7.5
Solution B : Volume 550 mL; pH = 6.5

$$pH = -\log_{10}[H^+]$$

For solution A:

$$-\log_{10}[H^+] = 7.5$$
$$\Rightarrow \quad [H^+]_A = 10^{-7.5}\text{ M}$$

For solution B:

$$-\log_{10}[H^+] = 6.5$$
$$\Rightarrow \quad [H^+]_B = 10^{-6.5}\text{ M}$$

Now,
$$[H^+]_{mix} = \frac{V_A \cdot [H^+]_A + V_B \cdot [H^+]_B}{V_A + V_B}$$

$$= \frac{450 \times 10^{-7.5} + 550 \times 10^{-6.5}}{450 + 550} = 1.88 \times 10^{-7}\text{ M}$$

$\therefore$
$$(pH)_{mix} = -\log_{10}[H^+]_{mix}$$
$$= -\log_{10}(1.88 \times 10^{-7}) = 7 - \log_{10}(1.88) = 6.72$$

Hence, pH of the mixture, $(pH)_{mix} = 6.72$.

Note: '*M*' refers to moles/*l*

5. Water Treatment

5.1 **A rapid sand filter is proposed for water supply treatment plant for a town with a population of 75000. The average rate of water supply is 150 litre/capita/day and rate of filtration is to be 100 litre/min/sq.m. Design:**

(i) Size and number of filter beds

(ii) Manifold lateral under drainage system

(iii) Wash water discharge required, if the rate of washing is 45 cm rise/min.

[18 marks : 1999]

Solution:

(i) Size and number of filter beds

The maximum water demand per day

$$= \text{Population} \times \text{Maximum daily rate of water supply}$$

$$= \text{Population} \times 1.8 \times \text{Average daily rate of water supply}$$

$$= 75000 \times 1.8 \times 150 = 20.25 \times 10^6 \text{ litres per day} = 20.25 \text{ MLD}$$

Assuming that 4% of filtered water is required for washing of the filter every day.

$$\text{Total filtered water required per day} = \frac{20.25}{0.96} = 21.09 \text{ MLD}$$

Also, assuming that 0.5 hour is lost every day in washing the filter.

$$\text{Filtered water required per hour} = \frac{21.09}{23.5} \times 24 = 21.538 \text{ MLD}$$

$$\frac{21.538}{24} = 0.897 \text{ MLH}$$

$$\text{Given rate of filtration} = 100 \text{ L/min/m}^2$$

$$\text{Rate of filtration} = \frac{\text{Filtered water required per hour}}{\text{Area of filter}}$$

$$\Rightarrow \quad 100 \times 60 = \frac{0.897 \times 10^6}{\text{Area of filter}}$$

$$\Rightarrow \quad \text{Area of filter} = 149.5 \text{ m}^2$$

Number of units (filter beds) may be roughly estimated by the equation developed by Morell and Wallace. It states that

$$N = 1.22\sqrt{Q}$$

where N is number of units Q is plant capacity in MLD

$$\therefore \quad N = 1.22\sqrt{21.09} = 5.6$$

Thus, providing 7 filter units = 6 operational + 1 standby

$$\therefore \quad \text{Area of each filter unit} = \frac{149.5}{6} = 24.92 \text{ m}^2 \Rightarrow 25 \text{ m}^2$$

Assuming the length of filter bed (L) as 1.5 times the width of the filter bed (B)

Now, $L \times B = 25$

$\Rightarrow \quad 1.5B \times B = 25$

$\Rightarrow \quad B = 4.1 \text{ m}$

and $L = 1.5 \times B = 1.5 \times 4.1 = 6.15 \text{ m}$

Take length of filter (L) say 6.3 m.

$$\therefore \quad B = \frac{6.3}{1.5} = 4.2 \text{ m}$$

Hence, adopting 6 filter units each of dimensions 6.3 m × 4.2 m

(ii) Design of manifold lateral under drainage system

Let a manifold and lateral system be provided below the filter bed, for receiving the filtered water and to allow backwashing for cleaning the filter. This consists of a central manifold pipe with laterals having perforations at their bottom.

To design this system, let us assume area of perforations to be 0.2% of filter area.

$$\therefore \quad \text{Total area of perforations} = 6.3 \times 4.2 \times \frac{0.2}{100} = 0.05292 \text{ m}^2$$

Now assuming the area of each lateral = four times the area of perforations in it

$$\therefore \quad \text{Total area of laterals} = 4 \times 0.05292 = 0.21168 \text{ m}^2$$

Now assuming the area of manifold to be twice the area of laterals, we have

$$\text{Area of manifold} = 2 \times 0.21168 = 0.4234 \text{ m}^2$$

$\therefore$Diameter of manifold (d) is given by

$$\frac{\pi}{4} d^2 = 0.4234$$

$$\Rightarrow \quad d = 0.734 \text{ m} = 73.4 \text{ cm}$$

Hence, using a 75 cm diameter manifold pipe laid lengthwise along the centre of the filter bottom. Laterals running perpendicular to the manifold (i.e. widthwise) emanating from the manifold may be laid at a spacing of say 15 cm (max. 30 cm). The number of laterals is then given as

$$\text{Number of laterals on each side} = \frac{\text{Length of filter bed}}{\text{Spacing between laterals}}$$

$$\therefore \quad \text{Number of laterals on each side} = \frac{6.3 \times 100}{15} = 42$$

Hence, use 84 laterals in all, in each filter unit. The diameter of the laterals is adopted as 6 mm.

$$\text{Now, length of each lateral} = \frac{\text{Width of filter}}{2} - \frac{\text{Diameter of manifold}}{2}$$

$$= \frac{4.2}{2} - \frac{0.75}{2} = 1.725 \text{ m}$$

Now, total number of perforations in all 84 laterals × Area of each lateral

$$= \text{Total area of perforations}$$

$$\therefore \text{Total number of perforations in all 84 laterals} \times \frac{\pi}{4} \times \left(\frac{6}{1000}\right)^2 = 0.05292$$

$\Rightarrow$Total number of perforations in all 84 laterals = 1871.66 ≈ 1872

$$\therefore \quad \text{Number of perforations in each lateral} = \frac{1872}{84} = 22.28 \text{ say } 23$$

$$\therefore \quad \text{Area of perforations per lateral} = 23 \times \frac{\pi}{4} \times (0.6)^2 = 6.5 \text{ cm}^2$$

$$\text{Now area of each lateral} = 4 \times \text{area of perforations in it}$$

$$= 4 \times 6.5 = 26 \text{ cm}^2$$

$$\therefore \quad \text{Diameter of each lateral} = \sqrt{\frac{26 \times 4}{\pi}} = 5.75 \text{ cm}$$

Hence, use 84 laterals each of 5.75 cm dia @ 15 cm c/c, each having 23 perforations of 6 mm size with 75 cm diameter manifold.

Check: $\dfrac{\text{Length of each lateral}}{\text{Diameter of lateral}} = \dfrac{1.725 \times 100}{5.75} = 30 < 60$ (Hence OK)

(iii) Wash water discharge

Given rate of washing of filter = 45 cm rise/min

$\therefore$ Wash water discharge $= \dfrac{0.45 \times 6.3 \times 4.2}{60} = 0.198 \text{ m}^3\text{/s}$

$\therefore$ Velocity of flow in the lateral for wash water $= \dfrac{0.198}{84 \times \dfrac{\pi}{4} \times \left(\dfrac{5.75}{100}\right)^2} = 0.91$ m/sec

Similarly, Velocity of flow in the manifold $= \dfrac{\text{Discharge}}{\text{Area}} = \dfrac{0.198}{\dfrac{\pi}{4} \times (0.75)^2} = 0.448$ m/sec

Thus velocity of flow is less than 1.8 to 2.4 m/s limit (Hence OK).

5.2 **Draw the flow sheet showing sequence of various treatment units of a typical water treatment plant with perennial river as the source of water. List these treatment units sequentially.**

[10 marks : 2000]

Solution:

Surface water sources (perennial rivers, lakes, ponds, etc.) often contain a wider variety of contaminants than ground water and treatment processes may be more complex. Most surface water contain turbidity in excess of drinking water standards. Although fast moving streams may carry larger material in suspension, most of the solids will be colloidal in size and will require chemical coagulation for removal. Depending on the geological formation of the water shed, hardness may or may not be a problem in surface waters. If low levels of colour and other organic material are present, adsorption onto surface active material, a process not significant in natural water systems, may be necessary. A wide variety of microorganisms, some of which may be pathogenic are also common constituents of surface waters. The flow sheet showing sequence of various treatment units in treating surface waters are shown in figure.

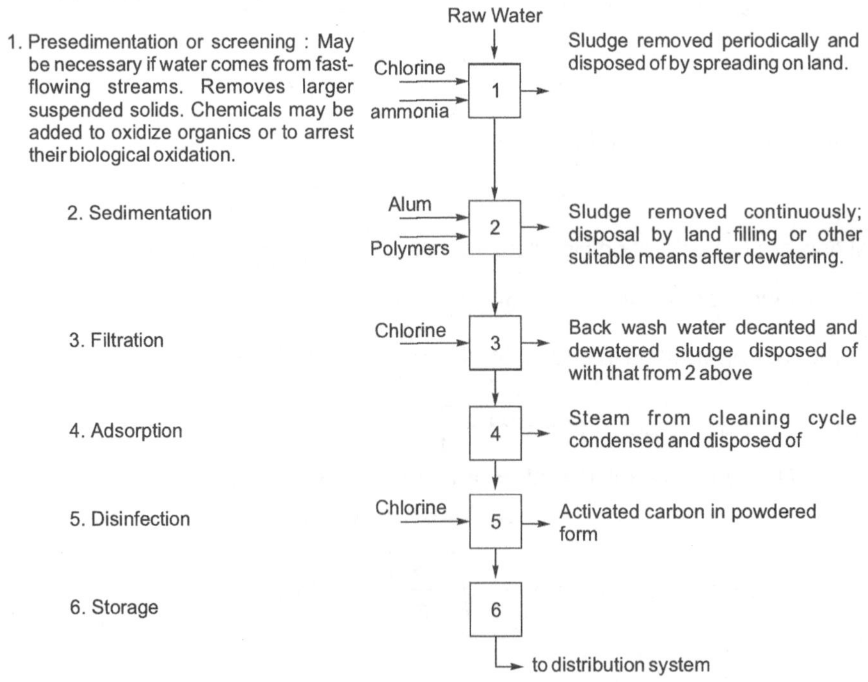

Raw Water

1. Presedimentation or screening : May be necessary if water comes from fast-flowing streams. Removes larger suspended solids. Chemicals may be added to oxidize organics or to arrest their biological oxidation.
2. Sedimentation
3. Filtration
4. Adsorption
5. Disinfection
6. Storage

Chlorine
ammonia
1
Sludge removed periodically and disposed of by spreading on land.

Alum
Polymers
2
Sludge removed continuously; disposal by land filling or other suitable means after dewatering.

Chlorine
3
Back wash water decanted and dewatered sludge disposed of with that from 2 above

4
Steam from cleaning cycle condensed and disposed of

Chlorine
5
Activated carbon in powdered form

6
to distribution system

Presedimentation or Screening: It may be necessary if water comes from fast flowing streams. It removes larger suspended solids. Some chemicals may be added to oxidize organics or to arrest their biological oxidation.

Sedimentation: It removes turbidity by coagulating colloids and settling them out. It may also remove colour caused by large organic molecules.

Filtration: It polishes to remove remaining turbidity. Any disinfectant may be added to prevent biological growth on filter medium.

Adsorption: It may be necessary if water contains dissolved organics. It may consist of activated carbon columns or activated carbon may be added in powdered form.

Disinfection: It destroys pathogens enough added to provide residual in the distribution system.

Storage: It provides contact time for disinfection and stores water for peak demand.

5.3 **Design the approximate dimensions of a set of two rapid gravity filters for treating water required for a population of 50,000; the rate of supply being 180 lit/day/person. The filters are rotated to work 5000 lit/hr/sq.m. Assume maximum daily demand to be 1.8 times the average daily demand. Assume any other data, if necessary.**

[10 marks : 2000]

Solution:

The maximum water demand per day = Population × maximum daily rate of supply

= Population × 1.8 × average daily rate of supply

= 50,000 × 1.8 × 180 = 16.2 MLD

Water demand per hour (neglecting time lost in washing the filter)

$$= \frac{16.2 \times 10^6}{24} = 675 \times 10^3 \text{ lit/hr}$$

Rate of filtration = 5000 lit/hr/m^2

$$\therefore \quad \text{Area of filter beds required} = \frac{\text{Water demand per hr}}{\text{Rate of filtration}} = \frac{675 \times 10^3}{5000} = 135 \text{ m}^2$$

Since two units are required to be designed, therefore area of each unit = $\frac{135}{2} = 67.5$ m^2

Assuming length of filter bed (L) as 1.5 times the width of filter bed (B)

$$\therefore \quad L \times B = 67.5$$

$$\Rightarrow \quad 1.5B \times B = 67.5$$

$$\Rightarrow \quad B = 6.71 \text{ m}$$

$$\therefore \quad L = 1.5B$$

$$= 1.5 \times 6.71$$

$$= 10.065 \text{ m}$$

Adopting 6.75 m width and 10 m length. Hence, two units of size 10 m × 6.75 m are required. One additional unit as standby may also be provided for breakdowns, repairs or cleaning operations.

5.4 **Explain how will you prepare the sand to be used in the Rapid Sand Filters from the available stock sand in case of a large water supply scheme.**

[15 marks : 2001]

Solution:

Traditionally, silica sand has been the most commonly used in slow sand filters as well as rapid sand filters. The important properties of such a material are:

(a) **Size:** The smaller the size of sand, smaller the pore openings through which the water must pass. Small pores increases filteration efficiency but increases the head loss resulting in a demised flow rule.

(b) **Gradation:** The proper gradation helps in arresting the impurities.

(c) **Density of medium:** When two or more materials with different specific gravities are used, the lighter material is located above the denser material of the same size.

Procedure to prepare the sand to be used in rapid sand filters:

- The filtering media consists of sand layers of depth about 60 to 90 cm.
- The effective size (D_{10}) of the sand various from 0.35 mm to 0.55 mm and uniformity coefficient $\left(\frac{D_{60}}{D_{10}}\right)$ generally ranges from 1.2 to 1.8.
- The larger size grains are kept on top with media of progressively decreasing size below.
- The larger pores on top retain large suspended particles while smaller pores retain succeeding finer material.
- As such the entire depth of filter functions efficiently.

5.5 **Defining coagulation, explain the different mechanisms of coagulation and its usefulness in removal of particles.**

A flocculation chamber 30 m long, 12 m wide and 4.5 m deep is to treat 75 MLD of water. It is equipped with 12 m long, 0.3 m wide paddles supported parallel to and moved by four horizontal shafts which rotate at a speed of 2.5 rpm. The centre line of paddles is 1.8 m from the shaft which is at mid depth of tank. Two paddles are mounted on each shaft, one opposite to another. If the mean velocity of water is 1/4 th velocity of paddles, find (i) Power consumption (ii) Time of flocculation and (iii) Value of *G*. Kinematic viscosity of water = 1.31×10^{-6} m²/s

[10 + 10 = 20 marks : 2002]

Solution:

Colloidal suspensions that do not agglomerate naturally are called stable. The most important factor contributing to the stability of colloidal suspensions is the excessively large surface to volume ratio resulting from their very small size. Surface phenomenon predominates over mass phenomenon. The most important surface phenomenon is the accumulation of electrical charges at the particle surface. The surface charge on colloidal particles gives them long term stability and hence the particles which might otherwise settle or coalesce are mutually repelled by their like charges. Thus coagulation is a chemical technique which is directed towards the destabilization of the charged colloidal particles. Flocculation on the other hand is the slow mixing technique which promotes the agglomeration of the stabilized particles.

For all practical purposes, however the entire process of addition of chemicals (coagulants) and mixing (flocculation) is usually referred to as coagulation.

In water treatment plants, chemical coagulation is usually accomplished by the addition of trivalent metallic salts such as $Al_2(SO_4)_3$ and $FeCl_3$.

There are four mechanisms which are thought to occur. These include:

1. **Ionic layer compression:** The quantity of ions present in the water surrounding a colloid has an effect on the decay function of the electrostatic potential

As shown in figure (a) high ionic concentration compresses the layers composed predominantly of counter ions towards the surface of the colloid. If this layer is sufficiently compressed, than the Vander Waals force will be predominant across the entire area of influence, so that the net force will be attractive and no energy barriers will exist. An example of ionic layer compression occurs in nature when a turbid stream flows in the ocean and deltas are formed from material which was originally so small that it could not have settled without coagulation.

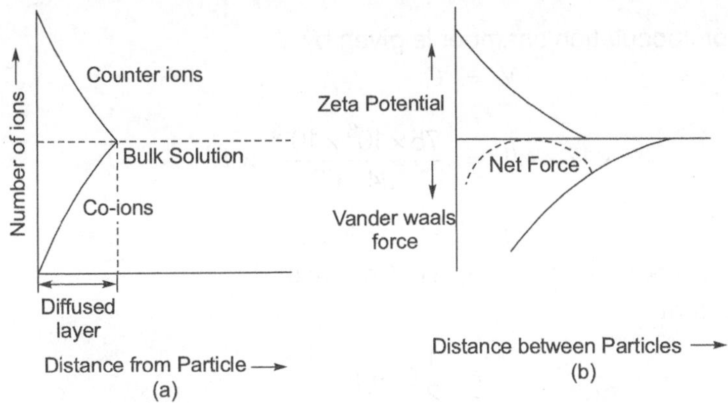

2. **Adsorption and charge neutralization:**
   - The nature rather than quantity of the ions is of prime importance in the theory of adsorption and charge neutralization.
   - The ionisation of $Al_2(SO_4)_3$ and $FeCl_3$ in water produces SO_4^{2-} and Cl^- along with Al^{3+} and Fe^{3+}. The Al^{3+} and Fe^{3+} cations react immediately with water to form a variety of aquometallic ions and hydrogen.

     $Al^{3+} + H_2O \longrightarrow Al(OH)_3 + 3H^+$

     $Fe^{3+} + H_2O \longrightarrow Fe(OH)_3 + 3H^+$
   - Tho aquometallic ions thus formed become part of the ionic cloud surrounding colloid and because they have a great affinity for surfaces are adsorbed onto the surface of the colloid where they neutralize the surface charge.
   - Once the surface charge has been neutralized, the ionic cloud dissipates and the electrostatic potential disappears so that contact occurs freely.

3. **Sweep Coagulation**
   - The last product formed in the hydrolysis of alum is $Al(OH)_3$.
   - The $Al(OH)_3$ forms in amorphous, gelatinous flocs that are heavier than water and settle by gravity.
   - Colloids may become entrapped in a floc as it is formed, or they may become enmeshed by its sticky surface as the flocs settle.
   - The process by which colloids are swept from suspension in this manner is known as sweep coagulation.

4. **Interparticle bridging:**
   - Large molecules may be formed when aluminium or ferric salts dissociate in water.
   - Synthetic polymers may also be used instead of, or in addition to, metallic salts which may be linear or branched and are highly surface reactive.

- Several colloids may become attached to one polymer and several of the polymer-colloid groups may become enmeshed resulting in a settleable mass.
- In addition to the adsorption forces, charges on the polymer may assist in the coagulation process.
- Metallic polymers formed by addition of aluminium or ferric salts are positively charged while synthetic polymers may carry positive or negative charges or may be neutral.
- Judicious choice of appropriate charges may do much to enhance the effectiveness of coagulation.

(i) Volume of flocculation chamber is given by

$$V = Qt$$

$$\Rightarrow \quad 30 \times 12 \times 4.5 = \frac{75 \times 10^6 \times 10^{-3}}{24 \times 60} \times t$$

$$\Rightarrow \quad t = 31.10 \text{ minutes}$$

Thus time of flocculation, $t = 31.10$ minutes

(ii) We know that

$$P = \frac{1}{2} C_D A_P \rho v_P^3$$

where P = Power input in watts

C_D = Dimensionless coefficient of drag, 1.8 for flat blades

ρ = Density of water in kg/m^3

A_P = Area of Paddles in m^2

v_P = Relative velocity of paddles in m/s which is about 75% of the actual paddle speed

Actual velocity of paddles = $\frac{2\pi rN}{60}$ [a minimum distance of 0.3 m should be maintained between paddle tips and all other structures in the flocculator]

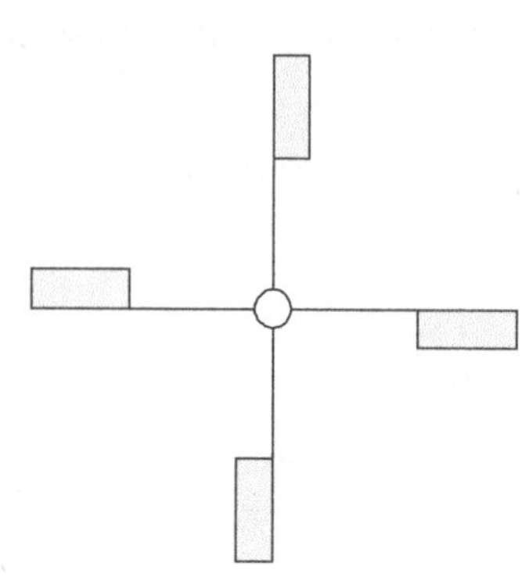

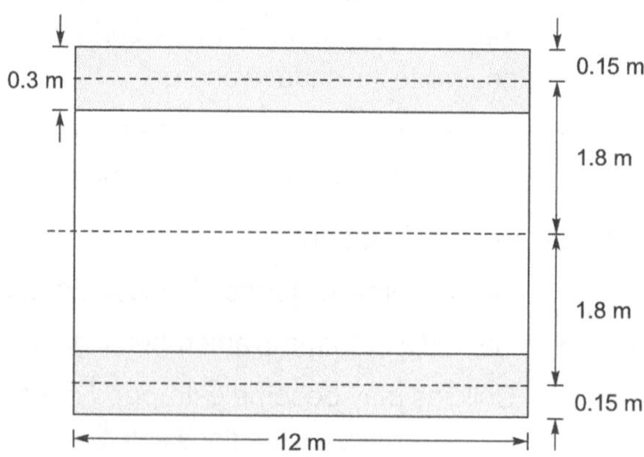

Thus $r = 1.8$

$$\therefore \text{Actual velocity of paddles} = \frac{2\pi \times 1.8 \times 2.5}{60} = 0.4712 \text{ m/s}$$

Velocity of water is one-fourth of that of paddles.

$$\therefore \quad v_P = 0.75 \times 0.4712 = 0.353 \text{ m/s}$$

$$A_P = 4 \times 2 \times 12 \times 0.3 = 28.8 \text{ m}^2$$

$$\therefore \quad P = \frac{1}{2} C_D A_P \rho v_P^3 = \frac{1}{2} \times 1.8 \times 28.8 \times 1000 \times (0.353)^3 \times 10^{-3}$$

$$P = 1.140 \text{ kW}$$

(iii)
$$G = \left(\frac{P}{V\mu}\right)^{1/2} = \left(\frac{1.140 \times 10^3}{30 \times 12 \times 4.5 \times 1.31 \times 10^{-6} \times 10^3}\right)^{1/2}$$
$$= 23.177 \text{ sec}^{-1}$$

5.6 **Why is sedimentation considered a necessity for treating raw water? What are the factors that affect sedimentation and how is sedimentation done? What is the purpose of coagulation in treatment of raw water? Give a neat sketch of 'Baffle type of mixing basin' for coagulation.**

[15 marks : 2003]

Solution:

Most of the suspended impurities present in water do have a specific gravity greater than that of water. In still water, these impurities will therefore tend to settle down under gravity, although in normal raw supplies, they remain in suspension because of the turbulence in water. Hence, as soon as turbulence is retarded by offering storage to the water, these impurities tend to settle down at the bottom of the storage tank. This is the principle behind sedimentation.

The settlement of a particle in water brought to rest, is opposed by the following factors:

(i) The velocity of flow which carries the particle horizontally. The greater the flow area, the lesser is the velocity and hence more easily the particle will settle down.

(ii) The viscosity of water in which the particle is travelling. The viscosity varies inversely with temperature. Warm water is less viscous and therefore offer less resistance to settlement. However, the temperature of water cannot be controlled to any appreciable extent in water purification processes and hence this factor is generally ignored.

(iii) The size, shape and specific gravity of particle. The greater is the specific gravity more readily the particle will settle. The size and shape of the particle also affect the settling rate.

The clarification of water by the process of sedimentation can be effected by providing conditions under which the suspended material present in water can settle out. The most important parameter among the above mentioned factors the velocity of flow and the shape and size of the particles are tried to be controlled in special basins called sedimentation basins. The viscosity is left uncontrolled as the same is not practically possible. The velocity of flow can be reduced by increasing the length of travel and by detaining the particles for a longer time in the sedimentation basin.

The size and shape of the particles can be altered by adding certain chemicals in water. These chemicals are called coagulants and they make the sedimentation quite effective by leading to the settlement of even very fine and colloidal particles. This is known as sedimentation aided with coagulation.

Mixing basin with baffle walls: The baffle type mixing basins are rectangular tanks which are divided by baffle walls. The baffles may either be provided in such a way as the water flows horizontally around their ends or they may be provided as to make the water move vertically over and under the baffles. The hindrances and the disturbances created by the provision of baffles in the path of flow, give it sufficient agitation, as to cause necessary mixing to develop the floc.

The brief sketch of a baffle type mixing basin is given below:

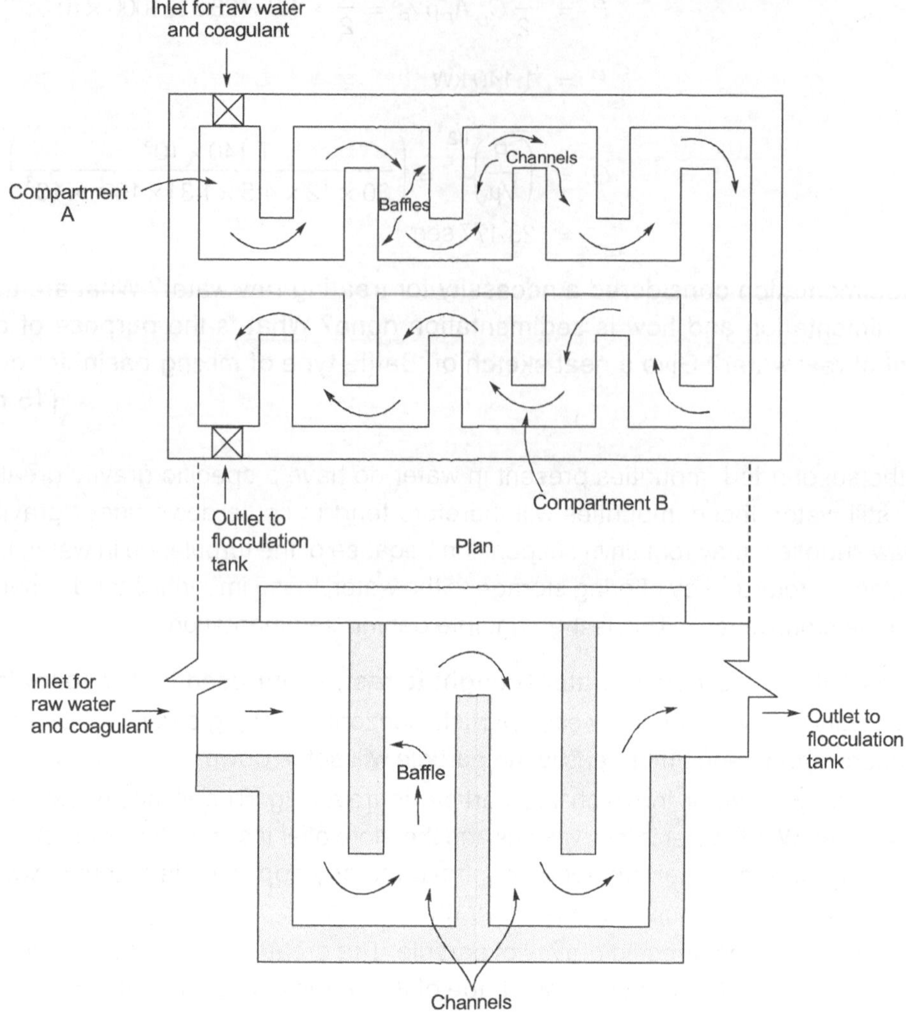

5.7 A Rapid Sand Filter is to be provided in a water treatment plant, to process the water for a town with a population of 275000. The water demand is 200 litre/capita/day. The rate of filtration is 15 $m^3/m^2/hr$. Allow 5% of filtered water for storage to meet the backwash requirements. Each backwashing period is of 30 min. Determine the number of filters required allowing one as a standby unit. The available surface area configuration of filter unit is 10 m × 4 m. Also compute the up-flow velocity and head-loss to expand the bed to 0.66 m The porosity of the bed is 0.50. Specific gravity is 2.5. The average particle size is 0.6 mm. The drag coefficient is 25.02 and the kinematic viscosity is 0.10136 × 10-5 m^2/s The flow is assume to be transitional flow.

[15 marks : 2004]

Solution:

$$\begin{aligned}\text{Maximum water demand per day} &= \text{Population} \times \text{maximum daily rate of water supply}\\ &= 275000 \times 1.8 \times \text{Average daily rate of water supply}\\ &= 275000 \times 1.8 \times 200 = 99 \text{ MLD}\end{aligned}$$

5% of the filtered water is stored to meet the backwash requirements.

$\therefore$ Total filtered water required per day $= 99 + \dfrac{5}{100} \times 99 = 103.95$ MLD

It is given that 30 min (0.5 hour) are required for backwashing.

$$\therefore \text{Filter water required per hour} = \frac{103.95}{23.5} = 4.4234 \times 10^6 \text{ litres per hour}$$

$$= 4423.4 \text{ m}^3 \text{ per hour}$$

$$\text{Area of filter} = \frac{\text{Filtered water required per hour}}{\text{Rate of filtration}}$$

$$= \frac{4423.40}{15} = 294.89 \text{ m}^2$$

The available surface area configuration of filter unit is 10 m × 4 m

$$\therefore \text{Number of filter units required} = \frac{294.89}{10 \times 4} = 7.37 \text{ units}$$

Hence, providing 8 units and 1 unit as a standby unit respectively.

Backwashing of granular medium filters is accomplished by reversing the flow and forcing clean water upward through the media. To clean the interior of the bed, it is necessary to expand it so that the granules are no longer in contact with each other, thus exposing all surfaces for cleaning. To hydraulically expand a porous bed, the head loss must be at least equal to the buoyant weight of the particles in the fluid. For a unit area of filter this is expressed by

$$h_{fb} = L(1-e)\frac{\rho_m - \rho_w}{\rho_w}$$

where h_{fb} is head loss required to initiate expansion in m, L is depth of bed in m, e is porosity of the medium, ρ_m is density of medium in kg/m^3 and ρ_w is density of water in kg/m^3

Assuming, depth of bed, $L = 0.60$ m

$$\rho_m = G\rho_w = 2.5 \times 1000 = 2500 \text{ kg/m}^3$$

$$\rho_w = 1000 \text{ kg/m}^3$$

$$e = 0.5$$

$$\therefore \quad h_{fb} = \frac{0.6(1-0.5)\times(2500-1000)}{1000}$$

$$\Rightarrow \quad h_{fb} = 0.45 \text{ m}$$

Now, the head loss through an expanded bed is essentially unchanged because the total buoyant weight of the bed is constant.

$$\therefore \quad \text{Weight of packed bed} = \text{Weight of bed fluidized}$$

$$L(1-e)\frac{(\rho_m - \rho_w)}{\rho_w} = L_{fb}(1-e_{fb})\frac{(\rho_m - \rho_w)}{\rho_w}$$

$$\therefore \quad L_{fb} = L\frac{(1-e)}{(1-e_{fb})}$$

where L_{fb} is depth of expanded bed and e_{fb} = porosity of expanded bed

$$\therefore \quad 0.66 = 0.6\frac{(1-0.5)}{(1-e_{fb})}$$

$$\Rightarrow \quad e_{fb} = 1 - \frac{0.6 \times 0.5}{0.66}$$

$$\Rightarrow \quad e_{fb} = 0.5454$$

Now this quantity e_{fb} is a function of the terminal settling velocity of the particles and the backwash velocity (up-flow velocity). An increase in the backwash velocity will result in a greater expansion of the bed. The expression commonly used to relate the bed expansion to backwash velocity and particle settling velocity is

$$e_{fb} = \left(\frac{v_b}{v_t}\right)^{0.22}$$

where v_b is the backwash velocity and v_t is terminal settling velocity.
Assuming flow to be transitional, then

$$C_D = \frac{24}{\text{Re}} + \frac{3}{\sqrt{\text{Re}}} + 0.34$$

$$\Rightarrow \quad 25.02 - 0.34 = \frac{24}{\text{Re}} + \frac{3}{\sqrt{\text{Re}}}$$

$$\Rightarrow \quad 24.68 = \frac{24}{\text{Re}} + \frac{3}{\sqrt{\text{Re}}}$$

Solving by **hit and trial method**, we get Re = 1.1 > 1 (Transitional)

$$\therefore \quad v_t^2 = \frac{4}{3}\frac{(\rho_m - \rho_w)gd}{C_D\rho_w}$$

$$\Rightarrow \quad v_t^2 = \frac{4 \times 9.81 \times (2500 - 1000) \times 0.6 \times 10^{-3}}{3 \times 25.02 \times 1000}$$

$$\Rightarrow \quad v_t^2 = 4.705 \times 10^{-4} \text{ m/s}$$

$$\Rightarrow \quad v_t = 0.022 \text{ m/s}$$

Now, we know that

$$e_{fb} = \left(\frac{v_b}{v_t}\right)^{0.22}$$

$$(e_{fb})^{1/0.22} = \frac{v_b}{v_t}$$

$$\Rightarrow \quad v_b = 0.022 \times (0.5454)^{1/0.22}$$

$$\Rightarrow \quad v_b = 1.39 \times 10^{-3} \text{ m/s}$$

5.8 **Explain the phenomenon of 'negative head' in filter. Why it should be avoided? How it is avoided in filter operation?**

[10 marks : 2005]

Solution:

Negative Head in Filter

- The motion of water percolating through the filter is opposed by the resistance offered by the sand grains and the impurities arrested by them.
- This causes the water to loose a portion of the available head and this head loss goes an increasing the time passes.
- A stage is ten reached when the frictional resistance offered by the filter media exceeds the static head of water above sand bed. Hence leads to a negative head in the bottom portion of filter media.

Reasons to Avoid Negative Head

- Due to negative head, water is sucked through filter media rather than getting flittered through it.
- The negative pressure so developed replaces air dissolved in water and leads to air binding which seriously affected the working of filter.

Steps to Avoid 'Negative Head' In filter

- Filters are to be frequently cleared as soon as the loss of head exceeds an optimum allowable value. The loss of head is generally limited to 2.5 m to 3.5 m and negative head to 1.2 m.
- The depth of water during filter operation may also be increased by 15-20 cm so as to increase the static head and the reduce negative head.

5.9 Derive equations for settling of a discrete particle in liquid.

[15 marks : 2006]

Solution:

When a discrete particle settles down in water, its downward settlement is opposed by the drag force offered by the water. The effective weight of the particle (actual weight - buoyancy) causes the particle to accelerate in the beginning, till it attains a sufficient velocity (v_s) at which the drag force becomes equal to the effective weight of the particle. After attaining that velocity (v_s), the particle falls down with that constant velocity

F_B = Buoyancy force

F_D = Drag force

W

Now, we know that

$$\text{Effective weight of the particle} = \text{Total weight} - \text{Buoyancy}$$

$$= \frac{4}{3}\pi r^3 \times \gamma_s - \frac{4}{3}\pi r^3 \times \gamma_w \quad \text{...(i)}$$

where r is radius of particle, γ_s is unit weight of particle and γ_w is unit weight of water

also,

$$\text{Drag force} = C_D \times A \times \rho_w \times \frac{V^2}{2} \quad \text{...(ii)}$$

where C_D is coefficient of drag, A is area of particle, ρ_w is density of water and v = Velocity of fall

Now, when v becomes equal to v_s, the drag force becomes equal to the effective weight of the particle

$$\therefore \quad C_D A \rho_w \frac{v_s^2}{2} = \frac{4}{3}\pi r^3 (\gamma_s - \gamma_w) \quad [\because A = \pi r^2]$$

$$\Rightarrow \quad C_D \,.\, \pi r^2 \,.\, \rho_w \frac{v_s^2}{2} = \frac{4}{3}\pi r^3 (\gamma_s - \gamma_w)$$

$$\Rightarrow \quad v_s^2 = \frac{4}{3} \times \frac{2r(\gamma_s - \gamma_w)}{\rho_w \,.\, C_D}$$

$$\Rightarrow \quad v_s^2 = \frac{4}{3} \times \frac{d(\gamma_s - \gamma_w)}{\rho_w \,.\, C_D} \quad [\because \gamma_s = \rho_s \times g \text{ and } \gamma_w = \rho_w \times g]$$

$$\Rightarrow \quad v_s^2 = \frac{4}{3} \times \frac{d(\rho_s g - \rho_w g)}{\rho_w \times C_D} = \frac{4}{3} \times d \times \rho_w g \left(\frac{\rho_s}{\rho_w} - 1\right) \times \frac{1}{\rho_w \,.\, C_D}$$

$$\Rightarrow \quad v_s^2 = \frac{4}{3} \times g \times d \times (S_s - 1) \times \frac{1}{C_D} \quad \text{...(iii)}$$

where d is the diameter of particle and S_s is the specific gravity of the particle.

Now for laminar flows and for small particles coefficient of drag, $C_D = \frac{24}{\text{Re}}$, where Re is particle Reynolds number

$$Re = \frac{v_s - d}{\nu}, \text{ where } \nu \text{ is the kinematic viscosity}$$

$$\therefore \quad v_s^2 = \frac{4}{3} \times g \times d \times (S_s - 1) \times \frac{1}{24/\text{Re}}$$

$$\Rightarrow \quad v_s^2 = \frac{4}{3} \times g \times d \times (S_s - 1) \times \frac{v_s \times d}{\nu \times 24}$$

$$\Rightarrow \quad v_s = \frac{g}{18}(S_s - 1) \times \frac{d^2}{\nu} \qquad \text{... (iv)}$$

Equation (iii) is the general equation for settling of a particle. Equation (iv) is valid for laminar flow and when particle diameter is less than 0.1 mm.

5.10 Differentiate between a rapid sand filter and a slow sand filter in respect of the following:
(i) Sand specifications through NUC and ES.
(ii) Method of collection of filtered water.
(iii) Media arrangement
(iv) Penetration and removal of impurities.

[8 marks : 2007]

Solution:

Comparison of slow sand and rapid gravity filters

S.No.	Item	Slow sand filters	Rapid gravity filters
1.	Sand specification	The effective size of filter sand ranges between 0.2 to 0.4 mm and uniformity coefficient varies between 1.8 to 2.5 or 3.0.	The effective size of the filter sand ranges between 0.35 to 0.55 and uniformity coefficient between 1.2 to 1.8
2.	Methods of collection of filters water	The "laterals" collect the filtered water which are open jointed pipe drains and they (laterals) discharge the filtered water into the "main drain", which leads the water to the "filtered water well". Main covered drain Inlet well Outlet well Open jointed lateral drains	The filtered water is collected by two methods (a) Collection of filtered water by perforated pipe laterals which discharges into central manifold pipe. (b) Collection of filtered water by strainers placed on the lateral drains which discharge into central manifold pipe.
3.	Media arrangement	The gravel base supports the sand. The grain size distribution is generally uniform throughout the depth of filter media, except that the to 10 to 15 cm may be laid of finer variety.	The gravel base supports the sand and also distributes the wash water uniformly on the surface of sand. The sand is laid in layer with smallest grain size at the top and coarsest grain size of the bottom.
4.	Penetration and removal of impurities	As of sand size is smaller hence impurities do not penetrate deeper. Very efficient in removing bacteria (98 to 99 percent) but less efficient in removing colour. They cannot handle turbid water containing turbidities more than 50 mg/*l*.	As of sand size is larger hence impurities penetrate deeper. Less efficient in removing bacteria (80 to 90 percent) but very efficient in removing in colour. They can handle very turbid waters.

5.11 What are different types of settling encountered when particles settle inliquid?

[10 marks : 2008]

Solution:

Depending upon the concentration of suspended matter and the characteristics of particles, settling can be classified in following four categories.

(i) **First Type:** In this particle have little or no tendency to flocculate in a dilute suspension. Such particles are known as discrete particles. Such type of settling is known as discrete settling.
- They settle as individual entities and there is no significant interaction with neighboring particles.
- The process is as in case of inorganic sand particles.

(ii) **Second Type:** In the second type, there in settling of flocculent particles in a dilute suspension.
- Particles agglomerate during settling which causes increase in size shape and density resulting in settlement at faster rate.
- This type of settling is known as flocculent settling.

(iii) **Third Type:** The third type of settling occurs when flocculent suspensions settle in an intermediate concentration.
- Due to proximity of particles with each other, they tend to remain in fixed positions with respect to each other and settle as a large mass (zone) rather than as individuals.
- This type of settling is known as zone or hindered settling.

(iv) **Fourth Type:** The fourth type of settling of flocculent particles in a very high concentration.
- Due to high concentration, particles come into physical contact and form a structure and further settling occur due to compression of the structure.
- This type of settling is also known as "compression settling."

5.12 What is Coagulation? Discuss the mechanisms of coagulation.

[10 marks : 2008]

Solution:

Coagulation is a chemical technique which is directed towards the destabilization of the charged colloidal particles. Flocculation on the other hand, is the slow mixing technique which promotes the agglomeration of the stabilized particles. For all practical purposes, however, the entire process of addition of chemicals (coagulants) and mixing (flocculation) is usually referred to as coagulation. In water treatment plants chemical coagulation usually accomplished by the addition of trivalent metallic salts such $Al_2(SO_4)_3$ or $FeCl_3$. The mechanisms of coagulation which are thought to occur are ionic layer compression, adsorption and charge neutralization, sweep coagulation and interparticle bridging.

(i) **Ionic layer compression:**
- The quantity of ions in the water surrounding a colloid has an effect on the decay function of the electrostatic potential.
- A high ionic concentration compresses the layers composed predominantly of counter ions towards the surface of colloid.
- If this layer is sufficiently compressed, then the vander Waals force will be predominant across the entire area of influence, so that the net force will be attractive and no energy barriers will exist.
- Although coagulants such as aluminium and ferric salts used in water treatment ionise, at the concentration commonly used they would not increase the ionic concentration sufficiently to affect ionic layer compression.

(ii) **Adsorption and charge neutralization:**
- The nature rather than quantity of the ions is of prime importance in the theory of adsorption and charge neutralization.
- The ionisation of $Al_2(SO_4)_3$ and $FeCl_3$ in water produces SO_4^{2-} and Cl^- along with Al^{3+} and Fe^{3+}. The Al^{3+} and Fe^{3+} cations react immediately with water to form a variety of aquometallic ions and hydrogen.

  $Al^{3+} + H_2O \longrightarrow Al(OH)_3 + 3H^+$

  $Fe^{3+} + H_2O \longrightarrow Fe(OH)_3 + 3H^+$

- The aquometallic ions thus formed become part of the ionic cloud surrounding colloid and because they have a great affinity for surfaces are adsorbed onto the surface of the colloid where they neutralize the surface charge.
- Once the surface charge has been neutralized, the ionic cloud dissipates and the electrostatic potential disappears so that contact occurs freely.

(iii) Sweep Coagulation

- The last product formed in the hydrolysis of alum is $Al(OH)_3$.
- The $Al(OH)_3$ forms in amorphous, gelatinous flocs that are heavier than water and settle by gravity.
- Colloids may become entrapped in a floc as it is formed, or they may become enmeshed by its sticky surface as the flocs settle.
- The process by which colloids are swept from suspension in this manner is known as sweep coagulation.

(iv) Interparticle bridging:

- Large molecules may be formed when aluminium or ferric salts dissociate in water.
- Synthetic polymers may also be used instead of, or in addition to, metallic salts which may be linear or branched and are highly surface reactive.
- Several colloids may become attached to one polymer and several of the polymer-colloid groups may become enmeshed resulting in a settleable mass.
- In addition to the adsorption forces, charges on the polymer may assist in the coagulation process.
- Metallic polymers formed by addition of aluminium or ferric salts are positively charged while synthetic polymers may carry positive or negative charges or may be neutral.
- Judicious choice of appropriate charges may do much to enhance the effectiveness of coagulation.

5.13 Design slow sand filters for a population of 40000 with an average rate of water supply of 150 litres per capita per day.

[5 marks : 2009]

Solution:

$$\text{Population} = 40000$$

$$\text{Per capita demand} = 150 \text{ litres per day}$$

$$\therefore \quad \text{Average daily demand} = \text{Population} \times \text{per capita demand}$$

$$= 40000 \times 150 = 6 \times 10^6 \text{ litres per day}$$

Assuming maximum demand as 1.8 times the average daily demand.

$$\therefore \quad \text{Maximum daily demand} = 1.8 \times \text{Average daily demand}$$

$$= 1.8 \times 6 \times 10^6 = 10.8 \times 10^6 \text{ litres per day}$$

Assuming rate of filtration as 180 litres per hour per m^2 of filter area.

$$\therefore \quad \text{Total surface area of filters required} = \frac{\text{Maximum daily demand}}{\text{Rate of filtration per day}} = \frac{10.8 \times 10^6}{180 \times 24} = 2500 \text{ m}^2$$

When surface area is more than 1200 m^2, then six slow sand filters are designed and one unit out of them will be kept as standby.

$$\therefore \quad \text{Area of each filter unit} = \frac{2500}{5} = 500 \text{ m}^2$$

Assuming length of each unit as twice its breadth.

$$\therefore \quad L \times B = 500$$

$$\Rightarrow \quad 2B \times B = 500$$

$$\Rightarrow \quad B = 15.81 \text{ m; say } 16 \text{ m}$$

$$\therefore \quad L = 2 \times 16 = 32 \text{ m}$$

Hence, **6 filter units will be used with one unit as standby, each unit of size 32 m × 16 m, arranged in series with 3 units on either side.**

5.14 **Design a rectangular grit chamber for a flow of 40 MLD. Specific gravity = 2.65 and size to be removed is 0.2 mm.**
Find the
(i) settling velocity of 0.2 mm particles,
(ii) critical horizontal velocity of flow, and
(iii) size of the grit chamber.
Assume kinematic viscosity of the liquid = 1.0 × 10^{-2} cm²/s.

[10 marks : 2009]

Solution:

(i) 1st iteration

Let the flow in the grit chamber be laminar. Thus, the settling velocity may be calculated by Stoke's equation i.e.

$$v_t = \frac{g(S_s - 1)d^2}{18\nu} = \frac{9.81 \times (2.65 - 1) \times (0.2 \times 10^{-3})^2}{18 \times 10^{-2} \times (10^{-2})^2} = 0.036 \text{ m/s}$$

$$Re = \frac{v_t d}{\nu} = \frac{0.036 \times 0.2 \times 10^{-3}}{10^{-2} \times (10^{-2})^2} = 7.2$$

The value of Reynolds number is greater than 1 but less than 10^4. Hence, the flow is transitional.

$$\therefore \quad C_D = \frac{24}{Re} + \frac{3}{(Re)^{1/2}} + 0.34 = \frac{24}{7.2} + \frac{3}{(7.2)^{1/2}} + 0.34 = 4.8$$

The general formula for the calculation of settling velocity is given by

$$v_t^2 = \frac{4}{3} \times g \times \frac{(S_s - 1)d}{C_D}$$

$$\Rightarrow \quad v_t^2 = \frac{4}{3} \times 9.81 \times \frac{(2.65 - 1) \times 0.2 \times 10^{-3}}{4.8}$$

$$\Rightarrow \quad v_t = 0.03 \text{ m/s}$$

2nd iteration

Again,
$$Re = \frac{v_t d}{\nu} = \frac{0.03 \times 0.2 \times 10^{-3}}{10^{-2} \times (10^{-2})^2} = 6$$

$$\therefore \quad C_D = \frac{24}{Re} + \frac{3}{(Re)^{1/2}} + 0.34 = \frac{24}{6} + \frac{3}{(6)^{1/2}} + 0.34 = 5.565$$

$$\therefore \quad v_t^2 = \frac{4}{3} \times 9.81 \times \frac{(2.65 - 1) \times 0.2 \times 10^{-3}}{5.565}$$

$$\Rightarrow \quad v_t = 0.028 \text{ m/s}$$

3rd iteration

$$\Rightarrow \quad Re = \frac{v_t d}{\nu} = \frac{0.028 \times 0.2 \times 10^{-3}}{10^{-2} \times (10^{-2})^2} = 5.6$$

$$\therefore \quad C_D = \frac{24}{Re} + \frac{3}{(Re)^{1/2}} + 0.34 = \frac{24}{5.6} + \frac{3}{(5.6)^{1/2}} + 0.34 = 5.893$$

$$\therefore \quad v_t^2 = \frac{4}{3} \times 9.81 \times \frac{(2.65-1) \times 0.2 \times 10^{-3}}{5.893}$$

$$\Rightarrow \quad v_t = 0.027 \text{ m/s} \approx 0.028 \text{ m/s} \quad \text{(Hence OK)}$$

Thus, the settling velocity of the 0.2 mm particles is 0.027 m/s.

(ii) Critical horizontal flow velocity can be calculated by modified Shield's formula as

$$v_h = 4.5\sqrt{gd(S_s-1)} = 4.5\sqrt{9.81 \times 0.2 \times 10^{-3} \times (2.65-1)} \simeq 0.26 \text{ m/s}$$

(iii) Let the length, width and depth of grit chamber be L, B and D respectively.

Quantity of flow, $Q = 40 \text{ MLD} \approx \dfrac{40 \times 10^6 \times 10^{-3}}{24 \times 60 \times 60} \text{ m}^3\text{/s} = 0.463 \text{ m}^3\text{/s}$

Now, we know that, $Q = v_h \times A$

$$\Rightarrow \quad A = \frac{0.463}{0.26} = 1.78 \text{ m}^2$$

Assuming depth of tank (D) as 1 m, then

$$D \times B = A$$

$$\Rightarrow \quad 1 \times B = 1.78$$

$$\Rightarrow \quad B = 1.78 \text{ m say } 1.8 \text{ m}$$

$$\text{Detention time} = \frac{\text{Depth of basin}}{\text{Settling velocity}} = \frac{D}{v_t} = \frac{1}{0.027} = 37 \text{ seconds}$$

$\therefore$ Lenth of tank, L = Critical horizontal flow velocity × Detention time = 0.26 × 37 = 9.62 m

Thus, the dimensions of the tank will be 9.62 m × 1.8 m × 1 m

5.15 **If the alum dose required to coagulate is 10 ppm, find out the amount of alum (in quintals) needed to treat 10 MLD of water.**

[2 marks : 2010]

Solution:

Quantity of water to be treated = 10×10^6 litres/day

Alum dose required = 10 ppm i.e. 10 mg/L

Amount of alum required per day = $10 \times 10^6 \times 10$ mg

$= 10^8 \times 10^{-6}$ kg $= 10^2 \times 10^{-2}$ quintal

= 1 quintal

5.16 **How much will be the settling velocity of a spherical particle A (having specific gravity of 2.65) of diameter 10×10^{-3} cm? Determine the size for a floating spherical particle having a specific gravity of 0.80 (rising with the same velocity as that of particle A).**

Assume kinematic viscosity of water as 1.012×10^{-2} cm²/sec.

[8 marks : 2010]

Solution:

Settling velocity of a particle for laminar flow as per Stoke's law is given by

$$v_S = \frac{g}{18}(G-1) \times \frac{d^2}{\nu}$$

For Particle A, velocity of settling will be

$$v_{SA} = \frac{981}{18} \times (2.65-1) \times \frac{(10\times 10^{-3})^2}{1.012\times 10^{-2}} \simeq 0.89 \text{ cm/sec}$$

Reynolds number, $Re = \dfrac{v_{SA}d}{\nu} = \dfrac{0.89\times 10\times 10^{-3}}{1.012\times 10^{-2}} \simeq 0.88 < 1$ OK

Thus the assumption of the flow being laminar is correct.

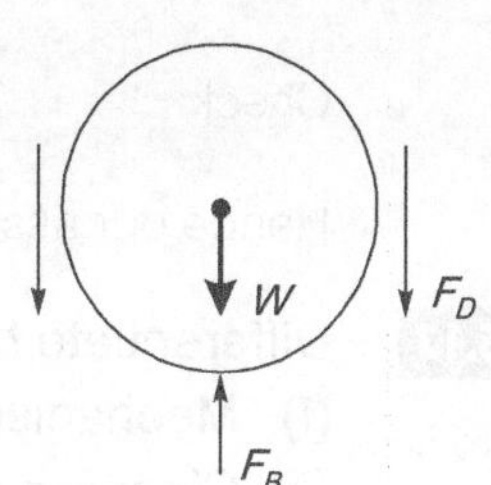

Now another particle (say *B*) is rising with the same velocity as that of particle *A* in water. When the spherical particle *B* rises the force due to gravity acts in downward direction and the buoyant force in upward direction.

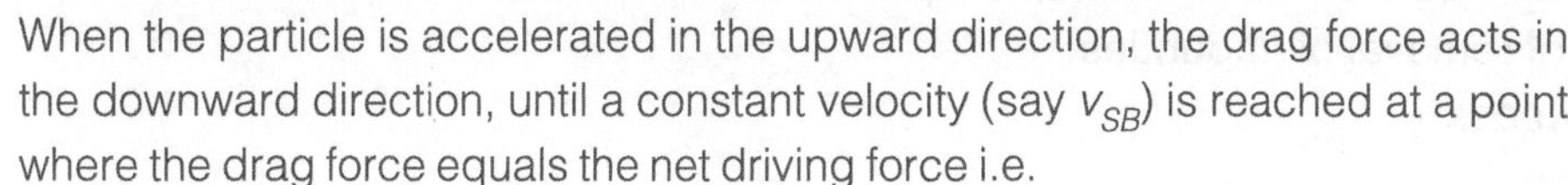

When the particle is accelerated in the upward direction, the drag force acts in the downward direction, until a constant velocity (say v_{SB}) is reached at a point where the drag force equals the net driving force i.e.

$$F_D + W = F_B$$

$$\Rightarrow \frac{C_D\times\rho_w A_{SB}v_{SB}^2}{2} + \rho_B g V_{SB} = \rho_w g V_{SB}$$

$$\Rightarrow \frac{24}{\text{Re}}\times\frac{\rho_w}{2}\times A_{SB}\times v_{SB}^2 = (\rho_w-\rho_B)gV_{SB} \qquad \left[\because C_D = \frac{24}{\text{Re}}\right]$$

$$\Rightarrow \frac{12}{v_{SB}d/\nu}\times\rho_w\times A_{SB}\times v_{SB}^2 = (1-G_B)\,\rho_w g V_{SB}$$

$$\Rightarrow \frac{12\nu}{d_B}\times\frac{\pi d_B^2}{4}\times v_{SB} = (1-G_B)g\times\frac{4}{3}\times\pi\left(\frac{d_B}{2}\right)^3$$

$$\Rightarrow 3\pi\nu\times d_B\times v_{SB} = (1-0.8)\times g\times\frac{4}{3}\times\frac{\pi}{8}\times d_B^3$$

$$\Rightarrow d_B^2 = \frac{90\nu\times v_{SB}}{g} \qquad [\text{But } v_{SB} = v_{SA}]$$

$$\Rightarrow d_B^2 = \frac{90\times 1.012\times 10^{-2}\times 0.89}{981}$$

$$\Rightarrow d_D = 2.87\times 10^{-2}\text{ cm } \textbf{(Ans.)}$$

$$R_e = \frac{vd_B}{\nu} = \frac{0.89\times 2.87\times 10^{-2}}{1.012\times 10^{-2}} = 2.52$$

Since $R_e = 2.52 > 1$

Hence the flow is transitional

$$\therefore \quad C_d = \frac{24}{R_e} + \frac{3}{(R_e)^{1/2}} + 0.34 = \frac{24}{2.52} + \frac{3}{2.52^{1/2}} + 0.34 = 11.75$$

$$F_D = F_B - W$$

$$\frac{1}{2}C_d\rho_w A_{SB}v_{SB}^2 = (\rho_w - P_B)gv_{SB}$$

$$\frac{1}{2}\times 11.75\times\frac{\pi}{4}\times d_B^2\times v_{SB}^2 = (1-0.8)\times g\times\frac{4}{3}\pi\left(\frac{d_B}{2}\right)^3$$

$$\frac{1}{2}\times 11.75\times\frac{\pi}{4}\times d_B^2\times\left(0.89\times 10^{-2}\right)^2 = 0.2\times 9.81\times\frac{4}{3}\times\frac{\pi}{8}\times d_B^3$$

$$d_B = \frac{\frac{1}{2}\times 11.75\times\frac{\pi}{4}\times\left(0.89\times 10^{-2}\right)^2}{0.2\times 9.81\times\frac{4}{3}\times\frac{\pi}{8}} = 3.55\times 10^{-4}\text{ m} = 3.55\times 10^{-2}\text{ cm}$$

Check: $$R_e = \frac{vd_B}{\nu} = \frac{0.89\times 3.55\times 10^{-2}}{1.012\times 10^{-2}} = 3.13 > 1$$

Hence our assumption of flow to be transitional is true. and $d_B = 3.55 \times 10^{-2}$ cm.

5.17 **Differentiate between a slow sand filter and rapid sand filter in respect to**
(i) Mechanism of removal of impurities.
(ii) Cleaning process
(iii) Effluent quality
(iv) Bed size in qualitative term.

[8 marks : 2010]

Solution:

Comparison of slow sand and rapid garvity filters.

S.No.	Item	Slow sand filters	Rapid gravity filters
1.	Mechanism of removal of impurities.	Certain micro-organisms and bacteria are generally present in the voids of filter. These micro-organism require organic impurities as their food for their survival. Hence micro-oganisms utilise such organic impurities and convert them into harmless compounds.	As the size of suspended particles present in water is bigger than the size of sand, hence these impurities cannot pass through these voids and get arrested in them. Even small size of suspended particles also arrested in these voids because void spaces act like tiny coagulation-sedimentation tanks.
2.	Cleaning process	Scrapping and removing to 1.5 to 3 cm thick layer and wasting down by hoses.	Agitating the sand grains and back washing with or without compressed air.
3.	Effluent quality	Very efficient in removing bacteria (98 to 99%) but less efficient in removing colour.	Less efficient in removing bacteria (80 to 90%) but very efficient in removing colour.
4.	Bed size	It consist of 30 to 75 cm thick gravels of different sizes placed in layers. Generally three to four layer each of 15-20 cm depth are used. Size of each unit is large such as (30 m × 60 m). The area varying from 100 to 2000 sq.m. of filter area.	It consists of sand layers about 60 to 90 cm in depth and placed over a gravel support. Size of each unit is small, such as 5 m × 8 m. The area varying from10 to 80 sq.m.

5.18 **In a water treatment plant, raw water undergoes the following treatment processes viz. coagulation, flocculation, sedimentation and filtration. A plant of capacity to treat 3 × 10^4 m^3/d of raw water is installed for a township. The raw water source is having a suspended solid concentration of 400 ppm. Alum [$Al_2(SO_4)_3$ 14 H_2O] is used as a primary coagulant at a dose of 45 ppm. Compute the daily production of sludge if 95% of the total suspended solids are removed by the treatment processes. Neglect aluminium hydroxide quantity formed due to Alum dose.**

[10 marks : 2011]

Solution:

Neglecting $Al(OH)_3$ formation and computing sludge produced due to suspended solid only.

Suspended solids concentration in raw water = 400 ppm = 400 mg/l

Suspended solids removed from raw water = $\frac{95}{100}\times 400$ = 380 mg/l

Total dry sludge produced per day = (3 × 10^4 × 10^3) × 380 = 1140 × 10^7 mg/day = 11400 kg/day

5.19 **Calculate the size of a rectangular tank to treat 2.0 million liters of water per day. The detention period may be assumed as 3 hours and overflow rate less than 40,000 lit. per sq.m. of the surface area per day.**

[10 marks : 2012]

Solution:

Supply of water to be treated per day = 2×10^6 litres

Supply of water to be treated during the detention period of 3 hours i.e., the capacity of the tank

$$= \frac{2\times10^6}{24}\times3 = 0.25\times10^6 \text{ liters} = 250 \text{ m}^3$$

Let us assume the velocity of flow to be maintained through the tank

= 20 cm/minute = 0.2m/minute

∴ The length of tank required = velocity of flow × detention period = (0.2 × 3 × 60) = 36 m

Cross-sectional area of the tank required

$$= \frac{\text{Capacity of tank}}{\text{Length of tank}} = \frac{250}{36} = 6.94 \text{ m}^2$$

Assuming an overflow rate of 36000 litres/day/m²

$$\therefore \quad \frac{Q}{BL} = 36000 \ l/\text{day/m}^2$$

$$\Rightarrow \quad \frac{Q}{BL} = 1500 \ l/\text{hr/m}^2$$

But

$$Q = \frac{2\times10^6}{24}\text{litres/hr} = 0.0833\times10^6 \text{ litres/hr}$$

$$\therefore \quad BL = \frac{Q}{1500} = \frac{0.0833\times10^6}{1500} = 55.56 \text{ m}^2$$

or

$$B = \frac{55.56}{36} = 1.543 \text{ m}$$

$$\therefore \quad \text{The depth} = \frac{250}{36\times1.543} = 4.5 \text{ m}$$

Using a free board of 0.5 m, the overall depth

= 0.5 + 4.5 = 5 m

Hence, a rectangular sedimentation tank with an overall size of 36 m x 1.543 x 5 m can be used.

5.20 **What is type-I settling? How analysis of type-I settling is done? Find the settling velocity of a discrete particle in water having diameter 4×10^{-3} cm and specific gravity 2.65. Take kinematic viscosity of water = 1.02×10^{-2} cm²/sec.**

[10 marks : 2013]

Solution:

Type-I Settling: Particles whose shape, size, specific gravity do not change with time are called as "DISCRETE PARTICLES" and particles whose surface properties are such that they coalesce/combine with other particles upon contact thereby changing shape, size and specific gravity of particles are called "FLOCCULATING PARTICLES" settling of discrete particles in dilute suspension is called as Type-I settling.

When a particle is suspended in water, initially it has only two forces acting upon it viz.

1. **Force of gravity** = $F_g = \rho_p V_p g$

   Where ρ_p and V_p are density and volume of particles respectively.

2. Buoyant force = $F_B = \rho_w V_p g$
Where ρ_w is the density of water.
Now $F_g = F_B$ is $\rho_p = \rho_w$ and no acceleration of the particles will take place.
If $\rho_p \neq \rho_w$ which usually always happens, a net force acts on the particles and particle accelerates in the direction of net force (F_{net}).
Thus $F_{net} = (\rho_p - \rho_w) g V_p$ = Driving force for acceleration.
Once motion of particles has started, a third force come into play due to viscous friction. This force is called as "DRAG FORCE" (F_D), given by

F_B, F_D, $W = F_g$

$$F_D = \frac{1}{2} C_D A_P \rho_w V_P^2$$

Where C_D = Drag Coefficient

$$\therefore \quad (\rho_p - \rho_w) g V_p = \frac{1}{2} C_D A_P \rho_w V_P^2$$

For spherical particles, $$\frac{V_P}{A_P} = \frac{\frac{\pi}{6} d^3}{\frac{\pi}{4} d^2} = \frac{2}{3} d$$

$$\therefore \quad V_P^2 = \frac{4}{3} g \frac{(\rho_P - \rho_w) d}{C_D \rho_w} \quad \text{...(i)}$$

Drag coeff. $$(C_D) = \begin{cases} \frac{24}{Re} \text{ for laminar flow.} \\ 0.4 \text{ for turbulent flow.} \end{cases}$$

Here Re = Reynold's no. = $\frac{v \rho_w d}{\mu}$

Substuting 'C_D; in (i), $$v = \frac{g d^2 \rho_w (G-1)}{18\mu}$$

G = Specific gravity of particles.
$d = 4 \times 10^{-3}$ cm/s $= 4 \times 10^{-5}$ m/s
$G = 2.65$

Assuming laminar flow, $$v = \frac{9.81(16 \times 10^{-10})(2.65-1)}{18 \times 1.02 \times 10^{-6}} = 14.106 \times 10^{-4} \text{ m/s}$$

$$\therefore \quad Re = \frac{14.106 \times 10^{-4} \times 4 \times 10^{-5}}{1.02 \times 10^{-6}} = 55.318 \times 10^{-3} < 1$$

⇒ Assumptions of laminar flow is true.

5.21 **A square filter box is to be designed for a filtration rate of 2.8 L/m²s. What are the required surface area and side dimension of the unit if the flow rate is 6 ML/d? If the filter is backwashed once a day for 12 minutes at a rate of 10 L/m²s, what percentage of the total flow is used for cleaning the filter?**
[8 marks : 2014]

Solution:

Given: Filtration rate = 2.8 L/m²s
Flow rate, Q = 6 MLD
Filtered water is used for backwashing

Filtered water required per day + water required for backwashing= water filtered at a rate of 2.8 L/m^{2}s in a day
Let the dimension of filter be 'B'

$$6 \times 10^6 + 10 \times B^2 \times 12 \times 60 = 2.8 \times B^2 \times 24 \times 60 \times 60$$

$$\Rightarrow \quad 6 \times 10^6 = [2.8 \times 86400 - 7200] \times B^2$$

$$\Rightarrow \quad B^2 = \frac{6 \times 10^6}{234720} = 25.56 \text{m}^2$$

$$\therefore \quad B \simeq 5.06 \text{ m}$$

The percentage of the total flow rate used for cleaning the filter

$$= \frac{10 \times B^2 \times 12 \times 60}{6 \times 10^6 + 10 \times B^2 \times 12 \times 60} \times 100$$

$$= \frac{184345.92}{6184345.92} \times 100 = 2.98\%$$

5.22 **A water treatment plant is designed to treat a flow of 20000 m^3/day. The chlorine dosage is 1 mg/L. What size of containers should be used in this plant? What is the minimum weight of chlorine which should be kept in hand?**

[8 marks : 2014]

Solution:

Given: Q = 20000 m^3/day

Chlorine dosage = 1 mg/L

Generally in chlorination contact period is of about 10 minutes

$\therefore$ Detention period, t_d = 10 min

$$\text{Volume of container used in plant, } V = Q \times t_d = \frac{20000}{24 \times 60} \times 10 = 138.89 \text{m}^3$$

Minimum weight of chlorine which should be kept in hand

$$= V \times \text{chlorine dosage} = 138.89 \times 10^3 \times 1 \times 10^{-3} \text{g} = 138.89 \text{ g}$$

5.23 **Explain the significance of alkalinity in coagulation practice and in lime-soda softening.**

[8 marks : 2015]

Solution:

The coagulation of metallic salts release hydrogen ion (H^+). These H^+ neutralise alkalinity and if initial alkalinity of water is low, coagulation will result in further reduction of pH.

Proper pH values must be maintained for best coagulation and since alkalinity must be present for hydroxide floc formation, low alkalinity water must be added with lime [$Ca(OH)_2$] or soda ash [Na_2CO_3] to increase the alkalinity.

If alkalinity is insufficient in water, lime [$Ca(OH)_2$] or soda ash (Na_2CO_3) may be added in water.

$$Al_2(SO_4)_3 \cdot 18H_2O + 3Ca(OH)_2 \longrightarrow \underset{\text{Permanent Hardness}}{3\,CaSO_4} + \underset{\text{ppt}}{2Al(OH)_3 \downarrow} + 18\,H_2O$$

$$Al_2(SO_4)_3 \cdot 18H_2O + 3Na_2CO_3 \longrightarrow 3\,Na_2SO_4 + \underset{\text{ppt}}{2\,Al(OH)_3 \downarrow} + 3CO_2 + 15\,H_2O$$

When softening and coagulation process is combined i.e. when lime soda is added with alum, lesser quantity of coagulant is required. Due to increased alkalinity by lime soda process, it reduces the corrosion of pipes. Its also help in killing bacteria because of high causticity. Its also help in removing upto some extent of Fe and Mn but the process produces huge quantity of sludge.

5.24 **In a treatment plant, two tanks of rapid sand filtration are proposed after sedimentation. Each filter bed has a surface area of 3 m × 2 m. The design flow rate to each bank of filters is 0.044 m³/sec. The design loading rate to each bank of filters is 150 m³/day/m². Determine the number of filter beds in each bank of filters. Determine the loading rate when one filter is out of service. The maximum acceptable loading rate is 235 m³/day/m².**

[8 marks : 2015]

Solution:

$$\text{Area of each filter bed} = 3 \times 2 = 6\,\text{m}^2$$

$$\text{Design flow rate to each bank of filter} = 0.044\,\text{m}^3/\text{s}$$

$$\text{Design loading rate to each bank of filter} = 150\,\text{m}^3/\text{day/m}^2$$

$$\text{Total area of filter beds required} = \frac{\text{Design flow rate}}{\text{Design loading rate}} = \frac{0.044 \times 3600 \times 24}{150} = 25.344\,\text{m}^2$$

$$\text{Number of filter beds required} = \frac{\text{Total area of filter beds required}}{\text{Area of one filter bed}}$$

$$= \frac{25.344}{6} = 4.224$$

Provide 5 filter beds.

When one gets out of service,

$$\text{Area of filter beds} = 4 \times 2 \times 3 = 24\,\text{m}^2$$

$$\text{Loading rate} = \frac{0.044 \times 3600 \times 24}{24} = 158.4\,\text{m}^3/\text{day/m}^2$$

Which is less than maximum acceptable loading rate (235 m³/day/m²).

Hence, design is safe even if one filter bed gets out of service.

5.25 **With help of a sketch, explain the reactions that happen at different stages of the breakpoint chlorination curve. Also, state the importance of breakpoint chlorination in water treatment.**

[20 marks : 2017]

Solution:

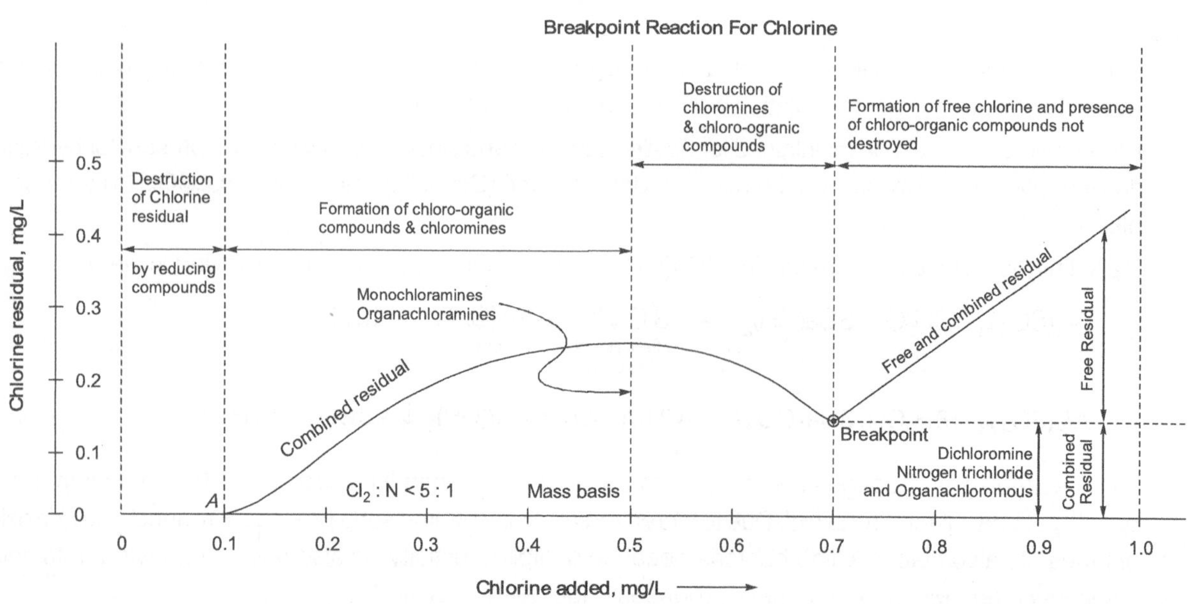

Chlorine is added to the water supply in two ways. It is mostly added as a gas i.e., $Cl_2(g)$. However, it can also be added as a salt such as sodium hypochlorite (NaOCl) or bleach. Chlorine gas dissolves in water following Henry's law.

$$Cl_2\,(g) \rightarrow Cl_2\,(aq) \qquad k_H = 6.2 \times 10^{-2}$$

Once dissolved, the following reaction occurs forming hypochlorous acid (HOCl),

$$Cl_2(aq) + H_2O \rightarrow HOCl + H^+ + Cl^-$$

Hypochlorous acid and hypochlorite ion collectively is called the free chlorine residual. These free chlorine compounds can react with many organic and inorganic compounds to form chlorinated compounds. If the products of these reactions posses oxidizing potential, they are considered the combined chlorine residual. A common compound in drinking water systems that reacts with chlorine to form combined residual is ammonia. Reaction between ammonia and chlorine form chloramines, which is mainly monochloromine (NH_2Cl), although some dichloramines and trichloramine (NCl_3) also can form. Many drinking water utilities use monochloramine as a dissinfactant. The overall reactions of the free chlorine and nitrogen can be represented by two simplified reactions as follows:

Monochloramine formation reaction: The reaction occurs rapidly when ammonia nitrogen is combined with free chlorine upto molar ratio of 1 : 1.

$$HOCl + NH_3 \rightarrow NH_2Cl + H_2O$$

Breakpoint Reaction: When excess free chlorine is added beyond the 1 : 1 initial molar ratio, the monochloramine is removed as follows:

$$2NH_2Cl + HOCl \rightarrow N_{2(g)} + 3H^+ + 3Cl^- + H_2O$$

The formation of chloramines and the breakpoint reaction creates a unique relationship between chlorine dose and amount of chlorine.

5.26 **A water treatment plant is to treat water at the rate of 6000 m³/day. If there are two rectangular sedimentation tanks (27 × 5 × 3.8 m) with total weir length of 50 m, determine**
(i) Detention time (ii) Overflow rate (iii) Weir loading

[6 marks : 2018]

Solution:

Given: $Q = 6000\ m^3/day,\ L = 27\ m,\ B = 5\ m,\ H = 3.8\ m$

Total weir length = 50 m

Flow in each tank, $$Q' = \frac{Q}{2} = 3000\ m^3/day$$

(i) $$\text{Detention time} = \frac{V}{Q'} = \frac{27 \times 5 \times 3.8}{3000}\ \text{day} = 0.171\ \text{day} = 4.104\ \text{hrs}$$

(ii) $$\text{Overflow rate} = \frac{Q'}{BL} = \frac{3000}{27 \times 5} = 22.22\ m^3/day/m^2$$

(iii) $$\text{Weir loading} = \frac{Q'}{\text{Length of weir}} = \frac{3000}{50/2} = 120\ m^3/day/m$$

5.27 **A water contains 110 mg/L carbonate ion and 80 mg/L bicarbonate ion at a pH of 10. Calculate the alkalinity exactly at 25°C. Approximate the alkalinity by ignoring hydroxide and hydrogen ion. What is the percentage error in approximation?**

[12 marks : 2019]

Solution:

Data given: $\left[CO_3^{2-}\right] = 110$ mg/*l*

$\left[HCO_3^{-}\right] = 80$ mg/*l*

As pH = 10

$\therefore$ pOH = 4

$\therefore$ $[OH^-] = 10^{-4}$ mol/*l*

$$= 10^{-4} \times 17 \frac{gm}{l} = 1.7 \frac{gm}{l}$$

$$[H^+] = 10^{-10} \text{ mol}/l$$

$$= 10^{-10} \times \frac{1 \times 10^3}{l} \frac{mg}{l}$$

$$= 10^{-7} \frac{mg}{l}$$

$\therefore$ Equivalent weight of $\left[CO_3^{2-}\right] = 30$

Equivalent weight of $\left[HCO_3^{-}\right] = 61$

Equivalent weight of $\left[OH^{-}\right] = 17$

$$\therefore \text{ Total alkalinity in term of } CaCO_3 = \frac{\left[CO_3^{2-}\right]}{\text{Equivalent weight of } \left[CO_3^{2-}\right]} \times \text{equivalent weight of } CaCO_3$$

$$+ \frac{\left[HCO_3^{-}\right]}{\text{Equivalent weight of } \left[HCO_3^{-}\right]} \times \text{equivalent weight of } CaCO_3$$

$$+ \frac{\left[OH^{-}\right]}{\text{Equivalent weight of } \left[OH^{-}\right]} \times \text{equivalent weight of } CaCO_3$$

$$- \frac{\left[H^{+}\right]}{\text{Equivalent weight of } \left[H^{+}\right]} \times \text{equivalent weight of } [H^+]$$

$$= \frac{110}{30} \times 50 + \frac{80}{61} \times 50 + \frac{1.7}{17} \times 50 - \frac{10^{-7}}{1} \times 50$$

$= 253.907$ mg/*l* as $CaCO_3$

Approximate alkalinity by ignoring hydroxide ion and hydrogen ion is given as

$$\text{Approximate alkalinity} = \frac{110}{30} \times 50 + \frac{80}{61} \times 50$$

$= 248.90$ mg/*l* as $CaCO_3$

$$\therefore \quad \text{Percentage error in approximation} = \frac{253.907 - 248.907}{253.907} \times 100 = 1.969\%$$

5.28 A bed of uniform sand, having particle size 0.65 mm diameter and specific gravity 2.66, porosity 0.48 and depth 75 cm is to be washed hydraulically. Compute

(a) Backwash rate so that expansion will be 50 percent.

(b) Head loss at this rate.

Take kinematic viscosity of water as 1.3×10^{-2} cm²/sec and 24 assume $C_D = \frac{24}{R}$.

[10 marks : 2020]

Solution:

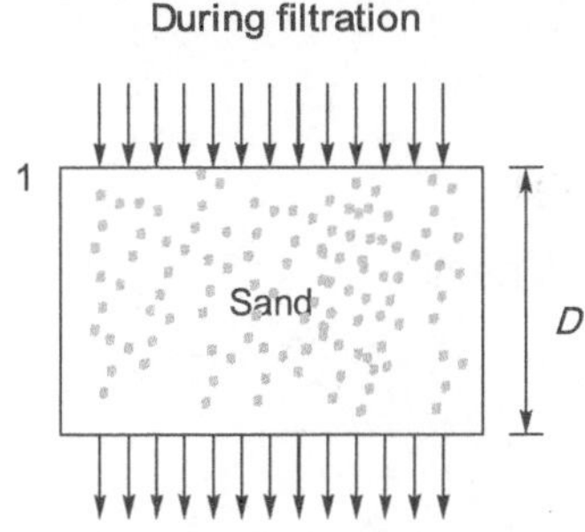

D = Depth of sand bed = 0.75 m
n = Porosity of sand bed = 0.48

During backwashing

2

Sand

D'

D' = Depth of expanded sand bed = 1.5 D = 1.5 × 0.75

n' = Porosity of expanded sand bed

Given, Particle size (d) = 0.65 mm

Specific gravity (G_s) = 2.66

To hydraulically expand the filter medium during backwashing loss of head through the medium is equal to buoyant weight of sand particles. The analysis is done by assuming no impurity is present and the loss of head is due to sand particles only

i.e $$h_L = D \times (1-n) \times (G_s - 1) \qquad \text{...(i)}$$

The buoyant weight during backwashing as well as filtration is constant hence, loss of head through the bed is constant.

$$h_{L1} = h_{L2}$$

$$D(1-n)(G_s-1) = D'(1-n')(G_s-1)$$

$$\Rightarrow \quad 0.75 \times (1-0.48) = 1.125\,(1-n')$$

$$\Rightarrow \quad n' = 0.653$$

(a) Porosity in the expanded condition is a function of backwash velocity (V_b) and terminal settling velocity as :

$$n' = \left(\frac{V_b}{V_s}\right)^{0.22} \qquad \text{...(ii)}$$

$$V_s = \sqrt{\frac{4}{3} \cdot \frac{gd(G_s-1)}{C_D}}$$

$$C_D = \frac{24}{\text{Re}} = \frac{24}{\left(\frac{V_s . d}{\nu}\right)}$$

$$V_s^2 = \frac{4}{3}\frac{gd(G_s - 1)}{24}\cdot\frac{V_s.d}{\nu}$$

$$V_s = \frac{g}{18}(G_s - 1)\frac{d^2}{\nu}$$

Given, kinematic viscosity (ν) = 1.3×10^{-2} cm^2/s

$$\therefore \quad V_s = \frac{9.81}{18}(2.66-1)\frac{(0.65\times10^{-3})^2}{1.3\times10^{-6}} = 0.294 \text{ m/s}$$

Now, from equation (i)

$$0.653 = \left(\frac{V_b}{0.294}\right)^{0.22}$$

$$\Rightarrow \quad V_b = (0.653)^{1/0.22} \times 0.294$$

$$= 0.0423 \text{ m/s} = 2.538 \text{ m/min}$$

Hence, Backwash rate = 2.538 m/min

(b) Head loss

$$h_L = D(1-n)(G_s - 1)$$

$$= 0.75(1-0.48)(2.66-1)$$

$$= 0.647 \text{ m} \simeq 0.65 \text{ m}$$

5.29 Briefly explain various factors affecting bactericidal efficiency of chlorine in water treatment process.

[10 marks : 2020]

Solution:

Various factors affecting bactericidal efficiency of chlorine in water treatment process are as following:

1. If the turbidity in the water increases the bactericidal efficiency decreases.
2. Metallic compound such as (iron, manganese) use chlorine in oxidising themselves thereby reducing bactericidal efficiency.
3. Ammonia combines with chlorine to form chloramines which reduces efficiency of chlorine.
4. pH of water also influences bactericidal efficiency. pH in the range of 5-7 is most effective. Beyond this range, bactericidal efficiency decreases.
5. The requirement of chlorine increases with decrease in temperature and increase in pH beyond 7.
6. More the time of contact of chlorine more will be the efficiency.

5.30 A water softener has 0.05 m^3 of ion exchange resin with an ion exchange capacity of 40 kg/m^3. The water requirement of users is 1000 litres per day. If the water contains 230 mg/l of hardness as $CaCO_3$ and users want to soften it to 100 mg/l as $CaCO_3$, how much water should bypass the softener daily? What is the time between regeneration cycles if it is assumed that the complete saturation of the resin occurs before regenerating?

[20 Marks : 2021]

Solution:

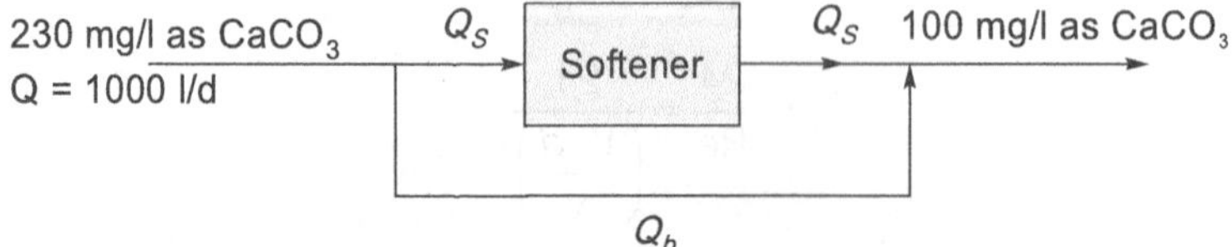

Q_S - discharge passing through the softener
Q_b - bypass discharge
Q - total discharge
H_s = Hardness of water went through softener
H_b = Hardness of bypass water

Assuming ion exchange resin works at an efficiency of 100% as long as it has the ions to exchange.

$$\text{Hardness of mix water} = \frac{Q_b H_b + Q_s H_s}{Q_b + Q_s}$$

$$\therefore\ 100\ \text{mg}/l \text{ as } CaCO_3 = \frac{Q_S \times O + Q_b \times 230 \text{ mg/l as } CaCO_3}{1000 \text{ l/d}}$$

$$Q_b = 434.782\ l/\text{d}$$

Now, the time of regeneration will come when the bed is exhausted.

$$\text{Volume of resin} = 0.05 \text{ m}^3$$

Total quantity of ions exchanged

$$= 40 \text{ kg/m}^3 \times 0.05 \text{ m}^3 = 2 \text{ kg}$$

$$\text{Hardness loading rate} = Q_s \times (\text{Initial hardness})$$
$$= (1000 - 434.782) \times 230 \times 10^{-6} \text{ kg/d}$$
$$= 0.13 \text{ kg/d}$$

$$\text{Time of regeneration} = \frac{\text{Total quantity of ions exchanged}}{\text{Hardness loading rate}}$$

$$= \frac{2}{0.13} \text{ kg/days} = 15.385 \text{ days}$$

5.31 **A drinking water treatment plant has a circular sedimentation basin to treat 13 million litres of river water per day. After storms occur upstream, the river often carries 0.010 mm silt particles with an average density of 2300 kg/m^3, and the silt must be removed before the water can be used. The sedimentation basin is 3.5 m deep and 21.0 m in diameter. The water is at 15°C. Answer the following:**

(i) What is the hydraulic detention time of the basin?

(ii) Will the sedimentation basin (clarifier) remove all of the silt particles from the river water? Justify your answer with appropriate calculations.

(Take density of water (ρ) = 999.1 kg/m^3 and viscosity of water (μ) = 0.00114 kg/m.s at 15°C)

[20 marks : 2022]

Solution:

(i) Given circular tank with side water depth,

$$H = 3.5 \text{ m}$$

and Diameter, $D = 21$ m

Hydraulic detention time,

$$D_t = \frac{\text{Volume of the tank}}{\text{Discharge passing through the tank}}$$

$$D_t = \frac{D^2(0.785H + 0.011D)}{Q} = \frac{21^2(0.785 \times 3.5 + 0.011 \times 21)\text{m}^3}{13 \times 10^3 \text{ m}^3/\text{d}}$$

$$= 0.101039 \text{ d} \times 24\text{h/d} = 2.425 \text{ h}$$

(ii) Surface overflow rate,

$$V_0 = \frac{Q}{\pi\frac{D^2}{4}} = \frac{13\times10^3\ \text{m}^3/\text{d}}{(\pi\times21^2/4)\text{m}^2} = 37.533\ \text{m}^3/\text{m}^2/\text{d}$$

The settling velocity can be found at by Stoke's law.

$$V_s = \frac{418(G_s-1)(3T+70)d^2}{100}$$

$$= \frac{418(2.3-1)(3\times15+70)}{100}\times(0.01)^2 = 0.0625\ \text{mm/s}$$

$$V_0 = 37.533\ \text{m/d} = 0.4344\ \text{mm/s}$$

$$V_s = \frac{(G-1)\gamma_w d^2}{18\mu} = \frac{(2.3-1)\times9.81\times999.1\times(0.010\times10^{-3})^2}{18\times0.00114}$$

$$= 0.0621\ \text{mm/s}$$

$\therefore$ $$V_s = \text{minimum}\{0.0624, 0.0621\}$$

$$= 0.0621\ \text{mm/s}$$

$\therefore$ Settling velocity is less than surface overflow rate and therefore all silt particles will not be removed.

5.32 **A PST 15 m long, 6 m wide and 3 m deep treats water for a town with a population of 20,000 supplied with 100 lpcd. The raw water sample indicated suspended solids conc. as 60 ppm. The PST worked with efficiency of 70% SS removal and the average specific gravity of the deposit in PST was 2.6. Compute**

(i) Detention Time

(ii) Horizontal Velocity

(iii) Rate of dry solids deposited

(iv) Overflow rate

[12 marks : 2023]

Solution:

Given: Length of tank, L = 15 m, Width of tank, B = 6 m, Height of tank, H = 3 m

Population of city, P = 20000, Per capita demand, q = 100 lpcd

Suspended solid concentration, SS_0 = 60 mg/L

Efficiency of removal of suspended solid, η = 70%

Specific gravity of deposit, G = 2.6

Now, Average discharge, $Q_{avg} = 20000\times100$

$= 2\times10^6$ Litre/day

Maximum daily discharge, $Q = 1.8\times2\times10^6$ Litre/day

$= 3.6\times10^6$ Litre/day

Now, Volume of tank, V = LBH

$= 15\ \text{m}\times6\ \text{m}\times3\ \text{m} = 270\ \text{m}^3$

(i) Detention time, $$t_d = \frac{\text{Volume of tank}}{\text{Maximum discharge}}$$

$$= \frac{270\text{m}^3}{3600\text{m}^3/\text{day}} = 0.075\ \text{day} = 1.8\ \text{hours}$$

(ii) Horizontal velocity, $$V_f = \frac{Q}{BH}$$

$$= \frac{3600 \text{m}^3/\text{day}}{(6 \times 3)\text{m}^2} = 200 \text{ m/day}$$

$$= 2.31 \times 10^{-3} \text{ m/s} = 2.31 \text{ mm/sec}$$

(iii) Suspended solids coming in tank = $Q \times SS_0$

$$= 3600 \text{ m}^3/\text{day} \times 60 \text{ mg/L}$$

$$= 216 \text{ kg/day}$$

Dry solids deposits in tank = $Q \times SS_0 \times \eta_0$

$$= 216 \times 0.7 = 151.2 \text{ kg/day}$$

(iv) Overflow rate, $V_0 = \frac{Q}{BL}$

$$= \frac{3600 \text{m}^3/\text{day}}{(15 \times 6)\text{m}^2}$$

$$= 40 \text{ m}^3/\text{day/m}^2$$

6. Design of Sewer

6.1 **Design a sewer to serve a population of 36,000; the daily per capita water supply allowance being 135 litres of which 80% finds its way into the sewer. The slope available for the sewer to be laid is 1 in 625 and the sewer should be designed to carry four times the dry weather flow when running full. What would be the velocity of flow in the sewer when running full? Assume n = 0.012 in Manning's formula.**

[15 marks : 2000]

Solution:

Average water supply = Population × daily per capita water supply

$$= 36{,}000 \times 135 = 4.86 \text{ MLD}$$

$\therefore$ Dry weather flow $= \frac{0.80 \times 4.86 \times 10^6}{10^3 \times 24 \times 60 \times 60} = 0.045 \text{ m}^3/\text{s}$

But design discharge, $Q = 4 \times$ Dry weather flow when running full $= 4 \times 0.045 = 0.18 \text{ m}^3/\text{s}$

If the diameter of the sewer is D, and hydraulic mean depth is R, then $R = \frac{D}{4}$ when sewer is running full.

As per Manning's equation the velocity of flow is given by

$$V = \frac{1}{n} R^{2/3} S^{1/2}$$

$$\therefore \quad Q = \frac{1}{n} A R^{2/3} S^{1/2}$$

$$\Rightarrow \quad 0.18 = \frac{1}{0.012} \times \frac{\pi}{4} \times D^2 \times \left(\frac{D}{4}\right)^{2/3} \times \left(\frac{1}{625}\right)^{1/2}$$

$$= \frac{1}{0.012} \times \frac{\pi}{4} \times \left(\frac{1}{4}\right)^{2/3} \times \left(\frac{1}{625}\right)^{1/2} \times (D)^{8/3}$$

$$\Rightarrow \quad (D)^{8/3} = \frac{0.18 \times 0.012 \times 4 \times \sqrt[3]{16} \times \sqrt{625}}{\pi} = 0.1732$$

$$\Rightarrow \quad D = 0.5182 \simeq 0.52 \text{ m}$$

$\therefore$ Velocity, $$V = \frac{Q}{A} = \frac{0.18}{\frac{\pi}{4} \times (0.52)^2} = 0.85 \text{ m/s}$$

6.2 **A combined sewer of circular section is to be laid to serve an area of 100 ha with a population of 90000 supplied with water at 200 *LPCD*. Assuming an impermeability factor of 0.50 and time of concentration of rainfall '*t*' as 20 minutes, calculate the size of the sewer when it has to run full with a velocity of 0.3 m/s. Assume suitable coefficients for '*a*' and '*b*' in the equation for '*R*', the intensity of rainfall relating to '*t*', the time of concentration.**

[10 marks : 2003]

Solution:

Given data: Population = 90000, Daily water supply per capita = 200 litres

Area (A) = 100 ha = 100×10^4 m² = 10^6 m², Impermeability factor or runoff coefficient = 0.50

Time of concentration (t) = 20 minutes

$\therefore$ The intensity of a rain is inversely proportional to the duration of the rain, an intensity duration curve can be represented by a generalised equation, of the form

$$R = \frac{a}{t+b}$$

where R = intensity of rainfall in cm/hr

t = time of concentration in minutes and a and b are constant.

When the value of t is between 5 to 20 minutes, $a = 75$ and b = 10 and $a = 100$ and $b = 20$ for $20 < t < 100$.

$$\therefore \quad R = \frac{75}{20+10} = 2.5 \text{ cm/hr}$$

Now average water supply = Population × average per capita water supply

= 90000 × 200 = 18 MLD

Assuming 80% of the water is converted into sewage

Volume of sewage = 0.80 × 18 = 14.4 MLD

$$\text{Average dry weather flow} = \frac{14.4 \times 10^6}{10^3 \times 24 \times 60 \times 60} = 0.1666 \text{ m}^3/\text{s}$$

$\therefore$ Maximum dry weather flow, $Q = 3 \times$ Average dry weather flow $= 3 \times 0.1666 = 0.5$ m³/s

Now, By Kuichling formula, we know that

$$Q_p = KRA$$

where Q_p is the peak rate of runoff in m³/s

$$\therefore \quad Q_p = 0.50 \times \frac{2.5 \times 10^{-2}}{60 \times 60} \times 10^6$$

$$\Rightarrow \quad Q_p = 3.47 \text{ m}^3/\text{s}$$

Thus, for the combined sewer, the discharge

$= Q + Q_p = 0.5 + 3.47 = 3.97$ m³/sec

We know that

Discharge in the sewer = Area of sewer × Velocity in sewer

$$3.97 = \frac{\pi}{4} \times D^2 \times 0.3$$

$$\Rightarrow \quad D^2 = \frac{3.97 \times 4}{0.3 \times \pi}$$

$$\Rightarrow \quad D = 4.10 \text{ m}$$

6.3 **A combined sewer of circular section is to be designed in a sewage system for a city with a population of 100000 in an area of 100 hectares. The mean flow of sewage from the city is 250 litre/capita/day. The rainfall intensity in the area is 4 cm/hr. The coefficient of runoff of the area is 0.48. The ratio of peak to average sewage flow is 2.0. The Manning's roughness coefficient is 0.012 and the Hazen-William's coefficient is 85. Using Manning's equation and Hazen William's expression, determine the gradient of the sewer to carry the peak flow with a velocity of 1.2 m/s.**

[15 marks : 2004]

Solution:

$$\therefore \quad \text{Average sewage flow} = \text{Population} \times \text{Mean flow of sewage}$$

$$= 100000 \times \frac{250 \times 10^{-3}}{24 \times 60 \times 60} = 0.289 \text{ m}^3/\text{s}$$

$\therefore$ Peak sewage flow, $Q = 2.0 \times 0.289 = 0.579 \text{ m}^3/\text{s}$

Area, $A = 100 \text{ ha} = 100 \times 10^4 \text{ m}^2 = 10^6 \text{ m}^2$

Coefficient of runoff, $K = 0.48$

Rainfall intensity, $p = 4 \text{ cm/hr}$

$\therefore$ Peak rate of runoff, $Q_P = KpA = 0.48 \times \dfrac{4 \times 10^{-2}}{60 \times 60} \times 10^6 = 5.33 \text{ m}^3/\text{s}$

$\therefore$ Total flow through the sewer $= Q + Q_P = 0.579 + 5.33 = 5.909 \text{ m}^3/\text{s}$

If the diameter of the combined sewer is D, then

$$\frac{\pi}{4} \times D^2 = \frac{5.909}{1.2} \qquad [\because Q = AV]$$

$$\Rightarrow \quad D^2 = \frac{5.909 \times 4}{1.2 \times \pi}$$

$$\Rightarrow \quad D = 2.5 \text{ m}$$

(i) As per Manning's equation

$$V = \frac{1}{n} R^{2/3} S^{1/2}$$

Given $V = 1.2 \text{ m/s}$

$$R = \frac{D}{4} = \frac{2.5}{4} = 0.625 \text{ m}$$

$$n = 0.012$$

$$\therefore \quad 1.2 = \frac{1}{0.012} \times (0.625)^{2/3} \times S^{1/2}$$

$$\Rightarrow \quad S^{1/2} = \frac{1.2 \times 0.012}{(0.625)^{2/3}}$$

$$\Rightarrow \quad S = \frac{1}{2577}$$

(ii) As per Hazen-William's formula

$$V = 0.85\, C_H R^{063} S^{0.54}$$

where $V = 1.2 \text{ m/sec}$

$$C_H = 85$$

$$R = 0.625 \text{ m}$$

$$\therefore \quad 1.2 = 0.85 \times 85 \times (0.625)^{0.63} (S)^{0.54}$$

$$\Rightarrow \quad (S)^{0.54} = \frac{1.2}{0.85 \times 85 \times (0.625)^{0.63}}$$

$$\Rightarrow \quad S = \frac{1}{1141.62}$$

6.4 **Explain the importance of self cleansing velocity and how it affects the design of sewer?**

[5 marks : 2005]

Solution:

The silting of sewers can be avoided by generating such just enough velocities that would not permit the solids to settle down i.e. the velocity should be such as to cause automatic self-cleansing effect. Such a self-cleansing velocity i.e. the velocity which will even scour the deposited particles of a given size, must be developed in the sewers, at least once a day, so as not to allow any deposition in the sewers. If certain deposition takes place and is not removed, it will obstruct free flow, causing further deposition and finally leading to the complete blocking of the sewer.

Hence, while finalising the sizes and gradients of the sewers, they must be checked for the minimum velocity that would be generated at minimum discharge (about 1/3 of average discharge).

The minimum velocity, generated in the sewers will not only ensure the adequate transportation of the suspended solids, but will also help in keeping the sewer size under control and preventing the sewage from getting stale and decomposed by moving it faster, thereby preventing evolution of foul gases.

6.5 **What is self-cleansing velocity? A 20 cm diameter sewer with invert slope of 1 in 500 is running full. Calculate the rate of flow in the sewer. Compare the velocity with self-cleansing velocity. Assume Manning's $n = 0.12$.**

[8 marks : 2007]

Solution:

Self cleansing velocity is that minimum velocity which should be generated in sewer to scour the deposited particles of a given size, at least once in a day. The silting of sewers can be avoided by generating such high velocities that would not permit the solids to settle down i.e. the velocity should be such as to cause automatic self cleansing effect.

The velocity in the sewer may be calculated by using Manning's equation i.e.

$$V = \frac{1}{n} R^{2/3} S^{1/2}$$

Given that $\quad n = 0.12$ (This value should be 0.012)

$$S = \frac{1}{500}; \; D = 20 \text{ cm} = 0.2 \text{ m}$$

The sewer is running full, hence $\quad R = \frac{D}{4} = \frac{0.2}{4} = 0.05 \text{ m}$

$$\therefore \quad V = \frac{1}{0.012} \times (0.05)^{2/3} \times \left(\frac{1}{500}\right)^{1/2} = 0.51 \text{ m/s}$$

The rate of flow in the sewer is given by

$$Q = AV = \frac{\pi}{4} \times (0.2)^2 \times 0.51 = 0.016 \text{ m}^3/\text{s}$$

The self-cleansing velocity for 1 mm size inorganic sand particles and 5 mm size organic particles is about 0.45 m/s. Hence the velocity in the sewer is greater than self cleansing velocity.

Comment: The value of n = 0.012, but it is given as 0.12 which is not correct.

6.6 Work out the hydraulic characteristics of a circular sewer sections running partially full.

[8 marks : 2007]

Solution:

Let us denote the properties of the section when running partially full by small letters and when running full by capital letters.

When sewer is running full, then

Area of cross section, $A = \frac{\pi D^2}{4}$

Wetted perimeter, $P = \pi D$

Hydraulic mean depth, $R = \frac{A}{P} = \frac{\pi D^2/4}{\pi D} = \frac{D}{4}$

When sewer is running partially full at a depth '*d*', then

$$d = \frac{D}{2} - \frac{D}{2}\cos\frac{\alpha}{2}$$

$$\Rightarrow \quad \frac{d}{D} = \frac{1}{2}\left(1 - \cos\frac{\alpha}{2}\right)$$

Area of cross-section, $a = \frac{\pi D^2}{4} \times \frac{\alpha}{360°} - \frac{D}{2}\cos\frac{\alpha}{2} \times \frac{D}{2}\sin\frac{\alpha}{2} = \frac{\pi D^2}{4}\left[\frac{\alpha}{360°} - \frac{\sin\alpha}{2\pi}\right]$

$$\left[\because \sin\alpha = 2\sin\left(\frac{\alpha}{2}\right) \times \cos\left(\frac{\alpha}{2}\right)\right]$$

$$\frac{a}{A} = \frac{\alpha}{360} - \frac{\sin\alpha}{2\pi}$$

Wetted perimeter, when running partially full,

$$p = \pi D \times \frac{\alpha}{360°}$$

$$\Rightarrow \quad \frac{p}{P} = \frac{\alpha}{360°}$$

Hydraulic mean depth, while running partially full, $r = \frac{a}{p}$

$$\Rightarrow \quad r = \frac{A\left[\frac{\alpha}{360°} - \frac{\sin\alpha}{2\pi}\right]}{P\left[\frac{\alpha}{360°}\right]} = R\left[1 - \frac{360° \sin\alpha}{2\pi\alpha}\right]$$

$$\Rightarrow \quad \frac{r}{R} = 1 - \frac{360 \sin\alpha}{2\pi\alpha}$$

Now, **velocity of flow** is given by Manning's equation, as

$$v = \frac{1}{n} r^{2/3} S_0^{1/2}$$

and

$$V = \frac{1}{n} R^{2/3} S_o^{1/2}$$

$$\therefore \quad \frac{v}{V} = \left(\frac{r}{R}\right)^{2/3} \quad [\because \text{For the same sewer value of n and } S_o \text{ do not change}]$$

$$\Rightarrow \quad \frac{v}{V} = \left[1 - \frac{360 \sin\alpha}{2\pi\alpha}\right]^{2/3}$$

Proportionate discharge,
$$\frac{q}{Q} = \frac{av}{AV} = \frac{a}{A} \times \frac{v}{V}$$

$$\frac{q}{Q} = \left[\frac{\alpha}{360^\circ} - \frac{\sin\alpha}{2\pi}\right] \times \left[1 - \frac{360^\circ \sin\alpha}{2\pi\alpha}\right]^{2/3}$$

6.7 **A 1200 m long storm sewer collects wastewater from a catchment area of 50 hectare, where 35% area is covered by roof (I = 0.9), 20% area is covered by pavements (I = 0.8) and 45% area is covered by open land (I = 0.18). Determine the average I, and diameter of storm sewerline assuming (i) the time of entry = 3 min (ii) velocity of full flow = 1.5 m/s (iii) n = 0.013 and slope = 0.001 I = runoff ratio.**

[10 marks : 2009]

Solution:

The total area A of the catchment can be considered to be made of smaller areas like roof (A_1), pavements (A_2) and open land (A_3) having run off ratios as I_1, I_2 and I_3 respectively. Then the runoff ratio I for the entire area may be computed by using

$$I = \frac{I_1A_1 + I_2A_2 + I_3A_3}{A_1 + A_2 + A_3}$$

$$\Rightarrow \quad I = \frac{I_1A_1 + I_2A_2 + I_3A_3}{A}$$

$$\therefore \quad I_1A_1 = 0.9 \times \frac{35}{100} \times 50 = 15.75$$

$$I_2A_2 = 0.8 \times \frac{20}{100} \times 50 = 8$$

$$I_3A_3 = 0.18 \times \frac{45}{100} \times 50 = 4.05$$

$$\therefore \quad I = \frac{15.75 + 8 + 4.05}{50} = 0.556$$

From Manning's equation, we have

$$V = \frac{1}{n} R^{2/3} S^{1/2}$$

The hydraulic mean depth (R) is given by

$$R = \frac{\text{Area of flow}}{\text{Wetted perimeter}}$$

$$\Rightarrow \quad R = \frac{\pi D^2/4}{\pi D} = \frac{D}{4}$$

Given that $V = 1.5$ m/s; $n = 0.013$; $S = 0.001$

$$\therefore \quad 1.5 = \frac{1}{0.013} \times \left(\frac{D}{4}\right)^{2/3} \times (0.001)^{1/2}$$

$$\Rightarrow \quad D^{2/3} = 1.5538$$

$$\Rightarrow \quad D = 1.94 \text{ m}$$

6.8 **Using $n = 0.015$ in Manning's formula, design a sewer running half-full at a flow rate of 650 litres per second and laid at an invert slope of 0.0001.**

[8 marks : 2010]

Solution:

Given that n = 0.015, Q = 650 litres per second, S = 0.0001

Let the diameter of sewer be D.

The sewer is running half full, therefore the depth of flow will be half the diameter of sewer i.e.

$$d = \frac{D}{2}$$

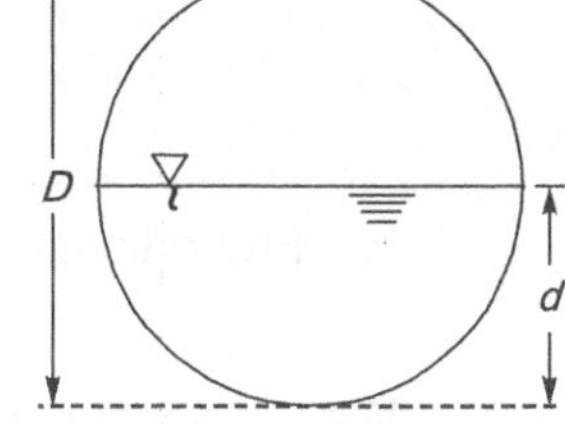

Area of flow, $$a = \frac{1}{2}\times\frac{\pi D^2}{4} = \frac{\pi D^2}{8}$$

Wetter perimeter, $$p = \frac{1}{2}\times\pi D = \frac{\pi D}{2}$$

Hydraulic mean depth, $$r = \frac{a}{p} = \frac{\pi D^2/8}{\pi D/2} = \frac{\pi D^2\times 2}{\pi D\times 8} = \frac{D}{4}$$

As per **Manning's equation**, the velocity of flow is given by

$$v = \frac{1}{n}r^{2/3}S^{1/2}$$

But $$q = av$$

$$\Rightarrow \quad q = \frac{1}{n}\times a\times r^{2/3}\times S^{1/2}$$

$$\Rightarrow \quad 650\times 10^{-3} = \frac{1}{0.015}\times\frac{\pi D^2}{8}\times\left(\frac{D}{4}\right)^{2/3}\times(0.0001)^{1/2}$$

$$\Rightarrow \quad D^{8/3} = \frac{650\times 10^{-3}\times 0.015\times 8\times(4)^{2/3}}{\pi\times(0.0001)^{1/2}}$$

$$\Rightarrow \quad D^{8/3} = 6.256$$

$$\Rightarrow \quad D = 1.99 = 2\text{ m}$$ **(Ans.)**

6.9 **Why a circular section is most widely adopted for sewer pipes? Discuss the hydraulic characteristics of circular sewer sections running partially full.**

[8 marks : 2013]

Solution:

For sewer pipes, circular section is the most suitable since the minimum flow velocity required to prevent sitting of sewers can be maintained even at very low sewer discharges.

Hydraulic characteristics of circular sewer running partially full:

Let there be a circular sewer of diameter 'D' running partially full at depth 'd', there by subtending an angle 'α' at the centre by the free surface of the flowing sewage.

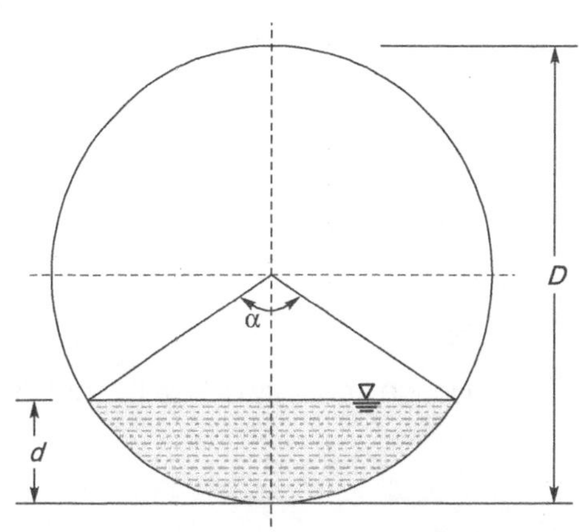

When pipe is running full

Area, $$A = \frac{\pi}{4}D^2$$

Wetted perimeter, $$P = \pi D$$

$\therefore$ HMD, $$R = \frac{A}{P} = \frac{D}{4}$$

When pipe is running partially full at depth '*d*'

$$d = \frac{D}{2} - \frac{D}{2}\cos\left(\frac{\alpha}{2}\right)$$

$\therefore$ Proportionate depth $= \frac{d}{D} = \frac{1}{2}\left[1-\cos\left(\frac{\alpha}{2}\right)\right]$

Area, $a = \frac{\pi}{4}D^2\frac{\alpha}{360} - \frac{D}{2}\cos\left(\frac{\alpha}{2}\right)\frac{D}{2}\sin\left(\frac{\alpha}{2}\right)$

$$= \frac{\pi}{4}D^2\left[\frac{\alpha}{360} - \frac{\sin\alpha}{2\pi}\right] \qquad \left[\because \sin\alpha = 2\sin\frac{\alpha}{2}.\cos\frac{\alpha}{2}\right]$$

$\therefore$ Proportionate area $\left(\frac{a}{A}\right) = \left[\frac{\alpha}{360} - \frac{\sin\alpha}{2\pi}\right]$

Wetted perimeter. $p = \pi D\frac{\alpha}{360}$

$\therefore$ Proportionate perimeter $\left(\frac{p}{P}\right) = \frac{\alpha}{360}$

Hydraulic Mean Depth (HMD), $r = \frac{a}{p} = \frac{D}{4}\left[1 - \frac{360^\circ \sin\alpha}{2\pi\alpha}\right]$

$\therefore$ Proportionate HMD $= \frac{r}{R} = \left[1 - \frac{360^\circ \sin\alpha}{2\pi\alpha}\right]$

6.10 **A 500 mm diameter sewer pipe has been installed with an invert slope of 0.008. Determine the capacity of flow when the sewer pipe is flowing half-full. The number of sewer pipes used is 2. Manning's roughness coefficient is 0.012. Assume that flow in the sewer pipe is uniform.**

[4 marks : 2016]

Solution:

Given: Diameter of sewer (D) = 0.5 m, Bed slope (S) = 0.008, Manning's roughness coefficient (N) = 0.12

Number of sewer = 2

Flow through sewer when sewer is running full,

$$Q = AV$$

$$Q = \frac{\pi D^2}{4} \times \frac{1}{N} R^{2/3} S^{1/2}$$

$$Q = \frac{\pi D^2}{4} \times \frac{1}{N}\left(\frac{D}{4}\right)^{2/3} S^{1/2}$$

$$Q = \frac{\pi \times 0.5^2}{4} \times \frac{1}{0.012} \times \left(\frac{0.5}{4}\right)^{2/3} (0.008)^{1/2}$$

$$Q = 0.37 \text{ m}^3/\text{sec}$$

When sewer is running half full, $\frac{q}{Q} = \frac{1}{2}$

$\Rightarrow$ $q = \frac{Q}{2} = 0.185 \text{ m}^3/\text{sec}$

Hence capacity of flow through two sewers when it is running half full

$$= 2q$$

$$= 0.37 \text{ m}^3/\text{sec} = 370 \ l/\text{sec}$$

6.11 **A rectangular sewer with width twice its depth is hydraulically equivalent to a circular sewer. Find the relation between the width of the rectangular sewer and the diameter of the circular sewer assuming that sewer is running completely full.**

[12 marks : 2020]

Solution:

Let B and D_1 represents the width and depth of the rectangular sewer, respectively.

$$\therefore \quad B = 2D_1$$

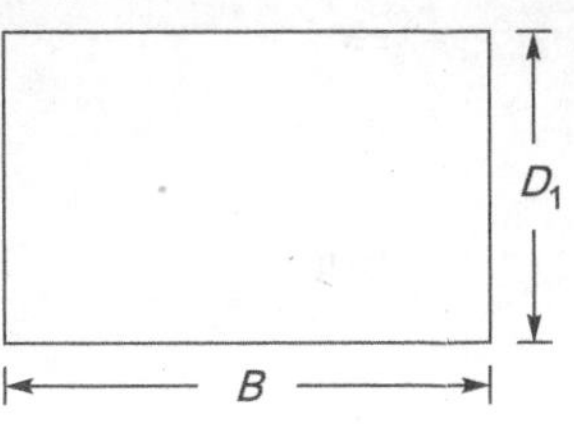

Now, when this rectangular sewer is running completely full.

The area of flow, $\quad A = B \cdot D_1 = 2D_1 \cdot D_1 = 2D_1^2$

The wetted perimeter, $\quad P = 2(B + D_1)$

As it is mentioned in the question that the rectangular sewer is running completely full, thus wetted perimeter is

$$P = 2(2D_1 + D_1) = 6 \cdot D_1$$

$\therefore$ Hydraulic radius, $\quad R = \dfrac{A}{P} = \dfrac{2D_1^2}{6D_1} = \dfrac{D_1}{3}$

Discharge capacity of rectangular sewer while running full

$$Q = \frac{1}{N} \cdot AR^{2/3} \cdot S^{1/2}$$

$$= \frac{1}{N} \cdot (2D_1^2) \cdot \left(\frac{D_1}{3}\right)^{2/3} \cdot S^{1/2} \qquad \text{...(1)}$$

Now, let D be the diameter of circular sewer.

When this circular sewer is running completely full,

The area of flow, $\quad A = \dfrac{\pi}{4} D^2$

The wetted perimeter, $\quad P = \pi D$

$\therefore$ Hydraulic radius, $\quad R = \dfrac{A}{P} = \dfrac{\frac{\pi}{4}D^2}{\pi D} = \dfrac{D}{4}$

Discharge capacity of rectangular sewer while running full

$$Q = \frac{1}{N} \cdot AR^{2/3} \cdot S^{1/2}$$

$$= \frac{1}{N} \cdot \left(\frac{\pi}{4} D^2\right) \cdot \left(\frac{D}{4}\right)^{2/3} \cdot S^{1/2} \qquad \text{...(2)}$$

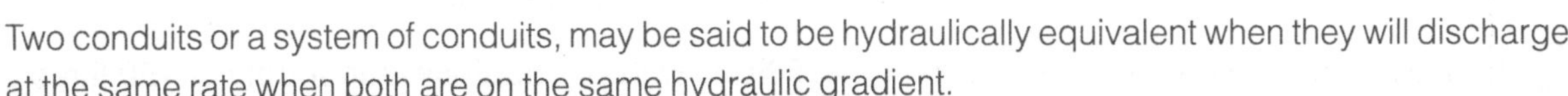

Two conduits or a system of conduits, may be said to be hydraulically equivalent when they will discharge at the same rate when both are on the same hydraulic gradient.

Note: Since sewers are designed to ensure flow under gravity, this implies, for the sewer, bed slope is same if they are hydraulically equivalent.

$\therefore$ From equation (1) and equation (2)

$$Q_1 = Q_2$$

$$\Rightarrow \quad \frac{1}{N} \cdot (2D_1^2) \cdot \left(\frac{D_1}{3}\right)^{2/3} \cdot S^{1/2} = \frac{1}{N} \cdot \left(\frac{\pi}{4} D^2\right) \cdot \left(\frac{D}{4}\right)^{2/3} \cdot S^{1/2}$$

$$\Rightarrow \quad \frac{2}{(3)^{2/3}} \cdot D_1^{8/3} = \frac{\pi}{(4)^{5/3}} \cdot D^{8/3}$$

$$\Rightarrow \quad D_1^{8/3} = \frac{\pi \cdot (3)^{2/3}}{2 \cdot (4)^{5/3}} \cdot D^{8/3}$$

$$\Rightarrow \quad D_1 = \left(\frac{\pi \cdot (3)^{2/3}}{2 \cdot (4)^{5/3}}\right)^{3/8} \cdot D$$

$$D_1 = 0.655\,D \qquad ...(3)$$

But, $B = 2D_1$

$$\therefore \quad D_1 = \frac{B}{2} \qquad ...(4)$$

$\therefore$ from equation (3) and equation (4)

$$\frac{B}{2} = 0.655D$$

$$\Rightarrow \quad B = 1.31D$$

This is the required relation between width of rectangular sewer (B) and diameter of circular sewer (D).

6.12 **A city has its catchment area of 7500 hectares. If the population density of the city is 200 persons per hectare and the water is supplied at the rate of 175 litres per capita per day, what would be the design flow for a combined sewer? Take intensity of rainfall equal to 30 mm/hour, average runoff coefficient equal to 0.50 and only 75% of water supplied contributes to the sewage. Also, peak discharge factor should be taken as 3.0.**

[10 Marks : 2021]

Solution:

Total water supplied = 175 l/c/d × 200 c/ha × 7500 ha = 262.5×10^6 l/d

Average sewage production = $0.75 \times 262.5 \times 10^6$ l/d

= 196.875×10^6 l/d = 2.278 m^3/s

$$\text{Total volume of runoff} = ciA = \frac{30 \times 10^{-3}}{3600} \times 0.5 \times 7500 \times 10^4 = 312.5\ m^2/s$$

Peak discharge for sewage = 3 × 2.278 = 6.834 m^3/s

Combined sewer is designed on the basis of peak discharge

$= (Q_{peak})_{sewage} + Q_{runoff}$

$= 6.834 + 312.5 = 319.334\ m^3/s$

6.13 **Design a sewer to serve a population of 50,000 with per capita water supply of 150 litres per day. Assume that the sewer should run 0.7 times full at the maximum discharge. The slope available for the sewer to be laid is 1 in 500 and the sewer should be designed with a peaking factor of 3.0. Assume Manning's rugosity coefficient value (N) = 0.012. Also perform check for self-cleansing velocity. The following table may be used if required:**

Proportionate depth	Proportionate velocity	Proportionate discharge
0.20	0.615	0.088
0.30	0.776	0.196

[20 marks : 2022]

Solution:

Design discharge of the sewer,

$$q = 3 \times (150 \text{ l/c/d} \times 0.8) \times 50000 \text{ c}$$
$$= 18 \times 10^6 \text{ l/d}$$

(A factor of 0.8 is taken for assuming 80% of water supplied is converted to sewage).

$$\therefore \quad q = 0.2083 \text{ m}^3\text{/s}$$

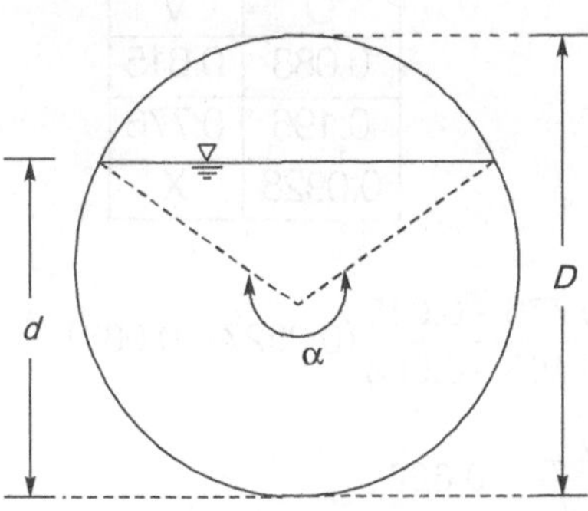

Given: $$\frac{d}{D} = 0.7$$

$$\therefore \quad 0.7 = \frac{1}{2}\left(1 - \cos\frac{\alpha}{2}\right)$$

$$\Rightarrow \quad 1.4 = 1 - \cos\frac{\alpha}{2}$$

$$\Rightarrow \quad \alpha = 227.156°$$

Now, $$\frac{q}{Q} = \left(\frac{\alpha}{360°} - \frac{\sin\alpha}{2\pi}\right)\left(\frac{\frac{\alpha}{360°} - \frac{\sin\alpha}{2\pi}}{\frac{\alpha}{360°}}\right)^{2/3}$$

$$= 0.748 \times \frac{(0.748)^{2/3}}{(0.631)^{2/3}} = 0.838$$

For assumed full flow

$$Q = \frac{q}{0.838} = \frac{0.2083 \text{ m}^3\text{/s}}{0.838}$$

$$\Rightarrow \quad 0.249 = \left(\frac{1}{0.012}\left(\frac{D}{4}\right)^{2/3}\sqrt{\frac{1}{500}}\right)\frac{\pi D^2}{4}$$

$$\Rightarrow \quad D = 0.561 \text{ m}$$

Check for self cleansing velocity:

Self cleansing velocity must be generated at minimum flow.

Assuming the minimum flow in the sewer to be $\frac{1}{3}^{rd}$ the average flow.

$$q_{min} = (150 l\text{/c/d} \times 0.8 \times 50000c) \times \frac{1}{3}$$

$$= 2 \times 10^6 \, l/d = 0.0231 \text{ m}^3/\text{s}$$

$$\therefore \quad \frac{q_{min}}{Q} = \frac{0.0231}{0.249} = 0.0928$$

For $\frac{q_{min}}{Q} = 0.0928$, from the given table, by interpolation, $\frac{V_{min}}{V}$ and $\frac{d_{min}}{D}$ can be found out.

$\frac{q_{min}}{Q}$	$\frac{V_{min}}{V}$
0.088	0.615
0.196	0.776
0.0928	X

$$X - 0.615 = \frac{0.776 - 0.615}{0.196 - 0.088}(0.0927 - 0.088)$$

$$\Rightarrow \quad X = \frac{V_{min}}{V} = 0.622$$

$$\therefore \quad V_{min.} = 0.622 \times \frac{1}{n} R^{2/3} S^{1/2}$$

$$= 0.622 \times \frac{1}{0.012} \times \left(\frac{0.561}{4}\right)^{2/3} \sqrt{\frac{1}{500}}$$

$$= 0.626 \text{ m/s}$$

Usually, minimum velocity greater than 0.6 m/s is satisfactory as per clause 3.15.1 of GOI Manual on Sewage and Sewage Treatment. Hence, OK.

7. Quality and Characteristics of Sewage

7.1 **Explain the procedure of estimating BOD of a given sample of sewage.**
The five day BOD at 20°C of a given sample is 450 mg/L. Calculate the ultimate BOD at 35°C given the deoxygenation constant at 20°C as 0.1 per day.

[5 + 10 = 15 marks : 2005]

Solution:

The standard 5 day BOD written as BOD_5 or simply BOD is determined in the laboratory by mixing or diluting a known volume of a sample of waste water with a known volume of aerated pure water and then calculating the DO of this diluted sample. The diluted sample is then incubated for 5 days at 20°C. The *DO* of the dilute sample after this period of incubation is again calculated. The difference between the initial *DO* value and the final *DO* value will indicate the oxygen consumed by the sewage sample in 5 days. The BOD in ppm is then calculated by using the equation.

$$BOD_5 = [\text{Initial } DO - \text{Final } DO] \times \frac{\text{Volume of diluted sample}}{\text{Volume of undiluted sewage sample}}$$

Given

$$BOD_5 = Y_t = Y_5 = 450 \text{ mg/L}$$

$$t = 5 \text{ day}$$

$$k_D \text{ at } 20° = 0.1 \text{ per days}$$

$$\text{Ultimate } BOD = L_o \text{ in mg/L}$$

Note: *Here the deoxygenation constant value = 0.1 falls in the range corresponding to base 10.*

We know that $Y_t = L_0(1-10^{-k_D t})$

$\Rightarrow$ $450 = L_0(1-10^{-0.1\times 5})$

$\Rightarrow$ $L_0 = \dfrac{450}{0.684}$

$\Rightarrow$ $L_0 = 658.11 \text{ mg/L}$

Ultimate BOD does not change with temperature, hence ultimate BOD at 35°C is 658.11 mg/L

7.2 **For a BOD test, raw sewage (3.0 mL) was diluted to 300 mL (capacity of a BOD bottle). The diluted sewage was observed for its dissolved oxygen at the beginning and end of 5 days incubation at 20°C. The respective values were 8.6 mg/L and 4.6 mg/L. Determine the BOD of the raw sewage.**

[2 marks : 2010]

Solution:

Volume of sample of raw sewage = 3.0 mL

Volume of diluted sample = 300 mL

$$\therefore \quad \text{Dilution factor} = \frac{\text{Volume of diluted sample}}{\text{Volume of undiluted sample}} = \frac{300}{3} = 100$$

The BOD of the raw sewage is given as

$$BOD_5 = [\text{Initial DO} - \text{Final DO}] \times \text{Dilution factor}$$
$$= (8.6 - 4.6) \times 100 = 400 \text{ mg/L}$$

7.3 **The BOD rule constant (k) for a rivers BOD assimilation was determined to be 2.0 day^{-1} (base e). The BOD of this river after leaving a heavily populated town was determined to be 50 mg/L. Determine the distance after which the river's BOD would become 4 mg/L. When the average velocity of the river was 1 m/sec. What would have been the k value if this distance would have been 300 km and state what k manifests.**

[8 marks : 2010]

Solution:

1 m/sec

P Q

Given, $k = 2 \text{ day}^{-1} (e)$

$(BOD_5)_P = 50 \text{ mg/L}$

$(BOD_5)_Q = 4 \text{ mg/L}$

Velocity, $v = 1 \text{ m/sec}$

Let ultimate BOD at P and Q be L_P and L_Q.

$(BOD_5)_P = 50 = L_P(1 - e^{-kt})$ ...(i)

Similarly $(BOD_5)_Q = 4 = L_Q(1 - e^{-kt})$ ...(ii)

From eq. (i) and (ii) $\dfrac{L_P}{L_Q} = \dfrac{50}{4}$

Now, $L_Q = P_P e^{-kt}$ (ultimate BOD will also decay at rate k)

Hence $4 = 50\, e^{-2\times t}$

$t = 1.2628 \text{ days}$

Hence distance, $D = vt = 1 \times 1.2628 \times 24 \times 60 \times 60 = 109.11 \times 10^3 \text{ m}$

$D = 109.11 \text{ km}$

If this distance D = 300 km

then $$t = \frac{D}{V} = \frac{300 \times 10^3}{1 \times 86400} = 3.471 \text{ days}$$

$$L_Q = L_P \, e^{-kt}$$

$$4 = 50 \, e^{-k \times 3.4+1}$$

$$k = 0.7274 \text{ day}^{-1} \text{ (base } e)$$

k signifies the rule of oxidation of organic matter and depends upon the nature of organic matter and temperature.

7.4 **Define BOD. Explain how BOD is determined for wastewater. For a BOD test carried out in the laboratory, 5 mL of wastewater having no initial DO was taken in a 300 mL BOD bottle and mixed with dilution water of 9.2 mg/L DO concentration. After incubating the bottle for 5 days at 20°C, the DO of the mixture was found to be 5.0 mg/L. Compute the BOD_5 (at 20°C) of wastewater.**

[8 marks : 2014]

Solution:

"Bio-chemical oxygen demand (BOD) is used as a measure of the quantity of oxygen required for oxidation of bio-degradable organic matter present in wastewater sample by aerobic bio-chemical action".

Determination of BOD_5: The standard 5 day BOD (BOD_5) is determined in the laboratory by mixing a known volume of a sample of wastewater with known volume of pure water and calculating the dissolved oxygen (D.O.) of this diluted sample. The diluted sample is then incubated for 5 days at 20°C. The dissolved oxygen (D.O.) of the diluted sample, after this period of incubation is again calculated. **Then BOD_5 in mg/*l* is calculated as**

$$BOD_5 = (D.O_i - D.O_f) \times \frac{\text{Vol. of the diluted sample}}{\text{Vol. of the undiluted sewage sample}}$$

where, $D.O_i$ = initial D.O. of diluted sample

$D.O_f$ = Final D.O. of diluted sample after 5 days incubation at 20°C

Give: Vol. of Waste water = 5 mL

DO of waste water = 0 mg/L

Vol. of pure water = 300 – 5 = 295 mL

DO of pure water = 9.2 mg/L

$$\text{Initial DO of diluted sample, } DO_i = \frac{V_{waste} \times BOD_5 + V_{pure} \times BOD_5}{V_{waste} + V_{pure}} = \frac{5 \times 0 + 295 \times 9.2}{300} = 9.0467 \text{ mg/L}$$

After incubating the bottle for 5 day, DO of mixture was found 5.0 mg/L

$\therefore$ DO_f = 5.0 mg/L

$$\therefore \quad BOD_5 = (DO_i - DO_f) \times \frac{\text{Vol. of diluted sample}}{\text{Vol. of undiluted sewage sample}}$$

$$= (9.0467 - 5.0) \times \frac{300}{5} = 242.8 \text{ mg/L}$$

7.5 A town on the bank of river Ganga discharges 18000 m^3 day^{-1} of treated wastewater into the river. The treated wastewater has a BOD_5 of 20 mg L^{-1} and a BOD decay constant of 0.12 day^{-1} at 20°C. The river has a flow rate of 0.43 m^3 sec^{-1} and an ultimate BOD of 5.0 mg L^{-1}. The DO of the river is 6.0 mg L^{-1} and the DO of the wastewater is 0.4 mg L^{-1}. Compute the DO and initial ultimate BOD in the river, immediately after mixing.

[8 marks : 2015]

Solution:

Discharge of treated wastewater, $Q_t = 18000 \text{ m}^3/\text{day}$

5 day BOD of treated wastewater, $BOD_{5t} = 20 \text{ mg}/l$

Ultimate BOD of treated wastewater, $BOD_t = \frac{BOD_{5t}}{(1-10^{-kt})} = \frac{20}{(1-10^{-0.12\times5})} = 26.709 \text{ mg}/l$

DO of treated wastewater, $DO_t = 0.4 \text{ mg}/l$

Discharge of river Ganga, $Q_R = 0.43 \text{ m}^3/\text{s}$

$$= 0.43 \times 3600 \times 24 \text{ m}^3/\text{day}$$

$$= 37152 \text{ m}^3/\text{day}$$

Ultimate BOD of river, $BOD_R = 5.0 \text{ mg}/l$

DO of water, $DO_R = 6.0 \text{ mg}/l$

Ultimate BOD of the mix, $BOD_{mix} = \frac{Q_R BOD_R + Q_t BOD_t}{Q_R + Q_t}$

$$= \frac{37152 \times 5.0 + 18000 \times 26.709}{37152 + 18000} = 12.085 \text{ mg}/l$$

DO of the mix, $DO_{mix} = \frac{Q_R DO_R + Q_t DO_t}{Q_R + Q_t}$

$$= \frac{37152 \times 6.0 + 18000 \times 0.4}{37152 + 18000} = 4.17 \text{ mg}/l$$

7.6 **Differentiate between BOD and COD in waste water.**
20 m*l* of sewage sample was added directly into a 300 mL BOD incubation bottle. The initial DO of the diluted sample was 7.8 mg/*l* and final DO after 5 day was 1.3 mg/*l*. The corresponding initial and final DO of the seeded dilution water were 8.1 and 6.9 mg/*l* respectively. What is the BOD_5 of the sewage?

[8 marks : 2016]

Solution:

BOD: Bio-chemical Oxygen Demand is the amount of oxygen required by the aerobic organisms to breakdown the organic matter present in water. BOD of water depends on temperature and nutrient concentration. BOD of polluted water is much higher than fresh water.

COD: Chemical Oxygen Demand is the (stoichiometric) measurement of oxygen required for oxidizing all the chemicals present in water. This includes both biodegradable and non-biodegradable matter. Magnitude wise COD > BOD.

True sample (*TS*)	+	Sewage water (*SW*)	→	Diluted sample (*DS*)
20 m*l*		280 m*l*		300 m*l*

Let, Initial DO of TS be DO_i

Final DO of TS be DO_f

$$BOD_5 = DO_i - DO_f$$

Then, $DO_{iDS} = \frac{V_{TS} DO_i + V_{SW} DO_{iSW}}{V_{TS} + V_{SW}}$

$$DO_{fDS} = \frac{V_{TS} DO_f + V_{SW} DO_{fSW}}{V_{TS} + V_{SW}}$$

$$DO_{iDS} - DO_{fDS} = \frac{V_{TS}(DO_i - DO_f) + V_{SW}(DO_i - DO_f)_{SW}}{V_{TS} + V_{SW}}$$

$$7.8 - 1.3 = \frac{20BOD_5 + 280(8.1 - 6.9)}{20 + 280}$$

$$BOD_5 = 80.7 \text{ mg}/l$$

7.7 **A wastewater sample was diluted 10% with tap water for conducting BOD Test. The DO content in wastewater sample was 1.2 mg/*l* and water was 6 mg/*l*. After 5 days incubation at 20°C, the DO of the sample was 0.5 mg/*l*. Calculate**

(i) 5 day BOD of wastewater and

(ii) ultimate BOD. Assume K_{20} = 0.12/day.

[12 marks : 2017]

Solution:

10% dilution means out of 100 mL, 10 mL is the wastewater (original sample) and 90 mL is the tapwater.

$$\therefore \text{ Initial dissolved oxygen in diluted sample} = \frac{(DO)_{waste} \times 10 + (DO)_{water} \times 90}{100}$$

$$= \frac{1.2 \times 10 + 6 \times 90}{100} = 5.52 \text{ mg}/l$$

Final DO of diluted sample after 5 days incubation at 20°C = 0.5 mg/*l*

(i) BOD_5 of wastewater = (Initial DO – Final DO) × Dilution factor

$$= (5.52 - 0.50) \times \frac{100}{10} = 50.2 \text{ mg}/l$$

(ii) We know, $BOD_5 \text{ of 5 days} = L_0\left(1 - 10^{-k_D t}\right)$

Where, L_0 = ultimate BOD in mg/*l*

k_0 = Deoxygenation constant = k_{20} = 0.12/day

t = time of incubation in days

$$\Rightarrow \quad 50.2 = L_0\left(1 - 10^{-k_{20} \times 5}\right)$$

$$\Rightarrow \quad L_0 = \frac{50.20}{\left(1 - 10^{-0.12 \times 5}\right)}$$

$$\Rightarrow \quad L_0 = 67.04 \text{ mg}/l$$

7.8 **With the help of data given below, estimate the per capita solid waste generated per day in a colony with 20,000 houses.**

Volume of waste /truck load = 20 m^3

Volume of waste/tractor load = 2 m^3

Volume of waste/small truck = 8 m^3

No. of truck loads = 2/day

No. of tractor loads = 1/day

No. of small truck loads = 3/day

Avg. no. of people/house = 4

Specific weight of waste in (kg/m^3) = 500 for truck; 225 for small truck; 50 for tractor.

[12 marks : 2017]

Solution:

Given: Volume of waste /truck load = 20 m^3

Volume of waste/tractor load = 2 m^3

Volume of waste/small truck = $8\ m^3$
No. of truck loads = 2/day
No. of tractor loads = 1/day
No. of small truck loads = 3/day
Avg. no. of people/house = 4

(i) Volume of waste, with truck load = $20 \times 2 = 40\ m^3/day$
Specific weight of waste = $500\ kg/m^3$
$\therefore$ Weight of waste with truckload = (40×500) kg/day
= 20000 kg/day ...(i)

(ii) Volume of waste with tractor load = $(2 \times 1)\ m^3/day = 2\ m^3/day$
Specific weight of waste = $50\ kg/m^3$
Weight of waste with tractor load = $50 \times 2 = 100$ kg/day ...(ii)

(iii) Volume of waste with small truck = $(8 \times 3) m^3/day = 24\ m^3/day$
Specific weight of waste = $225\ kg/m^3$
Weight of waste with small tractor = $225 \times 24 = 5400$ kg/day ...(iii)

Total weight of waste per day from colony = $(20000 + 100 + 5400) = 25500$ kg/day
Total number of people in colony = $4 \times$ Number of houses = $4 \times 20000 = 80000$

$$\therefore \text{Per capita solid waste generated per day} = \frac{25500}{80000} = 0.31875 \text{ kg/capita/day} \simeq 0.32 \text{ kg/capita/day}$$

7.9 **Compute the theoretical oxygen demand of 108.75 mg/*l* of glucose.**

[4 marks : 2020]

Solution:

Given, concentration of glucose ($C_6H_{12}O_6$) = 108.75 mg/*l*
To find theoretical oxygen demand (ThOD)

$$\underset{180\text{ g}}{C_6H_{12}O_6} + \underset{192\text{ g}}{6O_2} \longrightarrow 6CO_2 + 6H_2O$$

$\Rightarrow$ 1 part of glucose requires $\left(\frac{192}{180}\right)$ part of oxygen for oxidation to carbon di-oxide and water.

$\therefore$ 108.75 mg/*l* of glucose will require $\left(\frac{192}{180} \times 108.75\right)$ oxygen = 116 mg/*l*

$\therefore$ Theoretical oxygen demand (ThOD) = 116 mg/*l*

8. Disposing of the Sewage Effluents

8.1 **A river before its entry into a town had a discharge of 100 litres per seconds and 20 mg/L as the concentration of a conservative parameter. The town's waste water outfall having 200 mg/L concentration of the same conservative parameter raised the river concentration of 50 mg/L after a complete mix with the river water. Determine the dilution ratio resulting from the discharge of the said waste water outfall.**

[8 marks : 2007]

Solution:

Discharge of river, Q_r = 100 litres per second

Concentration of conservative parameter in river water, C_r = 20 mg/L

Concentration of conservative parameter in waste water, C_s = 200 mg/L

Concentration of conservative parameter in mix, $C_{mix} = 50$ mg/L

Discharge of waste water, $Q_s = ?$

The concentration of the conservative parameter after mixing of waste water with river water is given by

$$C_{mix} = \frac{Q_r C_r + Q_s C_s}{Q_r + Q_s}$$

$$\Rightarrow \quad 50 = \frac{100 \times 20 + Q_s \times 200}{100 + Q_s}$$

$$\Rightarrow \quad 5000 + 50\,Q_s = 2000 + 200\,Q_s$$

$$\Rightarrow \quad Q_s = \frac{3000}{150} = 20 \text{ litres per second}$$

$$\therefore \quad \text{Dilution ratio} = \frac{\text{Total discharge after mixing}}{\text{Discharge of waste water}} = \frac{Q_r + Q_s}{Q_s} = \frac{100+20}{20} = 6$$

8.2 **A large stream has a reoxygenation constant of 0.4 per day. At a velocity of 0.85 m/s; and at the point at which an organic pollutant is discharged, it is saturated with oxygen at 10 mg/L ($D_o = 0$). Below the outfall, the ultimate demand for oxygen is found to be 20 mg/L and the deoxygenation constant is 0.2 per day. What is the dissolved oxygen 48.3 km downstream?**

[10 marks : 2009]

Solution:

Given data: Reoxygenation constant (k_R) = 0.4 per day, Velocity of flow in large stream (v) = 0.85 m/s

Initial saturation DO = 10 mg/L, Initial oxygen deficit (D_o) = 0, Ultimate BOD (L_o) = 20 mg/L

Deoxygenation constant (k_D) = 0.2 per day, Location of critical DO = 48.3 km downstream

The time required for a certain amount of DO at 48.3 km downstream is given by

$$t = \frac{\text{Distance downstream}}{\text{Velocity of flow in large stream}}$$

$$= \frac{48.3 \times 10^3}{0.85} \times \frac{1}{60 \times 60 \times 24} = 0.6577 \text{ days}$$

The dissolved oxygen deficit after a time t is given by

$$D_t = \frac{k_D L_o}{k_R - k_D}\left[10^{-k_D t} - 10^{-k_R t}\right] + \left[D_o \times 10^{-k_R t}\right]$$

$$= \frac{0.2 \times 20}{0.4 - 0.2}\left[10^{-0.2 \times 0.6577} - 10^{-0.4 \times 0.6577}\right] + 0 = 3.86 \text{ mg/L}$$

But $\quad D_t$ = Initial saturation DO – DO at 48.3 km downstream

$\Rightarrow \quad$ DO 48.3 km downstream = 10 – 3.86 = 6.14 mg/L

8.3 **What are different zones of pollution in river stream? Explain the importance of re-oxygeneration, de-oxygeneration and oxygen deficit in problems of stream sanitation.**

[10 marks : 2013]

Solution:

Zones of pollution in a river stream: A river/stream undergoing self purification can be divided into following **four zones of pollution viz.**

(i) **Zone of degradation:** This zone exists upto a certain length beyond the point just below the sewage out fall.

Here in this zone, water becomes dark and turbid along with foundation of sludge deposits at the bottom. Dissolved oxygen gets reduced to about 40% of the saturation value ($\simeq$ 7.6 mg/lit at 30°C).

Rate of de-oxygenation is much higher than re-oxygenation rate.

Due to such un-favourable conditions, aquatic life such as algae etc. dies out.

(ii) **Zone of active decomposition:** This is the highly polluted zone.

Here water becomes greyish and much darker. Dissolved Oxygen (DO) level falls down to almost zero. Anaerobic conditions get set in thereby resulting in evolution of gases like CH_4, CO_2, H_2S etc. Due to decomposition of organic matter, the upper part of this zone is dominated by anaerobic bacteria while the lower part is dominated by aerobic bacteria.

(iii) **Zone of recovery:** Here the stream recourse from its degraded condition. Water becomes clear and algae reappears. Dissolved Oxygen (DO) level rises and organic matter gets mineralized to form nitrates, sulphates, phosphates and carbonates.

(iv) **Zone of cleaner water:** Here the river attains its original condition, DO level reaches to saturation value and usual aquatic life prevails.

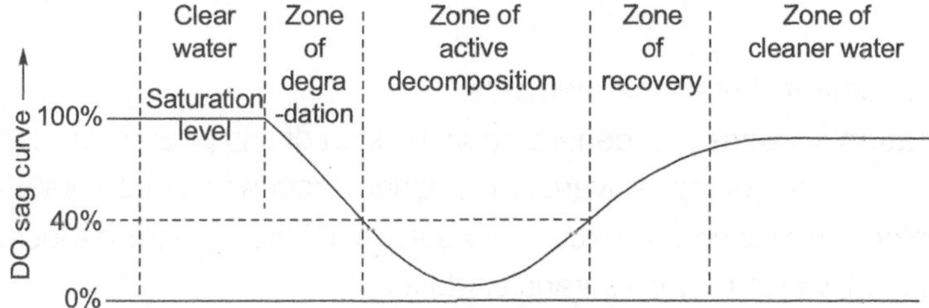

De-oxygeneration: Due to decomposition of organic matter in a polluted stream, DO level goes on decreasing. Rate of de-oxygenation follows the first order kinetics w.r.t. amount of organic matter remaining to be oxidized at prevalent temperature.

Re-oxygenation: In order to make-up the uses of DO in de-oxygenation process, atmosphere supplies oxygen to water and this process is referred to as re-oxygenation. Rate of re-oxygenation depends on the water body condition i.e., running on static water body, oxygen deficit and depth of receiving water.

Oxygen Deficit: In an running polluted stream, both de-oxygenation and re-oxygenation process goes simultaneously.

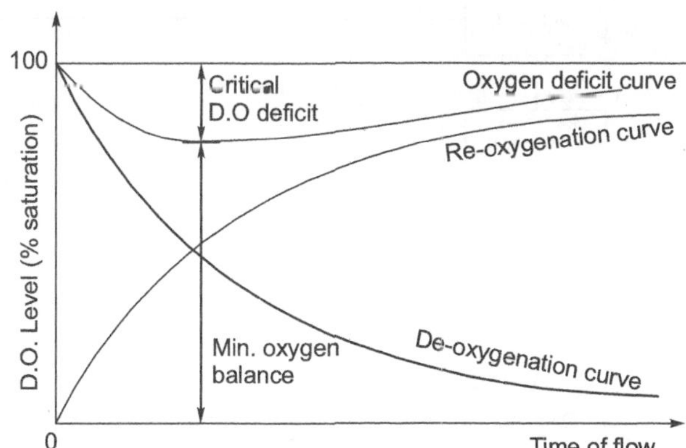

If R_D = rate of deoxygenation

and R_R = rate of reoxygenation

Then oxygen deficit occurs if

$$R_D > R_R$$

Net oxygen deficit = Algebraic sum of deoxygenation and reoxygenation curves.

8.4 Discuss thermal stratification and its importance in temperature of lakes.

[4 marks : 2013]

Solution:

Thermal Stratification: Heat gain and heat loss is a very slow process in water than on land or in air.

Temperature of water body remains nearly constant and changes gradually with season. In temperate zones of reservoir where the influence of turbulence and current is very low, there heat transfer is controlled by a phenomenon called thermal stratification.

In warm seasons, reservoir water gets divided into an upper warm layer called "**EPILIMNION**" and a lower relatively cool layer as "**HYPOLIMNION**". These two layers/zones are separated by a layer called "**THERMOCLINE**" where sharp temperature gradient exists.

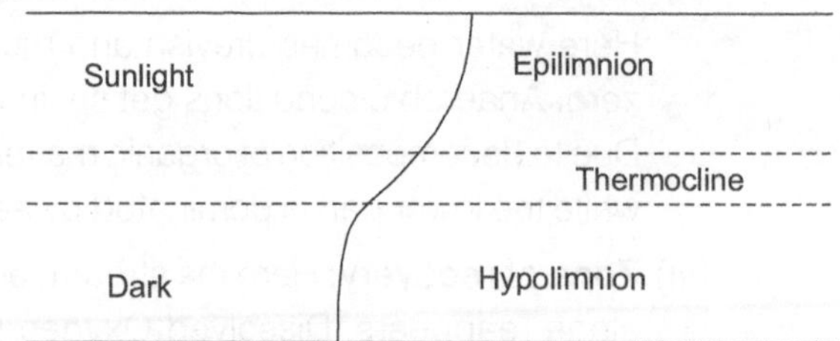

8.5 **Present a flow chart indicating various elements in municipal solid waste management and explain salient aspects of each element.**

[20 marks : 2017]

Solution :

Solid waste management : Functional elements.

(i) **Waste generation :** Wastes are generated at the start of any process and thereafter at every stage as the raw materials are being converted into various goods for consumption pruposes.

(ii) **Waste collection :** This includes collection of wastes and hauling them to a location where the collection vehicle is emptied, which may be a transfer station.

(iii) **Waste storage :** Storage is a key functional element because collection of wastes never takes place at the source or at time of their generation.

(iv) **Transfer and Transport :** This functional element involves the transfer of wastes from smaller collection vehicles where necessary to overcome the problem of narrow access lanes and subsequent transport of wastes usually over long distance to disposal sites.

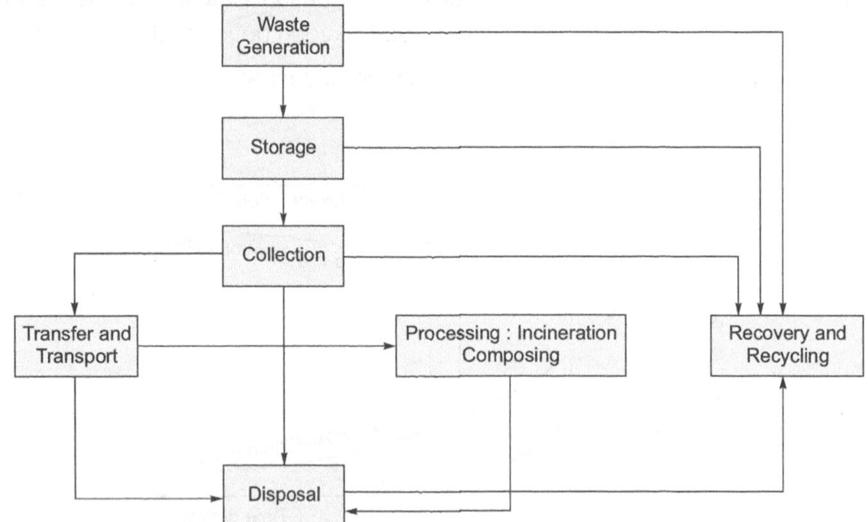

(v) **Processing :** Processing is required to alter the physical and chemical characteristics of wastes for energy and resource recovery and recycling. The important techniques include compaction, thermal volume reduction, manual separation of waste components, incineration and composting.

(vi) **Recovery and Recycling :** This includes various techniques, equipment facilities used to improve both the efficiency of disposal system and recovery of usable material and energy. Recovery involves the separation of valuable resources from the mixed solid wastes, delivered at transfer locations or processing plants.

(vii) **Waste disposal :** Disposal is the ultimate fate of all solid wastes, be it residential wastes, semisolid wastes from municipal and industrial treatment plants, incinerator residues, composts or other substances that have no further use to society.

8.6 **A treatment plant disposes 50 MLD of treated effluent into a river. The river flow rate is 20 m³/s and its DO is 8 mg/L before the mixing point. If the BOD of the effluent is 50 mg/L, find the BOD and DO of the river water at the d/s of mixing point. Assume BOD of river water as 0 and DO in WW effluent as 0.**

[6 marks : 2018]

Solution:

Subscript used:

'*W*' for wastewater (effluent), '*R*' for river, '*m*' for mix.

So, given data:

$$Q_w = 50\text{ MLD} = \frac{50\times10^6}{24\times60\times60\times10^3}$$

$$\Rightarrow \quad Q_w = 0.5787\text{ m}^3/\text{s}$$

$$Q_R = 20\text{ m}^3/\text{s}$$

$$DO_R = 8\text{ mg/L}$$

$$BOD_W = 50\text{ mg/L}$$

Also $$DO_W = 0 \text{ and } BOD_R = 0$$

Now, $$BOD_m = \frac{(BOD_w\times Q_W)+(BOD_R\times Q_R)}{Q_W+Q_R}$$

$$= \frac{(50\times0.5787)+(0\times20)}{0.5787+20} = 1.406\text{ mg/L (Ans)}$$

And $$DO_m = \frac{(DO_W\times Q_W)+(DO_R\times Q_R)}{Q_W+Q_R}$$

$$= \frac{(0\times0.5787)+(8\times20)}{0.5787+20} = 7.775\text{ mg/L (Ans.)}$$

8.7 **Explain the phenomena responsible for self purification in rivers. Discuss the factors influencing the self purification capacity.**

[20 marks : 2018]

Solution:

The various natural forces of purification which helps in effective self-purification process are:

1. Physical forces
   - Dilution and dispersion
   - Sedimentation
   - Sunlight (acting through biochemical forces)
2. Chemical forces (aided by biological forces)
   - Oxidation
   - Reduction

(i) **Dilution and dispersion:** When the putrescible organic matter is discharged into a large volume of water contained in the river stream, it gets rapidly dispersed and diluted. The action results in diminishing the concentration of organic matter, thus reduces the potential nuisance of sewage.

(ii) **Sedimentation:** The settleable solids, if present in sewage effluent, will settle down into the bed of the river, thus helping in the self purification process.

(iii) **Sunlight:** The sunlight has a bleaching and stabilizing effect of bacteria. If also helps certain microorganisms to derive energy from it and convert themselves into food for other forms of life, thus absorbing carbon dioxide and releasing oxygen by process known as photosynthesis. The evolution of oxygen will help in achieving self purification.

(iv) Oxidation: The oxidation of matter present in sewage effluent, will start as soon as the sewage outfalls into the river water containing dissolved oxygen. The deficiency of oxygen so created will be filled up by the atmospheric oxygen. The process of oxidation will continue till the organic matter has been completely oxidized.

(v) Reduction: Reduction occurs due to hydrolysis of organic matter settled at the bottom either chemically or biologically, anaerobic bacteria will help in splitting the complex organic constituents of sewage into liquids and gases.

Factors affecting the self purification capacity:

- **Temperature:** The temperature affects the rate of biological and chemical activities, which are enhanced at higher temperature and depressed at lower temperature.
- **Turbulence:** The turbulence in the water body helps in breathing the surface of stream or lake and helps in rapid re-aeration from the atmosphere. Thus it helps in maintaining aerobic conditions in the river stream and keeping it clean. However, too much turbulence erodes the bed of river.
- **Hydrography:** It affects the velocity and surface expanse of river stream.
- **Amount of dissolved oxygen:** The larger the amount of DO present in water, the better and earlier the self-purification will occur.
- **Rate of aeration:** The rate at which the DO deficiency in replenished, will considerably govern the self purification process. The greater is this rate, the quicker will be self-purification.

8.8 **Explain in detail the various process parameters required to control the aerobic composting of solid waste. Discuss the relevance of each parameter also.**

[10 marks : 2019]

Solution:

Disposal of Refuse by Composting

- Composting of refuse is a biological method of decomposing solid wastes.
- This decomposition can be affected either under aerobic conditions or under anaerobic condition or both.
- The final end product is a manure, called the compact or humus.
- If the organic materials, excluding plastics, rubber and leather are separated from municipal solid wastes and are subjected to bacterial decomposition, the end product remaining after dissimilatory and assimilatory bacterial activity is humus or compost. The entire process involving both separation and bacterial conversion of the organic solid wastes is known as composting.
- Decomposition of the organic solid wastes may be accomplished either aerobically or anaerobically depending on the availability of oxygen.
- Basically, composting is considered to be an aerobic process, because it involves piling up of refuse and its regular turning, either manually or by mechanical devices, so as to ensure sufficient supply of air and oxygen during its decomposition by bacteria, fungi and other microorganisms, like antinomycetes.
- Initially, the process starts with the mesophilic bacteria, which oxidises the organic matter (in the refuse) to carbon dioxide and liberate heat. The temperature rise about 45°C and at this point, the thermophilic bacteria take over and continue the decomposition. During this phase, the temperature further rises about 60°C which has to be maintained for at least 3 days in order to destroy pathogenic bacteria. This temperature control is crucial, because optimal decomposition occurs between 55 and 60°C, but if the temperature exceeds 60°C, decomposition slows down.

- Complete stabilisation occurs after the compact is allowed to cure for another 2 to 8 weeks. During the active early decomposition phase, the thermophilic bacteria are more active during the curing stage.
- The entire composting, thus gets completed in about 3-4 months time. Volume reductions of the original organic material of up to 50% are achieved under ideal conditions. The finally produced compost usually, has earthy smell and a dark brown colour.
- Moisture content of the compost mass should, however be controlled to ensure optimum aerobic decomposition, because excessive moisture will make it difficult to maintain aerobic conditions, while deficient moisture inhabits biological life. A moisture content of about 55% should be established, so that aerobic biological activity may proceed at an optimum rate.
- Most composting operation involve three basic steps:
  1. Preparation of the solid wastes.
  2. Decomposition of the soil wastes
  3. Product preparation and marketing
- Receiving, sorting, separation, size reduction and moisture and nutrient addition are part of the preparation step. Several techniques have been developed to accomplish the decomposition step. Once the solid wastes have been converted to a humus, they are ready for the third step, product preparing and marketing. This step include fine grinding, blending with various additives, granulation, bagging, storage, shipping and in some case direct marketing.
- The principal design consideration associated with the biological decomposition of prepared solid wastes are presented in table below:

Important design considerations for aerobic composting processes

Item	Comment
Particle Size	For optimum results the size of solid wastes should be between 25 and 75 mm (1 and 3 in).
Seeding and Mixing	Composting time can be reduced by seeding with partially decomposed solid wastes to the extent of about 1 to 5 percent by weight. Sewage sludge can also be added to prepared solid wastes. Where sludge is added, the final moisture content is the controlling variable.
Mixing or turning	To prevent drying, caking, and air channeling, material in the process of being composted should be mixed or turned on a regular schedule or as required. Frequency of mixing or turning will depend on the type of composting operation.
Air requirements	Air with at least 50 percent of the initial oxygen concentration remaining should reach all parts of the composting material for optimum results, especially in mechanical systems.
Total Oxygen Requirements	The theoretical quantity of oxygen required can be estimated.
Moisture Content	Moisture content should be in the range between 50 and 60 percent during the composting process. The optimum value appears to be about 55 percent.
Temperature	For best results, temperature should be maintained between 323 and 328 K (50 and 55°C) for the first few days and between 328 and 333 K (55 and 60°C) for the remainder of the active composting period. If temperature goes beyond 339 K (66°C), biological activity is reduced significantly.
Carbon-nitrogen Ratio	Initial carbon-nitrogen ratios (by mass) between 30 and 50 are optimum for aerobic composting. At lower ratios, ammonia is given off. Biological activity is also impeded at lower ratios. At higher ratios, nitrogen may be a limiting nutrient.
pH	To minimize the loss of nitrogen in the form of ammonia gas, pH should not rise above about 8.5.
Control of pathogens	If properly conducted, it is possible to kill all the pathogens, weeds and seeds during the composting process. To do this, the temperature must be maintained between 333 and 343 K (60 and 70°C) for 24 h.

8.9 **A waste water treatment plant discharge 1.0 m³/sec of effluent having an ultimate BOD of 45.0 mg/*l* into a river flowing at 10.0 m³/sec. Just upstream from the discharge point, the river water has an ultimate BOD of 4.0 mg/*l*. The deoxygenation constant (k_d) is 0.22/day. Answer the following:**

(i) Assuming complete and instantaneous mixing, what would be the ultimate BOD of the mixture of waste and river just downstream from the outfall?

(ii) Assuming a constant cross-sectional area for the river equal to 60 m², what ultimate BOD would you expect to find at a point 10.0 km downstream?

[12 Marks : 2021]

Solution:

k_D = 0.22/day.

Q_s - discharge of sewage

$(BOD_u)_s$ - ultimate BOD of sewage

Q_R - discharge of river

$(BOD_U)_R$ - ultimate BOD of river.

$(BOD_{u,\ mix})$ - Ultimate BOD of the mix

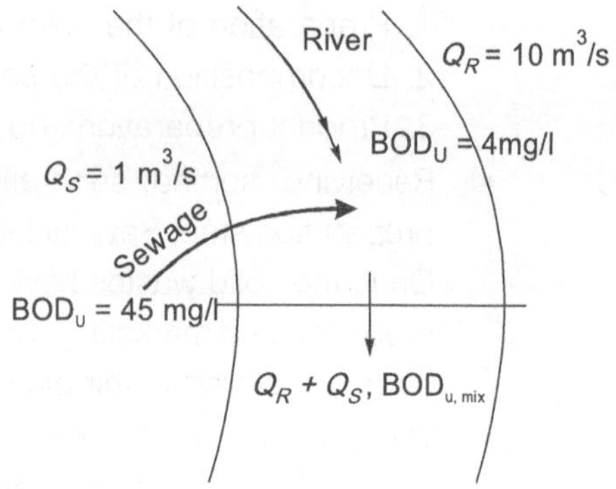

(i)
$$(BOD_{u,\ mix}) = \frac{(BOD_U)_S Q_s + (BOD_U)_R Q_R}{Q_S + Q_R}$$

$$= \frac{45 \times 1 + 4 \times 10}{11} mg/l = 7.727\ mg/l$$

(ii) Avg velocity of the mix,

$$V = \frac{Q_S + Q_R}{Area} = \frac{11 m^3/s}{60 m^2} = 0.183\ m/s$$

Time required to travel 10 km,

$$= \frac{10 \times 10^3 m}{0.183\ m/s \times 86400\ s/day} = 0.6324\ day$$

Hence, BOD at point (p), 10 km downstream from mixing point, i.e. after 0.6324 days.

$$(BOD)_t = 7.727 mg/l \left(1 - e^{-0.22 d^{-1} \times 0.6324 d}\right) = 1\ mg/l$$

(Assuming the k_D given is w.r.t base '*e*')

$$\text{BOD Remaining} = 7.727 - 1 \approx 6.727\ mg/l$$

8.10 **Discuss the impact of heavy metals in industrial wastewater when disposed into surface water. With the help of sketches, explain the working principle of the two methods used for removal of heavy metal from industrial wastewater.**

[12 marks : 2023]

Solution:

The disposal of industrial waste water containing heavy metals into surface water can have significant environmental and human health impacts. Heavy metals, such as, lead, mercury, cadmium, chromium and arsenic are toxic substances that can persist in the environment and accumulate in living organisms, leading to variety of adverse effects.

1. Environmental impact:

- **Water pollution**: Heavy metals can contaminate surface water bodies, leading to water pollution. This contamination can disrupt aquatic ecosystem by affecting the health and survival of aquatic organisms, including fish, amphibians etc.

- **Bioaccumulation and Biomagnification**: Heavy metals tend to accumulate in the tissues of organisms and can magnify up the food chain through a process called **biomagnification**. Predatory species at the top of the food chain, including human, can be exposed to high levels of heavy metals through consumption of contaminated organisms.
- **Eutrophication**: Some heavy metals can contribute to eutrophication, a process in which excess nutrients in water bodies promote the growth of algae. This can lead to oxygen depletion and negatively impact other aquatic organisms.

2. Human Health Impact:
    - **Water Contamination**: Heavy metals in surface water can contaminate drinking water sources posing a risk to human health. Ingesting or coming into contact with water containing high levels of heavy metals can lead to various health issues including neurological disorder, kidney damage, liver damage etc.
    - **Occupational exposure**: Industrial worker involved in the handling and disposal of wastewater can face direct exposure to heavy metals, increasing the risk of occupational health hazards.

The methods used for the removal of heavy metals from industrial wastewater are: precipitation and ion exchange.

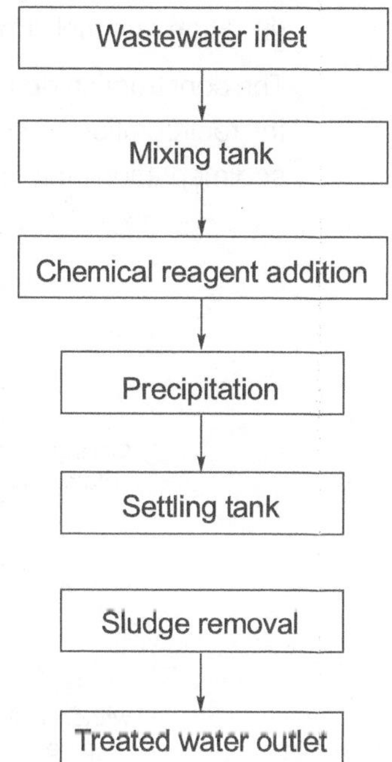

1. **Precipitation method**:
   This method involves the addition of chemical reagent to the wastewater that forms an insoluble precipitate with the heavy metal ions, allowing them to settle and be separated from the water.
    - The industrial water enters the mixing tank through the inlet.
    - A chemical reagent such as lime on sodium hydroxide is added to the wastewater. The reagent reacts with the heavy metal ions, forming insoluble metal hydroxide precipitates.
    - The mixture then flows into a settling tank where the precipitates settle at the bottom due to their increased weight and decreased solubility.
    - The settled precipitates form a sludge which is periodically removed from the bottom of the tank.
    - The treated water, with reduced heavy metal concentrations, is collected and discharged through the treated water outlet.
2. **Ion exchange method**:
   This method involves the use of specially designed resin material that can selectively absorb heavy metal ions from the waste water, replacing them with less harmful ions.

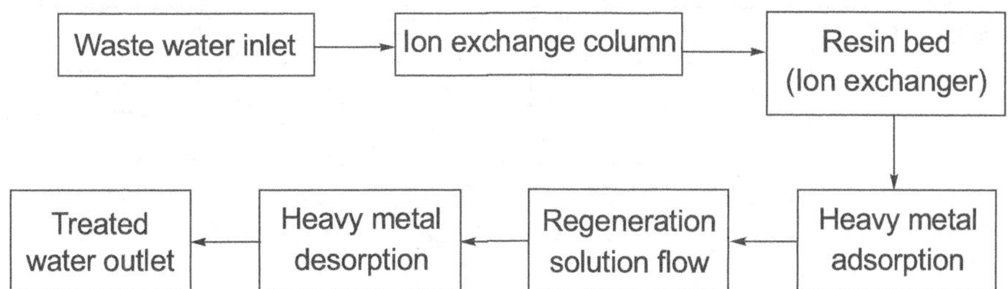

- The industrial wastewater enters the ion exchange column through the inlet.
- Inside the column there is a resin bed having functional groups that attract and bind heavy metal ions.
- As the wastewater flows through the resin bed, heavy metal ions in the water are adsorbed by the resin replacing other less harmful ions.
- After certain period, the resin becomes saturated with heavy metals ion and need regeneration. A regeneration solution such as acid or salt solutionis introduced to the column to desorb the heavy

metal ions from the resin.

- The desorbed heavy metal ions, along with the regeneration solution, are collected separately for further treatment or disposal.
- The treated water, with reduced heavy metal concentrations, passes through the resin bed and is collected at the treated water outlet.

9. Treatment of Sewage

9.1 **Compare 'Trickling Filter' and 'Recirculation with Bio-filters'.**
Describe with a neat sketch the working of a 'Sludge digestion tank with floating cover'.
[7 + 8 = 15 marks : 1999]

Solution:

Trickling filters are classified in two types:

(i) Standard rate trickling filter
(ii) High rate trickling filter

The construction details of these filters are same, except that provision is made in high rate trickling filters for recirculation of sewage through the filter, by pumping a part of the filtered effluent to the primary sedimentation tank, and repassing through it and the filter.

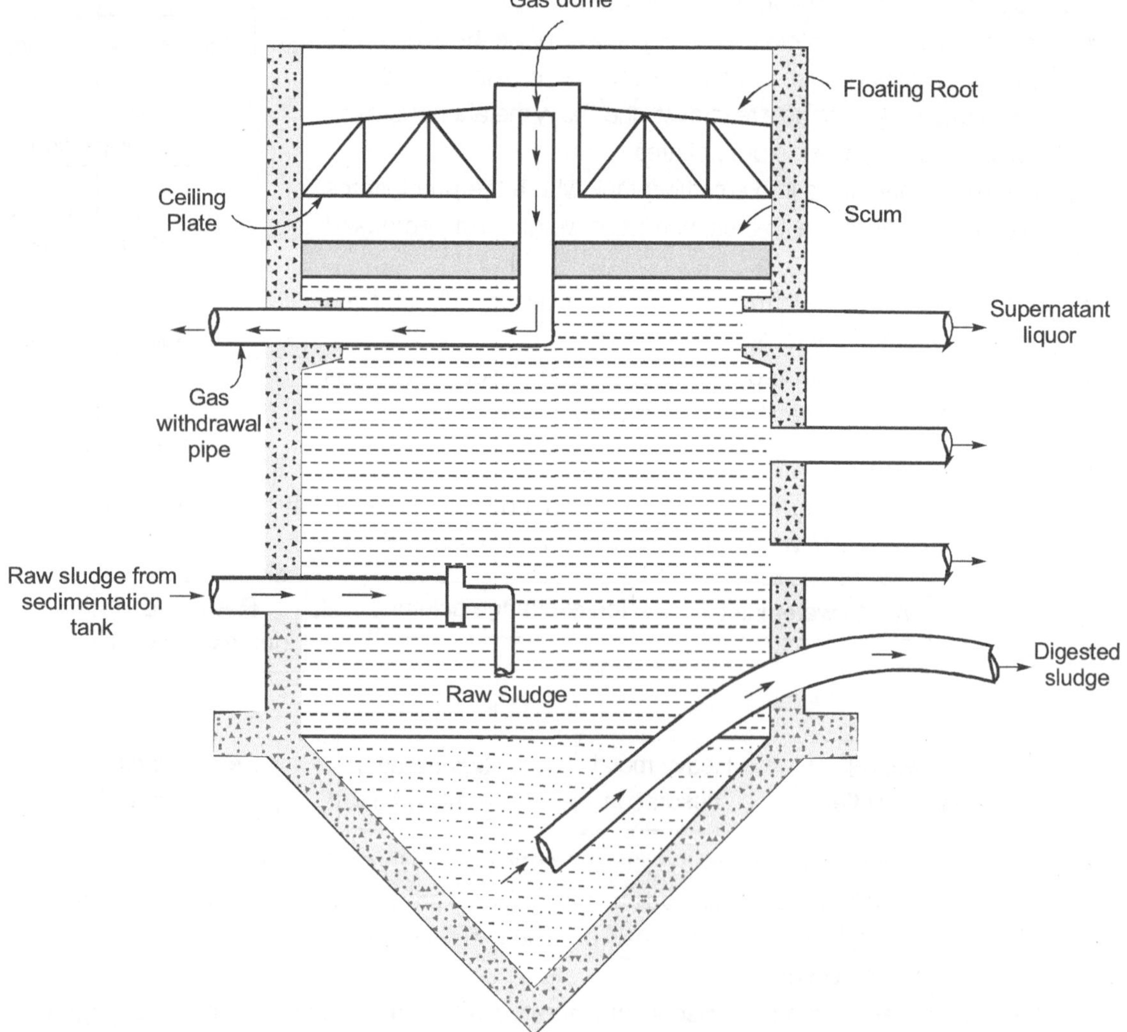

The high rate trickling filters make it possible to pass sewage at greater loadings, thus require lesser space and lesser filter media. The value of hydraulic loading for conventional (standard rate) trickling filters may vary between 22 to 44 M litre/ha/day. The hydraulic loading can still be increased to about 110 to 330 M litre/ha/day in high rate trickling filters. The value of organic loading for conventional filters may vary between 900 to 2200 kg of BOD_5 per ha-m. This organic loading value can be further increased to about 6000 to 18000 kg of BOD_5 per ha-m in high rate trickling filters. The depth of high rate trickling filters varies between 1.2 to 1.8 m.

Bio filters are essentially high rate filters, but comparatively shallower than trickling filters. The depth varies between 1.2 m to 1.5 m (the depth is kept less on the consideration that the main action of treatment is involved in the upper surface layer of the filter). The filter utilizes recirculation of a portion of the filter effluent to the PST for a second passage through the filter. If additional treatment is necessary to lower the BOD content in the effluent, such as in the case of strong sewages, a second stage filter may be provided. Also the quality of the final effluent can also be changed by altering the loading rate and the recirculation ratio. These filters are capable of giving any degree of treatment by appropriate selection of flow diagram and recirculation ratio. Organic loading normally ranges between 9000 to 11000 kg of BOD_5 per ha-m per day. The total hydraulic loading may range between 110 to 330 million litres per day per hectare.

A sludge digestion tank consists of a circular RCC tank with hoppered bottom and having a floating roof over its top. The raw sludge is pumped into the tank, and when the tank is first put into operation, it is seeded with the digested sludge from another tank. A screw pump with an arrangement for circulating the sludge from bottom to top of the tank or vice versa is commonly used for stirring the sludge. Sometimes, power driven mechanical devices may be used for stirring the sludge. A typical sludge digestion tank is shown above.

In cold countries, the tank may have to be provided with heating coils through which hot water is circulated in order that the temperature inside the tank is maintained at optimum digestion temperature level.

The gases of decomposition (chiefly methane and carbon dioxide) are collected in a gas dome. The digested sludge which settles down to the hoppered bottom of the tank is removed under hydrostatic pressure, periodically once a weak or so. The supernatant liquor lying between the sludge and the scum is removed at suitable elevations, through a number of withdrawal pipes. The supernatant liquor, being higher in BOD and suspended solids content is sent back for treatment along with the raw sewage in the treatment plant. The scum formed at the top surface of the supernatant liquor is broken by the recirculating flow or through the mechanical rockers called scum breakers.

9.2 **How do imhoff tanks differ from septic tanks in principle of operation? Mention the advantages and disadvantages of imhoff tanks.**

[10 marks : 2000]

Solution:

An imhoff tank is an improvement over septic tank, in which the incoming sewage is not allowed to get mixed up with the sludge produced and the outgoing effluent is not allowed to carry with it large amount of organic load as in the case of a septic tank. An imhoff tank is infact a two storeyed tank, which is sometimes also known as two storey digestion tank.

The advantages of imhoff tanks are:

(i) Detention period of imhoff tanks is very less compared to septic tanks.

(ii) Imhoff tanks combine the advantages of both the septic as well as sedimentation tanks and as such, they find use in case of small treatment plants requiring only primary treatment.

(iii) They are quite economical and do not require skilled attention during operations.

(iv) The results obtained are quite good, with 60 to 65% removal of solids, and 30 to 40% removal of BOD.

(v) There is no problem of sludge disposal, as in the case of sedimentation tanks.

Imhoff tanks have, however certain disadvantages:

(i) Depth of tank is more which may make the constructions costlier, especially in hard rocks or quick sands. At such places, these tanks may become uneconomical.

(ii) They may give out offensive odours, when improperly operated.

(iii) They are unsuitable and do not function properly where sewage is highly acidic in nature.

(iv) These tanks have a tendency to foam or boil. This may cause the scum to go up to the top of the tank and it may also force the sludge particles to enter the sedimentation chamber through the slot. The foaming may thus, adversely affect the quality of effluent.

9.3 **A trickling filter plant treats 1500 cum per day of sewage with a BOD_5 of 220 mg/L and a suspended solid concentration of 250 mg/L. Estimate the total solid production assuming that primary clarification removes 30% of BOD and 60% of influent solids. Take the solid production in the trickling filter as a 0.5 kg/kg of the applied BOD.**

[10 marks : 2001]

Solution:

Solids removed in primary clarification units of the trickling filter plant

$$= 60\% \text{ of influent suspended solids} = \frac{60}{100} \times 250 = 150 \text{ mg/L}$$

$$= \frac{150 \times 10^{-6}}{10^{-3}} \text{ kg/m}^3 = 150 \times 10^{-3} \text{ kg/m}^3$$

$\therefore$ Solids removed per day $= 150 \times 10^{-3} \times 1500 = 225$ kg

BOD_5 removed in primary clarification = 30%

$\therefore$ BOD applied to filters $= 100 - 30 = 70\%$

$$\therefore \text{Total BOD applied to filters} = 0.70 \times \frac{220 \times 10^{-6}}{10^{-3}} \times 1500 = 231 \text{ kg/day}$$

Solid production in filters $= 0.5$ kg/kg of BOD applied $= 0.5 \times 231 = 115.5$ kg/day

$\therefore$ **Total solid production** = Solid removed in primary clarification + Solids produced in filters

$= 225 + 115.5 = 340.5$ kg/day

9.4 **A waste water plant produces 1000 kg of dry solids per day at a moisture content of 95%. The solids are 70% volatile with a specific gravity of 1.05 and the remaining are non-volatile with a specific gravity of 2.5. Find the sludge volume after digestion, which reduces volatile solids content by 50% and decreases the moisture content to 90%.**

[15 marks : 2001]

Solution:

Total solids produced = 1000 kg per day

Volatile solids = 70% of total solids produced

$$= \frac{70}{100} \times 1000 = 700 \text{ kg}$$

$\therefore$ Non-volatile solids $= 1000 - 700 = 300$ kg

Now, volatile solids removed in digestion = 50% of volatile solids $= \frac{50}{100} \times 700 = 350$ kg

$\therefore$ Volatile solids left in the digested sludge $= 700 - 350 = 350$ kg

Non-volatile solids in the digested sludge = 300 kg (as in original sludge)

We know that 90% moisture content of digested sludge indicates that in a 100 kg digested sludge, 90 kg is water and remaining 10 kg are solids.

Total amount of solids in the digested sludge = 350 + 300 = 650 kg

$\therefore$ Mass of water in the digested sludge = $650 \times \frac{90}{10} = 5850 \text{ kg} = M_w$

where, M_w = Mass of water

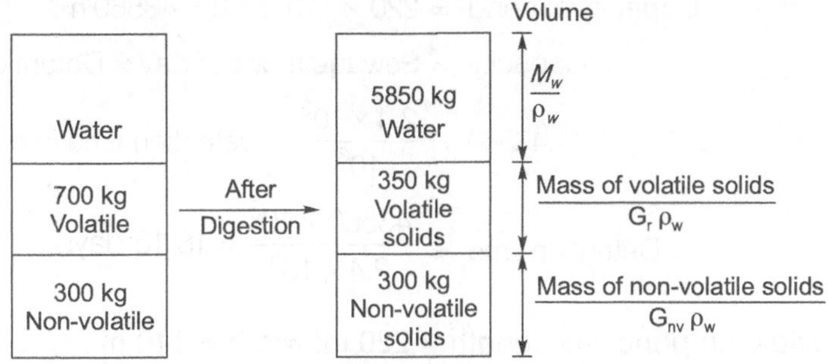

$$\text{Density of volatile solids} = G_v \times \rho_w = 1.05 \times 1000 = 1050 \text{ kg/m}^3$$

Density of non-volatile solids = $G_{nv} \times \rho_w = 2.5 \times 1000 = 2500 \text{ kg/m}^3$

Volume of volatile solids in the digested sludge

$$= \frac{\text{Mass}}{\text{Density}} = \frac{350}{1050} = 0.333 \text{ m}^3$$

Volume of non-volatile solids in the digested sludge = $\frac{300}{2500} = 0.12 \text{ m}^3$

Volume of water in digested sludge = $\frac{5850}{1000} = 5.85 \text{ m}^3$

$\therefore$ **Sludge volume after digestion** = $0.333 + 0.12 + 5.85 = 6.303 \text{ m}^3$

9.5 **Population of a town is 20,000 with an assured water supply of 150 litre per head per day. BOD of the waste water is 150 mg/L. Design the most suitable waste water treatment system (without power supply) for the town.**

Which symbiotic system is applied in waste water treatment? Explain its principle and under what conditions the process may be found suitable for a community.

[8 + 7 = 15 marks : 2002]

Solution:

For the effective removal of BOD either trickling filter or oxidation pond may be designed. The waster water treatment plant should be without any power supply as given in the problem. Hence, oxidation pond may be designed as given below:

The quantity of water supplied per day = $20000 \times 150 = 3 \times 10^6$ litres

Assuming 80% of the water supplied is converted in sewage

$\therefore$ Quantity of sewage produced = $0.80 \times 3 \times 10^6 = 2.4 \times 10^6$ litres

BOD content of waste water = $2.4 \times 10^6 \times 150 \times 10^{-6} = 360$ kg/day

Assuming the organic loading in the pond is 150 kg/ha/day.

$$\therefore \quad \text{Surface area required} = \frac{\text{BOD content}}{\text{organic loading}} = \frac{360}{150} = 2.4 \text{ ha} = 24000 \text{ m}^2$$

Assuming length of the pond (L) as twice of its width (B).

$$\therefore \quad L \times B = 24000$$

$$\Rightarrow \quad 2 \times B \times B = 24000$$

$$\Rightarrow \quad B = 109.5 \text{ m say } 110 \text{ m}$$

$$\therefore \quad L = 2 \times B = 220 \text{ m}$$

Using a pond with effective depth as 1.8 m, we get

$$\text{Capacity of pond} = 220 \times 110 \times 1.8 = 43560 \text{ m}^3$$

But

$$\text{Capacity} = \text{Sewage flow per day} \times \text{Detention time in days}$$

$$\therefore \quad 43560 = \frac{2.4 \times 10^6}{10^3} \times \text{Detention time in days}$$

$$\Rightarrow \quad \text{Detention time} = \frac{43560 \times 10^3}{2.4 \times 10^6} = 18.15 \text{ days}$$

Hence using an oxidation pond with length = 220 m; width = 110 m and over all depth = 1.8 + 1 = 2.8 m and a detention period of 18.15 days.

Design of inlet and outlet pipe:

Assuming an average velocity of sewage as 0.9 m/s and daily flow for 8 hrs only

$$\therefore \quad \text{Discharge} = \frac{2.4 \times 10^6}{10^3 \times 8 \times 60 \times 60} = 0.0833 \text{ m}^3/\text{sec}$$

$$\therefore \quad \text{Area of inlet pipe required} = \frac{\text{Discharge}}{\text{Velocity}} = \frac{0.0833}{0.9} = 0.09 \text{ m}^2$$

$$\therefore \quad \text{Diameter of inlet pipe} = \sqrt{\frac{4 \times 0.09 \times 10^4}{\pi}} = 33.85 \text{ cm} \approx 34 \text{ cm}$$

Diameter of outlet pipe may be takes as 1.5 times that of inlet i.e. 51 cm.

Algal symbiosis or algal photosynthesis is applied in waste water treatment. The principle is that in a total aerobic pond, the stabilization of waste is brought about by aerobic bacteria, which flourish in the presence of oxygen.

The oxygen demand of such bacteria in such a pond is met by the combined action of algae and other micro-organisms such as bacteria and protozoa. This is called as algal symbiosis. In this symbiosis, the algae growing in the presence of sunlight produce oxygen by action of photosynthesis and this oxygen is utilized by the bacteria for oxidising the waste organic matter. The end products of the process are carbon dioxide, ammonia and phosphates which are required by algae to grow and continue to produce oxygen.

This process can be applied in treating sewage and biodegradable waste waters.

9.6 **Design an oxidation pond with inlet and outlet pipes for treating sewage from a residential colony with 500 persons contributing sewage at 120 litres/capita/day. The 5-day BOD of sewage is 300 ppm. Assume velocity of sewage flow as 0.9 m/s, specific gravity of organic load as 1, and daily flow for 8 hrs organic loading in the pond at 300 kg/ha/day.**

[15 marks : 2003]

Solution:

Quantity of sewage to be treated per day = $500 \times 120 = 0.06 \times 10^6$ litres

$$\text{BOD content} = 0.06 \times 10^6 \times 300 \times 10^{-6} = 18 \text{ kg/day}$$

$$\text{Organic loading in the pond} = 300 \text{ kg/ha/day (given)}$$

$$\therefore \quad \text{Surface area required} = \frac{\text{BOD content}}{\text{Organic loading}} = \frac{18}{300} = 0.06 \text{ ha} = 600 \text{ m}^2$$

Assuming length of tank (L) as twice of its width (B),

$$L \times B = 600$$
$$\Rightarrow \quad 2B^2 = 600$$
$$\Rightarrow \quad B = 17.32 \text{ m say } 17.5 \text{ m}$$
$$\therefore \quad L = 2B = 2 \times 17.5 = 35 \text{ m}$$

Using a tank with effective depth as 1.5 m, we get

$$\text{Capacity of tank} = 35 \times 17.5 \times 1.5 = 918.75 \text{ m}^3$$

$$\therefore \quad \text{Detention time in days} = \frac{\text{Capacity of tank}}{\text{Sewage flow per day}} = \frac{918.75 \times 10^3}{0.06 \times 10^6} = 15.31 \text{ days}$$

Hence, using a oxidation pond with length = 35 m, width = 17.5 m

Overall depth = 1.5 + 1 = 2.5 m and a detention period of 15.31 days.

Design of inlet pipe and outlet pipe

Velocity of sewage flow = 0.9 m/s (given)

Time of daily flow = 8 hrs (given)

$$\therefore \quad \text{Discharge} = \frac{0.06 \times 10^6}{10^3 \times 8 \times 60 \times 60} = 2.083 \times 10^{-3} \text{ m}^3/\text{sec}$$

$$\therefore \quad \text{Area of inlet pipe required} = \frac{2.083 \times 10^{-3}}{0.9} = 2.3144 \times 10^{-3} \text{ m}^2 = 23.14 \text{ cm}^2 \simeq 23.15 \text{ cm}^2$$

$$\therefore \quad \text{Diameter of inlet pipe} = \sqrt{\frac{23.15 \times 4}{\pi}} = 5.43 \approx 6 \text{ cm}$$

Diameter of outlet pipe may be assumed as 1.5 times that of inlet pipe i.e. 9 cm.

9.7 **Find out the volume of an anaerobic digestion tank for 5 MLD of domestic waste water treatment plant having 60% suspended solid removal efficiency of primary clarifier and 250 mg/litre. Suspended solids in waste water based on sludge volume reduction in digester. Moisture content of influent sludge is 96%, initial volatile solids content in sludge is 70%, volatile solids destroyed is 65%, digested sludge solid concentration is 80%, specific gravity of primary sludge is 1.03, specific gravity of digested sludge is 1.04, density of water is 1000 kg/m³, mean cell residence time in days is 15.**

[20 marks : 2006]

Solution:

Average sewage flow = 5 MLD

Mass of suspended solids in 5 M litres of waste water = $250 \times 10^{-6} \times 5 \times 10^6$ = 1250 kg per day

Suspended solid removal efficiency of primary clarifier = 60% (given)

$$\therefore \text{ Mass of solids removed in primary clarifier} = \frac{60}{100} \times 1250 = 750 \text{ kg per day}$$

Now moisture content of influent sludge = 96% (given)

i.e. 4 kg of dry solids will make 100 kg of wet sludge

$\therefore$ 750 kg of dry solids will make $= \frac{100}{4} \times 750 = 18750$ kg of wet sludge per day

Specific gravity of primary (wet) sludge = 1.03 (given)

Density of water = 1000 kg/m^3 (given)

$\therefore$ Density of primary (wet) sludge = 1.03 × 1000 = 1030 kg/m^3

$\therefore$ Volume of primary sludge produced is given by

$$V_1 = \frac{18750}{1030} \text{ m}^3/\text{day} = 18.20 \text{ m}^3/\text{day}$$

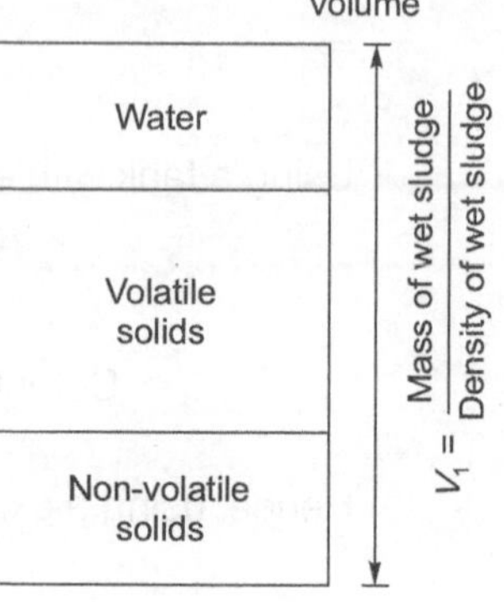

Initial volatile solids content in sludge = 70% (given)

Mass of volatile solids in sludge $= \frac{70}{100} \times 750 = 525$ kg

$\therefore$ Mass of non-volatile solids in sludge = 750 – 525 = 225 kg

Volatile solids destroyed = 65% (given)

Mass of digested volatile solids $= \frac{65}{100} \times 525 = 341.25$ kg

$\therefore$ Mass of undigested volatile solids = 525 – 341.25 = 183.75 kg

Mass of solids in digested sludge = 225 + 183.75 = 408.75 kg

Digested sludge solid concentration = 8.0% (given)

$\because$ 8 kg of dry solids will make 100 kg of digested sludge

$\therefore$ 408.75 kg of dry solids will make $= \frac{100}{8} \times 408.75$

= 5109.375 kg of digested sludge per day

Specific gravity of digested sludge = 1.04 (given)

$\therefore$ Density of digested sludge = 1.04 × 1000 = 1040 kg/m^3

$\therefore$ Volume of digested sludge produced is given by

$$V_2 = \frac{5109.375}{1040} = 4.91 \text{ m}^3/\text{day}$$

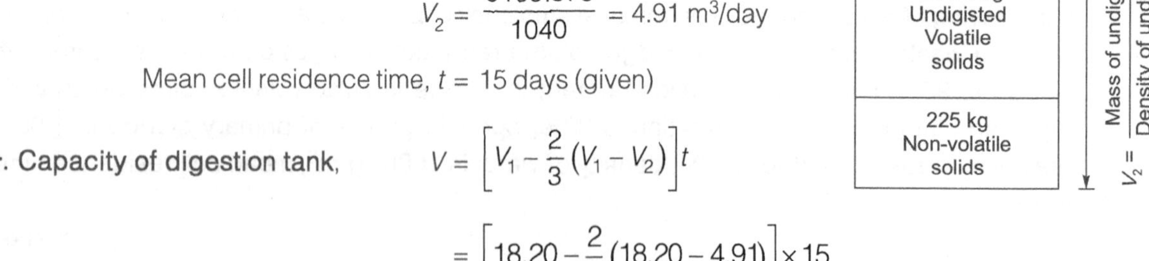

Mean cell residence time, t = 15 days (given)

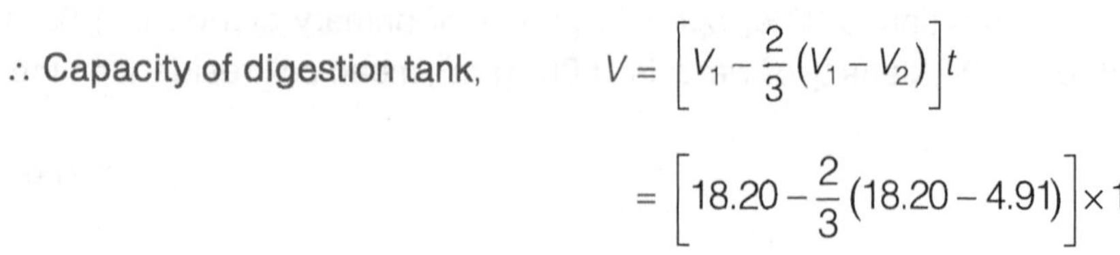

$\therefore$ **Capacity of digestion tank,**

$$V = \left[V_1 - \frac{2}{3}(V_1 - V_2)\right]t$$

$$= \left[18.20 - \frac{2}{3}(18.20 - 4.91)\right] \times 15$$

$$= 140.10 \text{ m}^3$$

9.8 Mention the operational troubles of a standard rate trickling filter and their remedies.

[8 marks : 2007]

Solution:

The major operational troubles and their remedies of a standard rate trickling filter are :

(i) Fly Nuisance:

- The filter fly psychoda which develops in the filter particles, may prove to be nuisance, as the same may be carried away into the habitation.

- This problem may be controlled by flooding the filter with sewage for 24 hours or more. The flooding will destroy the larvae, and usually interferes slightly with the results of operation.
- Another method of controlling fly nuisance is by using insecticides such as DDT, Chlorodane and BHC in the filter plant.

(ii) Odour Nuisance:

- Odours generally do not prevail in trickling filters using rotary distributors; but however when fixed nozzles are used, H_2S and other odorous gases are frequently released from the sprays into the atmosphere.
- The usual remedy is to chlorinate the sewage to prevent formation of H_2S or to neutralize that already formed.

(iii) Ponding trouble:

- Sometimes, the voids in the filter media gets clogged due to heavy growth of fungi and algae. This may result in ponding of the sewage over the filter bed.
- This trouble can be controlled by chlorinating the sewage which kills the algae, thus causing unloading of the accumulated material.
- Another method of controlling algae is to add $CuSO_4$ to the sewage and resting the bed for sometime.

9.9 Design a septic tank for 200 persons with a water supply of 125 litre per capita per day. Assume any other data and mention it.

[10 marks : 2008]

Solution:

Quantity of water supplied = Population × rate of water supply per capita
= 200 × 125 = 25000 litres per day

Assuming that 80% of the water supplied is converted into sewage.

$$\text{Quantity of sewage produced} = 25000 \times \frac{80}{100} = 20000 \text{ litres per day}$$

Assuming the detention time to be 24 hours.

$$\text{Quantity of sewage produced during detention period} = 20000 \times \frac{24}{24} = 20000 \text{ litres}$$

Assuming rate of accumulation of sludge = 30 litres per capita per year

Also, assuming the period of cleaning = 1 year

∴ Volume of sludge deposited = 200 × 30 × 1 = 6000 litres

∴ Capacity of the septic tank required = Volume of sewage produced + Volume of sludge deposited
$= 20000 + 6000 = 26000 \text{ litres} = 26000 \times 10^{-3} \text{ m}^3 = 26 \text{ m}^3$

Assuming the depth of the tank = 1.5 m

$$\therefore \quad \text{Surface area of tank} = \frac{\text{Volume of tank}}{\text{Depth of tank}} = \frac{26}{1.5} = 17.33 \text{ m}^2$$

Assuming the ratio of length to width as 3 : 1, we get

$$3B \times B = 17.33$$

$$\Rightarrow \quad B = 2.4 \text{ m}$$

$$\therefore \quad L = 3 \times 2.4 = 7.2 \text{ m} \simeq 7.3 \text{ m}$$

Thus, the dimensions of the septic tank will be
7.3 m × 2.4 m × (1.5 + 0.3) m overall depth 0.3 m is used as free board
Hence a tank of size 7.3 m × 2.4 m × 1.8 m is provided.

9.10 **A mixed liquor with 2000 mg/L of suspended solids has the settled volume of 200 mL from a litre of this mixed liquor. Calculate its sludge volume index. Is it safe?**

[2 marks : 2010]

Solution:

Settled sludge volume, $V_{ob} = 200\text{ mL}$

Concentration of suspended solids in mixed liquor,

$$X_{ob} = 2000\text{ mg/L}$$

Sludge volume Index is given by $SVI = \frac{V_{ob}}{X_{ob}} \times 1000\text{ mL/g} = \frac{200}{2000} \times 1000\text{ mL/g} = 100\text{ mL/g}$

The usual adopted range of *SVI* is between 50 – 150 mL/g and such a value indicates good settling sludge. In this case *SVI* is 100 mL/g, thus it is safe as it is a good settling sludge.

9.11 **On the basis of a detention period of 24 hours, determine the size (assuming length to width ratio of around 2 and depth of waste water about 1 m) of a septic tank required for a large house dwelling 100 persons. The flow into the tank may be assumed at the rate of 70 LPCD. What will be the surface loading and equivalent weir loading of the tank?**

[8 marks : 2010]

Solution:

Quantity of sewage supplied = Population × Rate of sewage supply per capita
= 100 × 70 = 7000 litres per day

Detention time = 24 hours

Quantity of sewage produced during detention period $= 7000 \times \frac{24}{24} = 7000$ litres

Assuming rate of accumulation of sludge = 30 litres per capita per year

Also, assuming the period of cleaning = 1 year

∴ Volume of sludge deposited = 100 × 30 × 1 = 3000 litres

∴ Capacity of the septic tank required = volume of sewage produced + volume of sludge deposited
$= 7000 + 3000 = 10000 \text{ litres} = 10\text{ m}^3$

Depth of waste water = 1 m

Hence assuming the depth of tank as 1.5 m

$$\therefore \quad \text{Surface area of tank} = \frac{\text{Volume of tank}}{\text{Depth of tank}} = \frac{10}{1.5} = 6.67\text{ m}^2$$

Given that length to width ratio i.e. $\frac{L}{B} = 2 \Rightarrow L = 2B$

$$\therefore \quad L \times B = 6.67$$

$$\Rightarrow \quad 2B \times B = 6.67$$

$$\Rightarrow \quad B = 1.826\text{ m} \approx 1.83\text{ m}$$

$$\therefore \quad L = 2B = 2 \times 1.83 = 3.66\text{ m} \approx 3.7\text{ m}$$

Thus the dimensions of the septic tank will be 3.7 m × 1.83 m × (1.5 + 0.3) m

0.3 m is used as freeboard.

Hence a septic tank of size 3.7 m × 1.83 m × 1.8 m is provided

$$\textbf{Surface overloading} = \frac{Q}{BL} = \frac{7000}{1.83 \times 3.7} = 1033.821\text{ litres/m}^2\text{/day}$$

$$\textbf{Equivalent weir loading rate} = \frac{Q}{B} = \frac{7000\,l\text{/day}}{1.83\text{ m}} = 3825.14\ l\text{/m/day} = 3.83\text{ m}^2\text{/m/day}$$

9.12 **Explain the following terms generally adopted in waste water collection system :**

A. (1) Sewage (2) Sewers (3) Sewerage system. Briefly mention the principles involved in 'Combined sewer design'.

B. A site adjacent to a river is considered for the location of a municipal waste water treatment plant having an average inflow of 2500 m³/day. The selected treatment process sequence is as follows:

(1) grit channel (2) primary sedimentation (3) activated sludge and (4) secondary sedimentation. Design the grit channels assuming maximum hydraulic loading as twice the average inflow rate. Assume other appropriate design parameters, if required.

[5+5=10 marks : 2011]

Solution:

A. (1) **Sewage:** The mixture of water and waste products of society including the human excreta is called sewage. It is classified into two types.

(a) **Domestic Sewage:** Liquid wastes originating from the residential, commercial or institutional buildings.

(b) **Industrial Sewage:** Liquid wastes originating from the industrial processes of various industries, such as Dyeing, paper making, Brewing etc.

(2) **Sewers:** Sewers are underground pipes that is used to carry sewage from the source of origin to treatment units. Sewers are designed to flow under gravity.

(3) **Sewerage System:** It is a network of sewer pipes laid in order to collect, treat and finally dispose of the sewage. The network of sewers may consist of house sewers, lateral sewers, branch sewers, main sewers, and outfall sewers etc.

Principles involved in 'Combined Sewer Design' are:

(i) A combined system requires laying only one set of bigger sized conduit, thus making the system cheaper.

(ii) The sewer pipes in the combined system are liable to frequent silting during the non-monsoon season (when the flows in them are quite less) so they are laid at sufficiently steeper slopes, which in turn, makes them deeper, requiring more excavation.

(iii) In a combined system, the less-foul drainage water gets mixed with the highly foul sewage water, thus necessitating the treatment of the entire flow, needing more capacity for the treatment plant.

(iv) A combined sewer is laid deeper and carries more water therefore adequate pumping requirements are to be provided.

B.

$$\text{Maximum hydraulic loading} = 2 \times 2500 = 5000 \text{ m}^3/\text{day} = \frac{5000}{24 \times 60 \times 60} = 0.058 \text{ m}^3/\text{sec}$$

Assuming the flow velocity through the tank as

$$V_H = 0.2 \text{ m/sec} \text{ and detention period of 2 minutes.}$$

Length of the grit chamber,

$$L = \text{Velocity} \times \text{Detention time}$$

$$= 0.2 \times (2 \times 60) = 24 \text{ m}$$

$$\text{Cross sectional area required} = \frac{\text{Discharge}}{V_H} = \frac{0.058}{0.2} = 0.29 \text{ m}^2$$

Assuming the water depth in the chamber to be 1 m.

$$\text{Width} = \frac{\text{Cross-sectional area}}{\text{Depth}} = \frac{0.29}{0.1} = 0.29 \text{ m} = 29 \text{ cm}$$

Providing a free board of 0.3 m.

Total dimension of grit channel is 24 × 0.29 × 1.3 m

9.13 **Explain the process of Anaerobic sludge digestion. Name the experiments to be performed in the laboratory to determine the digestibility of sludge.**

[10 marks : 2012]

Solution:

Anaerobic digestion is by far the most common process for dealing with wastewater sludges containing primary sludge.

The process of anaerobic sludge digestion takes place following three stages:

(a) Hydrolysis: In this stage the complex organic matter like proteins, cellulose etc. into soluble simple organic matters.

(b) Acid fermentation: In this stage, the soluble organic matter formed in the first stage of hydrolysis is converted into acids and alcohol of low molecular weight. This done with help of acid formers.

(c) Methane formation: The methane formers consist of strict anaerobic bacteria that convert acid and alcohol, along with hydrogen and carbon dioxide, to methane.

The experiments to be performed in the laboratory to determine the digestibility of sludge are:

(a) BOD measurement

(b) Test for total, suspended and volatile solids determination.

(c) pH test

(d) Temperature

(e) Determination of Alkalinity as $CaCO_3$

9.14 **Given the following data:**

Plant inflow = 9000 m^3/day

Recycle flow = 6000 m^3/day

Mixed liquid suspended solids (MLSS) = 4000 mg/L

Sludge volume withdrawn = 580 m^3/day

Suspended solids = 3800 kg day

Solid loading rate of 150 kg/m^2-day at average flow

Obtain the final settling tank size for an extended aeration plant.

[8 marks : 2014]

Solution:

Given: Q_0 = 9000 m^3/day, Q_R = 6000 m^3/day, MLSS, X = 4000 mg/L

and Solid loading rate = 150 kg/m^2-day

Assuming over flow rate for 2° extended aeration tank as 25 m^3/m^2/day for average depth of 4.0 m

∴ Overflow rate = 25 m^3/m^2/day

Depth of tank = 4.0 m

The surface adopted should be maximum of surface are a obtained from overflow rate and solid loading rate.

(i) $$A = \frac{Q_0}{\text{over flow rate}} = \frac{9000}{25} = 360\,\text{m}^2$$ ...(Based on over flow rate)

(ii) $$A = \frac{(Q_0 + Q_R)X}{\text{solid loading rate}}$$ ...(Based on solid loding rate)

(ii) $$A = \frac{(9000 + 6000) \times 10^3 \times 4000 \times 10^{-6}}{150} = 400\,\text{m}^2$$

Hence, adopted surface area of settling tank is 400 m^2.

Taking
$$\frac{L}{B} = 2, \text{ then}$$
$$L \times B = 400$$
$$2B^2 = 400$$
$$B = 14.14 \simeq 14.15 \text{ m}$$
$$\therefore \quad L = 28.3 \text{ m}$$

Thus, the final size of settling tank is 28.3 m × 14.15 m × 4 m.

9.15 **A wastewater plant produces 1000 kg of dry solids/day at a moisture content of 95%. The solids are 65% volatile with a specific gravity of 1.05 and the specific gravity of the non-volatile portion is 2.5. Determine the sludge volume:**

(*i*) as produced;

(*ii*) after digestion reduced the volatile solids by 50% and decreases moisture content by 90%.

(*iii*) after dewatering to 75% moisture;

(*iv*) after drying to 10% moisture;

(*v*) after incineration (only non-volatile solids remain).

[8 marks : 2014]

Solution:

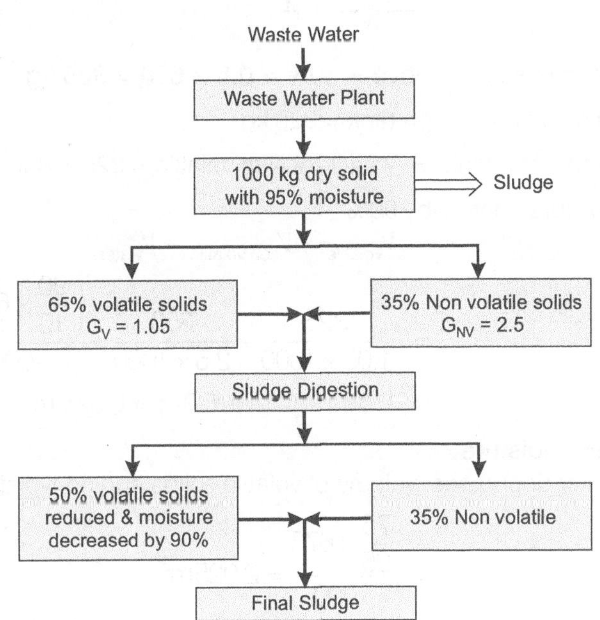

$$\text{Total solid produced} = 1000 \text{ kg}$$
$$\text{Volatile solid} = 65\% \text{ of total produced} = 0.65 \times 1000 = 650 \text{ kg}$$
$$\text{Non volatile solid} = 35\% \text{ of total solid produced} = 0.35 \times 1000 = 350 \text{ kg}$$

(i) Sludge volume produced just after treatment

$$V = V_{volatile} + V_{non\ volatile} + V_{water}$$
$$V_{volatile} = \frac{650 \text{kg}}{G_V \times 1000} = \frac{650}{1.05 \times 1000} = 0.619 \text{m}^3$$
$$V_{non\ volatile} = \frac{350}{G_{NV} \times 1000} = \frac{350}{2.5 \times 1000} = 0.14 \text{m}^3$$

Moisture content of sludge is 95%. It mean 5 kg of dry solid needed 95 kg of water to make 100 kg of sludge.

$\therefore$ 1000 kg of dry solids will have $\frac{95}{5} \times 1000\,\text{kg}$ of water

Hence, $$V_{water} = \frac{\left(95 \times \frac{1000}{5}\right)}{1000} = 19\,\text{m}^3$$

Hence volume of sludge, $V = 0.619 + 0.14 + 19 = 19.759\ \text{m}^3$

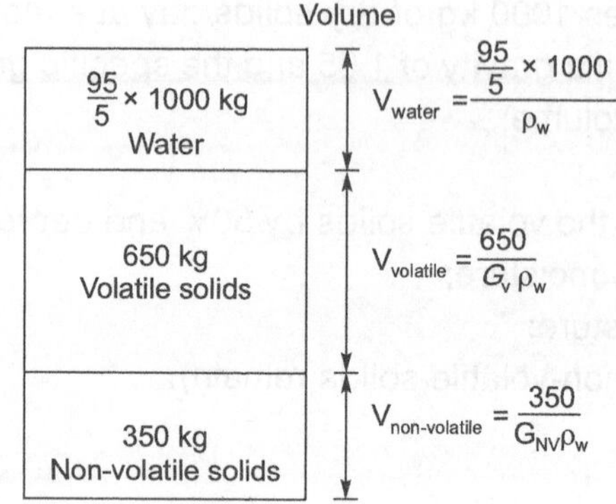

(ii) After digestion volatile solid remove = 50% = 0.5 × 650 = 325 kg

Volatile solid remaining in digestion tank = 325 kg

Total solid in degestion tank = Volatile + Non volatile = 325 + 350 = 675 kg

and decreases moisture content by 90%

Now, $$V = V_{volatile} + V_{non\ volatile} + V_{water}$$

$$= \frac{325}{1.05 \times 1000} + \frac{350}{2.5 \times 1000} + \frac{\left(\frac{90}{10} \times 675\right)}{1000}$$

$$= 0.309 + 0.14 + 6.075 = 6.524\ \text{m}^3$$

(iii) Dewatering to 75% moisture

(Only volume of water decreases, volume of volatile solids and nonvolatile remains same)

$$V_{water} = \frac{\frac{75}{25} \times 675}{1000} = 2.025\,\text{m}^3$$

$$\therefore \quad V = 0.309 + 0.14 + 2.025 = 2.474\ \text{m}^3$$

(iv) Drying to 10% moisture

(Volume of volatile and nonvolatile voids remains same)

$$V_{water} = \frac{\frac{10}{90} \times 675}{1000} = 0.075\,\text{m}^3$$

$$\therefore \quad V = 0.309 + 0.14 + 0.075 = 0.524\ \text{m}^3$$

(v) After incineration,

$$V_{volatile} = 0 \text{ and } V_{water} = 0$$

$$\therefore \quad V = 0.14\,\text{m}^3$$

9.16 **What is grit? Why should grit be removed from wastewater? What is the basic principle behind the design of grit chambers? What is the reason to have constant velocity of flow in grit chamber [conventional horizontal flow] and how is it achieved?**

[8 marks : 2015]

Solution:

Grit chambers or Grit channels, as they are usually called, are the sedimentation basins placed usually after the fine screens and certainly, before the primary sedimentation tank. The grit chamber removes the inorganic grit such as sand, gravel, and other mineral matter that has a nominal diameter of 0.15 to 0.20 mm or more.

Grit removal basins, such as Grit chamber or Grit channels or Detritus tank are the sedimentation basins placed in front of the wastewater treatment plant to remove the inorganic particles such as sand, gravel, grit, shells, bones and other non-putresible materials that may clog channels or damage pumps due to abrasion and to prevent their accumulation in sludge digestors.

Grit chambers are infact, nothing but like sedimentation tanks designed to separate the intended heavier inorganic materials by the process of sedimentation due to gravitation forces and to pass forward the lighter organic materials.

The basic principle in the design of the grit chambers is that the flow velocity should neither be too low as to cause the settling of lighter matter organic matter, nor should it be so high as not to cause the settlement of the entire silt and grit present in sewage.

If there are large variations in discharge, then the grit chamber is designed for generating optimum velocity at peak discharge and a velocity control section, such as a properly designed modified weir, called a proportional flow weir or a sutro weir, is provided at the lower end of the rectangular grit channel, which helps in varying the flow area of the section in direct proportion to the flow, and thus, helps to maintain a constant velocity in the channel (within the permissible limits of ± 5 to 10% over the designed value), even at varying discharges.

9.17 **A wastewater treatment plant operating primary settling tank gets an inflow of 12.96 million liters per day with influent suspended solids concentration of 280.0 mg/L. The primary settling tank has a removal efficiency of 59% with sludge concentration of 5%. If the volatile solids concentration in settled sludge is 60% and the specific gravity of volatile solids is 0.990; and the fixed solids concentration is 40% and the specific gravity of fixed solids is 2.65, determine the daily sludge production.**

[8 marks : 2015]

Solution:

Total suspended solid produced $= 12.96 \times 10^6 \times 280 \times 10^{-6} = 3628.8$ kg

$\because$ PST removal efficiency is 59%

$\therefore$ Total produced sludge in PST $= 0.59 \times 3628.8 = 2140.992$ kg

Volatile solid concentration $= 0.6 \times 2140.992 = 1284.595$ kg

Volume of volatile solid concentration,

$$V_1 = \frac{1284.595}{0.990 \times 1000} = 1.2975 \text{m}^3$$

Fixed solid concentration $= 0.4 \times 2140.992 = 856.3968$ kg

$\therefore$ Volume of fixed solid cocentration,

$$V_2 = \frac{856.3968}{2.65 \times 1000} = 0.3231 \text{m}^3$$

Raw sludge has 95% moisture and remaining 5% as solid content.

It means 5 kg of dry solid needed 95 kg of water to make 100 kg of sludge.

Hence 2140.992 kg produced solids will have water

$$= \frac{95}{5} \times 2140.992 = 40678.848 \text{ kg}$$

$\therefore$ Volume of water in raw sludge, $V_3 = \dfrac{40678.848}{1000} = 40.679\,m^3$

Hence, **volume of daily sludge production** $= V_1 + V_2 + V_3$

$= 1.2975 + 0.3231 + 40.679$

$= 42.2996\,m^3$

9.18 **Design a circular primary settling tank for a sewage treatment plant. Given data are:**

Flow rate = 27000 m^3/d Retention time = 2.5 hr Weir length = 30 m

Total number of tanks = 2 Surface loading rate = 30 $m^3/m^2/d$

[4 marks : 2016]

Solution:

Flow rate, $Q_0 = 27000\,m^3/day$

Retention time, $t_d = 2.5\,hr$

Volume of tank, $V = Q_0 t_d = 27000 \times \dfrac{2.5}{24}$

$V = 2812.5\,m^3$

Volume of each tank, $\dfrac{V}{2} = \dfrac{2812.5}{2} = 1406.25\,m^3$

Surface Loading Rate, $SLR = 30\,m^3/m^2/d$

Surface area of one tank $= \dfrac{Q_0/2}{SLR}$

$\Rightarrow \quad \dfrac{\pi d^2}{4} = \dfrac{27000}{2 \times 30} = 450\,m^2$

$\therefore \quad d = 23.9 \simeq 24\,m$

Depth of each tank $= \dfrac{V}{\text{Surface area}} = \dfrac{1406.25}{450} = 3.125\,m$

Hence $d = 24\,m$, $H = 3.125\,m$ } for 2 Tanks

9.19 **Design a high rate trickling filter for treating wastewater with BOD of 160 mg/*l*. The rate of wastewater generated is 5 MLD. The effluent BOD required is 25 mg/*l*. Adopt recirculation ratio of 1.5.**

[10 marks : 2017]

Solution :

Total BOD in raw sewage (wastewater) $= 160\,mg/l \times 5 \times 10^{+6} \times 10^{-6} = 800\,kg/day$

Assume, BOD removed in primary clarifier = 25%

$\therefore$ BOD entering per day in the filter units $= 0.75 \times 800 = 600\,kg$

Permissible BOD concentration in the effluents = 25 mg/*l*

$\therefore$ BOD allowed to go into the effluent $= 25\,mg/l \times 5 \times 10^6\,l = 125\,kg$

$\therefore$ BOD removed by filter per day $= 600\,kg - 125\,kg = 475\,kg$

Hence, Efficiency of filter $= \dfrac{\text{BOD removed}}{\text{Total BOD applied}} \times 100$

$$= \frac{475}{600} \times 100 = 79.2\%$$

But, we know that

$$\eta = \frac{100}{1 + 0.0044\sqrt{\frac{Y}{VF}}}$$

where Y = Total BOD applied to the filter per day in kg
V = Filter volume in ha.m
F = Recirculation factor

Given, Recirculation ratio, $\frac{R}{I} = 1.50$

$$\therefore \quad F = \frac{1 + \left(\frac{R}{I}\right)}{\left[1 + 0.1\left(\frac{R}{I}\right)\right]^2} = \frac{[1 + 1.5]}{[1 + 0.1 \times 1.5]^2} = 1.89$$

$$\therefore \quad 79.2 = \frac{100}{1 + 0.0044\sqrt{\frac{600}{V \times 1.89}}}$$

$$\Rightarrow \quad 1 + 0.0044\sqrt{\frac{317.46}{V}} = 1.263$$

$$\Rightarrow \quad V = 0.08886 \text{ ha.m} = 888.6 \text{ m}^3$$

Using 1.5 m depth of filter, we get

$$\text{Area required} = \frac{888.6}{1.50} = 592.4 \text{ m}^2$$

$$\therefore \quad \text{Diameter of filter tank required} = \sqrt{\frac{4 \times 592.4}{\pi}} \simeq 27.464 \text{ m} \simeq 27.50 \text{ m}$$

$\therefore$ Provide a filter unit of diameter 27.5 m and depth 1.5 m.

9.20 **What is sludge and how is it classified? With the help of flow chart, explain different processes/ operations involved in sludge treatment and indicate the objectives of each process/operation.**

[12 marks : 2018]

Solution:

Sludge in a mixture of liquids and solids, which may be volatile or non-volatile.

The sludge which is deposited in primary sedimentation tank is called raw sludge and the sludge which is deposited in a secondary clarifier is called the secondary sludge.

Raw sludge is odourous, contains highly putrescible matter and is thus very objectionable. Secondary sludge is also putrescible, through a little less objectionable.

Treatment Process:

1. **Thickening:** Thickening is often the first step in a sludge treatment process. Sludge from primary or secondary clarifier may be stirred to form large, more rapidly settling aggregates.
2. **Dewatering:** Water contents of sludge may be reduced by centrifugation, filtration and or evaporation to reduce transportation cost of disposal or to improve suitability for composting.

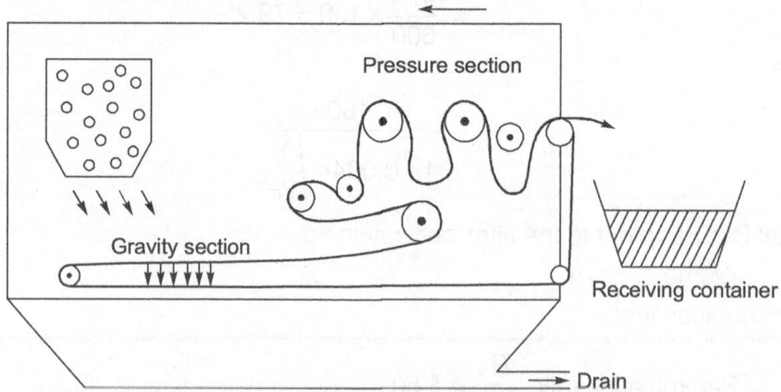

Schematic diagram of belt filter less to dewater the sludge.

3. **Digestion:** Sludges are treated using variety of techniques of digestion, the purpose of which is to decrease the amount of organic matter present in the solids.

   The most common being the anaerobic digestion, aerobic digestion and composting. Sludge digestion offers significant cost advantages by reducing the sludge quantity nearly by 50% and providing biogas as valuable energy source.

   Stages in sludge digestion process:
   - Acid fermentation
   - Acid regression
   - Alkaline fermentation

4. **Sludge stabilization**

   It means to stabilize the sludge and decreasing the pathogenic bacterial content.

5. **Sludge drying**

   Digested sludge from digestion tank is dewatered or dried up before further disposal. In India, water of sludge is removed by drying sludge on drying beds.

9.21 **A wastewater treatment plant consists of primary treatment clarifier followed by an activated sludge treatment unit. The primary and secondary sludge are mixed, thickened in a gravity thickener and sent to further treatment. Wastewater, treatment plant and sludge characteristics are as follows:**
- Influent SS = 220 mg/L; primary clarifier diameter = 25 mm
- Influent BOD = 250 mg/L; aerator volume = 3000 m^3
- Effluent BOD = 30 mg/L; MLSS in aerator = 3000 m^3/L
- Flow = 20000 m^3/L; solids in thickener supernatant = negligible
- Primary sludge = 5% solids; secondary sludge = 0.75% solids and thickened sludge = 4% solids
- Efficiency of primary clarifier for SS and BOD removal are 58% and 32% respectively.
- Biomass conversion factor in aerator = 0.35

Determine:

(i) Solids loading in kg/day to the sludge disposal facilities;

(ii) $\frac{F}{M}$ ratio is aerator;

(iii) percent volume reduction by the thickener.

[20 marks : 2019]

Solution:

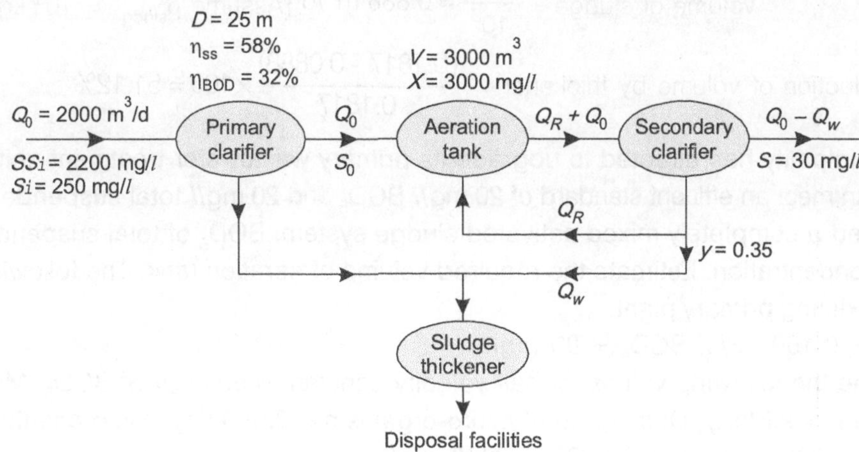

(i) Solids entering into PST = $Q_0 SSi$ = 20000 × 220 × 10^{-6} = 4.4 kg/d

Solids settling into PST = $\eta_{SS} Q_0 SSi = 0.58 \times 4.4 = 2.552$ kg/d ...(a)

BOD entering into PST = $Q_0 Si$ = 20000 × 250 × 10^{-6} = 5 kg/d

BOD entering into AT = $(1-\eta_{BOD})Q_0 Si = Q_0 S_o = (1-0.32)5 = 3.4$ kg/d

BOD leaving from AT = $Q_0 S = 20000 \times 30 \times 10^{-6} = 0.6$ kg/d

BOD removed in AT = (BOD entering – BOD leaving) from AT

= 3.4 – 0.6 = 2.8 kg/d

Biomass in sludge obtained from SST = 0.35 × 2.8 = 0.98 kg/d ...(b)

Total solids entering = Solids from PST (a) + Solids from SST (b) into sludge thickener = 2.552 + 0.98 = 3.532 kg/d

Since there is negligible solids in thickener supernatant solids to sludge disposal facilities = Solids entering into sludge thickener

= 3.532 kg/d

(ii)

$$\frac{F}{M} = \frac{Q_0 S_0}{VX} = \frac{3.4}{3000 \times 3000 \times 10^{-6}} = 0.377 \text{ kg/kg/d}$$

(iii)

$$\text{Vol. of sludge entering into thickener} = \left[\begin{bmatrix}\text{Vol. of sludge}\\ \text{from PST}\end{bmatrix} + \begin{bmatrix}\text{Vol. of Sludge}\\ \text{from SST}\end{bmatrix}\right]$$

For primary sludge

$$\text{Mass of sludge} = \frac{100}{5} \times 2.552 = 51.04 \text{ kg/d}$$

$$\text{Volume of sludge} = \frac{51.04}{10^3} = 0.05104 \text{ m}^3/\text{d}$$ [Assume ρ_{sludge} = 10^3 kg/m³]

For secondary sludge

$$\text{Mass of sludge} = \frac{100}{0.75} \times 0.98 = 130.66 \text{ kg/d}$$

$$\text{Volume of sludge} = \frac{130.66}{10^3} = 0.13066 \text{ m}^3/\text{d}$$ [Assume ρ_{sludge} = 10^3 kg/m³]

Volume of sludge entering = 0.05104 + 0.13066 = 0.1817 m³/d into thickener

For thickened sludge

$$\text{Mass of sludge} = \frac{100}{4} \times 3.552 = 88.8 \text{ kg/d}$$

$$\text{Volume of sludge} = \frac{88.8}{10^3} = 0.888 \text{ m}^3/\text{d} \text{ [Assume } \rho_{sludge} = 10^3 \text{ kg/m}^3]$$

$$\% \text{ Reduction of volume by thickener} = \frac{(0.1817 - 0.0888)}{0.1817} \times 100 = 51.12\%$$

9.22 **A municipality has directed to upgrade its primary wastewater treatment unit to a secondary unit that can meet an effluent standard of 20 mg/*l* BOD$_5$ and 20 mg/*l* total suspended solids. They have selected a completely mixed activated sludge system. BOD$_5$ of total suspended solids is 63% of TSS concentration. Estimate the required volume of aeration tank. The following data is available from existing primary plant:**
Flow = 0.150 m^3/s, BOD$_5$ = 80 mg/*l*.
Assume the following values for half velocity constant = 95 mg/*l* of BOD$_5$; Maximum growth rate constant = 2.5/day; Decay rate of micro-organism = 0.050/day; Yield coefficient = 0.50 mg VSS per mg BOD$_5$ removed; MLVSS = 2000 mg/*l*.

[20 marks : 2020]

Solution:

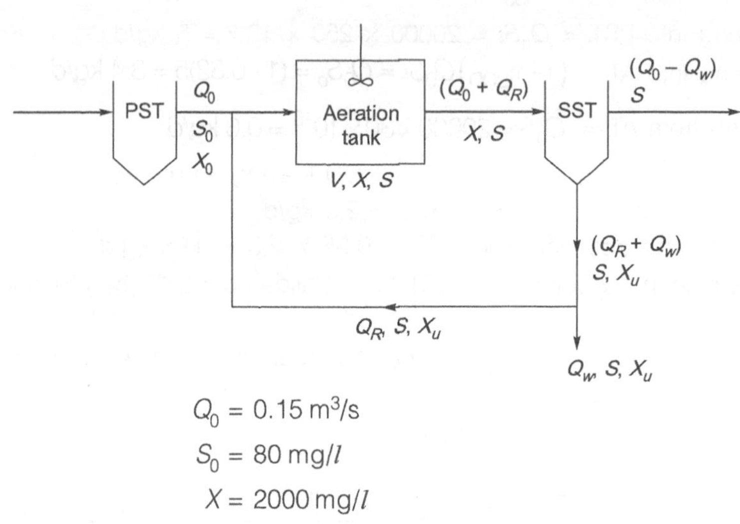

$$Q_0 = 0.15 \text{ m}^3/\text{s}$$

$$S_0 = 80 \text{ mg}/l$$

$$X = 2000 \text{ mg}/l$$

$$y = 0.5 \begin{Bmatrix} \text{fraction of food} \\ \text{mass converted to biomass} \end{Bmatrix}$$

For completely mix process,

Biomas in + Biomass growth= Biomass out

$$Q_0 \cdot X_0 + (k - k_d) X \cdot V = (Q_0 - Q_w) X_e + Q_w X_u \quad \text{...(1)}$$

Assuming influent and effluent biomass concentration are negligible compared to biomass at other points (i.e. X_0 and X_e) are negligible .

∴ Equation (1) becomes,

$$(k - k_d) X \cdot V = Q_w X_u$$

$$k - k_d = \frac{Q_w X_u}{VX}$$

$$k = \frac{Q_w X_u}{VX} + k_d$$

$$k = \frac{1}{\theta_c} + k_d \qquad \because \left\{ \text{Sludge age } (\theta_c) = \frac{VX}{Q_w X_u} \right\}$$

$$k = \frac{k_0 S}{k_s + S} = \frac{1}{\theta_c} + k_d \text{ (Monod's equation)}$$

k_0 = Maximum growth rate constant = 2.5 d^{-1}

S = Concentration of limiting food (in mg/l)

= 20 – 0.63 × 20 = 7.4 mg/l

k_s = Half saturation constant = 95 mg/l

Now,
$$\frac{2.5 \times 7.4}{95 + 7.4} = \frac{1}{\theta_c} + 0.05$$

$$\frac{1}{\theta_c} = \frac{669}{5120} \text{ days}^{-1}$$

$$\Rightarrow \quad \theta_c = 7.65 \text{ days}$$

Also,
$$VX = \frac{y \cdot Q_0 (S_0 - S) \cdot \theta_c}{1 + k_d \cdot \theta_c}$$

$$V \times 2000 = \frac{0.5 \times 0.15 \times 86400 \times (80 - 7.4) \times 7.65}{1 + (0.05 \times 7.65)}$$

$$\Rightarrow \quad V = 1301.6 \text{ m}^3$$

$\therefore$ Volume of aeration tank = 1301.6 m^3

9.23 **Explain the factors that cause sludge bulking in activated sludge process for wastewater treatment.**

[10 marks : 2020]

Solution:

Sludge bulking: A sludge that exhibits poor settling characteristics is called a bulking sludge. It may be due to either.

(a) The growth of filamentous micro-organisms which do not allow desirable compaction.

(b) Due to the production of non-filamentous highly hydrated biomass.

- There are many reasons for sludge bulking, the presence of toxic substances in influent, lowering of temperature, insufficient aeration and shock loading can also cause sludge bulking.
- Proper supply of air and proper design to maintain endogenous growth phase of metabolism will not produce bulking of sludge.
- The sludge bulking can be controlled by restoring proper air supply, eliminating shock loading to the aerator, or by increasing temperature of the wastewater or by small hypochlorite dosing to the return sludge line to avoid the growth of filamentous hygroscopic micro-organisms.

9.24 **Determine the size and depth of a high rate single stage trickling filter for treating the following domestic and industrial wastewater of a town of 50,000 population. Also determine the efficiency of filter and BOD_5 of the effluent.**

- **Domestic sewage @ 150 litres per capita per day having 225 mg/l of BOD_5, and**
- **Industrial wastewater @ 2,00,000 litres per day having 650 mg/l of BOD_5.**

Assume the following data:

— **BOD_5 removal in primary clarifier = 35%**

— **Permissible organic loading of trickling filter = 900 g/day/m^3 (excluding recirculated sewage)**

— **Recirculation ratio = 1.0**

— **Permissible hydraulic loading = 20 m^3/day/m^2 (including recirculated sewage)**

[20 Marks : 2021]

Solution:

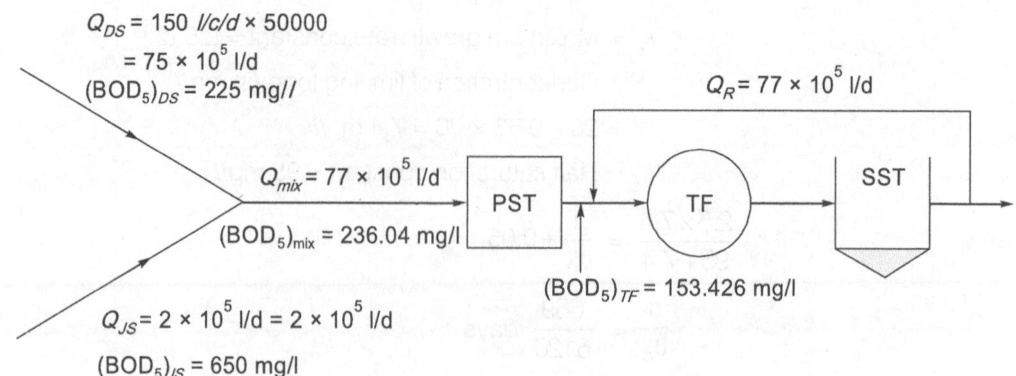

BOD_5 of the sewage after mixing,

$$(BOD_5)_{mix} = \frac{\left(75\times10^5 \text{ l/d} \times 225\text{mg/l}\right)+\left(2\times10^5 \text{ l/d}\times 650 \text{ mg/l}\right)}{77\times10^5 \text{ l/d}}$$

$$= 236.04 \text{ mg/l.}$$

Influent BOD, into the trickling filter $(BOD)_{i,\,TF} = (BOD_5)_{mix} \times 0.65 = 236.04 \times 0.65$

$$= 153.426 \text{ mg/l}$$

$$R = 1$$

$$\therefore \quad F = \frac{1+R}{\left[1+(1-f)R\right]^2};$$

$f \rightarrow$ treatability factor

f = 0.9 (for sewage)

F = Recirculation factor

$$F = \frac{1+1}{(1.1\times1)^2} = 1.653$$

$$\text{OLR} = 900 \text{ g/d/m}^3 = \frac{\text{Qty. of BOD entering the filter}}{\text{Volume of the filter medium}}$$

Volume of the filter medium,

$$V = \frac{153.426 \text{ mg/l} \times 77\times10^5 \text{ l/d}}{900 \text{ g/d/m}^3 \times 10^3 \text{ mg/g}} = 1312.64 \text{ m}^3$$

$$\text{HLR} = 20 \text{ m}^3\text{/d/m}^2 = \frac{\text{Discharge passing through the filter}}{\text{Surface area of the filter}}$$

$\therefore$ Surface area of the filter,

$$SA = \frac{\left(77\times10^2 + 77\times10^2\right)\text{m}^3\text{/d}}{20 \text{ m}^3\text{/m}^2\text{/d}} = 770 \text{ m}^2$$

$$\therefore \quad \frac{\pi D^2}{4} = 770 \text{ m}^2 \text{ ; where D-diameter of the filter}$$

$$D = 31.31 \text{ m}$$

$\because$ Maximum diameter permitted is 60 m, hence, OK.

$$V = \frac{\pi D^2}{4} \times H \text{ ; where H - height of the filter medium}$$

$$H = \frac{1312.64}{770} = 1.7\text{m}$$

∵ H is between 0.9 m – 2.5 m.

Final dimensions:

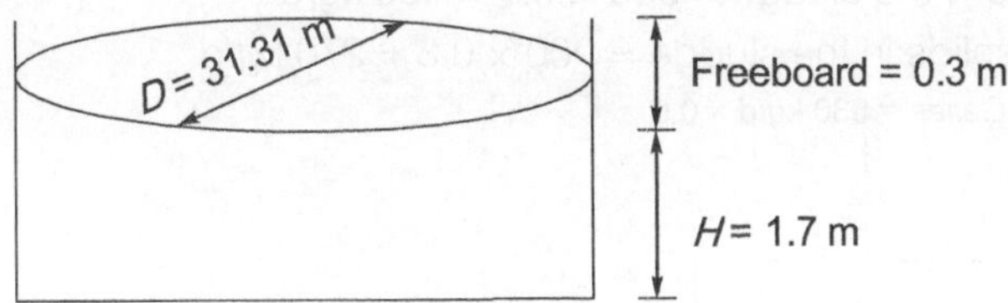

Now, efficiency of single stage high rate trickling filter:

$$\eta = \frac{100}{1+0.44\sqrt{\frac{OLR}{F}}} = \frac{100}{1+0.44\sqrt{\frac{0.9}{1.653}}} = 75.49 \approx 75.5\%$$

Now, BOD_5 of the effluent = (1 – 0.755) × 153.426 = 37.589 ≈ 37.6 mg/l.

9.25 **What do you understand by aerobic digestion? What are its advantages and disadvantages?**

[10 Marks : 2021]

Solution:

The sludge withdrawn from the primary sedimentation tank, which is less biological in nature can be digested aerobically i.e., decomposing the organic matter under aerobic conditions and then disposed of suitably after drying on drying beds.

This process of stabilization of primary sludge is aerobically sludge digestion.

Advantages:

1. As incoming sludge has less BOD concentration, so supernatant liquid will also have very less BOD concentration.
2. The end product is odourless, humus which are quite stable and can be used for sewage farming.
3. Operation is relatively easier than anaerobic digestion.
4. Capital cost is also comparatively lower.

Disadvantages:

1. In aerobic digestion as external food source is not supplied, so aerobic digestion is an endogenous respiration 'Processing' which organism are forced to metabolize. Now as dead cell mass of microorganism retains moisture, so dewatering is difficult.
2. It is associated with high power cost, as supply of oxygen required.
3. As incoming sludge is less biological in nature, so no useful by product like CH_4 is recovered.
4. The process is significantly by temperature, location and type of tank material etc.

9.26 **A wastewater treatment plant produces 900 kg of dry solids per day at a moisture content of 96 percent. The solids are 70 percent volatile with a specific gravity of 1.0 and 30 percent non-volatile with a specific gravity of 2.5. What would be the sludge volume under the following conditions:**

(i) After digestion process, which reduces volatile solids content by 50 percent and decreases the moisture content to 91 percent.

(ii) After dewatering process to 72 percent moisture.

[20 marks : 2022]

Solution:

Quantity of dry sludge produced = 900 kg/d

$$\text{Total mass of sludge} = \frac{900}{0.04} = 900\,\text{kg/d} \times \frac{100}{4} = 22500\ \text{kg/d}$$

Quantity of volatile solids in the sludge = 900 × 0.7 = 630 kg/d

Quantity of non-volatile solids in the sludge = 900 × 0.3 = 270 kg/d

(i)

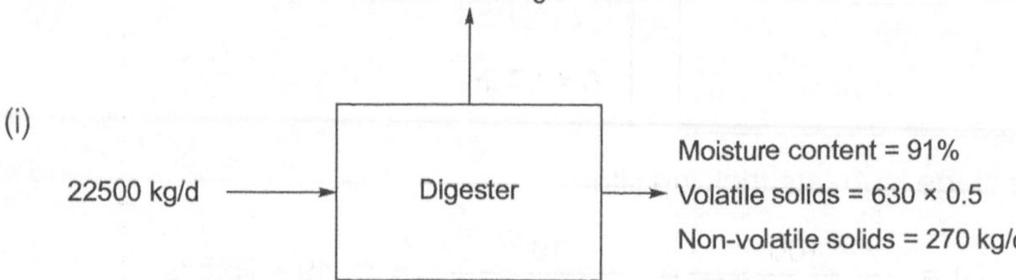

Total solids present in the digested sludge

$$= (630 \times 0.5) + 270 = 585\ \text{kg/d}$$

These solids are 9% of total mass

$$\therefore \quad \text{Total mass of sludge} = 585\,\text{kg/d} \times \frac{100}{9} = 6500\,\text{kg/d}$$

$$\text{Mass of water in the sludge} = 6500 - 585 = 5915\ \text{kg/d}$$

$$\text{Volume of water in the digested sludge} = \frac{5915\,\text{kg/d}}{1000\,\text{kg/m}^3} = 5.915\,\text{m}^3/\text{d}$$

Specific gravity of volatile solids = 1

$$\therefore \text{Volume of volatile solids in the digested sludge} = \frac{(630 \times 0.5)\,\text{kg/d}}{1 \times 1000\ \text{kg/m}^3} = 0.315\,\text{m}^3/\text{d}$$

Specific gravity of non-volatile solids = 2.5

$$\therefore \text{Volume of non-volatile solids in the digested sludge} = \frac{270\,\text{kg/d}}{2.5 \times 1000\,\text{kg/m}^3} = 0.108\,\text{m}^3/\text{d}$$

$$\therefore \text{Total volume of digested sludge} = V_{water} + V_{volatile} + V_{non\text{-}volatile}$$
$$= (5.915 + 0.315 + 0.108)\text{m}^3/\text{d}$$
$$= 6.338\ \text{m}^3/\text{d}$$

(ii)

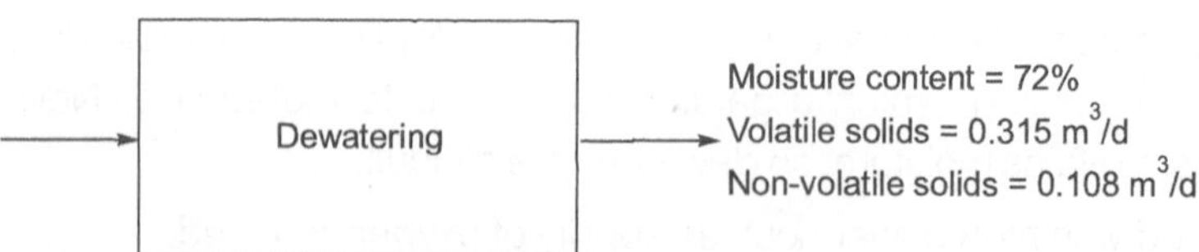

Dewatering is assumed to reduce volume by explusion of water only.

Mass of solids in the dewatered sludge = 315 + 270 = 585 kg/d

$$\text{Now, mass of moisture in the dewatered sludge} = 585\,\text{kg/d} \times \frac{72}{28} = 1504.286\,\text{kg/d}$$

$$\text{Volume of moisture in the dewatered sludge} = \frac{1504.286\,\text{kg/d}}{1000\,\text{kg/m}^3} = 1.504\,\text{m}^3/\text{d}$$

Thus, volume of dewatered sludge,

$$\text{Volume} = (1.504 + 0.315 + 0.108)\ \text{m}^3/\text{d} = 1.927\ \text{m}^3/\text{d}$$

9.27 **Differentiate between Symbiosis and Parasitism relationship. Explaining the working principles of oxidation ponds, discuss the importance of Algal-Bacteria symbiosis relationship in oxidation ponds.**
[20 marks : 2023]

Solution:

The comparison between symbiosis and parasitism relationships are tabulated below:

Parameter	Symbiosis	Parasitism
Definition	A close and long term interaction between two species, where at least one species benefits.	In this type of relationship, one species (parasite) benefits at the expense of another species (host).
Interaction	Mutualistic, commensalistic or parasitic.	Predatory or parasitic.
Benefit to host	Both species can benefit or one species benefits while the other is unaffected.	The host is harmed or negatively affected.
Dependency	Both species are dependent on each other for survival or reproduction.	The parasite is dependent on the host for resources or reproduction. The host is not dependent on the parasite.
Examples	Lichen (algae and fungi), coral reefs (corals and zooxanthellae), legume plants and nitrogen fixing bacteria.	Fleas on dogs, ticks on deer, tapeworms in human.
Control	The interaction is usually well balanced and regulated to prevent over-exploitation or harm to either species.	The parasite may exert control over the host, often causing harm or disease. The host may evolve defences to minimise the impact of the parasite.

Oxidation ponds, also known as stabilization ponds or waste stabilization ponds are shallow, man-made bodies of water used for wastewater treatment. They rely on the interaction between algae and bacteria to remove pollutants and purify the water.

The key roles of the algae-bacterial symbiosis relationship in oxidation ponds are:

1. **Oxygenation**: Algae photosynthesize and release oxygen, creating an oxygen rich environment that supports the growth of aerobic bacteria. This oxygenation is crucial for the activity of aerobic bacteria, which are responsible for the breakdown of organic matter and the removal of pollutants.
2. **Nutrient cycling**: Algae and bacteria work together to cycle and remove nutrients from the wastewater. Algae take up nitrogen and phosphorus during their growth, reducing their concentration in the water. Bacteria then decompose the organic matter provided by the algae, releasing nutrients that can be reused by the algae. This symbiotic relationship helps to maintain a balanced nutrient cycle and prevent excessive nutrient buildup in the water.
3. **Pollutant removal**: The combined activity of algae and bacteria leads to the degradation and removal of organic compounds, including pollutants and pathogens from the wastewater. Algae provide a source of organic matter for bacterial decomposition, while bacteria break down complex organic molecules into simpler, less harmful forms.
4. **Algae-Bacteria floc formation**: Algae and bacteria contribute to the formation of floc or clumps of organic matter in the oxidation pond. These flocs settle to bottom, facilitating the removal of suspended solids and some pollutants through sedimentation. The floc formation is a result of the mutual interactions between algae and bacteria, enhancing the efficiency of solid removal in the poind.

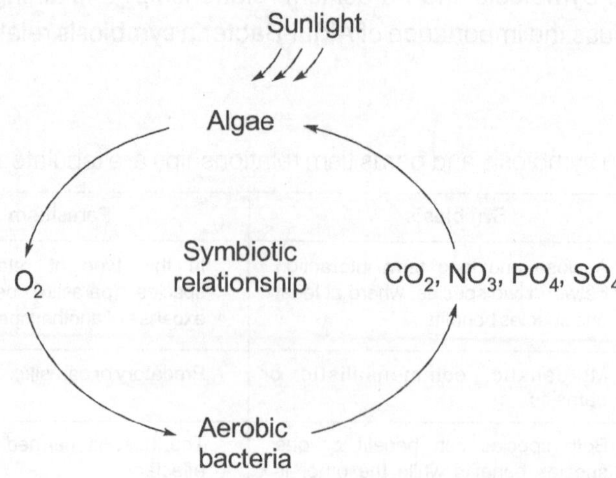

9.28 **The MLSS concentration in an aeration tank of ASP system was 3000 mg/L. Sludge volume was 180 mL after 30 minutes of settling in 1000 mL graduated measuring jar. For the above sample, find**

(i) SVI

(ii) SDI

(iii) Return sludge ratio required

(iv) SS conc. in recirculated sludge

[20 marks : 2023]

Solution:

Given, MLSS, $X = 3000$ mg/l

Sludge volume, $V = 180$ mL (Settled after 30 minutes)

(i) Sludge volume index

In 1 litre jar, 3000 mg of suspended solid has occupied 180 mL in SVI jar.

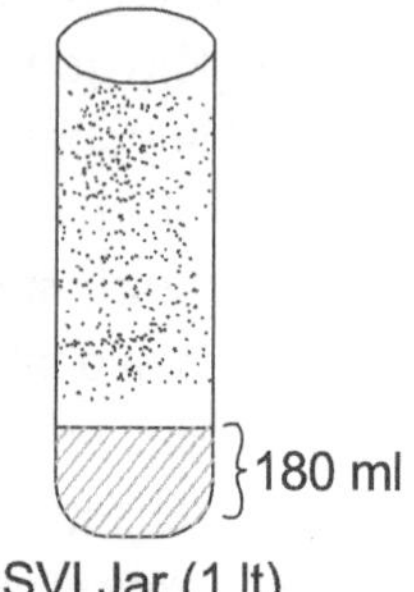

SVI Jar (1 lt)

$$\Rightarrow \quad \text{SVI} = \frac{V(\text{mL})}{X(g)} = \frac{180\text{mL}}{3000\text{mg}}$$

$$= \frac{180\text{mL}}{3\text{g}} = 60 \text{ mL/g}$$

(ii) $$\text{Sludge density index} = \frac{100}{SVI} = \frac{100}{60} = 1.67 \text{ g/mL}$$

(iii) $$\text{Return sludge ratio, } R = \frac{Q_R}{Q_0} = \frac{X}{X_u - X}$$

where, $$X_u = \frac{10^6}{SVI} = \frac{10^6}{60} = 16.67 \times 10^3 \text{ mg/}l$$

$$\therefore \quad R = \frac{3000}{(16.67-3)\times 10^3} = 0.219$$

(iv) S.S concentration in recirculated sludge,

$$X_u = \frac{10^6}{SVI} = \frac{10^6}{60} = 1.67 \times 10^4 \text{ mg/l}$$

10. Air, Sound, Land Pollution and EIA

10.1 **Explain the term 'refuse' and give its composition and classification. Describe briefly the various methods employed for the collection and disposal of the refuse.**

[10 marks : 2000]

Solution:

Refuse represents the dry wastes or solid wastes of the society except human excreta and sullage. It includes garbage, ashes, rubbish, dust, etc.

(i) **Garbage** includes all sorts of putrescible organic wastes obtained from kitchens, hotels, restaurants, etc. All waste food articles, vegetable peelings, fruit peelings, etc., are thus included in this term.

(ii) **Ashes denote** the incombustible waste products from hearths and furnaces, and houses or industries.

(iii) **Rubbish** includes all non-putrescible wastes except ashes. It, thus includes all combustible and non-combustible wastes such as rags, paper pieces, broken pieces of glass and furniture, card boards, broken crockery, etc.

Besides the technical classification based on the type of wastes, the refuse may also be classified, depending on its source, as: (i) House refuse; (ii) Street refuse and (iii) Trade refuse.

Collection of refuse: The refuse is generally collected in individual houses in small containers and from there, it is collected by sweepers in small hand driven lorries/carts and then dumped into the masonry chambers constructed along roadsides by municipalities. The refuse is finally carted away by municipal trucks, for further disposal during some day time.

Disposal of refuse: The refuse can be disposed of by various methods, such as

(a) **Sanitary land filling:** In this method of refuse disposal, refuse is carried and dumped into the low lying area under an engineered operation, designed and operated according to the acceptable standards, as not to cause any nuisance or hazards to public health or safety. Area method and trench method are two methods of land filling which are usually adopted.

(b) **Burning or incineration:** Burning of refuse at high temperatures in furnaces called incinerators is quite a sanitary method of refuse disposal and is widely adopted in USA where the refuse is of high calorific value and hence quite suitable for burning. The heat produced during burning of the refuse is used in the form of steam power for running turbines to generate electricity.

(c) **Barging it out into the sea:** This method may be used to dispose of refuse by throwing it away into the sea, after carrying it at reasonable distance from the coast. The sea depth at such disposal point should not be less than 30 m or so, and the direction of the currents should be as not to bring it back towards the shore.

(d) **Pulverization:** In this method refuse is pulverized in grinding machines, so as to reduce its volume and to change its physical character. The grinded or pulverized refuse becomes practically odourless and unattractive to the insects.

(e) **Composting:** It is a biological method of decomposing solid wastes. This decomposition can be effected either under aerobic or anaerobic conditions or both. The final end product is manure called compost or humus, which is in great demand as fertilizer for farms.

10.2 **Explain briefly noise pollution, its causes, effects and remedies.**

[10 marks : 2000]

Solution:

Noise can be defined as that unwanted sound pollutant which produces undesirable physiological and psychological effects in an individual, by interfering with one's social activities like work, rest, recreation, sleep, etc.

The unwanted sound certainly produces several undesirable effects on our body health, and it can therefore be called a pollutant. The basic reason why noise has so far not been regarded as a killer pollutant is the fact that so far noise has generally not exceeded to such levels as to cause drastic adverse impacts on human beings. Except in industrial areas and big cities noise pollution is not prominent and hence has generally remained ignored from being treated as an environmental pollutant. However, with the increasing industrialization and commercialisation of our society, and with consequential increase in noise producing automobiles and machines, noise is becoming a slow poison and is becoming more and more prominent as a pollutant.

Causes:

(i) Traffic including air traffic, road traffic and sea shore and inland water traffic.

(ii) Industrialization

(iii) Human activities such as blaring of loud speaker and sirens, shouting of hawkers and playing of children, etc.

Effects: Noise of sufficient intensity and duration can induce health problems like temporary and sometimes permanent hearing loss, besides causing several other diseases like general annoyance, irritation, disturbance, headaches, insomnia, fatigue, mental torture, nausea, high blood pressure, high pulse rate, greater perspiration, etc.

Remedies:

(i) By enforcing laws and ordinances, the sounds from loudspeakers, sirens, etc., can be controlled.

(ii) By use of good technology in vehicles and others machines which produces noise.

(iii) By proper town planning techniques e.g. the residential complexes should be separated from commercial and industrial complexes. Parks, playgrounds should be provided for children separately.

(iv) Noise produced by trains and automobiles can be abated by construction of walls on both sides of roads and railway lines.

10.3 For a medium sized town, organic waste is to be disposed off by the method of composting using trenches. Explain the salient points of this method.

[10 marks : 2001]

Solution:

Composting of refuse is a biological method of decomposing solid wastes. This decomposition can be effected either under aerobic or anaerobic conditions or both. The final end product is a manure called compost or humus, which is in great demand in European countries as fertilizer for farms.

Basically composting is considered to be an aerobic process, because it involves piling up of refuse and its regular turning either manually or by mechanical devices, so as to ensure sufficient supply of air and oxygen during its decomposition by bacteria, fungi and other micro organisms, like antinomycetes. Initially the process starts with mesophilic bacteria, which oxidise the organic matter to carbon dioxide and liberate heat. The temperature rises to about 45°C and at the point, the thermophilic bacteria take over and continue the decomposition. During this phase temperature further rises to about 60°C. After 3 weeks, the compost is stabilized and this is shown by appreciable fall in the temperature of the compost mass. The final compost should have an earthy smell and a dark brown colour. Moisture content of the compost mass should be controlled because excessive moisture will make it difficult to maintain aerobic conditions while deficient moisture inhibits biological life. It may sometimes, becomes necessary to add water to the compost mass during its turning, for maintaining satisfactory moisture content. Two methods which are generally adopted in composting are Indore Method which is an aerobic method and Bangalore method which is an anaerobic method.

10.4 **What is Environmental Impact Assessment? Briefly describe different methodology adopted for Environmental Impact Assessment.**

[10 marks : 2002]

Solution:

Assessing the nature and extent of degradation of environment due to human interference is called **Environmental Impact Assessment.**

Engineering projects involving development of thermal power, mining operations and even river valley projects have been found to be causing certain adverse and negative impacts on our surrounding environment, which has forced us to make it compulsory to evaluate these adverse impacts in details, well before the project is cleared for execution. With this end in view, all project clearance cells do evaluate and examine the detailed environmental assessment report, which is prepared and submitted along with Detailed Project Report (DPR) of every such project. Submission of such EIAs have been made compulsory by the GOI for all projects, which are likely to cause harm to our surrounding environment.

All such impact assessments should thoroughly examine and discuss the various possible environmental damages, whether pertaining to water pollution, air pollution, ground pollution, noise pollution, or any other kind of environmental pollution and their suggested remedial measures to prevent or to mitigate such hazardous environmental effects.

Methodology adopted for EIA: Environmental Impact Assessment essentially involve three steps:

(i) **Impact identification:** It may be carried out with the help of checklists, matrices or networks. Checklist merely present a list of environmental parameters to be investigated for possible impacts. Matrices are two dimensional checklists in which cause-effect relationships are established by listing possible project activities along one axis and potentially impacted environmental characteristics of conditions along the other. Network illustrates cause condition effect linkages as also temporal dimensions and therefore provide the most comprehensive methodology for impact identification.

(ii) **Prediction of environmental impacts:** It requires the greatest degree of scientific application. This step involves projecting the baseline environmental setting into future, with and without the project, and then performing the necessary calculations for predicting real impacts of the proposed development.

(iii) **Evaluation of Impacts:** It calls for conversion of the predicted values for various environmental parameters to a comparable set of units using some system of normalisation. Ideally, the environmental impacts should be expressed in monetary units for easy and objective comparison with other costs and benefits of the project. The major problem arises when while assigning monetary values to intangible parameters. Methods involving numerical rating and ranking or weighting and scaling of environmental impacts are therefore commonly used.

10.5 **Explain the components of an ecosystem. Give a pictorial example of structural components of ecosystem.**

[10 marks : 2002]

Solution:

The four major elements i.e. land, water, air and living organisms (plants and animals) together constitutes what is known as ecosystem.

Ecosystem can be further subdivided into:

1. Physical environment

2. Biological environment

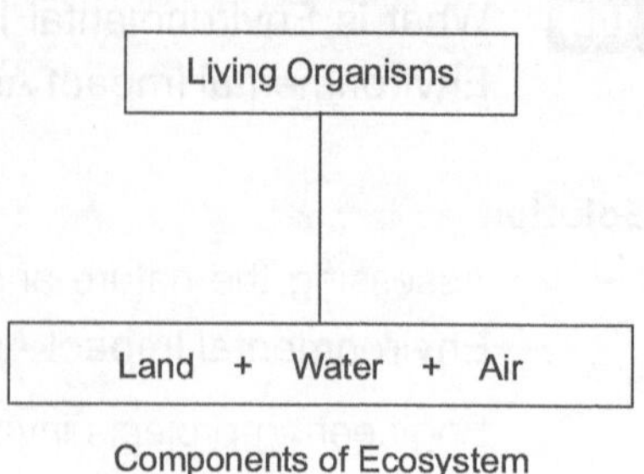

Land, water and air together, in fact, forms one group of environment called physical environment whereas the living organisms form another group of environment called biological environment. While the physical environment (land, water and air) is essential for existence of life in various forms, the biological environment provides the necessary food, so very essential for the sustenance of man on earth. Man as a matter of fact cannot survive on the earth without plant and animal life.

10.6 **Compute the effective stack height of a coal burning power plant with physical stack height of 200 m and stack diameter of 0.8 m, stack gas exit velocity of 18.3 m/sec, temperature of gas 140°C, when the ambient air temperature is 8°C, atmospheric pressure is 1000 millibars and average wind speed is 4.5 m/sec. Also compute the emission rate of sulphur dioxide of the plant assuming burning of 24,000 tonnes of coal per day with a sulphur content of 4.2%.**

[15 marks : 2003]

Solution:

As per Holland's equation, the plume height can be given as

$$\Delta h = \frac{V_s D}{u}\left[1.5 + 2.68 \times 10^{-3} P.D\left(\frac{T_s - T_a}{T_s}\right)\right]$$

where Δh is rise of plume above stack height in m, V_s is stack gas velocity in m/s, D is exit diameter of stack in m, u is wind speed in m/s, P is atmospheric pressure in millibars, T_s is stack gas temperature in °K and T_a is ambient air temperature in °K

Given V_s = 18.3 m/s; D = 0.8 m; u = 4.5 m/s; P = 1000 millibars; h = 200m

$$T_s = 140 + 273 = 413\ °K$$

$$T_a = 8 + 273 = 281\ °K$$

$$\therefore \quad \Delta h = \frac{V_s D}{u}\left[1.5 + 2.68 \times 10^{-3} P.D\left(\frac{T_s - T_a}{T_s}\right)\right]$$

$$= \frac{18.3 \times 0.8}{4.5}\left[1.5 + 2.68 \times 10^{-3} \times 10^3 \times 0.8\left(\frac{413 - 281}{413}\right)\right] = 7.11 \text{ m}$$

Effective height of stack = Physical stack height (h) + Δh = 200 + 7.11 = 207.11m

Now, Sulphur content of coal = 4.2%

Coal burnt per day = 24000 t

$$\therefore \quad \text{Sulphur produced per day} = \frac{4.2}{100} \times 24000 = 1008 \text{ tonnes}$$

Now, this sulphur will combine with oxygen to produce SO_2

$$\underset{32}{S} + \underset{32}{O_2} \longrightarrow \underset{64}{SO_2}$$

32 t of sulphur produce 64 t of sulphur dioxide

$$\therefore \quad \text{1008 t of sulphur will produce} = 1008 \times \frac{64}{32} = 2016 \text{ tonnes of } SO_2 \text{ per day}$$

$$\therefore \quad \textbf{Emission rate of } SO_2 = \frac{2016 \times 1000}{24 \times 60 \times 60} = 23.33 \text{ kg/sec}$$

10.7 **Classify the solid wastes, giving suitable example for each of them. Also explain the different methods of disposal of solid wastes.**

[10 marks : 2004]

Solution:

Refuse represents the dry wastes or solid wastes of the society except human excreta and sullage. It includes garbage, ashes, rubbish, dust, etc.

(i) Garbage includes all sorts of putrescible organic wastes obtained from kitchens, hotels, restaurants, etc. All waste food articles, vegetable peelings, fruit peelings, etc., are thus included in this term.

(ii) Ashes denote the incombustible waste products from hearths and furnaces, and houses or industries.

(iii) Rubbish includes all non-putrescible wastes except ashes. It, thus includes all combustible and non-combustible wastes such as rags, paper pieces, broken pieces of glass and furniture, card boards, broken crockery, etc.

Besides the technical classification based on the type of wastes, the refuse may also be classified, depending on its source, as: (i) House refuse; (ii) Street refuse and (iii) Trade refuse.

Collection of refuse: The refuse is generally collected in individual houses in small containers and from there, it is collected by sweepers in small hand driven lorries/carts and then dumped into the masonry chambers constructed along roadsides by municipalities. The refuse is finally carted away by municipal trucks, for further disposal during some day time.

Disposal of refuse:

The refuse can be disposed of by various methods, such as

(a) **Sanitary land filling:** In this method of refuse disposal, refuse is carried and dumped into the low lying area under an engineered operation, designed and operated according to the acceptable standards, as not to cause any nuisance or hazards to public health or safety. Area method and trench method are two methods of land filling which are usually adopted.

(b) **Burning or incineration:** Burning of refuse at high temperatures in furnaces called incinerators is quite a sanitary method of refuse disposal and is widely adopted in USA where the refuse is of high calorific value and hence quite suitable for burning. The heat produced during burning of the refuse is used in the form of steam power for running turbines to generate electricity.

(c) **Barging it out into the sea:** This method may be used to dispose of refuse by throwing it away into the sea, after carrying it at reasonable distance from the coast. The sea depth at such disposal point should not be less than 30 m or so, and the direction of the currents should be as not to bring it back towards the shore.

(d) **Pulverization:** In this method refuse is pulverized in grinding machines, so as to reduce its volume and to change its physical character. The grinded or pulverized refuse becomes practically odourless and unattractive to the insects.

(e) **Composting:** It is a biological method of decomposing solid wastes. This decomposition can be effected either under aerobic or anaerobic conditions or both. The final end product is manure called compost or humus, which is in great demand as fertilizer for farms.

10.8 **Explain different local and global effects of air pollution. Support your answer with suitable examples.**

[10 marks : 2004]

Solution:

Air as we all know is important for life. This life supporting important natural element, sometimes becomes our bitter enemy, as and when it gets polluted, since it causes number of diseases in our body. The harmful effects caused to human body-by the polluted air depends upon the type and concentration of the pollutants present in the air. The polluted air is not only harmful to man, but is also harmful to all types of life, including plants, animals and birds. It is also harmful to non living materials likes metals, marbles and other stones, woods, paints, papers, etc., which gets spoiled by the contact with polluted air, either due to mere physical corrosive action of polluted air or/and due to the chemical attack of the pollutants on such materials.

Three global issues which threatens the earth's atmosphere are:

(i) **Ozone depletion:** Ozone gas (O_3) is found to exist in upper natural environment (stratosphere) and helps life on earth by absorbing harmful ultraviolet radiations released by sunlight. The sun light is thus filtered out by the ozone molecules, before the same reaches us on earth.

The shortage of ozone in the stratosphere, causes the harmful UV radiations to reach our planet, causing several adverse impacts on human, animal as well as plant life existing on earth. Chlorofluorocarbons (CFCs) emitted by various industries especially aviation and refrigeration industry are depleting the ozone layer.

The most dreaded adverse impact of UV radiations is to cause mutation of DNA, leading to skin cancers including the deadly melanoma so much so that a 10% overall depletion of ozone is estimated to cause 26% increase in the incidence of non-melanoma skin cancers world wide.

(ii) **Acid rains:** Natural rainfalls through unpolluted environments are very slightly acidic, with pH just less than 7. However, when the environment is polluted with primary pollutants like SO_2 and NO_x, the resultant rainfalls tend to become more and more acidic, with the increasing concentrations of these pollutants.

The acidity in the rainwater is caused by the formation of secondary pollutants like H_2SO_4, HNO_3, HCl, etc., due to the reaction of water vapour with SO_2, NO_x and HCl gas. It has been specified that when the pH of rain water falls below 5.6, the rain is specifically termed as acidic.

Acid rain has the potential to convert forests into wastelands by damaging trees, plants and crops; acidify fresh waters of lakes, rivers, reservoirs, springs, wells and underground sources; induce respiratory illness in humans; poison wild life and inflict irreversible damages on buildings to deface the old treasured monuments.

In Germany, Switzerland, Sweden and Norway considerable portion of natural forests are dead or dying. 80% of all lakes in Norway and 20% in Sweden are acidified. As far as India is concerned, acid rains so far occurred in highly industrialised localities like Chembur in Maharashtra.

(iii) **Global warming:** During the past few years, earth's climate has been noticed to be getting hotter and hotter, year after year. The cause of this rising in temperature is the green house effect. It is the heating effect caused by green house gases since these gases in the lower atmosphere act like the glass of a green house, trapping some of the heat as it radiates back into the space. Excessive quantities of these gases may cause excessive heat to make like impossible.

The level of CO_2 in the earth's environment has increased by about 25% during the last decade along with a 1% increase in release of methane due to large scale industrial and developmental activities throughout the world leading to global warming and melting of polar icecaps and its consequential adverse impacts of rising of sea levels, etc.

10.9 **What arrangements are made in a landfill to collect and control the movement of leachate produced? What preventive measures to be taken to prevent contaminating ground water from the landfills?**

[8 marks : 2005]

Solution:

Following arrangements are made in a landfill to control the movement of leachate.

(i) **Natural liners:** These refer to compacted clay or shale, bitumen or soil sealants etc. and are generally less permeable, resistant to chemical attack and have good sorption properties. These generally do not act as true containment barriers, because sometimes leachate migrates through them.

(ii) **Synthetic (geo-membrane) liners:** These are typically made up to high or medium density polyethylene and are generally less permeable, easy to install, relatively strong and have good deformation characteristics. Note that natural and geo-membrane liners are often combined to enhance the overall efficiency of the containment systems.

Some of the leachate collection system include impermeable linear, granular material, collection piping, leachate storage tanks.

10.10 Discuss different processes for composting of solid waste. Explain important design considerations for one of the composting processes.

[10 marks : 2006]

Solution:

There are two methods of composting of solid wastes. They are:

(i) Ordinary composting

(ii) Mechanised composting

Ordinary composting of solid wastes can be performed by **Indore Process (aerobic)** or **Bangalore Process (anaerobic).**

Indore Process: It uses manual turning of piled up mass (refuse + night soil) for its decomposition under aerobic conditions. In this method, layers of vegetable waste and night soil are alternatively piled in depths of about 7.5 to 10 cm each to a total depth of about 1.5 m in a trench or above the ground to form a mound called a wind row. The mixture is kept aerobic by turning regularly for 2 to 3 months. This compost is then left for another one to one and half months without any turning after which the compost becomes ready for use.

Bangalore process: It is a method primarily anaerobic in nature and does not involve any turning or handling of the mass and hence is more clean than Indore method. The mixture of refuse and night soil is kept in a underground trench (about 10 m × 1.5 m × 1.5 m). This mass is covered at its top by layer of earth of about 15 cm depth and is finally left over for decomposition. Within 2 to 3 days of burial, intensive biological action starts and as a result a considerable amount of heat is generated which raises the temperature of the composting mass to about 75°C. After 4 to 5 months, the refuse gets fully stabilized and changes into a brown coloured odourless innocuous powdery mass called humus. This humus is recovered from trenches and served on 12.5 mm sieve and then sold out in the market as manure.

Mechanised composting involve mechanised receipt, segregation and pulverization of refuse before its digestion in closed digestors.

The refuse is received at the plant at the rate of 2 to 6 tonnes per vehicle. Dry refuse is then allowed to move along a belt conveyor, whereby the ferrous (iron) matter is removed by magnetic separators and other non-compostable organic matter are separated out for recycling and reuse. The separation of such materials can be done manually as well as by rotary and shaker screens. The remaining refuse is now pulverised by using different equipments like hammer mill, rusping machine, dano grinder, etc. The above refuse is finally digested in mechanical digestors under controlled conditions of temperature and moisture. Digestion period in such digestors normally vary between 2 to 5 days for refuse containing low cellulose or low C-N ratio and 7-9 days for refuse having more quantities of cellulose or high C-N ratio.

10.11 Write about the plume behaviour coming out from a stack.

[10 marks : 2006]

Solution:

Dispersion is the process of spreading out pollution emission over a large area and thus reducing their concentration. Wind speed and environmental lapse rates directly influence the dispersion pattern. **Seven classifications of plume behaviour** which may occur under some commonly encountered metrological conditions are discussed in the following sections.

(i) Looping Plume:

Under super-adiabatic condition, both upward and downward movement of the plume is possible. Large eddies of a strong wind cause a looping pattern. Although the large eddies tend to disperse pollutants over a wide region, high ground level concentrations may occur close to the stack.

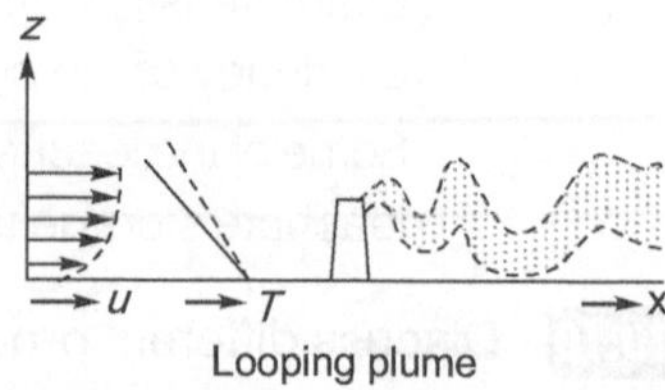

Looping plume

(ii) Neutral Plume:

Neutral plume is the upward vertical rise of the plume from the stack, which occurs when the environmental lapse rate is equal to or very near to the adiabatic lapse rate. The upward lifting of the plume will continue till it reaches an air of density similar to that of the plume itself.

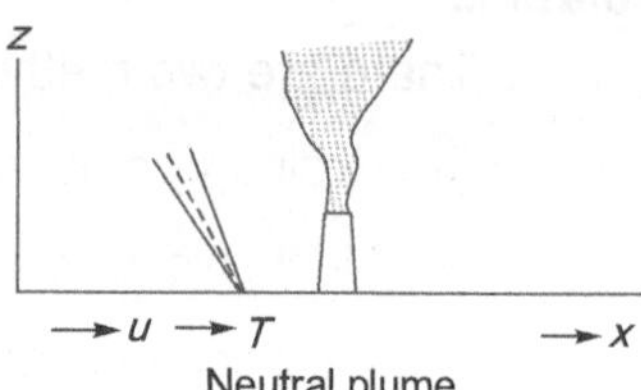

Neutral plume

(iii) Coning Plume:

A coning plume occurs under essential neutral stability when environmental lapse rate is equal to adiabatic lapse rate and moderate to strong winds occur. The plume enlarges in the shape of a cone. A major part of pollution may be carried fairly far downwind before reaching ground.

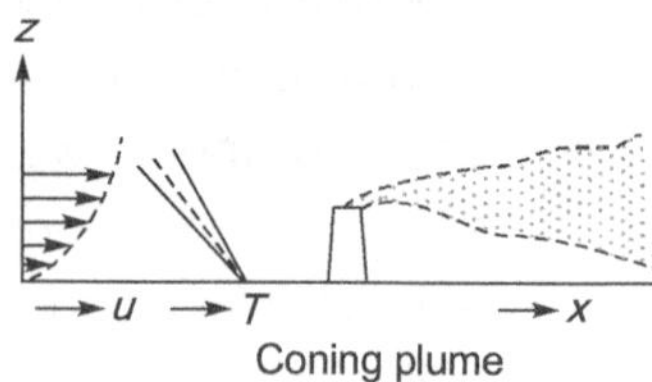

Coning plume

(iv) Fanning Plume:

A fanning plume occurs in the presence of a negative lapse rate when vertical dispersion is restricted. The pollutants disperse at the stack height, horizontally in the form of a fanning plume.

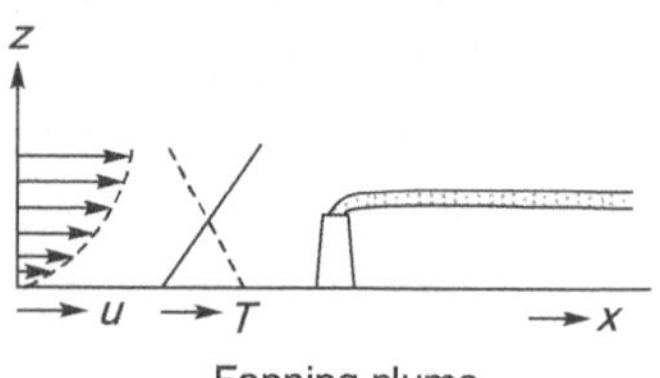

Fanning plume

(v) Lofting Plume:

When the stack is sufficiently high and the emission is above an inversion layer mixing in the upward direction is uninhibited, but downward motion is restricted. Such lofting plumes do not result in any significant concentration at ground level. However, the pollutants are carried hundreds of kilometers from the source.

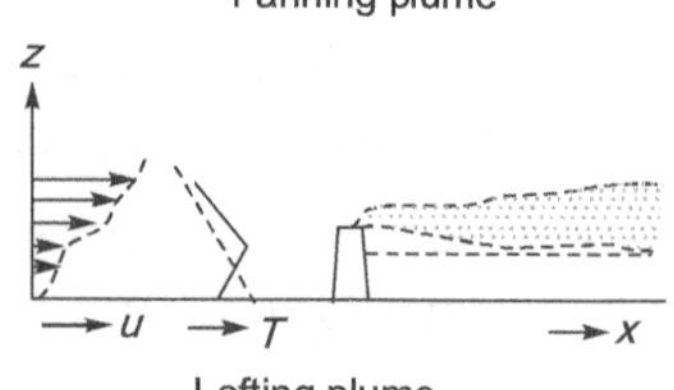

Lofting plume

(vi) Fumigating Plume:

When the emission from the stack is under an inversion layer, the movement of the pollutants in the upward direction is restricted. The pollutants move downwards. The resultant fumigation can lead to a high ground level concentration downwind of the stack.

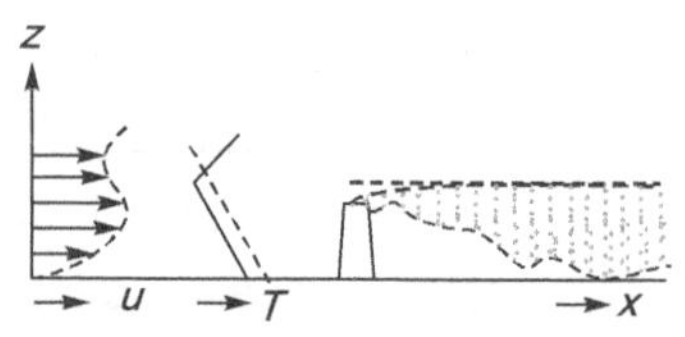

Fumigating plume

(vii) Trapping Plume:

When inversion layers exist above the emission source, as well as below the source, then naturally, the emitted plume will neither go up, nor will it go down, and would remain confined between the two inversions. Such a plume is called a trapping plume, and is considered bad condition for dispersion, as the dispersion cannot go above a certain height.

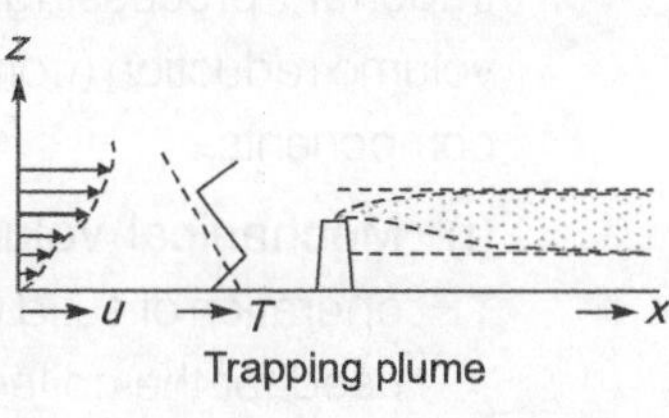

Trapping plume

10.12 What are different methods used for land filling in dry areas? Discuss them.

[8 marks : 2007]

Solution:

The principal methods used for landfilling in dry areas may be classified as, Area method, Trench method and Depression method.

(i) **Area method:** The area method is used when the terrain is unsuitable for the excavation of trenches in which to place the solid wastes. The filling operation usually is started by building an earthen levee against which wastes are placed in thin layers and compacted. Each layer is compacted as the filling progresses until the thickness of the compacted wastes reaches a height varying from 2 to 3 m. At that time, and at the end of each day's operation, a 150 to 300 mm layer of cover material is placed over the completed fill. The cover material must be hauled in by truck or earth-moving equipment from adjacent land or from borrow pit areas. In some newer landfill operations, the daily cover material is omitted. A completed lift, including the cover material, is called a **cell**. Successive lifts are placed on top of one another until the final grade called for in the ultimate development plane is reached. A final layer of cover materials is used when the fill reaches the final design height.

(ii) **Trench method:** The trench method of landfilling is ideally suited to areas where an adequate depth of cover material is available at the site and where the water table is well below the surface. To start the process (for a small landfill), a portion of the trench is dug with a bulldozer and the dirt is stockpiled to form an embankment behind the first trench. Wastes are then placed in the trench, spread into thin layers and compacted. The operation continues until the desired height is reached. Cover material is obtained by excavating an adjacent trench or continuing the trench that is being filled. In large landfills, a dragline and one or more scrapers are used to excavate a deep rectangular pit.

(iii) **Depression method:** At locations where natural or artificial depressions exist, it is often possible to use them effectively for landfilling operations. Canyons, ravines, dry borrow pits and quarries have all been used for this purpose. The techniques to place and compact solid wastes in depression landfills vary with the geometry of the site, the characteristics of the cover material, the hydrology and geology of the site, and the access to the site. In a canyon site, filling starts at the head end of the canyon and ends at the mouth. This practice prevents the accumulation of water behind the landfill. Wastes usually are deposited on the canyon floor and from there are pushed up against the canyon face at a slope of about 2 to 1. In this way, a high degree of compaction can be achieved.

10.13 Why is processing of solid waste required for solid waste management? Discuss various processing techniques used in solid waste management.

[8 marks : 2007]

Solution:

Processing techniques are used in solid waste management systems,

(i) to improve the efficiency of solid-waste disposal systems,
(ii) to recover resources (usable materials) and
(iii) to prepare materials for the recovery of conversion products and energy.

Important processing techniques used routinely in municipal solid waste systems include mechanical volume reduction (compaction), thermal volume reduction (incineration) and manual separation of waste components.

(i) **Mechanical volume reduction:** It is perhaps the most important factor in the development and operation of solid waste management systems. Vehicles equipped with compaction mechanisms are used for the collection of most municipal solid wastes. To increase the useful life of landfills, wastes are compacted. When compacting a broad range of municipal solid wastes, it has been found that the final density (typically about 1100 kg/m^3) is essentially the same regardless of the starting density and applied pressure.

(ii) **Thermal Volume reduction:** The volume of municipal wastes can be reduced by more than 90 per cent by incineration. In the past, incineration was quite common. However, with more restrictive air pollution control requirements necessitating the use of expensive cleanup equipment only a limited number of municipal incinerators are currently in operation.

(iii) **Manual components separation:** The manual separation of solid waste components can be accomplished at the source where solid wastes are generated, at a transfer station, at a centralised processing station, or at the disposal site. Manual sorting at the source of generation is the most positive way to achieve the recovery and reuse of materials. The number and types of components salvaged or sorted (e.g., cardboard and high quality paper, metals and wood) depend on the location, the opportunities for recycling and the resale market.

10.14 Estimate the moisture content of a solid waste sample with the following composition :

Component	Food wastes	Paper	Cardboard	Plastics	Garden Trimmings	Wood	Tin cans
Percent by mass	20	40	10	5	5	5	5
Moisture (%)	70	6	5	2	60	20	3

[10 marks : 2008]

Solution:

Let the total mass of the solid waste sample be W.

Thus, the summation of the mass of all the components of the sample should be W.

$$\therefore \quad \frac{20}{100}\times W+\frac{40}{100}\times W+\frac{10}{100}\times W+4\times\frac{5}{100}\times W = 0.9\,W$$

$\therefore$ The difference of total mass of sample and the mass of components will be

$$= W-0.9\,W = 0.1\,W$$

Let us assume that this difference is equal to the mass of water in the sample.

Component	Mass	Moisture
Food wastes	0.2W	$\frac{70}{100}\times 0.2\text{W} = 0.14\text{W}$
Paper	0.4W	$\frac{6}{100}\times 0.4\text{W} = 0.024\text{W}$
Card board	0.1W	$\frac{5}{100}\times 0.1\text{W} = 0.005$

Plastics	0.05 W	$\frac{2}{100} \times 0.05\,W = 0.001W$
Garden trimmings	0.05 W	$\frac{60}{100} \times 0.05\,W = 0.03\,W$
Wood	0.05 W	$\frac{20}{100} \times 0.05\,W = 0.01W$
Tin cans	0.05 W	$\frac{3}{100} \times 0.05\,W = 0.0015\,W$
Water	0.1W	$\frac{100}{100} \times 0.1W = 0.1W$
		Total = 0.3115 W

$$\therefore \quad \text{Moisture content of sample} = \frac{\text{Summation of moisture of all components}}{\text{Total mass of sample}}$$

$$= \frac{0.3115\text{ W}}{\text{W}} \times 100 = 31.15\%$$

10.15 Discuss the impact of air on

(i) Monuments (ii) Animals (iii) Plants (iv) Climate

[8 marks : 2010]

Solution:

(i) **Effect of air pollution on monuments:**

- Acid rain results when gaseous emissions of sulphur oxides (SO_x) and nitrogen oxides (NO_x) interact with water vapour and sunlight and are chemically converted to strong acidic compounds such as sulphuric acid (H_2SO_4) and Nitric acid (HNO_3). These chemicals deposited on earth surface during rain hence decolourisation and change of colour of monuments takes place.
- Due to acid rain corrosion of metallic surface takes place.

(ii) **Effect of air pollution of animals:**

- The adverse health effects of air pollutants on farm animal occurs when animals eat the plants, grasses, fodder and other vegetation which has been contaminant by the air pollutant.
- The cattle grazing on vegetation that has been contaminated with fluorine may develop fluorosis.
- The cattle grazing on vegetation that has been contaminated with arsenic and lead that may cause arsenic poisoning and lead poisoning respectively.

(iii) **Effect of air pollution on plants:**

- The most prominent air pollutant fluorine, which is emitted from factories manufacturing aluminium, glass, phosphate, fertilizers etc., if it's concentration is in excess of about 0.3 $\mu g/m^3$ cause photo-toxicological effects on plants.
- The damage caused by air pollutant like SO_2, H_2S, H_2N, Cl_2, PAN, O_3, NH_3, Smog etc. to the plant and vegetation occurs in the leaf structure. The pollutants clog the stomata of the leaf, thereby reducing the intake of CO_2, which adversely affects the photosynthesis.

(iv) **Effect of air pollution on climate:**

- Due to air pollution like increase in concentration of CO_2 may lead to phenomenon of global warming. So as a result of increases temperature, the polar ice caps may melt vigorously and there would be rise in ocean level threatening the population near the coastal zones.

- Ozone layer depletion is the most dreaded aspect of air pollution. This ozone layer cut off short wave length radiations (called ultraviolet radiation) from reaching the earth surface. This process serve as a protective shield to human life against the adverse effects of UV like skin cancer. It is obvious that any deplection of stratosphere ozone would be harmful to life on this earth.

10.16 Describe photochemical oxidants. What are the sources of photochemical oxidants and how the detection and analysis for the same is done?

[12 marks : 2012]

Solution:

Photochemical Oxidants: Oxidants or total oxidants, two terms used to describe levels of photochemical oxidants, generally indicate the net oxidizing ability of the ambient air. Ozone (O_3), the major photochemical oxidant, makes up approximately 90 percent of the oxidant pool. Other photochemical oxidants of concern in air pollution monitoring include nascent oxygen (O), excited molecular oxygen (O_2), peroxy-acetyl nitrate (PAN), peroxy-propinol nitrate (PPN), peroxy-butyl nitrate (PBN), nitrogen dioxide (NO_2), hydrogen peroxide (H_2O_2), and alkyl nitrates.

Sources of Oxidants: Ozone, the photochemical oxidant of major concern in air pollution, is produced in the upper atmosphere by solar radiation, and small concentrations of this gas diffuse downwards. Also, small concentrations are produced by lightning and forest fires. Background concentrations of ozone range from 39 to 78 $\mu g/m^3$ (0.02 to 0.04 ppm). However, background concentrations can reach the federal ambient-air-quality standard of 235 $\mu g/m^3$ (0.12 ppm). Anthropogenic precursors are usually involved where oxidant levels in rural areas have exceeded ambient standards. Ozone concentrations often are higher in suburban and even rural areas than in urban areas because O_3 in urban areas is lost to reactions with NO.

The photochemical systems can be reproduced in the laboratory, and the relationship of oxidant levels to hydrocarbon and oxides-of-nitrogen concentrations can be developed. However, because of the complex systems involved in atmospheric mixing, these laboratory relationships cannot be extrapolated to the ambient air.

Detection and Analysis: The photochemical oxidant pool consists of both organic and inorganic gases capable of oxidizing specific reagents. The reagent used most frequently is a neutral-phosphate-buffered potassium iodide solution which is calibrated by bubbling through a known concentration of ozone. The oxidizing intensity of ambient air can be monitored by measuring the amount of iodine released from the potassium iodide solution.

$$O_3 + 2KI + H_2O \longrightarrow O_2 + 2KOH + I_2$$

In the past, the buffered KI method was recommended as the reference method for calibration. Ultraviolet photometry is currently recommended as the primary calibration method.

Monitoring for oxidants where reducing agents such as SO_2 and H_2S are present will reduce the iodine released from the potassium iodide solution and thus reduce the oxidant readings. "Corrected" or "adjusted" oxidant concentrations indicate measurements which have been corrected for sulfur dioxide, nitrogen dioxide, and hydrogen sulfide.

Other devices used for monitoring ozone are chemiluminescence and coulometric analysis. Long-path infrared spectroscopy and gas chromatography have been used to monitor PAN.

10.17 Discuss various methods of land filling and its operation. Also discuss control of gas movement and leachate movement in landfill sites.

[10 marks : 2013]

Solution:

Land Filling: The following two methods are most commonly adopted for land filling viz.

(i) Area method (ii) Trench method

Area Method of Land Filling: It is used when it is not possible to excavate further at the chosen land fill site usually in the conditions of high ground water table. Here in the entire land fill site is divided into subareas by constructing embankments and roads. Roads over the embankments act as service roads for bringing solid waste from city to land fill site. Waste is simply brought and dumped on the ground, spread in layers of about 500 mm thickness and compacted.

Trench Method of Land Filling: When there is a possibility to excavate the land fill site, this trench method of land filling is employed. Here a trench is excavated of sufficient width to allow easy movement of vehicles on it. Daily this trench is filled with waste. Earth excavated from the trench can be used to provide daily cover as well as find cover on waste fillings.

Control of Gas Movement in Land Fill Sites:

(i) **By Vents and Barriers:** The latest movement of gas produced in a land fill can be managed by providing vents. These vents are generally constructed from gravels. Gravel layer of cell vent is placed directly above the daily cell cover.

(ii) **By Gas Recovery:** In very large sized landfills, the gas evolved due to decomposition of waste in collected through gas recovery wells. The recovered gases can be used for meeting the daily energy demands such as for cooling, power generation etc.

Control of Leachate Movement in Land Fill Site: Usually leachate is found at the bottom of land fill and from there its downward and lateral movement occurs. As leachate moves downward, many of the chemical and biological constituents are removed by filtering and adsorptive action of soil. However due to potential contamination of GLO by leachate, it is always desirable either to eliminate the production of leachate or to collect and treat it separately. For this, geotaxtiles are used along with clay lining.

Also by the use of impervious clay layer over the top of landfill with proper side slopes and adequate drainage, surface infiltration can be controlled from the land fill.

10.18 **What are various secondary air pollutants? Discuss sources, origin and hazardous effect of sulphur dioxide and carbon monoxide on human body.**

[10 marks : 2013]

Solution:

Secondary air pollutants: The large no. of primary air pollutants present in the air often react with one-another aided by sun-light and water vapours to give rise to totally new types of pollutants called as secondary air pollutants.

Various secondary air pollutants are:

(i) Sulphuric acid $SO_2 + H_2O \rightarrow H_2SO_4$

(ii) Ozone $3O_2 \rightarrow 2O_3$

(iii) Formaldehydes

(iv) Peroxy-acyl-nitrate (PAN)

Sulphur Dioxide (SO_2): SO_2 when inhaled, affects the mucous membranes of human body. It is a primary pollutant which in combination with water vapours forms a much hazardous secondary air pollutant viz. sulphuric acid (H_2SO_4). SO_2 increases the breathing rate and causes oxygen deficit in human body. SO_2 usually originates from oil refineries, chemical factories and combustion of fuels. Thermal power plants also emit SO_2 in huge proportions.

Carbon Monoxide (CO): CO mainly originates from combustion of fuels in vehicles. Incomplete combustion of organic matter in limited supply of air gives rise to CO.

CO has a very high affinity towards blood haemoglobin (Hb) rather than with oxygen. Thus when inhaled, it replaces oxygen and forms carboxy-haemoglobin which is of no use for the human respiratory system. Thus when nearly half of the blood haemoglobin gets used up to focus carboxy-haemoglobin, death results.

10.19 A family of four people generates solid waste at the rate of 0.45 kg/cap/day and the bulk density of refuse in a typical garbage cans is about 120 kg/m^3. If collection is once a week, how many 120 liters garbage cans will be needed for the household?

[8 marks : 2014]

Solution:

Given data:

Solid waste generated at the rate of = 0.45 kg/cap/day

$$\text{Bulk density } (\rho) = 120 \text{ kg/m}^3$$

In a week total solid waste generated $= 0.45 \times 7 \times 4 = 12.6$ kg

$$\text{Volume of solid waste generated in a week} = \frac{12.6}{120} = 0.105 \text{ m}^3 = 105 \text{ litres}$$

$$\text{Numbers of 120 liters garbage cans} = \frac{105}{120} = 0.875 \simeq 1$$

Hence numbers of 120 liters garbage can is one for one week.

10.20 Determine the area required for a new landfill site with a projected life of 20 years for a population of 150000 generating 25 kg per household per week. Assume that the density of waste is 500 kg/m^3. A planning restriction limits the height of the landfill to 10 m.

[8 marks : 2014]

Solution:

Garbage produced per household per week = 25 kg

According to 2011 census, the household size is below 4 persons in cities and it generally between 4 to 6 persons depending on the state.

Landfill site are generally designed for rural area, so we will be taking 5 persons as the optimum size of the household.

$$\text{Population} = 150000$$

$$\text{Number of household} = \frac{150000}{5} = 30000$$

$$\text{Projected life} = 20 \text{ years}$$

Total garbage expected to produced in 20 years,

$$\text{Total garbage produced} = \frac{25}{7} \times 365 \times 30000 \times 20 \text{ kg} = 782,\ 142,857.1 \text{ kg}$$

$$\text{Volume of garbage,} \quad V = \frac{M}{\rho} = \frac{782,142,857.1 \text{kg}}{500 \text{kg/m}^3} = 1564285.7 \text{m}^3$$

Since height of landfill is restricted to 10 m, take $H = 10$ m and let 'A' be area required for new landfill site then

$$V = A \times H$$

$$1564285.7 = A \times 10$$

$$\therefore \quad A = 156428.57 \text{ m}^2 \approx 15.6429 \text{ ha}$$

10.21 Traffic noise data are given in the table below. Compute L_{eq}.

Time (in s)	10	20	30	40	50	60	70	80	90	100
Noise (dBa) $L(t)$	71	75	70	78	80	84	76	74	75	74

[8 marks : 2014]

Solution:

L_{eq} is defined as the constant noise level, which, over a given time, expands the same amount of energy, as is expanded by the fluctuating levels over the same time.

The L_{eq} is calculated as
$$L_{eq} = 10\log_{10}\sum_{i=1}^{i=n}(10)^{\frac{Li}{10}} \times t_i$$

where, L_i = The noise level of any i^{th} sample

t_i = Time duration of ith sample expressed as fraction of total sample time

Here, Total sample time = 100 sec

Time (in s)	10	20	30	40	50	60	70	80	90	100
Noise (dBA) L(t)	71	75	70	78	80	84	76	74	75	74
$t_i = \frac{t}{100}$	0.1	0.1	0.1	0.1	0.1	0.1	0.1	0.1	0.1	0.1
$\frac{L_i}{10}$	7.1	7.5	7.0	7.8	8.0	8.4	7.6	7.4	7.5	7.4
$\Sigma(10)^{\frac{L_i}{10}} \times t_i$	59016763.06									

$$\therefore \quad L_{eq} = 10\log_{10}\sum_{1}^{10}(10)^{\frac{L_i}{10}} \times t_i$$

$$= 10 \times \log_{10}(59016763.06) = 77.71 \text{ dBA}$$

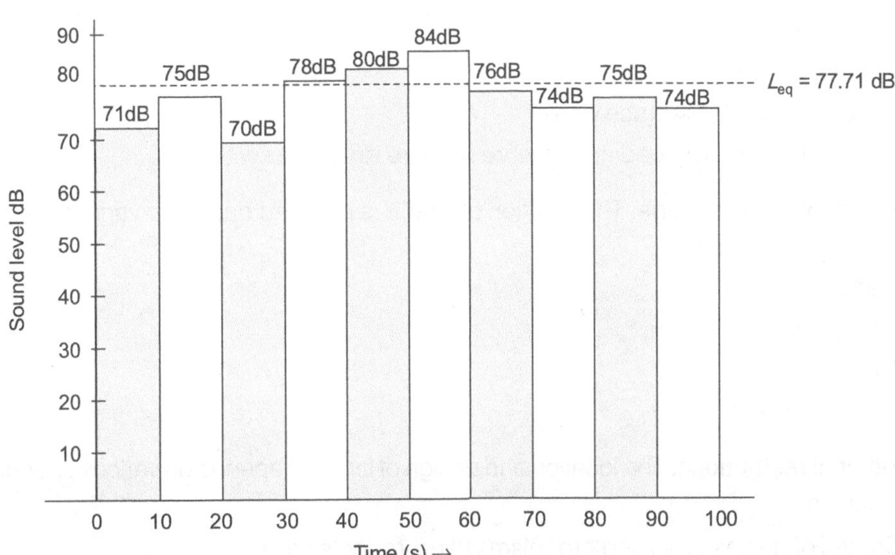

10.22 A town of population of 50000 generates municipal solid waste of 500 g per person per day. How many trucks would be needed to collect the waste twice weekly? The trucks have a capacity of 4.5 t (metric tonnes) each and operate 5 days per week. Assume that the trucks average two load per day at 75% capacity. If 20% of the waste generated is recycled, determine the weight of MSW that enters the landfill. If the density of the waste is 280 kg/m^3, what is the volume of MSW?

[8 marks : 2015]

Solution:

Population of town = 50000

Municipal solid waste (MSW) per person = 500 g/day

Total MSW generated in week $= 50000 \times 500 \times 7 \times 10^{-6}\, t = 175\, t$

Waste is to be collected two times in a week.

$\therefore$ Waste to be collected at one time $= \frac{175}{2} = 87.5t$

Loading capacity of each truck $= 0.75 \times 4.5t = 3.375\, t$

When 20% generated waste is recycled, MSW that enters landfill

$= 0.8 \times 175t = 140t$ per week

$$\text{Number of trucks required} = \frac{\text{Waste to be collected}}{\text{Capacity of one truck}} = \frac{87.5}{3.375}$$

$$= 25.93 \simeq 26 \text{ trucks}$$

$$\text{Volume of MSW per week} = \frac{\text{Weight of MSW that enters landfill}}{\text{Density}} = \frac{140}{280} \times 1000 = 500\text{m}^3$$

10.23 Describe briefly the environmental factors considered in designing a landfill.

[4 marks : 2016]

Solution:

Environmental factors affecting the design of landfill are following:

(i) **Rainfall intensity:** Rainfall intensity affects the landfill design to a significant extent. The problems associated with too much of rainfall are:
- Slope failure
- Leachate seepage
- Adverse impact of surface water
- Too much of erosion leading to failure of entire drainage system

(ii) **Environmental restrictions:** The location of landfill is decided based on various restrictions imposed in the locality such as:
- Wetlands
- Floodplains
- Airports
- Fault areas, etc.

(iii) **Operational restrictions:** The location and design of landfill depends on various operational restrictions also such as:
- Control of disease causing organisms (flies, rodents etc.)
- Protection of surface water

- Restriction on liquid disposal
- Ban on open burning
- Control of explosive gases, etc.

10.24 Define air pollution as per OECD definition. Differentiate between 'natural' and 'anthropogenic' air pollution sources, with examples. What are the effects of ambient ozone, PM_{10}, $PM_{2.5}$ and lead on human health?

[8 marks : 2016]

Solution:

Air pollution: As per OECD (Organisation for Economic Co-operation and Development), it is defined as: "*Air pollution is the presence of contaminant or pollutant substances in the air that do not disperse properly and that interfere with human health or welfare, or produce other harmful environmental effects.*"

Natural air pollution is caused due to natural forces and are not caused by human activities. Some of the sources of natural air pollutants are as follows:

- Dust originating from eruption of volcanoes
- Dust storms
- Forest fires
- Volatile organic compounds emitted from plants that are converted to aerosols

Anthropogenic air pollution is caused by the human activities. Some of the sources of anthropogenic air pollutants are:

- Powerplant emission
- Open burning of wastes
- Dust originating from construction sites and mining
- Industrial emissions
- Vehicular emissions

Effect of ambient ozone on human health: Inhalation of ozone causes various respiratory disorders like inflammation of air ducts, dis-functioning of lungs etc. Its symptoms include coughing, throat irritation, discomfort in chest while breathing etc.

Effect of PM_{10} on human health: PM_{10} has the ability to reach the lower regions of respiratory tract. It may cause damage to lung tissues, influenza, cancer and pre-mature death.

Effect of $PM_{2.5}$ on human health: $PM_{2.5}$ causes cardio-vascular disorders which increases mortality. It also increases the risk of cancer.

Effect of lead on human health: Lead is highly dangerous for children health. It causes learning disabilities, damages nervous system and kidneys, causes speech and lung impairment, reduces bone growth etc.

In adults, lead causes high blood pressure, digestion disorders, nervous disorders, muscle and joint pain etc.

10.25 Discuss in brief the essentials of rural sanitation. Explain how you dispose off (i) dry refuse, (ii) sullage and (iii) excretal waste in rural area.

[8 marks : 2016]

Solution:

Essentials of rural sanitation: As far as rural sanitation is concerned, **the following points must be observed:**

(i) Rural population is generally of low economic level and thus sanitation method in rural areas must be cost effective both in terms of construction and maintenance.

(ii) In the absence of proper waste treatment plants in rural areas, the sanitation method must aim to dispose off the waste near to its source of generation, like providing privy in houses etc.
(iii) Any rural sanitation method must require low maintenance.
(iv) The sanitation method must not be temperature susceptible i.e. it should work in all weathers round the year.
(v) Like in any other sanitation method, rural sanitation must not contaminate the soil and the ground water.

Disposal of dry refuse in rural area: Dry refuse includes vegetable wastes, paper wastes etc. This type of waste does not give offensive odours. It can be disposed off by collecting them in dustbins followed by collection at one place which is done by sweepers. From here, it is lifted up in waste collecting lorries. Also it can be dumped in low lying areas to raise the average level of ground. It can also be disposed off by burning but this method pollutes the environment and is banned in many cities.

Disposal of sullage in rural area: Sullage (domestic wastewater coming out from kitchens and bathrooms) contains chemicals used in household cleaning and washing. Sullage can be used for gardening purpose particularly in arid areas where water availability is sparse. Sullage can also be allowed to flow into the septic tank if already built.

Disposal of excretal waste in rural area: Rural excretal waste can be disposed off through *privy method.* Privy method helps to decompose the excreta as soon as it is deposited and gets converted to odourless and stable product. This decomposition of excreta may take time varying from few days in carefully controlled composting to several months like in pit privy. The end products of decomposition contain nutrients useful for soil and can be used as manure.

10.26 Describe the method of sampling suspended particulate matter by high volume sampler. Also explain how the mass concentration of particulates is determined.

[8 marks : 2016]

Solution:

Sampling suspended particulate matter by high volume sampler:

- High volume sampler can sample more than 1500 m^3 of air in 24 hours which is much higher than low volume sampler.
- This high volume sampler is used to collect large volume of air for 24 hours through a pre-weighed filter.
- The filter used in high volume sampler is generally made of glass fiber with collection efficiency upto 99%.
- Many a times, cellulose is also used as the filter material.
- Sampling of air for suspended particulate matter may either be *continuous type or time averaged* sampling.
- Location of sampling plays a very crucial role. Sampling must be done at the breathing level height if objective is to study the health hazards. For vegetation purpose, sampling must be done at foliage level. For determining the background concentrations, the location of sampler must be away from the source of pollution.
- The number of samples depends on the availability of concentration in the area to be investigated.
- The period, frequency and duration of sampling depend on the purpose of study.

Determination of mass concentration of particulates:

- After sampling, the filter is weighed again. The difference in the two weights of filter gives the mass of particulate matter collected on the filter.

- Dividing this mass of particulate matter with the volume of air sampled gives concentration of suspended particulate matter.
- This collected particulate matter can be further analyzed to determine the various pollutants present in it like lead and other heavy metals.

10.27 **Present classification of natural ecosystems and describe the components of ecosystem.**

[10 marks : 2017]

Solution:

Classification of natural ecosystems: Natural ecosystems operate under natural conditions without any major interference by man.

1. **Terrestrial Ecosystem :** Forest grassland and desert etc.
2. **Aquatic Ecosystem :**
   - *Freshwater* : Lotic (running water like spring, stream or rivers) or lentic (standing water as lake, pond, pools etc.
   - *Marine water* : Such as deep bodies as ocean or shallow ones as sea or an estuary.
   - *Saline water*

Components of ecosystem : An ecosystem is comprised of biotic and abiotic organisms. Energy cycles operate through abiotic and biotic organisms to maintain the balance within the ecosystem. The abiotic parts of the ecosystem include the non-living components such as air, water and the basic components of the environment.

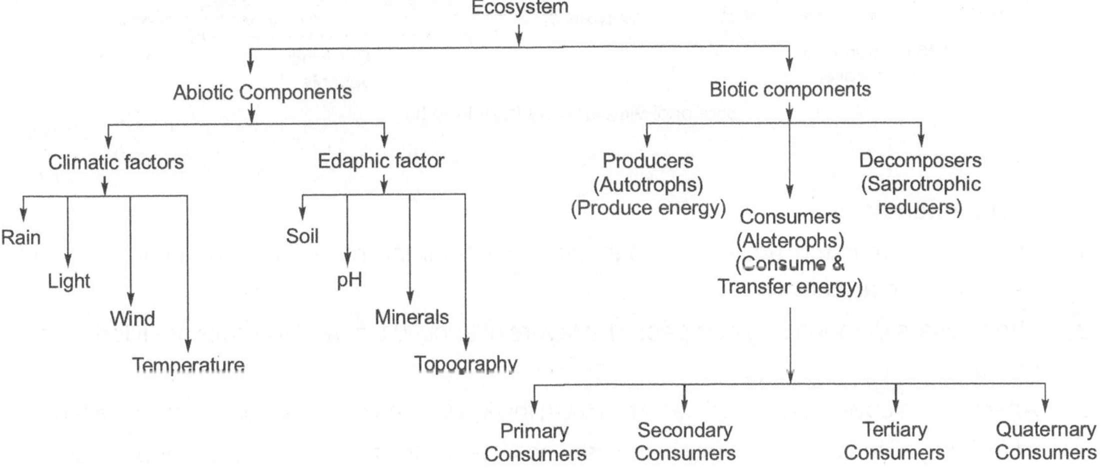

10.28 **Explain in detail how landfills are environmentally safe over open dumps. Explaining landfill operations, state the problems associated with landfills.**

[10 marks : 2018]

Solution:

Landfills are an environmentally safer alternative as compared to open dumps because:

- In this method of refuse disposal, refuse in carried and dumped into low-lying area under an engineered operation, designed and operated in an environmentally sound manner, as not to cause any public nuisance or hazards to public health or safety.
- The refuse is well compacted with bulldozers, trucks, rollers, etc., and is well covered with good earth, it doesn't cause any trouble like mosquito and insects breeding.
- Problem of evolution of unpleasant odours and four smells, as may be the case with open dumps, can be tackled easily.

- The refuse in managed landfills, may usually get stabilized within a period of 2-4 months and settle by 20-40% of its original volume/height. The filled up land can be used for developing some green land, parks or other recreational spots.
- The movement of gas in landfills can be controlled with properly designed engineering methods such as:
  - constructing vents and barriers
  - by gas recovery

Hence landfills are better than open dumps.

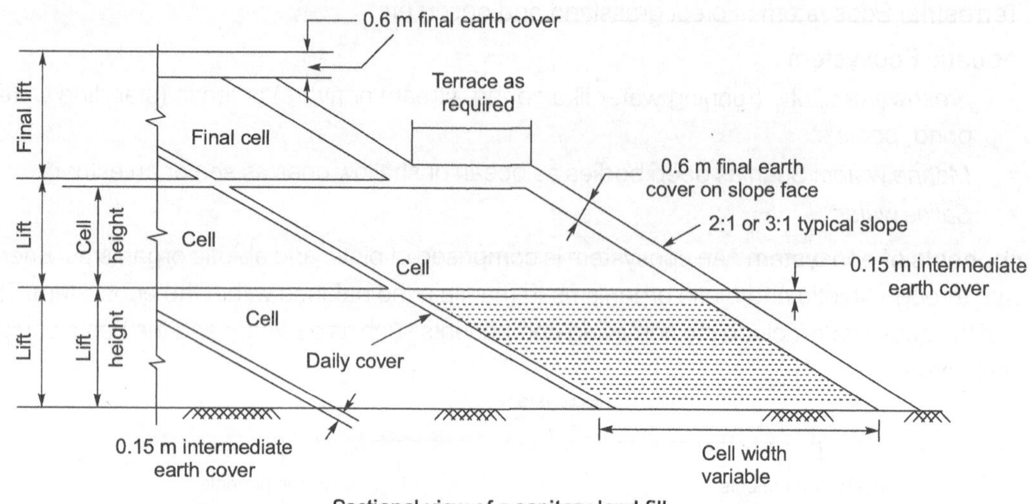

Sectional view of a sanitary land fill

Landfilling operation:

1. The filling of refuse is actively done in sanitary land filling by dividing the entire area into smaller portion called as cells.
2. The refuse is dumped and compacted in layers of about 0.5 m and the cells are filled upto the height of 1.5 m.
3. After this it is covered by good earth of 15 cm thickness. The cover of good earth is called daily cover.
4. After filling all the cells with first lift, the second lift is laid in about 1.5 m height and covered with good earth cover of about 0.15 m depth, called the intermediate cover.
5. The process will continue till the top most lift is piled up, over which the final cover of good earth of about 0.6 m depth shall be laid and well compacted.

Problem associated with landfills:

1. Low-lying depressions or dumping sites may not always be available.
2. There is continuous evolution of foul smells near badly maintained landfills and sometimes these gases are explosive in nature.
3. Evolution of greenhouse gases is also a major area for concern.
4. Since, the dumped garbage may contain harmful and sometimes carcinogenic non-biodegradable substances such as plastics unused medicines, paints, insecticides, sanitary napkins etc., they may start troubling on a later date, especially during rainy season, as leachate.
5. This leachate is highly polluted and hence very poisonous and may seep to the ground water table and contaminate it.

10.29 **With the help of line sketches, explain in working principles of gravity settlers and electrostatic precipitators used for air pollution control. Also, list their applications and limitations.**

[10 marks : 2018]

Solution:

(a) **Gravity settlers:**

Working principle: The emitted smokes, when made to pass through a settling chamber, drop some of their larger sized particles in the chamber under Stoke's law.

The largest size particle, d, that can be removed with 100% efficiency in a chamber of length L and height H is given by

$$d = C\sqrt{\frac{18\mu V_h H}{gL\rho_p}}$$

where, V_h = Horizontal velocity of gas, ρ_p = Density of particles
μ = Viscosity of air, C = Correction factor

Particulate Emission Control Devices

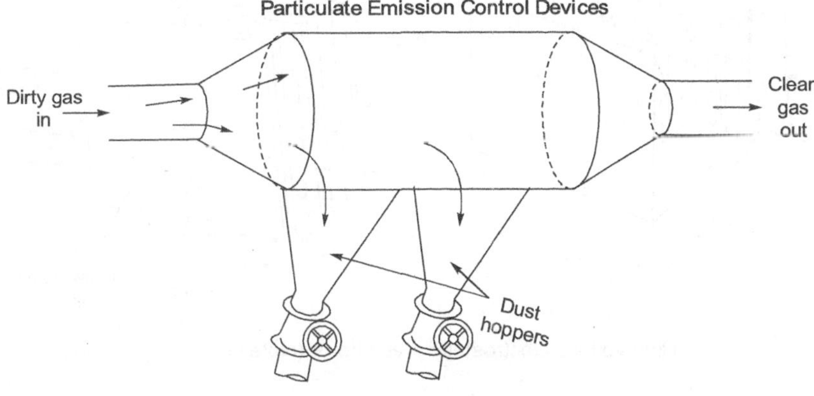

Gravitational Settling Chamber

Advantages	–	Simple to design
	–	Easy to maintain
	–	Low pressure loss
Limitations	–	Requires larger space
	–	Collection efficiency is low
	–	Only larger sized particles are separated out
Minimum particles size	>	50 μm
Efficiency	<	50%

(b) **Electrostatic Precipitators:**

Working principle: In electrostatic precipitator, the flue gas is made to pass through highly ionized zone, where the particles get electrically charged and are separated out from gas, with the help of electrostatic forces in the powerful electric field.

Advantage:	–	Particles may be collected wet or dry
	–	99% + efficiency
	–	Even small particles can be removed
	–	Maintenance is nominal, unless corrosive and adhesive materials are present in flue gas
	–	Can be operated at high temperatures upto 400°-500°C

Limitations
- Higher initial costs
- Sensitive to variable dust loadings
- Risk to personal safety, because they are high voltage
- Collection efficiency reduces with time

Minimum particle size that can be removed > 1 μm

Efficiency 95-99%

Use: Widely used in thermal power plants, paper and pump industries, mining and metallurgical industries iron and steel plants etc.

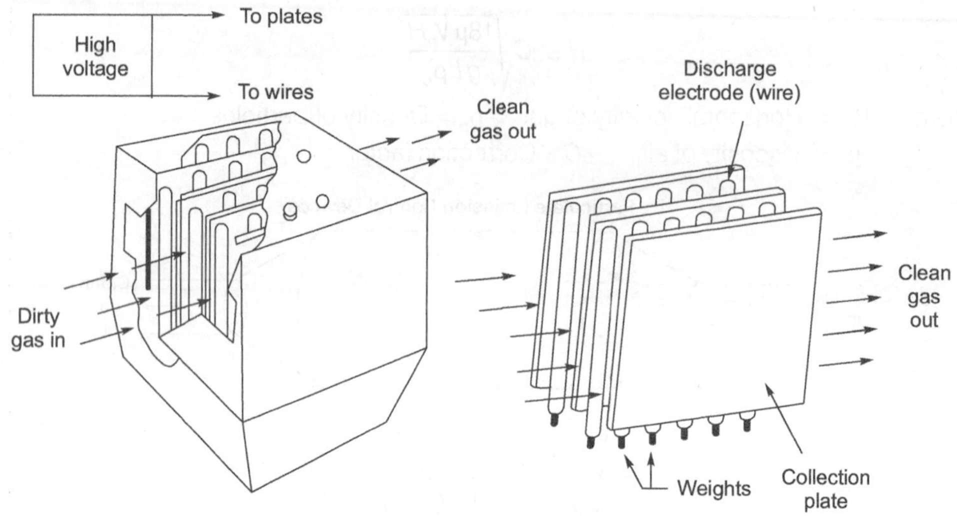

High voltage electrostatic precipitator (plate type)

10.30 Compute the average sound pressure level from the following sound pressure readings:

(1) 39 dBA (2) 52 dBA
(3) 67 dBA (4) 77 dBA

[5 marks : 2019]

Solution:

Average sound level is given as

$$\bar{L}_P = 20\log\frac{1}{N}\sum_{n=1}^{N}(10)^{L_n/20}$$

$$= 20\log\frac{1}{4}\left[10^{39/20} + 10^{52/20} + 10^{67/20} + 10^{77/20}\right] = 67.78 \text{ dB}$$

10.31 Discuss the isokinetic sampling process of flue gas stack sampling and explain with the help of diagram, how the results will be affected if the sampling is not done isokinetically.

[10 marks : 2019]

Solution:

Isokinetic sampling is a technique used to collect particles in a moving stream which moves at the same velocity in the sampling nozzle as elsewhere in the stream. This can increase the accuracy and reliability of results. It is used for activities like monitoring pollution in factory stacks, taking general air samples in an area of interest, and checking equipment for dust and other concerns.

In an isokinetic sampling procedure, the sampling nozzle is set up to allow the sample stream to enter without changing speed. This reduces the risk of concentrating larger or smaller particles.

If sampling is done other than at isokinetic velocity, then it will cause errors for two reasons. First sampling at greater or less than isokinetic rates tends to cause respectively larger or a smaller volume to be withdraw from the flue gases than accounted for by the cross-section area of the probe. Also, particles greater than 3.5μ size have sufficient inertia so that particle motion may deviate significantly from the gas flow stream line pattern, Thus, if sampling is not done isokinetically, then we will be unable to obtain representative sample.

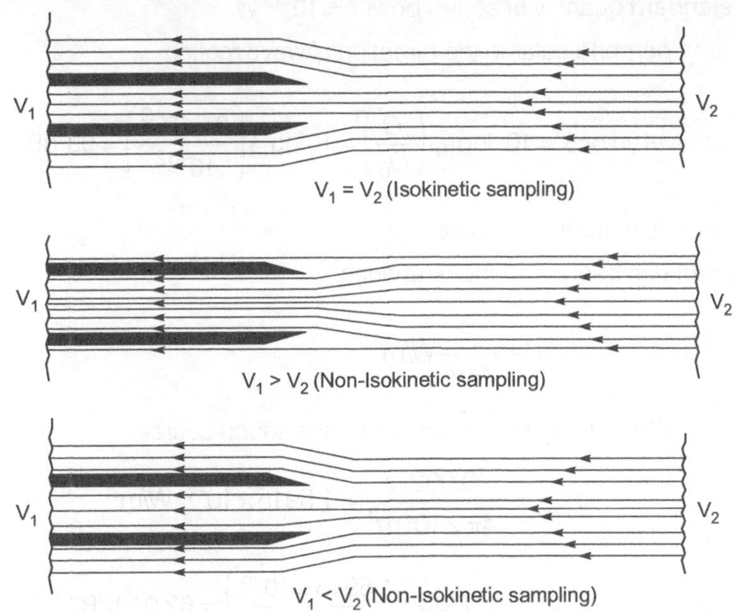

10.32 **A town with a population of 3 lakh produces solid waste at a rate of 2.5 kg/capita/day. If the waste is compacted to a density of 1500 kg/m³, how much volume of landfill site is needed in a year? Assuming that the ratio of solid waste to cover is 4 : 1, what volume of cover soil is needed in a year? What type of soil would you recommend as a cover?**

[4 marks : 2020]

Solution:

Given, Population, $(P) = 300000$

Rate of productivity $= 2.5$ kg/capita/day

$\therefore$ Total solid waste produced $= 2.5 \times 3 \times 10^5 = 7.5 \times 10^5$ kg/day

Density of waste $= 1500$ kg/m³

$$\therefore \quad V_{\text{solid waste}} = \frac{7.5 \times 10^5}{1500} = 500 \text{ m}^3/\text{day}$$

Ratio of solid waste to cover = 4 : 1

$$\therefore \quad V_{\text{cover soil}} = \frac{V_{\text{solid waste}}}{4} = \frac{500}{4} = 125 \text{ m}^3/\text{day}$$

$\therefore$ Total volume of landfill site required (per day)

$$= V_{\text{solid waste}} + V_{\text{cover soil}} = 500 + 125 = 625 \text{ m}^3/\text{day}$$

Volume of landfill site required in a year $= 625 \times 365 = 228125$ m³

Volume of cover soil required in a year $= 125 \times 365 = 45625$ m³

For cover purposes, it is recommended by GOI manual on MSWM, 2016 to use clay after thorough compaction.

10.33 **The sound power from a voice shouting is 0.002 *W*. What is the Sound Power Level? What are the Sound Intensity, Sound Intensity Level, the Sound Pressure and the Sound Pressure Level at a distance of 10 metres from the source? Assume that sound radiates from the source in all directions.**

[8 marks : 2020]

Solution:

Q = Measured quantity of sound power = 0.002 W

Q_0 = Reference standard quantity of sound power = 10^{-12} W

Assuming a unit area perpendicular to the direction of wave motion.

$$\therefore \quad \text{Sound power level (in dB)} = 10 \cdot \log_{10} \cdot \left(\frac{Q}{Q_0}\right) = 10\log_{10}\left(\frac{2\times 10^{-3}}{10^{-12}}\right) = 93 \text{ dB}$$

Now, at a distance of 10 m from the source,

Since, the sound radiates freely in all the directions,

$$\text{Sound intensity,} \quad I = \frac{P}{4\pi r^2} \text{ W/m}^2$$

where, r is the radius of imaginary sphere enclosing the sound source.

$$\therefore \quad I_{10\text{ m}} = \frac{0.002W}{4\pi \times 10^2 \text{ m}^2} = 1.5915\times 10^{-6} \text{ W/m}^2$$

$$\text{Sound intensity level,} \quad L_I = 10\log_{10}\left(\frac{1.5915\times 10^{-6}}{10^{-12}}\right) = 62.018\text{ dB}$$

Sound pressure level is measured in rms (root mean square).

$$62.018 \text{ dB} = 20\log_{10}\left(\frac{p_{rms}}{p_{rms_0}}\right)$$

p_{rms_0} is reference standard rms pressure

$$p_{rms_0} = 20\ \mu\text{Pa}$$

$$\therefore \quad 62.018 \text{ dB} = 20\log_{10}\left(\frac{p_{rms}}{20\mu Pa}\right)$$

$$\Rightarrow \quad p_{rms} = 25230\ \mu\text{Pa}$$

10.34 **In a factory, coal is burnt at a rate of 1 kg/second. Analysis of the coal reveals a sulphur content of 3 percent. The sulphur in the ash is 5 percent of the input sulphur. What is the annual rate of emission of sulphur dioxide?**

[10 marks : 2020]

Solution:

For a second, $\quad$ Coal burnt = 1 kg

$$\text{Sulphur content in coal} = 3\% = \frac{3}{100} \text{ kg}$$

$$\text{Sulpher in ash} = \frac{5}{100}\times\left(\frac{3}{100}\right) = 0.0015 \text{ kg}$$

Amount of sulphur which reacted with O_2 to become SO_2

$$= 0.03 - 0.0015 = 0.0285 \text{ kg} = 28.5 \text{ g}$$

Chemical reaction $S + O_2 \longrightarrow SO_2$

Using mole concept

1 mole of sulphur produces 1 mole of SO_2

$$n_S = \frac{28.5}{32} = 0.89 \left[\text{No. of moles} = \frac{\text{Weight (g)}}{\text{Molecular weight}}\right]$$

$$n_{SO_2} \text{ required} = 0.89$$

$$(wt)_{SO_2} = 0.89 \times 64 = 57 \text{ g} = 0.057 \text{ kg}$$

In one second 0.057 kg of SO_2 is emitted

In a year SO_2 emission = $0.057 \times 86400 \times 365 = 1797.552\ t$

10.35 Describe various functional elements of a solid waste management system.

[10 marks : 2020]

Solution:

Functional elements of a solid waste management system: The activities involved with the management of solid wastes from the point of generation to final disposal have been grouped into six functional elements.

1. Waste generation
2. On-site handling, storage and processing
3. Collection
4. Transfer and transport
5. Processing and recovery
6. Disposal

Table : Description of the functional elements of a solid waste management system

Component	Description
Waste generation	Those activities in which materials are identified as no longer being of value and are either thrown away or gathered together for disposal
On site handing. storage, and processing	Those activities associated with the handling, storage and processing of solid wastes at or near the point of generation.
Collection	Those activities associated with the gathering of solid wastes and the hauling of wastes after collection to the location where the collection vehicle is emptied
Transfer and transport	Those activities associated with (1) the transfer of wastes from the smaller collection vehicle to the larger transport equipment and (2) the subsequent transport of the wastes, usually over long distance, to the disposal site.
Processing and recovery	Those techniques equipment and facilities used to improve the efficiency of the other functional elements and to recover usable materials, conversion products or energy from solid wastes
Disposal	Those activities associated with ultimate disposal of solid wastes, including those wastes collected and transported directly to a landfill site, semisolid wastes (sludge) from wastewater treatment plants, incinerator residue, compost or other substances from the various solid-waste processing plants that are of no further use.

10.36 Differentiate and compare anaerobic digestion process and composting process used for solid waste treatment.

[10 marks : 2020]

Solution:

Anaerobic digestion process and composting process used for solid waste treatment

Anaerobic Digestion:

- It is common process for dealing with wastewater sludge containing primary sludge, because it contains large amounts of readily available organics that would induce a rapid growth of biomass if treated anaerobically.
- The anaerobic decomposition produces considerably less biomass than aerobic process.
- The principal function of anaerobic digestion, therefore is to convert as much of the sludge as possible to end products such as liquids and gases while producing little residual biomass as possible.
- In anaerobic process, the organisms are broadly classified as
  (a) acid formers
  (b) methane formers

Acid formers:

(i) They consist of facultative and anaerobic bacteria and include organisms that solubilize the organic solids through hydrolysis.

(ii) The soluble products are then fermented to acids and alcohols of low molecular weight.

Methane formers:

(i) They consist of strict anaerobic bacterias that convert the acids and alcohols, along with the hydrogen and carbon dioxide to methane.

(ii) Methane former act in the pH range of 6.5 - 7.5 and they are very delicate. A shock loading can be disastrous to such anaerobic bacteria.

(iii) Acid former respond quickly to food supply and hence acid increase quickly. The methane formers do not respond so quickly and hence pH may reduce.

(iv) As pH level falls down below "pH tolerance level" of the methane former, methane formation ceases and the pH may fall to even toxic level. To control this lime is added.

Advantage of Anaerobic Digestion:

1. Recovery the energy in form of Methane
2. Anaerobically digested sludge contains nutrients and organic matter that can improve the fertility of soil.
3. Pathogens in the sludge die off during relatively long detention period.

Disadvantage of Anaerobic Digestion:

1. Close process. Control is required to prevent upsets.
2. Supernatant liquid from an anaerobic disaster have high O_2 demand.

The digested sludge is dewatered, dried up and used as fertilizer; while the gases produced are also used for fuel or for driving gas engines.

Composting:

- Composting may be defined as a method of solid waste treatment in which the organic component of the solid waste stream is biologically decomposed under controlled aerobic conditions to a state in which it can be easily and safely handled, stored and applied to the land without adversely affecting the environment.
- Thus, composting is a controlled or engineered biological system.
- It is often a cost effective and environmentally alternative for the stabilisation and ultimate disposal of wastewater sludge.

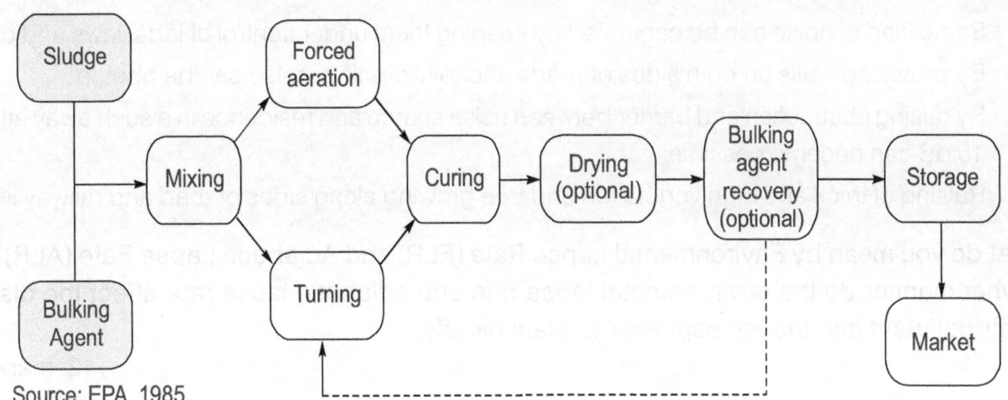

- The three major types of composting systems used are the aerated static pile, window and invessel system.
- Majority are aerated static pile followed by windows and by invessel.

10.37 Define 'noise' and explain as to why and how it should be regarded as an environmental pollutant. Also explain briefly the major factors and actions that may help in noise abatement in a modern society.

[12 Marks : 2021]

Solution:

Sound in the environment is caused by vibrations in the air (or some other medium) that reach human ears and stimulate a sensation of hearing. When the sound becomes loud, or disagreeable, or unwanted, it becomes noise. Since the unwanted sound (i.e., noise) certainly produces severs undesirable effects on our body health, it can be termed as an environmental pollutant. The air (prevention and control of Pollution) Act 1981, includes noise as one of the air pollutant.

Noise can, therefore, also be defined as that unwanted sound pollutant, which produces undesirable physiological and psychological effects in the individual, by interfering with one's social activities like work, rest, recreation, sleep, etc.

Harmful effects of noise include:

- **Noise induces annoyance:** One of the most important effect of noise on humans is annoyance and irritation due to disturbance.
- **Noise induced disease:** Noise may produce several undesirable physiological and psychological diseases in human beings. The disease caused may include: anxiety, tenses, nervousness, headaches, fatigue, nausea, insomnia, high blood pressure, high pulse rate, greater perspiration, gastric secretions, etc.
- **Sleeplessness:** The noise may induce sleep disturbances including shorter sleep durations, more frequent awakenings, etc.
- **Communication interference:** Noise can badly disturb communication when a person is speaking on telephone, or when individuals are talking face to face.
- **Noise induced hearing loss:** Exposure of human ears to intense noise for a long enough duration may cause damage to the inner ear, thereby decreasing ones ability to hear.
- **Effect of noise on wild life:** Wildlife, like humans, is also badly affected by noise. Health of several zoo-animals, particularly those of deers, rhinos, lions, etc. are adversely affected by noise. Several migratory birds have stopped resting in habitates close to noise cities.

Control of Noise Source by Design

(a) Some kind of noise can be controlled by keeping them under control of legal laws and ordinance.

(b) By providing walls on both sides of roads and railways line, noise can be abated.

(c) By raising obstruction and barrier between noise source and residence in a such away attenuation of 15 dB can become possible.

(d) Raising of thick and high vegetation and tree growing along sides of road and railway lines.

10.38 What do you mean by Environmental Lapse Rate (ELR) and Adiabatic Lapse Rate (ALR)? How and in what manner do the environmental lapse rate and adiabatic lapse rate affect the dispersion of an air pollutant into the atmosphere? Explain clearly.

[12 marks : 2022]

Solution:

Lapse rate: In the troposphere, the temperature of the ambient air usually decreasea with an increase in attitude. This rate of temperature change is called as lapse rate. This rate will differ from place to place and from time to time even at the same place.

Environmental lapse rate (ELR)/Ambient lapse rate: The ELR can be determined by sending up a balloon equipped with a thermometer. The balloon moves through the air and the temperature gradient of ambient air, which the rising balloon measures is called the ambient lapse rate, the environmental lapse rate or the prevailing lapse rate.

Adiabatic Lapse Rate: When a parcel of air, which is hotter and lighter than the surrounding air is released, then naturally it tends to rise up, until it reaches to a level at which its own temperature and density becomes equal to that of the surrounding using the law of conservation of energy and gas law. This rate of decrease of temperature with height is called adiabatic lapse rate. Dry air expanding adiabatically, cools at a rate of 9.8°C per kilometer and it is called dry adiabatic lapse rate. In wet adiabatic process a saturated parcel of air rises and cools. Temperature changes of the parcel are due to liberation of latent heat as well as of expansion of the air. Wet adiabatic lapse rate (6°C/km) is thus lesser than dry adiabatic lapse rate. Since a rising parcel of emitted smokes will normally, neither be fully dry nor fully saturated, the actual adiabatic lapse rate (ALR), representing cooling of the emitted smokes will be between the dry adiabatic rate and wet adiabatic rate. The three major relative position of ELR line with reference to ALR line are as follows:

(a) **Super Adiabatic Lapse Rate:** When the ELR is more than the ALR then ambient lapse rate is super adiabatic and the environment is said to unstable.

Dispersion of pollutants will be rapid due to rapid vertical mixing of the air making the environment unstable.

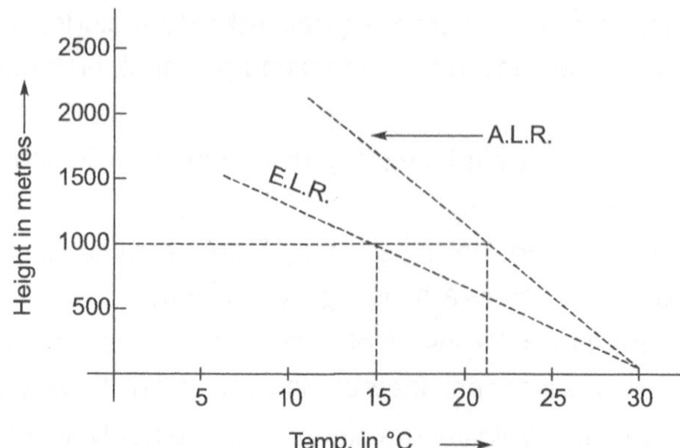

(b) **Sub-adiabatic Lapse Rate:** When ELR is less than the ALR the environment set to be stable and environmental lapse rate is called the sub-adiabatic lapse rate.

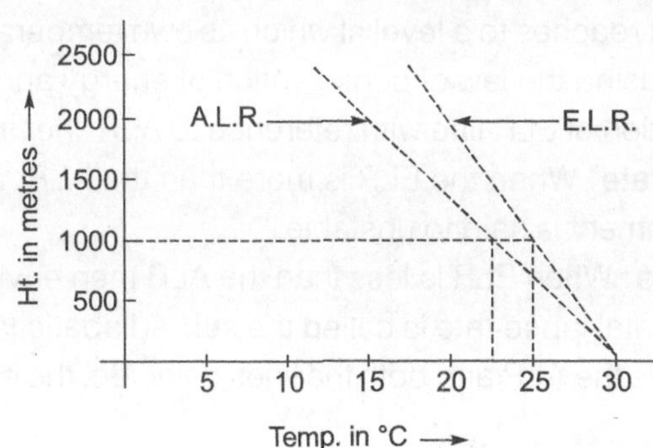

Under such condition, the atmosphere is said to be stable.

(c) **Neutral**: When ELR equals the ALR and both the lines coincide, the environment in such a case is called neutral.

10.39 **A community of 50,000 people uses a 12 ha landfill site that can be filled to an average depth of 20 m. If the municipal solid waste is generated at the rate of 25 N per person per day, and its compacted unit weight in the fill is 8 kN/m³ and the municipal solid waste to cover ratio is 4 : 1, what is the anticipated useful life of the landfill site?**

[12 marks : 2022]

Solution:

Volume of landfill site available $= 12 \text{ ha} \times 20 \text{ m} = 240$ ha-m

Total MSW generated $= 50000 \times 25 \text{ N/d} = 1250000 \text{ N/d}$

$$\frac{MSW}{\text{Cover}} = \frac{4}{1}$$

$$\therefore \quad \text{Weight of cover} = \frac{1}{4} \times 1250000 \text{ N/d} = 312500 \text{ N/d}$$

Total landfill weight filled per day = 1250000 + 312500 = 1562500 N/d

$$\text{Volume of compacted landfill filled per day} = \frac{1562500\, N/d}{8000 \text{ N/m}^3} = 195.31 \text{ m}^3/\text{d}$$

$$= \frac{240 \times 10^4 \text{ m}^3}{195.31 \text{ m}^3/\text{d}} = 12288.16 \text{ days} = 33.67 \text{ yrs.}$$

10.40 **With the help of sketches, explain the effect of lapse rate on plume behaviour indicating the possible plume shapes and dispersion conditions.**

[20 marks : 2023]

Solution:

Lapse Rate: With the increase in altitude, temperature of normal atmosphere decreases above the earth surface (in troposphere). The lapse rate is defined as the rate at which temperature changes with height in the atmosphere. This rate will differ from place to place, and from time to time even at the same place.

Types of lapse rate:

(i) **Environmental lapse rate (ELR)**: The ELR is determined by sending up a balloon equipped with a thermometer. The balloon moves through the air, not with it, and the temperature gradient of ambient air, which the rising balloon measures is called the Ambient lapse rate or the Environmental lapse rate or the prevailing lapse rate.

(ii) **Adiabatic lapse rate (ALR)**: When a parcel of air, which is hotter and lighter than the naturally it tends to rise up, until of course, it reaches to a level at which its own temperature and density becomes equal to that of the surrounding using the law of conservation of energy and gas law.

The three major relative position of ELR line with reference to ALR line are discussed below:

(a) **Super adiabatic lapse rate**: When the ELR is more than the ALR, then ambient lapse rate is super adiabatic and the environment is said to unstable.

(b) **Sub-adiabatic lapse rate**: When ELR is less than the ALR then environment is said to be stable and this prevailing environmental lapse rate is called the sub-adiabatic lapse rate.

(c) **Neutral**: When ELR equals the ALR and both the lines coincide, the environment in such a condition is called neutral.

Plume behaviour:

1. **Looping Plume:** Looping plume has wavy character and occurs in super adiabatic environments which produces highly unstable atmosphere, because of rapid mixing.

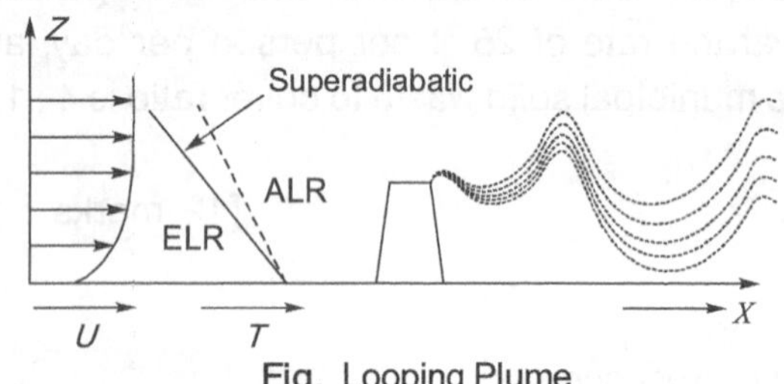

Fig. Looping Plume

2. **Neutral Plume:** Neutral plume is the upward vertical rise of the plume from the stack, which occurs when the ELR is equal to ALR.

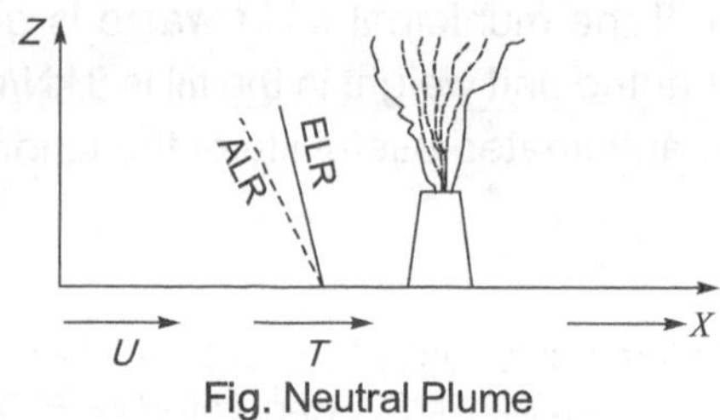

Fig. Neutral Plume

3. **Conning Plume:**
   - Conning plume occurs under sub-adiabatic conditions.
   - The plume dispension is known as conning because the plume makes a cone like shape about the plume line.

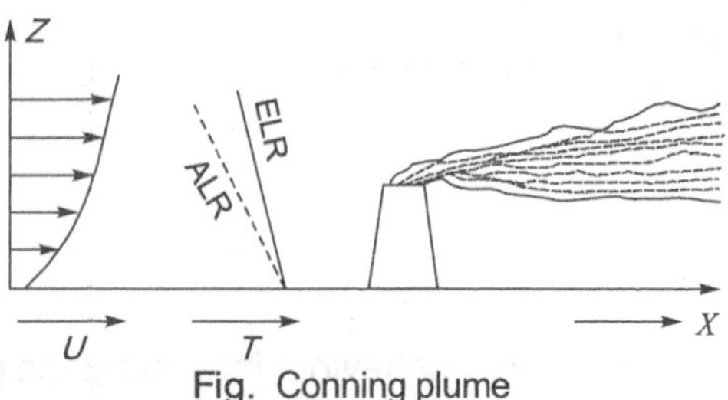

Fig. Conning plume

4. **Fanning plume:** There will be no vertical mixing and the plume will extend horizontally over large distances.

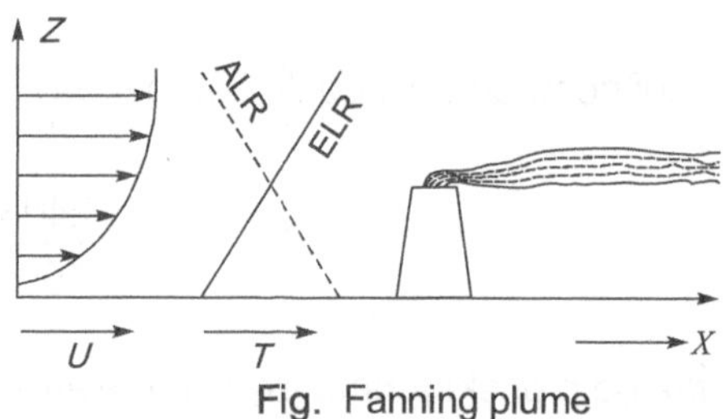

Fig. Fanning plume

5. **Lofting plume:** Where there exists a strong super adiabatic lapse rate above a surface inversion, then the plume is said to be lofting.

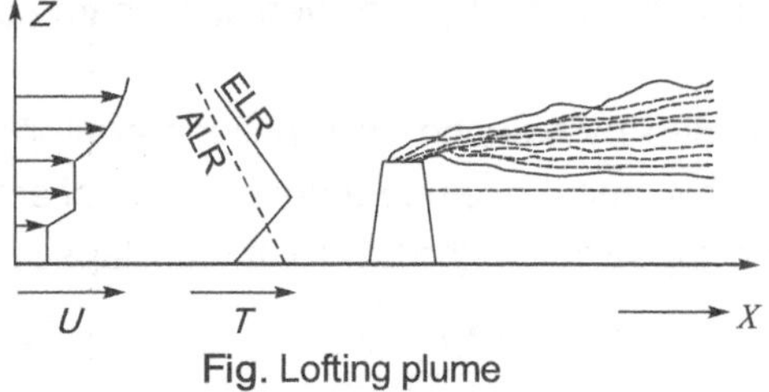

Fig. Lofting plume

6. **Fumigation plume:** When an inversion layer occurs at a short distance above the top of the stack and super adiabatic conditions prevail below the stack, then the plume is said to be fumigation.

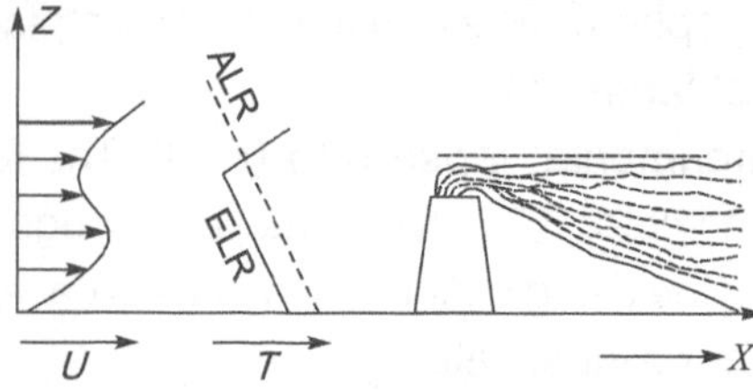

Fig. Fumigating plume

7. **Trapping plume:** When inversion layer occurs above the emission source, as well as below the source, then naturally, the emitted plume will neither go up nor will it go down and would remain confined between the two inversion.

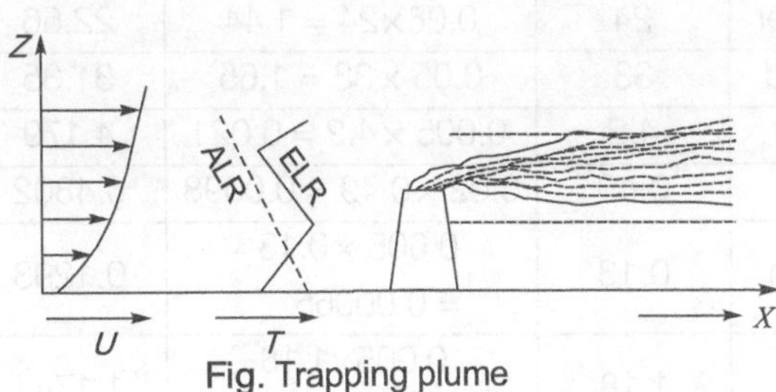

Fig. Trapping plume

10.41 **Using the data given below, find the moisture content (Wet and dry basis) of the municipal solid waste. Also estimate the as-discarded density. If the compaction is 3, find the size of collection vehicle required for 1000 kg of MSW.**

S.No.	Waste component	Mass(%)	MC(%)	Density (kg/m^3)
1.	Newspaper	15	6	85
2.	Other paper	24	6	85
3.	Cardboard	33	5	50
4.	Glass	4.2	0.5	195
5.	Plastic	0.49	2	65
6.	Aluminium	0.13	0.5	160
7.	Iron	1.18	0.5	320
8.	Non-ferrous	0.35	0.5	160
9.	Yard wastes	17.97	60	105
10.	Food wastes	1.67	60	290
11.	Soil and dust	2.01	8	480

[20 marks : 2023]

Solution:

Moisture content of a solid waste is expressed in two ways:

1. **Wet mass method**

   In this method, moisture in a sample is expressed as a percentage of wet mass of material.

   So, $\text{moisture content(\%)} = \left(\frac{a-b}{a}\right) \times 100$

   where, a = Initial mass of sample

   b = Mass of sample after drying

2. **Dry mass method**

   In this method, moisture in a sample is expressed as a percentage of dry mass of material.

   So, $\text{moisture content (\%)} = \left(\frac{a-b}{b}\right) \times 100$

   where, a = Initial mass of sample

   b = Mass of sample after drying

   Now, for the given solid waste sample, calculations are done below in table. Let the mass of given sample is 100 kg.

S.No.	Waste component	Mass(kg)	MC(kg)	Dry mass (kg)	Density (kg/m^3)	Volume (m^3)
1.	Newspaper	15	0.06×15=0.9	14.1	85	0.1765
2.	Other paper	24	0.06×24 = 1.44	22.56	85	0.2824
3.	Cardboard	33	0.05 ×33 = 1.65	31.35	50	0.66
4.	Glass	4.2	0.005 ×4.2 = 0.021	4.179	195	0.0215
5.	Plastic	0.49	0.02 ×0.49 = 0.0098	0.4802	65	0.0075
6.	Aluminium	0.13	0.005 ×0.13 = 0.00065	0.1293	160	0.00081
7.	Iron	1.18	0.005×1.18 = 0.0059	1.1741	320	0.00368
8.	Non-ferrous	0.35	0.005 ×0.35 = 0.00175	0.34825	160	0.00218
9.	Yard wastes	17.97	0.6×17.97 = 10.782	7.188	105	0.1711
10.	Food wastes	1.67	0.6×1.67=1.002	0.668	290	0.00576
11.	Soil and dust	2.01	0.08×2.01 =0.1608	1.8492	480	0.00419
				84.02605		1.3356

NOTE: In table, dry mass is calculated by subtracting moisture from given mass of individual component and volume is calculated by dividing the given mass of component by its density.

So,
$$a = 100 \text{ kg}$$
$$b = 84.02605 \text{ kg}$$

$$\text{So, moisture content (on wet basis)} = \left(\frac{a-b}{a}\right)\times 100$$
$$= \left(\frac{100-84.02605}{100}\right)\times 100 \simeq 15.97\%$$

$$\text{Moisture content (on dry basis)} = \left(\frac{a-b}{b}\right)\times 100$$
$$= \left(\frac{100-84.02605}{84.02605}\right)\times 100 = 19.01\%$$

(ii) As-discarded density

Now, volume of 100 kg sample = 1.3356 m^3

So,
$$\text{Density} = \frac{\text{Mass}}{\text{Volume}} = \frac{100\text{kg}}{1.3356\text{m}^3} = 74.87 \text{ kg/m}^3$$

(iii) Size of collection vehicle,

$$\text{Compaction ratio} = \frac{\text{Density of MSW in vehicle}}{\text{As-discarded density of MSW}}$$

$$\Rightarrow \quad 3 = \frac{\text{Density of MSW in vehicle}}{74.87}$$

$$\Rightarrow \quad \text{Density of MSW in vehicle} = 74.87 \times 3 = 224.61 \text{ kg/m}^3$$

$$\text{Now, volume of MSW in vehicle} = \frac{\text{Mass of MSW}}{\text{Density of MSW}}$$
$$= \frac{1000\text{kg}}{224.61\text{kg/m}^3} = 4.45 \text{ m}^3$$

■■■■

5 Soil Mechanics and Foundation Engineering

Revised Syllabus of ESE : *Geo-technical Engineering:* *Soil exploration - planning & methods, Properties of soil, classification, various tests and inter-relationships; Permeability & Seepage, Compressibility, consolidation and Shearing resistance, Earth pressure theories and stress distribution in soil; Properties and uses of geo-synthetics.*

Foundation Engineering: *Types of foundations & selection criteria, bearing capacity, settlement analysis, design and testing of shallow & deep foundations; Slope stability analysis, Earthen embankments, Dams and Earth retaining structures: types, analysis and design, Principles of ground modifications.*

---

1. Properties of Soils

1.1 The void ratio and specific gravity of a sample of clay are 0.73 and 2.7 respectively. If the voids are 92% saturated, find the bulk density, the dry density and the water content.
What would be the water content for complete saturation, the void ratio remaining the same?

[10 marks : 1999]

Solution:

Given data: $e = 0.73$, $G = 2.7$, $S = 92\%$

We know that $Se = Gw$

$$\Rightarrow \quad w = \frac{Se}{G}$$

$$\Rightarrow \quad w = \frac{0.92 \times 0.73}{2.7}$$

$$\Rightarrow \quad w = 0.2487 \text{ or } 24.87\%$$

Now,

$$\gamma = \frac{(G + Se)\gamma_w}{1+e}$$

$$\Rightarrow \quad \gamma = \left[\frac{2.7 + (0.92 \times 0.73)}{1+0.73}\right] \times 9.81$$

$$\Rightarrow \quad \gamma = 19.12 \text{ kN/m}^3$$

But Bulk density,

$$\rho = \frac{\gamma}{g} = \frac{19.12 \times 10^3}{9.81} = 1949.0316 \text{ kg/m}^3$$

Dry density,

$$\rho_d = \frac{\rho}{1+w} = \frac{1949.0316}{1+0.2487} = 1560.85 \text{ kg/m}^3$$

When $S = 100\%$

then $$Se = Gw$$

$\Rightarrow$ $$w = \frac{Se}{G} = \frac{1\times 0.73}{2.7}$$

$\Rightarrow$ $$w = 0.2703 \text{ or } 27.03\%$$

1.2 A sample of sand above water table was found to have a natural moisture content of 15% and a unit weight of 18.84 kN/m³. Laboratory tests on a dried sample indicated values of 0.5 and 0.85 for minimum and maximum void ratios respectively for densest and loosest states. Calculate the degree of saturation and relative density. Assume $G = 2.65$

[10 marks : 2000]

Solution:

Given data: $G = 2.65$, $w = 15\%\gamma = 18.84$ kN/m³, $e_{max} = 0.85$, $e_{min} = 0.5$

$$\gamma = \left(\frac{G+Se}{1+e}\right)\times\gamma_w \qquad [\because Se = Gw]$$

$\Rightarrow$ $$18.84 = \frac{(2.65+Gw)}{1+e}\times 9.81$$

$\Rightarrow$ $$18.84 = \left[\frac{2.65+(2.65\times 0.15)}{1+e}\right]\times 9.81$$

$\Rightarrow$ $$e = 1.5868 - 1 = 0.5868 \approx 0.59$$

Now, $$S = \frac{Gw}{e} = \frac{2.65\times 0.15}{0.59}\times 100 = 67.37\%$$

Relative Density, $$I_D = \left[\frac{e_{max}-e}{e_{max}-e_{min}}\right]\times 100 = \left[\frac{0.85-0.59}{0.85-0.5}\right]\times 100 = 74.28\%$$

The value of I_D gives the relative compactness of a natural soil deposit. Thus, the above value shows that given soil deposit is a dense soil deposit.

1.3 Earth is required to be excavated from borrow pits for building an embankment of height 6 m, top width 2 m and side slopes 1 : 1. The unit weight of undisturbed soil in wet condition is 18 kN/m³ and its natural water content is 8 per cent. The dry density required in the embankment is 20 kN/m³ with a water content of 10%. The specific gravity of soil solids is 2.70. Estimate the quantity of earth required to be excavated in the borrow area to construct one meter length of the embankment. If each truck has a capacity to carry 80 kN per trip, what is the number of truck loads required per meter length of embankment? What are the values of porosity and degree of saturation on the embankment?

[15 marks : 2003]

Solution:

Given data: $\gamma_1 = 18$ kN/m³, $w_1 = 8\%$, $\gamma_{d_2} = 20$ kN/m³ $w_2 = 10\%$, $G = 2.70$, $V_1 = ?$

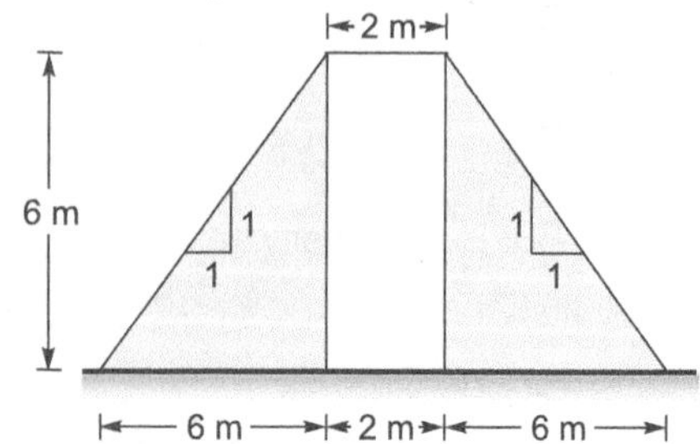

Area of embankment = $\left(\frac{2+14}{2}\right)\times 6 = 48$ m²

Volume of embankment = Area × Length = 48 × 1 = 48 m³ = V_2

Using subscript 1 and 2 for borrow pit and embankment respectively. All the symbols have their usual meanings

$\therefore$ $$\gamma_{d_2} = \frac{G\gamma_w}{1+e_2}$$

$$\Rightarrow \quad 20 = \frac{2.7 \times 9.81}{1+e_2}$$

$$\Rightarrow \quad e_2 = \frac{2.7 \times 9.81}{20} - 1 = 0.32$$

Now, porosity of the embankment soil,

$$n_2 = \frac{e_2}{1+e_2} = \frac{0.32}{1+0.32} = 0.2424$$

Degree of saturation of embankment soil can be given as

$$S_2 e_2 = w_2 G$$

$$\Rightarrow \quad S_2 = \frac{w_2 G}{e_2}$$

$$\Rightarrow \quad S_2 = \frac{0.10 \times 2.70}{0.32} \times 100$$

$$\Rightarrow \quad S_2 = 84.37\%$$

Now, if V_1 is the volume of soil to be excavated from the borrow pit,

then $$\frac{V_1}{1+e_1} = \frac{V_2}{1+e_2} = V_s$$

But to calculate V_1, we require 'e_1' which can be calculated as

$$\gamma_{d_1} = \frac{G\gamma_w}{1+e_1} \quad \text{... (i)}$$

$$\gamma_{d_1} = \frac{\gamma_1}{1+w_1} \quad \text{... (ii)}$$

From (i) and (ii), we get $\dfrac{G\gamma_w}{1+e_1} = \dfrac{\gamma_1}{1+w_1}$

$$\Rightarrow \quad \frac{2.70 \times 9.81}{1+e_1} = \frac{18}{1+0.08}$$

$$\Rightarrow \quad e_1 = \frac{2.70 \times 9.81 \times 1.08}{18} - 1 = 0.589$$

Now, we have

$$\frac{V_1}{1+e_1} = \frac{V_2}{1+e_2}$$

$$\Rightarrow \quad V_1 = \left(\frac{1+e_1}{1+e_2}\right) \times V_2 = \left(\frac{1+0.589}{1+0.32}\right) \times 48 = 57.78 \text{ m}^3$$

Hence, 57.78 m^3 of earth is to be excavated in the borrow area.

Total weight of Borrow pit soil $= V_1 \times \gamma_2 = 57.78 \times 18 = 1036.8$ kN

Number of truck loads required $= \dfrac{\text{Total weight of embankment soil}}{\text{Capacity of truck per trip}}$

$= \dfrac{1036.8}{80} = 12.96 \approx 13$ number of truck loads, for construction of per meter length of embankment.

1.4 Explain and discuss the use of Liquidity Index, Activity number, Thixotropy and Sensitivity of soil.
[10 marks : 2008]

Solution:

Liquidity index:

- The index that is used to indicate the consistency of undisturbed soils is called as the liquidity index or water plasticity ratio.
- It is expressed as $I_L = \dfrac{w_n - w_p}{I_P}$

  where w_n is the natural moisture content of soil in undisturbed state
  w_p is the plastic limit
  I_P is the plasticity index
- The liquidity index indicates the state of the soil in the field.
- If natural moisture content of the soil is closer to the liquid limit, the soil is considered to be soft.
- The soil is stiff if the natural moisture content is closer to plastic limit.
- Soils such as montmorillonite group have natural moisture contents greater than the liquid limits. The soil of this type when disturbed by vibration flow like a liquid. The liquidity index of such soils is greater than one.

Activity number:

- The activity of clay is expressed by an activity number,

$$A_c = \frac{I_P}{\text{Per cent of clay particles finer than 2 micron}}$$

- The clay soil which has an activity number greater than 1.4 can be considered as belonging to swelling type.
- Clays can be grouped into three qualitative categories depending upon their activity viz. Inactive (<0.75), Normal ($0.75 - 1.4$) and active (> 1.4).

Thixotropy:

- It is the property of certain clays by virtue of which they regain a part of the strength lost due to remoulding at unaltered moisture content, if left alone for a sufficient time.
- The increase in strength with passage of time is attributed to the tendency of cohesive soils to regain their chemical equilibrium, with the reorientation of water molecules in the adsorbed water layer.
- Thixotropy has important application in connection with pile driving operations and other types of construction in which disturbance of natural clay formations is inevitable.

Sensitivity:

- It is defined as the ratio of the unconfined compressive strength of an undisturbed specimen of the soil to the unconfined compressive strength of a specimen of the same soil after remoulding at unaltered water content.
- It is expressed as $S_t = \dfrac{q_u \text{ of undisturbed specimen}}{q_u \text{ of remoulded specimen}}$
- It is a measure of the loss of strength of soils as a result of remoulding.
- It is significant for clays.
- The clays that have sensitivity greater than 8 should be treated with care during construction operations because disturbance tends to transform them at least temporarily, into viscous fluids. Such clays belong to the montmorillonite group and possess flocculent structure.

1.5 **An oven dry soil sample of volume 225 cm³ weighs 390 g. If the specific gravity is 2.72, determine the void ratio and shrinkage limit. What will be the water content which will fully saturate the sample and cause an increase in volume equal to 8% of the original dry volume?**

[10 marks : 2011]

Solution:

Given: Volume of dry sample of soil = $V = 225$ cm³, Weight of dry sample of soil = $W = 390$ g

$$G = 2.72$$

Dry density of sample,
$$\gamma_d = \frac{W}{V} = \frac{390}{225} = 1.73 \text{ g/cm}^3$$

Using
$$\gamma_d = \frac{G\gamma_w}{1+e}$$

$$\Rightarrow \quad 1.73 = \frac{2.72 \times 1}{1+e}$$

$$\Rightarrow \quad e = 0.57$$

$$\text{Shrinkage limit} = \frac{e}{G} = \frac{0.57}{2.72} = 20.95\%$$

Water content for 100% saturation and increase in volume corresponding to 8%

$$V = 225 \text{ cm}^3$$

$$V_V + V_S = 225 \text{ cm}^3$$

$$e = \frac{V_V}{V_S} = 0.57$$

$$V_V = 81.7 \text{ cm}^3$$

$$V_S = 143.3 \text{ cm}^3$$

Increase in volume is 8%, hence $V_V = 225 \times 0.08 + 81.7 = 99.7$ cm³

Now,
$$e = \frac{V_V}{V_S} = \frac{99.7}{143.3} = 0.696$$

$$S \times e = W \times G$$

$$1 \times 0.696 = W \times 2.72$$

$$W = 25.6\%$$

1.6 **What do you understand by "index properties of soil"? Explain and list the properties under different categories?**

[4 marks : 2012]

Solution:

Index properties are those properties which are indicative of engineering behaviour of the soil.

The index properties for coarse grained soils are:

(a) Grain size distribution

- The percentage of various sizes of soil grains presents in a given dry soil sample.
- It is carried out by sieve analysis.

(b) Grain shape

- The grain shape can be angular, rounded, flaky etc.

(c) Relative density (I_D)

- It represents the degree of decreases of natural deposits of coarse grained soils

$$I_D = \frac{e_{max} - e_{nat}}{e_{max} - e_{min}} \times 100$$

The index properties for fine grained soils are:

Consistency and Atterberg limit:

- Consistency represents the relative ease with which soil can be deformed.
- Atterberg classified the consistency in four stages
  - (a) Solid stage
  - (b) Semisolid stage
  - (c) Plastic stage
  - (d) Liquid stage

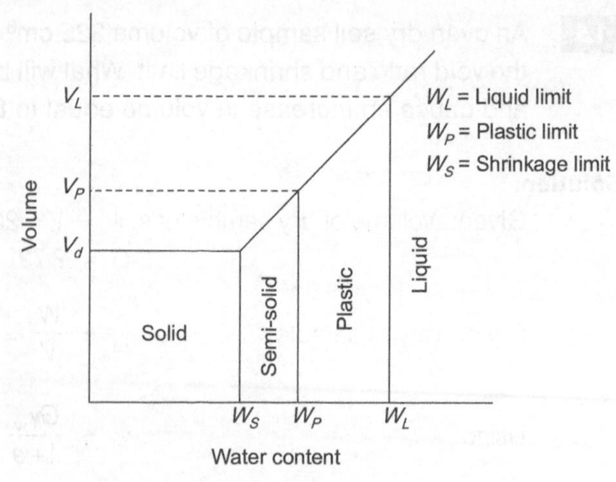

Atterberg limits

1.7 **A solid sample has a porosity of 40%. The specific gravity of solids is 2.7. Calculate the (i) void ratio (ii) dry density (iii) unit weight if the soil is 50% saturated and (iv) unit weight if the solid is completely saturated.**

[5 marks : 2012]

Solution:

Given data: Porosity of soil sample (n) = 40%, Specific gravity of soil solids (G_s) = 2.7

Now, as we know

(i)
$$n = \frac{e}{1+e}$$

$$\therefore \quad e = \frac{n}{1-n} = \frac{0.4}{1-0.4} = 0.667$$

(ii) Also
$$\gamma_d = \left(\frac{G_s}{1+e}\right)\gamma_w$$

Taking
$$\gamma_w = 9.81 \text{ kN/m}^3$$

$$\gamma_d = \frac{2.7}{1.667} \times 9.81$$

$$\Rightarrow \quad \gamma_d = 15.889 \text{ kN/m}^3$$

(iii) When soil is 50% saturated

$$\gamma = \left(\frac{G_s + Se}{1+e}\right)\gamma_w$$

where
$$S = 0.5$$

$$\therefore \quad \gamma = \left(\frac{2.7 + 0.5 \times 0.667}{1+0.667}\right) \times 9.81 = 17.852 \text{ kN/m}^3$$

(iv) When the soil is completely saturated, $S = 1$

$$\gamma_{sat} = \left(\frac{2.7 + 1 \times 0.667}{1+0.667}\right) \times 9.81 = 19.814 \text{ kN/m}^3$$

1.8 **A liquid limit test conducted on a soil sample in the cup device gave the following results:**

Number of blow	10	19	23	27	40
Water content (%)	60.00	45.20	39.80	36.50	25.20

Two determinations for the plastic limit gave water content of 20.30% and 20.80%. Determine:

(i) the liquid limit and plastic limit.

(ii) the plasticity index

(iii) the liquidity index if the natural water content is 27.40% and

(iv) the void ratio at the liquid limit, if the sp. gravity, $G_s = 2.7$. If the soil were to be loaded to failure, would you expect a brittle failure.

[10 marks : 2013]

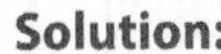
Solution:

Corresponding to $N = 25$ blows,

liquid limit is 38.15%

Taking the average value,

$$\text{plastic limit} = \frac{20.3 + 20.8}{2} = 20.55\%$$

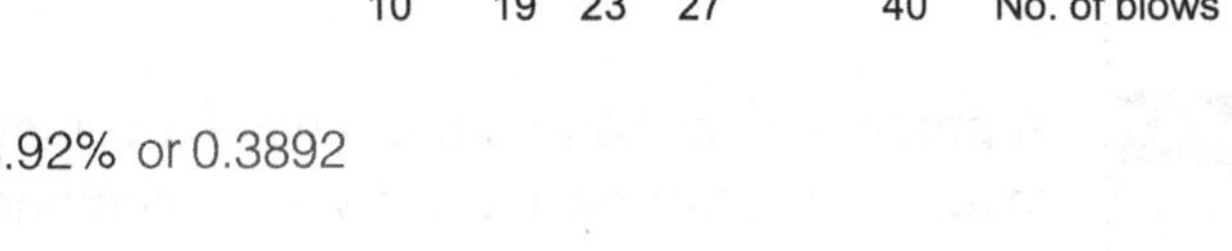

$$\text{Plasticity index } (I_p) = w_L - w_P$$

$$= 38.15 - 20.55 = 17.6\%$$

$$\text{Liquidity index} = \frac{w - w_P}{I_P} \times 100$$

$$= \frac{27.4 - 20.55}{17.6} \times 100 = 38.92\% \text{ or } 0.3892$$

Void ratio $$e = \frac{wG}{S} = 0.3815\,(2.7) = 1.03005 = 1.03 \qquad (\because S = 1)$$

As liquidity index is in between 0.25 to 0.50 hence, it is in medium stiff state. It will go plastic failure, it will not go Brittle failure.

1.9 **The mass of saturated soil sample is 150 gm and its mass when oven dried is 90 gm, find the water content. Suppose that the sample. Used for triaxial test, has a diameter of 38 mm and the height of 76 mm, find the void ratio.**

[4 marks : 2013]

Solution:

Given:

Mass of saturation soil sample, $M = 150$ m

Mass of oven dried sample, $M_s = 90$ g

$\therefore$ Mass of water, $M_w = 150 - 90 = 60$ g

$$\therefore \quad \text{Water content of sample} = \frac{60}{90} = \frac{2}{3} = 66.67\%$$

$$\text{Volume of sample} = \frac{\pi}{4}(38)^2\, 76\text{mm}^3 = 86192.74 \text{ mm}^3 = 86.193 \text{ cm}^3$$

Now, $$\text{Volume of water} = \frac{\text{Mass of water}}{\text{Density}} = \frac{60\text{g}}{1\text{g/cc}} = 60 \text{ cm}^3$$

For satuated soil sample,

Volume of voids = Volume of water

$\therefore$ Volume of solids $= 86.193 - 60 = 26.193$ cm^3

$$\therefore \quad \text{Void ratio,} \quad e = \frac{60}{26.193} = 2.29$$

1.10 **A soil deposit has a void ratio of 0.9. Its void ratio is reduced to 0.6 by compaction. Determine the percent reduction of volume by this compaction.**

[4 marks : 2014]

Solution:

Let volume of solid in soil is V_s.

Voids V_v

Solid V_s

Initially void ratio, $e_1 = 0.9$

$$\frac{V_v}{V_s} = 0.9$$

Total volume of soil, $V_1 = V_v + V_s = 0.9\,V_s + V_s = 1.9\,V_s$

Void ratio after compaction, $e_2 = 0.6$

$$\frac{V_v}{V_s} = 0.6$$

∴ Total volume of soil after compaction,

$$V_2 = V_v + V_s = 0.6V_s + V_s = 1.6V_s$$

% reduction of volume, $\Delta V = \dfrac{V_1 - V_2}{V_1} \times 100 = \dfrac{1.9V_s - 1.6V_s}{1.9V_s} \times 100 = 15.79\%$

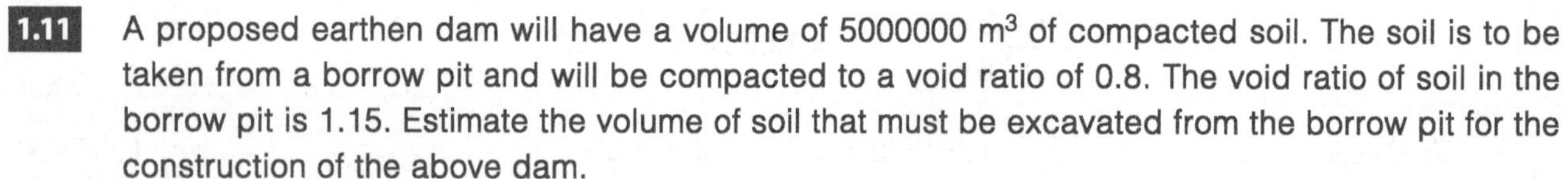

1.11 **A proposed earthen dam will have a volume of 5000000 m³ of compacted soil. The soil is to be taken from a borrow pit and will be compacted to a void ratio of 0.8. The void ratio of soil in the borrow pit is 1.15. Estimate the volume of soil that must be excavated from the borrow pit for the construction of the above dam.**

[4 marks : 2015]

Solution:

Given:

Volume of soil required in dam, $V_1 = 5 \times 10^6\ \text{m}^3$

Void ratio of soil in dam, $e_1 = 0.8$

Void ratio of soil in pit, $e_2 = 1.15$

Let volume of soil to be excavated $= V_2$

$$e = \frac{V_V}{V_S}$$

$$\Rightarrow \quad e + 1 = \frac{V_V + V_S}{V_S} = \frac{V}{V_S}$$

$$\Rightarrow \quad V_S = \frac{V}{1+e}$$

As volume of solids in dam and pit is equal,

$$\frac{V_1}{1+e_1} = \frac{V_2}{1+e_2}$$

$$\Rightarrow \quad V_2 = V_1\left(\frac{1+e_2}{1+e_1}\right) = 5 \times 10^6\left(\frac{1+1.15}{1+0.8}\right)$$

$$\therefore \quad V_2 = 5.972 \times 10^6\ \text{m}^3$$

1.12 **A soil has a void ratio of 0.70, degree of saturation 50% and G_S = 2.7. Find the water content, porosity, bulk density and dry density. By how much can the water content be increased without changing γ_d?**

[12 marks : 2017]

Solution :

Given: Void ratio (e) = 0.70, Degree of saturation (S) = 0.50, Specific gravity (G_s) = 2.70

(i) We know, $Se = wG_s$

$$\Rightarrow \quad w = \frac{Se}{G_s} = \frac{0.50 \times 0.70}{2.70} = 0.1296$$

$\therefore$ Water content, $w = 12.96\%$

(ii) Bulk density,

$$\gamma_t = \left(\frac{G_s + Se}{1+e}\right)\gamma_w$$

$$= \left(\frac{2.70 + 0.50 \times 0.70}{1+0.70}\right) \times 9.81 = 17.6 \text{ kN/m}^3$$

(iii) Dry density,

$$\gamma_d = \frac{\gamma_t}{1+w} = \frac{17.6}{1+0.1296} = 15.58 \text{ kN/m}^3$$

(iv) Porosity,

$$n = \frac{e}{1+e} = \frac{0.70}{1+0.70} = 0.4118 = 41.18\%$$

In order to have γ_d = constant

e = constant

We know, $Se = wG_s$

Taking, S = 100% [Maximum possible degree of saturation is 100% for w_{max}]

$$\therefore \quad \frac{100}{100} \times e = w_{max} \times G_s$$

$$\Rightarrow \quad w_{max} = \frac{e}{G_s} = \frac{0.70}{2.70} = 0.2593 = 25.93\%$$

$\therefore$ Increase in water content = (25.93 – 12.96)% = 12.97%

Therefore, water content may be increased by 12.97% without changing γ_d value.

1.13 **Liquid limit (LL) and plastic limit (PL) tests were carried out on a soil sample as per Indian Standard method. The values were 60% and 36% respectively for LL and PL. What is the type of soil based on the above test data as per Indian Standard Classification System? Justify your answer.**

[10 marks : 2019]

Solution:

Given: Liquid limit = 60%

Plastic limit = 36%

As per IS classification I_P of A-line = 0.73 (LL – 20)

= 0.73 (60 – 20) = 29.2%

Now, I_P of soil = 60 – 36 = 24%

$\because$ I_P of A-line > I_P of soil

Therefore, the soil will lie below A-line. Also LL > 50%, therefore soil should be highly plastic (highly compressible).

So soil type is MH or OH.

1.14 **The mass specific gravity of a fully saturated specimen of clay having a water content of 29% is 2.00. On oven drying, the mass specific gravity drops to 1.60. Calculate the specific gravity of clay assuming the void ratio to remain unchanged.**

[10 Marks : 2021]

Solution:

(i) Given that:

At $$w = 29\%$$

$$G_m = \frac{\gamma}{\gamma_w} = \frac{G\gamma_w}{1+e}\left(\frac{1+w}{\gamma_w}\right) = \frac{G(1+w)}{1+e} = 2$$

∴ Soil is saturated.

∴ $$e = wG$$

$$\frac{G(1+w)}{1+wG} = 2$$

$$\frac{G(1+0.29)}{1+0.29(G)} = 2$$

$$G = 2.82$$

1.15 **An embankment is to be constructed using sandy clay compacted to dry unit weight of 18 kN/m³. The sandy clay transported to the site from a borrow pit. The bulk unit weight of the sandy clay in the borrow pit is l6 kN/m³ and its natural water content is 11%. Calculate the volume of sandy clay from the borrow pit required for 1 cubic metre of finished embankment. Assume that the soil swells by 10% due to excavation and during transportation. You can take $G_s = 2.7$.**

[12 marks : 2023]

Solution:

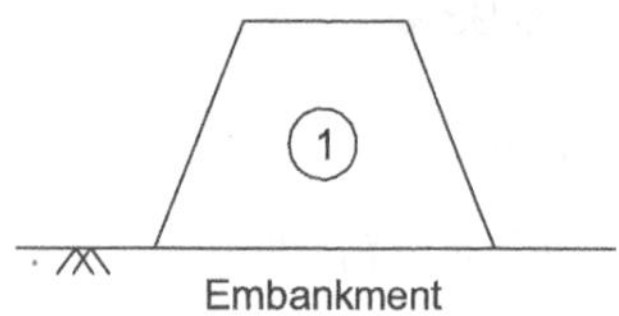

Borrow pit

2

For Embankment,

Dry unit weight, $(\gamma_{dry})_1 = 18\ \text{kN/m}^3$

Total volume, $(V_T)_1 = 1\ \text{m}^3$

Swell = 10%

For Borrow pit,

Bulk unit weight, $(\gamma_{bulk})_2 = 16\ \text{kN/m}^3$

Water content, $w_2 = 11\%$

Now, volume of solids remain same in both the cases.

Dry unit weight in Borrow pit, $$(\gamma_{dry})_2 = \frac{(\gamma_{bulk})_2}{1+w} = \frac{16}{1+0.11} = 14.41\ \text{kN/m}^2$$

and $$(\gamma_{dry})_2 = \frac{G\cdot\gamma_w}{1+e_2}$$ where e_2 is void ratio in Borrow pit

$$14.41 = \frac{2.7\times 9.81}{1+e_2}$$

$$\therefore \quad e_2 = 0.84$$

Also $$(\gamma_{dry})_1 = \frac{G \cdot \gamma_w}{1+e_1} \text{ where } e_1 \text{ is void ratio in Embankment}$$

$$18 = \frac{2.7 \times 9.81}{1+e_1}$$

$$e_1 = 0.47$$

Now, $$(V_S)_1 = (V_S)_2$$

where $(V_S)_1$ and $(V_S)_2$ are volume of solids in embankment and borrow pit respectively.

$$\frac{(V_T)_1}{1+e_1} = \frac{(V_T)_2}{1+e_2}$$

where $(V_T)_2$ is total volume in borrow pit.

$$\frac{1}{1+0.47} = \frac{(V_T)_2}{1+0.84}$$

$$(V_T)_2 = 1 \times \frac{1.84}{1.47} = 1.25 \text{ m}^3$$

Volume of soil excavated from Borrow pit = $(V_T)_2 \times$ Swell factor

$$= 1.25 \times 1.1 = 1.375 \text{ m}^3$$

1.16 **A clay has plastic limit and liquid limit of 18 and 39 percent, respectively. What water content would correspond to its liquidity index of –0.1? Comment on the consistency of this clay.**

[10 marks : 2023]

Solution:

Given: Liquid limit, w_L = 39%, Plastic limit, w_P = 18%, Liquidity index, I_L = – 0.1

Let, Natural water content = w_n

Now, $$I_L = \frac{w_n - w_P}{w_L - w_P}$$

$$\Rightarrow \quad -0.1 = \frac{w_n - 0.18}{0.39 - 0.18}$$

$$\Rightarrow \quad w_n = 0.159 \text{ or } 15.9\%$$

So, natural water content (w_n) corresponding to (I_L = –0.1) is 15.9%

Since, liquidity index (I_L) is less than zero, hence soil is in semi solid or very hard state of consistency.

2. Effective Stresses, Permeability and Seepage Analysis

2.1 **A trench is to be excavated 240 m away from a river and in a confined aquifer 4.5 m thick. The trench has to run parallel to the river and is to be 300 m long. Water level in the trench is to be maintained 6 m above the lower confining layer and 3 m below the water level in the river. Determine the rate at which water should be pumped from the trench, if the hydraulic conductivity of the aquifer is 4.5 m/day. Assume there is no contribution to flow from the land side of the trench. State the condition when the assumption can be valid.**

[12 marks : 1999]

Solution:

As there is no contribution of flow from land side of the trench, the hydraulic gradient will vary linearly between the river and trench, otherwise the flow could have been radial.

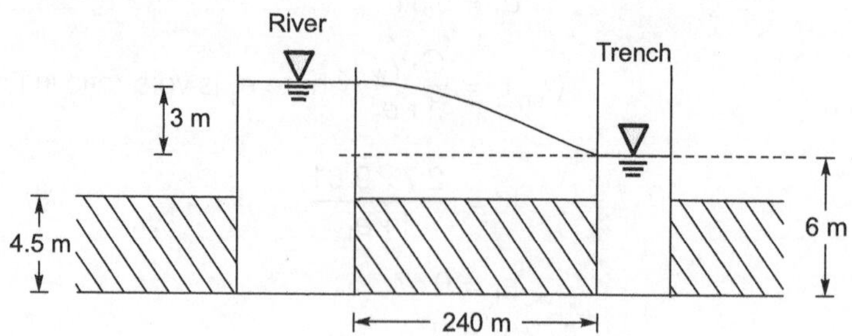

Given data:

Hydraulic conductivity (k) = 4.5 m/day, Length of trench (L) = 300 m, Thickness of aquifer (B) = 4.5 m

Hydraulic gradient between river and trench, $i = \frac{(6+3)-6}{240} = \frac{3}{240}$

Let the rate of discharge be Q, then

$$Q = kiA = kiLB = 4.5 \times \frac{3}{240} \times 300 \times 4.5 = 75.94 \text{ m}^3/\text{day}$$

where A is area through which water flows in the trench.

The above given assumption can only be valid when the flow is laminar i.e. **Darcy's law** is valid. It also means that the soil to be dewatered is homogeneous and isotropic.

2.2 **Write basic principles used in the following:**
(i) Electrical analogy method
(ii) Terzaghi's filter design approach

[10 marks : 2001]

Solution:

(i) **Electrical Analogy Method:** This method is quite extensively used for obtaining flownets. This method is based on the fact that the Darcy's law which govern the flow of water through soils is analogous to the Ohm's law governing the flow of electricity in a conducting medium. In the analogy the current being proportional to the voltage drop is similar to seepage being proportional to head dissipated. Thus if an electrical model is constructed such that its boundary conditions are similar to the boundary conditions governing seepage, the pattern of flow of electricity that is obtained in the electrical model, will be similar to the flow pattern during seepage through soils.

(ii) **Terzaghi's Filter Design Approach:** The use of filters prevents erosion and reduces uplift pressure. A protective filter consists of one or more layers of coarse grained, free draining material placed over a less pervious soil called the base. A filter would prevent the migration of finer particles but without inhibiting the flow of seepage water. So there is hardly any loss of head. This ensure that within the filter itself, seepage forces are reduced.

If the voids in the filter layers are much larger than the finest grains of the protected material (base), these grains are likely to be washed into the voids of the filter material and would ultimately obstruct the free flow. On the other hand, if the voids in the filter are too small, seepage forces are likely to develop to unacceptable levels. Both these situations have to be avoided. To achieve this, the filter material must have grain sizes that satisfy certain requirements. Terzaghi laid down certain criteria for protective filters.

They are based primarily on the grain size distributions of the filter material and the protected material. These are:

(i) $\frac{D_{15}(\text{filter})}{D_{85}(\text{Protected material})} < 5$ (ii) $4 < \frac{D_{15}\,(\text{filter})}{D_{15}\,(\text{Protected material})} < 20$ (iii) $\frac{D_{50}(\text{filter})}{D_{50}(\text{Protected material})} < 25$

D_{15}, D_{50} and D_{85} refer to the particle sizes from the grain size distribution curves. The first specification ensures that no significant invasion of particles from the protected soil to the filter shall take place. The first part of the second criterion will ensure that sufficient head is lost in flow through the filters without a build-up of seepage pressure. The third specification and the second part of second criterion are additional guidelines for the selection of filter material.

2.3 **What will be the ratio of average permeability in the horizontal direction to that in the vertical direction for a soil deposit consisting of three horizontal layers if the thickness and permeability of the second layer are twice those of the first and those of the third layer twice those of the second?**

[10 marks : 2002]

Solution:

Given data:

For 1st layer H_1, k_1

For 2nd layer $H_2 = 2H_1;\ k_2 = 2k_1$

For 3rd layer $H_3 = 2H_2 = 4H_1;\ k_3 = 2k_2 = 4k_1$

When the flow is in the horizontal direction, the hydraulic gradient is same through each layer but the discharge and velocity of flow through each layer is different.

$$\therefore \quad k_H = \frac{k_1H_1 + k_2H_2 + k_3H_3}{H_1 + H_2 + H_3} = \frac{k_1H_1 + (2k_1 \times 2H_1) + (4k_1 \times 4H_1)}{H_1 + 2H_1 + 4H_1}$$

$$= \frac{k_1H_1 + 4k_1H_1 + 16k_1H_1}{7H_1} = \frac{21k_1H_1}{7H_1}$$

$$\therefore \quad k_H = 3k_1$$

Now when the flow is vertical, then the discharge and velocity of flow remains same but hydraulic gradient changes in each layer

$$\therefore \quad k_V = \frac{H_1 + H_2 + H_3}{\frac{H_1}{k_1} + \frac{H_2}{k_2} + \frac{H_3}{k_3}} = \frac{H_1 + 2H_1 + 4H_1}{\frac{H_1}{k_1} + \frac{2H_1}{2k_1} + \frac{4H_1}{4k_1}} = \frac{7H_1}{3H_1/k_1} = \frac{7}{3}k_1$$

$$\therefore \quad \frac{k_H}{k_V} = \frac{3k_1}{\frac{7}{3}k_1} = \frac{3 \times 3}{7} = \frac{9}{7}$$

2.4 **A laboratory constant head permeability test was conducted on a silty sand specimen of void ratio 0.45. The cylindrical specimen had a diameter of 7.3 cm and a height of 16.8 cm. The head during the test was 75 cm. After 1 minute of testing at room temperature of 20°C, a total 775.6 gm of water was collected. Compute the coefficient of permeability in metres per sec. If the void ratio changes to 0.38, what would be the change in permeability.**

[10 marks : 2003]

Solution:

Given data: $e_1 = 0.45$, $e_2 = 0.38$, Diameter of specimen (d) = 7.3 cm, Length of specimen (L) = 16.8 cm
Head during the test (h) = 75 cm, Time of testing (t) = 1 min = 60 sec, Weight of water collected = 775.6 gm

Area of specimen, $$A = \frac{\pi}{4}d^2 = \frac{\pi}{4} \times (7.3)^2 = 41.85 \text{ cm}^2$$

Volume of water collected, $$Q = \frac{\text{Weight of water collected}}{\text{Density of water}} = \frac{775.6}{1} = 775.6 \text{ cm}^3$$

We know that $$k_1 = \frac{QL}{hAt} = \frac{775.6 \times 16.8}{75 \times 41.85 \times 60} \times 10^{-2} \text{ m/sec}$$

$$\Rightarrow \quad k_1 = 6.92 \times 10^{-4} \text{ m/sec}$$

We also know that $\quad k_1 : k_2 = \dfrac{c_1 e_1^3}{1+e_1} : \dfrac{c_2 e_2^3}{1+e_2}$ (By Kozny Karman Equation)

where coefficients c_1 and c_2 are shape factors which depend upon the manner of packing of grains and the shape characteristics of the pores. For sands 'c' changes very slightly with the void ratio i.e. for all practical purposes $c_1 = c_2$

$$\therefore \quad k_1 : k_2 = \frac{e_1^3}{1+e_1} : \frac{e_2^3}{1+e_2}$$

$$\Rightarrow \quad \frac{6.92\times 10^{-4}}{k_2} = \frac{(0.45)^3/(1+0.45)}{(0.38)^3/(1+0.38)} = \frac{(0.45)^3 \times 1.38}{(0.38)^3 \times 1.45}$$

$$\Rightarrow \quad k_2 = \frac{6.92\times 10^{-4} \times (0.38)^3 \times 1.45}{(0.45)^3 \times 1.38} = 4.38 \times 10^{-4} \text{ m/sec}$$

2.5 **A deposit of sand has porosity of 35% and $G = 2.7$. The soil is dry in top 1.5 m depth, it has 15% moisture content in next 1.8 m depth and it is submerged below it.**

(i) Find effective pressure at a depth of 8 m below the ground level.

(ii) Find change in effective pressure if the water table suddenly drops to a level of 6 m below the ground level.

(iii) Find shear strength of the soil on horizontal plane at 8 m depth for both the positions of ground water table if $\phi = 30°$.

[10 marks : 2004]

Solution:

For 1.5 m depth:

We know that $\quad e = \dfrac{n}{1-n} = \dfrac{0.35}{1-0.35} = 0.54$

For depth upto 1.5 m, S = 0, since sand is dry

$$\therefore \quad \gamma_1 = \frac{(G+Se)\gamma_w}{1+e}$$

$$\Rightarrow \quad \gamma_1 = \frac{[2.7+(0\times 0.54)]\times 9.81}{1+0.54}$$

$$\Rightarrow \quad \gamma_1 = \frac{2.7\times 9.81}{1.54} = 17.20 \text{ kN/m}^3$$

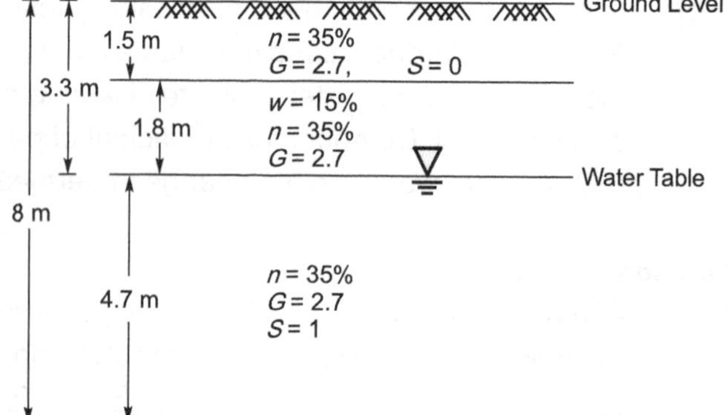

For a depth of 1.8 m below 1.5 m:

$$\gamma_2 = \frac{(G+Se)\gamma_w}{1+e}$$

$$\therefore \quad \gamma_2 = \frac{(G+Gw)\gamma_w}{1+e} \quad [\because Se = Gw]$$

$$\Rightarrow \quad \gamma_2 = \frac{(1+w)\,G\gamma_w}{1+e} = \frac{(1+0.15)\times 2.7\times 9.81}{1+0.54} = 19.78 \text{ kN/m}^3$$

$$\gamma_3 = \frac{(G+Se)\gamma_w}{1+e} = \frac{[2.7+(1\times 0.54)]\times 9.81}{1+0.54} = 20.64 \text{ kN/m}^3 \qquad [\because S = 1]$$

(i) **Total pressure at 8 m below *GL*,**

$$\sigma = (1.5\times 17.20) + (1.8\times 19.78) + (4.7\times 20.64)$$

$$\Rightarrow \qquad \sigma = 158.41 \text{ kN/m}^2$$

Pore water pressure at 8 m below *GL*,

$$u = 4.7\times 9.81 = 46.11 \text{ kN/m}^2$$

Effective pressure at 8 m below *GL*,

$$\bar{\sigma} = \sigma - u = 158.41 - 46.11 = 112.30 \text{ kN/m}^2$$

(ii) **When the water table drops to 6 m from ground level,** the soil will remain saturated for sometime; Thus, the effective pressure will be reduced only due to reduction in pore water pressure.

$\therefore$ For 1.5 m depth from *GL*, $\gamma_1 = 17.20 \text{ kN/m}^3$

For next 1.8 m after it, $\gamma_2 = 19.78 \text{ kN/m}^3$

For next 4.7 m after it, $\gamma_3 = 20.64 \text{ kN/m}^3$

Hence total pressure at 8 m below *GL*,

$$\sigma = (1.5\times 17.20) + (1.8\times 19.78) + (4.7\times 20.64)$$

$$\Rightarrow \qquad \sigma = 158.41 \text{ kN/m}^2$$

Pore water pressure, $u = 2\times 9.81 = 19.62 \text{ kN/m}^2$

Effective pressure at 8 m depth,

$$\bar{\sigma} = \sigma - u = 158.41 - 19.62 = 138.79 \text{ kN/m}^2$$

$\therefore$ Change in effective stress $= 138.79 - 112.30 = 26.49 \text{ kN/m}^2$

(iii) **Shear strength,**

$$\tau = c + \bar{\sigma}\tan\theta$$

Given soil strata is sand, hence $c = 0$, $\phi = 30°$

$$\therefore \qquad \tau = \bar{\sigma}\tan\theta$$

Now for effective pressure, $\bar{\sigma} = 112.30 \text{ kN/m}^2$ and $\phi = 30°$

$$\tau = 112.30\tan 30° = 64.84 \text{ kN/m}^2$$

and for effective pressure, $\bar{\sigma} = 138.79 \text{ kN/m}^2$ and $\phi = 30°$

$$\tau = 138.79\tan 30° = 80.13 \text{ kN/m}^2$$

2.6 **Water flows in a confined aquifer from a fully penetrating river with a piezometric head of 12 m towards a river with a piezometric head of 10 m located 400 m away. If the aquifer hydraulic conductivity and effective porosity are 3 m/day and 0.15 respectively, estimate the seepage velocity in the aquifer. Determine the draw-down and piezometric head 150 m from an upstream river.**

[12 marks : 2005]

Solution:

Given data: $h_1 = 12$ m; $h_2 = 10$ m; $L = 400$ m; $k = 3$ m/day; $n = 0.15$; $v_s = ?$

From **Darcy's law,** we have $v = ki$

where 'v' is discharge velocity, 'k' is hydraulic conductivity and 'i' is hydraulic gradient

$$i = -\frac{(h_2 - h_1)}{L} = -\frac{(10-12)}{400} = \frac{2}{400} = \frac{1}{200}$$

Now,

$$v = ki = 3\times\frac{1}{200} = \frac{3}{200} \text{ m/day}$$

Seepage velocity, $v_s = \frac{v}{n} = \frac{3/200}{0.15} = 0.1$ m/day

Now, for determining drawdown and piezometric head at 150 m from the upstream river we have,

$h_1 = 12$ m; $L = 150$ m; $v = \frac{3}{200}$ m/day; $k = 3$ m/day; $h_2 = ?$

Again, we have $v = ki$

$$\Rightarrow \quad v = -k \times \left(\frac{h_2 - h_1}{L}\right)$$

$$\Rightarrow \quad \frac{3}{200} = -3 \times \left(\frac{h_2 - 12}{150}\right)$$

$$\Rightarrow \quad \frac{15}{20} = 12 - h_2$$

$$\Rightarrow \quad h_2 = 12 - 0.75$$

$$\Rightarrow \quad h_2 = 11.25 \text{ m}$$

But drawdown, $s = h_1 - h_2 = 12 - 11.25 = 0.75$ m

∴ Piezometric head at 150 m away from upstream river = 11.25 m and drawdown = 0.75 m

2.7 What do you understand by seepage pressure and quick sand conditions?

[10 marks : 2008]

Solution:

Seepage Pressure: Flow will occur between two points in a soil mass only when there exists some difference in the potentials of those two points. The difference between the potentials of these two points, say '*h*' represents the head difference between these two points and is called the available potential or initial potential. Infact, this head gets lost or dissipated in enabling the flow of water between the two considered points. If the two considered points are at *L* distance apart, then certainly the dissipated head per unit length of flow will be represented by $\frac{h}{L}$, and this is called hydraulic gradient. This available potential lost (*h*) in flow between two points is infact consumed by the friction offered by the soil to the movement of water through it. This frictional drag pressure would be equal to $h\gamma_w$. This is counter balanced by the flow force, which is called the seepage pressure and may be represented by p_s. The seepage pressure thus can be defined as the pressure exerted by water on the soil through which it percolates. The seepage pressure always acts in the direction of flow. For vertical flows, the effective stress may therefore get decreased or increased by this seepage pressure. The effective stress in the soil mass, subjected to seepage pressure is therefore, given by

$$\bar{\sigma} = \gamma' L \pm p_s$$

If the flow occurs in the downward direction, the vertical effective pressure caused in the soil mass is increased and hence + ve sign will be used. Similarly, if the flow occurs in upward direction, then the vertical effective pressure is decreased and hence – ve sign will be used.

Quicksand Condition:

Let there be an arrangement as shown below:

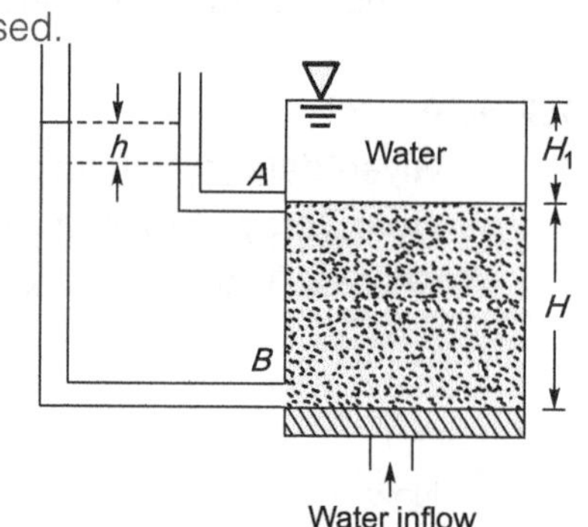

Total pressure, $\sigma_B = H_1 \gamma_w + \gamma_{sat} H$

Pore water pressure, $u_B = (H_1 + H + h)\gamma_w$

∴ Effective pressure at B, $\bar{\sigma}_B = \sigma_B - u_B$

$$= H_1 \gamma_w + \gamma_{sat} H - (H_1 + H + h)\gamma_w$$

$$= H_1 \gamma_w + \gamma_{sat} H - H_1 \gamma_w - H\gamma_w - h\gamma_w$$

$$= H(\gamma_{sat} - \gamma_w) - h\gamma_w$$

$$\Rightarrow \qquad \bar{\sigma}_B = H\gamma' - h\gamma_w$$

It is clear from the above expression that by increasing total head difference '*h*' it is possible to reach a condition, when effective stress in the soil becomes zero, i.e.

$$H\gamma' - h\gamma_w = 0 \qquad \left[\because \gamma' = \frac{(G-1)\gamma_w}{1+e}\right]$$

$$\Rightarrow \qquad \frac{(G-1)\gamma_w}{1+e} \times H - \frac{h}{H} \times H\gamma_w = 0$$

$$\Rightarrow \qquad H\gamma_w \times \left[\frac{G-1}{1+e} - i\right] = 0 \qquad \text{where } i = \frac{h}{H}$$

$$\Rightarrow \qquad i = i_{cr} = \frac{G-1}{1+e}$$

where i_{cr} is called critical hydraulic gradient. When upward flow takes place at the critical hydraulic gradient, a soil such as sand loses all its shearing strength and it cannot support any load. The soil is said to have become 'alive' or 'quick' and boiling will occur. This phenomenon is called quick sand. It is not a type of sand but only a hydraulic condition.

In case of sands, when seepage is upward, net effective pressure becomes zero when $i = i_{cr}$, hence sand particles become weightless and since these are not bound by cohesion, they may flow with water. In order to prevent this condition i should be less than i_{cr}. In case of clay this condition doesnot exist because clay particles have binding tendency i.e. cohesion.

2.8 **A deposit of fine sand has a porosity of 45%. Estimate the critical hydraulic gradient to develop quicksand condition if the specific gravity of grain is 2.7.**

[5 marks : 2009]

Solution:

Given data: Porosity of sand, n = 45%, Specific gravity of sand grains, G = 2.7

The critical hydraulic gradient to develop **quicksand condition** is given by

$$i_c = \frac{G-1}{1+e}$$

But void ratio,

$$e = \frac{n}{1-n} = \frac{0.45}{1-0.45} = 0.82$$

$$\therefore \qquad i_c = \frac{2.7-1}{1+0.82} = 0.9341$$

2.9 **A flow net is plotted for a homogeneous earthen dam of 30.0 m height with a free board of 5.0 m. If $k = 6 \times 10^{-4}$ cm/sec, Number of flow channels = 4, Number of potential drops = 10, calculate the discharge per metre run of dam.**

[2 marks : 2010]

Solution:

Quantity of seepage or discharge per meter run of dam is given by

$$q = k \times \frac{N_f}{N_d} \times H = 6 \times 10^{-4} \times 10^{-2} \times \frac{4}{10} \times (30-5) \text{ m}^3/\text{sec/m}$$

$$= 6 \times 10^{-5} \text{ m}^3/\text{sec/m} = 60 \text{ m}l/\text{sec/m}$$

2.10 **A 1.5 m layer of soil is subjected to an upward seepage head of 1.95 m. What depth of coarse sand will be required above this soil to provide a factor of safety of 1.5 against piping. Coarse sand and soil have specific gravity 2.67 and porosity as 30%.**

[8 marks : 2010]

Solution:

Given that seepage head, $h = 1.95$ m, Factor of safety, $F = 1.5$, Specific gravity, $G = 2.67$ and porosity, $n = 30\%$

Void ratio of the soil is given by

$$e = \frac{n}{1-n} = \frac{0.3}{1-0.3} = \frac{0.3}{0.7} = \frac{3}{7} = 0.43$$

Critical hydraulic gradient is given by

$$i_c = \frac{G-1}{1+e} = \frac{2.67-1}{1+0.43} = 1.1678$$

With a factor of safety of 1.5 against piping, the gradient will be

$$i = \frac{i_c}{F} = \frac{1.1678}{1.5} = 0.7785$$

But, $$i = \frac{h}{L}$$

$$\Rightarrow \quad L = \frac{h}{i} = \frac{1.95}{0.7785} = 2.5 \text{ m}$$

Available flow path i.e. thickness of existing soil = 1.5 m

∴ Depth of coarse sand required = $L - 1.5 = 2.5 - 1.5 = 1$ m

2.11 **Estimate the value of coefficient of permeability for a uniform graded sand of size $D_{10} = 0.15$ mm obtained from sieve analysis. $G = 2.67$.**

[2 marks : 2010]

Solution:

As per **Allen Hazen's equation**, the coefficient of permeability for sands is given by

$$k = 100 D_{10}^2$$

where k is in cm/s and D_{10} effective size of particles in cm

$$\therefore \quad k = 100 \times \left(\frac{0.15}{10}\right)^2 = 2.25 \times 10^{-2} \text{ cm/s}$$

2.12 **Draw the total stress, pore pressure and effective stress diagrams for a sand stratum having a thickness of 10 m. The water table is at a depth of 2 m below the ground level and there is a capillary rise of 1 m above the water table. Assume the dry and saturated unit weights of the sand as 17 kN/m³ and 21 kN/m³ respectively.**

[10 marks : 2011]

Solution:

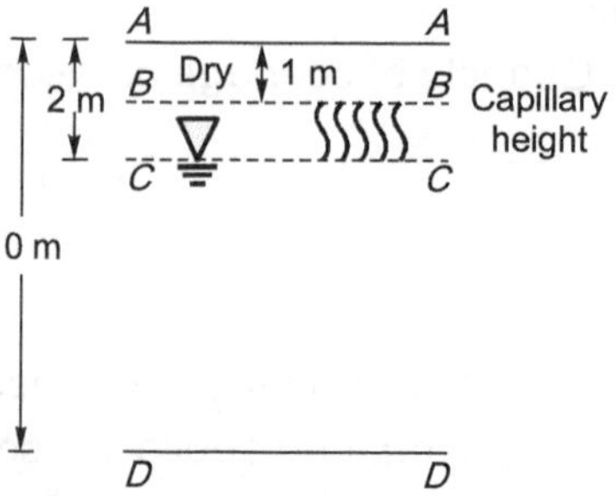

Given:

$$\gamma_d = 17 \text{ kN/m}^3$$
$$\gamma_{sat} = 21 \text{ kN/m}^3$$

A A
2 m B Dry 1 m B Capillary height
C C
10 m
D D

Section AA: $\sigma = u = \bar{\sigma} = 0$

Section BB (Top layer):

Total stress $= \sigma = \gamma_d \times 1 = 17$ kN/m²

Pore water pressure $= u = 0$

Effective stress $= \bar{\sigma} = \sigma - u = 17$ kN/m²

Section *BB* (Bottom layer) :

$$\text{Total stress} = \sigma = \gamma_d \times 1 = 17 \text{ kN/m}^2$$

$$\text{Pore water pressure} = u = -\gamma_w \times 1 = -10 \text{ kN/m}^2$$

$$\text{Effective stress} = \bar{\sigma} = \sigma - u = 17 - (-10) = 27 \text{ kN/m}^2$$

Section *CC* :

$$\text{Total stress} = \sigma = \gamma_d \times 1 + \gamma_{sat} \times 1 = 17 \times 1 + 21 \times 1 = 38 \text{ kN/m}^2$$

$$\text{Pore water pressure} = u = 0$$

$$\text{Effective stress} = 38 \text{ kN/m}^2$$

Section *DD*:

$$\text{Total stress} = \gamma_d \times 1 + \gamma_{sat} \times 9 = 17 \times 1 + 21 \times 9 = 206 \text{ kN/m}^2$$

$$\text{Pore water pressure} = u = \gamma_w \times 8 = 80 \text{ kN/m}^2$$

$$\text{Effective stress} = 206 - 80 = 126 \text{ kN/m}^2$$

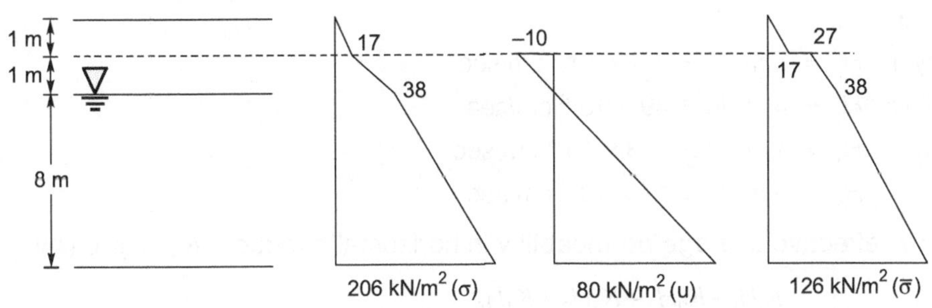

2.13 What is the intrinsic permeability of a water saturated medium that has hydraulic conductivity of 15.24 m/day? Assume the ground water is at atmospheric pressure at 20°C and has a density of 998.2 kg/m³ and viscosity of 1.002 × 10⁻³ kg/ms.

[4 marks : 2011]

Solution:

Hydraulic conductivity, $K = 15.24$ m/day

$$\rho_w = 998.2 \text{ kg/m}^3$$

$$\mu = 1.002 \times 10^{-3} \text{ kg/ms}$$

Intrinsic permeability, $K' = K\dfrac{\mu}{\gamma_w}$

$$\Rightarrow \quad K' = \frac{15.24}{(24 \times 60 \times 60)} \times \frac{(1.002 \times 10^{-3})}{(998.2) \times 9.81} = 1.8 \times 10^{-11} \text{ m}^2$$

2.14 A homogeneous earthdam 42 m high is built to store water for irrigation water requirement with a freeboard of 2 m. A horizontal filter of 30 m length is provided at its downstream end. The coefficient of permeability of the material of the dam is given as 2 × 10⁻³ cm/sec. A flow net was constructed through the body of the dam and the following result were obtained :

No. of potential drops = 25

No. of flow channels = 4

Calculate the seepage flow per meter length of the dam in *l*ps.

[6 marks : 2012]

Solution:

Quantity of seepage or seepage flow per meter length of dam is given by

$$q = k \times \frac{N_f}{N_d} \times H$$

Given:

$$k = 2 \times 10^{-3} \text{ cm/s} = 2 \times 10^{-5} \text{ m/s}$$
$$N_f = 4$$
$$N_d = 25$$
$$H = 42 - 2 = 40 \text{ m}$$

$$\therefore \quad q = 2\times10^{-5}\times\frac{4}{25}\times40$$
$$q = 12.8 \times 10^{-5} \text{ m}^3\text{/s} = 12.8 \times 10^{-5} \times 10^3 \text{ lps} = 0.128 \text{ lps}$$

2.15 **A horizontal stratified deposit consists of four layers each uniform in itself. The permeabilities of the layers are 7.5 x 10⁻⁴ cm/sec, 49 x 10⁻⁴ cm/sec, 13 x 10⁻⁴ cm/sec and 17 x 10⁻⁴ cm/sec and their thicknesses are 5 m, 4 m, 17 m and 6 m respectively. Find the effective average permeabilities of the deposit in horizontal and vertical directions.**

[5 marks : 2012]

Solution:

Given data:

For 1st layer $H_1 = 5$ m, $K_1 = 7.5 \times 10^{-4}$ cm/sec

For 2nd layer $H_2 = 4$ m, $K_2 = 49 \times 10^{-4}$ cm/sec

For 3rd layer $H_3 = 17$ m, $K_3 = 13 \times 10^{-4}$ cm/sec

For 4th layer $H_4 = 6$ m, $K_4 = 17 \times 10^{-4}$ cm/sec

As we know, **effective average permeability in horizontal direction,** K_H is given by

$$\Rightarrow \quad K_H = \frac{K_1H_1+K_2H_2+K_3H_3+K_4H_4}{H_1+H_2+H_3+H_4}$$

$$\Rightarrow \quad K_H = \frac{7.5\times10^{-4}\times5+49\times10^{-4}\times4+13\times10^{-4}\times17+17\times10^{-4}\times6}{5+4+17+6}$$

$$\Rightarrow \quad K_H = \frac{10^{-4}\times(7.5\times5+49\times4+13\times17+17\times6)}{32}$$

$$\Rightarrow \quad K_H = 17.39\times10^{-4} \text{ cm/sec}$$

and **effective permeability in the vertical direction,** K_V is given by

$$K_V = \frac{H_1+H_2+H_3+H_4}{\frac{H_1}{K_1}+\frac{H_2}{K_2}+\frac{H_3}{K_3}+\frac{H_4}{K_4}}$$

$$\Rightarrow \quad K_V = \frac{5+4+17+6}{\frac{5}{7.5\times10^{-4}}+\frac{4}{49\times10^{-4}}+\frac{17}{13\times10^{-4}}+\frac{6}{17\times10^{-4}}} = \frac{32}{2.4089\times10^{4}}$$

$$\Rightarrow \quad K_V = 13.284\times10^{-4} \text{ cm/sec}$$

2.16 **An earth dam is built on an impervious foundation with a horizontal filter at the base near the toe. The permeability of the soil in the horizontal and vertical directions are 3 x 10⁻² mm/sec and 1 x 10⁻² mm/sec respectively. The full reservoir level is 30 m above the filter. A flow net constructed for the transformed section of the dam, consists of 4 flow channels and 16 head drops. Estimate the seepage loss per metre length of the dam.**

[4 marks : 2013]

Solution:

Given: Horizontal permeability of soil (k_H) = 3 × 10⁻² mm/sec, Vertical permeability of soil (k_v) = 1 × 10⁻² mm/sec

$\therefore$ Equivalent permeability, $k = \sqrt{k_H . k_v} = \sqrt{3 \times 10^{-2} \times 1 \times 10^{-2}} = 1.732 \times 10^{-2}$ mm/s

Head above filter, $H = 30$ m

In flow net, no. of flow channels, $N_f = 4$

and no. of head drops, $N_d = 16$

$\therefore$ **Seepage per unit length of dam,** $Q = kH\dfrac{N_f}{N_d} = (1.732 \times 10^{-2} \times 10^{-3})30\left(\dfrac{4}{16}\right) = 0.13 l/\text{sec/m}$

2.17 **To estimate the seepage loss through a cofferdam foundation, the flow nets were constructed. The results of the flow net study gave $N_f = 7$, $N_d = 17$. The head of the water lost during seepage was 6 m. If $K = 5 \times 10^{-5}$ m/minute, compute the seepage loss per meter length of dam per day.**

[4 marks : 2014]

Solution:

Given: $N_f = 7$, $N_d = 17$

Head loss, $h = 6$ m

Coefficient of permeability, $K = 5 \times 10^{-5}$ m/minute

The seepage loss per metre length of dam is given by

$$q = K \cdot h \cdot \left(\frac{N_f}{N_d}\right)$$

$$= 5 \times 10^{-5} \times 6 \times \left(\frac{7}{17}\right) \text{ per minute}$$

$$= 5 \times 10^{-5} \times 6 \times \left(\frac{7}{17}\right) \times 60 \times 24 \text{ per day} = 0.178 \text{ m}^3\text{/day/m}$$

2.18 **A layer of sand 8.0 m thick lies above a layer of clay. The water table is at a depth of 1.0 m below the ground surface. The saturated unit weight of sand is 20.0 kN/m³ and its dry unit weight is 17.0 kN/m³. Plot the total stress, neutral stress and effective stress diagram with depth up to 8.0 m. If the sand above the water table gets saturated due to capillary moisture, what will be the changes in stress diagram?**

[8 marks : 2014]

Solution:

Assuming unit weight of water, $\gamma_w = 10$ kN/m³

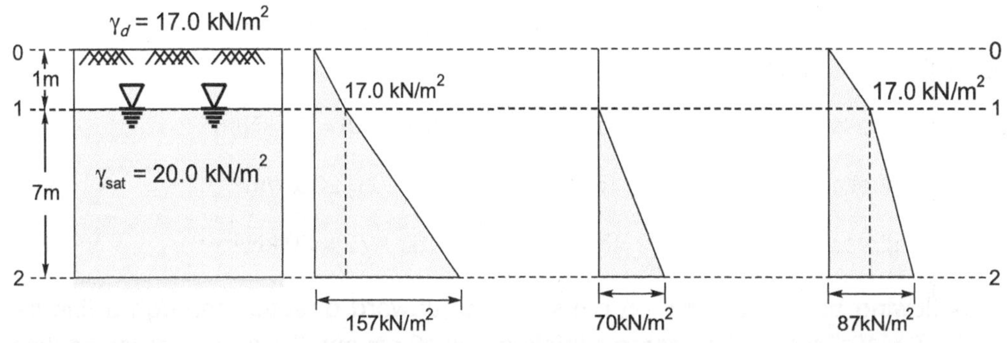

Fig (i) Total stress *Fig (ii) Neutral stress* *Fig (iii) Effective stress*

Case I: When soil above GWT is dry

(i) **Total stress (σ):**

Total stress at (0 – 0), $\sigma_0 = 0$

Total stress at (1 – 1), $\sigma_1 = \gamma_d \times 1 = 17 \times 1 = 17 \text{ kN/m}^2$

Total stress at (2 – 2), $\sigma_2 = \gamma_d \times 1 + \gamma_{sat} \times 7 = 17 \times 1 + 20 \times 7 = 157 \text{ kN/m}^2$

(ii) **Neutral stress (u)**

Neutral stress at (0 – 0), $\sigma_0 = 0$

Neutral stress at (1 – 1), $\sigma_1 = 0$ [∵ soil is dry]

Neutral stress at (2 – 2), $u_2 = \gamma_w \times 7 = 10 \times 7 = 70 \text{ kN/m}^2$

(iii) **Effective stress ($\bar{\sigma}$):**

Effective stress at (0 – 0), $\bar{\sigma}_0 = \sigma_0 - u_0 = 0$

Effective stress at (1 – 1), $\bar{\sigma}_1 = \sigma_1 - u_1 = 17 - 0 = 17 \text{ kN/m}^2$

Effective stress at (2 – 2), $\bar{\sigma}_2 = \sigma_2 - u_2 = 157 - 70 = 87 \text{ kN/m}^2$

Case II: When soil above GWT is capillary saturated

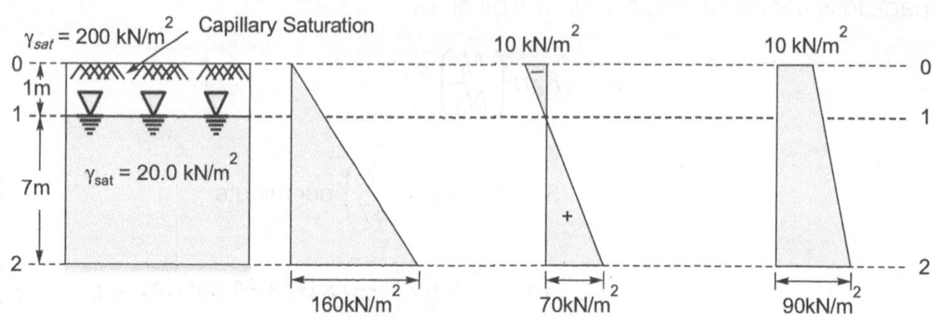

Fig (i) Total stress *Fig (ii) Neutral stress* *Fig (iii) Effective stress*

(i) **Total stress (σ):**

Total stress at (0 – 0), $\sigma_0 = 0$

Total stress at (1 – 1), $\sigma_1 = \gamma_{sat} \times 1 = 20.0 \times 1 = 20 \text{ kN/m}^2$

Total stress at (2 – 2), $\sigma_2 = \gamma_{sat} \times 8 = 20 \times 8 = 160 \text{ kN/m}^2$

(ii) **Neutral stress (u):**

Neutral stress at (0 – 0), $u_0 = -\gamma_w \times 1 = -10 \times 1 = -10 \text{ kN/m}^2$

Neutral stress at (1 – 1), $u_1 = 0$

Neutral stress at (2 – 2), $u_2 = \gamma_w \times 7 = 10 \times 7 = 70 \text{ kN/m}^2$

(iii) **Effective stress ($\bar{\sigma}$):**

Effective stress at (0 – 0), $\bar{\sigma}_0 = \sigma_0 - u_0 = 0 - (-10) = 10 \text{ kN/m}^2$

Effective stress at (1 – 1), $\bar{\sigma}_1 = \sigma_1 - u_1 = 20 - 0 = 20 \text{ kN/m}^2$

Effective stress at (2 – 2), $\bar{\sigma}_2 = \sigma_2 - u_2 = 160 - 70 = 90 \text{ kN/m}^2$

2.19 Water is flowing at the rate of 0.05 cm^3s^{-1} in the upward direction through a fine sand sample whose $k = 2 \times 10^{-3}$ cm s^{-1}. The sample thickness is 12 cm and the area of cross-section is 50 cm^2. Determine the effective pressure at the middle and bottom sections of the sample if the saturated unit weight of the sand is 19.4 kN/m^3.

[8 marks : 2014]

Solution:

Given: $q = 0.05$ cm³/s, $k = 2 \times 10^{-3}$ cm/s

Thickness of sample = 12 cm

Area cross section = 50 cm²

Saturated unit weight of sand = 19.4 kN/m³

According to Darcy's law,

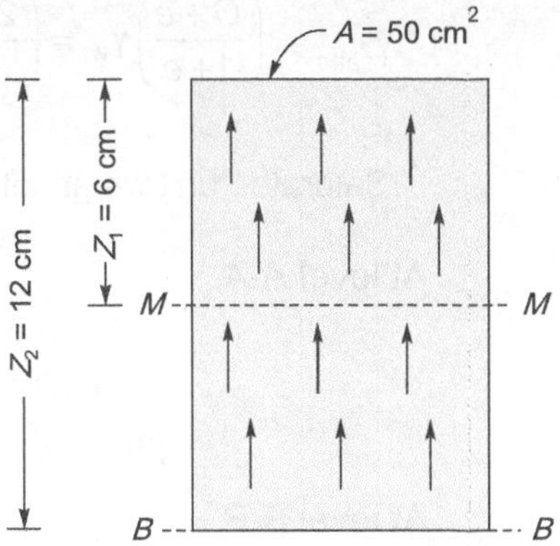

$$i = \frac{Q}{kA} = \frac{0.05}{2\times10^{-3}\times50} = 0.5$$

Effective stress in Hydrodynamic flow is given by

$$\bar{\sigma}_{dy} = \bar{\sigma}_{static} \pm iz\gamma_w$$

where, +ve for downward flow

–ve for upward flow

Assuming unit weight of water, $\gamma_w = 10$ kN/m³

∴ **Effective stress at mid of sample,**

$$\bar{\sigma}_M = \gamma' z_1 - iz_1\gamma_w = (19.4 - 10) \times 6 \times 10^{-2} - 0.5 \times 6 \times 10^{-2} \times 10 = 0.564 - 0.3 = 0.264 \text{ kN/m}^2$$

Effective stress at bottom of sample,

$$\bar{\sigma}_B = \gamma' z_2 - iz_2\gamma_w = (19.4 - 10) \times 12 \times 10^{-2} - 0.5 \times 12 \times 10^{-2} \times 10 = 1.128 - 0.6 = 0.528 \text{ kN/m}^2$$

2.20 The soil at a site consists of two layers of thickness H each. The coefficient of permeability of the soil of 1st layer is k_1 in both horizontal and vertical directions, whereas for the 2nd layer, it is $k_1/2$. What will be the equivalent permeability of the two-layered soil in horizontal and vertical directions?

[4 marks : 2015]

Solution:

Equivalent horizontal permeability,

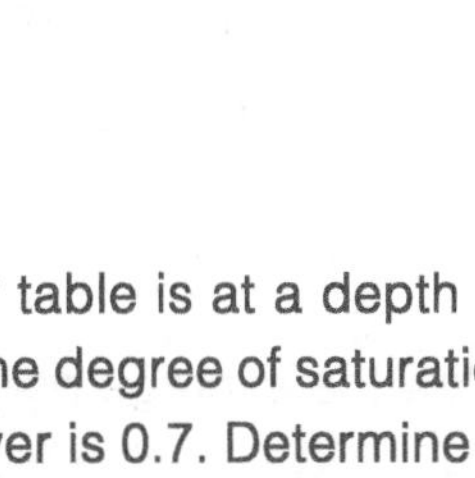

$$k_H = \frac{\Sigma kH}{\Sigma H} = \frac{k_1 H + \frac{k_1}{2}H}{H+H}$$

$$k_H = \frac{3}{4}k_1$$

Equivalent vertical permeability,

$$k_V = \frac{\Sigma H}{\Sigma H / K} = \frac{H+H}{\frac{H}{k_1} + \frac{2H}{k_1}} = \frac{2}{3}k_1$$

2.21 A layer of sand 6.0 m thick lies above a layer of clay soil. The water table is at a depth of 2.0 m below the ground surface. The void ratio of the sand layer is 0.6 and the degree of saturation of the sand layer above the water table is 40%. The void ratio of the clay layer is 0.7. Determine the total stress, neutral stress and effective stress at a point 10 m below the ground surface. Assume specific gravity of the sand and clay soil respectively as 2.65 and 2.7.

[8 marks : 2015]

Solution:

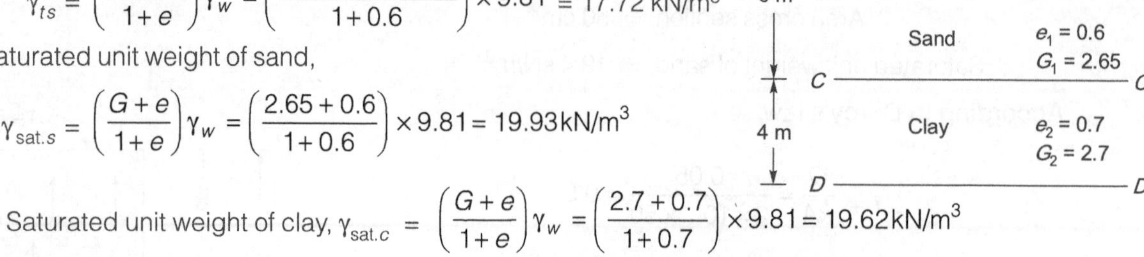

Bulk unit weight of sand above water table

$$\gamma_{ts} = \left(\frac{G+Se}{1+e}\right)\gamma_w = \left(\frac{2.65+0.4\times0.6}{1+0.6}\right)\times 9.81 = 17.72 \text{ kN/m}^3$$

Saturated unit weight of sand,

$$\gamma_{sat.s} = \left(\frac{G+e}{1+e}\right)\gamma_w = \left(\frac{2.65+0.6}{1+0.6}\right)\times 9.81 = 19.93 \text{kN/m}^3$$

Saturated unit weight of clay, $\gamma_{sat.c} = \left(\frac{G+e}{1+e}\right)\gamma_w = \left(\frac{2.7+0.7}{1+0.7}\right)\times 9.81 = 19.62 \text{kN/m}^3$

At level *A-A*,

Total stress, $\sigma_A = 0$

Neutral stress, $u_A = 0$

Effective stress, $\bar{\sigma}_A = 0$

At level *B-B*,

Total stress, $\sigma_B = \gamma_{ts}\times 2 = 17.72\times 2 = 35.44 \text{ kN/m}^2$

Neutral stress, $u_B = 0$

Effective stress, $\bar{\sigma}_B = \sigma_B - u_B = 35.44 \text{ kN/m}^2$

At level *C-C*,

Total stress, $\sigma_C = \gamma_{ts}\times 2 + \gamma_{sat.s}\times 4 = 17.72\times 2 + 19.93\times 4 = 115.16 \text{ kN/m}^2$

Neutral stress, $u_C = \gamma_w \times 4 = 9.81\times 4 = 39.24 \text{ kN/m}^2$

Effective stress, $\bar{\sigma}_C = \sigma_C - u_C = 75.92 \text{ kN/m}^2$

At level *D-D*,

Total stress, $\sigma_D = \gamma_{ts}\times 2 + \gamma_{sat.s}\times 4 + \gamma_{sat.c}\times 4$

$= 17.72\times 2 + 19.93\times 4 + 19.62\times 4 = 193.64 \text{ kN/m}^2$

Neutral stress, $u_D = \gamma_w\times 8 = 9.81\times 8 = 78.48 \text{ kN/m}^2$

Effective stress, $\bar{\sigma}_D = \sigma_D - u_D = 115.16 \text{ kN/m}^2$

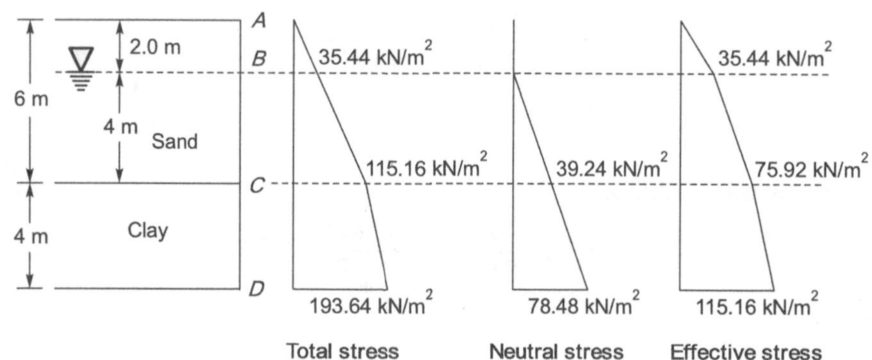

2.22 How the average permeability of a soil deposit consisting of a number of layers is determined?

[20 marks : 2017]

Solution:

Permeability of stratified soils: When a subsoil in field consists of number of strata which have different permeabilities, calculation of seepage through such a stratified deposit requires computation of average values of *k* applicable for whole deposit.

There are two such average values:

(i) Average horizontal coefficient of permeability k_H.

(ii) Average vertical coefficient of permeability, k_V.

(a) Horizontal flow (parallel to bedding planes) : When flow is horizontal it is taking place through all the layers at the same time and therefore, the hydraulic gradient is same in each layer that is $i_1 = i_2 = i_n = i$. But velocity of flow is different in different layers.

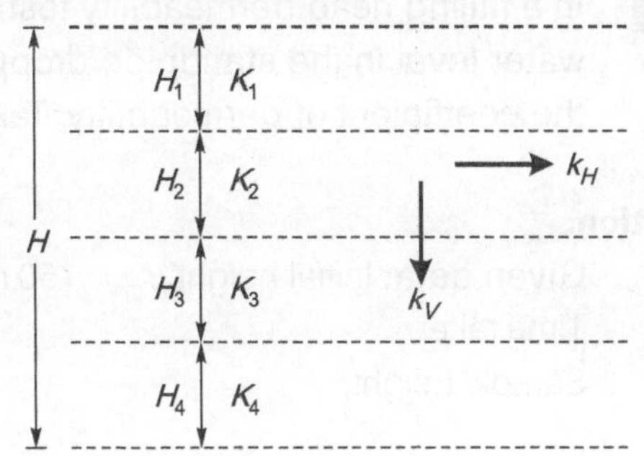

Average coefficient of permeability in layered soils

Considering unit width in the plane normal to the plane of cross-section

$$Q = kiA$$

$$Q = Q_1 + Q_2 + Q_3 + \ldots + Q_n$$

$$k_H iA = k_1 iA_1 + k_2 iA_2 \ldots + k_n iA_n$$

$$k_H i\,[H_1 + H_2 + H_3 + \ldots + H_n] \times 1 = k_1\,[H_1 \times 1]\,i + k_2\,[H_2 \times 1]\,i + \ldots + k_n\,[H_n \times 1]\,i$$

$$k_H = \frac{k_1 H_1 + k_2 H_2 + \ldots + k_n H_n}{H_1 + H_2 + \ldots + H_n};$$

where $H = H_1 + H_2 + H_3 \ldots + H_n$

(b) Vertical flow (normal to bedding planes): In this case, the flow is taking place in one layer after another and continuity of flow requires velocity of flow in each of the layers to be same. The hydraulic gradients on the other hand, change from layer to layer.

Let the hydraulic gradients in different layers be denoted by $i_1, i_2 \ldots i_n$ and total head loss in flow over the total thickness H be denoted by h made up of head losses in each of the layer equal to $h_1, h_2 \ldots h_n$.

The constant velocity of flow is

$$v = k_V \frac{h}{H} = k_1 i_1 = k_2 i_2 = \ldots k_n i_n$$

$$\Rightarrow \quad v = k_V \frac{h}{H} = k_1 \frac{h_1}{H_1} = k_2 \frac{h_2}{H_2} = \ldots \frac{k_n h_n}{H_n}$$

$$h_1 = k_V \frac{h}{H} \times \frac{H_1}{k_1}$$

$$h_2 = k_V \frac{h}{H} \times \frac{H_2}{k_2}$$

$$\vdots$$

$$h_n = k_V \frac{h}{H} \times \frac{H_n}{k_n}$$

Loss of head in each layer, $h = h_1 + h_2 + \ldots + h_n$

$$\because \quad = \frac{k_v h}{H} \cdot \frac{H_1}{k_1} + \frac{k_v h}{H} \cdot \frac{H_2}{k_2} + \ldots + \frac{k_v h}{H} \cdot \frac{H_n}{k_n}$$

$$\therefore \quad k_v = \frac{H}{\frac{H_1}{k_1} + \frac{H_2}{k_2} + \ldots \frac{H_n}{k_n}} = \frac{H_1 + H_2 + \ldots + H_n}{\frac{H_1}{k_1} + \frac{H_2}{k_2} + \ldots + \frac{H_n}{k_n}}$$

Where $H = H_1 + H_2 + \ldots + H_n$

2.23 In a falling head permeability test on a 120 mm high and 100 mm diameter cylindrical sample, the water level in the standpipe dropped from a height of 750 mm to 250 mm in one hour. Determine the coefficient of permeability. Take the internal diameter of the standpipe equal to 6 mm.

[10 marks : 2018]

Solution:

Given data: Initial height, $h_1 = 750$ mm, Final height, $h_2 = 250$ mm

Time taken, $t = 1$ hr

Sample height, $h = 120$ mm

Sample area, $A = \frac{\pi}{4}D^2,\ D = 100$ mm

Stand pipe area, $a = \frac{\pi}{4}d^2,\ d = 6$ mm

Now, coefficient of permeability

$$k = \frac{a}{A}\frac{L}{t}\ln\left(\frac{h_1}{h_2}\right) = \frac{\pi/4\times 6^2}{\pi/4\times 100^2}\times\frac{120}{1}\times\ln\left(\frac{750}{250}\right)$$

$\Rightarrow$ $k = 0.475$ mm/hr

2.24 A pit of 6.4 m deep is to be excavated in a fine sand stratum completely saturated up to the ground surface. The saturated unit weight of the sand was obtained as 20.3 kN/m^3. To stabilize the bottom of the excavation (prevent boiling), it was decided to drive steel sheet piles to act as cutoff walls that encircle the excavation. Determine the total length of sheet pile wall to provide a factor of safety of 1.5 against sand boiling. Assume specific gravity of soil, $G_s = 2.7$ and unit weight of water, $\gamma_w = 9.81$ kN/m^3.

[12 marks : 2019]

Solution:

Depth of excavation = 6.4 m (given)

$\gamma_{sat} = 20.3$ kN/m^3

FOS = 1.5

$G_S = 2.7$

$\gamma_W = 9.81$ kN/m^3

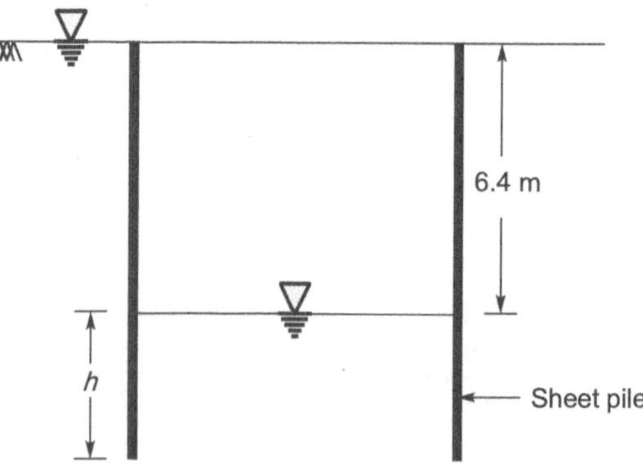

$\because$ $$\gamma_{sat} = \frac{G+e}{1+e}\cdot\gamma_w$$

$\therefore$ $$20.3 = \frac{2.7+e}{1+e}\times 9.81$$

$\Rightarrow$ $e = 0.5899$

$$i_{cr} = \frac{G-1}{1+e} = \frac{2.7-1}{1+0.5899} = 1.0692$$

$$i = \frac{i_{cr}}{FOS} = \frac{1.0692}{1.5} = 0.713$$

$$i = \frac{\Delta H}{L} \quad \text{where } \Delta H = 6.4 \text{ m},\quad L = h$$

$\therefore$ $$\frac{6.4}{h} = 0.713$$

$\Rightarrow$ $h = 8.976$ m

Total length of sheet pile= 6.4 + 8.976 = 15.376 m

2.25 The soil profile in a particular site consists of 7 m thick sandy layer overlain by a layer of clay. The water table is at 1 m below the ground surface. Above the water table, the sand is saturated with capillary moisture. The dry unit weight of sand is 17 kN/m^3 and its saturated unit weight is 20 kN/m^3. Plot the total stress, neutral stress and effective stress with depth up to a depth of 7 m.

[12 marks : 2020]

Solution:

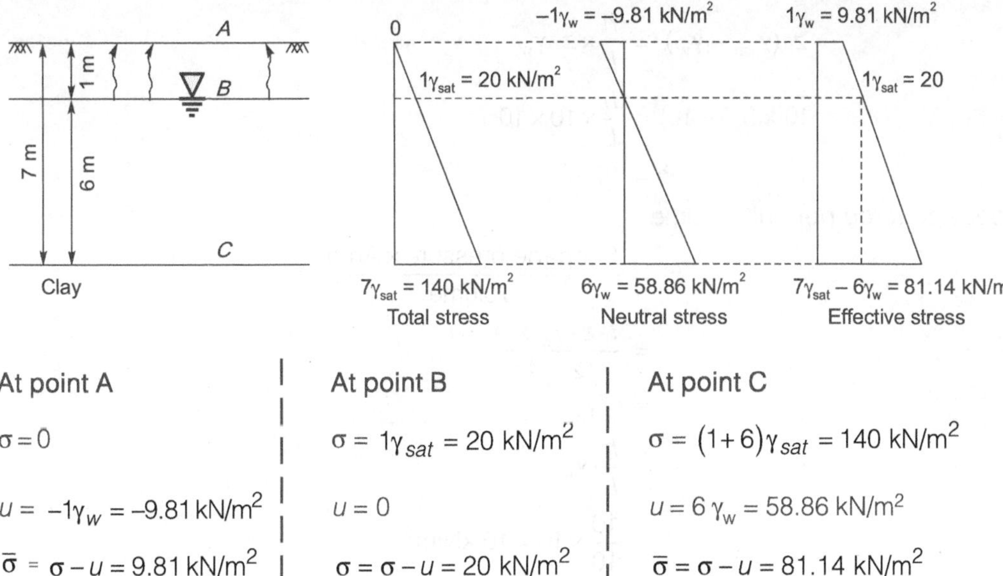

At point A	At point B	At point C
$\sigma = 0$	$\sigma = 1\gamma_{sat} = 20$ kN/m^2	$\sigma = (1+6)\gamma_{sat} = 140$ kN/m^2
$u = -1\gamma_w = -9.81$ kN/m^2	$u = 0$	$u = 6\,\gamma_w = 58.86$ kN/m^2
$\bar{\sigma} = \sigma - u = 9.81$ kN/m^2	$\sigma = \sigma - u = 20$ kN/m^2	$\bar{\sigma} = \sigma - u = 81.14$ kN/m^2

2.26 Soil sample and flow conditions are shown in the following figure:

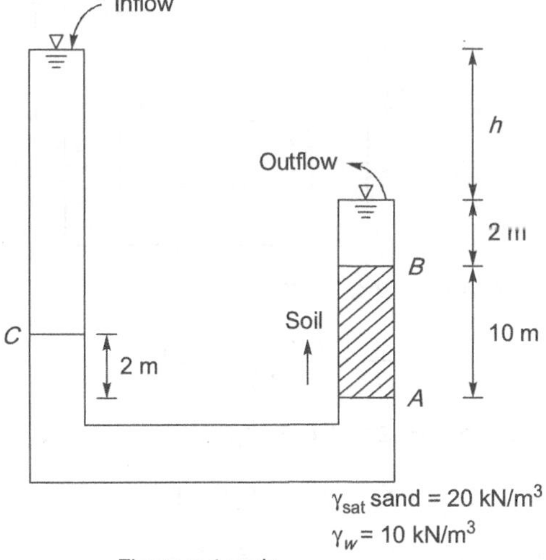

Figure not scale

(i) Find the head, h required to cause quick condition.

(ii) Compute the seepage force per unit volume at quick condition. Use cross-section area of tube as 1 m^2.

(iii) A student accidentally broke the left hand riser tube to the point C at an elevation 2 m above point A. Assuming that the water level is now maintained at C, compute the new hydraulic gradient, effective stress at elevation A and seepage force at elevation A.

[20 marks : 2023]

Solution:

(i) **For quick sand to occur:** Shear strength at A must be zero and for that effective stress at A, $\bar{\sigma}_A$ must be zero.

As upward flow is there, $\bar{\sigma}_A = 0$

$$z \cdot \gamma' - iz \cdot \gamma_w = 0$$

where γ' is submerged unit weight of soil and i is hydraulic gradient.

$$z \cdot (\gamma_{sat} - \gamma_w) = \frac{h}{L} \times z \cdot \gamma_w$$

$$10 \times (20 - 10) = \frac{h}{L} \times 10 \times 10$$

$$h = 10 \text{ m}$$

(ii) **Seepage force per unit volume:**

$$F_S = \frac{\text{Seepage pressure} \times \text{Area}}{\text{Volume}}$$

$$= \frac{i \cdot z \cdot \gamma_w \times A}{z \times A}$$

$$= i \cdot \gamma_w$$

$$= \frac{h}{L} \cdot \gamma_w$$

$$= \frac{10}{10} \times 10 = 10 \text{ kN/m}^3$$

(iii)

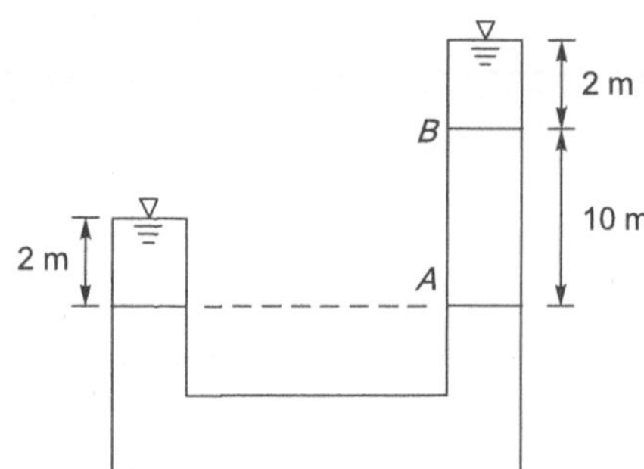

Now in this condition, downward flow is there.

$$\text{Hydraulic gradient (i)} = \frac{h_L}{L} = \frac{2 + 10 - 2}{10} = 1$$

$$\text{Effective stresses at } A,\ \bar{\sigma}_A = z\gamma' + i \cdot z \cdot \gamma_w$$

$$\bar{\sigma}_A = 10 \times (20 - 10) + 1 \times 10 \times 10 = 200 \text{ kN/m}^2$$

$$\text{Seepage force at } A,\ F_S = (i \cdot z \cdot \gamma_w) \times A$$

$$= 1 \times 10 \times 100 \times 1 = 100 \text{ kN}$$

3. Compressibility and Consolidation

3.1 **Undisturbed soil sample 30 mm thick got 50% consolidated in 20 minutes with drainage allowed at top and bottom in the laboratory. If the clay layer from which the sample was obtained is 3 m thick in field conditions, estimate the time it will take to consolidate 50% with (i) double surface drainage (ii) single surface drainage if in both cases consolidation pressure is uniform.**

[10 marks : 2001]

Solution:

We know that $T_v = \frac{C_v t}{d^2}$

where T_v is time factor, C_v is coefficient of consolidation, t is time to complete any stage of consolidation and d is length of drainage path

Also $T_v = \frac{\pi}{4}U^2$ where U is degree of consolidation in fraction and this relation is valid for $U \le 60\%$

$\therefore$ For $U = 50\%,\ T_V = \frac{\pi}{4}(0.5)^2$

$\Rightarrow$ $T_v = 0.1963$

$$d = \frac{30}{2} = 15 \text{ mm (double drainage)}$$

t = 20 minutes for 50% consolidation

$$\therefore \quad C_v = \frac{T_v d^2}{t} = \frac{0.1963 \times (15)^2}{20} = 2.2084 \text{ mm}^2\text{/min}$$

(i) **Time required for 50% consolidation with double drainage i.e.**

$$d = \frac{3}{2} = 1.5 \text{ m} = 1500 \text{ mm}$$

$$\therefore \quad t = \frac{T_v d^2}{C_v} = \frac{0.1963 \times (1500)^2}{2.2084} \times \frac{1}{60 \times 24} = 138.89 \text{ days}$$

(ii) **Time required for 50% consolidation with single drainage i.e. d = 3 m = 3000 mm**

$$\therefore \quad t = \frac{T_v d^2}{C_v} = \frac{0.1963 \times (3000)^2}{2.2084} \times \frac{1}{24 \times 60} = 555.55 \text{ days}$$

3.2 **In a consolidation test, the void ratio of the specimen which was 1.068 under the effective pressure of 214 kN/m², changed to 0.994 when the pressure was increased to 429 kN/m². Calculate the coefficient of compressibility, compression index and the coefficient of volume compressibility. Find the settlement of foundation resting on above type of clay if thickness of layer is 8 m and the increase in pressure is 10 kN/m².**

[15 marks : 2002]

Solution:

Given data: $e_1 = 0.994$; $e_0 = 1.068$; $\bar{\sigma}_1 = 429$ kN/m²; $\bar{\sigma}_0 = 214$ kN/m²; $H_0 = 8$ m; $\Delta\bar{\sigma}_0 = 10$ kN/m²

Coefficient of compressibility, $a_v = -\frac{\Delta e}{\Delta \bar{\sigma}}$

where Δe is change in void ratio $= e_1 - e_0$

and $\Delta\bar{\sigma}$ is change in effective pressure $= \bar{\sigma}_1 - \bar{\sigma}_0$

$$\therefore \quad a_v = -\left(\frac{0.994 - 1.068}{429 - 214}\right) = 3.44 \times 10^{-4} \text{ m}^2\text{/kN}$$

Coefficient of volume compressibility,

$$m_v = \frac{a_v}{1 + e_0} = \frac{3.44 \times 10^{-4}}{1 + 1.068} = 1.66 \times 10^{-4} \text{ m}^2\text{/kN}$$

Compression index, $$C_c = -\frac{\Delta e}{\log_{10}\left(\frac{\bar{\sigma}_1}{\bar{\sigma}_0}\right)} = -\frac{(0.994 - 1.068)}{\log_{10}\left(\frac{429}{214}\right)} = 0.245$$

Settlement of foundation, $\Delta H = m_v H \Delta\bar{\sigma}_0 = 13.28$ mm

3.3 **Explain the following: (i) Initial compression (ii) Primary consolidation (iii) Secondary consolidation (iv) Primary compression ratio.**

[8 marks : 2004]

Solution:

(i) **Initial Compression:** If soil is partially saturated, then immediately after the application of load, the volume decreases due to expulsion of air as well as due to compression of pore air which is called initial compression. At the end of initial compression of soil, it becomes fully saturated if load is sufficiently large. The result of initial compression is the immediate settlement which is usually determined by using elastic theory, even though the deformation itself is not truely elastic. Computation of immediate settlement has to be made in the design of shallow foundations.

(ii) **Primary Consolidation:** After the initial compression, soil is fully saturated and further decrease in volume occurs due to the expulsion of pore water and compression of pore water (water is incompressible, hence volume change due to compression of pore water is negligible). It is a time dependent phenomenon which depends upon permeability of soil and magnitude of load applied. The rate of flow is controlled by pore pressure, the permeability and compressibility of soil with the passage of time as the pore pressure dissipates, the rate of flow decreases and eventually, flow ceases altogether, leading to a condition of constant effective stress. This signifies the end of primary consolidation.

(iii) **Secondary Consolidation:** After the completion of primary settlement when pore water pressure ceases to zero at the top surface, and no decrease in volume may be expected hence forth. But in practice there is always a certain decrease in volume after primary consolidation after a long time which is called secondary consolidation. The actual cause for secondary consolidation is not well established but it may be attributed to plastic readjustment of soil solids. The secondary settlement in granular soils is insignificant but in highly plastic soils it is 10-20% of total consolidation.

(iv) **Primary Compression Ratio:** It is the ratio of primary compression to total compression and is denoted by r_p.

$$\therefore \quad r_p = \frac{\text{Primary compression}}{\text{Total compression}}$$

Primary compression ratio and other ratios like initial compression ratio and secondary compression ratios are used while determining the coefficient of consolidation (C_v) in laboratory. The two methods namely Casagrande logarithm of time fitting method and Taylor square root of time method use these ratios effectively. In these methods dial gauge readings which give compression of sample are plotted against logarithm of time or square root of time for various degree of consolidation. Thus

$$r_p = \frac{10}{9} \times \frac{(R_0 - R_{90})}{(R_i - R_f)} = \frac{R_0 - R_{100}}{R_i - R_f}$$

Where R_i = initial dial gauge reading

R_f = final dial gauge reading

R_0 = dial gauge reading at zero per cent consolidation

R_{90} = dial gauge reading at 90% consolidation

R_{100} = dial gauge reading at 100% consolidation

3.4 **Under a certain loading, a layer of clay is expected to undergo full settlement of 18 cm. Also it is expected to settle by 5 cm in the period of first 2 months of loading. Find the time required for the clay layer to settle by 10 cm.**

[7 marks : 2004]

Solution:

We know that **degree of consolidation** is given by

$$U = \frac{\Delta h}{\Delta H} \times 100$$

Where Δh is settlement at any stage and ΔH is total settlement

Given $\Delta H = 18$ cm, $\Delta h_1 = 5$ cm, $t_1 = 2$ months, $\Delta h_2 = 10$ cm, $t_2 = ?$

Now $$U_1 = \frac{5}{18} \times 100 = 27.78\% = 0.2778$$

$$U_2 = \frac{10}{18} \times 100 = 55.55\% = 0.5555$$

Now, we know that, $$T_v = \frac{C_v t}{d^2}$$

where T_v is time factor $= \frac{\pi}{4} U^2$

C_v is coefficient of consolidation which is constant for a particular soil.

d is length of drainage path

t is time required to complete any stage of consolidation

$\therefore \quad t \propto T_v$ [But $T_v \propto U^2$ where U is in fraction]

$\Rightarrow \quad t \propto U^2$

$$\Rightarrow \quad \frac{t_1}{t_2} = \left(\frac{U_1}{U_2}\right)^2$$

$$\Rightarrow \quad t_2 = \left(\frac{U_2}{U_1}\right)^2 \times t_1 = \left(\frac{0.5555}{0.2777}\right)^2 \times 2 = 8 \text{ months}$$

3.5 Write the significance of pre-consolidation pressure in soil. How would you determine the pre-consolidation pressure using Casagrande method.

[10 marks : 2006]

Solution:

Soils tend to retain the effect of stress changes that have taken place in their geological history, in the form of their structure. When a soil is stressed to a level greater than maximum stress to which it was ever subjected in the past, perhaps some kind of a break down in the soil structure occurs, resulting in a much higher compressibility indicated by a steep void ratio-effective stress curve $(e-\bar{\sigma})$

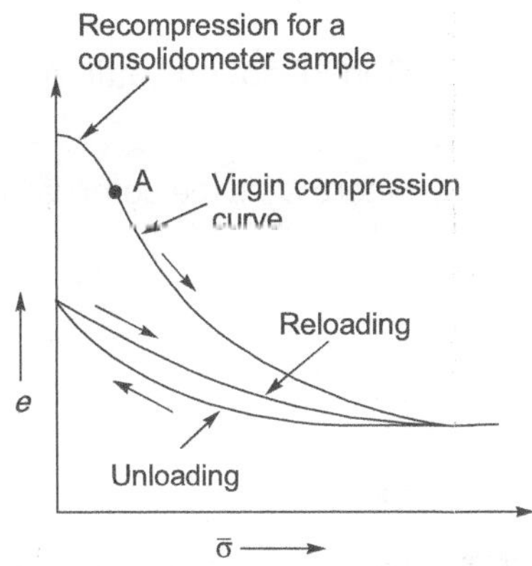

The initial flatter portion of the $e-\bar{\sigma}$ curve is called recompression curve and the steeper portion after the break in the curve (attributed to a breakdown in a structure) is called virgin compression curve, because the soil is experiencing first time the increased stresses in this part. Some where between these two parts of the curve lies the (point A).

Corresponding to the maximum value of stress the soil has ever experienced, called the preconsolidation pressure or stress $\bar{\sigma}_C$.

A soil is said to be preconsolidated or over consolidated , if the existing effective stress is less than the preconsolidation stress, that is $\bar{\sigma} < \bar{\sigma}_C$. Here comes into picture the term OCR (over consolidation ratio) which is the ratio of the preconsolidation stress to the present vertical effective stress; that OCR $= \frac{\bar{\sigma}_C}{\bar{\sigma}}$

For normally consolidated soil OCR = 1 and for a preconsolidated soil OCR > 1.

Determination of preconsolidation stress by casagrande method:

Casagrande studied the shapes of several e vs log $\bar{\sigma}$ curves and gave the following procedure as illustrated in figure.

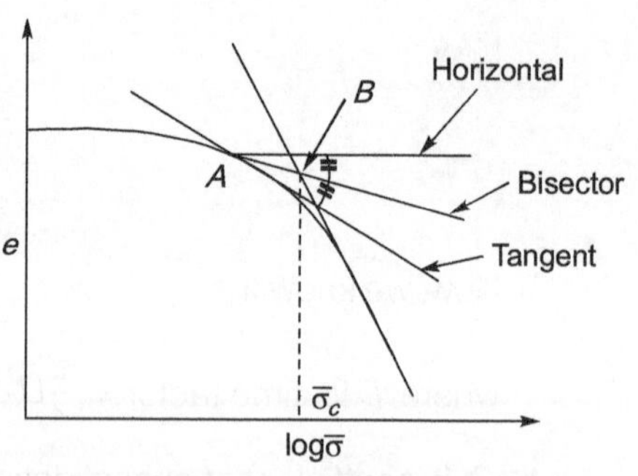

(i) By judgement of eye, point A, the point of maximum curvature on the consolidation curve is located.

(ii) At A, a horizontal line is drawn.

(iii) A tangent is drawn to the curve at A.

(iv) The angle obtained by steps (ii) and (iii) is bisected.

(v) The straight line part of the curve is extended back to meet the bisector line obtained in step (iv). The point of intersection of these lines (point B) gives the preconsolidation stress $\bar{\sigma}_c$.

3.6 **A soil of specific gravity 2.67 has a moisture content of 20% when fully saturated 2.0 cm thick sample of this soil tested in a consolidometer shows a compression of 0.050 cm when the load is increased from 45 kN/m² to 90 kN/m². Compute the compression index of the soil.**

[10 marks : 2008]

Solution:

Given data: Specific gravity of soil (G) = 2.67, Moisture content (w) = 20%, Degree of saturation (S) = 100%, Initial thickness of sample (H_0) = 2.0 cm

Change in thickness or settlement, $\Delta H = 0.050$ cm

Initial overburden pressure, $\bar{\sigma}_0 = 45$ kN/m²

Final overburden pressure, $\bar{\sigma}_1 = 90$ kN/m²

$\because \quad S \times e = G \times w$ (where e is the void ratio)

$$\therefore \quad e = \frac{G \times w}{100} = \frac{2.67 \times 20}{100} = 0.534$$

Consolidation settlement,

$$\Delta H = \frac{C_c \times H_0}{1+e} \times \log_{10}\left[\frac{\bar{\sigma}_0 + \Delta\bar{\sigma}}{\bar{\sigma}_0}\right]$$

where C_c is the compression index of soil

$$\Delta\bar{\sigma} = \bar{\sigma}_1 - \bar{\sigma}_0 = 90 - 45 = 45 \text{ kN/m}^2$$

$$\therefore \quad 0.05 = \frac{C_c \times 2}{1+0.534} \times \log_{10}\left[\frac{45+45}{45}\right]$$

$$\Rightarrow \quad C_c = 0.1274$$

3.7 **How many days would be required by a clay stratum 5 m thick, draining at both ends with coefficient of consolidation = 50 × 10⁻⁴ cm²/sec to attain 50% of its ultimate settlement? Given : T_{50} = 0.197.**

[10 marks : 2009]

Solution:

Given data: Thickness of clay layer, $H = 5$ m

Coefficient of consolidation, $C_v = 50 \times 10^{-4}$ cm²/sec

Degree of consolidation, $U = 50\%$

Time factor, $T_{50} = 0.197$

Since the clay stratum is draining at both ends, the length of drainage path, $d = \frac{H}{2} = \frac{5}{2} = 2.5\text{ m}$

From Taylor's formula, we have

$$T_v = \frac{C_v t}{d^2}$$

$$\Rightarrow \quad T_{50} = \frac{C_v t_{50}}{d^2}$$

$$\Rightarrow \quad t_{50} = \frac{T_{50} d^2}{C_v} = \frac{0.197 \times (2.5)^2}{50 \times 10^{-4} \times 10^{-4} \times 60 \times 60 \times 24} \text{ days}$$

$$= 28.5 \text{ days}$$

3.8 **In a consolidation test done in laboratory a sample of 20 mm thick consolidated 50% in 15 minutes with double drainage. How much time a 5.0 m thick layer of same soil will consolidate 50% and 30%? If the soil layer has a rock below, how much time it will take to consolidate 50% and 30%?**

[8 marks : 2010]

Solution:

As per Taylor's formula, time factor is given by

$$T_v = \frac{C_v t}{d^2}$$

where C_v is coefficient of consolidation, t is time to complete any stage of consolidation and d is length of drainage path.

When degree of consolidation i.e. $U \leq 60\%$; $T_v = \frac{\pi}{4} U^2$

$\therefore$ For $U = 50\%$, $$T_v = \frac{\pi}{4} \times (0.5)^2 = 0.1963$$

$$d = \frac{20}{2} = 10 \text{ mm (double drainage)}$$

$$t = 15 \text{ minutes for 50\% consolidation}$$

$$\therefore \quad C_v = \frac{T_v d^2}{t} = \frac{0.1963 \times (10)^2}{15} = 1.31 \text{ mm}^2/\text{min}$$

For $U = 30\%$, $$T_v = \frac{\pi}{4} \times (0.3)^2 = 0.0707$$

(i) **For $U = 50\%$,** $$d = \frac{5}{2} = 2.5 \text{ m} = 2500 \text{ mm}$$

$$t = \frac{T_v d^2}{C_v} = \frac{0.1963 \times (2500)^2}{1.31} \times \frac{1}{60 \times 24} = 650.38 \text{ days} = 1.78 \text{ years}$$

For $U = 30\%$, $$t = \frac{T_v d^2}{C_v} = \frac{0.0707 \times (2500)^2}{1.31} \times \frac{1}{60 \times 24} = 234.24 \text{ days}$$

(ii) **For $U = 50\%$,** $$d = 5 \text{ m} = 5000 \text{ mm}$$

$$t = \frac{T_v d^2}{C_v} = \frac{0.1963 \times (5000)^2}{1.31} \times \frac{1}{60 \times 24} \times \frac{1}{365} = 7.13 \text{ years}$$

For $U = 30\%$ $D = 5\text{ m} = 5000\text{ mm}$

$$t = \frac{T_v d^2}{C_v} = \frac{0.0707 \times (5000)^2}{1.31} \times \frac{1}{60 \times 24} \times \frac{1}{365} = 2.57 \text{ years}$$

3.9 **A 20 mm thick laboratory soil sample reaches 60% consolidation in 32.5 seconds under double drainage condition. How much time will be required for a 10 m thick layer in the field to reach the same degree of consolidation, if the clay layer is sandwiched between a sandy layer and an impermeable layer?**

[10 marks : 2011]

Solution:

Given: Thickness of soil sample = 20 mm, $U = 60\%$ consolidation, $t = 32.5$ sec

Let coefficient of consolidation of soil = c_v

Using, $$c_v = \frac{T_v d^2}{t} = \frac{T_v\left(\frac{20}{2}\right)^2}{32.5}$$ $\left[\because \text{ For double drainage } d = \frac{\text{total thickness}}{2}\right]$

or, $$c_v = \frac{100\,T_v}{32.5} \quad \text{...(i)}$$

For 10 m thick layer in field, d = 10 m for single drainage

T_v for 60% consolidation = Same as for soil sample

c_v = Same as for soil sample

$$\Rightarrow \quad c_v = T_v \times \frac{(10 \times 10^3)^2}{t} \quad \text{...(ii)}$$

Equating (i) and (ii) $$\frac{100\,T_v}{32.5} = \frac{10^8 T_v}{t}$$

$$\Rightarrow \quad t = 32.5 \times 10^6 \text{ sec} = 376.15 \text{ days}$$

3.10 **At a vertical stress of 200 kPa, the void ratio of a saturated soil sample tested in an oedometer is 1.52 and lies on the normal consolidation line. An increment of vertical stress of 150 kPa in the second stage compress the sample to a void ratio of 1.43.**

(i) Determine the compression index C_c of the soil.

(ii) The sample was unloaded to a vertical stress of 200 kPa and the void ratio increased to 1.45. Determine the slope of the recompression index C_r.

(iii) What is the over consolidation ratio of the soil at second state?

(iv) If the soil were reloaded to a vertical stress of 500 kPa, what void ratio would be attended?

[10 marks : 2013]

Solution:

Given: $\bar{\sigma}_0 = 200$ kPa, $S = 1$, $e_1 = 1.52$

$$\Delta\bar{\sigma} = 150 \text{ kPa}$$

$$e_2 = 1.43$$

$$\therefore \quad \Delta e = (e_2 - e_1) = -0.09$$

$\therefore$ Compression index $$C_C = \frac{-\Delta e}{\log\left(\frac{\bar{\sigma}_0 + \Delta\bar{\sigma}}{\sigma_0}\right)} = \frac{0.09}{\log\left(\frac{200+150}{200}\right)} = 0.37$$

Recompression index $$C_r = \frac{-(1.43-1.45)}{\log\left(\frac{150+200}{200}\right)} = 0.0823$$

Over Consolidation Ratio, $$(OCR) = \frac{350}{200} = 1.75$$

$$\Delta e = C_R \log\left(\frac{350}{200}\right) + C_c \log\left(\frac{500}{350}\right) = 0.02 + 0.0573 = 0.077$$

$$\therefore \quad e = 1.45 - 0.077 = 1.372$$

3.11 **For a flexible foundation with sides 2 m and 3 m and an axial load of 750 kN, determine the elastic settlement under one of the corners. The soil has properties ν = 0.37 and E = 9.8 MPa.**

[4 marks : 2013]

Solution:

Given: Length of footing (L) = 3 m, Width of footing (B) = 2 m, Axial load (P) = 750 kN

∴ Uniformly distributed load, $$q = \frac{750}{3 \times 2} = 125 \text{ kN/m}^2$$

$$E_s = 9.8 \times 10^6 \text{ N/m}^2 = 9800 \text{ kN/m}^2$$

$$\frac{L}{B} = \frac{3}{2} = 1.5$$

For flexible foundations with $\frac{L}{B}$ = 1.5, influence factor (I) for corner location is 0.68.

∴ **Elastic settlement,** $$S_i = qB\left(\frac{1-\mu^2}{E_s}\right)I = 125(2)\left(\frac{1-\mu^2}{9.8 \times 10^3}\right)0.68 \text{ m}$$

$$= 14.972 \times 10^{-3} \text{ m} \simeq 15 \text{ mm}$$

3.12 **A footing is to be constructed 1.8 m below the ground surface as shown in fig. The base of the footing is 2.7 m x 2.7 m and it carries a total load of 1800 kN which includes the column load and weight of the footing. Compute the total expected settlement of the footing (consider only consolidation settlement):**

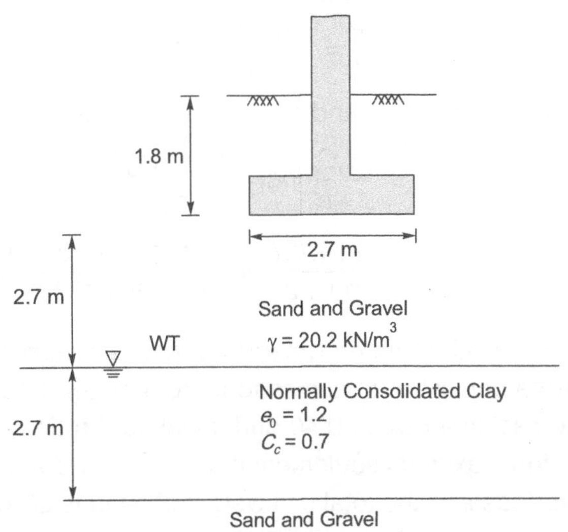

[8 marks : 2015]

Solution:

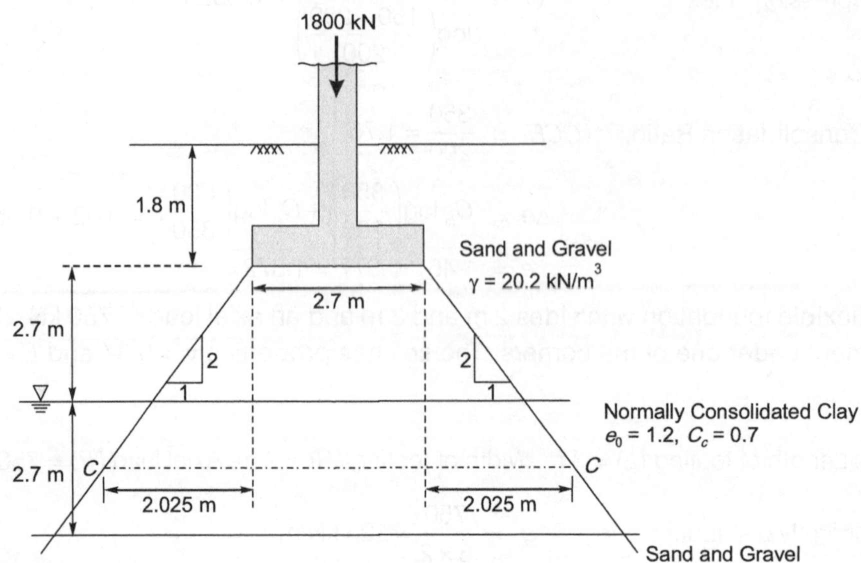

As load distribution is not given, let us consider load distribution to be 2 vertical to 1 horizontal.

Stress area at the centre of clay layer,

$$A = (2.7 + 2 \times 2.025)^2 = 45.5625 \text{ m}^2$$

Assuming specific gravity of solids of clay layer,

$$G = 2.7$$

$$\gamma_{sat_{clay}} = \left(\frac{G+e}{1+e}\right) \times \gamma_w = \left(\frac{2.7+1.2}{1+1.2}\right) \times 9.81 = 17.39 \text{ kN/m}^2$$

Initial effective overburden pressure at centre of clay layer,

$$\bar{\sigma}_0 = (1.8 + 2.7) \times \gamma + (\gamma_{sat} - \gamma_w) \times \frac{2.7}{2}$$

$$= (1.8 + 2.7) \times 20.2 + (17.39 - 9.81) \times \frac{2.7}{2}$$

$$\Rightarrow \quad \bar{\sigma}_0 = 101.133 \text{ kN/m}^2$$

Increase in effective overburden pressure at centre of clay layer

$$\Delta\bar{\sigma}_0 = \frac{P}{A} = \frac{1800}{45.5625} = 39.51 \text{ kN/m}^2$$

Assuming consolidation settlement of sand and gravel layer to be negligible,

$$\Delta H = \frac{H_0 C_c}{1+e_0} \log_{10}\left(\frac{\bar{\sigma}_0 + \Delta\sigma_0}{\bar{\sigma}_0}\right)$$

$$= \frac{2.7 \times 0.7}{1+1.2} \log_{10}\left(\frac{101.133 + 39.51}{101.133}\right) = 0.123 \text{ m} = 12.3 \text{ cm}$$

3.13 **The base of a 20 m × 50 m raft rests on ground surface and applies a stress of 100 kPa to the subsoil. The ground strata consist of dense sand up to 4 m depth followed by a clay layer of 2 m thickness having compression index of 0.40 and initial void ratio of 1.0. The clay is followed by more dense sand. Find the maximum settlement that will occur due to consolidation of clay. Water table is at the ground surface and the total unit weight of sand is 20 kN/m³ and of clay is 18 kN/m³. Calculations should be based on two equal divisions. Use 2 : 1 stress distribution.**

[8 marks : 2016]

Solution:

Let us consider two division of clay layer of 1 m each.

Layer-1

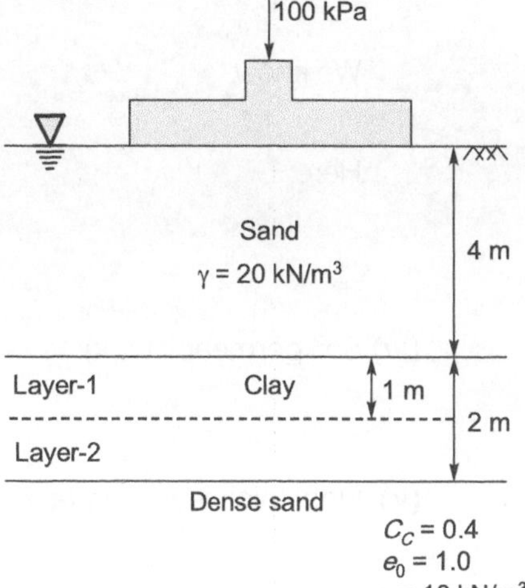

$$H_0 = 1 \text{ m}$$

$\bar{\sigma}_0$ at centre of layer 1

$$\bar{\sigma}_0 = (\gamma_{sand} - \gamma_w) \times 4 + (\gamma_{clay} - \gamma_w) \times 0.5$$
$$= (20 - 9.81) \times 4 + (18 - 9.81) \times 0.5$$
$$= 44.855 \text{ kN/m}^2$$

$\Rightarrow$
$$\Delta\bar{\sigma} = \frac{q(B \times L)}{(B + 2nz)(L + 2nz)}$$
$$z = 4.5 \quad n = 0.5$$
$$\Delta\bar{\sigma} = \frac{100 \times 20 \times 50}{\left(20 + 2 \times \frac{1}{2} \times 4.5\right)(50 + 4.5)}$$
$$\Delta\bar{\sigma} = 74.89 \text{ kN/m}^2$$

Layer-2

$$H_0 = 1 \text{ m}$$

$\bar{\sigma}_0$ at centre of layer-2

$$\bar{\sigma}_0 = (\gamma_{sand} - \gamma_w) \times 4 + (\gamma_{clay} - \gamma_w) \times 1.5$$
$$= (20 - 9.81) \times 4 + (18 - 9.81) \times 1.5 = 53.045 \text{ kN/m}^3$$
$$\Delta\bar{\sigma} = \frac{100 \times 20 \times 50}{(20 + 5.5)(50 + 5.5)} = 70.658 \text{ kN/m}^2$$
$$\Delta H = \Sigma \frac{H_0 C_C}{1 + e_0} \log \frac{\bar{\sigma}_0 + \Delta\bar{\sigma}}{\bar{\sigma}_0} = \frac{1 \times 0.4}{1 + 1}\left[\log\left(\frac{44.855 + 74.89}{44.855}\right) + \log\left(\frac{53.045 + 70.658}{53.045}\right)\right]$$
$$\Delta H = 0.1588 \text{ m} = 158.8 \text{ mm}$$

3.14 **A clay soil specimen 25 mm thick was tested in the laboratory allowing drainage both sides. The specimen showed a decrease in void ratio from 0.92 to 0.78 when the pressure was increased from 60 kN/m² to 120 kN/m². Calculate coefficient of compressibility, coefficient of volume change and coefficient of consolidation. If time for 50% consolidation was 5 minutes determine the soil permeability in cm/sec, and calculate final thickness of the specimen at the end of the test.**

[20 marks : 2017]

Solution :

(i) Coefficient of compressibility (a_v);

Change in void ratio, $\Delta e = e_0 - e = 0.92 - 0.78 = 0.14$

Change in pressure, $\Delta\bar{\sigma} = \bar{\sigma}_2 - \bar{\sigma}_1 = 120 - 60 = 60 \text{ kN/m}^2$

$$\therefore \quad a_v = \frac{\Delta e}{\Delta\bar{\sigma}} = \frac{0.14}{60} = 2.333 \times 10^{-3} \text{ m}^2/\text{kN}$$

(ii) Coefficient of volume change (m_v):

$$m_v = \frac{a_v}{1 + e_0} = \frac{2.333 \times 10^{-3}}{1 + 0.92} = 1.215 \times 10^{-3} \text{ m}^2/\text{kN}$$

(iii) Coefficient of consolidation (c_v);

$\because$ Degree of consolidation, $U = 50\%$

$\therefore$ $$T_V = \frac{\pi}{4}(0.50)^2 = 0.196$$

We know, $$T_V = \frac{C_V t}{H^2}$$

Here, $$H = \frac{25}{2} = 12.5 \text{ mm} = 0.0125 \text{ m} \quad \text{(double drainage)}$$

$\Rightarrow$ $$C_V = \frac{T_V H^2}{t} = \frac{0.196 \times (0.0125)^2}{5 \times 60} = 1.021 \times 10^{-7} \text{ m}^2/\text{sec}$$

(iv) Soil permeability (k);

$$k = C_v m_v \gamma_w = 1.021 \times 10^{-7} \times 1.215 \times 10^{-3} \times 9.81$$
$$= 1.217 \times 10^{-9} \text{ m/sec}$$

(v) Final thickness of specimen, H_f

$$\Delta H = \left(\frac{\Delta e}{1+e_0}\right) \times H_0 = \frac{0.14}{1.92} \times 25 = 1.823 \text{ mm}$$

$\therefore$ $$H_f = H - \Delta H = 25 - 1.823 = 23.18 \text{ mm}$$

3.15 **An 8 m thick saturated clay layer lies above a permeable dense sand layer. The clay settles by 40 mm in 2 years when subjected to a widespread load of 50 kN/m² at its surface.**

(i) What will be the ultimate settlement of the clay?

(ii) Calculate the compression index, C_c of the clay.

Assume that the water table is below the dense sand layer and will not influence the settlement. Properties of clay: Average bulk unit weight, γ_{bulk} = 18 kN/m³, Coefficient of consolidation, C_v = 1 m²/yr and Specific gravity, G_s = 2.65. Unit weight of water is equal to 9.81 kN/m³.

[20 marks : 2018]

Solution:

(i)

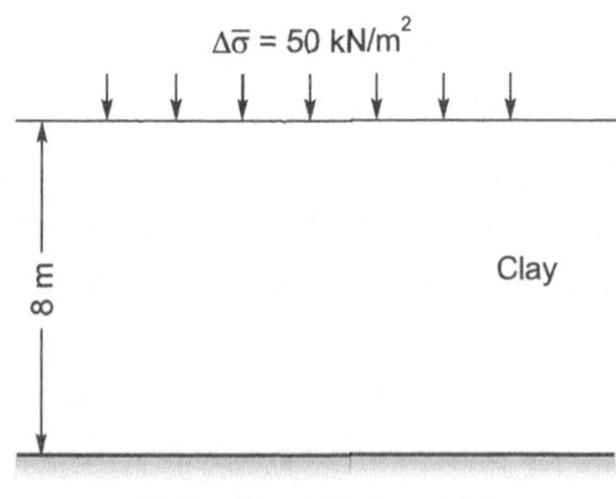

$H_0 = 8$ m, $\Delta h = 40$ mm in $t = 2$ yeas, $\gamma_b = 18$ kN/m³, $C_v = 1$ m²/yr, $G_s = 2.65$, $\gamma_w = 9.81$ kN/m³

Drainage path, $d = 8$ m (double drained)

Now, $$T_v = C_v \frac{t}{d^2} = \frac{\pi}{4} U^2$$

$\Rightarrow$ $$C_v \frac{t}{d^2} = \frac{\pi}{4}\left(\frac{\Delta h}{\Delta H}\right)^2$$ ΔH = Ultimate settlement

$$\Rightarrow \quad \frac{1\times 2}{4^2} = \frac{\pi}{4}\left(\frac{40}{\Delta H}\right)^2$$

$$\Rightarrow \quad \Delta H = 100.265 \text{ mm}$$

(ii) $C_c = ?$

$$\gamma_b = 18 \text{ kN/m}^3 = \left(\frac{G+e}{1+e}\right)\gamma_w$$

$$\Rightarrow \quad 18 = \left(\frac{2.65+e}{1+e}\right)\times 9.81$$

$$\Rightarrow \quad e = 0.976$$

$$\bar{\sigma}_0 = 4\times 18 = 72 \text{ kN/m}^2$$ (Stress at center of clay layer)

$$\Delta H = \frac{H_0 C_c}{1+e_0}\log\left(\frac{\bar{\sigma}_0+\Delta\bar{\sigma}}{\bar{\sigma}_0}\right)$$

$$\Rightarrow \quad 100.265 = \frac{8000C_c}{1.976}\log\left(\frac{72+50}{72}\right)$$

$$\Rightarrow \quad C_c = 0.108$$

3.16 **The figure given below represents time-consolidation relationship for a 30 mm thick clay sample subjected to a given pressure range under double drainage condition. Determine the coefficient of consolidation, C_v, for the clay sample. How long will it take (in days) to reach 50% consolidation for the same soil if it was 2.5 m thick and drained in one direction only? Given**

$$T = \frac{\pi}{4}\left(\frac{U\%}{100}\right)^2, U \le 60\%$$

$$T = 1.781 - 0.9331 \log(100 - U\%), U \le 60\%$$

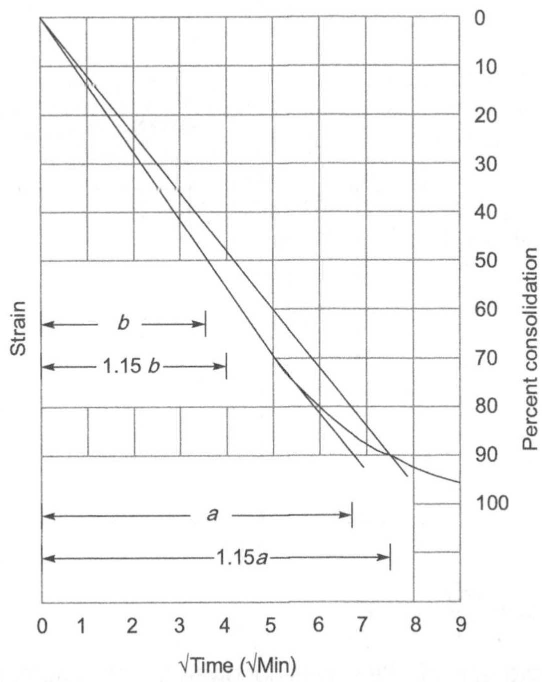

[15 marks : 2019]

Solution:

Given: Thickness of clay layer = 30 mm

$$T_{90} = 1.781 - 0.933 \log(100 - 90) = 0.848$$

From the graph

$$\sqrt{t_{90}} = 7.5$$

$$t_{90} = 56.25 \text{ min}$$

$$0.848 = \frac{C_V \times t_{90}}{H^2} \left(H = \frac{30}{2} = 15 \text{ mm as double drainage condition is given} \right)$$

$$\Rightarrow \quad \frac{0.848 \times 15^2}{56.25} = C_V$$

$$C_V = 3.392 \text{ mm}^2/\text{min} = 4.88 \times 10^{-3} \text{ m}^2/\text{day}$$

Now,

$$T_{50} = \frac{\pi}{4}\left(\frac{50}{100}\right)^2 = 0.196$$

$$0.196 = \frac{C_V \times t}{H^2} \quad \text{(Here, } H = 2.5 \text{ m, One way drainage is given)}$$

$$\therefore \quad 0.196 = \frac{4.88 \times 10^{-3} \times t}{2.5^2}$$

$$0.196 = \frac{4.88 \times 10^{-3} \times t}{2.5^2} = 251.02 \text{ days}$$

3.17 A square footing (2 m × 2 m) founded at a depth 1 m below GL has to support a column load of 400 kN. The soil profile consists of fine sand ($\gamma = 17$ kN/m^3) up to a depth of 3 m, followed by a 4 m thick layer of silty clay ($\gamma = 15$ kN/m^3, NMC = 92%, $C_c = 1.05$).

This is followed by dense sandy layer up to 12 m. The WT is at 2 m below the GL. Compute the possible consolidation settlement and state whether it is within permissible limits.

[12 marks : 2020]

Solution:

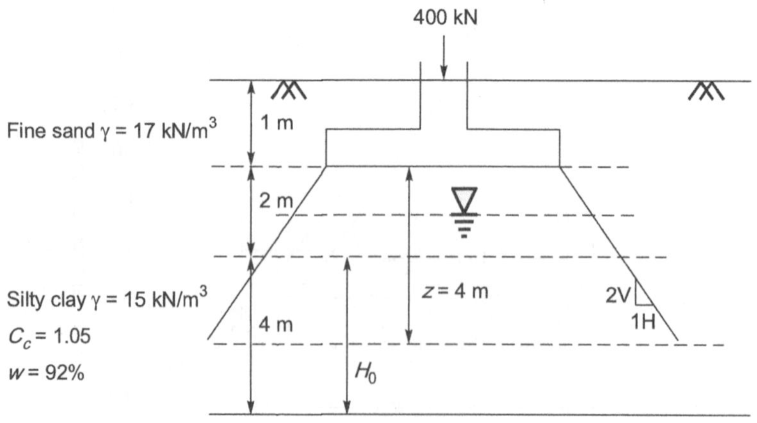

$$e_o = \frac{wG}{S} = 2.484$$

Note : Because G and void ratio of silty clay is not given hence assuming $G = 2.7$.

Note : Assuming γ and γ_{sat} same for fine sand because saturated unit weight of fine sand is not given.

Step 1 : H_o (Thickness of compressible layer) = 4 m

Step 2 : $\bar{\sigma}_o$ at c/c of compressible layer.

$$\bar{\sigma}_o = \sigma - u = (3\gamma + 2\gamma_{sat}) - 3\gamma_w$$

$$= 3 \times 17 + 2 \times 15 - 3 \times 9.81 = 51.57 \text{ kN/m}^2$$

Step 3 : Assuming load distribution as 2 V : 1 H

$$\Delta\bar{\sigma} = \frac{Q}{(B+2nz)^2} = \frac{400 \text{ kN}}{\left(2+2\times\frac{1}{2}\times4\right)} = 11.111 \text{ kN/m}^2$$

Step 4 :

$$\Delta H = \frac{H_o C_c}{1+e_o}\log\left(\frac{\bar{\sigma}_o + \Delta\bar{\sigma}}{\bar{\sigma}_o}\right)$$

$$= \frac{4\times1.05}{1+2.484}\log\left(\frac{51.57+11.111}{51.57}\right)$$

$$\Delta H = 0.102 \text{ m} = 102 \text{ mm}$$

Note : Settlement 102 mm is out of permissible limit hence there is risk of settlement failure.

Permissible limit as per IS Code :

A. For isolated footing on clay = 75 mm
B. For isolated footing on sand = 50 mm
C. For raft foundation on clay = 100 mm
D. For raft foundation on sand = 75 mm

3.18 **Soft saturated clay has a thickness of 6 m. After one year, when the clay consolidated by 50%, the observed settlement was to the tune of 10 cm. For an identical clay and loading condition, what will be the magnitude of settlement at the end of one year and five years if the thickness of the clay layer was 25 m?**

[20 Marks : 2021]

Solution:

Case-1:

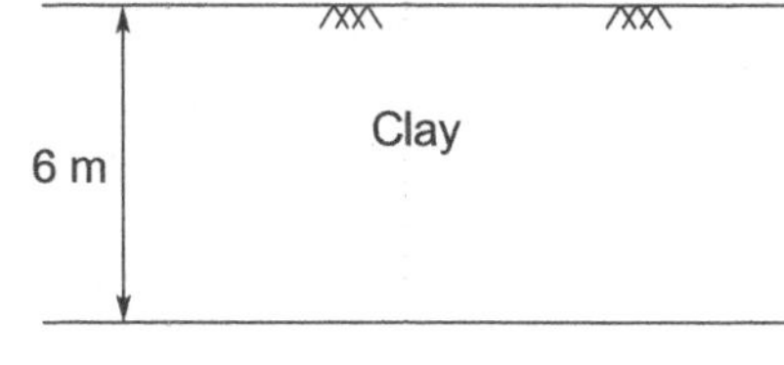

After 1year,

$$U = 50\%$$
$$\Delta h = 10 \text{ cm}$$
$$U = \frac{\Delta h}{\Delta H}$$
$$0.5 = \frac{100}{\Delta H}$$
$$\Delta H = 200 \text{ mm}$$

Case-2:

For 25 m clay layer

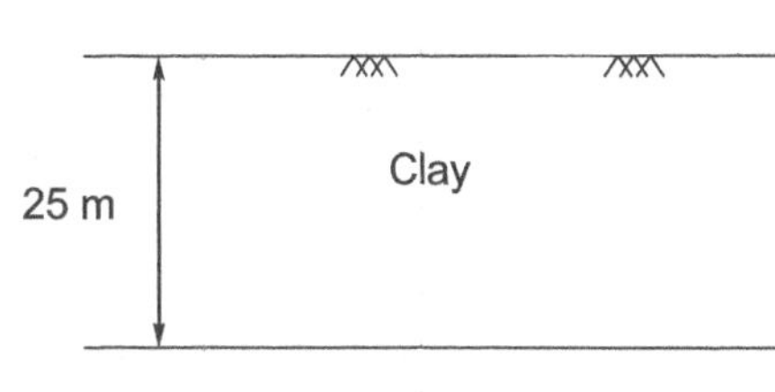

$$\Delta H = \frac{200}{6}\times 25 = 833.33 \text{ mm}$$

(Δh) After 1 year = ?

(Δh) After 5 years = ?

For identical clay and loading condition

$$T_v = \frac{C_v t}{d^2}$$

For $U < 50\%$

$$T_v \propto U^2 \propto \left(\frac{\Delta h}{\Delta H}\right)^2 \qquad \left(\because T_v = \frac{\pi}{4}U^2\right)$$

$\Rightarrow$ $t \propto d^2U^2$

After 1 year

$$\frac{t_2}{t_1} = \frac{d_2^2U_2^2}{d_1^2U_1^2}$$

$$\Rightarrow \quad \frac{1}{1} = \frac{(25)^2 \times \left(\frac{\Delta h_2}{833.33}\right)^2}{6^2 \times (0.5)^2}$$

$$\Delta h_2 = 99.99 \text{ mm} \simeq 100 \text{ mm}$$

After 5 years

$$\frac{5}{1} = \frac{(25)^2 \times \left(\frac{\Delta h_2}{833.33}\right)^2}{6^2 \times (0.5)^2}$$

$$\Delta h_2 = 223.6 \text{ mm}$$

$\therefore$ In 25 m clay layer

Settlement after 1 year = 100 mm

After 5 years settlement = 223.6 mm

3.19 **An oedometer test on a 2 cm thick clay sample took 15 minutes time to attain 50% consolidation under a loading with double drainage condition. How many days will it take in the field to achieve the same degree of consolidation for the same clay soil 4 m thick? Consider similar loading and draiange conditions for laboratory and field.**

[10 marks : 2022]

Solution:

Lab test sample: Thickness, $H = 2$ cm

Double drainage condition,

$$\therefore \quad T_v = C_v \frac{t}{[H/2]^2}$$

$$\text{For 50\% consolidation, } T_v = \frac{\pi}{4}(0.5)^2 = 0.196$$

$$t = 15 \text{ min}$$

$$\therefore \quad 0.196 = C_v \times \frac{15 \text{ min}}{\left[\frac{2 \times 10^{-2}}{2}\right]^2}$$

$$\Rightarrow \quad C_v = 1.31 \times 10^{-6} \text{ m}^2/\text{min}$$

Field conditions: Thickness, $H = 4$ m

$$[T_v]_{50\%} = C_v \times \frac{t}{[H/2]^2}$$

$$\Rightarrow \quad 0.196 = 1.31 \times 10^{-6} \times \frac{t}{[4/2]^2}$$

$$\Rightarrow \quad t = 415.6 \text{ days}$$

4. Compaction of Soils

4.1 **Work out the theoretical maximum dry density for a soil sample having specific gravity of 2.7 and OMC = 16%. Also explain the difference in OMC values in case of Proctor Test and Modified Proctor Test for cohesive soils and granular soils.**

[10 marks : 2001]

Solution:

Given data: $G = 2.7$, OMC $= w = 16\%, \gamma_w = 9.81$ kN/m^3

We know that

$$\gamma_d = \frac{G\gamma_w}{1+e}$$

But $$Se = Gw$$

$$\Rightarrow \quad e = \frac{Gw}{S}$$

$$\therefore \quad \gamma_d = \frac{G\gamma_w}{1+\frac{Gw}{S}}$$

where γ_d is dry unit weight of soil and all other symbols have their usual meanings.

Now, for a given water content, the theoretical maximum value of dry unit weight for a compacted soil is obtained corresponding to the situation where no air voids are left i.e. when the degree of saturation becomes equal to 100% i.e. $S = 1$

$$\therefore \quad \gamma_{d,\max} = \frac{G\gamma_w}{1+\frac{Gw}{1}}$$

$$\Rightarrow \quad \gamma_{d,\max} = \frac{G\gamma_w}{1+Gw}$$

$$= \frac{2.7\times 9.81}{1+2.7\times 0.16}$$

$$= 18.49 \text{ kN/m}^3$$

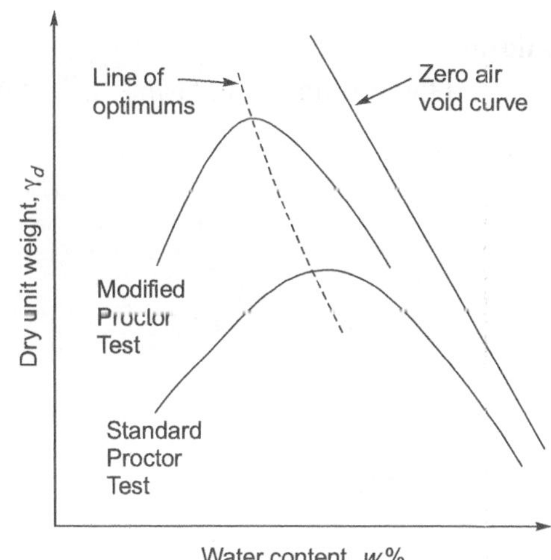

The only difference between **Proctor Test and Modified Proctor Test** is that the latter requires more compactive effort than the former. The dry density increases with the compactive effort but the water content decreases along with it. The compaction curve for the Modified Proctor Test shifts to the top and to the left w.r.t. the compaction curve for Proctor Test.

There is a difference between OMC values of cohesive and granular soils as well. Cohesive soils have large value of OMC than granular soils.

4.2 **For compacting which types of soils;**
(i) sheep foot rollers, and (ii) vibratory rollers are most effective?
Explain giving reasons.

[6 marks : 2004]

Solution:

(i) **Sheep Foot Rollers:** These are most effective for cohesive type of soils such as clays. These are suitable for compacting all fine grained soils, but is generally not suitable for use on cohesionless granular materials. The pads on the sheep foot drum penetrate through the top lift and actually compact the lift below. When the drum rotates the pads pull out of the soil, they kick up or fluff the material because of their shape.

Since the sheep foot rollers tends to aerate the soil as it compacts, it is ideally suited for working soils that have moisture contents above the acceptable moisture range. The sheep foot roller does not adequately compact the upper 2 to 3 inch in a lift and it should therefore be followed by a lighter pneumatic tyred or steel wheeled roller if no succeeding lift is to be placed.

(ii) **Vibratory Rollers:** These are most effective for cohesionless type of soils such as sands and gravels. The soils respond quite well to compaction produced by a combination of pressure and vibration. When these soils are vibrated, the particles shift their positions and nestle more closely with adjacent particles to increase the density of the mass.

The vibration has two measurements - amplitude, which give the measurement of movement or throw and frequency which is the rate of movement, or number of vibrations per second. The amplitude controls the effective area or depth to which the vibration is transmitted into the soil, while the frequency determines the number of flows or oscillations that are transmitted in a period of time.

The impacts imparted by the vibrations produce pressure waves that set the soil particles in motion producing compaction. In compacting granular material, frequency is usually the critical parameter as opposed to amplitude.

4.3 **In a Proctor compaction test, the soil specimen of one of the observations had a bulk density of 19 kN/m^3 with a moisture content of 15%. Find,**

(i) degree of saturation of the specimen if G = 2.7

(ii) additional moisture content required for saturating the soil specimen.

[7 marks : 2004]

Solution:

Given data γ = 19 kN/m^3, w = 15%, G = 2.7

(i)
$$\gamma = \frac{G+Se}{1+e} \times \gamma_w \qquad [\because Se = Gw]$$

$$\Rightarrow \quad \gamma = \frac{G+Gw}{1+e} \times \gamma_w$$

$$\Rightarrow \quad 1+e = \frac{(1+w)\,G\gamma_w}{\gamma}$$

$$\Rightarrow \quad e = \frac{(1+0.15)\times 2.7\times 9.81}{19} - 1 = 0.603$$

But
$$S = \frac{Gw}{e} = \frac{2.7\times 0.15}{0.603}$$
$$= 0.6716 = 67.16\%$$

Thus, degree of saturation is 67.16%

(ii) **For degree of saturation = 100%, S = 1**

But
$$Se = Gw$$

$$w = \frac{Se}{G}$$

$$\Rightarrow \quad w = \frac{1\times 0.603}{2.7} \times 100 = 22.33\%$$

$\therefore$ Additional water content required = 22.33 – 15 = 7.33%

4.4 State and discuss different factors influencing compaction of soil in the field.

[15 marks : 2006]

Solution:

There are four main factors which influence compaction and they are as follows:

(i) **Water Content:** There are two theories to explain the typical water content-dry unit weight relationship. They are the Lubrication theory by Proctor and the Electrical Double Layer theory by Lambe.

According to Lubrication theory at lower water contents, the soil is stiff and the soil grains offer more resistance to compaction. As the water content increases, the dry density increases and air voids are decreased till the optimum water content is reached, a stage when lubrication effect is maximum. With further increase in moisture content, however the water starts to replace the soil particles and since $\gamma_w << \gamma_s$, the dry unit weight starts decreasing.

Lambe uses concept of soil structure and the Electrical Double Layer theory to explain the effect of water content on dry unit weight. In case of cohesive soils, there is an attractive force namely the Van-der Waal's forces which acts between two soil particles and a repulsive force which is due to double layers of adsorbed water tending to come into contact with each other. While the attractive forces remains same in magnitude, the repulsive force is directly related to the size of double layers. If the net force between the particles is attractive, flocculated structure is the result; if it is repulsive, the particles tends to move away - 'disperse'. At low water contents attractive forces are predominant which makes it difficult for the particles to move about when compactive effort is applied. A low dry unit weight is the consequence. As the water content is increased, the double layer expands and inter particle repulsive forces increase. The particles easily slide over one another and get packed more closely, resulting in higher dry unit weight.

The maximum expansion of the double layer is at the OMC, beyond that, the addition of water does not add any further to the expansion of double layer but the water tends to occupy space which otherwise would have been occupied by soil particles. Hence a decrease in unit weight. It also explains why the shape of the compaction curve is not the typical inverted V shape in the case of soils which are not cohesive and plastic in nature.

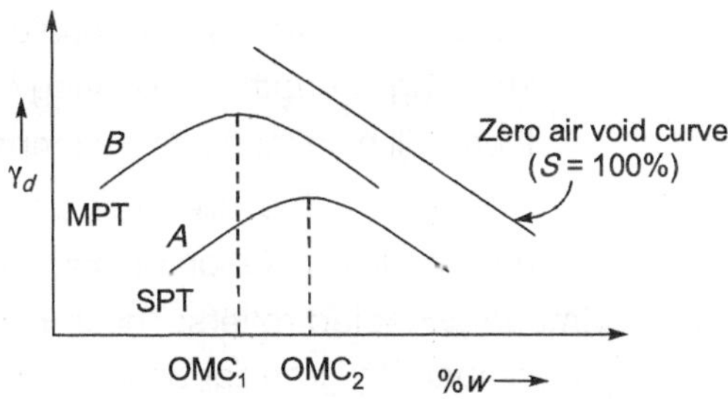

(ii) **Compactive Effort:** For a given type of compaction, the higher the compactive effort, the higher the maximum dry unit weight and lower the OMC. In the above figure compaction curve *B* corresponds to the higher compactive effort in a MPT, comparing it with the compaction curve *A* for SPT, one can see the compaction curve shifts to the top and to the left when compactive effort is increased. However, the margin of increase becomes smaller and smaller even on the dry side of the OMC while on the wet side of OMC, there is hardly any increase at all. If the peaks of compaction curves for different compactive efforts are joined together a 'line of optimums' is obtained which is nearly parallel to zero air void line. This brings out the fact that even a higher compactive effort does not result in a higher efficiency of compaction.

(iii) **Types of Soil:**

(a) Coarse grained, well graded soils compact to high dry unit weight especially if they contain some fines.

(b) Poorly graded sands lead to lowest dry unit weight values.

(c) In clay soils, the maximum dry unit weight tends to decrease as plasticity increases.

(d) Cohesive soils have generally high values of OMC.

(e) Heavy clays with high plasticity have very low maximum dry density and very high OMC.

(iv) Methods of Compaction:

Ideally speaking, the laboratory test must reproduce a given field compaction procedure, because the mode of compaction does influence somewhat the shape and the position of the 'γ_d' vs 'w' plot. Since the field compaction is essentially a kneading type compaction or rolling type compaction and the laboratory tests use the dynamic impact type compaction, one must expect some divergence in the OMC and $\gamma_{d(max)}$ in the two cases.

4.5 Explain the objects of compaction and effect of inadequate compaction. Discuss the advantages and application of various compacting equipment for construction of subgrade and embankment.

[8 marks : 2007]

Solution:

Compaction is the process by which the soil particles are artificially rearranged and packed together into a state of closer contact by mechanical means in order to decrease its porosity and thereby increase its dry density. This is usually achieved by dynamic means such as tamping, rolling or vibration. The process of compaction involves the expulsion of air only.

The following are the important objects of compaction:

(i) To increase the dry density of soil, thus increasing its shear strength and bearing capacity through an increase in frictional characteristics.
(ii) To decrease the tendency for settlement of soils
(iii) To bring about a low permeability of the soil
(iv) To reduce the compressibility of the soil
(v) To control undesirable volume changes through swelling and shrinkage

If the compaction is inadequate then all the above mentioned objects cannot be achieved. Thus the shearing strength and bearing capacity of the soil cannot be increased. The soil will settle more and there will be a increase in compressibility and permeability of the soil.

Smooth wheeled rollers and pneumatic tyred rollers are the most suitable compacting equipments for construction of subgrade and embankment.

Smooth-wheeled rollers: These impart static compression to the soil. These appear to be more suitable for compacting granular base courses and paving mixtures for highway and airfield work rather than for compacting earth fill. The relatively smooth surface obtained acts as a sort of seal at the end of a days work and drains off rain water very well. These are suitable for crushed rocks, gravels and sands.

4.6 (i) Explain the term "optimum moisture content". How is it affected by compacting effort?
(ii) State the factors affecting field compaction of soil.

[5 + 5 marks : 2009]

Solution:

(i) Moisture plays an important role in compaction. In fact, when moisture (water) is added to a dry soil, the soil grains get surrounded by a film of absorbed water. If more moisture is added, then this film becomes thicker and soil grains surrounded by such films can slide over each other more easily. The grains thus become capable to occupy closer spacing more easily on application of external loading. The moisture thus acts only as a lubricant in the compaction process and a closer packing of soil grains is achieved on rolling by the expulsion of air from the voids.

If more water is continued to be added, then a stage comes when moisture becomes excess and starts occupying the space which otherwise could have been occupied by soil grains under compaction. Evidently, moisture then starts hindering the closer packing of grains. In the limiting stage, there occurs a particular amount of water which causes maximum lubrication without becoming excess to cause hinderance in compaction by occupying space for its own molecules. This limiting moisture which is most useful content, the soil would be compacted maximum possible and would have the maximum density or unit weight i.e.. the weight of soil grains in a unit volume of compacted soil mass would be maximum.

It may however be noted that maximum dry unit weight so obtained is only for a given compactive effort and method of compaction. It is also not necessarily the maximum dry unit weight that can be obtained in the field.

For a given type of compactive effort, the higher the maximum dry unit weight and lower the OMC. However, as the moulding water content increases, the influence of compactive effort on dry unit weight tends to diminish. Again the maximum dry unit weight does not go on increasing as the compactive effort is increased. The margin of increase becomes smaller and smaller even on the dry side of OMC, while on the wet side of OMC, there is hardly any increase at all. If the peaks of compaction curves for different compactive efforts are joined together, a line of optimums is obtained which is nearly parallel to the zero air void curve.

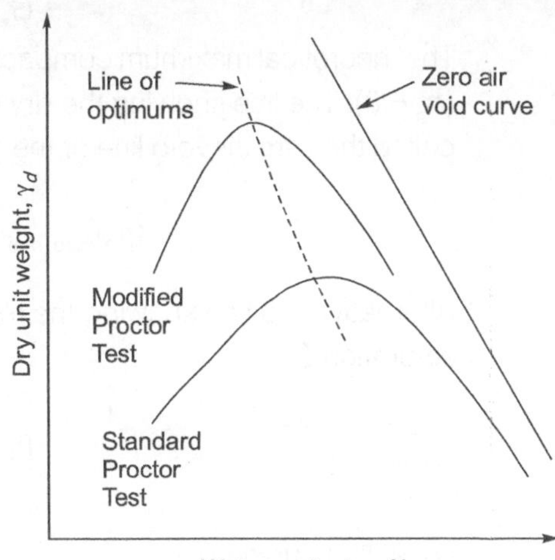

This brings out the fact that even a higher compactive effort does not result in a higher efficiency of compaction. The degree of saturation remains almost the same in all cases.

(ii) The factors affecting field compaction of soil are:

(a) Water content
(b) Compactive effort
(c) Type of soil
(d) Method of compaction

4.7 **Indian Standard (IS) Light Compaction test is usually conduced in 3 layers in a 1000 cc mould. How many blows per layer would be necessary if the above test is conducted in a 2250 cc mould?**

[1 marks : 2018]

Solution:

For 1000 cc mould, No. of layer = 3

No. of blows per layer = 25

When 2250 cc mould is used,

Total compaction energy per unit volume is same

$$\therefore \quad \frac{25 \times 3}{1000} = \frac{n \times 3}{2250}$$

$$\Rightarrow \quad n = 56.25 \quad (n = \text{no. of blows per layer for 2250 cc mould})$$

4.8 **With respect to a compaction curve, explain how one can plot the zero air voids line, 90% saturation line and 10% air voids line.**

[10 marks : 2020]

Solution:

Zero Air Void Line : A line which shows the water content and dry density relation for the compacted soil containing a constant percentage air voids is known as an air-voids line,

$$\rho_d = \frac{(1-n_a)\cdot G_s \cdot \rho_w}{1+w.G_s}$$

n_a = Percent air voids

w = Water content

ρ_d = Dry density corresponding to water content w

G_s = Specific gravity

The theoretical maximum compaction for any given water content corresponds to zero air voids condition ($n_a = 0$). The line showing the dry density as a function of water content for soil containing no air voids is called the zero air void line or the 100% saturation line.

$$(\rho_d)_{\text{theo, max}} = \frac{G_s \cdot \rho_w}{1+w.G_s} \quad ...(i)$$

Alternatively, a line showing the relation between water content and dry density for a constant degree of saturation S is

$$\rho_d = \frac{G_s \cdot \rho_w}{1+\left(\frac{w.G_s}{S}\right)}$$

For 90% saturation

$$\rho_d = \frac{G_s \cdot \rho_w}{1+\left(\frac{w.G_s}{0.9}\right)} \quad ...(a)$$

Instead of drawing lines corresponding to different degrees of saturation, it is sometimes more convenient to draw lines corresponding to different percentage air voids (n_a)

$$\rho_d = \frac{(1-n_a)\cdot G_s \cdot \rho_w}{1+w.G_s}$$

For theoretical maximum density

$$n_a = 0$$

$$(\rho_d)_{\text{theo, max}} = \frac{G_s \cdot \rho_w}{1+w.G_s}$$

Thus, the zero air-void line and 100% saturation line are identical.

The line for other percentage of air voids such as 10%.

$$\rho_d = \frac{0.90(G_s\rho_w)}{1+wG_s} \quad ...(b)$$

Thus, from equation (a) and (b), it may be noted that 10% air void line and 90% saturation lines are not identical.

4.9 **A 1.2 m thick embankment of loose sand is to be compacted using a Vibratory Roller. If the void ratio decreases from 1.2 to 0.8 due to compaction, calculate the final thickness of the embankment.**

[8 marks : 2023]

Solution:

Given:

Thickness of embankment, $H_o = 1.2$ m

Initial void ratio, $e_o = 1.2$

Final void ratio, $e_f = 0.8$

Now, $$\frac{\Delta H}{H_o} = \frac{\Delta e}{1+e_o} = \frac{e_o - e_f}{1+e_o}$$

where ΔH is change in thickness of embankment.

$$\Delta H = \frac{e_o - e_f}{1+e_0} \times 40 = \frac{1.2-0.8}{1+1.2} \times 1 \cdot Q = 0.218 \text{ m}$$

$$\therefore \quad H_{final} = 1.2 - 0.218 = 0.982 \text{ m}$$

5. Stress Distribution in Soil

5.1 **A water tower is constructed on a circular raft 5 m in diameter at a depth of 2 m below ground level in sand having bulk unit weight of 20 kN/m^3. The gross intensity of pressure at the base of the raft is 140 kN/m^2. What will be the increase in vertical stress due to raft at a point 4 m beneath the centre of the raft?**

[10 marks : 2002]

Solution:

Given data: Diameter of raft = 5 m, Radius of raft, $r = \frac{5}{2} = 2.5$ m, Depth of raft below ground level, $D_f = 2$ m, Bulk unit weight of soil, $\gamma = 20$ kN/m^3

Gross intensity of pressure at the base of raft, $q_g = 140$ kN/m^2

$\therefore$ Net intensity of pressure at the base of raft, $q_{net} = q_g - \gamma D_f = 140 - 20 \times 2 = 100$ kN/m^2

Depth below the centre of raft, $z = 4$ m

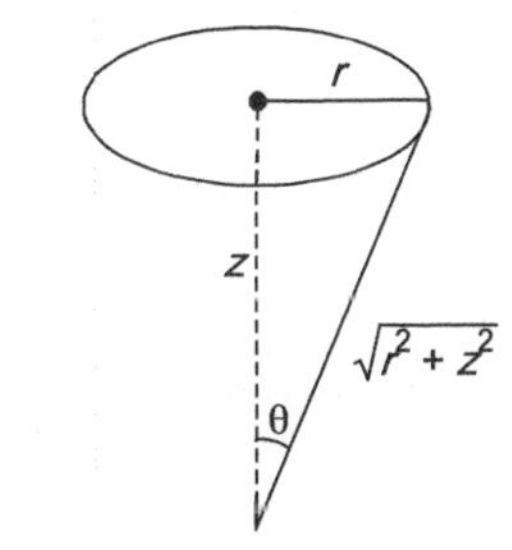

$$\therefore \quad \cos\theta = \frac{z}{\sqrt{z^2+r^2}} = \frac{4}{\sqrt{4^2+2.5^2}} = 0.848$$

Now increase in vertical stress is given by

$$\sigma_z = q_{net}(1-\cos^3\theta) = 100\,[1-(0.848)^3]$$
$$= 100 \times 0.3902 = 39.02 \text{ kN/m}^2$$

5.2 **The 1.0 m × 1.5 m footing shown in figure below supports a vertical load of 475 kN. Compute the magnitude of σ_z at a point P, considering both geostatic and induced stress.**

[6 marks : 2005]

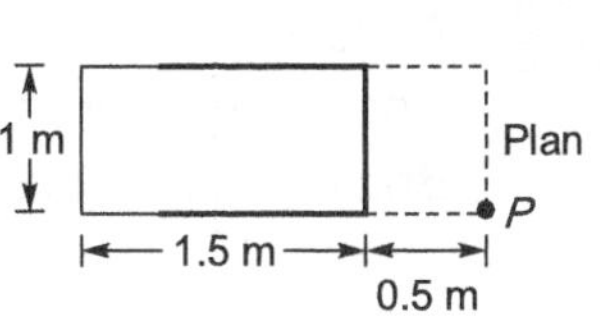

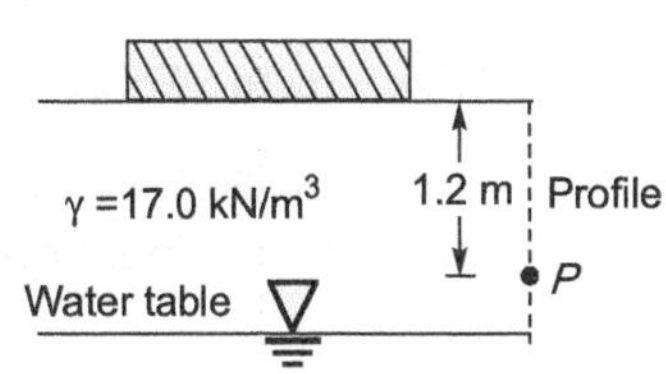

Solution:

The total stress at point P is the sum of geostatic stress and induced stress due to the load.

Total stress = Geostatic stress + Induced stress

Geostatic stress = $\gamma \times z = 17 \times 1.2 = 20.4$ kN/m^2

Load intensity, $$q = \frac{\text{Load}}{\text{Area}} = \frac{475}{1.0 \times 1.5} = 316.67 \text{ kN/m}^2$$

Vertical stress below the corner of a uniformly loaded rectangular area is given by

$$\sigma_z = \frac{q}{4\pi}\left[\left(\frac{2mn\,(m^2+n^2+1)^{1/2}}{m^2+n^2+1+m^2n^2}\right) \times \left(\frac{m^2+n^2+2}{m^2+n^2+1}\right) + \tan^{-1}\left(\frac{2mn(m^2+n^2+1)^{1/2}}{m^2+n^2+1-m^2n^2}\right)\right]$$

where $m = \frac{L}{z}$ and $n = \frac{B}{z}$

(i) When $L = 2.0$ m and B = 1 m and z = 1.2 m

$$m = \frac{L}{z} = \frac{2}{1.2} = 1.67$$

$$n = \frac{B}{z} = \frac{1}{1.2} = 0.83$$

$$\therefore \quad \sigma_{z_1} = \frac{316.67}{4\pi}\left[\left(\frac{2\times1.67\times0.83\times(1.67^2+0.83^2+1)^{1/2}}{1.67^2+0.83^2+1+1.67^2\times0.83^2}\right)\times\left(\frac{1.67^2+0.83^2+2}{1.67^2+0.83^2+1}\right) + \tan^{-1}\left(\frac{2\times1.67\times0.83\times(1.67^2+0.83^2+1)^{1/2}}{1.67^2+0.83^2+1-1.67^2\times0.83^2}\right)\right]$$

$$\Rightarrow \quad \sigma_{z_1} = \frac{316.67}{4\pi}[1.1214 + 1.1598] = 57.48 \text{ kN/m}^2$$

(ii) When $L = 0.5$ m and $B = 1$ m and z = 1.2 m

$$m = \frac{L}{z} = \frac{0.5}{1.2} = 0.42$$

$$n = \frac{B}{z} = \frac{1}{1.2} = 0.83$$

$$\therefore \quad \sigma_{z_2} = \frac{316.67}{4\pi}\left[\left(\frac{2\times0.42\times0.83\times(0.42^2+0.83^2+1)^{1/2}}{0.42^2+0.83^2+1+0.42^2\times0.83^2}\right)\times\left(\frac{0.42^2+0.83^2+2}{0.42^2+0.83^2+1}\right) + \tan^{-1}\left(\frac{2\times0.42\times0.83\times(0.42^2+0.83^2+1)^{1/2}}{0.42^2+0.83^2+1-0.42^2\times0.83^2}\right)\right]$$

$$\Rightarrow \quad \sigma_{z_2} = \frac{316.67}{4\pi}[0.7362 + 0.4998]$$

$$\Rightarrow \quad \sigma_{z_2} = 31.15 \text{ kN/m}^2$$

∴ Induced stress,

$$\sigma_z = \sigma_{z_1} - \sigma_{z_2} = 57.48 - 31.15 = 26.33 \text{ kN/m}^2$$

∴ **Total stress = 20.4 + 26.33 = 46.76 kN/m²**

5.3 **A rectangular foundation 6.0 m × 3.0 m in size exerts a pressure of 20 kN/m² to the soil underneath. Compute the increase of vertical stress at a point 0.5 m below the centre of foundation. Use Boussinesq's equation.**

[8 marks : 2010]

Solution:

Given data : Foundation size = 6 m × 3 m, Pressure (q) = 20 kN/m², Depth (Z) = 0.5 m

The increase of vertical stress at a point 0.5 m below the centre of foundation can be found by dividing the foundation in four equal parts and then finding the vertical stress below the corner of each part (σ_z).

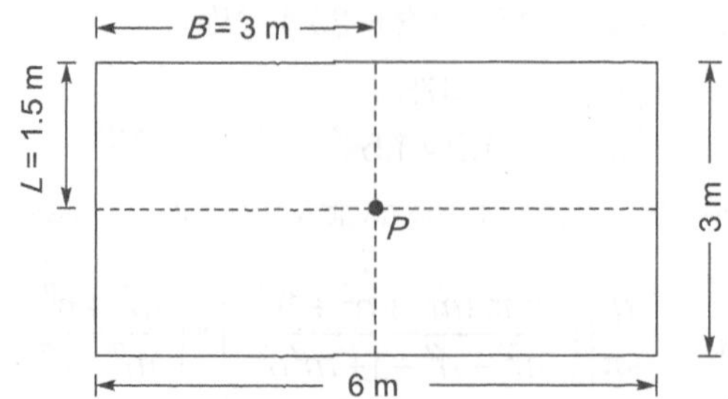

Here, $m = \frac{B}{Z} = \frac{3}{0.5} = 6$

$n = \frac{L}{Z} = \frac{1.5}{0.5} = 3$

Let influence factor I_N

Hence, $\sigma_z = qI_n$

$$I_n = \frac{1}{4\pi}\left[\frac{2mn(m^2+n^2+1)^{1/2}}{m^2+n^2+m^2n^2+1}\times\frac{m^2+n^2+2}{m^2+n^2+1}+\tan^{-1}\frac{2mm(m^2+n^2+1)^{1/2}}{m^2+n^2+1-m^2n^2}\right]$$

$$I_n = \frac{1}{4\pi}\left[\frac{2\times6\times3(6^2+3^2+1)^{1/2}}{6^2+3^2+6^2\times3^2+1}\times\frac{6^2+3^2+2}{6^2+3^2+1}+\tan^{-1}\frac{2\times6\times3(6^2+3^2+1)^{1/2}}{6^2+3^2+1-6^2\times3^2}\right]$$

$$I_n = \frac{1}{4\pi}\left[0.6742+\tan^{-1}(-0.8783)\right]$$

[The $\tan^{-1}$ term here is in degrees and negative, for correct solution convert the angle to positive and radians]

$$\tan^{-1}(-0.8783) = -41.29° \text{ [}\tan(180+\theta)=\tan\theta\text{]}$$

$\therefore$ $$= 180-41.29 = 138.7° = 138.7\times\frac{\pi}{180} = 2.421$$

Hence, $$I_n = \frac{1}{4\pi}[0.6742+2.421] = 0.2463$$

$$\sigma_z = 20\times0.2463 = 4.926 \text{ kN/m}^2$$

Hence increased vertical stress 0.5 m below centre of foundation

$$= 4\sigma_z = 4\times4.926 = 19.704 \text{ kN/m}^2$$

5.4 **An elevated structure with a total weight of 10,000 kN is supported on a tower with 4 legs. The legs rest on piers located at the corners of a square 6 m on a side. What is the vertical stress increment due to this loading at a point 7 m beneath the centre of the structure?**
(Assume that the load be approximated to a point load acting at the corners of a square of 6 m side)
[5 marks : 2012]

Solution:

Given data Total weight supported on a tower with 4 legs = 10,000 kN

$\therefore$ Weight transferred to each pier = 2500 kN

The horizontal distance, r from each of the load is equal to

$$r = \sqrt{3^2+3^2} = \sqrt{18} = 4.243 \text{ m}$$

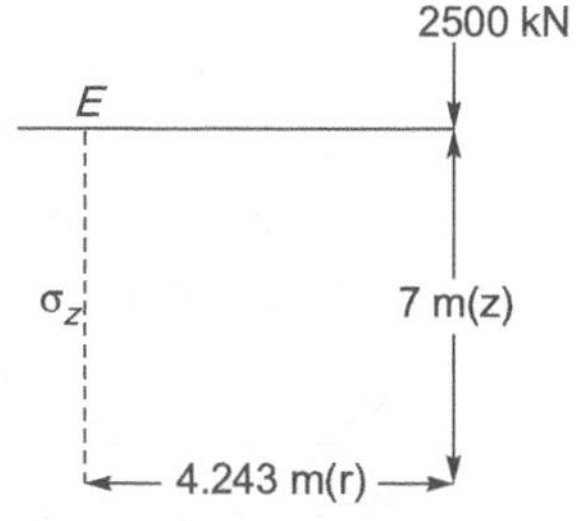

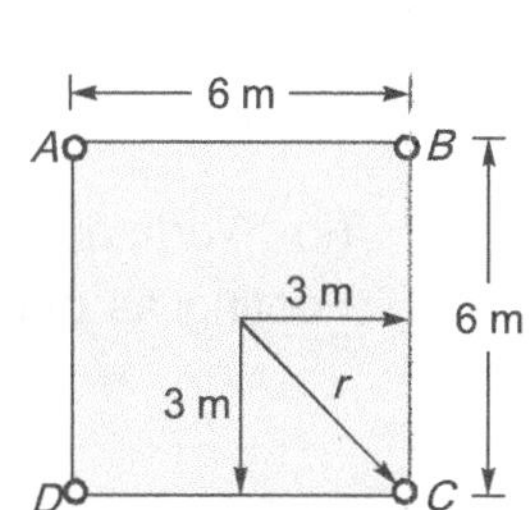

The vertical stress, at a horizontal distance 'r' and vertical depth 'z' is given by

$$\sigma_z = \frac{3Q}{2\pi z^2}\left[\frac{1}{1+(r/z)^2}\right]^{5/2}$$

$$Q = 2500 \text{ kN}, z = 7 \text{ m}, r = 4.243 \text{ m}$$

Hence, **vertical stress increment due to the total weight** is equal to

$$= \frac{4 \times 3 \times 2500}{2\pi \times 7^2}\left[\frac{1}{1+(4.243/7)^2}\right]^{5/2}$$

$$= 4 \times 11.141 = 44.564 \text{ kN/m}^2$$

5.5 **Two square footings with equal contact pressure of 250 kPa are at 5 m apart (centre-to-centre). The size of the one footing (A) is 2 m × 2 m and the other one (B) is 2.5 m × 2.5 m. Determine the vertical stress at 2 m vertically below (i) the footing (A), (ii) the footing (B) and (iii) the midpoint between the footings. Use Boussinesq's point load formula.**

[10 marks : 2019]

Solution:

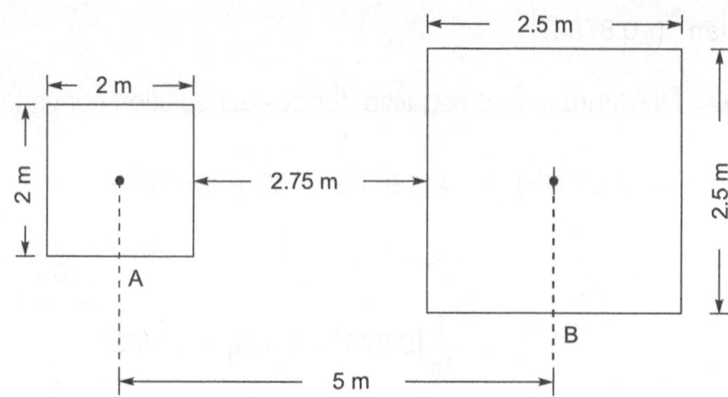

Contact pressure = 250 kPa (given) equal for both the footings.

(i) Vertical stress at 2 m vertically below 2 m × 2 m

Footing as per Boussinesq's point load formula.

$$\sigma_z = \left[\frac{3Q}{2\pi(z)^2}\left(\frac{1}{1+\left(\frac{r}{z}\right)^2}\right)^{5/2}\right]_A + \left[\frac{3Q}{2\pi(z)^2}\left(\frac{1}{1+\left(\frac{r}{z}\right)^2}\right)^{5/2}\right]_B$$

$$= \left[\frac{3 \times 250 \times 2^2}{2\pi(2)^2}\left(\frac{1}{1+0^2}\right)^{5/2}\right]_A + \left[\frac{3 \times 250 \times 2.5^2}{2\pi(2)^2}\left(\frac{1}{1+\left(\frac{5}{2}\right)^2}\right)^{5/2}\right]_B$$

$$\therefore \quad \sigma_z = 119.366 + 186.5096\,(7.0657 \times 10^{-3})$$

$$= 120.683 \text{ kN/m}^2$$

(ii) Vertical stress at 2 m vertically below 2.5 m × 2.5 m

Footing as per Boussinesq's point load formula

$$\sigma_z = \left[\frac{3Q}{2\pi(z)^2}\left(\frac{1}{1+\left(\frac{r}{z}\right)^2}\right)^{5/2}\right]_A + \left[\frac{3Q}{2\pi(z)^2}\left(\frac{1}{1+\left(\frac{r}{z}\right)^2}\right)^{5/2}\right]_B$$

$$= \left[\frac{3 \times 250 \times 2.5^2}{2\pi(2)^2}\left(\frac{1}{1+0^2}\right)^{5/2}\right]_B + \left[\frac{3 \times 250 \times 2^2}{2\pi(2)^2}\left(\frac{1}{1+\left(\frac{5}{2}\right)^2}\right)^{5/2}\right]_A$$

$$\therefore \quad \sigma_z = 186.5096 + 119.366\,(7.0657 \times 10^{-3})$$
$$= 187.353 \text{ kN/m}^2$$

(iii) Mid point between the footings

$$\sigma_z = \left[\frac{3\times 250\times 2\times 2}{2\pi\times(2)^2}\times\left(\frac{1}{1+\left(\frac{2.5}{2}\right)^2}\right)^{5/2}\right]_A + \left[\frac{3\times 250\times 2.5\times 2.5}{2\pi\times(2)^2}\times\left(\frac{1}{1+\left(\frac{2.5}{2}\right)^2}\right)^{5/2}\right]_B$$

$$\sigma_z = (119.366 \times 0.095) + (186.581 \times 0.095) = 29.064 \text{ kN/m}^2$$

5.6 **An electric power transmission pole is embedded 3 m into the ground. The pole weighs 30 kN and has base diameter of 450 mm. If the pole is assumed to transmit the load as point load in the soil, determine the stress increase at a depth 1 m below the base, and :**

(i) Along the centre, and

(ii) 1 m from the centre.

[12 marks : 2023]

Solution:

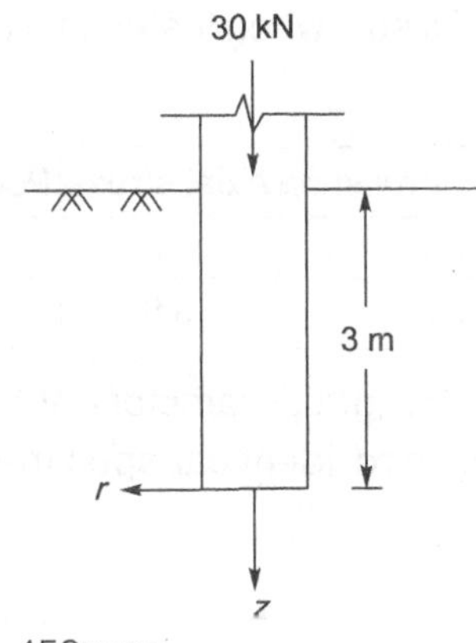

Given: Diameter of pole = 450 mm

Depth, z = 1 m

(i) **Stress increase at $r = 0$ and $z = 1$ m:** As the load is transferred as point load only, so stresses due to side friction is neglected.

As per Boussinesq's equation,

$$\sigma_z = \frac{3Q}{2\pi z^2}\left[\frac{1}{1+\left(\frac{r^2}{z^2}\right)}\right]^{5/2}$$

$$= \frac{3\times 30}{2\pi\times 1^2}\left[\frac{1}{1+\left(\frac{0}{1^2}\right)}\right]^{5/2}$$

$$= \frac{3\times 30}{2\pi\times 1^2} = 14.32 \text{ kN/m}^2$$

(ii) **Stress increase at $z = 1$ m and $r = 1$ m:**

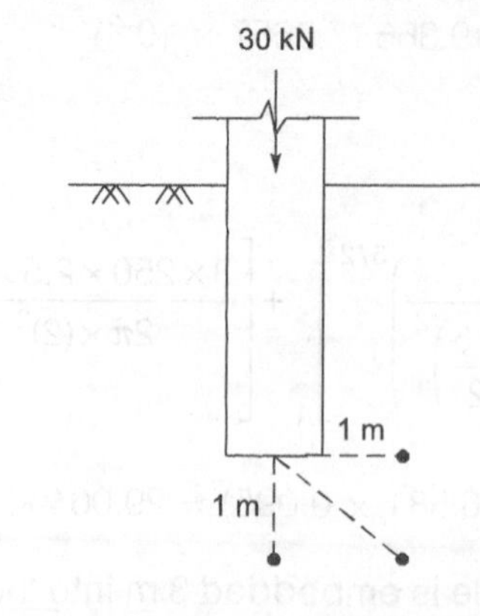

$$\sigma_z = \frac{3\times 30}{2\pi\times 1^2}\left[\frac{1}{1+\left(\frac{1}{1}\right)^2}\right]^{5/2} = 2.532\ \text{kN/m}^2$$

6. Shear Strength of Soil

6.1 Two identical specimens of a back fill soil, when tested in a triaxial compression test, gave following stresses (in kg/cm^2) at failure.

Sample No.	All round pressure	Axial stress	Pore water pressure
1.	1.0	2.2	0.6
2.	2.0	3.6	1.1

(i) Find total and effective shear strength parameters of the soil.

(ii) Find axial stress at which, the third identical specimen of the same soil will fail if all-round pressure is 2.8 kg/cm^2.

[15 marks : 2004]

Solution:

(i) For sample 1,

$$\sigma_3 = 1\ \text{kg/cm}^2,\ \sigma_d = 2.2\ \text{kg/cm}^2,\ u = 0.6\ \text{kg/cm}^2$$

$$\sigma_1 = \sigma_3 + \sigma_d = 1 + 2.2 = 3.2\ \text{kg/cm}^2$$

$$\therefore \quad \sigma_3' = \sigma_3 - u = 1 - 0.6 = 0.4\ \text{kg/cm}^2$$

$$\sigma_1' = \sigma_1 - u = 3.2 - 0.6 = 2.6\ \text{kg/cm}^2$$

If c′ is the cohesion and ϕ′ is the angle of shearing resistance, then

$$\sigma_1' = \sigma_3' \tan^2\left(45^\circ + \frac{\phi'}{2}\right) + 2c'\tan\left(45^\circ + \frac{\phi}{2}\right)$$

$$\Rightarrow \quad 2.6 = 0.4\tan^2\left(45^\circ + \frac{\phi'}{2}\right) + 2c'\tan\left(45^\circ + \frac{\phi'}{2}\right) \quad \text{...(i)}$$

Similarly for sample 2

$$\sigma_3 = 2.0\ \text{kg/cm}^2,\ \sigma_d = 3.6\ \text{kg/cm}^2,\ u = 1.1\ \text{kg/cm}^2$$

$$\sigma_1 = \sigma_3 + \sigma_d = 2 + 3.6 = 5.6\ \text{kg/cm}^2$$

$$\therefore \quad \sigma_1' = \sigma_1 - u = 5.6 - 1.1 = 4.5\ \text{kg/cm}^2$$

$$\sigma_3' = \sigma_3 - u = 2 - 1.1 = 0.9\ \text{kg/cm}^2$$

$\therefore \quad 4.5 = 0.9\tan^2\left(45° + \frac{\phi'}{2}\right) + 2c'\tan\left(45° + \frac{\phi'}{2}\right)$ ...(ii)

Subtracting (i) from (ii) we get

$$0.5\tan^2\left(45° + \frac{\phi'}{2}\right) = 1.9$$

$$\Rightarrow \quad \tan\left(45° + \frac{\phi'}{2}\right) = \sqrt{3.8}$$

$$\Rightarrow \quad 45° + \frac{\phi'}{2} = \tan^{-1}(1.9494)$$

$$\Rightarrow \quad \frac{\phi'}{2} = 17.84°$$

$$\Rightarrow \quad \phi' = 35.68°$$

Putting $\phi' = 35.68°$ in (i) we get

$$0.4\tan^2(45° + 17.84°) + 2c'\tan(45° + 17.84°) = 2.6$$

$$c' = \frac{2.6 - 1.52}{2\tan 62.84°} = 0.28 \text{ kg/cm}^2$$

Thus the effective shear strength parameters are $c' = 0.28$ kg/cm² and $\phi' = 35.68°$.

(ii) Let the total shear strength parameters are c and ϕ,

For Sample 1:

$\sigma_3 = 1$ kg/cm², $\sigma_d = 2.2$ kg/cm², $\sigma_1 = \sigma_3 + \sigma_d = 1 + 2.2 = 3.2$ kg/cm²

$$\sigma_1 = \sigma_3\tan^2\left(45° + \frac{\phi}{2}\right) + 2c\tan\left(45° + \frac{\phi}{2}\right)$$

$3.2 = 1 \times \tan^2\left(45° + \frac{\phi}{2}\right) + 2c\tan\left(45° + \frac{\phi}{2}\right)$ ...(i)

For Sample 2:

$\sigma_3 = 2.0$ kg/cm², $\sigma_d = 3.6$ kg/cm², $\sigma_1 = \sigma_3 + \sigma_d = 2 + 3.6 = 5.6$ kg/cm²

$\therefore \quad 5.6 = 2\tan^2\left(45° + \frac{\phi}{2}\right) + 2c\tan\left(45° + \frac{\phi}{2}\right)$ ...(ii)

Subtracting (i) from (ii) we get

$$\tan^2\left(45° + \frac{\phi}{2}\right) = 2.4$$

$$\Rightarrow \quad \tan\left(45° + \frac{\phi}{2}\right) = \sqrt{2.4}$$

$$\Rightarrow \quad 45 + \frac{\phi}{2} = 57.16$$

$$\Rightarrow \quad \frac{\phi}{2} = 12.16$$

$$\Rightarrow \quad \phi = 24.32°$$

Putting $\phi = 24.32°$ in (i), we get

$\tan^2(45° + 12.16°) + 2c\tan(45° + 12.16°) = 3.2$

$$\Rightarrow \quad c = \frac{0.8}{2\tan 57.16} = 0.26 \text{ kg/cm}^2$$

Thus the total shear strength parameters are c = 0.26 kg/cm² and φ = 24.32°

(ii) For third specimen

$$\sigma_3 = 2.8 \text{ kg/cm}^2$$

$$\sigma_1 = \sigma_3 \tan^2\left(45° + \frac{\phi}{2}\right) + 2c\tan\left(45° + \frac{\phi}{2}\right)$$

$$= 2.8\tan^2\left(45° + \frac{24.32}{2}\right) + 2 \times 0.26\tan\left(45° + \frac{24.32}{2}\right)$$

$$= 7.527 \text{ kg/cm}^2$$

Axial stress for failure of third specimen

$$= \sigma_1 - \sigma_3$$

$$= 7.527 - 2.8$$

$$= 4.727 \text{ kg/cm}^2$$

6.2 **What is the significance of pore water pressure in the triaxial shear test? Explain the pore water pressure parameters as given by A.W. Skempton.**

[10 marks : 2006]

Solution:

When the applied stresses on a soil mass are increased, the full load is not immediately transferred to the soil grains, but a part of it is transferred to the water, standing in the soil pores. This increased hydrostatic pressure in the soil pores is nothing but the increased pore pressure. This pore pressure reduces the effective shear strength of the soil hence it becomes necessary to assess the value of pore pressure. This pore water pressure can be measured by a pore pressure measuring apparatus connecting to the pressure line of the triaxial cell after closing the valve in the drainage line. The pore water pressure is measured under conditions of no flow either into the specimen or out of the specimen. In a saturated soil , when the test is undrained , there will be no volume change in the soil sample.

If it is not possible to determine pore water pressure practically, it can be calculated by using theoretical approach given by A.W. Skempton. The pore pressure parameter represent the response of pore pressure due to change in the total stress under undrained conditions. Such parameters are *A* and *B*.

The pore water pressure can be classified in two stages:

(i) Consolidation stage or cell pressure stage

(ii) Shear stage or deviator stress stage

The value of B can be given as

$$B = \frac{\Delta u_c}{\Delta \sigma_c} = \frac{\Delta u_c}{\Delta \sigma_3}$$

Here B is the ratio of pore pressure developed to the change in confining pressure. Δu_c represents increase in pore pressure due to increase in cell pressure, $\Delta\sigma_c$. For a fully saturated soil $B = 1$ and for a fully dried soil $B = 0$. The value of *B* can also be determined by using soil properties. Thus, the value of *B* can be given as

$$B = \frac{1}{1 + n\dfrac{C_v}{C_c}}$$

Where n is porosity, C_v is coefficient of consolidation, C_c is coefficient of compression

A is defined as $\bar{A} = AB$

where $\bar{A}$ is related to change in pore water pressure due to change in deviator stress i.e.

$$\bar{A} = \frac{\Delta u_d}{\Delta \sigma_d} = \frac{\Delta u_d}{\Delta \sigma_1 - \Delta \sigma_3}$$

where Δu_d is change in pore water pressure due to change in deviator stress.

Now $\Delta u = \Delta u_c + \Delta u_d = B \cdot \Delta\sigma_3 + \bar{A} \cdot \Delta\sigma_d = B \cdot \Delta\sigma_3 + AB \cdot \Delta\sigma_d = B \cdot \Delta\sigma_3 + AB(\Delta\sigma_1 - \Delta\sigma_3)$

$\therefore$ $\Delta u = B[\Delta\sigma_3 + A(\Delta\sigma_1 - \sigma_3)]$

The value of A may be as high as 2 to 3 for saturated fine sand and may be as low as –0.5 for heavily over consolidated clays. For a given soil A depends upon the strain, anisotropy, sample disturbance and the over consolidation ratio.

6.3 **What are the application of pore pressure coefficients?**

[5 marks : 2008]

Solution:

The effective stresses rather than total stresses, control the shear strength of a soil. For a given total stress, the effective stress is directly related to the magnitude of the pore pressure that is built up in a soil. Pore pressure coefficients are empirical coefficients which are used to express the response of pore pressure to changes in total stress under undrained conditions. These are very useful in field problems where pore pressures, that are induced consequent to change in the total stress may have to be computed. Construction of an earth embankment over a soft clay deposit is an example of such a problem. If the rate of construction of the embankment is such that the pore water pressure in the foundation soil cannot dissipate, undrained loading condition will prevail. If excessive pore pressures are allowed to develop, the stability of the foundation soil is endangered. Computation of pore pressure using pore pressure coefficients will help to estimate the chances of such an eventuality.

6.4 **Samples of a dry sand are to be tested in triaxial and direct shear test. In the triaxial test the specimen fails when the major and minor principal stresses are 960 kN/m² and 260 kN/m² respectively. What shear strength would be expected in direct shear test when the normal stress is 230 kN/m²?**

[5 marks : 2008]

Solution:

Given data: Major principal stress at failure, $\sigma_{1f} = 960$ kN/m^2

Minor principal stress at failure, $\sigma_{3f} = 260$ kN/m^2

Normal stress, $\sigma = 230$ kN/m^2

If ϕ is the angle of internal friction, then

$$\sin\phi = \frac{\sigma_{1f} - \sigma_{3f}}{\sigma_{1f} + \sigma_{3f}}$$

$$\Rightarrow \quad \sin\phi = \frac{960 - 260}{960 + 260}$$

$$\Rightarrow \quad \phi = \sin^{-1}(0.5738)$$

$$\Rightarrow \quad \phi = 35°$$

The shear strength of a soil is given by

$$\tau = c + \sigma \tan\phi \qquad \text{[where for sand } c = 0\text{]}$$

$$\Rightarrow \quad \tau = 0 + 230 \times \tan 35°$$

$$\Rightarrow \quad \tau = 161.05 \text{ kN/m}^2$$

6.5 **Discuss pore pressure parameters.**

[3 marks : 2010]

Solution:

If it is not possible to determine pore water pressure practically, it can be calculated by using theoretical approach given by **A.W. Skempton**. The pore pressure parameter represent the response of pore pressure due to change in the total stress under undrained conditions. Such parameters are A and B.

The pore water pressure can be classified in two stages:

(i) Consolidation stage or cell pressure stage

(ii) Shear stage or deviator stress stage

The value of B can be given as

$$B = \frac{\Delta u_c}{\Delta \sigma_c} = \frac{\Delta u_c}{\Delta \sigma_3}$$

Here B is the ratio of pore pressure developed to the change in confining pressure. Δu_c represents increase in pore pressure due to increase in cell pressure, $\Delta \sigma_c$. For a fully saturated soil $B = 1$ and for a fully dried soil $B = 0$. The value of B can also be determined by using soil properties. Thus, the value of B can be given as

$$B = \frac{1}{1 + n\frac{C_v}{C_c}}$$

Where n is porosity, C_v is coefficient of consolidation, C_c is coefficient of compression

A is defined as

$$\bar{A} = AB$$

where $\bar{A}$ is related to change in pore water pressure due to change in deviator stress i.e.

$$\bar{A} = \frac{\Delta u_d}{\Delta \sigma_d}$$

$$\bar{A} = \frac{\Delta u_d}{\Delta \sigma_1 - \Delta \sigma_3}$$

where Δu_d is change in pore water pressure due to change in deviator stress.

Now $$\Delta u = \Delta u_c + \Delta u_d = B \cdot \Delta \sigma_3 + \bar{A} \cdot \Delta \sigma_d$$

$$= B \cdot \Delta \sigma_3 + AB \cdot \Delta \sigma_d = B \cdot \Delta \sigma_3 + AB(\Delta \sigma_1 - \Delta \sigma_3)$$

$\therefore$ $$\Delta u = B[\Delta \sigma_3 + A(\Delta \sigma_1 - \sigma_3)]$$

The value of A may be as high as 2 to 3 for saturated fine sand and may be as low as –0.5 for heavily over consolidated clays. For a given soil A depends upon the strain, anisotropy, sample disturbance and the over consolidation ratio.

6.6 **In a laboratory vane shear test a vane 100 mm long and 60 mm diameter was pressed into the soft cohesive soil (G = 2.72). A torque of 40 kN-mm was required to achieve the failure. Same soil when remoulded required 15 kN-mm to achieve failure. Calculate the cohesion in both cases and value of sensitivity. Take the void ratio of soil as 30%.**

[5 marks : 2010]

Solution:

Given that: $d = 60$ mm, $h = 100$ mm, $G = 2.72$, $e = 30\%$, $T_1 = 40$ kN-mm, $T_2 = 15$ kN-mm

$\because$ $$s = c_u + \bar{\sigma}_n \tan\phi = \frac{T}{\pi d^2 \left(\frac{h}{2} + \frac{d}{6}\right)}$$

Since $\phi = 0$ in a cohesive soil, thus

$$s = c_u = \frac{T}{\pi d^2\left(\frac{h}{2}+\frac{d}{6}\right)}$$

$$\Rightarrow \quad T = c_u \pi d^2\left[\frac{h}{2}+\frac{d}{6}\right]$$

Assuming that both upper and lower end of the vane shear the soil.

In undisturbed condition, $$T_1 = c_u \pi d^2\left[\frac{h}{2}+\frac{d}{6}\right]$$

$$\Rightarrow \quad 40\times10^{-3} = c_u \times \pi \times (0.06)^2 \times\left[\frac{0.1}{2}+\frac{0.06}{6}\right]$$

$$\Rightarrow \quad c_u = \frac{40\times10^{-3}}{6.786\times10^{-4}} = 58.95 \text{ kN/m}^2$$

Since the soil is soft clay, the unit undrained shear resistance will be the shear strength of the soil. Therefore shear strength of soil in undisturbed state is 58.95 kN/m^2.

In remoulded condition,

$$T_2 = c_{ur} \pi d^2\left[\frac{h}{2}+\frac{d}{6}\right]$$

$$\Rightarrow \quad 15\times10^{-3} = c_{ur} \times \pi \times (0.06)^2 \times\left[\frac{0.1}{2}+\frac{0.06}{6}\right]$$

$$\Rightarrow \quad c_{ur} = \frac{15\times10^{-3}}{6.786\times10^{-4}} = 22.1 \text{ kN/m}^2$$

Thus shear strength of soft clay in remoulded condition is 22.1 kN/m^2

$$\text{Sensitivity of soil} = \frac{\text{Undisturbed soil strength}}{\text{Remoulded soil strength}} = \frac{58.95}{22.1} = 2.67$$

Thus the soil is normal as the value of sensitivity lies between 1 and 4

6.7 **Consolidated undrained triaxial tests were performed on two identical specimens of saturated clay with pore pressure measurements. The observations are**

Cell pressure at failure (kPa)	Deviator stress at failure (kPa)	Pore pressure at failure (kPa)
250	179	101
350	242	145

Determine the shear strength parameters in terms of both total and effective stresses (by analytical method).

[10 marks : 2011]

Solution:

The calculation for the total and effective stress is shown below in the table.

Sample	Cell Pressure σ_3 (kN/m^2)	σ_1 kN/m^2 $= \sigma_3 + \sigma_d$	Pore Pressure (u)kN/m^2	$\overline{\sigma}_3$	$\overline{\sigma}_1$
1.	250	250 + 179 = 429	101	250 – 101 = 149	429 – 101 = 328
2.	350	350 + 242 = 592	145	350 – 145 = 205	592 – 145 = 447

Minor principal stress = cell pressure = σ_3

$$\text{Major principal stress} = \sigma_1 = \sigma_3 + \sigma_d$$

$$\text{Effective stresses : } \bar{\sigma}_3 = \sigma_3 - u$$

$$\bar{\sigma}_1 = \sigma_1 - u$$

Calculation of shear strength parameters in terms of total stress:

$$\sigma_1 = \sigma_3 \tan^2\left(45 + \frac{\phi}{2}\right) + 2c\tan\left(45 + \frac{\phi}{2}\right)$$

For sample 1

$$\sigma_1 = 429 \text{ kN/m}^2$$

$$\sigma_2 = 250 \text{ kN/m}^2$$

$$429 = 250\tan^2\left(45 + \frac{\phi}{2}\right) + 2c\tan\left(45 + \frac{\phi}{2}\right) \quad \text{...(i)}$$

For sample 2

$$\sigma_1 = 592 \text{ kN/m}^2$$

$$\sigma_3 = 350 \text{ kN/m}^2$$

$$592 = 350\tan^2\left(45 + \frac{\phi}{2}\right) + 2c\tan\left(45 + \frac{\phi}{2}\right) \quad \text{...(ii)}$$

On solving (i) and (ii) equation,

$$529 - 429 = (350 - 250)\tan^2\left(45 + \frac{\phi}{2}\right)$$

$$\tan^2\left(45 + \frac{\phi}{2}\right) = 1.63$$

$$\tan\left(45 + \frac{\phi}{2}\right) = 1.2767$$

$$\phi = 13.86°$$

From equation (i),

$$429 = 250 \times 1.63 + 2c \times 1.2767$$

$$c = 8.42 \text{ kN/m}^2$$

Hence, shear strength parameters in terms of total stress are

$$c = 8.42 \text{ kN/m}^2$$

$$\phi = 13.86°$$

Calculation of shear strength parameters in terms of effective stress:

$$\bar{\sigma}_1 = \bar{\sigma}_3 \tan^2\left(45 + \frac{\phi'}{2}\right) + 2c'\tan\left(45 + \frac{\phi'}{2}\right)$$

For sample 1,

$$\bar{\sigma}_1 = 328 \text{ kN/m}^2$$

$$\bar{\sigma}_3 = 149 \text{ kN/m}^2$$

$$328 = 149\tan^2\left(45 + \frac{\phi'}{2}\right) + 2c'\tan\left(45 + \frac{\phi'}{2}\right) \quad \text{...(ii)}$$

For sample 2,

$$\bar{\sigma}_1 = 447 \text{ kN/m}^2$$

$$\bar{\sigma}_3 = 205 \text{ kN/m}^2$$

$$447 = 205\tan^2\left(45 + \frac{\phi'}{2}\right) + 2c'\tan\left(45 + \frac{\phi'}{2}\right) \quad \text{...(iii)}$$

On solving (iii) and (iv) equation,

$$447 - 328 = (205 - 149)\tan^2\left(45 + \frac{\phi'}{2}\right)$$

$$\tan^2\left(45 + \frac{\phi'}{2}\right) = 2.125$$

$$\tan\left(45 + \frac{\phi'}{2}\right) = 1.4577$$

$$\phi' = 21.1°$$

From eq. (iii)

$$328 = 149 \times 2.125 + 2c' \times 1.4577$$

$$c' = 3.9 \text{ kPa}$$

Hence shear strength parameters in terms of effective stress are

$$c' = 3.9 \text{ kPa}$$

$$\phi' = 21.1°$$

6.8 **A specimen of fine dry sand, when subjected to tri-axial compression test, failed at a deviator stress of 400 kN/m². It failed with a pronounced failure plane with an angle of 24° to the axis of the sample. Compute the lateral pressure to which the specimen would have been subjected.**

[5 marks : 2012]

Solution:

Given data:

$$\sigma_d = \Delta\sigma = 400 \text{ kN/m}^2$$

As we know

$$\sigma_1 = \sigma_3 \tan^2\left(45° + \frac{\phi}{2}\right) + 2c\tan\left(45° + \frac{\phi}{2}\right) \quad ...(i)$$

Angle of failure = 24° to the axis of sample

$$\therefore \quad \theta_c = 90° - 24° = 66°$$

$$\therefore \quad \theta_c = 45° + \frac{\phi}{2} = 66°$$

$$c = 0 \quad \text{(For fine dry sand)}$$

Now,

$$\sigma_1 = \Delta\sigma + \sigma_3 = 400 + \sigma_3$$

Therefore from eq. (i)

$$\therefore \quad 400 + \sigma_3 = \sigma_3 \tan^2 66°$$

$$\Rightarrow \quad \sigma_3 = \frac{400}{\tan^2 66° - 1} = 98.90 \text{ kN/m}^2$$

$\therefore$ **Lateral pressure** $\sigma_3 = 98.90 \text{ kN/m}^2$

6.9 **When an unconfined compression test was conducted on a cylindrical soil sample, it failed under an axial stress of 120 kN/m². The failure plane makes an angle of 50° with the horizontal. Determine the cohesion and the angle of internal friction of the soil.**

[4 marks : 2014]

Solution:

Unconfined compression results:

Confining pressure, $\sigma_c = \sigma_3 = 0$

Deviator stress, $\sigma_d = \sigma_1 = 120 \text{ kN/m}^2$

Angle of failure plane, $\theta_c = 45° + \frac{\phi}{2} = 50°$

Where, ϕ = angle of internal friction

$$\Rightarrow \quad \phi = 10°$$

Using,

$$\sigma_1 = \sigma_3 \tan^2\left(45° + \frac{\phi}{2}\right) + 2c\tan\left(45° + \frac{\phi}{2}\right)$$

$$\Rightarrow \quad \sigma_1 = \sigma_3 \tan^2\theta_c + 2c\tan\theta_c$$

$$120 = 0 + 2c\tan 50°$$

$$\Rightarrow \quad c = 50.35 \text{ kN/m}^2$$

6.10 **Determine the axial stress at failure for a dry dense sand in triaxial loading if σ_3 = 300 kN/m². A previous test had given σ_3 = 150 kN/m²; σ_1 = 735 kN/m² at failure.**

[4 marks : 2014]

Solution:

For dense sand: Triaxial test result are:

$$\sigma_3 = 150 \text{ kN/m}^2$$

$$\sigma_1 = 735 \text{ kN/m}^2$$

Let angle of internal friction of sand is ϕ

$$\therefore \quad \sigma_1 = \sigma_3 \tan^2\left(45° + \frac{\phi}{2}\right) + 2c\tan\left(45° + \frac{\phi}{2}\right)$$

$$\Rightarrow \quad 735 = 150\tan^2\left(45° + \frac{\phi}{2}\right) + 0 \qquad [\because c = 0, \text{ for pure sand}]$$

$$\Rightarrow \quad \tan\left(45° + \frac{\phi}{2}\right) = 2.213$$

$$\Rightarrow \quad 45° + \frac{\phi}{2} = 65.68°$$

$$\Rightarrow \quad \frac{\phi}{2} = 20.69°$$

$$\Rightarrow \quad \phi = 41.38°$$

If triaxial test is repeated with cell pressure σ_3 = 300 kN/m², then **axial stress at failure (σ_1)** may be given by

$$\sigma_1 = \sigma_3 \tan^2\left(45° + \frac{\phi}{2}\right) + 2c\tan\left(45° + \frac{\phi}{2}\right)$$

$$= 300\tan^2(45° + 20.69°) + 0 = 1470.2 \text{ kN/m}^2$$

6.11 **A sample of normally consolidated clay was subjected to a consolidated undrained tri-axial compression test that was carried out until the specimen failed at a deviator stress of 50 kN/m². The pore water pressure at failure was recorded to be 20 kN/m² and confining pressure of 50 kN/m² was used in the test. Determine the consolidated undrained friction angle ϕ_{CU} and drained friction angle ϕ_{CD}.**

[8 marks : 2015]

Solution:

Since only one observation is given hence assume c = 0 in consolidated undrained test.

(i) Undrained condition.

$$\sigma_d = 50 \text{ kN/m}^2$$

$$u = 20 \text{ kN/m}^2$$

$$\sigma_3 = 50 \text{ kN/m}^2$$

$$\sigma_1 = \sigma_d + \sigma_3 = 100 \text{ kN/m}^2$$

$$\sin\phi_{cu} = \frac{\sigma_1 - \sigma_3}{\sigma_1 + \sigma_3} = \frac{100 - 50}{100 + 50} = \frac{1}{3}$$

$$\Rightarrow \quad \phi_{cu} = 19.47°$$

(ii) Drained condition

$$\bar{\sigma}_1 = \sigma_1 - u$$

$$= 100 - 20 = 80 \text{ kN/m}^2$$

$$\bar{\sigma}_3 = \sigma_3 - u = 50 - 20 = 30 \text{ kN/m}^2$$

$$\sin \phi_{CD} = \frac{\bar{\sigma}_1 - \bar{\sigma}_3}{\bar{\sigma}_1 + \bar{\sigma}_3} = \frac{80 - 30}{80 + 30} = \frac{50}{110}$$

$$\Rightarrow \quad \phi_{CD} = 27.04°$$

6.12 **Draw a neat sketch of a vane shear apparatus. Also derive the relationship between undrained shear strength, C_u and torque, T.**

[4 marks : 2016]

Solution:

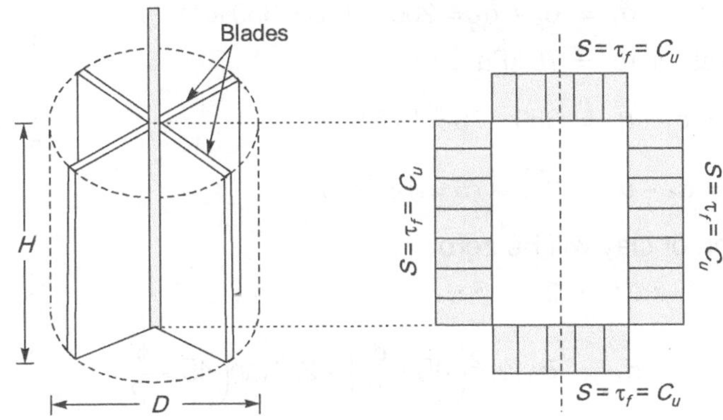

In undrained conditon, $S = C_u + \bar{\sigma}_n \tan\phi^{\nearrow 0} = C_u$

Relationship between undrained shear strength (C_u) and Torque (T)

Let T_1 = Resisting torque offered at the sides of cylindrical hole

T_2 = Resisting torque offered by the soil at the top and bottom

$$\therefore \quad T_1 = (\pi DHC_u)\frac{D}{2}$$

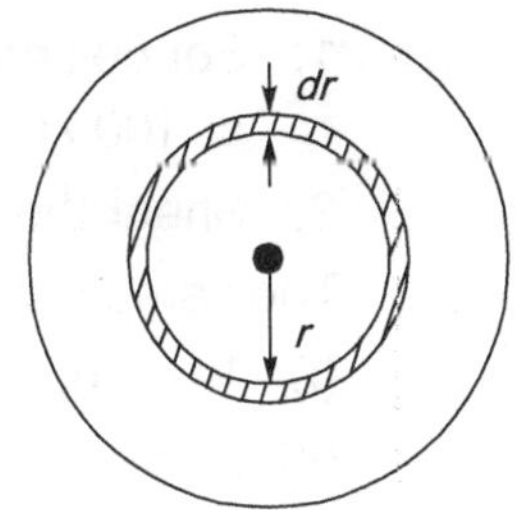

$$T_2 = 2\int_0^{D/2} (2\pi r dr)C_u \cdot r$$

$$= 4\pi C_u\left[\frac{r^3}{3}\right]_0^{D/2} = \frac{4\pi}{3}C_u\left[\frac{D^3}{8}\right] = \frac{\pi C_u D^3}{6}$$

$$\therefore \quad T = T_1 + T_2$$

$$\Rightarrow \quad T = \pi DHC_u \cdot \frac{D}{2} + \pi C_u \frac{D^3}{6}$$

$$\Rightarrow \quad C_u = \frac{T}{\pi D^2\left(\frac{H}{2} + \frac{D}{6}\right)}$$

If shear vane apparatus does not fully penetrate into the soil then

$$T_2 = \frac{1}{2}\left(\pi C_u \frac{D^3}{6}\right) = \left(\pi C_u \frac{D^3}{12}\right)$$

$$\therefore \quad C_u = \frac{T}{\pi D^2\left(\frac{H}{2}+\frac{D}{12}\right)}$$

6.13 In a consolidated undrained test with pore pressure measurement on a normally consolidated clay, a sample consolidated under a stress of 200 kPa failed at an additional axial stress of 150 kPa. The pore pressure at failure was 75 kPa. Determine analytically the shear strength parameters in terms of effective stresses.

[8 marks : 2016]

Solution:

$$\sigma_3 = 200\text{ kPa} \quad \sigma_d = 150\text{ kPa}$$

$$\sigma_1 = \sigma_3 + \sigma_d = 200 + 150 = 350\text{ kPa}$$

$$\text{Pore pressure at failure} = 75\text{ kPa}$$

$$\bar{\sigma}_1 = \sigma_1 - u = 350 - 75 = 275\text{ kPa}$$

$$\bar{\sigma}_3 = \sigma_3 - u = 200 - 75 = 125\text{ kPa}$$

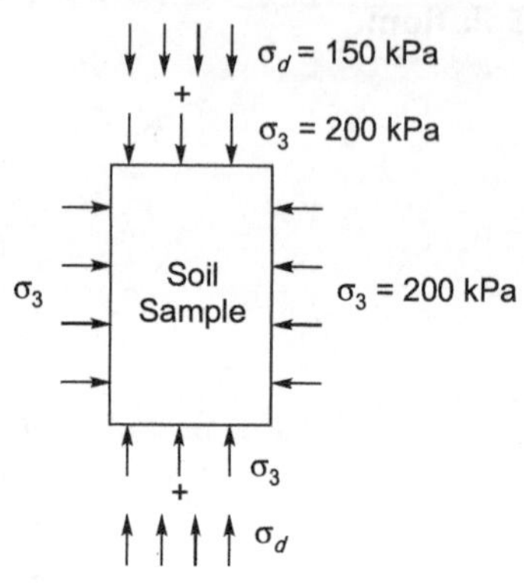

In CU test, cohesion of clay will be zero.

$$C' = 0$$

$$\therefore \quad \bar{\sigma}_1 = \bar{\sigma}_3 \tan^2\left(45+\frac{\phi'}{2}\right) + 2C'\tan\left(45+\frac{\phi'}{2}\right)$$

$$275 = 125\tan^2\left(45+\frac{\phi'}{2}\right)$$

$$\phi' = 22.02^\circ$$

$\therefore$ Shear strength parameters are $C' = 0$ and $\phi' = 22.02°$.

6.14 A conventional consolidated drained (CD) triaxial test was conducted on saturated clean and sample by using the following steps:

1. Set cell pressure to 250 kPa and allow the sample to consolidate with its drainage valve open at 100 kPa back pressure.
2. Shear the sample without any change in the drainage condition.

The sample failed when the deviator stress reached 300 kPa. Use analytical solution to determine:

(i) The slope of failure envelope in degrees. Assume c′ to be zero.

(ii) Slope of the failure plane in degrees.

(iii) Shear stress and normal stress on the failure plane (in kPa).

(iv) The maximum shear stress at failure (in kPa)

[12 marks : 2018]

Solution:

Given data: Cell pressure, $\sigma_3 = 250$ kPa, Back pressure, $u = 100$ kPa, Deviator stress, $\sigma_d = 300$ kPa

So, $\bar{\sigma}_3 = \sigma_3 - u = 250 - 100$

$\Rightarrow$ $\bar{\sigma}_3 = 150\text{ kPa}$

and $\bar{\sigma}_1 = \bar{\sigma}_3 + \sigma_d = 150 + 300$

$\Rightarrow$ $\bar{\sigma}_1 = 450\text{ kPa}$

(i)
$$\bar{\sigma}_1 = \bar{\sigma}_3 \tan^2\left(45+\frac{\phi'}{2}\right)+2c'\tan\left(45+\frac{\phi'}{2}\right)$$

$$\Rightarrow \quad 450 = 150\tan^2\left(45+\frac{\phi'}{2}\right) \quad ...(c' = 0)$$

$$\Rightarrow \quad \phi' = 30^\circ \quad \text{(slope of failure envelope)}$$

(ii) Slope of failure plane

$$\theta_c = 45^\circ+\frac{\phi}{2} = 45^\circ+\frac{30^\circ}{2}$$

$$\Rightarrow \quad \theta_c = 60^\circ \quad \text{(with major principal plane)}$$

(iii) On failure plane,
$$\theta_c = 60^\circ \quad \text{(with major principal plane)}$$
$$\theta = \text{Inclination with vertical axis} = 30^\circ$$

(a) Shear stress

$$\tau = \left(\frac{\bar{\sigma}_1-\bar{\sigma}_3}{2}\right)\sin 2\theta = \left(\frac{450-150}{2}\right)\sin(2\times 60^\circ)$$

$$\Rightarrow \quad \tau = 129.9\text{ kPa}$$

(b) Normal stress

$$\sigma_n = \frac{\sigma_1+\sigma_3}{2}+\left(\frac{\sigma_1-\sigma_3}{2}\right)\cos(2\theta_c)$$

$$= \frac{450+100}{2}+\frac{450-100}{2}\cos(2\times 60)$$

$$\Rightarrow \quad \sigma_n = 225\text{ kPa}$$

(iv) Maximum shear stress at failure

$$\tau_{max} = \left(\frac{\bar{\sigma}_1-\bar{\sigma}_3}{2}\right) = \frac{(450-150)}{2}$$

$$\Rightarrow \quad \tau_{max} = 150\text{ kPa}$$

6.15 **Consolidated undrained type Triaxial tests were carried out to failure on two identical specimens of silty clay with pore water pressure measurements, as given below:**

Sl. No.	Confining pressure (kPa)	Deviator stress (kPa)	Pore pressure (kPa)
1	100	150	40
2	200	220	75

Determine the shear strength parameters, if

(i) Construction is done at a faster rate,

(ii) Construction is done slowly.

[20 marks : 2020]

Solution:

(i) Construction is done at a faster rate i.e., soil will be in undrained condition hence total stress should be used.

(ii) Construction is done slowly i.e., soil will be in drained condition hence effective stress should be used.

SNo.	$\sigma_3(i)$	$\sigma_d(ii)$	$u(iii)$	$\sigma_1 = \sigma_3+\sigma_d$ $(iv) = (i+ii)$	$\bar{\sigma}_1 = \sigma_1 - u$ $= (iv)-(iii)$	$\bar{\sigma}_3 = \sigma_3 - u$ $= (i)-(iii)$
1	100	150	40	250	210	60
2	200	220	75	420	345	125

(i) Undrained/total shear parameters (construction is done at a faster rate)

$$\sigma_1 = \sigma_3 \tan^2\left(45+\frac{\phi}{2}\right)+2C\tan\left(45+\frac{\phi}{2}\right)$$

$$250 = 100\tan^2\left(45+\frac{\phi}{2}\right)+2C\tan\left(45+\frac{\phi}{2}\right) \quad \text{...(i)}$$

$$420 = 200\tan^2\left(45+\frac{\phi}{2}\right)+2C\tan\left(45+\frac{\phi}{2}\right) \quad \text{...(ii)}$$

$$170 = 100\tan^2\left(45+\frac{\phi}{2}\right)$$

$$\phi = 15.026°$$

From eq. (i)

$$C = 30.68 \text{ kN/m}^2$$

(ii) Drained/effective shear parameters (construction is done slowly)

$$\bar{\sigma}_1 = \bar{\sigma}_3 \times \tan^2\left(45+\frac{\phi}{2}\right)+2C'\tan\left(45+\frac{\phi'}{2}\right)$$

$$210 = 60\tan^2\left(45+\frac{\phi'}{2}\right)+2C'\tan\left(45+\frac{\phi'}{2}\right) \quad \text{...(iii)}$$

$$345 = 125\tan^2\left(45+\frac{\phi'}{2}\right)+2C'\tan\left(45+\frac{\phi'}{2}\right) \quad \text{...(iv)}$$

$$135 = 65\tan^2\left(45+\frac{\phi'}{2}\right)$$

∴ $\phi' = 20.487°$

From eq. (iii)

$$C' = 29.63 \text{ kN/m}^2$$

6.16 **With a neat labelled sketch of stress conditions and failure envelope in a triaxial compression test, derive the relation between major and minor principal stress in terms of cohesion and angle of internal friction.**

$\sigma_1 = \sigma_3 N_\phi + 2c\sqrt{N_\phi}$ — **σ_1 = Major principal stress**

σ_3 = Minor principal stress — **c = Cohesion**

ϕ = Angel of internal fiction

[12 Marks : 2021]

Solution:

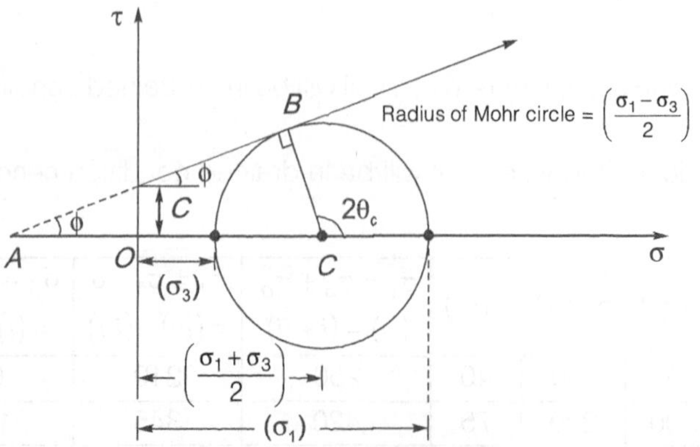

In ΔABC, $$\sin\phi = \frac{BC}{AC} = \frac{BC}{AO+OC} = \frac{\left(\frac{\sigma_1-\sigma_3}{2}\right)}{c\cot\phi + \left(\frac{\sigma_1+\sigma_3}{2}\right)} \qquad \left(\because \tan\phi = \frac{C}{AO}\right)$$

$$\frac{\sigma_1-\sigma_3}{2} = c\cot\phi\sin\phi + \left(\frac{\sigma_1+\sigma_3}{2}\right)\sin\phi$$

$$\sigma_1-\sigma_3 = 2c\cos\phi + (\sigma_1+\sigma_3)\sin\phi$$

$$\sigma_1(1-\sin\phi) = \sigma_3(1+\sin\phi) + 2c\cos\phi$$

$$\sigma_1 = \sigma_3\left(\frac{1+\sin\phi}{1-\sin\phi}\right) + 2c\times\frac{\cos\phi}{1-\sin\phi} \qquad \text{...(i)}$$

$\because$ $$\cos\phi = \sqrt{1-\sin^2\phi}$$

$\therefore$ $$\cos\phi = \sqrt{(1-\sin\phi)(1+\sin\phi)}$$

$\therefore$ $$\frac{\cos\phi}{1-\sin\phi} = \frac{\sqrt{(1-\sin\phi)(1+\sin\phi)}}{1-\sin\phi} = \sqrt{\frac{1+\sin\phi}{1-\sin\phi}}$$

Put in eq. (i)

$\therefore$ $$\sigma_1 = \sigma_3\left(\frac{1+\sin\phi}{1-\sin\phi}\right) + 2c\sqrt{\frac{1+\sin\phi}{1-\sin\phi}} \qquad \left\{\because \tan^2\left(45^\circ+\frac{\phi}{2}\right) = \frac{1+\sin\phi}{1-\sin\phi}\right\}$$

$$\sigma_1 = \sigma_3\tan^2\left(45^\circ+\frac{\phi}{2}\right) + 2c\tan\left(45^\circ+\frac{\phi}{2}\right) \qquad \text{(At plastic equilibrium)}$$

7. Retaining Wall/Earth Pressure Theories

7.1 A retaining wall 6 m high supports earth with its face vertical. The earth is cohesionless with particle specific gravity 2.69, angle of internal friction 35° and porosity 40.5%. The earth surface is horizontal and level with the top of the wall. Determine the earth thrust and its line of action on the wall if the earth is water logged to level 2.5 m below the top surface. Neglect wall friction. Draw the pressure diagrams.

[15 marks : 1999]

Solution:

Given data: Height of retaining wall (H) = 6 m, Specific gravity of soil particles (G) = 2.69

Angle of internal friction (ϕ) = 35°, Porosity (n) = 40.5% = 0.405

Depth of water table below top surface = 2.5 m

We know that: $$n = \frac{e}{1+e}$$

$\therefore$ $$e = \frac{n}{1-n} = \frac{0.405}{1-0.405} = 0.681$$

$$\gamma_{sat} = \left(\frac{G+e}{1+e}\right)\gamma_w = \left(\frac{2.69+0.681}{1+0.681}\right)\times 9.81 = 19.67\ \text{kN/m}^3$$

$$\gamma_{dry} = \frac{G\gamma_w}{1+e} = \frac{2.69\times 9.81}{1+0.681} = 15.7\ \text{kN/m}^3$$

$$\gamma' = \gamma_{sat} - \gamma_w = 19.67 - 9.81 = 9.86\ \text{kN/m}^3.$$

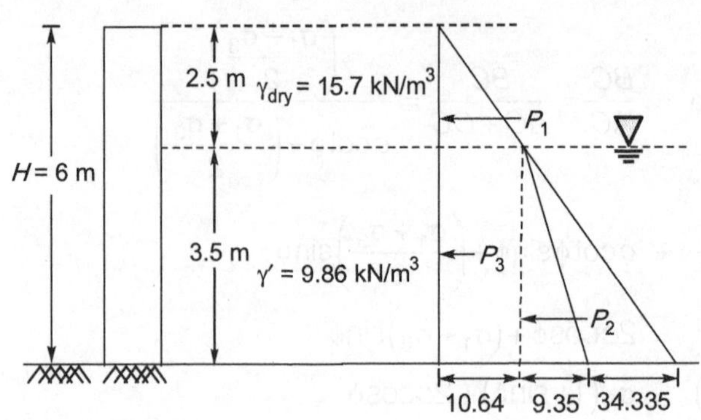

$$K_a = \frac{1-\sin\phi}{1+\sin\phi} = \frac{1-\sin 35°}{1+\sin 35°} = 0.271$$

For $z = 2.5$ m

$$P_v = K_a\gamma z - 2c\sqrt{K_a} \qquad [\because \text{for cohesionless soil } c = 0]$$

$$P_v = K_a\gamma z = 0.271 \times 15.7 \times 2.5 = 10.64 \text{ kN/m}^2$$

∴ Total active thrust for z = 2.5 m is given by

$$P_1 = \frac{1}{2} \times P_v \times 2.5$$

$$= \frac{1}{2} \times 10.64 \times 2.5 = 13.3 \text{ kN/m}$$

For $z = 3.5$ m

$$P_{v_1} = K_a\gamma' z = 0.271 \times 9.86 \times 3.5 = 9.35 \text{ kN/m}^2$$

$$P_{v_2} = \gamma_w z = 9.81 \times 3.5 = 34.335 \text{ kN/m}^2$$

$$P_{v_1} + P_{v_2} = 9.35 + 34.335 = 43.685 \text{ kN/m}^2$$

∴ Total active thrust for z = 3.5 m is given by

$$P_2 + P_3 = \frac{1}{2} \times (43.685) \times 3.5 + 10.64 \times 3.5 = 76.45 + 37.24 = 113.69 \text{ kN/m}$$

∴ Total active thrust on the retaining wall = $P_1 + P_2 + P_3$ = 13.3 + 76.45 + 37.24 = 126.99 kN/m

$$\text{Location of total thrust from base} = \frac{13.3 \times \left(3.5 + \frac{2.5}{3}\right) + 37.24 \times \left(\frac{3.5}{2}\right) + 76.45 \times \left(\frac{3.5}{3}\right)}{126.99} = 1.67 \text{ m}$$

7.2 For an earth retaining structure shown in figure below, construct earth pressure diagram for active state and find the total thrust per unit length of the wall.

[15 marks : 2000]

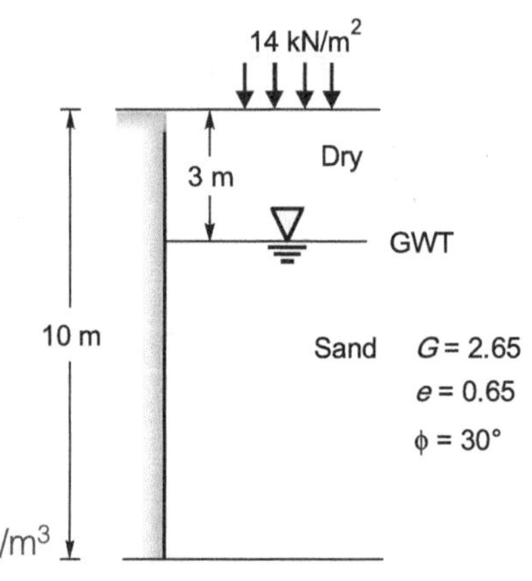

Solution:

Given data: $G = 2.65$; $e = 0.65$; $\phi = 30°$

$$\therefore \quad \gamma_d = \frac{G\gamma_w}{1+e} = \frac{2.65 \times 9.81}{1+0.65} = 15.76 \text{ kN/m}^3$$

$$\gamma' = \left(\frac{G-1}{1+e}\right)\gamma_w = \left(\frac{2.65-1}{1+0.65}\right) \times 9.81 = 9.81 \text{ kN/m}^3$$

also $$K_a = \frac{1-\sin\phi}{1+\sin\phi} = \frac{1-\sin 30^\circ}{1+\sin 30^\circ} = 0.333$$

Earth pressure diagram

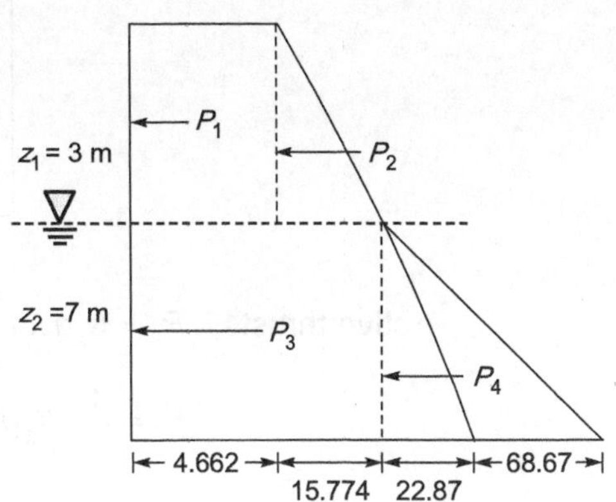

$$P_{a_1} = K_a q + K_a \gamma_d z_1$$

$$= 0.333 \times 14 + 0.333 \times 15.76 \times 3$$

$$= 4.662 + 15.744 = 20.406 \text{ kN/m}^2$$

$$P_{a_2} = K_a \gamma' z_2 = 0.333 \times 9.81 \times 7$$

$$= 22.87 \text{ kN/m}^2$$

$$P_{a_3} = \gamma_w z_2 = 9.81 \times 7 = 68.67 \text{ kN/m}^2$$

Total thrust per unit length

$$P_1 = 4.662 \times 3 = 13.986 \text{ kN/m}$$

$$P_2 = \frac{1}{2} \times 15.744 \times 3 = 23.616 \text{ kN/m}$$

$$P_3 = 20.406 \times 7 = 142.842 \text{ kN/m}$$

$$P_4 = \frac{1}{2}(22.87 + 68.67) \times 7 = 320.39 \text{ kN/m}$$

$$\therefore \quad P_1 + P_2 + P_3 + P_4 = 13.986 + 23.616 + 142.842 + 320.39 = 500.83 \text{ kN/m}$$

Location of resultant thrust from base

$$= \frac{13.986 \times \left(7 + \frac{3}{2}\right) + 23.616 \times \left(7 + \frac{3}{3}\right) + 142.842 \times \left(\frac{7}{2}\right) + 320.39 \times \left(\frac{7}{3}\right)}{500.83}$$

$$= 3.106 \text{ m}$$

7.3 A retaining wall 2 m in height has a smooth vertical surface. The backfill has a horizontal levelled surface with the top of retaining wall. The density of the backfill is 1.8 t/m^3, shearing resistance angle of 30° and cohesion zero. A uniformly distributed surcharge load of 3 t/m^2 intensity is acting on the backfill.

(i) Calculate the magnitude and point of application of active earth pressure per meter length of the retaining wall.

(ii) If during rainy season water table rises behind the wall to a height of 1 m above the base of the retaining wall, workout the effect on the value of active earth pressure if there is no change in the angle of shearing resistance. Submerged unit weight of backfill is 1.25 t/m^3.

[20 marks : 2001]

Solution:

Given data: $H = 2$ m; $\gamma = 1.8$ t/m^3; $\phi = 30°$; $c = 0$

Surcharge, $q = 3$ t/m^2

$$K_a = \frac{1-\sin\phi}{1+\sin\phi} = \frac{1-\sin 30^\circ}{1+\sin 30^\circ} = \frac{0.5}{1.5} = \frac{1}{3}$$

(i)

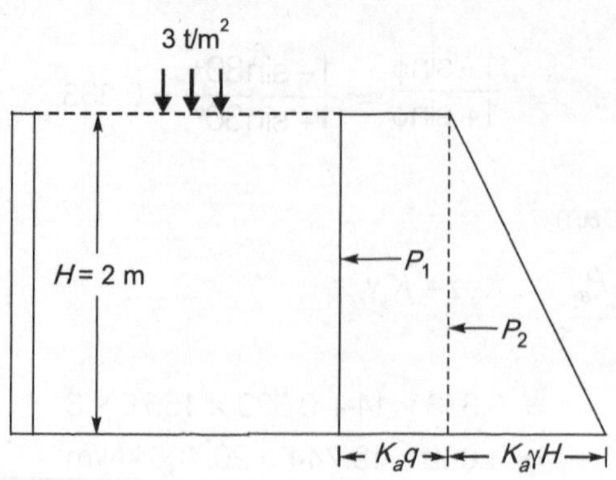

Active thrusts, $P_1 = K_a q \times H = \frac{1}{3} \times 3 \times 2 = 2$ t/m at $\frac{H}{2}$ from base

$$P_2 = \frac{1}{2} \times K_a \gamma H \times H = \frac{1}{2} \times \frac{1}{3} \times 1.8 \times 2^2 = 1.2 \text{ t/m at } \frac{H}{3} \text{ from base}$$

$\therefore$ Location of resultant thrust from base $= \dfrac{2 \times \frac{2}{2} + 1.2 \times \frac{2}{3}}{2 + 1.2} = 0.875$ m

Total active earth pressure per unit length of wall = 2 + 1.2 = 3.2 t/m

(ii)

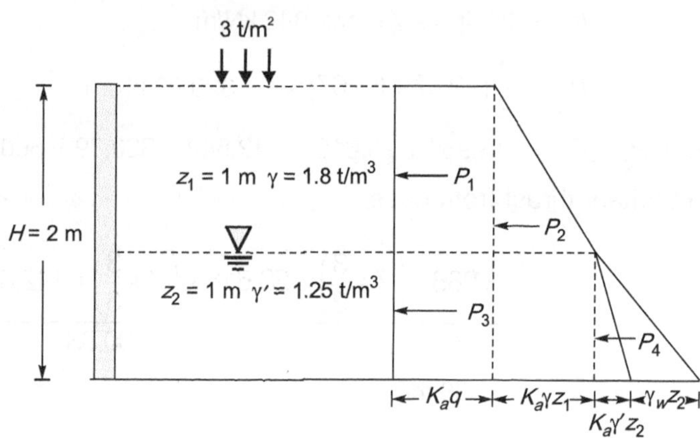

Active Pressure, $P_1 = K_a q z_1 = \frac{1}{3} \times 3 \times 1 = 1$ t/m at $\left(1 + \frac{1}{2}\right)$m from base

$$P_2 = \frac{1}{2} K_a \gamma z_1^2 = \frac{1}{2} \times \frac{1}{3} \times 1.8 \times 1^2 = 0.3 \text{ t/m at } \left(1 + \frac{1}{3}\right)\text{m from base}$$

$$P_3 = (K_a q + K_a \gamma z_1) z_2 = \left(\frac{1}{3} \times 3 + \frac{1}{3} \times 1.8 \times 1\right) \times 1 = 1.6 \text{ t/m at } \frac{1}{2} \text{ m from base}$$

$$P_4 = \frac{1}{2} K_a \gamma' z_2^2 + \frac{1}{2} \times \gamma_w z_2^2 = \frac{1}{2} \times \frac{1}{3} \times 1.25 \times 1^2 + \frac{1}{2} \times 1 \times 1^2 = 0.71 \text{ t/m at } \frac{1}{3} \text{ m from base}$$

Total active earth pressure per unit length

$$= P_1 + P_2 + P_3 + P_4 = 1 + 0.3 + 1.6 + 0.71 = 3.61 \text{ t/m}$$

Location of resulant thrust from base $= \dfrac{(1 \times 1.5) + \left(0.3 \times \frac{4}{3}\right) + (1.6 \times 0.5) + \left(0.71 \times \frac{1}{3}\right)}{3.61} = 0.813$ m

Change in total active thrust due to rise in water table = 3.61 – 3.2 = 0.41 t/m

7.4 **A smooth retaining wall 6 m high retains dry granular backfill weighing 16 kN/m³ to its level surface. The active thrust on the wall is 96 kN/m of wall. What will be the total active thrust if the water table comes up to backfill surface? Take specific gravity of backfill = 2.65.**

[15 marks : 2002]

Solution:

Given data: $H = 6$ m; $\gamma = 16$ kN/m³; $P = 96$ kN/m; $G = 2.65$

When backfill is dry, the active earth pressure diagram is

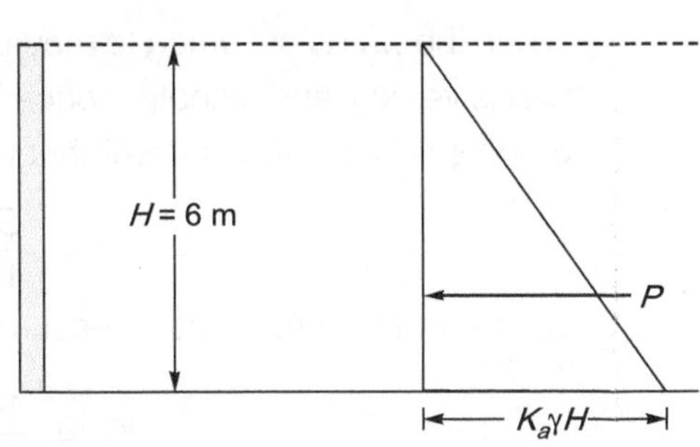

$$\therefore \quad P = \frac{1}{2}K_a\gamma H \times H$$

$$\Rightarrow \quad 96 = \frac{1}{2} \times K_a \times 16 \times 6 \times 6$$

$$\Rightarrow \quad K_a = \frac{1}{3}$$

But $$\frac{1-\sin\phi}{1+\sin\phi} = K_a$$

$$\Rightarrow \quad \frac{1-\sin\phi}{1+\sin\phi} = \frac{1}{3}$$

$$\Rightarrow \quad 3 - 3\sin\phi = 1 + \sin\phi$$

$$\Rightarrow \quad \sin\phi = \frac{1}{2}$$

$$\Rightarrow \quad \phi = 30°$$

When the water table comes upto the surface of backfill, the active earth pressure is increased.

Submerged density, $$\gamma' = \left(\frac{G-1}{1+e}\right)\gamma_w \quad \text{...(i)}$$

also $$\gamma_d = \frac{G\gamma_w}{1+e} \quad \text{...(ii)}$$

From (i) and (ii) equating values of $1 + e$, we get

$$\frac{(G-1)\gamma_w}{\gamma'} = \frac{G\gamma_w}{\gamma}$$

$$\left(\frac{G-1}{G}\right)\gamma = \gamma'$$

$$\therefore \quad \gamma' = \left(\frac{2.65-1}{2.65}\right) \times 16 = 9.96 \text{ kN/m}^3$$

Total active earth pressure per unit length of wall

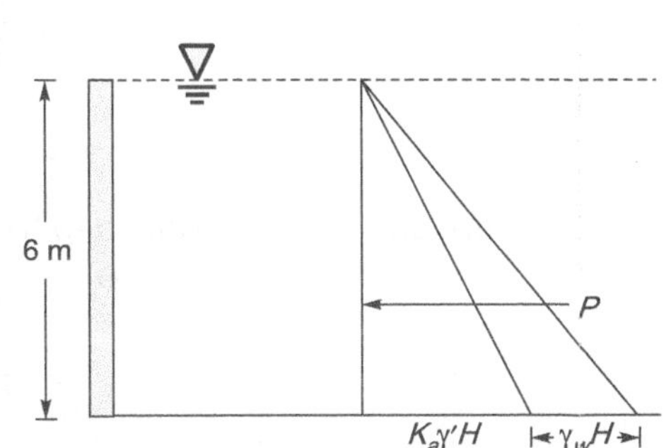

$$= \frac{1}{2}(K_a\gamma' H + \gamma_w H)H$$

$$= \frac{1}{2}\left(\frac{1}{3} \times 9.96 \times 6 + 9.81 \times 6\right)6$$

$$= 236.34 \text{ kN}$$

Location of resultant thrust from base $= \frac{6}{3} = 2$ m

7.5 **A retaining wall 10 m high is proposed to hold dry sand of void-ratio of 0.6. The value of angle of internal friction ϕ = 30° and specific-gravity of soil grains is 2.7. The back face of wall is vertical and smooth. Top surface of backfill is horizontal. Calculate the magnitude of the total active earth thrust against the wall assuming the wall is free to move. Also show the distribution of earth pressure and point of application of the resultant. Assume unit weight of water = 10 kN/m³.**

[10 marks : 2009]

Solution:

Given Data: Height of retaining wall (H) = 10 m, Void ratio (e) = 0.6, Angle of internal friction (ϕ) = 30°
Specific gravity of soil grains (G) = 2.7, Unit weight of water (γ_w) = 10 kN/m³

When the wall is assumed to move freely, the active earth pressure is developed. Given that back face of wall is vertical and smooth, hence Rankine's theory can be applied.
Since soil is dry, the unit weight of soil,

$$\gamma = \frac{G\gamma_w}{1+e} = \frac{2.7\times 10}{1+0.6} = 16.875 \text{ kN/m}^3$$

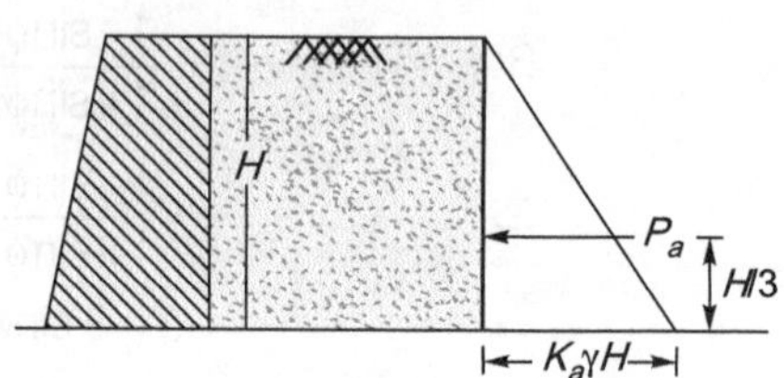

Coefficient of active earth pressure,

$$K_a = \frac{1-\sin\phi}{1+\sin\phi} = \frac{1-\sin 30°}{1+\sin 30°} = \frac{1}{3}$$

Pressure at the bottom of retaining wall = $K_a\gamma H$

Thus, the distribution of the active earth pressure with depth is linear. For a total height of H of the wall, the total active earth thrust per unit length of the wall is given by

$$P_a = \frac{1}{2}K_a\gamma H^2 = \frac{1}{2}\times\frac{1}{3}\times 16.875\times 10^2 = 281.25 \text{ kN}$$

This may be taken to act at a height of $\frac{H}{3}$ i.e. $\frac{10}{3}$ m above the base, through the centroid of the pressure distribution diagram.

7.6 **Calculate the active earth pressure at a depth of 3.6 m in a sandy soil with angle of internal friction as 30° and having a density of 1.9 gm/cc.**

[2 marks : 2010]

Solution:

Given that H = 3.6 m, ϕ = 30°, ρ = 1.9 gm/cc
Coefficient of active earth pressure is given as

$$K_a = \frac{1-\sin\phi}{1+\sin\phi} = \frac{1-\sin 30°}{1+\sin 30°} = \frac{1}{3}$$

$$\gamma = \rho g = \frac{1.9\times 10^{-3}}{10^{-6}}\times 9.81 = 18639 \text{ N/m}^3 = 18.639 \text{ kN/m}^3$$

Active earth pressure at a depth of 3.6 m is given by

$$p_a = k_a\gamma H$$

$$= \frac{1}{3}\times 18.639\times 10^3\times 3.6 = 22366.8 \text{ N/m}^2 = 22.37 \text{ kN/m}^2$$

7.7 **A 6.0 m high retaining wall is to support a soil with unit weight γ = 17.4 kN/m³, ϕ = 26° and c' = 14.36 kN/m². Determine the Rankine active force per unit length of wall before the tensile crack occurs. Find the critical depth.**

[8 marks : 2010]

Solution:

Given that: $H = 6.0$ m, $\gamma = 17.4$ kN/m^3, $\phi = 26°$, $c' = 14.36$ kN/m^2

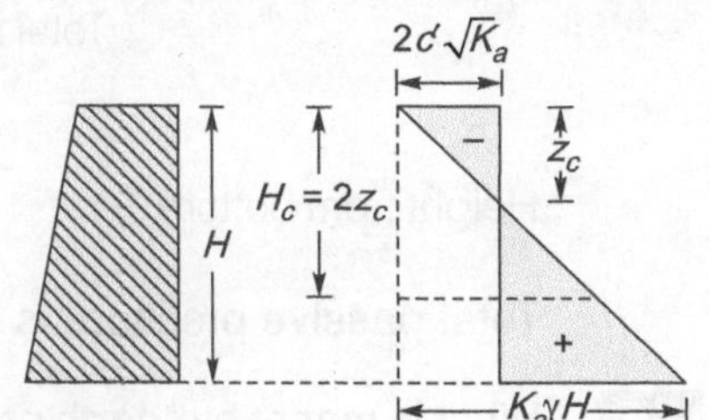

The lateral pressure distribution is given here.

Rankine coefficient of active earth pressure,

$$K_a = \frac{1-\sin\phi}{1+\sin\phi} = \frac{1-\sin 26°}{1+\sin 26°} = 0.3905$$

The Rankine active force per unit length of wall ignoring tensile cracks is given by

Rankine active force before tension crack appear,

$$P_a = \frac{1}{2}K_a\gamma H^2 - 2c'H\sqrt{K_a}$$

$$= \frac{1}{2}\times 0.3905 \times 17.4 \times (6)^2 - 2\times 14.36\times 6\times\sqrt{0.3905}$$

$$= 122.3046 - 107.6828 = 14.62 \text{ kN}$$

The critical depth is given by

$$H_c = \frac{4c'}{\gamma\sqrt{K_a}} = \frac{4\times 14.36}{17.4\sqrt{0.3905}} = 5.28 \text{ m}$$

7.8 **A retaining wall 6 m high, with a smooth vertical back is pushed against a soil mass having $c' = 40$ kN/m^2, $\phi' = 15°$ and $\gamma = 19$ kN/m^3. Using Rankine's theory, compute the total pressure and the point of application of the resultant thrust, if the horizontal soil surface carries a uniform surcharge load of 50 kN/m^2.**

[10 marks : 2011]

Solution:

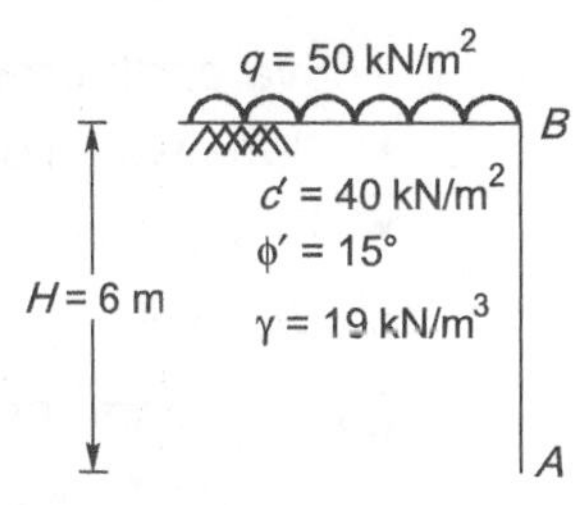

Pushed against soil mass means passive pressure is acting

Passive earth pressure coefficient = $K_p = \frac{1+\sin 15°}{1-\sin 15°} = 1.7$

At point A:

Pressure due to weight of soil = $K_p\gamma\cdot H$

$$P_1 = 1.7 \times 19 \times 6 = 193.8 \text{ kN/m}^2$$

Pressure due to surcharge load, $P_2 = q\cdot K_p$

$\Rightarrow$ $P_2 = 50 \times 1.7 = 85$ kN/m^2

Pressure due to cohesion, $P_3 = 2c'\sqrt{K_p}$

$\Rightarrow$ $P_3 = 2\times 40\sqrt{1.7} = 104.31$ kN/m^2

Total passive pressure at A = $P_1 + P_2 + P_3$

$= 193.8 + 85 + 104.31 = 383.11$ kN/m^2

Pressure diagram for each component is shown below:

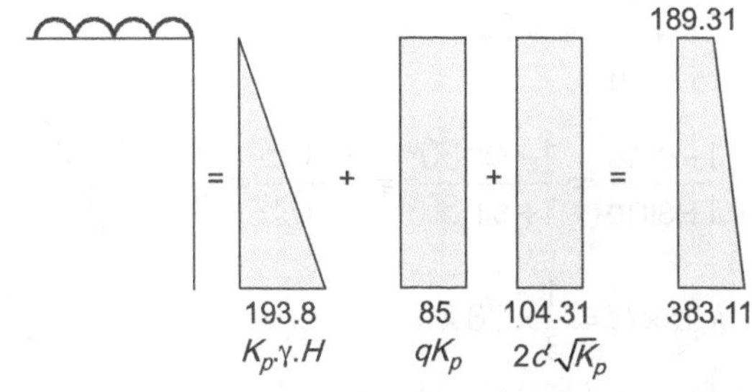

$$\text{Total pressure} = \frac{1}{2}\times 193.8\times 6+85\times 6+104.31\times 6 = 1717.26 \text{ kN/m}$$

Height from bottom, $$\bar{h} = \frac{\frac{1}{2}\times 193.8\times 6\times 2+85\times 6\times 3+104.31\times 6\times 3}{1717.26} = 2.66 \text{ m}$$

Total passive pressure is 1717.26 kN/m and it acts at a height of 2.66 m from bottom.

7.9 **What is meant by "earth pressure at rest", active earth pressure and passive earth pressure? Explain the difference in terms of wall movement.**

[4 marks : 2012]

Solution:

Structures which are used to hold back a soil mass are called retaining structure. Retaining walls, sheet pile walls, crib walls, basement walls etc. are examples of retaining.

The soil that is retained at a slope steeper than it can sustain by virtue of its shearing strength, exerts a force on the retaining wall. This force is called the earth pressure and the material that is retained by the wall is referred as

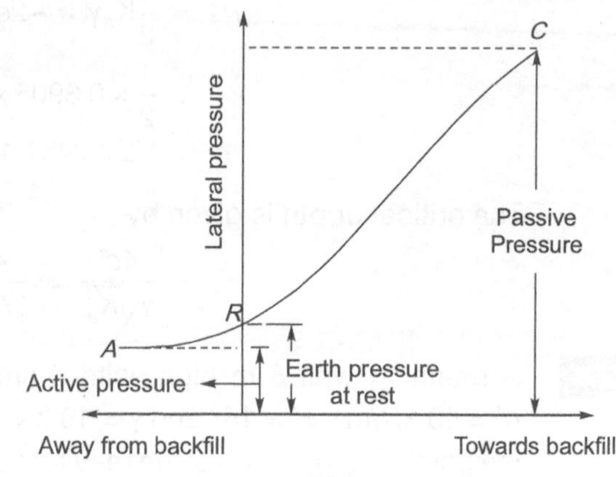

Earth pressure at rest:

- When the wall is rigid and unyielding, the soil mass is at rest.
- The earth pressure corresponding to this state is called earth pressure at rest.

Active earth pressure:

- When wall moves away from soil, the block of soil have the tendency to move downwards and outwards.
- The pressure developed on the wall when the soil is on the verge of failure is called active earth pressure.

Passive earth pressure:

- When the wall moves towards the soil, the block of soil has tendency to move up and inside.
- The pressure on the wall increases beyond the earth pressure at rest condition and this is called passive earth pressure.

7.10 **A retaining wall 4 m high, has a smooth vertical back. The backfill has a horizontal surface in level with the top of the wall. There is a uniformly distributed surcharge load of 36 kN/m^2 intensity over the backfill. The unit weight of backfill soil is 18 kN/m^3 with angle of shearing resistance, ϕ of 30° and cohesion is zero. Determine the magnitude and point of application of active pressure per meter length of wall.**

[16 marks : 2012]

Solution:

Given data: $H = 4$ m; $\gamma = 18$ kN/m^3; $\phi = 30°$; $c = 0$

Surcharge $q = 36$ kN/m^2

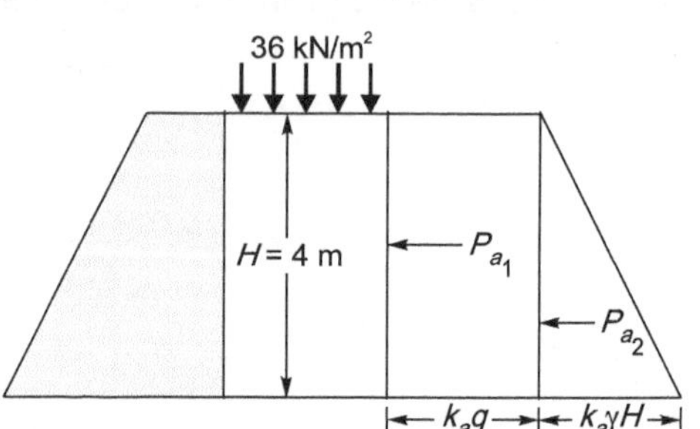

$$k_a = \frac{1-\sin\phi}{1+\sin\phi} = \frac{1-\sin 30°}{1+\sin 30°} = \frac{1-0.5}{1+0.5} = \frac{1}{3}$$

Active pressure $$P_{a_1} = k_a q\times H = \frac{1}{3}\times 36\times 4$$

$$= \quad 48 \text{ kN/m at } \frac{H}{2} \text{ from base}$$

$$P_{a_2} = \frac{1}{2}k_a\gamma H \times H = \frac{1}{2} \times \frac{1}{3} \times 18 \times 4 \times 4$$

$$= \quad 48 \text{kN/m at } \frac{H}{3} \text{ from base}$$

∴ Point of application of active pressure from base

$$= \frac{P_{a_1} \times \frac{H}{2} + P_{a_2} \times \frac{H}{3}}{P_{a_1} + P_{a_2}} = \frac{48 \times \frac{4}{2} + 48 \times \frac{4}{3}}{48 + 48} = 1.667 \text{ m from base}$$

Total active earth pressure per meter length of wall

$$= 48 + 48 = 96 \text{ kN/m}$$

7.11 A retaining wall with a smooth vertical back is 9 m high and retains a two-layer sand backfill with the following properties:

0-3 m depth : $c' = 0.0$, $\phi' = 30°$, $\gamma = 18$ kN/m^3

3-9 m depth : $c' = 0.0$, $\phi' = 35°$, $\gamma = 20$ kN/m^3

Show the active earth pressure distribution and determine the total active thrust on the wall. Assume that water table is well below the base of the wall.

[8 marks : 2015]

Solution:

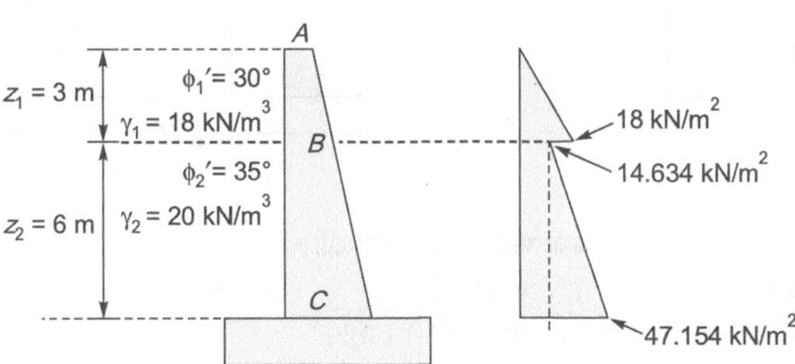

For upper layer of sand, $k_{a1} = \dfrac{1-\sin\phi_1'}{1+\sin\phi_1'} = \dfrac{1-\sin 30°}{1+\sin 30°} = \dfrac{1}{3}$

Active earth pressure,

At *A*, $p_A = 0$

At *B*, $p_B = k_{a_1}\gamma_1 z_1 = \dfrac{1}{3} \times 18 \times 3 = 18 \text{kN/m}^2$

For lower layer of sand, $k_{a2} = \dfrac{1-\sin\phi_2'}{1+\sin\phi_2'} = \dfrac{1-\sin 35°}{1+\sin 35°} = 0.271$

Active earth pressure

At *B*, $p_B = k_{a_2}\gamma_1 z_1 = 0.271 \times 18 \times 3 = 14.634 \text{ kN/m}^2$

At *C*, $p_C = k_{a_2}\gamma_1 z_1 + k_{a_2}\gamma_2 z_2 = 0.271 \times (18 \times 3 + 20 \times 6)$

$= 47.154 \text{ kN/m}^2$

$$\text{Total active thrust on wall} = \frac{1}{2}k_{a_1}\gamma_1 z_1^2 + k_{a_2}\gamma_1 z_1 z_2 + \frac{1}{2}k_{a_2}\gamma_2 z_2^2$$

$$= \frac{1}{2}\times\frac{1}{3}\times 18\times 3^2 + 0.271\times 18\times 3\times 6 + \frac{1}{2}\times 0.271\times 20\times 6^2$$

$$= 212.364 \text{ kN/m}$$

7.12 **A vertical wall 8.0 m high with a smooth back has sand behind it. The level of sand is horizontal. Water table is far below the ground surface. Determine the distribution of active earth pressure, total lateral force on the wall per metre length and the level at which the resultant lateral force acts.**

[10 marks : 2017]

Solution:

Active pressure on vertical retaining wall with cohesionless backfill :

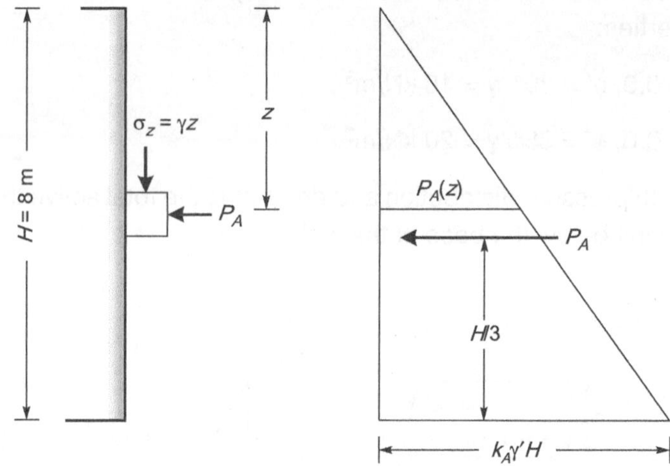

The back of the retaining wall is assumed to be smooth.

From Rankine's theory, $p_A = k_A\gamma z$

where γ is unit weight of the soil above the water table.

At the base of the wall, where $z = H$

$$p_A = k_A\gamma H$$

The distribution of earth pressure is shown above. The force per unit length of wall due to active pressure distribution is the total active thrust P_A and is given by the area of the active pressure distribution diagram and acts through the centre of the area.

Hence, $$P_A = \frac{1}{2}k_A\gamma H\times H = \frac{1}{2}k_A\gamma H^2$$

This acts at a distance $\frac{H}{3}$ above the base of the wall.

Given: $$H = 8\text{ m}$$

∴ $$P_A = \frac{1}{2}\times k_A\gamma(8)^2 = 32k_A\gamma$$

and acts at $\frac{8}{3} = 2.667$ m above base of the wall.

7.13 A 6 m deep excavation in sand is supported by a smooth vertical wall. The backfill is horizontal and supports a surcharge of 10 kN/m^2 on its surface.

(i) Determine the active thrust on the wall if the water table is far below the bottom of the excavation.

(ii) What should be the embedment depth if the wall in the above case is cantilever sheet pile wall?

Properties of the sand are: $c' = 0$, $\phi' = 30$ degree and $\gamma_{bulk} = 16$ kN/m^3

[20 marks : 2018]

Solution:

(i)

$$\phi' = 30°, \gamma_b = 16 \text{ kN/m}^3$$

$$k_a = \frac{1-\sin\phi'}{1+\sin\phi'} \quad \Rightarrow \quad k_a = \left(\frac{1}{3}\right)$$

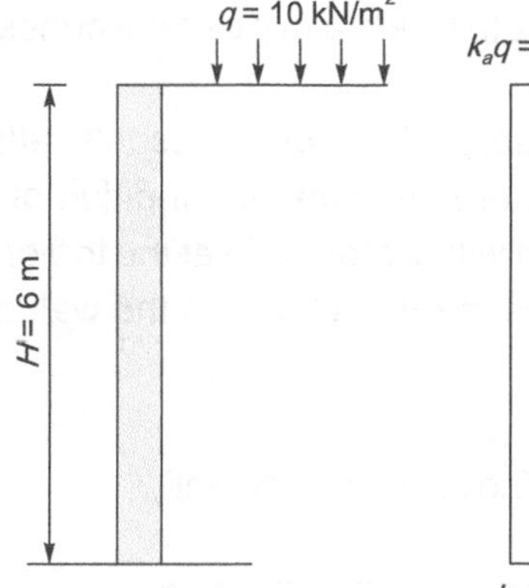

At top, $p_a = k_a q = 3.33$ kN/m^2

At bottom $p_b = k_a q + k_a \gamma H = 35.33$ kN/m^2

Active thrust, $P_a = \frac{1}{2} \times (3.33 + 35.33) \times 6$ ..(i.e., area of pressure diagram)

$\Rightarrow$ $P_a = 116$ kN/m

(ii)

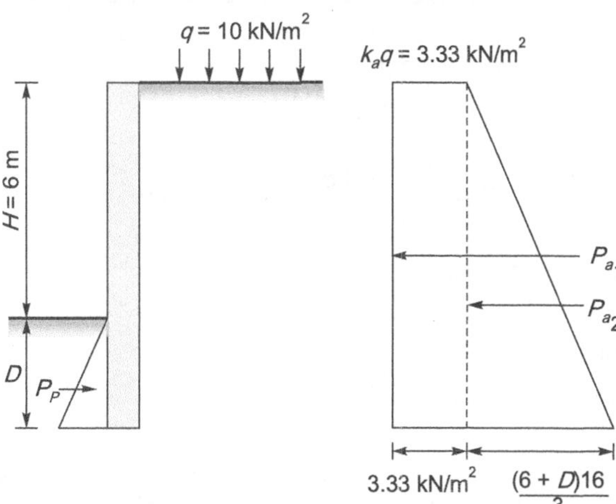

$$k_p = \left(\frac{1+\sin\phi'}{1-\sin\phi'}\right) = 3$$

Let D = depth of embankment

Using approximate analysis,

(Pressure distribution extends upto bottom of wall)

Taking moments about the bottom of wall

$$P_p \times \frac{D}{3} = P_{a_1}\left(\frac{H+D}{2}\right) + P_{a_2}\left(\frac{H+D}{3}\right)$$

$$\Rightarrow \quad k_p\gamma\frac{D^2}{2}\frac{D}{3} = k_a q(H+D)\left(\frac{H+D}{2}\right) + \frac{k_a\gamma}{2}(H+D)^2\left(\frac{H+D}{3}\right)$$

$$\Rightarrow \quad 3\times 16\times\frac{D^3}{6} = \frac{10(6+D)^2}{6} + \frac{16}{3}\frac{(6+D)^3}{6}$$

$$\Rightarrow \quad 48D^3 = 10(6+D)^2 + \frac{16}{3}(6+D)^3$$

On solving $D = 6.106$ m

7.14 **A 5 m high rigid vertical retaining wall has to retain a dry backfill cohesionless soil with the following properties:**
Void ratio (e) = 0.74, specific gravity of soil (G) = 2.68, Poisson's ratio of soil (μ) = 0.4.
The wall needs to be designed for no lateral movement condition of earth pressure. In monsoon, the dry backfill becomes fully submerged with water table at the top of the backfill surface. Estimate the percentage change in the total lateral thrust acting on the wall during monsoon.

[12 marks : 2022]

Solution:

(i) For design of rigid vertical retaining wall (no lateral movement),

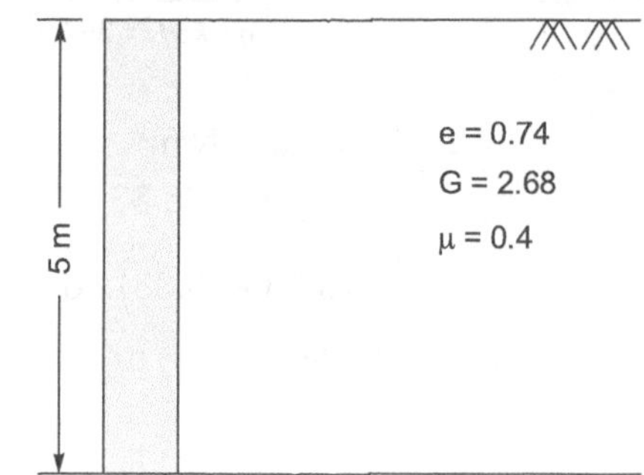

$$k_0 = \frac{\mu}{1-\mu} = \frac{0.4}{1-0.4} = 0.67$$

$$\gamma_d = \frac{G\gamma_w}{1+e} = \frac{2.68\times 9.81}{1+0.74} = 15.10 \text{ kN/m}^3$$

1. $\sigma_v = \gamma z = 15.1 \times z$ (at any depth z)
2. $p_0 = k_0\sigma_v = 0.67 \times 15.1 \times z$
3. for $z = 0$, $p_0 = 0$
   for $z = 5$ m, $p_0 = 50.585$ kN/m^2

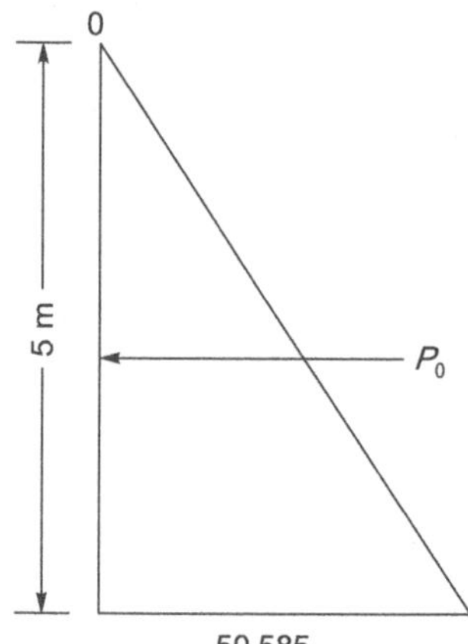

4. Total earth pressure, $P_0 = \frac{1}{2}k_0\cdot\gamma\cdot H^2 = \frac{1}{2}\times 0.67\times 15.1\times 5^2$

$$= 126.46 \text{ kN/m}$$

(ii) When the backfill is submerged:

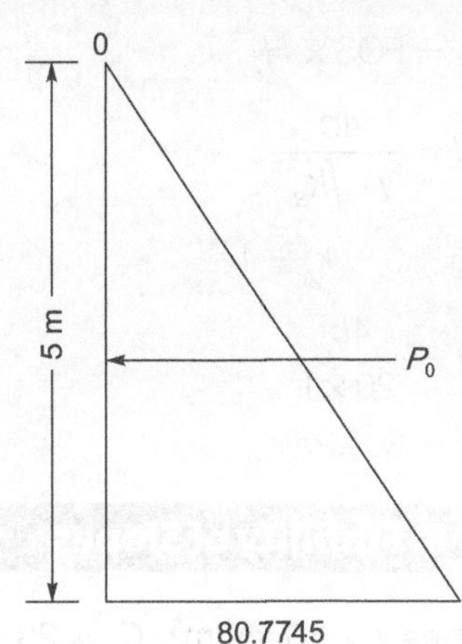

$$\gamma' = \left[\frac{G-1}{1+e}\right]\gamma_w = \frac{2.68-1}{1+0.74}\times 9.81 = 9.47 \text{ kN/m}^3$$

1. $\bar{\sigma}_v = \gamma' xz = 9.47 \times z$, $u = \gamma_w \times z = 9.81z$ [at any depth z]

2. $p_0 = k_0\bar{\sigma}_v + u = 0.67 \times 9.47z + 9.81z$

3. at $z = 0$, $p_0 = 0$
   at $z = 5$ m, $p_0 = 80.7745$ kN/m²

4. Total earth pressure, $P_0 = \frac{1}{2}\times 80.7745 \times 5 = 201.9$ kN/m

$$\text{Percentage change} = \frac{201.9 - 126.46}{126.46}\times 100 = 59.66\%$$

7.15 **A long trench with vertical sides is to be excavated in soft saturated clay deposits ($\phi_u = 0$) to lay a sewage pipeline. If the maximum depth of the trench is 2 m, what should be the approximate undrained cohesion of the clay (C_u) to maintain a minimum safety factor of 3? Assume that the clay has a unit weight of 20 kN/m³ and that the groundwater table is sufficiently below the excavation depth.**

[10 marks : 2023]

Solution:

Given: Factor of safety (FOS) = 3, Maximum depth of trench (H) = 2 m,
Unit weight of clay (γ) = 20 kN/m³

At critical depth, total active earth pressure is zero.

At depth z, earth pressure is zero.

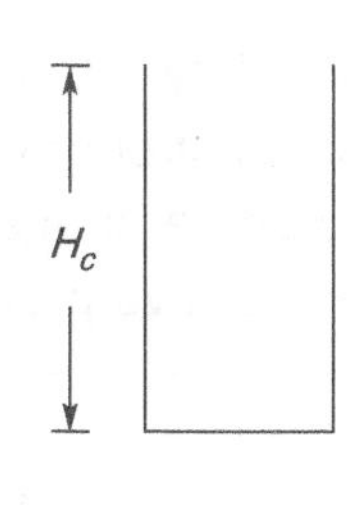

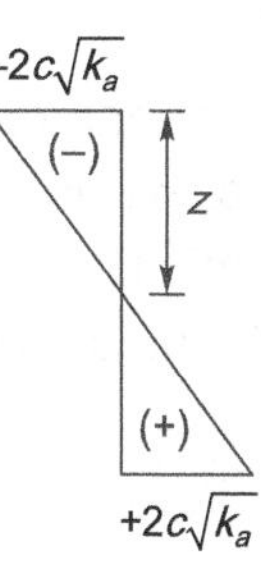

$$\Rightarrow \quad k_a \cdot \gamma \cdot z - 2c\sqrt{k_a} = 0$$

where c is cohesion

$$\Rightarrow \quad z = \frac{2c}{\gamma\sqrt{k_a}}$$

So, Critical depth,

$$H_c = 2z = \frac{4c}{\gamma \cdot \sqrt{k_a}}$$

Since,

$$\text{FOS} = \frac{H_c}{H}$$

$$\Rightarrow \quad H_c = \text{FOS} \times H$$

$$\Rightarrow \quad \text{FOS} \times H = \frac{4c}{\gamma \cdot \sqrt{k_a}}$$

for, $\phi = 0, \quad k_a = 1$

$$\Rightarrow \quad 3 \times 2 = \frac{4c}{20 \times 1}$$

So, Undrained cohesion, $c = 30$ kPa

8. Stability of Slopes

8.1 **A cut is to be made in a soil that has $\gamma = 16$ kN/m³, $C' = 28$ kN/m² and $\phi' = 20°$. The side of the cut slope will make an angle of 45° with the horizontal. What should be the depth of the cut that will have a factor of safety F_s of 3.5?**

[10 marks : 2013]

Solution:

For c-ϕ soils

Shear strength of soil, $\tau_f = c' + \bar{\sigma}\tan\phi'$

$$\tau_f = c' + \left(\gamma z \cos^2\beta\right)\tan\phi'$$

Shear stress at depth, $\tau = \gamma z \cos\beta \times \sin\beta$

Factor of safety, $\text{FOS} = \dfrac{\tau_f}{\tau}$

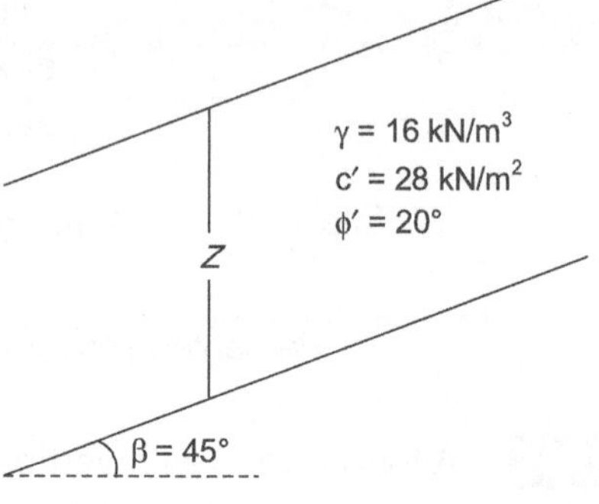

$$\text{FOS} = \frac{c' + \gamma z \cos^2\beta \times \tan\phi'}{\gamma z \cos\beta \times \sin\beta}$$

$$3.5 = \frac{28 + 16 \times Z \cos^2 45° \tan 20°}{16 Z \cos 45° \times \sin 45°}$$

$$\frac{28 + 2.912Z}{8Z} = 3.5$$

$$28 + 2.912\,Z = 28\,Z$$

$$25.088\,Z = 28$$

$$Z = 1.116 \text{ m}$$

Hence the soil can be excavated upto 1.116 m to have a factor of safety of 3.5 against shear failure.

8.2 **Calculate the factors of safety with respect to average shearing strength, cohesion and internal friction of a soil whose shear strength parameters obtained from laboratory tests are $\bar{C} = 40$ KN/m² and $\bar{\phi} = 28°$. The expected parameters of mobilised shearing resistance are $\bar{C}_m = 30$ KN/m² and $\bar{\phi}_m = 20°$. The average effective pressure on the failure plane is 100 kN/m².**

[10 marks : 2017]

Solution:

Given: Shear strength parameters from laboratory tests;

$$\bar{c} = 40 \text{ kN/m}^2 \text{ and } \bar{\phi} = 28°$$

Mobilised shear strength parameters

$$\bar{c}_m = 30 \text{ kN/m}^2 \text{ and } \bar{\phi}_m = 20°$$

Average shearing strength on the failure surface;

$$\tau_f = \bar{c} + \bar{\sigma}_f \tan\bar{\phi} = 40 + 100\tan 28°$$

$$\Rightarrow \qquad \tau_f = 93.171 \text{ kN/m}^2$$

Average value of shearing stress on the failure surface;

$$\tau = \bar{c}_m + \bar{\sigma}_f \tan\bar{\phi}_m = 30 + 100 \tan 20°$$

$$\Rightarrow \qquad \tau = 66.397 \text{ kN/m}^2$$

Thus, factor of safety with respect to shearing strength

$$F_s = \frac{\tau_f}{t} = \frac{93.171}{66.397} = 1.403$$

Factor of safety with respect to cohesion,

$$F_c = \frac{\bar{c}}{\bar{c}_m} = \frac{40}{30} = 1.333$$

Factor of safety with respect to friction,

$$F_\phi = \frac{\tan\bar{\phi}}{\tan\bar{\phi}_m}$$

$$\Rightarrow \qquad F_\phi = \frac{\tan 28°}{\tan 20°} = 1.461$$

8.3 **A long natural slope in overconsolidated fissured clay of saturated unit weight 22 kN/m³, is inclined at 14 degrees to the horizontal. The water table is at the surface and seepage is approximately parallel to the slope. A slip surface has developed on a plane parallel to the surface at a depth of 4 m. Determine the factor of safety along the slip plane using the residual strength parameter, ϕ'_r = 20 degrees.**

[10 marks : 2018]

Solution:

Given data: $\gamma_{sat} = 22 \text{ kN/m}^3$

$i = 14°, \quad z = 4 \text{ m}, \quad \phi = 20°$

Taking cohesion, $c = 0$ for over consolidated fissured clay.

Factor of safety, $$F = \frac{\gamma'}{\gamma_{sat}} \frac{\tan\phi}{\tan i}$$

$$= \left(\frac{22 - 9.81}{22}\right) \times \frac{\tan 20°}{\tan 14°} = 0.809$$

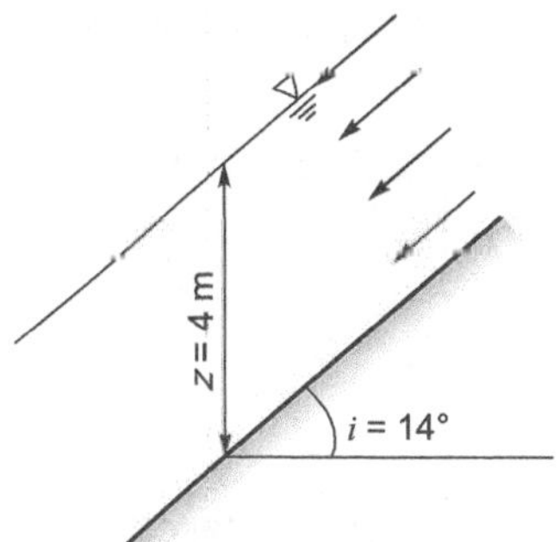

8.4 **An unsupported cut as shown in the figure below was made at a site for which unit weight of soil, γ_s = 18.2 kN/m³, cohesion, C = 25 kN/m² and angle of internal friction, ϕ = 10°. Determine the lateral stress at**

(i) the top of the excavation;

(ii) the bottom of the excavation;

(iii) the maximum depth of potential tension crack for the excavation.

What is the maximum depth up to which the excavation can be carried out safely without any support?

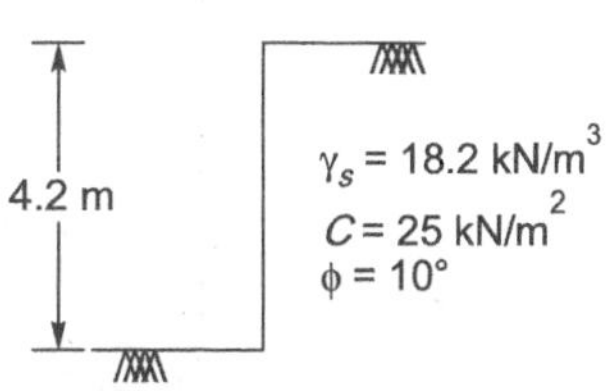

[12 marks : 2019]

Solution:

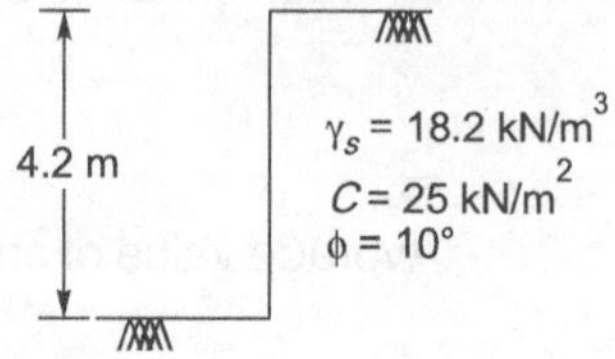

$$K_a = \tan^2\left(45 - \frac{\phi}{2}\right) = \tan^2(45-5) = 0.704$$

4.2 m

$\gamma_s = 18.2$ kN/m^3
$C = 25$ kN/m^2
$\phi = 10°$

(i) At top of the excavation

$$p_a = K_a\gamma_s Z - 2c\sqrt{K_a}$$

$\because$ $Z = 0$ at top

$$p_a = -2\times 25\times\sqrt{0.704} = -41.95 \text{ kN/m}^2$$

(ii) At the bottom of excavation (Z = 4.2 m)

$$p_a = 0.704\times 18.2\times 4.2 - 2\times 25\times\sqrt{0.704} = 11.86 \text{ kN/m}^2$$

(iii) Maximum depth of potential tension crack

$$0.704\times 18.2\times Z - 2\times 25\times\sqrt{0.704} = 0$$

$$\Rightarrow \quad Z = \frac{2\times 25\times\sqrt{0.704}}{0.704\times 18.2} = 3.27 \text{ m}$$

Maximum depth up to which the excavation can be carried out safely without any support $2Z = 6.54$ m.

8.5 **An infinite dry sandy slope is just stable at a slope angle of 35°. Unit weight of sand = 20 kN/m³. In monsoon, water starts flowing through the sand down the slope. At what inclination of slope will it be stable in such condition?**

[12 marks : 2022]

Solution:

Given: Slope angle, $\beta = 35°$

$$\gamma_{sand} = 20 \text{ kN/m}^3$$

In dry infinite slope, for the slope to be just stable,

$$\text{FOS} = 1$$

$$\Rightarrow \quad \frac{\tan\phi}{\tan\beta} = 1$$

$$\Rightarrow \quad \frac{\tan\phi}{\tan 35} = 1$$

$$\therefore \quad \phi = \beta = 35°$$

When flow takes place at ground slope surface,

$$\text{FOS} = \frac{\gamma'}{\gamma_{sat}}\times\frac{\tan\phi}{\tan\beta} \simeq \frac{1}{2}\frac{\tan\phi}{\tan\beta}$$

$$\Rightarrow \quad 1 = \frac{1}{2}\frac{\tan\phi}{\tan\beta}$$

$$\Rightarrow \quad \beta = \tan^{-1}\left[\frac{1}{2}\tan\phi\right] = \tan^{-1}\left[\frac{1}{2}\tan(35)\right]$$

$$\Rightarrow \quad \beta = 19.29°$$

9. Shallow Foundation and Bearing Capacity

9.1 A concrete strip footing rectangular in cross-section is located at ground level and extends 1.2 m below the ground level. It carries UDL of 15000 kg/m. The soil profile consists of homogeneous clay 6 m thick overlying rock. The clay properties are as under:

Saturated unit bulk weight = 1750 kg/m^3

Shear strength (undrained) = 8500 kg/m^2

Compressibility = 1 × 10^{-4} m^2/100 kg

Determine,

(i) width of footing for factor of safety, F = 2

(ii) ultimate consolidation settlement

Assume bulk unit weight of concrete = 2500 kg/m^3

Neglect the spread of load beneath the footing and any side cohesion on the foundation.

[15 marks : 1999]

Solution:

(i) As per **Terzaghi's bearing capacity equation** for strip footing on clayey soil,

$$q_u = 5.7\, c_u + \gamma D_f$$

Now, the net ultimate bearing capacity of the soil is

$$q_{nu} = 5.7\, c_u$$

Net safe bearing capacity,

$$q_{ns} = \frac{q_{nu}}{F}$$

and thus safe bearing capacity,

$$q_s = q_{ns} + \gamma D_f = \frac{q_{nu}}{F} + \gamma D_f$$

Given γ = 1750 kg/m^3; c_u = 8500 kg/m^2; m_v = 1 × 10^{-4} m^2/100 kg

Factor of safety, F = 2

Unit weight of concrete = 2500 kg/m^3

Depth of footing, D_f = 1.2 m

Now UDL = 15000 kg/m

Self weight of concrete = 1.2 × B × 2500 = 3000 B kg/m

Total load = 15000 + 3000 B

Thus, total load carried by the footing should be equal to the safe bearing capacity of soil

$$\therefore \quad 15000 + 3000\, B = \left(\frac{q_{nu}}{F} + \gamma_{sat} D_f\right) \times B$$

$$\Rightarrow \quad 15000 + 3000\, B = \left(\frac{5.7 c_u}{F} + \gamma_{sat} D_f\right) B$$

$$\Rightarrow \quad 15000 + 3000\, B = \left[\frac{5.7 \times 8500}{2} + (1750 \times 1.2)\right] \times B$$

$$\Rightarrow \quad B = 0.643 \text{ m}$$

(ii) Initial effective overburden pressure at the level of C-C,

$$\bar{\sigma}_1 = (1750 - 1000) \quad 3.6 = 2700 \text{ kg/m}^2$$

Final effective overburden pressure at the level C-C,

$$\bar{\sigma}_2 = 1.2 \times 2500 + 2.4 \times 1750 + \frac{15000}{B} - 3.6 \times 1000$$

$$= 3000 + 4200 + \frac{15000}{0.643} - 3600 = 26928.14 \text{ kg/m}^2$$

∴ Change in effective overburden pressure,

$$\Delta\bar{\sigma} = \bar{\sigma}_2 - \bar{\sigma}_1 = 26928.14 - 2700 = 24228.15 \text{ kg/m}^2$$

Ultimate consolidation settlement,

$$\Delta H = m_v H_0 \Delta\bar{\sigma}$$

$$= \frac{1 \times 10^{-4}}{100} \times 4.8 \times 24228.15 = 0.1163 \text{ m} = 9116.3 \text{ mm}$$

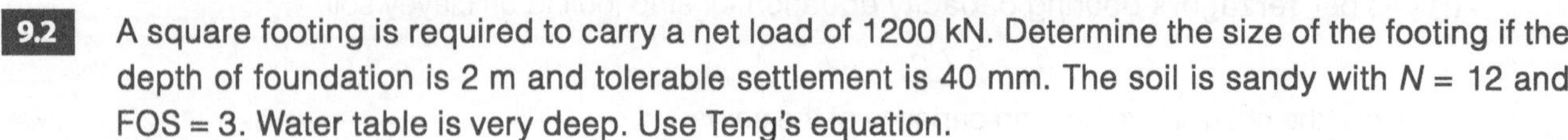

9.2 **A square footing is required to carry a net load of 1200 kN. Determine the size of the footing if the depth of foundation is 2 m and tolerable settlement is 40 mm. The soil is sandy with N = 12 and FOS = 3. Water table is very deep. Use Teng's equation.**

[15 marks : 2000]

Solution:

Given Data : Net load = 1200 kN, Depth of foundation (D_f) = 2 m, Tolerable settlement (S) = 40 mm N = 12; Factor of safety = 3

We know that net safe bearing capacity of soil as per Teng's equation is

$$q_{ns} = 1.4\,(N - 3)\left[\frac{B + 0.3}{2B}\right]^2 S\, C_w C_D$$

where N is corrected standard penetration value

B is width of footing in metres

S is tolerable settlement in mm

C_w is water table correction factor $= \frac{1}{2}\left(1 + \frac{D_{w_2}}{B}\right)$

C_D is depth factor $= \left(1 + \frac{D_f}{B}\right) \le 2$

D_{w_2} is depth of water table below the base of foundation

Since water table is very deep, hence $C_w = 1$

$$C_D = 1 + \frac{D_f}{B} = 1 + \frac{2}{B}$$

∴

$$q_{ns} = 1.4\,(12 - 3)\left[\frac{B + 0.3}{2B}\right]^2 \times 40 \times 1 \times \left(1 + \frac{2}{B}\right)$$

Also

$$\text{Net load} = \frac{q_{ns} \times B^2}{FOS}$$

$$\Rightarrow \quad \frac{1200 \times FOS}{B^2} = q_{ns}$$

$$\Rightarrow \quad \frac{1200 \times 3}{B^2} = 1.4\,(12-3) \times \left[\frac{B+0.3}{2B}\right]^2 \times 40 \times 1 \times \left(1+\frac{2}{B}\right)$$

$$\Rightarrow \quad 3600 = 126 \times (B+0.3)^2 \times \left(1+\frac{2}{B}\right)$$

$$\Rightarrow \quad 28.57 = (B^2 + 0.09 + 0.6B)\left(1+\frac{2}{B}\right)$$

$\Rightarrow$ $(B^2 + 0.6B + 0.09)\,(B + 2) - 28.57B = 0$

$\Rightarrow$ $B^3 + 0.6\,B^2 + 0.09\,B + 2B^2 + 1.2\,B + 0.18 - 28.57\,B = 0$

$\Rightarrow$ $B^3 + 2.6\,B^2 - 27.28\,B + 0.18 = 0$

$\Rightarrow$ $B = 4.08$ m

Thus the size of square footing is 4.08 m × 4.08 m

9.3 **A square footing located at a depth 1.5 m from the ground surface carries a column load of 150 kN. The soil is submerged having an effective unit weight of 11 kN/m^3 and an angle of shearing resistance of 30°. Find the size of footing using Terzhagi's theory if $F_s = 3$; for $\phi = 30°$, $N_q = 10$; $N_\gamma = 6.0$.**

[15 marks : 2000]

Solution:

Given Data: Column load (Q) = 150 kN, Depth of footing (D_f) = 1.5 m, Submerged unit weight (γ') = 11 kN/m^3, Angle of shearing resistance (ϕ) = 30°, Factor of safety (FOS) = 3

$$N_q = 10; \; N_\gamma = 6.0$$

We know that ultimate bearing capacity for square footing is given by

$$q_u = 1.3\,cN_c + \gamma' D_f N_q + 0.4\gamma' B N_\gamma$$

Since the value of c is not given, hence assuming the soil to be sandy.

Thus $\quad c = 0$

$\therefore \quad q_u = \gamma' D_f N_q + 0.4\gamma' B N_\gamma$

$\Rightarrow \quad q_u = 11 \times 1.5 \times 10 + 0.4 \times 11 \times B \times 6$

$\Rightarrow \quad q_u = 165 + 26.4B$

$\therefore \quad q_{nu} = q_u - \gamma' D_f = 165 + 26.4B - 11 \times 1.5$

$\Rightarrow \quad q_{nu} = 148.5 + 26.4B$

and $\quad q_{ns} = \dfrac{q_{nu}}{FOS} = \dfrac{148.5 + 26.4B}{3}$

$$q_{ns} \geq \frac{Q}{B^2}$$

(Assuming the column load to be net external load and equating to new safe bearing capacity)

$$\Rightarrow \quad \frac{148.5 + 26.4B}{3} \geq \frac{150}{B^2}$$

$\Rightarrow \quad 26.4\,B^3 + 148.5\,B^2 - 450 = 0$

$\Rightarrow \quad B = 1.54$ m

Here $D_f > B$ which is against the assumption of Terzaghi's theory. In strict sense, Terzaghi's theory is not applicable here.

9.4 A footing 2 m × 2 m in plan is founded at 1m depth below *GL* in sand having angle of shearing resistance of 36°. Compute the allowable load at the base of the footing if (i) factor of safety against shear is 3 and (ii) the maximum settlement of footing is not to exceed 50 mm. The water table is 1 m below GL. The allowable soil pressure for 50 mm settlement with water table at very great depth is 50 t/m². Saturated and dry unit weights of sand are 2 t/m³ and1.6 t/m³.
For $\phi = 36°$, $N_c = 50$; $N_q = 42$ and $N_\gamma = 46$
If the load on the footing is eccentric with $e_x = e_y = 0.25$ m, what will be the allowable load? Take soil above water table as dry.

[15 marks : 2002]

Solution:

(i) The ultimate bearing capacity for a square footing is given by

$$q_u = 1.3cN_c + \gamma D_f N_q + 0.4\gamma B N_\gamma$$

Given soil is sand, hence $c = 0$

$$q_u = \gamma D_f N_q + 0.4\gamma B N_\gamma$$

Since the water table is at the base of footing, therefore unit weight of sand above water table is taken as dry unit weight i.e. 1.6 t/m³ in the first term and unit weight of sand for second term is taken as submerged unit weight i.e. $\gamma_{sat} - \gamma_w = 2 - 1 = 1$ t/m³

$$\therefore \quad q_u = 1.6 \times 1 \times 42 + 0.4 \times 1 \times 2 \times 46 = 104 \text{ t/m}^2$$

$$q_{nu} = q_u - \gamma D_f = 104 - 1.6 \times 1 = 102.4 \text{ t/m}^2$$

$$q_{ns} = \frac{q_{nu}}{FOS} = \frac{102.4}{3} = 34.13 \text{ t/m}^2$$

$$q_s = q_{ns} + \gamma D_f = 34.13 + 1.6 \times 1 = 35.73 \text{ t/m}^2$$

∴ Allowable load, $Q_a = q_s \times B^2 = 35.73 \times 2^2 = 142.92 \text{ t} = 1402.04 \text{ kN}$

(ii) **The allowable soil pressure for 50 mm settlement is 50 t/m²**

$$\therefore \quad q_s = 50 \text{ t/m}^2$$

∴ Allowable load, $Q_a = q_s \times B^2 = 50 \times 2^2 = 200 \text{ t} = 1962 \text{ kN}$

If the load on the footing is eccentric with $e_x = e_y = 0.25$ m, then

$$q_{max} = \frac{Q}{A}\left(1 + \frac{6e_x}{L} + \frac{6e_y}{B}\right)$$

$$q_{min} = \frac{Q}{A}\left(1 - \frac{6e_x}{L} - \frac{6e_y}{B}\right)$$

Where q_{max} and q_{min} are maximum and minimum pressures; *A* is area of footing; *L* is length of footing; *B* is width of footing

In case of square footing, $A = B^2$, and $L = B$

and $q_{max} = q_s = 35.73$ t/m²

$$\therefore \quad 35.73 = \frac{Q}{2 \times 2}\left(1 + \frac{6 \times 0.25}{2} + \frac{6 \times 0.25}{2}\right)$$

$$Q = 57.17 \text{ t} = 560.84 \text{ kN}$$

9.5 Find the settlement of a footing of width 1.2 m with the following plate load test results for (i) Sandy soil, (ii) Clayey soil
Settlement of the plate 15 mm, width of the plate 400 mm.

[10 marks : 2003]

Solution:

Given data: Width of footing (B_f) = 1.2 m, Width of plate (B_p) = 400 mm = 0.4 m, Settlement of plate (S_p) = 15 mm, Settlement of footing (S_f)= ?

(i) For sandy soil, $S_f = S_p \times \left[\frac{B_f(B_p+0.3)}{B_p(B_f+0.3)}\right]^2 = 15\times\left[\frac{1.2(0.4+0.3)}{0.4\,(1.2+0.3)}\right]^2 = 29.4$ mm

(ii) For clayey soil, $S_f = S_p \times \frac{B_f}{B_p} = 15 \times \frac{1.2}{0.4} = 45$ mm

9.6 When a vertical face excavation was made in a deposit of clay, it failed at a depth of 2.8 m of excavation. Find the shear strength parameters of the soil if its bulk density is 17 kN/m³.
In this deposit, at some other location, a plate load test was conducted with 30 cm square plate, placed at a depth of 1 m below the ground level. The ultimate load was 13.5 kN. Water table was at a depth of 4 m below the ground level. Calculate the net safe bearing capacity for a 1.5 m wide strip footing to be founded at a depth of 1.5 m in this soil. Take factor of safety as 3. Use Terzaghi's bearing capacity theory.

[13 marks : 2004]

Solution:

Given Data: For clay (ϕ) = 0°, Bulk density (γ) = 17 kN/m³, Critical Height (H_c) = 2.8 m

The critical height of an unsupported cut in cohesive soil is given by

$$H_c = \frac{4c}{\gamma}\tan\left(45° + \frac{\phi}{2}\right) \text{ where c is cohesion in soil}$$

$$\Rightarrow \quad 2.8 = \frac{4c}{17}\tan(45° + 0°)$$

$$\Rightarrow \quad c = 11.9 \text{ kN/m}^2$$

Depth of water table below the test level = 4 – 1 = 3 m

Since this depth is much more than the width of footing, there will be no influence of water table.

For plate load test on square plate

$$\text{Size of plate} = 30 \text{ cm} \times 30 \text{ cm}$$

$$\text{Load on plate} = 13.5 \text{ kN}$$

$$\text{Ultimate bearing capacity of soil under plate} = \frac{13.5}{0.3\times0.3} = 150 \text{ kN/m}^2$$

As per Terzaghi's bearing capacity equation

$$q_u = 1.3\,cN_c + \gamma D_f N_q + 0.4\,\gamma B N_\gamma$$

For clay, $\phi = 0°$, $N_c = 5.7$, $N_q = 1$ and $N_\gamma = 0$

$$\therefore \quad q_u = 1.3 \times 5.7c + \gamma D_f$$

$$\Rightarrow \quad q_u = 7.41c + \gamma D_f$$

$$\Rightarrow \quad 150 = 7.41c + 17 \times 1$$

$$\Rightarrow \quad c = \frac{150-17}{7.41} = 17.95 \text{ kN/m}^2$$

Hence the lesser of the two values of cohesion is used i.e. c = 11.9 kN/m²

For strip footing

$$q_u = cN_c + \gamma D_f N_q + 0.5\gamma B N_\gamma$$

For clay, $\phi = 0°$, $N_c = 5.7$, $N_q = 1$, $N_\gamma = 0$

$\therefore$ $q_u = 5.7c + \gamma D_f = 5.7 \times 11.9 + 17 \times 1.5 = 93.33 \text{ kN/m}^2$

$\therefore$ Net ultimate bearing capacity, $q_{nu} = q_u - \gamma D_f = 93.33 - 17 \times 1.5 = 67.83 \text{ kN/m}^2$

$\therefore$ Net safe bearing capacity, $q_{ns} = \dfrac{q_{nu}}{\text{Factor of safety}} = \dfrac{67.83}{3} = 22.61 \text{ kN/m}^2$

9.7 **State difference between plastic deformation and elastic deformation. Give examples.**

[4 marks : 2005]

Solution:

Elastic deformation: At such a stage, large deformation is possible before absolute failure or rupture takes place.

Plastic deformation: At such a stage, the deflection increases very fast at constant load. The collapse of a structure can be partial, complete or over complete.

9.8 **A strip foundation is founded at a depth of 1.4m below ground level. Water table in the area is at the ground level and the soil is cohesionless. The footing is to be designed to carry a net safe load of 460 kN/m² with a factor of safety as 3. Find the width of the footing under Terzaghi's general failure criteria. Assume saturated unit weight γ_{sat} as 21.0 kN/m³ and angle of internal friction of soil as 36°. For ϕ = 36°, take N_q = 43.2, N_γ = 42.6.**

[8 marks : 2007]

Solution:

Given data:

Depth of footing, $D_f = 1.4$ m

Net safe bearing pressure, $q_{na} = 460 \text{ kN/m}^2$

Factor of safety, $F = 3$

Saturated unit weight, $\gamma_{sat} = 21.0 \text{ kN/m}^3$

Angle of internal friction, $\phi = 36°$

Bearing capacity factor, $N_q = 43.2$ and $N_\gamma = 42.6$

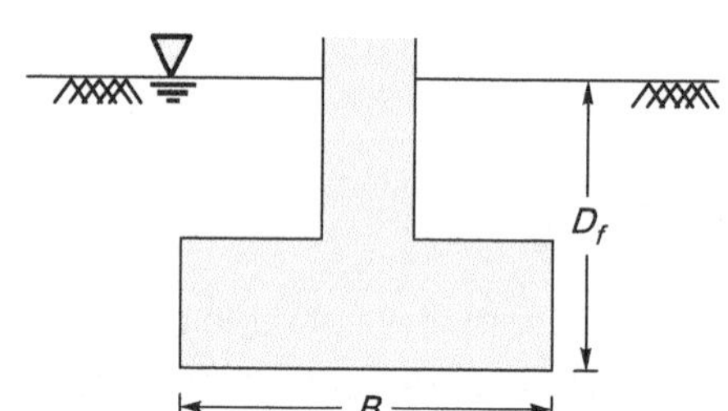

Let the width of strip footing be B.

The ultimate bearing capacity (q_u) in cohesionless soil for a strip footing under Terzaghi's general shear failure criteria is given by

$$q_u = \gamma D_f N_q + \frac{1}{2}\gamma B N_\gamma$$

But

$$\frac{q_u - \gamma D_f}{F} = q_{ns}$$

$$\therefore \quad q_{ns} \times F = \gamma D_f (N_q - 1) + \frac{1}{2}\gamma B N_\gamma$$

But the water table is at the ground level, therefore the unit weight of soil is taken as

$$\gamma_{sub} = \gamma_{sat} - \gamma_w = 21.0 - 10 = 11 \text{ kN/m}^3$$

$$\therefore \quad 460 \times 3 = 11 \times 1.4 \times (43.2 - 1) + \frac{1}{2} \times 11 \times B \times 42.6$$

$$\Rightarrow \quad B = \frac{1380 - 649.88}{234.3} = 3.12 \text{ m}$$

Thus, the width of the strip footing is 3.12 m.

9.9 With the help of neat sketches, explain different types of shallow foundations.

[10 marks : 2008]

Solution:

The different types of shallow foundations are explained below:

(i) Spread footing:

- It is basically a pad used 'spread out loads from walls or columns over a sufficiently large area of foundation soil.
- Spread footing required to support a wall is known as a continuous, wall or strip footing while that required to support a column is known as an individual or isolated footing.
- Spread footings are generally square, rectangular or circular in section.

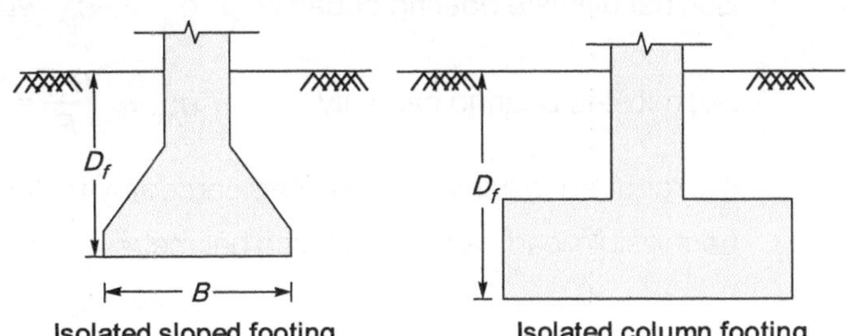

Isolated sloped footing Isolated column footing

(ii) Combined footing:

- A combined footing supports two or more columns in a row when the areas required for individual footings are such that they come very near each other.
- They are also preferred in situations of limited space on one side owing to the existence of the boundary line of private property.
- The plan shape of the footing may be rectangular or trapezoidal; the footing will then be called 'rectangular combined footing' or 'trapezoidal combined footing, as the case may be.

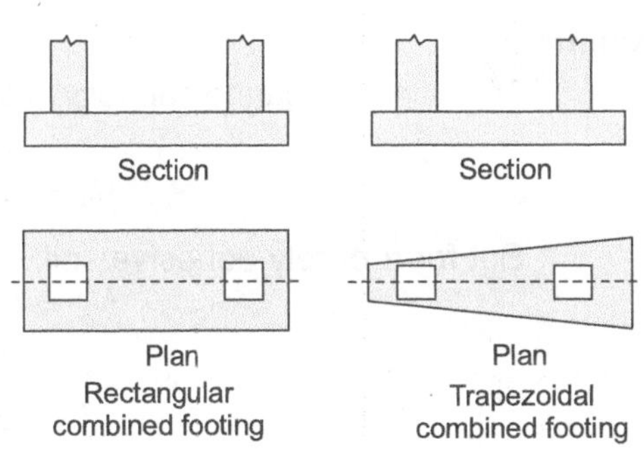

Rectangular combined footing Trapezoidal combined footing

(iii) Raft or mat foundations:

- These foundations covers the entire area of a structure, transmitting the entire structural load or load from several columns.
- These are used for a soil whose bearing capacity is low.
- It is adopted when individual column footings would tend to be too close or tend to overlap.

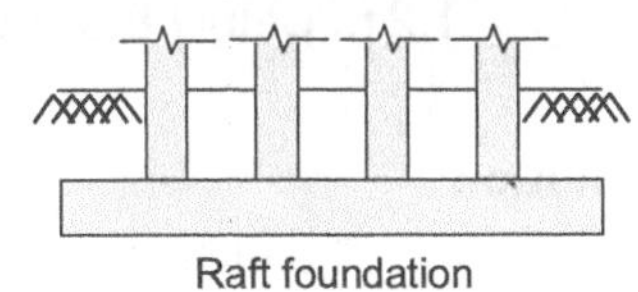
Raft foundation

(iv) Strap Foundations:

- It comprises two or more footings connected by a beam called strap.
- It is also called a 'cantilever footing' or 'pump-handle foundation'.
- It may be required when the footing of an exterior column cannot extend into an adjoining private property.

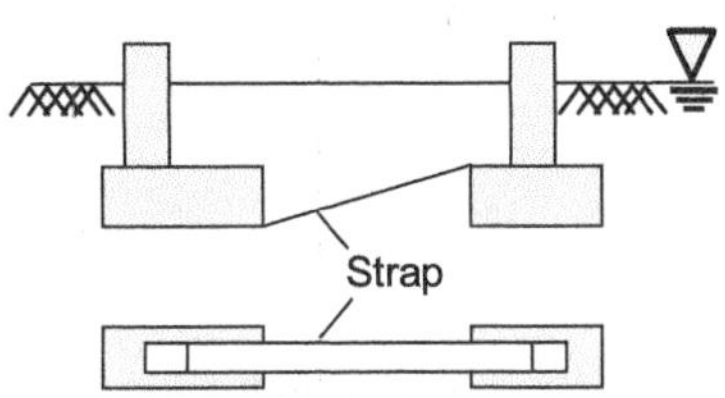

9.10 A 2 m wide strip footing is located at a depth of 2 m in a stiff clay of saturated unit weight of 20 kN/m³ and having $\phi_u = 0°$ and $c_u = 120$ kN/m². Using Terzaghi's bearing capacity equation, compute the safe load carried by footing per metre length with factor of safety $F = 3$ with respect to shear failure. Given : $N_c = 5.7$.

[10 marks : 2009]

Solution:

Given data: Width of strip footing (B) = 2 m, Depth of strip footing (D_f) = 2 m, Unit weight of clay (γ) = 20 kN/m^3, Factor of safety (F) = 3

$$N_c = 5.7;\ \phi_u = 0;\ c_u = 120\ \text{kN/m}^2$$

As per **Terzaghi's bearing capacity equation** for strip footing the ultimate bearing capacity of stiff clay is given by

$$q_u = cN_c + \gamma D_f$$

But, net ultimate bearing capacity, $q_{nu} = q_u - \gamma D_f = cN_c + \gamma D_f - \gamma D_f = cN_c$

But net safe bearing capacity, $q_{ns} = \dfrac{q_{nu}}{F} = \dfrac{cN_c}{F} = \dfrac{120 \times 5.7}{3} = 228\ \text{kN/m}^2$

It is required to calculate safe load carried by footing hence,

permissible load carried by footing per metre length

$$= q_{ns} \times B = 228 \times 2 = 456\ \text{kN/m length of footing.}$$

9.11 **Using Terzaghi theory find the ultimate bearing capacity for a square footing of 2.0 m × 2.0 m placed at depth of 1.2 m below the ground on a pure cohesive soil having density 18 kN/m^3. N_c = 5.7. Use local shear failure conditions. c = 40 kN/m^2.**

[2 marks : 2010]

Solution:

The ultimate bearing capacity of a square footing in local shear as per Terzaghi theory is given by

$$q_u = \frac{2}{3} \times 1.3\, cN_c + \gamma D_f N_q + 0.4 \gamma B N_\gamma$$

But for a purely cohesive soil,

$$N_c = 5.7; N_q = 1 \text{ and } N_\gamma = 0$$

$$\therefore \quad q_u = \frac{2}{3} \times 1.3\, cN_c + \gamma D_f = \frac{2}{3} \times 1.3 \times 40 \times 5.7 + 18 \times 1.2 = 219.2\ \text{kN/m}^2$$

9.12 **What will be the gross and net safe bearing capacity of sand having ϕ = 30° and density 2.1 t/m^3 below (a) 1.0 m wide strip footing (b) 1.0 m × 1.0 m square footing placed at a depth of 1.2 m below the ground. Take factor of safety as 2.5. Take N_c = 30.14, N_q = 18.4, N_γ = 22.4.**

[8 marks : 2010]

Solution:

(a) The ultimate bearing capacity of sand below **strip footing** as per **Terzaghi's equation** is given as

$$q_u = cN_c + \gamma D_f N_q + 0.5 \gamma B N_\gamma$$

For sands $c = 0$

$$\therefore \quad q_u = \gamma D_f N_q + 0.5 \gamma B N_\gamma$$

$$= 2.1 \times 1.2 \times 18.4 + 0.5 \times 2.1 \times 1 \times 22.4$$

$$= 69.888\ \text{t/m}^2 = 69.888 \times 9.81\ \text{kN/m}^2 = 685.60\ \text{kN/m}^2$$

Net safe bearing capacity, $q_{ns} = \dfrac{q_u - \gamma D_f}{F} = \dfrac{685.60 - 2.1 \times 9.81 \times 1.2}{2.5} = 264.35\ \text{kN/m}^2$

Gross safe bearing capacity, $q_s = q_{ns} + \gamma D_f = 264.35 + 2.1 \times 9.81 \times 1.2 = 289.07\ \text{kN/m}^2$

(b) The ultimate bearing capacity of sand below **Square footing** as per **Terzaghi's equation** is given as

$$q_u = 1.3\, cN_c + \gamma D_f N_q + 0.4 \gamma B N_\gamma$$

For sands $c = 0$

$$\therefore \quad q_u = \gamma D_f N_q + 0.4\gamma B N_\gamma$$

$$= 2.1 \times 1.2 \times 18.4 + 0.4 \times 2.1 \times 1 \times 22.4$$

$$= 65.184 \text{ t/m}^2 = 65.184 \times 9.81 \text{ kN/m}^2 = 639.46 \text{ kN/m}^2$$

Net safe bearing capacity, $q_{ns} = \dfrac{q_u - \gamma D_f}{F} = \dfrac{639.46 - 2.1 \times 9.81 \times 1.2}{2.5} = 245.89 \text{ kN/m}^2$

Gross safe bearing capacity, $q_s = q_{ns} + \gamma D_f = 245.89 + 2.1 \times 9.81 \times 1.2 = 270.61 \text{ kN/m}^2$

9.13 **The subsoil at a building site consists of medium sand with γ = 18 kN/m³, c' = 0, ϕ' = 32° and water table at the ground surface. A 2.5 m square footing is to be placed at 1.5 m below the ground surface. Compute the safe bearing capacity of the footing.**
What will be the safe bearing capacity, if the water table goes down to 3 m below the ground surface?
[For ϕ = 32°, N_q = 20.3 and N_γ = 19.7]

[10 marks : 2011]

Solution:

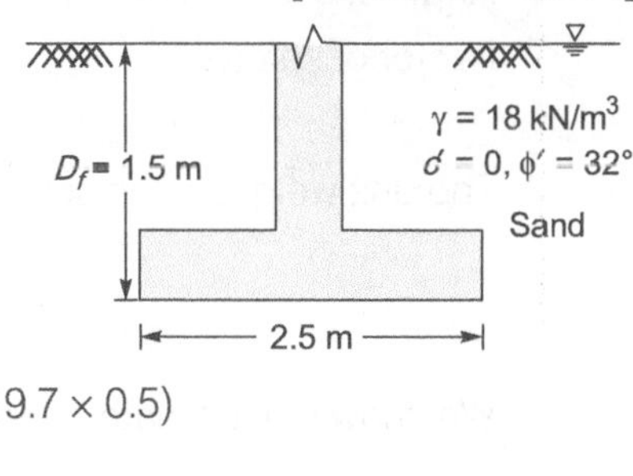

For square footing in a cohesionless soil

$$q_{ult} = \gamma D_f N_q R_{w1} + 0.4 B\gamma N_\gamma \cdot R_{w2}$$

For water table at ground surface,

$$R_{w1} = R_{w2} = 0.5$$

$$\therefore \quad q_{ult} = (18 \times 1.5 \times 20.3 \times 0.5) + (0.4 \times 2.5 \times 18 \times 19.7 \times 0.5)$$

$$= 451.35 \text{ kN/m}^2$$

$$q_{netult} = q_{ult} - \gamma d_f$$

$$= 451.35 - (18 - 10) \times 1.5 = 439.35 \text{ kN/m}^2$$

Considering factor of safety

$$= \text{FOS} = 3$$

$$q_{netsafe} = \frac{q_{netult}}{\text{FOS}} = 146.45 \text{ kN/m}^2$$

$$q_{safe} = q_{netsafe} + \gamma'.D_f = 146.45 + (18 - 10) \times 1.5 = 158.45 \text{ kN/m}^2$$

Safe bearing capacity when water table goes down to 3 m below the ground surface.

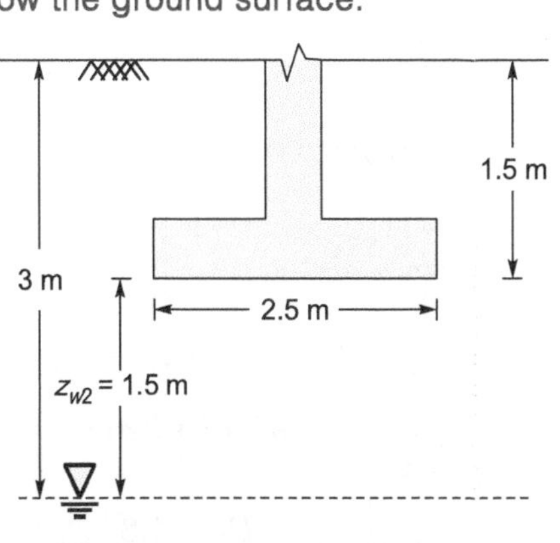

$$q_u = \gamma D_f N_q \cdot R_{w1} + 0.4 B\gamma \cdot N_\gamma R_{w2}$$

$$R_{w1} = 1$$

$$R_{w2} = \frac{1}{2}\left(1 + \frac{z_{w2}}{B}\right) = \frac{1}{2}\left(1 + \frac{1.5}{2.5}\right) = 0.8$$

$$\therefore \quad q_u = (18 \times 1.5 \times 20.3 \times 1) + (0.4 \times 2.5 \times 18 \times 19.7 \times 0.8)$$

$$= 831.78 \text{ kN/m}^2$$

$$q_{netult} = 831.78 - (18) \times 1.5 = 804.78 \text{ kN/m}^2$$

$$q_{netsafe} = \frac{804.78}{3} = 268.26 \text{ kN/m}^2$$

$$q_{safe} = 268.26 + (18) \times 1.5 = 295.26 \text{ kN/m}^2$$

9.14 **A square footing located at a depth of 1.3 m below the ground surface has to carry a safe load of 800 kN. Find the size of the footing if the desired factor of safety is 3. The soil has the following properties:**
Void ratio, $e = 0.55$
Degree of saturation, $S_r = 50\%$
Specific gravity = 2.67
$C = 8$ kN/m²
For $\phi = 30° - N_c = 37.2$, $N_q = 22.5$, $N_\gamma = 19.7$
(Bearing capacity factors)

[10 marks : 2012]

Solution:

Given data Safe column load (Q) = 800 kN, Depth of footing (D_f) = 1.3 m, Void ratio (e) = 0.55, Degree of saturation (S_r) = 50%, Specific gravity (G_s) = 2.67

$$c = 8 \text{ kN/m}^2$$

Angle of shearing resistance, $\phi = 30°$

Factor of safety, FOS = 3

$$N_c = 37.2;\ N_q = 22.5;\ N_\gamma = 19.7$$

The unit weight of the soil is given by

$$\gamma = \left(\frac{G_s + Se}{1+e}\right)\gamma_w = \left(\frac{2.67 + 0.5 \times 0.55}{1+0.55}\right) \times 9.81 = 18.639 \text{ kN/m}^3$$

We know that ultimate bearing capacity of square footing as given by Terzaghi is

$$q_u = 1.3\, cN_c + \gamma D_f N_q + 0.4\, \gamma B N_\gamma$$

$$\Rightarrow \quad q_u = 1.3 \times 8 \times 37.2 + 18.639 \times 1.3 \times 22.5 + 0.4 \times 18.639 \times B \times 19.7$$

$$\Rightarrow \quad q_u = 932.07 + 146.875\, B$$

$$\therefore \quad q_{nu} = q_u - \gamma D_f$$

$$\Rightarrow \quad q_{nu} = 932.07 + 146.875B - 18.639 \times 1.3 = 907.84 + 146.875B$$

and

$$q_{ns} = \frac{q_{nu}}{FOS} = \frac{907.84 + 146.875B}{3}$$

Thus,

$$q_s = q_{ns} + \gamma D_f$$

$$\Rightarrow \quad q_s = \frac{907.84 + 146.875B}{3} + 18.639 \times 1.3 = 326.84 + 48.958B$$

Since q_s should be greater than $\frac{Q}{B^2}$

$$\therefore \quad q_{safe} \geq \frac{Q}{B^2}$$

$$\therefore \quad 326.84 + 48.958\,B \geq \frac{800}{B^2}$$

$$\Rightarrow \quad 48.958B^3 + 326.84B^2 - 800 \geq 0$$

$$\Rightarrow \quad B = 1.42 \text{ m}$$

Note: Here $\frac{D_f}{B} = \frac{1.3}{1.42} < 1$

Hence, Terzaghi's theory is applicable here.

9.15 **A footing, 2 m square, rests on a soft clay with its base at a depth of 1.5 m from ground surface. The clay stratum is 3.5 m thick and is underlain by a firm sand stratum. The clay soil has following properties:**

$$W_L = 30\%, W_n = 40\%, G_s = 2.70$$
$$\phi_u = 0°, C_u = 0.5 \text{ kg/cm}^2$$

It is known that the clay stratum is normally consolidated. Using Skempton's equation determine the net safe bearing capacity of footing. Natural water table is close to the ground surface.

[10 marks : 2012]

Solution:

Size of square footing $= B \times B = 2 \text{ m} \times 2 \text{ m}$

$$C_u = 0.5 \text{ kg/cm}^2 = \frac{0.5 \times 9.81}{1000 \times 10^{-4}} \text{kN/m}^2 = 49.05 \text{ kN/m}^2$$

Depth of square footing, $D_f = 1.5$ m

We know that for clay, $\phi_u = 0$

and for such a condition, the net ultimate bearing capacity given by skempton is

$$q_{nu} = C_u N_c$$

where N_c is bearing capacity factor.

As per **Skempton's recommendations**, the value of N_c is given by

$$N_c = 6\left(1 + 0.2\frac{D_f}{B}\right) = 6\left(1 + \frac{0.2 \times 1.5}{2}\right) = 6.9\ (< 9)\ (\text{Ok})$$

$$\therefore \quad q_{nu} = 49.05 \times 6.9 = 338.445 \text{ kN/m}^2$$

Now, **net safe bearing capacity of footing,**

$$q_{ns} = \frac{q_{nu}}{FOS}$$

Assuming a FOS of 3 we have, $q_{ns} = \frac{338.445}{3} = 112.815 \text{ kN/m}^2$

9.16 **A circular concrete pier of 3 m diameter carries a gross load of 3500 kN. The supporting soil is a clayey sand having the following properties:**

$c = 5$ kN/m². $\phi = 30°$ and $\gamma = 18.5$ kN/m³. Find the depth at which the pier is to be located such that a factor of safety of 3.0 is assured. The bearing capacity factor for $\phi = 30°$ are $N_c = 30.1$, $N_q = 18.4$ and $N_\gamma = 22.4$.

[4 marks : 2013]

Solution:

$$q_{\text{gross press}} = \frac{3500}{\frac{\pi \times 3^2}{4}} \frac{\text{kN}}{\text{m}^2} = 495.15 \text{ kN/m}^2$$

$$q_{\text{net press}} = q_{\text{gross press}} - \gamma D_f = 495.15 - 18.5\, D_f \text{ kN/m}^2$$

$$q_{\text{net press}} \le q_{ns} = \frac{q_{nu}}{FOS}$$

$$(495.15 - 18.5\, D_f) \ge \frac{q_{nu}}{FOS = 3} \rightarrow \text{For circular footing}$$

$$3[495.15 - 18.5\, D_f] \le q_{nu} = 1.3\, cN_c + q(N_q - 1) + 0.3\, B\gamma N_\gamma$$

$$3[495.15 - 18.5\,D_f] \le 1.3(5)\,(30.1) + (18.5\,D_f)\,(18.4 - 1) + (0.3)(3)(18.5)(22.4)$$
$$3[495.15 - 18.5\,D_f] \le (195.65) + (321.9\,D_f) + 372.96$$

On solving above equation

$$D_f \ge 2.43 \text{ m which is less then size of footing i.e., 3 m}$$

9.17 **A wall footing carrying a load of 152 kN/meter length of wall is to be constructed at a depth of 1.2 m. Subsoil consists of a uniform deposit stiff clay with unit weight, γ = 18.8 kN/m^3 and unconfined compressive strength, q_u = 150 kN/m^2. Determine the width of the footing using a factor of safety of 3.**

[8 marks : 2014]

Solution:

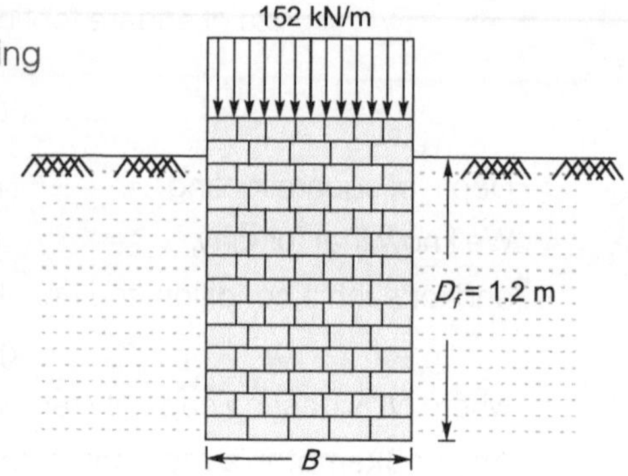

As per Terzaghi's bearing capacity: Theory the ultimate bearing capacity for strip footing on clayey may be given as

$$q_u = 5.7\,C_u + \gamma D_f$$

152 kN/m

D_f = 1.2 m

B

Since unconfined compressive strength,

$$q_u = 150 \text{ kN/m}^2$$

$$\therefore \quad C_u = \frac{q_u}{2} = 75 \text{ kN/m}^2$$

and

$$\gamma = 18.8 \text{ kN/m}^3$$

The net ultimate bearing capacity of soil

$$q_{nu} = q_u - \gamma D_f = 5.7\,C_u + \gamma D_f - \gamma D_f$$

$$\therefore \quad q_{nu} = 5.7\,C_u$$

If factor of safety is F, then net safe bearing capacity

$$q_{ns} = \frac{q_{nu}}{F} = \frac{5.7\,C_u}{F}$$

$$\text{Total weight of footing} = q_{ns} \times B$$

$$152 \text{ kN/m} = \frac{5.7 C_u}{\gamma} \times B$$

$$152 = \frac{5.7 \times 75}{3} \times B$$

$$B = \frac{152}{142.5} = 10.67 \text{ m}$$

9.18 **For a general c-ϕ soil, the cohesion c is 50 kPa. The total unit weight is 20 kN/m^3 and the bearing capacity factors are N_c = 8, N_q = 3, N_γ = 2. Using Terzaghi's formula, calculate the net ultimate bearing capacity for a strip footing of width 2 m at a depth of 1 m. Considering shear failure only, estimate the safe total load on a 10 m long, 2 m wide strip footing using a factor of safety of 3.**

[8 marks : 2014]

Solution:

Given: Cohesion (C) = 50 kN/m^2, γ=20 kN/m^2, Width of footing (B) = 2 m, D_f = 1 m

Bearing capacity factors are N_c = 8, N_q = 3 and N_γ = 2.

Assuming the water table is very deep

As per Terzarghi's theory, ultimate bearing capacity in C - ϕ soil is

$$q_u = CN_C + \gamma D_f N_q + \frac{1}{2}\gamma B N_r$$

$$= 50 \times 8 + 20 \times 1 \times 3 + \frac{1}{2} \times 20 \times 2 \times 2 = 500 \text{ kN/m}^2$$

Net ultimate bearing capacity, $q_{nu} = q_u - \gamma D_f = 500 - 20 \times 1 = 480 \text{ kN/m}^2$

If factor of safety, $F = 3$, then Net safe bearing capacity is given by

$$q_{ns} = \frac{q_{nu}}{F} = \frac{480}{3} = 160 \text{kN/m}^2$$

Gross safe bearing capacity, $q_s = q_{ns} + \gamma D_f = 160 + 20 \times 1 = 180 \text{ kN/m}^2$

Safe load on 10m long and 2m wide strip footing is

$$= q_s \times B \times L = 180 \times 2 \times 10 = 3600 \text{ kN}$$

9.19 **A wall footing carrying a load of 150 kN/m length of the wall is to be constructed at a depth 1.2 m below the ground surface. Subsoil consists of uniform deposit of stiff clay with unit weight, γ = 18.8 kN/m^3 and unconfined compressive strength = 160 kN/m^2. Determine the width of the footing using Terzaghi's theory. Use a factor of safety against bearing capacity failure = 3.0.**

[8 marks : 2015]

Solution:

Given: Load on wall = 150 kN/m length of wall, Depth of footing (D_f) = 1.2 m, Unit weight (γ) = 18.8 kN/m^2

Unconfined compression strength,

$$q_u = 160 \text{ kN/m}^2$$

Cohesion, $c_u = \frac{q_u}{2} = \frac{160}{2} = 80 \text{kN/m}^2$

Factor of safety, $F = 3.0$

Ultimate bearing capacity of clay according to Terzaghi's theory ona wall footing (strip footing)

$$q_u = cN_c + qN_q$$

Net safe bearing capacity, $q_{ns} = \frac{cN_c}{F} = \frac{80 \times 5.7}{3} = 152 \text{ kN/m}^2$

Let width of footing = B m

External pressure intensity at the base of footing = $\frac{150}{B} \text{kN/m}^2$

For safety, external pressure intensity ≤ safe bearing capacity.

$$\frac{150}{B} \leq 152$$

$$\Rightarrow \quad B \geq \frac{150}{152} \geq 0.987 \text{ m}$$

9.20 **A plate load test was conducted using a square plate of 30 cm side, at 2 m depth on loose sand. The plate settled by 8 mm at load intensity of 12 t/m^2. Determine the settlement of 2 m × 2 m square footing if it has to carry the same load intensity at 2 m depth at this site. Ignore the effect of embedment.**

[10 marks : 2018]

Solution:

Square plate:

Width, B_p = 30 cm, Settlement, S_P = 8 mm

The following corrections are required for N-values obtained :

(i) **Overburden Pressure Correction :** In granular soils, the overburden pressure affects the penetration resistance. If the two soils having same relative density but different confining pressures are tested, the one with a higher confining pressure gives a higher penetration number. As the confining pressure in cohesionless soils increases with the depth, the penetration number for soils at shallow depths is underestimated and that at greater depth is over-estimated. For uniformity, the N-values obtained from field tests under different effective overburden pressure are corrected to a standard effective overburden pressure.

$$N_0 = \text{Observed value of S.P.T.}$$

$$N_1 = N_0 \times \left(\frac{350}{\bar{\sigma}_o + 70}\right)$$

where, $\bar{\sigma}_o$ = Effective stress at test level (kN/m^2)

N_1 = Corrected N-value after overburden correction.

Overburden correction will not be applied if $\bar{\sigma}_o > 280$ kN/m^2

(ii) **Dilatancy Correction:**

- It is applied to the already corrected N-values for overburden pressure. Dilatancy correction is required only if ($N_1 > 15$) in saturated fine sand silt. (i.e., when water table is above test level).
- $N_1 > 15$, basically represents the dense sand which will have the tendency to dilate under rapid loading (undrained condition) and negative pore water pressure will develop. Hence, observed N-value will be more because shear resistance will increase.

$$N_2 = 15 + \frac{1}{2}(N_1 - 15)$$

N_2 = Corrected N-value after dilatancy correction.

9.23 **What is the basis for classifying foundations into shallow and deep? Briefly explain the situations in which different types of shallow foundations are adopted.**

[8 marks : 2020]

Solution:

According to Terzaghi :

- Foundation can be classified as shallow foundation when the depth of footing is less than or equal to width of footing, i.e., $\frac{D_f}{B} \leq 1$.
- In shallow foundation, only base resistance is considered and in deep foundation, skin resistance as well as base resistance is considered.

According to Skempton :

- Foundation is shallow if depth of footing is less than 2.5 times the base of the footing.

S.No.	Type of Soil and Loading Condition	Suitable Foundation
1.	Structural load is less and soil is medium to dense	Shallow foundation
2.	Structural load is heavy and soil is medium to loose.	Either raft or deep footing
3.	Footing area is 40% or more than plinth area.	Either raft or combined footing
4.	For individual houses and light building 2 to 4 storey.	Strip or isolated.

9.24 **Comment on the statement "The net bearing capacity of a shallow foundation in clayey soil is unaffected by the position of water table, whereas in sandy soil, it is very much affected".**

[5 marks : 2020]

Solution:

As per Terzaghi's for strip footing :

Net ultimate bearing capacity (q_{nu}) :

$$q_{nu} = CN_C + \gamma_t \cdot D_f(N_q - 1) + 0.5B\gamma_t N_\gamma \quad ...(1)$$

Case-1 : In clayey soil,

$$N_C = 5.7\text{m},\ N_q = 1,\ N_\gamma = 0$$

$$\therefore \quad q_{nu} = CN_C$$

From above, it is clear that net ultimate bearing capacity of a clayey soil is unaffected by the position of water table.

Case-2 : In sandy soil,

$$q_{nu} = \gamma_t \cdot D_f(N_q - 1) + 0.5B\gamma_t N_\gamma$$

If water table rises from great depth,

$$q_{nu} = \gamma_{sub} \cdot D_f(N_q - 1) + 0.5B \cdot \gamma_{sub} \cdot N_\gamma$$

As

$$\gamma_{sub} \simeq \frac{\gamma_t}{2}$$

Hence, the bearing capacity will almost reduced to half, i.e., in sandy soil, position of water table much affected the net ultimate bearing capacity of soil.

9.25 **A square footing 2 m wide in resting on soft saturated clay. The depth of foundation is 1.5 m from existing ground level. The thickness of clay layer is 4.0 m and is underlain by a firm sand stratum. The properties of clay are as under : Liquid limit 30%, natural moisture content 40%, specific gravity 2.7, angle of internal friction zero and undrained cohesion 0.5 kg/cm². The clay stratum is normally consolidated. Determine the net safe bearing capacity using the Vesic equation. Compute the settlement that would result if the load intensity were allowed to act on the footing. Natural water table is at ground level. Assume load spread 2 : 1.**

[20 Marks : 2021]

Solution:

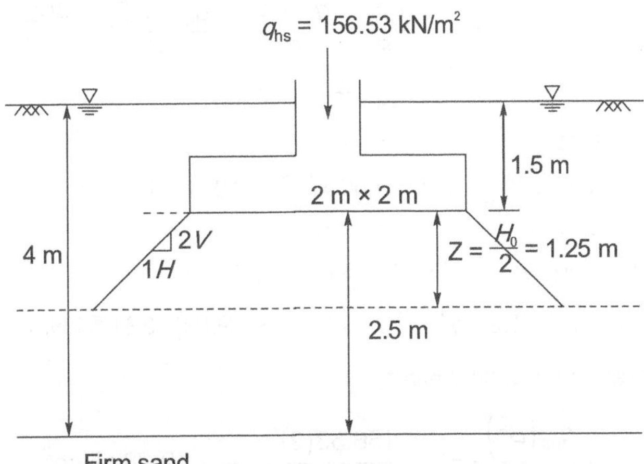

$$w_L = 30\%,\quad \omega_n = 40\%,\quad G = 2.7,\ \phi = 0,\quad S = 1$$

$$c_u = 0.5\text{ kg/cm}^2 = \frac{0.5\times 9.81\times 10^4}{10^3} = 49.05\text{ kN/m}^2$$

$$e = \frac{wG}{S} = \frac{0.4\times 2.7}{1} = 1.08$$

$$\gamma = \frac{(G+Se)\gamma_w}{1+e} = \frac{(2.7+1.08)\times 9.81}{1+1.08} = 17.83\text{ kN/m}^3$$

Calculation of net safe bearing capacity:

Vesic's theory:

For $\phi = 0$

$$N_c = 5.14$$

$$N_q = 1$$

$$N_\gamma = 0$$

$$S_c = 1+\frac{N_q}{N_c} = 1+\frac{1}{5.14} = 1.194$$

$$S_q = 1$$

$$S_\gamma = 0.6$$

$$d_c = 1+0.4\frac{D_f}{B} = 1+0.4\left(\frac{1.5}{2}\right) = 1.3$$

$$d_q = 1+2\tan\phi(1-\sin\phi)\frac{D_f}{B} = 1$$

$$d_\gamma = 1$$

$$i_c = i_q = i_\gamma = 1$$

Ultimate bearing capacity

$$q_u = CN_c\left(S_c d_c i_c\right)+\gamma' D_f N_q\left(S_q d_q i_q\right)+0.5B\gamma' N_\gamma\left(S_\gamma d_\gamma i_\gamma\right)$$

$$q_u = 49.05\times 5.14(1.194\times 1.3\times 1)+(17.83-9.81)(1.5)(1\times 1\times 1)$$

$$q_u = 403.36\text{ kN/m}^2$$

Net safe bearing capacity (Assuming FOS = 2.5)

$$q_{ns} = \frac{q_{nu}}{F} = \frac{q_u-\bar{\sigma}}{F} = \frac{q_u-\gamma' D_f}{F} = \frac{403.36-(17.83-9.81)\times 1.5}{2.5}$$

$$q_{ns} = 156.53\text{ kN/m}^2$$

Computation of settlement:

$$C_c = 0.009\,(w_L-10) = 0.009\,(30-10) = 0.18$$

1. $H_0 = 2.5\text{ m}$
2. At mid-clay layer" Initial effective stress:

$$\bar{\sigma}_0 = (1.5+1.25)\gamma' = 2.75\times(17.83-9.81) = 22.05\text{ kN/m}^2$$

3. Increase in stress at mid clay layer:

$$\Delta\bar{\sigma} = \frac{q_{hs}\left(B^2\right)}{(B+2nz)^2} = \frac{156.53(2)^2}{\left(2+2\times\frac{1}{2}\times 1.25\right)^2} = 59.28\text{ kN/m}^2$$

4. Settlement,

$$\Delta H = \frac{H_0 C_c}{1+e_0}\log\left(\frac{\bar{\sigma}_0 + \Delta\bar{\sigma}}{\bar{\sigma}_0}\right) = \frac{2.5\times 0.18}{1+1.08}\log\left(\frac{22.05+59.28}{22.05}\right)$$

$$= 0.1226 \text{ m} = 12.26 \text{ cm}$$

9.26 **A footing 2 m × 2 m in plan, transmits a pressure of 100 kN/m² on a cohesive soil layer having E = 6 × 10⁴ kN/m² and μ =0.50. Determine the immediate settlement of the footing at the centre, assuming it to be (i) a flexible footing, and (ii) a rigid footing.**

Influence Factor I_f for Vertical Displacement

Shape	I_f Flexible Foundation			I_f Rigid Foundation
	Centre	Corner	Average	
Circular	1.00	0.64	0.85	0.86
Square	1.12	0.56	0.95	0.82
Rectangular				
L/B = 1.5	1.36	0.68	1.20	1.06
L/B = 2	1.52	0.76	1.30	1.20
L/B = 5	1.10	1.05	1.83	1.70
L/B = 10	2.52	1.26	2.25	2.10
L/B = 100	3.38	1.69	2.96	3.40

[10 Marks : 2021]

Solution:

Given: Size of footing 2 m × 2 m

$$L/B = 1.0$$
$$q = 100 \text{ kN/m}^2$$
$$E = 6\times 10^4 \text{ kN/m}^2,\ \mu = 0.5$$

Immediate settlement, $$S_i = \frac{qB(1-\mu^2)}{E} I_f$$

Immediate settlement, at centre

(i) For flexible footing, $I_f = 1.12$ (From table given)

$$S_i = \frac{100\times 2\times(1-0.5^2)}{6\times 10^4}\times 1.12\times 1000 \text{ mm} = 2.8 \text{ mm}$$

(ii) For rigid footing, $I_f = 0.82$ (From table given)

$$S_i = \frac{100\times 2\times(1-0.5^2)}{6\times 10^4}\times 0.82\times 1000 \text{ mm} = 2.05 \text{ mm}$$

9.27 Two different contractors carried out plate load test at the same site location as per IS code. Details are as follows:

Parameters	Contractor 1	Contractor 2
Size of plate	45 cm × 45 cm	75 cm × 75 cm
Load	100 kN	175 kN
Settlement at above load	10 mm	10 mm

Estimate the maximum load which a footing of size 3 m × 3 m can carry at the settlement of 10 mm at the same site location. Consider the depths of both tests and that of proposed footing are same.

[10 marks : 2022]

Solution:

Plate load test:

Parameter	Contractor 1	Contractor 2
Area of plate	0.2025 m^2	0.5625 m^2
Perimeter of plate	1.8 m	3 m
Ultimate load	100 kN	175 kN

As per Housel's approach,

$$Q_1 = A_1 m + P_1 n$$

$$Q_2 = A_2 m + P_2 n$$

$$\Rightarrow \quad 100 = 0.2025\,\text{m} + 1.8n \quad \text{...(1)}$$

$$175 = 0.5625\,\text{m} + 3n \quad \text{...(2)}$$

On solving (1) and (2), we get

$$m = 37.04$$

$$n = 51.389$$

Now, for footing of size 3 × 3m:

$$Q_f = A_f m + P_f n$$

$$\Rightarrow \quad Q_f = (3 \times 3) \times 37.04 + [4 \times 3] \times 51.389$$

$$= 950.028 \text{ kN}$$

Ultimate bearing capacity of footing:

$$q_u = \frac{Q_f}{A_f} = \frac{950.028}{3 \times 3} = 105.56 \text{ kN/m}^2$$

∴ Maximum load of 950.028 kN can be carried by the footing.

9.28 A 6 m high pier rests on a 2 m × 2 m square footing at 1.5 m depth from the surface at a site having uniform clayey soils. The unconfined compressive strength of the clay is 100 kPa and its bulk unit weight is 20 kN/m^3. The pier carries a vertical load of 80 kN at the centre including its self-weight. A resultant horizontal load of 15 kN also acts on one side of the pier at 1.5 m above the surface. Determine the factor of safety with respect to the pier's net ultimate bearing capacity as per IS 6403 recommendations.

[12 marks : 2023]

Solution:

Given: Size of footing = 2 m × 2 m

Depth of footing = 1.5 m

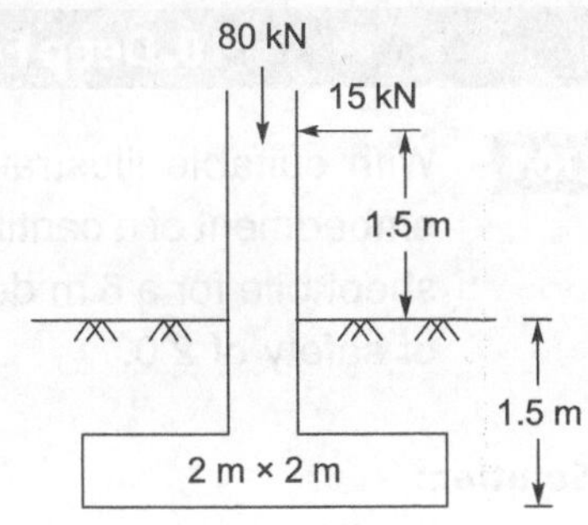

Unconfined compressive strength, $q_u = 100$ kPa

The horizontal load will have the effect of introducing both inclination and eccentricity of loading. The resultant of horizontal load and vertical load will be inclined at an angle α to the vertical.

where, $$\tan\alpha = \frac{\text{Horizontal load}}{\text{Vertical load}} = \frac{15}{80}$$

$$\Rightarrow \quad \alpha = \tan^{-1}\frac{15}{80} = 10.62°$$

Now, Height of horizontal load above the base of footing = 1.5 + 1.5 = 3 m

Eccentricity of the resultant load, e can be calculated as,

$$\frac{e}{3} = \tan\alpha$$

$$e = 3 \times \tan\alpha = 3 \times \tan 10.62° = 0.5625 \text{ m}$$

Now, reduced width B' on account of eccentricity of loading is given as

$$B' = B - 2e$$

$$= 2 - 2 \times 0.5625 = 0.875 \text{ m}$$

So, Reduced area, $$A' = B' \times L$$

$$= 0.875 \times 2 = 1.75 \text{ m}^2$$

Now, as per IS 6403,

for $\phi = 0$, $N_c = 5.14$, $N_q = 1$, $N_\gamma = 0$

$$N_\phi = \tan^2\left(45°+\frac{\phi}{2}\right) = \tan^2(45°+0°) = 1$$

$$S_c = \left[1+0.2\frac{B'}{L}\right] \times N_\phi = \left[1+0.2\times\frac{0.875}{2}\right]\times 1 = 1.0875$$

$$S_q - 1$$

$$d_c = 1+0.2\frac{D}{B'}\sqrt{N_\phi}$$

$$= \left[1+0.2\times\frac{1.5}{0.875}\right]\times\sqrt{1} = 1.34$$

$$d_q = 1$$

$$d_\gamma = \left[1+0.1\times\frac{D}{B'}\right] = \left[1+0.1\times\frac{1.5}{0.875}\right] = 1.17$$

$$i_c = i_q = \left(1-\frac{\alpha}{90°}\right)^2 = \left[1-\frac{10.62°}{90°}\right]^2 = 0.78$$

Now, Net ultimate bearing capacity, $q_{net} = [c\,N_c\,S_c\,d_c\,i_c + q\,(N_q - 1)\,S_q\,d_q\,i_q]$ $[\because N_\gamma = 0]$

$$= 50 \times 5.14 \times 1.0875 \times 1.34 \times 0.78 + 0 \qquad [\because N_q = 1]$$

$$= 292.12 \text{ kN/m}^2$$

Net ultimate load, $$Q_{net} = q_{net} \times A$$

$$= 292.12 \times 0.875 \times 2 = 511.21 \text{ kN}$$

$$\text{Factor of safety} = \frac{511.21}{80} = 6.39$$

10. Deep Foundation, Sheet Pile Walls and Machine Foundation

10.1 With suitable illustrations describe the simplified analysis method for designing the depth of embedment of a cantilever sheet pile in granular soils. Find the depth of embedment of the cantilver sheet pile for a 6 m deep excavation in a sandy soil layer for γ = 18 kN/m^3 and ϕ = 35° for a factor of safety of 2.0.

[15 marks : 2003]

Solution:

Let D be the depth of embedment and H be the height of the cantilever sheet pile above the dredge level. Assuming a concentrated force R acting at the foot of the pile. For equilibrium, the moments of active pressures on the right and passive resistance on the left about the point of reaction R must be balanced.

$$\text{FOS} = \frac{\text{Resisting moment}}{\text{Overturning moment}}$$

$$\text{FOS} = \frac{P_p \times \frac{D}{3}}{P_a \times \left(\frac{H+D}{3}\right)}$$

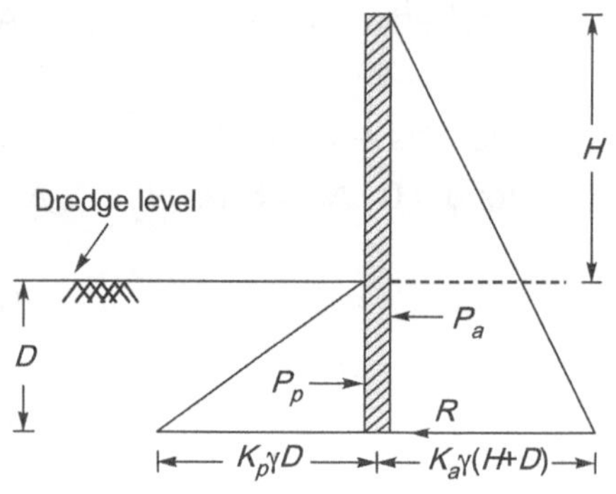

$$\therefore \quad \frac{P_p}{\text{FOS}} \times \frac{D}{3} - P_a \times \left(\frac{H+D}{3}\right) = 0$$

But
$$P_p = \frac{1}{2}K_p\gamma D^2$$

and
$$P_a = \frac{1}{2}K_a\gamma(H+D)^2$$

$$\Rightarrow \quad \frac{1}{2}\frac{K_p\gamma D^2}{\text{FOS}} \times \frac{D}{3} - \frac{1}{2}K_a\gamma(H+D)^2 \times \left(\frac{H+D}{3}\right) = 0$$

$$\Rightarrow \quad \frac{K_p D^3}{2} - K_1(H+D)^3 = 0$$

$$\Rightarrow \quad \frac{K_p D^3}{2} - K_a[H^3 + D^3 + 3HD(H+D)] = 0$$

$$\Rightarrow \quad \left(\frac{K_p}{2} - K_a\right)D^3 - K_a H^3 - 3HDK_a(H+D) = 0$$

$$\Rightarrow \quad \left(\frac{K_p}{2} - K_a\right)D^3 - 3HK_a D^2 - 3H^2K_a D - K_a H^3 = 0 \qquad \text{...(i)}$$

Given that H = 6 m; γ = 18 kN/m^3; ϕ = 35°; Factor of safety = 2

$$\therefore \quad K_a = \frac{1-\sin 35°}{1+\sin 35°} = 0.271$$

and
$$K_p = \frac{1}{K_a} = \frac{1}{0.271} = 3.69$$

Thus, **the depth of embedment is given by (i) i.e.**

$$\left(\frac{K_p}{2} - K_a\right)D^3 - 3HK_a D^2 - 3H^2K_a D - K_a H^3 = 0$$

$$\Rightarrow \left(\frac{3.69}{2} - 0.271\right) D^3 - 3 \times 6 \times 0.271 \times D^2 - 3 \times 6^2 \times 0.271 \times D - 0.271 \times 6^3 = 0$$

$$\Rightarrow 1.574D^3 - 4.878\, D^2 - 29.268\, D - 58.536 = 0$$

$$D = 6.7 \text{ m}$$

∴ Hence depth of emnbedment of sheet pile = 6.7 m

10.2 **Define pile group efficiency.**

[2 marks : 2005]

Solution:

The ratio of the ultimate load capacity of the pile group (Q_{ug}) to the sum of the individual load capacities of all the piles in the group, is called the group efficiency (η). It can be written as

$$\eta = \frac{Q_{ug}}{nQ_u}$$

Where n is the number of piles in the group and Q_u is the load capacity of one pile.

The group efficiency depends on the spacing between piles, type of soil in which the piles are installed and the manner of pile installation, that is, driven, bored, cast in situ, etc.

10.3 **What is negative friction?**

[3 marks : 2005]

Solution:

Piles installed in freshly placed fills of soft compressible deposits are subjected to a downward drag, a consequence of the consolidation of the strata after the piles are installed. This downward drag on the pile surface, when the soil moves down relative to the pile, adds to the structural loads and is called negative skin friction. This is in contrast to the usual shaft friction which is mobilised when the pile moves down relative to the soil. Thus, negative skin friction has an effect of reducing the allowable load on the pile.

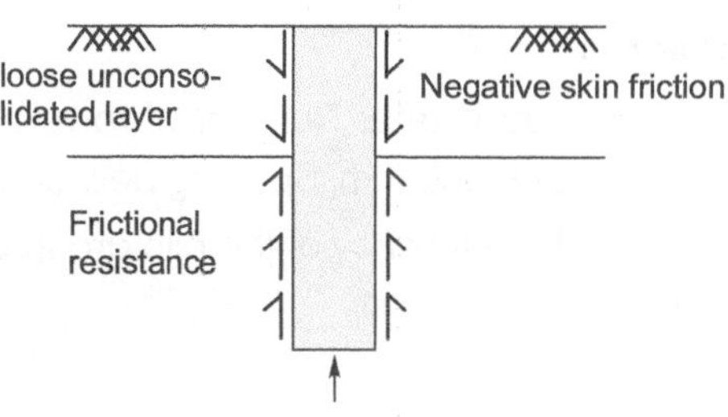

10.4 **Design a square pile group to carry 400 kN in clay with an unconfined compression strength of 60 kN/m². The piles are 300 mm diameter and 6 m long. Adhesion factor may be taken as 0.6.**

[10 marks : 2005]

Solution:

Given: UCS = 60 kN/m^2

So, $$C = \frac{60}{2} = 30 \text{ kN/m}^2$$

Assuming $$\eta > 100\%$$

$$Q_{design} = nQ_{up}$$

$$Q_{up} = \left[9C \times \left(\frac{\pi}{4} D^2\right) \times \alpha C \times \pi DL\right]$$

$$Q_{up} = \left[\left(9 \times 30 \times \frac{\pi}{4} \times 0.3^2\right) + 0.6 \times 30 \times \pi \times 0.3 \times 6\right]$$

$$Q_{up} = 120.87 \text{ kN}$$

So, $$n = \frac{Q_{design}}{Q_{up}} = \frac{400}{120.87} = 3.309 \simeq 4$$

So provide pile group of 2 × 2.

Now check for group efficiency.

Spacing assumed as 2.5d

$$= 2.5 \times 0.3 = 0.75 \text{ m}$$

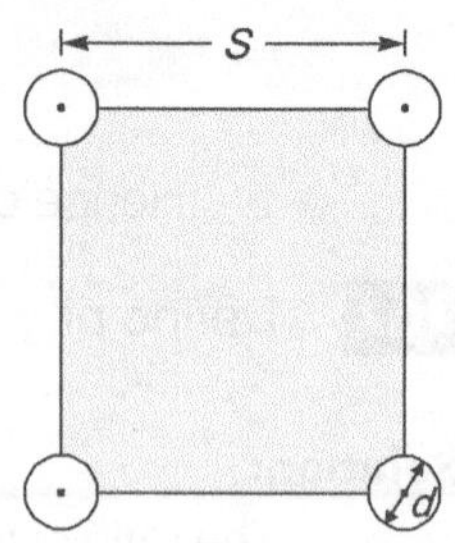

So, $B = S + d = 0.75 + 0.3 = 1.05$

Now, $Q_{ug} = 9C \times B^2 + \alpha C(4BL)$ $(\because \alpha = 1)$

$$= 9 \times 30 \times 1.05^2 + 1 \times 30 \times 4 \times 1.05 \times 6$$

$$= 1053.675 \text{ kN}$$

Now, efficiency, $\eta = \frac{Q_{ug}}{nQ_{up}} \times 100 = \frac{1053.675}{4 \times 120.87} \times 100$

$\eta = 217.935\% > 100\%$ (OK)

10.5 **A 4 × 5 pile group has the following details:**
Diameter of each pile, d = 300 mm
C/C spacing of pile = 900 mm
Capacity of a single pile = 500 kN
Determine the efficiency and capacity of free standing pile group.

[15 marks : 2006]

Solution:

Given data: Number of rows of piles (n) = 4, Number of columns of piles (m) = 5, Diameter of pile (d) = 300 mm, Spacing between piles (S) = 900 mm, Capacity of single pile (Q_u) = 500 kN

The efficiency of the pile group can be given by **Converse Labarre formula as**

$$\eta_g = 1 - \frac{\theta[(n-1)m + (m-1)n]}{90\,mn}$$

where $\theta = \tan^{-1}\left(\frac{d}{S}\right) = \tan^{-1}\left(\frac{300}{900}\right) = 18.43°$

$\therefore$ $\eta_g = 1 - \frac{18.43 \times [(4-1)5 + (5-1)4]}{90 \times 5 \times 4} = 0.6825 = 68.25\%$

Capacity of the free standing pile group,

$$Q_{ug} = \eta_g \times (mn\,Q_u) = 0.6825 \times 4 \times 5 \times 500 = 6825 \text{ kN}$$

10.6 **State and explain the Engineering News Formula for drop hammer. Write its application in the field.**

[5 marks : 2006]

Solution:

The **Engineering News Formula proposed by A.M. Wellington,** editor of the American Journal, Engineering News, is the simplest and the most popular of the dynamic pile load formulas. For an allowable load Q_a it may be given as

$$Q_a = \frac{WH\eta_H}{F(S+C)}$$

where Q_a is allowable load in kg or kN
W is weight of hammer in kg or kN
H is height of fall of hammer in cm
S is final penetration in cm per blow (which is termed as set)
F is Factor of safety usually taken as 6

C is Empirical constant which is equal to 2.5 cm for a drop hammer.
0.25 cm for a steam hammer.

Thus, allowable load, Q_a will be

$$Q_a = \frac{WH}{6(S+2.5)}$$

The detailed investigations carried out by Vesic on deep foundations in granular soils indicated that the ENF applicable to drop hammers gave pile loads as low as 44% of the actual loads. These formula could be used with more confidence in freely draining materials such as coarse sand. If the pile is driven to saturated loose fine sand and silt, there is every possibility of development of liquefaction which reduces the bearing capacity of the pile.

10.7 **A precast concrete pile of size 50 cm × 50 cm is to be driven into clay strata whose unconfined compressive strength is 110 kN/m². Compute the length of pile required to carry safe working load of 450 kN with factor of safety of 2.5. Assume the adhesion factor α (alpha) as 0.6**

[8 marks : 2007]

Solution:

Given data: Size of pile = 50 cm × 50 cm, Unconfined compression strength of clay (q_u) = 110 kN/m²
Safe working load (Q_a) = 450 kN, Factor of Safety (F) = 2.5, Adhesion factor (α) = 0.6

Let the length of the pile upto which it is embedded in the clay strata be L.

The ultimate load bearing capacity of the pile is given by

$$Q_u = c_b N_c A_b + A_s \alpha \bar{c}_u$$

Now,
$$Q_a = \frac{Q_u}{F}$$

$\Rightarrow$
$$Q_u = Q_a \times F = 450 \times 2.5 = 1125 \text{ kN}$$

Undrained cohesion at the base of pile,

$$c_b = \frac{q_u}{2} = \frac{110}{2} = 55 \text{ kN/m}^2$$

N_c = Bearing capacity factor for a deep foundation. For circular and square piles, N_c value of 9 is used as proposed by Skempton

A_b = Area at the base of pile = 0.50 × 0.50 = 0.25 m²

A_s = Surface area of pile embeded below ground surface = 4 × 0.5 × L = 2L

$\bar{c}_u$ = Average undrained cohesion in the embedded length of pile = c_b = 55 kN/m²

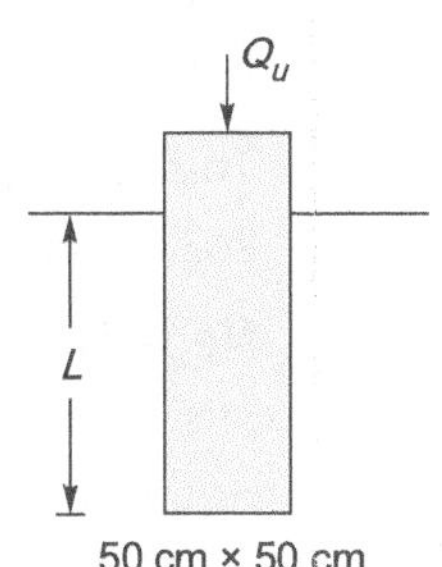

$\therefore$
$$Q_u = c_b N_c A_b + A_s \alpha \bar{c}_u$$

$\Rightarrow$
$$1125 = 55 \times 9 \times 0.25 + 2L \times 0.6 \times 55$$

$\Rightarrow$
$$L = \frac{1125 - 123.75}{66} = 15.17 \text{ m}$$

Thus, the length of the pile is 15.17 m

10.8 **How would you design the foundation of a structure in swelling soils?**

[10 marks : 2008]

Solution:

The rational methods of foundation design which are being used now to reduce or prevent the effects of swelling can be grouped into three categories, namely,

(a) isolating the structure from the swelling soil, (b) designing a structure to withstand the effects of swelling and (c) preventing the swelling.

Belled piers which are popular in USA for use in expansive soil conditions are isolated structures on swelling soils. The underreamed pile-beam construction is similar to that of belled piers and are designed for use under Indian conditions. The principal involved is to transfer the load of a building through the piles to a depth beyond the zone of seasonal variation in moisture content. The underreamed piles are bored cast in situ piles to a depth beyond the zone of seasonal variation in moisture content. The underreamed piles are bored cast in situ piles with their lower portions enlarged or reamed in the form of a bulb. The piles are connected at their top by plinth beams of reinforced concrete which support the super structure.

Designing a structure which is strong enough and rigid enough to withstand the effects of swelling may prove to be highly uneconomical except in the case of very small structures where even if the loads are supported by a central area or peripheral area much smaller than the plan area, the bearing pressures are within limits. Obviously, it is futile to attempt designing a structure that will remain undamaged in spite of swelling, because the cost of providing the structure with extra strength and rigidity would be much more as compared to the cost of underreamed pile or piers foundation.

Swelling can often be controlled, if not eliminated, by providing an impervious apron around the structure. By providing an apron, the moisture gradient between the centre of the structure and its edges is minimized and hence the differential swelling is controlled.

Elimination of possible swelling can theoretically be brought about by (a) pre-wetting the ground to a moisture content equal to the equilibrium moisture content, (b) making downward loads large enough to exceed swelling pressures and (c) chemical stabilization.

10.9 What is "negative skin friction" and its significance in the design of pile foundation?

[5 marks : 2009]

Solution:

Piles installed in freshly placed fills of soft compressible deposits are subjected to a downward drag, a consequence of the consolidation of the strata after the piles are installed. This downward drag on the pile surface, when the soil moves down relative to the pile, adds to the structural loads, and is called negative skin friction.

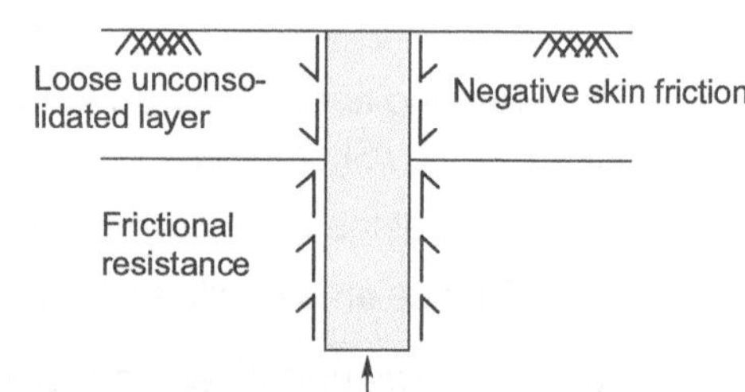

This is in contrast to the usual shaft friction which is mobilised when the pile moves down relative to the soil. Thus, negative skin friction has an effect of reducing the allowable load on the pile.

Negative skin friction may also occur by the lowering of ground water which increases the effective stress inducing consolidation and consequent settlement of the soil surrounding the pile. Some-times negative skin friction may develop even in the zone of the fill, if the fill itself is settling under its self weight.

It is necessary to subtract negative skin friction force from the total load that the pile can support. In such a case the factor of safety will be modified as follows:

$$\text{Factor of safety} = \frac{\text{Ultimate pile load capacity}}{\text{Working load + Negative skin friction force}}$$

Sometimes it may also be written as

$$\text{Factor of safety} = \frac{\text{Ultimate pile load capacity} - \text{Negative skin friction force}}{\text{Working load}}$$

Values of negative skin friction force are computed in just the same way as positive skin friction.

10.10 **A 450 mm × 450 mm concrete pile 20.0 m long is driven into sand deposits with γ = 17 kN/m³ and ϕ' = 30°. Find the ultimate load i.e. point load Q_P by Meyerhoff's method and Janbu method. Meyerhoff's $N_{q'}$ = 55, Atmospheric pressure = 100 kN/m², Janbu's $N_{q'}$ = 18.4**

[8 marks : 2010]

Solution:

(i) Meyerhoff's method

Point load capacity of a pile in cohesionless soils is given by

$$Q_p = \gamma D_f N_{q'} A_P$$

Given that: $\gamma = 17$ kN/m³, $D_f = 20.0$ m, $N_{q'} = 55$, $p_a = 100$ kN/m²

$$A_p = 0.45 \times 0.45 = 0.2025 \text{ m}^2$$

$$\therefore \quad Q_p = 17 \times 20 \times 55 \times 0.2025 = 3786.75 \text{ kN}$$

But the value of Q_p should not exceed the limiting load carrying capacity at pile point given as $0.5\, p_a N_{q'} \tan\phi' A_p$ i.e.

$$Q_p \leq 0.5\, p_a N_{q'} \tan\phi' A_p$$

$$\Rightarrow \quad Q_p \leq 0.5 \times 100 \times 55 \times \tan 30° \times 0.2025$$

$$\Rightarrow \quad Q_p \leq 321.51 \text{ kN}$$

Therefore the ultimate load carrying capacity of pile as per Meyerhoff's method is 321.51 kN

(ii) Janbu's Method

Point load capacity of pile in cohesionless soils is given by

$$Q_p = \gamma D_f N_{q'} A_p = 17 \times 20 \times 18.4 \times 0.2025 = 1266.84 \text{ kN}$$

10.11 **Nine piles of 300 mm dia and 8 m length are arranged in a square pattern for the foundation for a column in a uniform deposit of medium stiff clay (q_u = 100 kN/m²). If the centre to centre spacing of piles is 900 mm and adhesion factor = 0.9, calculate the capacity of the pile group assuming a F.S. of 2.5.**

[10 marks : 2011]

Solution:

Given:

Dia. of Pile = D = 300 mm = 0.3 m

Length of pile = L = 8 m

Adhesion factor = α = 0.9

$$c = \frac{q_u}{2} = 50 \text{ kN/m}^2$$

S = 0.9 m
B = 2.1 m
2.1 m

Ultimate load of pile group acting together.

$$Q_{g1(u)} = q_p A_g + c(4\,BL)$$
$$= cN_c \times B^2 + c(4\,BL)$$
$$= (50 \times 9) \times (2.1)^2 + 50 \times 4 \times 2.1 \times 8 = 5344.5 \text{ kN}$$

Ultimate load of pile group with each pile acting individually

$$Q_{g2(u)} = NQ_u \qquad \text{[where } N \text{ is no. of piles]}$$
$$= 9[cN_c A_p + \alpha c(\pi DL)]$$

$$= 9\left[50\times 9\times \frac{\pi \times 0.3^2}{4} + 0.9\times 50\times \pi \times 0.3\times 8\right] = 3339.9 \text{ kN}$$

As the ultimate load for individual pile failure case is less than the pile group load, safe load is given by.

$$Q_a = \frac{3339.9}{\text{FOS}} = \frac{3339.9}{2.5} = 1335.96 \text{ kN}$$

10.12 **Design a friction pile group to carry a load of 3000 kN including the weight of the pile cap at a site where the soil is uniform clay to a depth of 20 m underlain by rock. Average unconfined compressive strength of clay is 70 kN/m². The clay may be assumed to be of normal sensitivity and normally loaded with liquid limit of 60%. A factor of safety of 3 is required against shear failure. (Assume length of pile as 10 m).**

[10 marks : 2012]

Solution:

Given : $\frac{q_u}{2} = \frac{70}{2} = 35 \text{ kN/m}^2$, Permissible, $C = \frac{35}{3} \text{ kN/m}^2$

$\because$ Length of pile = 10 m

Let the diameter of pile = 0.6 m

Spacing of pile = $3d = 3 \times 0.6 \times 100 = 180$ cm

Let the number of pile = n

Considering the piles to act individually, the load at failure is given by

$$Q_{ug} = n C \pi d L$$

$$3000 = n\times \frac{35}{3}\times \pi \times 0.6\times 10$$

$$n = \frac{3000\times 3}{35\times \pi \times 0.6\times 10} = 13.64$$

For square arrangement, keep $n = 16$

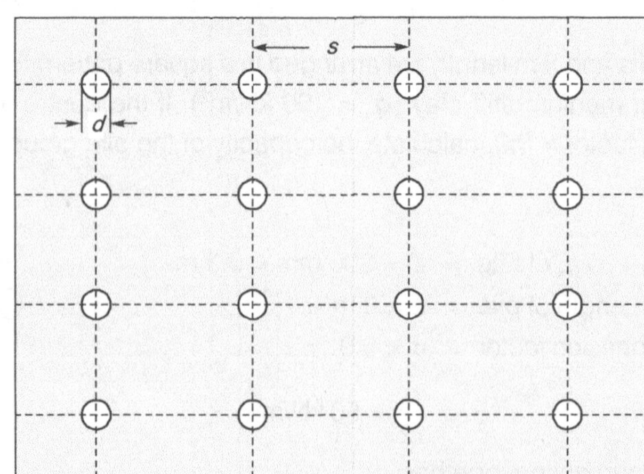

Check for the group action

$$B = 35 + df = 3 \times 180 \times 60 = 600 \text{ cm} = 6 \text{ m}$$

$\therefore$ Load taken by group action = $4BLC + A_p C N_C$

$$A_b = B\times B = 6^2 = 36 \text{ m}^2;\ N_C = 9$$

$$\therefore \quad Q_{ug} = 4\times 6\times 10\times \frac{35}{3} + 36\times \frac{35}{3}\times 9$$

$$= 6580 \text{ kN} > 3000 \text{ kN}$$ Hence safe

10.13 A square pile group consists of nine friction piles driven in cohesive soil. The diameter of each pile is 0.3 m and center-to-center spacing is 1.2 m. The ultimate capacity of each pile is 300 kN. Estimate the design capacity of the pile group.

[8 marks : 2014]

Solution:

Given: Diameter of pile (D) = 0.3 m, Ultimate capacity of pile (Q_{up}) = 300 kN, Centre to centre spacing (S) = 1.2 m, For friction piles in cohesive soil (Q_{up}) = $\alpha \bar{c} A_s$

Assume adhesion factor $\alpha = 0.9$, also assume length of pile and average cohesion are L and $\bar{c}$ respectively.

$$300\text{ kN} = 0.9\bar{c}(\pi DL)$$

$$\Rightarrow \quad \bar{c}L = \frac{300}{0.9 \times 0.3 \times \pi} = 353.68$$

Let size of pile group is $B \times B$

where, $B = 2s + D = 2 \times 1.2 + 0.3 = 2.7$ m

Ultimate load carrying capacity of pile group,

$$Q_{ug} = \alpha \bar{c} A_s$$

For pile group, $\alpha = 1$

$$\therefore \quad Q_{ug} = 1 \times \bar{c}(4 \times B \times L) = 1 \times 4 \times 2.7 \times (\bar{c}L)$$

$$= 1 \times 4 \times 2.7 \times 353.68 = 3819.74 \text{ kN}$$

Ultimate load capacity of pile group is taken smaller of the following

(i) $Q_{ug} = 3819.71$ kN

(ii) $nQ_{up} = 9 \times 300 = 2700$ kN

Hence ultimate load capacity of pile group is minimum of above two i.e., 2700 kN.

Asume FOS = 2.5

$$\text{Then safe load for pile group} = \frac{2700}{\text{FOS}} = \frac{2700}{2.5} = 1080 \text{ kN}$$

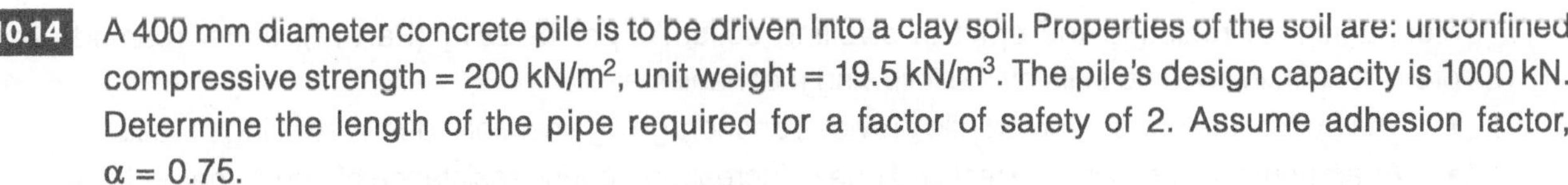

10.14 A 400 mm diameter concrete pile is to be driven into a clay soil. Properties of the soil are: unconfined compressive strength = 200 kN/m^2, unit weight = 19.5 kN/m^3. The pile's design capacity is 1000 kN. Determine the length of the pipe required for a factor of safety of 2. Assume adhesion factor, $\alpha = 0.75$.

[8 marks : 2015]

Solution:

Given: d = 400 mm = 0.4 m, q_u = 200 kN/m^2, γ = 19.5 kN/m^3, p_d = 1000 kN, F = 2, α = 0.75

$$\text{Cohesion of soil,} \quad C_u = \frac{q_u}{2} = \frac{200}{2}\text{kN/m}^2 = 100\text{kN/m}^2$$

Let length of pile = L m

$$\text{Load carrying capacity of pile,} \quad Q_u = q_b A_b + q_s A_s = (\alpha C_u)(\pi dL) + (9C_u) \times \left(\frac{\pi}{4}d^2\right)$$

$$= 0.75 \times 100 \times \pi \times 0.4 \times L + 9 \times 100 \times \frac{\pi}{4} \times 0.4^2 = 30\pi L + 36\pi$$

$$\textbf{For safety,} \quad P_u = \frac{Q_u}{F}$$

$$1000 \text{ kN} = \frac{30\pi L + 36\pi}{2}$$

$$\Rightarrow \quad L = \frac{1000 \times 2 - 36\pi}{30\pi} = 20\text{m}$$

10.15 **Draw a neat diagram of under-reamed pile foundation. Label different parts and give typical dimensions of different parts. Also explain the mechanism involved in resisting the uplift in swelling soils.**

[8 marks : 2016]

Solution:

Under reamed pile foundation: As per **IS 2911 (Part - III)**, the various components of *bored cast in-situ under reamed pile foundation* is shown below:

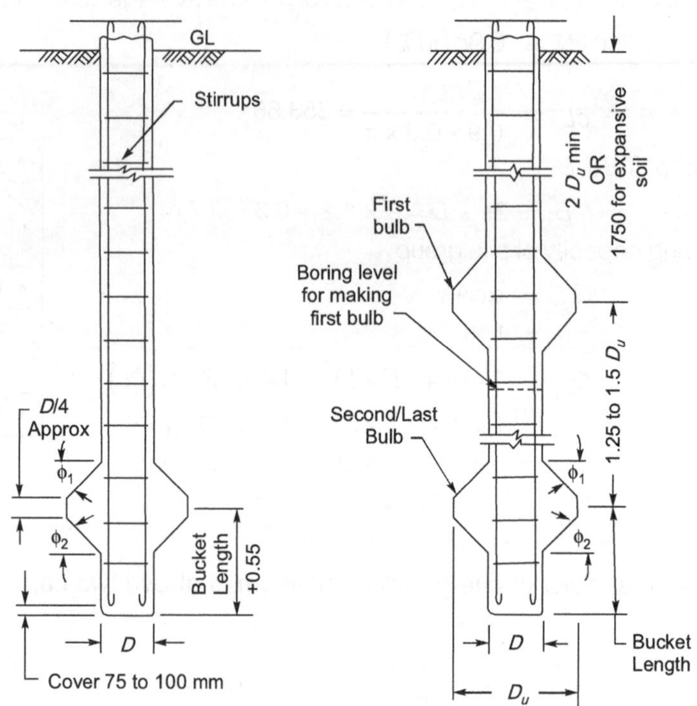

Mechanism of resisting the uplift in swelling soils: As per **IS 2911 (Part-III)**, under reamed piles prevent uplift in swelling soils by the following mechanisms:

- The bulb of the under reamed pile provides anchorage against uplift due to swelling pressure.
- Apart from this, uplift is also resisted due to increased bearing resistance of under-reamed pile.

10.16 **State which type of foundation you would adopt in the following cases and then give the reasons:**
(i) Transmission line tower.
(ii) Four-storey hospital building on erratic soil.
(iii) Two-storey residential building of load-bearing walls resting on medium dense sand.
(iv) Twenty-storey commercial building.

[8 marks : 2016]

Solution:

(i) **Transmission line tower:** For transmission line tower, pile foundation is best suited because of possible uplift due to wind forces.

(ii) **Four storey hospital building on erratic soil:** For this type of building in erratic soil, slab-beam (strap) foundation can be provided for columns in a row. The individual strap footings can be tied with a beam at plinth level.

(iii) **Two storey residential building with load bearing walls on medium dense sand:** For this type of building, a continuous strip footing is required below the load bearing walls.

(iv) **Twenty storey commercial building:** For this type of building, pile foundation is best suited as due to large height of the structure, lateral forces will be significant that need to be carried to the foundation safely. Pile foundation can transmit these large lateral forces to the soil efficiently.

10.17 What is negative skin friction? What is its effect on pile capacity?

[6 marks : 2017]

Solution:

Negative skin friction and its effect on pile capacity: Piles installed in freshly placed soft compressible deposits are subjected to a downward drag, a consequence of the consolidation of the strata after the piles are installed. This downward drag on the pile surface, when the soil moves down relative to the pile adds to the structural loads and is called negative skin friction.

Thus, negative skin friction has an effect of reducing the allowable load on the pile. Negative skin friction may also develop if the fill material is a loose sand deposit. It can also occur due to the lowering of the ground water table which increases the effective stress thus causing consolidation of the soil with the resultant down-drag on piles. A small relative movement between the soil and the pile of the order of about 10 mm may be sufficient for the full negative skin friction to materialise.

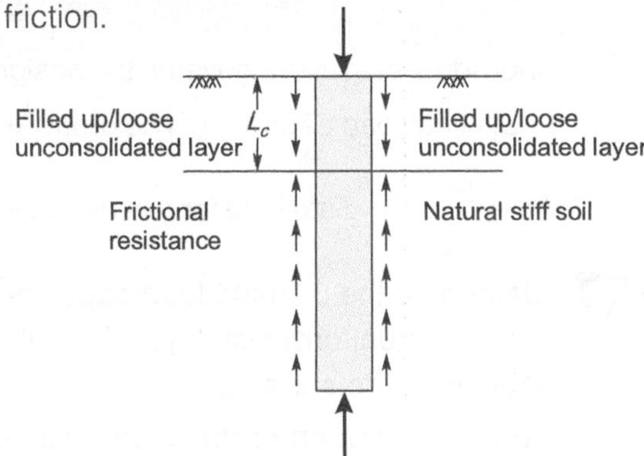

In bearing piles where the settlement of the pile is negligible, negative skin friction becomes a pile capacity problem. In compressible soils, the problem of negative skin friction should be considered a settlement problem.

10.18 A pile group consists of 9 piles of 30 cm diameter and 12 m long driven in clay. The centre to centre spacing of piles is 0.9 m. The clay has unconfined compressive strength of 80 kN/m² and γ = 20 kN/m³.
Determine the efficiency of pile group and the safe load for the pile group with factor of safety = 2.5. Take N_C = 9.0, adhesion factor = 0.6.

[14 marks : 2017]

Solution:

Unconfined compressive strength, $q_u = 80 \text{ kN/m}^2$

$$\therefore \quad C_u = \frac{q_u}{2} = 40 \text{ kN/m}^2$$

Given: $N_c = 9$ and adhesion factor, $\alpha = 0.60$

Individual pile failure:

Ultimate capacity of single pile is given as

$$Q_u = C_u N_c A_b + \alpha C_u A_s$$

$$= 40 \times 9 \times \frac{\pi}{4}(0.30)^2 + 0.6 \times 40 \times \pi \times 0.30 \times 12$$

$$= 25.45 + 271.43 = 296.88 \text{ kN} \quad \text{...(i)}$$

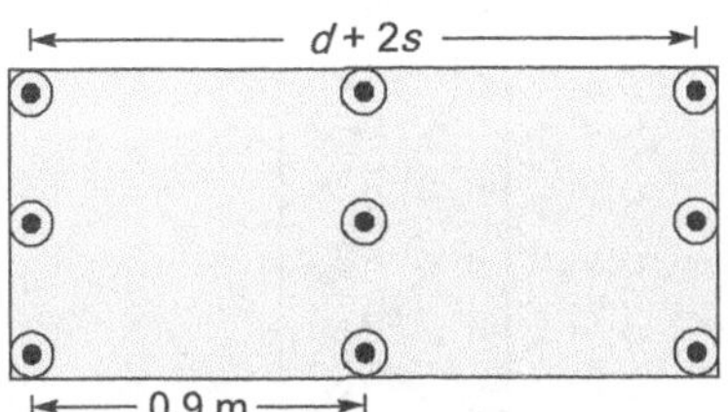

Size of pile group $= 2s + d$

$= 2 \times 0.9 + 0.3 = 2.1$ m

Pile group failure:

Ultimate load capacity of pile group by block failure is

$$Q_{ug} = C_u N_c A_g + P_b L C_u$$
$$= 40 \times 9 \times 2.1 \times 2.1 + 4 \times 2.1 \times 12 \times 40$$
$$= 1587.60 + 4032 = 5619.60 \text{ kN} \quad \text{...(ii)}$$

$$\text{Efficiency of pile group} = \frac{Q_{ug}}{nQ_u} = \frac{5619.60}{9 \times 296.88} = 2.1$$

Ultimate load capacity of pile group of nine piles on basis of individual action = nQ_u

$$= 9 \times 296.88 = 2671.92 \text{ kN} \quad \text{...(iii)}$$

Individual pile failure governs the design.

Ultimate load capacity of pile group = 2671.92 kN (smaller of equation (ii) and (iii))

$$\text{Safe load for pile group} = \frac{2671.92}{\text{FOS}} = \frac{2671.92}{2.50} = 1068.77 \text{ kN}$$

10.19 Determine the ultimate load capacity of a 600 mm diameter and 12.5 m long concrete pile driven through a uniform clay layer. The pile is made to rest on top of dense silica sand layer. The water table is at the surface.

What will happen to the above ultimate load capacity of the water table level is lowered by 5 m without any change in the soil property?

The properties of the two soil layers are:

Clay: Undrained shear strength = 40 kN/m^2

Unit weight = 20 kN/m^3, Adhesion factor = 0.8, N_c = 0.9.

Dense silica sand: ϕ' = 36°, Unit weight = 18 kN/m^3. Use N_q = 88.

Assume that the unit weight of water is 10 kN/m^3.

[12 marks : 2018]

Solution:

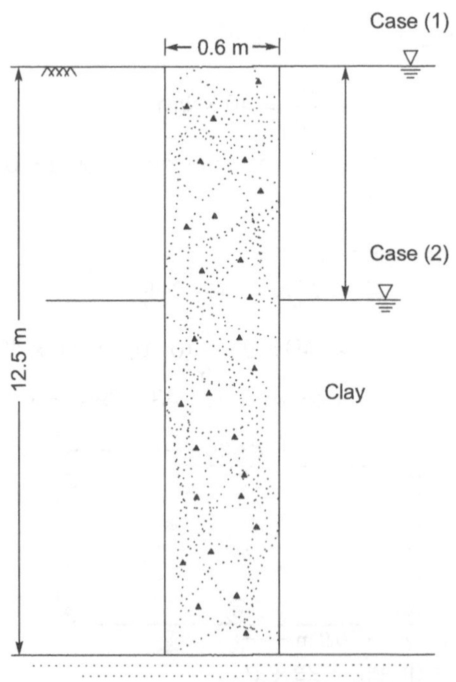

Case-1: Water table is at the surface ultimate load capcity Q_{up}

$$Q_{up} = A_b\bar{\sigma}N_q + A_s\alpha C$$

Here,

$$\bar{\sigma} = \text{Effective stress at bottom}$$
$$= \gamma' L$$
$$= (\gamma_{sat} - \gamma_w)L$$
$$= 125 \text{ kN/m}^2$$
$$A_b = \frac{\pi}{4}\times 0.6^2 = 0.28274 \text{m}^2$$
$$N_q = 88$$
$$A_s = \pi dl = 23.562 \text{ m}^2$$
$$\alpha = 0.8$$
$$C = \text{Undrained shear strength i.e.,} = 40 \text{ kN/m}^2$$

$$\therefore \quad Q_{up} = (0.28274 \times 125 \times 88) + (23.562 \times 0.8 \times 40) = 3864.124 \text{ kN}$$

Case-2: Water table level lowered by 5 m.

$$\bar{\sigma} = 5\gamma_{sat} + 7.5\,(\gamma_{sat} - \gamma_w) = 175 \text{ kN/m}^2$$

Rest all parameters remain the same.

$$Q_{up} = (0.28274 \times 175 \times 88) + (23.562 \times 0.8 \times 40)$$

$$\% \text{ increase} = \frac{5108.18\text{-}3864.12}{3864.12}\times 100 = 32.19\% = 5108.18 \text{ kN}$$

There is an increase of 32.19% due to lowering of water table.

Note: *Skin friction resistance remain the same, while end bearing/point resistance increases.*

10.20 **A pile group consists of four friction piles in cohesive soil. The unit weight and unconfined compressive strength of the soil are respectively 20.2 kN/m³ and 200 kPa. The diameter of each pile is 300 mm, length is 12.0 m and centre-to-centre spacing between the piles is 750 mm. Assuming an adhesion factor of 0.6, determine (i) load capacity of the group based on the individual pile failure, (ii) load capacity of the group based on the block failure and (iii) design load capacity of the group. Assume a factor of safety of 2.0 for individual pile failure and 3 for block failure.**

[20 marks : 2019]

Solution:

Given: $\gamma = 20.2 \text{ kN/m}^3$

Unconfined compressive strength = 200 kPa

Diameter of pile = 0.3 m

Length = 12 m

$\alpha = 0.6$

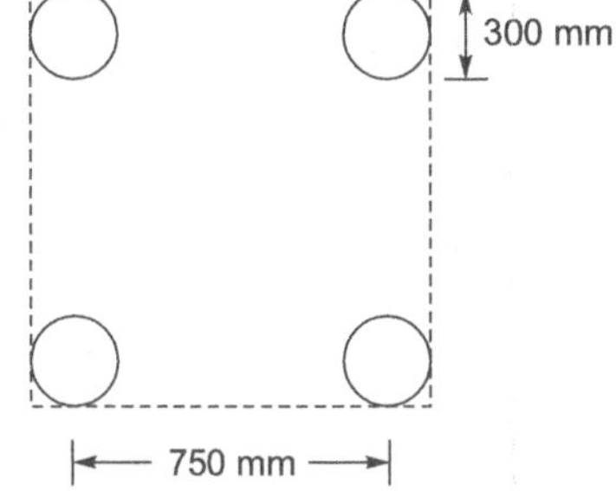

(i) Load capacity on individual pile failure

$$Q_{us} = q_s \cdot A_s + q_b \cdot A_b$$

$$q_s = \alpha\bar{c} = 0.6\times\frac{200}{2} = 60 \text{ kN/m}^2$$

$$A_s = \pi dL = \pi\times 0.3\times 12 = 11.3097 \text{ m}^2$$

q_bA_b term is neglected as the piles are friction piles.

$$Q_{us} = 60 \times 11.3097 = 678.24 \text{ kN}$$

$$Q_{us\ Total} = n \cdot Q_s = 4 \times 678.24 = 2712.96 \text{ kN}$$

$$Q_{an} = \frac{Q_{us,total}}{FOS} = \frac{2712.96}{2} = 1356.48 \text{ kN}$$

(ii) Load capacity of the group based on the block failure.

$$Q_{ug} = q_s A_s$$

$$Q_{ag} = \frac{q_s As}{FOS} = \frac{100 \times 4 \times 1.05 \times 12}{3} = 1680 \text{ kN}$$

(iii) Design load capacity (Q_{safe}) = Minimum of (Q_{an}, Q_{ag}) = 1356.48 kN

10.21 **The soil profile in a particular site consists of a 1.5 m thick filled up soil (N = 3, γ = 17 kN/m^3) followed by 2 m thick very soft clay layer (N = 0, C_u = 5 kN/m^2, γ = 15 kN/m^3). This is followed by 6 m thick sandy layer (av. N value = 8 and γ = 17 kN/m^3), which is followed by 11 m thick stiff clay layer (av. cohesion = 25 kN/m^2, γ = 15 kN/m^3). This is followed by dense sand upto 30 m (av. N value = 50, γ = 19 kN/m^3). The water table is at 1.5 m below GL. Calculate the safe load that a 25 m long 600 mm dia bored cast in situ pile can carry.**

Take for N = 3, ϕ = 24°; N = 8, ϕ = 28°

for N = 50, ϕ = 41°, N_q = 140 and N_γ = 152.

[20 marks : 2020]

Solution:

For 1.5 m filled up soil : Filled-up soil is not compacted, hence assuming negative skin friction in soil.

$$Q_{nf} \text{ using static method } = \frac{1}{2} K\gamma L \tan\delta \cdot A_s = K\bar{\sigma}_V \tan\delta \cdot A_s$$

Note : Pile is bored pile $K = K_o = 1 - \sin\phi = 0.593 = 0.6$

As per IS 2911, take $\delta = \phi$

$$Q_{nf1} = 0.6\left[\frac{0 + 17 \times 1.5}{2}\right] \tan 24(\pi \times 0.6 \times 1.5) = 9.63 \text{ kPa}$$

For 2 m very soft clay : Clay is very soft and N = 0. Hence assuming negative friction in soft clay.

$$Q_{nf2} = \alpha\bar{C}(\pi dL) = 0.7(5) \times \pi \times 0.6 \times 2 = 13.19 \text{ kPa}$$

Note : For very soft clay assume α = 0.7.

For 6 m sandy layer :

$$Q_{sf} \text{ using static method } = K(\bar{\sigma}_V)_{avg} \tan\delta \cdot A_s$$

Note : Assume $K = K_o = 1 - \sin\phi = 0.53$ for bored pile.

As per IS-2911, $\delta = \phi = 28°$

$$(\bar{\sigma}_V)_{avg} = \left[\frac{(1.5 \times 17 + 2 \times 5.19) + (1.5 \times 17 + 2 \times 5.18 + 6 \times 7.19)}{2}\right] = 57.45 \text{ kPa}$$

$$Q_{sf3} = 0.53 \times 57.45 \tan 28(\pi \times 0.6 \times 6) = 183.101 \text{ kPa}$$

For 11 m stiff clay :

$$Q_{sf4} = \alpha\bar{C}(\pi dL)$$

Note : Assuming α = 0.4 for stiff clay.

$$Q_{sf4} = 0.4 \times 25(\pi \times 0.6 \times 11) = 207.34 \text{ kPa}$$

For 4.5 m thick dense sand :

$$Q_{eb} = (\bar{\sigma}_V)_{\text{base}} \cdot N_q \cdot A_b + \left\{\frac{1}{2} B\gamma N_\gamma\right\} A_b$$

$$(\bar{\sigma}_V)_{\text{base}} = 1.5 \times 17 + 2 \times (5.19) + 6 \times 7.19 + 11 \times 5.19 + 4.5 \times 9.19 = 177.465 \text{ kPa}$$

$$Q_{eb} = 177.465 \times 140 \left\{\frac{\pi}{4}(0.6)^2\right\} + \frac{1}{2} \times 0.6 \times 19 \times 152 \times \frac{\pi}{4}(0.6)^2$$

$$= 7024.786 + 244.96 = 7269.754 \text{ kN}$$

Note : For bored pile end bearing can be assumed as $\frac{1}{2}$ to $\frac{2}{3}$ of driven pile.

$$Q_{eb} = \frac{1}{2}(7269.754) = 3634.87 \text{ kN}$$

$$Q_{sf5} = K(\bar{\sigma}_V)\tan\delta A_s = 0.344(\bar{\sigma}_V)_{\text{avg}} \tan\phi(\pi \times 0.6 \times 4.5)$$

$$(\bar{\sigma}_V)_{\text{avg}} = \left(\frac{136.11 + 177.465}{2}\right) = 156.7875 \text{ kPa}$$

$$Q_{sf5} = 0.344 \times 156.787 \tan(41)(\pi \times 0.6 \times 4.5) = 397.69 \text{ kPa}$$

Using Static Method :

$$Q_{up} = Q_{eb} + (-Q_{nf})_1 + (-Q_{nf})_2 + Q_{sf3} + Q_{sf4} + Q_{sf5}$$

$$= 3634.87 - 9.63 - 13.19 + 183.101 + 207.34 + 397.69$$

$$= 4400.188 \text{ kN}$$

Safe Load Carrying Capacity : (Assuming F.O.S. = 2.5)

$$Q_{\text{safe}} = \frac{Q_{up}}{F} = \frac{4400.188}{2.5} = 1760.07 \text{ kN}$$

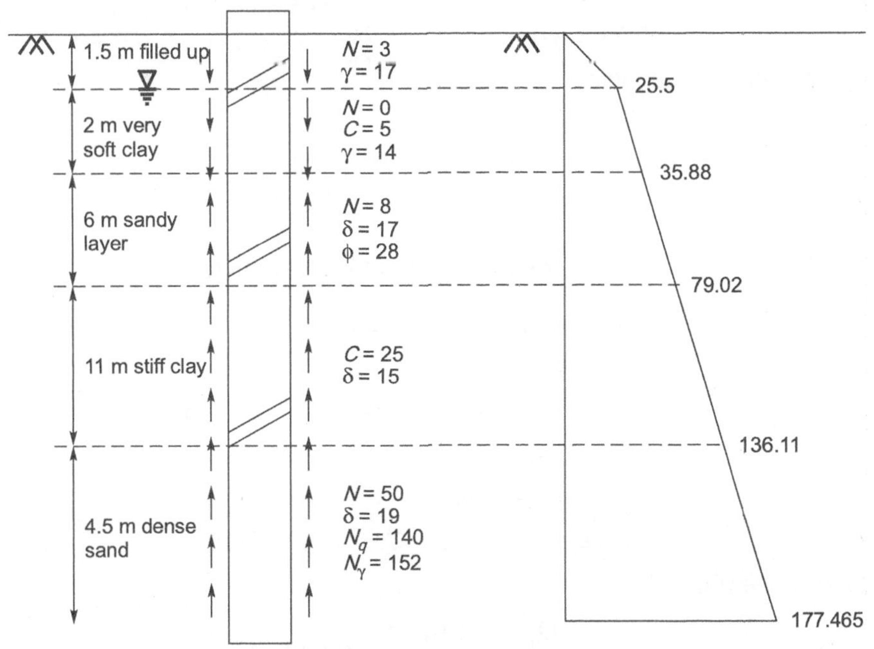

Note:

Consistency	N value	Bored piles	Driven cast in situ piles
Soft to very soft	<4	0.7	1.0
Medium	4-8	0.5	0.7
Stiff	8-15	0.4	0.4
Stiff to hard	>15	0.3	0.3

Values of α

10.22 A concrete pile 450 mm in diameter and 20 m long is driven through a system of layered cohesive soil. The top layer is 8 m thick and comprises of soft clay with cohesion of 30 kN/m^2 and adhesion factor of 0.90. The middle layer which is medium stiff clay has a thickness of 6 m and undrained cohesion of 50 kN/m^2 with adhesion factor of 0.75. The bottom-most layer which is stiff strata extends to a great depth with undrained cohesion of 105 kN/m^2 and adhesion of 0.50. Compute the ultimate and allowable capacity of pipe if the factor of safety assumed is 3.0. The water table is observed to be at ground level.

[12 Marks : 2021]

Solution:

From the data given in question:

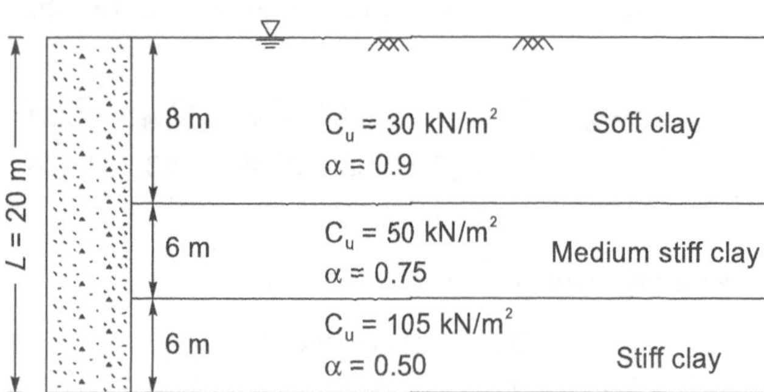

Dia. of pile, $D = 450$ mm

$FOS = 3.0$

To determine ultimate load capacity:

$$Q_{up} = Q_{eb} + Q_{fs}$$

Q_{eb} = Ultimate bearing capacity

$$= C_u N_c A_b$$

$$= 105 \times 9 \times \frac{\pi}{4} \times (0.45)^2 = 150.3 \text{ kN}$$

Q_{fs} = Skin friction resistance

$$= \Sigma \alpha C_u A_s$$

$$= 0.9 \times 30 \times \pi \times 0.45 \times 8 + 0.75 \times 50 \times \pi \times 0.45 \times 6 + 0.50 \times 105 \times \pi \times 0.45 \times 6$$

$$= 1068.77 \text{ kN}$$

$\therefore$ $$Q_{up} = 150.3 + 1068.77 = 1219.07 \text{ kN}$$

Allowable load on the pile,

$$Q_{ap} = \frac{Q_{up}}{FOS} = \frac{1219.07}{3.0} = 406.35 \text{ kN}$$

10.23 A pile group with 12 piles, each having a diameter of 0.5 m and 30 m long, supports a raft foundation. The piles are arranged in 3 rows and spaced at 1.25 m c/c. The properties of the foundation soil are as follows:

$$\gamma' = 11 \text{ kN/m}^3,\ C_u = 40 \text{ kN/m}^2,\ \phi = 0°.$$

Take $\alpha = 0.85$ and F.O.S = 2.5.

Determine the capacity of the pile group.

20 marks : 2022]

Solution:

Given:

Number of piles = 12

Diameter of each pile = 0.5 m

Length of each pile = 30 m

Spacing = 1.25 m c/c

$\gamma' = 11 \text{ kN/m}^3$

$C_u = 40 \text{ kN/m}^2$

$\phi = 0$

$\alpha = 0.85$

FOS = 2.5

s

2.5 + d = 3 m

3.75 + d = 4.25 m

Group action:

$$Q_{ug} = 9 \cdot c \cdot A_b + \alpha \cdot C \cdot A_s$$
$$= 9 \times 40 \times [4.25 \times 3] + 1 \times 40 \times [2(3 + 4.25) \times 30]$$
$$= 21990 \text{ kN}$$

Individual action:

$$NQ_{up} = n[9 \cdot C \cdot A_b + \alpha \cdot C \cdot A_s]$$
$$= 12\left[9 \times 40 \times \frac{\pi}{4} \times 0.5^2 + 0.85 \times 40 \times \pi \times 0.5 \times 30\right]$$
$$= 20074.78 \text{ kN}$$

Now,

$$Q_{safe} = \left\{\begin{matrix} \dfrac{Q_{ug}}{FOS} \\ \dfrac{nQ_{up}}{FOS} \end{matrix}\right\}_{\text{minimum}} = \left\{\begin{matrix} \dfrac{21990}{2.5} = 8796 \text{ kN} \\ \text{OR} \\ \dfrac{20074.78}{2.5} = 8029.9 \text{ kN} \end{matrix}\right\}_{\text{minimum}}$$

$$Q_{safe} = 8029.9 \text{ kN}$$

10.24 A twenty-storeyed building carries a load of 10 kN/m^2 at each floor level. A fully compensated (buoyant) raft foundation is proposed for such building at a soft clay soil site with unit weight of clay as 15 kN/m^3. Find the depth at which the raft foundation needs to be placed.

[10 marks : 2022]

Solution:

Load acting on each floor = 10 kN/m^2

Number of floors = 20

$\therefore$ Total load = 20 × 10 = 200 kN/m^2

For floating/buoyant raft:

$$\text{Total load} = \gamma \times D_f$$
$$\Rightarrow \quad 200 = 15 \times D_f$$
$$\Rightarrow \quad D_f = 13.33 \text{ m}$$

$\therefore$ Depth at which raft is to be placed is 13.33 m.

10.25 A prestressed concrete pipe of 0.5 m diameter is driven in medium dense sand up to 10 m depth. The groundwater table level is at 3 m depth below the surface. The properties of the sand are:

Angle of internal friction of sand, ϕ'	30 degrees
Angle of wall friction between pile and sand, δ	20 degrees
Post-driving horizontal earth pressure co-efficient	1
Saturated unit weight of sand	19 kN-m^3
Unit weight of sand above groundwater table	17 kN-m^3

If the unit shaft resistance reaches a limiting value at 15 D, where D is the diameter of the pile, estimate the skin friction resistance of the pile.

[20 marks : 2023]

Solution:

Given: Diameter of pile = 0.5 m, $\phi' = 30°$, $\delta = 20°$, $k = 1$

Saturated unit weight of sand, $\gamma_{sat} = 19 \text{ kN/m}^3$

Bulk unit weight of sand, $\gamma_{bulk} = 17 \text{ kN/m}^3$

It is given that maximum value of shaft resistance reaches at critical depth (L_c) of $15D = 15 \times 0.5 = 7.5$ m. Therefore, effective stresses will be uniform below 7.5 m depth.

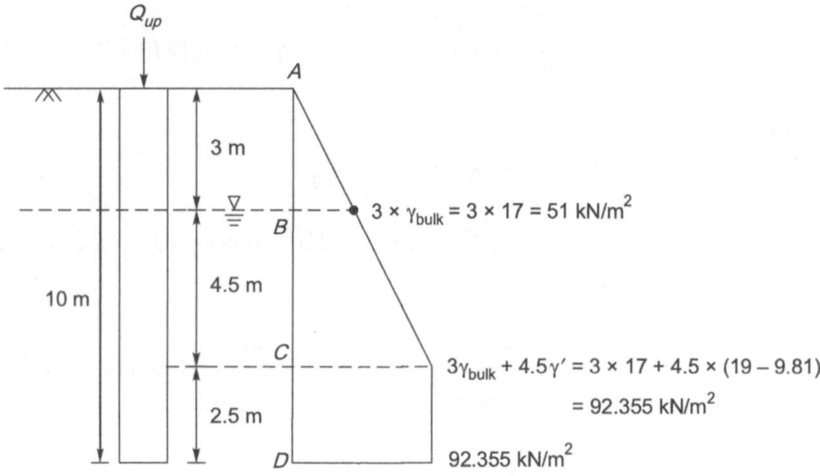

As only skin friction is asked,

$$Q_{up} = Q_{Sf_{AB}} + Q_{Sf_{BC}} + Q_{Sf_{CD}}$$

Now,
$$Q_{Sf_{AB}} = q_{s_1} \cdot A_{s_1}$$

where
$$q_{s_1} = k[\bar{\sigma}_V]_{avg.} \tan\delta$$

$$= 1 \times \left[\frac{0+51}{2}\right] \times \tan 20°$$

$$= 9.281 \text{ kN/m}^2 \leq 100 \text{ kN/m}^2 \quad \text{(o.k.)}$$

$\therefore$
$$Q_{Sf_{AB}} = q_{s_1} \cdot A_{s_1}$$

$$= 9.281 \times \pi \times 0.5 \times 3 = 43.74 \text{ kN}$$

Also,
$$Q_{Sf_{BC}} = q_{s_2} \cdot A_s$$

$$q_{s_2} = 1 \times \left[\frac{51+92.355}{2}\right] \times \tan 20°$$

$$= 26.09 \text{ kN/m}^2 < 100 \text{ kN/m}^2 \quad \text{(O.K.)}$$

$\therefore \quad Q_{Sf_{BC}} = 26.09 \times \pi \times 0.5 \times 4.5 = 184.42 \text{ kN}$

Also, $\quad Q_{Sf_{CD}} = q_{s_3} \cdot A_{s_3}$

$q_{s_3} = 1 \times 92.355 \times \tan 20°$

$= 33.61 \text{ kN/m}^2 \leq 100 \text{ kN/m}^2$ (O.K.)

$\therefore \quad Q_{Sf_{CD}} = 33.61 \times \pi \times 0.5 \times (7 - 4.5) = 132 \text{ kN}$

$\therefore$ Total skin friction $= Q_{Sf_{AB}} + Q_{Sf_{BC}} + Q_{Sf_{CD}}$

$= 43.74 + 184.42 + 132 = 360.16 \text{ kN}$

10.26 **A 600 mm diameter pile is installed up to the bottom of a 16 m thick stiff clayey soil. The pile rests on dense gravelly strata. The average undrained shear strength of the clay is 60 kPa ($\phi_u = 0$) and its saturated unit weight is 18 kN/m³. If the pile has an enlarged base of diameter 1.2 m, determine its ultimate uplift capacity. Assume that the groundwater level is at the ground surface. Ignore the benefit due to the weight of the pile. Take adhesion factor $\alpha = 0.8$ and friction coefficient in uplift $K = 0.5$.**

[12 marks : 2023]

Solution:

(i) Given:

Saturated unit weight, $\gamma_{sat} = 18 \text{ kN/m}^3$, $\phi = 0$, Diameter of pile, $d = 600$ mm

Diameter of bulb, $d_u = 1200$ mm, Adhesion factor, $\alpha = 0.8$

Coefficient of uplift, $K = 0.5$

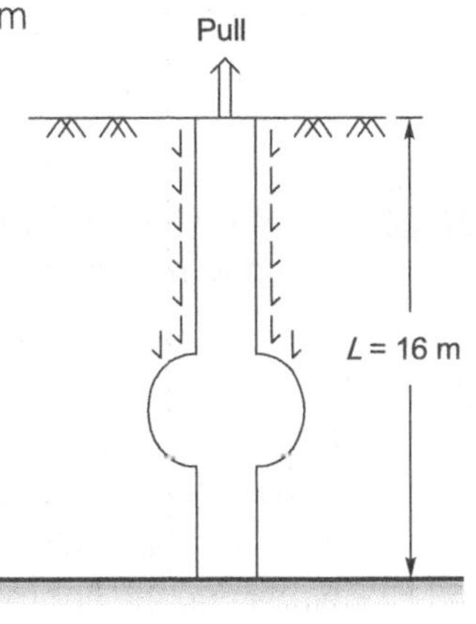

Now, Pullout capacity $= Q_{eb} + Q_{sf}$

$$= 9 \cdot c \cdot \frac{\pi}{4}\left[d_u^2 - d^2\right] + \alpha c \pi dL$$

where c is cohesion equal to undrained shear strength

$$= 9 \times 60 \times \frac{\pi}{4}\left[1.2^2 - 0.6^2\right] + 0.8 \times 60 \times \pi \times 0.6 \times 16$$

$= 1905.7 \text{ kN}$

$\therefore$ Ultimate uplift capacity $= K \times$ Pull capacity

$= 0.5 \times 1905.7 = 952.85 \text{ kN}$

11. Soil Stablization and Soil Exploration

11.1 **Explain with a neat sketch the Resistivity Method of Soil Exploration.**

[10 marks : 1999]

Solution:

The electrical resistivity method is based on the measurement and recording of changes in the mean resistivity or apparent specific resistance of various soils. The resistivity, ρ is usually defined as the resistance between the opposite faces of a unit cube of the material. Significant variations in resistivity can be detected between different types of soil strata or above and below the water table or between unfissured rocks and soils or between voids and soil / rock.

The test is carried out by driving four metal spikes to serve as electrodes into the ground along a straight line at equal distances. Current (I) from a battery flow through the soil between the two outer electrodes, producing an electrical field within the soil. The potential difference E between the two inner electrodes is then measured. The apparent resistivity, ρ is given by the equation

$$\rho = \frac{2\pi DE}{I}$$

It is customary to express D in cm, E in volts, I in amperes and ρ in ohm-cm. The arrangement of the devices in the method is shown in the figure below.

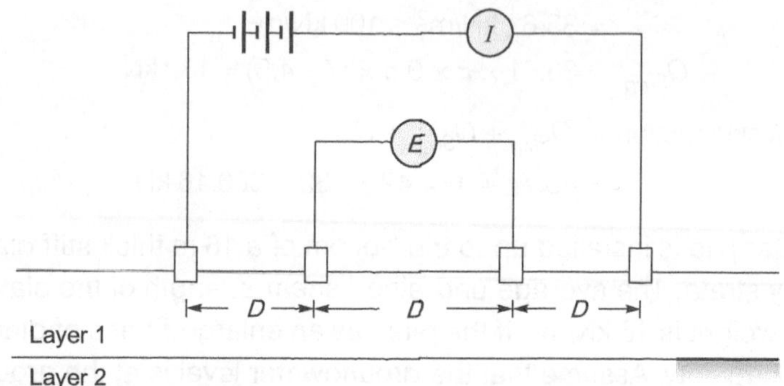

The apparent resistivity is the weighted average of true resistivity upto a depth D in a large volume of soil, the soil close to the surface being more heavily weighted than the soil at greater depths. If a stratum of low resistivity overlies a stratum of higher resistivity, the current is forced to flow closer to the ground surface, resulting in a higher voltage drop and hence a higher value of apparent resistivity. It would be the opposite if a stratum of high resistivity lies above a stratum of low resistivity. The method of sounding is used when the variation of resistivity with depth is required. The method of profiling is used when the lateral variation of soil strata is to be investigated. The electrical resistivity method is not as reliable as the seismic method.

11.2 **Write basic principles used in the following:**
Seismic refraction method of soil exploration.

[5 marks : 2001]

Solution:

Seismic Refraction Method of Soil Exploration: This method is based on the fact that seismic waves have different velocities in different types of soil or rock. Further the waves are refracted when they cross the boundary between different types of soil. The method enables the determination of the general soil types and the approximate depth of boundaries of strata or the bed rock.

The method consists of inducing impact or generating shock by exploding a small charge at or near the ground surface. The radiating shock waves are recorded by a device called geophone which records the time of travel of the waves. The geophones are installed at suitable known distances on the ground in a line from the source of shock or the same is moved away from the geophone to produce shock waves at given intervals. Some of the waves, termed direct or primary waves travel directly from the shock source along the ground surface in the direction of the geophones. Other waves travel in a downward direction at various angles to the horizontal and will be refracted if they pass into a stratum of different seismic velocity. If the underlying layer is denser, the refracted waves travel much faster. As the distance between the shock source and the geophone increases, the refracted waves reach the geophone earlier than the direct waves.

11.3 **Explain with a neat sketch how dynamic cone penetration test is carried out in the field.**

[10 marks : 2005]

Solution:

- In this test, a cone which has an apex angle of 60° is driven into the soil by blows of a hammer of 65 kg falling freely from a height of 750 mm.
- The blow count for every 100 mm penetration of the cone is continuously recorded. The cone is driven till refusal or upto the required depth.

- The number of blows required for 300 mm penetration is noted as the dynamic cone resistance (N_{cd}).
- The test are performed either by using a 50 mm diameter cone without bentonite slurry or a 65 mm diameter cone with bentonite slurry.
- The test is quick and helps to cover a larger area rather economically, then SPT.
- However no sample can be obtained in this test.
- It helps in identifying the uniformlity or the variability of the subsoil profile.

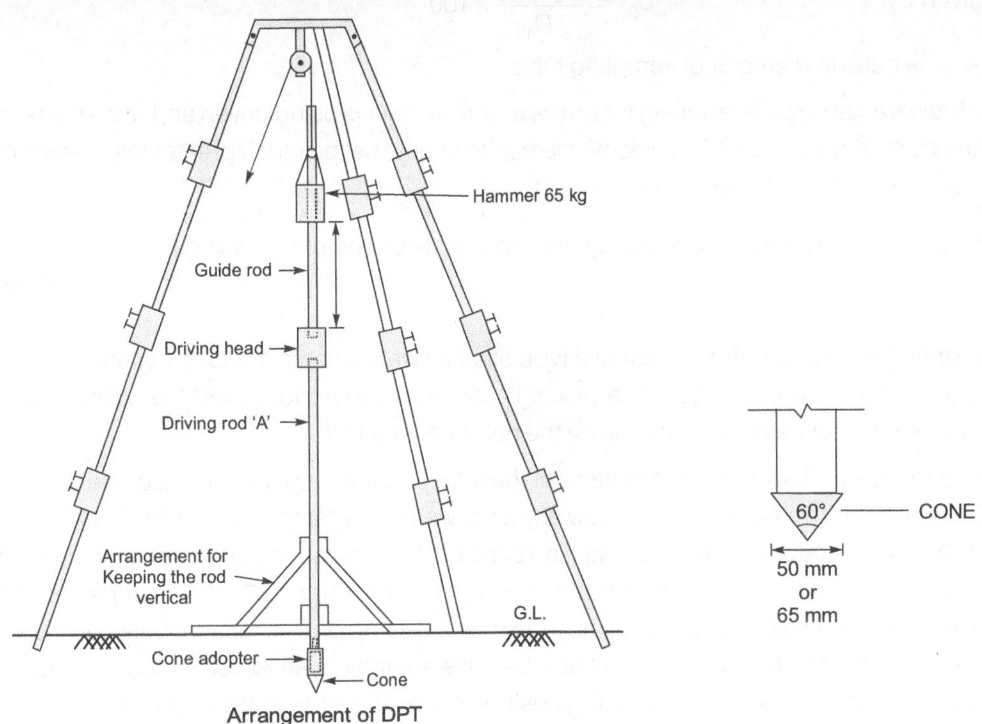

Arrangement of DPT

11.4 **With the help of a diagram, explain the terms with reference to soil sampler: area ratio, inside clearance, outside clearance.**

[6 marks : 2006]

Solution:

Area Ratio: It represents the volume of the soil displaced by the sampler as a ratio of the sample volume. The larger is the value of this ratio, the larger is the sample disturbance. So long as area ratio is less than 10%, the sample disturbance is presumed to be small. However, area ratio value upto 25% is even considered to be good. In order to keep area ratio less, the thickness of the cutting shoe must be reduced. Thin cutting edge is therefore preferred in a soil sampler. It is denoted by A_r.

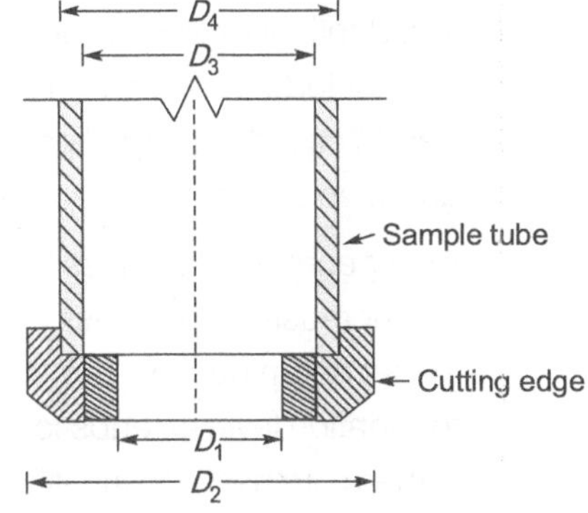

$$A_r = \frac{D_2^2 - D_1^2}{D_1^2} \times 100$$

where D_1 is inside diameter of cutting edge and D_2 is outside diameter of cutting edge

Inside Clearance:

It is given by
$$C_i = \frac{D_3 - D_1}{D_1} \times 100$$

where D_3 is inside diameter of sampling tube

The inside clearance is meant to reduce friction between the soil sample and the sampler when the soil enters the tube, by allowing for elastic expansion. If the inside clearance is too large, there will be too much of lateral expansion. The IS code recommends that the inside clearance C_i should be from 1 to 3%

Outside clearance:

It is given by
$$C_0 = \frac{D_2 - D_4}{D_4} \times 100$$

where D_4 is outside diameter of sampling tube

Outside clearance helps in reducing friction while the sampler is being driven and when it is being withdrawn after the sample is collected. The outside clearance should not be much greater than the inside clearance. Its value usually lies between 0 to 2 per cent.

11.5 Describe stationary piston sampler giving special features of this sampler.

[4 marks : 2006]

Solution:

A stationary piston sampler is a special type of a thin walled sampler, not provided with air relief hole or valve, but provided with a piston. The piston is connected to a rod, called the piston rod which passes through the sampler head and runs inside the hollow boring rod.

Keeping this piston lowered to the bottom of the tube and keeping them locked, the sampler is lowered into the bore hole. This piston will not allow any air or water to enter the tube while pushing the sampler into the soil below the bottom of the bore hole. The piston is made to touch the ground, and only tube is pushed down into the soil, until the sampler head meets the top of the piston. The sampler is then taken out, keeping the piston locked at the top of the tube over the sample. The vacuum between the piston and the sample would help to retain the soil in the tube. This sampler is an excellent tool for obtaining very fine undisturbed samples especially in soft sensitive clays. It can also be used for silts and silty sands, having some cohesion.

11.6 If you are in-charge of subsoil exploration of important structures, how would you decide the depth of exploration? List out the factors you will consider and their importance.

[10 marks : 2008]

Solution:

The depth of exploration in any soil investigation should extend through any unsuitable foundation material and include all soil strata which are likely to undergo settlement due to the load of the structure. This depth, called the "significant depth of exploration" depends on the type of structure its weight size, shape and disposition of loaded areas, intensity of loading, soil profile and soil properties.

As a general rule, unless bed rock is encountered, boring should be carried out to such a depth that the net increase in soil stress due to the load of the structure is less than 5% of the effective stress in the soil at that depth due to overburden. In case of square loaded areas, the isobar of 10% intensity of loading at foundation level extends to a depth of about twice the width of foundation below the base of foundation. It is therefore useful to investigate the subsoil to a depth of atleast twice the width of the anticipated largest size foundation below the base of foundation.

The depth to which seasonal variations affect the soil should be regarded as the minimum depth for the exploration of the sites. But where industrial processes affect the soil characteristics, this depth may be more.

The depth of exploration at the start of the work may be decided as given in table below which be modified as exploration proceeds, if required.

Depth of exploration (IS : 1892: 1979)

S.N.	Type of foundation	Depth of exploration
1.	Isolated spread footings or raft or adjacent footings with clear spacing equal or greater than four times the width	One and half times the width
2.	Adjacent footing with clear spacing less than twice the width	One and half times the length
3.	Adjacent rows of footings	
	(i) With clear spacing between rows less than twice the width	Four and half times the width
	(ii) With clear spacing between rows greater than twice the width	Three times the width
	(iii) With clear spacing between rows greater than or equal to four times the width	One and half times the width
4.	Pile and well foundations	One and half times the width of structure from bearing level (toe of pile or bottom of well)
5.	Road cuts	Equal to the bottom width of the cut
6.	Fill	Two metres below the ground level or equal to the height of the fill whicheve is greater

11.7 Describe 'differential free swell' test of soil.

[5 marks : 2009]

Solution:

Differential free swell test: Two samples of dried soil weighing 10 g each, passing through 425 μ sieve are taken. One is put in a 50 cc graduated glass cylinder containing kerosene oil. The other sample is put in a similar cylinder containing distilled water. Both the samples are left undisturbed for 24 hours and then their volumes are noted.

The differential free swell is expressed as

$$\text{DFS} = \frac{\text{Soil volume in water} - \text{Soil volume in kerosene}}{\text{Soil volume in kerosene}}$$

The table of degree of expansiveness and DFS is given below:

Degree of expansiveness	DFS(%)
Low	Less than 20
Moderate	20 to 35
High	35 to 50
Very high	Greater than 50

In areas where the soils have high or very high DFS, conventional shallow foundations will not be adequate.

11.8 For which geotechnical application does the static cone penetration test have distinct advantage over any other field tests? Why?

[4 marks : 2016]

Solution:

Static cone penetration test is used for various environmental site characterization and ground water monitoring activities. It is particularly used for soft clays and silts and fine to medium sand deposits. It gives a continuous record of variation of both cone resistance and friction resistance with depth. It measures the static resistance of the soil.

Static cone penetration test does not yield any sample. It increases accuracy, speed of deployment, more continuous soil profile and reduce cost over other soil testing methods.

11.9 **Describe any two laboratory methods to measure 'swelling pressure' of soils.**

[8 marks : 2016]

Solution:

Apparatus Required (Common for both methods):

Consolidometer: A device to hold the sample in a ring either fixed or floating with porous stones (or ceramic discs) on each face of the sample. A consolidometer shall also provide means for submerging the sample, for applying a vertical load and for measuring the change in the thickness of the specimen.

Dial gauge: To record the vertical expansion of the specimen precisely.

Specimen Diameter: The specimen shall be 60 mm in diameter (specimens of diameters 50, 70 and 100 mm may also be used in special case) precisely.

Specimen Thickness: The specimen shall be at least 20 mm thick in all cases. However, the thickness shall not be less than 10 times the maximum diameter of the grain in the soil specimen. The diameter to thickness ratio shall be a minimum of 3.

Porous Stones: The stones shall be of silicon carbide or aluminium oxide and of medium grade. It shall have a high permeability compared to that of the soil being tested.

Water Reservoir: To keep the soil sample submerged.

1. **Consolidometer Method:**

   Preparation of specimen from undisturbed soil sample: The container ring shall be cleaned and weighed empty. From one end of the undisturbed soil sample about 30 mm, or more if desired, of the soil sample shall be cut off and rejected. The consolidation ring should be gradually inserted in the sample by pressing with hands and carefully removing the material around the ring. The soil specimen so cut shall project as far as 10 mm on either side of the ring.

   The specimen shall then be trimmed smooth and flush with the top and bottom of the ring. Any voids in the specimen caused due to removal of gravel or limestone pieces, shall be filled back by pressing lightly the loose soil in the voids, care being taken to see that the specimen is not affected. The container ring shall be wiped clear of any soil sticking to the outside and weighed again with the soil. The whole process should be quick to ensure minimum loss of moisture and if possible shall be carried cut in the moisture room. Three representative specimens from the soil trimming shall be taken in moisture content cans and their moisture content determined.

   Preparation of specimen from disturbed soil sample: In case where it is necessary to use disturbed soil samples the soil sample shall be compacted to the desired (field) density and water content in a standard compaction proctor mould. Samples of suitable sizes are cut from it as given above.

   Procedure: All surfaces of the consolidometer which are to be enclosed shall be moistened. The porous stones shall be saturated by boiling in distilled water for at least 15 minutes. The consolidometer shall be assembled with the soil specimen (in the ring) and porous stones at top and bottom of the specimen, providing a filter paper rendered wet between the soil specimen and the porous stone. The loading block shall then be positioned centrally on the top porous stone.

   An initial setting load of 50 gram-force/cm^2 (this includes the weight of the porous stone and the loading pad) shall be placed on the loading hanger and the initial reading of the dial gauge shall be noted.

The system shall be connected to a water reservoir with the level of water in the reservoir being at about the same level as the soil specimen and water allowed to flow in the sample. The soil shall then be allowed to swell.

The free swell readings shown by the dial gauge under the seating load of 5 kN/m^2 (0.05 kgf/cm^2) shall be recorded at different time intervals.

The dial gauge readings shall be taken till equilibrium is reached. This is ensured by making a plot of swelling dial reading versus time in hours, which plot becomes asymptotic with abscissa (time scale). The equilibrium swelling is normally reached over a period of 6 to 7 days in general for all expansive soils.

The compression dial readings shall be recorded till the dial readings attain a steady state for each load applied over the specimen. The consolidation loads shall be applied till the specimen attains its original volume.

2. **Constant Volume Method:**

The consolidation specimen ring with the specimen shall be kept between two porous stones saturated in boiling water providing a filter paper between the soil specimen and the porous stone. The loading block shall then be positioned centrally on the top of the porous stone.

This assembly shall then be placed on the plate of the loading unit as shown in figure. The load measuring proving ring tip attached to the load frame shall be placed in contact with the consolidation cell without any eccentricity. A direct strain measuring dial gauge shall be fitted to the cell. The specimen shall be inundated with distilled water and allowed to swell.

Procedure: The initial reading of the proving ring shall be noted. The swelling of the specimen with increasing volume shall be obtained in the strain measuring load gauge. To keep the specimen at constant volume, the platen shall be so adjusted that the dial gauge always show the original reading. This adjustment shall be done at every 0.1 mm of swell or earlier. The duration of test shall conform to the requirements given in previous method. The assembly shall then be dismantled and the soil specimen extracted from the consolidation ring to determine final moisture content in accordance with IS : 2720 (Part IX)-1973.

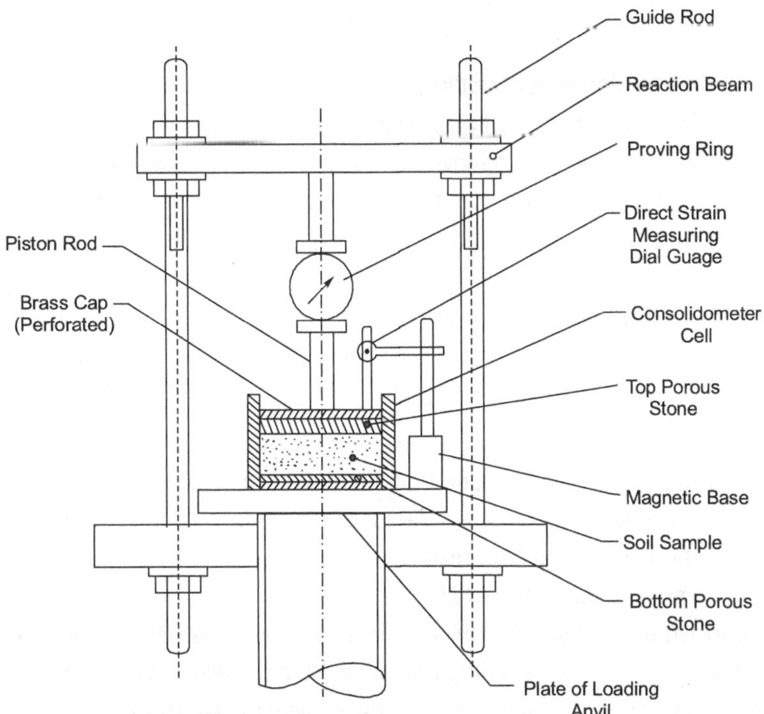

Set up for Measuring Swelling Pressure in the Constant Volume Method

11.10 **When is the dilatancy correction not applied to the N-value in a standard penetration test and why?**

[4 marks : 2016]

Solution:

Dilatancy correction in N value: In case soil is other than silty fine sand or fine sand, dilatancy correction is not required in the Standard Penetration Test. This correction is also not required when the recorded blow count (*N* value) is less than 15 because in this case negative pore water pressure developed is negligible, so don't require to apply the dilatancy correction in case of coarse sand and when *N* value is less than 15.

11.11 **Explain briefly the following factors associated with soil sampler**

(i) Area Ratio (A_r)

(ii) Inside clearance

(iii) Design of non-return valve

[12 marks : 2017]

Solution:

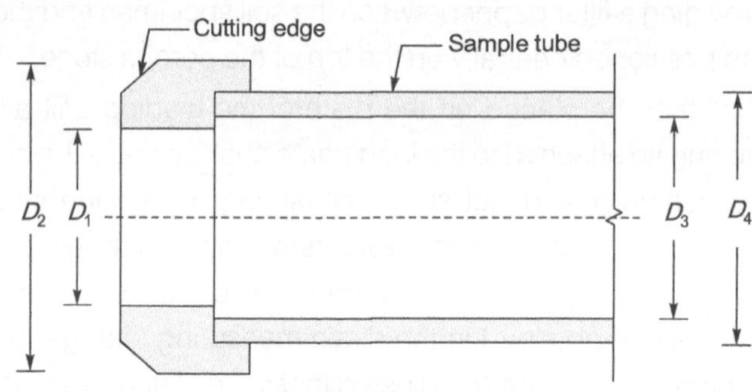

Section of a sampler

(i) **Area ratio, A_r :** The area ratio is one of the factors that determine the mechanical disturbance of the soil being the ratio of the volume of the soil displaced by the sampler tube in proportion to the volume of the sample. It is given as

$$A_r = \frac{D_2^2 - D_1^2}{D_1^2} \times 100\%$$

where, D_1 = Inside diameter of cutting edge

D_2 = Outside diameter of cutting edge

The area ratio, A_r should be kept as low as possible consistent with the strength requirements of the drive shoe and the sampling tube so as to reduce sample disturbance.

Its value should not be greater than about 20 percent for stiff formations whereas for soft sensitive clays, an area ratio of 10% or less is preferred.

(ii) **Inside clearance ratio:** It is defined as

$$C_i = \left(\frac{D_3 - D_1}{D_1}\right) \times 100 \text{ percent}$$

D_3 = Inside diameter of sampling tube

The inside clearance is meant to reduce friction between the soil sample and the sampler when the soil enters the tube, by providing an allowance for elastic expansion. If the inside clearance is too large, there will be too much of lateral expansion.

Inside clearance should be from 1 to 3%.

(iii) **Design of Non-return valves:** A non-return valve allows the escape of air or water as the sample enters the tube and assists in retaining it when the tube is withdrawn from the ground.

The non-return valve should have a large orifice to allow the air and water to escape quickly and easily when driving the sampler.

11.12 List four major uses of geotextiles. What is the difference between permittivity and transmissivity? [6 marks : 2018]

Solution:

Four major used of geotextiles are as follows:

1. **Geotextiles as separators**

   Main use of geotextiles as separator is in construction of highways on clayey soils. As particles size of granular base of highway is much larger than that of the subgrade (clay), a geotextile sheet is used between subgrade and the base course to prevent migration of clay particles into the bases course.

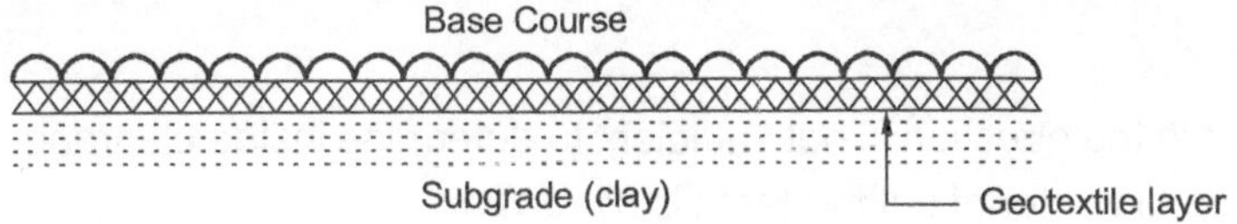

2. **Geotextiles as filter**

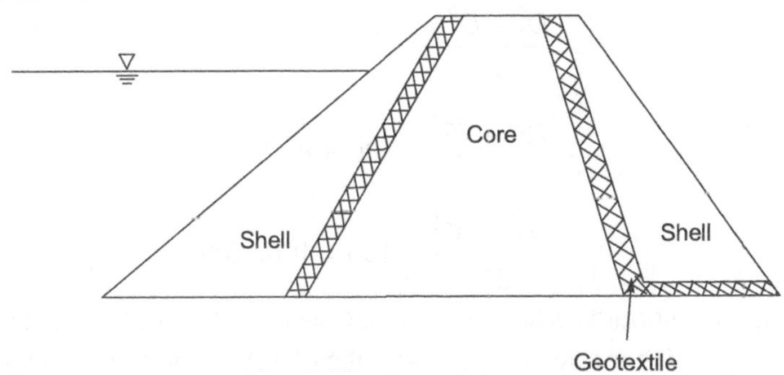

   Geotextile can be used as filters instead of conventional filters, When the silt laden turbid water passes through the textile, the silt particles are prevented from movement by the textile.

3. **Geotextile as drain**

   As geotextile are previous, they can effectively function as drain. The figure shows a typical application where geotextile is used for drainage behind a retaining wall.

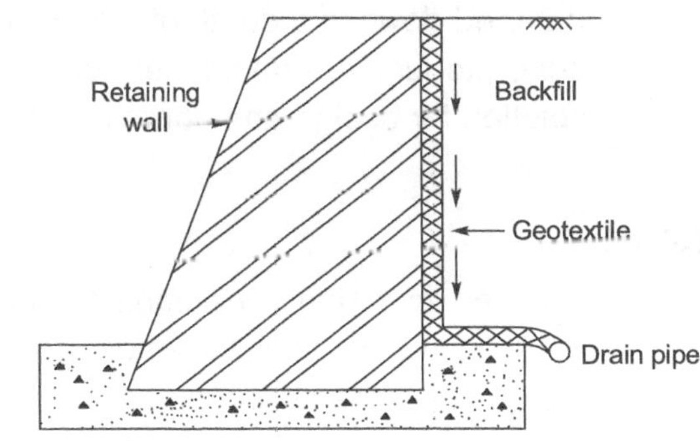

4. **Geotextile as reinforcement for strengthening soil**

   Geotextiles have high tensile strength. These can be used to increase the load carrying capacity of the soil.

   Geotextile are used as reinforcement in the soil, which is poor in tension but good in compression.

   The action is similar to that of steel bars in RC slab.

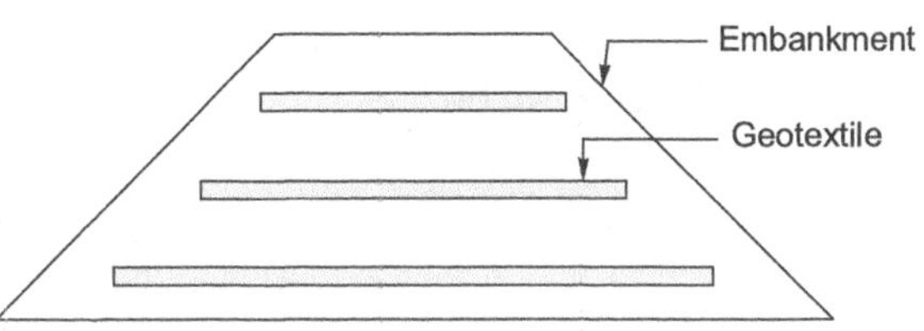

Permittivity is the volumetric flow of water per unit area through the soil or geotextile normal to the plane, as $m^3/m^2/sec$.

Transmissivity is the flow of water within the plane of geosynthetic. It is used for drainage application where water flows within, not just through the textile. It is quoted as m^2/s.

11.13 **The following two size of sampling tubes are available in the market:**

Parameters	Sampling Tube 1	Sampling Tube 2
Outer Dia (mm)	75	50
Inner Dia (mm)	72	35
Length (mm)	600	600

To obtain undisturbed soil sample from borehole, which sampling tube needs to be selected and why?

[12 marks : 2022]

Solution:

The data provided in the question is not sufficient to determine inside clearance, outside clearance and area ratio, because diameter of cutting teeth is not given.

So, as per given data:

$$A_r = \frac{D_0^2 - D_i^2}{D_i^2}$$

For sampling tube-1, $$A_r = \frac{75^2 - 72^2}{72^2} \times 100 = 8.5\%$$

For sampling tube-2, $$A_r = \frac{50^2 - 35^2}{35^2} \times 100 = 104.08\%$$

As per IS1892, area ratio should be kept as low as possible. It's value should not be greater than about 20% for stiff soil and for soft sensitive clays, area ratio of 10% or less should be preferred.

Hence Sample-1 should be selected.

11.14 **A single-storeyed structure is to be constructed at a site in which construction debris has been dumped down to a depth of 3 m over a period of time. The debris is in loose state and consists of concrete lumps, broken tiles and brickbats mixed with soil. Describe how to proceed to find a solution for design and construction of foundation without basement.**

[8 marks : 2023]

Solution:

The filled up soil must be compacted well before undertaking any compaction. Necessary soil filling must be done in order to fill the voids. Concrete blocks may be used to compact the soil sufficiently. After adequate compaction, grade slab may be casted in order to have a firm surface. On this grade slab, singly storey structure can be built.

Grade slab may or may not be provided with reinforcement. However, nominal reinforcement will be beneficial. Minimum thickness of this grade slab must be 100 mm. Before laying the grade slab, a layer of bitumen on the soil can be laid.

■■■■

6 Surveying & Geology

Revised Syllabus of ESE : *Surveying:* *Classification of surveys, various methodologies, instruments & analysis of measurement of distances, elevation and directions; Field astronomy, Global Positioning System; Map preparation; Photogrammetry; Remote sensing concepts; Survey Layout for culverts, canals, bridges, road/ railway alignment and buildings, Setting out of Curves.*
Geology: *Basic knowledge of Engineering geology & its application in projects.*

---

1. Fundamental Concepts of Surveying and Linear Measurements

1.1 Distinguish the pair of the following:
Cumulative error and compensating error in chain surveying.

[2 marks : 2004]

Solution:

Cumulative error and compensating error in chain surveying:

Cumulative errors or systematic errors follow some well defined mathematical or physical law or system. These errors have some well defined pattern. The magnitude and the sign of systematic errors can be determined and a suitable correction can be made to the measured quantity. A systematic error will always have the same sign and magnitude under the same conditions. The sign of correction is always opposite to the error. These errors are cumulative in nature, therefore, they are also known as cumulative errors. Cumulative errors in chaining are due to standardisation and slope.

Accidental errors are random in nature. These errors do not follow any fixed pattern or law. These errors can be positive or negative. Accidental errors are generally of small magnitude and they tend to distribute themselves equally on both sides of the true value. These errors tend to cancel themselves in a series of measurements and are therefore also called compensating errors. For linear measurements, the random (compensating) errors are proportional to $\sqrt{L}$ where L is the length of the line. On the other hand, the systematic (cumulative) errors are proportional to L.

Random errors in chaining generally occurs due to

(a) Incorrect holding of the chain
(b) Incorrect setting of the arrow
(c) Incorrect plumbing
(d) Incorrect reading or interpolation
(e) Incorrect calibration
(f) Variation in temperature and pull

2. Compass Surveying, Theodolites and Traverse Surveying

2.1 What are the errors induced in theodolite survey?

[10 marks : 2000]

Solution:

The errors which occur in theodolite survey may be classified into three types depending upon their sources:

(i) **Instrumental errors:** These may occur due to imperfect adjustments of the theodolite, due to constructional defects in the instrument or due to wear of the various components.

The errors due to imperfect adjustments are most common and some of them are as follows:

(a) Errors due to imperfect adjustments of the plate level
(b) Error due to line of collimation not being perpendicular to horizontal axis.
(c) Error due to horizontal axis not being perpendicular to the vertical axis.
(d) Error due to eccentricity of inner and outer axes.
(e) Error due to eccentricity of verniers.
(f) Error due to imperfect graduations on the horizontal scale.
(g) Vertical index error.
(h) Vertical cross hair not perpendicular to the horizontal axis.
(i) Error due to defective tripod.

(ii) **Personal errors:** These can be classified into two categories:

(a) **Errors in manipulation:** These can occur due to inaccurate centring, inaccurate levelling, slip in screws and improper use of tangent screws.

(b) **Errors in sighting and reading:** These errors occur due to inaccurate vernier reading, inaccurate sighting and parallax.

(iii) **Errors due to natural causes:** These errors occur due to temperature effect, wind effect, refraction effect and settlement of tripod.

2.2 A closed traverse has the following lengths and bearings: The length *CD* could not be measured due to some obstruction to chaining. The bearing of AB could not be taken as station *A* is badly affected by local attraction. Find the exact bearing of the side *AB* and calculate the length *CD*.

[15 marks : 2000]

Line	Length	Bearing
AB	200.0 m	Roughly east
BC	98.0 m	178°
CD	Not obtained	270°
DA	86.4 m	1°

Solution:

Let the length of line *CD* be x and the bearing of line *AB* be θ.

The closed traverse can be shown as below:

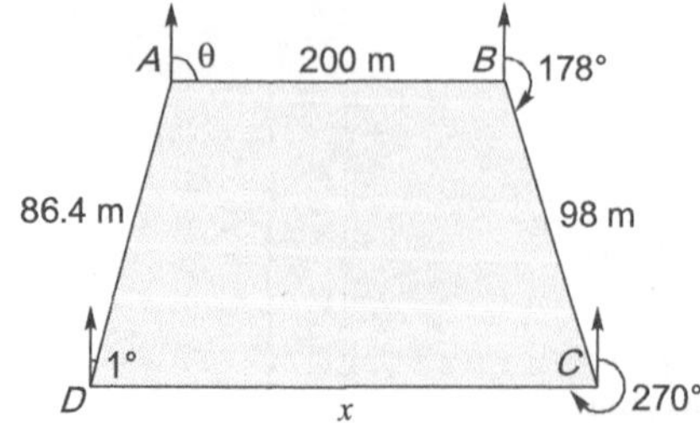

The latitude and departures of each line can be calculated and tabulated below as:

Line	Length	Bearing	Latitude	Departure
AB	200	θ	200 cosθ	200 sinθ
BC	98	178°	–97.94	3.42
CD	x	270°	x cos 270° = 0	x sin 270° = $-x$
DA	86.4	1°	86.39	1.51

We know that for a closed traverse:

(i) Sum of latitudes = 0

(ii) Sum of departures = 0

$$\therefore \quad 200\cos\theta - 97.94 + x\cos 270° + 86.39 = 0$$

$$\Rightarrow \quad 200\cos\theta = 11.55$$

$$\Rightarrow \quad \cos\theta = \frac{11.55}{200}$$

$$\Rightarrow \quad \theta = 86°\,41'\,21.58''$$

$$\text{also} \quad 200\sin\theta + 3.42 + x\sin 270° + 1.51 = 0$$

$$\Rightarrow \quad 200\sin 86°\,41'\,21'' + 3.42 + (-1)x + 1.51 = 0$$

$$\Rightarrow \quad x = 204.596\text{ m}$$

Thus the length of line *CD* is 204.596 m and bearing of *AB* of 86° 41′ 21.58″.

2.3 **The following readings were noted in a closed traverse:**

Line	*AB*	*BC*	*CD*	*DE*	*EA*
FB	32°	77°	112°	122°	265°
BB	212°	262°	287°	302°	85°

At which stations do you suspect local attraction? Find correct bearings of lines? What will be the true for-bearings (as reduced bearings) of lines, if the magnetic declination was 12°W?

[10 marks : 2002]

Solution:

We know that difference of fore bearings and back bearings is 180°. Based on this criteria, the calculations are tabulated below:

Line	*FB*	*BB*	Error	Corrected *FB*	Corrected *BB*
AB	32°	212°	0	32°	212°
BC	77°	262°	+5	77°	257°
CD	112°	287°	+5	107°	287°
DE	122°	302°	0	122°	302°
EA	265°	85°	0	265°	85°

True bearing = Magnetic bearing – Declination (west)

In the above table it is clear that only station C has an error of 5° due to local attraction.

Line	Corrected *FB*	Corrected *BB*	Declination	True corrected FB (WCB)	True corrected BB (WCB)	True corrected FB (RB)
AB	32°	212°	12°W	32°– 12° = 20°	212°– 12° = 200°	N 20° E
BC	77°	257°	12°W	65°	245°	N 65° E
CD	107°	287°	12°W	95°	275°	S 85° E
DE	122°	302°	12°W	110°	290°	S 7° E
EA	265°	85°	12°W	253°	73°	S17° W

2.4 Explain the following terms in the context of surveying:
(i) Least count (ii) Closing error (iii) Arithmetic check (iv) Local attraction (v) Whole to the part
[10 marks : 2003]

Solution:

(i) **Least Count:** It is the smallest possible value of one division of any scale. For a vernier scale, least count is thus equal to the value of the smallest division on the main scale divided by the total number of divisions on the vernier.

If S is smallest division of main scale and n is smallest division of vernier scale

$$\text{Then, least count} = \frac{S}{n}$$

(ii) **Closing error:** In a complete circuit, the sum of north latitudes must be equal to that of south latitudes and the sum of eastings must be equal to that of westings, if linear as well as angular measurements of the traverse along with their computation are correct. If not, the distance between the starting station and the position obtained by calculation is known as closing error.

The closing error is generally expressed as a ratio of the closing error(e) and the perimeter of the traverse (P). In that form it is called relative closing error. Thus,

$$\text{Relative closing error} = \frac{\text{Error of closure}}{\text{Perimeter of traverse}} = \frac{e}{P}$$

It is conventional to express the relative closing error with the numerator as unity.

Thus,
$$\text{Relative closing error} = \frac{1}{P/e}$$

(iii) **Arithmetic check:** These are simple arithmetic calculations which are made to check the correctness of entries in height of instrument method or rise and fall method. The arithmetic calculations can be checked by using the following equation:

$$\Sigma\text{Back sight} - \Sigma\text{Fore sight} = \text{Last}RL - \text{First}RL = \Sigma\text{Rise} - \Sigma\text{fall}$$

Thus, if the calculations on left hand side as well as on right hand side are equal, then the calculations made in filling the entries are correct.

(iv) **Local attraction:** Local attraction is the attraction of magnetic needle to a local magnetic field other than earth's magnetic field. The local magnetic field is caused by iron fences, iron pipes, steel bars, vehicles, steel doors and windows, iron deposits, etc. Even small items made of iron or steel such as the wrist watch case, pen, belt buckle, tapping arrows and steel tapes cause local attraction. DC power lines also develop a local magnetic field. A freely suspended magnetic needle takes the direction of the earth's magnetic field only if these is no local attraction. The magnetic needle will deviate from the magnetic meridian under the local magnetic field (forces).

(v) **Whole to the part:** It is the first principle of surveying. The surveyor should first establish accurately large main framework consisting of widely spaced control points. Between the large main framework, subsidiary small frameworks can be established by relatively less accurate surveys. The errors in small framework are thus localised and are not magnified and the accumulation of errors is controlled. In the reverse process of working from the part to whole, the small frameworks will be expanded to the large framework and the errors will get magnified.

2.5 **Distinguish the pair of the following:**
Loose needle method and fast needle method of theodolite traversing.

[2 marks : 2004]

Solution:

Loose needle method and fast needle method of theodolite traversing:

In loose needle method, the direction of the magnetic meridian is established at each traverse station and the direction of the line is determined with reference to the magnetic meridian. In other words, the magnetic bearing of each line is determined directly.

In fast needle method, the magnetic meridian is established only at the starting station. The magnetic bearing of the first line is measured directly. The magnetic bearings of all other lines are determined indirectly from the magnetic bearing of the first line and included angles.

2.6 **To find out the included angles in a closed survey *PQRSTP*, the following observations were made with the compass. Calculate the included angles after correcting for local attractions.**

Line	Fore bearing (FB)	Back bearing (BB)
PQ	N62°45′E	S62°15′W
QR	N21°00′W	S20°45′E
RS	N71°30′W	S71°30′E
ST	S39°00′W	N38°00′E
TP	S54°30′E	N53°15′W

[15 marks : 2005]

Solution:

Let us convert Quadrantal Bearings into Whole Circle Bearings.

Line	*FB*	*BB*	*FB*(WCB)	*BB*(WCB)	Difference	Corrected FB(WCB)	Corrected BB(WCB)
PQ	N62°45′E	S62°15′W	62°45′	242°15′	−179°30′	62°30′	242°30′
QR	N21°00′W	S20°45′E	339°	159°15′	179°45′	339°15′	159°15′
RS	N71°30′W	S71°30′E	288°30′	108°30′	180°	288°30′	108°30′
ST	S39°00′W	N38°00′E	219°	38°	181°	219°	39°
TP	S54°30′E	N53°15′W	125°30′	306°45′	−181°15′	126°30′	306°30′

The difference of *FB* and *BB* of each line is calculated and the difference is exactly 180° for line *RS*. Therefore station (point) *R* and *S* are free from local attraction. Also, the *BB* of *QR* and *FB* of *ST* are correct.

(i) Correct *FB* of *QR* $= 159^\circ\, 15' + 180^\circ = 339^\circ\, 15'$

Error at *Q* $= 339^\circ - 339^\circ\, 15' = -15'$

Correction at *Q* $= +15'$

(ii) Correct *BB* of *PQ* $= 242^\circ\, 15' + 15' = 242^\circ\, 30'$

Correct *FB* of *PQ* $= 242^\circ\, 30' - 180^\circ = 62^\circ\, 30'$

$$\text{Error at } P = 62^\circ 45' - 62^\circ 30' = 15'$$
$$\text{Correction at } P = -15'$$

(iii) **Correct *BB* of *TP*** $= 306^\circ 45' - 15' = 306^\circ 30'$

$$\text{Correct } FB \text{ of } TP = 306^\circ 30' - 180^\circ = 126^\circ 30'$$
$$\text{Error at } T = 125^\circ 30' - 126^\circ 30' = -1^\circ$$
$$\text{Correction at } T = +1^\circ$$

(iv) **Correct *BB* of *ST*** $= 38^\circ + 1^\circ = 39^\circ$

$$\text{Correct } FB \text{ of } ST = 39^\circ + 180^\circ = 219^\circ$$
$$\text{Error at } S = 219^\circ - 219^\circ = 0$$

(Hence OK)

The closed traverse can be drawn as shown below with the corrected *FB* and *BB* of respective lines.

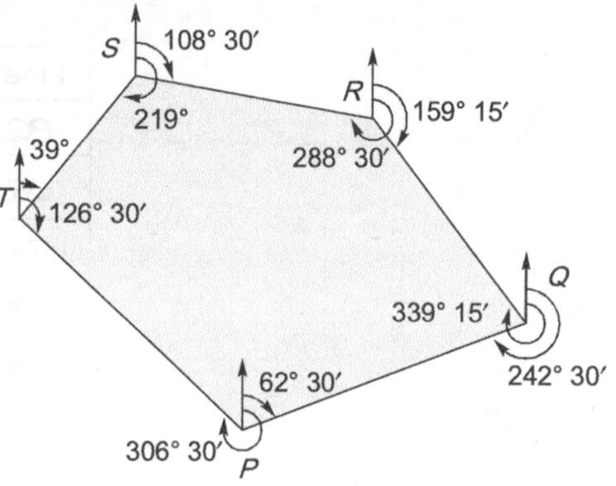

$\angle T = FB \text{ of } TP - BB \text{ of } ST = 126^\circ 30' - 39^\circ = 87^\circ 30'$

$\angle S = FB \text{ of } ST - BB \text{ of } RS = 219^\circ - 108^\circ 30' = 110^\circ 30'$

$\angle R = FB \text{ of } RS - BB \text{ of } QR = 288^\circ 30' - 159^\circ 15' = 129^\circ 15'$

$\angle Q = FB \text{ of } QR - BB \text{ of } PQ = 339^\circ 15' - 242^\circ 30' = 96^\circ 45'$

$\angle P = FB \text{ of } PQ - BB \text{ of } TP = 62^\circ 30' - 306^\circ 30' = -244^\circ$

The negative sign indicates, this angle is exterior.

$\therefore$ Interior $\angle P = 360^\circ - 244^\circ = 116^\circ$

Check:

Sum of all included angles = $(2n-4) \times 90$ where n is the number of lines in the traverse.

$\therefore$ $87^\circ 30' + 110^\circ 30' + 129^\circ 15' + 96^\circ 45' + 116^\circ = (2 \times 5 - 4) \times 90$

$\Rightarrow$ $540^\circ = 540^\circ$ (Hence OK)

2.7 **Discuss the five steps involved in the setting up of a Theodolite.**

[5 marks : 2011]

Solution:

Five steps involved in the setting up of a theodolite are:

1. **Setting Over the Station:** It involves two process centering of the instrument and approximate levelling. Centring of the instrument over the station mark is done by a plumb bob or by optical plummet. Approximate levelling is done either with reference to a small circular bubble provided on tribrach or is done by eye judgement.
2. **Levelling:** After having centred and approximately levelled the instrument, accurate levelling is done with the help of foot screws and with reference to the plate levels. The pupose of the levelling is to make the vertical axis truly vertical.
3. **Elimination of Parallax:** Parallax can be eliminated in two steps: (i) by focusing the eye-piece for distinct vision of the cross-hairs, and (ii) by focusing the objective to bring the image of the object in the plane of cross-hairs.
4. **Adjustment of Line of Sight:** The line of sight should coincide with the optical axis of the telescope. The object of the adjustment is to place the intersection of the cross-hair in the optical axis. Thus, both horizontal as well as vertical hair are to be adjusted.

5. **Adjustment of the Horizontal Axis:** The horizontal axis should be perpendicular to the vertical axis. The object of the adjustment is to make the horizontal axis perpendicular to the vertical axis so that it is perfectly horizontal when the instrument is levelled.

2.8 **In a city property survey, following observations were made. Calculate the latitude, departures and closing error, prior to computation of the area closed in the transverse. Adjust the coordinates using Bowditch's rule.**

[10 marks : 2013]

Line	Length in m	Whole Circle Bearing
AB	89.31	45°10′
BC	219.76	72°05′
CD	151.18	161°52′
DE	159.10	228°43′
EA	232.26	300°42′

Solution:

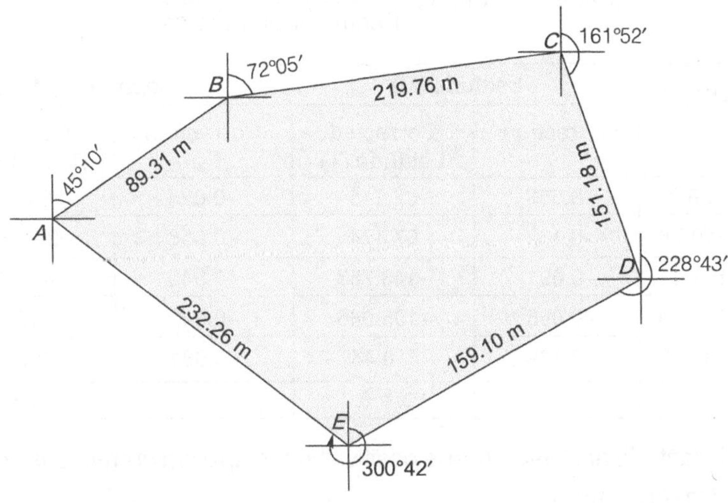

$$\text{Latitude} = h\cos\theta$$

$$\text{Departure} = h\sin\theta$$

Where, L = length of traverse line

θ = Bearing of traverse line

$$\text{Perimeter of traverse} = (89.31 + 219.76 + 151.18 + 159.10 + 232.26)\text{ m}$$

$$= 851.61\text{ m}$$

Traverse Line	Latitude (M)		Departure (M)	
	+	–	+	–
AB	62.968		63.335	
BC	67.606		209.103	
CD		143.672	47.052	
DE		104.971		119.557
EA	118.579			199.709
Σ	**249.153**	**–248.643**	**319.49**	**–319.266**

$$\text{Sum of latitudes} = 249.153 - 248.643 = 0.51\text{ m}$$

$$\text{Sum of departures} = 319.49 - 319.266 = 0.224\text{ m}$$

$\therefore$ Total error in latitude = +0.51 m

and Total error in departure = +0.224 m

$$\text{Closing error} = \sqrt{0.51^2 + 0.224^2} = 0.557 \text{ m}$$

Correction:

(i) Correction in the latitude of any particular line

$$= \text{Total in latitude} \times \frac{\text{Length of that line}}{\text{Perimeter of traverse}}$$

By Bowditch rule

(ii) Correction inthe depature of any particular line

$$= \text{Total error in departure} \times \frac{\text{Length of that line}}{\text{Perimeter of traverse}}$$

Line	Length in m	Latitude (M)		Departure (M)	
		Correction	Corrected Latitude (T)	Correction	Corrected Departure (T)
AB	89.31	–0.053	62.915	–0.023	63.312
BC	219.76	–0.132	67.474	–0.058	209.045
CD	151.18	–0.091	–143.763	–0.040	47.012
DE	159.10	–0.095	–105.066	–0.042	–119.599
EA	232.26	–0.139	118.44	–0.061	–199.770
			$\Sigma = 0$		$\Sigma = 0$

Note: Here error in latitude and departure is positive hence applied corrections are negative.

Computation of Traverse Area

Meridian $M_1 = \dfrac{63.312}{2} = 31.656$ m

$$M_2 = M_1 + \frac{D_1}{2} + \frac{D_2}{2}$$

$= 31.656 + 31.656 + 104.523$

$= 167.835$ m

$$M_3 = 167.835 + \frac{D_2}{2} + \frac{D_3}{2}$$

$= 167.835 + 104.523 + 23.506$

$= 295.864$ m

$M_4 = 295.864 + 23.506 + (-59.7995)$

$= 259.571$ m

$M_5 = 259.571 - 59.7995 - 99.885 = 99.8865$ m

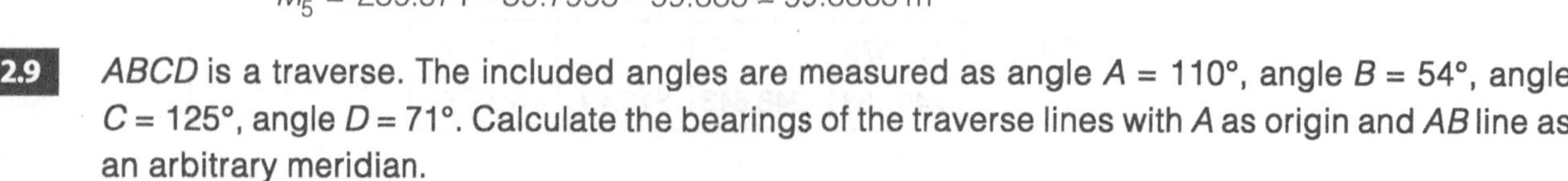

2.9 *ABCD* is a traverse. The included angles are measured as angle *A* = 110°, angle *B* = 54°, angle *C* = 125°, angle *D* = 71°. Calculate the bearings of the traverse lines with *A* as origin and *AB* line as an arbitrary meridian.

[8 marks : 2014]

Solution:

Take line *AB* as horizontal and draw given traverse. Hence bearing of line *AB* = 0°

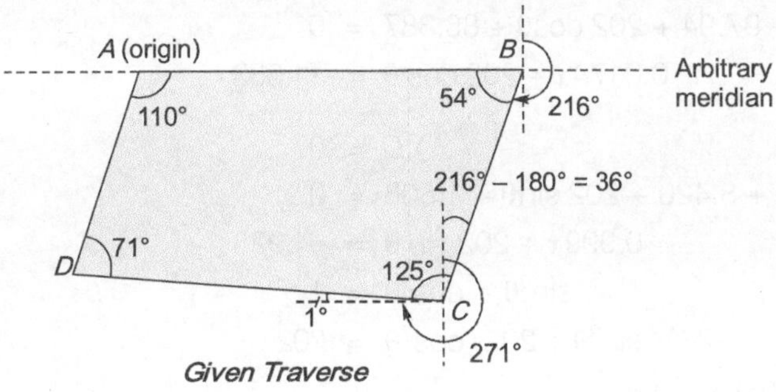

Given Traverse

Bearing of line BC = 270° – 54° = 216° (clockwise)
Bearing of line CD = 360° – 125° + 36° = 271° (clockwise)

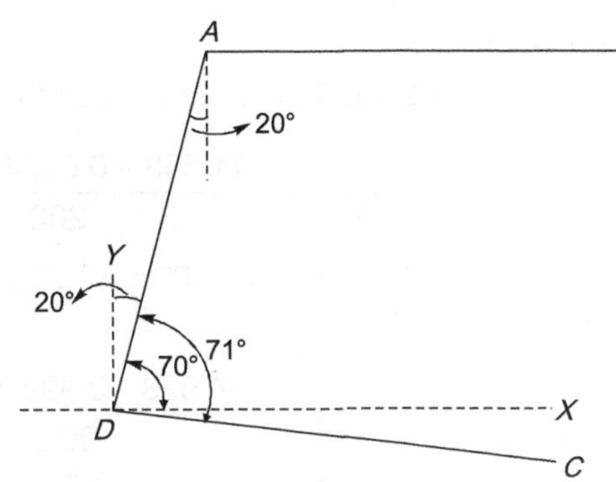

Bearing of line DA = 20° (clockwise)

2.10 During a compass traverse surveying of a closed traverse *PQRS*, the following readings were obtained:

Line	Length (m)	Whole circle bearing
PQ	Could not be measured	89°
QR	98.0	178°
RS	202.0	Roughly west local attraction
SP	86.4	1°

Determine the exact bearing of *RS* and length of *PQ*.

[4 marks : 2014]

Solution:

Latitude $L = l\cos\theta$

Departure, $D = l\sin\theta$

Line	Length	Bearing	Latitude $L = l\cos\theta$	Departure $D = l\sin\theta$
PQ	x	89°	$0.0174x$	$0.999x$
QR	98.0	178°	–97.940	3.420
RS	202.0	θ°	$202\cos\theta$	$202\sin\theta$
SP	86.4	1°	86.387	1.508

For a close traverse,

$$\Sigma\text{Latitude} = 0 \text{ and } \Sigma\text{Departure} = 0$$

(i)
$$\Sigma L = 0$$
$$0.0174x - 97.94 + 202\cos\theta + 86.387 = 0$$
$$0.0174x + 202\cos\theta = 11.553 \quad ...(i)$$

(ii)
$$\Sigma D = 0$$
$$0.999x + 3.420 + 202\sin\theta + 1.508 = 0$$
$$0.999x + 202\sin\theta = -4.928 \quad ...(ii)$$

We know, $\sin^2\theta + \cos^2\theta = 1$

Also $202^2\sin^2\theta + 202^2\cos^2\theta = 202^2 \quad ...(iii)$

From (i), (ii) and (iii), we get
$$(-4.928 - 0.999x)^2 + (11.553 - 0.0174x)^2 = 202^2$$
$$24.285 + 9.846x + 0.998x^2 + 133.472 - 0.4020x + 3.028 \times 10^{-4}x^2 = 40804$$
$$0.9983x^2 + 9.442x - 40646.243 = 0$$
$$x = 197.11 \text{ m}$$

Substituting value of x in equation (i),
$$202\cos\theta = 11.553 - 0.0174x$$
$$\cos\theta = \frac{11.553 - 0.0174 \times 197.11}{202} = 0.0402$$
$$\theta = 87.69° \simeq 87.7° \text{ as } \theta \text{ comes out to be positive.}$$

Substituting value of x in equation (ii)
$$\sin\theta = \frac{-4.928 - 0.999 \times 197.11}{202} = -0.999$$
$$\theta = -87.73° \simeq 87.7°$$

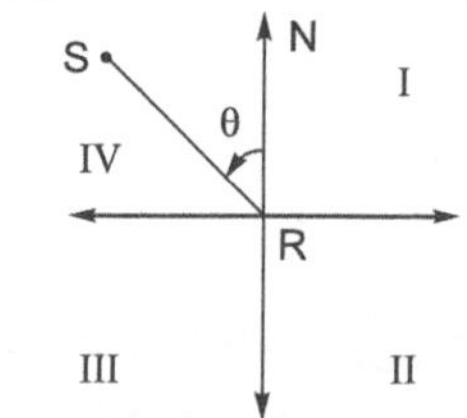

As we get value of sin is –ve and value of cos is +ve hence orientation of line is in (iv) quadrant.

Hence fore bearing of line RS is = 360° – 87.7° = 272.3°

2.11 **Four cars started from station O at the same time. After an hour, these cars reached at stations S1, S2, S3 and S4 respectively. Determine the correct interior angles in a closed compass traverse O-S1-S2-S3-S4-O formed by these cars and original station O. The traversing was done in the clockwise direction. The following bearings were observed in the closed traverse:**

Line	FB	BB
O-S1	S 42° 15′ E	N 42° 15′ W
S1-S2	S 48° 30′ W	N 49° 30′ E
S2-S3	N 76° 45′ W	S 75° 00′ E
S3-S4	N 15° 00′ E	S 16° 00′ W
S4-O	N 63° 45′ E	S 62° 30′ W

[8 marks : 2015]

Solution:

Converting the given bearings to whole circle bearings and applying corrections for local attraction.

Line	FB	BB	Difference	Error	Corrected Bearings	
					FB	BB
O-S1	137°45′	317°45′	–180°	–	137°45′	317°45′
S1-S2	228°30′	49°30′	179°	1°	228°30′	48°30′
S2-S3	283°15′	105°00′	178°15′	1°45′	284°15′	104°15′
S3-S4	15°00′	196°00′	–181°	1°	15°45′	195°45′
S4-O	63°45′	242°30′	–178°45′	1°15′	63°30′	243°30′

As difference in *FB* and *BB* of *O-S*1, is 180° therefore stations *O* and S1 are free from local attraction and all other angles are calculated with respect to them.

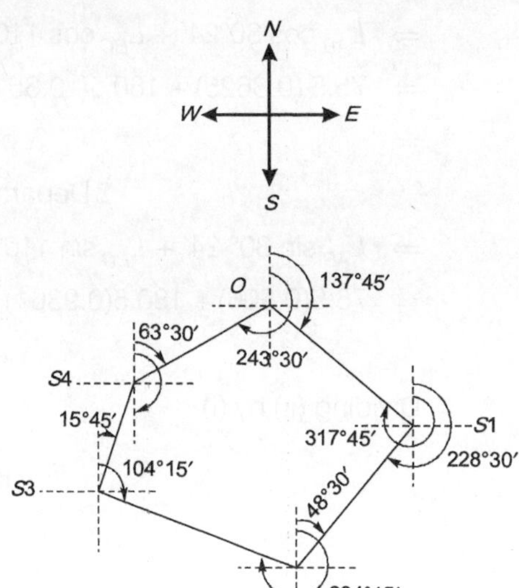

$$\angle O = BB \text{ of } S4O - FB \text{ of } OS1$$
$$= 243°30' - 137°45' = 105°45'$$
$$\angle S1 = BB \text{ of } OS1 - FB \text{ of } S1S2$$
$$= 317°45' - 228°30' = 89°15'$$
$$\angle S2 = BB \text{ of } S1S2 - FB \text{ of } S2S3$$
$$= 48°30' - 284°15' = -235°45'$$
$$= 360° - 235°45' = 124°15'$$
$$\angle S3 = BB \text{ of } S2S3 - FB \text{ of } S3S4$$
$$= 104°15' - 15°45' = -88°30'$$
$$\angle S4 = BB \text{ of } S3S4 - FB \text{ of } S4O$$
$$= 195°45' - 63°30' = -132°15'$$

2.12 **What are meant by 'face left' and 'face right' of theodolite? How would you change face? What instrumental errors are eliminated by 'face left' and 'face right' observations?**

[4 marks : 2016]

Solution:

Face left: When the vertical circle of the theodolite is on the left of the observer then the corresponding observation is called as *face left observation* and the theodolite is said to be in face left condition.

Face right: When the vertical circle of the theodolite is on the right of the observer then the corresponding observation is called as face right observation and the theodolite is said to be in face right condition.

The face of the theodolite is changed by rotating the telescope in the vertical plane through 180 degrees.

The mean of these two observations i.e. face left and face right readings eliminates the error due to collimation.

2.13 **The following records are obtained in a traverse survey where the length and bearing of the last line were not recorded:**

Compute the length and bearing of line DA.

[8 marks : 2016]

Line	Length (in m)	Bearing
AB	75.50	30°24′
BC	180.50	110°36′
CD	60.25	210°30′
DA	?	?

Solution:

Let the length of line DA be x m and whole circle bearing be θ.

Line	Length (in m)	Bearing
AB	75.50	30°24′
BC	180.50	110°36′
CD	60.25	210°30′
DA	x	θ

For a closed traverse,

$$\Sigma \text{ Departures} = \Sigma \text{ Latitudes} = 0$$

$$\because \quad \Sigma \text{ Latitudes} = 0$$

$\Rightarrow \; L_{AB} \cos 30°24' + L_{BC} \cos 110°36' + L_{CD} \cos 210°30' + L_{DA} \cos\theta = 0$

$\Rightarrow \; 75.5\,(0.8625) + 180.5(-0.3518) + 60.25(-0.8616) + x\cos\theta = 0$

$\Rightarrow \quad x\cos\theta = 50.293$ ...(i)

$\because \quad \Sigma\,\text{Departures} = 0$

$\Rightarrow \; L_{AB} \sin 30°\,24' + L_{BC} \sin 110°36' + L_{CD} \sin 210°\,30' + L_{DA} \sin\theta = 0$

$\Rightarrow \; 75.5(0.506) + 180.5(0.9361) + 60.25(-0.5075) + x\sin\theta = 0$

$x\sin\theta = -176.592$ ...(ii)

Dividing (ii) by (i)

$$\tan\theta = \frac{176.592}{50.293} = 3.5113$$

(Neglecting the sign of sinθ and cosθ in the calculation of bearing)

$\therefore \quad \theta = 74.1°$

Now cos θ is +ve and sin θ is –ve and thus θ will lie in II quadrant i.e.

$$\theta = 360° - 74.1°$$
$$= 285.9° = 285°\,54'$$

Substituting θ in any of (i) and (ii),

$$x\cos 285°54' = 50.293$$

$\Rightarrow \quad x = 183.58 \text{ m}$

Length of side *DA* = 183.58 m and bearing of side *DA* = 285°54′.

2.14 **What is repetition method in theodolite surveying? What are different instrumental errors which can be eliminated by the repetition method?**

[6 marks : 2019]

Solution:

Methods of Repetition:

This method is used to measure horizontal angle to a finer degree of accuracy obtainable with the least count of Vernier.

Following errors are eliminated by this method:

1. **Errors due to eccentricity** of Vernier's and centers are eliminated by taking both Vernier readings.
2. **Error due to adjustment** of line of collimation and the trunnion axis are eliminated by taking both face readings.
3. **Error due to in accurate graduations** are eliminated by taking the readings at different parts of the circle.

2.15 **The following internal angles and length of sides are observed for a closed traverse ABCDA (in anti-clockwise direction) :**

Angle	Observed value	Side	Measured length (m)
DAB	92°38′	AB	27.15
ABC	104°33′	BC	52.16
BCD	70°46′	CD	41.96
CDA	92°07′	DA	46.73

Adjust the internal angles for closing error. Also adjust the traverse by Bowditch method and calculate the consecutive coordinate of points A, B, C and D. Assume line AD in north direction.

[20 marks : 2019]

Solution:

$$\text{Sum of internal angles} = \angle A + \angle B + \angle C + \angle D$$
$$= 92°38' + 104°33' + 92°38' + 104°33'$$
$$= 360°04'$$
$$\text{Error} = 360°04' - 360° = 0°04'$$
$$\text{Correction} = -0°04'$$
$$\text{Correction per angle} = \frac{-0°04'}{4} = -0.01'$$

$\therefore$ Correct interior angles, $\angle A = 92°38' - 0°01' = 92°37'$
$$\angle B = 104°33' - 0°01' = 104°32'$$
$$\angle C = 70°46' - 0°01' = 70°45'$$
$$\angle D = 92°07' - 0°01' = 92°06'$$

FB of AB = 92°37′
FB of BC = 104°32′ + 2°37′ − 90° = 17°09′
FB of CD = 17°09′ + 70°45′ − 180° = 267°54′
FB of DA = 180°

Line	Length (L′) (in m)	Fore bearing	Latitude (L)	Correction	Corrected latitude	Departure (D)	Correction	Correction departure
AB	27.15	92°37′	−1.239	−0.054	−1.293	27.1216	−0.09218	27.029
BC	52.16	17°09′	49.84	−0.104	49.736	15.3806	−0.17709	15.203
CD	41.96	267°54′	−1.537	−0.083	−1.62	−41.9318	−0.14246	−42.074
DA	46.73	180°	−46.73	−0.093	−46.823	0	−0.15865	−0.15865
	$\Sigma L' = 168$ m		$\Sigma L = 0.334$	$\Sigma C = 0.334$		$\Sigma D = 0.5704$	$\Sigma C' = -0.5704$	

$$e_x = \Sigma D = 0.5704$$
$$e_y = \Sigma L = 0.334$$
$$\text{Closing error} = \sqrt{(0.5704)^2 + (0.334)^2} = 0.661$$
$$\theta = \tan^{-1}\left(\left|\frac{e_x}{e_y}\right|\right) = 59.65°$$

2.16 Determine the correct magnetic bearings of the lines of closed traverse having the following bearings as observed:

Line	AB	BC	CD	DE	EA
FB	81°05′	100°20′	171°35′	210°50′	300°50′
BB	260°20′	282°35′	351°45′	30°05′	121°10′

[20 marks : 2020]

Solution:

No any line creating difference of 180°.

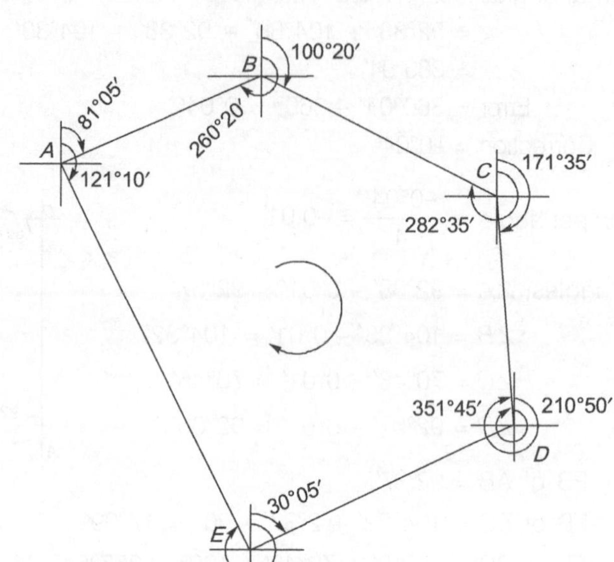

Calculation of internal angle

:

$$\angle A = 121°\ 10' - 81°\ 05' = 40°\ 5'$$
$$\angle B = 260°\ 20' - 100°\ 20' = 160°$$
$$\angle C = 282°35' - 171°35' = 111°$$
$$\angle D = 351°\ 45' - 210°\ 50' = 140°\ 55'$$
$$\angle E = 360° - (300°\ 50' - 30°\ 05') = 89°\ 15'$$

Designated sum of interior angle

$$= (2n - 4) \times 90$$
$$= (2 \times 5 - 4) \times 90$$
$$= 540$$

Total error in internal angle $= 541°\ 15' - 540° = +1°15'$

Total correction $= -1°\ 15'$

Correction per angle $= \frac{1}{5}(1°15') = -15'$

		Correction	Corrected angle
	40°5′	−15′	39°50′
	160°	−15′	159°45′
	111°	−15′	110°45′
	140°55′	−15′	140°40′
	89°15′	−15′	89°
Total	541°15′	−1°15′	540°

Correction for reference:

Line with minimum deviation will have minimum error of local attraction.

Line	$\Delta = \lvert FB - BB \rvert$	Deviation $\lvert 180 - \Delta \rvert$
AB	179°15′	45′
BC	182°15′	2°15′
CD	180°10′	10′
DE	180°45′	45′
EA	179°40′	20′

CD is the line of minimum deviation.

Hence bearings of line *CD* will be corrected.

$$FB \text{ of line } CD = 171°35' + \frac{10'}{2} = 171°40'$$

$$BB \text{ of line } CD = 351'45' - \frac{10'}{2} = 351°40'$$

Now corrected beam:

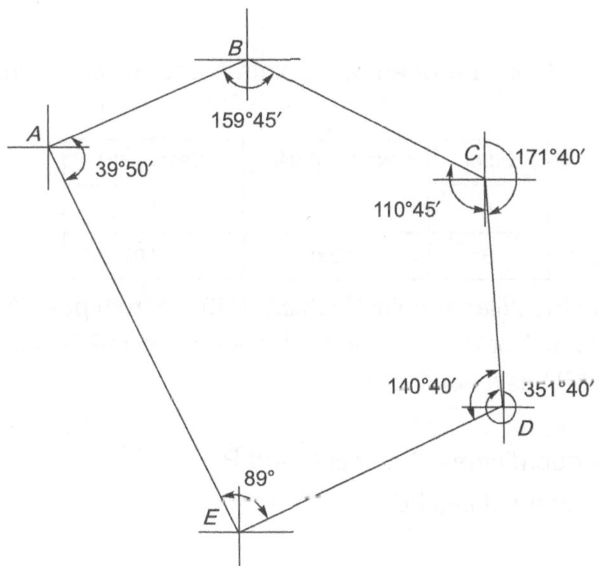

Calculation for corrected value

$$D\begin{bmatrix} BB \text{ of } CD = 351°40' \\ FB \text{ of } DE = 351°40° - 140°40' = 211° \end{bmatrix}$$

140°40′ 351°40′

89° 31°

$$E\begin{bmatrix} BB \text{ of } DE = 211° - 180° = 31° \\ FB \text{ of } EA = 360° - (89° - 31°) = 302° \end{bmatrix}$$

$$A\begin{bmatrix} BB \text{ of } EA = 302° - 180° = 122° \\ FB \text{ of } AB = 122° - 39°50' = 82°10' \end{bmatrix}$$

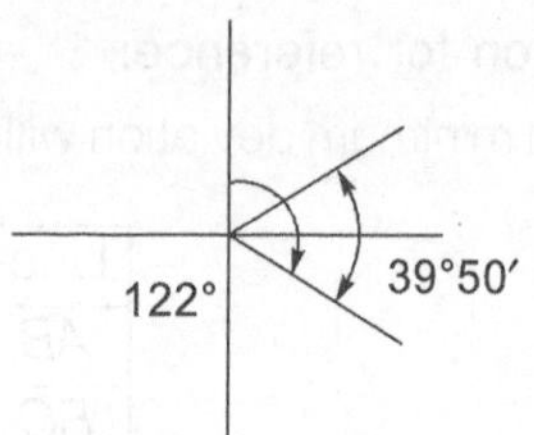

$$B\begin{bmatrix} BB \text{ of } AB = 180° + 82°10' = 262°10' \\ FB \text{ of } BC = 262°10' - 159°45' = 102°25' \end{bmatrix}$$

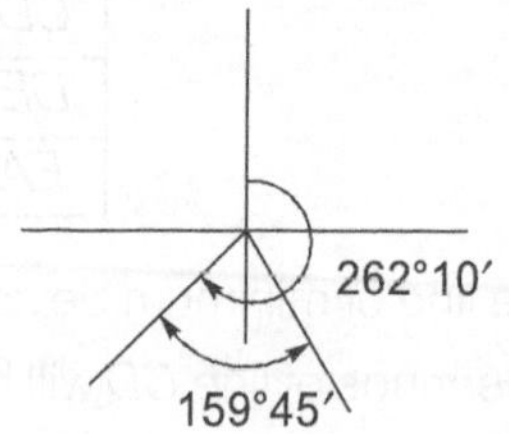

$$C\begin{bmatrix} BB \text{ of } BC = 102°25' + 180° = 282°25' \\ FB \text{ of } CD = 282°25' - 110°45' = 171°40' \end{bmatrix}$$

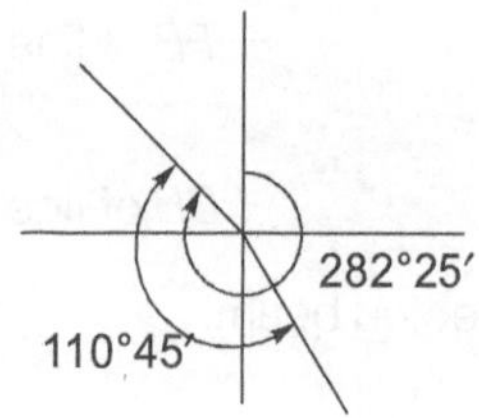

2.17 A straight bridge is set out between two points A and B, whose independent coordinates are given below:

Point	Northing (N)	Easting (E)
A	0	0
B	1200	100

It is required to set out the pillar at point C which is 400 m from point A. It is not possible to set the instrument either at point A or B. To set point C, another point P is selected at a horizontal distance of 600 m from A. Line AP has a bearing of 45°.

Calculate:

(i) The independent coordinates of points C and P.

(ii) The length and bearing of line PC.

[20 marks : 2023]

Solution:

In figure, $\tan\theta = \dfrac{100}{1200} = 4.7636°$

So, Bearing of line AB = 4.7636°

So, Length of line AB = $\sqrt{1200^2 + 100^2}$

= 1204.16 m

Now, Length of line AP = 600 m

Bearing of line AP = 45°

Now, Latitude of AP = 600 cos 45°

= 424.26 m

Departure of AP = 600 sin 45°

= 424.26 m

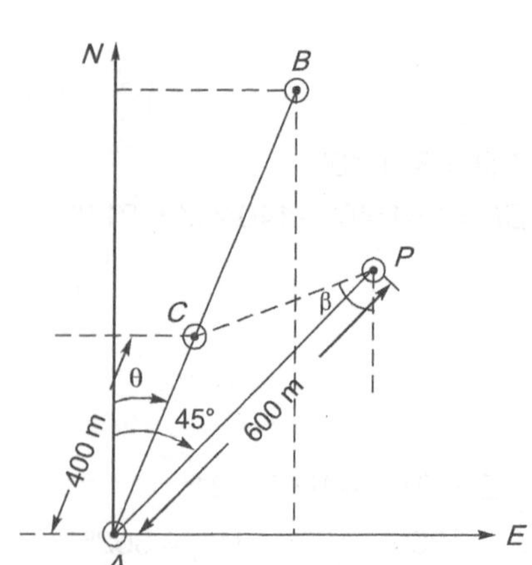

Latitude of AC = 400 cos 4.7636°

= 398.62 m

Departure of AC = 400 sin 4.7636° = 33.22 m

(i) So, Independent coordinate of P = 424.26 N, 424.26 E

So, Independent coordinate of C = 398.62 N, 33.22 E

(ii) For line PC

$$\text{Length of line PC} = \sqrt{(424.26-398.62)^2+(424.26-33.22)^2} = 391.88 \text{ m}$$

As APC is a closed traverse,

So, Departure of PC, ΔD = 424.26 − 33.22 = 391.04 m

So, Latitude of PC, ΔL = 424.26 − 398.62 = 25.64 m

So, $$\beta = \tan^{-1}\left(\frac{\Delta D}{\Delta L}\right) = \tan^{-1}\left(\frac{391.04}{25.64}\right) = 86.26°$$

From the figure, it can be seen that the bearing is S 86.26° W.

3. Levelling, Contouring and Plane Table Surveying

3.1 **The following observations were made in running fly levels from a bench mark of *RL* 60.65 m.**

Back sight	Fore sight
0.964	0.948
1.632	1.153
1.105	1.984
0.850	

Five pegs at 20 m intervals are to be set on a falling gradient of 1 in 100 m from the last position of the instrument. The first peg is to be at *RL* 60 m. Work out the staff readings required for setting the pegs and prepare the page of the level book.

[15 marks : 1999]

Solution:

The page of the level book can be prepared as follows:

Station	*BS*	*IS*	*FS*	*HI*	*RL*	Remarks
1.	0.964	-	-	61.614	60.65	*BM*
2.	1.632	-	0.948	62.298	60.666	-
3.	1.105	-	1.153	62.250	61.145	-
4.	0.850	-	1.984	61.116	60.266	-

Now pegs are to be set on a falling gradient of 1 in 100 at 20 m interval.

∴ Difference in elevation between two consecutive pegs $= \frac{1}{100} \times 20 = 0.2$ m

Staff readings:

Station	*BS*	*IS*	*FS*	*HI*	*RL*	Remarks
4	0.850	-	-	61.116	60.266	*BM*
P_1	-	-	1.116	-	60.00	-
P_2	-	-	1.316	-	59.80	-
P_3	-	-	1.516	-	59.60	-
P_4	-	-	1.716	-	59.40	-
P_5	-	-	1.916	-	59.20	-

3.2 In order to determine the elevation of top Q of a signal on a hill, observations were made from two stations P and R. The stations P, R and signal Q were on the same plane. If the angles of elevation of the top Q of the signal measured at P and R were 25°35′ and 15°5′ respectively, determine the elevation of the foot of the signal if the height of the signal above its base was 4 m. The staff readings upon the Bench Mark (RL 105.42 m) were respectively 2.755 m and 3.855 m when the instrument was at P and at R. The distance between P and R was 120 m.

[10 marks : 2001]

Solution:

The statement of the given problem can be represented through a diagram as shown below:

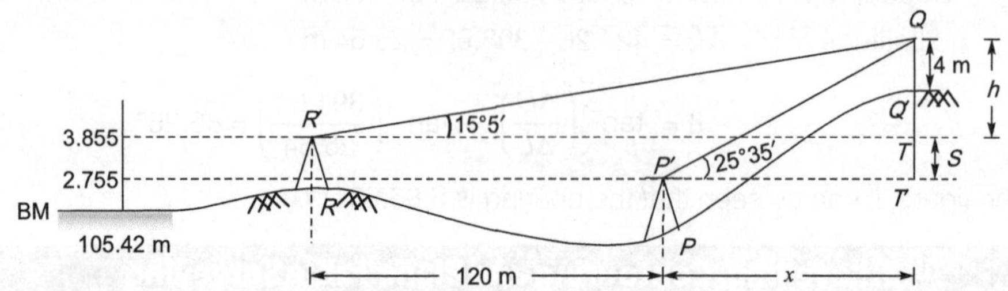

Let 'h' be the elevation of the signal from the line of sight of instrument at R, S be the difference of elevation of instrument at P and R and x be the horizontal distance between instrument at P and the signal.

$$\therefore \quad S = 3.855 - 2.755 = 1.1 \text{ m}$$

In $\Delta P'QT'$, we have

$$\tan 25° 35' = \frac{h+S}{x}$$

$$\Rightarrow \quad x = \frac{h+1.1}{\tan 25° 35'} \quad \text{... (i)}$$

In $\Delta R'QT$, we have

$$\tan 15° 5' = \frac{h}{120+x}$$

$$\Rightarrow \quad x = \frac{h}{\tan 15° 5'} - 120 \quad \text{...(ii)}$$

Equating (i) and (ii), we get

$$\frac{h+1.1}{\tan 25° 35'} = \frac{h}{\tan 15° 5'} - 120$$

$$\Rightarrow \quad h + 1.1 = \frac{\tan 25° 35'}{\tan 15° 5'} \times (h - 120 \tan 15° 5')$$

$$\Rightarrow \quad h + 1.1 = 1.78\, h - 57.45$$

$$\Rightarrow \quad h = \frac{58.55}{0.78} = 75.06 \text{ m}$$

But $\quad h = QQ' + Q'T$

$$\Rightarrow \quad 75.06 = 4 + Q'T$$

$$\Rightarrow \quad Q'T = 71.06 \text{ m}$$

Elevation of foot of signal = BM reduced Level + Staff reading when instrument is at R + $Q'T$

= 105.42 + 3.855 + 71.06 = 180.335 m

3.3 Define a bench mark. What are the different kinds?

A line of levels was run from a Bench Mark of *RL* 51.450 m and ended on a *BM* of *RL* 63.500 m. The sum of the back sights and fore sights were 87.755 and 73.725 respectively. What was the closing error of the work?

[10 marks : 2002]

Solution:

Bench mark is a fixed point of known elevation above the datum. Any point whose elevation is definitely known can be used as a bench mark. However, a relatively permanent object having a marked point whose elevation is known is usually taken as the bench mark. A triangle is generally cut or pointed around the bench mark to indicate the location.

The following types of bench marks are used depending upon the permanency and precision.

(i) **GTS bench mark:** The great trigonometrical survey (GTS) bench marks are established by the Survey of India throughout the country. The levels of the GTS bench marks are determined very accurately with respect to the mean sea level at Bombay Port.

(ii) **Permanent Bench mark:** The permanent bench marks are established at a closer interval between widely spaced GTS bench marks. These bench marks are established by the Survey of India or some other government agency such as PWD. The permanent bench marks are usually established on relatively permanently natural or artificial points such as isolated rocks, culverts, gate pillars of public buildings. These are generally less accurate than GTS bench marks.

(iii) **Temporary bench marks:** These are bench marks established temporarily whenever required. These are generally the points at which a days work is closed and from which next days work is started.

(iv) **Arbitrary bench marks:** These are bench marks whose elevations are arbitrary assumed for levelling of a small area. The elevations assumed do not refer to any fixed datum such as MSL.

Given data:

$$\text{First } RL = 51.45\text{ m; Last } RL = 63.50\text{ m; } \Sigma BS = 87.755\text{ m; } \Sigma FS = 73.725\text{ m}$$

We know that, arithmetic check for entries in a field level book is given by

$$\Sigma BS - \Sigma FS = \text{Last } RL - \text{First } RL$$

If LHS and RHS are equal, then entries are correct, otherwise the difference between LHS and RHS is the closing error of the work. Thus

$$\Sigma BS - \Sigma FS = 87.755 - 73.725 = 14.03\text{ m}$$

$$\text{Last } RL - \text{First } RL = 63.50 - 51.45 = 12.05\text{ m}$$

$$\therefore \quad \text{Closing error} = 14.03 - 12.05 = 1.98\text{ m.}$$

3.4 The field level book readings from a flylevel are as follows:

Staff station	*RL*	*BS*	*FS*	Remarks
BM-1	100.000	3.635	-	
A	×	×	2.375	
B	104.150	4.220	1.030	
C	106.650	3.990	×	
BM-2	108.000	×	×	

Find out the missing values marked (×) and perform the arithmetic check(s).

[10 marks : 2003]

Solution:

The entries in the field level book are tabulated below:

Staff station	*BS*	*FS*	*HI*	*RL*
BM - 1	3.635	-	103.635	100.000
A	3.920	2.375	105.180	101.260
B	4.220	1.030	108.370	104.150
C	3.990	1.720	110.640	106.650
BM - 2	-	2.640	-	108.000

Arithmetic Check:

$$\Sigma BS = 3.635 + 3.920 + 4.220 + 3.990 = 15.765$$

$$\Sigma FS = 2.375 + 1.030 + 1.720 + 2.640 = 7.765$$

$$\text{First } RL = 100.000$$

$$\text{Last } RL = 108.000$$

$$\Sigma BS - \Sigma FS = \text{Last } RL - \text{First } RL$$

$$\Rightarrow \quad 15.765 - 7.765 = 108.000 - 100.000$$

$$\Rightarrow \quad 8.000 = 8.000 \text{ (Hence OK)}$$

3.5 **Distinguish the pair of the following:**

(i) Height of collimation method and rise and fall method of computing levels.

(ii) Basic principle of two point problem and three point problem.

[2 + 2 = 4 marks : 2004]

Solution:

(i) Height of collimation method and rise and fall method of computing levels:

In height of collimation method, the reduced levels of points are calculated by using height of instrument. Whenever the instrument is shifted, the height of instrument is calculated, and then the *RL* are calculated by deducting the fore sight from it.

In rise and fall method of levelling two consecutive readings from the same instrument position give the difference of levels, which may be a rise from the preceding station or a fall. Thus, the rise or fall of each point is determined with respect to the preceding station.

(ii) Basic principle of two point problem and three point problem:

Three point problem can be used, if a surveyor wants to set up the plane table at a station *P* towards which no rays had been drawn when the plane table occupied other stations. Thus three point problem can be used if three well defined points *A*, *B* and *C* whose positions *a*, *b* and *c* had already been plotted on the map are visible from the plane table station *P*.

Principle of two point problem: The two-point problem consists of "locating the position of a plane table station on the drawing sheet by observation of two well-defined points, whose positions have already been plotted on plan."

Principle of three point problem: The three point problem consists of "locating the position of a plane table station on the drawing sheet by observation of three well defined points, whose positions have already plotted on plan."

3.6 **The top of a stack was sighted from two stations *A* and *B*, which are 125 m apart and are in the same vertical plane with the top of the stack. The angle of elevation of the top of the stack observed from instrument station *A* is 35°20′ and that observed from instrument station *B* is 22°28′. The angle of elevation from *B* to a vane 1.75 m above the foot of the staff held at station *A* is 16°15′. The height of instrument at *A* and *B* were 1.856 m and 1.565 m respectively. The *RL* of station *B* is 100.0 m. Find out the *RL* of the top of the stack.**

[20 marks : 2004]

Solution:

It is clear from the statement of the problem that the difference in level of instruments at *A* and *B* is large. That is why the height of instrument at *A* is calculated indirectly by keeping a vane at *A*.

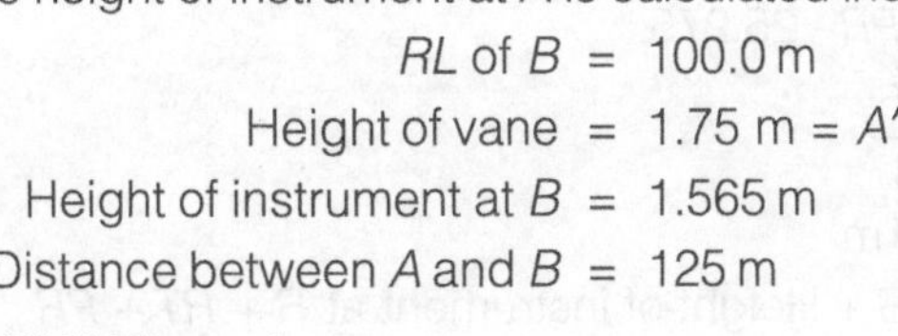

$$RL \text{ of } B = 100.0 \text{ m}$$

$$\text{Height of vane} = 1.75 \text{ m} = A'A$$

$$\text{Height of instrument at } B = 1.565 \text{ m}$$

$$\text{Distance between } A \text{ and } B = 125 \text{ m}$$

∴ In $\Delta B'A'A''$, we have

$$\tan 16° 15' = \frac{A'A + AA''}{A''B'}$$

$$\Rightarrow \quad \tan 16° 15' = \frac{1.75 + AA''}{125}$$

$$\Rightarrow \quad AA'' = 125 \tan 16° 15' - 1.75$$

$$\Rightarrow \quad AA'' = 34.684 \text{ m}$$

$$\therefore \quad RL \text{ of } A = RL \text{ of } B + \text{Height of instrument at } B + AA''$$

$$= 100.00 + 1.565 + 34.684 = 136.249 \text{ m}$$

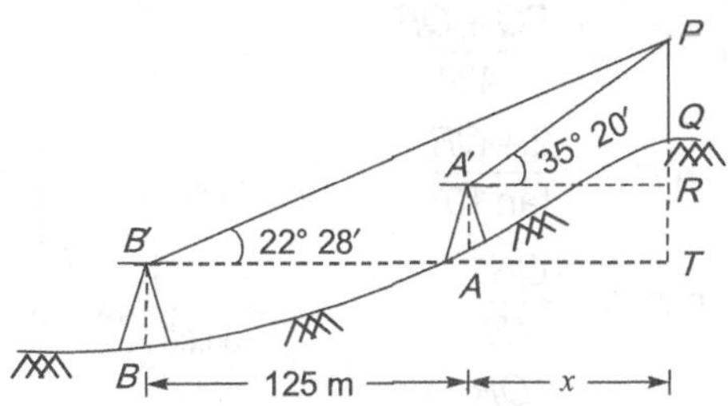

Let x be the horizontal distance between A' and R

$$RL \text{ of } B' = RL \text{ of } B + \text{Height of instrument at } B = 100.00 + 1.565 = 101.565 \text{ m}$$

$$RL \text{ of } A' = RL \text{ of } A + \text{Height of instrument at } A = 136.249 + 1.856 = 138.105 \text{ m}$$

∴ Difference in levels of A' and B' can be given as

$$RT = RL \text{ of } A' - RL \text{ of } B' - 138.105 - 101.565 = 36.54 \text{ m}$$

In $\Delta A'PR$,

$$\tan 35° 20' = \frac{PR}{x}$$

$$\Rightarrow \quad x = \frac{PR}{\tan 35° 20'} \quad \text{...(i)}$$

In $\Delta B'PT$

$$\tan 22° 28' = \frac{PR + RT}{125 + x}$$

$$\Rightarrow \quad 125 + x = \frac{PR + 36.54}{\tan 22° 28'}$$

$$\Rightarrow \quad x = \frac{PR + 36.54 - 125 \tan 22° 28'}{\tan 22° 28'}$$

$$\Rightarrow \quad x = \frac{PR - 15.152}{\tan 22° 28'} \quad \text{...(ii)}$$

Equating (i) and (ii), we get

$$\frac{PR}{\tan 35° 20'} = \frac{PR - 15.152}{\tan 22° 28'}$$

$$\Rightarrow \quad PR = \frac{\tan 35° 20'}{\tan 22° 28'} \times (PR - 15.152)$$

$$\Rightarrow \quad PR = 1.714\, PR - 25.975$$

$$\Rightarrow \quad PR = \frac{25.975}{0.714}$$

$$\Rightarrow \quad PR = 36.379 \text{ m}$$

$$\therefore \quad RL \text{ of top of stack} = RL \text{ of } B + \text{Height of instrument at } B + RT + PR$$

$$= 100.00 + 1.565 + 36.54 + 36.379 = 174.484 \text{ m}$$

3.7 **A flag post of height 2 m was erected on top of a building. Find the *RL* of the top of the flag post, if the vertical angles to the bottom and top of it were measured using a theodolite as 7° and 10° respectively. A staff reading of 1.245 m was taken over a bench mark of + 100.000 with a vertical angle of 0°0′0″.**

[10 marks : 2005]

Solution:

The statement of the problem can be represented through the diagram shown below:

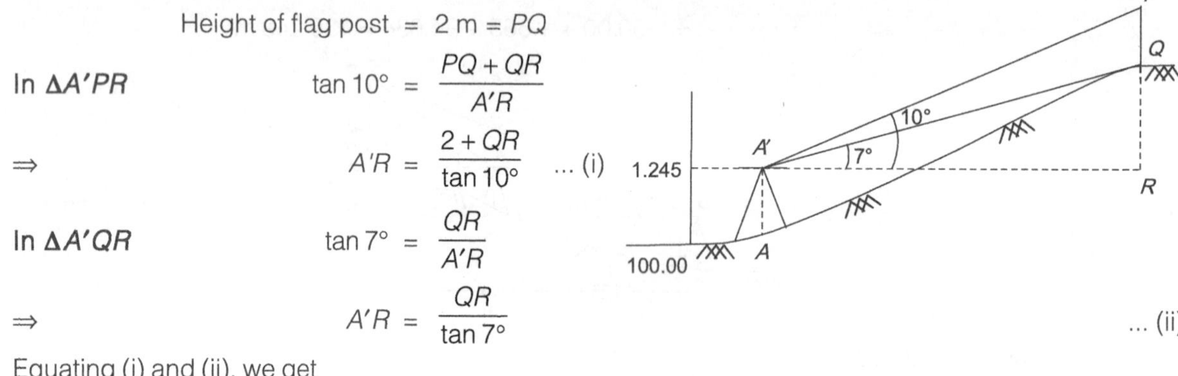

Height of flag post = 2 m = PQ

In $\Delta A'PR$

$$\tan 10° = \frac{PQ + QR}{A'R}$$

$$\Rightarrow \quad A'R = \frac{2 + QR}{\tan 10°} \quad \text{... (i)}$$

In $\Delta A'QR$

$$\tan 7° = \frac{QR}{A'R}$$

$$\Rightarrow \quad A'R = \frac{QR}{\tan 7°} \quad \text{... (ii)}$$

Equating (i) and (ii), we get

$$\frac{2 + QR}{\tan 10°} = \frac{QR}{\tan 7°}$$

$$\Rightarrow \quad 2 + QR = 1.436\, QR$$

$$\Rightarrow \quad QR = 4.587 \text{ m}$$

$$\therefore \quad RL \text{ of top of flag post} = RL \text{ of Bench mark} + \text{Staff reading at } BM + QR + PQ$$

$$= 100.00 + 1.245 + 4.587 + 2 = 107.832 \text{ m}$$

3.8 **In setting up the plane table at a station *P*, the corresponding plotted point '*p*' was not accurately centred over *P*. If the displacement of *P* was 10 cm in a direction at right angles to the ray, determine the displacement of the point from its rule position on the plan if the scale of the plot was 1 cm = 1 m.**

[8 marks : 2007]

Solution:

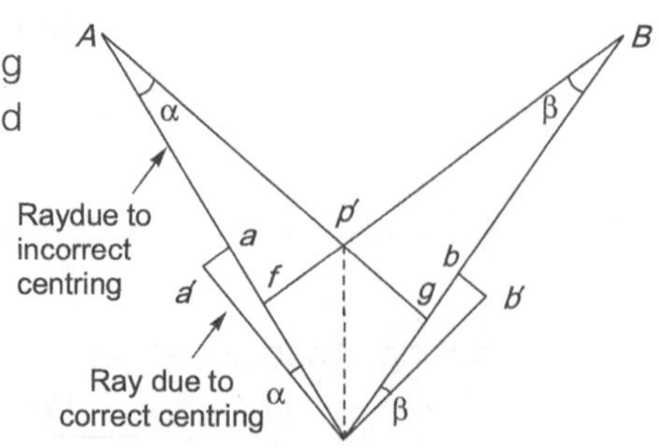

Let 'p' be the plotted position of P, while the position of exact centring should have been p', so that linear error in centring is $e = pp'$ and angular error in centring is $APB - apb = \alpha + \beta$

Dropping perpendicular form p' to ap and bp at f and g respectively

$$\therefore \quad p'f = AP \sin\alpha \text{ and } p'g$$

$$= Bp \sin\beta$$

$$\Rightarrow \qquad \sin\alpha = \frac{p'f}{AP} \approx \alpha \text{ as } \alpha \text{ is small}$$

Similarly $$\sin\beta = \frac{p'g}{BP} \approx \beta \text{ as } \beta \text{ is small}$$

Let us find out the error in plotting of a and *b*. Let us say that *a'* and *b'* are the positions of *A* and *B* for correct centring. Then the error in the position of *A* and *B* will be *aa'* and *bb'* respectively.

$$aa' = pa \times \alpha = pa \times \frac{p'f}{AP}$$

$$bb' = pb \times \beta = pb \times \frac{p'f}{BP}$$

Let $p'f = p'g = e$ and S = fractional scale (RF)

$\therefore$ $pa = PA \times S$ and $pb = PB \times S$

Thus $$aa' = pa \times \frac{p'f}{AP} = PA \times S \times \frac{e}{AP} = S \times e$$

Similarly $bb' = S \times e$

Given data:

Scale is given as 1 cm = 1 m

$$\therefore \quad S = \frac{1}{100}$$

$$e = 10 \text{ cm}$$

$$\therefore \quad aa' = S \times e = \frac{1}{100} \times 10 = 0.1 \text{ cm} = 1 \text{ mm}$$

3.9 **In running fly levels from a *BM* of *RL* 120.750, the following readings were obtained:**

BS	0.850	1.285	1.182	0.965	0.490
FS	0.555	1.150	1.945	1.755	-

From the last position of the instrument, seven pegs at 10.0 m interval are to be set out on a uniform descending gradient of 1 in 50. The first peg is to have a *RL* of 120.00 m. Work-out the staff readings required for setting the tops of pegs on the desired gradient. Enter the results and rule out a page of the level book. Carry out the arithmetic checks.

[10 marks : 2008]

Solution:

The table below shows various entries:

Station	Distance (m)	*BS*	*IS*	*FS*	*HI*	*RL*	Remarks
1	-	0.850	-	-	121.600	120.750	-
2	-	1.285	-	0.555	122.330	121.045	-
3	-	1.182	-	1.150	122.362	121.180	-
4	-	0.965	-	1.945	121.382	120.417	-
5	-	0.490	-	1.755	120.117	119.627	-
6	0	-	0.117	-	-	120.00	Peg1
7	10	-	0.317	-	-	119.800	Peg2
8	20	-	0.517	-	-	119.600	Peg3
9	30	-	0.717	-	-	119.400	Peg4
10	40	-	0.917	-	-	119.200	Peg5
11	50	-	1.117	-	-	119.000	Peg6
12	60	-	-	1.317	-	118.800	Peg7
		$\Sigma BS = 4.772$		$\Sigma FS = 6.722$			

Arithmetic check:

$$\Sigma BS - \Sigma FS = \text{Last } RL - \text{First } RL$$
$$4.772 - 6.722 = 118.800 - 120.750$$
$$-1.95 = -1.95 \quad \text{(OK)}$$

In the first part of the question, fly levelling was done, the computations can be carried out as in the table above. For the last setting of the instrument, when a backsight is taken on station 5, the height of instrument comes out to be 120.117. The *RL* of the first peg is to be 120.00.

$\therefore$ Staff reading $= HI - \text{staff point elevation} = 120.117 - 120.00 = 0.117$

This reading is entered in *IS* column. The *RL* of the next peg at the falling gradient of 1 in 50 will be

$$120.00 - 10 \times \frac{1}{50} = 119.80$$

Its staff reading will be 120.117 – 119.80 = 0.317

Similarly, the staff readings for other pegs are calculated and entered in the table given above.

3.10 **Find the difference in level between two points *A* and *B* in a reciprocal levelling with the following data:**

(i) $R \sin 1'' = 30.876$ m, $\sin 1'' = \frac{1}{206265}$
(ii) Horizontal distance between *A* and *B* = 6882.38 m
(iii) Angle of elevation from *A* to *B* = 1°50′20″
(iv) Angle of depression from *B* to *A* = 1°51′10″
(v) Height of signal at *A* = 4.145 m
(vi) Height of signal at *B* = 3.597 m
(vii) Height of instrument at *A* = 1.463 m
(viii) Height of instrument at *B* = 1.554 m

[8 marks : 2010]

Solution:

Axis signal correction at *A*, $\delta_1 = \dfrac{S_2 - h_1}{D \sin 1''}$

where S_2 = height of signal at B = 3.597 m
h_1 = Height of instrument at A = 1.463 m
D = Horizontal distance between A and B = 6882.38 m

$$\therefore \quad \delta_1 = \frac{3.597 - 1.463}{6882.38 \sin 1''} = \frac{2.134 \times 206265}{6882.38} = 63.956''$$

Axis signal correction at *B*, $\delta_2 = \dfrac{S_1 - h_2}{D \sin 1''}$

where S_1 = Height of signal at A = 4.145 m
h_2 = Height of instrument at B = 1.554 m

$$\therefore \quad \delta_2 = \frac{4.145 - 1.554}{6882.38 \sin 1''} = \frac{2.591 \times 206265}{6882.38} = 77.65''$$

Angle corrected for axis signal correction at *A* is given by

$\alpha_1 = \alpha - \delta_1$ where α is angle of elevation from *A* to *B*

$$\Rightarrow \quad \alpha_1 = 1°50'20'' - 0°0'63.956'' = 1°49'16.04''$$

Angle corrected for axis signal correction at B is given by

$$\beta_1 = \beta + \delta_2 \text{ where } \beta \text{ is angle of depression from } B \text{ to } A$$

$$\Rightarrow \quad \beta_1 = 1°51'10'' + 0°0'77.65'' = 1°52'27.65'' = 6747.65''$$

Taking curvature effect in consideration, we get

$$\theta = \frac{D}{R \sin 1''} = \frac{6882.38}{30.876} = 222.904''$$

$$\Rightarrow \quad \frac{\theta}{2} = \frac{222.904''}{2} = 111.452''$$

$$\therefore \quad \frac{\alpha_1 + \beta_1}{2} = \frac{6556.04'' + 6747.65''}{2} = 6651.845''$$

$$\Rightarrow \quad \frac{\alpha_1 + \beta_1}{2} + \frac{\theta}{2} = 6651.845'' + 111.452'' = 6763.297''$$

Thus the difference in level between points A and B is given by

$$H = \frac{D \sin\left(\frac{\alpha_1 + \beta_1}{2}\right)}{\cos\left[\left(\frac{\alpha_1 + \beta_1}{2}\right) + \frac{\theta}{2}\right]} = \frac{6882.38 \sin 6651.845''}{\cos 6763.297''} = 222.03 \text{ m}$$

H is positive, thus B is higher than A.

3.11 **The following observations were made to determine the sensitivity of two bubble tubes. Determine which bubble tube is more sensitive. The distance of the staff from the instrument was 100 m and the length of one division of both bubble tubes is 2 mm.**

[5 marks : 2011]

Bubble Tube		Bubble Reading LHS	Bubble Reading RHS	Staff Reading
A	(i)	13	05	1.681
	(ii)	08	12	1.767
B	(i)	15	03	1.635
	(ii)	06	14	1.788

Solution:

Bubble tube A:

When bubble is at extreme left division

$$\text{Bubble reading on LHS} = l_1 = 13$$

$$\text{Bubble reading on RHS} = r_1 = 5$$

$$\text{Staff reading} = s_1 = 1.681$$

When bubble is at extreme right division

$$\text{Bubble reading on LHS} = l_2 = 8$$

$$\text{Bubble reading on RHS} = r_2 = 12$$

$$\text{Staff reading} = s_2 = 1.767$$

Total number of divisions for which bubble is moved

$$n = \frac{(l_1 - r_1) - (l_2 - r_2)}{2} = \frac{(13-5)-(8-12)}{2} = 6 \text{ div.}$$

$$\text{Staff intercept} = S = s_2 - s_1 = 1.767 - 1.681 = 0.086 \text{ m}$$

Sensitivity of the bubble tube is given by

$$\alpha' = \frac{S}{nD} \times 206265 \text{ sec}$$

D = distance of the staff from the instrument = 100 m

$$\alpha' = \frac{0.086}{6 \times 100} \times 206265 = 29.56 \text{ sec}$$

Bubble Tube *B:*

Total number of divisions for which bubble is moved

$$= \frac{(15-3)-(6-14)}{2} = 10$$

$$\text{Staff intercept} = 1.788 - 1.635 = 0.153 \text{ m}$$

$$\text{Sensitivity} = \alpha' = \frac{0.153}{10 \times 100} \times 206265 = 31.56 \text{ sec}$$

Bubble tube *A* is more sensitive, because bubble tube *A* can measure smaller angle for one division of movement.

3.12 **A road is to be constructed with a uniform rising gradient of 1 in 100. Determine the staff readings required for setting the tops of the two pegs on the given gradient at 30 meters interval from the last position of the instrument. The *RL* of the first peg is 384.500 m. A fly levelling was carried out from a BM of *RL* 387.000 m. The following observations (in m) were recorded.**

Backsight	1.625	2.345	2.045	2.955
Foresight	1.315	3.560	2.355	

[8 marks : 2015]

Solution:

B.S.(m)	*F.S.(m)*	*H.I.(m)*	*R.L.(m)*
1.625		388.625	387.0
2.345	1.315	389.655	387.31
2.045	3.56	388.14	386.095
2.955	2.355	388.74	385.785

H.I. at last position of instrument = 388.74 m

R.L. of first peg = 384.50 m

Therefore, staff reading at first peg = 388.74 − 384.5 = 4.24 m

$$\text{Staff reading at second peg} = 4.24 + 30 \times \frac{1}{100} = 4.54 \text{ m}$$

3.13 **While levelling between two points *P* and *Q* on opposite banks of a river, the level was first set up near point *P* and the staff readings observed at point *P* and *Q* were 1.385 m and 3.005 m respectively. After this level was set up near point *Q* and the observed readings on point *P* and *Q* were 0.750 m and 2.320 m respectively. If the reduced level of point *Q* was 120.000 m, determine the reduced level of point *P*.**

[6 marks : 2017]

Solution :

Instrument Station	Staff Reading on *P*	Staff Reading on *Q*	
P	1.385	3.005	RL of *Q* = 120.00 m
Q	0.750	2.320	

Instrument near *P*

Incorrect level difference between *P* & *Q* = 3.005 − 1.385 = 1.620 m

Instrument near *Q*

Incorrect level difference between *P* & *Q* = mean of two incorrect differences

$$= \frac{1.620 + 1.570}{2} = 1.595 \text{ m} \qquad \text{(fall from } P \text{ to } Q\text{)}$$

∴ True reduced level of point *P* = (RL of *Q* + 1.595) = 120.00 + 1.595 = 121.595 m

3.14 **The following consecutive readings were taken with auto level and a staff of length 4 m along the centre line of under constructed road having continuously sloping ground at a constant interval of 20 metres:**

3.105, 2.120, 1.115, 0.410, 3.655, 2.310, 1.275,0.310, 3.310, 2.310, 1.200, 0.430

The reduced level of last station was 200.200 m. Rule out the page of level book and enter the readings.

Calculate the reduced level of the points by rise and fall method. Also, find the gradient of the line joining the first and last points.

[15 marks : 2018]

Solution:

Distance (m)	B.S. (m)	I.S. (m)	F.S. (m)	Rise (m)	Fall (m)	R.L. (m)	Remarks
0	3.105					191.28	
20		2.12		0.985		192.265	
40		1.115		1.005		193.27	
60	3.655		0.41	0.705		193.975	C.P.
80		2.31		1.345		195.32	
100		1.275		1.035		196.355	
120	3.31		0.31	0.965		197.32	C.P.
140		2.31		1.00		198.32	
160		1.2		1.11		199.43	
180			0.43	0.77		200.20	
	$\Sigma BS = 10.07$		$\Sigma FS = 1.15$	ΣRise = 8.92	ΣFall = 0		

Check:

$$\Sigma \text{BS} - \Sigma \text{FS} = \Sigma \text{Rise} - \Sigma \text{Fall} = \text{Last RL} - \text{First RL}$$

$$= 8.92 \text{ m}$$

$$\text{Gradient} = \frac{8.92}{180} = \frac{1}{20.18} \text{ or } 4.96\%$$

3.15 A level was set up between two stations A and B. The distance of level from station A was 520 m and the reading on the staff held at A was 1.620 m. The distance of level from station B was 720 m and the reading on the staff held at B was 2.120 m. The reduced level (RL) of point A was 100.000 m. What is the RL of point B? Assume that there is no error in the instrument.

[6 marks : 2019]

Solution:

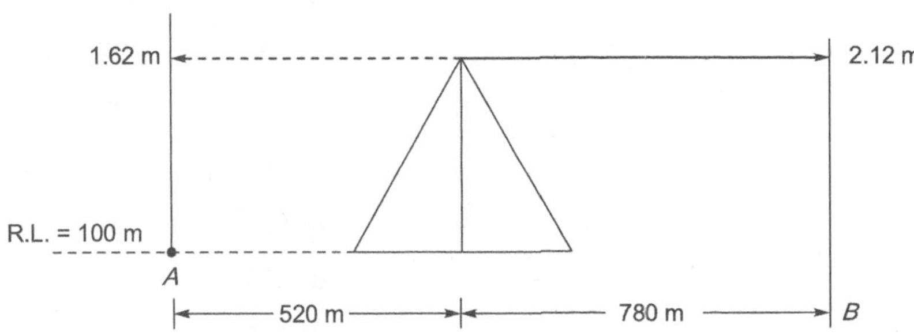

Given: RL of A = 100.00 m

Correct staff reading at station A = $1.62 - 0.0673\ (0.52)^2$

Correct staff reading at station B = $2.12 - 0.0673 \times 0.78^2 = 2.079$ m

Difference in elevation between A and B

$$\Delta h = 2.079 - 1.6018 = 0.4772 \text{ m}$$

$$\text{RL of } B = 100 - 0.4772 = 99.52 \text{ m}$$

3.16 In a running fly level from a benchmark of RL 187.215, the following readings were obtained.

BS	1.115	2.135	1.880	2.725
FS	0.865	3.930	0.880	–

From the last position of the instrument, five pegs at 20 m intervals are to be set out on a uniformly falling gradient of 1 in 40. The first peg is to have an RL of 185.670, Work out the staff readings required for setting the tops of the pegs on the given gradient.

[12 marks : 2020]

Solution:

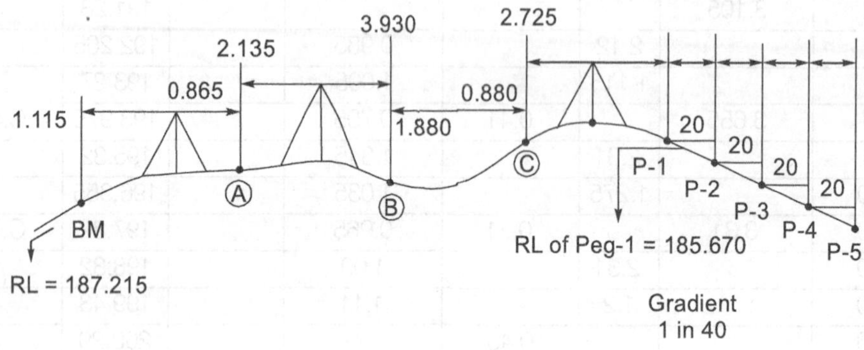

Gradient:
1 in 40

$\Rightarrow$ 1 m fall in 40 m distance

Hence for every 20 m distance fall = 0.5 m

$$\text{RL of Peg-1} = 185.670 \text{ m}$$
$$\text{RL of Peg-2} = 185.67 - 0.5 = 185.17 \text{ m}$$
$$\text{RL of Peg-3} = 185.17 - 0.5 = 184.67 \text{ m}$$
$$\text{RL of Peg-4} = 184.67 - 0.5 = 184.17 \text{ m}$$
$$\text{RL of Peg-5} = 184.17 - 0.5 = 183.67 \text{ m}$$

Level Book:

Station	BS	IS	FS	HI	RL	Remarks
BM	1.115			188.33	187.215	BM
A	2.135		0.865	189.60	187.465	CP
B	1.880		3.930	187.55	185.67	CP
C	2.725		0.880	189.395	186.67	CP
P-1		3.725			185.670	
P-2		4.225			185.170	
P-3		4.725			184.670	
P-4		5.225			184.170	
P-5			5.725		183.670	

Check:

$$\Sigma BS - \Sigma FS = (7.855 - 11.4) = -3.545$$
$$\text{Last RL} - \text{First RL} = (183.67 - 187.215) = -3.545 \qquad \text{(OK)}$$

3.17 Two points *A* and *B* were selected 80 m apart from each other for testing permanent adjustment of a dumpy level using two peg test. Following staff readings were observed at both the points while keeping dumpy level at two different locations:

Instrument Location	Staff Reading (in m)	
	A	*B*
Midway betwen A and B	1.430	1.780
At P (10 m from A and 70 m from B)	1.500	1.950

Calculate the following:

(I) Level difference between point *A* and point *B*.

(II) Inclination of line of sight, if any.

(III) Corrected staff readings at point *A* and point *B* when dumpy level was set at *P* and having no error in its permanent adjustment.

[5 marks : 2022]

Solution:

Assume, inclination error = α (downward)

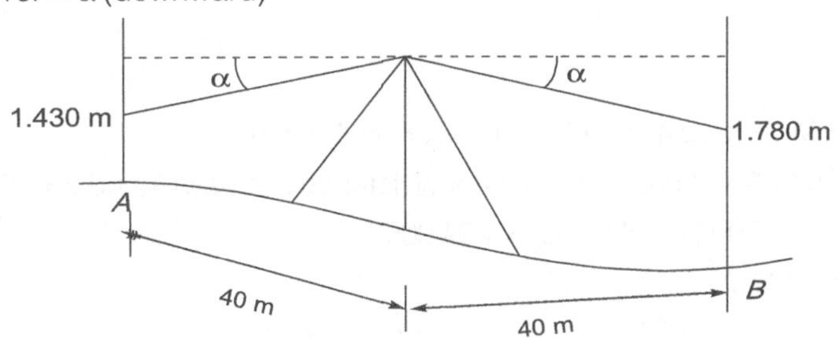

When instrument is kept midway,

$$\text{Collimation error} = 40\tan\alpha$$

Corrected staff reading at A = $1.430 + 40\tan\alpha$

Corrected staff ready at B = $1.780 + 40\tan\alpha$

1. Level difference between A and B

$$H = |\text{Corrected staff reading at } A - \text{Corrected staff reading at } B|$$

$$= |1.430 + 40\tan\alpha - (1.780 + 40\tan\alpha)| = 0.35 \text{ m}$$

2. Inclination of line of sight

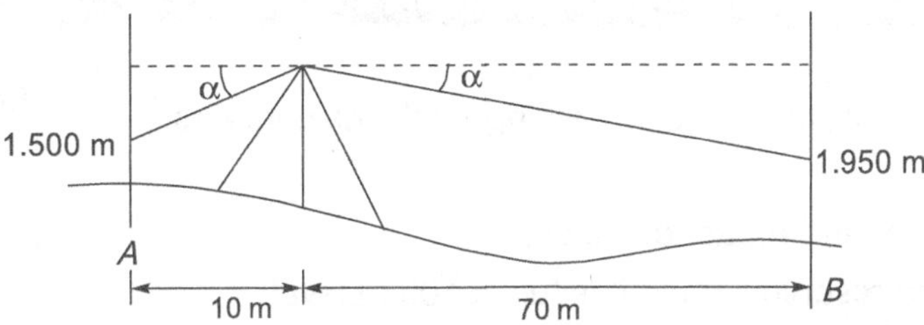

Corrected staff reading at A

$$= 1.5 + 10\tan\alpha$$

Correcterd staff reading at B

$$= 1.95 + 70\tan\alpha$$

$$\text{Level difference} = |\text{Corrected reading at } A - \text{Corrected reading at } B|$$

$$\Rightarrow \quad 0.35 = |1.95 + 70\tan\alpha - 1.5 - 10\tan\alpha|$$

$$\Rightarrow \quad 0.35 = 0.45 + 60\tan\alpha$$

$$\Rightarrow \quad \alpha = -0°5'43.77''$$

It means our assumption was incorrect and inclination error is upward.

3. Corrected staff reading at A

$$= 1.5 - 10\tan\alpha$$

$$= 1.5 - 10\tan 0°5'43.77'' = 1.483 \text{ m}$$

Corrected staff reading at B

$$= 1.95 - 70\tan\alpha$$

$$= 1.95 - 70\tan(0°5'43.77'') = 1.833 \text{ m}$$

3.18 **In running fly levels from a benchmark of reduced level 212.40 m, a surveyor took an intermediate sight of 0.420 m with the staff held on a benchmark of reduced level 264.005 m. The sum of Back Sights and Fore Sights from the start to second BM is 75.205 m and 23.450 m, respectively. What is the closing error on the second benchmark? If the distance between the first BM and second BM is 30 km, comment whether the work is satisfactory for ordinary levelling for location and construction survey or not.**

[12 marks : 2023]

Solution:

Given: R.L. of B.M.$_1$ = 212.4 m, R.L. of B.M.$_2$ = 264.005 m

Σ Back sight = 75.205, Σ Fore sight = 23.450, Distance between 1 and 2 = 30 km

Now, True R.L of B.M.$_2$ = 264.005

Measured R.L of B.M.$_2$

$$\Sigma BS - \Sigma FS = \text{Last RL} - \text{First RL}$$

$$75.205 - 23.450 = B.M._2 - 212.4$$

So, Measured R.L. of B.M.$_2$ = 264.155

Now, Closing error = Measured value – True value

$$= 264.155 - 264.505 = 0.15 \text{ m}$$

$$\text{Permissible error for ordinary levelling} = \pm 0.025\sqrt{d} \quad \text{where } d \text{ is in km}$$

$$= \pm 0.025\sqrt{30} = 0.1369 \text{ m}$$

Since actual error is more than permissible error, work is not satisfactory.

4. Calculation of Area, Volume and Theory of Errors

4.1 In a proposed reservoir, the areas containing within the contours are:

Contour in (m)	Area in (ha)
100	32
95	26
90	24
85	18
80	15
75	13
70	7
65	2

Using the method of end areas, calculate:

(i) Capacity of the reservoir when it is full at 100 m level.

(ii) Elevation of the water level when it is 60% full.

Ignore the volume below RL 65 m. Give your remarks abut the method of *'end areas'*.

[15 marks : 2006]

Solution:

Let us arrange the data in increasing order of the contours.

Contour (m)	65	70	75	80	85	90	95	100
Area (ha)	2	7	13	15	18	24	26	32

(i) Volume of reservoir as per Trapezoidal Rule is given by

$$V = h\left[\frac{A_1 + A_8}{2} + (A_2 + A_3 + A_4 + A_5 + A_6 + A_7)\right]$$

$$= 5\left[\frac{2+32}{2} + (7 + 13 + 15 + 18 + 24 + 26)\right] = 600 \text{ ha-m}$$

Volume of reservoir as per Prismoidal Rule is given by

$$V = \frac{h}{3}\,[A_1 + A_n + 4(A_2 + A_4 + A_6 + ...) + 2(A_3 + A_5 + ...)]$$

But it is applicable only when n is an odd number. Thus applying prismoidal formula from *RL* 65 to 95

$$V = \frac{5}{3}\left[(2 + 26) + 4(7 + 15 + 24) + 2(13 + 18)\right]$$

$$= \frac{5}{3}\,[28 + 184 + 62] = 456.67 \text{ ha-m}$$

Volume of reservoir from *RL* 95 to 100 = $\left(\frac{26+32}{2}\right) \times 5 = 145$ ha-m

$\therefore$ Total Volume = 456.67 + 145 = 601.67 ha-m

Hence, capacity of the reservoir when it is full can be taken as 601.67 ha-m.

(ii) Capacity of reservoir when it is 60% full = $\frac{60}{100}$ **× 601.67 = 361 ha-m**

Capacity of reservoir upto *RL* of 85 m = $\frac{5}{3}\left[(2 + 18) + 4(7 + 15) + (2 \times 13)\right]$

$$= \frac{5}{3}\,[20 + 88 + 26] = 223.33 \text{ ha-m}$$

Capacity of reservoir from *RL* of 85 m to 90 m = $\left(\frac{18+24}{2}\right) \times 5 = 105$ ha-m

$\therefore$ Total capacity of reservoir upto *RL* of 90 m = 223.33 + 105 = 328.33 ha-m

Total capacity of reservoir upto *RL* of 95 m = 456.67 ha-m (calculated above)

Now, for an elevation of 90 m, capacity of reservoir = 328.33 ha-m

$\therefore$ **Elevation for which capacity of reservoir becomes 361 ha-m will be**

$$= 90 + \frac{95 - 90}{(456.67 - 328.33)} \times (361 - 328.33)$$

$$= 90 + 1.273 = 91.273 \text{ m}$$

Method of end areas gives the capacity of reservoir on higher side which is good for design perspective. The results are more authentic than the trapezoidal method.

4.2 The following are the observed values of an angle and their weightage :
Find :

(i) Probable error of single observation of unit weight.
(ii) Probable error of weighted arithmetic mean.
(iii) Probable error of single observation of weight 3.

Angle	Weightage
30° 24′ 20″	2
30° 24′ 18″	2
30° 24′ 19″	3

[10 marks : 2009]

Solution:

Since the error is in seconds only the degrees and minutes of the quantities have not been included in the tabulation. The computations are arranged in the tabular form below:

S.No.	Value	Weight(w)	V	V^2	wV^2
1.	20″	2	1″	1	2
2.	18″	2	−1″	1	2
3.	19″	3	0	0	0
		$\Sigma w = 7$			$\Sigma wV^2 = 4$

Weighted arithmetic mean of the seconds readings of the observed angles

$$= \frac{20'' \times 2 + 18'' \times 2 + 19'' \times 3}{2+2+3} = 19''$$

$\therefore$ Weighted arithmetic mean of the angle $= 30°\,24'\,19''$

$$\therefore \quad V_1 = 20'' - 19'' = 1''$$
$$V_2 = 18'' - 19'' = -1''$$
$$V_3 = 19'' - 19'' = 0$$

(i) Probable error of single observation unit weight is given by

$$E_s = \pm 0.6745\sqrt{\frac{\Sigma wV^2}{n-1}} = \pm 0.6745\sqrt{\frac{4}{3-1}} = \pm 0.95'' \qquad \text{...(i)}$$

(ii) Probable error of the arithmetic mean is given by

$$E_m = \pm 0.6745\sqrt{\frac{\Sigma wV^2}{\Sigma w(n-1)}} = \pm 0.6745\sqrt{\frac{4}{7(3-1)}} = \pm 0.36''$$

(iii) Probable error of single observation of weight 3 is given by

$$E_w = \frac{E_s}{\sqrt{w}} = \pm 0.6745\sqrt{\frac{\Sigma wV^2}{w(n-1)}}$$

$$= \pm \frac{0.95}{\sqrt{3}} \text{ (substuting } E_s \text{ from (i))} = \pm 0.55''$$

4.3 Using Prismoidal Rule, calculate the volume of a 5 m deep pit whose top and bottom dimensions are respectively 10 m × 20 m and 20 m × 40 m.

[2 marks : 2010]

Solution:

Area of top portion, $A_1 = 10 \text{ m} \times 20 \text{ m}$

Area of bottom portion, $A_2 = 20 \text{ m} \times 40 \text{ m}$

Area of mid portion, $A_m = \left[\frac{(20-10)}{5} \times 2.5 + 10\right] \times \left[\left(\frac{40-20}{5}\right) \times 2.5 + 20\right] = 15 \text{ m} \times 30 \text{ m}$

Total depth of pit, $d = 5 \text{ m}$

As per the Prismoidal Rule, the volume of the pit will be given as

$$V = \frac{d}{6}[A_1 + A_2 + 4A_m] = \frac{5}{6}[(10 \times 20) + (20 \times 40) + (4 \times 15 \times 30)]$$

$$= 2333.33 \text{ m}^3$$

5. Tacheometric, Curve, Hydrographic Survey, Tides and Triangulation

5.1 **Determine the gradient from a point *A* to *B* from the following observations made with a fixed hair tacheometer fitted with an anallatic lens, the constant of the instrument being 100.**

	Bearing	Reading on Stadia Hairs		Reading on Axial Hair	Vertical Angle
To *A*	345°	0.750	2.120	1.435	+15°
To *B*	75°	0.625	3.050	1.835	+15°

[10 marks : 2012]

Solution:

Let the instrument station be held at point *P*

Case-I: Observation from *P* to *A*

$$s = 2.120 - 0.750 = 1.37$$

Assuming that the staff is held vertical

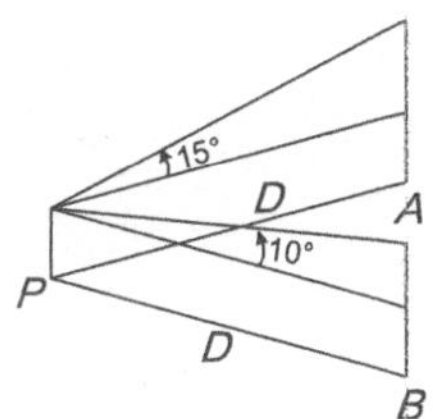

$$\therefore \quad D = ks\cos^2\theta \qquad (\text{As } C = 0)$$

$$\Rightarrow \quad D = 100 \times 1.37 \times \cos^2 15°$$

$$= 127.823 \text{ m}$$

$$V = ks\frac{1}{2}\sin 2\theta$$

$$= \frac{1}{2} \times 100 \times 1.37 \times \sin 30° = 34.25 \text{ m}$$

Hence, difference in elevation between *A* and instrument axis

$= 34.25 - 1.435 = 32.815$ m (A being higher)

Case-II: Observation from *P* to *B*

$$s = 3.050 - 0.625 = 2.425 \text{ m}$$

$$D = ks\cos^2\theta = 100 \times 2.425 \times \cos^2 10° = 235.188 \text{ m}$$

$$V = ks\frac{1}{2}\sin 2\theta = 100 \times 2.425 \times \frac{1}{2} \times \sin 20° = 41.47 \text{ m}$$

$\therefore$ Difference in elevation between *B* and instrument axis

$= 41.47 - 1.835 = 39.635$ m (*B* being higher)

As seen in the figure

$$\angle APB = 360° - 345° + 75° = 90°$$

$\therefore$ $$AB = \sqrt{AP^2 + BP^2}$$

Distance $AP = 127.823$ m

Distance $BP = 235.188$ m

$\therefore$ $$AB = \sqrt{(127.823)^2 + (235.18)^2} = 267.672 \text{ m}$$

Difference in elevation between A and B

$$= (39.635 - 32.815) = 6.82 \text{ m (B being higher)}$$

$\therefore$ Gradient from A to B = $\dfrac{6.82}{267.672}$ = 1 in 39.25 (rising)

5.2 **Two straight AB and BC falling to the right at gradients 12% and 6% respectively, are to be connected by a vertical parabolic curve of length 240 m. Chainage and reduced level of point B are 3000 m and 60.00 m respectively. Calculate the chainage and reduced level of the first three and last three points of the curve by tangent correction method. Take peg interval as 20 m.**

[15 marks : 2018]

Solution:

Two straight AB and BC falling to right.

$$N = |n_1 - n_2| = 6\%$$

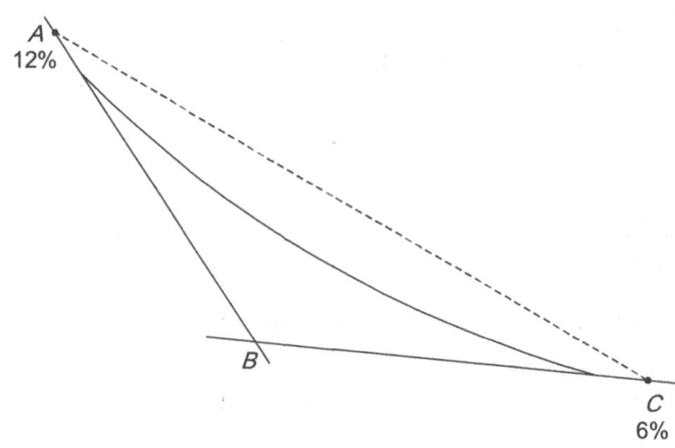

- As per the figure, it will be a sag curve.
- Length of curve AC = 240 m

  No. of chord on one side of apex = $\dfrac{120}{20} = 6$

  Given chainage of B = 3000 m

  RL of B = 60 m
- Chainage of point A = 3000 – (6 × 20) = 2880 m

  Chainage of point C = 3000 + (6 × 20) = 3120 m
- Elevation of tangent AB falls at 12% rate

$\therefore$ Fall per peg interval $= -\frac{12}{100} \times 20 \text{ m} = -2.4 \text{ m}$

Elevation of A = Elevation of B – (– Rate of fall × No. of chords to A)

$= 60 + 2.4 \times 6 = 74.4 \text{ m}$

- Elevation of tangent BC falls at 6% rate

$\therefore$ Fall per peg interval $= \frac{-6}{100} \times 20 = -1.2 \text{ m}$

Elevation of C = Elevation of B + (–Rate of fall × No. of chords till C)

$= 60 - 1.2 \times 6 = 52.8 \text{ m}$

We know, tangent correction is given by

$$y = kx^2$$

To here,

$$k = \left(\frac{e_1 - e_2}{4N}\right) = \left(-\frac{2.4-(-1.2)}{24}\right) = -0.05 \text{ m}$$

$$y = -0.05x^2$$

y will be additive to tangent elevation to get elevation on the curve.

- Elevation of points on curve = Tangent Elevation + Tangent correction

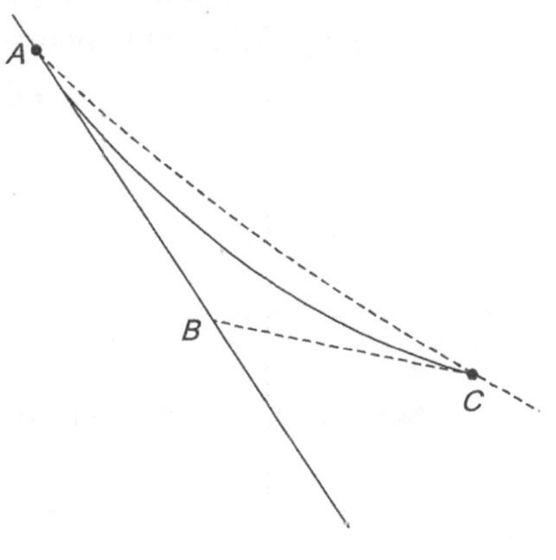

Tangent elevations:

Point-1: $74.4 - 2.4 \times 1 = 72 \text{ m}$

Point-2: $74.4 - 2.4 \times 2 = 69.6 \text{ m}$

Point-3: $74.4 - 2.4 \times 3 = 67.2 \text{ m}$

Point-10: $74.4 - 2.4 \times 10 = 50.4 \text{ m}$

Point-11: $74.4 - 2.4 \times 11 = 48 \text{ m}$

Point-12: $74.4 - 2.4 \times 12 = 45.6 \text{ m}$

Tangent corrections:

$$y = kx^2 = 0.05x^2$$

$$y_1 = 0.05 \times 1 = 0.05 \text{ m}$$

$$y_2 = 0.05 \times 2^2 = 0.2 \text{ m}$$

$$y_3 = 0.05 \times 3^2 = 0.45 \text{ m}$$

$$y_4 = 0.05 \times 4^2 = 0.80 \text{ m}$$

$$y_{10} = 0.05 \times 10^2 = 5.0 \text{ m}$$

$$y_{11} = 0.05 \times 11^2 = 6.05 \text{ m}$$

$$y_{12} = 0.05 \times 12^2 = 7.2 \text{ m}$$

$\therefore$ Elevation of point on curve = Tangent elevation + Tangent correction

Point	Elevation	Chainge
A	74.4m	2880
1.	72.05m	2990
2.	69.8m	2920
10.	55.4m	3080
11.	54.05m	3100
C.	52.80m	3120

5.3 A simple circular curve of radius 30 chain length has been set out to connect two tangents with external deflection angle of 30°. The chainage of point of tangency is 300 chains. On further inspection, it is proposed to alter the radius of curve and point of tangency for revised curve. Also calculate the length of long chord for revised curve. Also calculate the length of long chord for revised curve. [Chain length = 20 m]

[10 Marks : 2019]

Solution:

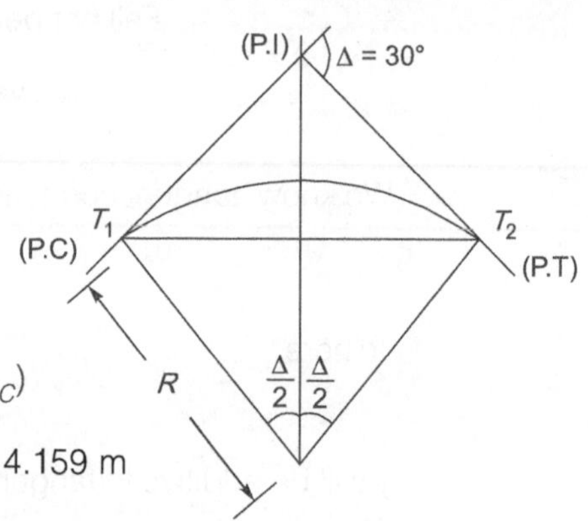

$R = 30$ chains

$= 30 \times 20 = 600$ m

Chainage of point of tangency,

$T_2 = 300$ chains

$= 300 \times 20 = 6000$ m

Also, chainage of T_2 = Chainage of T_1 + Length of curve (L_C)

$$L = 2\pi R \times \left(\frac{\Delta}{360°}\right) = \frac{2 \times \pi \times 600 \times 30°}{360°} = 314.159 \text{ m}$$

$\therefore$ Chainage of $T_1 = 6000 - 314.159 = 5685.84$ m

Chainage of point intersection (PI)

= Chainage of T_1 + Tangent length (T)

$$= 5685.84 + R\tan\frac{\Delta}{2}$$

$$= 5685.84 + 600\tan\left(\frac{30}{2}\right)$$

$$= 5685.84 + 160.77 = 5846.609 \text{ m}$$

Now, revised radius of curve $R' = 45 \times 20 = 900$ m

$$\text{Chainage of } T_1 = R'\tan\frac{\Delta}{2} = 5846.609 - 900\tan\left(\frac{30}{2}\right)$$

$$= 5846.609 - 241.154 = 5605.454 \text{ m}$$

$$\therefore \quad \text{Chainage of } T_2 = \text{Chainage of } T_1 + 2\pi R' \times \left(\frac{\Delta}{360}\right)$$

$$\Rightarrow \quad 5605.454 + 2\pi \times 900 \times \left(\frac{30}{360}\right) = 6076.692 \text{ m}$$

$$\text{Length of long chord } (T_1T_2) = 2R'\sin\frac{\Delta}{2} = 2 \times 900 \times \sin\left(\frac{30}{2}\right) = 465.874 \text{ m}$$

5.4 Two sets of tacheometric readings were taken from an instrument station A (R.L. = 100.50 m) to a staff station B as given below:

Instrument	P	Q
Multiplying constant	95	90
Additive constant	0.25	0.35
Height of instrument	1.35 m	1.40 m
Staff held	Vertical	Normal

Instrument	Instrument station	Staff station	Vertical angle	Stadia readings (m)		
P	*A*	*B*	8°12′	0.905	1.305	1.555
Q	*A*	*B*	8°12′	?	?	?

Determine:

(i) The distance between instrument station and staff station.

(ii) The reduced level of staff station B.

(iii) The stadia readings with instrument Q.

[20 Marks : 2021]

Solution:

Case: 1 Instrument P at station A and staff held vertical at *B*.

$$\left.\begin{array}{l}K=95\\C=0.25\end{array}\right| H.I.=1.35\text{ m},\ S=(1.555\text{-}0.905)=0.65\text{ m}$$

$$L = KS\cos\theta + C$$

$$= 95\times 0.65\cos 8^\circ 12' + 0.25 = 61.368\text{m}$$

$$RL_A = 100.5\text{ m}$$

Distance $$D = L\cos\theta = 61.368\cos 8^\circ 12' = 60.74\text{ m}$$

Elevation, $$V = L\sin\theta = 61.368\sin 8^\circ 12' = 8.75\text{ m}$$

$$RL_B = (RL_A + \text{H.I} + V - S_2) = (100.5 + 1.35 + 8.75 - 1.305)$$

$$= 109.295\text{ m}$$

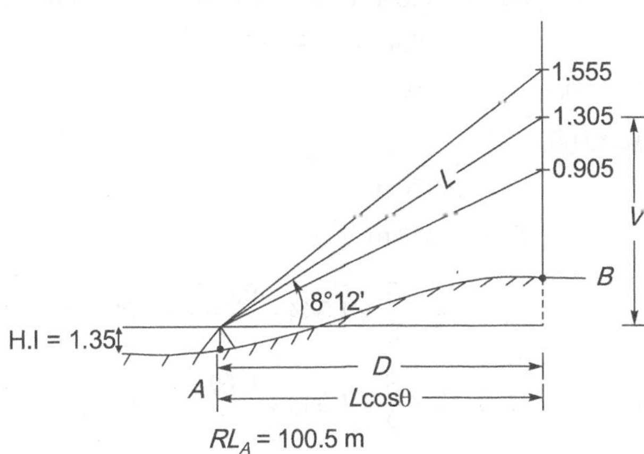

Case: 2 Instrument *Q* at station *A* and staff held normal at *B*:

$$\left.\begin{array}{l}K=90\\C=0.35\end{array}\right| H.I=1.40\text{ m}$$

Staff intercept = S

Axial reading = S_2

$$L = KS + C$$

$$L = (90S + 0.35)$$

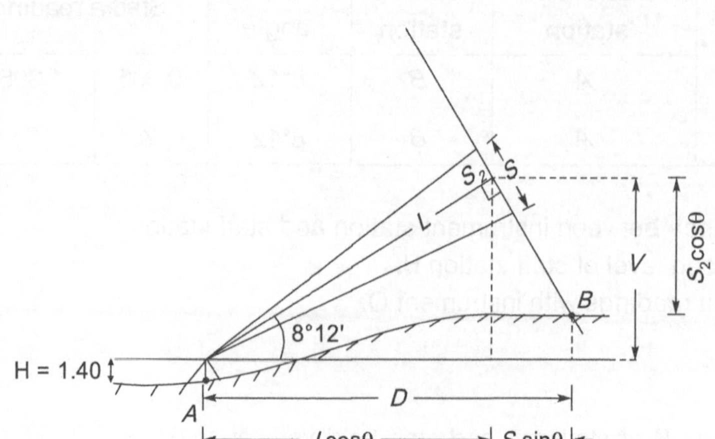

Vertical distance, $V = L \cdot \sin\theta$

Horizontal distance, $D = L\cos\theta + S_2 \sin\theta$

$$\Rightarrow \quad D = (90S + 0.35)\cos 8^\circ 12' + S_2 \sin 8^\circ 12'$$

$$\Rightarrow \quad 60.74 = (90\,S + 0.35)\cos 8^\circ 12' + S_2 \sin 8^\circ 12'$$

$$\Rightarrow \quad 60.74 = 89.08\,S + 0.346 + S_2(0.1426)$$

$$\Rightarrow \quad 89.08S + 0.1426\,S_2 = 60.394 \qquad \text{...(i)}$$

$$RL_B = RL_A + H.I + V - S_2\cos\theta$$

$$\Rightarrow \quad 109.295 = 100.5 + 1.40 + [90S + 0.35]\sin 8^\circ 12' - S_2 \cos 8^\circ 12'$$

$$\Rightarrow \quad 109.295 = 100.5 + 1.40 + 12.836\,S + 0.0499 - S_2 \times 0.989$$

$$\Rightarrow \quad 7.345 = 12.836\,S - 0.989\,S_2$$

$$\Rightarrow \quad 12.836\,S - 0.989\,S_2 = 7.345 \qquad \text{...(ii)}$$

From equation (i) and (ii)

$$S = 0.676\text{ m}$$

$$S_2 = 1.344\text{ m}$$

Upper stadia reading, $S_1 = S_2 - \dfrac{S}{2} = 1.344 - \dfrac{0.676}{2} = 1.006\text{ m}$

Lower stadia reading, $S_3 = S_2 + \dfrac{S}{2} = 1.344 + \dfrac{0.676}{2} = 1.682\text{ m}$

∴ **Stadia readings are:** $\begin{cases} S_1 = 1.006\text{m} \\ S_2 = 1.344\text{m} \\ S_3 = 1.682\text{m} \end{cases}$

(i) Distance between instrument and staff station : 60.74 m

(ii) RL of B = 109.295 m

(iii) Stadia readings at B $\begin{cases} S_1 = 1.006\text{m} \\ S_2 = 1.344\text{m} \\ S_3 = 1.682\text{m} \end{cases}$

5.5 The top point *P* of a tower having reduced level of 200 m was sighted using the theodolite from two stations *A* and *B* which were 100 m apart and were on the same side of the tower. All the three points *A*, *B* and *P* were in the same plane. The angles of elevation of point P from instrument stations A and B were 10°30′ and 16° 20′ respectively. The horizontal axis of instrument at point A was 2.5 m below the horizontal axis of instrument at point B and 1.5 m above the base of the tower. Calculate the:

(i) horizontal distance between point *A* and the top of the tower.
(ii) height of the tower.
(iii) reduced level of station *A*, if the height of instrument at *A* was 1.5 m.

[12 marks : 2022]

Solution:

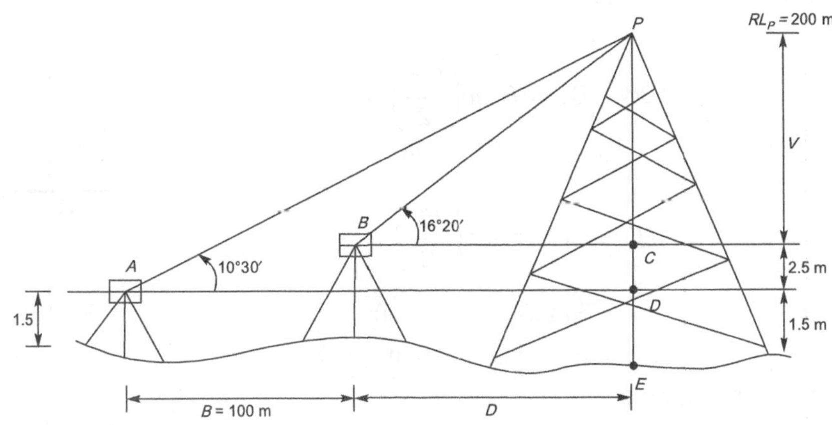

From ΔAPD:

$$\tan 10°30' = \frac{V+2.5}{D+100}$$

$$\Rightarrow \quad V + 2.5 = D\tan 10°30' + 100\tan 10°30' \quad ...(1)$$

From ΔBPC,

$$\tan 16°20' = \frac{V}{D}$$

$$\Rightarrow \quad V = D\tan 16°20' \quad ....(2)$$

$$\Rightarrow \quad V = 0.293\, D$$

From (1) and (2), $D = 148.93$ m

$V = 43.64$ m

(i) Horizontal distance between (*A*) and tower:

$$= (D + 100) = 149.93 + 100 = 248.93 \text{ m}$$

(ii) Height of tower $PE = (V + 2.5 + 1.5)$

$$= (43.64 + 2.5 + 1.5) = 47.64 \text{ m}$$

(iii) RL of ground station *A*: $= RL_P - V - 2.5 - 1.5$

$$= (200 - 43.64 - 2.5 - 1.5) = 152.36 \text{ m}$$

5.6 A simple circular curve is to set out in a National Highway touching the three lines having following details :

Line	Reduced Bearing	Length (m)
AB	N 90° E	–
BC	S 0° E	170
CD	S 70° W	–

If the chainage of point B is 700 m, calculate the radius of curve and chainage of all the tangent points in the curve.

[20 marks : 2023]

Solution:

In figure, R is radius of required curve

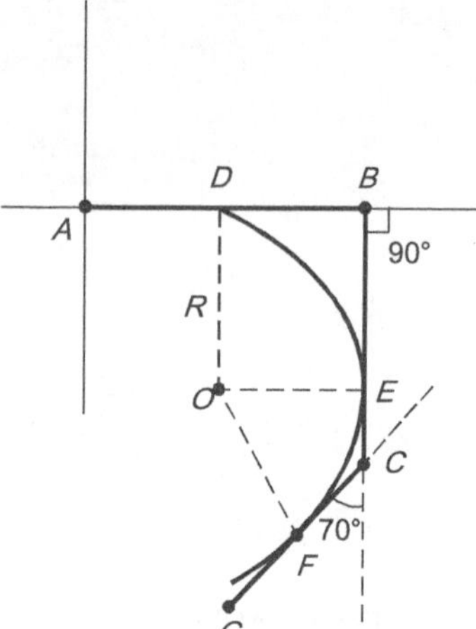

Now, $$DB = BE = R\tan\left(\frac{90°}{2}\right)$$

$$= R \quad ...(i)$$

Also, $$EC = CF = R\tan\left(\frac{70°}{2}\right)$$

$$= R\tan 35° \quad ...(ii)$$

Now, $$BC = BE + EC = 170 \text{ m} \quad ... (iii)$$

Putting (i) and (ii) in (iii), we get

$$R + R\tan 35° = 170$$

$$= R \approx 100 \text{ m}$$

Now, Tangent length $BD = R = 100$ m

Also, $BE = 100$ m

$$\text{Length of curve from D to E} = \frac{\pi R \times 90°}{180°} = \frac{\pi \times 100 \times 90°}{180°} = 157.08 \text{ m}$$

$$\text{Length of curve from E to F} = \frac{\pi R \times 70°}{180°} = \frac{\pi \times 100 \times 70°}{180°} = 122 \text{ m}$$

Now, chainage of point B = 700 m

So, chainage of point D = Chainage of B – BD

= 700 – 100 = 600 m

Chainage of point E = Chainage of point D + Length of curve from D to E

= 600 + 157.08 = 757.08 m

Chainage of point F = Chainage of point E + Length of curve from E to F

= 757.08 + 122 = 879.08 m

6. Field Astronomy and Photogrammetric Survey

6.1 Explain the following terms in brief:

(i) Ecliptic
(ii) Sensible horizon
(iii) Equation of time
(iv) Local mean time

[10 marks : 1999]

Solution:

(i) Ecliptic:

Because of the annual motion of the earth about the sun, the apparent path of the sun is in the plane which passes through the centre of celestial sphere and which intersects it in a great circle called the ecliptic. The angle between the plane of celestial equator and the ecliptic is called the obliquity of ecliptic. The points of intersection of the ecliptic with the celestial equator are called the equinoctial points.

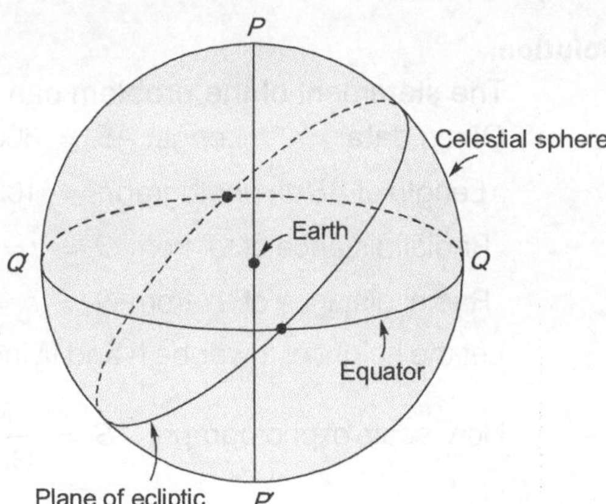

(ii) Sensible horizon:

The sensible horizon at a point on the surface of the earth is a circle in which a plane passing through the point of observation and tangential to the earth's surface intersects the celestial sphere. The line of sight of an accurately levelled telescope of a theodolite indicates the sensible horizon at the point.

(iii) Equation of time:

The local mean time (*LMT*) and the local apparent time (*LAT*) at a place are related as:

$$LMT = LAT - ET$$

$$\Rightarrow \quad ET = LAT - LMT$$

Thus, the difference between local apparent time and local mean time at any instant is known as the equation of time. Because the mean sun is a fictitious celestial body, it cannot be observed in the field.

The equation of time is positive when the apparent solar time is greater than the mean solar time. This can be understood by the diagram below:

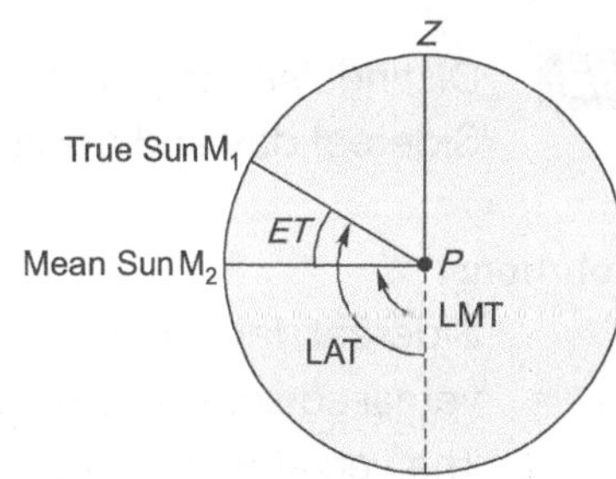

The value of equation of time varies from zero to 16 minutes throughout the year. The equation of time is zero four times in a year. On these four days *LAT* and *LMT* are same.

(iv) Local mean time:

The apparent solar time is variable and since it is inconvenient to work in terms of the variable apparent solar time. The astronomers invented a fictitious sun, called the mean sun which moves at a uniform rate along the celestial equator. It makes one complete revolution from west to east in one year. The mean solar day commences when the mean sun is at the antipode $\bar{Z}$ of the meridian. The time interval as measured by one daily revolution of the mean sun is called a mean solar day. Thus, when the mean sun is at the meridian *PZ* at *Z*, the local mean time is 12 hour. In the case of mean solar time the equation becomes

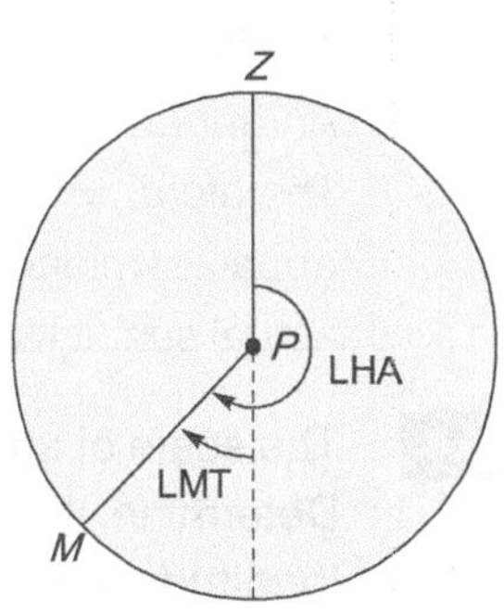

$$LMT = \text{LHA of mean sun} + 12$$

6.2 **A section line AB 300 m long on a flat terrain measures 102.4 mm on the vertical photograph. A radio tower also appears on the photograph. The distance measured from the principal points to the image of the bottom and top of the radio tower found to be 7 cm and 8 cm respectively. The average elevation of the terrain was 553 m. Determine the height of the tower. Take f = 152.4 mm.**

[10 marks : 2001]

Solution:

The statement of the problem can be understood with the help of diagram shown below:

Given data: Length AB = 300 m; h_{av} = 553 m

Length of AB on photograph = 102.4 mm

Radial distance of 'a' from O = $r = 8$ cm

Radial distance of 'b' from O = $r_0 = 7$ cm

Let the height of tower be h and flying height be H.

Now, scale of photograph, $$S = \frac{\text{Map distance}}{\text{Ground distance}}$$

$$= \frac{102.40}{300 \times 1000} = \frac{1}{2929.69}$$

But $$S = \frac{f}{H - h_{av}}$$

$$\Rightarrow \quad \frac{1}{2929.69} = \frac{152.4 \times 10^{-3}}{H - 553} \qquad [\because f = 152.4 \text{ mm}]$$

$$\Rightarrow \quad H = 553 + (2929.69 \times 0.1524) = 999.48 \text{ m}$$

Also, relief displacement, $$d = r - r_0 = 8 - 7 = 1 \text{ cm}$$

$\therefore$ **Height of tower,** $$h = \frac{d(H - h_{av})}{r} = \frac{1 \times (998.48 - 553)}{8} = 55.81 \text{ m}$$

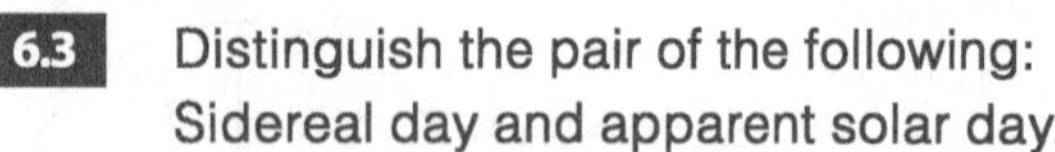

6.3 **Distinguish the pair of the following:**
Sidereal day and apparent solar day.

[2 marks : 2004]

Solution:

Sidereal day and apparent solar day: A sidereal day on any particular meridian commences when the vernal equinox is at the upper transit of that meridian. At that instant the local sidereal time is zero hour. The vernal equinox moves clock-wise when observed from the top of the celestial sphere. The vernal equinox returns again to the upper transit of the meridian just 24 sidereal hours after the previous transit. The time interval between two successive transits of the vernal equinox is equal to one sidereal day.

An apparent solar day commences when the actual sun is at the antipode of the observer's meridian. Therefore, when the true sun is at the antipode, the local apparent time is zero hour. It is also called local apparent midnight. An apparent solar day is the time interval between two successive upper transits of the actual sun. It may be noted that apparent solar day begins at midnight.

6.4 **The scale of an aerial photograph is 1 cm = 150 m and the size of the photograph is 20 cm × 20 cm. Determine the number of photographs in each strip to cover an area of 15 km × 15 km if the longitudinal overlap is 70% and side overlap 30%.**

[2 marks : 2010]

Solution:

The number of photographs in each strip can be expressed as

$$N_1 = \frac{L_1}{(1 - p_l) \times \frac{1}{S} \times L} + 1$$

where L_1 is the length of ground, l is the length of photograph, p_l is the longitudinal overlap and S is the scale of photograph.

Scale of photograph $S = 1 \text{ in } 150 \times 100 = 1 \text{ in } 15000$

$$\therefore \quad N_1 = \frac{15 \times 10^3}{(1-0.7) \times 15000 \times 0.2} + 1 = 17.67 \approx 18$$

Number of flight lines required is given by

$$N_2 = \frac{W_1}{(1-p_w) \times \frac{1}{S} \times w} + 1$$

where W_1 is the width of the ground, w is the width of photograph, p_w is the side overlap

$$\therefore \quad N_2 = \frac{15 \times 10^3}{(1-0.30) \times 15000 \times 0.2} + 1 = 8.14 \approx 9$$

$\therefore$ **Total number of photographs** $= N_1 \times N_2 = 18 \times 9 = 162$

6.5 **The observed altitude of β-ursae Minoris at lower and upper culminations are 29° 58′15″ and 60° 45′3″. Find the latitude of the place of observation assuming the correction for refraction to be equal to 57″ × tangent of apparent zenith distance.**

[8 marks : 2010]

Solution:

Observed altitude of star at lower culmination,

$$\alpha = 29°58'15''$$

Observed altitude of star at upper culmination,

$$\beta = 60°45'3''$$

True altitude of star at lower culmination is given by

$$\alpha_1 = \text{Observed altitude} - \text{Refraction correction}$$
$$= 29°58'15'' - 57'' \times \tan z$$
$$= 29°58'15'' - 57'' \times \cot \alpha$$
$$= 29°58'15'' - 57'' \times \frac{1}{\tan \alpha} = 29°58'15'' - 57'' \times \frac{1}{\tan(29°58'15'')}$$
$$= 29°58'15'' - 57'' \times 1.734 = 29°58'15'' - 98.4'' = 29°56'36.6''$$

True altitude of star at upper culmination is given by

$$\beta_1 = \beta - 57'' \times \frac{1}{\tan\beta} = 60°45'3'' - 57'' \times \frac{1}{\tan(60°45'3'')}$$
$$= 60°45'3'' - 31.92'' = 60°44'31.08''$$

$\therefore$ **Latitude of the place of observation is given by**

$$\theta = \frac{\alpha_1 + \beta_1}{2} = \frac{29°56'36.6'' + 60°44'31.08''}{2} = 45°20'33.84''$$

6.6 **Briefly explain how Global Positioning System (GPS) is useful in surveying work.**

[4 marks : 2015]

Solution:

The advantages of GPS in surveying work are as follows:

- Surveys becomes independent of weather conditions as a result of using radio frequencies to transmit the signals.
- Unlike conventional surveying procedures, there is no need for intervisibility between stations.

- Use of same field and data reduction procedures results in position accuracy which are independent of network shape or geometry.
- The points can be established wherever they are required and need not be located at evenly distributed sites atop mountains to satisfy intervisibility, or network geometry criteria.
- GPS surveying is more efficient, more flexible and less time consuming positioning technique than using conventional terrestrial survey technologies.
- GPS can be used to obtain high accuracy three dimensional (3D) information, analytical and anytime.

6.7 **(i) Define remote sensing and briefly explain the principle of remote sensing. Explain the terms 'passive remote sensing' and 'active remote sensing'.**

(ii) Explain the concept of global positioning system.

[8 marks : 2016]

Solution:

(i) **Remote Sensing:** It is the art of collecting information about an object or area or a phenomenon from a distance without being getting into physical contact with them. The human visual system is an example of a remote sensing system in the general sense.

Active remote sensing: In active remote sensing, the source has its own electromagnetic energy. This energy is reflected back by the object and the returned energy is measured.

Passive remote sensing: It uses the natural energy from the Sun and records the energy that is directly reflected or radiated from the objects.

(ii) **Global Positioning System:** Global Positioning System (GPS) uses the satellite data to calculate the accurate position on earth. It provides location and time information on or near to the earth's surface where there is unobstructed line of sight with four or more satellites. GPS receiver calculates the position by precisely timing the signals sent by the satellites high above the earth. All Global Positioning Systems work on the same principle with the only difference that is the number of satellites they can simultaneously communicate with. Most GPS receivers can communicate with 12 satellites with some modern receivers capable of communicating with 14 - 20 satellite.

6.8 **An area of 10 km × 20 km is to be surveyed using aerial photogrammetry. Average scale of photograph is 1 : 10,000 at ground elevation of 400 m above the datum. Focal length of camera used is 20 cm and size of photographs are 23 cm × 23 cm. The speed of aircraft is 270 kmph. The forward lap in photographs is 70% and side lap is 30%. Determine the flying height, exposure interval and number of photographs required to complete the survey.**

[15 marks : 2017]

Solution :

Given: Size of photograph = 23 cm × 23 cm Average scale of photograph $(S) = 1 : 10000$

i.e., 1 cm = 100 m

Average elevation of terrain = 400 m

Longitudinal overlap, $P_l = 70\%$

Side overlap, $P_w = 30\%$

Ground speed of aircraft, $V = 270 \text{ kmph} = 270 \times \frac{5}{18} = 75 \text{ m/s}$

Focal length of camera, $f = 20 \text{ cm} = 200 \text{ mm}$

Flying height

Scale, $$S = \frac{f}{(H - h_{av})} = \frac{1}{10000} = \frac{0.20}{(H - 400)}$$

$\Rightarrow$ $H = 2400 \text{ m}$

Theoretical ground spacing of flight lines $W = (1-P_w)\frac{W}{S} = (1-0.3)\times 23\times 10^{-2}\times\frac{1}{1/10000} = 1610 \text{ m}$

Number of flight lines required, $N_2 = \frac{L_2}{w}+1 = \frac{20\times 10^3}{1610}+1 = 13.42 = 14$

Ground distance between exposure, $L = (1-P_l)\frac{l}{S} = (1-0.7)\times\frac{23\times 10^{-2}}{1/10000} = 690 \text{ m}$

$$\text{Exposure interval} = \frac{L}{V} = \frac{690}{75} = 9.2 \text{ sec}$$

$$\simeq 9.0 \text{ sec (Least count} = 0.5 \text{ sec)}$$

Adjusted ground distance between exposures $= 75 \times 9 = 675$ m

Number of photographs per flight line, $N = \frac{10\times 10^3}{675}+1 = 15.81 = 16$

Hence, total number of photographs required $= N_1 \times N_2 = 16 \times 14 = 224$

6.9 Briefly explain the various sources of errors in Global Positioning System (GPS).

[6 marks : 2017]

Solution:

GPS error sources : The error sources can be classified into three groups, namely

- Satellite related errors {Systematic errors}
- Propagation-medium related errors {Systematic errors}
- Receiver-related errors {Systematic errors}

However, sometimes errors are introduced intentionally known as selected availability.

Systematic Errors (Natural Errors):

1. **Satellite Errors :** GPS satellites are equipped with very accurate atomic clocks. But as good as these clocks are, they are not perfect. Slight inaccuracies in their timekeeping ultimately lead to inaccuracies in our position measurements. GPS satellites are placed into very high orbits and so are relatively free from the perturbing effects of the earth's upper atmosphere but even then they still drift slightly from their predicted orbits contributing to the errors.

2. **Signal propagation errors:** GPS satellites transmit their timing information by radio and that is another source of error because radio signals in the earth's atmosphere do not behave as predictably desired.

3. **Receiver errors:** The receivers are also not perfect. They can introduce their own errors which usually stem from their clock or internal noise multipath, antenna face centre variation etc.

Multipath: As the GPS signal arrives at the surface of the earth it may get reflected by local obstructions and gets to the receiver's antenna via more than one path. This form of error is called the multipath errors, because in a sense the signal is getting to the antenna by multiple paths.

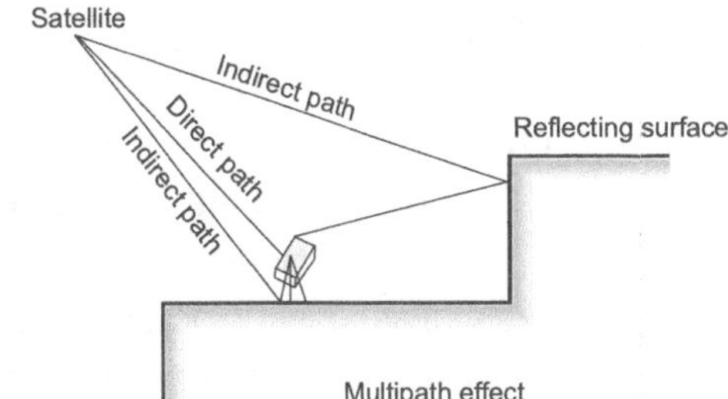

Multipath effect

Selective availability: Sometimes noise is introduced intentionally into the GPS satellite clocks, which reduces their accuracy.

When implemented it is the largest source of error in the GPS system. Selective availability is also known as intentional error. For example: For civil and military purposes.

6.10 A pair of overlap aerial photographs was taken with a camera from an altitude of 3000 m above datum. The focal length of camera was 120 mm and the mean distance between two principal points, both of which lie on the datum, was 71.20 mm. In the common overlap area, a tall chimney 200 m high with its base in the datum surface was observed. Find out the difference of parallax for the top and bottom of the chimney.

[6 marks : 2018]

Solution:

Given data: $H = 3000$ m, f = 120 m

Distance between principal points, $b = 71.2$ mm

B = Aircraft base

Chimney height, $h = 200$ m

Now, scale of photograph for datum elevation,

$$S = \frac{f}{H} = \frac{120/1000}{3000} = \left(\frac{1}{25000}\right)$$

Also we have,

$$\frac{f}{H} = \frac{b}{B}$$

$$\Rightarrow \quad B = \frac{bH}{f} = \frac{71.2}{120} \times 3000 = 1780 \text{ m}$$

Now, parallax at bottom

$$h_{avg} = 0$$

So,

$$p_b = \frac{fB}{H} = \frac{120 \times 1780}{3000} = 71.2 \text{ mm}$$

Now, parallax at top

$$h_{avg} = h = 200 \text{ m}$$

So,

$$p_t = \frac{fB}{H-h} = \frac{120 \times 1780}{3000-200} = 76.29 \text{ mm}$$

So, difference in parallax, $p_t - p_b = 76.29 - 71.2 = 5.086$ mm

6.11 Discuss the various method of relative positioning in GPS.

[5 marks : 2018]

Solution:

GPS receivers use timing signal from atleast four satellites, and any number of errors or delays can occur during the signals, transit to earth.

DGPs (Differential Global Positioning System) is an enhancement to GPS that was developed to correct these errors and inaccuracies in the GPS system, allowing for more accurate positioning information.

There are basically three methods of relative (differential) positioning in GPS:

(i) **Differential (Relative) Positioning:**

It requires at least two receivers set up at two stations (usually one is known) to collect satellite date simultaneously in order to determine coordinate differences. This method will position the two stations relative to each other and can provide the accuracies required for basic land surveying and hydrographic surveying.

(ii) **Differential Positioning (Code Pseudo-Range Tracking)**

It results in absolute coordinates of the user on the earth's surface.

Errors in range are directly reflected in resultant coordinates errors.

Differential positioning is not so concerned with the absolute position of the user but with the relative difference between the user positions, which are simultaneously observing the same satellites.

(iii) **Differential Positioning (Carrier Phase Tracking)**

The process becomes somewhat more complex when carrier-signals are tracked such that range changes are measured by phase resolution.

It provides for a more accurate range resolution due to short wave length and the ability of the receiver to resolve the carrier phase down to 2 mm.

So, this method has primary application in engineering, topographic, and geodetic surveying.

6.12 **Explain briefly about the spatial and spectral resolution in Remote Sensing.**

[6 marks : 2018]

Solution:

Spatial resolution:

- Spatial resolution describes how much detail in a photographic image is visible to human eye.
- The ability to "resolve" or separate, small details is one way of describing what we call spatial resolution.
- Spatral resolution of images acquired by satellite sensor systems is usually expressed in meters.

Spactral resolution:

- Sensor's spactral resolution specifies the number of spectral bands in which the sensors can collect reflected radiance.
- The choice or number of spactral bands required will depend on the application of use.
- The spectral reflectance curves or special signatures of different types of ground targets provide the knowledge base for information extraction.
- Reflectance measurement helps in revealing the mineral content of rocks, the moisture of soil, health of vegetation and other invisible details.

6.13 **What is spectral reflectance curve (spectral signature) in remote sensing? Explain any four applications of remote sensing in civil engineering?**

[10 Marks : 2019]

Solution:

Electromagnetic energy incidont on the surface features are partially reflected, absorbed or transmitted through it. The fractions that are reflected absorbed or transmitted vary with material type and the condition of the feature. It also varies with the wavelength of the incident energy. Majority of the remote sensing systems operate in the region in which the surface features mostly reflect the incident energy. The reflectance characteristics of the surface features are represented using spectral reflectance curves.

Applications of Remote Sensing in Civil Engineering

1. **Agriculture**
   (i) Early season estimation of total cropped area
   (ii) Monitoring crop condition using crop growth profile.
   (iii) Identification of crops and their coverage estimation in multi-cropped regions.
   (iv) Crop yield modelling
   (v) Cropping system/crop rotation studies
   (vi) Command area management
   (vii) Detection of moisture stress in crops and quantification of its effect on crop yield
   (viii) Detection of crop violations
   (ix) Zoom cultivation–desertification
2. **Land use and soils**
   (i) Mapping land use/cover (level-III) at 1 : 25000 scale or better

(ii) Change detection
(iii) Identification of degraded lands/erosion prone areas
(iv) Soil categorization

3. **Geology**
(i) Lithological and structural mapping
(ii) Geomorphological mapping
(iii) Ground water exploration
(iv) Engineering geological
(v) Geo-environmental studies
(vi) Drainage analysis
(vii) Mineral exploration
(viii) Coal fire mapping
(ix) Oil field detection

4. **Urban Land use**
(i) Urban land use level IV mapping
(ii) Updating of urban transport network
(iii) Monitoring urban sprawl
(iv) Identification of unauthorised structures

6.14 Discuss how the sensors are classified in Remote Sensing and briefly explain their salient features.
[10 marks : 2020]

Solution:

The instrument that measures the properties of electromagnetic radiations leaving a surface/medium due to scattering or emission, is called a remote sensor.

Remote sensors are of two types :
1. Passive sensors
2. Active sensors

1. **Passive Sensors :** The remote sensors, which sense natural radiations either emitted or reflected from the earth, are called passive sensors.
2. **Active Sensors :** The remote sensors which produce their own electromagnetic radiation of specific wavelength or a band of wavelength, as a part of the sensor system, are called, active remote sensors.

As the technology involved in developing sensors throughout the electromagnetic spectrum, is not the same, the remote sensors, are also classified as under :

(i) The remote sensors which operate in the optical infrared region.

(ii) The remote sensors which operate in the optical microwave region.

Both the optical infrared and microwave sensors may be either imaging or non-imaging sensors. The imaging remote sensors give a two-dimensional spatial distribution of the emitted or reflected intensity of radiation (such as a photographic camera). The non-imaging sensors measure the intensity of radiation, within the field of view. Vertical temperature profiling radiometer (VTPR) is a type of non-imaging sensor.

Sensor on Board Satellite Scanning the Ground surface : Sensors mounted on aircraft or satellite platforms measure the amounts of energy reflected from or emitted by earth's surface. The sensor measures the ground surface along the one dimensional profile on the ground below the platform on either side of the ground track of the platform. The sensor scans the ground below the satellite platform. Movement of the platform in forward direction, enables the sensor to build an image of the earth's surface.

The satellite sensor scans the ground surface along line. Each scan line of a remotely sensed image is a digital or numerical record of radiance measurements made at regular intervals along the line. Consecutive scan lines form an image. Depending upon the direction of principal axis of the sensor, the sensors are called Nadir looking or side looking sensor.

A Nadir looking sensor images the ground area on either side of the satellite platform. Whereas a side-looking sensor images the earth's surface lying on one side of the satellite track.

6.15 Write a detailed note on ideal and real remove-sensing system by highlighting their six components as well as their six shortcomings.

[20 Marks : 2021]

Solution:

Ideal and Real Remote-Sensing Systems:

An ideal remote-sensing system is wherein electromagnetic energy of all wavelengths and of known uniform intensity is produced by an ideal source; the energy propagates from the source without loss to a homogeneous target; and the energy of various wavelengths selectively interacts with the target, resulting in a return signal of reflected and emitted energy.

The returned signal propagates without loss to a sensor that responds linearly to energy of all wavelengths of any intensity. In real time, an intensity versus wavelengths response is recorded, processed into an interpretable format, and recognized as being unique to the particular target in its particular chemical state. The information obtained without the particular target is made readily available in a useful form to the users.

The basis components of an ideal remote sensing system include the following:

1. **Uniform energy source: This** source would provide energy over all wavelengths, at a constant, known, high level of output, irrespective of time and place.
2. **Non-interfering atmosphere:** This would be an atmosphere that would not modify the energy from the source in any manner, whether that energy were on its way to the earth's surface of coming form it.
3. **Series of unique energy/matter interactions at the earth:** These interactions would generate reflected and/or emitted signals that are not only selective with respect to wavelength, but are also known, invariant, and unique to each and every earth surface features type and subtype of interest.
4. **Super sensor:** This would be a sensor, highly sensitive to all wavelengths, yielding spatially detailed data on the absolute brightens (or radiance) from a scene as a function of wavelength, throughout the spectrum. This super sensor would be simple and reliable, require virtually no power or space, and be accurate and economical to operate.
5. **Real-time data handling system:** In this system, the instant the radiance versus wavelength response over a terrain element were generated, it would be processed into an interpretable format and recognized as being unique to the particular terrain element from which it came.
6. **Multiple data users:** These people would have knowledge of great depth, both of their respective disciplines and of remote sensing data acquisition and analysis technique.

Unfortunately, an ideal remote sensing system does not exist. The elements of ideal system discussed above have the following general shortcomings:

1. **Energy source:** All passive remote sensing system rely energy that is either reflected and /or emitted from earth surface features.
2. **Atmosphere:** To some extent, the atmosphere always modifies the strength and spectral distribution of energy received by the sensor.
3. **Energy/matter interaction at the earth surface:** Remote sensing would be simple if every material reflected and/or emitted energy in a unique, known way. Although spectra response patterns (signatures) play a central role in detecting identifying and analyzing earth surface materials, the spectral world is full of ambiguity.

4. **Sensor:** No single sensor is sensitive to all wavelengths. All real sensors have fixed limit of spectral sensitivity. They also have a limit on how small an object on the earth's surface can be "seen" by a sensor being separate from its surroundings.
5. **Data handling system:** Processing sensor data into an interpretable format can be—and often is—an effort entailing considerable thought, instrumentation, time experience and reference data.
6. **Multitype data users:** Central to the successful application to any remote sensing system is the person (or persons) using the remote sensor data from that system.

6.16 **A star of declination 40°30′30″ S is to be observed at lower and upper transit at a place of latitude 75°30′30″ S. Draw a neat labelled figure depicting declination, latitude, altitude in upper transit, altitude in lower transit and find the approximate apparent altitude at which the star should be sought.**

[12 Marks : 2021]

Solution:

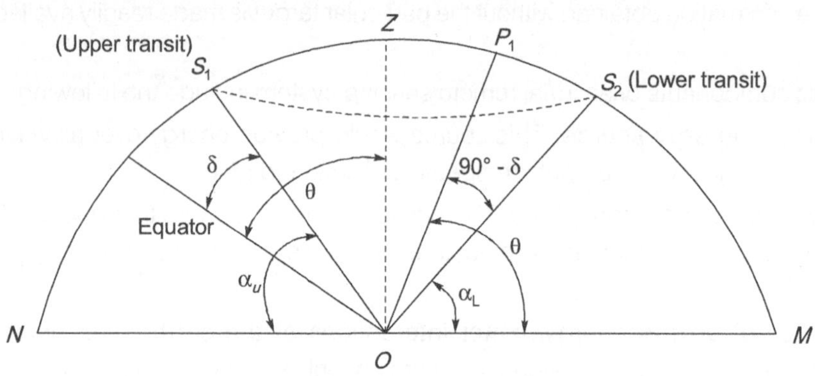

Given: $\delta = 42° \; 30' \; 30''S$;

$\theta = 75° \; 30' \; 30'' \; S$; α = corrected altitude

At lower transit:

$$\alpha_L = \theta - (90° - \delta)$$

$$= 75° \; 30' \; 30'' - (90° - 42° \; 30' \; 30'') = 28° \; 1' \; 0''$$

Hence apparent altitude $= 28° \; 1' \; 0'' + 58'' \times \cot(28° \; 1'0'') = 28° \; 2' \; 49{\cdot}01''S$

At upper transit:

$$\alpha_u = 90° - \theta + \delta$$

$$= 90° - 75° \; 30' \; 30'' + 42° \; 30' \; 30'' = 57° \; 0' \; 0''$$

Apparent altitude $= 57°0'0'' + 58'' \cot(57°0'0'') = 57°0'37.67'' \; N$

6.17 **(i) Briefly explain the working principle of Global Poistioning System (GPS). Write any three advantages and any three limitations of GPS in surveying.**

(ii) A city of site 25 km × 50 km is to be surveyed for aerial photogrammetry using camera having focal length of 150 mm mounted on an airplane flying at the height of 1500 m above the ground level. The longitudinal and side lap will be 80% and 30% respectively. The size of the photograph will be 250 mm × 250 mm. It is planned to set the exposure interval of camera at 10 seconds. Calculate the following:

(I) The number of photographs required to cover the area.

(II) The flying speed of the airplane.

(III) The total time required to capture the photographs.

[10 marks : 2022]

Solution:

(i) **Working Principle**: The global positioning system is a space based navigation system. It consists of 24 satellites nominally orbiting the earth in medium earth orbit.

- Each GPS satellite transmits data that indicates its location and the current time.
- All GPS satellites synchronize operations so that these repeating signals are transmitted at the same instant.
- The signals, moving at the speed of light, arrive at a GPS receiver at slightly different times because some satellites are further away than others.
- The distance to the GPS satellites can be determined by estimating the amount of time it takes for their signals to reach the receiver.
- When the receiver estimates the distance to at least four GPS satellites, it can calculate its position in three dimensions.

Advantages:

1. Positional accuracy irrespective of network geometry acts as a function for two or more different station distance.
2. Comparatively more flexible, less time consuming and also more effective.
3. High accuracy with 3-dimensional geographic information irrespective of place and time.

Limitations of GPS:

1. Clear and fine visibility of sky and no obstructions through any obstacles e.g. branches etc.
2. Not easily universally acceptable.
3. Highly skilled personnels are required.

(ii)
$$\text{Area} = [25 \text{ km} \times 50 \text{ km}]$$
$$f = 150 \text{ mm}$$
$$(H - h_{avg}) = 1500 \text{ m}$$
$$\text{Scale} = \left(\frac{f}{H - h_{avg}}\right) = \frac{150 \times 10^{-3}}{1500} = \frac{1}{10000}$$
$$p_l = 80\%, \quad p_b = 30\%$$

Size of photo given is 250 mm × 250 mm.

$$T = 10 \text{ sec.}$$

1. Number of photos per strip,

$$n_L = \frac{L}{ls(1-p_e)} + 1 = \frac{50}{250 \times 10000 \times 10^{-6}(1-0.8)} + 1 = 101$$

Number of strips, $$n_s = \frac{B}{lb(1-p_b)} + 1 = \frac{25}{250 \times 10000 \times 10^{-6}(1-0.3)} + 1$$
$$= 15.286 \simeq 16$$

Number of photos, $$n = (n_L \times n_s) = 101 \times 16 = 1616 \text{ photos}$$

2. $$T = \frac{3600L}{v}$$
$$L = ls(1 - p_e)$$
$$= 250 \times 10000 \times (1 - 0.8) \times 10^{-6} = 0.5 \text{ km}$$
$$10 \text{ sec} = \frac{3600 \times 0.5}{v}$$
$$\Rightarrow \quad v = 180 \text{ kmph}$$

3. Time taken to cover one complete strip = 10 × (100) = 1000 s

$$\text{Total time} = 1000 \times 16 = 1600 \text{ s}$$

6.18 **What are the different types of resolutions in Remote Sensing? Briefly explain the significance of each resolution in the field of Civil Engineering.**

[5 marks : 2022]

Solution:

Resolution in Remote Sensing: Resolution is the ability to obtain the finer details of an image. It determines the details of the surface features that can be distinguished. Four types of resolutions are generally considered-spatial resolution, spectral resolution, temporal resolutions, and radiometric resolution.

1. **Spatial resolution**: This refers to the ability of the remote sensing system to record spatial details. The concept of spatial resolution is different for photographs and digital images. In the case of aerial photographs, spatial resolution is defined as the sharpness of the image. In the case of optical scanning systems, resolution depends upon the focal length of the optical system, the height of the imaging platform, and the size of the sensor element. Resolution is usually expressed in terms of the instant field of view.
2. **Spectral resolution**: This refers to the wavelengths to which the remote sensing system is sensitive. Aerial photography uses only a narrow band of the spectrum (visible light). The spectral resolution of aerial photographs thus happens to be low. The resolution depends upon the number of wavelength bands as well as on the width of each band. For higher resolutions, a large number of bands of narrower bandwidth are used.
3. **Temporal resolution**: This refers to the frequency with which images are obtained in terms of time. Temporal resolution is important for predicting environmental changes, vegetation cover, and crop patterns.
4. **Radiometric resolution**: This refers to the sensitivity of the system to small changes in the radiation (wavelength). In photography, this referred to as contrast or the number of grey tones in the image.

7. Geology

7.1 **Dust, gravel and intact pieces of rock are produced during rock coring with core advance of 2 m. The lengths of the rock pieces are 150 mm, 200 m, 90 mm, 300 mm, 60 mm, 250 mm, 120 mm, 170 mm, 80 mm, 210 mm and 75 mm.**

Calculate *RQD* and comment on the rock quality based on *RQD*.

[20 marks : 2022]

Solution:

Total length of rock pieces greater than 100 mm,

$$l = 150 + 200 + 300 + 250 + 120 + 170 + 210 = 1400 \text{ mm}$$

$$\text{Total core run length} = 2000 \text{ mm}$$

$$\therefore \quad RQD = \frac{1400}{2000} \times 100 = 70\%$$

For RQD value between 51 – 75%, rock quality is considered to be fair i.e. moderate weathered. So, rock quality is fair.

■■■■

7 Highway Engineering

Revised Syllabus of ESE : *Highways-Planning & construction methodology, Alignment and geometric design; Traffic Surveys and Controls; Principles of Flexible and Rigid pavements design.*

1. Highway Geometric Design

1.1 **A highway is provided with a horizontal curve of radius 300 m in certain locality. Calculate the super elevation needed to maintain the design speed of 90 kmph. Also calculate the maximum allowable speed, if the super elevation is limited to 0.07 and safe limit of transverse coefficient of friction is 0.15. Comment on the result.**
Explain the terms:
(i) Ruling Minimum Radius and
(ii) Absolute Minimum Radius of horizontal curve.

[15 marks : 1999]

Solution:

Given data: $V = 90$ kmph; $R = 300$ m; $e = 0.07$; $f = 0.15$

We know that super elevation, $e = \dfrac{V^2}{127R}$

where V is in kmph and R is radius of curve in m. Calculation of super elevation for mixed traffic condition is a complex problem, as different vehicles ply on the road with a wide range of speeds. To super elevate the pavement upto the maximum limit so as to counteract the centrifugal force fully, neglecting the lateral friction is safer for fast moving vehicles. But for slow moving vehicles this may be quite inconvenient. On the contrary to provide lower value of super elevation, thus relying more on the lateral friction would be unsafe for fast moving vehicles. As a compromise and from practical considerations it is suggested that the super elevation should be provided to fully counteract the centrifugal force due to 75% of the design speed, (by neglecting lateral friction developed) and limiting the maximum superelevation to 0.07.

$\therefore$ $$e = \frac{(0.75V)^2}{127R} = \frac{V^2}{225R} = \frac{90^2}{225 \times 300} = 0.12 > 0.07$$

Hence a superelevation, $e = 0.07$ is provided.

Check: $$f = \frac{V^2}{127R} - 0.07 = \frac{90^2}{127 \times 300} - 0.07 = 0.143 < 0.15 \text{ (Hence Safe)}$$

Now, we know that $$e + f = \frac{V_a^2}{127R}$$

$\therefore$ $$\frac{V_a^2}{127 \times 300} = 0.07 + 0.15$$

$$\Rightarrow \qquad V_a^2 = 8382$$

$$\Rightarrow \qquad V_a = 91.55 \text{ kmph}$$

If the allowable speed as calculated above is higher than design speed, then design is adequate. Thus, the maximum allowable speed is 91.55 kmph.

(i) **Ruling minimum radius:** For a certain speed of a vehicle the centrifugal force is dependent on the radius of the horizontal curve. To keep the centrifugal ratio within a low limit, the radius of the curve is kept correspondingly high. The centrifugal force which is counteracted by the superelevation and lateral friction is given by

$$e + f = \frac{V^2}{127R}$$

$$\Rightarrow \qquad R = \frac{V^2}{127(e+f)}$$

This radius is called ruling minimum radius of curve for the ruling design speed of the road.

(ii) **Absolute minimum radius:** When the minimum design speed is adopted instead of ruling design speed, the corresponding radius is called absolute minimum radius. This length of radius must be provided to counteract the centrifugal forces.

1.2 **An ascending gradient 1 in 60 meets a descending gradient of 1 in 50. Find the length of summit vertical curve for a stopping sight distance of 180 m.**

[15 marks : 2002]

Solution:

The length of summit vertical curve is given by

$$L = \frac{NS^2}{\left(\sqrt{2H} + \sqrt{2h}\right)^2} \qquad \text{when } L > \text{SSD}$$

where L = length of summit curve in m
S = stopping sight distance in m
H = height of eye level of driver above roadway surface in m
h = height of object above the pavement surface in m
N = deviation angle equal to algebraic difference in grades, radians or tangent of the deviation angle.

Given $\qquad n_1 = +\frac{1}{60}; \quad n_2 = -\frac{1}{50}; \quad S = 180 \text{ m}$

$$\therefore \qquad N = \frac{1}{60} - \left(-\frac{1}{50}\right) = \frac{1}{60} + \frac{1}{50} = \frac{11}{300}$$

As per IRC $H = 1.2$ m and $h = 0.15$ m

$$\therefore \qquad L = \frac{\frac{11}{300} \times (180)^2}{\left(\sqrt{2 \times 1.2} + \sqrt{2 \times 0.15}\right)^2}$$

$$\Rightarrow \qquad L = 270.18 \text{ m} > 180 \text{ m (Hence OK)}$$

1.3 **Calculate the extra width of the pavement required on a horizontal curve of radius 700 m on a two lane highway, the design speed being 80 kmph. Assume wheel base length of 6 m.**

[8 marks : 2005]

Solution:

The extra widening of pavement on horizontal curves is divided into two parts (i) mechanical widening and (ii) psychological widening

Mechanical Widening: The widening required to account for the off tracking due to rigidity of wheel base is called mechanical widening (W_m) and may be calculated as given below:

Let, R_1 = radius of path traversed by the outer rear wheel in m

R_2 = radius of the path traversed by the outer front wheel in m

W_m = off tracking or mechanical widening in m

l = length of wheel base in m

$$W_m = OC - OA = OB - OA = R_2 - R_1$$

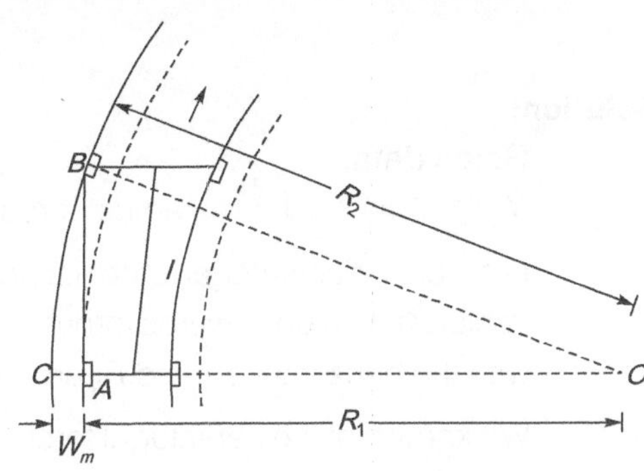

From Δ OAB, we know that

$$OA^2 = OB^2 - BA^2$$

$$\Rightarrow \quad R_1^2 = R_2^2 - l^2$$

But $$R_1 = R_2 - W_m$$

$$\therefore \quad (R_2 - W_m)^2 = R_2^2 - l^2$$

$$\Rightarrow \quad R_2^2 + W_m^2 - 2R_2W_m = R_2^2 - l^2$$

$$\Rightarrow \quad l^2 = W_m(2R_2 - W_m)$$

$$\Rightarrow \quad W_m = \frac{l^2}{2R_2 - W_m}$$

$$\Rightarrow \quad W_m = \frac{l^2}{2R} \text{ (approx.)} \qquad \left[\because W_m \text{ is very less compared to radius of curve}\right]$$

where R is the mean radius of the curve. The mechanical widening calculated above is required for one vehicle negotiating a horizontal curve along one traffic lane. Hence in a road having n traffic lanes as 'n' vehicles can travel simultaneously, the total mechanical widening required is given by

$$W_m = \frac{nl^2}{2R}$$

Here given data: $n = 2, \quad l = 6 \text{ m}, \quad R = 700 \text{ m}$

$$\therefore \quad W_m = \frac{nl^2}{2R} = \frac{2\times(6)^2}{2\times 700} = 0.0514 \text{ m}$$

Psychological Widening: Extra width of pavement is also provided for psychological reasons such as to provide for greater maneuverability of steering at higher speeds, to allow for the extra space requirements for the overhangs of vehicles and to provide greater clearance for crossing and overtaking on the curves. Psychological widening is therefore important in pavements with more than one lane. An empirical formula has been recommended by IRC to find the psychological widening 'W_{ps}' which is given as

$$W_{ps} = \frac{V}{9.5\sqrt{R}}$$

where V is design speed in kmph and R is radius of horizontal curve in m

Here V = 80 kmph and R = 700 m

$$\therefore \quad W_{ps} = \frac{80}{9.5\sqrt{700}} = 0.3183 \text{ m}$$

Hence total widening $W_e = W_m + W_{ps} = 0.0514 + 0.3183 = 0.3697$ m

1.4 **A truck with CG at $X = 1.4$ m and $Y = 1.7$ m is travelling on curved road of radius 200 m and super elevation as 0.05. Determine the maximum safe speed to avoid both slipping and overturning, assuming that the coefficient of sliding friction is 0.15. Sketch, explain and derive the expression for the conditions when overturning is likely to happen.**

[20 marks : 2006]

Solution:

Given data:

$X = 1.4$ m; $Y = 1.7$ m, radius of road, $R = 200$ m, super elevation, $\tan\theta = 0.05$; $f = 0.15$

Let P be the centrifugal force acting horizontally outwards through the centre of gravity; W be the weight of vehicle acting downwards through centre of gravity; θ be the angle of inclination of outer edge of pavement with the horizontal; F_A and F_B are the frictional forces developed on wheel A and B respectively.

We know that the centrifugal force is opposed by corresponding value of the friction developed and by a component of the force of gravity due to the super elevation provided.

The figure below shows the cross section of a pavement with all the forces acting on the vehicle resolved parallel and perpendicular to the inclined road surface.

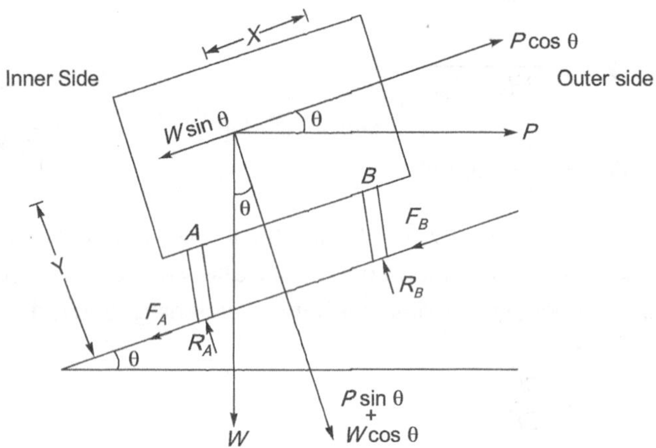

Analysis for skidding:

To avoid skidding, the component of centrifugal force P parallel to the plane must be less than or equal to the combined effect of component of gravity parallel to the plane and the friction force F_A and F_B.

$$\therefore \quad P\cos\theta \leq W\sin\theta + F_A + F_B$$

$$\Rightarrow \quad P\cos\theta \leq W\sin\theta + f(R_A + R_B)$$

where f is the coefficient of lateral friction and R_A and R_B are the normal reactions at A and B respectively.

$$\therefore \quad P\cos\theta \leq W\sin\theta + f(W\cos\theta + P\sin\theta)$$

$$\Rightarrow \quad P\cos\theta - Pf\sin\theta \leq W\sin\theta + fW\cos\theta$$

Dividing both sides by $W\cos\theta$, we get

$$\frac{P}{W} - \frac{P}{W}f\tan\theta \leq \tan\theta + f$$

$$\Rightarrow \quad \frac{P}{W} \leq \frac{f + \tan\theta}{1 - f\tan\theta} \quad \text{...(i)}$$

Analysis for overturning:

Let X and Y be the distance of centre of gravity from wheel B.

To avoid overturning, the moments of all the resolved components of forces about B should be greater than or equal to zero, if the moments are taken positive in counter clockwise sense.

$\therefore \quad W\cos\theta \times X + W\sin\theta \times Y + P\sin\theta \times X - P\cos\theta \times Y \geq 0$

$\Rightarrow \quad P(X\sin\theta - Y\cos\theta) + W(X\cos\theta + Y\sin\theta) \geq 0$

$\Rightarrow \quad P(Y\cos\theta - X\sin\theta) - W(X\cos\theta + Y\sin\theta) \leq 0$

$$\Rightarrow \quad \frac{P}{W} \leq \frac{X\cos\theta + Y\sin\theta}{Y\cos\theta - X\sin\theta}$$

$$\Rightarrow \quad \frac{P}{W} \leq \frac{X + Y\tan\theta}{Y - X\tan\theta} \text{ [Dividing numerator and denominator by } \cos\theta] \quad \text{...(ii)}$$

We know that $\frac{P}{W} = \frac{V^2}{gR}$

(a) Safe speed as per skidding

$$\frac{V^2}{gR} \leq \frac{f + \tan\theta}{1 - f\tan\theta} \quad \text{[from (i)]}$$

$$\Rightarrow \quad V^2 \leq 9.81 \times 200 \times \left[\frac{0.15 + 0.05}{1 - (0.15)\times 0.05}\right]$$

$$\Rightarrow \quad V \leq 19.88 \text{ m/s}$$

(b) Safe speed as per overturning

$$\frac{V^2}{gR} \leq \frac{X + Y\tan\theta}{Y - X\tan\theta} \quad \text{[from (ii)]}$$

$$\Rightarrow \quad V^2 \leq 9.81 \times 200 \times \left[\frac{1.4 + (1.7\times 0.05)}{1.7 - (1.4\times 0.05)}\right]$$

$$\Rightarrow \quad V \leq 42.28 \text{ m/s}$$

Hence skidding will occur prior to overturning, hence the safe speed will be 19.88 m/s.

1.5 **An express way of four lane divided passing through a flat terrain has a horizontal curve of radius equal to ruling minimum radius. If the design speed is 120 kmph calculate the following elements of the curve: (i) Ruling minimum radius (ii) Super elevation (iii) Extra widening, and (iv) Length of the transition curve.**

Suggest the best suitable shape of the transition curve.

[10 marks : 2007]

Solution:

(i) As per IRC

Maximum allowable superelevation (e) = 0.07

Design coefficient of lateral friction (f) = 0.15

The ruling minimum radius is given by

$$R_{\text{ruling}} = \frac{V^2}{127(e+f)} = \frac{(120)^2}{127(0.07 + 0.15)} = 515.39 \text{ m}$$

(ii) The superelevation for 75% of design speed is calculated neglecting the friction

$$\therefore \quad e = \frac{(0.75\,V)^2}{127\,R} \qquad [\because R_{ruling} = R]$$

$$\Rightarrow \quad e = \frac{(0.75 \times 120)^2}{127 \times 515.39}$$

$$\Rightarrow \quad e = 0.124 > 0.07$$

Hence, the maximum superelevation provided will be 0.07.

(iii) The extra widening is given as

$$E_w = \frac{nl^2}{2R} + \frac{V}{9.5\sqrt{R}}$$

As per IRC, length of wheel base (l) is taken as 6.1 m

$$\therefore \quad E_w = \frac{4 \times (6.1)^2}{2 \times 515.39} + \frac{120}{9.5\sqrt{515.39}} = 0.701 \text{ m}$$

(iv) Length of the transition curve

(a) As per IRC, the length of transition curve for a flat terrain is given as

$$L_s = \frac{2.7\,V^2}{R} = \frac{2.7 \times (120)^2}{515.39} = 75.44 \text{ m}$$

(b) As per rate of change of acceleration,

$$L_s = \frac{v^3}{CR}$$

where v is in m/s and C is the rate of change of acceleration.

As per IRC, $\quad C = \frac{80}{75 + V}$ where V is in kmph

Maximum and minimum values of C are 0.8 and 0.5 respectively

$$\therefore \quad C = \frac{80}{75 + 120} = 0.41$$

$$\therefore \quad \text{Adopt } C = 0.5$$

Also $\quad L_s = \frac{v^3}{CR} = \frac{(0.278 \times 120)^3}{0.5 \times 515.39} = 144.07 \text{ m}$

(c) As per rate of change of super elevation

$L = 150x$ in plain and rolling terrain as per IRC where x is raise of outer surface of pavement

If the pavement is rotated about the centre line, then

$$x = \left[\frac{W + E_w}{2}\right] e$$

As per IRC maximum super elevation, $e = 0.07$ for plain and rolling terrain and width of pavement, $W = 14$ m for a four lane expressway.

$$\therefore \quad x = \left[\frac{14 + 0.701}{2}\right] \times 0.07 = 0.5145$$

$$\therefore \quad L_s = 150\,x = 150 \times 0.5145 = 77.18 \text{ m}$$

Hence the length of transition curve will be the maximum of (a), (b) and (c) i.e. 144.07 m. Also spiral transition curve is the most ideal choice for Highways.

1.6 **A valley curve of a State Highway is formed by a descending gradient of 1 in 20 meeting an ascending gradient of 1 in 30. Design the length of a valley curve to fulfil both comfort condition and head light sight distance requirement for a design speed of 80 kmph. Assume allowable rate of change of centrifugal acceleration C = 0.60 m/sec^2. Suggest the best suitable shape of the valley curve. Consider reaction time t = 2.5 sec and coefficient of longitudinal friction f = 0.35.**

[10 marks : 2008]

Solution:

Given data: $n_1 = -\frac{1}{20}$; $n_2 = +\frac{1}{30}$; v = 80 kmph = 22.22 m/sec; C = 0.60 m/sec^3; t = 2.5 sec; f = 0.35

Deviation angle, $$N = n_1 - n_2 = -\frac{1}{20} - \frac{1}{30} = \frac{-3-2}{60} = -\frac{1}{12}$$

Length of valley curve for comfort condition is given by

$$L = 2\left[\frac{N v^3}{C}\right]^{1/2} = 2\left[\frac{1}{12} \times \frac{(22.22)^3}{0.6}\right]^{1/2} = 78.07 \text{ m}$$

Neglecting the ascending and descending gradients at the valley curve and calculating sight distance as per the equation given below, we get

$$S = vt + \frac{v^2}{2gf} = 22.22 \times 2.5 + \frac{(22.22)^2}{2 \times 9.81 \times 0.35} = 127.45 \text{ m}$$

If L > SSD, then

$$L = \frac{NS^2}{1.5 + 0.035S} = \frac{1}{12} \times \frac{(127.45)^2}{1.5 + 0.035 \times 127.45} = 227.09 \text{ m}$$

As this value is higher than SSD of 127.45 m, the assumption is correct. The valley curve length based on headlight sight distance being higher than that based on comfort condition, the design length of the valley curve is 227.09 m or say 228 m.

The valley curve is assumed to be of parabolic shape when its length is calculated as per the head light sight distance criterion.

1.7 **A national-highway passing through rolling-terrain in heavy rainfall area has a horizontal curve of radius 550 meter. Design the length of transition curve, assuming design-speed of 80 kmph, normal pavement width = 7 meter, allowable rate of change of centrifugal acceleration between 0.5 to 0.8 and allowable rate of introduction of super-elevation is 1 in 150.**

[10 marks : 2012]

Solution:

Given data: Radius of curve (R) = 550 m, Width of pavement (W) = 7 m

∴ No. of lanes, $$n = \frac{7}{3.5} = 2$$

Design speed, V = 80 kmph

The length of transition curve is calculated for three cases and maximum of the three is adopted.

(i) Length of transition curve by rate of change of centrifugal acceleration;

$$L_s = \frac{v^3}{CR}$$

where V is in m/s and C is rate of change of acceleration in m/s^3

$$C = \frac{80}{75+V} = \frac{80}{75+80} = 0.516$$

This value is between 0.5 and 0.8 and hence accepted

$$\therefore \quad L_s = \frac{(0.278 \times 80)^3}{0.516 \times 550} = 38.76\text{m}$$

(ii) Length of transition curve by rate of introduction of superelevation

Superelevation rate $e = \frac{V^2}{225R} = \frac{80^2}{225 \times 550} = 0.0517 < 0.07$ (OK)

We know that,

Extra widening at curve, $E_w = \frac{nl^2}{2R} + \frac{V}{9.5\sqrt{R}}$

where l = length of wheel base in m

As per IRC it is taken to be 6.1 m

$$\therefore \quad E_w = \frac{2 \times 6.1^2}{2 \times 550} + \frac{80}{9.5\sqrt{550}} = 0.427\text{m}$$

Total raise of outer edge of pavement with respect to the centre line;

$$x = \frac{E}{2} = \left[\frac{W+E_w}{2}\right]e = \left[\frac{7+0.427}{2}\right] \times 0.0517 = 0.192\text{m}$$

$$\therefore \quad L = 150\,x = 150 \times 0.192 = 28.80 \text{ m}$$

(iii) As per empirical formula

As per IRC, the length of transition curve for a plain and rolling terrain is given by

$$L = \frac{2.7V^2}{R} = \frac{2.7 \times 80^2}{550} = 31.42\text{m}$$

Adopt the highest value of the three i.e., 38.76 or say 39 m as the design length of transition curve.

1.8 A descending gradient of 1 in 30 meets an ascending gradient of 1 in 40 to form a valley curve. Find the length of the curve if the stopping sight distance is 120 m.

[5 marks : 2013]

Solution:

$$n_1 = -\frac{1}{30} \text{ (Descending gradient)}$$

$$n_2 = \frac{1}{40} \text{ (Ascending gradient)}$$

Stopping sight distance, SSD = 120 m

Let L = Length of valley curve.

Let L > SSD

(i) Length of valley curve as per headlight sight distance,

$$\therefore \quad L = \frac{NS^2}{(1.5+0.035S)}$$

Where, $N = \left|\frac{-1}{30} - \frac{1}{40}\right| = \left|\frac{-4-3}{120}\right| = \frac{7}{120}$

$$\therefore \quad L = \frac{\left(\frac{7}{120}\right)120^2}{(1.5+0.035\times120)}$$

and $$S = SSD = 120 \text{ m}$$

$$= \frac{7\times120}{5.7} = 147.368 \text{ m} > 120 \text{ m (SSD)} \quad ...(i)$$

(ii) Length of valley curve as per comfort condition,

$$L = 2\sqrt{\frac{Nv^3}{C}} = 2\sqrt{\frac{\frac{7}{120}\times\left(120\times\frac{5}{18}\right)^3}{0.6}} = 120.014 \text{ m} \quad ...(ii)$$

$\therefore$ Length of curve = 147.368 m (maximum of (i) and (ii)).

1.9 **Calculate the safe stopping distance for a vehicle travelling at 90 kmph, on an upward gradient of 3 per cent, given total reaction time is 2.5 seconds and the coefficient of friction for the road surface as 0.35.**

[4 marks : 2013]

Solution:

Given: Total reaction time, $t = 2.5$ sec

Coefficient of friction, $f = 0.35$

$v = 90$ km/hr = 25 m/sec

Upward gradient $(n) = 3\%$

$\therefore$ Stopping Sight Distance, $$\text{SSD} = vt + \frac{v^2}{2g(f+n)} = 25(2.5) + \frac{25^2}{2\times9.81(0.35+0.03)} = 146.33 \text{ m}$$

Thus stopping sight distance is 146.33 m

1.10 **Determine the minimum non-passing sight distance that should be provided for a vehicle coming down a 6 per cent gradient, using the following data:**

1. **Design speed, $V = 56$ kmph**
2. **Reaction time of the driver = 2 seconds**
3. **Coefficient of friction between Tyre and Road surface = 0.5**

[5 marks : 2013]

Solution:

$$S = vt + \frac{v^2}{2g(f-n)}$$

$$S = (0.278\times56)(2) + \frac{(0.278\times56)^2}{2(0.5-0.06)(9.81)}$$

$$S = 31.136 + 28.075 = 59.211 \text{ m}$$

1.11 **While aligning a highway in a built-up area, it was necessary to provide a horizontal circular curve of radius 300 meters. Design the following geometric features:**

(*i*) Superelevation

(*ii*) Extra widening of pavements

(*iii*) Length of transition curve

[8 marks : 2014]

Solution:

Given: Radius of curve, $R = 300$ m

Assuming design speed for Highway in a built-up area to be 80 kmph

(i) Super Elevation:

For mixed traffic conditions, *SE* is calculated for 75% of design speed neglecting the friction.

$$\therefore \qquad e = \frac{(0.75V)^2}{127R} = \frac{V^2}{225R} = \frac{80^2}{225 \times 300} = 0.095 > 0.07(\text{max})$$

As the value of SE is greater than the maximum SE of 0.07. Hence the actual SE to be provided is restricted to 0.07.

Check for coefficient of lateral friction at full speed

$$e_{max} + f = \frac{V^2}{127R}$$

$$f = \frac{80^2}{127 \times 300} - 0.07 = 0.0979 < 0.15\ (\text{max}) \qquad \text{(OK)}$$

(ii) Extra widening of Pavements:

Assuming highway pavement width, $W = 7$ m

$\therefore$ No. of lanes, $n = 2$

Also assuming wheel base of vehicle,

As per IRC, $l = 6.1$ m

Extra widening,
$$E_W = \frac{nl^2}{2R} + \frac{V}{9.5\sqrt{R}} = \frac{2 \times 6.1^2}{2 \times 300} + \frac{80}{9.5\sqrt{300}} = 0.124 + 0.486 = 0.61 \text{ m}$$

(iii) Length of transition curve:

$$e = \frac{V^2}{225R} = \frac{80^2}{225 \times 300} = 0.095 > 0.07$$

$$\therefore \qquad e = 0.07$$

If pavement is rotated about inner edge

Rise of outer edge,
$$x = (W + E_w) \times e$$
$$= (7 + 0.61) \times 0.07$$
$$= 0.5327 \text{ m}$$

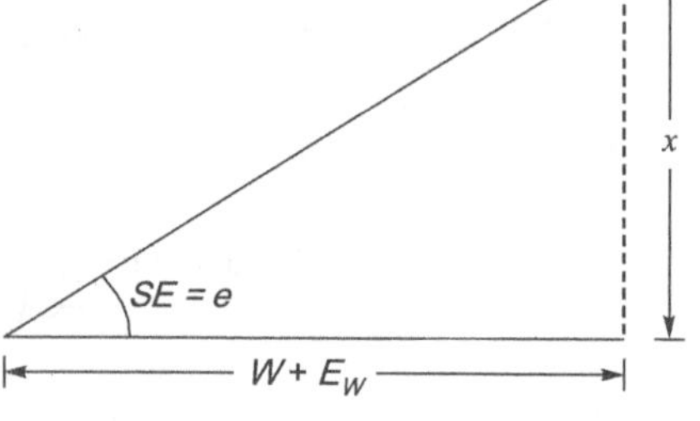

Hence length of transition curve in built-up area

$$L = 100x$$
$$= 100 \times 0.5327 = 53.27 \text{ m}$$

1.12 On a two-lane two-way highway, a car A was following a truck B and both were travelling at a speed of 40 kmph. While looking for an opportunity to overtake the truck, the driver of car *A* saw another car *C* coming from the opposite direction. At that moment, the distance between *A* and *C* was 450 m. After an initial hesitation period of 2 seconds, the driver of car *A* started the overtaking operation. The distance between *A* and *B* at that instant was 30 m. *A* overtook *B* by accelerating at a uniform rate of 1.20 ms^{-2}. When the overtaking operation was completed, there was a distance of 25 m between *B* and *A*. Determine the distance between different vehicles given as measured from the front bumper of the one vehicle to the front bumper of the another vehicle. Design speed of the highway is 80 kmph.

[8 marks : 2014]

Solution:

Before Overtaking

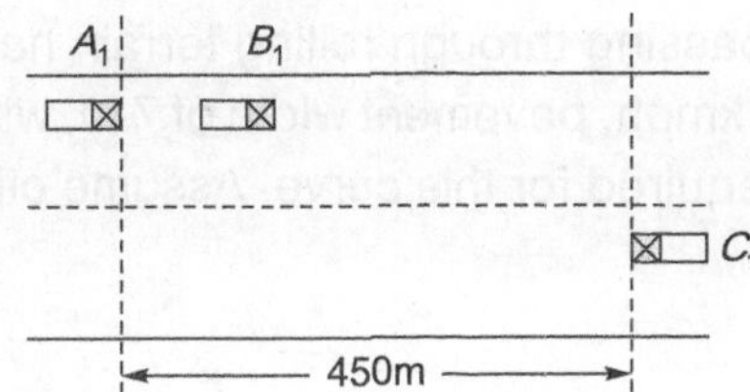

Assume, length of a vehicle = 6 m

Overtaking operation

s_1 = 30 m

s_2 = 25 m

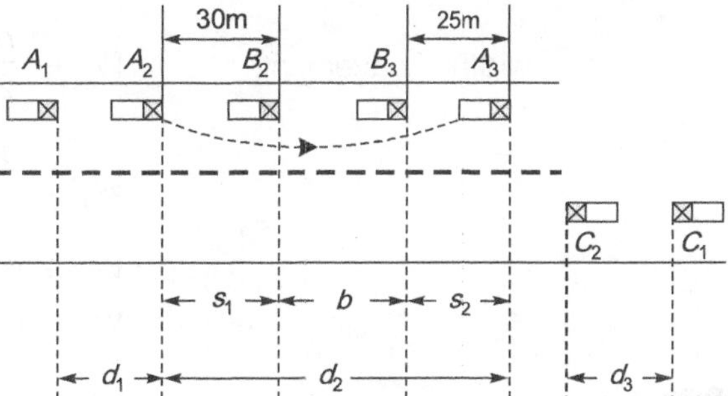

Before overtaking and overtaking operation is as shown above in the figures.

Total distance travelled by A during the overtaking operation,

$$D_A = d_1 + d_2$$

where, d_1 = distance travelled in reaction time

d_2 = distance travelled in overtaking

Now, reaction time, $t_R = 2$ sec

$\therefore$ $d_1 = V_A \times t_R$

$$= \left(40 \times \frac{5}{18}\right) \times 2 \qquad \ldots V_A = V_B = 40 \text{ kmph}$$

$$d_1 = 22.22 \text{ m}$$

Time taken during overtaking operation

$$T = \sqrt{\frac{2(s_1 + s_2)}{a}}$$

where, a = acceleration = 1.2 m/s^2

Now,

$$T = \sqrt{\frac{2(30+25)}{1.2}}$$

$$T = 9.5743 \text{ sec}$$

Now,

$$d_2 = s_1 + b + s_2 = 30 + V_B \times T + 25 = 55 + \left(40 \times \frac{5}{18}\right) \times 9.5743$$

$$d_2 = 161.38 \text{ m}$$

Thus, $D_A = d_1 + d_2 = 22.22 + 161.38 = 183.6$ m

During overtaking distance travelled by vehicle C

$$d_3 = V_c(t_R + T)$$

$$d_3 = 80 \times \frac{5}{18}(2 + 9.5743)$$

$$d_3 = 257.2 \text{ m}$$

Hence remaining distance between front bumpers of vehicles A & C

$$= 450 - (d_1 + d_2 + d_3)$$

$$= 450 - (183.6 + 257.2) = 9.2 \text{ m}$$

and remaining distance between front bumpers at vehicles B & C

$$= s_2 + 9.2 = 25 + 9.2 = 34.2 \text{ m}$$

1.13 **A two-lane national highway passing through rolling terrain has horizontal curve of 450 m radius. Assuming design speed of 90 kmph, pavement width of 7 m, with a wheel base of 6.1 m, determine extra width of the pavement required for this curve. Assume other parameters as per relevant IRC recommendations.**

[4 marks : 2016]

Solution:

Given: Radius of curve (R) = 450 m, Design speed (V) = 90 kmph, Wheel base (l) = 6.1 m, Pavement width (W) = 7 m, Number of lanes (n) = 2

Extra width of pavement, $$W_e = \frac{nl^2}{2R} + \frac{V}{9.5\sqrt{R}}$$

$$W_e = \frac{2 \times (6.1)^2}{2 \times 450} + \frac{90}{9.5\sqrt{450}}$$

$$W_e = 0.083 + 0.446$$

$$W_e = 0.529 \text{ m}$$

1.14 **Two straight roads intersect at chainage of 15,000 m and have external deflection angle equal to 100°. The road bend is to be designed for maximum speed of 80 kmph, maximum centrifugal ratio of 1/4 and a maximum rate of the change of acceleration of 0.3 m/sec³.**
The curve consists of a circular curve and two cubical parabola.
Calculate :
(i) the radius of circular curve
(ii) length of transition curve
(iii) chainage of all junction points

[20 marks : 2017]

Solution :

Given: External deflection angle (Δ) = 100°, Maximum centrifugal ratio = $\frac{1}{4}$

Maximum speed, $V = 80$ kmph
Maximum rate of change of acceleration, $\alpha = 0.3 \text{ m/sec}^3$
Chainage of intersection, $I = 15000$ m

(i) $\because$ $$V = 80 \text{ kmph} = 80 \times \frac{5}{18} = 22.22 \text{ m/sec}$$

$$\text{Centrifugal ratio} = \frac{V^2}{gR} = \frac{1}{4}$$

Where, R = Radius of the circular curve

$\therefore$ $$R = \frac{4V^2}{g} = \frac{4 \times (22.22)^2}{9.81} = 201.32 \text{ m}$$

(ii) Length of the transition curve, $$L = \frac{V^3}{\alpha R} = \frac{(22.22)^3}{0.3 \times 201.32} = 181.65 \text{ m}$$

(iii) Spiral Angle, $$\phi_s = \frac{L}{2R} \text{radians} = \frac{181.65}{2 \times 201.32} \times \frac{180}{\pi} = 25.85°$$

$$\text{Central Angle} = \Delta - 2\phi_s = 100° - 2 \times 25.85° = 48.30°$$

$$\text{Length of circular curve} = \frac{\pi \times 201.32 \times 48.30°}{180°} = 169.71 \text{ m}$$

Length of tangent, $$IT_1 = (R+S)\tan\left(\frac{\Delta}{2}\right) + \frac{L}{2}$$

Shift of curve, $$S = \frac{L^2}{24R} = \frac{(181.65)^2}{24 \times 201.32} = 6.83 \text{ m}$$

∴ Tangent length, $$T = IT_1$$

$$= (201.32 + 6.83)\tan\left(\frac{100}{2}\right) + \left(\frac{181.65}{2}\right) = 338.89 \text{ m}$$

Chainage of tangent point T_1 (First tangent point)

$$= \text{Chainage of } I - IT_1 = (15000 - 338.89) = 14661.11 \text{ m}$$

Chainage of first junction point = 14661.11 + L = 14661.11 + 181.65 = 14842.76 m

Chainage of second junction point = 14842.76 + 169.71 = 15012.47 m

Chainage of second tangent point = 15012.47 + 181.65 = 15194.12 m

1.15 **The speed of overtaking vehicle is 90 km/hr and acceleration is 3.0 km/hr/sec. Calculate the safe passing sight distance for one way and two way traffic. Calculate the length of overtaking zone. Also draw the same.**

[12 marks : 2017]

Solution:

Design speed = Speed of overtaking vehicle

⇒ $$V = 90 \text{ kmph} = 90 \times \frac{5}{18} = 25 \text{ m/sec}$$

∴ Speed of overtaken vehicle, $V_b = V - 16 = 90 - 16 = 74$ kmph = 20.56 m/sec

Reaction time,

Adopt t = 2 secs

∴ $$d_1 = 0.28\, V_b \times t = 0.28 \times 74 \times 2 = 41.44 \text{ m}$$

$$d_2 = 0.28\, V_b T + 2s$$

$$s = 0.2\, V_b + 6 = 0.2 \times 74 + 6 = 20.8 \text{ m}$$

$$T = \sqrt{\frac{14.4s}{A}} = \sqrt{\frac{14.4 \times 20.8}{3}} = 9.99 \text{ sec} \simeq 10 \text{ secs}$$

∴ $$d_2 = 0.28 \times 74 \times 10 + 2 \times 20.8 = 248.80 \text{ m}$$

$$d_3 = 0.28\, VT = 0.28 \times 90 \times 10 = 252 \text{ m}$$

O.S.D. on one-way traffic road = $d_1 + d_2$ = 41.44 + 248.80 = 290.24 ≃ 291 m

Minimum length of overtaking zone = 3 × OSD = 3 × 291 = 873 m

Desirable length of overtaking zone = 5 × OSD = 5 × 291 = 1455 m

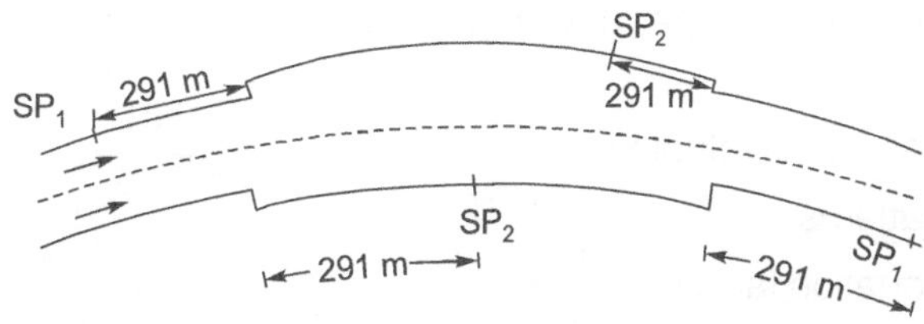

O.S.D. on two-way traffic road = $d_1 + d_2 + d_3$ = 41.44 + 248.80 + 252 = 542.24 m ≃ 543 m

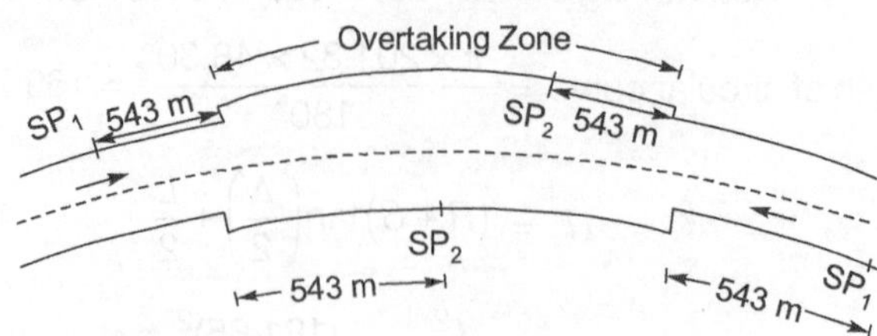

SP_1 = Sign post 'overtaking zone ahead'

SP_2 = Sign post 'end of overtaking zone'

Minimum length of overtaking zone = 3 (OSD) = 3 × 542.24 m = 1626.72 m

Desirable length of overtaking zone = 5 × OSD = 5 × 542.24 = 2711.2 m

1.16 **Briefly describe planning surveys for highways. How are these used and interpreted?**

[12 marks : 2018]

Solution:

Highway planning phase includes:

(i) Assessment of road length requirement for an area.

(ii) Preparation of master plan showing the phasing of plan in annual and or five year plans.

The planning surveys consists of following studies, the details are as follows:

(a) Economic studies

In order to estimate the economics involved and to give economic justification for each plan, economic studies are to be undertaken. Details to be collected include:

(i) Population and its distribution.

(ii) Trend of population growth.

(iii) Agricultural and industrial products, areawise.

(iv) Existing communication, recreation and education facilities.

(v) Per capita income.

(b) Financial studies

Financial studies are essential to study the financial aspects like sources of income and manner in which funds for the project may be mobilized. Following details should be collected:

- Sources of income and estimated revenue from taxation on road transport.
- Living standards.
- Resources at local level, toll taxes, vehicle registration and fines etc.
- Future trends in financial aspects.

(c) Traffic or road use studies

Traffic surveys should be carried out in order to collect the following details:

- Traffic volume in vehicle per day, annual average daily traffic, peak and design hourly traffic volume.
- Origin and destination studies.
- Traffic flow patterns.
- Mass transportation facilities.
- Accident analysis.
- Future trends and growth in traffic

(d) Engineering studies

The studies include:

- Topographic surveys.
- Soil surveys.
- Location and classification of existing roads.
- Road life studies.
- Traffic studies.
- Special problems in drainage, construction and maintenance of roads.

1.17 **Design the length of a valley curve which is formed by a descending grade of 1 in 30 meeting an ascending grade 1 in 40 for a design of 100 kmph. Assume: Driver's reactions time = 2.5 sec, Coefficient of friction = 0.35 and Allowable Rate of Change of Centrifugal Acceleration = 0.6 m/sec^3.**

[15 marks : 2018]

Solution:

Given data:

Descending grade = 1 in 30

Ascending grade = 1 in 40

Design speed, $V = 100$ kmph = 27.78 m/s

Reaction time, $t = 2.5$ sec

Coefficient of friction, $f = 0.35$

Rate of change of centrifugal acceleration,

$C = 0.6$ m/sec^3

(i) Based on comfort criteria

Length of valley curve, $$L = 2\sqrt{\frac{NV^3}{C}}$$

$$N = \frac{-1}{30} - \left(\frac{1}{40}\right) = \left(\frac{-7}{120}\right)$$

$$\Rightarrow \quad L = 2\sqrt{\frac{\frac{7}{120}\times\left(\frac{100\times5}{18}\right)^3}{0.6}} = 91.3 \text{ m}$$

(ii) Based on head light sight distance criteria

Assume $L > SSD$

$$L = \frac{NS^2}{2h + 2S\tan\alpha} \qquad (h = 0.75 \text{ m}, \alpha = 1°)$$

S = SSD = Lag distance + Braking distance

$$= Vt + \frac{V^2}{2gf} = (27.78\times2.5) + \left(\frac{27.78^2}{2\times9.81\times0.35}\right) = 181.83 \text{ m}$$

$$\therefore \quad L = \frac{\frac{7}{120}\times181.83^2}{(2\times0.75) + (2\times181.83\times\tan1°)}$$

$\Rightarrow$ $L = 245.76$ m > SSD

$\therefore$ Assumption is correct

So, length of valley curve is maximum of above two values.

$\Rightarrow$ Length of valley curve = 245.76 m

1.18 A two-lane, two-way highway is designed for design speed of 80 km/hr. A vertical curve is to be provided at intersection of downward gradient of 1 in 50 with another downward gradient of 1 in 20. Calculate the length of the vertical curve fulfilling the requirement of stopping sight distance and overtaking sight distance. The coefficient of longitudinal friction and the acceleration may be taken as 0.35 and 3.6 km/hr/sec respectively.

[25 marks : 2019]

Solution:

$$N_1 = -\frac{1}{50}$$

$$N_2 = -\frac{1}{20}$$

Deviation angle, $N = N_1 - N_2$

$$= -\frac{1}{50} - \left(-\frac{1}{20}\right)$$

$$N = \frac{3}{100}$$

$N_1 = \frac{1}{50}$, $N_2 = \frac{1}{20}$

Design speed (v) = 80 kmph = 22.22 m/s

(I) Stopping sight distance (SSD) calculation:

$$\text{SSD} = vt + \frac{v^2}{2gfb}, \text{ [Consider brake efficiency } b = 100\%]$$

Coefficient of longitudinal friction, $f = 0.35$

Assume reaction time = 2.5 seconds

$$\therefore \quad \text{SSD} = 22.22 \times 2.5 + \frac{(22.22)^2}{2 \times 9.81 \times 0.35} = 127.448 \text{ m}$$

(II) Overtaking sight distance (OSD) calculation:

Since speed of overtaken vehicle V_b is not given,

$$\therefore \quad V_b = (V - 4.5) \text{ m/s} = 17.72 \text{ m/s}$$

OSD for two lane, two-way highway = $d_1 + d_2$

$$d_1 = V_b \times t_r$$

[Assume, reaction time of driver, t_r = 2 sec]

$$\therefore \quad d_1 = 17.72 \times 2 = 35.44 \text{ m}$$

$$d_2 = b + 2s,$$

where $S = 0.7V_b + 6$

$$\therefore \quad S = 0.7 \times 17.72 + 6 = 18.404 \text{ m}$$

Also, $b = V_b \times T$, where $T = \sqrt{\frac{4S}{a}}$

a = 3.6 km/hr/sec (given)

$$\therefore \quad T = \sqrt{\frac{4 \times 18.404}{3.6 \times \frac{5}{18}}} = 8.5799 \text{ sec}$$

$$\therefore \quad b = 17.72 \times 8.5799 = 152.037 \text{ m}$$

$$\therefore \quad d_2 = b + 2s = 152.037 + 2 \times 18.404 = 188.845 \text{ m}$$

$$d_3 = vT = 22.22 \times 8.5799 = 190.645 \text{ mm}$$

Hence, $\text{OSD} = d_1 + d_2 + d_3$

$= 35.44 + 188.845 + 190.645$

$= 414.93 \text{ m} \simeq 415 \text{ m}$

Case 1 : Assume (L_C > SSD)

$$L_C = \frac{NS^2}{4.4} = \frac{\left(\frac{3}{100}\right) \times 127.448^2}{4.4}$$

$$= 110.747 \text{ m} < 127.448 \text{ m}$$

Hence, our assumption is wrong.

$\therefore \quad L_C < \text{SSD}$

$$L_C = 2S - \frac{4.4}{N} = 2 \times 127.448 - \frac{4.4}{\left(\frac{3}{100}\right)}$$

Length of vertical curve based on SSD;

$$L_C = 108.23 \text{ m}$$

Case 2 : Assume, (L_C > OSD)

$$L_C = \frac{NS^2}{9.6} = \frac{\left(\frac{3}{100}\right) \times 415^2}{9.6} = 538.20 \text{ m}$$

Hence, our assumption is correct.

Length of vertical curve based on OSD = 538.20 m

1.19 **Design the length of transition curve to be provide on a horizontal curve of radius 484 m on a National Highway with double lane passing through heavy rainfall area. Following design data is given:**

Ruling design speed = 80 kmph

Type of terrain = Rolling terrain

Rate of introduction of superelevation = 1 in 150

Wheel base of design vehicle = 6 m

[25 marks : 2020]

Solution:

Given : $R = 484$ m, $V = 80$ kmph (Rolling terrain), $N = 150$, $l = 6$ m

Case (i) : Based on rate of change of centrifugal acceleration

$$C = \frac{80}{75 + V} = \frac{80}{75 + 80} = 0.516 \text{ m/s}^3$$

$$L_T = \frac{V^3}{CR} = \frac{\left(\frac{5}{18} \times 80\right)^3}{0.516 \times 484} = 43.94 \text{ m}$$

Case (ii) : As per rate of introduction of superelevation.

Design of superelevation

$$e = \frac{V^2}{225R} = \frac{80^2}{225 \times 484} = 0.0587 < 0.07 \qquad \text{(Safe)}$$

$$e = 5.87\%$$

Assume rotation about centreline
(For radius $R > 300$ m extra widening is not required for 2 lane road)

$$L_T = \frac{1}{2}eN(W + W_e)$$

$$= \frac{1}{2} \times 0.0587 \times 150 \times 7 = 30.81 \text{ m}$$

Case (iii) : Minimum length

$$L_T = 2.7\frac{V^2}{R} = 2.7 \times \frac{80^2}{484} = 35.70 \text{ m}$$

Length of tranection curve, $$L_T = \max\begin{cases}43.94\\30.81 = 43.94 \text{ m}\\35.70\end{cases}$$

1.20 **Clearly state the nine goals and objectives of National Transport Plan of India.**

[12 Marks : 2021]

Solution:

The following are the broad goals and objectives for National Transport Plan:

1. Balanced development for the total road network comprising three functional groups viz., the primary system (National Highway (NH) and expressway(s)), secondary system (State Highways and Major District Roads) and rural roads.
2. Development of roads to be considered an integral part of the total transport system supplementing other modes, integrating the development plans with railways and other modes of transport.
3. Completion of the National Highways Development Project comprising the golden Quadrilateral and the North-South and East-West corridors.
4. Phased removal of deficiencies in the existing NH network in tune with traffic for the next 10-15 years with emphasis on four-laning of high-density corridors.
5. To plan and take preliminary action for expressways to be built in future in those sections where these can be economically justified.
6. To make long distance travel safer and faster so as to give a boost to the economy.
7. Priority is to be accorded to areas like overloading of trucks, control of encroachments and unplanned ribbon development, energy conservation and environment protection.
8. Greater attention to be paid to rehabilitation and reconstruction of weak/dilapidated bridges for traffic safety.
9. Special attention is to be paid to the development of roads in the North-Eastern region.

1.21 **Design a summit curve for a National Highway for a stopping sight distance of 200 m at the junction of a rising gradient of 1 in 60 and a falling gradient of 1 in 30. Set out the curve with a chord 25 m long. Determine the R.L. of the point immediately below the intersection point of the grade line and also the R.L. of the highest point on the curve. Assume sight distance is less than the length of vertical curve, R.L. at the start f the curve as 10.0. h_1 and h_2 as per current Indian practice IRC 66-1976.**

[20 Marks : 2021]

Solution:

Given: Stopping sight distance,

$$S = 200 \text{ m}$$

$$n_1 = +\frac{1}{60},\ n_2 = -\frac{1}{30}$$

$$N = \left|\frac{1}{60} + \frac{1}{30}\right| = \frac{1}{20}$$

Since,

$$L_S > S$$

$$L_S = \frac{NS^2}{2\left(\sqrt{h_1} + \sqrt{h_2}\right)^2}$$

As per IRC 66 – 1976, h_1 = Hight of driver = 1.2 m

h_2 = Hight of object = 0.15 m

$$L_s = \frac{NS^2}{4.4} = \frac{\left(\frac{1}{20}\right) \times 200^2}{4.4}$$

$$= 454.54 \text{ m} > 200 \text{ m (OK)}$$

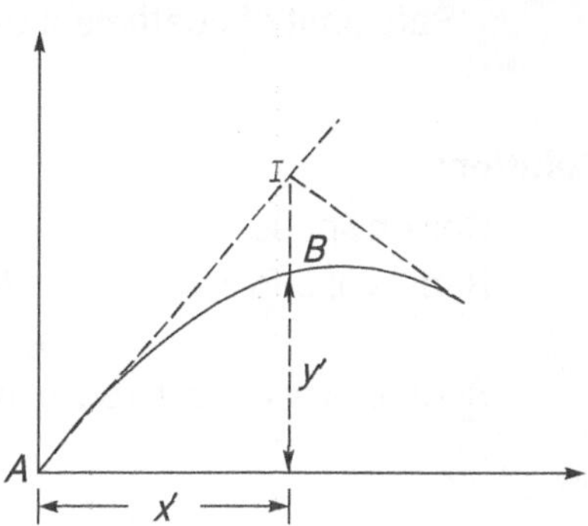

(a) **RL of point just below VPI (i.e. RL_B)**

we know that, *VPI* exist at $x = \frac{L_S}{2}$

$$RL_B = RL_A + y' @ x = \frac{L}{2} \qquad \left(\because y' = \frac{-Nx^2}{2L_s} + x_1 x\right)$$

$$\therefore \quad RL_B = RL_A + \left(-\frac{N}{2L_S}\right) \times \frac{L_S^2}{4} + n_1 \frac{L_S}{2}$$

$$= RL_A + n_1 \frac{L_S}{2} - N\frac{L_S}{8}$$

$$= 10 + \left(\frac{1}{60}\right) \times \frac{454.54}{2} - \frac{1}{20} \times \frac{454.54}{8} = 10.947 \text{ m}$$

(b) **RL of highest point on the curve**:

Position of summit point, from origin exist at $x = \frac{n_1}{N} L_s$

$$\frac{\frac{1}{60}}{\frac{1}{20}} \times 454.54 = 151.51 \text{ m}$$

RL of highest point

$$= RL_A + y' @ x = 151.51 \qquad \left(\because y' = \frac{-Nx^2}{2L_s} + n_1 x\right)$$

$$\therefore \quad RL_{\text{summit}} = 10 + \left[\frac{-N}{2L_S}\right] x^2 + n_1 x$$

$$= 10 + \left(\frac{-\left(\frac{1}{20}\right)}{2 \times 454.54}\right) \times 151.51^2 + \frac{1}{60} \times 151.51 = 11.263 \text{ m}$$

1.22 The design speed of a highway is 80 km/h. There is a horizontal curve of radius 200 m in a certain locality. What should be the superelevation required to maintain this design speed? If the maximum superelevation of 0.07 is not to be exceeded, what should be the maximum allowable speed on this curve? Also determine the extra widening required and length of transition curve using the following data:

Length of wheel base of the largest vehicle = 6.1 m.

Pavement width = 7.2 m.

Number of lanes = 2.

Rate of introduction of superelevation = 1 in 200.

Type of terrain = Plain.

Safe limit of coefficient of friction = 0.15.

[20 marks : 2022]

Solution:

Design speed, $V = 80$ kmph

Radius of curve, $R = 200$ m

$e_{max} = 7\%$

Superelevation required to maintain design speed,

$$e = \frac{V^2}{225R} = \frac{80^2}{225 \times 200} = 0.1422 > 0.07$$

Provide, $e = 7\%$ and check for friction

$$f = \left(\frac{V^2}{127R} - e_{max}\right) = \left(\frac{80^2}{127 \times 200} - 0.07\right) = 0.1819 \not< 0.15$$

Maximum allowable speed,

$$V_{max} = \sqrt{127R(e_{max} + f)} = \sqrt{127 \times 200 \times (0.07 + 0.15)} = 74.75 \text{ kmph}$$

Extra widening,

$$W_e = \left(\frac{nl^2}{2R} + \frac{V}{9.5\sqrt{R}}\right) = \frac{2 \times 6.1^2}{2 \times 200} + \frac{80}{9.5\sqrt{200}} = 0.781 \text{ m}$$

Length of transition curve:

Criterion : 1 As per rate of change of centrifugal acceleration,

$$C = \frac{80}{75 + V} = \frac{80}{75 + 80} = 0.516 \text{ m/s}^3$$

$$L_T = \frac{0.0215V^3}{CR} = \frac{0.0215 \times 80^3}{0.516 \times 200} = 106.67 \text{ m}$$

Criterion: 2 As per rate of attainment of superelevation,

Assume rotation about @ inner edge,

$$L_T = eN(W + W_e) = 0.07 \times 200 \times (7.2 + 0.781) = 111.734 \text{ m}$$

Criterion: 3 As per minimum length, IRC method for plain terrain.

$$L_T = 2.7\frac{V^2}{R} = 2.7 \times \frac{80^2}{200} = 86.4 \text{ m}$$

Minimum length of transition curve is maximum of all the above criteria i.e. 111.734 m.

1.23 A car moving with a speed of 80 kmph has to overtake another car moving at a speed of 64 kmph in the two-lane one-way highway. If the reaction time of the driver is 2.5 s and acceleration of overtaking car is 0.95 m/s², calculate the safe overtaking sight distance.

[12 marks : 2023]

Solution:

Given: Two-lane one-way road

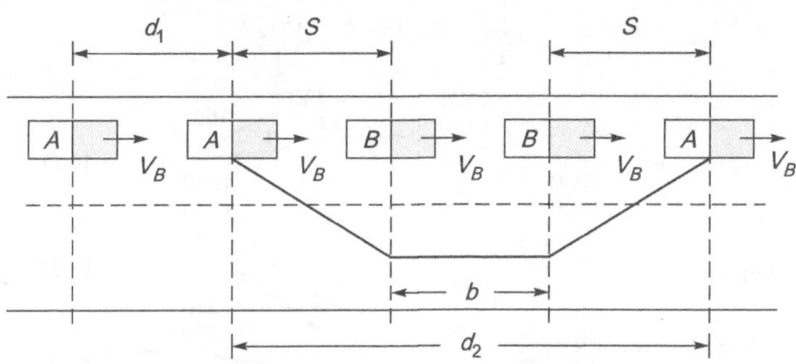

Speed of A, V_A = 80 kmph, Speed of B, V_B = 64 kmph, Reaction time, t_R = 2.5 sec

Acceleration, a = 0.95 m/s²,

$$d_1 = 0.278 \times V_B \times t_R$$
$$= 0.278 \times 64 \times 2.5 = 44.48 \text{ m}$$
$$d_2 = 2s + 0.278 \times V_B \times T$$

where,
$$s = 0.2 \times V_b + 6$$
$$= 0.2 \times 64 + 6 = 18.8 \text{ m}$$

and
$$T = \sqrt{\frac{4s}{a}} = \sqrt{\frac{4 \times 18.8}{0.95}} = 8.89 \text{ sec}$$

$$\therefore \quad d_2 = 2 \times 18.8 + 0.278 \times 64 \times 8.89 = 195.8 \text{ m}$$
$$\therefore \quad \text{OSD} = d_1 + d_2$$
$$= 44.48 + 195.8 = 240.28 \text{ m}$$

2. Traffic Engineering

2.1 Four legs in a rotary intersection are designated as 1, 2, 3 and 4. The traffic volumes (V_{ij}) in terms of PCU per hour are given as

V_{12}	150	V_{21}	310	V_{31}	1520	V_{41}	30
V_{13}	450	V_{23}	200	V_{32}	570	V_{42}	1080
V_{14}	412	V_{24}	1090	V_{34}	240	V_{43}	600

What is the proportion of weaving traffic to total traffic in the weaving section between legs 2 and 3? What is the use of this value?

[10 marks : 2001]

Solution:

Traffic flow leaving leg 1 $= V_{12} + V_{13} + V_{14} = 150 + 450 + 412 = 1012$

Traffic flow approaching leg 1 $= V_{21} + V_{31} + V_{41} = 310 + 1520 + 30 = 1860$

Traffic flow leaving leg 2 $= V_{21} + V_{23} + V_{24} = 310 + 200 + 1090 = 1600$

Traffic flow approaching leg 2 $= V_{12} + V_{32} + V_{42} = 150 + 570 + 1080 = 1800$

Traffic flow leaving leg 3 = $V_{31} + V_{32} + V_{34}$ = 1520 + 570 + 240 = 2330

Traffic flow approaching leg 3 = $V_{13} + V_{23} + V_{43}$ = 450 + 200 + 600 = 1250

Traffic flow leaving leg 4 = $V_{41} + V_{42} + V_{43}$ = 30 + 1080 + 600 = 1710

Traffic flow approaching leg 4 = $V_{14} + V_{24} + V_{34}$ = 412 + 1090 + 240 = 1742

The traffic assigned to the network is shown below:

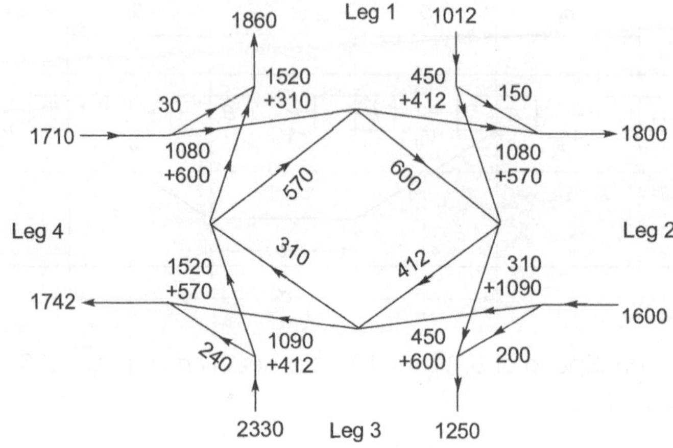

Weaving traffic between legs 2 and 3 = (310 + 1090) + (450 + 600) = 2450

Total traffic between legs 2 and 3 = (310 + 1090) + (450 + 600) + 412 + 200 = 3062

$\therefore$ Proportion of weaving traffic to total traffic between legs 2 and 3 = $\frac{2450}{3062} = 0.8$

This value is also known as weaving ratio. It is useful in calculating the practical capacity of the rotary intersection. Its value should lie between 0.4 and 1.0

2.2 **A road intersection has five legs designated as 1, 2, 3, 4 and 5. Leg 1 is in N-S direction and others are marked clockwise. The traffic volumes in terms of PCU(V_{ij}) per hour during peak period are given below as:**

V_{12}	37	V_{31}	466	V_{41}	182	V_{51}	182
V_{13}	303	V_{32}	122	V_{42}	54	V_{52}	54
V_{14}	64	V_{34}	47	V_{43}	18	V_{53}	18
V_{15}	52	V_{35}	657	V_{45}	116	V_{54}	116

Find the weaving ratio between the legs 1 and 2. What is the use of this value? Draw a sketch showing the traffic volumes between legs 1 and 2.

[10 marks : 2006]

Solution:

Traffic flow leaving leg 1 = $V_{12} + V_{13} + V_{14} + V_{15}$ = 37 + 303 + 64 + 52 = 456

Traffic flow leaving leg 2 = 0

Traffic flow leaving leg 3 = $V_{31} + V_{32} + V_{34} + V_{35}$ = 466 + 122 + 47 + 657 = 1292

Traffic flow leaving leg 4 = $V_{41} + V_{42} + V_{43}$ I V_{45} = 182 + 54 + 18 + 116 = 370

Traffic flow leaving leg 5 = $V_{51} + V_{52} + V_{53} + V_{54}$ = 45 + 132 + 62 + 15 = 254

Traffic flow approaching leg 1 = $V_{31} + V_{41} + V_{51} = 466 + 182 + 45 = 693$

Traffic flow approaching leg 2 = $V_{12} + V_{32} + V_{42} + V_{52} = 37 + 122 + 54 + 132 = 345$

Traffic flow approaching leg 3 = $V_{13} + V_{43} + V_{53} = 303 + 18 + 62 = 383$

Traffic flow approaching leg 4 = $V_{14} + V_{34} + V_{54} = 64 + 47 + 15 = 126$

Traffic flow approaching leg 5 = $V_{15} + V_{35} + V_{45} = 52 + 657 + 116 = 825$

The traffic assigned to the network is shown below:

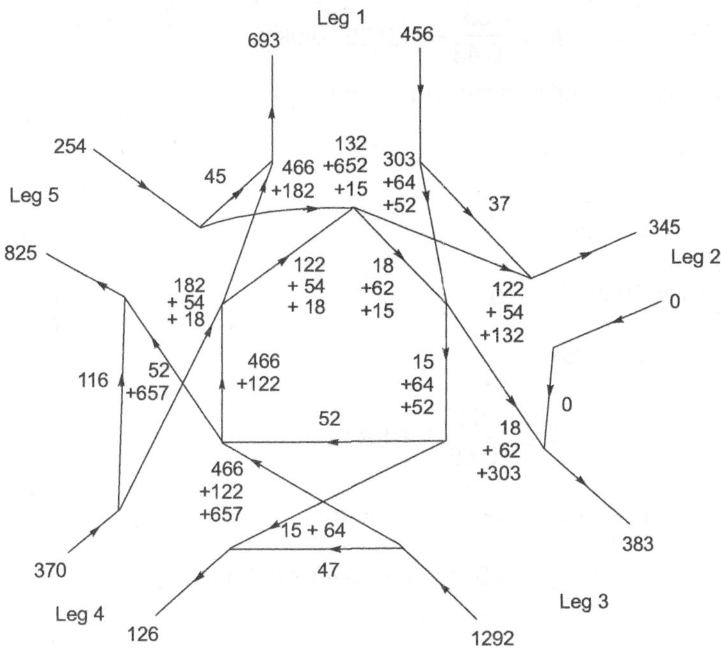

Weaving traffic between leg 1 and leg 2 = (303 + 64 + 52) + (122 + 54 + 132) = 727

Total traffic between leg 1 and 2 = (303 + 64 + 52) + (122 + 54 + 132) + (18 + 62 + 15) + 37 = 859

$\therefore$ Weaving ratio between legs 1 and 2 = $\frac{727}{859} = 0.85$

This value is useful in calculating the practical capacity of the rotary intersection. It should lie between 0.4 and 1.0.

2.3 **For a given section of road, relationship between speed and density is obtained as $u = 53 - 0.43k$, where 'u' is speed in kmph and 'k' is density in vehicle/km. Estimate the maximum flow, mean speed and mean density. Sketch the plots of three flow relationships and mark the values as obtained. State the significance of these values.**

[10 marks : 2006]

Solution:

We know that

Capacity of road, C = Traffic volume or flow in veh/hr

But, Traffic volume = Traffic density × Speed = $\frac{\text{veh}}{\text{km}} \times \frac{\text{km}}{\text{hr}}$

(i) Plot of relationship between speed and density

$$\therefore \quad C = ku$$

$$\Rightarrow \quad C = k(53 - 0.43k)$$

$$\Rightarrow \quad C = 53k - 0.43k^2$$

Now if $\quad C = 0$

$$\Rightarrow \quad 53k - 0.43k^2 = 0$$

$$\Rightarrow \quad k(53 - 0.43k) = 0$$

$$\Rightarrow \quad \text{Either } k = 0$$

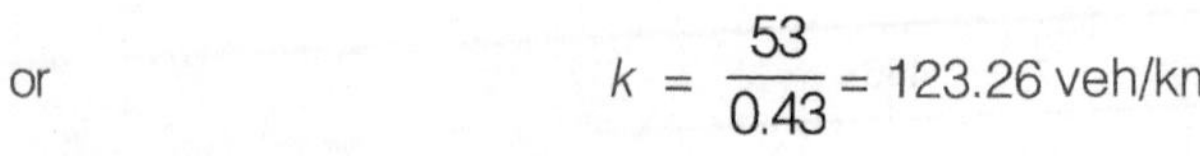

$$\text{or} \quad k = \frac{53}{0.43} = 123.26 \text{ veh/km}$$

(ii) Plot of relationship between density and volume

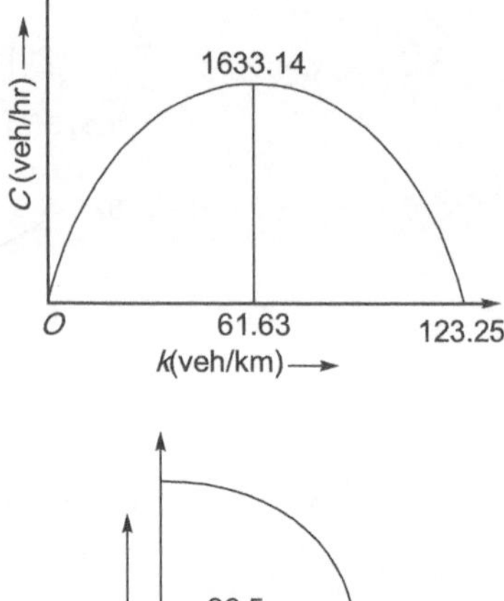

Now for maximum value of C,

$$\frac{dC}{dk} = 0$$

$$\Rightarrow \quad \frac{d}{dk}(53k - 0.43k^2) = 0$$

$$\Rightarrow \quad 53 - 0.86k = 0$$

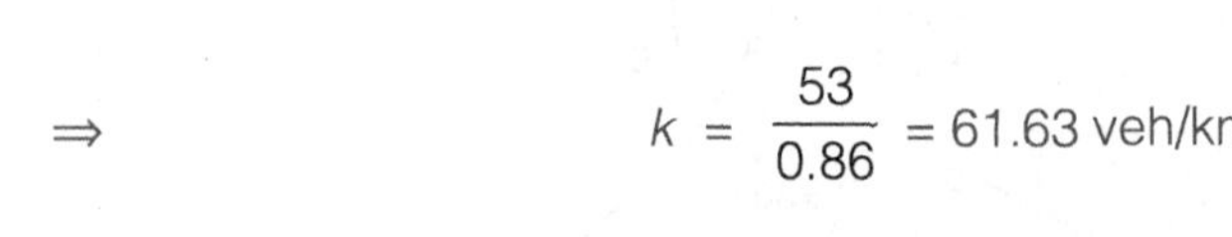

$$\Rightarrow \quad k = \frac{53}{0.86} = 61.63 \text{ veh/km}$$

(iii) Plot of relationship between speed and volume

$$\therefore \quad C_{max} = 53 \times 61.63 - 0.43 \times (61.63)^2$$

$$= 1633.14 \text{ veh/hr}$$

$$u = 53 - 0.43k$$

$$\therefore \quad U_{mean} = 53 - (0.43 \times 61.63) = 26.5 \text{ km/hr.}$$

The maximum flow is 1633.14 veh/hr; mean speed is 26.5 km/hr and the mean density is 61.63 veh/km.

Density and flow relationship is shown in the graph. As per the graph the capacity of the given section of road is zero when the density is zero. Capacity increases with density and attains a maximum value of 1633.14 veh/hr when density is 61.63 veh/km. The capacity decreases, if the density increase from this value and it becomes zero at a density of 123.26 veh/km.

2.4 Traffic flow in an urban area at right angled intersection of two major roads in the design year are given below:

Approach Road	Traffic Flow in PCU /hr		
	Left Turning	Straight ahead	Right Turning
North	415	650	300
East	300	550	250
South	350	400	225
West	400	500	300

Both roads have a carriage way width of 15 m. Design and draw a Rotary Intersection and check for its practical capacity, making suitable assumption as per IRC recommendations.

[15 marks : 2007]

Solution:

The traffic in PCU per hour is depicted in figure below:

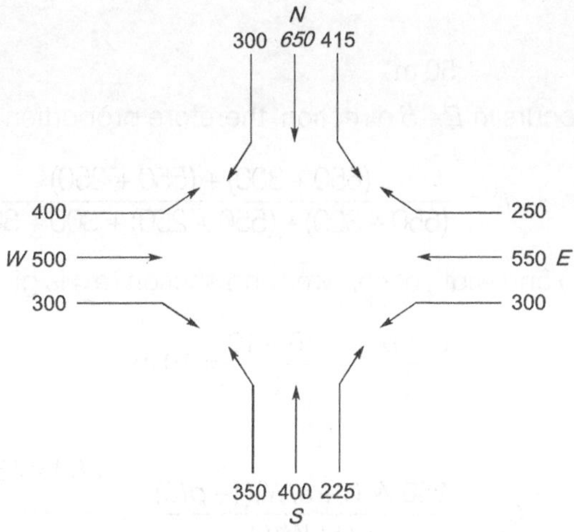

The traffic in PCU per hour on each leg is depicted below:

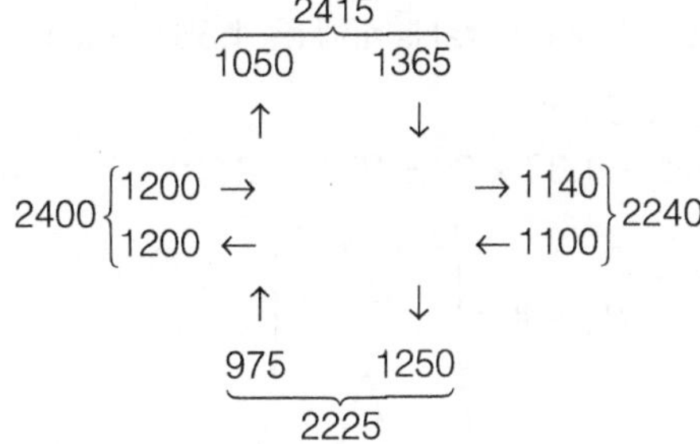

The traffic assigned to the network is shown in figure below:

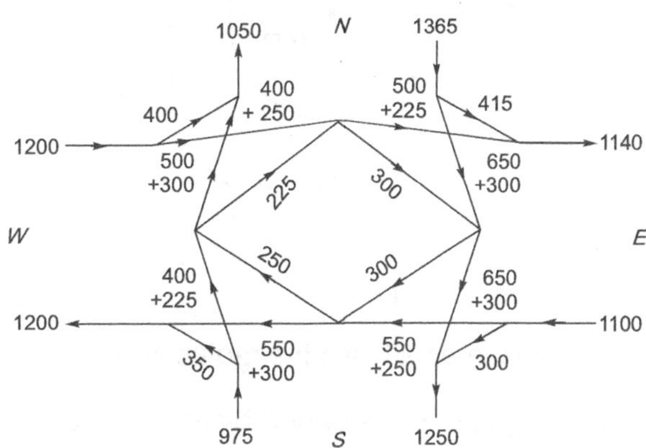

The maximum 2 way flow in the intersection leg (North) is 2415 PCU/hr and the maximum in one direction is 1365 PCU/hr.

For a carriageway width of 15 m, width of carriageway at entry and exit is taken as 10.0 m. The width of non-weaving section is also kept as 10.0 m

Width of weaving section, $W = \frac{e_1 + e_2}{2} + 3.5 = \frac{10+10}{2} + 3.5 = 13.5\ m$

As per IRC, the weaving length L is taken between 30 m to 60 m for 30 kmph design speeds so that $W/L = 0.12$ to 0.4

∴ Let us take $L = 50$ m

The maximum weaving occurs in $E - S$ direction, therefore proportion of weaving traffic is given by

$$p = \frac{(650+300)+(550+250)}{(650+300)+(550+250)+300+300} = 0.745$$

Average width of entry (e_1) and width of non-weaving section (e_2) is given by

$$e = \frac{e_1 + e_2}{2} = \frac{10+10}{2} = 10 \text{ m}$$

$$\therefore \quad Q_p = \frac{280\,W\,(1+e/W)(1-p/3)}{(1+W/L)} = \frac{280 \times 13.5 \times \left(1+\frac{10}{13.5}\right)\left(1-\frac{0.745}{3}\right)}{1+\frac{13.5}{50}}$$

$$= 3894.46 \text{ PCU per hour } (> 2415;\ \text{Hence OK})$$

IRC recommends radius at entry as 15 to 25 m for 30 kmph speed.

IRC recommends minimum radius of central island as 1.33 times the radius at entry.

Let us assume radius at entry as 20 m

∴ Radius of central island = 1.33 × 20 = 26.6 = 27 m

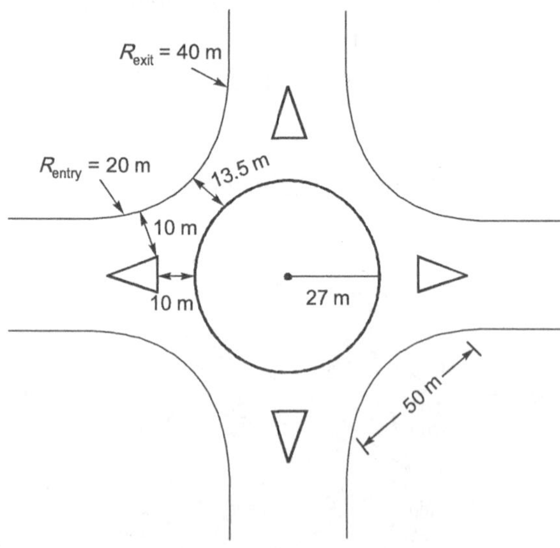

The radius of rotary at exit is taken as 2 times the radius at entry i.e. 2 × 20 = 40 m

2.5 **A vehicle moving at 65 kmph speed was stopped by applying brake and the length of skid mark was 25.50 m. If the average skid resistance of the pavement surface is known to be 0.70, determine the brake efficiency of the test vehicle.**

[5 marks : 2008]

Solution:

Given data: Initial speed (u) = 65 kmph = 18.06 m/sec, Braking distance (L) = 25.50 m, Average skid resistance of surface (f) = 0.70

But we know that $L = \frac{u^2}{2gf'}$ where f' is the average skid resistance developed.

$$\therefore \quad f' = \frac{u^2}{2gL} = \frac{(18.06)^2}{2 \times 9.81 \times 25.50} = 0.6519$$

$$\therefore \quad \text{Brake efficiency} = \frac{\text{Average skid resistance developed}}{\text{Average skid resistance of surface}} \times 100$$

$$= \frac{f'}{f} \times 100 = \frac{0.6519}{0.70} \times 100 = 93.129\%$$

2.6 Spot Speed Studies were conducted at a certain stretch of a highway and consolidated data collected are given below:

Speed Range, kmph	No. of vehicles observed
0 – 10	10
10 – 20	40
20 – 30	50
30 – 40	50
40 – 50	150
50 – 60	250
60 – 70	300
70 – 80	90
80 – 90	40
90 – 100	20

Determine :

(i) The upper and lower values of speed limits for regulation of mixed traffic flow.

(ii) The design speed for checking the geometric design elements of the highway.

[10 marks : 2008]

Solution:

The different data for solving the problem are tabulated below:

Speed Range kmph	Mid Speed kmph	Frequency f	Frequency %	Cumulative frequency
0 – 10	5	10	1	1
10 – 20	15	40	4	5
20 – 30	25	50	5	10
30 – 40	35	50	5	15
40 – 50	45	150	15	30
50 – 60	55	250	25	55
60 – 70	65	300	30	85
70 – 80	75	90	9	94
80 – 90	85	40	4	98
90 – 100	95	20	2	100
		$\Sigma f = 1000$	100	

Using the values of mid-speed and cumulative frequency % of the above table, cumulative speed distribution curve is plotted. From this graph, the following results are obtained:

(i) Upper speed limit for regulation = 85th percentile speed = 65 kmph

Lower speed limit for regulation = 15th percentile speed = 35 kmph

(ii) Speed to check design element = 98th percentile speed = 85 kmph

2.7 **Differentiate between time mean speed and space mean speed.**

[2 marks : 2009]

Solution:

Time mean speed	Space mean speed
• It is the average of spot speeds of vehicles at any point over a duration of time.	• It is the average speed of vehicles over a certain road length at any time is called space mean speed.
• It is the arithmatic mean of the speed of vehicles, at a certain section, during an interval oftime	• It is the harmonic mean of speed of vehicles, at the particular instant of time, over a certain road length.
• Time mean sped $V_t = \frac{\sum_{i=1}^{n} n_i v_i}{n}$	• Space mean speed $V_s = \frac{n}{\sum_{i=1}^{n} \frac{1}{V_i}}$

2.8 **The speeds of overtaking and overtaken vehicles are 80 kmph and 60 kmph respectively. If the acceleration of the overtaking vehicle is 2.5 kmph per second, calculate the safe passing sight distance for (i) single lane one way traffic (ii) three lane both way traffic. Assume perception time of driver = 2 sec.**

[8 marks : 2010]

Solution:

Given that $V_A = 80$ kmph and $V_B = 60$ kmph, $A = 2.5$ kmph/sec $t = 2$ seconds.

(a) The overtaking vehicle is forced to reduce its speed to the speed of overtaken vehicle and moves behind it allowing a space 's' till there is an opportunity for safe overtaking operation. The distance travelled by the overtaking vehicle during this time is given by

$$d_1 = 0.278\ V_B t = 0.278 \times 60 \times 2 = 33.36 \text{ m}$$

(b) After travelling distance d_1, the overtaking vehicle starts accelerating shifts to the adjoining lane, overtakes the other vehicle and shifts back to its original lane ahead of overtaken vehicle in position in time T seconds. The distance travelled by the overtaking vehicle in this operation is given by

$$d_2 = 0.278\ V_B T + 2s$$

where $T = \sqrt{\frac{14.4s}{A}}$ and $s = 0.2V_B + 6$

$\therefore$ $s = 0.2 \times 60 + 6 = 18$ m

$$T = \sqrt{\frac{14.4 \times 18}{2.5}} = 10.18 \text{ seconds}$$

$\therefore$ $d_2 = 0.278 \times 60 \times 10.18 + 2 \times 18 = 205.80$ m

(c) The distance travelled by any vehicle coming from opposite direction moving at speed of overtaking vehicle during the overtaking operation of the overtaking vehicle during time T is given by

$$d_3 = 0.278 V_A T = 0.278 \times 80 \times 10.18 = 226.40 \text{ m}$$

(i) The safe passing sight distance for single lane one way traffic will be given as

$$OSD = d_1 + d_2 = 33.36 + 205.80 = 239.16 \text{ m}$$

(ii) The safe passing sight distance for three lane both way traffic will be given as

$$OSD = d_1 + d_2 + d_3 = 33.36 + 205.81 + 226.4 = 465.56 \text{ m}$$

2.9 Calculate the design traffic in million standard axles (MSA) and commercial vehicles per day (CVD) required for design of flexible pavement as per current Indian procedure (revised IRC method) for the following data:

(i) Annual daily commercial vehicles at last count (May 2008) = 200

(ii) Rate of traffic growth = 6%

(iii) Design life = 10 years

(iv) Vehicle damage factor = 2

(v) The road is proposed to be completed in May 2011.

[8 marks : 2010]

Solution:

Design traffic in million standard axles as per revised IRC method is given as

$$N = \frac{365 \times [(1+r)^n - 1]}{r} \times A \times D \times F$$

where A is initial traffic after construction in CVD, n is design life in years, r is traffic growth rate, F is vehicle damage factor and D is lane distribution factor.

Initial traffic after construction in CVD is given by

$$A = P(1 + r)^x$$

where P is annual daily commercial vehicles at last count and x is number of years between the last count and the year of completion of construction.

$\therefore$ $A = 200\,(1 + 0.06)^3 = 238.20$ CVD **(Ans.)**

Designing the road as a single lane road or two lane road.

$D = 1$ for single lane road with traffic in both directions

$D = 0.75$ for two lane road with traffic in both directions

$$\therefore \quad N = \frac{365 \times [(1+0.06)^{10} - 1]}{0.06} \times 238.20 \times 2 \times D \times 10^{-6} \text{ MSA (Million Standard Axles)}$$

$= 2.3\,D$ MSA

For $D = 1$, $N = 2.3 \times 1 = 2.3$ MSA **(Ans.)**

For $D = 0.75$, $N = 2.3 \times 0.75 = 1.725$ MSA **(Ans.)**

2.10 Current Truck-traffic Volume (AADTT) for a Six-lane freeway are shown in the table below. A bituminous pavement with a 20 years design life is to be designed according to the Hveem method. Determine the design Esal for lane-3 of this freeway, if it is assumed that traffic-volumes for each truck classification will grow linearly by 25% over the next 20 years.

Vehicle - type	Current Traffic Volume (AADTT)
2 axle trucks	1000
3 axle trucks	500
4 axle trucks	250
5 axle trucks or more	1200

Assume :

(i) Lane distribution factor for Lane-3 is 0.80.

(ii) ESAL constant as

Vehicle - type	ESAL constant
2 axle trucks	1380
3 axle trucks	3680
4 axle trucks	5880
5 axle trucks or more	13780

[5 marks : 2012]

Solution:

Given: Traffic grows linearly by 25% over 20 years.

Taking the average increase i.e.

$$\frac{1+1.25}{2} = 1.125$$

$\therefore$ Expanded volume will be 1.125 times the current volume

Vehicle type	Current Volume	Expansion Factor	Expanded volume
2 axle trucks	1000	1.125	1125
3 axle trucks	500	1.125	562.5
4 axle trucks	250	1.125	281.25
5 axle or more trucks	1200	1.125	1350

Design ESAL is calculated by multiplying the expanded volumes by 20 year ESAL constant

Vehicle type	Expanded Volume	ESAL Constant	20 year ESAL
2 axle trucks	1125	1380	1,552,500
3 axle trucks	562.5	3680	20,70,000
4 axle trucks	281.25	5880	16,53,750
5 axle or more	1350	13780	18,603,000
			2,38,79,250

Now, lane distribution factor for lane 3 is 0.80

$\therefore$ **Design ESAL for lane 3 = 0.8 × 2,38,79,250 = 1,91,03,400**

2.11 What are advantages of Traffic-Rotary?

[5 marks : 2012]

Solution:

A rotary intersection or traffic rotary is an enlarged road intersection where all converging vehicles are forced to move round a large central island in one direction (clock wise direction) before they can weave out of traffic flow into their respective directions radiating from the central island. The main objects of providing a rotary are to eliminate the necessity of stopping even for crossing streams of vehicles and to reduce the area of conflict. The crossing of vehicles is avoided by allowing all vehicles to merge into the streams around the rotary and then to diverge out to the desired radiating road. Thus the crossing conflict is eliminated and converted into weaving manoeuvre or a merging operation from the right and a diverging operation to the left.

Various Advantages of Rotary are:

(i) Crossing manoeuvre is converted into weaving or merging and diverging operations. Hence there is no necessity of any of the vehicles, even those which have to go in cross directions, to stop and proceed within a traffic rotary. Thus the journey is more consistent and comfortable when compared with any other intersection at grade.

(ii) All traffic including those turning right or going straight across the rotary have equal opportunity as those turning left.

(iii) The variable cost of operation of automobile is less at a traffic rotary than at a signalized intersection where the vehicles have to stop and proceed. Though the distance to be traversed by vehicles which are to turn to the right or proceed straight across is higher in rotarary, still the fuel consumed in the process of crossing the rotary intersection is likely to be less.

(iv) There is no necessity of traffic police or signal to control the traffic as the traffic rotary could function by itself as a traffic controlled intersection and is the simplest of all controls. The maintenance cost is hence almost nil.

(v) The possible number of accidents and the severity of accidents are quite low because of low relative speed. Further weaving, merging and diverging manoeuvres are easier and less dangerous operation than crossing. Check on speed of vehicles is automatically enforced by proper design.

(vi) Rotaries can be constructed with advantage when the number of intersecting roads is between four and seven.

(vii) The capacity of the rotary intersection is the highest among all other intersections at grade. The rotary can accommodate a total traffic upto about 3000 vehicles per hour and enable radial streets to carry traffic almost to their full capacity.

2.12 Briefly explain

1. **Why AADT may be considered better parameter to ADT in representing average daily traffic and**
2. **Why thirtieth highest hourly volume may generally be considered as the hourly volume for design of traffic facilities.**

[4 marks : 2015]

Solution:

1. The AADT, which is $\frac{1}{365}$th of the total annual flow, is a common measure of flow utilized in geometric standards for highways, improvement of existing facilities and standards for pavement design and maintenance. If the flow is not measured for all the 365 days, but only for a few days the average flow is known by the term ADT, Average Daily Traffic.
2. The 30th highest hourly volume is the hourly volume that will be exceeded only 29 times in a year and all other hourly volumes of the year will be less than this value. The highest or peak hourly volume of the year will be too high that it will not be economical to design the facilities according to this volume. The highway facilities designed with capacity for 30th highest hourly traffic volume in the assumed year is found to be satisfactory from both facility and economic considerations. This is because the cost will be less when compared to the peak hourly volume and hence reasonable. There will be congestion only during 29 hours in the year. Thus the thirtieth highest hourly volume is generally taken as the hourly volume for design.

2.13 **A traffic survey conducted on four-lane single carriageway road reported traffic of 1400 CVPD (in both directions). Assuming growth rate of 5% design life of 20 years and vehicle damage factor of 3.5, calculate design traffic to be used in pavement design in terms of million standard axles (msa). As per IRC, lane distribution factor = 0.75.**

[8 marks : 2016]

Solution:

Design traffic, $$N_S = \frac{365A\left[\left(1+\frac{r}{100}\right)^n - 1\right]\times LDF \times VDF}{\frac{r}{100}}$$

1400 1400

Four Lane-two way Road [Undivided]

Given, $LDF = 0.75$, $VDF = 3.5$, $r = 5\%$, $n = 20$ years

$\therefore$ $A = 1400$ CVPD [For 2 Lane]

$$\therefore \quad N_S = \frac{365\times 1400\left[\left(1+\frac{5}{100}\right)^{20} - 1\right]\times 0.75\times 3.5}{\frac{5}{100}}\times 10^{-6}$$

$$N_S = 44.35 \text{ msa}$$

2.14 **An isolated three-phase signal is being designed for optimum cycle length of 90 s. The total lost time from all the three phases is 15 s. The critical flow ratios for phase I and II are 0.2 and 0.3 respectively. Determine the critical flow ratio for phase III as per Webster's method of traffic signal design. Given that saturation flow for volume in phase III is 500 pcu.**

[8 marks : 2016]

Solution:

Total lost time per cycle, $L = 15$ sec

But $L = 2n + R$

where, n = Number of phase = 3, R = All Red Time

$\Rightarrow$ $15 = 2(3) + R$

$\Rightarrow$ $R = 9$ sec

Critical flow ratio for Phase-I (y_1) = 0.2

Critical flow ratio for Phase-II (y_2) = 0.3

Let critical flow ratio for Phase-III = y_3

$\therefore$ $Y = y_1 + y_2 + y_3 = 0.2 + 0.3 + y_3 = 0.5 + y_3$

Optimum cycle length (C_0) = 90 sec.

$$\therefore \quad C_0 = \frac{1.5L+5}{1-Y}$$

$$\Rightarrow \quad 90 = \frac{1.5(15)+5}{1-Y}$$

$\Rightarrow$ $Y = 0.6944$

But $Y = 0.5 + y_3$

$\Rightarrow$ $0.6944 = 0.5 + y_3$

$\Rightarrow$ $y_3 = 0.1944$

2.15 **Define optimum signal cycle time. Design two phase traffic signal with pedestrian crossing by Webster's method for an average normal flow of traffic on cross roads A and B during design hour as 480 PCU and 250 PCU per hour, the saturation flows on roads A and B are given as 1200 PCU and 1000 PCU per hour respectively. All red time required for pedestrian crossing is 12 seconds and amber times of 2 seconds for clearance in each phase is to be provided.**

[12 marks : 2020]

Solution:

Optimum Signal Cycle Time: It is the time just enough to pass all the arrived traffic in same cycle at the intersection. There is a minimum cycle time of 25 sec fixed from safety considerations while a maximum cycle time of 120 sec is generally considered desirable.

Given data:

Normal flow	**Saturation flow**
q_A = 480 PCU/hr	S_A = 1200 PCU/hr
q_B = 250 PCU/hr	S_B = 1000 PCU/hr

All red time, R = 12 sec, n = 2, A = 2 sec

Total lost time = $(2n + R) = 2 \times 2 + 12 = 16$ sec

Sum of critial flow ratio

$$Y = y_A + y_B$$

$$y_A = \frac{q_A}{S_A} = \frac{480}{1200} = 0.4$$

$$y_B = \frac{q_B}{S_B} = \frac{250}{1000} = 0.25$$

$$Y = (0.4 + 0.25) = 0.65$$

Optimum cycle time,

$$C_0 = \frac{1.5L + 5}{1 - Y} = \frac{1.5 \times 16 + 5}{1 - 0.65} = 82.857 \text{ sec}$$

Green time,

$$G_A = \frac{y_a}{Y}(C_0 - L) = \frac{0.4}{0.65} \times (82.857 - 16) = 41.14 \text{ sec}$$

$$G_B = \frac{y_b}{Y}(C_0 - L) = \frac{0.25}{0.65} \times (82.857 + 10) = 25.71 \text{ sec}$$

Now providing amber time of 2 sec and 12 sec of all red time.

Total cycle time = (41.14 + 25.71 + 12 + 2 + 2)

= 82.85 sec

2.16 **List the five traffic surveys and elaborate all of them.**

[20 Marks : 2021]

Solution:

Traffic studies or surveys are carried out to analyze the traffic characteristics.

(a) Traffic volume study

Volume of traffic is a very important variable and is essentially the quantity of movement per unit of time at a specified location. It is expressed as vehicles per day and vehicles per hour

Traffic volume,

$$q = \frac{n \times 3600}{T}$$

Volume studies are basically useful to establish:

(i) the relative importance of any route
(ii) the fluctuations in flow
(iii) the distribution of traffic on the road system
(iv) the trends in the road use

(b) Speed studies

Speed is the rate of travel expressed in kmph or in m/s. Over a particular route, the actual speed of vehicle may vary. Speed of a vehicle depends upon several factors such as geometric features, traffic conditions, time, place, environment and driver.

Types of Speed

(i) **Spot Speed**: It is the instantaneous speed of a vehicle at any specified point. The spot speeds are affected by physical features of road such as pavement width, curve, sight distance, gradient, pavement unevenness and road side developments.

(ii) **Average Speed:** It is the average of spot speed of vehicles at a particular section or location. There are two types of average speed or mean speed:

(a) **Time Mean Speed:** It is defined as the average speed of all the vehicles passing a point on a highway over some specified time interval.

$$\text{Time mean speed,} \quad V_t = \frac{\sum_{i=1}^{n} V_i}{n}$$

(b) **Space Mean Speed:** It is the average speed of all vehicle at a certain road length over some specified time period. It is obtained from the observed travel time of the vehicles over a relatively long stretch of the road. Space mean speed is the harmonic mean of the speed of the vehicles passing a point on a highway during a particular interval of time.

$$\text{Space mean speed,} \quad V_s = \frac{3.6\,dn}{\sum_{i=1}^{n} t_i} = 3.6\left[\frac{n}{\frac{t_1}{d_1}+\frac{t_2}{d_2}+\ldots\frac{t_n}{d_n}}\right]$$

$$= 3.6\left[\frac{n}{\frac{1}{V_1}+\frac{1}{V_2}+\ldots\frac{1}{V_n}}\right]$$

(iii) **Running Speed:** It is the average speed maintained by a vehicle over a particular stretch of road, while the vehicle is in motion.

$$\text{Running speed} = \frac{\text{Total distance travelled}}{\text{Total running time}}$$

(iv) **Travel Speed or Journey Speed:** It is defined as the total distance travelled upon total time taken including all stoppage and delay.

(c) Origin and destination study

This study is generally carried out to

(i) Plan the road network and other facilities for vehicular traffics.
(ii) Plan the schedule of different modes of transportation for the trip demand.

(iii) To judge the adequacy of existing routes and to use in planning new network to roads.
(iv) To locate expressway or major routes along the desires lines.
(v) To establish preferential routes for various categories to vehicle including by pass.
(vi) To locate intermediate stops of public transport.
(vii) It is also used for mass rapid transit system.

There are number of methods for collecting the *O* and *D* data:

(i) Road Side Interview Method
(ii) License Plate Method
(iii) Return Post Car Method
(iv) Tag on Car Method
(v) Home Interview Method
(vi) Work Spot Interview Method

(d) Traffic flow characteristics and studies

Traffic flow characteristics are divided under two categories:

1. **Macroscopic characteristics:** Traffic flow theory assumes that there is a fundamental relationship among the three principle variables of traffic flow, speed, and density as follows:

$$q = k \times U$$

2. **Microscopic characteristics:**

Time Headway: The time interval between the passage of successive vehicles moving in the same lane and measured from head to head as they pass a point on the road is known as the time headway.

Space Headway: The distance between successive vehicles moving in the same lane measured from head at any instance is the space headway.

(e) Traffic capacity studies

Some important related terms which are often used are:

(i) **Traffic Volume (q):** It is the number of vehicles moving in a specified direction on a given lane or roadway that pass a given point during specified unit of time. It is expressed as vehicles per hour or vehicles per day.

(ii) **Traffic Density (k):** It is defined as the number of vehicles occupying a unit length of lane of roadway at a given instant. It is expressed in vehicles per kilometre.

(iii) **Traffic Capacity:** It is the ability of a roadway to accommodate traffic volume. It is expressed as vehicles per hour per lane. The capacity of roadway depends on a number of prevailing roadway and traffic conditions. Traffic capacity is always greater than or equal to traffic volume.

(iv) **Basic Capacity:** It is the maximum number of vehicles that can pass a given point on a lane or roadway during one hour under the most nearly ideal roadway and traffic conditions which can possibly be at trained.

(v) **Possible Capacity:** It is the maximum number of vehicles that can pass a given point on a lane or roadway during one hour under prevailing roadway and traffic conditions. The value of possible capacity varies between zero to basic capacity.

(vi) **Practical Capacity:** It is the maximum number of vehicle that can pass a given point on a lane or roadway during one hour, without traffic density being so great as to cause unreasonable delay, hazard or restriction to the drivers freedom to manoeuvre under the prevailing roadway and traffic conditions. This is also known as design capacity.

2.17 **What do you mean by spot speed, running speed, space-mean speed and time-mean speed? Explain them with appropriate examples. Also discuss the main purposes of spot speed studies.**

[12 marks : 2022]

Solution:

1. Spot speed is the instantaneous speed of a vehicle at a specified section or location.
2. Running speed is the average speed maintained by a vehicle over a particular stretch of road, while the vehicle is in motion. This is obtained by dividing the distance covered by the time during which the vehicle is actually in motion.
3. Space-mean speed represents the average speed of vehicles in a certain road length at any time. This is obtained from the observed travel time of the vehicles over a relatively long stretch of the road. Space-mean speed is calculated from:

$$V_s = \frac{3.6dn}{\sum_{i=1}^{n} t_1}$$

where V_s = space-mean speed, kmph
d = length of road considered, (in m)
n = number of individual vehicle observations
t_1 = observed travel time (sec) for the vehicle to travel distance d, m

The average travel time of all the vehicles is obtained from the reciprocal of space mean speed.

4. Time-mean speed represents the speed distribution of vehicles at a point on the roadway and it is the average of instantaneous speeds of observed vehicles at the spot. Time-mean speed is calculated from:

$$V_t = \frac{\sum_{i=1}^{n} V_i}{n}$$

where, V_t = time-mean speed, kmph
V_i = observed instantaneous speed of ith vehicles, kmph
n = number of vehicles observed

Thus space-mean speed is slightly lower than time mean speed under typical speed conditions on rural highway.

Purpose of spot speed study:

(a) Planning traffic control and in traffic regulations.
(b) Geometric design-for redesigning existing highways.
(c) In accident studies.
(d) Study the traffic capacity.
(e) Decide the speed trends.
(f) Compare diverse types of drivers and vehicles under specified conditions.

2.18 **Spot speed study was carried out to redesign the stretch of major district road. The data collected during the study is given below:**

Speed range kmph	Frequency of Vehicles		
	Two wheelers	Cars	Others
0 - 10	5	0	0
10 - 20	20	6	4
20 - 30	24	12	4
30 - 40	20	30	5
40 - 50	30	60	30
50 - 60	35	35	30
60 - 70	25	35	15
70 - 80	10	15	10
80 - 90	10	18	2
90 - 100	1	9	0

(i) What is the design speed for redesigning existing MDR?

(ii) What are upper and lower speed limits for mixed traffic?

(iii) What are the different measures to increase the spot speed of vehicles moving on the road?

(iv) Check whether the speed distribution is reasonably normal or not.

[20 marks : 2023]

Solution:

Speed range	Avg. speed	Total number of vehicles	Cumulative number of vehicles	Cumulative frequency (%)
0 - 10	5	5	5	1
10 - 20	15	30	35	7
20 - 30	25	40	75	15
30 - 40	35	55	130	26
40 - 50	45	120	250	50
50 - 60	55	100	350	70
60 - 70	65	75	425	85
70 - 80	75	35	460	92
80 - 90	85	30	490	98
90 - 100	95	10	500	100

(i) Design speed = 98th percentile speed = 85 kmph

(ii) Upper safe limit = 85th percentile speed = 65 kmph

Lower speed limit = 15th percentile speed = 25 kmph

(iii) In order to increase the spot speed of the vehicles, the variation in vehicles class must be reduced and as far as possible, only similar types of vehicles should ply on road thereby avoiding the mixed traffic flow. In the given speed range of 30 kmph to 70 kmph, maximum number of other types of vehicles are there which are occupying significant space on road and thus the number of other types of vehicles must be reduced on the road by providing alternate routes to them.

(iv) Speed distribution is reasonably normal because on either side of the mid-speed (i.e. 50 kmph), roughly equal number of vehicles are there thereby making the speed distribution relatively normal distribution.

3. Pavement Design

3.1 PBT conducted with 30 cm diameter plate on a soil subgrade yielded a pressure of 1 kg /cm² at 5 mm deflection. The test carried out over 18 cm base course yielded a pressure 5 kg/cm² at 5 mm deflection. Design the pavement section for wheel load of 4100 kg with a tyre pressure of 6 kg/cm² and allowable deflection of 5 mm. Use Burmister's method.

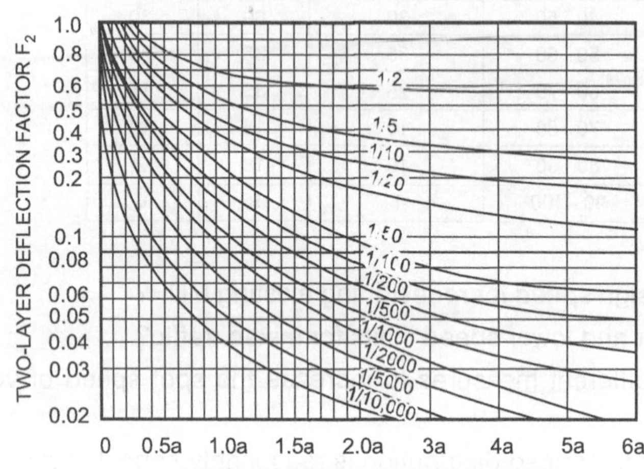

Figure : Relationship of F_2 and h in a Two-layer System (Burmister Method)

[15 marks : 1999]

Solution:

Case (i) Plate load test over soil subgrade

Diameter of plate = 30 cm

$\therefore$ Radius of contact area, $a = \frac{30}{2} = 15$ cm

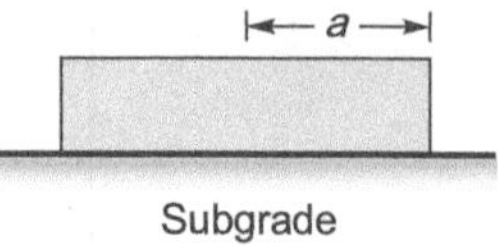

$$\Delta = 5 \text{ mm} = 0.5 \text{ cm}$$

$$p = 1 \text{ kg/cm}^2$$

This is a single layer system, as per **Burmister method**

$\therefore$ $h = 0$ and $F_2 = 1.0$

$\therefore$ For rigid plate,

$$\Delta = \frac{1.18\, pa}{E_s} \times F_2$$

$$\Rightarrow \quad 0.5 = \frac{1.18 \times 1 \times 15}{E_s} \times 1$$

$$\Rightarrow \quad E_s = 35.4 \text{ kg/cm}^2$$

Case (ii) For plate load test over 18 cm base course

$$\Delta = 0.5 \text{ cm}, p = 5 \text{ kg/cm}^2, h = 18 \text{ cm}$$

$$E_s = 35.4 \text{ kg/cm}^2$$

$$a = 15 \text{ cm}$$

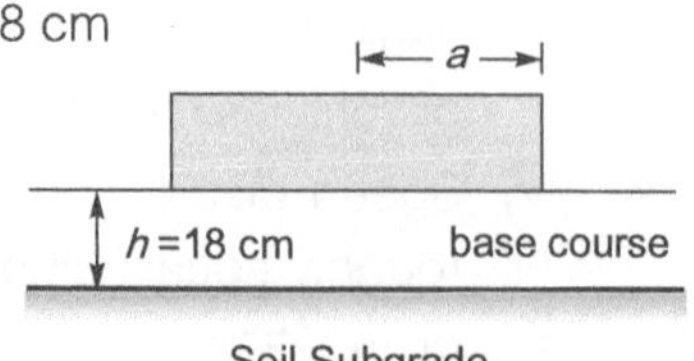

$\therefore$ For a rigid plate,

$$\Delta = \frac{1.18\, pa}{E_s} F_2$$

$$\Rightarrow \quad 0.5 = \frac{1.18 \times 5 \times 15}{35.4} \times F_2$$

$$\Rightarrow \quad F_2 = 0.2$$

and $$\frac{h}{a} = \frac{18}{15} = 1.2$$

From the graph in given problem, for a value of $F_2 = 0.2$ and $\frac{h}{a} = 1.2$ or $h = 1.2\,a$, the $\frac{E_s}{E_p}$ value lies between $\frac{1}{50}$ and $\frac{1}{100}$, but it is more close to $\frac{1}{100}$. Say $\frac{E_s}{E_p} = \frac{1}{90}$

$$\therefore \quad \frac{E_s}{E_p} = \frac{1}{90}$$

$$\Rightarrow \quad E_p = 90\,E_s = 90 \times 35.4 = 3186 \text{ kg/cm}^2$$

Design of Pavement:

$$P = 4100 \text{ kg}; \; p = 6 \text{ kg/cm}^2$$

$$\therefore \quad \text{Contact area} = \frac{P}{p} = \frac{4100}{6} = 683.33 \text{ cm}^2$$

$\therefore$ Radius of contact area, $$a = \sqrt{\frac{683.33}{\pi}} = 14.75 \text{ cm}$$

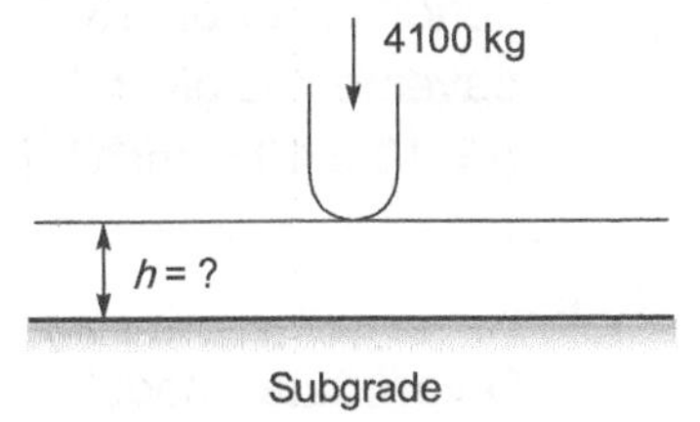

$$\Delta = 0.5 \text{ cm}$$

$$\frac{E_s}{E_p} = \frac{1}{90}$$

$\therefore$ For a flexible case, $$\Delta = \frac{1.5\,pa}{E_s} \times F_2$$

$$\Rightarrow \quad 0.5 = \frac{1.5 \times 6 \times 14.75}{35.4} \times F_2$$

$$\Rightarrow \quad F_2 = 0.133$$

$\therefore$ For a F_2 value of 0.133 and $\frac{E_s}{E_p} = \frac{1}{90}$, the value of h in terms of a can be found easily from the graph

$$\Rightarrow \quad h = 2.3\,a = 2.3 \times 14.75 = 33.93 \text{ cm}$$

Thus the pavement thickness = 33.93 cm = 34 cm

3.2 **Differentiate between flexible and rigid pavements.**

[7 marks : 2005]

Solution:

Some points of differences between flexible and rigid pavements are:

Flexible Pavements	Rigid Pavements
1. These have low flexural strength and are rather flexible in their structural action.	1. These possess noteworthy flexural strength or flexural rigidity.
2. The critical condition of stress in the flexible pavement is the distribution of compressive stresses.	2. The critical condition of stresses in the rigid pavements is the maximum flexural stress occurring in the slab due to wheel loads and temperature changes.
3. The flexible pavement layers transmit the stresses to the lower layers by grain to transfer through the points of contact in the granular structure.	3. The stresses are not transferred from grain to grain to the lower layers as in the case of flexible pavement layers.
4. A typical flexible pavement is made up of four components viz., soil subgrade, sub-base course, base course and surface course.	4. A typical rigid pavement is made up of three components viz. soil subgrade, base course and CC slab.
5. If the lower layer of a flexible pavement or soil subgrade is undulated, the flexible pavement surface also gets undulated.	5. The rigid pavements remain in their original position even if the soil subgrade is undulated.
6. Flexible pavements are commonly designed using empirical design charts or equations.	6. Rigid pavements are usually designed using the elastic theory.
7. Flexible pavements are made up of bituminous concrete and other granular materials with or without bituminous binder.	7. The rigid pavements are made up of portland cement concrete either plain, reinforced or prestressed.

3.3 Determine the warping stresses at interior edge and corner region in a 32 cm thick concrete pavement with transverse joints and longitudinal joints at 4.5 m and 3.6 m interval respectively. The modulus of subgrade reaction is 6 kg/cm²/cm. Assume temperature differential for a day condition to be 15°C and radius of contact area is 15 cm. Additional data for cement concrete pavement is given below:

$e = 10 \times 10^{-6}$ per°C; $E = 3 \times 10^5$ kg/cm²; $\mu = 0.15$; $C_x = 0.80$; $C_y = 0.45$

[8 marks : 2007]

Solution:

Given data: Modulus of elasticity of concrete (E) = 3×10^5 kg/cm², Thermal coefficient of concrete (e) = 10×10^{-6} per °C, Temperature difference between the top and bottom of slab (t) = 15°C, Poisson's ratio (μ) = 0.15, (C_x) = 0.80 and (C_y) = 0.45

The **warping stress at the interior region** may be given as

$$S_i = \frac{Eet}{2}\left[\frac{C_x + \mu C_y}{1-\mu^2}\right]$$

$$\Rightarrow \quad S_i = \frac{3\times10^5 \times 10\times10^{-6}\times 15}{2} \times \left[\frac{0.80+(0.15\times0.45)}{1-(0.15)^2}\right]$$

$$\Rightarrow \quad S_i = 19.97 \text{ kg/cm}^2 \textbf{ (Ans.)}$$

The **warping stress at the edge region** is given by

$$S_e = \left.\begin{matrix}\dfrac{C_x Eet}{2} \\ \dfrac{C_y\, Eet}{2}\end{matrix}\right\} \text{whichever is higher}$$

$$\frac{C_x\, Eet}{2} = \frac{0.80\times3\times10^5\times10\times10^{-6}\times15}{2} = 18 \text{ kg/cm}^2$$

$$\frac{C_y Eet}{2} = \frac{0.45\times3\times10^5\times10\times10^{-6}\times15}{2} = 10.125 \text{ kg/cm}^2$$

$$\therefore \quad S_e = 18 \text{ kg/cm}^2 \textbf{ (Ans.)}$$

The **warping stress at corner region** may be given as

$$S_c = \frac{Eet}{3(1-\mu)}\sqrt{\frac{a}{l}}$$

where, a = radius of contact = 15 cm, l = radius of relative stiffness

But
$$l = \left[\frac{Eh^3}{12k(1-\mu^2)}\right]^{1/4}$$

where, h = thickness of concrete pavement slab = 32 cm, k = modulus of subgrade reaction = 6 kg/cm^3

$$\therefore \quad l = \left[\frac{3\times10^5\times(32)^3}{12\times6(1-0.15^2)}\right]^{1/4} = 108.71 \text{ cm}$$

$$\therefore \quad S_c = \frac{Eet}{3(1-\mu)}\sqrt{\frac{a}{l}}$$

$$\Rightarrow \quad S_c = \frac{3\times10^5\times10\times10^{-6}\times15}{3(1-0.15)}\sqrt{\frac{15}{108.71}}$$

$$\Rightarrow \quad S_c = 6.56 \text{ kg/cm}^2 \textbf{ (Ans.)}$$

3.4 **Define Radius of Relative Stiffness and Equivalent Radius of Resisting Section. Calculate the wheel load stress at corner using Westergaard's corner load stress analysis from the following data:**

(i) Design wheel load = 4200 kg
(ii) '*E*' value of cement concrete = 2.8 × 10^5 kg/cm^2
(iii) Thickness of pavement slab = 20 cm
(iv) μ value of cement concrete = 0.15
(v) Modulus of subgrade reaction = 7.0 kg/cm^3
(vi) Radius of contact are = 14.0 cm

[10 marks : 2008]

Solution:

Radius of relative stiffness: A certain degree of resistance to slab deflection is offered by the subgrade. This is dependent upon the stiffness or pressure deformation properties of the subgrade material. The tendency of the slab to deflect is dependent upon its properties of flexural strength.

The resultant deflection of the slab which is also the deformation of subgrade is a direct measure of the magnitude of subgrade pressure. The pressure deformation characteristics of rigid pavement is thus a function of relative stiffness of slab to that of subgrade. Westergaard defined this term as radius of relative stiffness. Mathematically, it is given as

$$l = \left[\frac{Eh^3}{12k(1-\mu^2)}\right]^{1/4} = \left[\frac{2.8\times10^5\times(20)^3}{12\times7\times(1-0.15^2)}\right]^{1/4} = 72.27 \text{ cm}$$

Equivalent Radius of Resisting Section: Considering the case of interior loading, the maximum bending moment occurs at the loaded area and acts radially in all directions. With the load concentrated on a small area of the pavement, the question arises as to what sectional area of the pavement is effective in resisting the bending moment. According to Westergaard, the equivalent radius of resisting section is approximated, in terms of radius of load distribution and slab thickness.

$$b = \sqrt{1.6a^2 + h^2} - 0.675\,h$$

Given that a = 14.0 cm and h = 20 cm

$$\therefore \quad b = \sqrt{1.6(14.0)^2 + (20)^2} - 0.675 \times 20 = 13.21 \text{ cm} \qquad ...(< 1.724\,h)$$

The commonly used equation for theoretical computation of wheel load stress at corner region is

$$S_c = \frac{3P}{h^2}\left[1 - \left(\frac{a\sqrt{2}}{l}\right)^{0.6}\right] = \frac{3 \times 4200}{(20)^2}\left[1 - \left(\frac{14.0 \times \sqrt{2}}{72.27}\right)^{0.6}\right] = 17.02 \text{ kg/cm}^2$$

3.5 **In a locality where the rainfall is heavy, it is proposed to construct an ODR of WBM pavement (single lane), a two lane MDR of thin bituminous surface pavement. Taking IRC recommended camber and carriageway width, find out the height of the crown with respect to the edges.**

[2 marks : 2010]

Solution:

Camber for WBM pavement in area of heavy rainfall = 1 in 33 or 3.03%

Camber for thin bituminous pavement in area of heavy rainfall = 1 in 40 or 2.5%

Width of carriage way for single lane road = 3.75 m

Width of carriageway for two lane road without raised kerbs = 7.0 m

(i) Rise of crown with respect to edges for single lane road

$$= \frac{W}{2} \times \text{camber} = \frac{3.75}{2} \times \frac{1}{33} = 0.057 \text{ m}$$

(ii) Rise of crown with respect to edges for two lane road without raised kerbs

$$= \frac{W}{2} \times \text{camber} = \frac{7}{2} \times \frac{1}{40} = 0.0875 \text{ m or } 0.088 \text{ m}$$

3.6 **Design the pavement section by triaxial test method using the following data:**
Wheel load = 5100 kg
Radius of contact area = 15 cm
Traffic coefficient, X = 1.5
Rainfall coefficient, Y = 0.9
Design deflection, Δ = 0.25 cm
E-value of subgrade soil, E_s = 100 kg/cm²
E-value of base course material, E_b = 400 kg/cm²
E-value of 7.5 cm thick bituminous concrete surface course = 1000 kg/cm².
Also draw the pavement section with base course. Assume the pavement to consist of single layer of base course material.

[10 marks : 2011]

Solution:

Assuming the pavement to consist of single layer of base course material only; the pavement thickness is given by:

$$T_b = \left\{\sqrt{\left(\frac{3PXY}{2\pi E_s \Delta}\right)^2 - a^2}\right\}\left(\frac{E_s}{E_b}\right)^{1/3}$$

$$= \left\{\sqrt{\left(\frac{3\times 5100\times 1.5\times 0.9}{2\pi\times 100\times 0.25}\right)^2 - 15^2}\right\}\left(\frac{100}{400}\right)^{1/3}$$

$$= 130.63\times 0.63 \simeq 82.3 \text{ cm}$$

Let 7.5 cm bituminous concrete surface with E_c = 1000 kg/cm^2 be equivalent to the thickness t_b of base course. The equivalent replacement t_b is obtained from the relation:

$$\frac{t_b}{t_c} = \left(\frac{E_c}{E_b}\right)^{1/3} \quad \text{i.e.,} \quad \frac{t_b}{7.5} = \left(\frac{1000}{400}\right)^{1/3}$$

$$t_b = 7.5\times\left(\frac{1000}{400}\right)^{1/3} = 10.2 \text{ cm}$$

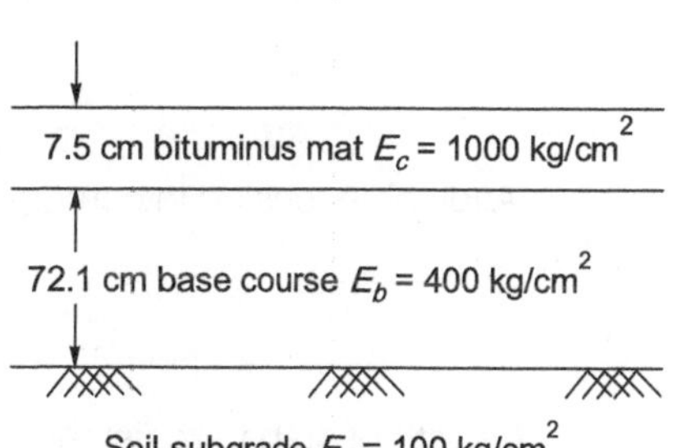

Therefore the required base course thickness

$$= 82.3 - 10.2 = 72.1 \text{ cm}$$

The pavement section consists of 72.1 cm thick WBM base course and 7.5 cm thick bituminous concrete surface course.

3.7 **The maximum increase in temperature is expected to be 26°C after the construction of a CC pavement. If the expansion joint gap is 2.2 m, design the spacing between the expansion and contraction joints. Assume coefficient of thermal expansion as 10 × 10^{-6} per °C, unit wt of concrete = 2400 kg/m^3, allowable stress in tension during curing = 0.8 kg/cm^2 and coefficient of friction of the interface =1.4.**

[12 marks : 2017]

Solution:

Spacing between expansion joints

Let δ′ = Half the joint width $= \frac{\delta}{2} = \frac{2.2}{2} = 1.1$ m

Let δ is the maximum expansion in a cement concrete slab of length of L_c with a temperature rise from T_1 and T_2.

$$\delta' = L_c C(T_2 - T_1)$$

$$1.1 \text{ m} = L_c \times 10\times 10^{-6}\times 26$$

where, C = Thermal expansion coefficient

$$L_c = 4230.77 \text{ m}$$

Here spacing between expansion joint is very high. It shows given data have some error.

$$\delta' = L_c C(T_2 - T_1)$$

$$\frac{2.2}{2} = L_c\times 10\times 10^{-6}\times 26$$

$$L_c = 4230.77 \text{ cm}$$

$$L_c = 42.3077 \text{ m}$$

$$L_c \simeq 42.31 \text{ m}$$

Therefore, provide spacing of expansion joints = 9 × 4.5 = 40.50 m

(As 10 × 4.5 = 45 m which is higher than 42.31 m)

Hence, expansion joints are provided after eight contraction joints or after ninth slab).

Spacing between contraction joints:

If L_e is the slab length in metre, the maximum stress occurs at half the length.

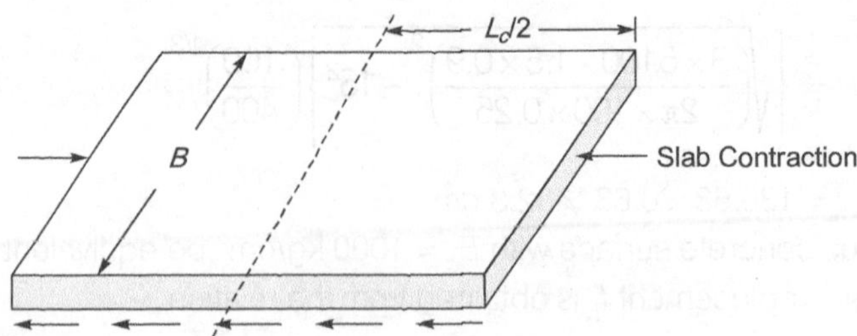

Total frictional resistance upto distance, $\frac{L_c}{2}$ will be

$$= W \times b \times \frac{L_c}{2} \times \frac{h}{100} \times f$$

Allowable tension in cement concrete

$$= S_c \times h \times b \times 100$$

$$\frac{WbL_chf}{200} = 100\, S_cbh$$

Length of slab to resist the frictional drag i.e. spacing of contraction joint.

$$\therefore \quad L_c = \frac{2S_c}{Wf} \times 10^4$$

where, L_c = Slab length or spacing between contraction joints, m

h = Slab thickness

W = Unit weight of cc, kg/m^3 = 2400 kg/m^3

S_c = Allowable stress in tension = 0.8 kg/m^2

f = Coefficient of friction = 1.40

$$\therefore \quad L_c = \frac{2 \times 0.8 \times 10^4}{2400 \times 1.4} = 4.76 \text{ m}$$

Maximum spacing suggested by IRC is 4.5 m for plain cement concrete pavement and so adopt L_c = 4.5 m.

3.8 **Design a flexible pavement for 7.0 m wide carriage way. Out of total present 600 CVPD, 200 CVPD have VDF of 2.5 and remaining vehicles have VDF of 3.5. Design the flexible pavement if effective CBR of the subgrade is 6.0%. Planning and construction period is 1.5 year and design life is 15 years. Assume necessary data suitably. Design template for 6.0% of CBR value is as under:**

Design Traffic	Wearing course (mm)	Binder course (mm)	Base (mm)	Sub base (mm)
2 msa	20 SDBC	50 DBMM	225	175
5 msa	25 SDBC	50 DBM	250	210
10 msa	40 BC	65 DBM	250	260
20 msa	40 BC	90 DBM	250	260

[20 marks : 2017]

Solution :

Given: Width of carriage way = 7.0 m, Number of lanes = 2.0, CBR value of subgrade = 6%

Design life of pavement, $n = 15$ years

We know,

Design traffic in million standard axles as per IRC method is given as:

$$N = \frac{365\left[(1+r)^n - 1\right]}{r} \times A \times D \times F$$

where, A = Initial traffic after construction in commercial vehicles per day CVPD, D = Lane distribution factor = 0.75 (for two lane), F = Vehicle damage factor, n = Design life in years = 15 years.

Initial traffic after construction in CVPD is given by

$$A = P(1 + r)^x$$

where, P = annual daily commercial vehicles at last count, x = Number of years between last count and year of completion of construction = 1.5 years.

Let us assume r = rate of traffic growth = 7.0%

$\therefore$ $$A_1 = 200\,(1 + 0.07)^{1.5} = 221.36 \text{ CVPD}$$

and $$A_2 = 400\,(1 + 0.07)^{1.5} = 442.73 \text{ CVPD}$$

$\therefore$ $$N = \frac{365\left[(1+0.07)^{15} - 1\right]}{0.07} \times [A_1F_1 + A_2F_2] \times D \times 10^{-6}$$

$$N = 9172.09 \times [221.36 \times 2.5 + 442.73 \times 3.5] \times 0.75 \times 10^{-6}$$

$\Rightarrow$ $$N = 14.466 \text{ msa}$$

From table for CBR = 6% and design traffic = 14.46 msa, we get

$$\text{Wearing course} = 40 \text{ BC, mm}$$

$$\text{Binder course} = 65 + \frac{90-65}{20-10}(20 - 14.466) = 78.835 \text{ DBM, mm}$$

$$\text{Base course} = 250 \text{ mm}$$

$$\text{Subbase course} = 260 \text{ mm}$$

3.9 **Explain single layer system in brief. Determine the total thickness of flexible pavement over subgrade having elastic modulus of 180 kg/cm².**

Assume following data:

Design wheel load = 5100 kg

Tyre pressure = 7 kg/cm²

Permissible deflection = 0.25 cm

[12 marks : 2018]

Solution:

Boussinesq's single layer system can be considered as special case of Bermister's layered system analysis. If layers of soil subgrade, subbase course and base course are assigned elastic modulii of E_s, E_{sb}, E_b. Then as per Boussinesq's single layer system $E_s = E_{sb} = E_b$, while in layered analysis it is taken as $E_b > E_{sb} > E_s$.

Triaxial method proposed by L.A. palmar and E.S. Barber based on Boussinesq's displacement equation for homogeneous elastic single layer:

$$\Delta = \frac{3Pa^2}{2E\left(a^2 + z^2\right)^{1/2}}$$

$$p = \frac{P}{\pi a^2}$$

$$\Delta = \frac{3P}{2\pi E\left(a^2 + z^2\right)^{1/2}}$$

$$\left(a^2 + z^2\right)^{1/2} = \frac{3P}{32\pi E\Delta}$$

$$a^2 + z^2 = \left(\frac{3P}{2\pi E\Delta}\right)^2$$

$$z = \sqrt{\left(\frac{3P}{2\pi E\Delta}\right)^2 - a^2}$$

Assuming, that pavement is incompressible, z becomes T, the thickness of pavement

$$T = \sqrt{\left(\frac{3P}{2\pi E\Delta}\right)^2 - a^2}$$

P = Wheel load, kg
E = Modulus of elasticity of subgrade, kg/cm^2
a = Radius of contact area
Δ = Design deflection(0.25 cm)

$$\therefore \quad T = \sqrt{\left(\frac{3\times 5100}{2\pi\times 180\times 0.25}\right)^2 - 15.23^2} = 51.95 \text{ cm}$$

Hence total thickness over subgrade = 51.95 cm $\simeq$ 52 cm.

3.10 **Write down the construction steps for Water Bound Macadam road. Also compare WBM construction with WMM construction.**

[20 marks : 2019]

Solution:

The construction of WBM roads may be divided into following steps:

(i) **Preparation of foundation for receiving the WBM course:** The foundation layer i.e. subgrade, subbase or base course is prepared to required grade and camber. On existing road surfaces, the depressions and pot holes are filled and corrugations are removed by scarifying and reshaping the surface to the required grade and camber.

(ii) **Provision of Lateral Confinement:** It may be done by constructing the shoulders to advance to a thickness equal to that of compacted *WBM* layer.

(iii) **Spreading of Coarse Aggregate:** The coarse aggregates are spread uniformly to proper profile to even thickness upon the prepared foundation.

(iv) **Rolling:** Rolling is started from the edges and then gradually shifted towards the centreline of the road.

(v) **Application of screenings:** After the coarse aggregates are rolled adequately, the dry screenings are applied gradually over the surface to fill the interstices in three or more applications.

(vi) **Sprinkling and Grouting:** After screeching, the surface is sprinkled with water, swept and rolled.

(vii) **Application of Binding Material**

(viii) **Setting and Drying**

Difference between WMM and WBM

Wet Mix Macadam (WMM)	Water Bound Macadam (WBM)
Stone aggregates are comparatively smaller in size (4.75-20 mm)	Stone aggregate are larger in size (45-90 mm)
Stone aggregates and binding materials are premixed in a batching plant	Stone aggregates, screening material and binders are overlaid one after another
Prepared by stone aggregates and binders	Prepared by stone aggregate, screening and binding material
Water is premixed	Water is sprinkled on dry material

3.11 **A rigid pavement of 25 cm thickness of M40 grade of concrete is supported over a subgrade having modulus of subgrade reaction as 8.0 kg/cm^3. If dowel bars are placed at centre-to-centre spacing of 30 cm, calculate the maximum load carried by a single dowel which is just below the wheel. Assume wheel load as 4100 kg, participation of dowel bars in load distribution up to 1.0 × radius of relative stiffness and load to be transferred by joint as 50%. Poisson's ratio of the concrete may be taken as 0.15.**

[20 marks : 2019]

Solution:

Given: Slab thickness (h) = 25 cm

Poisson's ratio of concrete (μ) = 0.15

Modulus of subgrade reaction (k) = 8 kg/cm^3

Center to center spacing of dowel bars (δ) = 30 cm

Load to be transferred by joint is 50% of the wheel load (P)

Distance upto which dowel bars are effective = 1.0 × radius of relative stiffness

Also, modulus of elasticity of cement concrete (E) = 3 × 10^5 kg/cm^2

$$\text{Radius of relative stiffness } (l_{cm}) = \left[\frac{Eh^3}{12k(1-\mu^2)}\right]^{1/4}$$

$$l = \left[\frac{3\times 10}{12k(1-\mu^2)}\right]^{1/4} = 84.069 \text{ cm}$$

∴ Distance upto which dowel bars are effective = 84.069 cm

$$\text{Load capacity} = 1+\frac{(1\times l-\delta)}{1\times l}+\frac{(1\times l-2\delta)}{1\times l}+\frac{(1\times l-3\delta)}{1\times l}+\ldots$$

[As long as numerator is positive]

$$\therefore \quad \text{Load capacity} = 1+\frac{(84.069-30)}{84.069}+\frac{(84.069-60)}{84.069} = 1.929$$

Maximum load carried by single dowel

$$P_s = \frac{0.5\times P}{1.929} = \frac{0.5\times 4100}{1.929} = 1062.73 \text{ kg}$$

3.12 **Plate bearing tests were conducted with 30 cm diameter plate on soil subgrade and over 15 cm base course. The pressure yielded at 0.25 cm deflection is 1.25 kg/cm^2 and 5 kg/cm^2 for subgrade and base course respectively. Design the thickness of flexible pavement for a wheel load of 4080 kg with tyre pressure of 4.5 kg/cm^2. Take rigidity factor 1.11.**

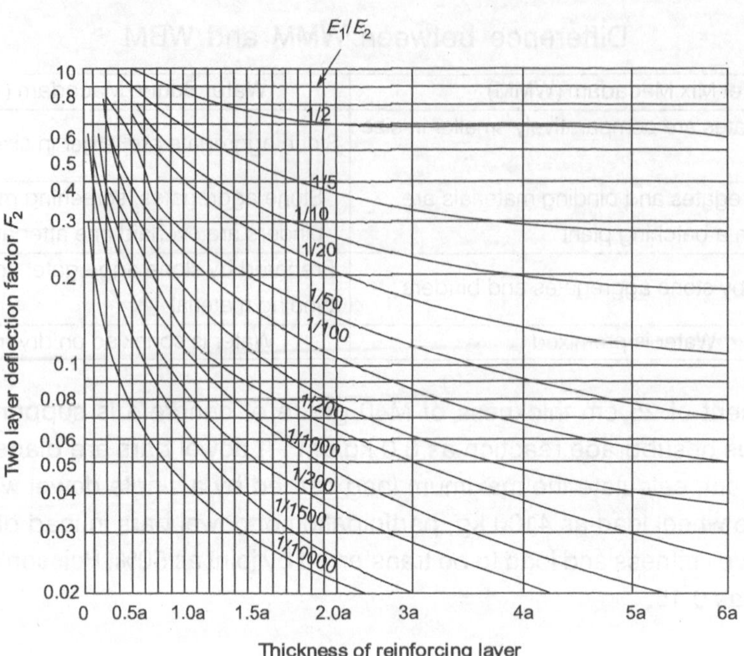

[20 marks : 2022]

Solution:

Plate bearing test conducted on subgrade,

$$a = \frac{30}{2}$$

$$p = 1.25 \text{ kg/cm}^2$$

$$\Delta = 0.25 \text{ cm} \qquad F_2 = 1$$

$$\Delta = 1.18\frac{p.a}{E_s} \times F_2$$

$$\Rightarrow \quad 0.25 = 1.18 \times \frac{1.25 \times 15}{E_s} \times 1$$

$$\Rightarrow \quad E_s = 88.5 \text{ kg/cm}^2$$

Plate bearing test conducted on 15 cm base,

$$h = 15 \text{ cm}, \ a = 15 \text{ cm}, \ \frac{h}{a} = 1$$

$$\Delta = 1.18\frac{p.a}{E_s} \times F_2$$

$$\Rightarrow \quad 0.25 = \text{m}1.18\frac{5 \times 15}{88.5} \times F_2$$

$$\Rightarrow \quad F_2 = 0.25$$

Using given curve, $\frac{h}{a} = 1$

$$\Rightarrow \quad h = 1.0a$$

$$F_2 = 0.25$$

$$\frac{E_1}{E_2} = \frac{1}{100}$$

$$\Rightarrow \quad E_2 = 100E_1$$

For pavement thickness:

$$P = 4080 \text{ kg}$$

$$p = 4.5 \text{ kg/cm}^2$$

$$p = \frac{P}{\pi a^2}$$

$$\Rightarrow \quad a = \sqrt{\frac{P}{\pi p}} = \sqrt{\frac{4080}{4.5 \times \pi}} = 16.99\,\text{cm}$$

$$\Delta = 1.5\frac{p.a}{E_s} \times F_2$$

$$\Rightarrow \quad 0.25 = 1.5 \times \frac{4.5 \times 16.99}{88.5} \times F_2$$

For $F_2 = 0.193$ and $\frac{E_1}{E_2} = \frac{1}{100} \Rightarrow \frac{h}{a} = 1.25$

From curve

For $F_2 = 0.193$

$$\frac{E_1}{E_2} = \frac{1}{100}, \quad \frac{h}{a} = 1.25$$

$$\therefore \quad h = 1.25 \times 16.99 = 21.24\text{ cm}$$

Thickness of pavement = 21.24 cm

4. Highway Materials, Maintenance and Properties

4.1 **Benkelman beam test was carried out on a stretch of flexible pavement of a National Highway, as per IRC guidelines. The mean value of deflections and standard deviations computed are 1.57 mm and 0.12 mm respectively. The other data are:**

(i) Temperature during the test : 25°C

(ii) Correction factor for subgrade moisture content = 1.2

(iii) Initial commercial traffic in the year of completion of construction of overlay : 1200/day

(iv) Annual growth rate of commercial vehicles :7.5%

(v) Design Life = 10 years

(vI) Vehicle damage factor = 3.5

Take 1 cm of bituminous macadam equal to 0.7 cm of DBM/AC/SDC.

Determine the thickness of asphaltic concrete overlay using the figure below.

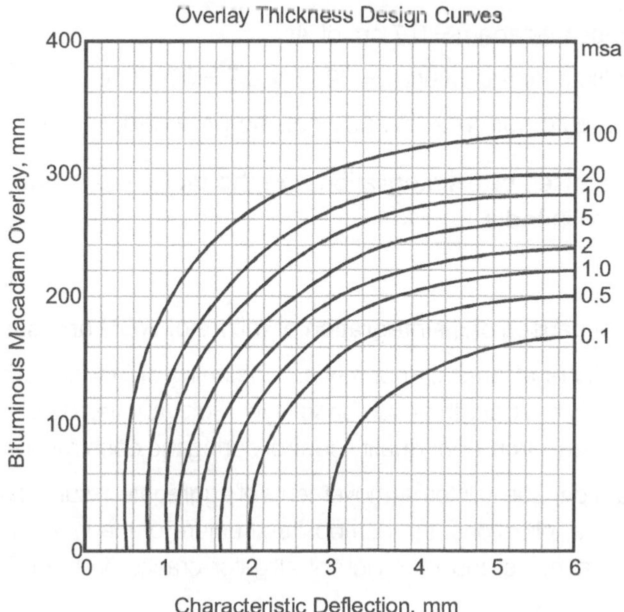

[15 marks : 2001]

Solution:

Given data: Test temperature = 25°C, Correction factor for subgrade moisture content = 1.2, Initial commercial traffic (P) = 1200 vehicle per day, Annual growth rate (r) = 7.5%, Design life = 10 years, Mean deflection ($\bar{D}$) = 1.57 mm, Standard deviation (σ) = 0.12 mm

$$\text{Design commercial traffic} = P\left[1+\frac{r}{100}\right]^{10} = 1200\left[1+\frac{7.5}{100}\right]^{10} \simeq 2474 \text{ vehicle per day}$$

$$\text{Characteristic deflection,} \quad D_c = \bar{D} + \sigma = 1.57 + 0.12 = 1.69 \text{ mm}$$

IRC has suggested a standard pavement temperature of 35°C and a correction factor of 0.0065 mm per °C to be applied for the variation from this standard pavement temperature. The correction will be negative when the pavement temperature is above 35°C and positive when it is lower.

∴ Deflection after temperature correction

$$= 1.69 - 0.0065\,(25-35) = 1.755 \text{ mm}$$

Corrected deflection for subgrade moisture

$$= 1.2 \times 1.755 \simeq 2.11 \text{ mm}$$

$$\text{As per IRC,} \quad h_o = 550 \log_{10}\left(\frac{D_c}{D_a}\right)$$

where h_o = thickness of granular or WBM overlay in mm

D_c = characteristic deflection after correction

D_a = allowable deflection which is equal to 1 mm for a design traffic between 1500 and 4500 vehicle per day

$$\therefore \quad h_o = 550 \log_{10}\left(\frac{2.11}{1}\right) = 178.36 \text{ mm}$$

When bituminous concrete or bituminous macadam with bituminous surface course is provided as the overlay, an equivalency factor of 2 is suggested by IRC to decide the actual overlay thickness required.

$$\therefore \text{Thickness of bituminous macadam} = \frac{178.36}{2} = 89.18 \text{ mm}$$

Also, 1 cm of bituminous macadam = 0.7 cm of *AC*

∴ Thickness of asphaltic concrete overlay = 89.18 × 0.7 = 62.43 mm

Hence 65 mm of asphaltic concrete overlay can be provided.

4.2 **Discuss the various causes of disintegration and the major faults occurring in *WBM* and surface treated (asphalt roads) in India.**

[15 marks : 2003]

Solution:

Following are the chief causes of disintegration or failure of WBM and surface treated roads in India:

(i) **Inadequate stability or strength**: Poor mix proportioning or inadequate thickness are main reasons for the lack of stability or strength of sub base layer. Soft varieties of stone aggregates also make the base course layer weak. Improper quality control during the construction results in poor base course.

(ii) **Loss of binding action**: Due to internal movements of aggregate in sub base or base course layers under the repeated stress applications, the composite structure of the layers gets disturbed. This results in loosening of the total mass and formation of alligator cracks on the bituminous surfacing of flexible pavements.

(iii) **Loss of base course materials**: The loss of base course materials is only possible when either the base course is not covered with a wearing course or the wearing course has completely worn out. Due to the fast moving vehicles plying on the road, there is a suction caused between the pneumatic tyres and the exposed base material. This causes removal of binding material in WBM base and the stone aggregates are left in a loose state.

(iv) **Inadequate wearing course**: Absence of wearing course or inadequate thickness or stability of wearing course exposes the base course to the damaging effects of climatic variations mainly due to rains, frost action and traffic. Bullock cart traffic affects the surface course. Pervious wearing course also permits the surface water to seep through and soften the base course thus weakening it.

(v) **Use of inferior materials**: Many failures mainly structural failures are attributed due to the use of inferior materials in the paving jobs. Some materials exhibit satisfactory characteristics initially, but show rapid deterioration due to weathering.

(vi) **Failure of wearing course**: Failure of wearing course is observed due to lack of proper mix design, improper gradation of aggregates, inadequate binder content and inferior type of binder result in a poor bituminous surfacing.

Some major faults occurring in flexible pavements are:

(a) **Alligator (map) cracking**: This is the most common type of failure and occurs due to relative movement of pavement layer materials. This may be caused due to repeated application of heavy wheel loads resulting in fatigue failure or due to moisture variations resulting in swelling and shrinkage of subgrade and other pavement materials.

(b) **Consolidation of pavement layers**: Formation of ruts are mainly attributed to the consolidation of one or more layers of pavement. The repeated application of loads along the same wheel path cause cumulative deformation resulting in consolidation deformation or longitudinal ruts. Shallow ruts on the surfacing course can also be due to wearing along the wheel path.

(c) **Shear failure and cracking**: Shear failures are associated with the inherent weakness of the pavement mixtures, the shearing resistance being low due to inadequate stability or excessively heavy loading. The shear failure causes upheaval of pavement materials by forming a fracture or cracking.

(d) **Longitudinal cracking**: Due to frost action and differential volume changes in subgrade, longitudinal cracking is caused in pavement traversing through the full pavement thickness. Settlement of fill and sliding of side slopes would also cause this type of failure.

(e) **Frost heaving**: Frost heaving is often misunderstood for shear or other types of failures. In shear failure, the upheaval of portion of pavement is followed with a depression. In the case of frost heaving, there is mostly a localized heaving up of pavement portion depending upon the ground water and climatic conditions.

(f) **Lack of binding with lower layer**: Slipping occurs when the surface course is not bound with the underlying base. This results in opening up and loss of pavement materials forming patches or pot holes. This condition is more pronounced when the prime/tack coat in between two layers is lacking.

(g) **Reflection cracking**: This type of cracking is observed in bituminous overlays provided over existing cement concrete pavements. The crack pattern as existing in cement concrete pavements are mostly reflected on bituminous surfacing in the same pattern. Structural action of the total pavement section is not much influenced by the presence of reflection cracks but since the cracks appear at the surface, these allow surface water to seep through and cause damage to the soil subgrade or result in mud pumping.

4.3 **Describe field method of finding CBR value of a subgrade. Also explain the method of design of a flexible pavement using the soaked CBR-value.**

[10 marks : 2004]

Solution:

The method of design of a flexible pavement using the soaked CBR-value is explained below:

(i) In order to design a pavement by CBR method, first the soaked CBR value of the soil subgrade is evaluated.

(ii) Then the appropriate design curve is chosen by taking the design wheel load or by taking the anticipated traffic into consideration.

(iii) Thus the total thickness of flexible pavement needed to cover the subgrade of the known CBR value is obtained.

(iv) In case there is a material superior than the soil subgrade, such that it may be used as sub-base course then the thickness of construction over this material could be obtained from the design chart knowing the CBR value of the sub-base.

(v) Thickness of the sub-base course is the total thickness minus the thickness over the sub-case.

4.4 Write short notes on:
(i) Reinforcement in cement concrete roads.
(ii) Processes in bituminous road construction.

[10 marks : 2006]

Solution:

(i) Reinforcement in cement concrete pavements:

The reinforcement used in cement concrete pavement is in the form of welded wire fabric or bar mats. The function of reinforcement is to hold the cracked slab portions together and thus not to allow them to open up any more. Sufficient load transfer across the crack is affected due to the reinforcement. It may be clearly understood that the quantity of reinforcement recommended for use, does not provide any added flexural strength to the pavement. The function of the reinforcement is more or less similar to the dowel bars in contraction joints and tie bars to withstand tensile forces due to the movement of slabs and the frictional resistance between the bottom fibre of the concrete and the subgrade.

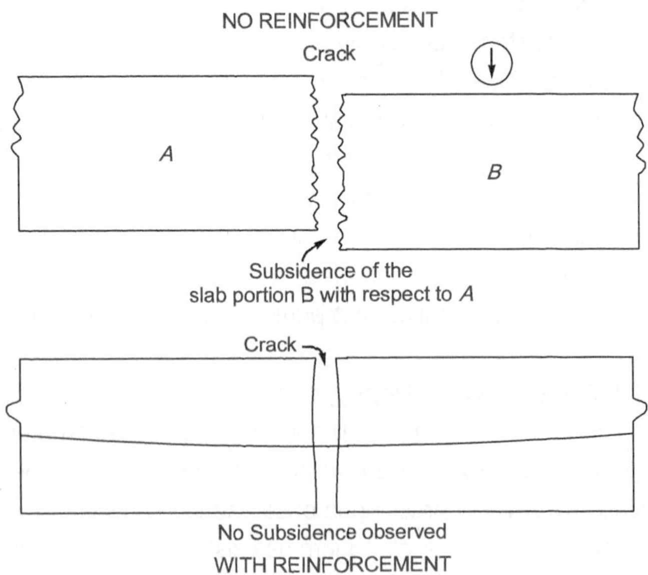

Experience shows that hair cracks often occur at frequent intervals in RCC pavement, but such cracks show no tendency to open up. The recommended values for providing reinforcement for moderate and heavy traffic are 3.5 kg/m^2 and 7.0 kg/m^2 respectively.

In order to obtain the maximum advantage, it is suggested that greater quantity of reinforcement should be placed in longitudinal direction. Further the reinforcement should either be placed in the mid depth or towards the top of the pavement for better functioning. Steel wire fabric should conform to **IS : 1556-1960** and bar mats should conform to **IS : 432-1960**.

(ii) Processes in bituminous road construction:

A number of processes are in use for bituminous pavement construction.

On the basis of method of construction they can be classified as:

(a) Interface treatments like prime coat and tack coat

(b) Surface dressing and seal coat

(c) Grouted or penetration type constructions

  (i) Penetration Macadam

  (ii) Built up spray Grout

(d) Premix methods:

  (i) Bituminous bound macadam

  (ii) Carpet c

  (iii) Bituminous concrete

  (iv) Sheet asphalt or rolled asphalt

  (v) Mastic asphalt

In the above methods, the choice of particular binder depends upon the type of construction, availability of materials and equipment, climatic conditions, etc. Bitumen and tar, require heating to bring them to a proper viscosity for their use. The construction technique using these materials is termed as hot technique. Cutback and emulsion do not normally require heating, therefore applied cold and the technique is known as cold mix.

Based on the technique of mixing and construction, the bituminous constructions are classified as road mix and central plant mix methods.

Depending upon the gradation, the premix constructions are classified as open graded and dense graded. The premixed carpet is open graded whereas the bituminous concrete is dense graded.

The methods of bituminous construction employed in base course are grouted or penetration macadam and bitumen bound macadam. The different bituminous construction used in surface course of pavements are surface dressing, grouted or penetration macadam with seal coat, and bituminous premix surfacing. The premix may be either carpet, bituminous concrete, sheet asphalt or mastic asphalt.

4.5 **Enumerate and explain the desirable properties of soil as a highway material. Classify and discuss the suitability of the soil as a subgrade material from the result of soil sample given below:**

(i) Soil portion passing through 0.074 mm sieve = 60%.

(ii) Liquid limit = 50%

(iii) Plastic limit = 30%

[8 marks : 2007]

Solution:

The desirable properties of soil as a highway material are:

(i) The soil should possess adequate stability or resistance to permanent deformation under loads and should possess resistance to weathering, thus retaining the desired subgrade support.

(ii) Incompressibility

(iii) Permanency of strength

(iv) Minimum changes in volume and stability under adverse conditions of weather and ground water. Minimum variation in volume will ensure minimum variation in differential expansion and differential strength values.

(v) It is essential to avoid excessive moisture retention and to reduce the potential frost action.

(vi) It ensures higher dry density and strength under particular type and amount of compaction.

Given data: $LL = 50; PL = 30\%$ and $P = 60\%$

Group Index is given by

$$GI = 0.2a + 0.005ac + 0.01\,bd$$

Plasticity index, $PI = LL - PL = 50 - 30 = 20\%$

$$a = P - 35 = 60 - 35 = 25 < 40 \text{ (OK)}$$

$$b = P - 15 = 60 - 15 = 45 > 40 \text{ adopt } b = 40$$

$$c = LL - 40 = 50 - 40 = 10 < 20$$

$$d = PI - 10 = 20 - 10 = 10 < 20$$

$$\therefore \quad GI = 0.2 \times 25 + 0.005 \times 25 \times 10 + 0.01 \times 40 \times 10 = 10.25$$

GI Value	Soil Suitability
0-1	Good Soil
2-4	Fair Soil
5-9	Poor Soil
10-20	Very Poor Soil

Here GI > 10, hence given subgrade material is very poor, hence not suitable as subgrade material.

4.6 **Discuss the specification of materials, plant and equipment and construction method to be followed as per IRC guidelines to ensure the quality control in the field for dense bituminous macadam. Write checklist for proper functioning of hot mixed plant.**

[8 marks : 2007]

Or

Explain briefly construction procedure and quality control measures for dense bituminous macadam (DBM).

[5 marks : 2009]

Solution:

The **Bituminuous Macadam (BM)** is a premix laid immediately after mixing and then compacted. It is an open graded construction suitable only as a base or binder course. When this layer is exposed as a surface course, at least a seal coat is necessary.

Specifications of Materials: The grades of bitumen used are 30/40, 60/70 and 80/100 penetration. Road tar RT-4, cutback and emulsion can also be used in cold mix construction technique. The binder content used varies from 3.0 to 4.5 per cent by weight of the mix.

Aggregates used are of low porosity fulfilling the following requirements for the base course.

Los Angeles abrasion value	50% max.
Aggregate impact value	35% max.
Flakiness index	15% max.
Stripping at 40°C after 24 hours immersion (CRRI test)	25% max.
Loss with sodium sulphate, 5 cycles.	12% max.

For binder course the specified maximum abrasion and impact values are 40 and 30 per cent respectively.

The quantity of aggregates required for 10 m^2 of bitumen bound macadam are 0.60 to 0.75 m^3 and 0.90 to 1.0 m^3 respectively, for 50 and 75 mm compacted thickness. The bitumen quantity would be determined based on the gradings adopted.

Plants and Equipment: Various plants and equipment required for the job include sprayer, mechanical mixer or improvised hand mixer, spreader (mechanical paver or finisher, grader, or manual methods) and roller.

There are different types of mixer in use. Mechanical mixers include travelling plant or central plant and are employed depending upon the job. Manual hand mixer is also used which is fixed vane type.

Construction Steps:

(i) **Preparation of existing layer:** The existing layer is prepared to a proper profile. Pot holes are patched and irregularities are made even. The surface is properly cleaned.

(ii) **Tack coat or prime coat application:** A tack coat is applied of thin layer of bitumen binder on the existing layer either using the sprayer or a pouring can. The quantity of application is 4.0 to 7.5 kg per 10 m^2 for black top layer and 7.5 to 10 kg per 10 m^2 for untreated WBM layer.

(iii) **Premix preparation:** The bitumen binder and aggregates as per recommended gradings are separately heated to the specified temperatures and are then placed in the mixer chosen for the job. The mixing temperature for each grading and the bitumen binder is also specified based on the laboratory results. A tolerance of ± 10°C is allowed. The mixing is done till a homogeneous mixture is obtained. The mixture is then carried to the site for its placement through a transporter or a wheel barrow.

(iv) **Placement:** The bituminous paving mixture is then immediately placed on the desired location and is spread with rakes to a pre-determined thickness. The camber profile is checked with a template. It may be stated here that a compacting temperature also influence the strength characteristic of the resulting pavement structure. It is therefore required that the minimum time is spent between the placement of the mix and the rolling operations.

(v) **Rolling and finishing the paving mix:** The rolling is done with 8 to 10 tonnes tandem roller. The rolling is commenced from the edges of the pavement construction towards the centre, and uniform overlapping is provided. The finished surface should not show separate lines of markings due to defective or imprópor rolling. The roller wheels are kept damp otherwise the paving mix may partly stick to the wheels and the finishing may not be good. A variation of 6 mm over 3 m length is allowed in the cross profile. The number of undulations exceeding 10 mm should be less than 30 in 300 m length of pavement.

4.7 **Explain briefly the construction steps of 'Built-up Spray Grout' for strengthening of existing flexible pavement structure.**

[5 marks : 2008]

Solution:

The construction steps of 'Built up spray Grout' are:

(i) **Preparation of existing base:** The depressions and pot holes are filled and the surface is brought to the required grade and camber. The surface is cleaned and a prime coat is applied and cured if the surface is porous.

(ii) **Tack coat:** Heated bituminous binder is applied at a rate of 7.5 to 10 kg on WBM surface and 6 to 8 kg on black top surface per 10 m^2 area.

(iii) **Spreading first layer:** Coarse aggregates are spread and hand packed at a rate of 0.5 m^3 per 10 m^2 area.

(iv) **Rolling:** The aggregate is rolled using 8 to 10 tonnes roller as usual.

(v) **Application of binder:** Bitumen binder is sprayed uniformly at a rate of 12.5 to 15 kg per 10 m^2 over the dry surface of the compacted aggregates.

(vi) **Spreading of second layer of coarse aggregates:** The second layer of coarse aggregates is spread at a rate of 0.5 m^3 per 10 m^2 area, immediately after applying the binder.

(vii) **Rolling:** This layer is again rolled thoroughly starting from the edges.

(viii) **Second application of binder:** Binder is applied at a rate of 12.5 to 15.0 kg per 10 m^2 area.

(ix) **Application of key aggregates:** Immediately after the application of the binder, key aggregates are spread at a rate of 0.13 m^3 per 10 m^2 area and then rolled.

(x) **Surface finish:** The surface unevenness is checked with 3 m straight edge. Longitudinal profile should not have undulations exceeding 12 mm and the number of undulations greater than 10 mm size should not exceed 30 in 300 m length of road. In cross profile the maximum allowable variation in surface is 8.0 mm.

(xi) **Opening to traffic:** Built up spray grout should not be exposed to traffic before providing surface course.

4.8 How are following bitumen binders classified?
(i) Penetration grade bitumen
(ii) Blown bitumen
(iii) Cutbacks
(iv) Emulsions

[5 marks : 2009]

Solution:

(i) **Penetration grade bitumen:** The bitumen grade is specified in terms of penetration value. 80-100 or 80/100 grade bitumen means that the penetration value of the bitumen is in the range 80 to 100 at standard test conditions. The penetration values of various types of bitumen used in pavement construction in India range between 20 and 225, 30/40 and 80/100 grade bitumen are more commonly used, depending on construction type and climatic conditions. In hot climates a lower penetration grade bitumen like 30/40 bitumen is preferred.

(ii) **Blown bitumen:** Blown bitumen or oxidised bitumen are produced by passing air through soft bitumen mixtures under controlled temperature conditions. Their classification is designated by two numbers representing the midpoints of their softening point and penetration ranges. They are designated like blown bitumen 85/25, blown bitumen 95/15, blown bitumen 100/15 etc.

(iii) **Cutbacks:** Cutbacks are classified as Rapid Curing (RC), Medium Curing (MC) and Slow Curing (SC). The classification is based on the rate of curing or hardening after the application. The grade of cutback or its fluidity is designated by a numerical figure following the initials. The cutback with the lowest viscosity is designated by numeral 0, e.g. RC-0, MC-0 and SC-0. The suffix numerals 0, 1, 2, 3, 4 and 5 designates progressively more viscous cutbacks and indicates a definite viscosity irrespective of type of cutbacks.

(iv) **Emulsions**: A bitumen emulsion is a liquid product in which a substantial amount of bitumen is suspended in a finely divided condition in an aqueous medium and stabilized by means of one or more suitable materials. An emulsion is a two phase system consisting of two immiscible liquids; the one being dispersed as fine globules in the other. Three types of bituminous emulsion are prepared, viz., (i) Rapid Setting (RS), (ii) Medium Setting (MS) and (iii) Slow Setting (SS) types. Rapid Setting type emulsion is suitable for surface dressing and penetration macadam type of construction. Medium

Setting type is used for premixing with coarse aggregates and Slow Setting type emulsion is suitable for fine aggregate mixes.

4.9 **Differentiate between hydrophilic aggregates and hydrophobic aggregate.**

[2 marks : 2009]

Solution:

Hydrophilic Aggregate	Hydrophobic Aggregate
• Hydrophilic aggregates have good affinity for water compared to the affinity for bitumen.	• Hydrophobic aggregates have good affinity for bitumen compared to the affinity for water.
• Hydrophilic aggregates in general acidic in nature, siliceous such as sand stone, quartz.	• Hydrophobic aggregates are usually basic in nature typically as calcareous such as limestone, solomite.
• These are usually negatively charged in presence of water.	• These get positively charged in presence of water.

4.10 **Explain the procedure of conducting impact test as per IS 2386 part IV. What are the limits for the material to be used for 1. Sub-base, 2. Base course, 3. Surface course.**

[5 marks : 2011]

Solution:

A test designed to evaluate the toughness of stone or the resistance of the aggregates to fracture under repeated impacts is called impact test.

The aggregate impact testing machine consists of a metal base and a cylindrical steel cup of internal diameter 10.2 cm and depth 5 cm in which the aggregate specimen is placed. A metal hammer of weight of 13.5-14.0 kg having a free fall from a height 38 cm is arranged to drop through vertical guides.

Aggregate specimen passing 12.5 mm sieve and retained on 10 mm sieve is filled in the cylindrical measure in 3 layers by tamping each layer by 25 blows. The sample is transferred from the measure to the cup of the aggregate impact testing machine and compacted by tamping 25 times. The hammer is raised to a height of 38 cm above the upper surface of the aggregate in the cup and is allowed to fall freely on the specimen. After subjecting the test specimen to 15 blows, the crushed aggregate is sieved on 2, 36 mm sieve. The aggregate impact value is expressed as the percentage of the fine formed in terms of the total weight of the sample.

The maximum aggregate impact value for materials to be used for

1. Base course: 35% for bituminous macadam and 40% for water bound macadam.
2. Surface course: 30%

4.11 **Differentiate between the following with reference to bituminous construction:**

1. Prime coat and Tack coat

2. Bituminous concrete and Bituminous macadam

[4 + 4 = 8 marks : 2015]

Solution:

1. **Prime Coat:** A prime coat is a coating applied directly to a prepared base before additional layers of support or coating are applied. Prime coat asphalt preparation is a vital element, as it directly affects the shear strength of the final asphalt product.

   A prime coat is mainly responsible for protecting the subgrade of a construction project before additional layers are added. In asphalt preparation, they can also act as a binder with secondary and tertiary compounds, creating a better adhesion between the layers. Prime coat asphalt acts as an initial sealer in the asphalt laying process to block the other layers from moisture, dust and debris before additional coating installations.

Important things to remember about installing a prime coat or prime coat asphalt are:

- It provides temporary waterproofing, but must have additional compound layers in order for the floor, wall, or ceiling coating to perform long-term waterproofing duties.
- The asphalt prime coat must completely penetrate the base surface in order to be effective.
- A prime coat must completely cure before an additional layer is incorporated.
- It is vital to be aware of weather conditions when apply an asphalt prime coat. In the case of wet weather, this step should be postponed to avoid runoff into the water supply.

Tack Coat: A tack coat is applied after the prime coat, to form an adhesive bond between the tack coat and the next layer of coating. The tack coat prevents slippage and may sometimes function as a more long-term sealer to protect the substrate from moisture and bacteria. For asphalt prime coat systems, the tack coat is one of the most vital parts of the process, as it connects the subsequent layers and forms the base of those layers strength.

Important things to remember about installing a tack coat are:

- It is vital that the prime coat is clean and dry before installing a tack coat.
- The higher the viscosity of the tack layer, the more shear strength the overall structure will have. Diluted, slow-set emulsions are the most typical style of applying a tack coat.

2. **Bituminous Macadam (BM)** or Bitumen Bound Macadam is a premixed construction method consisting of one or more courses of compacted crushed aggregates premixed with bituminous binder, laid immediately after mixing. The BM is laid in compacted thickness of 75 mm or 50 mm and three different gradations of aggregates have been suggested for each thickness to provide open graded and semi-dense constructions. The BM is essentially a base course or binder course and hence should be covered by a suitable surfacing course before exposing to traffic. BM base course is considered to be much superior than other types of base course materials such as WBM with respect to load dispersion characteristics and durability.

   Bituminous Concrete or Asphaltic Concrete (AC) is a dense graded premixed bituminous mix which is well compacted to form a high quality pavement surface course. The AC consists of a carefully proportioned mixture of coarse aggregates, fine aggregates, mineral filler and bitumen and the mix is designed by an appropriate method such as the Marshal method to fulfil the requirements of stability, density, flexibility and voids. The thickness of bituminous concrete surface course layer usually ranges from 40 to 75 mm. The IRC has provided specification for 40 mm thick AC surface course for highway pavements.

■■■■

8 Railway Engineering

Revised Syllabus of ESE : *Terminology, Planning, designs and maintenance practices; track modernization.*

1. Rail Joints, Welding of Rails and Signals

1.1 What are the functions of Railway Stations? Describe the factors that influence the selection of site for a railway station.

[10 marks : 2000]

Solution:

A railway station is the selected place on a railway line where trains halt for one or more of the following purposes:

(i) For exchange of passengers.

(ii) For exchange of goods.

(iii) For control of train movements.

(iv) To enable the following express trains to overtake.

(v) To enable the trains on a single line track to cross from opposite directions.

(vi) For taking diesel or coal and water for locomotives.

(vii) For detaching engines and running staff.

(viii) For detaching or attaching of compartments and wagons.

(ix) For sorting of bogies to form new trains, housing of locomotives in loco-sheds etc.

(x) In emergencies in case of dislocation of track due to rains or accidents etc.

For selecting the site for a railway station, following factors should be considered:

(i) **Acquisition of land:** Sufficient land area including land width for a single track or double track system and extra area to accommodate additional lines, station buildings, platforms, cabins, staff quarters and for future extension needs to be acquired in the beginning. This is necessary from economic consideration because once the land is acquired, the land values greatly increase due to start of development on adjoining land.

(ii) **Proximity to town or village:** The station site should be close to the town or village as far as possible to facilitate the transport of passengers and goods.

(iii) **Nature of land area:** The land area should preferably be a level ground for convenience and safety of marshalling and shunting wagons. Steep gradients may cause the wagons to move even when not in operation due to high wind velocities and may result in accidents.

(iv) **Approach roads to station site:** It is highly essential to encourage rail transport by providing approach roads connected a town or village to a station site.

(v) **Station site alignment:** Station site should never be located on curves and as far as possible should not be near the curves also, otherwise the signals with either be difficult to sight or out of sight.

(vi) **Site drainage:** The site should have quick drainage facilities and minimum grade of 1 in 1000 is recommended in station yards for this purpose.

(vii) **Station amenities:** Amenities like water supply, coal, diesel, electricity, etc., should be provided at the station site. A plentiful supply of drinking water is essential.

(viii) **Type of station and yard:** A station may be either a through type or terminal type station. The shape of the plot to be acquired for station site will depend upon location of station buildings, yards and other facilities at station site.

(ix) **Role of authorities:** The civil and military authorities, should be consulted before the final selection of site is made. This is necessary to execute the development scheme and for justification of railway project.

1.2 **(i) Define and explain the term "Breathing length of LWR".**

(ii) Determine the minimum theoretical length of LWR beyond which the central portion of 52 kg rail would not be subjected to longitudinal movement due to 30°C temperature variation. Use the following additional data:

(a) RAIL:

- **50 kg rail section**
- **Cross sectional area = 66.15 cm^2**
- **Modulus of elasticity of rail section = 2.1 x 10^6 kg/cm^2**
- **Coefficient of thermal expansion of rail section = 11.5 x 10^{-6}/°C**

(b) SLEEPER :

- **Sleeper spacing = 60 cm**
- **Average resisting force/sleeper/rail = 300 kg**

[15 marks : 2001]

Solution:

(i) A long welded rail continues to expand at its ends for a particular length till adequate forces gets developed towards the centre. A stage reaches when at a particular length of the rail from its ends, the resistance offered by the track structure accumulates and becomes equal to thermal forces created as a result of changes in temperature. There is no movement of rail beyond this point. The cumulative value of expansion or contraction of these portions of the rail at the ends (breathing length) is given by

$$\delta L = \frac{L\alpha t}{2}$$

where L is length of the portion of rail, α is coefficient of thermal expansion of rail and t is variation in temperature

Thus, the portion of the long welded rail in the end, which undergoes movement due to temperature variations, and absence of resisting forces of track is called the breathing length.

(ii) Suppose L is the length of rail required, then

$$\delta L = L\alpha t$$

$$\therefore \quad \text{Strain prevented} = \frac{\delta L}{L} = \frac{L\alpha t}{L} = \alpha t$$

Given that $\alpha = 11.5 \times 10^{-6}$/°C and $t = 30$°C

$$\therefore \quad \text{Strain prevented} = 11.5 \times 10^{-6} \times 30 = 345 \times 10^{-6} = 3.45 \times 10^{-4}$$

$$\therefore \quad \text{Stress developed} = \text{Strain prevented} \times E$$

$$= 3.45 \times 10^{-4} \times 2.1 \times 10^{6} = 724.5 \text{ kg/cm}^2$$

$$\therefore \quad \text{Force developed} = \text{Stress developed} \times \text{Cross sectional area of rail}$$

$$= 724.5 \times 66.15 = 47925.675 \text{ kg}$$

Given that average resisting force per sleeper per rail = 300 kg

∴ Number of sleepers required to prevent the force developed,

$$n = \frac{\text{Force developed}}{\text{Average resisting force per sleeper}}$$

$$= \frac{47925.675}{300}$$

$$= 159.75 \approx 160 \text{ (approx.)}$$

Given that sleeper spacing = 60 cm (say S)

We know that minimum length of welded rail

$$= (n-1)\,S = (160-1) \times 60 \times 10^{-2} = 95.4 \text{ m}$$

1.3 What are the key factors considered in the modernization of railway track with the aim to increase the average speed and capacity of the modernized section?

[12 marks : 2023]

Solution:

The following factors are to be considered for the modernization of railway track with aim to increase average speed and capacity of the section:

(i) Track geometry having proper alignment, appropriate super-elevation, smooth curves.

(ii) Upgrading track structure to accommodate higher speeds and heavy loads using stronger materials for sleepers, rails, ballast etc.

(iii) Focus on optimizing operational practices, such as efficient train scheduling, improved coordination between employee's at all hierarchy levels.

(iv) Implementation of track maintenance by improving regular inspections, timely repairs and the use of modern track inspection techniques.

(v) Electrification of railway track enables faster acceleration, higher speeds and increased capacity

(vi) Investing funds to upgrade tracks and signalling systems to allow higher speeds and reduce delays.

2. Ballast, Formation and Sleepers

2.1 What are the requirement of an ideal permanent way? Draw a dimensional cross section of a *BG* track in embankment on a straight track and mark the details.

[10 marks : 2002]

Solution:

Permanent track is regarded to be semi-elastic in nature. There is possibility of track getting disturbed by the moving wheel loads. The track should therefore, be constructed and maintained keeping the requirements of a permanent way, in view, so as to achieve higher speed and better riding qualities with less future maintenance.

Following are some of the basic requirements of a permanent way:

(i) The gauge should be correct and uniform.

(ii) The rails should be in proper level. Thus in a straight track, two rails must be at the same level. On curves, the outer rail should have proper super elevation and there should be proper transition at the junction of a straight and a curve.

(iii) The alignment should be correct, i.e. it should be free from kinks or irregularities.

(iv) The gradient should be uniform and as gentle as possible. Any change of gradient should be followed by a smooth vertical curve to give smooth riding quality.

(v) The track should be resilient and elastic in order to absorb shocks and vibrations of running track.

(vi) The track should have enough lateral strength, so that alignment is maintained even due to effects of side thrust on tangent lengths and centrifugal forces on curves.

(vii) The radii and super elevation on curves should be properly designed and maintained.

(viii) Drainage system must be perfect for enhancing safety and durability of track.

(ix) Joints including points and crossings which are regarded to be weakest points of the railway track, should be properly designed and maintained.

(x) The track structure should be strong, low in initial cost as well as maintenance cost.

The typical cross section of BG track in embankment on straight track is shown below:

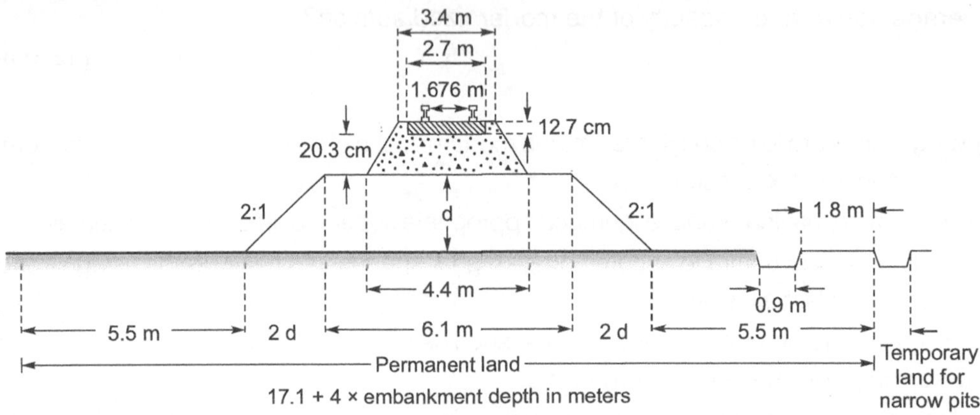

2.2 **State the reasons due to which the vertical, horizontal and transverse loadings are caused to the sleepers in a railway track. Why is ballast filled between the sleepers? Write the merits and demerits of concrete sleepers and explain the reasons of these being extensively used in India.**

[10 marks : 2006]

Solution:

Some of the reasons due to which the vertical, horizontal and transverse loadings (stresses are caused to the sleepers in a railway track are:

(i) Wheel load
(ii) Weight transfer from wheel to wheel on the same and different axles.
(iii) Irregularities of the rails
(iv) Speed
(v) Dynamic effect of wheels on rails
(vi) Elasticity of the rails
(vii) Efficiency of fastenings
(viii) Longitudinal forces in rails

Ballast is filled between the sleepers because it performs the following functions:

(i) It transfers the load from the sleeper to the subgrade and then distributes it uniformly over a larger area of the formation.

(ii) It holds the sleepers in position and prevents the lateral and longitudinal movement, due to dynamic loads and vibrations of moving trains.
(iii) It imparts some degree of elasticity to the track.
(iv) It provides easy means of maintaining the correct levels of the two lines of a track and for correcting track alignment.
(v) It provides good drained foundation immediately below the sleepers and helps to protect the top surface of the formation. This is achieved by providing coarse and rough aggregates with plenty of voids.

Merits of Concrete Sleepers:
(i) These sleepers are free from natural decay and attacks by vermin, insects, etc.
(ii) They have maximum life when compared to other sleepers, the life under normal conditions is 40 to 60 years.
(iii) These are not affected by moisture, chemical action of ballast, cinder and subsoil salt.
(iv) These have higher elastic modulus and hence can withstand the stresses induced by fast and heavy traffic.
(v) Concrete sleepers in the elastic fastenings offers an ideal track in respect of gauge, cross level and alignment.

Demerits of Concrete Sleepers:
(i) The weight of concrete sleeper is as high as 2.5 to 3 times of wooden sleeper, requiring the mechanical appliances for handling.
(ii) These sleepers require pads and plugs for spikes.
(iii) They damage the bottom edge during the packing.
(iv) The scrap value is almost nil.
(v) The damages to the concrete sleepers is very heavy in case of derailment.

2.3 **A BG track has a sleeper density of $n + 6$. If the track is laid with welded rails of 26 m length, find out the number of sleepers required for constructing a railway track of 1690 m.**

[2 marks : 2010]

Solution:

Length of each rail, $n = \frac{26}{2} = 13\text{ m}$

Sleeper density $= n + 6 = 13 + 6 = 19$

Total number of rails required $= \frac{1690}{13} = 130$

$\therefore$ Total number of sleepers = Number of rails × Sleeper density = 130 × 19 = 2470

2.4 **(a) Write a detailed note on track ballast. Also write the recommended ballast depth.**
(b) With a neat labelled sketch, write a note on points or turnouts.
(c) Write a detailed note on sleeper density for railways.

[5 + 5 + 5 = 15 marks : 2011]

Solution:

(a) Ballast is the granular material usually broken stone or brick, shingle or kankar, gravel or sand placed and packed below and around the sleepers to transmit load (due to the wheels of the train) from sleepers, to formation and at the same time allowing drainage of the track.

Functions of Ballast:

Ballast performs the following functions:
(i) It transfers the load from the sleeper to the subgrade and then distributes it uniformly over a larger area of the formation.

(ii) It holds the sleepers in position and prevents the lateral and longitudinal movement, due to dynamic loads and vibrations of moving trains.

(iii) It imparts some degree of elasticity to the track.

(iv) It provides easy means of maintaining the correct levels of the two lines of a track (i.e., level in straight portions and correct super-elevation on curves) and for correcting track alignment.

(v) It provides good drained foundation immediately below the sleepers and helps to protect the top surface of the formation.

The recommended ballast dimensions for various gauges are tabulated below:

	Dimensions	B.G.	M.G.	N.G.
1.	Width of ballast	3.35 m	2.25 m	1.83 m
2.	Depth of ballast	20 to 25 cm	15 to 20 cm	15 cm
3.	Quantity of stone ballast per metre length	1.036 m^3	0.71m^3	0.53 m^3

(b) Turnout is the simplest combination of points and crossings which enables one track either a branch line or a siding, to take off from another track. So the object of turnout is to provide facilities for safe movement of trains in either direction on both the tracks.

As shown in the figure following are the parts of a turnout:

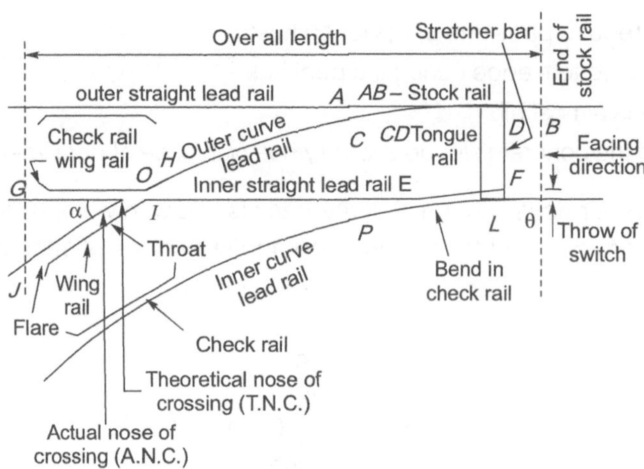

(i) A pair of points or switches (*ABCD* and *EFPQ*).

(ii) A pair of stockrails.

(iii) A Vee crossing (*GHIJ*).

(iv) Two check rails.

(v) Four lead rails.

(vi) 'Switch tie-plate' or gauge tie chair and crossing tie-plate.

(vii) Studs or stops.

(viii) Bearing plates; slide chairs, stretcher bars etc.

(ix) For operating the points–Rods, cranks, levers etc.

(x) For locking system-lock bar, plunger bar etc.

(c) Sleeper density is the number of sleepers per rail length and it is specified as ($M + x$ or $N + x$), where M is the length of the rail in metres (N is the length of rail in yards) and x is a number, varying according to the following factors and is fixed by the Railway Board for various axle loads.

The factors governing the sleeper density are:

(i) Axle load and speed

(ii) Type and section of the rails

(iii) Type of ballast and ballast cushion

(iv) Type and strength of sleepers, i.e., Bearing area of a particular sleeper on the ballast and

(v) Nature of foundation.

Thus, if M is the length of a rail in metres, the sleeper density is expressed as $M(M+1)$, $(M+2)$ and so on. For instance, if length of the rail is 13 m, 17 sleepers are expressed as $M+4$ (i.e., 13 $M+4$) is the density of sleepers.

Since sleeper also provides lateral stability to the track, so, more the number of sleepers more is the lateral stability. The number of sleepers, however, cannot be increased indefinitely as certain minimum space between sleepers is required for packing of ballast. In India, this minimum distance for manual packing of ballast is kept 30.5 cm to 35.5 cm except at joints.

2.5 **List and briefly discuss the geological factors affecting the construction of high speed rail project.**
[10 marks : 2022]

Solution:

Following geological factors can affect the construction of high speed rail:

1. Rock Types
2. Attitude of Rocks
3. Batter and Slope Angle
4. Surface and Groundwater
5. Joints, Shear Zones, Fault and Fault Zone

1. **Rock Types:** The overall behavior of rock mass will depend on its intrinsic properties such as density, porosity, strength, cohesion factor of mineral composition and texture of rocks as well as on the presence of joints and shear zones in a rock mass.
2. **Attitude of Rocks:** The strike and dip of layers in rocks such as volcanic flows, bedding, lamination and foliation planes with respect to slope cut and bench plays a very important role. Horizontal rocks will provide moderate to good conditions while vertical rocks will pose problems on cut slope and bench.
3. **Batter and Slope Angle:** The inclination and height of side slope are also important as they are going to control the space that would be available after cut and fill for laying the rail.
4. **Surface and Groundwater:** Water is the most important factor in aiding and triggering the landslides. Surface runoff caused by rainfall result into sudden mass transportation of debris on the slopes. Infiltration of water through pores and fractures result in to reduction of frictional resistance along fractures, increased pore pressure and swelling of clays all causing landslides.
5. **Joints, Shear Zones, Fault and Fault Zone:** Joints the most ubiquitous feature in rock mass will control the overall rock mass characteristics. Lesser the joint sets, farther the joints, closed joint apertures and high degree of asperities will make the rock slope stronger and stable.

3. Geometric Design of the Track

3.1 **What special measures and precautions are required in geometric features of a modern railway track?**

[10 marks : 1999]

Solution:

Special measures required in geometric features of a modern railway track are:

(i) **Gauge of track:** A wider, correct and uniform gauge is desirable for attaining high speeds. Always wider is the gauge better would be the stability and hence higher the speed. The gauge over crossings is kept 1 mm tight to improve running. The gauge should be laid and maintained tight by 3 mm on straight and on curves upto 4 mm for high speed routes.

(ii) **Alignment of track:** For high speed tracks, both horizontal and vertical alignment should be perfect. For horizontal alignment, the flat curves along with transition curves and adequate super elevation should be provided. The sharp curves in tracks should either be eliminated by diverting the track or replacing them by flat curves. For vertical alignment, the steep gradients should be eliminated either by replacing them by gentle gradients or again by diverting the track or tunnelling through hills. Adequate cant deficiency should be provided.

(iii) **Other geometric elements:** Other geometric elements like cant, radius of the curve, degree of the curve, length of transition curve, etc., for high speeds should be provided.

(iv) **Track centres:** The centre to centre distance between tracks in station yards is kept more for high speeds tracks as compared to a section between stations. This offers several advantages such as safety of staff, elimination of problems of loading gauge and safety margin and possibility of allowing trains at high speeds over crossovers to greater intermediate track length.

Precautions required are as follows:

(i) Maintenance of track shall be within permanent tolerances.

(ii) Rails shall be ultrasonically tested at least once in two years for detecting cracks or flaws.

(iii) All critical points, such as points and crossings, vicinity of water columns, old rails over 10 years, portions with past history of rail breakage etc., shall be tested more frequently.

(iv) After one year of the train introduction, track irregularities shall be recorded once in two months and recording by oscillograph car atleast once in four months.

(v) There must not be less than (M + 3) effective sleepers for 13 m rail length (BG track) on the entire rate where M is length of rail in metres.

3.2 **For a broad gauge track in a transition zone, in order to allow locomotives with a maximum permissible speed of 110 kmph, calculate the following:**

(i) Radius of curvature (ii) Degree of curvature (iii) Super elevation (iv) Length of transition

[10 marks : 2003]

Solution:

We know that, by **Martin's formula** for a *BG* track in a transition zone for speed > 100 kmph

$$V = 4.58\sqrt{R}$$

where *V* is maximum permissible speed in kmph and R is radius of curvature in meters

Now, $V = 110$ kmph (given)

$\therefore$ $110 = 4.58\sqrt{R}$

$\Rightarrow$ $R = 576.84$ m

Degree of curvature is given as

$$\frac{2\pi R}{360°} = \frac{30}{D}$$

$\therefore$ $D = 2.98° \approx 3°$

For $D = 3^\circ$,

$$R = \frac{1720}{D^\circ} = \frac{1720}{3^\circ} = 573.33 \text{ m}$$

Theoretical superelevation is given by, $e_{th} = \frac{GV^2}{127R}$ where G = Gauge distance

$$= \frac{(1.676 \times 100) \times 110^2}{127 \times 573.33} = 27.852 \text{cm}$$

Now, maximum cant deficiency, $D_{max} = 10$ cm

$\therefore$ Actual superelevation, $e_{act} = e_{th} - D_{max}$

$$= 27.852 - 10 = 17.852 \text{ cm}$$

(> 16.5 cm as V < 120 kmph) Not OK

$\therefore$ Now use Degree of curvature, $D = 2^\circ$

Radius of curvature is, $R = \frac{1720}{D} = \frac{1720}{2}$

$\therefore$ $R = 860$ m

Now, theoretical superelevation is given as,

$$e_{th} = \frac{GV^2}{127R} = \frac{(1.676 \times 100) \times 110^2}{127 \times 860}$$

$$e_{th} = 18.567 \text{ cm}$$

$$D_{max} = 10 \text{ cm}$$

$$e_{act} = e_{th} - D_{max}$$

$$= 18.567 - 10 = 8.567 \text{ cm}$$

Length of transition curve, $L = ?$

(i) $L = 7.2e = 7.2 \times 8.567 = 61.68$ m

(ii) $L = 0.073eV_{max} = 0.073 \times 8.567 \times 110 = 68.8$ m

(iii) $L = 0.073DV_{max} = 0.073 \times 10 \times 110 = 80.3$ m

Thus, length of transition curve = 80.3 ≈ 81 m (maximum of above 3).

3.3 **What is the equilibrium cant required on a 5° curve on a *BG* track for the train speed of 60 kmph?**
[10 marks : 2004]

Solution:

Given: Degree of curvature of curve, $D = 5^\circ$

For a *BG* track, $G = 1.676$ m

Speed of train $V = 60$ kmph

Radius of curvature, $R = \frac{1720}{D} = \frac{1720}{5} = 344$ m

Super elevation, $e = \frac{GV^2}{127R} = \frac{1.676 \times (60)^2}{127 \times 344} = 0.1381 \text{ m} = 13.81 \text{ cm}$

3.4 **What is the importance of track drainage? State the measures to be adopted with suitable illustrations.**
[10 marks : 2004]

Solution:

The bearing power (or strength and stability of soil i.e. resistance to shear of soils) are greatly reduced due to presence of excess moisture. The variation in bearing strength and stability depends upon the percentage of moisture, soil type and mode of stress application.

Track drainage is important due to following reasons:

(i) The modern-track embankments, which are subjected to heavy and fast loads, get settled due to presence of excess moisture.

(ii) The presence of excess sub-surface moisture reduces the track stability and results in ballast pockets, dirty ballast, low joints pumping sleepers, unstable formation and slipping and subsiding banks.

(iii) The presence of surface water and ground water if not properly drained results in recurrent soft spots, unstable banks and cuttings, bank slips and land slides.

(iv) The erosion of soil from the bank, slopes of embankment, cut and hill side is caused due to surface water.

(v) In the rainy season, the presence of a badly drained track formation is the main cause of accidents due to derailments.

The following remedial measures can be adopted to prevent sinking of ballast into wet soil:

(i) **Use of pervious cess:** If trouble in track performance due to formation of water pocket is not excessive, in such cases pervious cess can be provided as shown in figure.

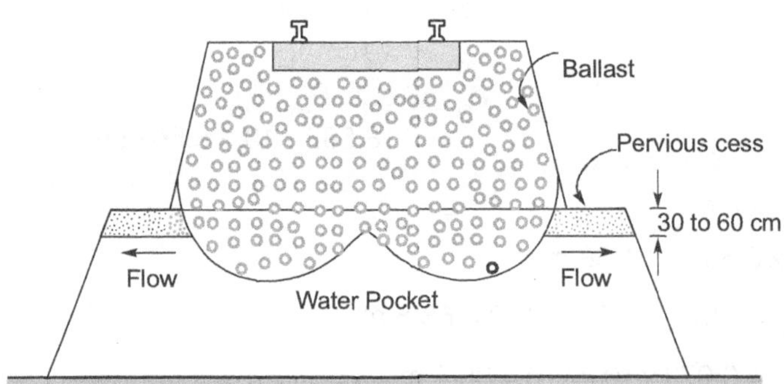

(ii) **Use of perforated pipes and trench drains:** Water can also be successfully drained by use of perforated pipes and trench drains. These perforated pipes are laid with loose joints so that water which surrounds them can enter inside after it has seeped to the rubble filled trench.

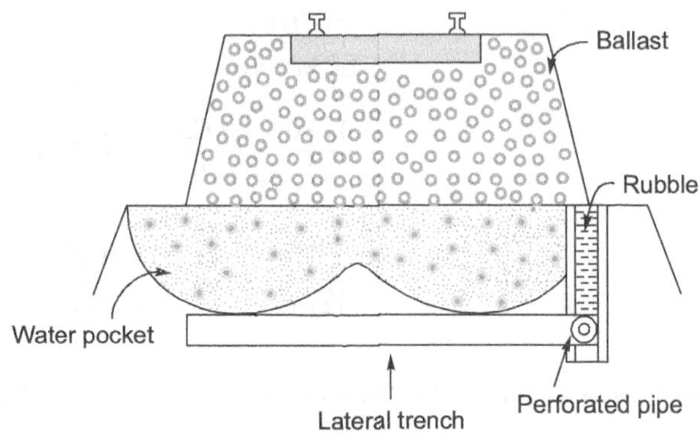

(iii) **Use of inverted filter blanket:** If water pocket formation causes excessive trouble then the method is to first screen the ballast, drain off water pockets and then provide 20 cm to 30 cm thick inverted filter blanket.

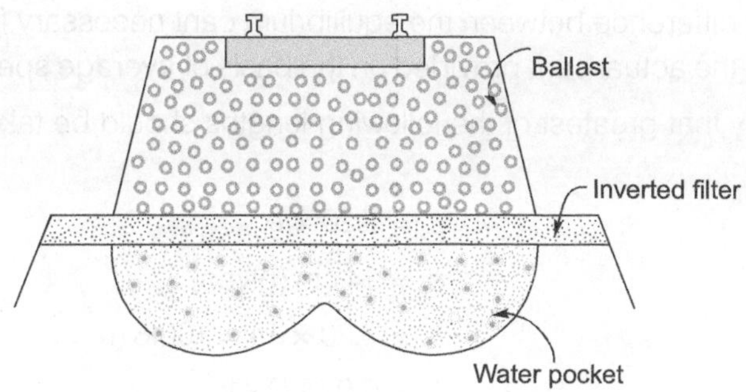

(iv) **Use of sand piles:** In case a pervious soil strata is existing below cohesive soil, the sand piles can be used. Water pocket is formed in cohesive soil and water from it can be collected in the sand piles which further passes it to pervious layer and formation gets stabilized.

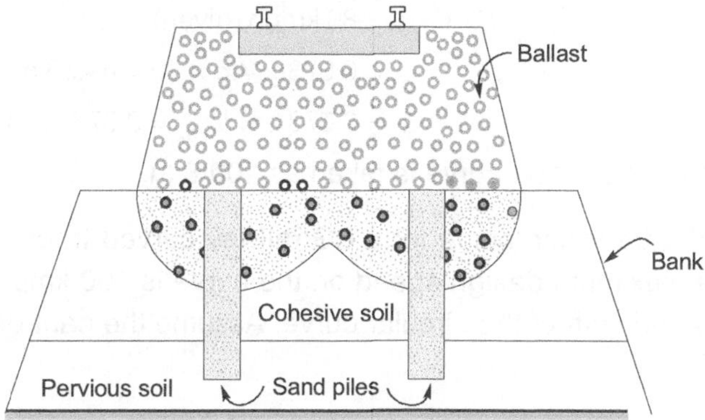

(v) **Use of capillary break:** If the trouble of water pocket is caused due to capillary use, then capillary break is the best remedy. But use of capillary break in embankment is not possible when embankment is built. So while constructing formation this should be borne in mind.

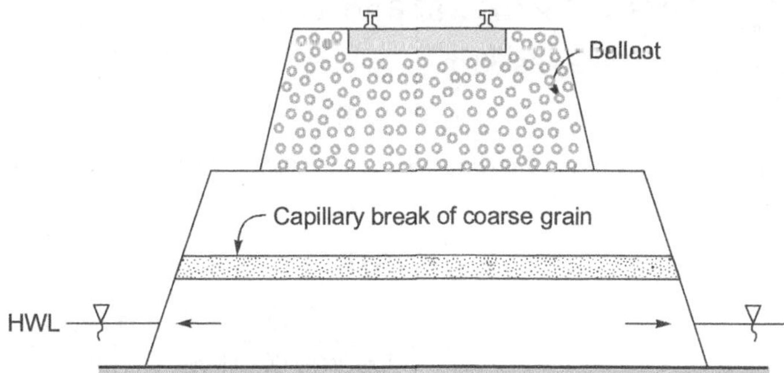

3.5 **Define cant and cant deficiency. Calculate the length of the transition curve for a broad gauge curved track having 5° deflection and a cant of 14 cm. The maximum permissible speed on the curved track is 80 kmph.**

[9 marks : 2007]

Solution:

When a train moves round a curve, it is subjected to a centrifugal force acting horizontally at the centre of gravity of each vehicle radially away from the centre of the curve. This increases the weight on the outer rail. To counteract the effect of centrifugal force, the level of the outer rail is raised above the inner rail by a certain amount to introduce the centripetal force. This raised elevation of outer rail above the inner rail at a horizontal curve is called super elevation or cant.

Cant deficiency is the difference between the equilibrium cant necessary for the maximum permissible speed on a curve and the actual cant provided on the basis of average speed of trains.

Indian railways specify that greatest of the following lengths should be taken as length of the transition curve:

(i) $L = 7.20 \times e$

$e = 14 \text{ cm (given)}$

$\therefore \quad L = 7.20 \times 14 = 100.8 \text{ m}$

(ii) $L = 0.073\, D \times V_{max}$

where D is cant deficiency in cm

For BG tracks and for speeds up to 100 kmph,

$D = 7.6 \text{ cm}$

$V_{max} = 80 \text{ kmph (given)}$

$\therefore \quad L = 0.073 \times 7.6 \times 80 = 44.38 \text{ m}$

(iii) $L = 0.073\, e \times V_{max} = 0.073 \times 14 \times 80 = 81.76 \text{ m}$

Hence the length of the transition curve is taken as 100.8 m.

3.6 Design the length of a transition curve for a BG railway curved track having 4° deflection and a 'cant' of 12 cm. The maximum design speed on the curve is 100 kmph. Also calculate offsets at every 15 m intervals and shift of the circular curve. Assume the cant deficiency as 7.6 cm.

[10 marks : 2008]

Solution:

As per Indian Railways the length of the transition curve is the greater out of the following three values

(i) $L = 7.20 \times e = 7.20 \times 12 = 86.4 \text{ m}$

(ii) $L = 0.073\, D \times V_{max} = 0.073 \times 7.6 \times 100 = 55.48 \text{ m}$

(iii) $L = 0.073\, e \times V_{max} = 0.073 \times 12 \times 100 = 87.6 \text{ m} \approx 88 \text{ m}$

Hence length of curve is to be taken as 88 m.

Now, $L = 90 \text{ m (say)}$

The equation of cubic parabola which is used as a transition curve is given by

$$y = \frac{x^3}{6RL}$$

But, $$R = \frac{1720}{\text{Degree of curve}} = \frac{1720}{4°} = 430 \text{ m}$$

Taking offsets at 15 m intervals, we get

$y = \dfrac{x^3}{6RL} \text{(m)}$

$$y_{15} = \frac{(15)^3 \times 100}{6 \times 430 \times 90} = 1.45 \text{ cm}$$

$$y_{30} = \frac{(30)^3 \times 100}{6 \times 430 \times 90} = 11.63 \text{ cm}$$

$$y_{45} = \frac{(45)^3 \times 100}{6 \times 430 \times 90} = 39.24 \text{ cm}$$

$$y_{60} = \frac{(60)^3 \times 100}{6 \times 430 \times 90} = 93.02 \text{ cm}$$

$$y_{75} = \frac{(75)^3 \times 100}{6 \times 430 \times 90} = 181.69 \text{ cm}$$

$$y_{90} = \frac{(90)^3 \times 100}{6 \times 430 \times 90} = 313.95 \text{ cm}$$

Chainage (m)	15	30	45	60	75	90
Offset (cm)	1.45	11.63	39.24	93.02	181.69	313.95

Shift of the curve, $S = \frac{L^2}{24R} = \frac{(90)^2}{24 \times 430} = 0.785 \text{ m}$

3.7 **Determine the length of transition curve and offsets at every 15 metres for BG curved track having 4° curvature and cant of 12 cm. The maximum permissible speed on curve is 85 kmph.**

[10 marks : 2009]

Solution:

Indian Railways specify that greatest of the following lengths should be taken as length of the transition curve:

(i) $L = 7.20 \times e$ where e is cant in cm

$\therefore$ $L = 7.20 \times 12 = 86.4 \text{ m}$

(ii) $L = 0.073\, D \times V_{max}$

where D is cant deficiency in cm and V_{max} is maximum permissible speed on the curve.

For *BG* tracks and for speeds upto 100 kmph, $D = 7.6$ cm

$V_{max} = 85$ kmph (given)

$\therefore$ $L = 0.073 \times 7.6 \times 85 = 47.158 \text{ m}$

(iii) $L = 0.073\, e \times V_{max} = 0.073 \times 12 \times 85 = 74.46 \text{ m}$

Hence the length of transition curve is 86.4 m, but it is taken as 90 m (being multiple of chain length).

Radius of curve, $R = \frac{1720}{D} = \frac{1720}{4} = 430 \text{ m}$

The equation of cubic parabola is

$$y = \frac{x^3}{6RL}$$

$$\text{Offset at 15 m} = \frac{(15)^3}{6 \times 430 \times 90} \times 100 = 1.45 \text{ cm}$$

$$\text{Offset at 30 m} = \frac{(30)^3}{6 \times 430 \times 90} \times 100 = 11.63 \text{ cm}$$

$$\text{Offset at 45 m} = \frac{(45)^3}{6 \times 430 \times 90} \times 100 = 39.24 \text{ cm}$$

$$\text{Offset at 60 m} = \frac{(60)^3}{6 \times 430 \times 90} \times 100 = 93.02 \text{ cm}$$

$$\text{Offset at 75 m} = \frac{(75)^3}{6 \times 430 \times 90} \times 100 = 181.69 \text{ cm}$$

$$\text{Offset at 90 m} = \frac{(90)^3}{6 \times 430 \times 90} \times 100 = 313.95 \text{ cm}$$

Chainage (m)	15	30	45	60	75	90
Offset (cm)	1.45	11.63	39.24	93.02	181.69	313.95

3.8 **What are the factors to be considered in selection of a gauge? Write advantages of uniformity of gauges?**

[8 marks : 2012]

Solution:

The gauge of a railway track is defined as the clear minimum perpendicular distance between the inner faces of the two rails as shown in the figure.

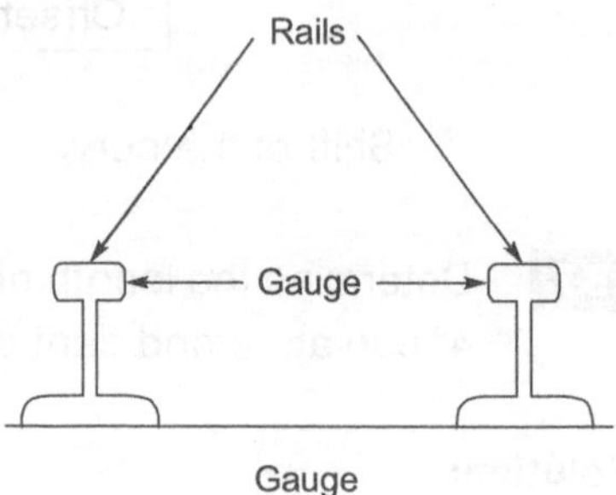

Following are the factors affecting the choice of a gauge:

1. **Traffic condition** : If the intensity of traffic on the track is likely to be more, a gauge wider than the standard gauge is suitable.
2. **Development of poor areas** : The narrow gauges are laid in certain parts of the world to develop a poor area and thus link the poor area with the outside developed world.
3. **Cost of track :** The cost of railway track is directly proportional to the width of its gauge. Hence, if the fund available is not sufficient to construct a standard gauge, a metre gauge or a narrow gauge is preferred rather than to have no railways at all. In India, the approximate costs of M.G. track and B.G. track per kilometre length are about Rs. 50 lakhs and Rs. 1 crore respectively. However there is no appreciable increase in cost of track, if a wider gauge is adopted initially because of the following facts:
   (i) There is only proportional increase in the costs of ballast, sleepers, rails, etc. for different gauges.
   (ii) The cost of acquisition of land for the permanent track is not much affected by the change of gauge width.
   (iii) The costs of culverts, bridges, tunnels, signals, staff quarters, level-crossings and other works associated with the permanent track are more or less same for all the gauges.
   (iv) The cost of rolling stock is independent of the gauge of track for carrying the same volume of traffic.
4. **Speed of movement:** The speed of a train is a function of the diameter of wheel which in turn is limited by the gauge. The wheel diameter is usually about 0.75 times the gauge width and thus the speed of a train is almost proportional to the gauge. If higher speeds are to be attained, the B.G. track is preferred to the M.G. or N.G. track.
5. **Nature of country:** In mountainous country, it is advisable to have a narrow gauge of the track since it is more flexible and can be laid to a smaller radius on the curves. This is the main reason why some important railways, covering thousands of kilometres, are laid with a gauge as narrow as 610 mm.

Following are the advantages of uniformity of gauges:

1. **No transport-bottlenecks:** In a unigauge system there is no transport bottleneck. It gives improved operational efficiency and fast movements of goods and passengers.
2. **No transhipment hazards:** In unigauge system there is no need of changing trains, hence no damage to goods and no inconvenience to the passengers.
3. **Improved utilization of track:** In unigauge system the track can be used to the full capacity, resulting in reduction of working expenses of the railway.

4. **Provision of alternate routes:** The unigauge system provides alternate routers for free movement of traffic, resulting in reduced pressure on the track.
5. **Better turn round:** The unigauge system provides better turn round of wagons and locomotives resulting in improved operating ratio of the railway system. This will result in immense benefits to the nation.
6. **Balanced economic growth:** Due to the unigauge system the development of all areas will be uniform resulting in balanced economic growth.
7. **No multiple tracking works:** The unigauge system eliminates the duplicate traffic facilities and tracking works, resulting in saving of resources.

3.9 **Determine the equilibrium speed and cant to be provided on a BG curve of 3 degree if the speeds of several trains running on the line are as follows:**

15 trains at a speed of 50 km/hr
12 trains at a speed of 60 km/hr
8 trains at a speed of 70 km/hr
3 trains at a speed of 80 km/hr

[8 marks : 2014]

Solution:

(i) Equilibrium speed:

15 trains at a speed of 50 km/hr
12 trains at a speed of 60 km/hr
8 trains at a speed of 70 km/hr
3 trains at a speed of 80 km/hr

Equilibrium speed, V_{avg} = Weighted average of speed

$$V_{avg} = \frac{(15\times 50)+(12\times 60)+(8\times 70)+(3\times 80)}{15+12+8+3} = 59.74 \text{ Kmph}$$

(ii) Cant Provided:

Given: degree of curve, $D = 3°$

$$\therefore \quad R = \frac{1720}{3} = 573.33 \text{ m} \qquad \left[\because D = \frac{1720}{R}\right]$$

Actual cant provided, $e_{act} = \frac{GV_{avg}^2}{127R}$ m

For *BG* track, $G = 1.676$ m

$$\therefore \quad e_{act} = \frac{1.676\times(59.74)^2}{127\times 573.33} = 0.0821 \text{ m or } 8.21 \text{ cm} < 16.5 \text{ cm (max)}$$

Thus, the actual cant provided is 8.21 cm.

3.10 (i) Differentiate between Loading gauge and Construction gauge.

(ii) A BG railway track is to be constructed in an area where ruling gradient permissible is 1 in 100. A curve of 3° is also required to be provided. Determine the gradient to be provided at this location so that the combined resistance due to gradient and curve should not exceed the resistance due to ruling gradient.

[4 + 4 = 8 marks : 2015]

Solution:

(i) **Loading gauge:** A loading gauge defines the maximum height and width for railway vehicles and their loads to ensure safe passage through bridges, tunnels and other structures. The loading gauge limits the size of passenger carriages, goods wagons (freight cars) and shipping containers that can be conveyed on a section of railway line and varies across the world and often within a single railway system. Over time there has been a trend towards larger loading gauges and more standardization of gauges with older lines having their structure gauges enhanced by raising bridges, increasing the height of tunnels and making other necessary alterations.

Construction gauge: Construction gauge is the gauge which is decided by adding necessary clearances to the loading gauge so that the vehicle can move safely without infringement. This is the condition which is to be provided in case of all the bridges or the tunnels or the pavements or the platforms which needs to be constructed or the shapes of all the platforms because there are certain clearances which are added to the loading gauge and then with those clearances this construction gauge is provided.

(ii) Let 'w' be the weight of the train and 1 in x be the required gradient.
Hence,

$$\text{Resistance due to curve} = 0.04\% \text{ per degree of curve} = \frac{0.04}{100} \times 3 \times w$$

$$\text{Resistance due to required gradient} = \frac{1}{x} \times w$$

$$\text{Resistance due to ruling gradient} = \frac{1}{100} \times w$$

Therefore, under given condition,

$$\frac{1}{x} \times w + \frac{0.04}{100} \times 3 \times w = \frac{1}{100} \times w$$

$$\Rightarrow \quad x = 113.63$$

∴ **Gradient can be provided 1 in 115.**

3.11 **From a BG main line curve of 2°, a 1 : 12 turnout takes off in contrary flexure for a branch line. The speed on the branch line is restricted to 42 km/hr. Determine the speed restriction on the main line. Assume permissible deficiency in cant as 75 mm.**

[20 marks : 2017]

Solution:

We have, Degree of 1 : 12 turn out curve = 4°

Given: Speed on branchline = 42 kmph

Degree of mainline curve = 2°

Thus, degree of branchline curve = 4° – 2° = 2° [In contrary flexure]

$$\text{Radius of branch line curve} = \frac{1720}{2} = 860 \text{ m}$$

$$\therefore \text{ Theoretical cant needed for branch line} = \frac{GV^2}{127R_e} = \frac{1.676 \times (42)^2}{127 \times 860} = 0.0271 \text{ m}$$

$$= 2.71 \text{ cm} = 27.1 \text{ mm}$$

$$e_{\text{theoritical}} = e_{\text{actual}} + \text{cant deficiency}$$

$$e_{\text{actual}} = e_{\text{theoritical}} - \text{cant deficiency}$$

$$= 27.1 - 75$$

$$e_{actual} \text{ (on branch line)} = -47.9 \text{ mm}$$

Hence,
$$e_{actual} \text{ (main line)} = 47.9 \text{ mm}$$
$$e_{theoritical} \text{ (main line)} = e_{actual} + \text{Cant deficficney}$$
$$= 47.9 + 75 = 122.9 \text{ mm} = 12.29 \text{ cm} < 16.5 \text{ cm} \quad \text{(OK)}$$

Hence,
$$e_{theoritical} = \frac{GV_{max}^2}{127R}$$
$$0.1229 = \frac{1.676 \times V_{max}^2}{127 \times 860}$$
$$V_{max} = 89.49 \text{ kmph}$$

From martin formula
$$V_{max} = 4.35\sqrt{R - 67}$$
$$= 4.35\sqrt{860 - 67} = 122.497 \text{ kmph}$$

Hence speed restriction on main line will be 89.49 kmph.

3.12 **Draw a typical diagram showing all important features of a crossover with intermediate portion straight and crossing angles equal for two parallel railway tracks. Find intermediate straight distance and overall length of crossover for Broad Gauge tracks of same crossing number 1 in 12. The distance between centres of tracks is 5 m.**

[20 marks : 2018]

Solution:

Crossover (between parallel tracks)

When two adjacent parallel or diverging tracks, which are straight or curved, are connected by two sets of turnouts with or without a straight length between them, the connecting line is known as crossover.

Important features:

- It consists of two pairs of switches, two acute angle crossing with reverse curves, with or without straight length in between and four check rails.
- Use of reverse curves between two crossing shortens the length of crossover but its use becomes limited for low speed siding.
- The length of straight portion depends on spacing D between the two adjacent tracks to be joined.
- Cross-overs are useful to change the track, when train are approaching from one direction only.

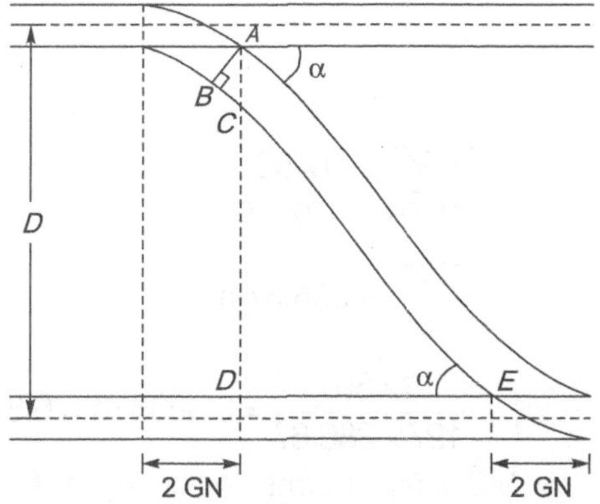

$$CD = D - \frac{G}{2} - \frac{G}{2} - AC$$

From ΔCDE

$$\tan\alpha = \frac{CD}{DE}$$

$$\tan\alpha = \frac{1}{N}$$

From ΔABC

$$\cos\alpha = \frac{G}{AC}$$

$$\cos\alpha = \frac{N}{\sqrt{1+N^2}}$$

$$AC = G\sec\alpha$$

$$\sec\alpha = \frac{\sqrt{1+N^2}}{N}$$

$\because$

$$DE = CD\ (\text{at } \alpha = (D - G - AC) \times N$$

$$= (D - G - G\sec\alpha)\,N$$

$$= \left(D - G - \frac{G\sqrt{1+N^2}}{N}\right)N$$

$$DE = N(D-G) - G\sqrt{1+N^2}$$

Straight intermediate portion (projected)

$$= N(D-G) - G\sqrt{1+N^2}$$

$$= 12(5-1.676) - 1.676\sqrt{1+12^2}$$

$$= 19.708 \text{ m}$$

Total cross over length $= 4\,GN + N(D-G) - G\sqrt{1+N^2}$

$$= 4 \times 1.676 \times 12 + 12 \times (5 - 1.676) - 1.676\sqrt{1+12^2}$$

$$= 100.156 \text{ m}$$

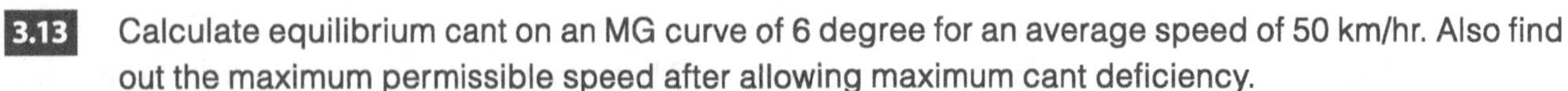

3.13 Calculate equilibrium cant on an MG curve of 6 degree for an average speed of 50 km/hr. Also find out the maximum permissible speed after allowing maximum cant deficiency.

[12 marks : 2019]

Solution:

Given Degree of curve = 6°

$$V = 50 \text{ kmph}$$

$$e_a = \frac{GV^2}{127R} = \frac{1 \times 50^2}{127 \times R}$$

$$R = \frac{1720}{6} = 286.67 \text{ m}$$

$$e_a = \frac{1 \times 50^2}{127 \times 286.67} = 0.0686 \text{ m} = 6.87 \text{ cm}$$

$\because$ For speed < 100 kmph, for MG limit of cant deficiency is equal to 5.1 cm.

$$e_{th} = e_a + e_d = 6.87 + 5.1 = 11.97 \text{ cm} = 0.1197 \text{ m}$$

Now, $$e_{th} = \frac{GV_{max}^2}{127R}$$

$$0.1197 = \frac{1 \times V_{max}^2}{127 \times 286.67}$$

$\Rightarrow$ $$V_{max} = 66.01 \text{ kmph}$$

By Martin's formula, $$V_{max} = 4.35\sqrt{(R-67)}$$

$$V_{max} = 4.35\sqrt{(286.67-67)} = 64.47 \text{ kmph}$$

So, maximum permissible speed = 64.47 kmph $\simeq$ 64 kmph.

3.14 **Write a note on ten requirements of track drainage system for typical B.G. railway line.**

[12 Marks : 2021]

Solution:

Water considered as greatest threat to railway track. A poor drainage system may lead to settlement of embankment, reduced bearing capacity of formation, shrinkage and cracking of banks and formation of ballast pockets etc.

So, to overcome these adverse conditions a good track drainage system should be planned in such a way that:

(a) Surface and underground water should be well away from track
(b) Surface water of adjoining land should be prevented to enter into track
(c) The side drains should have enough capacity and slope.
(d) Flow of surface water should not cause erosion of banks and embankment.
(e) Sub surface water should be drained off by subsurface drainage system.
(f) High water table area should be avoided
(g) The track alignment should be made to rest on pervious, naturally drained soil
(h) Water logged areas should be avoided
(i) The drains or pipe should be kept at closer spacing to keep water table well below formation to prevent capillary rise.
(j) Underground water pocket, water bearing strata on the sides, black cotton soil etc. should be avoided for track construction.

3.15 **(i) With a neat labelled sketch, derive the relation between superelevation (e), for a broad gauge railway with a speed 'V' (km/h) and radius of curvature (R).**

(ii) If the wheel base of a vehicle moving on BG track is 6 m, the diameter of wheel is 1.50 m and the depth of flanges below the top of rail is 3.20 cm, determine the extra width required to be provided on gauge, if the radius of the curve is 150 m.

[10 + 10 Marks : 2021]

Solution:

(i) Raising of outer track with respect to inner track to generate centripetal force against the centrifugal force which works outward from C.G, of train when it moves on a curve, is known as super elevation. Super elevation not only gives counteracting force for i.e. but also reduces centrifugal force.

Where, $\frac{mV^2}{R}\cos\theta$ = Centrifugal force at superelevated track

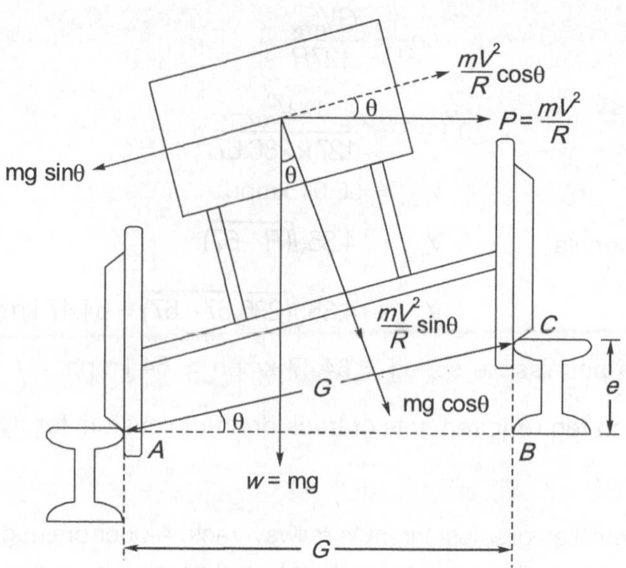

$$mg\sin\theta = \text{Counteracting centripetal force}$$
$$V = \text{Speed of train (kmph)}$$
$$R = \text{Radius of curvature (m)}$$
$$G = \text{Gauge distance (m)}$$
$$e = \text{Superelevation (m)}$$

From the figure shown above,

$$\Sigma F_x = 0$$

$$mg\sin\theta = \frac{mV^2}{R}\cos\theta$$

$$\tan\theta = \frac{V^2}{gR} \quad ...(i)$$

From geometry of track,

$$\sin\theta = \frac{e}{G} \quad ...(ii)$$

$\because$ θ is very small

$\therefore \sin\theta = \tan\theta = \theta = \frac{e}{G}$

Equating equation (i) and (ii)

$$\frac{e}{G} = \frac{V^2}{gR}$$

$$e = \frac{GV^2}{gR}$$

$$V = \text{Speed of train (m/s)}$$
$$g = 9.81 \text{ m/s}^2$$

If we put g value and correct V in kmph

$$e = \frac{GV^2}{127R}$$

(ii) Extra width, $d = \frac{13(B+L)^2}{R}$

Where given B = wheel base (m) i.e. 6 m, R = Radius of curve (m) i.e. 150 m

L = lap length of flange in (m) i.e. $L = 0.02\sqrt{h^2 + Dh}$

h = depth of flange below rail head in cm i.e. 3.2 cm (given)

D = diameter of wheel in cm i.e. 150 cm

$$\therefore \quad L = 0.02\sqrt{3.20^2 + 150 \times 3.20} = 0.4428 \text{ m}$$

$$\therefore \quad d = \frac{13(6 + 0.4428)^2}{150} = 3.597 \text{ cm}$$

3.16 **A 400 m radius curve is introduced between straight portions of a Broad Gauge (BG) railway line intersecting to form a deflection angle of 50 degrees. The speed for determining the equilibrium cant is fixed at 100 kmph and the maximum sectional speed is 120 kmph. Determine the equilibrium cant, the maximum permissible speed (considering the cant deficiency, cant excess) and desirable length of transition curve. The maximum permissible cant and cant deficiency are 165 mm and 100 mm respectively. Cant excess is restricted to 75 mm. The lowest speed of any train can be taken as 50 kmph.**

[20 marks : 2022]

Solution:

Given:
Radius of curve = 400 m
Equilibrium speed = 100 kmph
Maximum sectional speed = 120 kmph
Speed of lowest train = 50 kmph

$$\text{Cant required for lowest train} = \frac{GV^2}{127R} = \frac{1750 \times 50^2}{127 \times 400} = 86.12 \text{ mm}$$

$$e_{act} = e_{slow\ moving} + \text{cant excess}$$

Actual cant to be provided = 86.12 + 75 = 161.12 mm

Theoretical maximum permissible speed possible on given track

$$= 0.27\sqrt{(C_a + C_d)R}$$

$$= 0.27\sqrt{(161.12 + 100) \times 400} = 87.26 \text{ kmph}$$

Length of transion curve,

$$L = 0.72\, C_a = 0.72 \times 161.12 = 116 \text{ m}$$

$$L = 0.008\, C_a V_{max} = 0.008 \times 161.2 \times 87.26 = 112.5 \text{ m}$$

$$L = 0.008\, C_d V_{max} = 0.008 \times 100 \times 87.26 = 69.8 \text{ m}$$

Length of transition curve = 116 m

4. Points and Crossing

4.1 **Enumerate the various types of track junction in use. Discuss the suitability of each under different conditions.**

[15 marks : 2000]

Solution:

The common track junctions are as follows:

(i) **Turnouts:** This is the simplest combination of points and crossings. These provide facilities for turning of trains from one track to another. One turnout provides facilities for turning of vehicles or trains in one direction of the main track only. Depending upon the facilities provided for turning on right or left of main track, the turnouts are called right hand or left hand turnouts respectively.

(ii) **Symmetrical Split:** When a turnout is taken off from a curved track, it is termed as a split. When a straight track is split up in two different directions with equal radii, it is known as a symmetrical split. This provides facilities for turning of trains in both left and right directions of the main track. It is suitable for locations where turnout from a straight track would take too much of space.

(iii) **Three throw switches:** When two turnouts take off from the same point of a main straight track, it is called "Three throw" arrangement. Three throw switches may be used in congested areas where space is not enough and heavy traffic is not prevailing such as goods yards, entrances to locomotives yards, etc. This track junction is obsolete these days.

(iv) **Double turnout:** This arrangement is an improvement over three throw switch. This can be used on main lines with heavy traffic and are of special significance in the congested areas where economic consideration in space is of primary importance.

(v) **Diamond crossing:** When straight tracks or curved tracks of the same or different gauges cross each other at an angle less than 90°, a diamond shape is formed, so this crossing is called diamond crossing. They are provided on straight tracks only as they necessitate restriction on speed.

(vi) **Crossovers:** When two adjacent parallel and diverging tracks, which may be straight or curved, are connected by two sets of turnouts, with or without a straight length between them, the connecting line is a crossover. These are useful to change the track, when trains are approaching from one direction only.

(vii) **Single slip and double slip:** In a diamond crossing two tracks cross each other but the trains cannot be taken from one track to another but in slip arrangement the trains can move from one track to another.

(viii) **Gauntlet track and fixed point system:** Gauntlet track is that arrangement of the track which is adopted when a double track is to be narrowed over a short distance of the track. This is used to economize the cost of a double line bridge. This is also used when part of a double line bridge is under repair.

(ix) **Scissor cross over:** It is a combination of one crossover over the other cross over in the opposite direction to enable the trains to change the track from either direction along the main track. This arrangement is useful where enough space for two separat cross overs is not available and shunting operations are frequent.

(x) **Gathering lines or ladder tracks:** When a number of parallel tracks are branched off from the straight track in continuation of a turnout, it is called a gathering line or ladder track. This arrangement is known as "Herring bone Grid" and is used in marshalling yards and goods yards where sidings are almost of equal length.

(xi) **Temporary diversion:** A temporary diversion is required when the track has to be directed from its original position if some heavy or time consuming work required is to be done on original track or track has failed under floods etc.

(xii) **Triangle:** The triangles are required to change the direction of engines. Turn tables are also used to change the direction of engines but they are very costly. So, if enough area is available and railway sections are not important, then the installation of a turn table is not justified.

(xiii) **Double junctions:** These junctions become essential where two or more main tracks are running and where branches are taking off from main tracks.

4.2 **What is diamond crossing on a railway track? Give a sketch of such crossing for a *BG* track. Explain the salient features of different parts of the crossing.**

[10 marks : 2004]

Solution:

When straight tracks or curved tracks of the same or different gauges cross each other at an angle less than 90°, a diamond shape is formed. So this crossing is called a diamond crossing.

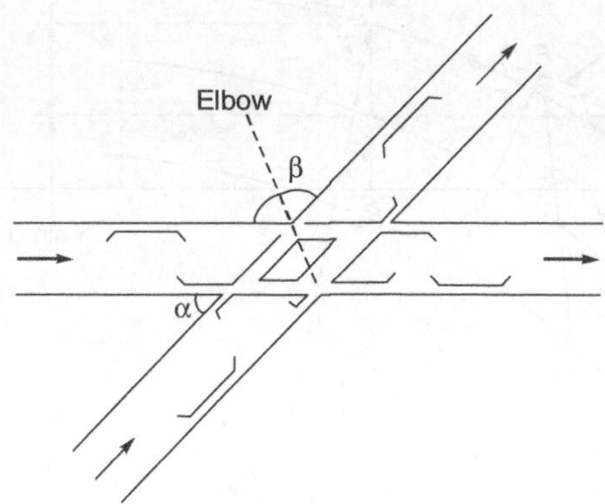

The salient features of diamond crossing are:

(i) It consists of two acute angle crossings (α and α) and two obtuse angle crossings (β and β) and four check rails.

(ii) The length of the gap between two noses of an obtuse crossing increases as the acute angle of crossing increases.

(iii) Indian Standards specify the limit of flattest diamond to be 1 in 10 for *BG* tracks and 1 in 8.5 for other tracks. Crossing of tracks at sharper angles to this limit is undesirable.

(iv) Diamond crossings should be avoided as far as possible on curves as they necessitate restriction on speed.

4.3 Calculate lead and radius of a turnout on a Broad Gauge railway track with the following data:
Heel divergence = 130 mm
Straight length between theoretical nose of crossing and tangent point of crossing = 1.3 m
Angle of crossing = 4° 45' 49"
Angle of switch = 1° 08′ 00″
Broad Gauge Width = 1.676 m
Show the values on a neat sketch of turnout.

[12 marks : 2020]

Solution:

Given for a BG track,

Heel divergence, $d = 130$ mm

Straight length between theoretical nose of crossing and tangent point of crossing

$x = 1.3$ m

Angle of crossing, $\alpha = 4° 45' 49''$

Angle of switch, $\beta = 1° 8' 00''$

Gauge distance, $G = 1.676$ m

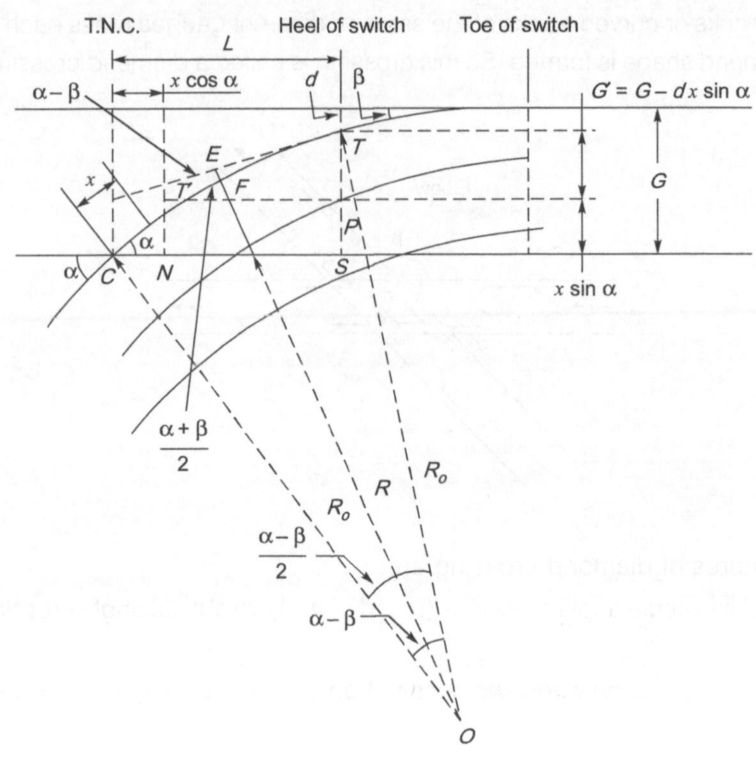

(i) Radius, R

$$R_o = \frac{G - d - x\sin\alpha}{\cos\beta - \cos\alpha}$$

$$= \frac{1.676 - 0.13 - 1.3 \times \sin(4°45'49'')}{\cos(1°08'00'') - \cos(4°45'49'')} = 441.31 \text{ m}$$

$\therefore$ Central radius,

$$R = R_o - \frac{G}{2} = 441.31 - \frac{1.676}{2} = 440.472 \text{ m}$$

(ii) Crossing Lead (L):

$$L = x\cos\alpha + (G - d - x\sin\alpha) \cdot \cot\left(\frac{\alpha + \beta}{2}\right)$$

$$= 1.3\cos(4°45'49'') + (1.676 - 0.13 - 1.3\sin(4°45'49'')$$

$$\times \cot\left(\frac{4°45'49'' + 1°08'00''}{2}\right)$$

$$= 1.2955 + 1.438 \times 19.415 = 29.214 \text{ m}$$

4.4 What is meant by crossing? What are the essential requirements of a good crossing? Discuss various types of crossing in use in Indian Railways.

[12 marks : 2022]

Solution:

A 'crossing' or a 'frog' is a device which provides two flangeways through which the wheels of the flanges may move, when two rails intersect each other at an angle.

The flanged wheel of the train jump over the gap from 'throat' or 'nose' of crossing and to check the wheel flanges from striking the nose, the opposite wheel flanges are guided by a use of 'check rails' inside the running rail.

Components of a crossing:

(i) A crossing or Vee piece.
(ii) Point and splice rails.
(iii) Wing rails.
(iv) Check rails.
(v) Chairs at crossing, at toe and at heel.
(vi) Throat of crossing

Requirements and characteristics of a good crossing:

(i) The assembly of a crossing has to be rigid to with stand against severe vibrations.and splice rails.
(ii) The wear on parts of the wing rails, opposite the nose and also of nose itself must be protected. The medium manganese steel (i.e., an alloy steel) is used where traffic is light whereas high manganese steel is used where traffic is heavy.
(iii) The crossing body should be as rigid as possible and as long as practicable.
(iv) Ramp the wing rails by 3 mm to 6 mm at the top from throat to nose and flushing it to the normal level of the chair.
(v) By use of the distance blocks closely touching the web.
(vi) The nose of crossing should have some thickness, varying from 6 mm to 18 mm.

Type of crossings: Crossing can be classified as below:

1. Acute angle crossing or "V" crossing or Frog.
2. Obtuse angle crossing or Diamond crossing.
3. Square crossing.

1. **Acute angle crossing**: It is obtained when a left-hand rail of one track crosses a right-hand rail of another track or vice versa. This crossing is used widely.
2. **Obtuse angle crossing**: This crossing is obtained when left-hand rail of one track crosses right hand rail of another track or vice versa at an obtuse angle.

   In diamond crossing, a pair of special crossing is used which is called "Obtuse crossing".
3. **Square crossing**: When two straight tracks cross each other at right angles, they give rise to square crossing. This type of crossing must be avoided on main lines because there is heavy wear and tear due to dynamic loads.

5. Track Stresses, Traction and Tractive Resistance

5.1 **A locomotive on BG track with four pairs of driving wheels each carrying axle load of 20 tonnes is required to haul a train at a speed of 80 kmph. The train is made to run on a level track with curvature of 2°. Calculate the maximum permissible load that can be pulled by the engine. Take hauling capacity as one-sixth of the load on driving wheels.**

[8 marks : 2010]

Solution:

Hauling capacity of locomotive, $H = \frac{1}{6} \times \text{load on driving wheels} = \frac{1}{6} \times 4 \times 20 = \frac{40}{3}$ tonnes

On a curved level track, train resistance is given by

$$T = 0.0016\,W + 0.00008\,WV + 0.0000006\,WV^2 + 0.0004\,WD$$

where W is weight of train in tonnes including weight of locomotive, V is speed of train in kmph and D is degree of curve.

$\therefore$ $T = 0.0016\,W + 0.00008\,W \times 80 + 0.0000006\,W \times (80)^2 + 0.0004\,W \times 2$

$\Rightarrow$ $T = 0.01264\,W$

But, Hauling capacity = Train resistance

$\Rightarrow$ $H = T$

$\Rightarrow$ $\frac{40}{3} = 0.01264\,W$

$\Rightarrow$ $W = 1054.85$

$\Rightarrow$ $W \approx 1055$ tonnes

5.2 **Discuss the cause and effects of "Creep of Rails".**

[4 marks : 2013]

Solution:

Creep of Rails: Creep of rails is the longitudinal movement of rails of the track in the direction of motion of locomotives. Creep is common to all railways and value of creep ranges from zero to about 6 inches or 16 cm.

Causes of Creep:

(i) Unequal contraction and expansion of rails due to variation in temperatures give rise to creep.

(ii) Creep develops in rails when a train starts on stops. While starting, train pushes the rail backwards and while stopping due to application of brakes, creep in rails occur in forward direction.

Effects of Creep of Rails:

(i) The most serious effect of creep is the buckling of track in lateral directions. If unattended and not properly removed them if causes derailments which leads to accident.

(ii) Sleepers move out of position which leads to the change in gauge and alignment of the track.

(iii) Rail joints are opened out of their limit in some case and stresses are set up in fish plates and bolts which leads to the breakage of the bolts.

(iv) Points and crossing get disturbed.

(v) Maintenance and replacement of tracks become difficult.

(vi) Smashing of fish plates and bolts, bending of bars, kinks at joint are other effect of creep.

5.3 **Calculate the maximum permissible load that a B.G. steam locomotive with three pairs of driving wheels with axle load of 22 tons each, on a straight level track at a speed of 80 kmph.**

Also calculate the reduction in speed, if the train has to run on a rising gradient of 1 in 200. What would be the further reduction in speed, if the train has to negotiate a 4° curve on the rising gradient? Assume, coefficient of friction as 0.20.

[10 marks : 2013]

Solution:

Hauling capacity of steam locomotive

$\Rightarrow$ $H = f \times$ total load of driving wheels

$= 0.2 \times 3 \times 22 = 13.2$ tonne

Train resistance is given by $\Rightarrow T$

$$T = 0.0016\,W + 0.00008\,WV + 0.000000\,6\,WV^2$$

where V = 80 kmph

$$T = 0.0016\,W + 0.0000\,8 \times W \times 80 + 0.000000\,6 \times W \times 80^2$$

$$T = 0.01184\,W$$

But Hauling capacity = Train resistance

$$H = T$$

$$13.2 = 0.01184\,W$$

$$W = 1114.865 \text{ tonnes}$$

Total resistance when gradient is introduced

$$T = 0.0016\,W + 0.00008\,WV + 0.0000006\,WV^2 + W\tan\theta$$

$$T = 0.0016 \times 1114.865 + 0.00008 \times 1114.865 \times V + 0.0000006 \times 1114.865 \times V^2 + 1114.865$$

$$T = 7.35811 + 0.0892\,V + 6.6892 \times 10^{-4}\,V^2$$

But Hauling capacity = Train resistance

$$H = T$$

$$13.2 = 7.35811 + 0.0892\,V + 6.6892 \times 10^{-4}\,V^2$$

$$6.6892 \times 10^{-4}\,V^2 + 0.0892\,V - 5.84189 = 0$$

$$V = 48.124 \text{ kmph}$$

Hence reduction in speed after introducing the gradient

$$= 80 - 48.124 = 31.876 \text{ kmph}$$

Total resistance when curvature is also introduces is

$$T = 0.0016\,W + 0.00008\,WV + 0.0000006\,WV^2 + W\tan\theta + 0.0004\,WD$$

$$T = 0.0016 \times 1114.865 + 0.00008 \times 1114.865 \times V + 0.0000006 \times 1114.865 \times V^2 + 1114.865 \times \frac{1}{200} + 0.0004 \times 1114.865 \times 4$$

$$T = 9.141893 + 0.0891892\,V + 6.68919 \times 10^{-4}\,V^2$$

But Hauling capacity = Train resistance

$$H = T$$

$$13.2 = 9.14183 + 0.0891892\,V + 6.68919 \times 10^{-4}\,V^2$$

$$6.68919 \times 10^{-4}\,V^2 + 0.0891892\,V - 4.05817 = 0$$

$$V = 35.8575 \text{ kmph}$$

Hence reduction in speed after introducing gradient and curvature

$$= 48.124 - 35.8575 = 12.2665 \text{ kmph}$$

■■■■

9 Airport, Dock, Harbour and Tunnelling Engineering

Revised Syllabus of ESE : *Airports:* *Layout, planning & design.*
Tunneling: *Alignment, methods of construction, disposal of muck, drainage, lighting and ventilation.*
Harbours: *Terminology, layouts and planning.*

1. Airport

1.1 **Explain the need for air traffic control. Name the various enroute aids for controlling the air traffic on air routes and various possible landing aids for aircrafts.**

[10 marks : 2000]

Solution:

The need for air traffic control is mainly for two reasons. Firstly to safeguard life and property and secondly to expedite the traffic movements. The control of air traffic deals with that phase of air transportation which ensures safe, convenient and economic movement of aircraft from one airport to another airport.

The aircraft flight from one airport to another involves the following basic actions:

(i) The aircraft takes off from an airport
(ii) It maintains a proper altitude in air
(iii) It navigates from point to point safely
(iv) It lands at the desired airport

The primary functions of air traffic control devices can therefore be summarised as follows:

Airport traffic control:
This deals with the following:

(i) To guide the aircraft, desiring to land or take off
(ii) To control the taxing of arriving and departing aircraft on the airfield between the apron and the runway.

Airway traffic control: This regulates the movement of aircraft along the air routes with adequate lateral and vertical separation to avoid collision. This is particularly essential when visibility is poor.

Airway communication: This deals with conveying of airway and weather information to the pilot during the flight.

General or non-airway traffic control: This presents a serious problem when personal flying is done by a large number of people. In such cases the movement of aircraft, not flying along the airway must be regulated to prevent interference to the main air traffic.

Various enroute aids are available to the pilot during his flight from one airport to another. They are as follows:

(i) Airway beacon
(ii) Low/Medium frequency (LF/MF) ratio range

(iii) Very high frequency omnidirectional range (VOR)
(iv) Air ground communication
(v) Tactical air navigation (TACAN)
(vi) Distance measuring equipment (DME)
(vii) Marker beacon
(viii) Direction finder
(ix) Air route surveillance radar (ARSR)

The following aids are available to any aircraft while landing:

(i) Instrument landing system (ILS)
(ii) Precision approach radar (PAR) or Ground approach control (GAC)
(iii) Airport surveillance radar (ASR)
(iv) Airport surface detection equipment (ASDE)
(v) Approach lights

1.2 **Enumerate and explain all possible factors which are to be assessed precisely and to be taken into account during the process of decision making for suitability of a potential site for a Civilian Airport.**
[10 marks : 2008]

Solution:

The factors listed below are for the selection of a suitable site for a major civilian airport installation:

(i) **Regional plan:** The site selected should fit well into the regional plan, thereby forming it an integral part of the national network of airport.

(ii) **Airport Use:** The selection of site depends upon the use of an airport i.e. whether for civilian or for military operations. However, during the emergency, civilian airports are taken over by the defence. Therefore, the airport site selected should be such that it provides natural protection to the area from air raids.

(iii) **Proximity to other airports:** The site should be selected at a considerable distance from the existing airports so that the aircraft landing in one airport does not interfere with the movement of aircraft at other airport.

(iv) **Ground Accessibility:** The site should be so selected that it is readily accessible to the users. The airline passenger is more concerned with his door-to-door time rather than the actual time in air travel. The time to reach the airport is, therefore, an important consideration specially for short-haul operations. The best location is a site adjacent to the main highway. This provides a quick access and minimises the cost of an entrance road.

(v) **Topography:** This includes natural features like ground contours, trees; streams etc. A raised ground e.g. a hill top, is usually considered to be an ideal site for an airport.

(vi) **Obstructions:** When aircraft is landing or taking off, it loses or gains altitude very slowly as compared to the forward speed. For this reason, long clearance areas are provided on either side of runway known as approach areas over which the aircraft can safely gain or lose altitude. The areas should be kept free of obstructions. The obstructions may consist of fences, trees, pole lines, building and other natural or man made objects.

(vii) **Visibility:** Poor visibility lowers the traffic capacity of the airport. The site selected should therefore be free from visibility reducing conditions, such as fog, smoke and haze. Fog generally settles in the area where wind blow is minimum, e.g. in a valley. Smoke and haze nuisance exist at sites nearer to the industrial areas.

(viii) **Wind:** Runway is so oriented that landing and take off is done by heading into the wind. Wind data, i.e., direction, duration and intensity of wind should be collected over a minimum period of about five years. This helps in proper orientation of runway and influences the shape of the site needed for the development of airport. The site should be located to windward direction of the city, so that a minimum smoke from the city is blown over the site.

(ix) **Noise Nuisance:** The extent of noise nuisance depends upon the limb-out path of aircraft, type of engine propulsion and the gross weight of aircraft. The problem becomes more acute with jet engine aircrafts. Therefore, the site should be so selected that the landing and take off paths of the aircrafts pass over the land which is free from residential or industrial development. Sometimes buffer zone may have to be provided between the take off end of a runway and a nearby residential area.

(x) **Grading, Drainage and Soil Characteristics:** Grading and drainage play an important role in the construction and maintenance of airport which in turn influences the site selection. The possibility of floods at the valley sites should be investigated. Sites with high water tables, which may require costly sub-soil drainage should be avoided. The cost of grading and drainage can often be reduced by selecting a site with favourable soil characteristics. The most desirable type of soil for airport construction is the one which contains a reasonable amount of previous material such as gravel, sand of decomposed granite combined with a suitable natural binder.

(xi) **Future Development:** Taking into consideration the anticipated future developments in the airport, larger area should be acquired initially. Zoning ordinates should be implemented to prevent the growth of undesirable structures within the area.

(xii) **Availability of Utilities from Town:** An airport has to be provided with facilities like water supply, sewer, telephone, electricity etc. In the selection of site, the availability of these utilities from the town should be given due consideration.

(xiii) **Economic Consideration:** The cost estimates should be prepared for the entire airport construction considering both initial and ultimate stages. The estimates should include land cost, clearing and grading of land, drainage, removal of hazards, paving, turfing, lighting, construction of buildings, access roads and automobile parking areas. Amongst the various alterate sites, one which is economical should be preferred.

1.3 **What are the assumed conditions of runway length for standard environment which decide the basic runway length?**

[5 marks : 2009]

Solution:

The assumed conditions at the airport are as follows:

(i) Airport altitude is at sea level.
(ii) Temperature at the airport is standard (15°C).
(iii) Runway is levelled in the longitudinal direction.
(iv) No wind is blowing on runway.
(v) Aircraft is loaded to its full loading capacity.
(vi) There is no wind blowing enroute to the destination.
(vii) Enroute temperature is standard.

1.4 **Determine the actual runway length after applying necessary corrections for elevation and temperature as per ICAO and gradient correction as per FAA specification for the data given below.**

(i) Basic runway length = 1800 metres
(ii) Elevation of airport site = 600 metres
(iii) Monthly mean of average daily temperature for the hottest month of the year = 15°C
(iv) Monthly mean of maximum daily temperature for the same month = 21.6°C
(v) Effective gradient = 0.6%

[5 marks : 2009]

Solution:

(i) **Correction for elevation:**

ICAO recommends that the basic runway length should be increased at the rate of 7 percent per 300 m rise in elevation above mean sea level.

$$\therefore \quad \text{Correction for elevation} = \frac{7}{100} \times 1800 \times \frac{600}{300} = 252 \text{ m}$$

Thus, Corrected length = 1800 + 252 = 2052 m

(ii) **Correction for temperature:**

Standard atmospheric temperature at the given elevation = 15° – 0.0065 × 600 = 11.1°C

$$\text{Airport reference temperature} = T_a + \frac{T_m - T_a}{3} = 15° + \frac{21.6° - 15°}{3} = 17.2°\text{C}$$

As per ICAO the basic runway length after having been corrected for elevation should be further increased at the rate of 1% for every 1°C rise of airport reference temperature.

Rise in temperature = 17.2 – 11.1 = 6.1°C

$$\text{Correction for temperature} = 2052 \times \frac{1}{100} \times 6.1 = 125.172 \text{ m}$$

Thus, corrected length = 2052 + 125.172 = 2177.172 m

Check for the total correction for elevation plus temperature:

$$\text{Total correction in percentage} = \frac{2177.172 - 1800}{1800} \times 100 = 20.95\% < 35\% \qquad \text{(OK)}$$

As per ICAO, the total correction for elevation plus temperature should not exceed 35% of the basic runway length.

(iii) **Correction for gradient:**

FAA recommends that the runway length after having been corrected for elevation and temperature should be further increased at the rate of 20% for every 1% of effective gradient.

$$\text{Correction for gradient} = \frac{20}{100} \times 2177.172 \times 0.6 = 261.261 \text{ m}$$

Thus, corrected length = 2177.172 + 261.261 = 2438.433 m

Rounding the above value to the nearest 10 m, the corrected runway length is 2440 m.

1.5 **Differentiatite between stop way and cleary way or runway length.**

[2 marks : 2009]

Solution:

Clear way (or runway length)	Stop way
It is an area extending beyond the runway end, centrally located about the extended centre line of runway, to facilitates completion of take off operation.	Stop way is the area at the end of take-off runway , centered upon extended center line of runway, to facilitates aeroplane during an aborted take-off without causing structural damage to the aeroplane.

1.6 **Determine the radius of a taxiway for a supersonic aircraft to negotiate the curve at a turning speed of 50 kmph. The wheel base is 35 m and the tread of main landing gear is 7.5 m. The airport is of type A as per ICAO. Assume coefficient of friction between tyre and pavement surface as 0.13.**

[8 marks : 2010]

Solution:

Given: Turning speed (V) = 50 kmph, Wheel base of aircraft (W) = 35 m, Tread of main landing gear = 7.5 m, Coefficient of friction between tyre and pavement surface (f) = 0.13, For type A airport as per ICAO width of taxiway, T = 22.5 m

(i) **Turning radius may be expressed as**

$$R = \frac{V^2}{125f} = \frac{(50)^2}{125 \times 0.13} = 153.85 \text{ m}$$

(ii) **Turning radius as per Horonjeff's equation is given by**

$$R = \frac{0.388\,W^2}{\frac{T}{2} - S}$$

where S is distance between midway point of tread of landing gear and the edge of taxiway pavement. Taking the distance of oleo strut of the landing gear as 6 m, we get

$$S = 6 + \frac{7.5}{2} = 9.75 \text{ m}$$

$$\therefore \quad R = \frac{0.388 \times (35)^2}{\frac{22.5}{2} - 9.75}$$

$$\Rightarrow \quad R = \frac{475.3}{1.5}$$

$$\Rightarrow \quad R = 316.87$$

$$\Rightarrow \quad R \approx 317 \text{ m}$$

(iii) **For supersonic aircrafts** minimum radius of taxiway is taken as 180 m.

Therefore the maximum value among the three cases discussed above will be the turning radius to be actually provided = 317 m

1.7 **Runway length required for landing at sea level in standard atmospheric condition is 2100 m. Runway length required for take off at a level site at sea level in standard atmospheric condition is 2500 m. Aerodrome elevation is 200 m and reference temperature is 24°C. Temperature in the standard atmosphere for 200 m is 15°C and runway slope is 0.5%. Determine the length of the runway after applying correction to runway length.**

[5 marks : 2011]

Solution:

(a) Corrections to runway take-off length.

(i) **Correction for elevation** $= \frac{7}{100} \times 2500 \times \frac{200}{300} = 116.67 \text{ m}$

Corrected length = 2500 + 116.67 = 2616.67 m

(ii) **Correction for temperature**

Rise of temperature = 24° – 15° = 9°C

$$\text{Correction} = \frac{1}{100} \times 9 \times 2616.67 = 235.5 \text{ m}$$

Corrected length = 2616.67 + 235.5 = 2852.17 m = 2852.2 m

Check for total correction for elevation plus temperature

$$\text{Total correction in percent} = \frac{2852.2 - 2500}{2500} \times 100 = 14.09\% < 35\% \qquad \text{(OK)}$$

which is less than 35% as prescribed by ICAO.

(iii) **Correction for gradient** $= \frac{20}{100} \times 2852.2 \times 0.5 = 285.22 \text{ m}$

Corrected length = 2852.2 + 285.22 = 3137.42 m ≈ 3138 m

(b) Correction to runway landing length

(i) Correction for elevation $= \frac{7}{100} \times 2100 \times \frac{200}{300} = 98$ m

Corrected length = 2100 + 98 = 2198 m

No corrections are need to landing length for temperature and gradient

Actual runway length to be provided would be greater of (a) and (b) = 3138 m

1.8 **Explain detail the eight requirements of airport pavement.**

[10 marks : 2011]

Solution:

The requirements of airport pavement are:

(i) Good rideability.
(ii) Good frictional characteristics.
(iii) High strength and stability to withstand the shear stresses induced by heavy wheel loads and high tyre pressures.
(iv) A durable hard-wearing weatherproof surface free from loose material and sharp edges that might endanger aircraft.
(v) Resistance to fuel spillage and jet blast.
(vi) Pavement should have rut resistance under channelled traffic or parked aircraft.
(vii) Light reflectance for excellent runway visibility for approaching aircraft.
(viii) Permanent surface texture or grooving to minimize potential of aircraft skidding.

1.9 **Determine the turning radius of the taxiway of a supersonic transport aircraft with a wheel-base of 30 m and tread of main landing gear as 6.0 m for a design turning speed of 50 kmph. Assume coefficient of friction between tyre and pavement surface as 0.13 and width of taxiway pavement as 22.5 m.**

[5 marks : 2012]

Solution:

Given data: Turning speed (V) = 50 kmph, Wheel base of aircraft (W) = 30 m, Tread of main landing gear = 6 m, Coefficient of friction between tyre and pavement surface, (f) = 0.13, Width of taxiway (T) = 22.5

(a) Turning radius may be expressed as

$$R = \frac{V^2}{125f} = \frac{(50)^2}{125 \times 0.13} = 153.85\text{m}$$

(b) Turning radius as per Horonjeff's equation is given by

$$R = \frac{0.388W^2}{\frac{T}{2} - S}$$

where, S is the distance between midway point of tread of landing gear and the edge of taxiway pavement.

Taking the distance of oleo strut of the landing gear as 6 m, we get

$$S = 6 + \frac{6}{2} = 9\text{m}$$

$$\therefore \quad R = \frac{0.388 \times 30^2}{\frac{22.5}{2} - 9} = \frac{349.2}{2.25} = 155.2 \simeq 156 \text{ m}$$

(c) For supersonic aircrafts minimum radius of taxiway is taken as 180 m.
Therefore, the maximum value amongst the three cases discussed above will be the turning radius to be actually provided = 180 m

1.10 **Sketch a typical cross-section of a runway and taxiway. Explain the importance of drainage of Airport.**

[4 marks : 2013]

Solution:

Typical cross-section of a runway

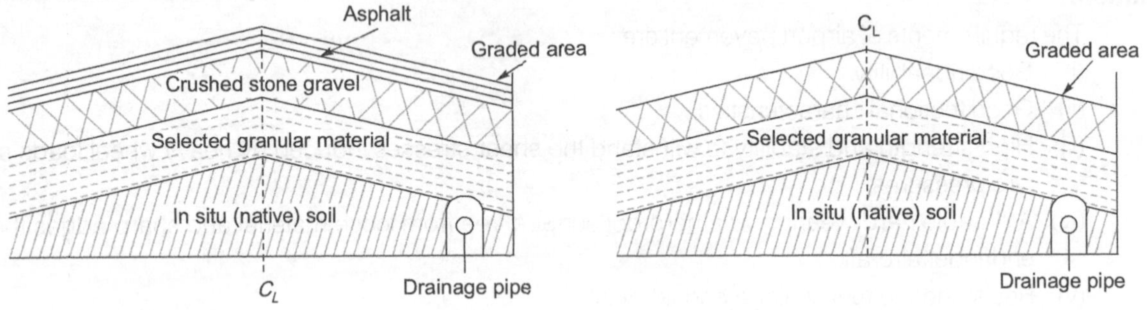

Asphaltic Concrete (AC) flexible pavement *Portland Cement Concrete (PCC) rigid pavement*

Typical Section of a Taxiway

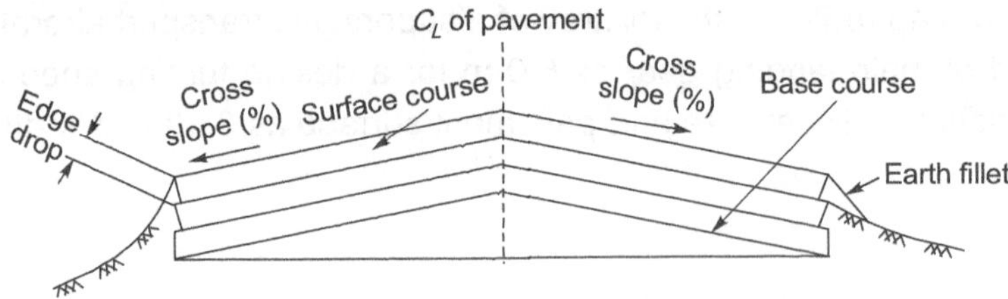

Airport Drainage: A well designed airport drainage system is a prime requisite for safe operation, efficiency and pavement durability. Improper drainage system may lead to huge damages due to flooding and may pose a serious prospective hazard for air traffic operations/activities.

Inadequate drainage also causes weakening of foundations, slope erosion etc.

Airport drainage system is characterized by vast flat areas and urgent need for prompt removal of surface water from pavements. The resulting runoff then must be removed from the airport by means of surface drains and underground storm drainage system.

Some of the drainage design aspects are:

(i) Estimation of runoff
(ii) Design of basic system for collection and disposal of runoff.
(iii) Adequate provisions of subsurface drainage.

1.11 **An exit taxiway is to be designed for Boeing 707 with turn off speed of 65 km/hr. Calculate the turning radius of the exit taxiway using the following data:**
Coefficient of lateral friction = 0.13
Wheel base = 18.0 m
Tread of main landing gear = 7.0 m
Width of taxiway = 22.5 m

[8 marks : 2014]

Solution:

Given: Turn off speed (V) = 65 Kmph, Coefficient of lateral friction (f) = 0.13, Wheel base (W) = 18.0 m, Tread of main landing gear = 7.0 m, Width of taxiway (T) = 22.5 m

(i) Turning radius base on lateral friction:

$$R = \frac{V^2}{125\,f}\ \text{m}$$

where, V = turn off speed (Kmph)

f = Coefficient of lateral friction

$$\therefore \qquad R = \frac{(65)^2}{125 \times 0.13} = 260\,\text{m}$$

(ii) Turning radius based on Horonjeff equation:

$$R = \frac{0.388\,W^2}{\frac{T}{2} - S}$$

where W = Wheel base of aircraft (m)

T = Width of taxiway pavement (m)

S = Distance between midway point of the main gear and the edge of the taxiway pavement (m)

Taking the distance of oleo struct of landing gear as 6 m,

$$S = 6 + \frac{7.0}{2} = 9.5\,\text{m}$$

$$\therefore \qquad R = \frac{0.388 \times (18)^2}{\left(\frac{22.5}{2}\right) - 9.5} \simeq 72\,\text{m}$$

(iii) For super sonic air crafts, minimum radius of taxiway is taken as 180 m.

Therefore, the radius of taxiway should be maximum value of the three cases discussed above.

$\therefore$ Taxiway turning radius (R) = 260 m

1.12 **An airport is to be constructed at a site 190 m above mean sea level and on a level ground. The runway length required under standard atmospheric condition at sea level for landing is considered as 2100 m and for takeoff as 1600 m respectively. Determine the actual runway length to be provided at this airport site. Airport reference temperature may be considered as 21°C.**

[8 marks : 2015]

Solution:

(a) Correction to runway landing length

$$\text{Correction for elevation} = \frac{7}{100} \times 2100 \times \frac{190}{300} = 93.1\,\text{m}$$

$$\text{Corrected length} = 2100 + 93.1 = 2193.1\ \text{m}$$

(b) (i) Corrections to runway take-off length

$$\text{Correction for elevation} = \frac{7}{100} \times 1600 \times \frac{190}{300} = 70.93\ \text{m}$$

$$\text{Corrected length} = 1600 + 70.93 = 1670.93\ \text{m}$$

(ii) **Correction for temperature:**

Standard atmospheric temperature at the given elevation = 15° – 0.0065 × 190 = 13.765°C

Airport reference temperature = 21°C

As per ICAO the basic runway length after having been corrected for elevation should be further increased at the rate of 1% for every 1°C rise of airport reference temperature.

$$\text{Rise in temperature} = 21 - 13.765 = 7.235°\text{C}$$

$$\text{Correction for temperature} = 1670.93 \times \frac{1}{100} \times 7.235 = 120.9 \text{ m}$$

Thus, $\quad \text{corrected length} = 1670.93 + 120.9 = 1791.83 \text{ m} \simeq 1792 \text{ m}$

Thus, actual runway length is 2193.1 m which is greater of (a) and (b).

1.13 **Determine the actual length of runway after applying necessary corrections for elevation, temperature and gradient (as per relevant ICAO, FAA recommendations) for the data given below:**

(i) Basic runway length = 2500 m

(ii) Elevation of airport = 500 m

(iii) Monthly mean of average daily temperature for the hottest month of the year = 15°C

(iv) Monthly mean of maximum daily temperature for the same month = 22°C

(v) Effective gradient = 0%

[8 marks : 2016]

Solution:

Correction for elevation:

$$\text{Correction for elevation} = \frac{7}{100} \times 2500 \times \frac{500}{300} \simeq 292 \text{ m}$$

$\therefore \quad \text{Corrected length} = 2500 + 292 = 2792 \text{ m}$

Correction for temperature:

Standard atmospheric temperature at the given elevation

$$= 15° - 0.0065 \times 500 = 11.75°\text{C}$$

$$\text{Airport reference temperature} = T_a + \frac{T_m - T_a}{3} = 15° + \frac{22° - 15°}{3} = 17.33°\text{C}$$

$$\text{Rise in temperature} = 17.33° - 11.75° = 5.58°\text{C}$$

$$\text{Correction for temperature} = 2792 \times \frac{1}{100} \times 5.58 = 155.79 \text{ m}$$

Thus, $\quad \text{corrected length} = 2792 + 155.79 = 2947.79 \text{ m}$

Check for the total correction for elevation and temperature:

$$\text{Total correction in percentage} = \frac{2947.79 - 2500}{2500} = 17.91\% \not> 35\% \qquad \text{(OK)}$$

As per ICAO, the total correction for elevation and temperature should not exceed 35% of the basic runway length.

Correction for gradient = 0.

Therefore, the corrected runway length is nearly 2950 m.

1.14 **Discuss instrumental landing system and air navigation aids.**

[4 marks : 2016]

Solution:

Instrumental landing system abbreviated as **ILS** is a ground based instrument approach system that guides the lateral and vertical movement of the aircraft while landing on a runway with the help of radio signals and high intensity light arrays. It is helpful for the pilot when the ground visibility is poor and in bad weather conditions.

Air navigation aids: The various air navigation aids work on similar principles like navigating the aircraft without getting lost or endangering the life of ground personnel and those on board etc. Some of the air navigation aids are GPS, NDB, VOR, INS and OMEGA.

1.15 **What do you understand by Wind Rose Diagram? How it is prepared? Discuss its significance in runway orientation.**

[20 marks : 2017]

Solution:

Wind rose diagram : The wind data pertaining to a particular area is presented in the form of geometrical diagram. It indicates the direction, intensity and duration of wind component. It is correlated with meteorological data to predict the feasibility of construction of the airport component. It gives warning of the bad weather and inform the aircrafts about the feasibility of the check landing or take off. The windrose diagram is shown below which shows the prevailing direction of wind.

Construction of wind rose diagram : Wind roses may be constructed from the data obtained over a given time period such as particular month or season or a year. The convention used is wind direction refers to the direction from which the wind is blowing. A line or bar extending to the north on the wind rose indicates the frequency of winds blowing from the north.

The wind rose diagram is prepared using an appropriate scale to represent percentage frequencies of wind directions and appropriate index shades, lines etc. to represent various speeds. Observations corresponding wind speed below 1 km/hr are recorded as calm.

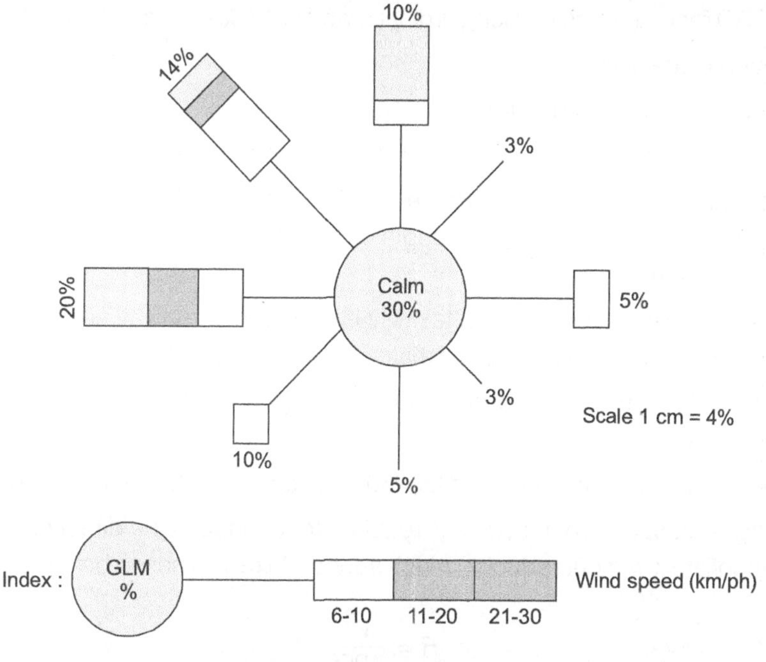

Wind Rose Construction

Significance in Runway Orientation: Runways are oriented in the direction of the prevailing wind. The reason behind that is to utilize it fully to get the maximum force of wind at the time of take off and landing of an aircraft. According to FAA standards, runways should be oriented so that the aircraft can take off and/or land atleast 95% of time without exceeding the allowable crosswinds.

Wind data is required in many dimensions i.e. in terms of duration, intensity of wind in the vicinity of the airport and this is required for development of windrose diagram which is finally used to identify the orientation of runway based on wind coverage area.

Wind direction is studied from wind rose diagram to examine whether the wind will attack aircraft from the headside or tailside or from lateral sides. Also direction of wind is not necessarily same throughout the year. It will keep on changing.

1.16 What are the elements that are considered in the geometric design of a taxiway? Explain them in brief. Design the turning radius of a taxiway for a class B airport handling supersonic aircrafts with a wheel base of 30 m and tread of main landing gear as 7.2 m. The aircraft design speed to negotiate the curve is given as 60 kmph and coefficient of friction as 0.13.

[20 marks : 2018]

Solution:

A taxiway is a path for aircraft at an airport connecting runways with aprons, hangars, terminals and other facilities.

Elements considered in the geometric design of a taxiway:

(i) **Length of taxiway:** It should be short so as to save fuel consumption.

(ii) **Width of taxiway:** It is much lower than that of runway because speed at taxiway is lower, so pilot can considerably manoeuvre the aircraft at smaller width.

(iii) **Width of safety area:** It includes taxiway pavement on either side plus the area that is graded and drained.

(iv) **Longitudinal gradient:**
- Steeper gradient demands greater fuel consumption.
- ICAO recommends that longitudinal gradient 1.5% for A and B and 3% for other types at airports.

(v) **Transverse gradient**
- Provided for quick drainage.
- It should not be greater than 1.5% for A, B and C type and 2% for D and E types of airport.

(vi) **Rate of change of longitudinal gradient:** ICAO recommends that rate of change of slope in longitudinal direction should be 1% per 30 m length of vertical curve for A, B and C types and 1.2% for D and E types for airports.

(vii) **Sight distance:** Since speed on taxiway is slower than runway so smaller sight distance is require comparatively. ICAO recommends that:
- For A, B, C type of airport, surface of taxiway should be visible from 3 m height for 300 m distance.
- Whereas, the height is 2.1 m for 250 m distance, for D, E type of airport.

(viii) **Turning radius:** Horizontal curve is required for change in the direction of taxiway such that aircraft can negotiate it without reducing much speed. So radius has to be large.

Radius of curve, $$R = \frac{V^2}{125f} \quad ...(i)$$

R = Radius in meter
V = Speed in kmph
f = Transverse coefficient of friction = 0.13

Also, $$R_{min} = 120 \text{ m (subsonic jet)} \quad \text{...(ii)}$$
$$R_{min} = 180 \text{ m (Supersonic jet)}$$

And, according to Horonjeff,

$$R = \frac{0.388W^2}{\frac{T}{2} - S} \quad \text{...(iii)}$$

W = Wheel base of aircraft
T = Width of taxiway pavement
S = Distance between midway point of the main gears and edge of taxiway pavement in meter

$$S = \text{Oleo distance} + \frac{T'}{2}$$

$$\Rightarrow \quad S = 6 + \frac{T'}{2}$$

For the given question, R will be taken as the maximum of values obtained by equation (i), (ii) and (iii)

1. $$R = \frac{V^2}{125f} \quad (V = 60 \text{ kmph}, f = 0.13)$$

$$\therefore \quad R = 221.54 \text{ m}$$

2. $$R_{min} = 180 \text{ m} \quad \text{(supersonic aircraft)}$$

3. $$R = \frac{0.388W^2}{\frac{T}{2} - S}$$

$$W = 30 \text{ m},\ T = 22.5 \text{ m},\ S = 6 + \frac{7-2}{2} = 9.6 \text{ m}$$

$$\therefore \quad R = \frac{0.388 \times 30^2}{\frac{22.5}{2} \quad 9.6} = 211.64 \text{ m}$$

So, R is maximum of (221.54, 180, 211.64)

$$\Rightarrow \quad R = 221.54 \text{ m}$$

1.17 How are runways oriented? Explain the terms 'wind coverage' and 'crosswind' component.

[12 marks : 2019]

Solution:

Runway Orientation

- The orientation of runway depends upon the direction of wind and to some extent area available for development.
- Runway is oriented in the direction of prevailing wind such that the direction of wind is opposite to the direction of landing and take off.
- The head wind provides greater lift force on the wings of aircraft while taking off, thus aircraft rises above the ground earlier i.e. in a shorter length of runway.
- During landing, head wind provides greater drag force so that it comes to hault in a shorter length of runway.

2. Dock and Harbour

2.1 Explain the function and working of a Floating Dry Dock with a neat sketch.

[10 marks : 1999]

Or

Explain giving a neat sketch, the use and working of a Floating Dry Dock.

[10 marks : 2002]

Solution:

Floating dry dock may be defined as a floating vessel which can lift a ship out of water and retain it above water by means of its own buoyancy. It is a hollow structure of steel or reinforced cement concrete consisting of 2 side walls and a floor, with the ends open. To receive a ship, the structure is sunk to required depth by ballasting its interior chambers with water, the ship is then floated into position and berthed. The dock is raised bodily with the berthed ship by unballasting the chambers by pumping out the water.

There are three important types of floating docks that have been developed viz. Rigid type floating docks, self docking type floating docks and self docking offshore type floating docks.

(i) **Rigid type floating dry docks:**
In this type, the side walls are rigidly fixed to the pontoon or bottom section. The floor portion is divided into a number of chambers, so as to assist in canting the dock if necessary to berth damaged ships, by partial unballasting of the chambers.

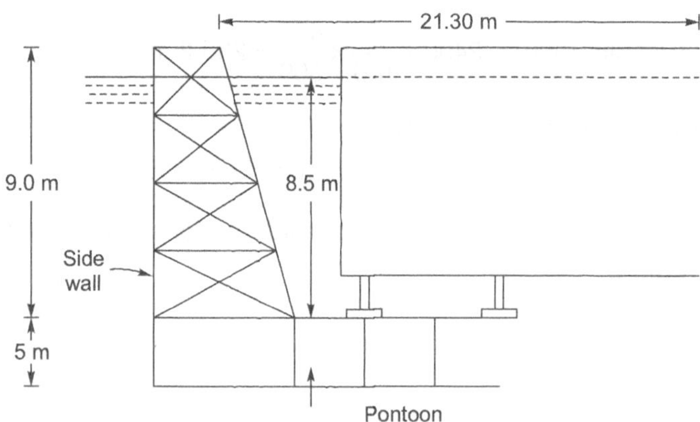

(ii) **Self docking type floating docks:**
Self docking refers to a type of floating dock, which is divided into sections longitudinally, and one of which is capable of being lifted and docked on the remainder of the dock for purposes of cleaning, painting or repairing. A typical self docking dry dock known as bolted sectional type is shown below:

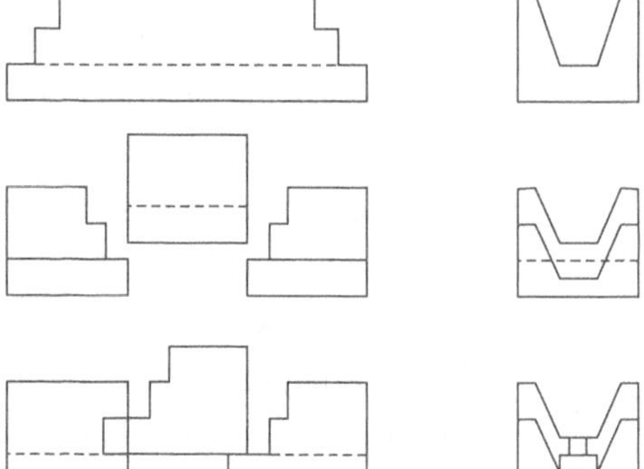

Firstly the whole dock (having minimum three sections) is shown assembled; secondly the centre section is shown detached and about to be docked on the two end sections and thirdly an end section is seen being docked on the other two.

This type is usually constructed in three equal sections, the two end sections having stepped ends to form landings during self docking. It combines the advantages of strength of the rigid type with self docking facility.

(iii) Self docking offshore type floating docks:
The offshore type floating dock has no side wall on the water side and has an L-shaped cross section. The side wall is connected to the shore by hinged parallel body capable of lifting or lowering the dock. The ship to be docked, could be brought on to the dock from either end or side ways.

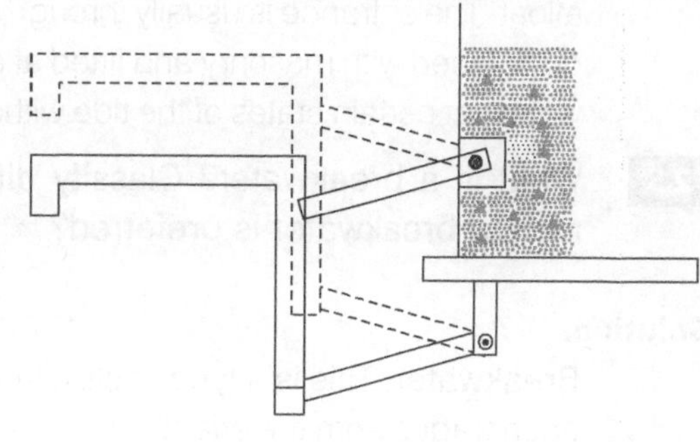

The dock is longitudinally made into two sections, so as to dock one half on the other. The dock and the self docking operation are illustrated in the figure. This type of dock is convenient in a sheltered situation and adoptable for being attached to river quays.

2.2 Enumerate the forces acting on an upright wall breakwater and explain their destructive effects.

[10 marks : 2000]

Solution:

The forces acting on an upright wall break water are:

1. **Shearing of bed joints due to horizontal pressure of the wave:** This can be prevented by the adhesive force of the mortar joint and frictional resistance.
2. **Overturning due to horizontal pressure of wave:** This is another aspect of the horizontal pressure and the design should provide for this like ordinary walls to keep the resultant of the weight and the horizontal pressure within the third of the base to avoid tension cracks in the foundation courses.
3. **Uplift force:** This force is generated by difference in water levels between outside and inside of harbour basin. The only force to eliminate this is the weight of the masonry.
4. **Fracture:** This does not result directly from water action. It may be caused by the dislocated blocks knocking against each other and breaking loose the joint. Such failures are usually avoided by proper binding in the masonry by joggles, etc.

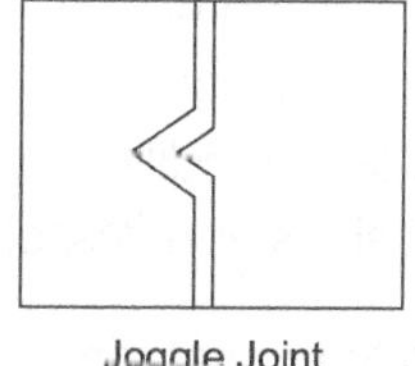
Joggle Joint

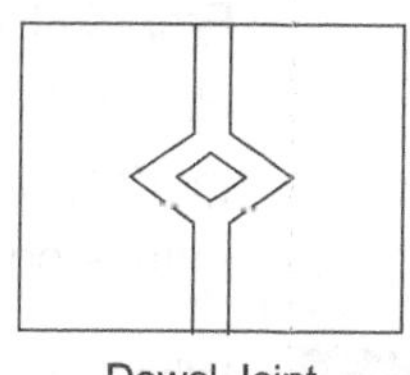
Dowel Joint

2.3 Differentiate between Wet-dock and Dry-dock.

[2 marks : 2009]

Solution:

Wet-docks	Dry-docks
Docks required for berthing of ships or vessels to facilitate the loading and unloading of passengers and cargo are called wet docks. These are also known as harbour docks.	The docks used for repairs of ships are known as dry docks.

2.4 What is Wet-dock? Write the uses of Wet-dock and explain operation of wet-dock using Lock and gate.

[5 marks : 2012]

Solution:

A Wet-dock or impounded dock is a variant in which the water is impounded either by dock gates or by a lock thus allowing ships to remain afloat at low tide in places with high tidal ranges. The level of water in the dock is maintained despite the rasing and lowering of the tide. This makes transfer of cargo easier. It works like a lock which controls the water level and allows passage of ships. It is basically an area of water which except for its entrance is enclosed with walls of masonry and excavated to a depth sufficient to take floating ships. The water is shut in by caissons or gates and is kept at a level sufficient to ensure that ships remain afloat. The entrance is usually through a lock. Lock is an excavated channel or approach to a basin or Wet-dock faced with masonry and fitted at each in with a caisson or gates. Ships may be moved to or from tidal waters at certain states of the tide without appreciably altering the level of water in the basin or Wet-dock.

2.5 **What is a breakwater? Classify different types of breakwaters. Under what condition a rubble mound breakwater is preferred?**

[2 marks : 2013]

Solution:

Breakwater: This is a type of structure constructed on coast as a part of coastal defence, to protect the anchorages from the effect of sea.

Types of breakwater:

1. **Cairson breakwater:** These have vertical sides and use self weight and mass of fill within to resist the overturing forces by waves.
2. **Rubble mound breakwater:** They use structural voids to dissipitate the wave energy.

2.6 **Describe Beanfort Scale.**

[2 marks : 2013]

Solution:

Beanfort Scale: The original Beanfort scale does not represent the actual velocity with which the air moves. It is merely a scale of wind conditions that was used by sailors to categorize different sailing conditions. With increasing no. of type of sailing ships and eventually the arrival of steam boats, exchanging and comparing the wind force information became a problem. Attempts have been made to express the Beanfort scale nos. into physical units by cumulating the Beanfort nos. for wind force with measured speed by anemometers. There are always exist the possibility to arrive at unexpected results while converting different anemometer readings i.e., knots as m/s to Beanfort numbers.

Francis Beanfort set up a scale of fourteen (14) wind force categories in 1806.

He numbered them as: $O \rightarrow$ calm $B \rightarrow$ storm

In 1838, a slightly modified scale at 13 wind categories (0 to 12) became mandatory for ship logbook entries on all ships and vessels of British Royal Navy.

2.7 **What are navigational aids? Why are they necessary?**

[2 marks : 2013]

Solution:

Navigational aids: Any sort of device/maker which helps the traveller in navigation. It is mostly used in sea travel and air travel. Such aids include lighthouses, buoys, fog signals etc. They are used to help the traveller in navigation, giving a rough estimate of direction etc.

2.8 **What is a dock? Classify the docks and explain the purpose of each type.**

[2 marks : 2013]

Solution:

Dock: It is a structure for holding the ships.

Types of Docks:

1. **Dry dock:** A basin that can be flooded and drained to allow a load to come to rest on a dry platform.
2. **Floating dock:** It is a type of jetty resting on pontoons.
3. **Marina dock:** It is a type of dock wherein different little docks are connected.
4. **Wharf:** It is a type of fixed platform, usually on piles, where ships are loaded and unloaded.

2.9 **What are the uses of wet docks in harbours?**

[4 marks : 2015]

Solution:

A Wet-dock or impounded dock is a variant in which the water is impounded either by dock gates or by a lock thus allowing ships to remain afloat at low tide in places with high tidal ranges. The level of water in the dock is maintained despite the rasing and lowering of the tide. This makes transfer of cargo easier. It works like a lock which controls the water level and allows passage of ships. It is basically an area of water which except for its entrance is enclosed with walls of masonry and excavated to a depth sufficient to take floating ships. The water is shut in by caissons or gates and is kept at a level sufficient to ensure that ships remain afloat. The entrance is usually through a lock. Lock is an excavated channel or approach to a basin or Wet-dock faced with masonry and fitted at each in with a caisson or gates. Ships may be moved to or from tidal waters at certain states of the tide without appreciably altering the level of water in the basin or Wet-dock.

2.10 **State the functions of docks. Compare floating dock and dry dock.**

[5 marks : 2018]

Solution:

A dock is the area of water between human-made structures that are involved in the handling of boats or ships.

Functions of docks are:

- Berting the ships to keep them afloat at a uniform level.
- This facilitates loading and unloading of cargo as well as passengers.
- Docks also allow repairing of ships.

Floating Dock	Dry Dock
(i) A floating vessel which can lift a ship out of water and retain it above water by means of its own buoyancy.	(i) It is a vessel that can be flooded to allow a load to be floated in, then drained to allow that load to come to rest on a dry platform.
(ii) Initial and working costs is cheaper.	(ii) Up keep and maintenance cost is lesser.
(iii) It can be transferred from point to point (mobility).	(iii) Advantage of mobility is not present.
(iv) It has no elaborate entrance or gate arrangement.	(iv) The open end of the chamber is provided with a gate and acts as the entrance to the dock.

2.11 **Classify wet docks and write advantages and disadvantages of each of them.**

[10 marks : 2020]

Solution:

Classification of Wet Docks :

1. **Wet docks in tidal basin :** These are docks located on the open sea coast protected by an outlying breakwater.

   Advantages :

   - Vessel can come in and berth or leave any time.
   - Costly arrangement like lock gates for entrance closing are not required.

   Disadvantages :

   - If tidal range is more, then operations of loading and unloading are seriously effected.
   - The fluctuating water levels causes rubbing of sides of ships against the berths.

2. Enclosed Wet Dock :
   - In enclosed wet dock, water level remains constant by lock gates so that cargo handling and commercial activities are easy to perform.
   - Major disadvantages include that it is costly and because of lock gates ships will take more time for entry and exit.

Advantages :

- These helps in effective handling of cargo.
- These prevent the rubbing of the ship sides.
- These are effective when there is increase in draft of the vessel, or where considerable silting takes place.

2.12 **Discuss the essential features of an ideal commercial harbour.**

[5 marks : 2023]

Solution:

Essential features of an ideal commercial harbour:

(a) A commercial harbour should have deep waterways to ensure safe navigation and accessibility for ships.

(b) The harbour should have proper berthing facilities.

(c) The harbour must have efficient cargo handling facilities.

(d) Harbour is the important transportation hub that facilitate goods movement.

(e) Commercial harbours require customs facilities to process import/export documentation, inspect cargo and collect duties and taxes.

(f) It should be accessible to various transportation networks so as to transport commodities easily and quickly.

3. Tunnelling

3.1 **What are the various methods of tunnelling in hard rocks? Write the advantages of Mole-tunnelling method.**

[5 marks : 2012]

Solution:

In hard rock, the tunnelling rigs have a larger field of application and offer several advantages such as : preformation without harming the surrounding rocks a smooth cunt surface that reduces the necessity of support and/or a concrete lining and less personnel, etc. Excavation with explosions is widely accepted method of tunnelling in hard rocks.

Following sequence of operations are generally adopted for driving the tunnel through rock:

1. **Setting up and drilling:** For driving a tunnel, number of holes are drilled as per drilling pattern in size and depth as decided depending upon the size of the tunnel and its formation. Drifters are generally used for drilling in the tunnels where in water is used to remove the cuttings from the holes instead of compressed air to reduce the amount of dust in the air. The modern equipment used for drilling blast holes in tunnels is 'drill jumbo'.
2. **Loading the holes with explosives and firing:** In tunnels the holes should be fired in the sequence as given in the drilling pattern. Drilling pattern when followed produces most economical and efficient breakage of rock for a given tunnel, and is determined by conducting tests using different patterns.
3. **Ventilating and removing the dust resulted by explosion.**
4. **Loading and hauling the muck:** Mucking is the operation of loading the broken rock or earth for removal from a tunnel. Mucking is done by hand in small tunnels, drifts, and pilot tunnels. Mucking operations can be performed by power shovels, tractor loader, mucking machines, etc.
5. **Removing groundwater from the tunnel, if necessary:** Ground water in a tunnel construction is controlled in two phases; namely preventing the excess water entering the tunnel, and removing the water which has entered the tunnel. Water enters the tunnel either from the ground through which the

tunnel is driven or that from the drilled holes which carry the cuttings to clean the hole. Estimation for the ground water can be done with the help of exploratory holes. If heavy inflow of ground water is noticed which is due to broken rocks, it is possible to grout off the heavy flow of water and so solidify the formation. The seepage water, and the water from the drill holes is collected in sumps, and from there it is pumped out either with air-driven centrifugal pumps or with electric-motor-driven centrifugal pumps.

6. **Erection of supports for the roof and sides, if necessary. This includes rock-bolting or steel supports:** Ground adjacent to the tunnel is supported until the permanent concrete lining is installed. These provide support against all shattered and loosened materials without further loosening the virgin rock. These temporary supports should be strong enough to resist the pressure transmitted to them by the ground, and should be able to prevent any moving of the surrounding rock and also to protect against the fall of loosened rock pieces. The supports are provided immediately after blasting but before proceeding to clear the muck.
7. **Placing reinforcement.**
8. **Placing the concrete lining:** After excavating the tunnel, it is required to be lined with concrete. The lining is done in two or three parts i.e., overt, invert and kerb. For lining of overt of the tunnel, travelling type of shutter forms of suitable length are used. In tunnels to be used for water, concrete lining is done to prevent seepage of water. For tunnels to be used for traffic, concreting is done in portions where rock formation needs additional support.
9. **Curing and shuttering removal:** Tunnel Boring Machines (TBMs) also known as Moles are the recent developments in the tunnel driving technique. The function of TBM is to loosen the earth or break the rock continuously in the entire section of the tunnel, into cuttings and convey to the rear of the machine, where it can be loaded into muck cars or dumpers or on to conveyor belts for the transportation to the ultimate disposal site. The Moles are transported to the site of work in parts and then assembled at the project site either outside the tunnel portal, if space for assembly is available or inside the tunnel by driving the tunnel, sufficient in length for enabling its assembly using conventional method.

   The conventional drilling and blasting method is very time consuming. In hydro-electric projects, construction of tunnels is one of the most critical activities and has led to delays in commissioning of projects. In drilling and blasting technique, it has been experienced that only 25 percent of explosive energy is utilized in actual rock breaking process and the rest is dissipated through surrounding medium. There is, therefore, a pressing need for employing tunnel boring machines for driving tunnels for speedy, economic and timely completion of hydro electric and other projects requiring tunnelling work. Use of tunnels boring machine have revolutionised the tunnel practices. The tunnels which would have been completed in years are executed in months.

3.2 **Why is efficient ventilation system required in tunnelling operation. What are the requirements of ventilation system? Explain combination of blowing and exhausting system of ventilation.**

[5 marks : 2012]

Solution:

Main objects of ventilation system in a tunnel are:

(i) To provide fresh air for workers.

(ii) To remove poisonous gases and fumes produced by explosion.

(iii) To remove the dust caused by drilling, blasting, mucking and other operations performed in the tunnel.

If a pilot or a drift is driven in the tunnel from portal or portal, sufficient natural ventilation shall be available. When natural ventilation is not adequate, a mechanical method is provided. Mechanical ventilation is provided by one or more electric-motor-driven fans, which may blow fresh air in the tunnel or exhaust the dust and foul gases from the tunnel. Air is blown into the tunnel thorough a light weight pipe or a fibre duct, whereas if the air is exhausted, the duct should be of sufficient strength so that it may not collapse under partial vacuum.

Tunnel may have both blower and exhaust system. When fresh air is blown into the tunnel, it is released near the working face, and as it flows towards the portal it carries the dust and fumes with it. While exhaust system is used, dust and foul air is drawn keeping the duct opening at about 50 to 60 m from working face. Thereby causing the fresh air to flow into the tunnel, from the portal. The system should be such that, it provides minimum of 6 to 8 cum m of fresh air per worker, while dust and fumes should be removed within 20 minutes of blasting. Air is generally supplied at a discharge velocity of about 9 m/min at the end of the duct. As tunnelling work progresses and length of air duct increases, blowing power of the electric fans should be increased.

3.3 **What is dredging? Classify different types of dredging work.**

[2 marks : 2013]

Solution:

Dredging: It is an excavation activity usually carried out atleast partially under water. It is done in shallow seas or in fresh water areas for collecting bottom sediments and disposing them off to a different location. This technique is often used to keep the water ways navigable.

Different types of Dredging Work:

1. **Construction dredging:** For construction projects that require creation of trenches for pipelines, tunnels, foundation for structures etc.
2. **Environmental dredging:** To treat contaminated areas to restore it at non-contaminated level. Hydraulic transfer of contaminated sediments has become the most cost effective and efficient way to restore contaminated areas.
3. **Mining dredging:** This is done to excavate sand, gravel, precious, stones, coal, phosphates etc. Most mining dredgers require a high level of throughput.
4. **Navigational dredging:** The deeper the channel and water ways to improve navigation. Mostly used in posts and harbours.
5. **Recreational dredging:** The expand, maintain and deeper the lakes that have progressively become shallow due to deposits and vegetation.
6. **Restoration dredging:** To restore wetlands, lakes, beaches, coastlines and other land reclamation projects.

3.4 **Briefly compare the construction of tunnels in rock strata and construction of tunnels in soft strata.**

[4 marks : 2015]

Solution:

(a) **Tunneling in hard rocks is carried by one of the following methods:**

1. **Full Face Method:** This method of tunneling is adopted when the length of the tunnel is more than 3 meters. Large sized tunnels in rocks are always driven by this method. With the development of drill carriage this method is becoming more and more popular. In this method vertical columns are fixed to the face of the tunnel to which a large number of drills may be mounted or fixed at any suitable height as shown in the figure below. A series of drill holes are drilled at about 120 cm center to center in any number of desired rows, preferably in two rows. The size of the holes may vary from 10 to 40 mm. These holes are then charged with explosives and ignited. The muck is removed before the next operation of drilling holes.
2. **Heading and Benching Method:** Tunnel cross section is divided into two parts, the top portion of the tunnel is known as the *heading* and the bottom portion as *bench*. Usually this method is adopted for railway tunnels. In this method of tunneling, top portion or heading will be about 3.70 to 9.6 m ahead of the bottom portion. In hard rock which may permit the roof to withstand without supports, the top heading generally is advanced by one round of bottom portion. If the rock is broken then heading may be driven well ahead of the bottom portion and after giving proper

support to the roof, the bottom portion is completed. In hard rock the heading is bored first and the holes are driven for the bench portion at the same time as the removal of the muck. This required less explosive than the full face method, but due to the development of the drill carriage or jumbo, the use of this method is decreasing.

3. **Drift Method:** Drift is a small tunnel, usually its size is 3 m × 3 m. In driving a large tunnel it has been found advantageous to drive a drift first through the full length or in a portion of the length of the tunnel prior to the excavating the full bore.

   The drift may be provided at the center, sides, bottom or top as desired. In this method after driving the drift, the drill holes are drilled all round the drift in the entire cross section of the tunnel, filled with explosives and ignited. The rock shatters, the muck removed and the tunnel expanded to the full cross section.

(b) **Driving tunnels in soft ground:** While tunneling in soft grounds, explosives are not used and tunneling is done with the hand tools such as pick-axes, shovels etc. During excavation operation supports for soil are required immediately depending upon the type of soil. In the old days, timber was the only material used for supporting soft ground till the introduction of the steel liner plates few years ago. As heavy supporting system is needed to support the roof and sides, there is more obstruction in the movement inside the tunnel, which reduces the progress of the work. Care should taken to ensure that all struts should be sufficiently strong to bear the pressure coming on them. The method to be adopted in the soft ground tunneling depends upon the type of ground.

Needle beam method, sequence of construction operations: This method is useful for tunneling in the soft ground whose roof soil can stand without support for few minutes. In this method 5 to 6 meters long R.S. joist or timber beams are required in addition to other timber boards and struts. This method requires large number of jacks which cause obstruction in the efficient working of the laborers. For tunneling in soft ground it is more economical than other methods.

3.5 Explain 'full-face' and 'drift' methods of driving tunnel in rocks.

[4 marks : 2016]

Solution:

Full face method of tunneling: In the 'full face' method of tunneling, the entire diameter of the tunnel is excavated at the same timo. This is suitable for tunnels passing through hard rocks and in firm soils where the excavation can hold itself for sufficient time. This allows sufficient time to have mocking and supporting operations.

Drift method of tunneling: In the 'drift' method of tunneling, small drifts are made at various locations and are expanded. In fact, drift is a small tunnel of size 3 m x 3 m. This drift is provided at the center, bottom, sides or at the top. After driving the drift at the desired locations, drill holes are drilled all-round the drift and filled with explosive. On igniting the explosive, the rock breaks and the muck is removed and tunnel gets expanded for the full cross-section.

3.6 Explain various geological considerations for selection of tunnel sites.

[5 marks : 2018]

Solution:

Geological consideration for selection of tunnel sites:

1. Self supporting strength of the rocks through which tunnel has to be made, should be properly analysed.
2. The rocks should be checked for suitability for blasting or boring.
3. In extra hard rocks, there is problem of stress concentration and may fail in an explosive manner called as rock burst due to their brittle nature and stress concentration.

4. In order to decide the method of tunneling, the strength and extent of lining and hence the cost of the project the information separating mineralogical composition, textures and structures of rocks through which the proposed tunnel has to pass is to great importance.
5. Geological structures–dip and strike are two quantitative properties of rock. They determine the attitude (disposition in space) of rocks and hence influence the design of excavation to a great extent.

3.7 **Describe tunnel lining and various materials used for it.**

[10 marks : 2020]

Solution:

The objects of providing a tunnel with permanent lining are manifold.

1. It gives correct section to the tunnel.
2. It withstands soil pressure when driven in soft soils.
3. It reduces losses in friction and erosive action, and ensures stream line motion, when the tunnel has to carry water by providing a smooth passage at good velocity, free from turbulence.
4. It forms a good protective covering to certain types of rocks prone to air slaking.
5. It keeps the inside of the tunnel free from water percolation.
6. It supports large slabs of rock which might have become loosened during blasting.

Materials for Lining :

Masonry : Brick masonry was the standard material for tunnel lining, but is now rapidly going out of use, except in the case of underground sewers, as bricks are more acid resisting and suitable to carry sewage. A great disadvantage in using brick lining is the difficulty in back packing the space between the tunnel roof and the extrodas of the arch which at best has to be handpacked and is imperfect. At a later stage this may cause uneven pressures on the arch lining. The packing material employed is usually spalls, sand and brick bats, well rammed. On account of so many indeterminate factors in design, a very heavy section may be necessary the construction of which becomes cumbersome and costly.

Stone Masonry : It has more or less the same disadvantages as brick lining and in addition is very heavy necessitating very strong centres. But is still used for lining the sides.

Cement concrete has become the standard material for tunnel lining in both rock and soft soils. Its main advantage lies in its plasticity which allows it to be well packed between the form and the soil. The waterproof qualities of cement concrete, makes for a first class watertight lining. It could be used to form an unbroken ring right round, forming a shell. If unusual soil pressures have to be reckoned with, the thickness could be controlled and reinforced, suitably.

Timber : It is one of the oldest lining materials though of late, it is slowly yielding place to concrete. It is used both as a temporary support during construction and as a permanent support later.

Modern practice is to use either timber for semi-permanent lining and cement concrete as a standard practice. An attempt will be made to describe these two types of linings in detail. Masonry linings more or less follow the same modus operandi as concrete, like erection of centering, construction of the arch, easing of centres etc., except that the material is different.

3.8 **(i) How do the methods of tunnel construction in hard rock differ from the methods of tunnel construction in soft ground? State the various operations involved in hard rock tunnelling and soft ground tunnelling.**

(ii) List the ten typical features of a harbour. Elaborate any three at length.

[10 + 10 Marks : 2021]

Solution:

(i) Rocks are strong materials that are difficult to cut. Tunnelling in rock involves using tunnel boring machines with special hardened steel drill tools to cut into rock and remove the spoils. Rock cutting may even require drilling and blasting. If the rock is strong, massive and without discontinuities, then tunnels can be permanently stable without support. Since such rock formations are impervious, water does not seep into tunnels. Presence of joints and fissures causes instability in the rock mass. The rock mass can slip along these discontinuities and water can gush in from them. Tunnels in such formations require support and waterproof lining.

In contrast, soils are soft materials that are easy to cut. Excavation in soils is much easier than in rocks. However, tunnels in soils are usually unstable and seepage of water is always a problem beneath water table. Hence side support and waterproof lining in invariably provided in them.

Actual operations in the construction of a tunnel depend upon the size of the tunnel, kind of formation encountered and method of attacking the heading etc. However , following operations may be carried out during construction of a tunnel.

1. Setting up and drilling
2. Loading holes with explosives and firing them
3. Ventilations and removing the dust after explosion
4. Loading and hauling muck
5. Removing ground water if necessity arises.
6. Erecting supports for sides and roof if necessary
7. Placing reinforcement
8. Placing concrete lining.

(ii) **Features of a harbour:**

Main function of a harbour is to provide safe and suitable accomodation for vessels seeking refuge, supplies, refuelling, repairs or the transfer of cargo and passengers. The following is the ten typical features of a harbour:

1. Harbour entrance
2. Approach channel
3. Turning basin
4. Sheltered basin
5. Break waters
6. Wharves and quays
7. Jetties and piers
8. Lock and locked basin
9. Dry docks and spillways
10. Ancillaries

1. **Approach channel:** Ideally, the depth of water naturally available in the entire harbour area should be sufficient for navigation of design vessel at all the times. When such ideal conditions do not prevail, a channel with sufficient depth and width must be dredged to provided for a passage of ships between the harbour entrance and the docks. The alignment and the dimensions of channel are determined after considering factors involved in channel design. The terminology approach channel is used for the dredged fairway through which ships proceed from the open sea to the harbour basin. The portion of channel which lies beyond the harbour entrance in the open sea is called an outer channel. The portion lying between the harbour entrance and harbour

basin is called an inner channel. The inner channel is protected from storms and waves by natural configurations or by breakwaters.

2. **Turning basin**: It is the area required for manoeuvring the ship when it goes to or leaves the berth, os that a ship can leave head-on. The size of the turning basin primarily depends on the design vessel. It should preferably be designed to have a ship turn under continuous headway without help of a tug. This means that the turning basin should be large enough to permit a free turning.
3. **Breakwaters**: The main function of a breakwater, or a system of breakwaters, is to protect the enclosed area of water from storm waves. Thus a breakwater helps achieve calm in the harbour and thereby contribute to the safety of the vessel within and its easy working. The monolithic structure usually provided at the tip of the breakwater is called the pierhead.

■■■■